the Astronomy Place

YOUR CONVENIENT ONLINE ACCESS TO THE MOST POPULAR ASTRONOMY STUDENT WEBSITE AVAILABLE

With your purchase of a new copy of *The Cosmic Perspective,* **Third Edition,** you should have received a Student Access Kit for **the Astronomy Place.** The kit contains instructions and a code for you to access this dynamic website. Your Student Access Kit looks like this:

DON'T THROW YOUR ACCESS KIT AWAY!
If you did not purchase a new textbook or cannot locate the **Student Access Kit** and would like to access the wealth of AstronomyPlace resources, you may purchase your subscription online with a major credit card. Go to www.astronomyplace.com, click on the textbook cover for *The Cosmic Perspective,* **Third Edition,** click Buy Now and follow the on-screen instructions.

W9-AWW-668

REQUIRED for access to o...

the Astronomy Place
www.astronomyplace.com

STUDENT ACCESS KIT FOR

the Astronomy Place

PEARSON
Addison Wesley

THE COSMIC PERSPECTIVE, THIRD EDITION
BENNETT, DONAHUE, SCHNEIDER, VOIT

WHAT IS Astronomy Place?

Astronomy Place is a dynamic website that features award-winning self-paced animated and interactive tutorials—each designed specifically to help you master key concepts throughout in the course. The site also features narrated and animated movies, chapter-specific quizzes, summaries and overviews, self-test quizzes, flashcards for reviewing, weblinks and more.

To log in to Astronomy Place
After you register using the instructions in the Student Access Kit or purchase access online, simply go to www.astronomyplace.com, click on your book cover and type your Login Name and Password (that you created during registration).

Minimum System Requirements
Windows: 250 MHz; OS 98, NT, 2000, XP
Macintosh: 233 MHz; OS 9.2, 10
Both:
- 64 RAM installed
- 1024 x 768 screen resolution
- Browsers: PC: Internet Explorer 5.5 & 6.0, Netscape 6.2.3; Mac: Internet Explorer 5.1, Netscape 6.2.3
Plug Ins: Macromedia Flash 5.0, 6.0; QuickTime 5.0, 6.0

Joining an online class (if available)
An online class may be available to you for the Astronomy Place website associated with this textbook. If your instructor chooses to include this as part of your coursework, you will be provided with a *Class ID.*

To participate in an online class, (after you have logged in to Astronomy Place for *The Cosmic Perspective,* **Third Edition**), click Join a Class and follow the on-screen instructions, providing the Class ID when asked. From then on, whenever you log in to this site with your login name and password, you will have access to this online class.

Technical Support
M-F 9am – 6pm Eastern (US & Canada)
www.aw.com/techsupport

WHAT STUDENTS ARE SAYING...

Enhanced pedagogy helps students navigate the cosmos and structure their studies

Besides the compelling narrative, *The Cosmic Perspective,* **Third Edition** includes a number of new pedagogical features designed to focus students' attention and enhance learning.

LEARNING GOALS

2.1 Patterns in the Sky
- What is a constellation?
- What is the celestial sphere?
- Why do we see a band of light called the "Milky Way" in our sky?

2.2 The Circling Sky
- What are the basic features of the local sky?
- How does the sky vary with latitude?
- Why are some stars above the horizon at all times?
- How does the night sky change through the year?

2.3 The Reason for Seasons
- What is the cause of the seasons on Earth?
- What are the solstices and equinoxes?
- Why are the warmest days typically a month after the beginning of summer?

2.4 Precession of Earth's Axis
- What is Earth's cycle of precession?

2.5 The Moon, Our Constant Companion
- Why do we see phases of the Moon?
- What conditions are necessary for an eclipse?
- Why were eclipses difficult for ancient peoples to predict?

2.6 The Ancient Mystery of the Planets
- Why do planets sometimes seem to move backward relative to the stars?
- Why did the ancient Greeks reject the idea that Earth goes around the Sun even though it offers a more natural explanation for observed planetary motion?

THE BIG PICTURE

Putting Chapter 1 into Context

In this first chapter, we developed a broad overview of our place in the universe. It is not yet necessary for you to remember the details—everything presented in this chapter will be covered in greater depth later in the book. However, you should understand enough so that all the following "big picture" ideas are clear:

- The Earth is not the center of the universe but instead is a planet orbiting a rather ordinary star in the Milky Way Galaxy. The Milky Way Galaxy, in turn, is one of billions of galaxies in our observable universe.

- We are "star stuff." The atoms from which we are made began as hydrogen and helium in the Big Bang and were later fused into heavier elements in massive stars. When these stars died, they released these atoms into space, where our galaxy recycled them. Our solar system formed from such recycled matter some 4.6 billion years ago.

- Cosmic distances are literally astronomical, but we can put them in perspective with the aid of scale models and other scaling techniques. When you think about these enormous scales, don't forget that every star is a sun and every planet is a unique world.

- We are latecomers on the scale of cosmic time. The universe was more than halfway through its history by the time our solar system formed, and then it took billions of years more before humans arrived on the scene.

- All of us are being carried through the cosmos on spaceship Earth. Although we cannot feel this motion in our everyday lives, the associated speeds are surprisingly high. Learning about the motions of spaceship Earth not only gives us a new perspective on the cosmos, but also helps us understand its nature and history.

- Throughout history, the development of astronomy has gone hand in hand with social and technological development. In this sense, astronomy touches

NEW! Learning goals are presented as key questions at the start of each chapter to help students focus on the most important concepts ahead. These are combined with the chapter table of contents to provide a clear study plan.

"The Big Picture" box at the end of each chapter helps students rethink what they've learned in the overall context of the astronomy course and their place in the cosmos.

NEW! End-of-chapter summaries wrap up the learning goals with answers that help reinforce the students' understanding of the key concepts presented in the chapter.

SUMMARY OF KEY CONCEPTS

2.1 Patterns in the Sky

- *What is a constellation?* A constellation is a region of the sky. The sky is divided into 88 official constellations.
- *What is the celestial sphere?* It is an imaginary sphere surrounding Earth upon which the stars, Sun, Moon, and planets appear to reside.
- *Why do we see a band of light called the "Milky Way" in our sky?* This band of light traces the galactic plane as it appears from our location in the Milky Way Galaxy.

2.2 The Circling Sky

- *What are the basic features of the local sky?* The horizon is the boundary between Earth and sky. The meridian traces a half-circle from due south on your horizon, through the zenith (the point directly overhead), to due north on your horizon. Any point in the sky can be located by its altitude and direction.
- *How does the sky vary with latitude?* As the celestial sphere appears to rotate around us each day, we see different portions of the paths of stars from different latitudes. The altitude of the celes-

the summer solstice, so that the Northern Hemisphere is entering winter and the Southern Hemisphere is entering summer. The spring and fall equinoxes mark the times midway between the solstices. Keep in mind that the two hemispheres experience opposite seasons and that the solstices and equinoxes bear the names that make sense for the Northern Hemisphere.

- *Why are the warmest days typically a month after the beginning of summer?* The summer solstice is usually considered the first day of summer in the Northern Hemisphere, but the warmest days come later because it takes time for the more direct sunlight to heat up the ground and oceans from the winter cold.

2.4 Precession of Earth's Axis

- *What is Earth's cycle of precession?* It is a roughly 26,000-year cycle over which Earth's axis sweeps out a circle as it gradually points to different places in space.

2.5 The Moon, Our Constant Companion

- *Why do we see phases of the Moon?* At any time,

THINK ABOUT IT

Hold your pinhead-size model Earth in your hand and look toward your grapefruit-size model Sun 15 meters away. Aside from the tiny Earth that you hold, there is nowhere else in the solar system—and nowhere we yet know of in the universe—where humans can survive outside the artificial environment of a spacecraft or spacesuit. Does holding the model Earth affect your perspective on human existence in any way? Does it affect your perspective on the Earth? Explain.

COMMON MISCONCEPTIONS

Light-Years

A recent advertisement illustrated a common misconception by claiming "It will be light-years before anyone builds a better product." This advertisement makes no sense, because light-years are a unit of *distance*, not a unit of time. If you are unsure whether the term *light-years* is being used correctly, try testing the statement by remembering that 1 light-year is approximately 10 trillion kilometers, or 6 trillion miles. The advertisement then reads "It will be 6 trillion miles before anyone builds a better product," which clearly does not make sense.

MEDIA EXPLORATIONS

FOR A COMPLETE LIST OF MEDIA RESOURCES AVAILABLE, GO TO WWW.ASTRONOMYPLACE.COM, AND CHOOSE CHAPTER 2 FROM THE PULLDDOWN

Astronomy Place Web Tutorials

Tutorial Review of Key Concepts

Use the following interactive Tutorials at www.astronomyplace.com to review key concepts from this chapter.

Seasons Tutorial
Lesson 1 Factors affecting seasonal changes
Lesson 2 The solstices and equinoxes
Lesson 3 The Sun's position in the sky

Phases of the Moon Tutorial
Lesson 1 The causes of lunar phases
Lesson 2 Time of day and horizons
Lesson 3 When the moon rises and sets

Eclipses Tutorial
Lesson 1 Why and when do eclipses occur?
Lesson 2 Types of solar eclipses
Lesson 3 Lunar eclipses

Phases of the Moon Tutorial, Lesson 3
1. When does a new moon rise and set?
2. In which phase does the moon set just after the Sun sets?
3. What factors influence our view of the Moon from Earth?

Eclipses Tutorial, Lesson 1
1. What happens during a solar eclipse?
2. What happens during a lunar eclipse?
3. What role does the Moon's orbit play in the appearance of a solar eclipse?

Exploring the Sky & Solar System

Of the many activities available in the *Voyager:SkyGazer* CD-ROM accompanying your book, use the following files to observe key phenomena covered in this chapter. Go to the File: Basics folder for the following demonstrations:

1. Wide Field Milky Way
2. Eclipse 1991–1992 Views
3. Winter Sky

Go to the File: Demo folder for the following demonstrations:

1. Russian Midnight Sun
2. Earth Orbiting the Moon

Give your students the benefit of the most popular astronomy student website available

www.astronomyplace.com

The Astronomy Place is a dynamic website that features award-winning self-paced interactive tutorials. **By popular demand, these tutorials can now be conveniently assigned as pre- or post- class activities, allowing you to monitor and grade your students' performance using a new built-in tracking grading system.** The site also features narrated animations, movies, chapter-specific quizzes, chapter summaries and overviews, questions, weblinks, and more. Access to this wealth of resources is included free with each new copy of the book.

"**These tutorials greatly increased my knowledge and confidence... I like... that you learn something and then you are able to apply it through questions and visual sequences.**"
—*James Tomlinson*

"**... one of the most awesome educational programs I have ever logged on to. I am very pleased with the understanding of phases of the moon that this program has provided me with.**"
—*David Richardson*

"**I was amazed how well I understood phases of the moon [after the tutorial on astronomyplace]... The Light and Spectroscopy tutorial was especially helpful for understanding absorption and emission line spectrums. I had a firm grasp of those concepts after I did the tutorial.**"
—*Cassalyn David*

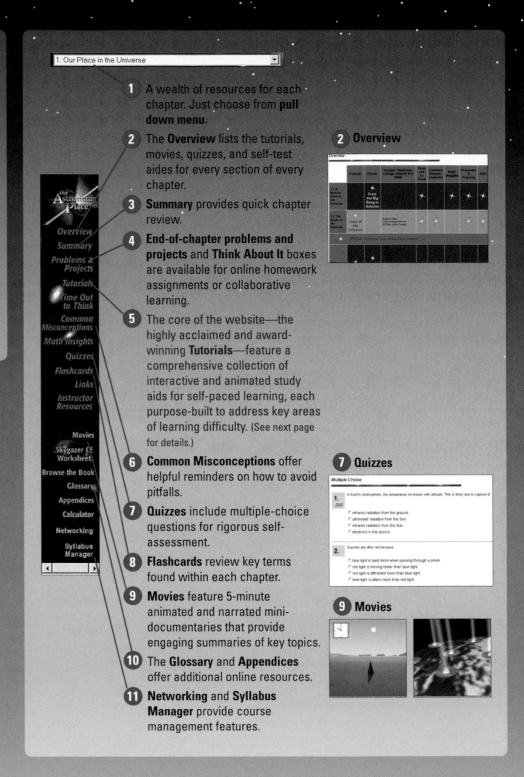

1: Our Place in the Universe

1. A wealth of resources for each chapter. Just choose from **pull down menu.**

2. The **Overview** lists the tutorials, movies, quizzes, and self-test aides for every section of every chapter.

3. **Summary** provides quick chapter review.

4. **End-of-chapter problems and projects** and **Think About It** boxes are available for online homework assignments or collaborative learning.

5. The core of the website—the highly acclaimed and award-winning **Tutorials**—feature a comprehensive collection of interactive and animated study aids for self-paced learning, each purpose-built to address key areas of learning difficulty. (See next page for details.)

6. **Common Misconceptions** offer helpful reminders on how to avoid pitfalls.

7. **Quizzes** include multiple-choice questions for rigorous self-assessment.

8. **Flashcards** review key terms found within each chapter.

9. **Movies** feature 5-minute animated and narrated mini-documentaries that provide engaging summaries of key topics.

10. The **Glossary** and **Appendices** offer additional online resources.

11. **Networking** and **Syllabus Manager** provide course management features.

2 Overview

7 Quizzes

9 Movies

The acclaimed Astronomy Place Tutorials focus on important and hard-to-grasp concepts and build understanding through carefully structured, step-by-step interactive pedagogy

These online tutorials encourage interactive learning as students manipulate tools, adjust slider bars, generate graphs, etc. The **Solar System Tutorial**, for example, requires students to explore each planet to compare and contrast properties and thereby answer specific questions.

The Astronomy Place Website is packed with purpose-built interactive tools, animations, and movies—all specifically designed to help students with key topics of difficulty. The library of award-winning tutorials now includes:

Multiple representations provide cues for students with different learning styles, helping them to translate between graphs, animations of internal and external properties, etc. Interactive questions actively develop these skills. The **Stellar Evolution Tutorial,** for example, takes the student through a sequence of interactive tools, animations and quizzes focusing on a star's evolution from a main-sequence star to a red giant to a white dwarf. In this way, students develop an understanding of how the star's view from space, the inner workings of its core, and its location on the H-R Diagram are connected.

NEW! Scale of the Universe

■ Seasons

■ Phases of the Moon

NEW! Eclipses

■ Orbits and Kepler's Laws

■ Light and Spectroscopy

NEW! Doppler Effect

NEW! Formation of the Solar System

NEW! Detecting Extrasolar Planets

■ Shaping Planetary Surfaces

■ Surface Temperature of Terrestrial Planets

■ Hertzsprung-Russell Diagram

■ Stellar Evolution

NEW! Measuring Cosmic Distances

■ Detecting Dark Matter in Spiral Galaxies

■ Hubble's Law

NEW! Fate of the Universe

This **Light and Spectroscopy Tutorial** builds a student's understanding of stellar spectra by exploring the properties of spectra from different elements, and later experimenting with what happens when the radiation passes through a hydrogen gas cloud.

These tutorials include:

• Nearly 60 interactive lessons comprising 18 award-winning tutorials

• 125 animations

• 100 interactive tools

• 450 tutorial questions

• 10 animated mini-documentaries

• 75 pre-built SkyGazer planetarium software activities

• More than 2,500 quizzes and questions in total

The **Surface Temperature of Terrestrial Planets Tutorial** explores such topics as the greenhouse effect on Venus. Students adjust a planet's albedo, orbital radius, and more to explore how these parameters effect the surface temperature and other associated questions.

The Cosmic Lecture Launcher CD-ROM puts a wealth of superior presentation resources in the palm of your hand

Do you want a powerful teaching tool that provides a wealth of presentation tools AND the means to integrate media seamlessly into your course? The **new Cosmic Lecture Launcher CD-ROM** contains all of the visual resources you need to complement your lecture presentation. The **Cosmic Lecture Launcher** is an extensive visual library of more than 1,200 high-resolution images, purpose-built interactive tools (applets) and animations, PowerPoint® slides, and more—saving you time looking for that missing slide or locating that latest web image, and giving you the ability to integrate and customize all these for your course. It's the most comprehensive library of purpose-built astronomy presentation materials available.

More than 100 purpose-built interactive tools

A wealth of interactive tools from the highly acclaimed Astronomy Place Tutorials are now available for you to use for in-class demonstrations. For instance, demonstrate how the Tully-Fisher relationship can be used to predict the distance to a galaxy using an interactive tool found in the **Galaxies** folder.

More than 150 animations

You'll find a wide range of superior, purpose-built animations from the Astronomy Place Tutorials along with additional movies from a variety of other well-known web sources. For instance, you can illustrate the process of planetary tectonics using one of the many animations found in the **Planetary Worlds** folder.

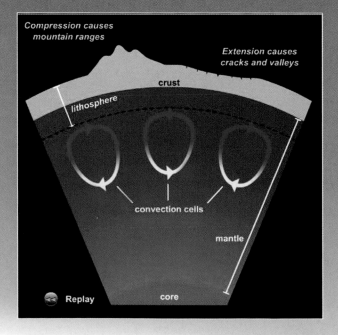

More than 800 photos and illustrative art from *The Cosmic Perspective*, Third Edition

A wealth of exceptional images from the highly acclaimed illustration program for *The Cosmic Perspective*, **Third Edition** are provided, digitized in high resolution for optimal classroom presentation and organized chapter by chapter.

The photos show how she would see the lunar phases from Earth as she turns to face the moon in each position as it orbits the Earth. Rise and set times are approximate.

Waning Crescent
Rise: 3 A.M.
Meridian: 9 A.M.
Set: 3 P.M.

Third Quarter
Rise: midnight
Meridian: 6 A.M.
Set: noon

To Sun

New Moon
Rise: 6 A.M.
Meridian: noon
Set: 6 P.M.

Waning Gibbous
Rise: 9 P.M.
Meridian: 3 A.M.
Set: 9 A.M.

Waxing Crescent
Rise: 9 A.M.
Meridian: 3 P.M.
Set: 9 P.M.

First Quarter
Rise: noon
Meridian: 6 P.M.
Set: midnight

Waxing Gibbous
Rise: 3 P.M.
Meridian: 9 P.M.
Set: 3 A.M.

Full Moon
Rise: 6 P.M.
Meridian: midnight
Set: 6 A.M.

More than 200 additional images from well-known astronomy resources

In addition to the art and photos from the text, a library of well-known and up-to-date images from other astronomical resources are provided—organized chapter by chapter—so you'll never have to surf the web again looking for the best image to illustrate your course, or try locating that missing slide.

The Celestial Sphere

north celestial pole

- The sky above looks like a dome…a hemisphere..
- If we imagine the sky around the entire Earth, we have the **celestial sphere**.
- This a 2-dimensional representation of the sky
← Because it represents our view from Earth, we place the Earth in the center of this sphere.

Copyright © Addison Wesley

2003, Pearson Education Inc., publishing as Addison Wesley

PowerPoint® presentations

Chapter-by-chapter PowerPoint® slides are provided with lecture/topic suggestions, images, and examples of embedded links using animations, tools, and simulations. You can use the presentations as they stand, or customize them to suit your particular lecture needs.

The Cosmic Lecture Launcher CD-ROM accompanies *The Cosmic Perspective*, **Third Edition** (including *The Solar System*, **Third Edition**, and *Stars, Galaxies, and Cosmology*, **Third Edition**) and is for the exclusive use of professors who adopt this text.

Inspire your students with the updated Cosmos series

Carl Sagan's landmark TV series **Cosmos** excited a generation, with more than 600 million viewers in more than 60 countries. This Peabody and Emmy award-winning series is arguably the most popular science television program ever produced. Inspire your students with the newly updated **Best of Cosmos** series (VHS), and the fully updated **Cosmos** series (VHS and DVD). These are available FREE to all adopting professors. Plus you'll get a lesson plan on how to integrate these episodes with *The Cosmic Perspective* in your classes.

SkyGazer planetarium software brings the universe to your students' fingertips

Purpose-built for students in introductory astronomy, **Voyager: SkyGazer, College Edition v3.4*** combines exceptional planetarium software with informative pre-packaged tutorials. Based on the popular Voyager software, **SkyGazer** makes it easy for students to learn more about their place in the universe through interactive exercises. This dual-platform CD-ROM is included in all new copies of *The Cosmic Perspective*, **Third Edition.** You'll find suggested **SkyGazer** Activities at the end of each chapter in the text, in the **Astronomy Media Workbook,** and online at The Astronomy Place (www.astronomyplace.com).

Voyager: SkyGazer, College Edition v3.4 comes with 75 pre-packaged demos. These are designed to help you quickly master all the features of **SkyGazer** while improving your understanding of key astronomical phenomena.

***SkyGazer, College Edition v3.4** is part of Carina Software's Voyager suite of products that includes Voyager, Voyager with SkyPilot, and Voyager: SkyGazer.

The new Voyager: SkyGazer "Explore" menu provides simulations of special astronomical events and familiar astronomical concepts

Explore features include:
- Sky Theater
- Satellite Map
- Star Calendar
- Seasons of the Earth
- Wobble of the Earth
- Shadows on the Earth
- Paths of the Planets
- Phases of the Planets
- The Solar System
- The Solar Neighborhood

Sky Theater QuickTime movies now include:
- Andromeda and Pegasus
- Aquila the Eagle
- Auriga the Charioteer
- Bootes the Herdsman
- Canis Major the Great Dog
- Cassiopeia and Perseus
- Comet Halley
- Crux and Centaurus
- Cygnus the Swan
- Gemini the Twins
- Hercules
- Hubble Views Mars
- Hubble Views Vesta
- Leo the Lion
- Lyra the Lyre
- Mission to Jupiter
- Orion the Hunter
- Scorpius and Sagittarius
- Taurus the Bull
- Ursa Major and Minor

THE COSMIC PERSPECTIVE

THIRD EDITION

About the Cover

The artwork used on the cover of this book is from Voyage: A Journey Through Our Solar System, an exhibition developed by Challenger Center for Space Science Education, the Smithsonian Institution, and NASA. Images created by ARC Science Simulations, 2002.

Voyage, an outdoor scale model of the solar system, depicts the sizes and distances between the planets at one ten-billionth actual size. The exhibition, which celebrates the human capacity to explore, is on permanent display along the National Mall in Washington, D.C. To learn more about the Voyage exhibition, see pages 10 and 11 in this book and visit the exhibition website: www.voyageonline.org. To learn more about Challenger Center, go to: www.challenger.org.

A Journey Through Our Solar System

The background photographs in this composite cover image are from Atlas Image mosaic, obtained as part of the Two Micron All Sky Survey (2MASS), a joint project of the University of Massachusetts and the Infrared Processing and Analysis Center/ California Institute of Technology, funded by the National Aeronautics and Space Administration and the National Science Foundation.

THE COSMIC PERSPECTIVE

THIRD EDITION

Jeffrey Bennett
University of Colorado at Boulder

Megan Donahue
Michigan State University

Nicholas Schneider
University of Colorado at Boulder

Mark Voit
Michigan State University

PEARSON

Addison
Wesley

San Francisco   Boston   New York
Capetown   Hong Kong   London   Madrid   Mexico City
Montreal   Munich   Paris   Singapore   Sydney   Tokyo   Toronto

Executive Editor:	*Adam Black, Ph.D.*
Market Developer:	*Susan Winslow*
Marketing Manager:	*Christy Lawrence*
Assistant Editor:	*Stacie Kent*
Developmental Editor:	*Patricia Brewer*
Production Coordination:	*Vivian McDougal*
Production:	*Mary Douglas, Rogue Valley Publications*
Photo Research:	*Myrna Engler*
Graphic Artists:	*John and Judy Waller/Scientific Illustrators, Joe Bergeron, John Goshorn/Techarts, Blakeley Kim, Emiko-Rose Koike/fiVth.com, Quade Paul/fiVth.com*
Copyeditor:	*Mary Roybal*
Text Designer:	*Mark Ong/Side By Side Studios*
Cover Designer:	*Blakeley Kim*
Composition:	*Thompson Type/Alma Bell, Lori Shranko, Janice Adamski*
Cover Printer:	*Phoenix Color*
Prepress Services:	*H&S Graphics, Inc./Tom Anderson, Lori Jewell*
Printer and Binder:	*Von Hoffmann Corp.*

For permission to use copyrighted material, grateful acknowledgment is made to the copyright holders on pp. C-1–C-3, which are hereby made part of this copyright page.

ISBN
Full book: 0-8053-8762-5
Solar System: 0-8053-8930-X
Stars and Galaxies: 0-8053-8931-8

2 3 4 5 6 7 8 9 10—VHC—05 04 03

We shall not cease from exploration
And the end of all our exploring
Will be to arrive where we started
And know the place for the first time.

T. S. Eliot

Dedication

TO ALL WHO HAVE EVER WONDERED about the mysteries
of the universe. We hope this book will answer some of
your questions—and that it will also raise new questions
in your mind that will keep you curious and interested in
the ongoing human adventure of astronomy.

And, especially, to the members of the "baby boom"
that has occurred among the authors and editors during
the writing of this book: Michaela, Emily, Rachel, Sebastian, Elizabeth, Nathan, Grant, Georgia, Brooke, Brian, and
Angela. The study of the universe begins at birth, and we
hope that you will grow up in a world with far less poverty,
hatred, and war so that all people will have the opportunity
to contemplate the mysteries of the universe into which
they are born.

Brief Contents

Detailed Contents

PART II

KEY CONCEPTS FOR ASTRONOMY

PART III

LEARNING FROM OTHER WORLDS

PART VI

GALAXIES AND BEYOND

Preface

We humans have gazed into the sky for countless generations, wondering how our lives are connected to the Sun, Moon, planets, and stars that adorn the heavens. Today, through the science of astronomy, we know that these connections go far deeper than our ancestors ever imagined. This book tells the story of modern astronomy and the new perspective—*The Cosmic Perspective*—with which it allows us to view ourselves and our planet. It is written for anyone who is curious about the universe, but it is designed primarily as a textbook for college students who do not intend to major in mathematics or science.

This book grew out of our experience teaching astronomy to both college students and the general public over the past 25 years. During this time, a flood of new discoveries fueled a revolution in our understanding of the cosmos but had little impact on the basic organization and approach of most astronomy textbooks. We felt the time had come to rethink how to organize and teach the major concepts in astronomy to reflect this renaissance in understanding. This book is the result.

Themes of *The Cosmic Perspective*

The Cosmic Perspective offers a broad survey of modern understanding of the cosmos and of how we have gained that understanding. Such a survey can be presented in a number of different ways. We have chosen to interweave a few key themes throughout the narrative—each selected to help make the subject more appealing to students who may never have taken any formal science courses and who may begin the course with little understanding of how science works. We built our book around the following five key themes:

- *Theme 1: We are a part of the universe and thus can learn about our origins by studying the universe.* This is the overarching theme of *The Cosmic Perspective*, as we continually emphasize that learning about the universe helps us understand what has made our existence possible. Studying the intimate connections between human life and the cosmos not only gives students a reason to care about astronomy but also deepens their appreciation of the unique and fragile nature of our planet and the life it supports.

- *Theme 2: The universe is comprehensible through scientific principles that can be understood by anyone.* The universe is comprehensible because the same physical laws appear to be at work in every aspect, on every scale, and in every epoch of the universe. Moreover, while the laws generally have been discovered by professional scientists, their fundamental features can be understood by anyone. Students can learn enough in one or two terms of astronomy to comprehend the basic reasons for all the phenomena they see around them, from seasonal changes and phases of the Moon to the most esoteric astronomical images that appear in the news.

- *Theme 3: Science is not a body of facts but rather a process through which we seek to understand the world around us.* Many students assume that science is just a body of facts, but the long history of astronomy clearly shows that science is a process through which we learn about our universe—a process that is not always a straight line to the "truth." That is why our ideas about the cosmos sometimes change as we learn more, as they did dramatically when we first recognized that Earth is a planet going around the Sun rather than the center of the universe. We continually emphasize the nature of science so that students can see how and why modern theories have gained acceptance and can understand why these theories may still be subject to change in the future.

- *Theme 4: A course in astronomy is the beginning of a lifelong learning experience.* Building upon the prior themes, we emphasize that what students learn in their astronomy course is not an end but a beginning. By remembering a few key physical principles and learning to appreciate the nature of science, students can follow astronomical developments for the rest of their lives. We therefore seek to motivate students enough so that they will continue to participate in the ongoing human adventure of astronomical discovery.

- *Theme 5: Astronomy affects each of us personally with the new perspectives it offers.* We all conduct the daily business of our lives with reference to some "world view"—a set of personal beliefs about our place and purpose in the universe—that we have developed through a combination of schooling, religious training, and personal thought. This world view shapes our beliefs and many of our actions. Although astronomy does not mandate a particular set of beliefs, it does provide perspectives on the architecture of the universe that can influence how we view ourselves and our world, and these perspectives can potentially affect our behavior. For example, someone who believes Earth to be at the center of the universe might treat our planet quite differently from someone who views it as a tiny and fragile world in a vast cosmos. In many respects, the role of astronomy in shaping world views may represent the deepest connections between the universe and the everyday lives of humans.

Pedagogical Principles of *The Cosmic Perspective*

No matter how an astronomy course is taught, it is very important to present material according to a clear set of pedagogical principles. The following list briefly summarizes the major pedagogical principles that we apply throughout the book. (The *Instructor's Guide* describes these principles in more detail.)

- *Stay focused on the "big picture."* Astronomy is filled with interesting facts and details, but they are meaningless unless they fit into a big picture view of the universe. We therefore take care to stay focused on the big picture (essentially the themes discussed above) at all times. A major benefit of this approach is that although students may forget individual facts and details after the course is over, the big picture framework should stay with them for life.

- *Always provide context first.* We all learn new material more easily when we understand why we are learning it. In essence, this is simply the idea that it is easier to get somewhere when you know where you are going. We therefore begin the book (Chapter 1) with a broad overview of modern understanding of the cosmos, so that students can understand what they will be learning in the rest of the book. We maintain this "context first" approach throughout the book and always tell students what they will be learning, and why, before diving into the details.

- *Make the material relevant.* It's human nature to be more interested in subjects that seem relevant to our lives. Fortunately, astronomy is filled with ideas that touch each of us personally. For example, the study of our solar system helps us better understand and appreciate our planet Earth, and the study of stars and galaxies helps us learn how we have come to exist. By emphasizing our personal connections to the cosmos, we make the material more meaningful, inspiring students to put in the effort necessary to learn it.

- *Emphasize conceptual understanding over "stamp collecting" of facts.* If we are not careful, astronomy can appear to be an overwhelming collection of facts that are easily forgotten when the course ends. We therefore emphasize a few key conceptual ideas that we use over and over again. For example, the laws of conservation of energy and conservation of angular momentum reappear throughout the book, and we find that the wide variety of features found on the terrestrial planets can be understood through just a few basic geological processes. Research shows that, long after the course is over, students are far more likely to retain such conceptual learning than individual facts or details.

- *Proceed from the more familiar and concrete to the less familiar and abstract.* It's well known that children learn best by starting with concrete ideas and then generalizing to abstractions later. In fact, the same is true for many adults. We therefore always try to build "bridges to the familiar"—that is, to begin with concrete or familiar ideas and then gradually draw more general principles from them.

- *Use plain language.* Surveys have found that the number of new terms in many introductory astronomy books is larger than the number of words taught in many first courses in foreign language. In essence, this means the books are teaching astronomy in what looks to students like a foreign language! Clearly, it is much easier for students to understand key astronomical concepts if they are explained in plain English—without resorting to unnecessary jargon. We have gone to great lengths to eliminate jargon as much as possible or, at minimum, to replace standard jargon with terms that are easier to remember in the context of the subject matter.

- *Recognize and address student misconceptions.* Students do not arrive as blank slates. Most students enter our courses not only lacking the knowledge we hope to teach but often holding misconceptions about astronomical ideas. Therefore, to teach correct ideas, we must also help students recognize the paradoxes in their prior misconceptions. We address this issue in a number of ways, the most obvious being the presence of many "Common Misconceptions" boxes. These summarize commonly held misconceptions and explain why they cannot be correct.

The Topical (Part) Structure of *The Cosmic Perspective*

The Cosmic Perspective is organized into six broad topical areas (the six "Parts" in the table of contents), each approached in a distinctive way designed to help maintain the focus on the themes discussed above. Here, we summarize the guiding philosophy through which we have approached each topic. We also highlight a few of the major changes in the third edition and list a few of the key new figures that illustrate the general improvements we have made in this new edition. (The *Instructor's Guide* describes the topical structure in much more detail.)

Part I Developing Perspective (Chapters 1–3, S1)

Guiding Philosophy: Introduce the big picture, the process of science, and the historical context of astronomy

The basic goal of these chapters is to give students a "big picture" overview and context for the rest of the book and to be sure they develop an appreciation for the process of science and how science has developed through history. Chapter 1 offers an overview of our modern understanding of the cosmos, thereby giving students perspective on the entire universe. Chapter 2 provides an introduction to basic sky phenomena, including seasons and phases of the Moon, and a perspective on how phenomena we experience every day are tied to the broader cosmos. Chapter 3 discusses the nature of science, offering an historical perspective on the development of science and giving students perspective on how science works and how it differs from nonscience. The supplementary Chapter S1 (optional) covers more detail about the sky, including celestial timekeeping and navigation.

New for the third edition Throughout the book, we have edited to improve the flow of ideas, improved art pieces, and added new illustrations to help students understand particularly tricky concepts. We have also reorganized these chapters in three major ways:

1. A much-enhanced discussion of seasons and precession now appears in Chapter 2 (moved from Chapter 1 in the previous edition).

2. The tour of the solar system that formerly appeared in Chapter 1 now appears in our new Chapter 8.

3. The discussion of the Copernican revolution now appears in Chapter 3 (moved from Chapter 5 in the previous edition).

In addition, we have expanded and completely revised both the discussion of the Greek role in the historical development of science (Section 3.3) and the discussion of the nature of science (Section 3.5).

Key new figures include: 1.3, 1.5, 2.15, 2.21, 2.23, 3.13

Part II Key Concepts for Astronomy (Chapters 4–7)

Guiding Philosophy: Bridges to the familiar

These chapters lay the groundwork for understanding astronomy through what is sometimes called the "universality of physics"—the idea that a few key principles governing matter, energy, light, and motion explain both the phenomena of our daily lives and the mysteries of the cosmos. We approach this material by following the principle of building "bridges to the familiar." Each chapter begins with a section on science in everyday life, in which we remind students how much they already know about scientific phenomena from their everyday experiences. We then build on this everyday knowledge to help students learn the formal principles of physics needed for the rest of their study of astronomy. Chapter 4 covers basic ideas of matter and energy—the types of energy (kinetic, potential, radiative), conservation of energy, the atomic structure of matter, phase changes, and energy levels in atoms. Chapter 5 covers Newton's laws and gravity, including discussions of the "why" of Kepler's laws, the origin of tides and tidal forces, and the many astronomical concepts that can be understood by considering orbital energy. Chapter 6 covers the nature of light and spectra, including the electromagnetic spectrum, the particle and wave nature of light, the formation of spectral lines, the laws of thermal radiation, and the Doppler effect. Chapter 7 covers telescopes and astronomical observing techniques.

New for the third edition Again, we have edited to improve the text flow, improved art pieces, and added new illustrations to help students understand key concepts. In Chapter 5, we have enhanced our discussion of the "why" of Kepler's laws (and, as noted previously, moved discussion of the Copernican revolution from this chapter to Chapter 3). Chapter 7 has been updated to reflect the ongoing revolution in astronomical observatories, and the section on spacecraft that formerly appeared in Chapter 7 has been moved to the new Chapter 8.

Key new figures include: 5.6, 6.3, 6.18

Part III Learning From Other Worlds (Chapters 8–14)

Guiding Philosophy: True comparative planetology

This set of chapters begins with a broad overview of the solar system in Chapter 8, including a 10-page tour that highlights some of the most important and interesting features of the Sun and each of the nine planets in turn. In Chapters 9–14, we explain these features through a true comparative planetology approach, in which the discussion emphasizes the processes that shape the planets rather than a "stamp collecting" of facts about them. Using the concrete features of the solar system presented in Chapter 8, Chapter 9 builds student understanding of the current theory of solar system formation and explores how the theory has been affected by discoveries of planets around other stars. Chapters 10 and 11 focus on the terrestrial planets, covering key ideas of geology and atmospheres, respectively. In both chapters, we use examples drawn from our own planet Earth to help students understand the types of features that are found throughout the terrestrial worlds and the fundamental processes that explain how these features came to be. We conclude each of these two chapters by summarizing how the various processes have played out on each individual world. Chapter 12 covers the jovian planets and their moons and rings. Chapter 13 covers small bodies in the solar system—asteroids, comets, and Pluto. It also covers cosmic collisions, including the impact linked to the extinction of the dinosaurs and a discussion of how seriously we should take the ongoing impact threat. Finally, Chapter 14 turns our attention back to Earth. Having already studied and understood all the ways in which Earth is similar to other worlds, this chapter covers how and why Earth is different, with emphasis on the role of life. *Note that Part III is essentially independent of Parts IV through VI, and thus can be covered either before or after them.*

New for the third edition We have added the new Chapter 8 to provide an overview of the solar system and a summary of each of its major worlds, one by one. To make our comparative planetology approach easier to teach and more accessible to students, we have completely rewritten the chapters that follow. We have also added numerous new, easy-to-understand summary diagrams that highlight all of the major concepts, and have added scale bars to most planetary images to give students a better sense of scale. In Chapters 10 and 11, we have placed increased emphasis on familiar features of Earth as a way to introduce key geological ideas. We have also enhanced and expanded our tours of the individual terrestrial worlds to make sure that the planets can be understood individually as well as comparatively. These changes have allowed us to focus Chapter 14 more clearly on the unique features of Earth, such as plate tectonics, atmospheric oxygen, climate stability, and the role of life. Discussion of life elsewhere, which formerly appeared in this chapter, now appears in the new Chapter 24.

Key new figures include: 8.1, 8.5, 8.6, 9.2, 9.13, 10.16, 10.22, 10.24, 11.20, 11.29, 12.35, 13.26, 14.16

Part IV—A Deeper Look at Nature (Chapters S2–S4)

Guiding Philosophy: Ideas of relativity and quantum mechanics are accessible to anyone.

Nearly all students have heard of things like the prohibition on faster-than-light travel, curvature of spacetime, and the uncertainty principle. But few, if any, students enter an introductory astronomy course with any idea of what these things mean, and they are naturally curious about them. Moreover, a basic understanding of ideas of relativity and quantum mechanics makes it possible to gain a much deeper appreciation of many of the most important and interesting topics in modern astronomy, including black holes, gravitational lensing, and the overall geometry of the universe. Thus, the three chapters of Part IV cover special relativity (Chapter S2), general relativity (Chapter S3), and key astronomical ideas of quantum mechanics (Chapter S4). The main thrust throughout is to demystify relativity and quantum mechanics by convincing students that they are capable of understanding the key ideas despite the reputation of these subjects for being hard or counterintuitive. ***These chapters are labeled "supplementary" because coverage of them is optional.*** Including them in your course will give students a deeper understanding of the topics that follow on stars, galaxies, and cosmology, but the later chapters are self-contained and may be covered without having covered Part IV at all.

Part V Stellar Alchemy (Chapters 15–18)

Guiding Philosophy: We are intimately connected to the stars.

These are our chapters on stars and stellar lifecycles. Chapter 15 covers the Sun in depth, so that it can serve as our concrete model for building an understanding of other stars. Chapter 16 describes the general properties of other stars, how we measure these properties, and how we classify stars with the HR diagram. Chapter 17 covers stellar evolution, tracing the birth-to-death lives of both low- and high-mass stars. Chapter 18 covers the end points of stellar evolution: white dwarfs, neutron stars, and black holes. Today, we know so much about stars that the primary challenge in teaching them is deciding what details can be skipped without losing the main points. We therefore have chosen to focus on those aspects of stars that support our themes—especially those aspects that reveal our intimate connection to the stars (for example, how stars have produced the elements from which we are made):

New for the third edition In addition to our overall editing and art program improvement, significant changes include an updated discussion of the solar neutrino problem and enhanced discussion of stellar birth.

Key new figures include: 15.8, 15.10, 16.1, 17.2, 17.4, 17.6, 17.15

Part VI Galaxies and Beyond (Chapters 19–24)

Guiding Philosophy: Present galaxy evolution in a way that parallels the teaching of stellar evolution, and integrate cosmological ideas in the places where they most naturally arise

These chapters cover galaxies and cosmology. These topics traditionally have been more difficult to teach than stars, largely because we knew so much less about them. Fortunately, the state of knowledge has improved dramatically in the past couple decades, making it possible for us to teach these topics in a much more coherent and tightly integrated fashion. We therefore cover topics with an organization that closely parallels the organization used for stars.

For example: Chapter 19 presents the Milky Way as a paradigm for galaxies in much the same way that Chapter 15 uses the Sun as a paradigm for stars; Chapter 20 presents the variety of galaxies and how we determine key parameters such as galactic distances, much as Chapter 16 presents the variety of stars and how we determine key stellar parameters; and Chapter 21 discusses the current state of knowledge regarding galaxy evolution, just as Chapter 17 covers stellar evolution. Throughout Part VI, we integrate cosmological ideas as they arise. For example, students first encounter dark matter when we discuss the rotation curve of the Milky Way in Chapter 19, and we go into depth on Hubble's law in Chapter 20 because of its importance to the cosmic distance scale and to our understanding of what we see when we look at distant galaxies. This approach also lays the groundwork for our discussion of dark matter and the fate of the universe in Chapter 22 and of the Big Bang in Chapter 23. *Note that the final Chapter 24, which covers life in the universe, is essentially independent of the rest of Parts IV through VI, so it can be covered either here or after Part III.*

New for the third edition In addition to our overall editing and art program improvement, we have updated or revised discussion of spiral arms and the center of the galaxy in Chapter 19. We have expanded our discussion of galaxy types and the Hubble tuning fork diagram in Chapter 20, updated discussion of possible acceleration of the expansion and the so-called dark energy in Chapter 22, and updated discussion of the cosmic microwave background to include implications of recent data from the WMAP satellite in Chapter 23. Chapter 24 is almost entirely new, offering a general discussion of issues related to the possibility of life beyond Earth.

Key new figures include: 19.22, 19.23, 19.24, 20.10, 21.6, 22.17, 22.18, 23.16

Pedagogical Features of *The Cosmic Perspective*

Alongside the main narrative, *The Cosmic Perspective* includes a number of pedagogical devices designed to enhance student learning. Here is a brief summary, beginning with features new to the third edition.

NEW ■ **Learning Goals** Presented as key questions at the start of each chapter, these goals help students focus their attention on the most important concepts ahead.

NEW ■ **Chapter Summary** The end-of-chapter summary offers concise answers to the learning goal questions, helping reinforce student understanding of key concepts from the chapter.

NEW ■ **Key Concept Figures** Dozens of new figures have been added and many more have been improved so that nearly every important concept in the book is now accompanied by a figure that summarizes it visually. Thus, students can get an overview of all the key chapter concepts by studying the illustrations (with their captions), then go back to read the chapter in detail.

NEW ■ **Wavelength/Observatory Icons** For astronomical photographs (or art that might be confused with photographs), simple icons identify the wavelength band of the photo or identify the figure as an art piece or computer simulation. Along with the wavelength icon for photos, another icon indicates whether the image came from ground-based or space-based observations.

NEW ■ **Media Explorations** Each chapter ends with a section or page of "Media Explorations" that highlight some of the many media resources available to aid students in studying the chapter material. These sections include suggested "Web projects" designed for independent research.

■ **The Big Picture** Every chapter narrative ends with this feature. It helps students put what they've learned in the chapter into the context of the overall goal of gaining a new perspective on ourselves and our planet.

■ **End-of-Chapter Questions** Each chapter includes an extensive set of exercises that can be used for study, discussion, or assignment.

■ **Think About It** This feature, which appears throughout the book as short questions integrated into the narrative, gives students the opportunity to reflect on important new concepts. It also serves as an excellent starting point for classroom discussions.

■ **Common Misconceptions** These boxes address and correct popularly held but incorrect ideas related to the chapter material.

■ **Mathematical Insights** These boxes contain most of the mathematics used in the book and can be covered or skipped, depending on the level of mathematics that you wish to include in your course.

■ **Special Topic Boxes** These boxes contain supplementary discussion topics related to the chapter material but not prerequisite to the continuing discussion.

■ **Cross-References** When a concept is covered in greater detail elsewhere in the book, we include a cross-reference, in brackets, to the relevant section (e.g., [Section 5.2]).

■ **Glossary** A detailed glossary makes it easy for students to look up important terms.

■ **Appendixes** The appendixes include a number of useful references and tables, including key constants (Appendix A), key formulas (Appendix B), key mathematical skills (Appendix C), and numerous data tables and star charts.

Resources and Supplements for *The Cosmic Perspective*

The Cosmic Perspective is much more than just a textbook. It is a complete package of resources designed to help both teachers and students. Here is a brief summary of the available resources and supplements.

■ **FREE with All New Books Astronomy Place (www.astronomyplace.com)** The Astronomy Place Web site offers a wealth of study resources for students, and many resources and course management tools for teachers. With more than 70,000 users per month, this is the most popular astronomy textbook Web site available to students. A subscription to the site is included free with every new book. Look for your personal access kit with your new book. If you did not receive an access kit with your book, you may purchase access online at www.astronomyplace.com. Among the many resources at the Astronomy Place, you'll find:

● **Interactive, educational tutorials** We now have a total of 18 full-length online tutorials, which together include nearly 60 individual tutorial lessons and hundreds of interactive tools and animations, each focused on a key concept. The text includes icons in section headers to point to relevant online tutorial lessons, plus suggested tutorial activities in the Media Explorations sections at the end of each chapter.

● **Online, multiple-choice chapter quizzes** New for the third edition, the Astronomy Place now has two quizzes for each chapter in the book. The first quiz focuses on basic definitions and ideas, while the second asks more conceptual questions.

● *Skygazer* **activity worksheets** These worksheets are designed to be used with *Voyager: SkyGazer, College Edition*—the planetarium software packaged with the textbook.

● **And much more** Animated movies, flash cards, study resources for individual chapters, and many other useful study aids can be found at the Astronomy Place Web site.

■ **FREE with All New Books** *Voyager: Skygazer, College Edition* Based on *Voyager III,* one of the world's most popular planetarium programs, *SkyGazer* makes it easy for students to learn constellations and explore the wonders of the sky through interactive exercises. The *Skygazer* CD is packaged free with all new copies of this book. It comes preloaded with

75 demos, and suggested activities appear in the Media Explorations section at the end of each chapter in the book.

■ **FREE with New Books The Addison Wesley Astronomy Tutor Center** This center provides one-on-one tutoring by qualified college instructors in any of four ways—phone, fax, email, and the Internet—during evening and weekend hours. Tutor center instructors will answer questions and provide help with examples and exercises from the text. Tutor center registration is free with new books only when the professor orders books with the special tutor center package. (Professors: Contact your local Addison Wesley sales representative if you wish to order this package.) Otherwise, it can be purchased separately. See www.aw.com/tutorcenter for more information.

■ **Astronomy Media Workbook** (ISBN 0-8053-8755-2) This supplementary workbook offers an extensive set of printed activities and more in-depth projects—suitable for labs or homework assignments—that use the Astronomy Place Web site tutorials and *Skygazer* software.

Several additional supplements are available for instructors only. Contact your local Addison Wesley sales representative to find out more about the following supplements:

■ **Cosmic Lecture Launcher CD** (ISBN 0-8053-8749-8) This CD provides a wealth of presentation tools to help prepare course lectures. It includes a set of Power-Point slides for every section in the textbook, a comprehensive collection of high-resolution figures from the book and other astronomical sources, and a library of more than 250 interactive applets and simulations.

■ **Instructor's Guide** (ISBN 0-8053-8748-X) This guide contains a detailed overview of the text, sample syllabi for courses of different emphasis and duration, suggestions on teaching strategies, answers or discussion points for all Think About It questions in the text, solutions to end-of-chapter problems, and a detailed reference guide summarizing media resources available for every chapter and section in the book.

■ **Carl Sagan's** *Cosmos* **(DVD or Video)** The *Best of Cosmos* and the complete, revised, enhanced, and updated *Cosmos* series are available free to qualified adopters of *The Cosmic Perspective.*

■ **Test Bank** Available in both computerized (ISBN 0-8053-8745-5) or printed (ISBN 0-8053-8746-3) form, the Test Bank contains a broad set of multiple-choice, true/false, and free-response questions for each chapter, including the questions from the online quizzes.

■ **Transparency Acetates** (ISBN 0-8053-8747-1) For those who use overhead projectors in lectures, this set contains more than 180 images from the text.

Acknowledgments

A textbook may carry author names, but it is the result of hard work by a long list of committed individuals. We could not possibly list everyone who has helped, but we would like to call attention to a few people who have played particularly important roles. First, we thank our editors and friends at Addison Wesley who have stuck with us through thick and thin, including Adam Black, Linda Davis, Stacie Kent, Christy Lawrence, Stacy Treco, Liana Allday, Nancy Benton, Vivian McDougal, Joan Marsh, Ben Roberts, Robin Heyden, Sami Iwata, and Bill Poole. Special thanks to our production team, especially Mary Douglas, Myrna Engler, Mary Roybal, Karen Stough, and Nancy Ball; our art and design team, Blakeley Kim, Mark Ong, Judy Waller, and John Waller; our supplements team, including Tom Fleming and Stacy Palen (work on the Cosmic Lecture Launcher CD), Jonathan Williams (work on the Test Bank), and Michael LoPresto (Media Workbook author); and our Web team, led by Claire Masson, Jim Dove, and Ian Shakeshaft.

We've also been fortunate to have an outstanding group of reviewers whose extensive comments and suggestions helped us shape the book. We thank all those who have reviewed drafts of the book in various stages, including:

Christopher M. Anderson, University of Wisconsin
Peter S. Anderson, Oakland Community College
John Beaver, University of Wisconsin at Fox Valley
Timothy C. Beers, Michigan State University
David Brain, University of Colorado, Boulder
Priscilla J. Benson, Wellesley College
David Branch, University of Oklahoma
Jean P. Brodie, UCO/Lick Observatory, University of California, Santa Cruz
Eric Carlson, Wake Forest University
Supriya Chakrabarti, Boston University
Dipak Chowdhury, Indiana University–Purdue University at Fort Wayne
Josh Colwell, University of Colorado
Christopher Crow, Indiana University Purdue University, Fort Wayne
John M. Dickey, University of Minnesota
Robert Egler, North Carolina State University at Raleigh
Robert A. Fesen, Dartmouth College
Sidney Freudenstein, Metropolitan State College of Denver
Martin Gaskell, University of Nebraska
Richard Gelderman, Western Kentucky University
Richard Gray, Appalachian State University
Kevin Grazier, Jet Propulsion Laboratory
David Griffiths, Oregon State University
David Grinspoon, University of Colorado
Bruce Gronich, University of Texas, El Paso
Jim Hamm, Big Bend Community College
Charles Hartley, Hartwick College
Joe Heafner, Catawba Valley Community College
Richard Holland, Southern Illinois University, Carbondale
Richard Ignace, University of Wisconsin

Bruce Jakosky, University of Colorado
Adam Johnston, Weber State University
Steve Kipp, University of Minnesota, Mankato
Kurtis Koll, Cameron University
John Kormendy, University of Texas, Austin
Kristine Larsen, Central Connecticut State University
Ana Marie Larson, University of Washington
Larry Lebofsky, University of Arizona
Nancy Levenson, University of Kentucky
Patrick Lestrade, Mississippi State University
David M. Lind, Florida State University
Michael LoPresto, Henry Ford Community College
William R. Luebke, Modesto Junior College
Marie Machacek, Massachusetts Institute of Technology
Marles McCurdy, Tarrant County College
Stacy McGaugh, University of Maryland
Steven Majewski, University of Virginia
Phil Matheson, Salt Lake Community College
Barry Metz, Delaware County Community College
Dinah Moche, Queensborough Community College of City University, New York
Zdzislaw E. Musielak, University of Texas, Arlington
Gerald H. Newsom, Ohio State University
Brian Oetiker, Sam Houston State University
John P. Oliver, University of Florida
Russell L. Palma, Sam Houston State University
Jorge Piekarewicz, Florida State University
Harrison B. Prosper, Florida State University
Monica Ramirez, Aims College, Colorado
Christina Reeves-Shull, Richland College
Elizabeth Roettger, DePaul University
Roy Rubins, University of Texas, Arlington
Rex Saffer , Villanova University
John Safko, University of South Carolina
James A. Scarborough, Delta State University
Joslyn Schoemer, Denver Museum of Nature and Science
James Schombert, University of Oregon
Gregory Seab, University of New Orleans
Paul Sipiera, William Harper Rainey College
Michael Skrutskie, University of Virginia
Mark H. Slovak, Louisiana State University
Dale Smith, Bowling Green State University
John Spencer, Lowell Observatory
Darryl Stanford, City College of San Francisco
John Stolar, West Chester University
Jack Sulentic, University of Alabama
C. Sean Sutton, Mount Holyoke College
Beverley A. P. Taylor, Miami University
Donald M. Terndrup, Ohio State University
David Trott, Metro State College
Darryl Walke, Rariton Valley Community College
Fred Walter, State University of New York, Stony Brook
James Webb, Florida International University
Mark Whittle, University of Virginia
Paul J. Wiita, Georgia State University
Jonathan Williams, University of Florida

J. Wayne Wooten, Pensacola Junior College
Arthur Young, San Diego State University
Min S. Yun, University of Massachusetts, Amherst
Dennis Zaritsky, University of California, Santa Cruz
Robert L. Zimmerman, University of Oregon

Historical Accuracy Reviewer—Owen Gingerich,
     Harvard–Smithsonian

In addition, we thank the following colleagues who helped us clarify technical points or checked the accuracy of technical discussions in the book:

Thomas Ayres, University of Colorado
Cecilia Barnbaum, Valdosta State University
Rick Binzel, Massachusetts Institute of Technology
Howard Bond, Space Telescope Science Institute
Humberto Campins, University of Florida
Robin Canup, Southwest Research Institute
Josh Colwell, University of Colorado
Mark Dickinson, Space Telescope Science Institute
Jim Dove, Metropolitan State College of Denver
Harry Ferguson, Space Telescope Science Institute
Andrew Hamilton, University of Colorado
Todd Henry, Georgia State University
Dave Jewitt, University of Hawaii
Hal Levison, Southwest Research Institute

Mario Livio, Space Telescope Science Institute
Mark Marley, New Mexico State University
Kevin McLin, University of Colorado, Boulder
Rachel Osten, University of Colorado, Boulder
Bob Pappalardo, Brown University
Michael Shara, American Museum of Natural History
Glen Stewart, University of Colorado
John Stolar, West Chester University
Dave Tholen, University of Hawaii
Nick Thomas, MPI/Lindau (Germany)
Dimitri Veras, University of Colorado
John Weiss, University of Colorado, Boulder
Don Yeomans, Jet Propulsion Laboratory

Finally, we thank the many people who have greatly influenced our outlook on education and our perspective on the universe over the years, including Tom Ayres, Fran Bagenal, Forrest Boley, Robert A. Brown, George Dulk, Erica Ellingson, Katy Garmany, Jeff Goldstein, David Grinspoon, Don Hunten, Bruce Jakosky, Catherine McCord, Dick McCray, Dee Mook, Cheri Morrow, Charlie Pellerin, Carl Sagan, Mike Shull, John Spencer, and John Stocke.

Jeff Bennett
Megan Donahue
Nick Schneider
Mark Voit

About the Authors

JEFFREY BENNETT

MEGAN DONAHUE

Jeffrey Bennett received a B.A. in biophysics from the University of California, San Diego (1981) and a Ph.D. in astrophysics from the University of Colorado, Boulder (1987). He currently spends most of his time as a teacher, speaker, and writer. He has taught extensively at all levels, including having founded and run a science summer school for elementary and middle school children. At the college level, he has taught more than fifty classes in subjects ranging from astronomy, physics, and mathematics, to education. He served two years as a visiting senior scientist at NASA headquarters, where he helped create numerous programs for science education. He also proposed the idea for and helped develop the *Voyage* Scale Model Solar System, which opened in 2001 on the National Mall in Washington, D.C. (He is pictured here with the model Sun.) In addition to this astronomy textbook, he has written college-level textbooks in astrobiology, mathematics, and statistics, and a book for the general public, *On the Cosmic Horizon* (Addison Wesley, 2001). He also recently completed his first children's book, *Max Goes to the Moon* (Big Kid Science, 2003). When not working, he enjoys participating in masters swimming and in the daily adventures of life with his wife Lisa, his children Grant and Brooke, and his dog, Max. You can read more about his projects on his personal Web site, www.jeffreybennett.com.

Megan Donahue is an associate professor in the Department of Physics and Astronomy of Michigan State University. Her current research is mainly on clusters of galaxies: their contents—dark matter, hot gas, galaxies, active galactic nuclei—and what they reveal about the contents of the universe and how galaxies form and evolve. She grew up on a farm in Nebraska and received a bachelor's degree in physics from MIT, where she began her research career as an X-ray astronomer. She has a Ph.D. in astrophysics from the University of Colorado, for a thesis on theory and optical observations of intergalactic and intracluster gas. That thesis won the 1993 Trumpler Award from the Astronomical Society for the Pacific for an outstanding astrophysics doctoral dissertation in North America. She continued post-doctoral research in optical and X-ray observations as a Carnegie Fellow at Carnegie Observatories in Pasadena, California, and later as an STScl Institute Fellow at Space Telescope. Megan was a staff astronomer at the Space Telescope Science Institute until 2003, when she joined the MSU faculty. Megan is married to Mark Voit, who is also a frequent collaborator of hers on many projects, including this textbook and the raising three children, Michaela, Sebastian, and Angela. Between the births of Sebastian and Angela, Megan qualified for and ran the 2000 Boston Marathon. She hopes to run another one soon.

Nicholas Schneider is an associate professor in the Department of Astrophysical and Planetary Sciences at the University of Colorado and a researcher in the Laboratory for Atmospheric and Space Physics. He received his B.A. in physics and astronomy from Dartmouth College in 1979 and his Ph.D. in planetary science from the University of Arizona in 1988. In 1991, he received the National Science Foundation's Presidential Young Investigator Award. His research interests include planetary atmospheres and planetary astronomy, with a focus on the odd case of Jupiter's moon Io. He enjoys teaching at all levels and is active in efforts to improve undergraduate astronomy education. Off the job, he enjoys exploring the outdoors with his family and figuring out how things work.

Mark Voit is an associate professor in the Department of Physics and Astronomy at Michigan State University. He earned his A.B. in astrophysical sciences at Princeton University and his Ph.D. in astrophysics at the University of Colorado in 1990. He continued his studies at the California Institute of Technology, where he was a research fellow in theoretical astrophysics, then moved on to Johns Hopkins University as a Hubble Fellow. Before coming to Michigan State, Mark worked in the Office of Public Outreach at the Space Telescope, where he developed museum exhibitions about the Hubble Space Telescope and was the scientist behind NASA's HubbleSite. His research interests range from interstellar processes in our own galaxy to the clustering of galaxies in the early universe. He is married to co-author Megan Donahue, and they try to play outdoors with their three children whenever possible, enjoying hiking, camping, running, and orienteering. Mark is also author of the popular book *Hubble Space Telescope: New Views of the Universe.*

How to Succeed in Your Astronomy Course

Using This Book

Each chapter in the book is designed to make it easy for you to study effectively and efficiently. To get the most out of each chapter, you might wish to use the following study plan:

- Begin by reading the Learning Goals to make sure you know what you will be learning about in each chapter.

- Before reading in depth, start by skimming the chapter, focusing only on the illustrations. Study each illustration and read the captions so that you will get an overview of the key chapter concepts.

- Next, read the chapter narrative. Try to answer the Think About It questions as you go along, but you may save the other boxed features (Common Misconceptions, Special Topics, Mathematical Insights) to read later.

- After reading the chapter once, go back through and read the boxed material. Also look for the tutorial icons that tell you when there is a relevant Web-based tutorial on the Astronomy Place (www.astronomyplace.com). If you are having difficulty with a concept, be sure you try the tutorial.

- Study the chapter's Summary of Key Concepts by first trying to answer the Learning Goals questions for yourself, then checking your understanding against the answers given in the summary.

- Check your understanding by trying the online quizzes at www.astronomyplace.com. Do the basic quiz first. Once you clear up any difficulties you have with the basic quiz, try the conceptual quiz.

The Key to Success: Study Time

The single most important key to success in any college course is to spend enough time studying. A general rule of thumb for college classes is that you should expect to study about 2 to 3 hours per week *outside* of class for each unit of credit. For example, based on this rule of thumb, a student taking 15 credit hours should expect to spend 30 to 45 hours each week studying outside of class. Combined with time in class, this works out to a total of 45 to 60 hours spent on academic work—not much more than the time a typical job requires, and you get to choose your own hours.

Of course, if you are working while you attend school, you will need to budget your time carefully.

As a rough guideline, your studying time in astronomy might be divided as shown in the table at the top of p. xxvii. If you find that you are spending fewer hours than these guidelines suggest, you can probably improve your grade by studying more. If you are spending more hours than these guidelines suggest, you may be studying inefficiently; in that case, you should talk to your instructor about how to study more effectively.

General Strategies for Studying

- Don't miss class. Listening to lectures and participating in discussions is much more effective than reading someone else's notes. Active participation will help you retain what you are learning.

- As you read, make notes to remind yourself of ideas you'll want to review in more detail later. The best way to do this is to make notes in the margins of the book. If you want to mark text for later review, don't highlight—underline! Using a pen or pencil to underline material requires greater care than highlighting and therefore helps keep you alert as you study. Be careful to underline selectively—it won't help you later if you've underlined everything.

- Budget your time effectively. One or 2 hours each day is more effective, and far less painful, than studying all night before homework is due or before exams.

- If a concept gives you trouble, do additional reading or studying beyond what has been assigned. And if you still have trouble, ask for help: You surely can find friends, colleagues, or teachers who will be glad to help you learn.

- Working together with friends can be valuable in helping you understand difficult concepts. However, be sure that you learn *with* your friends and do not become dependent on them.

- Be sure that any work you turn in is of *collegiate quality*: neat and easy to read, well organized, and demonstrating mastery of the subject matter. Although it takes extra effort to make your work look this good, the effort will help you solidify your learning and is

If Your Course Is:	Time for Reading the Assigned Text (per week)	Time for Homework Assignments (per week)	Time for Review and Test Preparation (average per week)	Total Study Time (per week)
3 credits	2 to 4 hours	2 to 3 hours	2 hours	6 to 9 hours
4 credits	3 to 5 hours	2 to 4 hours	3 hours	8 to 12 hours
5 credits	3 to 5 hours	3 to 6 hours	4 hours	10 to 15 hours

also good practice for the expectations that future professors and employers will have.

Preparing for Exams

- Study the review questions, and rework problems and other assignments; try additional questions to be sure you understand the concepts. Study your performance on assignments, quizzes, or exams from earlier in the term.

- Study the relevant online tutorials and chapter quizzes available at www.astronomyplace.com.

- Study your notes from lectures and discussions. Pay attention to what your instructor expects you to know for an exam.

- Reread the relevant sections in the textbook, paying special attention to notes you have made on the pages.

- Study individually *before* joining a study group with friends. Study groups are effective only if every individual comes prepared to contribute.

- Don't stay up too late before an exam. Don't eat a big meal within an hour of the exam (thinking is more difficult when blood is being diverted to the digestive system).

- Try to relax before and during the exam. If you have studied effectively, you are capable of doing well. Staying relaxed will help you think clearly.

PART I

DEVELOPING PERSPECTIVE

1 Our Place in the Universe

F ar from city lights on a clear night, you can gaze upward at a sky filled with stars. If you lie back and watch for a few hours, you will observe the stars marching steadily across the sky. Confronted by the seemingly infinite heavens, you might wonder how Earth and the universe came to be. With these thoughts, you will be sharing an experience common to humans around the world and in thousands of generations past.

Modern science offers answers to many of our fundamental questions about the universe and our place within it. We now know the basic content and scale of the universe. We know the age of Earth and the approximate age of the universe. And, although much remains to be discovered, we are rapidly learning how the simple constituents of the early universe developed into the incredible diversity of life on Earth.

In this first chapter, we will survey the content and history of the universe, the scale of the universe, and the motions of Earth in our universe. We'll develop a "big picture" perspective on our place in the universe that will provide a base on which we can build a deeper understanding in the rest of the book.

1.1 A Modern View of the Universe

If you observe the sky carefully, you can see why most of our ancestors believed that the heavens revolved about Earth. The Sun, Moon, planets, and stars appear to circle around our sky each day, and we cannot feel the constant motion of Earth as it rotates on its axis and orbits the Sun. Thus, it seems quite natural to assume that we live in an Earth-centered, or *geocentric*, universe.

Nevertheless, we now know that Earth is a planet orbiting a rather average star in a vast cosmos. (In astronomy, the term *cosmos* is synonymous with *universe*.) The historical path to this knowledge was long and complex, involving the dedicated intellectual efforts of thousands of individuals. In later chapters, we'll encounter many of these individuals and explore how their discoveries changed human understanding of the universe. We'll see that many ancient beliefs made a lot of sense and changed only when people were confronted by strong evidence to the contrary. We'll also see how the process of science has enabled us to acquire this evidence and thereby discover that we are connected to the stars in ways our ancestors never imagined.

First, however, it's useful to have at least a general picture of the universe as we know it today. This big picture will make it easier for you to understand the historical development of astronomy, the evidence for our modern ideas, and the mysteries that remain. Let's begin by examining what modern astronomy has to say about our cosmic location and origins.

Our Cosmic Address

Take a look at Figure 1.1. Going counterclockwise from Earth, this painting illustrates the basic levels of structure that describe what we might call our "cosmic address."

Earth is a planet in our **solar system**, which consists of the Sun and all the objects that orbit it: nine planets and their moons, the chunks of rock we call asteroids, the balls of ice we call comets, and countless tiny particles of interplanetary dust.

Our Sun is a star, just like the stars we see in our night sky. The Sun and all the stars we can see with the naked eye make up only a small part of a huge, disk-shaped collection of stars called the **Milky Way Galaxy**. A galaxy is a great island of stars in space, containing from a few hundred million to a trillion or more stars. The Milky Way Galaxy is relatively large, containing more than 100 billion stars. Our solar system is located a little over halfway from the galactic center to the edge of the galactic disk.

Some galaxies are fairly isolated, but many others congregate in groups. Our Milky Way, for example, is one of the two largest galaxies among about 40 galaxies in the **Local Group**. Groups of galaxies with more than a few dozen members are often called **galaxy clusters**.

This box summarizes a few key astronomical definitions introduced in this chapter and used throughout the book.

Basic Astronomical Objects

star Our Sun and other ordinary stars are large, glowing balls of gas that generate heat and light through nuclear fusion in their cores. (The term *star* is also applied to objects that are in the process of becoming true stars, such as protostars, and to the remains of stars that have died, such as neutron stars.)

planet A moderately large object that orbits a star. Planets may be rocky, icy, or gaseous in composition, and they shine primarily by reflecting light from their star. Astronomers sometimes disagree about what counts as a planet, because there are no official minimum or maximum sizes. For example, some astronomers argue that Pluto is too small to count as a planet. On the large side, astronomers disagree about whether an object a couple dozen times the size of Jupiter should be called a very large planet or a "failed star" (such as a *brown dwarf* [Section 17.2]).

moon (or **satellite**) An object that orbits a planet. The term *satellite* is also used more generally to refer to any object orbiting another object.

asteroid A relatively small and rocky object that orbits a star. Asteroids are sometimes called *minor planets* because they orbit much like planets but are smaller than anything we consider to be a true planet.

comet A relatively small and icy object that orbits a star.

Collections of Astronomical Objects

solar system Our solar system consists of the Sun and all the material that orbits it, including the planets. The term *solar system* technically refers only to our own star system (because *solar* means "of the Sun"), but it is sometimes applied to other star systems.

star system A star (sometimes more than one star) and any planets and other materials that orbit it. (Roughly half of all star systems contain two or more stars.)

galaxy A great island of stars in space, containing from a few hundred million to a trillion or more stars, all held together by gravity and orbiting a common center.

cluster (or *group*) **of galaxies** A collection of galaxies bound together by gravity. Small collections (up to a few dozen galaxies) are generally called *groups*, with the term *cluster* reserved for larger collections of galaxies.

supercluster A gigantic region of space where many individual galaxies and many groups and clusters of galaxies are packed closer together than elsewhere in the universe.

universe (or **cosmos**) The sum total of all matter and energy, that is, everything within and between all galaxies.

observable universe The portion of the entire universe that, at least in principle, can be seen from Earth. The observable universe is probably only a tiny portion of the entire universe.

Astronomical Distance Units

astronomical unit (AU) The average distance between Earth and the Sun, which is about 150 million kilometers. (More technically, 1 AU is the length of the semimajor axis of Earth's orbit.)

light-year The distance that light can travel in 1 year, which is about 9.46 trillion kilometers.

Terms Relating to Motion

rotation The spinning of an object around its axis. For example, Earth rotates once each day around its axis, which is an imaginary line connecting the North Pole to the South Pole (and passing through the center of Earth).

revolution (**orbit**) The orbital motion of one object around another. For example, Earth revolves (orbits) around the Sun once each year.

expansion (of the universe) We say that the universe is expanding because the average distance between galaxies is increasing with time. Note that while the universe as a whole is expanding, individual galaxies and their contents (as well as groups and clusters of galaxies) are *not* expanding.

On a very large scale, the universe appears frothlike, with galaxies and galaxy clusters loosely arranged in giant chains and sheets. The galaxies and galaxy clusters are more tightly packed in some places than in others, forming giant structures called **superclusters**. The supercluster to which our Local Group belongs is called, not surprisingly, the **Local Supercluster**. Between the vast groupings of galaxies lie huge voids containing few, if any, galaxies.

Finally, the **universe** is the sum total of all matter and energy, encompassing the superclusters and voids and everything within them. To review the different levels of structure in the universe, you might imagine how a faraway friend would address a postcard to Earth (Figure 1.2).

THINK ABOUT IT

Some people think that our tiny physical size in the vast universe makes us insignificant. Others think that our ability to learn about the wonders of the universe gives us significance despite our small size. What do *you* think?

Figure 1.1 This painting illustrates our cosmic address. Earth is one of nine planets orbiting the Sun in our solar system. Our solar system is one of more than 100 billion star systems in the Milky Way Galaxy. Our galaxy is one of the two largest of about 40 galaxies in the Local Group. The Local Group lies near the outskirts of the Local Supercluster. The Local Supercluster is one piece of the complex, large-scale structure traced by galaxies throughout the universe.

the Milky Way Galaxy

the Solar System
(not to scale)

the Local Group

the Local Supercluster

Earth

the Universe

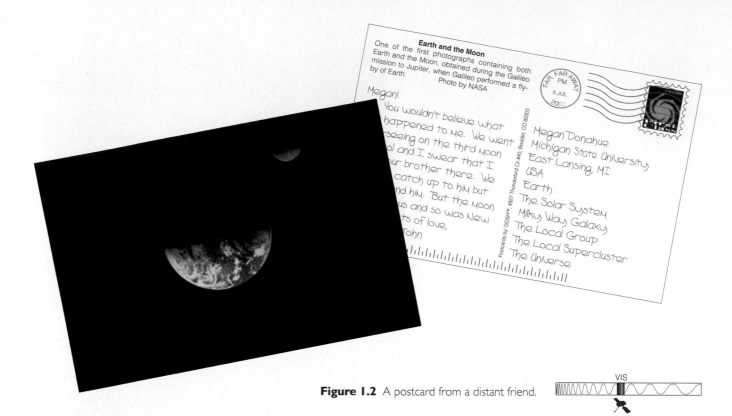

Figure 1.2 A postcard from a distant friend.

Our Cosmic Origins

How did we come to be? Much of the rest of this text discusses the scientific evidence concerning our cosmic origins, and we'll see that humans are newcomers in an old universe. For now, let's look at a quick overview of the scientific story of creation, as summarized in Figure 1.3.

As we'll discuss shortly, telescopic observations of distant galaxies show that the entire universe is *expanding*. That is, average distances between galaxies are increasing with time. If the universe is expanding, everything must have been closer together in the past. From the observed rate of expansion, astronomers estimate that the expansion started about 14 billion years ago. Astronomers call this beginning the **Big Bang**.

Expansion Versus Gravity The universe as a whole has continued to expand ever since the Big Bang, but on smaller size scales the force of gravity has drawn matter together. Structures such as galaxies and clusters of galaxies occupy regions where gravity has won out against the overall expansion That is, while the universe as a whole continues to expand, individual galaxies and their contents do *not* expand. Most galaxies, including our own Milky Way, probably formed within a few billion years after the Big Bang.

Within galaxies, gravity drives the collapse of clouds of gas and dust to form stars and planets. Stars are not living organisms, but they nonetheless go through "life cycles." After their birth in giant clouds of gas and dust, stars shine for millions or billions of years. The energy that makes stars shine comes from **nuclear fusion**, the process in which lightweight atomic nuclei smash together and stick

(or fuse) to make heavier nuclei. Nuclear fusion occurs deep in a star's core throughout its life. A star "dies" when it finally exhausts all its usable fuel for fusion.

In its final death throes, a star blows much of its content back out into space. In particular, massive (but short-lived) stars die in titanic explosions called *supernovae*. The returned matter mixes with other matter floating between the stars in the galaxy, eventually becoming part of new clouds of gas and dust from which new generations of stars can be born. Thus, galaxies function as cosmic recycling plants, recycling material expelled from dying stars into new generations of stars and planets. Our own solar system is a product of many generations of such recycling.

Star Stuff The recycling of stellar material has another, even more important, connection to our own existence. By studying stars of different ages, we have learned that the early universe contained only the simplest chemical elements: hydrogen and helium (and a trace amount of lithium). We and Earth are made primarily of "other" elements, such as carbon, nitrogen, oxygen, and iron. Where did these other elements come from? Astronomers have discovered that all these elements were manufactured by massive stars, either through the nuclear fusion that makes them shine or through nuclear reactions accompanying the explosions that end their lives.

The processes of heavy-element production and cosmic recycling had already been taking place for several billion years by the time our solar system formed, about 4.6 billion years ago. The cloud that gave birth to our solar

Within a few billion years after the Big Bang, gravity caused local concentrations of matter to collapse into galaxies even while the universe as a whole continued to expand.

A star forms at the center of a collapsing cloud of gas and dust, and planets may form in the spinning disk that surrounds the young star.

Stars shine with the energy produced by nuclear fusion in their cores; the fusion also creates heavier elements from lighter ones.

The universe has been expanding ever since its hot and dense beginning in the Big Bang. Each of the three cubes represents the same region of the universe, showing how the region expands with time.

Galaxies like the Milky Way act as cosmic recycling plants: stars are made from the material in clouds of gas and dust within the galaxy, and stars return material to interstellar space when they die.

Massive stars explode when they die, scattering the elements they've produced into space.

Figure 1.3 Our cosmic origins: All the matter and energy in the universe was created in the Big Bang. This sequence of paintings shows the progression of that matter and energy from the Big Bang to human life. Note that the elements from which we are made were produced in stars that shined long ago. These elements formed Earth through the recycling role played by our galaxy.

The Earth was built with elements produced in stars that lived and died in the Milky Way before our solar system formed.

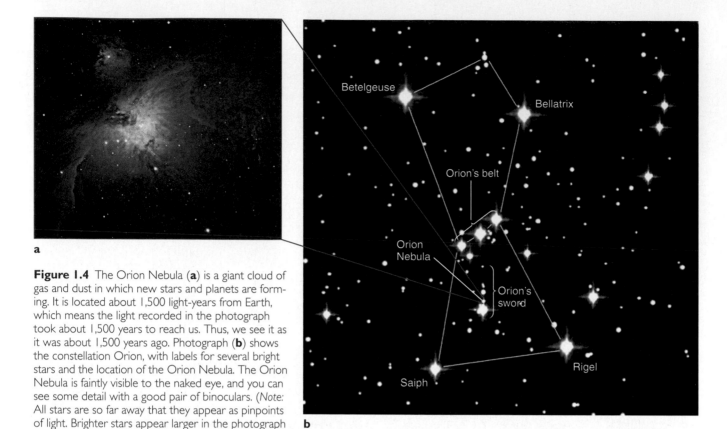

a

b

Figure 1.4 The Orion Nebula (**a**) is a giant cloud of gas and dust in which new stars and planets are forming. It is located about 1,500 light-years from Earth, which means the light recorded in the photograph took about 1,500 years to reach us. Thus, we see it as it was about 1,500 years ago. Photograph (**b**) shows the constellation Orion, with labels for several bright stars and the location of the Orion Nebula. The Orion Nebula is faintly visible to the naked eye, and you can see some detail with a good pair of binoculars. (*Note:* All stars are so far away that they appear as pinpoints of light. Brighter stars appear larger in the photograph only because they are overexposed. The crosses on the bright stars are an artifact of the telescope used to take the photograph.)

system was about 98% hydrogen and helium. The other 2% contained all the other chemical elements. The small rocky planets of our solar system, including Earth, were made from a small part of this 2%. We do not know exactly how the elements on the Earth's surface developed into the first forms of life, but it appears that microbial life was already flourishing on Earth more than 3.5 billion years ago. Biological evolution took over once life arose, leading to the great diversity of life on Earth today.

In summary, most of the material from which we and our planet are made was created inside stars that died before the birth of our Sun. We are intimately connected to the stars because we are products of stars. In the words of astronomer Carl Sagan (1934–1996), we are "star stuff."

Seeing into the Past

We study the universe by studying light from distant stars and galaxies. Light travels extremely fast by earthly standards: The speed of light is 300,000 kilometers per second. At this speed it would be possible to circle Earth nearly eight times in just 1 second. Nevertheless, even light takes a substantial amount of time to travel the vast distances in space.

For example, light takes about 1 second to reach Earth from the Moon and about 8 minutes to reach Earth from the Sun. Light from the stars takes many years to reach us, so we measure distances to the stars in units called **light-years**. One light-year is the distance that light can travel in 1 year—about 10 trillion kilometers, or 6 trillion miles. Note that a light-year is a unit of *distance,* not time.

The brightest star in the night sky, Sirius, is about 8 light-years from our solar system. This means it takes light from Sirius about 8 years to reach us. Thus, when we look at Sirius, we see light that left the star about 8 years ago.

The Orion Nebula, a star-forming region visible to the naked eye as a small, cloudy patch in the sword of the constellation Orion, lies about 1,500 light-years from Earth (Figure 1.4). Thus, we see the Orion Nebula as it looked about 1,500 years ago—about the time of the fall of the Roman Empire. If any major events have occurred in the Orion Nebula since that time, we cannot yet know about them because the light from these events would not yet have reached us.

Because light takes time to travel through space, we are led to a remarkable fact:

> **The farther away we look in distance, the further back we look in time.**

This fact allows us to see what parts of the universe looked like in the distant past. For example, if we look at a galaxy that is 1 billion light-years away, its light has taken 1 billion years to reach us—which means we are seeing it as it looked 1 billion years ago.*

---

*This assumes we have properly accounted for expansion during the billion years. More technically, in this and similar examples we are talking about *lookback time,* an idea we will discuss in Chapter 21.

We see this galaxy as it was 7 billion years ago, when the universe was only about half its current age.

We see this galaxy as it was 12 billion years ago—so if the universe is 14 billion years old today, we are seeing this galaxy as it looked when the universe was only 2 billion years old.

Light from this distance shows us how the universe looked very shortly after the Big Bang

If the universe is 14 billion years old and we try to look to a distance of, say, 15 billion light-years, we are trying to look to a time before the universe existed—which means we cannot see anything at this distance, even in principle.

7 billion light-years

12 billion light-years

14 billion light-years

Figure 1.5 Because light travels at a finite speed, looking farther away in space means looking further back in time. Thus, the age of the universe limits the extent of our observable universe. This figure assumes the universe is 14 billion years old.

Now assume the universe is 14 billion years old. In that case, if we look at a galaxy that is 7 billion light-years away, its light has taken 7 billion years to reach us—which means we are seeing it as it looked 7 billion years ago, when the universe was only half its current age. If we look at a galaxy that is 12 billion light-years away, we see it as it was 12 billion years ago, when the universe was only 2 billion years old. Thus, simply by looking to great distances, we can see what parts of the universe looked like when the universe was younger. The key limitation to this ability is the power of our telescopes. Modern telescopes are capable of seeing bright galaxies 12 billion or more light-years away, and astronomers eagerly await new telescopes that will allow us to see fainter objects at such great distances.

Because looking to great distances means looking into the past, the age of the universe imposes a fundamental limit on how far we can see (Figure 1.5). If the universe is 14 billion years old, we cannot possibly see anything more than 14 billion light-years away, because we'd be trying to look to a time before the universe existed. Thus, our **observable universe**—the portion of the entire universe that we can potentially observe—consists only of objects that lie within 14 billion light-years of Earth. This fact does not put any limit on the size of the *entire* universe, which may be far larger than our observable universe. We simply have

no hope of seeing or studying anything beyond the bounds of our observable universe.

It is amazing to realize that any "snapshot" of a distant galaxy or cluster of galaxies is a picture of both space and time. For example, the Great Galaxy in Andromeda, also known as M 31, lies about 2.5 million light-years from Earth. Figure 1.6 is therefore a picture of how M 31 looked about 2.5 million years ago, when early humans were first walking on Earth. Moreover, the diameter of M 31 is about 100,000 light-years, so light from the far side of the galaxy requires 100,000 years more to reach us than light from the near side. Thus, the picture of M 31 shows 100,000 years of time. This single photograph captured light that left the near side of the galaxy some 100,000 years later than the light it captured from the far side. When we study the universe, it is impossible to separate space and time.

THINK ABOUT IT

Suppose that, at this very moment, students are studying astronomy on planets somewhere in the Great Galaxy in Andromeda. What would they see as they look from afar at our Milky Way? Could they know that we exist here on Earth? Explain.

Mathematical Insight **1.1** **How Far Is a Light-Year?**

It's easy to calculate the distance represented by a light-year if you remember that

$$\text{distance} = \text{speed} \times \text{time}$$

For example, if you travel at a speed of 50 kilometers per hour for 2 hours, you will travel 100 kilometers. A light-year is the distance covered by light, traveling at a speed of 300,000 kilometers per second, in a time of 1 year. In the process of multiplying the speed and the time, you must convert the year to seconds in order to arrive at a final answer in units of kilometers.

$$1 \text{ light-year} = (\text{speed of light}) \times (1 \text{ yr})$$
$$= \left(300{,}000 \frac{\text{km}}{\text{s}}\right) \times \left(1 \text{ yr} \times \frac{365 \text{ days}}{1 \text{ yr}}\right.$$
$$\left. \times \frac{24 \text{ hr}}{1 \text{ day}} \times \frac{60 \text{ min}}{1 \text{ hr}} \times \frac{60 \text{ s}}{1 \text{ min}}\right)$$
$$= 9{,}460{,}000{,}000{,}000 \text{ km}$$

Thus, "1 light-year" is just an easy way of saying "9.46 trillion kilometers" or "almost 10 trillion kilometers."

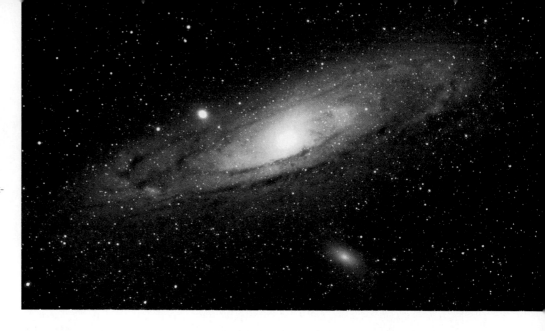

Figure 1.6 M 31, the Great Galaxy in Andromeda, is about 2.5 million light-years away, so this photo captures light that traveled through space for 2.5 million years to reach us. Because the galaxy is 100,000 light-years in diameter, the photo also captures 100,000 years of time in M 31. We see the galaxy's near side as it looked 100,000 years later than the time at which we see the far side.

The Meaning of a Light-Year

A recent advertisement illustrated a common misconception by claiming "It will be light-years before anyone builds a better product." This advertisement makes no sense, because a light-year is a unit of *distance*, not a unit of time. If you are unsure whether the term *light-years* is being used correctly, try testing the statement by remembering that 1 light-year is approximately 10 trillion kilometers, or 6 trillion miles. The advertisement then reads "It will be 6 trillion miles before anyone builds a better product," which clearly does not make sense.

 Scale of the Universe Tutorial, Lessons 1–3

1.2 The Scale of the Universe

The numbers we've given in our description of the size and age of the universe probably have little meaning for you—after all, they are literally astronomical. In this section, we will try to give meaning to incredible cosmic distances and times.

 Virtual Tour of the Solar System

A Walking Tour of the Solar System

One of the best ways to develop perspective on cosmic sizes and distances is to imagine our solar system shrunk down to a scale on which you could walk through it. The Voyage scale model solar system in Washington, D.C. (Figure 1.7) makes such a walk possible. The Voyage model shows the Sun and the planets, and the distances between them, at *one ten-billionth* of the actual sizes and distances. Figure 1.8a shows the planets at their scaled sizes, and Figure 1.8b shows a map

of the locations of the planets in the Voyage model. Table 1.1 lists both real and scaled sizes and distances. If you study the figures and table carefully, you should notice several key features of our solar system that are revealed on this scale:

- On the Voyage scale, the Sun is about the size of a large grapefruit. The planets range in size from dust-speck-size Pluto to marble-size Jupiter. Earth is about the size of the ball point in a pen. Most asteroids and comets are microscopic on this scale.

- The planets fall into two clear groups by distance. The four inner planets—Mercury, Venus, Earth, and Mars—all lie within just a few steps of the Sun on this scale. The outer planets—Jupiter, Saturn, Uranus, Neptune, and Pluto—are much more widely separated. For example, Jupiter is more than three times as far as Mars from the Sun.

- Compared to their small sizes, the distances between the planets are enormous.

- One of the most striking features of the solar system is its *emptiness* (perhaps that's why we call it *space*!). To show the complete orbits of the planets around the Sun, the Voyage model would require an area measuring over a kilometer (0.6 mile) on a side, equivalent to more than 300 football fields. Imagine this area, which is the size of a typical college campus. The only objects large enough to be seen by your naked eye would be the grapefruit-size Sun, the nine planets, and a few moons.

Earth is the only place in our solar system—and the only place we yet know of in the universe—where we could survive outside the artificial environment of a spacecraft or space suit. How does visualizing the Earth to scale (as a ball point orbiting a grapefruit at a distance of 15 meters) affect your perspective on human existence? How does it affect your perspective on our planet? Explain.

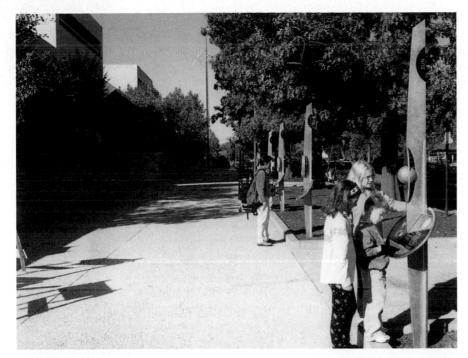

Figure 1.7 This photograph shows the pedestals for the Sun and the inner planets in the Voyage scale model solar system on the National Mall (Washington, D.C.). The model represents solar system sizes and distances at *one ten-billionth* of their actual values. The model Sun—the gold sphere visible on the nearest pedestal—is about the size of a large grapefruit on this scale. The model planets are too small to see in this photograph. For example, on this scale Earth is only about the size of the ball point in a pen. (The model planets are encased in the sidewalk-facing disks visible at about eye level on the planet pedestals.) The building at the left is the National Air and Space Museum. The planets of the outer solar system can be found farther along the walkway. Voyage was created for the National Mall and other locations around the world by the Challenger Center for Space Science Education, the Smithsonian Institution, and NASA.

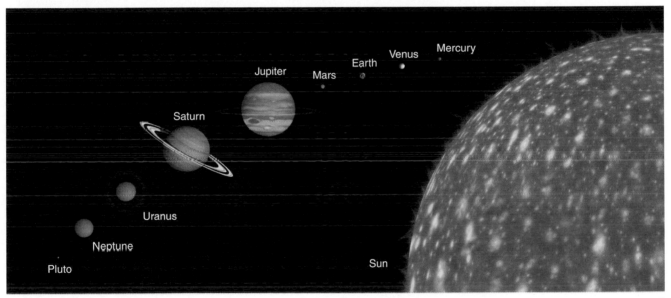

a This painting shows the planets and Sun at *one ten-billionth* of their actual sizes. (Distances are *not* to scale in this painting.)

b This map shows the locations of the Sun and planets in the Voyage model, which represent distances from the Sun at *one ten-billionth* of the actual distances. Although Voyage shows the planets in a line from the Sun, in reality each planet orbits the Sun independently, and a perfect alignment essentially never occurs. Nevertheless, the model gives a good sense of the distances between planetary orbits.

Figure 1.8 Planetary sizes and locations in the Voyage scale model solar system.

Seeing our solar system to scale also helps us put space exploration into perspective. The Moon is the farthest humans have traveled. Twelve astronauts walked on the Moon between 1969 and 1972 as part of the Apollo program (Figure 1.9). On the 1-to-10-billion Voyage scale, the Moon is a barely visible speck about two thumb-widths (4 centimeters) from Earth. If you place the model Earth in the center of your palm, the farthest place humans have ever reached also lies within your palm. Sending humans to Mars—a dream of many people—will require a journey some 200 times as far.

Our robotic spacecraft have gone much farther, visiting every planet in our solar system except Pluto. These trips stretch the limits of modern technology, and it can take years to reach the outer planets. For example, while in the Voyage model you can walk from Earth to Pluto in only about 10 minutes, plans for a real mission to Pluto envision a trip lasting almost a decade. Our solar system is a vast place with many worlds that we have barely begun to explore.

The Incredible Distances to the Stars

Imagine that you start at the Voyage model Sun in Washington, D.C., as pictured in Figure 1.8b. You walk the roughly 600-meter distance to Pluto (just over $\frac{1}{3}$ mile) and then decide to keep going to find the nearest star besides the Sun. How far would you have to go?

Amazingly, you would need to walk to California. That is, on the same scale that allows you to walk from the Sun to Pluto in minutes, even the nearest stars would be more than 4,000 kilometers (2,500 miles) away. If this answer seems hard to believe, you can calculate it for yourself. A light-year is about 10 trillion kilometers, which becomes 1,000 kilometers on the 1-to-10-billion scale (10 trillion ÷ 10 billion = 1,000). The nearest star system to our own,

called Alpha Centauri (Figure 1.10), is about 4.4 light-years away. Thus, Alpha Centauri's real distance of about 4.4 light-years becomes about 4,400 kilometers (2,700 miles) on the 1-to-10-billion scale, or roughly equivalent to the distance across the United States. (Alpha Centauri is actually a three-star system, which sometimes leads to confusion about the identity of the "nearest star." Proxima Centauri, the smallest and dimmest of the three stars, is over 0.1 light-year closer to us than the other two stars. Thus, while Alpha Centauri is the nearest star *system* to our own, Proxima Centauri is the nearest individual star.)

The tremendous distances to the stars give us some perspective on the technological challenge of astronomy. For example, because the largest star of the Alpha Centauri system is roughly the same size and brightness as our Sun, viewing it in the night sky is somewhat like being in Washington, D.C., and seeing a very bright grapefruit in San Francisco (neglecting the problems introduced by the curvature of the Earth). It may seem remarkable that we can see this star at all, but the blackness of the night sky allows the naked eye to see it as a faint dot of light. It looks much brighter through powerful telescopes, but we still cannot see any features of the star's surface.

Now, consider the difficulty of seeing *planets* orbiting nearby stars. It is equivalent to looking from Washington, D.C., and trying to see ball points or marbles orbiting grapefruits in California (or beyond). You probably won't be surprised to learn that we have not yet seen such planets directly. Indeed, the bigger surprise may be that we *have* discovered more than 100 extrasolar planets (planets around other stars) through indirect techniques. These techniques involve searching for signs that a planet is affecting the motion or the light of the star it orbits [Section 9.6]. All of the extrasolar planets detected as of 2003 are closer in size to Jupiter than to Earth. However, with planet-detecting technology improving rapidly, astronomers are hopeful

Table 1.1 Solar System Sizes and Distances, 1-to-10-Billion Scale

Object	Real Diameter	Real Distance from Sun (average)	Model Diameter	Model Distance from Sun
Sun	1,392,500 km	—	139 mm = 13.9 cm	—
Mercury	4,880 km	57.9 million km	0.5 mm	6 m
Venus	12,100 km	108.2 million km	1.2 mm	11 m
Earth	12,760 km	149.6 million km	1.3 mm	15 m
Mars	6,790 km	227.9 million km	0.7 mm	23 m
Jupiter	143,000 km	778.3 million km	14.3 mm	78 m
Saturn	120,000 km	1,427 million km	12.0 mm	143 m
Uranus	52,000 km	2,870 million km	5.2 mm	287 m
Neptune	48,400 km	4,497 million km	4.8 mm	450 m
Pluto	2,260 km	5,900 million km	0.2 mm	590 m

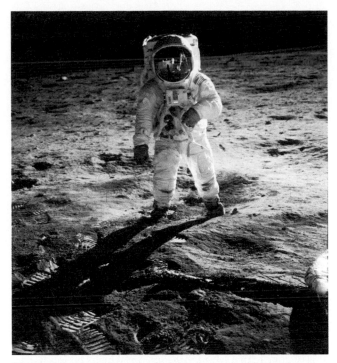

Figure 1.9 The Moon is the farthest place ever visited by humans, yet on the 1-to-10-billion scale of the Voyage model it is only about 4 centimeters (1.5 inches) from Earth. This famous photograph from the first moon landing (*Apollo 11* in July 1969) shows astronaut Buzz Aldrin, with his visor showing Neil Armstrong in the reflection. Armstrong was the first to step onto the Moon's surface, saying, "That's one small step for [a] man, one giant leap for mankind."

that the first discoveries of Earth-size planets around other stars will occur within a decade.

Our examination of stellar distances also offers a sobering lesson about the possibility of travel to the stars. Although science fiction shows like *Star Trek* and *Star Wars* may make interstellar travel seem easy, the reality is far different. Consider the *Voyager 2* spacecraft. Launched in 1977, *Voyager 2* flew by Jupiter in 1979, Saturn in 1981, Uranus in 1986, and Neptune in 1989. (Its trajectory did not take it near Pluto.) *Voyager 2* is now bound for the stars at a speed of close to 50,000 kilometers per hour—about 100 times as fast as a speeding bullet. Even at this speed, *Voyager 2* would take about 100,000 years to reach Alpha Centauri if it were headed in that direction (which it's not). Convenient interstellar travel remains well beyond our present technology [Section 24.5].

The Scale of the Milky Way Galaxy

The vast separation between our solar system and Alpha Centauri is typical of the separations among star systems here in the outskirts of the Milky Way Galaxy. Thus, the 1-to-10-billion scale is useless for modeling even just a few dozen of the nearest stars, because they could not all be spaced properly on the Earth's surface. Visualizing the entire galaxy requires a new scale.

Let's further reduce our solar system scale by a factor of 1 billion (making it a scale of 1 to 10^{19}). On this new

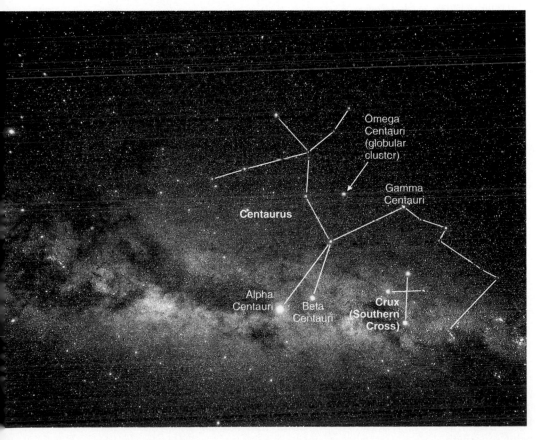

Figure 1.10 This photograph and diagram show the constellation Centaurus, which is visible only from tropical and southern latitudes. Note the location of Alpha Centauri, the nearest star system to our own. Its real distance is about 4.4 light-years, which is about 4,400 kilometers (2,700 miles) on the 1-to-10-billion Voyage scale. In other words, on the same scale on which Pluto is a short walk away, the distance to the nearest stars is equivalent to the distance across the United States.

Confusing Very Different Things

Most people are familiar with the terms *solar system* and *galaxy,* but people sometimes mix them up. Notice how incredibly different our solar system is from our galaxy. Our solar system is a *single* star system consisting of our Sun and the various objects that orbit it, including Earth and eight other planets. Our galaxy is a collection of some 100 billion star systems—so many that it would take thousands of years just to count them. Thus, confusing the terms *solar system* and *galaxy* represents a mistake by a factor of 100 billion—a fairly big mistake!

scale, each light-year becomes 1 millimeter, and the 100,000-light-year diameter of the Milky Way Galaxy becomes 100 meters, or about the length of a football field. Visualize a football field with a scale model of our galaxy centered over midfield. Our entire solar system is a microscopic dot located around the 20-yard line. The 4.4-light-year separation between our solar system and Alpha Centauri becomes just 4.4 millimeters on this scale—smaller than the width of your little finger. If you stood at the position of our solar system in this model, millions of star systems would lie within reach of your arms.

Another way to put the galaxy into perspective is to consider its number of stars—more than 100 billion. Imagine that tonight you are having difficulty falling asleep (perhaps because you are contemplating the scale of the universe). Instead of counting sheep, you decide to count stars. If you are able to count about one star each second, on average, how long would it take you to count 100 billion stars in the Milky Way? Clearly, the answer is 100 billion (10^{11}) seconds, but how long is that? Amazingly, 100 billion seconds turns out to be more than 3,000 years. (You can confirm this by dividing 100 billion by the number of seconds in 1 year.) Thus, you would need thousands of years just to *count* the stars in the Milky Way Galaxy, and this assumes you never take a break—no sleeping, no eating, and absolutely no dying!

The Number of Stars in the Universe

As incredible as the scale of our galaxy may seem, the Milky Way is only one of at least 100 billion galaxies in the observable universe. Just as it would take thousands of years to count the stars in the Milky Way, it would take thousands of years to count all the galaxies. Think for a moment about the total number of stars in all these galaxies. If we assume 100 billion stars per galaxy, the total number of stars in the observable universe is roughly 100 billion × 100 billion or 10,000,000,000,000,000,000,000 (10^{22}).

How big is this number? Visit a beach. Run your hands through the fine-grained sand. Imagine counting each tiny grain of sand as it slips through your fingers. Then imagine counting every grain of sand on the beach and continuing on to count *every* grain of dry sand on *every* beach on Earth. If you could actually complete this task, you would find that the number of grains of sand is similar to the number of stars in the observable universe (Figure 1.11).

Contemplate the fact that there may be as many stars in the observable universe as grains of sand on all the beaches on Earth and that each star is a potential sun for a system of planets. With so many possible homes for life, do you think it is conceivable that life exists only on Earth? Why or why not?

The Scale of Time

Now that we have developed some perspective on the scale of space, we can do the same for the scale of time. Imagine the entire history of the universe, from the Big Bang to the present, compressed into a single year. We can represent this history with a *cosmic calendar,* on which the Big Bang takes place at the first instant of January 1 and the present day is just before the stroke of midnight on December 31 (Figure 1.12). For a universe that is about 14 billion years old, each month on the cosmic calendar represents a little more than 1 billion years. (More precisely, an average month represents 1.17 billion years.)

On this scale, the Milky Way Galaxy probably formed sometime in February. Many generations of stars lived and died in the subsequent cosmic months, enriching the

Figure 1.11 The number of stars in the observable universe is similar to the number of grains of dry sand on all the beaches on Earth.

Jan. 1
The Big Bang

Feb.
The Milky Way forms.

JANUARY

FEBRUARY

MARCH

APRIL

MAY

JUNE

JULY

AUGUST

SEPTEMBER

OCTOBER

NOVEMBER

DECEMBER

Sept. 3
Earth forms.

Sept. 22
Earliest
evidence of
life on Earth

DECEMBER

S	M	T	W	T	F	S
1	2	3	4	5	6	7
8	9	10	11	12	13	14
15	16	17 The Cambrian explosion	18	19	20	21
22	23	24	25	26 Rise of the dinosaurs	27	28
29	30 (7:00 A.M.) Dinosaurs extinct	31				

59 seconds:
Kepler and Galileo
prove Earth orbits the Sun.

49 seconds:
Pyramids are built.

DECEMBER 31

Morning...
12:00 pm _____
1:00 pm _____
2:00 pm _____
3:00 pm _____
4:00 pm _____
5:00 pm _____
6:00 pm _____
7:00 pm _____
8:00 pm _____
9:00 pm Early hominids
10:00 pm _____
11:00 pm _____
11:58 pm Modern humans evolve.
11:59 pm _____
12:00 am _____

35 seconds:
Agriculture arises.

Figure 1.12 The cosmic calendar compresses the history of the universe into 1 year. This version assumes that the universe is 14 billion years old, so each month represents a little more than 1 billion years. Only within the last few seconds of the last day has human civilization taken shape. (This version of the cosmic calendar is adapted from one created by Carl Sagan.)

15

galaxy with the "star stuff" from which we and our planet are made.

Our solar system and our planet did not form until early September on this scale, or 4.6 billion years ago in real time. By late September, life on Earth was flourishing. However, for most of Earth's history, living organisms remained relatively primitive and microscopic in size. On the scale of the cosmic calendar, recognizable animals became prominent only in mid-December, with the period of diverse evolution that biologists call the Cambrian explosion [Section 14.5]. Early dinosaurs appeared on the day after Christmas. Then, in a cosmic instant, the dinosaurs disappeared forever—probably due to the impact of an asteroid or a comet [Section 13.6]. In real time, the death of the dinosaurs occurred some 65 million years ago, but on the cosmic calendar it was only yesterday. With the dinosaurs gone, small furry mammals inherited Earth. Some 60 million years later, or around 9 P.M. on December 31 of the cosmic calendar, early hominids (human ancestors) walked upright.

Perhaps the most astonishing thing about the cosmic calendar is that the entire history of human civilization falls into just the last half-minute. The ancient Egyptians built the pyramids only about 11 seconds ago on this scale. About 1 second ago, Kepler and Galileo proved that Earth orbits the Sun rather than vice versa. The average college student was born about 0.05 second ago, around 11:59:59.95 P.M. on the cosmic calendar. On the scale of cosmic time, the human species is the youngest of infants, and a human lifetime is a mere blink of an eye.

THINK ABOUT IT

Notice that, while life has existed for most of our planet's history, intelligent life is a very recent development. Some people use this fact to argue that even if life itself is common in the universe, intelligent life and civilizations will prove to be very rare. Explain the logic behind this argument. Then try to think of counterarguments to explain why it might still be possible for thousands or millions of civilizations to exist in our galaxy alone. Which arguments do you find most persuasive? Defend your opinion.

1.3 Spaceship Earth

The next step in our "big picture" overview is getting a sense of motion in the universe. Wherever you are as you read this book, you probably have the feeling that you're "just sitting here." Nothing could be further from the truth. In fact, you are being spun in circles as Earth rotates, you are racing around the Sun in Earth's orbit, and you are careening through the cosmos in the Milky Way Galaxy. In the words of noted inventor and philosopher R. Buckminster Fuller (1895–1983), you are a traveler on *spaceship Earth*.

Rotation and Orbit

The most basic motions of Earth are its **rotation** (spin) and its **orbit** (sometimes called *revolution*) around the Sun. Earth rotates once each day around its axis, an imaginary line connecting the North Pole to the South Pole (and passing through the center of Earth). Although we do not feel any obvious effects from Earth's rotation, the speed of rotation is substantial (Figure 1.13). Unless you live at very high latitude, you are whirling around Earth's axis at a speed of 1,000 kilometers per hour (600 miles per hour) or more—faster than most airplanes travel.

Earth rotates from west to east, which is counterclockwise as viewed from above the North Pole. As a result, the Sun (as well as the Moon and stars) *appears* to go around us in the opposite direction, from east to west. That is why the Sun rises in the east and sets in the west each day. (We will discuss the apparent motion of the sky in more detail in the next chapter.)

At the same time Earth is rotating, it is also orbiting around the Sun. It takes 1 year to complete each orbit. Again, while we don't feel any effects from the orbit, the speed is quite impressive: We and our planet are right now racing around the Sun at a speed in excess of 100,000 kilometers per hour (60,000 miles per hour).

Earth's orbital path defines a flat plane that we call the **ecliptic plane**. Earth's axis happens to be tilted by $23\frac{1}{2}°$ from a line *perpendicular* to the ecliptic plane (Figure 1.14). Keep in mind that this notion of tilt makes sense only in relation to the ecliptic plane. That is, the idea of "tilt" by itself has no meaning in space, where there is no absolute up or down. In space, "up" and "down" mean only away

0 km/hr
830 km/hr
1,275 km/hr
1,550 km/hr
1,650 km/hr
1,550 km/hr
1,275 km/hr

Figure 1.13 As Earth rotates, your speed around Earth's axis depends on your latitude. Unless you live at very high latitude, your speed is over 1,000 km/hr. Notice that Earth rotates counterclockwise as viewed from above the North Pole, so you are always rotating from west to east—which is why the Sun rises in the east and sets in the west.

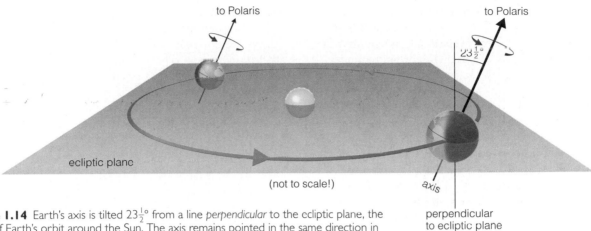

ecliptic plane

(not to scale!)

to Polaris

to Polaris

$23\frac{1}{2}°$

axis

perpendicular
to ecliptic plane

Figure 1.14 Earth's axis is tilted $23\frac{1}{2}°$ from a line *perpendicular* to the ecliptic plane, the plane of Earth's orbit around the Sun. The axis remains pointed in the same direction in space—toward Polaris, the North Star—at all times throughout the year. From far above the North Pole, you would notice that Earth rotates *and* orbits counterclockwise.

from the center of Earth (or another planet) and toward the center of Earth, respectively.

THINK ABOUT IT

If there is no up or down in space, why do you think nearly all globes have the North Pole on top and the South Pole on the bottom? Would it be equally correct to have the South Pole on top or to turn the globe sideways? Explain.

Earth's rotation and orbit exhibit many other features, some that are easy to notice and others that are quite subtle. For our purposes in this book, three other key features are important to understand and will come up in later discussions.

- **Earth's axis remains pointed in the same direction in space at all times throughout each year.** We'll see in Chapter 2 how this fact helps explain the seasons. The axis (going from south to north) happens to point very nearly in the direction of a star called Polaris, which is why Polaris is also known as the North Star.

- **Earth orbits the Sun in the same direction that it rotates on its axis.** That is, both rotation and orbit go counterclockwise as viewed from above the North Pole. This is not a coincidence but a consequence of how our planet was born. The giant cloud of gas and dust from which our solar system was born must also have been spinning [Section 9.2], and the direction of the cloud's spin is reflected in the directions of Earth's rotation and orbit.

- **Earth's orbit is not a perfect circle.** Rather, it is a slightly oval shape known as an *ellipse* [Section 3.4]. As a result, Earth's distance from the Sun varies slightly over the course of each year (Figure 1.15). The Earth's *average* distance from the Sun, which is about 150 million kilometers (93 million miles), is given a special name: an **astronomical unit**, or **AU**. (More technically,

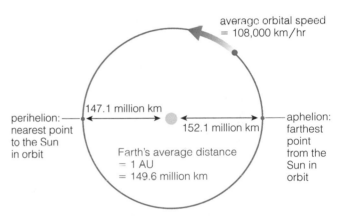

average orbital speed
= 108,000 km/hr

perihelion:
nearest point
to the Sun
in orbit

147.1 million km

152.1 million km

aphelion:
farthest
point
from the
Sun in
orbit

Earth's average distance
= 1 AU
= 149.6 million km

Figure 1.15 Earth's orbit is not quite a perfect circle, though it is close enough to circular to appear like a circle to the eye. This diagram is correctly scaled (you can measure to see that it is not a perfect circle) and shows Earth's minimum (perihelion) and maximum (aphelion) distances from the Sun. It also reminds us that Earth's average speed in its orbit is over 100,000 kilometers per hour (60,000 miles per hour) and that Earth's average distance from the Sun defines 1 astronomical unit (AU).

1 AU is the *semimajor axis* of the Earth's elliptical orbit.) Distances within our solar system are commonly described in astronomical units because these are easier to interpret than units in kilometers. For example, knowing that Mars is about 230 million kilometers from the Sun may not mean much to you, but knowing that it is 1.5 AU from the Sun immediately tells you that Mars is 1.5 times as far from the Sun as is Earth.

Traveling in the Milky Way Galaxy

Rotation and orbit are only a small part of the travels of spaceship Earth. In fact, our entire solar system is on a great journey within the Milky Way Galaxy.

Our Local Solar Neighborhood Let's begin with the motion of our solar system relative to nearby stars in what we call our *local solar neighborhood* (the region of the Sun and nearby stars). Figure 1.16 shows that stars within the local solar neighborhood move essentially at random relative to one another. It also offers an important reminder of the incredible scale of the galaxy. Imagine drawing the tiniest dot that you can make on the galaxy painting in Figure 1.16. Your dot will probably be about 10,000 times smaller than the picture of the galaxy as a whole—but it will cover a region representing more than 10 million stars! (The entire galaxy contains more than 100 billion stars, and 100 billion ÷ 10,000 = 10 million.) We usually think of our local solar neighborhood as an even smaller region of the galaxy including the nearest few thousand to few million stars.

The stars of the local solar neighborhood (or any other small region of the galaxy) generally move quite fast relative to one another. For example, we are moving relative to nearby stars at an average speed of about 70,000 kilometers per hour (40,000 miles per hour), about three times as fast as the Space Station orbits Earth. Given these high speeds, why don't we see nearby stars racing around our sky?

The answer lies in their vast distances from us. You've probably noticed that a distant airplane appears to move through your sky more slowly than one flying close overhead. If we extend this idea to the stars, we find that even at speeds of 70,000 kilometers per hour stellar motions would

be noticeable to the naked eye only if we watched them for thousands of years. That is why the patterns in the constellations seem to remain fixed. Nevertheless, in 10,000 years the constellations will be noticeably different from those we see today. In 500,000 years they will be unrecognizable. If you could watch a time-lapse movie made over millions of years, you *would* see stars racing across our sky.

Despite the chaos of motion in the local solar neighborhood over millions and billions of years, collisions between star systems are extremely rare. Explain why. (*Hint:* Consider the sizes of star systems, such as the solar system, relative to the distances between them.)

Galactic Rotation If you look closely at leaves floating in a stream, their motions relative to one another might appear random, just like the motions of stars in the local solar neighborhood. As you widen your view, you see that all the leaves are being carried in the same general direction by the downstream current. In the same way, as we widen our view beyond the local solar neighborhood the seemingly random motions of its stars give way to a simpler and even faster motion: The entire Milky Way Galaxy is rotating.

Stars at different distances from the galactic center take different amounts of time to complete an orbit. Our

Figure 1.16 This painting illustrates the motion of stars within our local solar neighborhood. The "zoom out" box is necessary because even the tiniest dot you could draw on the picture of the galaxy would cover a region representing more than 10 million stars.

The box represents stars and their motions in the local solar neighborhood.

solar system, located about 28,000 light-years from the galactic center, completes one orbit of the galaxy in about 230 million years (Figure 1.17). Even if you could watch from outside our galaxy, this motion would be unnoticeable to your naked eye. However, if you calculate the speed of our solar system as we orbit the center of the galaxy, you will find that it is close to 800,000 kilometers per hour (500,000 miles per hour).

The galaxy's rotation reveals one of the greatest mysteries in science—one that we will study in depth in Chapter 22. The speeds at which stars orbit the galactic center depend on the strength of gravity, and the strength of gravity depends on how mass is distributed throughout the galaxy. Thus, careful study of the galaxy's rotation allows us to determine the distribution of mass in the galaxy.

Such studies suggest that the stars in the disk of the galaxy represent only the "tip of the iceberg" compared to the mass of the entire galaxy (Figure 1.18). That is, most of the mass of the galaxy seems to be located outside the visible disk, in what we call the galaxy's *halo*. We don't know the nature of this mass. Because we have not detected any light coming from it, we call it **dark matter**. Studies of other galaxies suggest that they also are made mostly of dark matter. In fact, most of the mass in the universe seems to be made of this mysterious dark matter, but we do not yet know what it is.

The Expanding Universe

The billions of galaxies in the universe also move relative to one another. Within the Local Group (see Figure 1.1), some of the galaxies move toward the Milky Way Galaxy, some move away from it, and some move in more complex ways. For example, two small galaxies, known as the Large and Small Magellanic Clouds, apparently orbit the Milky Way. Again, the speeds are enormous by earthly standards. In fact, the Milky Way is moving toward the Great Galaxy in Andromeda (M 31) at about 300,000 kilometers per hour (180,000 miles per hour)—but this motion is unnoticeable to our eyes. Despite the high speed, we needn't worry about a collision anytime soon. Even if the Milky Way and Andromeda Galaxies are approaching each other head-on (which they might not be), it will be nearly 10 billion years before any collision begins.

When we look outside the Local Group, however, we find two astonishing facts that were first recognized in the 1920s by Edwin Hubble, for whom the Hubble Space Telescope was named:

1. Virtually every galaxy outside the Local Group is moving *away* from us.

2. The more distant the galaxy, the faster it appears to be racing away from us.

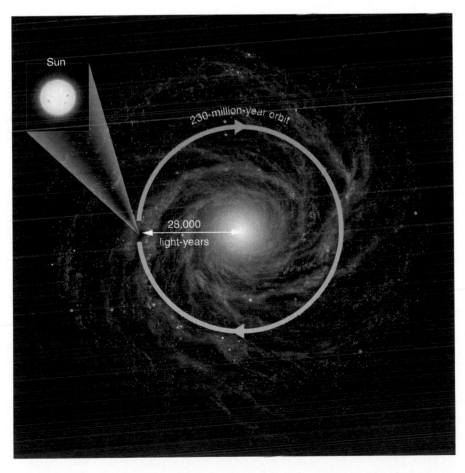

Sun

230-million-year orbit

28,000 light-years

Figure 1.17 This painting shows how the entire Milky Way Galaxy rotates (in a direction that would tend to wind up the spiral arms). Our Sun and solar system are located about 28,000 light-years from the galactic center. At this distance, each orbit around the galactic center takes about 230 million years. Although this orbital motion would be unnoticeable to the eye even from outside the galaxy, it is quite fast—about 1 million km/hr (600,000 mi/hr).

Most of the galaxy's light comes from stars and gas in the galactic disk...

...but most of the galaxy's mass lies above and below the disk in the *halo*.

Figure 1.18 This painting shows an edge-on view of the Milky Way Galaxy. Most visible stars reside within the galaxy's thin *disk*, which runs horizontally across the page in this figure. Careful study of galactic rotation suggests that most of the mass lies in the galactic *halo*—a large, spherical region that surrounds and encompasses the disk. Because this mass emits no light that we have detected, we call it dark matter. (The dark matter may extend quite far out; see Figure 22.1.)

Upon first hearing of these two facts, you might be tempted to conclude that our Local Group (which is held together by gravity) suffers a cosmic case of chicken pox. However, there is a natural explanation: *The entire universe is expanding.* We'll save details about this expansion for later in the book (Chapter 20), but you can understand the basic idea by thinking about a raisin cake baking in an oven.

Imagine that you make a raisin cake in which the distance between adjacent raisins is 1 centimeter. You place the cake in the oven, where it expands as it bakes. After 1 hour, you remove the cake, which has expanded so that the distance between adjacent raisins has increased to 3 centimeters (Figure 1.19). The expansion of the cake seems fairly obvious. But what would you see if you lived *in* the cake, as we live in the universe?

Pick any raisin (it doesn't matter which one), call it the Local Raisin, and identify it in the pictures of the cake both before and after baking. Figure 1.19 shows one possible choice for the Local Raisin, with three nearby raisins labeled. The accompanying table summarizes what you would see if you lived within the Local Raisin. Notice, for example, that Raisin 1 starts out at a distance of 1 centimeter before

baking and ends up at a distance of 3 centimeters after baking, which means it moves a distance of 2 centimeters away from the Local Raisin during the hour of baking. Hence, its speed as seen from the Local Raisin is 2 centimeters per hour. Raisin 2 moves from a distance of 2 centimeters before baking to a distance of 6 centimeters after baking, which means it moves a distance of 4 centimeters away from the Local Raisin during the hour. Hence, its speed is 4 centimeters per hour, or twice as fast as the speed of Raisin 1. Generalizing, the fact that the cake is expanding means that all raisins are moving away from the Local Raisin, with more distant raisins moving away faster.

Hubble's discovery that galaxies are moving in much the same way as the raisins in the cake, with most moving away from us and more distant ones moving away faster, implies that the universe in which we live is expanding much like the raisin cake. If you now imagine the Local Raisin as representing our Local Group of galaxies and the other raisins as representing more distant galaxies or clusters of galaxies, you have a basic picture of the expansion of the universe. Like the expanding dough between the raisins in the cake, *space* itself is growing between galaxies. More

Distances and Speeds As Seen from the Local Raisin

Raisin Number	Distance Before Baking	Distance After Baking (1 hour later)	Speed
1	1 cm	3 cm	2 cm/hr
2	2 cm	6 cm	4 cm/hr
3	3 cm	9 cm	6 cm/hr
⋮	⋮	⋮ ⋮	⋮

Figure 1.19 An expanding raisin cake illustrates basic principles of the expansion of the universe. From the outside, the raisin cake appears to expand uniformly. From the inside, anyone living in one of the raisins would find that all other raisins are moving away as the cake expands, with more distant raisins moving away faster. This analogy shows why the fact that more distant galaxies move away from us faster than nearer ones implies that our universe is expanding.

distant galaxies move away from us faster because they are carried along with this expansion like the raisins in the expanding cake. And, just as the raisins themselves do not expand, the individual galaxies and clusters of galaxies do not expand because they are bound together by gravity.

There's one important distinction between the raisin cake analogy and the universe: Because a cake is small in size, it has a center and edges that we can see. In contrast, our observable universe is probably just part of the entire universe, so we could not identify any center and edges even if the universe had them. The effects of expansion would appear basically the same from any place in the universe. Anyone living in any galaxy would see other galaxies moving away, with more distant ones moving faster. No place can claim to be any more "central" than any other place. Thus, unlike the cake, we say that our universe has no center.

Summary of Our Motion

Let's summarize the motions we have covered. We spin around Earth's axis as Earth orbits the Sun. Our solar system moves among the stars of the local solar neighborhood as this entire neighborhood orbits the center of the Milky Way Galaxy. Our galaxy, in turn, moves among the other galaxies of the Local Group as the Local Group is carried along with the overall expansion of the universe. Table 1.2 lists the motions and their associated speeds. Spaceship Earth is carrying us on a remarkable journey!

Table 1.2 The Motions of Spaceship Earth

Motion	Typical Speed
rotation	1,000 km/hr or more around axis, with one rotation taking 1 day
orbit of Sun	100,000 km/hr around Sun, with one orbit taking 1 year
motion within local solar neighborhood	70,000 km/hr relative to nearby stars
rotation of the Milky Way Galaxy	800,000 km/hr around galactic center, with one galactic rotation taking about 230 million years
motion within Local Group	300,000 km/hr toward Andromeda Galaxy
universal expansion	more distant galaxies moving away faster, with the most distant moving at speeds close to the speed of light

1.4 The Human Adventure of Astronomy

In a relatively few pages, we've laid out a fairly complete overview of our modern scientific ideas about the universe. But our goal in this book is not for you simply to be able to recite these ideas. Rather, it is to help you understand the evidence supporting them and the extraordinary story of how they developed.

Astronomy is a human adventure in the sense that it affects virtually everyone—even those who have never looked at the sky—because the development of astronomy has been so deeply intertwined with the development of civilization as a whole. Revolutions in astronomy have gone hand in hand with the revolutions in science and technology that have shaped modern life.

Witness the repercussions of the Copernican revolution, which changed our view of Earth from being the center of the universe to being just one planet orbiting the Sun. This revolution, which we will discuss further in Chapter 3, began when Copernicus published his idea of a Sun-centered solar system in 1543. Three subsequent figures—Tycho Brahe, Johannes Kepler, and Galileo—provided the key evidence that eventually led to wide acceptance of the Copernican idea. The revolution culminated with Isaac Newton's uncovering of the laws of motion and gravity. Newton's work, in turn, became the foundation of physics that helped fuel the industrial revolution.

More recently, the development of space travel and the computer revolution have helped fuel tremendous progress in astronomy. We've learned a lot about our solar system by sending probes to the planets, and many of our most powerful observatories, including the Hubble Space Telescope, reside in space. On the ground, computer design and control have led to tremendous growth in the size and power of telescopes, particularly in the past decade.

Many of these efforts, along with the achievements they spawned, have led to profound social change. The most famous example involved Galileo, whom the Vatican put under house arrest in 1633 for his claims that Earth orbits the Sun. Although the Church soon recognized that Galileo was right, he was formally vindicated only with a statement by Pope John Paul II in 1992. In the meantime, his case spurred great debate in religious circles and had a profound influence on both theological and scientific thinking.

As you progress through this book and learn about astronomical discovery, try to keep in mind the context of the human adventure. You will then be learning not just about a science, but about one of the great forces that have helped shape our modern world. This context will also lead you to think about how the many astronomical mysteries that remain—such as the makeup of dark matter, the events of the first instant of the Big Bang, and the question of life beyond Earth—may influence our future.

What would it mean to us if we were ever to learn the complete story of our cosmic origins? How would our view of Earth be changed if we came to learn that Earth-like planets are common or exceedingly rare? Only time may answer these questions, but the chapters ahead give you the foundation you need to understand how we changed from a primitive people looking at patterns in the night sky to a civilization capable of asking deep questions about our existence.

THE BIG PICTURE

Putting Chapter 1 into Context

In this first chapter, we developed a broad overview of our place in the universe. It is not yet necessary for you to understand the details—everything presented in this chapter will be covered in greater depth later in the book. However, you should understand enough so that the following "big picture" ideas are clear:

- Earth is not the center of the universe but instead is a planet orbiting a rather ordinary star in the Milky Way Galaxy. The Milky Way Galaxy, in turn, is one of billions of galaxies in our observable universe.

- We are "star stuff." The atoms from which we are made began as hydrogen and helium in the Big Bang and were later fused into heavier elements by massive stars. When these stars died, they released these atoms into space, where our galaxy recycled them into new stars and planets. Our solar system formed from such recycled matter, some 4.6 billion years ago.

- Cosmic distances are literally astronomical, but we can put them in perspective with the aid of scale models and other scaling techniques. When you think about these enormous scales, don't forget that every star is a sun and every planet is a unique world.

- We are latecomers on the scale of cosmic time. The universe was more than halfway through its history by the time our solar system formed, and then it took billions of years more before humans arrived on the scene.

- All of us are being carried through the cosmos on spaceship Earth. Although we cannot feel this motion in our everyday lives, the associated speeds are surprisingly high. Learning about the motions of spaceship Earth gives us a new perspective on the cosmos and helps us understand its nature and history.

- Throughout history, astronomy has developed hand in hand with social and technological development. Astronomy thereby touches all of us and is a human adventure that can be enjoyed by all.

1.1 A Modern View of the Universe

- *What is our physical place in the universe?* Earth is a planet in our solar system, which is one of some 100 billion star systems in the Milky Way Galaxy, which is one of about 40 galaxies in the Local Group, which is part of the Local Supercluster, which is part of the universe.

- *What are our cosmic origins and why do we say that we are made of "star stuff"?* The universe began in the Big Bang and has been expanding ever since, except in localized regions where gravity has caused matter to collapse into galaxies and stars. The Big Bang essentially produced only two chemical elements: hydrogen and helium. The rest have been produced by stars and recycled within galaxies from one generation of stars to the next, which is why we are "star stuff."

- *Why does looking into space mean looking back in time?* Light takes time to travel through space. A light-year is the distance light can travel in 1 year, which is about 10 trillion kilometers. Thus, when we look farther away, we see light that has taken a longer time to reach us.

1.2 The Scale of the Universe

- *What does our solar system look like when viewed to scale?* On a scale of 1 to 10 billion, the Sun is about the size of a grapefruit. Planets are much smaller, with Earth the size of a ball point and Jupiter the size of a marble on this scale. The distances between planets are huge compared to their sizes.

- *How far away and how numerous are the stars?* On the 1-to-10-billion scale, it is possible to walk from the Sun to Pluto in just a few minutes. On the same scale, the nearest stars besides the Sun are thousands of kilometers away. The rest of the Milky Way Galaxy must be viewed on a different scale, and there are so many stars in our galaxy that it would take thousands of years just to count them. The number of stars in the observable universe is about the same as the number of grains of dry sand on all the beaches on Earth.

- *How do human time scales compare to the age of the universe?* On a cosmic calendar that compresses the history of the universe into 1 year, human civilization is just a few seconds old.

1.3 Spaceship Earth

- *What are the basic motions of spaceship Earth?* Earth rotates on its axis once each day and orbits the Sun once each year, at an average distance of 1 AU and with its axis tilted by $23\frac{1}{2}°$ to a line perpendicular to the ecliptic plane. All stars are in motion, at surprisingly high speeds, but they are so far away that our eyes do not notice this motion. Our solar system orbits the center of the Milky Way Galaxy about once every 230 million years. Galaxies in the Local Group move relative to one another, while all other galaxies are moving away from us with the expansion of the universe.

- *How do we know that the universe is expanding?* We observe nearly all other galaxies to be moving away from us, with more distant ones moving faster.

1.4 The Human Adventure of Astronomy

- *How is astronomy interwoven with other aspects of human society?* Astronomy has been a crucial part of the development of human culture and technology.

❓ Does It Make Sense?

Decide whether each statement makes sense and explain why it does or does not.

Example: I walked east from our base camp at the North Pole.

Solution: The statement does not make sense because *east* has no meaning at the North Pole—all directions are south from the North Pole.

1. Our solar system is bigger than some galaxies.

2. The universe is about 14 billion light-years old.

3. It will take me light-years to complete this homework assignment!

4. Someday we may build spaceships capable of traveling at a speed of 1 light-minute per hour.

5. Astronomers recently discovered a moon that does not orbit a planet.

6. NASA plans soon to launch a spaceship that will leave the Milky Way Galaxy to take a photograph of the galaxy from the outside.

7. The observable universe is the same size today as it was a few billion years ago.

8. Photographs of distant galaxies show them as they were when they were much younger than they are today.

9. At a nearby park, I built a scale model of our solar system in which I used a basketball to represent Earth.

10. Because nearly all galaxies are moving away from us, we must be located at the center of the universe.

Problems

(Quantitative problems are marked with an asterisk.)

11. *Old and New Views.* What do we mean by a geocentric universe? In broad terms, contrast a geocentric universe with our modern view of the universe.

12. *Expanding Universe.* What do we mean when we say that the universe is expanding? Why does an expanding universe suggest a beginning in what we call the *Big Bang*?

13. *Observable Universe.* What do we mean by the *observable universe*? Is it the same as the entire universe? Explain what limits the extent of the observable universe.

14. *Distances to Stars.* How do the distances to the stars compare to distances within our solar system? Give a few examples to put the comparison in perspective.

15. *Galactic Perspective.* Describe at least two ways to put the scale of the Milky Way Galaxy in perspective.

16. *Universal Perspective.* How many galaxies are in the observable universe? How many stars are in the observable universe? Put these numbers in perspective.

17. *Raisin Cake Universe.* Suppose that all the raisins in a cake are 1 centimeter apart before baking and 4 centimeters apart after baking.

   a. Draw diagrams to represent the cake before and after baking.

   b. Identify one raisin as the Local Raisin on your diagrams. Construct a table showing the distances and speeds of other raisins as seen from the Local Raisin.

   c. Briefly explain how your expanding cake is similar to the expansion of the universe.

18. *Scaling the Local Group of Galaxies.* Both the Milky Way Galaxy and the Great Galaxy in Andromeda (M 31) have a diameter of about 100,000 light-years. The distance between the two galaxies is about 2.5 million light-years.

   a. Using a scale on which 1 centimeter represents 100,000 light-years, draw a sketch showing both galaxies and the distance between them to scale.

   b. How does the separation between galaxies compare to the separation between stars? Based on your answer, discuss the likelihood of galactic collisions in comparison to the likelihood of stellar collisions.

*19. *Distances by Light.* Just as a light-year is the distance that light can travel in 1 year, we define a light-second as the distance that light can travel in 1 second, a light-minute as the distance that light can travel in 1 minute, and so on. Following the method of Mathematical Insight 1.1, calculate the distance in kilometers represented by each of the following: 1 light-second; 1 light-minute; 1 light-hour, 1 light-day.

*20. *Driving to the Planets (and Stars).* Imagine that you could drive your car at a constant speed of 100 km/hr (62 mi/hr), even in space. (In reality, the law of gravity would make driving through space at a constant speed all but impossible.)

   a. How long would it take to drive all around the Earth? Assume you can drive across both land and ocean. (*Hint:* Use Earth's circumference of approximately 40,000 kilometers.)

   b. Suppose you started driving from the Sun. How long would it take to reach Earth? How long would it take to reach Pluto? Use the data in Table 1.1.

   c. How long would it take to drive the 4.4 light-years to Alpha Centauri? (*Hint:* You'll need to convert the distance to kilometers.)

*21. *Voyager's Trip to the Stars.* The *Voyager 2* spacecraft is traveling at about 50,000 km/hr. At this speed, how long would it take to reach Alpha Centauri (if it were headed in the right direction, which it is not)?

*22. *Speed Calculations.* Calculate each of the following speeds in both kilometers per hour and miles per hour. In each case, assume a circular orbit; recall that the formula for the circumference of a circle is $2 \times \pi \times$ radius. (*Hint:* Divide the distance traveled in the circular orbit by the time it takes to complete one orbit.)

   a. The speed of the Earth going around the Sun; use the average Earth–Sun distance of 149.6 million kilometers.

   b. The speed of our solar system around the center of our galaxy; assume that we are located 28,000 light-years from the center and that each orbit takes 230 million years.

Discussion Questions

23. *Vast Orbs.* The chapter-opening quotation from Carl Sagan seems to suggest that humans might be better behaved and less inclined to wage war if everyone appreciated Earth's place in the universe. Do you agree? Defend your opinion.

24. *Infant Species.* In the last few tenths of a second before midnight on December 31 of the cosmic calendar, we have developed an incredible civilization and learned a great deal about the universe, but we also have developed technology through which we could destroy ourselves. The midnight bell is striking, and the choice for the future is ours. How far into the next cosmic year do you think our civilization will survive? Defend your opinion.

25. *A Human Adventure.* How important do you think astronomical discoveries have been to our social development? Defend your opinion with examples drawn from your knowledge of history.

For a complete list of media resources available, go to www.astronomyplace.com and choose Chapter 1 from the pull-down menu.

 Astronomy Place Web Tutorials

Tutorial Review of Key Concepts

Use the interactive **Tutorial** at www.astronomyplace.com to review key concepts from this chapter.

Scale of the Universe Tutorial

Lesson 1 Distances Scales: The Solar System

Lesson 2 Distances Scales: Stars and Galaxies

Lesson 3 Powers of 10

Supplementary Tutorial Exercises

Use the interactive **Tutorial Lessons** to explore the following questions.

Scale of the Universe Tutorial, Lesson 2

1. With spaceships like those that we use to explore our solar system, could we explore planets around other stars? Why or why not?

2. How does the distance between galaxies in the Local Group compare to the sizes of the galaxies?

3. Why are collisions between galaxies more likely than collisions between stars within galaxies?

Scale of the Universe Tutorial, Lesson 3

1. Why are powers of 10 useful for describing and comparing distances in the universe?

2. If you begin from the scale of the observable universe, how many powers of 10 must you zoom in to see the orbits of the planets in our solar system?

3. How many powers of 10 separate the scale on which you can see all the planetary orbits from the scale on which you can see individual people?

 Exploring the Sky and Solar System

Of the many activities available on the *Voyager: SkyGazer* **CD-ROM** accompanying your book, use the following files to observe key phenomena covered in this chapter.

Go to the **File: Basics** folder for the following demonstrations.

1. Chicago 10000AD

2. Dragging the Sky

Go to the **Explore** menu for the following demonstrations.

1. Solar Neighborhood

2. Paths of the Planets

Movies

Check out the following narrated and animated short documentaries available on www.astronomyplace.com for a helpful review of key ideas covered in this chapter.

From the Big Bang to Galaxies Movie

Web Projects

Take advantage of the useful Web links on www.astronomyplace.com to assist you with the following projects.

1. *Astronomy on the Web.* The Web contains a vast amount of astronomical information. Starting from the links on the textbook Web site, spend at least an hour exploring astronomy on the Web. Write two or three paragraphs summarizing what you learned from your Web surfing. What was your favorite astronomical Web site, and why?

2. *Tour Report.* Take the virtual tour of the Voyage scale model solar system on the text Web site (www.astronomyplace.com). After completing it, imagine that a friend asks you the following questions. Answer each question in one paragraph.

    a. Is the Sun really much bigger than Earth?

    b. Is it true that the Sun uses nuclear energy?

    c. Would it be much harder to send humans to Mars than to the Moon?

    d. In elementary school, I heard that Neptune is farther from the Sun than Pluto. Is this true?

    e. I read that Pluto is not really a planet. What's the story?

    f. Why didn't they have any stars besides the Sun in the scale model?

    g. What was the most interesting thing you learned during your tour?

3. *NASA Missions.* Visit the NASA Web site to learn about upcoming missions that concern astronomy. Write a one-page summary of the mission you feel is most likely to give us new astronomical information during the time you are enrolled in your astronomy course.

2 Discovering the Universe for Yourself

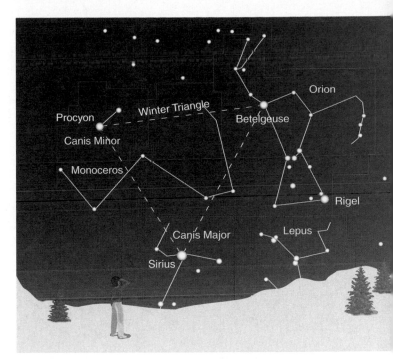

Figure 2.1 Red lines mark official borders of several constellations near Orion. Yellow lines connect recognizable patterns of stars within constellations. Sirius, Procyon, and Betelgeuse form a pattern spanning several constellations and called the Winter Triangle, which is easy to find on clear winter evenings.

This is an exciting time in the history of astronomy. A new generation of telescopes is probing the depths of the universe. Increasingly sophisticated space probes are collecting new data about the planets and other objects in our solar system. Rapid advances in computing technology allow scientists to analyze the vast amount of new data and to model the processes that occur in planets, stars, galaxies, and the universe.

One goal of this book is to help *you* share in the ongoing adventure of astronomical discovery. One of the best ways to become a part of this adventure is to discover the universe for yourself by doing what other humans have done for thousands of generations: Go outside, observe the sky around you, and contemplate the awe-inspiring universe of which you are a part. In this chapter, we'll discuss a few key ideas that will help you understand what you see in the sky.

2.1　Patterns in the Sky

Shortly after sunset, as daylight fades to darkness, the sky appears to fill slowly with stars. On clear, moonless nights far from city lights, as many as 2,000–3,000 stars may be visible to your naked eye. As you look at the stars, your mind might group them into many different patterns. If you observe the sky night after night or year after year, you will recognize the same patterns of stars.

People of nearly every culture gave names to patterns in the sky. The pattern that the Greeks named Orion, the hunter (see Figure 1.4), was seen by the ancient Chinese as a supreme warrior called *Shen*. Hindus in ancient India also saw a warrior, called *Skanda,* who as the general of a great celestial army rode a peacock. The three stars of Orion's belt were seen as three fishermen in a canoe by Aborigines of northern Australia. As seen from southern California, these three stars climb almost straight up into the sky as they rise in the east, which may explain why the Chemehuevi Indians of the California desert saw them

as a line of three sure-footed mountain sheep.[*] These are but a few of the many names, each accompanied by a rich folklore, that have been given to the pattern of stars we call Orion.

Constellations

The patterns of stars seen in the sky are usually called constellations. Astronomers, however, use the term **constellation** to refer to a specific *region* of the sky. Any place in the sky belongs to some constellation, and familiar patterns of stars merely help locate particular constellations. For example, the constellation Orion includes all the stars in the familiar pattern of the hunter plus the region of the sky in which these stars are found (Figure 2.1).

The official borders of the constellations were set in 1928 by members of the International Astronomical Union (IAU), an association of astronomers from around the world. The IAU divided the sky into 88 constellations (see Appendix I) whose borders correspond roughly to the star patterns recognized by Europeans. Thus, despite the wide variety of names given to patterns of stars by different cultures, the "official" names of constellations visible from the Northern Hemisphere can be traced back to the ancient

---

[*]These and other constellation stories are found in E. C. Krupp, *Beyond the Blue Horizon* (Oxford University Press, 1991).

Greeks and to other cultures of southern Europe, the Middle East, and northern Africa. No one knows exactly when these constellations were first named, although some names probably go back at least 5,000 years. The official names of the constellations visible from the Southern Hemisphere are primarily those given by seventeenth-century European explorers.

Learning your way around the constellations is no more difficult than learning your way around your neighborhood, and recognizing the patterns of just 20–40 constellations is enough to make the entire sky seem familiar. The best way to learn the constellations is to go out and view them, guided by the help of a few visits to a planetarium and the star charts in the back of this book (Appendix J). The *Skygazer* software that comes with this book can also help you learn constellations.

The Celestial Sphere

The stars in a particular constellation may appear to lie close to one another, but this is an illusion. Stars that appear right next to each other in the sky may be quite far apart in reality, because they may lie at very different distances from Earth (Figure 2.2). The illusion occurs because we lack depth perception when we look into space, a consequence of the fact that the stars are so far away. The ancient Greeks mistook this illusion for reality, imagining the Earth to be surrounded by a great **celestial sphere** on which the stars lay.

Today, the concept of a celestial sphere is still useful for our learning about the sky, even though we know that Earth does not really lie in the center of a giant ball of stars. We give names to special locations on the imaginary celestial sphere. As shown in Figure 2.2, the point directly over

Earth's North Pole is called the **north celestial pole**. The point directly over Earth's South Pole is called the **south celestial pole**. The **celestial equator** represents an extension of Earth's equator into space.

The stars form the patterns of the constellations on the celestial sphere, while the Sun, Moon, and planets appear to wander slowly among the stars. As we'll discuss in Section 2.6, the apparent motions of the Moon and the planets are fairly complex. The Sun, however, appears to circle the celestial sphere once each year on a simple path called the **ecliptic**. The ecliptic is the projection of the ecliptic plane (see Figure 1.14) onto the celestial sphere. A model of the celestial sphere typically shows the patterns of the stars, the borders of the 88 official constellations, the ecliptic, and the celestial equator and poles (Figure 2.3).

The Milky Way

As your eyes adapt to darkness at a dark site, you'll begin to see the whitish band of light called the *Milky Way*. Our Milky Way Galaxy gets its name from this band of light. You can see only part of the Milky Way at any particular time, but it stretches all the way around the celestial sphere. If you look carefully, you will notice that the Milky Way varies in width and has dark fissures running through it. The widest and brightest parts of the Milky Way are most easily seen from the Southern Hemisphere (Figure 2.4), which probably explains why the Aborigines of Australia

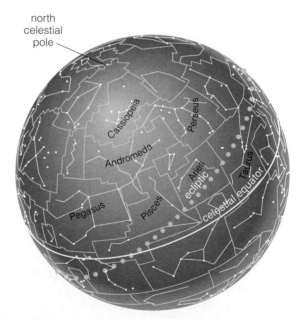

Figure 2.3 A model of the celestial sphere shows the patterns of the stars, the borders of the 88 official constellations, the ecliptic, and the celestial equator and poles. Because the celestial sphere represents the view from Earth, we imagine Earth to reside in the center of the sphere. Thus, when we look into the sky, we see patterns of stars as they would appear from *inside* this imaginary sphere (which means the patterns appear left-right reversed compared to what we see when we look at the model from the outside).

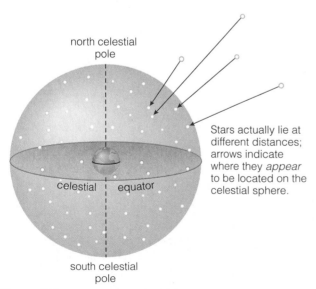

Figure 2.2 Our lack of depth perception when we look into space creates the illusion that Earth is surrounded by a *celestial sphere*. In reality, stars that appear very close together in our sky may actually lie at very different distances from Earth.

gave names to the patterns they saw within the Milky Way in the same way other cultures named patterns of stars.

The band of light called the Milky Way bears an important relationship to the Milky Way Galaxy: *It traces our galaxy's disk of stars—the galactic plane—as it appears from our location on the outskirts of the galaxy* (Figure 2.5). The Milky Way Galaxy is shaped like a thin pancake with a bulge in the middle. We view the universe from our location a little more than halfway out from the center of this "pancake." When we look in any direction *within* the plane of the galaxy, we see countless stars, along with interstellar gas and dust. These stars and glowing clouds of gas form the band of light we call the Milky Way. The dark fissures appear in regions where particularly dense interstellar clouds obscure our view of stars behind them. The central bulge of the galaxy makes the Milky Way wider in the direction of the galactic center, which is the direction of the constellation Sagittarius in our sky.

We see fewer stars when we look in directions pointing *away* from our location within the galactic plane. If we look in the direction of any of the white arrows in Figure 2.5, relatively little gas and dust will obscure our view of more distant objects. Thus, we have a clear view to the far reaches of the universe, limited only by what our

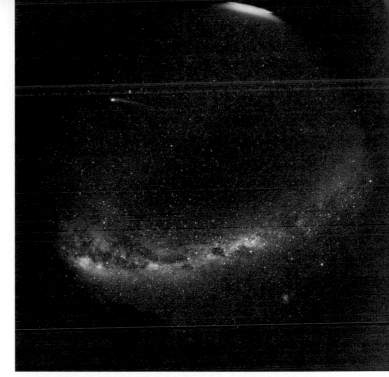

Figure 2.4 A "fish-eye" photograph of the Milky Way in the Australian sky. Near the upper left, Comet Hayakutake is visible in this 1996 photo.

Figure 2.5 Artist's conception of the Milky Way Galaxy from afar, showing how the galaxy's structure affects our view from Earth. When we look *into* the galactic plane in any direction, our view is blocked by stars, gas, and dust. Thus, we see the galactic plane as the band of light we call the Milky Way, stretching a full 360° around our sky (i.e., around the celestial sphere). We have a clear view to the distant universe only when we look *away from* the galactic plane.

Stars in the Daytime

Because we don't see stars in the daytime, some people believe that the stars vanish in the daytime and "come out" at night. In fact, the stars are always present. The reason your eyes cannot see stars in the daytime is that their dim light is overwhelmed by the bright daytime sky. You *can* see bright stars in the daytime with the aid of a telescope, and you may see stars in the daytime if you are fortunate enough to observe a total eclipse of the Sun. Astronauts can also see stars in the daytime. Above Earth's atmosphere, where no air is present to scatter sunlight through the sky, the Sun is a bright disk against a dark sky filled with stars. (However, because the Sun is so bright, astronauts must block its light and allow their eyes to adapt to darkness if they wish to see the stars.)

eyes or instruments allow. For example, if you are fortunate enough to have a very dark sky, you may see a fuzzy patch in the constellation Andromeda (Figure 2.6). Although this patch may look like nothing more than a small cloud, you are actually seeing the Great Galaxy in Andromeda—some 2.5 million light-years away.

Suppose our solar system were located on the opposite side of the galaxy at the same distance from the galactic center. Would we still be able to see the Great Galaxy in Andromeda? Would it still appear among the stars we recognize in the constellation Andromeda? Explain.

Figure 2.6 The location of the Great Galaxy in Andromeda—also known as M 31—among the constellations. From a dark site, you can see this galaxy with the naked eye as a small, fuzzy patch. This patch is the combined light of more than 100 billion stars, and its light has traveled through space for 2.5 million years to reach your eye. (See Figure 1.6 for a telescopic photo of this galaxy.)

2.2 The Circling Sky

If you spend a few hours out under a starry sky, you'll see stars (and the Moon and planets) rising and setting much like the Sun. In reality, we are the ones who are moving, not the stars. The Sun and stars appear to rise in the east and set in the west because Earth rotates in the opposite direction, from west to east (see Figure 1.13). In fact, Earth's rotation makes the entire celestial sphere appear to rotate around us each day. If you could view the celestial sphere from outside, the daily motion of the stars would appear as simple circles (Figure 2.7).

However, because we live on Earth, we see only *half* the celestial sphere at any one moment. The ground beneath us blocks our view of the other half. The particular half that we see depends on the time, the date, and our location on Earth. In this section, we'll discuss how to make sense of the sky as seen from Earth.

The Dome of the Local Sky

Picture yourself standing in a flat, open field. The sky appears to take the shape of a dome, making it easy to understand why people of many ancient cultures believed we live on a flat Earth lying under a great dome that encompasses the world. Today, we use the appearance of a dome to define the **local sky**—the sky as seen from wherever you happen to be standing (Figure 2.8). The boundary between Earth and sky is what we call the **horizon**. The point directly overhead is your **zenith**. Your **meridian** is an imaginary half-circle stretching from your horizon due south, through your zenith, to your horizon due north.

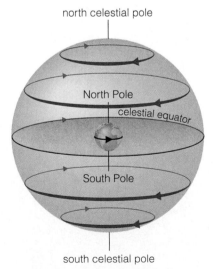

Figure 2.7 Earth rotates from west to east (black arrow), making the celestial sphere *appear* to rotate around us from east to west. The stars (and the Sun, Moon, and planets) therefore appear to make simple daily circles around us. The red circles represent the apparent daily paths of a few selected stars.

You can pinpoint the position of any object in your local sky by stating its **direction** along your horizon and its **altitude** above your horizon. For example, Figure 2.8 shows a person pointing to a star located in a southeasterly direction at an altitude of 60°. (The zenith has an altitude of 90° but no direction, because it is straight overhead.)

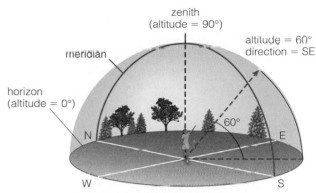

Figure 2.8 From any place on Earth, the local sky looks like a dome (hemisphere). This diagram shows key definitions in the local sky and how we can describe the position of any object in the local sky by its altitude and direction.

Angular Measures of Size and Distance

Because of our lack of depth perception in the sky, we cannot tell the true sizes of objects or the true distances between objects just by looking at them. For example, the Sun and the Moon look about the same size in our sky, but the Sun's diameter is actually about 400 times larger than the Moon's. We can, however, measure *angles* in the sky. The **angular size** of an object like the Sun or the Moon is the angle it appears to span in your field of view. The **angular distance** between a pair of objects is the angle that appears to separate them. For example, the angular size of the Moon is about $\frac{1}{2}°$ (Figure 2.9a), while the angular distance between the "pointer stars" at the end of the Big Dipper's bowl is about 5° (Figure 2.9b). You can use your outstretched hand to make rough estimates of angles in the sky (Figure 2.9c).

For more precise astronomical measurements, we subdivide each degree into 60 **arcminutes** and subdivide each arcminute into 60 **arcseconds**. Thus, there are 60 arcseconds in 1 arcminute, 60 arcminutes in 1°, and 360° in a full circle. We abbreviate arcminutes with the symbol ′ and arcseconds with the symbol ″. For example, we read 35°27′15″ as "35 degrees, 27 arcminutes, 15 arcseconds."

a The angular size of the Moon is about $\frac{1}{2}°$.

b The angular distance between the pointer stars of the Big Dipper is about 5°.

c You can estimate angular sizes or distances with your outstretched hand.

Stretch out your arm as shown here.

Figure 2.9 We measure *angular sizes* or *angular distances*, rather than actual sizes or distances, when we look at objects in the sky.

The Local Sky Varies with Latitude

If you stay in one place, you'll see the same set of stars following the same paths through your sky from one year to the next. But if you travel north or south, you'll notice somewhat different sets of stars moving through the sky on somewhat different paths. This observation convinced ancient scientists that there must be more to the sky than the simple dome visible from any one place.

By about 500 B.C., the famous mathematician Pythagoras was teaching that Earth is a sphere located at the center of a great celestial sphere [Section 3.3]. More than a century later, Aristotle (384–322 B.C.) cited observations of Earth's curved shadow on the Moon during lunar eclipses as evidence for a spherical Earth.

To understand why the sky changes with north–south travel, we must first review how we locate points on Earth (Figure 2.10a). **Latitude** measures positions north or south. Latitude is defined to be 0° at the equator, so the North Pole and the South Pole have latitude 90°N and 90°S, respectively. Note that "lines of latitude" are actually circles running parallel to the equator. **Longitude** measures east–west position, so "lines of longitude" are semicircles extending from the North Pole to the South Pole. The line of longitude passing through Greenwich, England, is defined to be longitude 0° (Figure 2.10b). This line is sometimes called the **prime meridian**. The decision to denote the line of longitude passing through Greenwich as the prime meridian was made by international treaty in 1884. Stating a latitude and a longitude pinpoints a location on Earth. For example, Figure 2.10a shows that Rome lies

at about 42°N latitude and 12°E longitude and that Miami lies at about 26°N latitude and 80°W longitude.

Paths of Stars Through the Local Sky

We have seen that Earth's daily rotation makes the celestial sphere appear to rotate around us (see Figure 2.7). However, because we see only half the celestial sphere from any

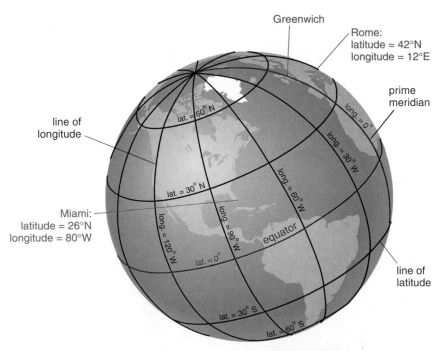

a Latitude measures angular distance north or south of the equator. Longitude measures angular distance east or west of the prime meridian, which passes through Greenwich, England.

b The entrance to the Old Royal Greenwich Observatory, near London. The line emerging from the door marks the prime meridian.

Figure 2.10 We can locate any place on Earth's surface by its latitude and longitude.

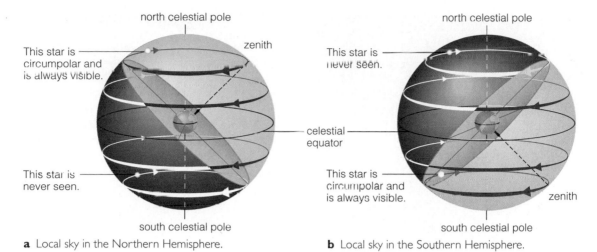

north celestial pole

zenith

This star is circumpolar and is always visible.

celestial equator

This star is never seen.

south celestial pole

a Local sky in the Northern Hemisphere.

north celestial pole

This star is never seen.

celestial equator

This star is circumpolar and is always visible.

zenith

south celestial pole

b Local sky in the Southern Hemisphere.

Figure 2.11 These diagrams show the local sky at two different latitudes—one in the Northern Hemisphere and one in the Southern Hemisphere. The horizon slices through different portions of the full celestial sphere at different latitudes. As a result, star paths in the local sky are tilted compared to the simple daily rotation of the celestial sphere. The diagrams are oriented with the North Pole up so that you can easily see how the sky differs at different latitudes. To follow star paths in each local sky, you should rotate the page so that the zenith for that sky points up.

location on Earth (at any one time), this simple motion looks more complicated in the local sky. We can understand why by comparing the full celestial sphere to the dome of the local sky.

Figure 2.11 compares the local sky to the celestial sphere at two sample latitudes on Earth, one in the Northern Hemisphere and one in the Southern Hemisphere. The diagrams are drawn with the North Pole facing up so that you can easily see why people at different latitudes view different portions of the celestial sphere. However, you'll be able to follow the star paths in each local sky better if you rotate the page until the zenith for that latitude points up. You'll immediately notice that the local horizon slices through the simple circles on the full celestial sphere. If you look more closely, you'll discover the following key facts about star paths through the local sky:

● In the Northern Hemisphere, stars relatively near the north celestial pole remain constantly above the horizon. We say that such stars are **circumpolar**. They never rise or set but instead make daily *counterclockwise* circles around the north celestial pole. In the Southern Hemisphere, circumpolar stars make daily *clockwise* circles around the south celestial pole. Figure 2.12 shows a beautiful photograph of the daily circles of stars.

● In the Northern Hemisphere, stars relatively near the south celestial pole remain constantly below the horizon and are never seen. In the Southern Hemisphere, stars near the north celestial pole are never visible. Thus, different sets of constellations are visible in northern and southern skies. Note that the set of constellations changes only with latitude, not with longitude.

● For both hemispheres, all other stars (those that appear above the horizon but are not circumpolar) daily rise in the east and set in the west. Stars located north of the celestial equator on the celestial sphere rise north of due east and set north of due west. Stars located on the celestial equator rise due east and set due west. Stars located south of the celestial equator rise south of due east and set south of due west.

The same ideas apply to the daily paths of the Sun, Moon, and planets through our sky. Although these objects all wander slowly among the constellations, on any particular day they appear essentially fixed among the stars on the celestial sphere. Thus, the paths of these objects follow the same rules stated above for star paths. For example, on a day when the Sun is located on the celestial equator (as it is on the days of the equinoxes, which we'll discuss shortly), it follows the same path through your local sky as a star on the celestial equator.

The Altitude of the Celestial Pole Equals Your Latitude

If you examine the geometry of the diagrams in Figure 2.11, you'll notice another key fact: *The altitude of the celestial pole in your sky is equal to your latitude.* For example, if the north celestial pole appears in your sky at an altitude of 40° above your north horizon, your latitude is 40°N. Similarly, if the south celestial pole appears in your sky at an altitude of 30° above your south horizon, your latitude is 30°S.

This feature is very useful for navigation, because it allows you to determine your latitude just by finding the celestial pole in your sky. Finding the north celestial pole is fairly easy, because it lies very close to the star Polaris (Figure 2.13a). In the Southern Hemisphere, you can find the

Figure 2.12 This time-exposure photograph, taken at Arches National Park in Utah, shows how Earth's rotation causes stars to trace daily circles around the sky. The north celestial pole lies at the center of the circles. Over the course of a full day, circumpolar stars trace complete circles, and stars that rise in the east and set in the west trace partial circles. Here we see only about one-quarter of each portion of the full daily path, because the time exposure lasted about 6 hours.

south celestial pole with the aid of the Southern Cross (Figure 2.13b). We'll discuss celestial navigation and how the sky varies with latitude in more detail in Chapter S1.

Answer the following questions for your latitude: Where is the north (or south) celestial pole in your sky? Where should you look to see circumpolar stars? What portion of the celestial sphere is never visible in your sky?

Annual Changes in the Night Sky

The basic patterns of motion in the sky remain the same from one day to the next. The Sun, Moon, planets, and stars trace daily circles around the sky. However, if you observe the sky night after night, you will notice changes that cannot be seen in a single night. For example, you may have noticed that the constellation Orion is prominent in the February evening sky but by September is visible only shortly before dawn.

The night sky changes through the year because of Earth's changing position in its orbit around the Sun. Figure 2.14 shows how this works. As we orbit the Sun over the course of a year, the Sun *appears* to move against the background of the distant stars in the constellations. We don't see the Sun and the stars at the same time, but if we could we'd notice the Sun gradually moving eastward along the ecliptic, completing one circuit each year. The constel-

lations along the ecliptic are called the constellations of the **zodiac**. (Tradition places 12 constellations along the zodiac, but the official borders include a wide swath of a thirteenth constellation, Ophiuchus.)

The Sun's apparent location along the ecliptic determines which constellations we see at night. For example, Figure 2.14 shows that in late August, Aquarius is visible on the meridian at midnight. If we could see stars in the daytime, the Sun would appear to be in Leo, which is opposite Aquarius on the celestial sphere. We therefore cannot see Leo in late August, because it moves with the Sun through the daytime sky. Six months later, in February, we see Leo at night, and Aquarius is above the horizon only in the daytime.

Based on Figure 2.14 and today's date, in what constellation does the Sun currently appear? What constellation of the zodiac will be on your meridian at midnight? What constellation of the zodiac will you see in the west shortly after sunset? Go outside at night to confirm your answers.

 Seasons Tutorial, Lessons 1–3

2.3 The Reason for Seasons

We have seen how Earth's rotation makes the sky appear to circle us daily and how the night sky changes as Earth orbits the Sun each year. The combination of Earth's rotation and its orbit also leads to the progression of the seasons. In this section, we'll explore the reason for seasons.

The Tilt of Earth's Axis Causes the Seasons

You know that we have seasonal changes, such as longer and warmer days in summer and shorter and cooler days in winter. But why do the seasons occur? The answer is that the tilt of Earth's axis causes sunlight to fall differently on Earth at different times of year, as shown in Figure 2.15.

Recall that Earth's axis remains pointed in the same direction in space (toward Polaris) throughout the year [Section 1.3]. Because Earth orbits the Sun, the orientation of the axis *relative to the Sun* changes over the course of

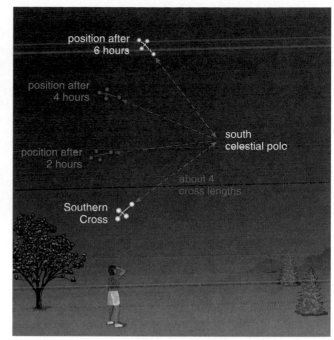

looking northward

looking southward

a In the Northern Hemisphere, the pointer stars of the Big Dipper point to Polaris, which lies within 1° of the north celestial pole. Note that the sky appears to turn *counterclockwise* around the north celestial pole.

b In the Southern Hemisphere, the Southern Cross points to the south celestial pole, which is not marked by any bright star. The sky appears to turn *clockwise* around the south celestial pole.

Figure 2.13 The altitude of the celestial pole in your sky is equal to your latitude.

Figure 2.14 This diagram shows why the Sun appears to move steadily eastward along the ecliptic, through the constellations of the zodiac. As Earth orbits the Sun, we see the Sun against the background of different zodiac constellations at different times of year. For example, on August 21 the Sun appears to be in the constellation Leo.

Spring Equinox
The Sun shines equally on both hemispheres. Northern Hemisphere is entering spring; Southern Hemisphere is entering fall.

Summer Solstice
Northern Hemisphere receives its most direct sunlight of the year (beginning of summer); Southern Hemisphere receives its least direct sunlight (beginning of winter).

Winter Solstice
Northern Hemisphere receives its least direct sunlight of the year (beginning of winter); Southern Hemisphere receives its most direct sunlight (beginning of summer).

Not to scale! On the scale the orbit is drawn, Earth would be too small to see (and the Sun would be a tiny dot).

Fall Equinox
The Sun shines equally on both hemispheres. Northern Hemisphere is entering fall; Southern Hemisphere is entering spring.

a Earth's axis points in the same direction (toward Polaris) throughout the year, causing its orientation *relative to the Sun* to change as Earth orbits the Sun.

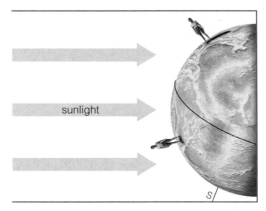

Summer Solstice: Midday sunlight strikes Earth more directly in the Northern Hemisphere—meaning the Sun is higher in the sky and casts smaller shadows—than in the Southern Hemisphere.

Winter Solstice: The situation is reversed from the summer solstice, with midday sunlight striking the Southern Hemisphere more directly and the Northern Hemisphere less directly.

b The shadows cast by the sunlight demonstrate the concentration of the Sun's energy. In the summer hemisphere, the sunlight is more intense because it is more direct and more concentrated. In the winter hemisphere, the less direct sunlight is spread over a larger area (as shown by the larger shadow) and is thus less intense.

Figure 2.15 The cause of the seasons. Note that the seasons are opposite in the Northern and Southern Hemispheres.

a year. For example, look at the left side of Figure 2.15a. The Northern Hemisphere is tipped toward the Sun, making it summer there, while the Southern Hemisphere is tipped away from the Sun, making it winter there. Half an orbit later (the right side of Figure 2.15a), the situation is reversed. The Northern Hemisphere is in winter because it is tipped away from the Sun, while the Southern Hemisphere is in summer because it is tipped toward the Sun. Notice that the axis has not changed the direction in which it is pointing. The change in which hemisphere is tipped toward the Sun occurs only because Earth moves between opposite sides of the Sun in its orbit. That is why the two hemispheres experience opposite seasons.

Figure 2.15b shows why the days are warmer in summer and cooler in winter. Notice that midday sunlight strikes the summer hemisphere at a steeper angle than it strikes the winter hemisphere. The Sun therefore follows a longer and higher path through the summer sky than through the winter sky. The steeper angle of summer sunlight also means that objects and people in the summer hemisphere cast smaller midday shadows than those in the winter hemisphere. In essence, sunlight is more concentrated in the summer hemisphere, and that is why summer tends to be warmer than winter. You can see the annual change in how high the Sun rises in your sky by observing the Sun's position at the same time each day (Figure 2.16).

Figure 2.16 This composite photograph shows images of the Sun, always from the same place and at the same time of day (mean solar time), snapped at 10-day intervals over an entire year. The three bright streaks show the path of the Sun's rise on three particular dates. This photograph looks east, so north is to the left and south is to the right. It was taken in the Northern Hemisphere. Notice that the Sun's altitude varies considerably. It is high in the summer and low in the winter. The sunrise position also changes. The Sun rises north of due east in the summer and south of due east in the winter. We'll discuss the reasons for the "figure 8" (called an *analemma*) in Chapter S1.

The Cause of Seasons

When asked what causes the seasons, many people mistakenly answer that the seasons are caused by variations in Earth's distance from the Sun. By knowing that the Northern and Southern Hemispheres experience opposite seasons, you'll realize that Earth's varying distance from the Sun *cannot* be the cause of the seasons. If it were, both hemispheres would have summer at the same time. Although Earth's distance from the Sun *does* vary slightly over the course of a year, this factor is greatly overwhelmed by the way the tilt of the rotation axis causes the Northern and Southern Hemispheres to alternately receive more or less direct sunlight. Earth's varying distance from the Sun has no noticeable effect on our seasons.

Solstices and Equinoxes

Figure 2.15 also shows where Earth is located in its orbit at four special times during the year: the two equinoxes and the two solstices. The **summer solstice**, which occurs around June 21 each year, is the day on which the Northern Hemisphere receives its most direct sunlight.* The **winter solstice**, which occurs around December 21, is the day on which the Northern Hemisphere receives its least direct sunlight. On the two equinoxes, Earth's orbital position is such that the Sun shines equally on both hemispheres. The

spring equinox (or *vernal equinox*) occurs around March 21 and marks the day on which the Northern Hemisphere switches from being tipped slightly away from the Sun to being tipped slightly toward the Sun. The **fall equinox** (or *autumnal equinox*) occurs around September 21 and marks the opposite change, when the Northern Hemisphere first starts to be tipped away from the Sun.

The names of the solstices and equinoxes reflect the northern seasons, which can make things sound strange when we talk about seasons in the Southern Hemisphere. For example, on the *summer* solstice it is *winter* in the Southern Hemisphere. This apparent injustice to people in the Southern Hemisphere arose because the solstices and equinoxes were named long ago by people living in the Northern Hemisphere. A similar injustice is inflicted on people living in equatorial regions. If you study Figure 2.15 carefully, you'll see that Earth's equator gets its most direct sunlight on the two equinoxes and its least direct sunlight on the solstices. Thus, people living near the equator don't experience four seasons in the same way as people living at mid-latitudes. Instead, equatorial regions generally have one portion of the year that tends to be rainier (often called the rainy or monsoon season) and one portion of the year that tends to be drier (the dry season), with the weather determined by global wind patterns.

On a related note, you are probably aware that seasonal variations become more extreme at high latitudes. For example, Alaska has much longer summer days and much longer winter nights than Florida. In fact, the Sun becomes circumpolar at very high latitudes (within the *Arctic* and *Antarctic Circles*) in the summer. The Sun never sets during these summer days in what we call the *land of the midnight Sun* (Figure 2.17). Of course, the name "land of noon darkness" would be more appropriate in the winter, when the Sun never rises above the horizon at these high latitudes.

---

*Historically, people thought of the summer solstice as occurring on a particular day of the year, and many ancient cultures built markers to help them know when this day came (see Chapter 3). Now that we understand Earth's orbit, we can pinpoint the precise *moment* of the summer solstice, which is why news reports tell both the date and the time at which it occurs. (The same is true for the equinoxes and the winter solstice.)

High Noon

When is the Sun directly overhead in your sky? Many people answer "at noon." It's true that the Sun reaches its *highest* point each day when it crosses the meridian, giving us the term "high noon" (though the meridian crossing is rarely at precisely 12:00 [Section S1.2]). However, unless you live in the Tropics (between latitudes 23.5°S and 23.5°N), the Sun is *never* directly overhead. In fact, any time you can see the Sun as you walk around, you can be sure it is *not* at your zenith. Unless you are lying down, seeing objects at the zenith requires tilting your head back into a very uncomfortable position. (To learn how to determine the Sun's maximum altitude in your sky, see Chapter S1.)

When Do the Seasons Begin?

You've probably heard it said that each equinox and solstice marks the first day of a season. For example, the summer solstice is usually said to mark the "first day of summer." However, if you think about what we've learned about the seasons so far, it might seem that the summer solstice should mark the middle rather than the beginning of summer. After all, the summer solstice is when the Northern Hemisphere is at its *maximum* tilt toward the Sun and has its "longest day" of the year—meaning the day with the most daylight and the least nighttime darkness. So why do we say that the summer solstice is the first day of summer?

Choosing the summer solstice to be the "first day of summer" is somewhat arbitrary. However, the choice is a fairly good one in at least two ways. First, it was much easier for ancient people to identify the day of the summer solstice than most other days of the year. Many prehistoric structures, such as Stonehenge, were used for this purpose [Section 3.2]. Second, the summer solstice comes fairly close to the beginning of the warmest three months of the year (in the Northern Hemisphere). Although the time around the summer solstice is when the Sun's path through the Northern Hemisphere sky is longest and highest, it is *not* usually the warmest time of the year. Instead, the warmest days tend to come about one to two months later. To understand why, think about what happens when you heat a pot of cold soup. Even though you may have the stove turned on high from the start, it takes a while for the soup to warm up. In the same way, it takes some time for sunlight to heat the ground and oceans from the cold of winter to the warmth of summer. Thus, "midsummer" in terms of weather comes in late July and early August, which makes the summer solstice a pretty good choice for the "first day of summer."

Why Orbital Distance Doesn't Matter

Recall that Earth's orbit is not a perfect circle, which means Earth's distance from the Sun varies over the course of each year [Section 1.3]. Yet distance variation plays no role in causing the seasons. The main reason why distance variation does not affect our seasons is that it is fairly small (see Figure 1.15). Earth is only about 3% farther from the Sun at its farthest point than at its nearest. The difference

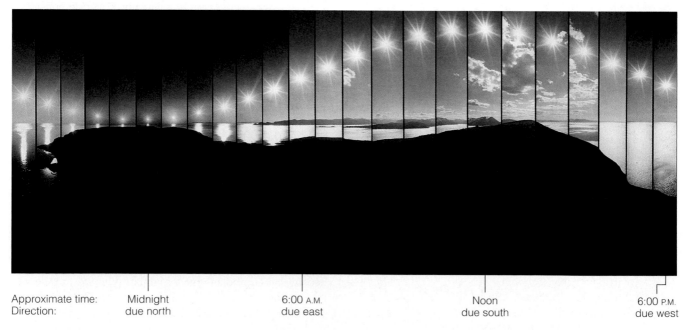

| Approximate time: | Midnight | 6:00 A.M. | Noon | 6:00 P.M. |
| Direction: | due north | due east | due south | due west |

Figure 2.17 This sequence of photos shows the progression of the Sun all the way around the horizon on the summer solstice at the Arctic Circle. Note that the Sun does not set but instead skims the northern horizon at midnight. It then gradually rises higher, reaching its highest point at noon, when it appears due south.

in the strength of sunlight due to this small change in distance is easily overwhelmed by the effects caused by the $23\frac{1}{2}°$ axis tilt.

Still, we might expect the varying orbital distance to make the seasons more extreme in one or the other hemisphere. For example, Earth is closest to the Sun (perihelion) in early January and farthest from the Sun (aphelion) in early July, which means that the Northern Hemisphere receives stronger winter sunlight and weaker summer sunlight than does the Southern Hemisphere. This "extra" winter sunlight should tend to make the Northern Hemisphere winter less cold, while the reduced summer sunlight should tend to make the Northern Hemisphere summer less hot. Thus, we would expect the Northern Hemisphere to have the more moderate seasons and the Southern Hemisphere to have the more extreme seasons. In fact, the opposite is true: The Northern Hemisphere seasons are more extreme than those of the Southern Hemisphere.

There are two reasons for this surprising fact. The more important one becomes obvious when you look at a map of Earth (Figure 2.18). Most of Earth's land lies in the Northern Hemisphere, with far more ocean in the Southern Hemisphere. As you can see at any beach, lake, or pool, water takes longer to heat or cool than soil or rock (largely because sunlight heats bodies of water to a depth of many meters while heating only the very top layer of land). The water temperature therefore remains fairly steady both day and night, while the ground can heat up and cool down dramatically. The Southern Hemisphere's larger amount of ocean moderates its climate. The North-ern Hemisphere, with more land and less ocean, heats up and cools down more easily, explaining why it has the more extreme seasons.

The second reason is the (much lesser) effect of Earth's orbital speed. Earth moves slightly faster in its orbit when it is closer to the Sun and slightly slower when it is farther from the Sun (a fact embodied in Kepler's second law [Section 3.4]). Thus, Earth is moving slightly slower during the Northern Hemisphere summer (when Earth is farther from the Sun), which makes the northern summer last about 2–3 days longer than the southern summer. A couple of extra days of more direct sunlight means more time to heat up.

Jupiter has an axis tilt of about 3°, small enough to be insignificant. Saturn has an axis tilt of about 27°, or slightly greater than that of Earth. Both planets have nearly circular orbits around the Sun. Do you expect Jupiter to have seasons? Do you expect Saturn to have seasons? Explain.

2.4 Precession of Earth's Axis

We have discussed both daily and annual changes in the sky. These changes are all we are likely to notice in our daily lives. However, a much longer cycle has been noticed over the centuries: a gradual change in the direction that Earth's axis points in space, known as **precession**. Although Earth's axis will remain pointed toward Polaris throughout our

equator

Figure 2.18 This map of Earth shows that most land lies in the Northern Hemisphere, while most ocean lies in the Southern Hemisphere. Because water heats and cools more slowly than land, the oceans tend to moderate temperature variations. Thus, the greater amount of ocean in the Southern Hemisphere makes its seasons less extreme than the Northern Hemisphere seasons.

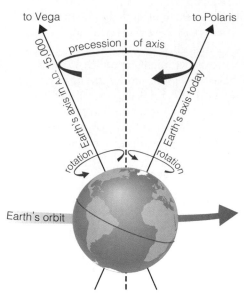

Figure 2.19 Precession affects the orientation, but not the tilt, of a spinning object's axis. The spinning top precesses because gravity tries to make it fall over. The rotating Earth precesses because gravitational pulls from the Sun and Moon try to "straighten out" Earth's axis. (Earth's axis precesses in the opposite direction from the top's—clockwise instead of counterclockwise—because its axis is being pulled toward the vertical, while the top's axis is being pulled toward the horizontal.)

a A spinning top wobbles, or *precesses*, more slowly than it spins.

b Earth's axis also precesses. Each precession cycle takes about 26,000 years. The axis tilt remains about the same ($23\frac{1}{2}°$) throughout the cycle, but the changing orientation of the axis means that Polaris is only a temporary North Star.

lifetimes, precession ensures that it has not always been pointed that way and will not always remain so.

Precession can occur with any rotating object. You can see it easily by spinning a top (Figure 2.19a). As the top spins rapidly, you'll notice that its axis also sweeps out a circle at a somewhat slower rate. We say that the top's axis *precesses*. Earth's axis precesses in much the same way, but far more slowly (Figure 2.19b). Each cycle of Earth's precession takes about 26,000 years. Today, the axis points toward Polaris, which makes it our North Star. In about 13,000 years, the axis will point nearly in the direction of the star Vega, making it the North Star at that time. During most of the precession cycle, the axis does not point toward any bright star. Note that the axis tilt remains close to $23\frac{1}{2}°$ throughout the cycle—it is only the axis orientation that changes.

Why does precession occur? It is caused by gravity's effect on a tilted, rotating object.* You have probably seen how gravity affects a top. If you try to balance a nonspinning top on its point, it will fall over almost immediately. This happens because the top will inevitably be leaning a little to one side. No matter how slight this lean, it is enough for gravity to tip the top over. But if you spin the top rapidly, it does not fall over so easily. The spinning top stays upright because rotating objects tend to keep spinning around the same rotation axis (a consequence of some-

thing known as the *law of conservation of angular momentum* [Section 5.2]). This tendency prevents gravity from immediately pulling the spinning top over, since falling over would mean a change in the spin axis from near-vertical to horizontal. Instead, gravity succeeds only in making the axis trace circles of precession. As friction slows the top's spin, the circles of precession get wider and wider, and ultimately the top falls over. If there were no friction to slow its spin, the top would spin and precess forever.

The rotating Earth precesses because of gravitational tugs from the Sun and Moon. These tugs try to "straighten

---

*A further requirement for precession is that the object *not* be a perfect sphere. Because the strength of gravity depends only on mass and not on an object's rate of spin, gravity can affect an object's tilt only if it is acting on some bulging mass that is not spherical. A top clearly is not spherical. Earth meets the requirement because, while it is close to a perfect sphere, it bulges somewhat around the equator. Thus, precession occurs because gravity acts on the Earth's "equatorial bulge."

Figure 2.20 The Moon phases demonstration. Your head represents Earth, and the ball represents the Moon. Hold the ball toward the Sun at arm's length. Then turn in a circle, always keeping the ball at arm's length and always looking directly at it. Be sure to swing the ball in a circle that carries it from being directly toward the Sun (when you are facing the Sun) to being directly opposite the Sun (when you are facing away from the Sun). As you turn, you will see the ball go through phases just like the phases of the Moon.

out" Earth's rotational axis, reducing its tilt. However, like any rotating object, the Earth tends to keep spinning around the same axis. The result is that gravity's attempt to reduce the Earth's axis tilt succeeds only in making the axis precess. That is why the tilt of the axis remains nearly constant (close to $23\frac{1}{2}°$) throughout each cycle of precession.

THINK ABOUT IT

You can observe and experiment with precession using an inexpensive toy *gyroscope*. Find a gyroscope and see how it works. How is its motion similar to that of the precessing Earth? How is it different? (*Hint:* Compare the time scales for the cycles of precession.) Can you use the idea that spinning objects tend to keep the same rotation axis to explain why it is much easier to ride a fast-moving bicycle than it is to balance on a non-moving bike?

 Phases of the Moon Tutorial, Lessons 1–3

2.5 The Moon, Our Constant Companion

Like all objects on the celestial sphere, each day the Moon rises in the east and sets in the west. Like the Sun, the Moon also moves gradually eastward through the constellations of the zodiac. However, it takes the Moon only about a month to make a complete circuit around the celestial sphere. If you carefully observe the Moon's posi-

tion relative to bright stars over just a few hours, you can notice the Moon's drift among the constellations.

As the Moon moves through the sky, both its appearance and the time at which it rises and sets change with the cycle of **lunar phases**. Each complete cycle from one new moon to the next takes about $29\frac{1}{2}$ days—hence the origin of the word *month* (think of "moonth"). The easiest way to understand the lunar phases is with a simple demonstration. Use a small ball to represent the Moon while your head represents Earth. If it's daytime and the Sun is shining, take your ball outside and notice how you see phases as you move the ball around your head (Figure 2.20). If it's dark or cloudy, you can place a flashlight a few meters away to represent the Sun. As you hold your ball at various places in its "orbit" around your head, you'll observe that phases result from just two basic facts:

1. At any particular time, half of the ball faces the Sun (or flashlight) and therefore is bright, while the other half faces away from the Sun and therefore is dark.

2. As you look at the ball, you see some combination of its bright and dark faces. This combination is the phase of the ball.

We see lunar phases for the same reason. Half of the Moon is always illuminated by the Sun, but the amount of this illuminated half that we see from Earth depends on the Moon's position in its orbit (Figure 2.21). The lunar phases

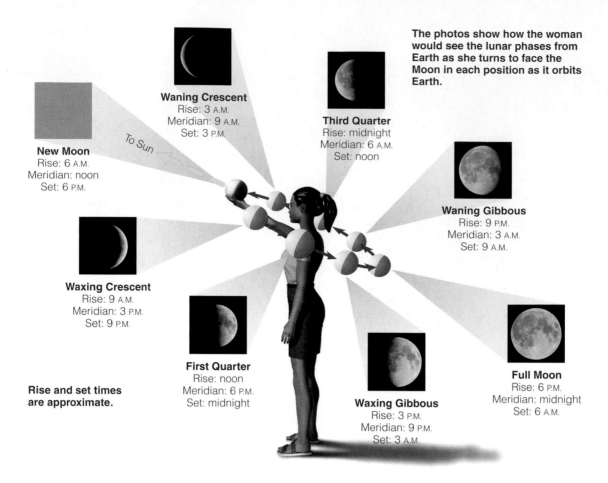

The photos show how the woman would see the lunar phases from Earth as she turns to face the Moon in each position as it orbits Earth.

Waning Crescent
Rise: 3 A.M.
Meridian: 9 A.M.
Set: 3 P.M.

Third Quarter
Rise: midnight
Meridian: 6 A.M.
Set: noon

New Moon
Rise: 6 A.M.
Meridian: noon
Set: 6 P.M.

To Sun

Waning Gibbous
Rise: 9 P.M.
Meridian: 3 A.M.
Set: 9 A.M.

Waxing Crescent
Rise: 9 A.M.
Meridian: 3 P.M.
Set: 9 P.M.

Rise and set times are approximate.

First Quarter
Rise: noon
Meridian: 6 P.M.
Set: midnight

Waxing Gibbous
Rise: 3 P.M.
Meridian: 9 P.M.
Set: 3 A.M.

Full Moon
Rise: 6 P.M.
Meridian: midnight
Set: 6 A.M.

Figure 2.21 As the Moon orbits Earth (represented here by the woman's head), the half facing the Sun is always illuminated, while the half facing away is always dark. But as we look at the Moon from Earth, we see some combination of the illuminated and dark portions, as shown in the photos of the lunar phases. (The new moon photo shows blue sky, because a new moon is always close to the Sun in the sky and hence hidden from view by the bright light of the Sun.) To understand the rise and set times (which are approximate and vary with latitude, time of year, and other factors), try the Phases of the Moon tutorial at astronomyplace.com.)

also determine the time of day during which the Moon is visible. For example, full moon occurs when the Moon is opposite the Sun in the sky, so the full moon rises around sunset, reaches the meridian at midnight, and sets around sunrise.

THINK ABOUT IT

Suppose you go outside in the morning and notice that the visible face of the Moon is half light and half dark. Is this a first-quarter or third-quarter moon? How do you know? (*Hint:* Study Figure 2.21.)

Although we see many *phases* of the Moon, we do not see many *faces*. In fact, from Earth we always see (nearly) the same face of the Moon.* This tells us that the Moon must rotate once on its axis in the same time that it makes a single orbit of Earth. You can observe this with another simple demonstration. Place a ball on a table to represent Earth while you represent the Moon. Start by facing the ball. If you

do not rotate as you walk around the ball, you'll be looking away from it by the time you are halfway around your orbit (Figure 2.22a). The only way you can face the ball at all times is by completing exactly one rotation while you complete one orbit (Figure 2.22b). (We'll learn why the Moon's periods of rotation and orbit are the same in Chapter 5.)

The Moon's appearance is affected to a lesser degree by its varying distance from Earth. Just as the Earth's orbit around the Sun is not a perfect circle, neither is the Moon's orbit around Earth. The Moon's distance from Earth varies during its orbit from a minimum of about 356,000 kilometers to a maximum of about 407,000 kilometers, with an average distance of 380,000 kilometers. When the Moon happens to be closer to Earth, it appears larger in angular

*Because the Moon's orbital speed varies while its rotation rate is steady, the Moon's visible face appears to wobble slightly back and forth as it orbits Earth. This effect, called *libration*, allows us to see a total of about 59% of the Moon's surface over the course of a month, even though we see only 50% of the Moon at any single time.

Moon in the Daytime

In traditions and stories, night is so closely associated with the Moon that many people mistakenly believe that the Moon is visible only in the nighttime sky. In fact, the Moon is above the horizon as often in the daytime as at night, though it is easily visible only when its light is not drowned out by sunlight. For example, a first-quarter moon is easy to spot in the late afternoon as it rises through the eastern sky, and a third-quarter moon is visible in the morning as it heads toward the western horizon (see rise and set times in Figure 2.21).

Another misconception appears in illustrations that show a star in the dark portion of the crescent moon (diagram below). A star in the dark portion appears to be in front of the Moon, which is impossible because the Moon is much closer to us than is any star.

This view, though common in art, can never occur because the star would have to be between Earth and the Moon.

Moon on the Horizon

You've probably noticed that the full moon appears to be larger when it is near the horizon than when it is high in your sky. However, this appearance is an illusion. If you measure the angular size of the full moon on a particular night, you'll find it is about the same whether it is near the horizon or high in the sky. (It actually appears slightly larger when overhead than when on the horizon, because you are viewing it from a position on Earth that is closer to the Moon by the radius of Earth.) In fact, the Moon's angular size in the sky depends only on its distance from Earth. Although this distance varies over the course of the Moon's monthly orbit, it does not change enough to cause a noticeable effect on a single night. (You can eliminate the illusion by viewing the Moon upside down between your legs when it is on the horizon.)

size. When it is farther from Earth, the Moon appears smaller in angular size.

The View from the Moon

A good way to solidify your understanding of the lunar phases is to imagine that you live on the side of the Moon that faces Earth. Look again at Figure 2.21. Note that at new moon you would be facing the day side of Earth. Thus, you would see *full earth* when people on Earth see new moon. Similarly, at full moon you would be facing the night side of Earth. Thus, you would see *new earth* when people on Earth see full moon. In general, you'd always see Earth in a phase opposite the phase of the Moon seen by people on Earth. Moreover, because the Moon always shows nearly the same face to Earth, Earth would appear to hang nearly

a If you do not rotate while walking around the model, you will not always face it.

b You will face the model at all times only if you rotate exactly once during each orbit.

Figure 2.22 The fact that we always see the same face of the Moon means that the Moon must rotate once in the same amount of time that it takes to orbit Earth once. You can see why by walking around a model of Earth while imagining that you are the Moon.

Figure 2.23 This illustration represents the ecliptic plane as the surface of a pond. The Moon's orbit is slightly tilted to the ecliptic plane. Thus, in this illustration, the Moon spends half of each orbit above the pond surface and half below the surface. The points at which the orbit crosses the surface represent the *nodes* of the Moon's orbit. Eclipses occur only when the Moon both is at a node (passing through the pond surface) *and* has a phase of either new moon or full moon—as is the case with the lower left and top right orbits shown. At all other times, new moons and full moons occur above or below the ecliptic plane, so no eclipse is possible.

Within the figure:
- Full and new moons not near nodes; no eclipses.
- Full and new moons occur near nodes; eclipses possible.
- Nodes are the points where the Moon's orbit crosses the ecliptic plane.
- Full and new moons not near nodes; no eclipses.
- Full and new moons occur near nodes; eclipses possible.
- The pond surface represents the ecliptic plane (the plane of Earth's orbit around the Sun).

stationary in your sky. In addition, because the Moon takes about a month to rotate, your "day" would last about a month. Thus, you'd have about 2 weeks of daylight followed by about 2 weeks of darkness as you watched Earth hanging in your sky and going through its cycle of phases.

Thinking about the view from the Moon clarifies another interesting feature of the lunar phases: The dark portion of the lunar face is not *totally* dark. Imagine that you are standing on the Moon when it is in a crescent phase. Because it's nearly new moon as seen from Earth, you would see nearly full earth in your sky. Just as we can see at night by the light of the Moon, the light of Earth would illuminate your night moonscape. (In fact, because Earth is much larger than the Moon, the full earth is much bigger and brighter in the lunar sky than the full moon is in Earth's sky.) This faint light illuminating the "dark" portion of the Moon's face is often called the *ashen light* or *earthshine*. This light enables us to see the outline of the full face of the Moon even when the Moon is not full.

 Eclipses Tutorial, Lessons 1–3

Eclipses

Look once more at Figure 2.21. If this figure told the whole story of the lunar phases, a new moon would always block our view of the Sun, and Earth would always prevent sunlight from reaching a full moon. More precisely, this figure makes it look as if the Moon's shadow should fall on Earth

during new moon and that Earth's shadow should fall on the Moon during full moon. Any time one astronomical object casts a shadow on another, it is called an **eclipse**. Figure 2.21 makes it look as if we should have an eclipse with every new moon and every full moon—but we don't.

The missing piece of the story in Figure 2.21 is that the Moon's orbit is inclined to the ecliptic plane by about 5°. To visualize this inclination, imagine the ecliptic plane as the surface of a pond, as shown in Figure 2.23. Because of the inclination of its orbit, the Moon spends most of its time either above or below this surface. It crosses *through* this surface only twice during each orbit: once coming out

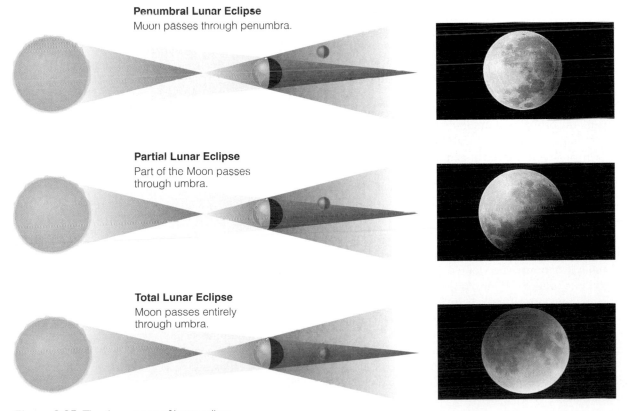

Figure 2.24 The shadow cast by an object in sunlight. Sunlight is fully blocked in the umbra and partially blocked in the penumbra.

and once going back in. The two points in each orbit at which the Moon crosses the surface are called the **nodes** of the Moon's orbit.

Figure 2.23 shows the position of the Moon's orbit at several different times of year. Note that the nodes are aligned the same way in each case (diagonally on the page). As a result, the nodes lie in a straight line with the Sun and Earth only about twice each year. (For reasons we'll discuss shortly, it is not *exactly* twice each year.) Because an eclipse can occur only when the Sun, Earth, and Moon lie along a straight line, two conditions must be met simultaneously for an eclipse to occur:

1. The nodes of the Moon's orbit must be nearly aligned with the Sun and Earth.

2. The phase of the Moon must be either new or full.

There are two basic types of eclipse. A **lunar eclipse** occurs when the Moon passes through Earth's shadow and therefore can occur only at *full moon*. A **solar eclipse** occurs when the Moon's shadow falls on Earth and therefore can occur only at *new moon*. But the full story of eclipse types is more complex, because the shadow of the Moon or of Earth consists of two distinct regions: a central **umbra**, where sunlight is completely blocked, and a surrounding **penumbra**, where sunlight is only partially blocked (Figure 2.24). Thus, an umbral shadow is totally dark, while a penumbral shadow is only slightly darker than no shadow.

Lunar Eclipses A lunar eclipse begins at the moment when the Moon's orbit first carries it into Earth's penumbra. After that, we will see one of three types of lunar eclipse (Figure 2.25). If the Sun, Earth, and Moon are nearly perfectly

Penumbral Lunar Eclipse
Moon passes through penumbra.

Partial Lunar Eclipse
Part of the Moon passes through umbra.

Total Lunar Eclipse
Moon passes entirely through umbra.

Figure 2.25 The three types of lunar eclipse.

aligned, the Moon will pass through Earth's umbra, and we will see a **total lunar eclipse**. If the alignment is somewhat less perfect, only part of the full moon will pass through the umbra (with the rest in the penumbra), and we will see a **partial lunar eclipse**. If the Moon passes *only* through Earth's penumbra, we will see a **penumbral lunar eclipse**.

Penumbral eclipses are the most common type of lunar eclipse, but they are difficult to notice because the full moon darkens only slightly. Partial lunar eclipses are easier to see because Earth's umbral shadow clearly darkens part of the Moon's face. (Note that Earth's umbra casts a curved shadow on the Moon, demonstrating that Earth is round.) A total lunar eclipse is particularly spectacular because the Moon becomes dark and eerily red during **totality** (the time during which the Moon is entirely engulfed in the umbra). The Moon is dark because it is in shadow, and it is red because Earth's atmosphere bends some of the red light from the Sun around Earth and toward the Moon.

Solar Eclipses We can also see three types of solar eclipse (Figure 2.26). If a solar eclipse occurs when the Moon is relatively close to Earth in its orbit, the Moon's umbra touches a small area of Earth's surface (no more than about 270 kilometers in diameter). Anyone within this area will see a **total solar eclipse**. Surrounding this region of totality is a much larger area (typically about 7,000 kilometers in diameter) that falls within the Moon's penumbral shadow. Anyone within this region will see a **partial solar eclipse**, in which only part of the Sun is blocked from view. If the eclipse occurs when the Moon is relatively far from Earth, the umbra may not reach Earth's surface at all. In that case, anyone in the small region of Earth directly behind the umbra will see an **annular eclipse**, in which a ring of sunlight surrounds the disk of the Moon. (Again, anyone in the surrounding penumbral shadow will see a partial solar eclipse.)

During any solar eclipse, the combination of Earth's rotation and the orbital motion of the Moon causes the circular umbral and penumbral shadows to race across the face of Earth at a typical speed of about 1,700 kilometers per hour (relative to the ground). As a result, the umbral (or annular) shadow traces a narrow path across Earth, and totality (or annularity) never lasts more than a few minutes in any particular place.

A total solar eclipse is a spectacular sight. It begins when the disk of the Moon first appears to touch the Sun. Over the next couple of hours, the Moon appears to take a larger and larger "bite" out of the Sun. As totality approaches, the sky darkens and temperatures fall. Birds head back to their nests, and crickets begin their nighttime chirping. During the few minutes of totality, the Moon completely blocks the normally visible disk of the Sun, allowing the faint *corona* to be seen (Figure 2.27). The surrounding sky takes on a twilight glow, and planets and bright stars become visible in the daytime. As totality ends, the Sun slowly emerges

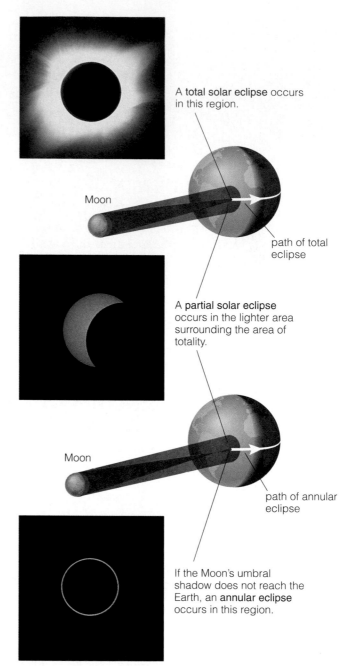

A **total solar eclipse** occurs in this region.

Moon

path of total eclipse

A **partial solar eclipse** occurs in the lighter area surrounding the area of totality.

Moon

path of annular eclipse

If the Moon's umbral shadow does not reach the Earth, an **annular eclipse** occurs in this region.

Figure 2.26 The three types of solar eclipse. (The photographs on the left are, from top to bottom, a total solar eclipse, a partial solar eclipse, and an annular eclipse.)

from behind the Moon over the next couple of hours. However, because your eyes have adapted to the darkness, totality appears to end far more abruptly than it began.

Predicting Eclipses Few phenomena have so inspired and humbled humans throughout the ages as eclipses. For many cultures, eclipses were mystical events associated with fate or the gods, and countless stories and legends surround eclipses. One legend holds that the Greek philosopher Thales (c. 624–546 B.C.) successfully predicted the year (but presumably not the precise time) that a total

Figure 2.27 This multiple-exposure photograph shows the progression of a total solar eclipse. Totality (central image) lasts only a few minutes, during which time we can see the faint corona around the outline of the Sun. This photo was taken July 22, 1990, in La Paz, Mexico.

eclipse of the Sun would be visible in the area where he lived, which is now part of Turkey. Coincidentally, the eclipse occurred as two opposing armies (the Medes and the Lydians) were massing for battle. The eclipse so frightened the armies that they put down their weapons, signed a treaty, and returned home. Because modern research shows that the only eclipse visible in that part of the world at about that time occurred on May 28, 585 B.C., we know the precise date on which the treaty was signed—the first historical event that can be dated precisely.

Much of the mystery of eclipses probably stems from the relative difficulty of predicting them. Look again at Figure 2.23. The two periods each year when the nodes of the Moon's orbit are nearly aligned with the Sun are called **eclipse seasons**. Each eclipse season lasts a few weeks, so

some type of lunar eclipse occurs during each eclipse season's full moon, and some type of solar eclipse occurs during its new moon.

If Figure 2.23 told the whole story, eclipse seasons would occur every 6 months, and predicting eclipses would be easy. For example, if eclipse seasons always occurred in January and July, eclipses would occur only on the dates of new and full moons in those months. But Figure 2.23 does not show one important thing about the Moon's orbit: The nodes slowly shift around the orbit. As a result, eclipse seasons actually occur slightly less than 6 months apart (about 173 days apart) and therefore do not recur in the same months year after year.

The combination of the changing dates of eclipse seasons and the $29\frac{1}{2}$-day cycle of lunar phases makes eclipses

---

SPECIAL TOPIC The Moon and Human Behavior

From myths of werewolves to stories of romance under the full moon, human culture is filled with claims that our behavior is influenced by the phase of the Moon. Can we say anything scientific about such claims?

The Moon clearly has important influences on Earth. Most visibly, the Moon is primarily responsible for the tides [Section 5.4]. However, the Moon's tidal force acts only over large distances and cannot directly affect objects as small as people.

If a physical force from the Moon cannot affect human behavior, could we be influenced in other ways? Certainly, anyone who lives near the oceans is influenced by the rising and falling of the tides. For example, fishermen and boaters must follow the tides. Thus, although the Moon does not directly influence their behavior, it does so indirectly through its effect on the oceans.

Many physiological patterns in many species appear to follow the lunar phases, and the average human menstrual cycle is so close in length to a lunar month that it is difficult to believe the similarity is mere coincidence. Nevertheless, aside from the obvious physiological cycles and the influence of tides on people who live near the oceans, claims that the lunar phase affects human behavior are difficult to verify scientifically. For example, although it is possible that the full moon brings out certain behaviors, it may also simply be that some behaviors are easier to exhibit when the sky is bright. A beautiful full moon may bring out your desire to walk on the beach under the moonlight, but there is no scientific evidence that the full moon would affect you the same way if you lived in a cave and couldn't see it.

Table 2.1   Total Lunar Eclipses 2003–2010

May 16, 2003
November 9, 2003
May 4, 2004
October 28, 2004
March 3, 2007
August 28, 2007
February 21, 2008
December 21, 2010

recur in a cycle of about 18 years $11\frac{1}{3}$ days. If a solar eclipse were to occur today, another would occur 18 years $11\frac{1}{3}$ days from now. This roughly 18-year cycle is called the **saros cycle**.

Astronomers in many ancient cultures identified the saros cycle and thus could predict *when* eclipses would occur. However, the saros cycle does not account for all the complications involved in predicting eclipses. If a solar eclipse occurred today, the one that would occur 18 years,

$11\frac{1}{3}$ days from now would not be visible from the same places on Earth and might not be of the same type. For example, one might be total and the other only partial. No ancient culture achieved the ability to predict eclipses in every detail.

Today, eclipses can be predicted because we know the precise details of the orbits of Earth and the Moon. Many astronomical software packages can do the necessary calculations. Table 2.1 lists upcoming total lunar eclipses, and Figure 2.28 shows paths of totality for upcoming total solar eclipses.

2.6   The Ancient Mystery of the Planets

Five planets are easy to find with the naked eye: Mercury, Venus, Mars, Jupiter, and Saturn. Mercury can be seen only infrequently, and then only just after sunset or just before sunrise because it is so close to the Sun. Venus often shines brightly in the early evening in the west or before dawn in the east. If you see a very bright "star" in the early evening

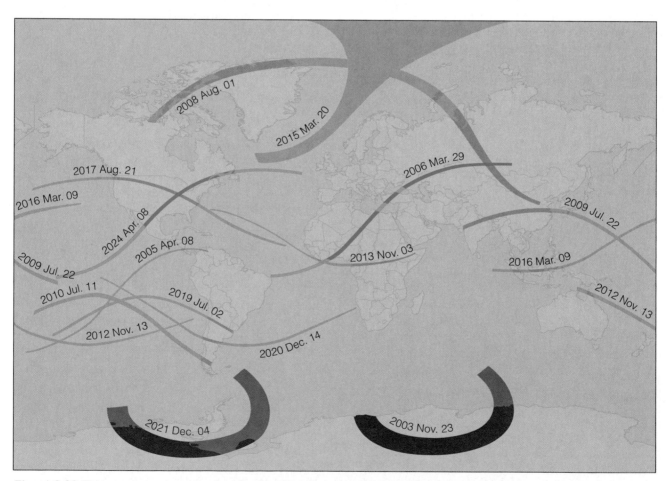

Figure 2.28 This map shows the paths of totality for solar eclipses from 2003 to 2025. Paths of the same color represent eclipses occurring in successive saros cycles, separated by 18 years 11 days. For example, the 2024 eclipse occurs 18 years 11 days after the 2006 eclipse (both shown in green).

or early morning it is probably Venus. Jupiter, when it is visible at night, is the brightest object in the sky besides the Moon and Venus. Mars is recognizable by its red color, but be careful not to confuse it with a bright red star. Saturn is also easy to see with the naked eye, but because many stars are just as bright as Saturn, it helps to know where to look. (It also helps to know that planets tend not to twinkle as much as stars.) Sometimes several planets may appear close together in the sky, offering a particularly beautiful sight (Figure 2.29).

Like the Sun and the Moon, the planets appear to move slowly through the constellations of the zodiac. (The word *planet* comes from the Greek for "wandering star.") However, while the Sun and the Moon always appear to move eastward relative to the stars, the planets occasionally reverse course and appear to move *westward* through the zodiac. A period during which a planet appears to move westward relative to the stars is called a period of **apparent retrograde motion** (*retrograde* means "backward"). Figure 2.30 shows a period of apparent retrograde motion for Jupiter.

Ancient astronomers could easily "explain" the daily paths of the stars through the sky by imagining that the celestial sphere was real and that it really rotated around Earth each day. But the apparent retrograde motion of the planets posed a far greater mystery: What could cause the planets sometimes to go backward? As we'll discuss in Chapter 3, the ancient Greeks came up with some very clever ways to explain the occasional backward motion of the planets, despite being wedded to the incorrect idea of an Earth-centered universe. However, their explanation was quite complex.

In contrast, apparent retrograde motion has a simple explanation in a Sun-centered solar system. You can demonstrate it for yourself with the help of a friend (Figure 2.31). Pick a spot in an open field to represent the Sun. You can represent Earth, walking counterclockwise around the Sun, while your friend represents a more distant planet (e.g.,

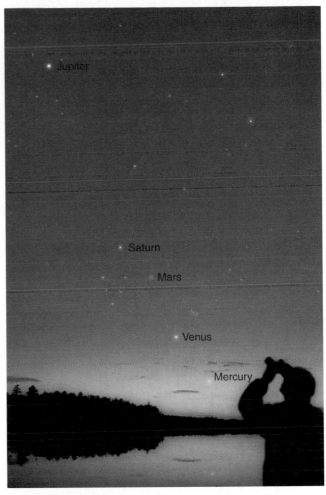

Figure 2.29 This photograph shows a rare planetary grouping in which all five planets that are easily visible to the naked eye appeared close together in the sky. It was taken near Chatsworth, New Jersey, just after sunset on April 23, 2002. The next such close grouping of these five planets in our sky will not occur until September 2040.

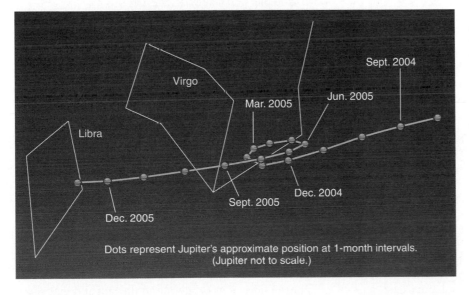

Dots represent Jupiter's approximate position at 1-month intervals. (Jupiter not to scale.)

Figure 2.30 This diagram shows Jupiter's approximate position among the stars in our sky during 2004–2005. Jupiter generally appears to drift eastward among the stars, but for about 4 months each year it turns back toward the west. Here, this apparent retrograde motion occurs between about February 2005 and June 2005. Note that Jupiter moves about one-twelfth of the way through the zodiac each year because it takes 12 years to complete one orbit of the Sun.

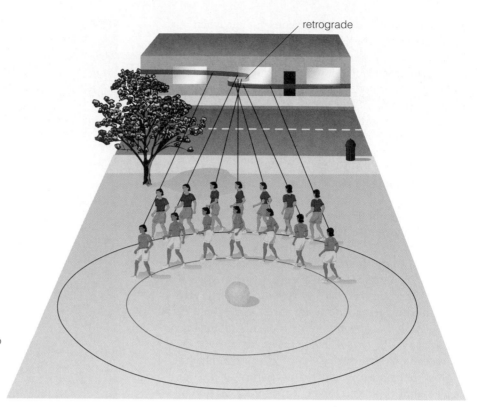

retrograde

Figure 2.31 The retrograde motion demonstration. Watch how your friend (in red) usually appears to you (in blue) to move forward against the background of the building in the distance but appears to move backward as you catch up to and pass him or her in your "orbit."

Mars, Jupiter, or Saturn) by walking counterclockwise around the Sun at a greater distance. Your friend should walk more slowly than you, because more distant planets orbit the Sun more slowly. As you walk, watch how your friend appears to move relative to buildings or trees in the distance. Although both of you always walk the same way around the Sun, your friend will *appear* to move backward against the background during the part of your "orbit" at which you catch up to and pass him or her. (To understand the apparent retrograde motions of Mercury and Venus, which are closer to the Sun than is Earth, simply switch places with your friend and repeat the demonstration.)

The apparent retrograde motion demonstration applies directly to the planets. For example, because Mars takes about 2 years to orbit the Sun (actually 1.88 years), it covers about half its orbit during the 1 year in which Earth makes a complete orbit. If you trace lines of sight from Earth to Mars from different points in their orbits, you will see that the line of sight usually moves eastward relative to the stars but moves westward during the time when Earth is passing Mars in its orbit (Figure 2.32). Like your friend in the demonstration, Mars never actually changes direction; it only *appears* to change direction from our perspective on Earth.

If the apparent retrograde motion of the planets is so readily explained by recognizing that Earth is a planet, why wasn't this idea accepted in ancient times? In fact, the idea that Earth goes around the Sun was suggested as early as 260 B.C. by the Greek astronomer Aristarchus and likely was debated many times thereafter. In a book published

in 1440, the German scholar Nicholas of Cusa wrote that Earth goes around the Sun. (Interestingly, although Galileo was punished by the Church for promoting the same belief two centuries later, Nicholas was ordained a priest in the year his book was published and later was elevated to cardinal.) The desire for a simple explanation for apparent retrograde motion was a primary motivation of Copernicus when he revived Aristarchus's idea in the early 1500s. (Copernicus was aware of the claim by Aristarchus but probably was not aware of the book by Nicholas of Cusa.)

Nevertheless, the idea that Earth goes around the Sun did not gain wide acceptance among scientists until the work of Kepler and Galileo in the early 1600s [Section 3.4]. Although there were many reasons for the historic reluctance to abandon the idea of an Earth-centered universe, perhaps the most prominent involved the inability of ancient peoples to detect something called **stellar parallax**.

Extend your arm and hold up one finger. If you keep your finger still and alternately close your left eye and right eye, your finger will appear to jump back and forth against the background. This apparent shifting, called *parallax,* occurs because your two eyes view your finger from opposite sides of your nose. If you move your finger closer to your face, the parallax increases. If you look at a distant tree or flagpole instead of your finger, you probably cannot detect any parallax by alternately closing your left eye and right eye. Thus, parallax depends on distance, with nearer objects exhibiting greater parallax than more distant objects.

If you now imagine that your two eyes represent Earth at opposite sides of its orbit around the Sun and that your

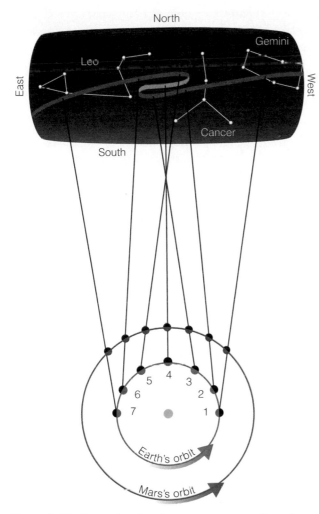

Figure 2.32 The explanation for apparent retrograde motion. Follow the lines of sight from Earth to Mars in numerical order. The period during which the lines of sight shift *westward* relative to the distant stars is the period during which we observe apparent retrograde motion for Mars.

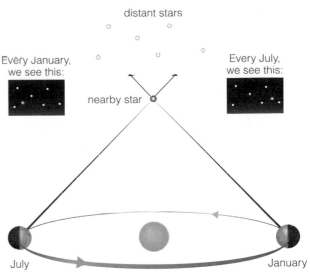

Figure 2.33 Stellar parallax is an apparent shift in the position of a nearby star as we look at it from different places in Earth's orbit. This figure is greatly exaggerated; in reality, the amount of shift is far too small to detect with the naked eye.

finger represents a relatively nearby star, you have the idea of stellar parallax. That is, because we view the stars from different places in our orbit at different times of year, nearby stars should *appear* to shift back and forth against distant stars in the background during the course of the year (Figure 2.33).

Because they believed that all stars lay on the same celestial sphere, the Greeks actually expected to see stellar parallax in a slightly different way. If Earth orbited the Sun, they reasoned that at different times of year we would be closer to different parts of the celestial sphere and thus would notice changes in the angular separations of stars. However, no matter how hard they searched, ancient

SPECIAL TOPIC Aristarchus

Until the early 1600s, nearly everyone believed that Earth was the center of the universe. Yet Aristarchus (c. 310–230 B.C.) had argued otherwise almost 2,000 years earlier.

Aristarchus proposed a Sun-centered system in about 260 B.C. Little of Aristarchus's work survives to the present day, so we cannot know why he made this proposal. It may have been an attempt to find a more natural explanation for the apparent retrograde motion of the planets. To account for the lack of detectable stellar parallax, Aristarchus suggested that the stars were extremely far away.

He further strengthened his argument by estimating the sizes of the Moon and the Sun. By observing the shadow of Earth on the Moon during a lunar eclipse, Aristarchus estimated the Moon's diameter to be about one-third of Earth's diameter—only slightly higher than the actual value. He then used a geometric argument, based on measuring the angle between the Moon and the Sun at first- and third-quarter phases, to conclude that the Sun must

be larger than Earth. (His measurements were imprecise, so he estimated the Sun's diameter to be about 7 times Earth's rather than the correct value of about 100 times.) His conclusion that the Sun is larger than Earth may have been another reason why he felt that Earth should orbit the Sun, rather than vice versa.

Like most scientific work, Aristarchus's work built upon the work of others. In particular, Heracleides (c. 388–315 B.C.) had previously suggested that Earth rotates. Aristarchus may have drawn on this idea to explain the apparent daily rotation of the stars in our sky. Heracleides also was the first to suggest that not all heavenly bodies circle Earth. Based on observations that Mercury and Venus always are close to the Sun in the sky, he argued that these two planets must orbit the Sun. Thus, in suggesting that all the planets orbit the Sun, Aristarchus was extending the ideas of Heracleides and others before him. Unfortunately, Aristarchus's arguments were not widely accepted in ancient times and were revived only with the work of Copernicus some 1,800 years later.

astronomers could find no sign of stellar parallax. They concluded that one of the following must be true:

1. Earth orbits the Sun but the stars are so far away that stellar parallax is undetectable to the naked eye.

2. There is no stellar parallax because Earth doesn't move; it is the center of the universe.

Unfortunately, with notable exceptions such as Aristarchus, ancient astronomers rejected the correct answer (1) because they could not imagine that the stars could be *that* far away. Today, we can detect stellar parallax with the aid of telescopes, providing direct proof that Earth really does orbit the Sun. Careful measurements of stellar parallax also provide the most reliable means of measuring distances to nearby stars [Section 16.2].

THINK ABOUT IT

How far apart are opposite sides of Earth's orbit? How far away are the nearest stars? Describe the challenge of detecting stellar parallax. It may help to visualize Earth's orbit and the distance to the stars on the 1-to-10-billion scale used in Chapter 1.

The ancient mystery of the planets drove much of the historical debate over Earth's place in the universe. In many ways, the modern technological society we take for granted today can be traced directly back to the scientific revolution that began because of the quest to explain the slow wandering of the planets among the stars in our sky.

THE BIG PICTURE
Putting Chapter 2 into Context

In this chapter, we surveyed the phenomena of our sky. Keep the following "big picture" ideas in mind as you continue your study of astronomy:

● You can enhance your enjoyment of learning astronomy by spending time outside observing the sky. The more you learn about the appearance and apparent motions of the sky, the more you will appreciate what you can see in the universe.

● From our vantage point on Earth, it is convenient to imagine that we are at the center of a great celestial sphere—even though we really are on a planet orbiting a star in a vast universe. We can then understand what we see in the local sky by thinking about how the celestial sphere appears from our latitude.

● Most of the phenomena of the sky are relatively easy to observe and understand. The more complex phenomena—particularly eclipses and apparent retrograde motion of the planets—challenged our ancestors for thousands of years and helped drive the development of science and technology.

SUMMARY OF KEY CONCEPTS

2.1 Patterns in the Sky

• *What is a constellation?* A constellation is a region of the sky. The sky is divided into 88 official constellations.

• *What is the celestial sphere?* It is an imaginary sphere surrounding Earth upon which the stars, Sun, Moon, and planets appear to reside.

• *Why do we see a band of light called the "Milky Way" in our sky?* This band of light traces the galactic plane as it appears from our location in the Milky Way Galaxy.

2.2 The Circling Sky

• *What are the basic features of the local sky?* The horizon is the boundary between Earth and sky. The meridian traces a half-circle from due south on your horizon, through the zenith (the point directly overhead), to due north on your horizon. Any point in the sky can be located by its altitude and direction.

• *How does the sky vary with latitude?* As the celestial sphere appears to rotate around us each day, we see different portions of the paths of stars from different latitudes. The altitude of the celestial pole (north or south) is the same as your latitude (north or south).

• *Which stars are above the horizon at all times?* All stars appear to make a daily circle. Circumpolar stars are those whose entire daily circles are above the horizon. Different sets of stars are circumpolar at different latitudes.

• *How does the night sky change through the year?* The visible constellations at a particular time of night depend on where Earth is located in its orbit around the Sun.

2.3 The Reason for Seasons

- *What is the cause of the seasons on Earth?* As Earth orbits the Sun, the tilt of its axis causes different portions of Earth to receive more or less direct sunlight at different times of year.

- *What are the solstices and equinoxes?* The summer and winter solstices are the times during the year when the Northern Hemisphere gets its most and least direct sunlight, respectively. The spring and fall equinoxes are the two times when both hemispheres get equally direct sunlight.

- *Why are the warmest days typically a month after the beginning of summer?* The summer solstice is usually considered the first day of summer in the Northern Hemisphere, but the warmest days come later because it takes time for the more direct sunlight to heat up the ground and oceans from the winter cold.

2.4 Precession of Earth's Axis

- *What is Earth's cycle of precession?* It is a roughly 26,000-year cycle over which Earth's axis sweeps out a circle as it gradually points to different places in space.

2.5 The Moon, Our Constant Companion

- *Why do we see phases of the Moon?* At any time, half the Moon is illuminated by the Sun and half is in darkness. The face of the Moon that we see is some combination of these two portions, determined by the relative locations of the Sun, Earth, and Moon.

- *What conditions are necessary for an eclipse?* An eclipse can occur only when the nodes of the Moon's orbit are nearly aligned with the Sun and Earth. When this condition is met, we can get a solar eclipse at new moon and a lunar eclipse at full moon.

- *Why were eclipses difficult for ancient peoples to predict?* There are three types of lunar eclipse and three types of solar eclipse. Although the pattern of eclipses repeats with the approximately 18-year saros cycle, it does not necessarily repeat with the same type of eclipse, and the eclipses are not necessarily visible from the same places on Earth.

2.6 The Ancient Mystery of the Planets

- *Why do planets sometimes seem to move backward relative to the stars?* A planet's period of apparent retrograde motion occurs over a few weeks to a few months as Earth passes the planet in its orbit (or as the planet passes Earth).

- *Why did the ancient Greeks reject the idea that Earth goes around the Sun even though it offers a more natural explanation for observed planetary motion?* A major reason was their inability to detect stellar parallax—the slight shifting of nearby stars against the background of more distant stars that occurs as Earth orbits the Sun. To most Greeks, it seemed unlikely that the stars could be so far away as to make parallax undetectable to the naked eye, even though that is, in fact, the case. They instead explained the lack of detectable parallax by imagining Earth to be stationary at the center of the universe.

❓ Does It Make Sense?

Decide whether the statement makes sense and explain why it does or does not. (For an example, see Chapter 1, "Does It Make Sense?")

1. If you had a very fast spaceship, you could travel to the celestial sphere in about a month.

2. The constellation Orion didn't exist when my grandfather was a child.

3. When I looked into the dark fissure of the Milky Way with my binoculars, I saw what must have been a cluster of distant galaxies.

4. Last night the Moon was so big that it stretched for a mile across the sky.

5. I live in the United States, and during my first trip to Argentina I saw many constellations that I'd never seen before.

6. Last night I saw Jupiter right in the middle of the Big Dipper. (*Hint:* Is the Big Dipper part of the zodiac?)

7. Last night I saw Mars move westward through the sky in its apparent retrograde motion. (*Hint:* How long does it take to notice apparent retrograde motion?)

8. Although all the known stars appear to rise in the east and set in the west, we might someday discover a star that will appear to rise in the west and set in the east.

9. If Earth's orbit were a perfect circle, we would not have seasons.

10. Because of precession, someday it will be summer everywhere on Earth at the same time.

Problems

(Quantitative problems are marked with an asterisk.)

11. *Dome of the Sky.* Why does the local sky look like a dome? Define *horizon, zenith,* and *meridian.* How do we describe the location of an object in the local sky?

12. *Angular Measures.* Explain why we can measure only *angular sizes* and *angular distances* for objects in the sky. What are *arcminutes* and *arcseconds?*

13. *No Axis Tilt.* Suppose Earth's axis had no tilt. Would we still have seasons? Why or why not?

14. *New Planet.* Suppose we discover a planet in another solar system that has a circular orbit and an axis tilt of 35°. Would you expect this planet to have seasons? If so, would you expect them to be more or less extreme than the seasons on Earth? If not, why not?

15. *View from Afar.* Describe how the Milky Way Galaxy would look in the sky of someone observing from a planet around a star in the Great Galaxy in Andromeda (M 31).

16. *Phases of the Moon.* Describe the Moon's cycle of *phases* and explain why we see phases of the Moon.

17. *Eclipses.* Why don't we see an *eclipse* at every new and full moon? Describe the conditions that must be met for us to see a *solar* or *lunar eclipse.*

18. *Stellar Parallax.* What is *stellar parallax?* Describe the role it played in making ancient astronomers believe in an Earth-centered universe.

19. *Your View.*

    a. Find your latitude and longitude, and state the source of your information.

    b. Describe the altitude and direction in your sky at which the north or south celestial pole appears.

    c. Is Polaris a circumpolar star in your sky? Explain.

    d. Describe the path of the meridian in your sky.

    e. Describe the path of the celestial equator in your sky. (*Hint:* Study Figure 2.11.)

20. *View from the Moon.* Suppose you lived on the Moon, near the center of the face that we see from Earth.

    a. During the phase of full moon, what phase would you see for Earth? Would it be daylight or dark where you live?

    b. On Earth, we see the Moon rise and set in our sky each day. If you lived on the Moon, would you see Earth rise and set? Why or why not? (*Hint:* Remember that the Moon always keeps the same face toward Earth.)

    c. What would you see when people on Earth were experiencing an eclipse? Answer for both solar and lunar eclipses.

21. *A Farther Moon.* Suppose the distance to the Moon were twice its actual value. Would it still be possible to have a total solar eclipse? An annular eclipse? A total lunar eclipse? Explain.

22. *A Smaller Earth.* Suppose Earth were smaller in size. Would solar eclipses be any different? If so, how? What about lunar eclipses? Explain.

23. *Observing Planetary Motion.* Find out what planets are currently visible in your evening sky. At least once a week, observe the planets and draw a diagram showing the position of each visible planet relative to stars in a zodiac constellation. From week to week, note how the planets are moving relative to the stars. Can you see any of the apparently "erratic" features of planetary motion? Explain.

24. *A Connecticut Yankee.* Find the book *A Connecticut Yankee in King Arthur's Court,* by Mark Twain. Read the portion that deals with the Connecticut Yankee's prediction of an eclipse (or read the entire book). In a one- to two-page essay, summarize the episode and how it helps the Connecticut Yankee gain power.

*25. There are 360° in a full circle.

    a. How many arcminutes are in a full circle?

    b. How many arcseconds are in a full circle?

    c. The Moon's angular size is about $\frac{1}{2}$°. What is this in arcminutes? In arcseconds?

Discussion Questions

26. *Earth-Centered Language.* Many common phrases reflect the ancient Earth-centered view of our universe. For example, the phrase "the Sun rises each day" implies that the Sun is really moving over Earth. We know that the Sun only *appears* to rise as the rotation of Earth carries us to a place where we can see the Sun in our sky. Identify other common phrases that imply an Earth-centered viewpoint.

27. *Flat Earth Society.* Believe it or not, there is an organization called the Flat Earth Society. Its members hold that Earth is flat and that all indications to the contrary (such as pictures of Earth from space) are fabrications made as part of a conspiracy to hide the truth from the public. Discuss the evidence for a round Earth and how you can check it for yourself. In light of the evidence, is it possible that the Flat Earth Society is correct? Defend your opinion.

For a complete list of media resources available, go to www.astronomyplace.com and choose Chapter 2 from the pull-down menu.

 Astronomy Place Web Tutorials

Tutorial Review of Key Concepts

Use the interactive **Tutorials** at www.astronomyplace.com to review key concepts from this chapter.

Seasons Tutorial

Lesson 1 Factors Affecting Seasonal Changes

Lesson 2 The Solstices and Equinoxes

Lesson 3 The Sun's Position in the Sky

Phases of the Moon Tutorial

Lesson 1 The Causes of Lunar Phases

Lesson 2 Time of Day and Horizons

Lesson 3 When the Moon Rises and Sets

Eclipses Tutorial

Lesson 1 Why and When Do Eclipses Occur?

Lesson 2 Types of Solar Eclipses

Lesson 3 Lunar Eclipses

Supplementary Tutorial Exercises

Use the interactive **Tutorial Lessons** to explore the following questions.

Seasons Tutorial, Lesson 1

1. What factors affect a planet's surface temperature?

2. Why does the tilt of Earth rather than Earth's orbit have a greater effect on Earth's seasons?

3. Why is the equator always hot and why are the poles always cold despite the passage of the seasons?

Phases of the Moon Tutorial, Lesson 3

1. When does a new moon rise and set?

2. In which phase does the Moon set just after the Sun?

3. What factors influence our view of the Moon from Earth?

Eclipses Tutorial, Lesson 1

1. What happens during a solar eclipse?

2. What happens during a lunar eclipse?

3. What role does the Moon's orbit play in the appearance of a solar eclipse?

 Exploring the Sky and Solar System

Of the many activities available on the *Voyager: SkyGazer* **CD-ROM** accompanying your book, use the following files to observe key phenomena covered in this chapter.

Go to the **File: Basics** folder for the following demonstrations.

1. Wide Field Milky Way

2. Eclipse 1991–1992 Views

3. Winter Sky

Go to the **File: Demo** folder for the following demonstrations.

1. Russian Midnight Sun

2. Earth Orbiting the Moon

3. Mars in Retrograde

Go to the **Explore** menu for the following demonstrations.

1. Shadows on Earth

2. Phases of the Planets

Web Projects

Take advantage of the useful Web links on www.astronomyplace.com to assist you with the following projects.

1. *Sky Information.* Search the Web for sources of daily information about sky phenomena (such as lunar phases, times of sunrise and sunset, or dates of equinoxes and solstices). Identify and briefly describe your favorite source.

2. *Constellations.* Search the Web for information about the constellations and their mythology. Write a short report about one or more constellations.

3. *Upcoming Eclipse.* Find information about an upcoming solar or lunar eclipse that you might have a chance to witness. Write a short report about how you could best witness the eclipse, including any necessary travel to a viewing site, and what you can expect to see. Bonus: Describe how you could photograph the eclipse.

3 The Science of Astronomy

We especially need imagination in science. It is not all mathematics, nor all logic, but is somewhat beauty and poetry.

Maria Mitchell (1818–1889), astronomer and first woman elected to American Academy of Arts and Sciences

Today we know that Earth is a planet orbiting a rather ordinary star, in a galaxy of a hundred billion or more stars, in an incredibly vast universe. We know that Earth, along with the entire cosmos, is in constant motion. We know that, on the scale of cosmic time, human civilization has existed only for the briefest moment. Yet we have acquired all this knowledge only recently in human history. How did we manage to learn these things?

It wasn't easy. Astronomy is the oldest of the sciences, with roots extending as far back as recorded history allows us to see. But while our current understanding of the universe rests on foundations laid long ago, the most impressive advances in knowledge have come in just the past few centuries.

In this chapter, we will trace how modern astronomy grew from its roots in ancient observations, including those of the Greeks. We'll pay special attention to the unfolding of the Copernican revolution, which overturned the ancient belief in an Earth-centered universe and laid the foundation for nearly all of modern science. Finally, we'll explore the nature of modern science and the scientific method.

3.1 Everyday Science

A common stereotype holds that scientists walk around in white lab coats and somehow think differently than other people. In reality, scientific thinking is a fundamental part of human nature.

Think about how a baby behaves. By about a year of age, she notices that objects fall to the ground when she drops them. She lets go of a ball—it falls. She pushes a plate of food from her high chair—it falls too. She continues to drop all kinds of objects, and they all plummet to Earth. Through powers of observation, the baby learns about the physical world: Things fall when they are unsupported.

Eventually, she becomes so certain of this fact that, to her parents' delight, she no longer needs to test it continually.

One day somebody gives the baby a helium balloon. She releases it, and to her surprise it rises to the ceiling! Her understanding of physics must be revised. She now knows that the principle "all things fall" does not represent the whole truth, although it still serves her quite well in most situations. It will be years before she learns enough about the atmosphere, the force of gravity, and the concept of density to understand *why* the balloon rises when most other objects fall. For now, she is delighted to observe something new and unexpected.

The baby's experience with falling objects and balloons exemplifies scientific thinking. In essence, it is a way of learning about nature through careful observation and trial-and-error experiments. Rather than thinking differently than other people, modern scientists simply are trained to organize this everyday thinking in a way that makes it easier for them to share their discoveries and employ their collective wisdom. This type of clear and organized thinking is at the heart of science. It is what has allowed us to acquire our present physical knowledge of the universe.

THINK ABOUT IT

When was the last time you used trial and error to learn something? Describe a few cases where you have learned by trial and error in cooking, participating in sports, fixing something, or any other situation.

Just as learning to communicate through language, art, or music is a gradual process for a child, the development of science has been a gradual process for humanity. Science in its modern form requires painstaking attention to detail, relentless testing of each piece of information to ensure its reliability, and a willingness to give up old beliefs that are not consistent with observed facts about the physical world. For professional scientists, these demands are the "hard work" part of the job. At heart, professional scientists are like the baby with the balloon, delighted by the unexpected and motivated by those rare moments when they—and all of us—learn something new about the universe.

3.2 The Ancient Roots of Science

We will discuss modern science shortly, but first we will explore how it arose from the observations of ancient peoples. Our exploration begins in central Africa, where people of many indigenous societies predict the weather with reasonable accuracy by making careful observations of the Moon. The Moon begins its monthly cycle as a crescent in the western sky just after sunset. Through long traditions of sky watching, central African societies learned that the

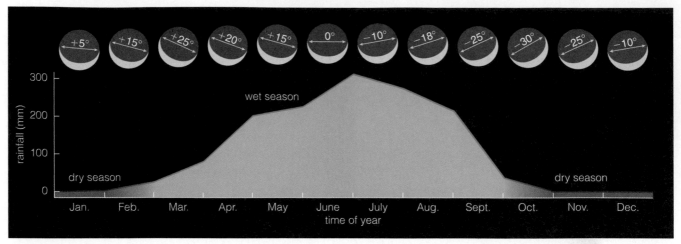

Figure 3.1 The roots of science lie in careful observation of the world around us. This diagram shows how central Africans used the Moon to predict the weather. The graph depicts the annual rainfall pattern in central Nigeria, characterized by a wet season and a dry season. The Moon diagrams represent the orientation of a waxing crescent moon relative to the western horizon at different times of year. The angle of the crescent "horns" allows observers to determine the time of year and hence the expected rainfall. (The orientation is measured in degrees, where 0° means that the crescent horns are parallel to the horizon.)

orientation of the crescent "horns" relative to the horizon is closely tied to rainfall patterns (Figure 3.1).

No one knows when central Africans first developed the ability to predict weather using the lunar crescent. The earliest-known written astronomical record comes from central Africa near the border of modern-day Congo and Uganda. It consists of an animal bone (known as the *Ishango* bone) etched with patterns that appear to be part of a lunar calendar, probably carved around 6500 B.C.

Why did ancient people bother to make such careful and detailed observations of the sky? In part, it was probably their inherent curiosity. In the daytime, they surely recognized the importance of the Sun to their lives. At night, without electric light, they were much more aware of the starry sky than we are today. Thus, it's not surprising that they paid attention to patterns of motion in the sky and developed ideas and stories to explain what they saw.

Astronomy also played a practical role in ancient societies by enabling them to keep track of time and seasons, a crucial skill for people who depended on agriculture for survival. This ability may seem quaint today, when digital watches tell us the precise time and date, but it required considerable knowledge and skill in ancient times, when the only clocks and calendars were in the sky.

Modern measures of time reflect their ancient astronomical roots. Our 24-hour day is the time it takes the Sun to circle our sky. The length of a month comes from the lunar cycle, and our calendar year is based on the cycle of the seasons. The days of the week are named after the seven naked-eye objects that appear to move among the constellations: the Sun, the Moon, and the five planets recognized in ancient times (Table 3.1). (The word *planet,* which means "wanderer," originally referred to the Sun and the Moon as well as the five visible planets. Earth was *not* considered a

planet in ancient times, since it was assumed to be stationary at the center of the universe.)

Determining the Time of Day

In the daytime, ancient peoples could tell time by observing the Sun's path through the sky. Many cultures probably used the shadows cast by sticks as simple sundials [Section S1.2]. The ancient Egyptians built huge obelisks, often inscribed or decorated in homage to the Sun, that probably also served as simple clocks (Figure 3.2).

At night, the Moon's position and phase give an indication of the time (see Figure 2.21). For example, a first-

Table 3.1 The Seven Days of the Week and the Astronomical Objects They Honor The correspondence between objects and days is easier to see in French and Spanish. In English, the correspondence becomes clear when we look at the names of the objects used by the Teutonic tribes who lived in the region of modern-day Germany.

Object	Teutonic Name	English	French	Spanish
Sun	Sun	Sunday	dimanche	domingo
Moon	Moon	Monday	lundi	lunes
Mars	Tiw	Tuesday	mardi	martes
Mercury	Woden	Wednesday	mercredi	miércoles
Jupiter	Thor	Thursday	jeudi	jueves
Venus	Fria	Friday	vendredi	viernes
Saturn	Saturn	Saturday	samedi	sábado

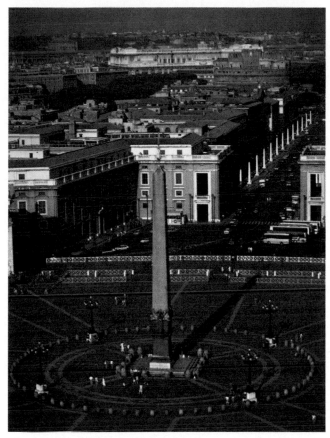

Figure 3.2 This ancient Egyptian obelisk, which stands 83 feet tall and weighs 331 tons, resides in St. Peter's Square at the Vatican in Rome. It is one of 21 surviving obelisks from ancient Egypt, most of which are now scattered around the world. Shadows cast by the obelisks may have been used to tell time.

quarter moon sets around midnight, so it is not yet midnight if the first-quarter moon is still above the western horizon. The positions of the stars also indicate the time if you know the approximate date. For example, in December the constellation Orion rises around sunset, reaches the meridian around midnight, and sets around sunrise. Hence, if it is winter and Orion is setting, dawn must be approaching. Most ancient peoples probably were adept at estimating the time of night, although written evidence is sparse.

Our modern system of dividing the day into 24 hours arose in ancient Egypt some 4,000 years ago. The Egyptians divided the daylight into 12 equal parts, and we still break the 24-hour day into 12 hours each of a.m. and p.m. (The abbreviations *a.m.* and *p.m.* stand for the Latin terms *ante meridiem* and *post meridiem*, respectively, which mean "before the middle of the day" and "after the middle of the day.")

The Egyptian "hours" were not a fixed amount of time because the amount of daylight varies during the year. For example, "summer hours" were longer than "winter hours," because one-twelfth of the daylight lasts longer in summer than in winter. Only much later in history did the hour become a fixed amount of time, subdivided into 60 equal minutes each consisting of 60 equal seconds.

The Egyptians also divided the night into 12 equal parts, and early Egyptians used the stars to determine the time at night. Egyptian *star clocks*, often found painted on the coffin lids of Egyptian pharaohs, essentially cataloged where particular stars appear in the sky at particular times of night and particular times of year. By knowing the date from their calendar and observing the positions of particular stars in the sky, the Egyptians could use the star clocks to estimate the time of night.

By about 1500 B.C., Egyptians had abandoned star clocks in favor of clocks that measure time by the flow of water through an opening of a particular size, just as hourglasses measure time by the flow of sand through a narrow neck.* These *water clocks* had the advantage of being useful even when the sky was cloudy. They eventually became the primary timekeeping instruments for many cultures, including the Greeks, Romans, and Chinese. Water clocks, in turn, were replaced by mechanical clocks in the 1600s and by electronic clocks in the twentieth century. Despite the availability of other types of clocks, sundials were in use throughout ancient times and remain popular today both for their decorative value and as reminders that the Sun and stars once were our only guides to time.

Determining the Time of Year

Many ancient cultures built structures to help them mark the seasons. One of the oldest standing human-made structures served such a purpose: Stonehenge in southern England, which was constructed in stages from about 2750 B.C. to about 1550 B.C. (Figure 3.3). Observers standing in its center see the Sun rise directly over the Heel Stone only on the summer solstice. Stonehenge also served as a social gathering place and probably as a religious site. No one knows whether its original purpose was social or astronomical. Perhaps it was built for both—in ancient times, social rituals and practical astronomy probably were deeply intertwined.

One of the most spectacular structures used to mark the seasons was the Templo Mayor in the Aztec city of Tenochtitlán, located on the site of modern-day Mexico City (Figure 3.4). Twin temples stood on top of a flat-topped, 150-foot-high pyramid. From the location of a royal observer watching from the opposite side of the plaza, the Sun rose directly through the notch between the twin temples on the equinoxes. Like Stonehenge, the Templo Mayor served important social and religious functions in addition to its astronomical role. Before it was destroyed by the Conquistadors, other Spanish visitors reported elaborate rituals, sometimes including human sacrifice, that took place at the Templo Mayor at times determined by astronomical observations.

---

*Hourglasses using sand were not invented until about the eighth century A.D., long after the advent of water clocks. Natural sand grains vary in size, so making accurate hourglasses required technology for getting uniform grains of sand.

a Stonehenge today.

Figure 3.3 Stonehenge helped its builders keep track of the seasons.

b A sketch showing how archaeologists believe Stonehenge looked when construction was completed around 1550 B.C. Note, for example, that observers standing in the center would see the Sun rise directly over the Heel Stone on the summer solstice.

Many ancient cultures aligned their buildings with the cardinal directions (north, south, east, and west), enabling them to mark the rising and setting of the Sun relative to the building orientation. Some cultures created monuments with a single special astronomical purpose. For example, someone among the ancient Anasazi people carved a spiral known as the Sun Dagger on a vertical cliff face near the top of a butte in Chaco Canyon, New Mexico (Figure 3.5). The Sun's rays form a dagger of sunlight that pierces the center of the carved spiral only once each year—at noon on the summer solstice.

Lunar Cycles and Calendars

Many ancient civilizations paid particular attention to the lunar cycle, often using it as the basis for lunar calendars. The months on lunar calendars generally have either 29 or 30 days, chosen to make the average agree with the approx-

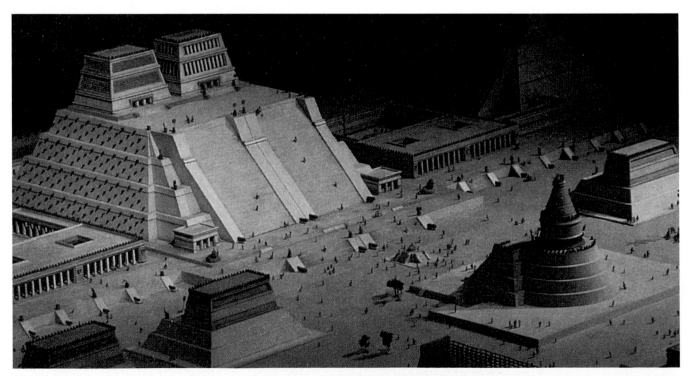

Figure 3.4 This scale model shows the Templo Mayor and the surrounding plaza as they are thought to have looked before Aztec civilization was destroyed by the Conquistadors. The structure was used to help mark the seasons.

a The Sun Dagger is an arrangement of rocks that shape the Sun's light into a dagger, with a spiral carved on a rock face behind them. The dagger pierces the center of the carved spiral each year at noon on the summer solstice.

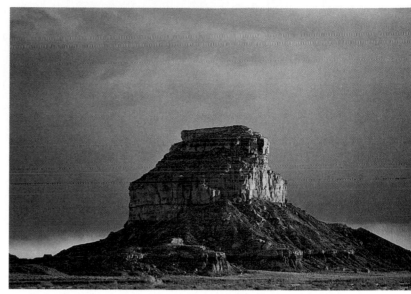

b The Sun Dagger is located on a vertical cliff face high on this butte in New Mexico.

Figure 3.5 The Sun Dagger, a summer solstice marker built by the Anasazi.

imately $29\frac{1}{2}$-day lunar cycle. Thus, a 12-month lunar calendar has only 354 or 355 days, or about 11 days fewer than a calendar based on the Sun. Such a calendar is still used in the Muslim religion. That is why the month-long fast of Ramadan (the ninth month) begins about 11 days earlier with each subsequent year.

Other lunar calendars take advantage of the fact that 19 years is almost precisely 235 lunar months. That is, every 19 years we get the same lunar phases on about the same dates. Because this 19-year cycle was recognized by the Greek astronomer Meton in 432 B.C., it is called the **Metonic cycle**. (However, the cycle was almost certainly known to Babylonian astronomers centuries before Meton.) A lunar calendar can be synchronized to the Metonic cycle by adding a thirteenth month to 7 of every 19 years (making exactly 235 months in each 19-year period), ensuring that "new year" comes on approximately the same date every nineteenth year.

The Jewish calendar follows the Metonic cycle, adding a thirteenth month in the third, sixth, eighth, eleventh,

---

Mathematical Insight **3.1** **The Metonic Cycle**

The Metonic cycle can be used to keep lunar calendars fairly closely synchronized with solar calendars. To see why, we need to know that each cycle of lunar phases takes an average of 29.53 days. Thus, a 12-month lunar calendar has $12 \times 29.53 = 354.36$ days, or about 11 days less than our 365-day solar calendar. If the new year on a 12-month lunar calendar occurs this year on October 1, next year it will occur 11 days earlier, on September 20. The following year it will occur 11 days before that, on September 9, and so on. Thus, on a lunar calendar that does not use the Metonic cycle, over many years the lunar new year would move all the way around the solar calendar.

The Metonic cycle is a period of 19 years on a solar calendar, which is almost precisely 235 months on a lunar calendar. To verify this fact, we can look up that a solar year is 365.25 days and then calculate that 19 solar years is equivalent to:

$$19 \text{ yr} \times \frac{365.25 \text{ days}}{1 \text{ yr}} = 6,939.75 \text{ days}$$

Similarly, we find that 235 lunar months is:

$$235 \text{ months} \times \frac{29.53 \text{ days}}{1 \text{ month}} = 6,939.55 \text{ days}$$

Notice that the difference between 19 solar years and 235 lunar months is only about 0.2 day, or about 5 hours. This near equivalence means that the dates of lunar phases repeat with each Metonic cycle. For example, there was a new moon on October 1, 1998, and we will have a new moon on October 1, 2017. (In general, the dates may be off by 1 day because of the 5-hour "error" in the Metonic cycle and also because 19 years on our calendar may be either 6,939 days or 6,940 days, depending on the leap-year cycle.)

For a lunar calendar to remain roughly synchronized with a solar calendar, it must have exactly 235 months in each 19-year period. Because a 12-month lunar calendar has a total of only $19 \times 12 = 228$ months in 19 years, the lunar calendar needs 7 extra months to reach the required 235 months. One way to accomplish this is to have 7 years with a thirteenth lunar month in every 19-year cycle, as is the case for the Jewish calendar.

fourteenth, seventeenth, and nineteenth years of each cycle. This explains why the date of Easter changes each year: The New Testament ties the date of Easter to the Jewish festival of Passover, which has its date set by the Jewish lunar calendar. In a slight modification of the original scheme, most Western Christians now celebrate Easter on *the first Sunday after the first full moon after March 21*. If the full moon falls on Sunday, Easter is the following Sunday. (Eastern Orthodox churches calculate the date of Easter differently, because they base the date on the Julian rather than the Gregorian calendar [Section S1.3].)

Some ancient cultures learned to predict eclipses by recognizing the 18-year saros cycle [Section 2.5]. In the Middle East, the ancient Babylonians achieved remarkable success in predicting eclipses more than 2,500 years ago. The most successful eclipse predictions prior to modern times were probably made by the Mayans in Central America. The Mayan calendar featured a sacred cycle that almost certainly was related to eclipses. This Mayan cycle, called the *sacred round*, lasted 260 days—almost exactly $1\frac{1}{2}$ times the 173.32 days between successive eclipse seasons. Unfortunately, we know little more about the extent of Mayan knowledge because the Spanish Conquistadors burned most Mayan writings.

The complexity of the Moon's orbit leads to other long-term patterns in the Moon's appearance. For example, the full moon rises at its most southerly point along the eastern horizon only once every 18.6 years, a phenomenon that may have been observed from the 4,000-year-old sacred stone circle at Callanish, Scotland (Figure 3.6).

Observations of Planets and Stars

Many ancient cultures also made careful observations of planets and stars. For example, Mayan observatories in Central America, such as the one still standing at Chichén Itzá (Figure 3.7), had windows strategically placed for observations of Venus.

The rising and setting of bright stars was sometimes used to track the seasons because the times at which stars are visible follow a simple annual pattern (see Figure 2.14). For example, early Central American people marked the beginning of their year when the group of stars called the Pleiades first rose in the east after having been hidden from view for 40 days by the glare of the Sun. On the other side of the world, southern Greeks also used the Pleiades. The Greek farmer Hesiod composed an epic poem (*Works and Days*) in about 800 B.C. that includes directions on planting and harvesting by the Pleiades:

> *When you notice the daughters of Atlas, the Pleiades, rising, start on your reaping, and on your sowing when they are setting. They are hidden from your view for a period of forty full days, both night and day, but then once again, as the year moves round, they reappear at the time for you to be sharpening your sickle.*

Some people built elaborate observation aids to mark the rising and setting of the stars. More than 800 lines, some stretching for miles, are etched in the dry desert sand of Peru between the Ingenio and Nazca Rivers. Many of these lines may simply have been well-traveled pathways. Others are aligned in directions that point to places where

Figure 3.6 The Moon rising between two stones of the 4,000-year-old sacred stone circle at Callanish, Scotland (on the Isle of Lewis in the Scottish Hebrides). The full moon rises in this position only once every 18.6 years.

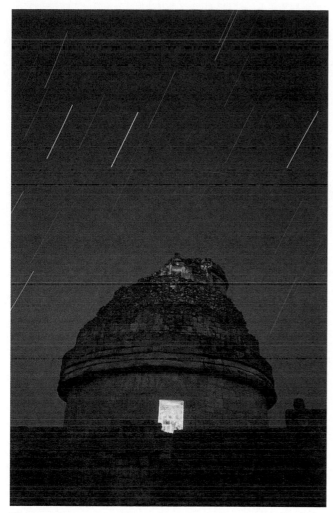

Figure 3.7 The ruins of the Mayan observatory at Chichén Itzá.

Figure 3.8 Hundreds of lines and patterns are etched in the sand of the Nazca desert in Peru. This aerial photo shows a large etched figure of a hummingbird.

bright stars or the Sun rose at particular times of year. In addition to the many straight lines, the desert features many large figures of animals. These may be representations of constellations made by the Incas who lived in the region (Figure 3.8).

THINK ABOUT IT

The animal figures show up clearly only when seen from above. As a result, some UFO enthusiasts argue that the patterns must have been created by aliens. What do you think of this argument? Defend your opinion.

The great Incan cities of South America clearly were built with astronomy in mind. Entire cities appear to have been designed so that particular arrangements of roads, buildings, or other human-made structures would point to places where bright stars rose or set, or where the Sun rose and set at particular times of year.*

---

*For more details about the layout of Incan cities and other ancient astronomy discussed in this chapter, see Anthony F. Aveni, *Ancient Astronomers* (Smithsonian Books, St. Remy Press and Smithsonian Institution, 1993).

Structures for astronomical observation also were popular in North America. Lodges built by the Pawnee people in Kansas featured strategically placed holes for observing the passage of constellations that figured prominently in their folklore. In the northern plains of the United States, Native American Medicine Wheels probably were designed for astronomical observations. The "spokes" of the Medicine Wheel at Big Horn, Wyoming, were aligned with the rising and setting of bright stars, as well as with the rising and setting of the Sun on the equinoxes and solstices (Figure 3.9). The 28 spokes of the Medicine Wheel probably relate to the month of the native Americans, which they measured as 28 days (rather than 29 or 30 days) because they did not count the day of the new moon.

Polynesian Navigators Perhaps the people most dependent on knowledge of the stars were the Polynesians, who lived and traveled among the many islands of the mid- and South Pacific. Because the next island in a journey usually was too distant to be seen, poor navigation meant becoming lost at sea. As a result, the most esteemed position in Polynesian culture was that of the Navigator, a person who had acquired the detailed knowledge necessary to navigate great distances among the islands.

The Navigators employed a combination of detailed knowledge of astronomy and equally impressive knowledge of the patterns of waves and swells around different islands (Figure 3.10). The stars provided their broad navigational sense, pointing them in the correct direction of their intended destination. As they neared a destination, the wave and swell patterns guided them to their precise landing point. A navigator memorized all his skills and passed them to the next generation through a well-developed program

Figure 3.9 Ground view of the Big Horn Medicine Wheel in Wyoming. Note the 28 "spokes" radiating out from the center. These spokes probably relate to the month of the Native Americans.

for training future Navigators. Unfortunately, with the advent of modern navigational technology, many of the skills of the Navigators have been lost.

From Observation to Science

Before a structure such as Stonehenge could be built, careful observations had to be made and repeated over and over to ensure their validity. Careful, repeatable observations also underlie modern science. To this extent, elements of modern science were present in many early human cultures.

The degree to which scientific ideas developed in different societies depended on practical needs, social and political customs, and interactions with other cultures. Because all of these factors can change, it should not be surprising that different cultures were more scientifically or technologically advanced than others at different times in history. The ancient Chinese, for example, began keeping remarkably detailed records of astronomical observations at least 5,000 years ago.

Despite this "head start," Chinese science and technology had clearly fallen behind that of Europe by the end

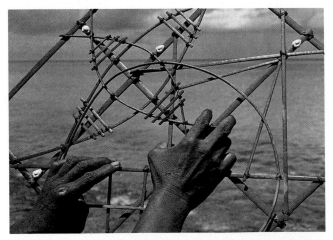

Figure 3.10 A traditional Polynesian navigational instrument.

Figure 3.11 This photo shows a model of the celestial sphere and other instruments on the roof of the ancient astronomical observatory in Beijing. The observatory was built in the 1400s; the instruments shown here were built later and show the influence of Jesuit missionaries.

of the 1600s (Figure 3.11). Some historians argue that a primary reason for this decline was that the Chinese tended to regard their science and technology as state secrets. This secrecy may have slowed Chinese scientific development by preventing the broad-based collaborative science that fueled the European advance beginning in the Renaissance.

In Central America, the ancient Mayans also were ahead of their time in many ways. For example, their system of numbers and mathematics looks distinctly modern. They invented the concept of zero some 500 years before its introduction in the Eurasian world (by Hindu mathematicians, around A.D. 600). The Aztecs, Incas, Anasazi, and other ancient peoples of the Americas may have been quite advanced in many other areas as well, but few written records survive to tell the tale.

It appears that virtually all cultures employed scientific thinking to varying degrees. If the circumstances of history had been different, any one of these many cultures might have been the first to develop what we consider to be modern science. In the end, however, history takes only one of countless possible paths. The path that led to modern science emerged from the ancient civilizations of the Mediterranean and the Middle East—and especially from ancient Greece.

3.3 Ancient Greek Science

By 3000 B.C., civilization was well established in two major regions of the Middle East: Egypt and Mesopotamia (Figure 3.12). Their geographical location placed these civilizations at a crossroads for travelers, merchants, and armies of Europe, Asia, and Africa. This mixing of cultures fostered creativity, and the broad interactions

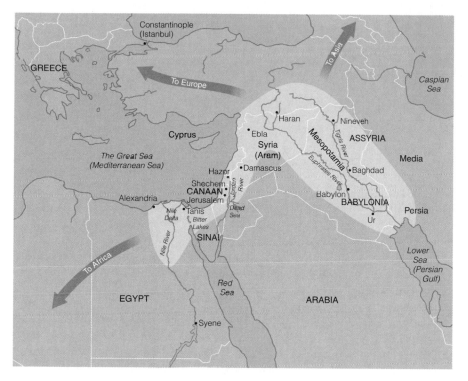

Figure 3.12 Map of Greece and the Middle East, with names spanning several centuries. Mesopotamia was the ancient Greek name for the region between the Tigris and Euphrates Rivers. Modern historians use the name for the entire region now called Iraq.

among peoples ensured that new ideas spread throughout the region.

Over the next 2,500 years, numerous great cultures arose. The ancient Egyptians built the Great Pyramids between 2700 and 2100 B.C., using their astronomical knowledge to orient the Pyramids with the cardinal directions. They also invented papyrus scrolls and ink-based writing. The Babylonians invented methods of writing on clay tablets and developed arithmetic to serve in commerce and later in astronomical calculations. Many more of our

---

SPECIAL TOPIC Eratosthenes Measures Earth

In a remarkable ancient feat, the Greek astronomer and geographer Eratosthenes estimated the size of Earth in about 240 B.C. He did it by comparing the altitude of the Sun on the summer solstice in the Egyptian cities of Syene (modern-day Aswan) and Alexandria.

Eratosthenes knew that the Sun passed directly overhead in Syene on the summer solstice. He also knew that in the city of Alexandria to the north the Sun came within only 7° of the zenith on the summer solstice. He therefore reasoned that Alexandria must be 7° of latitude to the north of Syene (see figure). Because 7° is $\frac{7}{360}$ of a circle, he concluded that the north-south distance between Alexandria and Syene must be $\frac{7}{360}$ of the circumference of Earth.

Eratosthenes estimated the north-south distance between Syene and Alexandria to be 5,000 stadia (the *stadium* was a Greek unit of distance). Thus, he concluded that:

$$\frac{7}{360} \times \text{circumference of Earth} = 5,000 \text{ stadia}$$

From this he found Earth's circumference to be about 250,000 stadia.

Today, we don't know exactly what distance a stadium meant to Eratosthenes. Based on the actual sizes of Greek stadiums, it must have been about $\frac{1}{6}$ kilometer. Thus,

Eratosthenes estimated the circumference of the Earth to be about $\frac{250,000}{6} = 42,000$ kilometers—remarkably close to the modern value of just over 40,000 kilometers.

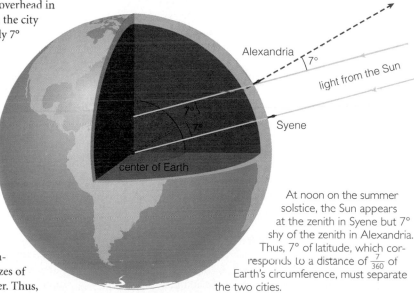

At noon on the summer solstice, the Sun appears at the zenith in Syene but 7° shy of the zenith in Alexandria. Thus, 7° of latitude, which corresponds to a distance of $\frac{7}{360}$ of Earth's circumference, must separate the two cities.

modern principles of commerce, law, and religion originated with the cultures of Egypt and Mesopotamia. But much of what we now call science came from ancient Greece, which rose as a power in the Middle East around 500 B.C.

Greek Philosophers Sought to Explain the Cosmos

We saw in Section 3.2 that many ancient societies recorded astronomical observations for religious and practical purposes. The Greeks stand out as the first people known to propose explanations for the motions of astronomical objects that relied on logic and geometry instead of supernatural forces. Many historians trace the origins of modern science directly back to Greek philosophers.

The Greeks tried to explain the architecture of the universe by constructing **models** of nature. The idea of modeling is central to modern science. Scientific models differ somewhat from the models you may be familiar with in everyday life. In our daily lives, we tend to think of models as miniature physical representations, such as model cars or airplanes. In contrast, a scientific model is a conceptual representation whose purpose is to explain and predict observed phenomena. For example, a model of Earth's climate uses logic and mathematics to represent what we know about how the climate works. Its purpose is to explain and predict climate changes, such as the changes that may occur with global warming. Just as a model airplane does not faithfully represent every aspect of a real airplane, a scientific model may not fully explain all our observations of nature. Nevertheless, even the failings of a scientific model can be useful, because they often point the way toward building a better model.

The Greek models of nature sought to explain things such as the properties of matter and the motions of the stars, Sun, Moon, and planets. Many of the scientific ideas discussed in this book originated with the Greeks. (See Figure 3.13 for a time line.) Although the Greek models may seem primitive from our modern perspective, they were an enormous step forward in scientific thinking. Let's look at how the Greeks developed models to explain the motions of celestial objects.

The Greek Geocentric Model

The Greek **geocentric model** of the cosmos, so named because it placed Earth at the center of the universe, was one of the greatest intellectual accomplishments of ancient times. We generally trace the origin of Greek science to the philosopher Thales (c. 624–546 B.C.; pronounced *thay-lees*). We encountered Thales earlier for his legendary prediction of a solar eclipse [Section 2.5]. Thales is said to have imagined Earth as a flat disk floating in a giant ocean of water.

His contemporary Anaximander (c. 610–546 B.C.) soon replaced this primitive model with a more sophisticated one. Anaximander suggested that Earth—which he imagined to be cylindrical in shape—floats in empty space surrounded by a sphere of stars and two separate rings along which the Sun and Moon travel. Thus, we credit him with inventing the idea of a celestial sphere [Section 2.2].

We do not know precisely who came up with the idea of a spherical Earth. The great mathematician Pythagoras (c. 560–480 B.C.) and his followers apparently imagined Earth as a sphere floating at the center of the celestial sphere. They may have based their belief in a spherical Earth in part on observations of how the positions of celestial objects change with latitude [Section 2.2]. However, much of their motivation was philosophical—they considered a sphere to be the most appropriate shape because it was geometrically perfect.

The supposed heavenly perfection of spheres influenced models of the cosmos for many centuries. Plato (428–348 B.C.), whose philosophy was based much more on pure thought than on observations, asserted that all heavenly objects move in perfect circles at constant speeds and therefore must reside on huge spheres encircling Earth (Figure 3.14). Greeks who took observations more seriously found this model problematic: The apparent retrograde motion of the planets [Section 2.6], already well known

---

SPECIAL TOPIC Aristotle

Aristotle (384–322 B.C.) is among the best-known philosophers of the ancient world. Both his parents died when he was a child, and he was raised by a family friend. In his 20s and 30s, he studied under Plato (428–348 B.C.) at Plato's Academy. He later founded his own school, called the Lyceum, where he studied and lectured on virtually every subject. Historical records tell us that his lectures were collected and published in 150 volumes. About 50 of these volumes survive to the present day.

Many of Aristotle's scientific discoveries involved the nature of plants and animals. He studied more than 500 animal species in detail, including dissecting specimens of nearly 50 species, and came up with a strikingly modern classification system. For example, he was the first person to recognize that dolphins should be classified with land mammals rather than with fish. In mathematics, he is known for laying the foundations of mathematical logic. Unfortunately, he was far less successful in physics and astronomy, areas in which many of his claims turned out to be wrong.

Despite his wide-ranging discoveries and writings, Aristotle's philosophies were not particularly influential until many centuries after his death. His books were preserved and valued by Islamic scholars but were unknown in Europe until they were translated into Latin in the twelfth and thirteenth centuries. Aristotle achieved his near-reverential status only after St. Thomas Aquinas (1225–1274) integrated Aristotle's philosophy into Christian theology. In the ancient world, Aristotle's greatest influence came indirectly, through his role as the tutor of Alexander the Great.

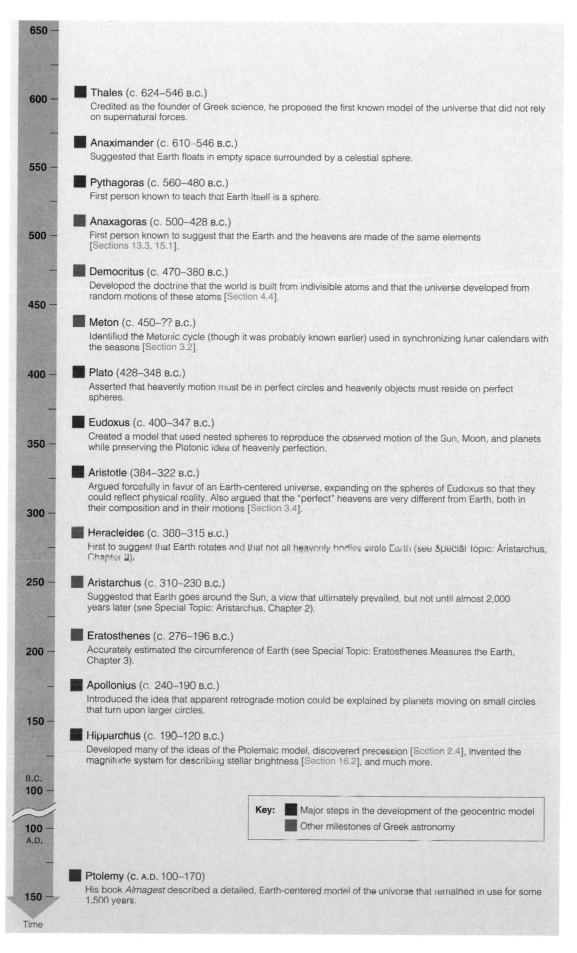

Thales (c. 624–546 B.C.)
Credited as the founder of Greek science, he proposed the first known model of the universe that did not rely on supernatural forces.

Anaximander (c. 610–546 B.C.)
Suggested that Earth floats in empty space surrounded by a celestial sphere.

Pythagoras (c. 560–480 B.C.)
First person known to teach that Earth itself is a sphere.

Anaxagoras (c. 500–428 B.C.)
First person known to suggest that the Earth and the heavens are made of the same elements [Sections 13.3, 15.1].

Democritus (c. 470–380 B.C.)
Developed the doctrine that the world is built from indivisible atoms and that the universe developed from random motions of these atoms [Section 4.4].

Meton (c. 450–?? B.C.)
Identified the Metonic cycle (though it was probably known earlier) used in synchronizing lunar calendars with the seasons [Section 3.2].

Plato (428–348 B.C.)
Asserted that heavenly motion must be in perfect circles and heavenly objects must reside on perfect spheres.

Eudoxus (c. 400–347 B.C.)
Created a model that used nested spheres to reproduce the observed motion of the Sun, Moon, and planets while preserving the Platonic idea of heavenly perfection.

Aristotle (384–322 B.C.)
Argued forcefully in favor of an Earth-centered universe, expanding on the spheres of Eudoxus so that they could reflect physical reality. Also argued that the "perfect" heavens are very different from Earth, both in their composition and in their motions [Section 3.4].

Heracleides (c. 388–315 B.C.)
First to suggest that Earth rotates and that not all heavenly bodies circle Earth (see Special Topic: Aristarchus, Chapter 2).

Aristarchus (c. 310–230 B.C.)
Suggested that Earth goes around the Sun, a view that ultimately prevailed, but not until almost 2,000 years later (see Special Topic: Aristarchus, Chapter 2).

Eratosthenes (c. 276–196 B.C.)
Accurately estimated the circumference of Earth (see Special Topic: Eratosthenes Measures the Earth, Chapter 3).

Apollonius (c. 240–190 B.C.)
Introduced the idea that apparent retrograde motion could be explained by planets moving on small circles that turn upon larger circles.

Hipparchus (c. 190–120 B.C.)
Developed many of the ideas of the Ptolemaic model, discovered precession [Section 2.4], Invented the magnitude system for describing stellar brightness [Section 16.2], and much more.

Key: ■ Major steps in the development of the geocentric model
      ■ Other milestones of Greek astronomy

Ptolemy (c. A.D. 100–170)
His book *Almagest* described a detailed, Earth-centered model of the universe that remained in use for some 1,500 years.

Time

Figure 3.14 This model represents the Greek idea of the heavenly spheres (c. 400 B.C.). Earth is a sphere that rests in the center. The Sun, the Moon, and each of the planets moves on its own sphere, and the outermost sphere holds the stars.

by this time, clearly showed that planets do not move at constant speeds around Earth.

An ingenious solution came from one of Plato's colleagues, Eudoxus (c. 400–347 B.C.). He created a model in which the Sun, the Moon, and the planets were each attached to their own sphere, which in turn was nested within several other spheres. Individually, the nested spheres turned in perfect circles. By carefully selecting rotation axis and rotation speed for each sphere, Eudoxus was able to make them work together in a way that re-created many of the observed motions of the Sun, Moon, and planets in our sky. Other Greeks refined the model by comparing its predictions to observations and adding more spheres to improve the agreement.

This is how things stood when Aristotle (384–322 B.C.) arrived on the scene. Whether Eudoxus and his followers thought of the nested spheres as real physical objects is not clear. Aristotle certainly did. In the cosmic model of Aristotle, all the spheres responsible for celestial motion were transparent and interconnected like the gears of a giant machine. Furthermore, Earth's position at the center of everything—as well as its composition and shape—was explained as a natural consequence of gravity. Aristotle argued that gravity pulled heavy things toward the center of the universe and allowed lighter things to float toward the heavens. Thus, all the dirt, rock, and water of the universe collected at the center, forming a spherical Earth. Of course, Aristotle was wrong about both gravity and Earth's location. However, largely because of his persuasive arguments for an Earth-centered universe, the geocentric view dominated Western thought for almost 2,000 years.

Ptolemy's Synthesis of the Geocentric Model

Greek modeling of the cosmos became much more quantitative after Aristotle, culminating in the work of Claudius Ptolemy (c. A.D. 100–170; pronounced *tol-e-mee*). Ptolemy's model still placed Earth at the center of the universe, but it differed in significant ways from the nested spheres of Eudoxus and Aristotle. We refer to Ptolemy's model as the **Ptolemaic model** to distinguish it from earlier geocentric models.

To explain the apparent retrograde motion of the planets, the Ptolemaic model applied an idea first suggested by Apollonius (c. 240–190 B.C.). This idea held that each planet moves around Earth on a small circle that turns upon a larger circle (Figure 3.15). (The small circle is sometimes called an *epicycle,* and the larger circle is called a *deferent.*) A planet following this circle-upon-circle motion traces a loop as seen from Earth, with the backward portion of the loop mimicking apparent retrograde motion.

Ptolemy also relied heavily on the work of Hipparchus (c. 190–120 B.C.), considered by many historians to have been the greatest Greek astronomer. Among his many accomplishments, Hipparchus developed the circle-upon-circle idea of Apollonius into a model that could predict planetary positions. To do this, Hipparchus had to add several features to the basic idea, such as adding even smaller circles that moved upon the original set of small circles and allowing the large circles to be positioned slightly off-center from Earth.

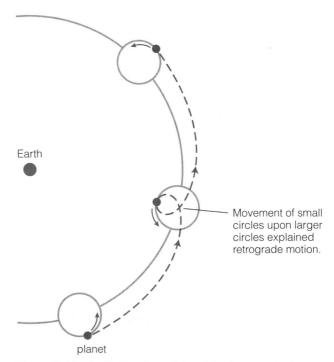

Earth

Movement of small circles upon larger circles explained retrograde motion.

planet

Figure 3.15 The Ptolemaic model explained apparent retrograde motion by supposing that each planet moved around Earth on a small circle that turned upon a larger circle. As shown by the dashed curve, this makes the planet trace a loop as seen from Earth. The backward portion of the loop represents apparent retrograde motion.

Ptolemy's great accomplishment was to adapt and synthesize these earlier ideas into a single system that agreed quite well with the astronomical observations available at the time. In the end, he created and published a model that could correctly forecast future planetary positions to within a few degrees of arc—roughly equivalent to holding your hand at arm's length against the sky. His model generally worked so well that it remained in use for the next 1,500 years. When his book describing the model was translated by Arabic scholars around A.D. 800, they gave it the title *Almagest,* derived from words meaning "the greatest compilation."

Toward a Scientific Renaissance

The Library of Alexandria One reason why Greek thought gained such broad influence was that the Greeks proved to be as adept at politics and war as they were at philosophy. In about 330 B.C., Alexander the Great (356–323 B.C.) began a series of conquests that expanded the Greek Empire throughout the Middle East, absorbing all the former empires of Egypt and Mesopotamia. Alexander was more than just a military leader. He also had a keen interest in science and education. As a teenager, Alexander's personal tutor had been none other than Aristotle.

Alexander encouraged the pursuit of knowledge and respect for foreign cultures. On the Nile delta in Egypt, he founded the city of Alexandria, which soon became a center of world culture. The heart of Alexandria was a great library and research center that opened in about 300 B.C. (Figure 3.16). The Library of Alexandria was the world's preeminent center of research for the next 700 years.

The library's end is closely tied to one of its last resident scholars, Hypatia (A.D. 370–415), probably the most prominent female scholar of the ancient world. In addition to her work at the library, Hypatia was the director of the observatory in Alexandria and one of the leading mathematicians and astronomers of her time. Unfortunately, she became a scapegoat during a time of rising sentiment against free inquiry and was gruesomely murdered by an anti-intellectual mob in A.D. 415. The final destruction of the Library of Alexandria took place not long after her death. At its peak, the Library of Alexandria held more than a half million books, handwritten on papyrus scrolls. Most of the scrolls probably were original manuscripts or the single copies of original manuscripts. When the library was destroyed, most of its storehouse of knowledge was lost forever.

Estimate the number of books you're likely to read in your life, and compare this number to the half million books once housed in the Library of Alexandria. Can you think of other ways to put into perspective the loss of ancient wisdom resulting from the destruction of the Library of Alexandria?

The Islamic Role Much more would have been lost if not for the rise of a new center of intellectual inquiry in Baghdad (present-day Iraq). While European civilization fell into the period of intellectual decline known as the Dark Ages, scholars of the new religion of Islam sought knowledge of mathematics and astronomy in hopes of better understanding the wisdom of Allah. During the eighth and ninth centuries A.D., scholars working in the Muslim empire centered in Baghdad translated and thereby saved many ancient Greek works.

Around A.D. 800, the Islamic leader Al-Mamun (A.D. 786–833) established a "House of Wisdom" in Baghdad with a mission much like that of the destroyed Library of Alexandria. Founded in a spirit of great openness and

a The Great Hall.

b A scroll room.

Figure 3.16 These renderings show an artist's reconstruction, based on scholarly research, of how the Great Hall and a scroll room of the Library of Alexandria might have looked.

tolerance, the House of Wisdom employed Jews, Christians, and Muslims, all working together in scholarly pursuits. Using the translated Greek scientific manuscripts as building blocks, these scholars developed algebra and many new instruments and techniques for astronomical observation. Most of the official names of constellations and stars come from Arabic because of the work of the scholars at Baghdad. If you look at a star chart, you will see that the names of many bright stars begin with *al* (e.g., Aldebaran, Algol), which simply means "the" in Arabic.

The Islamic world of the Middle Ages was in frequent contact with Hindu scholars from India, who in turn brought knowledge of ideas and discoveries from China. Hence, the intellectual center in Baghdad achieved a synthesis of the surviving work of the ancient Greeks and that of the Indians and the Chinese. The accumulated knowledge of the Arabs spread throughout the Byzantine Empire (the eastern part of the former Roman Empire). When the Byzantine capital of Constantinople (modern-day Istanbul) fell to the Turks in 1453, many Eastern scholars headed west to Europe, carrying with them the knowledge that helped ignite the European Renaissance.

3.4 The Copernican Revolution

The Greeks and other ancient peoples developed many important ideas of science, but what we now think of as science arose during the European Renaissance. Within a half-century after the fall of Constantinople, Nicholas Copernicus began the work that ultimately overturned the Earth-centered, Ptolemaic model. Over the next century and a half, philosophers and scientists (who were often one and the same) debated and tested his radical view of the cosmos. Ultimately, the new ideas introduced by Copernicus fundamentally changed how we perceive our place in the universe. This dramatic change, known as the **Copernican revolution**, spurred the development of virtually all modern science and technology.

Nicholas Copernicus: The Revolution Begins

Nicholas Copernicus was born in Torún, Poland, on February 19, 1473. His family was wealthy, and he received a first-class education in mathematics, medicine, and law. He began studying astronomy in his late teens. By that time, tables of planetary motion based on the Ptolemaic model were noticeably inaccurate. But few people were willing to undertake the difficult calculations required to revise the tables. Indeed, the best tables available had been compiled some two centuries earlier under the guidance of Spanish monarch Alphonso X (1221–1284). Commenting on the tedious nature of the work required to make these *Alfonsine Tables,* the monarch is said to have complained that "If I

had been present at the creation, I would have recommended a simpler design for the universe."

In his quest for a better way to predict planetary positions, Copernicus adopted the Sun-centered idea first suggested by Aristarchus some 1,800 years earlier [Section 2.6]. He was probably motivated in large part by the much simpler explanation for apparent retrograde motion offered by a Sun-centered system (see Figures 2.31 and 2.32). As he worked out the mathematical details of his model, Copernicus also discovered simple geometric relationships that allowed him to calculate each planet's true orbital period around the Sun (from its observed "synodic period" [Section S1.1]) and its true distance from the Sun in terms of Earth–Sun distance. (That is, he could find distances in astronomical units, but not in absolute units such as miles or kilometers.) The success of his model in providing a geometric layout for the solar system further convinced him that the Sun-centered idea must be correct.

Copernicus was hesitant to publish his work, fearing that his suggestion that Earth moved would be considered absurd. Nevertheless, he discussed his system with other scholars, generating great interest. At the urging of some of these scholars, including some high-ranking officials of the Catholic Church, he finally agreed to publish his work. Copernicus saw the first printed copy of his book, *De Revolutionibus Orbium Caelestium* ("Concerning the Revolutions of the Heavenly Spheres"), on the day he died—May 24, 1543.

Copernicus (1473–1543)

Early supporters of Copernicus were drawn to the aesthetic advantages of his model. However, it did not make substantially better predictions than Ptolemy's model, largely because he still believed that heavenly motion must be in perfect circles. Because the true orbits of the planets are *not* circles, Copernicus found it necessary to add circles upon circles to his system, just as in the Ptolemaic system. As a result, his complete model was no more accurate and no less complex than the Ptolemaic model, and the Sun-centered idea won relatively few converts in the 50 years after it was published. After all, why trade thousands of years of tradition for a new model that worked equally poorly?

Tycho Brahe: The Greatest Naked-Eye Observer of All Time

Part of the difficulty faced by astronomers who sought to improve either the Ptolemaic or the Copernican system was a lack of quality data. The telescope had not yet been invented, and existing naked-eye observations were not very accurate. In the late 1500s, Danish nobleman Tycho Brahe (1546–1601), usually known simply as Tycho, set about correcting this problem.

Tycho Brahe (1546–1601)

When Tycho was a young boy, his family discouraged his interest in astronomy. He therefore followed his passion in secret, learning the constellations from a miniature model of a celestial sphere that he kept hidden. As he grew older, Tycho was often arrogant about both his noble birth and his learned abilities. At age 20, he fought a duel with another student over which of them was the better mathematician. Part of his nose was cut off, and he designed a replacement piece made of silver and gold.

In 1563, Tycho decided to observe a widely anticipated alignment of Jupiter and Saturn. To his surprise, the alignment occurred nearly two days later than Copernicus had predicted. Resolving to improve the state of astronomical prediction, he set about compiling careful observations of stellar and planetary positions in the sky.

Tycho's fame grew after he observed what he called a *nova*, meaning "new star," in 1572 and proved that it was at a distance much farther away than the Moon. (Today, we know that Tycho saw a *supernova*—the explosion of a distant star.) In 1577, Tycho observed a comet and proved that it too lay in the realm of the heavens. Others, including Aristotle, had argued that comets were phenomena of Earth's atmosphere. King Frederick II of Denmark decided to sponsor Tycho's ongoing work, providing him with money to build an unparalleled observatory for naked-eye observations (Figure 3.17). After Frederick II died in 1588, Tycho moved to Prague, where his work was supported by German emperor Rudolf II.

Over a period of three decades, Tycho and his assistants compiled naked-eye observations accurate to within less than 1 arcminute—less than the thickness of a fingernail viewed at arm's length. Because the telescope was invented shortly after his death, Tycho's data remains the best set of naked-eye observations ever made. Despite the quality of his observations, Tycho never succeeded in coming up with a satisfying explanation for planetary motion. He was convinced that the *planets* must orbit the Sun, but his inability to detect stellar parallax [Section 2.6] led him to conclude that Earth must remain stationary. Thus, he advocated a model in which the Sun orbits Earth while all other planets orbit the Sun. Few people took this model seriously.

Figure 3.17 Tycho Brahe in his naked-eye observatory, which worked much like a giant protractor. He could sit and observe a planet through the rectangular hole in the wall as an assistant used a sliding marker to measure the angle on the protractor.

Although Tycho failed to explain the motions of the planets satisfactorily, he succeed in finding someone who could: In 1600, he hired the young German astronomer Johannes Kepler (1571–1630). Kepler and Tycho had a strained relationship, but Tycho recognized the talent of his young apprentice. In 1601, as he lay on his deathbed, Tycho begged Kepler to find a system that would make sense of the observations so "that it may not appear I have lived in vain."

Kepler's Reformation

Kepler was deeply religious and believed that understanding the geometry of the heavens would bring him closer to God. Like Copernicus, he believed that Earth and the other planets traveled around the Sun in circular orbits. He worked diligently to match circular motions to Tycho's data.

Kepler labored with particular intensity to find an orbit for Mars, which posed the greatest difficulties in matching the data to a circular orbit. After years of calculation,

Kepler found a circular orbit that matched all of Tycho's observations of Mars's position along the ecliptic (east-west) to within 2 arcminutes. However, the model did not correctly predict Mars's positions north or south of the ecliptic. Because Kepler sought a physically realistic orbit for Mars, he could not (as Ptolemy and Copernicus had done) tolerate one model for the east-west positions

Johannes Kepler (1571–1630)

and another for the north-south positions. He attempted to find a unified model with a circular orbit. In doing so, he found that some of his predictions differed from Tycho's observations by as much as 8 arcminutes.

Discarding Perfect Circles Kepler surely was tempted to ignore these discrepancies and attribute them to errors by Tycho. After all, 8 arcminutes is barely one-fourth the angular diameter of the full moon. But Kepler trusted Tycho's careful work, and the misses of his predictions by 8 arcminutes finally led him to abandon the idea of circular orbits—and to find the correct solution to the ancient riddle of planetary motion. About this event, Kepler wrote:

> *If I had believed that we could ignore these eight minutes [of arc], I would have patched up my hypothesis accordingly. But, since it was not permissible to ignore, those eight minutes pointed the road to a complete reformation in astronomy.*

Kepler's Laws of Planetary Motion Kepler's key discovery was that planetary orbits are not circles but instead are a special type of oval called an **ellipse**. You probably know how to draw a circle by putting a pencil on the end of a string, tacking the string to a board, and pulling the pencil around (Figure 3.18a). Drawing an ellipse is similar, except that you must stretch the string around *two* tacks (Figure 3.18b). The locations of the two tacks are called the **foci** (singular, **focus**) of the ellipse. By altering the distance between the two foci while keeping the same length of string, you can draw ellipses of varying **eccentricity**, a quantity that describes how much an ellipse deviates from a perfect circle (Figure 3.18c). A circle has zero eccentricity, and greater eccentricity means a more elongated ellipse.

Kepler summarized his discoveries with three simple laws that we now call **Kepler's laws of planetary motion**. He published the first two laws in 1610 and the third in 1618.

> **Kepler's first law (Figure 3.19):** *The orbit of each planet about the Sun is an ellipse with the Sun at one focus.* **(There is nothing at the other focus.)**

This law tells us that a planet's distance from the Sun varies during its orbit. It is closest at the point called **perihelion** and farthest at the point called **aphelion**. (*Helios* is Greek for the Sun, the prefix *peri* means "near," and the prefix *ap* [or *apo*] means "away." Thus, *perihelion* means "near the Sun" and *aphelion* means "away from the Sun.") The *average* of a planet's perihelion and aphelion distances is called its **semimajor axis**. We will refer to this simply as the planet's average distance from the Sun.

> **Kepler's second law (Figure 3.20):** *As a planet moves around its orbit, it sweeps out equal areas in equal times.*

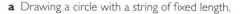

a Drawing a circle with a string of fixed length.

b Drawing an ellipse with a string of fixed length.

c *Eccentricity* describes how much an ellipse deviates from a perfect circle.

Figure 3.18 An ellipse is a special type of oval. These diagrams show how an ellipse differs from a circle and how different ellipses vary in their eccentricity.

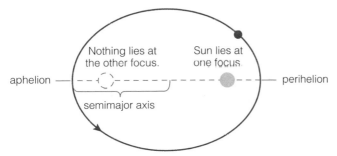

Figure 3.19 Kepler's first law: The orbit of each planet about the Sun is an ellipse with the Sun at one focus. (The eccentricity shown here is exaggerated compared to the actual eccentricities of the planets.)

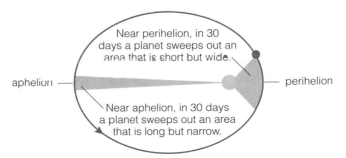

Figure 3.20 Kepler's second law: As a planet moves around its orbit, it sweeps out equal areas in equal times.

As shown in Figure 3.20, this means that the planet moves a greater distance when it is near perihelion than it does in the same amount of time near aphelion. That is, the planet travels faster when it is nearer to the Sun and slower when it is farther from the Sun.

> **Kepler's third law (Figure 3.21):** *More distant planets orbit the Sun at slower average speeds, obeying the following precise mathematical relationship:*
>
> $$p^2 = a^3$$
>
> *where p is planet's orbital period in years and a is its average distance from the Sun in astronomical units.*

Figure 3.21a shows the $p^2 = a^3$ law graphically. Notice that the square of each planet's orbital period (p^2) is indeed equal to the cube of its average distance from the Sun (a^3). Figure 3.21b shows that this law does indeed imply that more distant planets orbit the Sun more slowly.

The fact that more distant planets move more slowly led Kepler to suggest that planetary motion might be the result of a force from the Sun. He even speculated about the nature of this force, guessing that it might be related to magnetism. (This idea, shared by Galileo, was first suggested by William Gilbert [1544–1603], an early believer in the Copernican system.) Kepler was right about the existence of a force, but wrong in his guess about magnetism. More than a half-century later, Isaac Newton explained planetary motion as a consequence of the gravitational force attracting planets to the Sun [Section 5.3].

THINK ABOUT IT

Suppose a comet had a very eccentric orbit that brought it quite close to the Sun at its perihelion and beyond Mars at its aphelion, but with an average distance (semimajor axis) of 1 AU. According to Kepler's laws, how long would the comet take to complete each orbit of the Sun? Would it spend most of its time close to the Sun, far from the Sun, or somewhere in between? Explain.

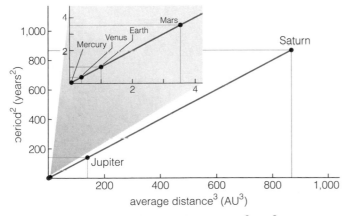

a The precise statement of Kepler's third law is $p^2 = a^3$, where p is a planet's orbital period in years and a is its average distance from the Sun in AU. The graph shows this relationship for the planets known in Kepler's time. The straight line tells us that p^2 (plotted along the vertical axis) is equal to a^3 (plotted along the horizontal axis).

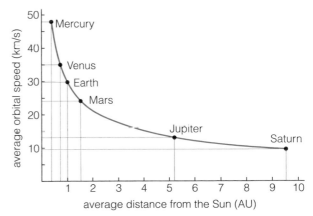

b Because Kepler's third law relates a planet's orbital distance to its orbital time (period), it can be used to calculate a planet's average orbital speed. This graph shows the result—more distant planets orbit the Sun more slowly. (Kepler knew the form of the relationship but could not determine speeds in km/s because the numerical value of the astronomical unit was not yet known.)

Figure 3.21 Kepler's third law tells us that more distant planets orbit the Sun at slower average speeds.

Galileo: The Death of the Earth-Centered Universe

The success of Kepler's laws in matching Tycho's data provided strong evidence in favor of Copernicus's placement of the Sun, rather than Earth, at the center of the solar system. Nevertheless, many scientists still voiced reasonable objections to the Copernican view. There were three basic objections, all rooted in the 2,000-year-old beliefs of Aristotle and other ancient Greeks.

- First, Aristotle had held that Earth could not be moving because, if it were, objects such as birds, falling stones, and clouds would be left behind as Earth moved along its way.

- Second, the idea of noncircular orbits contradicted Aristotle's claim that the heavens—the realm of the Sun, Moon, planets, and stars—must be perfect and unchanging.

- Third, as the ancients had argued, stellar parallax ought to be detectable if Earth orbits the Sun [Section 2.6].

Galileo (1564–1642)

Galileo Galilei (1564–1642), a contemporary and correspondent of Kepler, answered all three objections.

Galileo (nearly always known by only his first name) defused the first objection with experiments that almost single-handedly overturned the Aristotelian view of physics. In particular, he demonstrated that a moving object remains in motion *unless* a force acts to stop it (an idea now codified in Newton's first law of motion [Section 5.2]). This contradicted Aristotle's claim that the natural tendency of any moving object is to come to rest. Galileo concluded that objects such as birds, falling stones, and clouds that are moving with Earth should *stay* with Earth unless some force knocks them away. This same idea explains why passengers in an airplane stay with the moving airplane even when they leave their seats.

Tycho's supernova and comet observations already had challenged the validity of the second objection by showing that the heavens could change. Galileo shattered the idea of heavenly perfection after he built a telescope in late 1609. (Galileo did *not* invent the telescope. It was invented in 1608 by Hans Lippershey. However, Galileo took what was little more than a toy and turned it into a scientific instrument.) Through his telescope, Galileo saw sunspots on the Sun, which were considered "imperfections" at the time.

He also used his telescope to prove that the Moon has mountains and valleys like the "imperfect" Earth by noticing the shadows cast near the dividing line between the light and dark portions of the lunar face (Figure 3.22). If the heavens were in fact not perfect, then the idea of elliptical orbits (as opposed to "perfect" circles) was not so objectionable.

The third objection—the absence of observable stellar parallax—had been of particular concern to Tycho. Based on his estimates of the distances of stars, Tycho believed that his naked-eye observations were sufficiently precise to detect stellar parallax if Earth did in fact orbit the Sun.

Refuting Tycho's argument required showing that the stars were more distant than Tycho had thought and therefore too distant for him to have observed stellar parallax. Although Galileo didn't actually prove this fact, he provided strong evidence in its favor. In particular, he saw with his telescope that the Milky Way resolved into countless individual stars. His discovery helped him argue that the stars were far more numerous and more distant than Tycho had imagined.

Figure 3.22 The shadows cast by mountains and crater rims near the dividing line between the light and dark portions of the lunar face prove that the Moon's surface is not perfectly smooth.

In hindsight, the final nails in the coffin of the Earth-centered universe came with two of Galileo's earliest discoveries through the telescope. First, he observed four moons clearly orbiting Jupiter, *not* Earth (Figure 3.23). (By itself, this observation still did not rule out a stationary, central Earth. However, it showed that moons can orbit a moving planet like Jupiter, which overcame some critics' complaints that the Moon could not stay with a moving Earth.) Soon thereafter, he observed that Venus goes through phases in a way that proved that it must orbit the Sun and not Earth (Figure 3.24).

With Earth clearly removed from its position at the center of the universe, the scientific debate turned to the question of whether Kepler's laws were the correct model for our solar system. The most convincing evidence came in 1631, when astronomers observed a transit of Mercury across the Sun's face. Kepler's laws had predicted the transit with overwhelmingly better success than any competing model.

Although we now recognize that Galileo won the day, the story was more complex in his own time, when Catholic Church doctrine still held Earth to be the center of the universe. On June 22, 1633, Galileo was brought before a Church inquisition in Rome and ordered to recant his claim that Earth orbits the Sun. Nearly 70 years old and fearing for his remaining life, Galileo did as ordered. His life was spared. However, legend has it that as he rose from his knees he whispered under his breath, *Eppur si muove*—Italian for "And yet it moves." (Given the likely consequences if Church officials had heard him say this, most historians doubt the veracity of the legend.)

Galileo was not formally vindicated by the Church until 1992 [Section 1.4], but the Church gave up the argument long before that. Galileo's book, *Dialogue Concerning the Two Chief World Systems,* was removed from the Church's index of banned books in 1824. Today, Catholic scientists are at the forefront of much astronomical research, and official Church teachings are compatible not only with Earth's planetary status but also with the theories of the Big Bang and the subsequent evolution of the cosmos and of life.

Figure 3.23 A page from Galileo's notebook in 1610. His sketches show four "stars" near Jupiter (the circle) but in different positions at different times (and sometimes hidden from view). Galileo soon realized that the "stars" were actually moons orbiting the giant planet.

3.5 The Nature of Science

The story of how our ancestors gradually figured out the basic architecture of the cosmos exhibits many features of what today is considered "good science." For example, we have seen how models were formulated and tested against observations and were modified or replaced when they failed those tests. The story also illustrates some classic mistakes, such as the failure of anyone before Kepler to

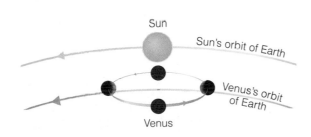

a In the Ptolemaic system, Venus follows a circle upon a circle that keeps it close to the Sun in our sky. Therefore, its phases would range only from new to crescent.

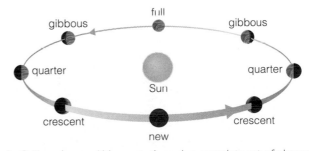

b Galileo observed Venus go through a complete set of phases and therefore proved that it orbits the Sun.

Figure 3.24 Galileo's telescopic observations of Venus proved that it orbits the Sun rather than Earth.

question the belief that orbits must be circles. The ultimate success of the Copernican revolution led scientists, philosophers, and theologians to reassess the various modes of thinking that played a role in the 2,000-year process of discovering Earth's place in the universe. The principles of modern science emerged from this reassessment.

Perhaps surprisingly, it turns out to be quite difficult to define the term *science* precisely. The word comes from

---

SPECIAL TOPIC And Yet It Moves

Although the evidence supporting the idea that Earth rotates and orbits the Sun was quite strong by the time of Galileo's trial in 1633, it was still indirect. Today, we have much more direct proof that Galileo was correct when he supposedly whispered of Earth *Eppur si muove*—"And yet it moves."

French physicist Jean Foucault provided the first direct proof of *rotation* in 1851. Foucault built a large pendulum that he carefully started swinging. Any pendulum tends to swing always in the same plane, but Earth's rotation made Foucault's pendulum appear to twist slowly in a circle. Today, *Foucault pendulums* are a popular attraction at many science centers and museums. A second direct proof that Earth rotates is provided by the *Coriolis effect*, first described by French physicist Gustave Coriolis (1792–1843). The Coriolis effect [Section 11.4], which would not happen if Earth were not rotating, is responsible for things such as the swirling of

hurricanes and the fact that missiles that travel great distances on Earth deviate from straight-line paths.

Direct proof that Earth orbits the Sun came from English astronomer James Bradley (1693–1762). To understand Bradley's proof, imagine that starlight is like rain, falling straight down. If you are standing still you should hold your umbrella straight over your head, but if you are walking through the rain you should tilt your umbrella forward, because your motion makes the rain appear to be coming down at an angle. Bradley discovered that observing light from stars requires that telescopes be tilted slightly in the direction of Earth's motion—just like the umbrella. This effect is called the *abberation of starlight*.

Stellar parallax also provides direct proof that Earth orbits the Sun. It was first measured by German astronomer Friedrich Bessel in 1838.

● Foucault pendulums are popular attractions at many science centers and museums.

the Latin *scientia,* meaning "knowledge," but not all knowledge is science. For example, you may know what music you like best, but your musical taste is not a result of scientific study. In this section, we'll explore in some detail the features that set modern science apart from other forms of knowledge.

Approaches to Science

One reason why science is difficult to define is that not all science works in the same way. For example, philosophers of science often distinguish between two primary scientific approaches: discovery science and hypothesis-driven science.

Discovery science involves going out and looking at nature in a general way in hopes of learning something new and unexpected. Once we gather the observations, we try to interpret and explain them. Often we find that they suggest new questions to be studied in greater depth. Galileo's first astronomical observations with a telescope are a prime example. Everywhere he pointed his revolutionary device, Galileo discovered unforeseen wonders that settled old scientific questions and sparked new ones.

Hypothesis-driven science involves proposing an idea and performing experiments or observations to test it. Galileo used this mode of inquiry as well. For example, he performed careful experiments to test his idea that all objects remain in motion unless some force acts to stop them. We often use this kind of thinking in our everyday lives.

Consider what you would do if your flashlight suddenly stopped working. You might question why it has stopped working, and you might *hypothesize* that the reason is that the batteries have died. In other words, you've created a tentative explanation, or **hypothesis,** for the flashlight's failure. A hypothesis is sometimes called an *educated guess*—in this case it is "educated" because you already know that flashlights need batteries. Your hypothesis then allows you to make a simple prediction: If you replace the batteries with new ones, the flashlight should work. You can test this prediction by replacing the batteries. If the flashlight now works, you've confirmed your hypothesis. If it doesn't, you must revise or discard your hypothesis, hopefully in favor of some other one that you can then test (such as that the bulb is dead). Figure 3.25 illustrates the basic flow of this process, often referred to as the "scientific method."

The scientific method is a useful idealization of scientific thinking, but science rarely progresses in such an orderly way. Most scientific progress involves a combination of discovery and hypothesis-driven science. Furthermore, scientists are human beings, and their intuition and personal beliefs inevitably influence their creation of new hypotheses. These aspects of human nature sometimes aid progress and sometimes impede it. Copernicus, for example, adopted the idea that Earth orbits the Sun not because he had carefully tested it but because he believed it made more sense than the prevailing view of an Earth-centered

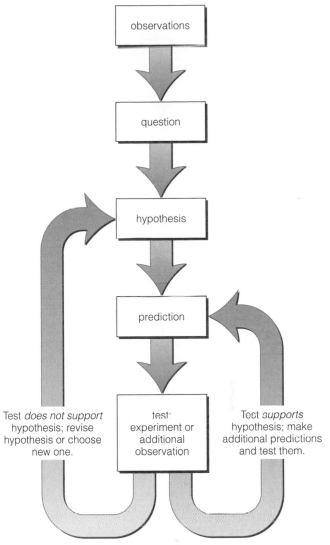

Figure 3.25 This diagram illustrates what we often call the scientific method.

universe. While his intuition guided him to the right general idea, he missed on the specifics because he still clung to Plato's ancient belief that heavenly motion must be in perfect circles. Only when Kepler reluctantly abandoned this belief did scientists finally realize the true nature of planetary motion.

Given that the idealized scientific method is an overly simplistic characterization of science, how do we decide whether something is scientific? To answer this question, we must look a little deeper at the distinguishing characteristics of scientific thinking.

Hallmarks of Science

One way to define scientific thinking is to list the criteria that scientists use when they judge competing models of nature. Historians and philosophers of science have examined (and continue to examine) this issue in great depth,

Figure 3.26 Hallmarks of science.

and different experts express somewhat different viewpoints on the details. Nevertheless, everything we now consider to be science shares the following three basic characteristics, which we will refer to as the "hallmarks" of science (Figure 3.26):

● Modern science seeks explanations for observed phenomena that rely solely on natural causes.

● Science progresses through the creation and testing of models of nature that explain the observations as simply as possible.

● A scientific model must make testable predictions about natural phenomena that would force us to revise or abandon the model if the predictions do not agree with observations.

Each of these hallmarks is evident in the story of the Copernican revolution. The first shows up in the way Tycho's exceptionally careful measurements of planetary motion motivated Kepler to come up with a better explanation for those motions. The second is evident in the way several competing models were compared and tested, most notably those of Ptolemy, Copernicus, and Kepler. The third shows up in the fact that each model could make precise predictions about the future motions of the Sun, Moon, planets, and stars in our sky. When a model's predictions failed, it was modified or ultimately discarded. Kepler's model gained acceptance in large part because its predictions matched Tycho's observations much better than Ptolemy's model.

Occam's Razor The simplicity feature of the second hallmark deserves further explanation. Remember that the original model of Copernicus did *not* match the data noticeably better than Ptolemy's model. Thus, a purely data-driven judgment based on the third hallmark might have led scientists to immediately reject the Sun-centered idea. Instead, many scientists found elements of the Copernican model

appealing, such as the simplicity of its explanation for apparent retrograde motion. They therefore kept it alive until Kepler found a way to make it work.

If agreement with data were the sole criterion for judgment, we could imagine a modern-day Ptolemy adding millions or billions of additional circles to the geocentric model in an effort to improve its agreement with observations. In principle, a sufficiently complex geocentric model could reproduce the observations with almost perfect accuracy—but it still would not convince us that Earth is the center of the universe. We would still choose the Copernican view over the geocentric view because its predictions would be just as accurate yet would follow from a much simpler model of nature. The idea that scientists should prefer the simpler of two models that agree equally well with observations is called *Occam's razor*, after medieval scholar William of Occam (1285–1349).

Theories in Science The most successful scientific models explain a wide variety of observations in terms of just a few general principles. When a powerful yet simple model makes predictions that survive repeated testing, scientists elevate its status and call it a **theory**. Some famous examples are Isaac Newton's theory of gravity, Charles Darwin's theory of evolution, and Albert Einstein's theory of relativity. Note that the use of the word *theory* in science contrasts with our everyday usage, which equates theories more closely with speculations or hypotheses. In everyday life, someone might get a new idea and say, for example, "I have a new theory about why people enjoy the beach." Without the support of a broad range of evidence that has been tested and confirmed by others, this idea is really only a hypothesis. Newton's theory of gravity qualifies as a scientific theory because it can be stated in simple mathematical terms and explains a great many observations and experiments.

Despite its success in explaining observed phenomena, a scientific theory can never be proved true beyond all doubt, because ever more sophisticated observations may eventually disagree with its predictions. However, anything that qualifies as a scientific theory must be supported by a large, compelling body of evidence.

THINK ABOUT IT

When someone claims that something is "only a theory," what do you think they mean? Does this meaning agree with the definition of a theory in science? Do scientists always use the word *theory* in its "scientific" sense? Explain.

Science and Pseudoscience

People often seek knowledge in ways that do not qualify as science. For example, suppose you are shopping for a car, learning to play drums, or pondering the meaning of life. In each case, you might make observations, exercise logic, and test hypotheses. Yet these pursuits clearly are not science, because they are not directed at developing testable explanations for observed natural phenomena. As long as nonscientific searches for knowledge make no claims about how the natural world works, they do not conflict with science.

However, you will often hear claims about the natural world that seem to be based on observational evidence but do not treat evidence in a truly scientific way. Such claims are often called **pseudoscience**, which literally means "false science." To distinguish real science from pseudoscience,

a good first step is to check whether a particular claim exhibits all three hallmarks of science. Consider the example of people who claim a psychic ability to "see" the future. If they truly can see the future, they should be able to make specific, testable predictions. In this sense, "seeing" the future sounds scientific, since we can test it. However, numerous studies have tested the predictions of "seers" and have found that their predictions come true no more often than would be expected by pure chance. If the "seers" were scientific, they would admit that this evidence undercuts their claim of psychic abilities. Instead, they generally make excuses. For example, they might say the predictions didn't come true because of some type of "psychic interference." Making testable claims but then ignoring the results of the tests marks the claimed ability to see the future as pseudoscience.

THINK ABOUT IT

Consider the claim that Earth is regularly visited by aliens from other worlds, traveling in "UFOs." Do you think this claim qualifies as science or pseudoscience? Defend your opinion.

Is Science Objective?

The boundaries between science and pseudoscience are sometimes blurry. In particular, because science is practiced by human beings, individual scientists bring their personal biases and beliefs to their scientific work. These biases can

---

SPECIAL TOPIC Logic and Science

In science, we attempt to acquire new knowledge through logical reasoning. The process of reasoning is carried out by constructing arguments in which we begin with a set of premises and try to draw appropriate conclusions. This concept of a logical argument differs somewhat from the definition of *argument* in everyday life. In particular, a logical argument need not imply any animosity or dissension.

There are two basic types of argument: *deductive* and *inductive*. Both are important in science.

The following is a simple example of a *deductive argument*:

PREMISE: All planets orbit the Sun in ellipses with the Sun at one focus.
PREMISE: Earth is a planet.
CONCLUSION: Earth orbits the Sun in an ellipse with the Sun at one focus.

As long as the two premises are true, the conclusion must also be true—this is the essence of deduction. Note the construction of a deductive argument: The first premise is a general statement that applies to all planets, whereas the conclusion is a specific statement that applies only to Earth. As this example suggests, deduction is often used to *deduce* a specific prediction from a more general theory. If the specific prediction proves to be false, then something must be wrong with the premises from which it was deduced. If it proves true, then we've acquired a piece of evidence in support of the premises.

Contrast the deductive argument above with the following example of an *inductive argument*:

PREMISE: Birds fly up into the air but eventually come back down.
PREMISE: People who jump into the air fall back down.
PREMISE: Rocks thrown into the air come back down.
PREMISE: Balls thrown into the air come back down.
CONCLUSION: What goes up must come down.

Imagine that you lived in the time of ancient Greece, when Earth was thought to be a realm distinct from the sky. Because each premise supports the conclusion, you might agree that this is a strong inductive argument and that its conclusion probably is true. However, no matter how many more examples you consider of objects that go up and come down, you could never *prove* that the conclusion is true—only that it seems likely to be true.

A single counterexample can prove the conclusion of an inductive argument to be false. In this case, we now know the conclusion is false because the spacecraft *Voyager 2* was launched from Earth and never fell back down.

Inductive arguments *generalize* from specific facts to a broader model or theory and hence are the arguments used to build scientific theories. That is why theories can never be proved true beyond all doubt—they can only be shown to be consistent with ever larger bodies of evidence. Theories *can* be proved false, however, if they fail to account for observed or experimental facts.

influence how a scientist proposes or tests a model. In some cases, scientists have been known to cheat—either deliberately or subconsciously—to obtain the result they desire. For example, in the late nineteenth and early twentieth centuries, some astronomers claimed to see artificial canals in their blurry telescopic images of Mars. They hypothesized that Mars was home to a dying civilization that used the canals to transport water to thirsty cities [Section 10.5]. No such canals actually exist. These astronomers apparently allowed their beliefs about extraterrestrial life to influence how they interpreted blurry images—in essence, a form of cheating, though certainly not intentional.

Bias can sometimes show up even in the thinking of the scientific community as a whole. Some valid ideas may not be considered by any scientist because the ideas fall too far outside the general patterns of thought, or **paradigm**, of the time. Einstein's theory of relativity provides an example. Many scientists in the decades before Einstein had gleaned hints of the theory but did not investigate them, at least in part because they seemed too outlandish.

The beauty of science is that it encourages continued testing by many people. Even if personal biases affect some results, tests by others will eventually uncover the mistakes. Similarly, when a new idea falls outside the accepted paradigm, sufficient testing and verification of the idea will eventually force a change in the paradigm. Thus, although individual scientists rarely follow the idealized scientific method, the collective action of many scientists over many years ensures that faulty ideas will eventually be identified

and improved. That is why, despite the biases of individual scientists, science as a whole is usually objective.

3.6 Astrology

We have discussed the development of astronomy and the nature of science in some depth. Now let's talk a little about a subject often confused with the science of astronomy: the similarly named subject of *astrology*. Today, astronomy and astrology are very different practices. In ancient times, however, astrology and astronomy often went hand in hand, and astrology played an important role in the historical development of astronomy.

In brief, the basic tenet of astrology is that human events are influenced by the apparent positions of the Sun, Moon, and planets among the stars in our sky. The origins of this idea are easy to understand. After all, there is no doubt that the position of the Sun in the sky influences our lives—it determines the seasons and hence the times of planting and harvesting, of warmth and cold, and of daylight and darkness. Similarly, the Moon determines the tides, and the cycle of lunar phases coincides with many biological cycles. Because the planets also appear to move among the stars, it seemed reasonable to ancient people that planets also influence our lives, even if these influences were much more difficult to discover.

Ancient astrologers hoped that they might learn *how* the positions of the Sun, Moon, and planets influence our lives. They charted the skies, seeking correlations with events on Earth. For example, if an earthquake occurred when Saturn was entering the constellation Leo, might Saturn's position have been the cause of the earthquake? If the king became ill when Mars appeared in the constellation Gemini and the first-quarter moon appeared in Scorpio, might it mean another tragedy for the king when this particular alignment of the Moon and Mars next recurred? Surely, the ancient astrologers thought, the patterns of influence would eventually become clear. The astrologers hoped that they might someday learn to forecast human events with the same reliability with which astronomical observations of the Sun could forecast the coming of spring.

Astrology's Role in Astronomical History

Because forecasts of the seasons and forecasts of human events were imagined to be closely related, astrologers and astronomers usually were one and the same in the ancient world. For example, in addition to his books on astronomy, Ptolemy published a treatise on astrology called *Tetrabiblios* that remains the foundation for much of astrology today. But Ptolemy himself recognized that astrology stood upon a far shakier foundation than astronomy. In the introduction to *Tetrabiblios*, Ptolemy compared astronomical and astrological predictions:

[Astronomy], which is first both in order and effectiveness, is that whereby we apprehend the aspects of the movements of sun, moon, and stars in relation to each other and to the earth. . . . I shall now give an account of the second and less sufficient method [of prediction (astrology)] in a proper philosophical way, so that one whose aim is the truth might never compare its perceptions with the sureness of the first, unvarying science. . . .

Other ancient scientists probably likewise recognized that their astrological predictions were far less reliable than their astronomical ones. Nevertheless, if there was even a slight possibility that astrologers could forecast the future, no king or political leader would dare to be without one. Astrologers held esteemed positions as political advisers in the ancient world and were provided with the resources they needed to continue charting the heavens and history. Much of the development of ancient astronomy was made possible through wealthy political leaders' support of astrology.

Throughout the Middle Ages and into the Renaissance, many astronomers continued to practice astrology. For example, Kepler cast numerous *horoscopes*—the predictive charts of astrology—even as he was discovering the laws of planetary motion. However, given Kepler's later description of astrology as "the foolish stepdaughter of astronomy" and "a dreadful superstition," he may have cast the horoscopes solely as a source of much-needed income. Modern-day astrologers also claim Galileo as one of their own, in part for his having cast a horoscope for the Grand Duke of Tuscany. However, whereas Galileo's astronomical discoveries changed human history, the horoscope he cast was just plain wrong: It predicted a long and fruitful life for the duke, who died just a few weeks later.

The scientific triumph of Kepler and Galileo in showing Earth to be a planet orbiting the Sun heralded the end of the linkage between astronomy and astrology. Nevertheless, astrology remains popular today. More people earn income by casting horoscopes than through astronomical research, and books and articles on astrology often outsell all but the most popular books on astronomy.

Scientific Tests of Astrology

Today, different astrologers follow different practices, which makes it difficult even to define *astrology*. Some astrologers no longer claim any ability to make testable predictions and therefore are practicing a form of *nonscience* that modern science can say nothing about. However, for most astrologers the business of astrology is casting horoscopes. Horoscopes often are cast for individuals and either predict future events in the person's life or describe characteristics of the person's personality and life. If the horoscope predicts future events, it can be evaluated by whether the predictions come true. If it describes the person's personality and life, the description can be checked for accuracy.

A *scientific* test of astrology requires evaluating many horoscopes and comparing their accuracy to what would be expected by pure chance. For example, suppose a horoscope states that a person's best friend is female. Because that is true of roughly half the population in the United States, an astrologer who casts 100 such horoscopes would be expected by pure chance to be right about 50 times. Thus, we would be impressed with the predictive ability of the astrologer only if he or she were right much more often than 50 times out of 100. In hundreds of scientific tests, astrological predictions have never proved to be accurate by a substantially greater margin than expected from pure chance. Similarly, in tests in which astrologers are asked to cast horoscopes for people they have never met, the horoscopes fail to match actual personality profiles more often than expected by chance. The verdict is clear: The methods of astrology are useless for predicting the past, the present, or the future.

What about newspaper horoscopes, which often appear to ring true? If you read them carefully, you will find that these horoscopes generally are so vague as to be untestable. For example, a horoscope that says "It is a good day to spend time with your friends" doesn't offer much for testing. Indeed, if you read the horoscopes for all 12 astrological signs, you'll probably find that several of them apply equally well to you.

THINK ABOUT IT

Look in a local newspaper for today's weather forecast and for your horoscope. Contrast the nature of their predictions. By the end of the day, you will know if the weather forecast was accurate. Will you know whether your horoscope was accurate? Explain.

Does It Make Sense?

In science, observations and experiments are the ultimate judge of any idea. No matter how outlandish an idea might appear, it cannot be dismissed if it successfully meets observational or experimental tests. The idea that Earth rotates and orbits the Sun at one time seemed outlandish, yet today it is so strongly supported by the evidence that we consider it a fact. The idea that the positions of the Sun, Moon, and planets among the stars influence our lives might sound outlandish today, but if astrology were to make predictions that came true, adherence to the principles of science would force us to take it seriously. However, given that scientific tests of astrology have never found any evidence that its predictive methods work, it is worth looking at its premises to see whether they make sense. Might there be a few kernels of wisdom buried within the lore of astrology?

Let's begin with one of the key premises of astrology: There is special meaning in the patterns of the stars in the constellations. This idea may have seemed quite reasonable in ancient times, when the stars were assumed to be fixed on an unchanging celestial sphere. Today we know that the patterns of the stars in the constellations are accidents of the moment. Long ago the constellations did not look the same, and they will look still different far in the future. Moreover, the stars in a constellation don't necessarily have any *physical* association. Because stars vary in distance, two stars that appear on opposite sides of our sky might well be closer together than two stars in the same constellation. Constellations are only *apparent* associations of stars, with no more physical reality than the water in a desert mirage.

Astrology also places great importance on the positions of the planets among the constellations. Again, this idea might have seemed quite reasonable in ancient times, when it was thought that the planets truly wandered among the stars. Today we know that the planets only *appear* to wander among the stars. In reality, the planets are in our own solar system, while the stars are vastly farther away. It is difficult to see how mere appearances could have profound effects on our lives.

Many other ideas at the heart of astrology are equally suspect. For example, most astrologers claim that a proper horoscope must account for the positions of *all* the planets. Does that mean that all horoscopes cast before the discovery of Pluto in 1930 were invalid? If so, why didn't astrologers notice that something was wrong with their horoscopes and predict Pluto's existence? Given that several moons in the solar system arc larger than Pluto, with two larger than Mercury, should astrologers also be tracking the positions of these moons? What about asteroids, which orbit the Sun like planets? What about planets orbiting other stars? Given seemingly unanswerable questions like these, there seems little hope that astrology will ever meet its ancient goal of being able to forecast human events.

THE BIG PICTURE

Putting Chapter 3 into Context

In this chapter, we focused on the scientific principles through which we have learned so much about the universe. Key "big picture" concepts from this chapter include the following:

- The basic ingredients of scientific thinking—careful observation and trial-and-error testing—are a part of everyone's experience. Modern science simply provides a way of organizing this everyday thinking to facilitate the learning and sharing of new knowledge.

- Although knowledge about the universe is growing rapidly today, each new piece rests upon foundations of older discoveries. The foundations of astronomy reach far back into history and are intertwined with the general development of human culture and civilization. The ancient Greeks played a particularly important role, including developing the idea that models can be used to explain and represent the architecture of the cosmos.

- The Copernican revolution, which overthrew the ancient Greek belief in an Earth-centered universe, did not occur instantaneously. It unfolded over a period of more than a century, during which many of the characteristics of modern science first appeared. Key figures in this revolution include Copernicus, Tycho, Kepler, and Galileo.

- Several key hallmarks distinguish science from other ways of gathering knowledge: Science seeks natural explanations for observed phenomena, it progresses through the creation and testing of models that explain the observations as simply as possible, and it revises or abandons models whose predictions disagree with observations.

- Although astronomy and astrology once developed hand in hand, today they represent very different things. The predictions of astrology either fail rigorous testing or are so vague as to be untestable. Astronomy is a science and is the primary means by which humans learn about the physical universe.

3.1 Everyday Science

- *How is scientific thinking similar to other everyday thinking?* Scientific thinking involves trial and error like much everyday thinking, but in a carefully organized way.

3.2 The Ancient Roots of Science

- *How is modern science rooted in ancient astronomical observations?* Ancient cultures observed the motions of the Sun, Moon, planets, and stars for religious and practical reasons. Science took root as cultures eventually sought to understand the patterns they discovered.

- *What did ancient civilizations achieve in astronomy?* Ancient accomplishments include Egyptian measurements of time; identification of the Metonic cycle; structures for observation, such as Templo Mayor, the Sun Dagger, Mayan observatories, and Medicine Wheels; Polynesian navigation; and Chinese record keeping of astronomical events.

3.3 Ancient Greek Science

- *How did the Greeks lay the foundations for modern science?* The Greeks developed models of nature and emphasized the importance of having the predictions of those models agree with observations of nature.

- *What was the Ptolemaic model?* Ptolemy's model was a synthesis of earlier Greek ideas about the geocentric universe that allowed prediction of planetary positions. This sophisticated geocentric model worked well enough to remain in use for nearly 1,500 years.

3.4 The Copernican Revolution

- *How did Copernicus, Tycho, Kepler, and Galileo change our view of the cosmos?* Copernicus created a Sun-centered model of the solar system designed to replace the Ptolemaic model, but it was no more accurate because he still used perfect circles. Tycho provided observations used by Kepler to refine the model by introducing orbits with the correct characteristics. Galileo's experiments and telescopic observations overcame remaining objections to the Copernican idea of Earth as a planet orbiting the Sun.

- *What are Kepler's three laws of planetary motion?* (1) The orbit of each planet is an ellipse with the Sun at one focus. (2) As a planet moves around its orbit, it sweeps out equal areas in equal times. (3) More distant planets orbit the Sun at slower average speeds, following a precise mathematical relationship ($p^2 = a^3$).

3.5 The Nature of Science

- *How can we distinguish science from nonscience?* Science generally exhibits these three hallmarks: (1) Modern science seeks explanations for observed phenomena that rely solely on natural causes. (2) Science progresses through the creation and testing of models of nature that explain the observations as simply as possible. (3) A scientific model must make testable predictions about natural phenomena that would force us to revise or abandon the model if the predictions do not agree with observations.

- *What is a theory in science?* A scientific theory is a model that explains a wide variety of observations in terms of just a few general principles and has survived numerous tests to verify its predictions and explanations.

3.6 Astrology

- *How were astronomy and astrology related in the past? Are they still related today?* Astronomy and astrology both grew out of ancient observations of the sky. However, astronomy developed into a modern science. Astrology, or the search for hidden influences of planets and stars on human lives, has never passed scientific tests and does not qualify as science.

❓ Does It Make Sense?

Decide whether the statement makes sense and explain why it does or does not. (For an example, see Chapter 1, "Does It Make Sense?")

1. If we defined hours as the ancient Egyptians did, we'd have the longest hours on the summer solstice and the shortest hours on the winter solstice.

2. The date of Christmas (December 25) is set each year according to a lunar calendar.

3. When navigating in the South Pacific, the Polynesians found their latitude with the aid of the pointer stars of the Big Dipper.

4. The Ptolemaic model reproduced apparent retrograde motion by having planets move sometimes counterclockwise and sometimes clockwise in their circles.

5. In science, saying that something is a theory means that it is really just a guess.

6. Ancient astronomers were convinced of the validity of astrology as a tool for predicting the future.

7. If the planet Uranus had been identified as a planet in ancient times, we'd probably have eight days in a week.

8. Upon its publication in 1543, the Copernican model was immediately accepted by most scientists because its predictions of planetary positions were essentially perfect.

Problems

9. *Ancient Accomplishments.* In as much depth as possible, describe one notable astronomical achievement of an ancient culture and why it was significant.

10. *Days of the Week.* How are the names of the seven days of the week related to astronomical objects?

11. *The Metonic Cycle.* What is the *Metonic cycle?* How does the Jewish calendar follow the Metonic cycle? How does this influence the date of Easter? Why does the Muslim fast of Ramadan occur earlier with each subsequent year?

12. *The Greek Model.* Briefly summarize the development of the Greek *geocentric model,* from Thales through Ptolemy.

13. *Unfolding of the Copernican Revolution.* What was the *Copernican revolution,* and how did it change the human view of the universe? Briefly describe major players and events in the Copernican revolution.

14. *Hallmarks of Science.* Briefly describe each of the three hallmarks of science and how they are useful.

15. *Kepler's Laws.* Clearly state each of Kepler's three laws of planetary motion. For each law, describe in your own words what it means in a way that could be understood by almost anyone.

16. *What Makes It Science?* Choose a single idea in the modern view of the cosmos as discussed in Chapter 1, such as "The universe is expanding," "The universe began with a Big Bang," "We are made from elements manufactured by stars," or "The Sun orbits the center of the Milky Way Galaxy once every 230 million years."

    a. Briefly describe how the idea you have chosen is rooted in each of the three hallmarks of science discussed in this chapter. (That is, explain how it is based on observations, how our understanding of it depends on a model, and how the model is testable.)

    b. No matter how strongly the evidence may support a scientific idea, we can never be certain beyond all doubt that the idea is true. For the idea you have chosen, describe an observation that might cause us to call the idea into question. Then briefly discuss whether you think that, overall, the idea is likely or unlikely to hold up to future observations. Defend your opinion.

17. *The Copernican Revolution.* Based on what you have learned about the Copernican revolution, write a one- to two-page essay about how you believe it altered the course of human history.

18. *Cultural Astronomy.* Choose a particular culture of interest to you, and research the astronomical knowledge and ac-

complishments of that culture. Write a two- to three-page summary of your findings.

19. *Astronomical Structures.* Choose an ancient astronomical structure of interest to you (e.g., Stonehenge, Nazca lines, Pawnee lodges) and research its history. Write a two- to three-page summary of your findings. If possible, also build a scale model of the structure or create detailed diagrams to illustrate how the structure was used.

20. *Venus and the Mayans.* The planet Venus apparently played a particularly important role in Mayan society. Research the evidence and write a one- to two-page summary of current knowledge about the role of Venus in Mayan society.

21. *Scientific Test of Astrology.* Find out about at least one scientific test that has been conducted to test the validity of astrology. Write a short summary of how the test was conducted and what conclusions were reached.

22. *Your Own Astrological Test.* Devise your own scientific test of astrology. Clearly define the methods you will use in your test and how you will evaluate the results. Then carry out the test and write a report on your methods and results.

Discussion Questions

23. *The Impact of Science.* The modern world is filled with ideas, knowledge, and technology that developed through science and application of the scientific method. Discuss some of these things and how they affect our lives. Which of these impacts do you think are positive? Which are negative? Overall, do you think science has benefited the human race? Defend your opinion.

24. *The Importance of Ancient Astronomy.* Why was astronomy important to people in ancient times? Discuss both the practical importance of astronomy and the importance it may have had for religious or other traditions. Which do you think was more important in the development of ancient astronomy, its practical or its philosophical role? Defend your opinion.

25. *Secrecy and Science.* The text mentions that some historians believe that Chinese science and technology fell behind that of Europe because of the Chinese culture of secrecy. Do you agree that secrecy can hold back the advance of science? Why or why not? For the past 200 years, the United States has allowed a greater degree of free speech than most other countries in the world. How great a role do you think this has played in making the United States the world leader in science and technology? Defend your opinion.

26. *Lunar Cycles.* We have now discussed four distinct lunar cycles: the $29\frac{1}{2}$-day cycle of phases, the 18-year–$11\frac{1}{3}$-day saros cycle, the 19-year Metonic cycle, and the 18.6-year cycle over which a full moon rises at its most southerly place along the horizon. Discuss how each of these cycles can be observed and the role each cycle has played in human history.

27. *Astronomy and Astrology.* Why do you think astrology remains so popular around the world even though it has failed all scientific tests of its validity? Do you think the popularity of astrology has any positive or negative social consequences? Defend your opinions.

For a complete list of media resources available, go to www.astronomyplace.com and choose Chapter 3 from the pull-down menu.

 Astronomy Place Web Tutorials

Tutorial Review of Key Concepts

Use the interactive **Tutorial** at www.astronomyplace.com to review key concepts from this chapter.

Orbits and Kepler's Laws Tutorial

Lesson 2 Kepler's First Law

Lesson 3 Kepler's Second Law

Lesson 4 Kepler's Third Law

Supplementary Tutorial Exercises

Use the interactive **Tutorial Lesson** to explore the following questions.

Orbits and Kepler's Laws Tutorial, Lesson 2

1. When is an ellipse a circle?

2. Use the ellipse tool in Lesson 2 to prove that the semimajor axis of an ellipse is a planet's average orbital radius:

   A: Orbital radius when the planet is closest to the Sun: perihelion = _____ AU

   B: Orbital radius when the planet is farthest from the Sun: aphelion = _____ AU

   C: Average of perihelion and aphelion = _____ AU

   D: Length of the major axis of the ellipse = _____ AU

   E: Length of the semimajor axis of the ellipse = _____ AU

   Are the average orbital radius (C) and the semimajor axis (E) equal?

 Exploring the Sky and Solar System

Of the many activities available on the *Voyager: SkyGazer* **CD-ROM** accompanying your book, use the following files to observe key phenomena covered in this chapter.

Go to the **File: Basics** folder for the following demonstrations.

1. Ptolemy on Venus

2. Phase of Mercury

3. Pluto's Orbit

Go to the **File: Demo** folder for the following demonstrations.

1. Hale–Bopp Path

2. Hyakutake nears Earth

3. Venus–Earth–Moon

Go to the **Explore** menu for the following demonstrations.

1. Solar System

2. Paths of the Planets

Web Projects

Take advantage of the useful Web links on www.astronomyplace.com to assist you with the following projects.

1. *Easter.* Find out when Easter is celebrated by different sects of Christianity. Then research how and why different sects set different dates for Easter. Summarize your findings in a one- to two-page report.

2. *Greek Astronomers.* Many ancient Greek scientists had ideas that, in retrospect, seem well ahead of their time. Choose one or more of the following ancient Greek scientists, and learn enough about their work in science and astronomy to write a one- to two-page "scientific biography."

Thales	Anaximander	Pythagoras
Anaxagoras	Empedocles	Democritus
Meton	Plato	Eudoxus
Aristotle	Callipus	Aristarchus
Archimedes	Eratosthenes	Apollonius
Hipparchus	Seleucus	Ptolemy
Hypatia		

3. *The Ptolemaic Model.* This chapter gives only a very brief description of Ptolemy's model of the universe. Investigate the model in greater depth. Using diagrams and text as needed, give a two- to three-page description of the model.

4. *The Galileo Affair.* In recent years, the Roman Catholic Church has devoted a lot of resources to learning more about the trial of Galileo and to understanding past actions of the Church in the Galileo case. Learn more about such studies, and write a short report about the current Vatican view of the case.

5. *Science or Pseudoscience.* Choose some pseudoscientific claim that has been in the news recently, and learn more about it and how scientists have "debunked" it. Write a short summary of your findings.

S1 Celestial Timekeeping and Navigation

Supplementary Chapter

Socrates: Shall we make astronomy the next study? What do you say?
Glaucon: Certainly. A working knowledge of the seasons, months, and years is beneficial to everyone, to commanders as well as to farmers and sailors.
Socrates: You make me smile, Glaucon. You are so afraid that the public will accuse you of recommending unprofitable studies.

Plato, Republic

In ancient times, the practical need for timekeeping and navigation was one of the primary reasons for the study of astronomy. The celestial origins of timekeeping and navigation are still evident. The time of day comes from the location of the Sun in the local sky, the month comes from the Moon's cycle of phases, and the year comes from the Sun's annual path along the ecliptic. The very name "North Star" tells us how it can be an aid to navigation.

Today, we can tell the time by glancing at an inexpensive electronic watch and navigate with hand-held devices that receive signals from satellites of the global positioning system (GPS). But knowing the celestial basis of timekeeping and navigation can still be useful, particularly for understanding the rich history of astronomical discovery. In this chapter, we will explore the apparent motions of the Sun, Moon, and planets in greater detail, enabling us to study the principles of celestial timekeeping and navigation.

S1.1 Astronomical Time Periods

When people began measuring time long ago, the only celestial motions that mattered were those that could be seen in the local sky. We can measure that local sky with much more precision now, and astronomers have identified and defined more than one kind of day, month, year.

Solar Versus Sidereal Day

We may think of our 24-hour day as the rotation period of Earth, but that's not quite true. Earth's rotation period is the time it takes Earth to complete one full rotation. Because Earth's daily rotation makes the celestial sphere appear to rotate around us (see Figure 2.7), we can measure the rotation period by timing how long it takes the celestial sphere to make one full turn through the local sky.

For example, we could start a stopwatch at the moment when a particular star is on our meridian (the semicircle stretching from due south, through the zenith, to due north [Section 2.2]), then stop the watch the next day when the same star again is on the meridian (Figure S1.1a). Measured in this way, Earth's rotation period is about 23 hours 56 minutes (more precisely $23^h56^m4.09^s$)—or about 4 minutes short of 24 hours. This time period is called a **sidereal day**, because it is measured relative to the apparent motion of stars in the local sky. *Sidereal* (pronounced *sy-dear-ee-al*) means "related to the stars."

Our 24-hour day, which we call a **solar day**, is based on the time it takes for the *Sun* to make one circuit around our local sky. We could measure this time period by starting the stopwatch at the moment when the Sun is on our meridian one day and stopping it when the Sun reaches the meridian the next day (Figure S1.1b). The solar day is indeed 24 hours on average, although it varies slightly (up to 25 seconds longer or shorter than 24 hours) over the course of a year.

A simple demonstration shows why a solar day is slightly longer than a sidereal day. Set an object on a table to represent the Sun, and stand a few steps away from the object to represent Earth. Point at the Sun and imagine that you also happen to be pointing toward some distant star that lies in the same direction. If you rotate (counterclockwise) while standing in place, you'll again be pointing at both the Sun and the star after one full rotation (Figure S1.2a). But because Earth orbits the Sun at the same time that it rotates, you can make the demonstration more realistic by taking a couple of steps around the Sun (counterclockwise) while you are rotating (Figure S1.2b). After one full rotation, you will again be pointing in the direction of the distant star, so this represents a sidereal day. However, because of your orbital motion, you'll need to rotate slightly more than once to be pointing again at the Sun. This "extra" bit of rotation makes a solar day longer than a sidereal day.

The only problem with this demonstration is that it exaggerates Earth's daily orbital motion. In reality, Earth moves about 1° per day around its orbit (because it makes a full 360° orbit in 1 year, or about 365 days). Because a single rotation means rotating 360°, Earth must actually rotate about 361° with each solar day (Figure S1.2c). The extra 1° rotation accounts for the extra 4 minutes by which the solar day is longer than the sidereal day. (To see why it takes 4 minutes to rotate 1°, remember that Earth rotates 360° in about 23 hours 56 minutes, or 1,436 minutes. Thus, 1° of rotation takes $\frac{1}{360} \times 1,436$ minutes $\approx$ 4 minutes.)

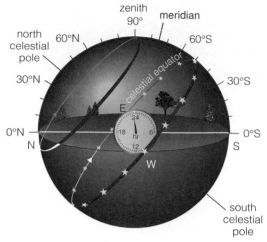

a A sidereal day is the time it takes any star to make a circuit of the local sky. It is about 23 hours 56 minutes.

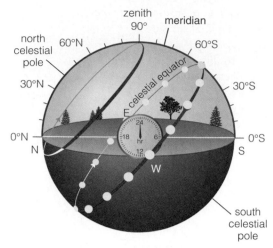

b A solar day is measured similarly, but by timing the Sun rather than a star. The length of the solar day varies over the course of the year but averages 24 hours.

Figure S1.1 Using the sky to measure the length of a day.

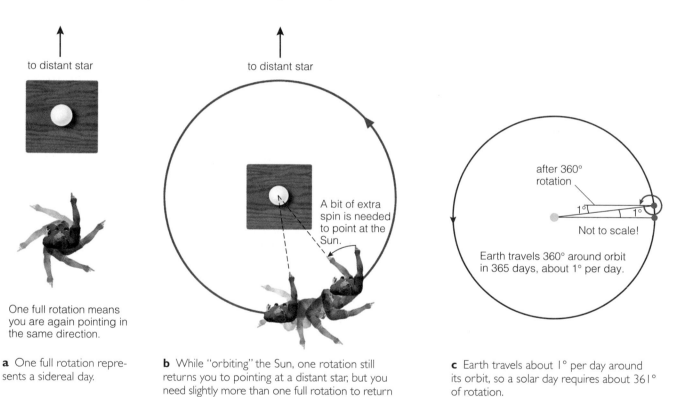

a One full rotation represents a sidereal day.

b While "orbiting" the Sun, one rotation still returns you to pointing at a distant star, but you need slightly more than one full rotation to return to pointing at the Sun.

c Earth travels about 1° per day around its orbit, so a solar day requires about 361° of rotation.

Figure S1.2 A demonstration showing why a solar day is slightly longer than a sidereal day.

Synodic Versus Sidereal Month

As we discussed in Chapter 2, our month comes from the Moon's $29\frac{1}{2}$-day cycle of phases (think "moonth"). More technically, the $29\frac{1}{2}$-day period required for each cycle of phases is called a **synodic month**. The word *synodic* comes from the Latin *synod*, which means "meeting." A synodic month gets its name because the Sun and the Moon "meet" in the sky with every new moon.

Just as a solar day is not Earth's true rotation period, a synodic month is not the Moon's true orbital period. Earth's motion around the Sun means that the Moon must complete more than one full orbit of Earth from one new moon to the next (Figure S1.3). The Moon's true orbital period, or **sidereal month**, is only about $27\frac{1}{3}$ days. Like the sidereal day, the sidereal month gets its name because it describes how long it takes the Moon to complete an orbit relative to the positions of distant stars.

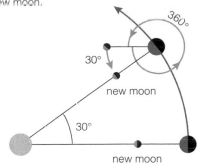

Earth travels about 30° per month around the Sun, so the Moon must orbit around Earth about 360° + 30° = 390° from new moon to new moon.

Figure S1.3 The Moon completes one 360° orbit in about $27\frac{1}{3}$ days (a sidereal month), but the time from new moon to new moon is about $29\frac{1}{2}$ days (a synodic month).

Tropical Versus Sidereal Year

A year is related to Earth's orbital period, but again there are two slightly different definitions for the length of the year. The time it takes for Earth to complete one orbit relative to the stars is called a **sidereal year**. But our calendar is based on the cycle of the seasons, which we can measure from the time of the spring equinox one year to the spring equinox the next year. This time period, called a **tropical year**, is about 20 minutes shorter than the sidereal year. A 20-minute difference might not seem like much, but it would make a calendar based on the sidereal year get out of sync with the seasons by 1 day every 72 years—a difference that would add up over centuries.

The difference between the sidereal year and the tropical year arises from Earth's 26,000-year cycle of axis precession [Section 2.4]. Precession not only changes the orientation of the axis in space but also changes the locations in Earth's orbit at which the seasons occur. Each year, the location of the equinoxes and solstices among the stars shifts about $\frac{1}{26,000}$ of the way around the orbit. And $\frac{1}{26,000}$ of a year is about 20 minutes, which explains the 20-minute difference between the tropical year and the sidereal year.

Planetary Periods (Synodic Versus Sidereal)

Planetary periods are not used in our modern timekeeping, but they were important to many ancient cultures. For example, the Mayan calendar was based in part on the apparent motions of Venus. Today, understanding planetary periods can help us make sense of what we see in the sky.

A planet's **sidereal period** is the time it takes to orbit the Sun. (As usual, it has the name *sidereal* because it is measured relative to distant stars.) For example, Jupiter's sidereal period is 11.86 years, so it takes about 12 years for Jupiter to make a complete circuit around the constellations of the zodiac. Thus, Jupiter appears to move through roughly one zodiac constellation each year. If Jupiter is currently in Leo (as it is for much of 2004), it will be in Virgo at this time next year and Libra the following year, returning to Leo in 12 years.

A planet's **synodic period** is the time between being lined up with the Sun in our sky one time and the next similar alignment. (As with the Moon, the term *synodic* refers to the planet's "meeting" the Sun in the sky.) Figure S1.4 shows that the situation is somewhat different for planets nearer the Sun than Earth (that is, Mercury and Venus) and planets farther away (all the rest of the planets).

Look first at the situation for the more distant planet in Figure S1.4. As seen from Earth, this planet will sometimes line up with the Sun in what we call a **conjunction**. At other special times, it will appear exactly opposite the Sun in our sky, or at **opposition**. We cannot see the planet during conjunction with the Sun because it is hidden by the Sun's glare and rises and sets with the Sun in our sky. At opposition, the planet moves through the sky like the full moon, rising at sunset, reaching the meridian at midnight, and setting at dawn. You can see that the planet is closest to Earth at opposition and hence appears brightest at this time.

Figure S1.4 shows that a planet *nearer* than Earth to the Sun has two conjunctions—an "inferior conjunction" between Earth and the Sun and a "superior conjunction" when the planet appears behind the Sun as seen from Earth—rather than one conjunction and one opposition.

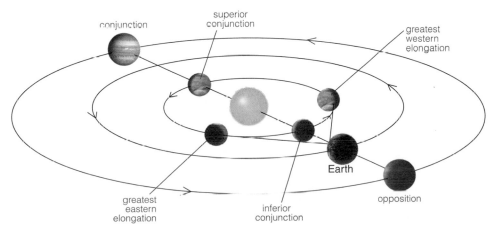

conjunction

superior conjunction

greatest western elongation

Earth

greatest eastern elongation

inferior conjunction

opposition

Figure S1.4 This diagram shows important positions of planets relative to Earth and the Sun. For a planet farther from the Sun than Earth (such as Jupiter), conjunction is when it appears aligned with the Sun in the sky, and opposition is when it appears on our meridian at midnight. Planets nearer the Sun (such as Venus) have two conjunctions and never get farther from the Sun in our sky than at their greatest elongations.

Mercury

Figure S1.5 NASA's *TRACE* satellite captured this image of a Mercury transit on November 15, 1999. The photograph was taken with ultraviolet light. The colors represent the intensity of this light. The structures seen with ultraviolet light are patches of hot gas just above the Sun's visible surface.

Mercury and Venus usually appear slightly above or below the Sun at inferior conjunction because they have slightly different orbital planes than Earth. Occasionally, however, these planets appear to move across the disk of the Sun during inferior conjunction, creating a **transit** (Figure S1.5). Mercury transits occur an average of a dozen times per century, with the next one coming on November 8, 2006. Venus transits typically come in pairs separated by a century or more. This century's pair of Venus transits occur on June 8, 2004, and June 6, 2012. If you miss them both, you'll have to wait 105 years for the next one.

The inner planets are best viewed when they are near their points of **greatest elongation**, that is, when they are farthest from the Sun in our sky. At its greatest eastern elongation, Venus appears about 46° east of the Sun in our sky, which means it shines brightly in the evening sky. Similarly, at its greatest western elongation, Venus appears about 46° west of the Sun in our sky, shining brightly in the predawn sky. In between the times when Venus appears in the morning sky and the times when it appears in the evening sky, Venus disappears from view for a few weeks with each conjunction. Mercury's pattern is similar, but because it is closer to the Sun it never appears more than about 28° from the Sun in our sky. Mercury is difficult to see, because it is almost always obscured by the glare of the Sun.

Based on Figure S1.4, explain why neither Mercury nor Venus can ever be on the meridian in the midnight sky.

Measuring a planet's synodic period is fairly easy. It simply requires observing its position relative to the Sun in the sky. In contrast, we must calculate a planet's sidereal period from the geometry of planetary orbits. Copernicus was the first to perform these calculations, and he found that more distant planets had longer sidereal periods. The simplicity of this pattern helped convince him that his idea of a Sun-centered solar system was correct [Section 3.4].

S1.2 Daily Timekeeping

Now that we have discussed astronomical time periods, we can turn our attention to modern measures of time. Our clock is based on the 24-hour solar day. You are already familiar with the basic principles, such as the idea that noon is around the time when the Sun is highest in the local sky. However, the precise details of timekeeping are somewhat subtler.

Apparent Solar Time

If we base time on the Sun's *actual* position in the local sky, as is the case when we use a sundial (Figure S1.6), we are measuring **apparent solar time**. Noon is the precise moment when the Sun is on the meridian and the sundial casts its shortest shadow. Before noon, when the Sun is rising upward through the sky, the apparent solar time is *ante meridiem* ("before the middle of the day"), or *a.m.* For example, if the Sun will reach the meridian 2 hours from now, the apparent solar time is 10 A.M. After noon, the apparent solar time is *post meridiem* ("after the middle of the day"), or *p.m.* If the Sun crossed the meridian 3 hours ago, the apparent solar time is 3 P.M. Note that, technically, noon and midnight are *neither* a.m. nor p.m. However, by

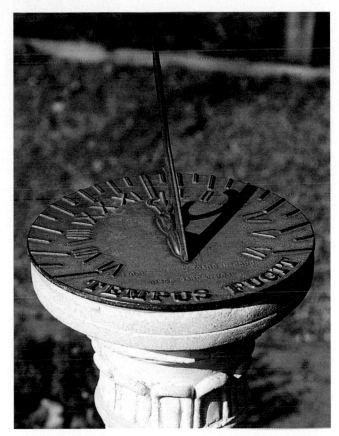

Figure S1.6 A basic sundial consists of a stick, or *gnomon*, that casts a shadow and a dial marked by numerals. Here, the shadow is on the Roman numeral III, indicating that the apparent solar time is 3:00 P.M. (The portion of the dial without numerals represents nighttime hours.) Because the Sun's path across the local sky depends on latitude, a particular sundial will be accurate only for a particular latitude.

convention we usually say that noon is 12 P.M. and midnight is 12 A.M.

THINK ABOUT IT

It is daytime or nighttime at 12:01 A.M.? 12:01 P.M.? Explain.

Mean Solar Time

Suppose you set a clock to read precisely 12:00 when a sundial reads noon today. If every solar day were precisely 24 hours, your clock would always remain synchronized with the sundial. However, while 24 hours is the *average* length of the solar day, the actual length of the solar day varies throughout the year. As a result, your clock will not remain perfectly synchronized with the sundial. For example, your clock is likely to read a few seconds before or after 12:00 when the sundial reads noon tomorrow, and within a few weeks your clock time may differ from the apparent solar time by several minutes. Your clock (assuming it is accurate) will again be synchronized with the Sun on the same date next year, since it keeps track of the average length of the solar day.

If we average the differences between the time a clock would read and the time a sundial would read, we can define **mean solar time** (*mean* is another word for *average*). A clock set to mean solar time reads 12:00 each day at the time that the sun crosses the meridian *on average*. The actual mean solar time at which the Sun crosses the meridian varies over the course of the year in a fairly complex way (see "Solar Days and the Analemma," p. 92). The result is that, on any given day, a clock set to mean solar time may read anywhere from about 17 minutes before noon to 15 minutes after noon (that is, from 11:43 A.M. to 12:15 P.M.) when a sundial indicates noon.

Although the lack of perfect synchronization with the Sun might at first sound like a drawback, mean solar time is actually more convenient than apparent solar time (the sundial time)—as long as you have access to a mechanical or electronic clock. Once set, a reliable mechanical or electronic clock can always tell you the mean solar time. In contrast, precisely measuring apparent solar time requires a sundial, which is useless at night or when it is cloudy.

Like apparent solar time, mean solar time is a *local* measure of time. That is, it varies with longitude because of Earth's west-to-east rotation. For example, clocks in New York are set 3 hours ahead of clocks in Los Angeles. If clocks were set precisely to local mean solar time, they would vary even over relatively short east-west distances. For example, mean solar clocks in central Los Angeles would be about 2 minutes behind mean solar clocks in Pasadena, because Pasadena is slightly farther east.

Standard, Daylight, and Universal Time

Clocks reading mean solar time were common during the early history of the United States. However, by the late 1800s, the growth of railroad travel made the use of mean solar time increasingly problematic. Some states had dozens of different "official" times, usually corresponding to mean solar time in dozens of different cities, and each railroad company made schedules according to its own "railroad time." The many time systems made it difficult for passengers to follow the scheduling of trains.

On November 18, 1883, the railroad companies agreed to a new system that divided the United States into four time zones, setting all clocks within each zone to the same time. That was the birth of **standard time**, which today divides the world into time zones (Figure S1.7). Depending on where you live within a time zone, your standard time may vary somewhat from your mean solar time. (In principle, the standard time in a particular time zone is the mean solar time in the *center* of the time zone so that local mean solar time within a 1-hour-wide time zone could never differ by more than a half-hour from standard time. However, time zones often have unusual shapes to conform to social, economic, and political realities, so larger variations between standard time and mean solar time sometimes occur.)

In most parts of the United States, clocks are set to standard time for only part of the year. Between the first Sunday in April and the last Sunday in October, most of the United States changes to **daylight saving time**, which is 1 hour ahead of standard time. Because of the 1-hour advance on daylight saving time, clocks read around 1 P.M. (rather than around noon) when the Sun is on the meridian.

As we will see, for purposes of navigation and astronomy it is useful to have a single time for the entire Earth. For historical reasons, this "world" time was chosen to be the mean solar time in Greenwich, England—the place that also defines longitude 0° (see Figure 2.10). Today, this *Greenwich mean time* (*GMT*) is often called **universal time** (**UT**). (Outside astronomy, it is often called universal coordinated

---

SPECIAL TOPIC   Solar Days and the Analemma

The average length of a solar day is 24 hours, but the precise length varies over the course of the year. Two effects contribute to this variation.

The first effect is due to Earth's varying orbital speed. Recall that, in accord with Kepler's second law, Earth moves slightly faster when it is closer to the Sun in its orbit and slightly slower when it is farther from the Sun. Thus, Earth moves slightly farther along its orbit each day when it is closer to the Sun. This means that the solar day requires more than the average amount of "extra" rotation (see Figure S1.2) during these periods—making these solar days longer than average. Similarly, the solar day requires less than the average amount of "extra" rotation when it is in the portion of its orbit farther from the Sun—making these solar days shorter than average.

The second effect is due to the tilt of Earth's axis, which causes the ecliptic to be inclined by $23\frac{1}{2}°$ to the celestial equator on the celestial sphere. Because the length of a solar day depends on the Sun's apparent *eastward* motion along the ecliptic, the inclination would cause solar days to vary in length even if Earth's orbit were perfectly circular. To see why, suppose the Sun appeared to move exactly 1° per day along the ecliptic. Around the times of the solstices, this motion would be entirely eastward, making the solar

day slightly longer than average. Around the times of the equinoxes, when the motion along the ecliptic has a significant northward or southward component, the solar day would be slightly shorter than average.

Together, the two effects make the actual length of solar days vary by up to about 25 seconds (either way) from the 24-hour average. Because the effects accumulate at particular times of year, the apparent solar time can differ by as much as 17 minutes from the mean solar time. The net result is often depicted visually by an **analemma** (Figure 1), which looks much like a figure-8. You'll find an analemma printed on many globes (Figure 2.16 shows a photographic version).

By using the horizontal scale on the analemma (Figure 1) you can convert between mean and apparent solar time for any date. (The vertical scale shows the declination of the Sun, which is discussed in Section S1.5.) For example, the dashed line shows that on November 10 a mean solar clock is about 17 minutes "behind the Sun," or behind apparent solar time. Thus, if the apparent solar time is 6:00 P.M. on November 10, the mean solar time is only 5:43 P.M. The discrepancy between mean and apparent solar time is called the **equation of time**. It is often plotted as a graph (Figure 2), which gives the same results as reading from the analemma.

The discrepancy between mean and apparent solar time also explains why the times of sunrise and sunset don't follow seasonal patterns perfectly. For example, the winter solstice around December 21 has the shortest daylight hours (in the Northern Hemisphere), but the earliest sunset occurs around December 7, when the Sun is still well "behind" mean solar time.

Figure 1 The analemma shows the annual pattern of discrepancies between apparent and mean solar time. For example, the dashed line shows that on November 10 a mean solar clock reads 17 minutes behind (earlier than) apparent solar time.

Figure 2 The discrepancies can also be plotted on a graph as the equation of time.

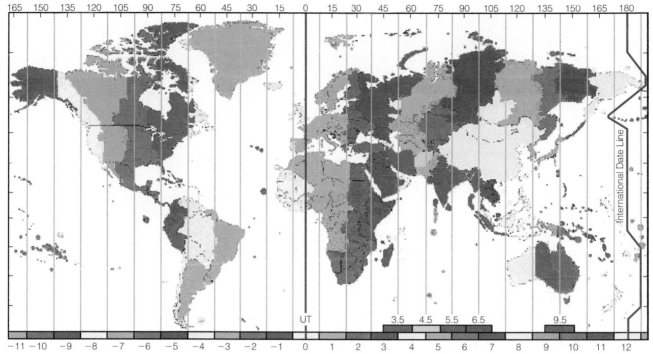

Figure S1.7 Time zones around the world. The numerical scale at the bottom shows hours ahead of (positive numbers) or behind (negative numbers) the time in Greenwich, England; the scale at the top is longitude. The vertical lines show standard time zones as they would be in the absence of political considerations. The color-coded regions show the actual time zones. Note, for example, that all of China uses the same standard time, even though the country is wide enough to span several time zones. Note also that a few countries use time zones centered on a half-hour, rather than an hour, relative to Greenwich time.

time [UTC]. Many airlines and weather services call it "Zulu time," because Greenwich's time zone is designated Z and "zulu" is a common way of phonetically identifying the letter Z.)

 Seasons Tutorial, Lesson 2

S1.3 The Calendar

Our modern calendar is based on the length of the tropical year, which is the amount of time from one spring equinox to the next. The origins of our calendar go back to ancient Egypt. By 4200 B.C., the Egyptians were using a calendar that counted 365 days in a year.

Because the tropical year actually is closer to $365\frac{1}{4}$ days, the Egyptian calendar slowly drifted out of phase with the seasons by about 1 day every 4 years. For example, if the spring equinox occurred on March 21 one year, 4 years later it occurred on March 20, 4 years after that on March 19, and so on. Over many centuries, the spring equinox moved through many different months. To keep the seasons and the calendar synchronized, Julius Caesar decreed the adoption of a new calendar in 46 B.C. This **Julian calendar** introduced the concept of **leap year**: Every fourth year has 366 days, rather than 365, so that the average length of the calendar year is $365\frac{1}{4}$ days.

The Julian calendar originally had the spring equinox falling around March 24. If it had been perfectly synchro-

nized with the tropical year, this calendar would have ensured that the spring equinox occurred on the same date every 4 years (that is, every leap-year cycle). It didn't work perfectly, however, because a tropical year is actually about 11 minutes short of $365\frac{1}{4}$ days. Thus, the moment of the spring equinox slowly advanced by an average of 11 minutes per year. By the late 1500s, the spring equinox was occurring on March 11.

In 1582, Pope Gregory XIII introduced a new calendar—the **Gregorian calendar**—designed to return the spring equinox to the same date after every 4-year cycle. The Gregorian calendar made two adjustments to the Julian calendar. First, Pope Gregory decreed that the day in 1582 following October 4 would be October 15. By eliminating the ten dates from October 5 through October 14, 1582, he pushed the date of the spring equinox in 1583 from March 11 to March 21. (He chose March 21 because it was the date of the spring equinox in A.D. 325, which was the time of the Council of Nicaea, the first ecumenical council of the Christian church.) Second, the Gregorian calendar added an exception to the rule of having leap year every 4 years: Leap year is skipped when a century changes (for example, in years 1700, 1800, 1900) *unless* the century year is divisible by 400. Thus, 2000 was a leap year because it is divisible by 400 (2,000 ÷ 400 = 5), but 2100 will *not* be a leap year. These adjustments make the average length of the Gregorian calendar year almost exactly the same as the actual length of a tropical year, which ensures that the spring equinox will occur on March 21 every fourth year for thousands of years to come.

Today, the Gregorian calendar is used worldwide for international communication and commerce. (Many countries still use traditional calendars, such as the Chinese, Islamic, and Jewish calendars, for cultural purposes.) However, as you might guess, the Pope's decree was not immediately accepted in regions not bound to the Catholic Church. For example, the Gregorian calendar was not adopted in England or in the American colonies until 1752, and it was not adopted in China until 1912 or in Russia until 1919.

S1.4 Mapping Locations in the Sky

We are now ready to turn our attention from timekeeping to navigation. The goal of celestial navigation is to use the Sun and the stars to find our position on Earth. Before we can do that, we need to understand the apparent motions of the sky. We'll begin in this section by discussing how we map the celestial sphere. The next section will use this map to help you understand motion in the local sky. Then, in the final section, we'll see how these ideas lead to the principles of celestial navigation.

A Map of the Celestial Sphere

For purposes of pinpointing objects in our sky, it's useful to think of Earth as being in the center of a giant celestial sphere (see Figure 2.3). From our point of view on Earth, the celestial sphere appears to rotate around us each day (see Figure 2.7).

We can use a model of the celestial sphere to locate stars or the Sun, much as we use a globe to locate places on Earth. The primary difference is that the celestial sphere models *apparent* positions in the sky, rather than true positions in space. As we discussed in Chapter 2, we can compare positions on the celestial sphere only by reference to the *angles* that separate them, not by actual distances. But aside from this important difference between a globe and the celestial sphere, we can use both as maps by identifying special locations (such as the equator and poles) and adding a system of coordinates (such as latitude and longitude).

We've already discussed the special locations we need for a map of the celestial sphere: the north and south celestial poles, the celestial equator, and the ecliptic. Figure S1.8 shows these locations on a schematic diagram. Earth is in the center because the celestial sphere represents the sky as we see it from Earth. The arrow along the ecliptic indicates the direction in which the Sun appears to move along it over the course of each year. It is much easier to visualize the celestial sphere if you make a model with a simple plastic ball. Use a felt-tip pen to mark the north and south celestial poles on your ball, and then add the celestial equa-

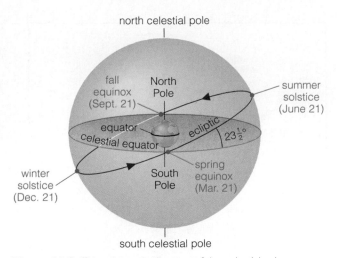

Figure S1.8 This schematic diagram of the celestial sphere, shown without stars, helps us create a map of the celestial sphere by showing the north and south celestial poles, the celestial equator, the ecliptic, and the equinoxes and solstices. As a study aid, you should use a plastic ball as a model of the celestial sphere, marking it with the same special locations.

tor and the ecliptic. Note that the ecliptic crosses the celestial equator on opposite sides of the celestial sphere at an angle of $23\frac{1}{2}°$ (because of the tilt of Earth's axis).

Equinoxes and Solstices

Remember that the equinoxes and solstices are special moments in the year that help define the seasons [Section 2.3]. For example, the *spring equinox,* which occurs around March 21 each year, is the moment when spring begins for the Northern Hemisphere and fall begins for the Southern Hemisphere. These moments correspond to positions in Earth's orbit (see Figure 2.15) and hence to apparent locations of the Sun along the ecliptic. As shown in Figure S1.8, the spring equinox occurs when the Sun is on the ecliptic at the point where it crosses from south of the celestial equator to north of the celestial equator. This point is also called the spring equinox. Thus, the term *spring equinox* has a dual meaning: It is the *moment* when spring begins and also the *point* on the ecliptic at which the Sun appears to be located at that moment.

Figure S1.8 also shows the points marking the summer solstice, fall equinox, and winter solstice, with the dates on which the Sun appears to be located at each point. Remember that the dates are approximate because of the leap-year cycle and because a tropical year is not exactly $365\frac{1}{4}$ days. (For example, the spring equinox may occur anytime between March 20 and March 23.)

Although no bright stars mark the locations of the equinoxes or solstices among the constellations, you can find them with the aid of nearby bright stars (Figure S1.9). For example, the spring equinox is located in the constellation Pisces and can be found with the aid of the four bright stars in the Great Square of Pegasus. Of course, when the Sun is located at this point around March 21, we cannot see

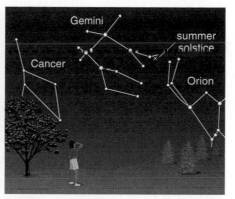

a The spring equinox

b The summer solstice

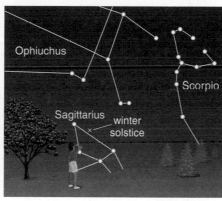

c The fall equinox

d The winter solstice

Figure S1.9 These diagrams show the locations among the constellations of the equinoxes and solstices. No bright stars mark any of these points, so you must find them by studying their positions relative to recognizable patterns. The time of day and night at which each point is above the horizon depends on the time of year.

Pisces or Pegasus because they are close to the Sun in our daytime sky.

THINK ABOUT IT

Using your plastic ball as a model of the celestial sphere (which you have already marked with the celestial poles, equator, and ecliptic), mark the locations and approximate dates of the equinoxes and solstices. Based on the dates for these points, approximately where along the ecliptic is the Sun on April 21? On November 21? How do you know?

Celestial Coordinates

We can complete our map of the celestial sphere by adding a coordinate system similar to the coordinates that measure latitude and longitude on Earth. This system will be the third coordinate system we've used in this book; Figure S1.10 reviews the three systems.

- Figure S1.10a shows the coordinates of *altitude* and *direction* (or *azimuth**) we use in the local sky.

- Figure S1.10b shows the coordinates of *latitude* and *longitude* we use on Earth's surface.

---

*Azimuth is usually measured clockwise around the horizon from due north. By this definition, the azimuth of due north is 0°, due east is 90°, due south is 180°, and due west is 270°.

- Figure S1.10c shows the system of **celestial coordinates** we use to pinpoint locations on the celestial sphere. These coordinates are called **declination (dec)** and **right ascension (RA)**.

Declination and Latitude If you compare Figures S1.10b and c, you'll see that declination on the celestial sphere is very similar to latitude on Earth:

- Just as lines of latitude are parallel to Earth's equator, lines of declination are parallel to the celestial equator.

- Just as Earth's equator has lat = 0°, the celestial equator has dec = 0°.

- Latitude is labeled north or south relative to the equator, while declination is labeled *positive* or *negative*. For example, the North Pole has lat = 90°N, while the north celestial pole has dec = +90°; the South Pole has lat = 90°S, while the south celestial pole has dec = −90°.

Right Ascension and Longitude The diagrams in Figures S1.10b and c also show that right ascension on the celestial sphere is very similar to *longitude* on Earth:

- Just as lines of longitude extend from the North Pole to the South Pole, lines of right ascension extend from the north celestial pole to the south celestial pole.

a We use altitude and direction to pinpoint locations in the local sky.

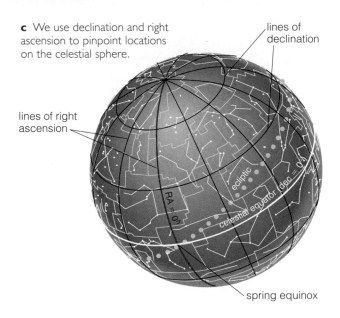

b We use latitude and longitude to pinpoint locations on Earth.

Greenwich

Rome:
latitude = 42°N
longitude = 12°E

prime meridian

line of longitude

line of latitude

Miami:
latitude = 26°N
longitude = 80°W

c We use declination and right ascension to pinpoint locations on the celestial sphere.

lines of declination

lines of right ascension

ecliptic

celestial equator

spring equinox

Figure S1.10 Celestial coordinate systems.

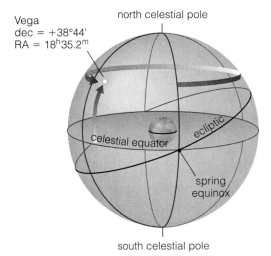

Vega
dec = +38°44'
RA = 18ʰ35.2ᵐ

north celestial pole

celestial equator

ecliptic

spring equinox

south celestial pole

Figure S1.11 This diagram shows how we interpret the celestial coordinates of Vega. Its declination tells us that it is 38°44' north of the celestial equator. We can interpret its right ascension in two ways: As an angle, it means Vega is about 279° (the angular equivalent of 18ʰ35ᵐ) east of the vernal equinox; as a time, it means Vega crosses the meridian about 18 hours 35 minutes after the spring equinox.

- Just as there is no natural starting point for longitude, there is no natural starting point for right ascension. By international treaty, longitude zero (the prime meridian) is the line of longitude that runs through Greenwich, England. By convention, right ascension zero is the line of right ascension that runs through the spring equinox.

- Longitude is measured in degrees east or west of Greenwich, while right ascension is measured in hours (and minutes and seconds) east of the spring equinox. A full 360° circle around the celestial equator goes through 24 hours of right ascension, so each hour of right ascension represents an angle of 360° ÷ 24 = 15°.

Example: Where's Vega? We can use celestial coordinates to describe the position of any object on the celestial sphere. For example, the bright star Vega has dec = +38°44' and RA = 18ʰ35ᵐ (Figure S1.11). The positive declination tells us that Vega is 38°44' *north* of the celestial equator. The right ascension tells us that Vega is 18 hours 35 minutes east of the spring equinox. Translating the right ascension from hours to angular degrees, we find that Vega is about 279° east of the spring equinox (because 18 hours represents $18 \times 15° = 270°$ and 35 minutes represents $\frac{35}{60} \times 15 \approx 9°$).

THINK ABOUT IT

On your plastic ball model of the celestial sphere, add a scale for right ascension along the celestial equator and also add a few circles of declination, such as declination 0°, ±30°, ±60°, and ±90°. Locate Vega on your model.

You may be wondering why right ascension is measured in units of time. The answer is that time units are convenient for tracking the daily motion of objects through the local sky. All objects with a particular right ascension cross the meridian at the same time. For example, all stars with RA = 0^h cross the meridian at the same time that the spring equinox crosses the meridian. For any other object, the right ascension tells us when it crosses the meridian in hours *after* the spring equinox crosses the meridian. Thus, for example, Vega's right ascension, 18^h35^m, tells us that on any particular day it crosses the meridian about 18 hours 35 minutes after the spring equinox. (This is 18 hours 35 minutes of *sidereal time* later, which is not exactly the same as 18 hours 35 minutes of solar time; see Mathematical Insight S1.1.)

Celestial Coordinates Change with Time The celestial coordinates of stars are not quite constant but rather change gradually with Earth's 26,000-year cycle of axis precession [Section 2.4]. The change occurs because celestial coordinates are tied to the celestial equator, which moves with precession relative to the constellations. (Axis precession does not affect Earth's orbit, so it does not affect the location of the ecliptic among the constellations.) Thus, the celestial coordinates of stars change even while the stars themselves remain fixed in the patterns of the constellations.

The coordinate changes are not noticeable to the naked eye, but precise astronomical work—such as aiming a telescope at a particular object—requires almost constant updating of celestial coordinates. Star catalogs therefore always state the year for which coordinates are given (for example,

---

Mathematical Insight **S1.1** **Time by the Stars**

The clocks we use in daily life are set to solar time, ticking through 24 hours for each day of mean solar time. In astronomy, it is also useful to have clocks that tell time by the stars, or **sidereal time**. Just as we define *solar time* according to the Sun's position relative to the meridian, *sidereal time* is based on the positions of stars relative to the meridian. We define the **hour angle** (**HA**) of any object on the celestial sphere to be the time since it last crossed the meridian. (For a circumpolar star, hour angle is measured from the *higher* of the two points at which it crosses the meridian each day.) For example:

- If a star is crossing the meridian now, its hour angle is 0^h.
- If a star crossed the meridian 3 hours ago, its hour angle is 3^h.
- If a star will cross the meridian 1 hour from now, its hour angle is -1^h or, equivalently, 23^h.

By convention, time by the stars is based on the hour angle of the spring equinox. That is, the **local sidereal time** (**LST**) is

$$\text{LST} = \text{HA}_{\text{spring equinox}}$$

For example, the local sidereal time is 00:00 when the spring equinox is *on* the meridian. Three hours later, when the spring equinox is 3 hours west of the meridian, the local sidereal time is 03:00.

Note that, because right ascension tells us how long after the spring equinox an object reaches the meridian, the local sidereal time is also equal to the right ascension (RA) of objects currently crossing your meridian. For example, if your local sidereal time is 04:30, stars with RA = 4^h30^m are currently crossing your meridian. This idea leads to an important relationship between any object's current hour angle, the current local sidereal time, and the object's right ascension:

$$\text{HA}_{\text{object}} = \text{LST} - \text{RA}_{\text{object}}$$

This formula will make sense to you if you recognize that an object's right ascension tells us the time by which it trails the spring equinox on its daily trek through the sky. Because the local sidereal time tells us how long it has been since the spring equinox was on the meridian, the difference $\text{LST} - \text{RA}_{\text{object}}$ must tell us the position of the object relative to the meridian.

Sidereal time has one important subtlety: Because the stars (and the celestial sphere) appear to rotate around us in one sidereal day (23^h56^m), sidereal clocks must tick through 24 hours of sidereal time in 23 hours 56 minutes of solar time. That is, a sidereal clock gains about 4 minutes per day over a solar clock. As a result, you cannot immediately infer the local sidereal time from the local solar time, or vice versa, without either doing some calculations or consulting an astronomical table. Of course, the easiest way to determine the local sidereal time is with a clock that ticks at the sidereal rate. Astronomical observatories always have sidereal clocks, and you can buy moderately priced telescopes that come with sidereal clocks.

Example 1: Suppose it is 9:00 P.M. on the spring equinox (March 21). What is the local sidereal time?

Solution: On the day of the spring equinox, the Sun is located at the point of the spring equinox in the sky. Thus, if the Sun is 9 hours past the meridian, so is the spring equinox. The local sidereal time is LST = 09:00.

Example 2: Suppose the local sidereal time is LST = 04:00. When will Vega cross your meridian?

Solution: Vega has RA = 18^h35^m. Thus, at LST = 04:00, Vega's hour angle is

$$\text{HA}_{\text{Vega}} = \text{LST} - \text{RA}_{\text{Vega}} = 4:00 - 18:35 = -14:35$$

Vega will cross your meridian in 14 hours 35 minutes, which also means it crossed your meridian 9 hours 25 minutes ago ($14^h35^m + 9^h25^m = 24^h$). (Note that these are intervals of sidereal time.)

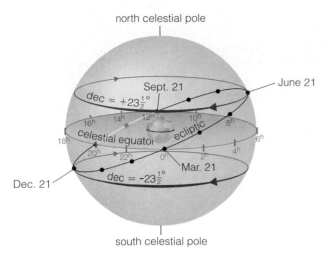

Figure S1.12 We can use this diagram of the celestial sphere to determine the Sun's right ascension and declination at monthly intervals.

"epoch 2000"). Astronomical software can automatically calculate day-to-day celestial coordinates for the Sun, Moon, and planets as they wander among the constellations.

Celestial Coordinates of the Sun

Unlike the Moon and the planets, which wander among the constellations in complex ways that require detailed calculations, the Sun moves through the zodiac constellations in a fairly simple way: It moves roughly 1° per day along the ecliptic. In a month, the Sun moves approximately one-twelfth of the way around the ecliptic, meaning that its right ascension changes by about $24 \div 12 = 2$ hours per month. Figure S1.12 shows the ecliptic marked with the Sun's monthly position and a scale of celestial coordinates. From this figure, we can create a table of the Sun's month-by-month celestial coordinates.

Table S1.1 starts from the spring equinox, when the Sun has declination 0° and right ascension 0^h. You can see in the shaded areas of the table that while RA advances steadily through the year, the Sun's declination changes much more rapidly around the equinoxes than around the solstices. For example, the Sun's declination changes from $-12°$ on February 21 to 12° on April 21, a change of 24° in just two months. In contrast, between May 21 and July 21, the declination varies only between $+20°$ and $+23\frac{1}{2}°$. This behavior explains why the daylight hours increase rapidly in spring and decrease rapidly in fall but stay long for a couple of months around the summer solstice and short for a couple of months around the winter solstice.

THINK ABOUT IT

On your plastic ball model of the celestial sphere, add dots along the ecliptic to show the Sun's monthly positions. Use your model to estimate the Sun's celestial coordinates on your birthday.

Table S1.1 The Sun's Approximate Celestial Coordinates at 1-Month Intervals

Approximate Date	RA	Dec
Mar. 21 (spring equinox)	0 hr	0°
Apr. 21	2 hr	+12°
May 21	4 hr	+20°
June 21 (summer solstice)	6 hr	$+23\frac{1}{2}°$
July 21	8 hr	+20°
Aug. 21	10 hr	+12°
Sept. 21 (fall equinox)	12 hr	0°
Oct. 21	14 hr	−12°
Nov. 21	16 hr	−20°
Dec. 21 (winter solstice)	18 hr	$−23\frac{1}{2}°$
Jan. 21	20 hr	−20°
Feb. 21	22 hr	−12°

 Seasons Tutorial, Lesson 3

S1.5 Understanding Local Skies

In Chapter 2, we briefly discussed how the daily circles of stars vary with latitude (see Figure 2.11). With our deeper understanding of the celestial sphere and celestial coordinates, we can now study local skies in more detail. We'll begin by focusing on star tracks through the local sky, which depend only on a star's declination. Then we'll discuss the daily path of the Sun, which varies with the Sun's declination and therefore with the time of year.

Star Tracks

The apparent daily rotation of the celestial sphere makes star tracks seem simple when viewed from the outside (see Figure 2.7). Local skies seem complex only because the ground always blocks our view of half of the celestial sphere. The half that is blocked depends on latitude. Let's first consider the local sky at Earth's North Pole, the easiest case to understand, and then look at the local sky at other latitudes.

The North Pole Figure S1.13a shows the rotating celestial sphere and your orientation relative to it when you are standing at the North Pole. Your "up" points toward the north celestial pole, which therefore marks your zenith. Earth blocks your view of anything south of the celestial equator, which therefore runs along your horizon. To make it easier for you to visualize the local sky, Figure S1.13b shows your horizon extending to the celestial sphere. Note

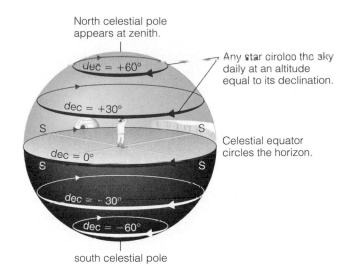

a The orientation of the local sky, relative to the celestial sphere, for the North Pole.

Figure S1.13 The sky at the North Pole.

b Extending the horizon to the celestial sphere makes it easier to visualize the local sky at the North Pole. (*Note:* To understand the extension, think of Earth in part (a) as being extremely small compared to the celestial sphere.)

that the horizon is marked with directions and that all directions are south from the North Pole. Thus, because the meridian is defined as running from north to south in the local sky, there is no meridian at the North Pole.

The daily circles of the stars keep them at constant altitudes above or below your horizon, and their altitudes are equal to their declinations. For example, a star with declination +60° circles the sky at an altitude of 60°, and a star with declination −30° remains 30° below your horizon at all times. As a result, all stars north of the celestial equator are circumpolar at the North Pole, never falling below the horizon. Similarly, stars south of the celestial equator never appear in the sky seen from the North Pole.

Notice that right ascension does not affect a star's path at all. It affects only the time of day and year at which a star is found in a particular direction along your horizon. If you are having difficulty visualizing the star paths, it may help you to watch star paths as you rotate your plastic ball model of the celestial sphere.

The Equator Next imagine that you are standing somewhere on Earth's equator (lat = 0°), such as in Ecuador, in Kenya, or on the island of Borneo. Figure S1.14a shows that "up" points directly away from (perpendicular to) Earth's rotation axis. Figure S1.14b shows the local sky more clearly by extending the horizon to the celestial sphere and rotating

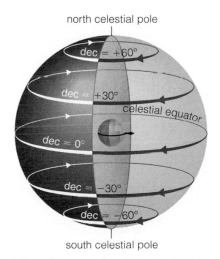

a The orientation of the local sky, relative to the celestial sphere, for Earth's equator.

Figure S1.14 The sky at the equator.

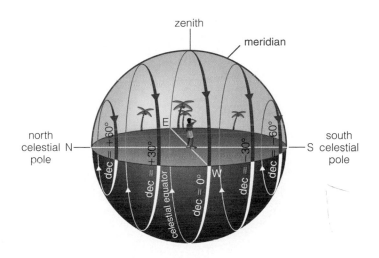

b Extending the horizon and rotating the diagram make it easier to visualize the local sky at the equator.

the diagram so the zenith is up. As everywhere except at the poles, the meridian extends from the horizon due south, through the zenith, to the horizon due north.

Look carefully at how the celestial sphere appears to rotate in the local sky. The north celestial pole remains stationary on your horizon due north. As we should expect, its altitude, 0°, is equal to the equator's latitude [Section 2.2]. Similarly, the south celestial pole remains stationary on your horizon due south. At any particular time, half of the celestial equator is visible, extending from the horizon due east, through the zenith, to the horizon due west. The other half lies below the horizon. As the equatorial sky appears to turn, all star paths rise straight out of the eastern horizon and set straight into the western horizon, with the following features:

- Stars with dec = 0° lie *on* the celestial equator and therefore rise due east, cross the meridian at the zenith, and set due west.

- Stars with dec > 0° rise north of due east, reach their highest point on the meridian in the north, and set north of due west. Their rise, set, and highest point depend on their declination. For example, a star with dec = +30° rises 30° north of due east, crosses the meridian 30° to the north of the zenith—that is, at an *altitude* of 90° − 30° = 60° in the north—and sets 30° north of due west.

- Stars with dec < 0° rise south of due east, reach their highest point on the meridian in the south, and set south of due west. For example, a star with dec = 50° rises 50° south of due east, crosses the meridian 50° to the south of the zenith—that is, at an *altitude* of 90° − 50° = 40° in the south—and sets 50° south of due west.

You can see that exactly half of any star's daily circle lies above the horizon. Thus, every star is above the horizon for exactly half of each sidereal day, or just under 12 hours, and below the horizon for the other half of the sidereal day.

THINK ABOUT IT

Visualize the daily paths of stars as seen from Earth's equator. Are any stars circumpolar? Are there stars that never rise above the horizon? Explain.

Other Latitudes We can use the same basic strategy to determine star tracks for other latitudes. Let's consider latitude 40°N, such as in Denver, Indianapolis, Philadelphia, or Beijing. First, as shown in Figure S1.15a, imagine standing at this latitude on a basic diagram of the rotating celestial sphere. Note that "up" points to a location on the celestial sphere with declination +40°. To make it easier to visualize the local sky, we next extend the horizon and rotate the diagram so the zenith is up (Figure S1.15b).

As we would expect, the north celestial pole appears 40° above the horizon due north, since its altitude in the local sky is always equal to the latitude. Half of the celestial equator is visible. It extends from the horizon due east, to the meridian at an altitude of 50° in the south, to the horizon due west. By comparing this diagram to that of the local sky for the equator, you can probably notice a general rule for the celestial equator at any latitude:

Exactly half the celestial equator is always visible, extending from due east on the horizon to due west on the horizon and crossing the meridian at an altitude of 90° minus the latitude.

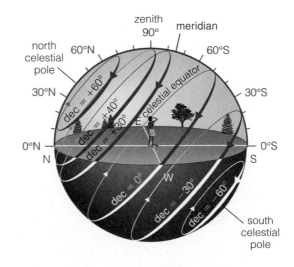

a The orientation of the local sky, relative to the celestial sphere, for latitude 40°N. Because latitude is the angle to Earth's equator, "up" points to the circle on the celestial sphere with declination +40°.

b Extending the horizon and rotating the diagram so the zenith is up make it easier to visualize the local sky. The blue scale along the meridian shows altitudes and directions in the local sky.

Figure S1.15 The sky at 40°N latitude.

The celestial equator runs through the southern half of the sky at locations in the Northern Hemisphere and through the northern half of the sky at locations in the Southern Hemisphere.

The diagram in Figure S1.15b also shows star tracks through the local sky for 40°N. You may find it easier to visualize them if you rotate your plastic ball model of the celestial sphere while holding it in the orientation of Figure S1.15b. As you study the tracks, note these key features:

● Stars with dec = 0° lie *on* the celestial equator and therefore follow the path of the celestial equator through the local sky. That is, for latitude 40°N, they rise due east, cross the meridian at altitude 50° in the south, and set due west.

● Stars with dec > 0° follow paths parallel to the celestial equator but farther north. Thus, they rise north of due east, cross the meridian north of where the celestial equator crosses it, and set north of due west. If they are within 40° of the north celestial pole on the celestial sphere (which means declinations greater than 90° − 40° = 50°), their entire circles are above the horizon, making them circumpolar. You can find the precise point at which a star crosses the meridian by adding its declination to the 50°S altitude at which the celestial equator crosses the meridian. For example, Figure S1.15b shows that a star with dec = +30° crosses the meridian at altitude 50° + 30° = 80° in the south and a star with dec = +60° crosses the meridian at altitude 70° in the north. (To calculate the latter result, note that the sum 50° + 60° = 110° goes 20° past the zenith altitude of 90°, making it equivalent to 90° − 20° = 70°.)

● Stars with dec < 0° follow paths parallel to the celestial equator but farther south. Thus, they rise south of due east, cross the meridian south of where the celestial equator crosses it, and set south of due west. If they are within 40° of the south celestial pole on the celestial sphere (which means declinations less than −90° + 40° = −50°), their entire circles are below the horizon, and thus they are never visible.

● The fraction of any star's daily circle that is above the horizon—and hence the amount of time it is above the horizon each day—depends on its declination. Because exactly half the celestial equator is above the horizon, stars on the celestial equator (dec = 0°) are above the horizon for about 12 hours per day. Stars with positive declinations have more than half their daily circle above the horizon and hence are above the horizon for more than 12 hours each day (with the range extending to 24 hours a day for the circumpolar stars). Stars with negative declinations have less than half their daily circle above the horizon and hence are above the horizon for less than 12 hours each day (with the range going to zero for stars that are never above the horizon).

We can apply the same strategy we used in Figure S1.15 to find star paths for other latitudes. Figure S1.16 shows the process for latitude 30°S. Note that the south celestial pole is visible to the south and that the celestial equator passes through the northern half of the sky. If you study the diagram carefully, you can see how star tracks depend on declination.

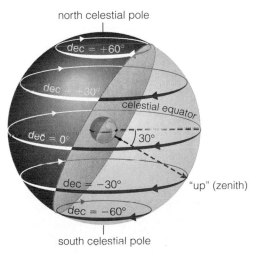

a The orientation of the local sky at latitude 30°S, relative to the celestial sphere. "Up" points to the circle on the celestial sphere with declination −30°.

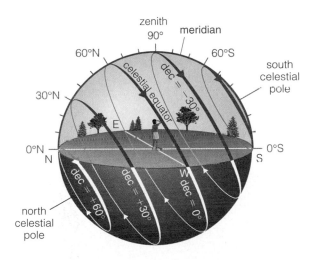

b Extending the horizon and rotating the diagram so the zenith is up make it easier to visualize the local sky. Note that the south celestial pole is visible at altitude 30° in the south, while the celestial equator stretches across the northern half of the sky.

Figure S1.16 The sky at 30°S latitude.

Study Figure S1.16 for latitude 30°S. Describe the path of the celestial equator. Explain how it obeys the 90° − latitude rule given above. Give a general description of how star tracks differ for stars with positive and negative declinations. What stars are circumpolar at this latitude?

The Path of the Sun

Just as a star's path through the sky depends only on its declination, the Sun's path through the sky on any particular day depends only on its declination for that day. For example, because the Sun's declination is $+23\frac{1}{2}°$ on the summer solstice, the Sun's path through the local sky on June 21 is the same as that of any star with declination $+23\frac{1}{2}°$. Thus, as long as we know the Sun's declination for a particular day, we can find the Sun's path at any latitude with the same local sky diagrams we used to find star tracks.

Figure S1.17 shows the Sun's path on the equinoxes and solstices for latitude 40°N. On the equinoxes, when the Sun is on the celestial equator (dec = 0°), the Sun's path follows the celestial equator: It rises due east, crosses the meridian at altitude 50° in the south, and sets due west. Like any object on the celestial equator, it is above the horizon for 12 hours. On the summer solstice, the Sun rises well north of due east,* reaches an altitude of $73\frac{1}{2}°$ when it crosses the meridian in the south, and sets well north of due west. The daylight hours are long because much more than half of the Sun's path is above the horizon. On the winter solstice, the Sun rises well south of due east, reaches

---

*Calculating exactly how far north of due east the Sun rises is beyond the scope of this book, but astronomical software packages will tell you exactly where—and at what time—the Sun rises and sets along the horizon for any location and any date.

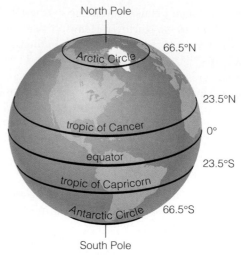

Figure S1.18 Special latitudes defined by the Sun's path through the sky.

an altitude of only $26\frac{1}{2}°$ when it crosses the meridian in the south, and sets well south of due west. The daylight hours are short because much less than half of the Sun's path is above the horizon.

We could make a similar diagram to show the Sun's path on various dates for any latitude. However, the $23\frac{1}{2}°$ tilt of Earth's axis makes the Sun's path particularly interesting at the special latitudes shown in Figure S1.18. Let's investigate these latitudes.

The North and South Poles

In Figure S1.13b, we saw that the celestial equator circles the horizon at the North Pole. Because the Sun appears *on* the celestial equator on the day of the spring equinox, the Sun circles the north polar sky *on the horizon* on March 21 (Figure S1.19), completing a full circle of the horizon in 24 hours (1 solar day).

Over the next 3 months, the Sun continues to circle the horizon, circling at gradually higher altitudes as its

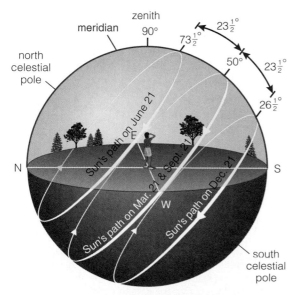

Figure S1.17 The Sun's daily paths for the equinoxes and solstices at latitude 40°N.

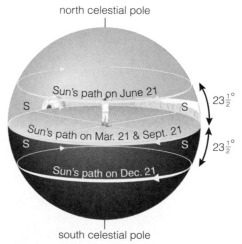

Figure S1.19 Daily paths of the Sun for equinoxes and solstices at the North Pole.

declination increases. It reaches its highest point on the summer solstice, when its declination of $+23\frac{1}{2}°$ means that it circles the north polar sky at an altitude of $23\frac{1}{2}°$. After the summer solstice, the daily circles gradually fall lower over the next 3 months, reaching the horizon on the fall equinox. Then, because the Sun's declination is negative for the next 6 months (until the following spring equinox), it remains below the north polar horizon. Thus, the North Pole essentially has 6 months of daylight and 6 months of darkness, with an extended twilight that lasts a few weeks beyond the fall equinox and an extended dawn that begins a few weeks before the spring equinox.

The situation is the opposite at the South Pole. Here the Sun's daily circles slowly rise above the horizon on the fall equinox to a maximum altitude of $23\frac{1}{2}°$ on the *winter* solstice, then slowly fall back to the horizon on the spring equinox. Thus, the South Pole has the Sun above the horizon during the 6 months it is below the north polar horizon.

Although we've correctly described the Sun's true position in the polar skies over the course of the year, two effects complicate what we actually see at the poles around the times of the equinoxes. First, the atmosphere bends light enough so that the Sun *appears* to be slightly above the horizon even when it is actually slightly below it. Near the horizon, this bending makes the Sun appear about 1° higher than it would in the absence of an atmosphere. Second, the Sun's angular size of about $\frac{1}{2}°$ means that it does not fall below the horizon at a single moment but instead sets gradually. Together, these effects mean that the Sun appears above each polar horizon for slightly longer (by several days) than 6 months each year.

The Equator At the equator, the celestial equator extends from the horizon due east, through the zenith, to the horizon due west. The Sun follows this path on each equinox, reaching the zenith at local noon (Figure S1.20). Following the spring equinox, the Sun's increasing declination means

that it follows a daily track that takes it gradually northward in the sky. It is farthest north on the summer solstice, when it rises $23\frac{1}{2}°$ north of due east, crosses the meridian at altitude $66\frac{1}{2}°$ in the north, and sets $23\frac{1}{2}°$ north of due west. Over the next 6 months, it gradually tracks southward until the winter solstice, when its path is the mirror image (across the celestial equator) of its summer solstice path.

Like all objects in the equatorial sky, the Sun is always above the horizon for half a day and below it for half a day. Moreover, the Sun's track is highest in the sky on the equinoxes and lowest on the summer and winter solstices. That is why equatorial regions do not have seasons like temperate regions [Section 2.3]. The Sun's path in the equatorial sky also makes it rise and set perpendicular to the horizon, making for a more rapid dawn and a briefer twilight than at other latitudes.

The Tropic Circles We've seen that, while the Sun reaches the zenith twice a year at the equator (on the spring and fall equinoxes), it never reaches the zenith at mid-latitudes (such as 40°N). The boundaries of the regions on Earth where the Sun sometimes reaches the zenith are the circles of latitude 23.5°N and 23.5°S. These latitude circles are called the **tropic of Cancer** and the **tropic of Capricorn**, respectively. (The region between these two circles is generally called the *tropics*.)

Figure S1.21 shows why the tropic of Cancer is special. The celestial equator extends from due east on the horizon to due west on the horizon, crossing the meridian in the south at an altitude of $90° - 23\frac{1}{2}°$ (the latitude) $= 66\frac{1}{2}°$. The Sun follows this path on the equinoxes (March 21 and September 21). As a result, the Sun's path on the summer solstice, when it crosses the meridian $23\frac{1}{2}°$ northward of the celestial equator, takes it to the zenith at local noon. Because the Sun has its maximum declination on the

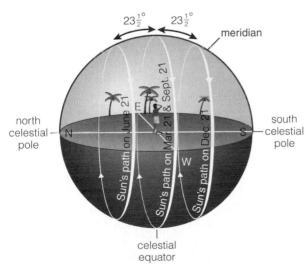

Figure S1.20 Daily paths of the Sun for the equinoxes and solstices at the equator.

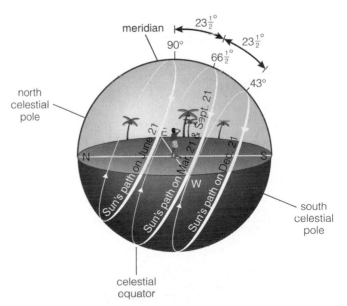

Figure S1.21 Daily paths of the Sun for the equinoxes and solstices at the tropic of Cancer.

summer solstice, the tropic of Cancer marks the northern-most latitude at which the Sun ever reaches the zenith. Similarly, at the tropic of Capricorn the Sun reaches the zenith at local noon on the winter solstice, making this the southernmost latitude at which the Sun ever reaches the zenith. Between the two tropic circles, the Sun passes through the zenith twice a year (the dates vary with latitude).

The names of the tropics of Cancer and Capricorn come from the zodiac constellations of the summer and winter solstices along the ecliptic—as they were about 2,000 years ago. Because of precession over the past 2,000 years, the summer solstice is currently in the constellation Gemini and the winter solstice is in the constellation Sagittarius (see Figure S1.9).

The Polar Circles At the equator, the Sun is above the horizon for 12 hours each day year-round. At latitudes progressively farther from the equator, the daily time that the Sun is above the horizon varies progressively more with the seasons. The special latitudes at which the Sun remains continuously above the horizon for a full day each year mark the polar circles: the **Arctic Circle** at latitude 66.5°N and the **Antarctic Circle** at latitude 66.5°S. Poleward of these circles, the length of continuous daylight (or darkness) increases beyond 24 hours, reaching the extreme of 6 months at the North and South Poles.

Figure S1.22 shows why the Arctic Circle is special. The celestial equator extends from due east on the horizon to due west on the horizon, crossing the meridian in the south at an altitude of $90° - 66\frac{1}{2}°$ (the latitude) $= 23\frac{1}{2}°$. As a result, the Sun's path is circumpolar on the summer solstice: It skims the northern horizon at midnight, rises through the eastern sky to a noon maximum altitude of 47° in the south, and then gradually falls through the western sky until it is back on the horizon at midnight (see the

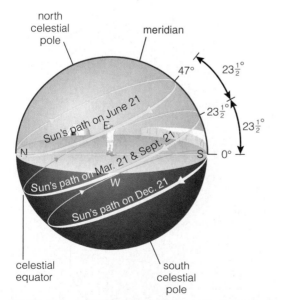

north
celestial
pole

meridian

47° $23\frac{1}{2}$

$23\frac{1}{2}$

$23\frac{1}{2}$

Sun's path on June 21

E

N   S   0°

Sun's path on Mar. 21 & Sept. 21

W

Sun's path on Dec. 21

celestial
equator

south
celestial
pole

Figure S1.22 Daily paths of the Sun for the equinoxes and solstices at the Artic Circle.

photograph of this path in Figure 2.17). At the Antarctic Circle, the Sun follows the same basic pattern on the winter solstice, except that it skims the horizon in the south and rises to a noon maximum altitude of 47° in the north.

However, as at the North and South Poles, what we actually see at the polar circles is slightly different from this idealization. Again, the bending of light by Earth's atmosphere and the Sun's angular size of about $\frac{1}{2}°$ make the Sun *appear* to be slightly above the horizon even when it is slightly below it. Thus, the Sun seems not to set for several days, rather than for a single day, around the summer solstice at the Arctic Circle (the winter solstice at the Antarctic Circle). Similarly, the Sun appears to peek above the horizon momentarily, rather than not at all, around the winter solstice at the Arctic Circle (the summer solstice at the Antarctic Circle).

S1.6 Principles of Celestial Navigation

Imagine that you're on a ship at sea, far from any landmarks. How can you figure out where you are? It's easy, at least in principle, if you understand the apparent motions of the sky discussed in this chapter.

Latitude

Determining latitude is particularly easy if you can find the north or south celestial pole: Your latitude is equal to the altitude of the celestial pole in your sky. In the Northern Hemisphere at night, you can determine your approximate latitude by measuring the altitude of Polaris. Because Polaris has a declination within 1° of the north celestial pole, its altitude is within 1° of your latitude. For example, if Polaris has altitude 17°, your latitude is between 16°N and 18°N.

If you want to be more precise, you can determine your latitude from the altitude of *any* star as it crosses your meridian. For example, suppose Vega happens to be crossing your meridian at the moment and appears in your southern sky at altitude 78°44′. Because Vega has dec = +38°44′ (see Figure S1.11), it crosses your meridian 38°44′ north of the celestial equator. As shown in Figure S1.23a, you can conclude that the celestial equator crosses your meridian at an altitude of precisely 40° in the south. Your latitude must therefore be 50°N because the celestial equator always crosses the meridian at an altitude of 90° minus the latitude. You know you are in the Northern Hemisphere because the celestial equator crosses the meridian in the south.

In the daytime, you can find your latitude from the Sun's altitude on your meridian if you know the date and have a table that tells you the Sun's declination on that date. For example, suppose the date is March 21 and the Sun crosses your meridian at altitude 70° in the north (Figure S1.23b). Because the Sun has dec = 0° on March 21,

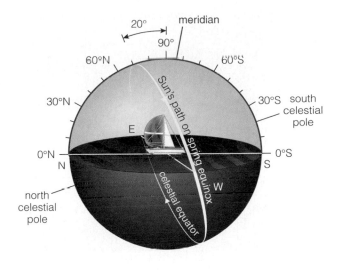

a Because Vega has dec = +38°44', it crosses the meridian 38°44' north of the celestial equator. From Vega's meridian crossing at altitude 78°44' in the south, the celestial equator must cross the meridian at altitude 40° in the south. Thus, the latitude must be 50°N.

b To determine latitude from the Sun's meridian crossing, you must know the Sun's declination, which you can determine from the date. The case shown is for the spring equinox, when the Sun's declination is 0° and hence follows the path of the celestial equator through the local sky. From the celestial equator's meridian crossing at 70° in the north, the latitude must be 20°S.

Figure S1.23 Determining latitude from a star and from the Sun.

you can conclude that the celestial equator also crosses your meridian in the north at altitude 70°. You must be in the Southern Hemisphere, because the celestial equator crosses the meridian in the north. From the rule that the celestial equator crosses the meridian at an altitude of 90° minus the latitude, you can conclude that you are at latitude 20°S.

Longitude

Determining longitude requires comparing the current positions of objects in your sky with their positions as seen from some known longitude. As a simple example, suppose you use a sundial to determine that the apparent solar time is 1:00 P.M. You immediately call a friend in England and learn that it is 3:00 P.M. in Greenwich (or you carry a clock that keeps Greenwich time). You now know that your local time is 2 hours earlier than the local time in Greenwich, which means you are 2 hours west of Greenwich. (An earlier time means you are *west* of Greenwich, because Earth rotates from west to east.) Each hour corresponds to 15° of longitude, so "2 hours west of Greenwich" means longitude 30°W.

At night, you can find your longitude by comparing the positions of stars in your local sky and at some known longitude. For example, suppose Vega is on your meridian and a call to your friend reveals that it won't cross the meridian in Greenwich until 6 hours from now. In this case, your local time is 6 hours later than the local time in Greenwich. Thus, you are 6 hours east of Greenwich, or at longitude 90°E (because 6 × 15° = 90°).

Celestial Navigation in Practice

Although celestial navigation is easy in principle, at least three practical considerations make it more difficult in practice. First, finding either latitude or longitude requires a tool for measuring angles in the sky. One such device, called an *astrolabe*, was invented by the ancient Greeks and significantly improved by Islamic scholars during the Middle Ages. The astrolabe's faceplate (Figure S1.24a) could be used to tell time, because it consisted of a rotating star map and horizon plates for specific latitudes. Today you can buy similar rotatable star maps, called *planispheres*. Most astrolabes contained a sighting stick on the back that allowed users to measure the altitudes of bright stars in the sky. These measurements could then be correlated against special markings under the faceplate (Figure S1.24a). Astrolabes were effective but difficult and expensive to make. As a result, medieval sailors often measured angles with a simple pair of calibrated perpendicular sticks, called a *cross-staff* or *Jacob's staff* (Figure S1.24c). A more modern device called a *sextant* allows much more precise angle determinations by incorporating a small telescope for sightings (Figure S1.24d). Sextants are still used for celestial navigation on many ships. If you want to practice celestial navigation yourself, you can buy an inexpensive plastic sextant at many science-oriented stores.

A second practical consideration is the need to know the celestial coordinates of stars and the Sun so that you can determine their paths through the local sky. At night, you can use a table listing the celestial coordinates of bright stars. In addition to knowing the celestial coordinates, you

a The faceplate of an astrolabe; many astrolabes had sighting sticks on the back for measuring the positions of bright stars.

b A copper engraving of Italian explorer Amerigo Vespucci (for whom America was named) using an astrolabe to sight the Southern Cross. The engraving by Philip Galle, from the book *Nova Reperta*, was based on an original by Joannes Stradanus in the early 1580s.

c A woodcutting of Ptolemy holding a cross-staff (artist unknown).

d A sextant.

Figure S1.24 Navigational instruments.

must either know the constellations and bright stars extremely well or carry star charts to help you identify them. For navigating by the Sun in the daytime, you'll need a table listing the Sun's celestial coordinates on each day of the year.

The third practical consideration is related to determining longitude: You need to know the current position of the Sun (or a particular star) in a known location, such as Greenwich, England. Although you could find this out

by calling a friend who lives there, it's more practical to carry a clock set to universal time (that is, Greenwich mean time). In the daytime, the clock makes it easy to determine your longitude. If apparent solar time is 1:00 P.M. in your location and the clock tells you that it is 3:00 P.M. in Greenwich, then you are 2 hours west of Greenwich, or at longitude 30°W. The task is more difficult at night, because you must compare the position of a *star* in your sky to its current position in Greenwich. You can do this with the aid of de-

Compass Directions

Most people determine direction with the aid of a compass rather than the stars. However, a compass needle doesn't actually point to true geographic north. Instead, the compass needle responds to Earth's magnetic field and points to *magnetic* north, which can be substantially different from true north. If you want to navigate precisely with a compass, you need a special map that shows local variations in Earth's magnetic field. Such maps are available at most camping stores. They are not perfectly reliable, however, because the magnetic field also varies with time. In general, celestial navigation is much more reliable than a compass for determining direction.

tailed astronomical tables that allow you to determine the current position of any star in the Greenwich sky from the date and the universal time.

Historically, this third consideration created enormous problems for navigation. Before the invention of accurate clocks, sailors could easily determine their latitude but not their longitude. Indeed, most of the European voyages of discovery beginning in the 1400s relied on little more than guesswork about longitude, although some sailors learned complex mathematical techniques for estimating longitude through observations of the lunar phases. More accurate longitude determination, upon which the development of extensive ocean commerce and travel depended, required the invention of a clock that would remain accurate on a ship rocking in the ocean swells. By the early 1700s, solving this problem was considered so important that the British government offered a substantial monetary prize for the solution. The prize was claimed in 1761 by John Harrison, with a clock that lost only 5 seconds during a 9-week voyage to Jamaica.*

The Global Positioning System

In the past decade, a new type of celestial navigation has supplanted traditional methods. It involves finding positions relative to a set of satellites in Earth orbit. These satellites of the **global positioning system** (**GPS**) in essence function like artificial stars. The satellite positions at any moment are known precisely from their orbital characteristics. The GPS currently involves about two dozen satellites orbiting Earth at an altitude of 20,000 kilometers. Each satellite transmits a radio signal that can be received by a small radio receiver—rain or shine, day or night. GPS receivers have a built-in computer that calculates your precise

*The story of the difficulties surrounding the measurement of longitude at sea and how the problem was finally solved by Harrison is chronicled in Dava Sobel, *Longitude* (Walker and Company, 1995).

position on Earth by comparing the signals received from several GPS satellites.

The United States originally built the GPS in the late 1970s for military use. Today, the many applications of the GPS include automobile navigation systems as well as systems for helping airplanes land safely, guiding the blind around town, and helping lost hikers find their way. The GPS has been used by geologists to measure *millimeter-*scale changes in Earth's crust.

With the rapid growth in the use of GPS navigation, the ancient practice of celestial navigation is in danger of becoming a lost art. Fortunately, many amateur clubs and societies are keeping the art of celestial navigation alive.

THE BIG PICTURE

Putting Chapter S1 into Context

In this chapter, we built upon concepts from the first three chapters to form a more detailed understanding of celestial timekeeping and navigation. We also learned how to determine paths for the Sun and the stars in the local sky. As you look back at what you've learned, keep in mind the following "big picture" ideas:

- Our modern systems of timekeeping are rooted in the apparent motions of the Sun through the sky. Although it's easy to forget these roots when you look at a clock or a calendar, the sky was the only guide to time for most of human history.

- The term *celestial navigation* sounds a bit mysterious, but it involves simple principles that allow you to determine your location on Earth. Even if you're never lost at sea, you may find the basic techniques of celestial navigation useful to orient yourself at night (for example, on your next camping trip).

- If you understand the apparent motions of the sky discussed in this chapter and also learn the constellations and bright stars, you'll feel very much "at home" under the stars at night.

S1.1 Astronomical Time Periods

- *Why isn't the Earth's rotation period exactly equal to the 24 hours in our day?* The 24-hour solar day is the average time between noon one day and noon the next day, which is longer than the sidereal day (rotation period) because of Earth's daily movement in its orbit around the Sun.

- *How are astronomical time periods based on the Moon, on our orbit, and on the planets?* Our month is based on the Moon's cycle of phases, or synodic period of $29\frac{1}{2}$ days. The Moon's sidereal period (about $27\frac{1}{3}$ days) is its true orbital period. Our calendar is based on the tropical year, which is the time from one spring equinox to the next. The tropical year differs slightly from Earth's true orbital period (sidereal year) because of precession. Planets have a synodic period from one opposition or conjunction to the next and a sidereal year based on their true orbital periods.

S1.2 Daily Timekeeping

- *What kind of time do our clocks tell?* Ordinary clocks tell standard or daylight saving time.

- *What is universal time (UT)?* It is the mean solar time in Greenwich, England.

S1.3 The Calendar

- *Why do we have leap years?* Leap years keep the calendar synchronized with the seasons.

- *Do we always have a leap year every 4 years?* No. We use the Gregorian calendar, which skips leap year in century years not divisible by 400.

S1.4 Mapping Locations in the Sky

- *How do we describe positions on the celestial sphere?* Positions on the celestial sphere are described with the coordinates of declination and right ascension.

- *How do the Sun's celestial coordinates change during the year?* The Sun's right ascension advances steadily by about 2 hours per month. The Sun's declination varies between $-23\frac{1}{2}°$ and $+23\frac{1}{2}°$.

S1.5 Understanding Local Skies

- *Why does the night sky vary with latitude?* Stars of different declination appear to rise, set, and cross the meridian at different altitudes depending on your latitude.

- *What is the path of the Sun through the sky at the North Pole?* Over 6 months, the Sun circles the horizon daily and gradually rises from the horizon to $23\frac{1}{2}°$ altitude and then falls back to the horizon. It then remains below the horizon for 6 months of night.

- *Where on Earth is the Sun sometimes directly overhead?* The Sun may be directly overhead only between the tropics of Capricorn and Cancer.

S1.6 Principles of Celestial Navigation

- *What must you know to measure your latitude?* You must know the declination of a star (or the Sun) that is crossing your meridian and its altitude as it crosses.

- *What must you know to measure your longitude?* You must know the position of an object in your sky and its position at the same time in the sky of Greenwich, England (or some other specific location). This is most easily done if you have a clock that tells universal time.

Does It Make Sense?

Decide whether each statement makes sense and explain why it does or does not. (*Hint:* For statements that involve coordinates—such as altitude, longitude, or declination—check whether the correct coordinates are used for the situation. For example, it does not make sense to describe a location on Earth by an altitude since altitude makes sense only for positions in the local sky.)

1. Last night I saw Venus shining brightly on the meridian at midnight.

2. The apparent solar time was noon, but the Sun was just setting.

3. My mean solar clock said it was 2:00 P.M., but a friend who lives east of here had a mean solar clock that said it was 2:11 P.M.

4. When the standard time is 3:00 P.M. in Baltimore, it is 3:15 P.M. in Washington, D.C.

5. The Julian calendar differed from the Gregorian calendar because it was based on the sidereal year.

6. Last night around 8:00 P.M. I saw Jupiter at an altitude of 45° in the south.

7. The latitude of the stars in Orion's belt is about 5°N.

8. Today the Sun is at an altitude of 10° on the celestial sphere.

9. Los Angeles is west of New York by about 3 hours of right ascension.

10. The summer solstice is east of the vernal equinox by 6 hours of right ascension.

11. If it were being named today, the tropic of Cancer would probably be called the tropic of Gemini.

12. Even though my UT clock had stopped, I was able to find my longitude by measuring the altitudes of 14 different stars in my local sky.

Problems

(Quantitative problems are marked with an asterisk.)

13. *Definition of a Day.* Briefly explain the difference between a *solar day* and a *sidereal day*.

14. *Definition of a Month.* Briefly explain the difference between a *synodic month* and a *sidereal month*.

15. *Length of the Year.* Why is the *tropical year* slightly shorter than the *sidereal year*?

16. *Planetary Periods.* What is the difference between a planet's *sidereal period* and its *synodic period*? Explain the meaning of *conjunction, opposition,* and *greatest elongation* for planetary orbits viewed from Earth.

17. *Telling Time.* What is *apparent solar time*? Why is it different from *mean solar time*? Also define *standard time, daylight saving time,* and *universal time (UT)*.

18. *Celestial Coordinates.* What are *declination* and *right ascension*? How are these *celestial coordinates* similar to latitude and longitude on Earth? How are they different?

19. *North Pole Sky.* Suppose you are standing at the North Pole. Where is the celestial equator in your sky? Where is the north celestial pole? Describe the daily motion of the sky. Do the same for the sky at the equator and at latitude 40°N.

20. *Solar Motion.* Describe the Sun's paths through the local sky on the equinoxes and on the solstices for latitude 40°N.

21. *Opposite Rotation.* Suppose Earth rotated in the opposite direction from its revolution; that is, suppose it rotated clockwise (as seen from above the North Pole) every 24 hours while revolving counterclockwise around the Sun each year. Would the solar day still be longer than the sidereal day? Explain.

22. *Fundamentals of Your Local Sky.* Answer each of the following for *your* latitude.

    a. Where is the north (or south) celestial pole in your sky?

    b. Describe the location of the meridian in your sky. Specify its shape and at least three distinct points along it (such as the points at which it meets your horizon and its highest point).

    c. Describe the location of the celestial equator in your sky. Specify its shape and at least three distinct points along it (such as the points at which it meets your horizon and crosses your meridian).

    d. Does the Sun ever appear at your zenith? If so, when? If not, why not?

    e. What range of declinations makes a star circumpolar in your sky? Explain.

    f. What is the range of declinations for stars that you can never see in your sky? Explain.

23. *Sydney Sky.* Repeat problem 22 for the local sky in Sydney, Australia (latitude 34°S).

24. *Path of the Sun in Your Sky.* Describe the path of the Sun through your local sky for each of the following days.

    a. The spring and fall equinoxes.

    b. The summer solstice.

    c. The winter solstice.

    d. Today. (*Hint:* Estimate the right ascension and declination of the Sun for today's date by using the data in Table S1.1).

25. *Sydney Sun.* Repeat problem 24 for the local sky in Sydney, Australia (latitude 34°S).

26. *Lost at Sea I.* During an upcoming vacation, you decide to take a solo boat trip. While contemplating the universe, you lose track of your location. Fortunately, you have some astronomical tables and instruments, as well as a UT clock. You thereby put together the following description of your situation:

    • It is the spring equinox.

    • The Sun is on your meridian at altitude 75° in the south.

    • The UT clock reads 22:00.

    a. What is your latitude? How do you know?

    b. What is your longitude? How do you know?

    c. Consult a map. Based on your position, where is the nearest land? Which way should you sail to reach it?

27. *Lost at Sea II.* Repeat problem 26, based on the following description of your situation:

    • It is the day of the summer solstice.

    • The Sun is on your meridian at latitude $67\frac{1}{2}°$ in the north.

    • The UT clock reads 06:00.

28. *Lost at Sea III.* Repeat problem 26, based on the following description of your situation:

    • Your local time is midnight.

    • Polaris appears at altitude 67° in the north.

    • The UT clock reads 01:00.

29. *Lost at Sea IV.* Repeat problem 26, based on the following description of your situation:

   - Your local time is 6 A.M.

   - From the position of the Southern Cross, you estimate that the south celestial pole is at altitude 33° in the south.

   - The UT clock reads 11:00.

*30. *Sidereal Time.*

   a. Suppose it is 4 P.M. on the spring equinox. What is the local sidereal time?

   b. Suppose the local sidereal time is 19:30. When will Vega cross your meridian?

   c. You observe a star that has an hour angle of +3 hours ($+3^h$) when the local sidereal time is 8:15. What is the star's right ascension?

Discussion Questions

31. *Northern Chauvinism.* Why is the solstice in June called the *summer solstice,* when it marks winter for places like Australia, New Zealand, and South Africa? Why is the writing on maps and globes usually oriented so that the Northern Hemisphere is at the top, even though there is no up or down in space? Discuss.

32. *Celestial Navigation.* Briefly discuss how you think the benefits and problems of celestial navigation might have affected ancient sailors. For example, how did they benefit from using the north celestial pole to tell directions, and what problems did they experience because of the difficulty in determining longitude? Can you explain why ancient sailors generally hugged coastlines as much as possible on their voyages? What dangers did this type of sailing pose? Why did the Polynesians become the best navigators of their time?

For a complete list of media resources available, go to www.astronomyplace.com and choose Chapter S1 from the pull-down menu.

 ## Astronomy Place Web Tutorials

Tutorial Review of Key Concepts

Use the following interactive **Tutorial** at www.astronomyplace.com to review key concepts from this chapter.

Seasons Tutorial

Lesson 2 The Solstices and Equinoxes

Lesson 3 The Sun's Position in the Sky

Supplementary Tutorial Exercises

Use the interactive **Tutorial Lessons** to explore the following questions.

Seasons Tutorial, Lesson 2

1. How are the solstices and equinoxes related to Earth's orbital position around the Sun?

2. What are day and night like at the North Pole?

3. How does the length of the day vary with the seasons at the Antarctic circle?

Seasons Tutorial, Lesson 3

1. When can you see the Sun directly over the equator?

2. Where can you see the Sun directly overhead on the winter solstice?

3. When is the Sun seen directly overhead at your latitude?

 ## Exploring the Sky and Solar System

Of the many activities available on the **Voyager: SkyGazer** **CD-ROM** accompanying your book, use the following files to observe key phenomena covered in this chapter.

Go to the **File: Basics** folder for the following demonstrations.

1. Analemma

2. Rubber Horizon

3. Three Cities

Go to the **File: Demo** folder for the following demonstrations.

1. Venus Transit of 1769

2. Celestial Poles

3. Russian Midnight Sun

Movies

Check out the following narrated and animated short documentaries available on www.astronomyplace.com for a helpful review of key ideas covered in this chapter.

The Celestial Sphere Movie

Time and Seasons Movie

Web Projects

Take advantage of the useful Web links on www.astronomyplace.com to assist you with the following projects.

1. *Sundials.* Although they are no longer necessary for timekeeping, sundials remain popular for their cultural and artistic value. Search the Web for pictures and information about interesting sundials around the world. Write a short report about at least three sundials that you find particularly interesting.

2. *The Analemma.* Learn more about the analemma and its uses from information available on the Web. Write a short report on your findings.

3. *Calendar History.* Investigate the history of the Julian or Gregorian calendar in greater detail. Write a short summary of some interesting aspect of the history you learn from your Web research. (For example, why did Julius Caesar allow one year to have 445 days? How did our months end up with 28, 30, or 31 days?)

4. *Global Positioning System.* Learn more about the global positioning system and its uses. Write a short report summarizing how you think the GPS will affect our lives over the next 10 years.

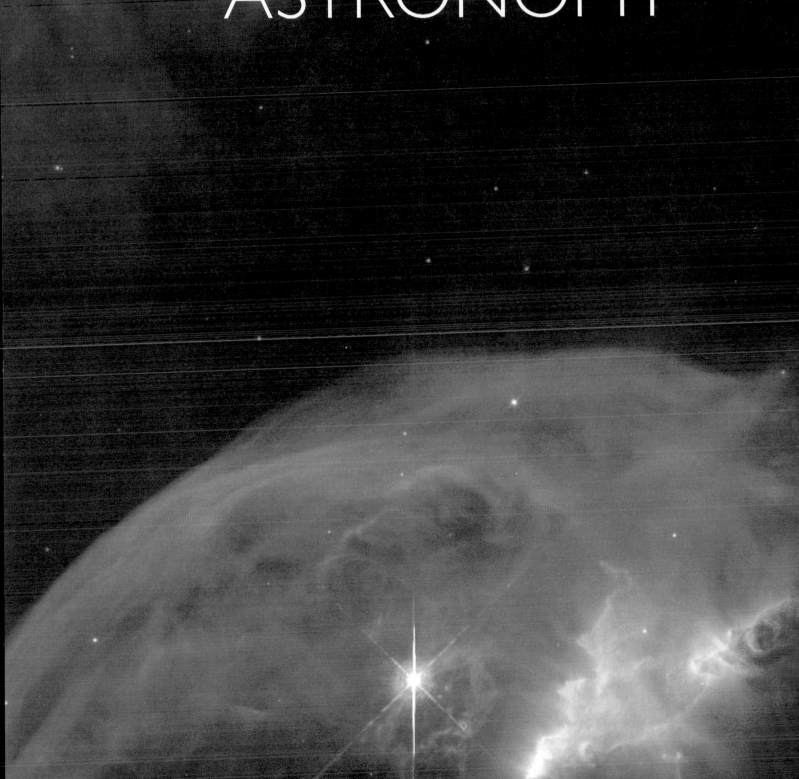

PART II

KEY CONCEPTS
FOR
ASTRONOMY

4 A Universe of Matter and Energy

In this and the next three chapters, we turn our attention to the scientific concepts that lie at the heart of modern astronomy. We begin by investigating the nature of matter and energy, the fundamental stuff from which the universe is made.

The history of the universe is essentially a story about the interplay between matter and energy since the beginning of time. Interactions between matter and energy began in the Big Bang and govern everything from the microscopic interactions of atoms to gargantuan collisions of galaxies that unfold over a billion or more years. Understanding the universe therefore depends on familiarity with how matter responds to the ebb and flow of energy.

The concepts of matter and energy presented in this chapter will enable you to understand most of the topics in this book. Some of the concepts and terminology may already be familiar to you. If not, don't worry. We will go over them again as they arise in various contexts, and you can refer back to this chapter as needed during the remainder of your studies.

4.1 Matter and Energy in Everyday Life

The meaning of **matter** is obvious to most people, at least on a practical level. Matter is simply material, such as rocks, water, or air. You can hold matter in your hand or put it in a box.

The meaning of **energy** is not quite as obvious, although we certainly talk a lot about it. We pay energy bills to the power companies, we use energy from gasoline to run our cars, and we argue about whether nuclear energy is a sensible alternative to fossil fuels. On a personal level, we often talk about how energetic we feel on a particular day. But what *is* energy?

Broadly speaking, energy is what makes matter move. For Americans, the most familiar way of measuring energy is in Calories, which we use to describe how much energy our bodies can draw from food. A typical adult uses about 2,500 Calories of energy each day. Among other things, this energy keeps our hearts beating and our lungs breathing, generates the heat that maintains our 37°C (98.6°F) body temperature, and allows us to walk and run.

Just as we can measure height in inches, feet, or meters, there are many alternatives to Calories for measuring energy. If you look closely at an electric bill, you'll probably find that the power company charges you for electrical energy in units called *kilowatt-hours*. If you purchase a gas appliance, its energy requirements may be labeled in *British thermal units*, or *BTUs*. In science and internationally, the favored unit of energy is the **joule**, which is equivalent to $\frac{1}{4,184}$ of a Calorie. Thus, the 2,500 Calories used daily by a typical adult are equivalent to about 10 million joules. For comparison, Table 4.1 lists various energies in joules.

Although energy can always be measured in joules, it has many different forms. We have already talked about food energy, electrical energy, and the energy of a beating heart. Fortunately, the many forms of energy can be grouped into three basic categories (Figure 4.1).

First, whenever matter is moving, it has energy of motion, or **kinetic energy** (*kinetic* comes from a Greek word meaning "motion"). Falling rocks, the moving blades on an electric mixer, a car driving down the highway, and the molecules moving in the air around us are all examples of objects with kinetic energy.

The second basic category of energy is **potential energy**, or energy being stored for possible later conversion into kinetic energy. A rock perched on a ledge has *gravitational* potential energy because it will fall if it slips off the edge. Gasoline contains *chemical* potential energy, which a car engine converts to the kinetic energy of the moving car. Power companies supply *electrical* potential energy, which we use to run dishwashers and other appliances.

The third basic category is energy carried by light, or **radiative energy** (the word *radiation* is often used as a synonym for *light*). Plants directly convert the radiative energy of sunlight into chemical potential energy through the process of *photosynthesis*. Radiative energy is fundamental to astronomy, because telescopes collect the radiative energy of light from distant stars.

THINK ABOUT IT

We buy energy in many different forms. For example, we buy chemical potential energy in the form of food to fuel our bodies. Describe several other forms of energy that you commonly buy.

Energy can change from one form to another. Looking at how energy changes can help us understand many common phenomena. For example, a diver standing on a 10-meter platform has gravitational potential energy owing to her height above the water and chemical potential energy stored in her body tissues. She uses the chemical potential energy to flex her muscles in such a way as to initiate her dive and then to help her execute graceful twists and spins (Figure 4.2). Meanwhile, her gravitational potential

Table 4.1 Energy Comparisons

Item	Energy (joules)
Average daytime solar energy striking Earth, per m² per second	1.3×10^3
Energy released by metabolism of one average candy bar	1×10^6
Energy needed for 1 hour of walking (adult)	1×10^6
Kinetic energy of average car traveling at 60 mi/hr	1×10^6
Daily energy needs of average adult	1×10^7
Energy released by burning 1 liter of oil	1.2×10^7
Energy released by fission of 1 kg of uranium-235	5.6×10^{13}
Energy released by fusion of hydrogen in 1 liter of water	7×10^{13}
Energy released by 1-megaton H-bomb	5×10^{15}
Energy released by major earthquake (magnitude 8.0)	2.5×10^{16}
Annual U.S. energy consumption	10^{20}
Annual energy generation of Sun	10^{34}
Energy released by supernova (explosion of a star)	10^{44}–10^{46}

Mathematical Insight **4.1** **Temperature Scales**

Three temperature scales are commonly used today (see the figure). In the United States, we usually use the **Fahrenheit** scale, defined so that water freezes at 32°F and boils at 212°F. Internationally, temperature is usually measured on the **Celsius** scale, which places the freezing point of water at 0°C and the boiling point at 100°C.

Scientists measure temperature on the **Kelvin** scale, which is the same as the Celsius scale except for its zero point. A temperature of 0 K is the coldest possible temperature, known as **absolute zero**, and 0 K is equivalent to −273.15°C. (The degree symbol ° is not used when writing temperatures on the Kelvin scale.) Thus, any particular temperature has a Kelvin value that is numerically 273.15 larger than its Celsius value. Using T to stand for temperature and a subscript to indicate the temperature scale, we can write the conversions between Celsius and Kelvin as follows:

$$T_{\text{Kelvin}} = T_{\text{Celsius}} + 273.15$$

$$T_{\text{Celsius}} = T_{\text{Kelvin}} - 273.15$$

To find the conversion between Fahrenheit and Celsius, we observe that the Fahrenheit scale has 180° (212°F − 32°F = 180°F) between the freezing and boiling points of water, whereas the Celsius scale has only 100° between these points. A temperature change of 1 Celsius degree therefore is equivalent to a temperature change of 1.8 Fahrenheit degrees. Furthermore, the freezing point of water is numerically 32 larger on the Fahrenheit scale than on the Celsius scale. Combining these two facts gives us the conversions between Fahrenheit and Celsius:

$$T_{\text{Celsius}} = \frac{T_{\text{Fahrenheit}} - 32}{1.8}$$

$$T_{\text{Fahrenheit}} = 32 + (1.8 \times T_{\text{Celsius}})$$

Example: Convert human body temperature of 98.6°F into Celsius and Kelvin.

Solution: First, we convert 98.6°F to Celsius:

$$T_{\text{Celsius}} = \frac{T_{\text{Fahrenheit}} - 32}{1.8} = \frac{98.6 - 32}{1.8} = 37.0°C$$

Next, we convert Celsius to Kelvin:

$$T_{\text{Kelvin}} = T_{\text{Celsius}} + 273.15 = 37.0 + 273.15 = 310.15 \text{ K}$$

Thus, human body temperature of 98.6°F is 37.0°C or 310.15 K.

● Three common temperature scales: Kelvin, Celsius, and Fahrenheit.

Energy can be converted between its three basic forms.

kinetic energy
(energy of motion)

radiative energy
(energy of light)

potential energy
(stored energy)

Figure 4.1 The three basic categories of energy. Energy can be converted from one form to another, but it can never be created or destroyed (an idea embodied in the law of conservation of energy [Section 4.2]).

energy becomes kinetic energy of motion as she falls toward the water.

Following how energy changes from one form to another also helps us understand the universe. For example, the particles in a collapsing cloud of interstellar gas convert gravitational potential energy into kinetic energy as they fall inward, and their motion generates heat that can eventually ignite a star. But before we study astronomical phenomena, we need quantitative ways to describe energy.

4.2 A Scientific View of Energy

In this section, we discuss a few ways to quantify kinetic and potential energy. We'll discuss radiative energy as part of our discussion of light in Chapter 6.

Kinetic Energy

We can calculate the kinetic energy of any moving object with a very simple formula:

$$\text{kinetic energy} = \tfrac{1}{2}\, mv^2$$

where m is the mass of the object and v is its speed (v for *velocity*). If we measure the mass in kilograms and the speed in meters per second, the resulting answer will be in joules. (That is, energy has units of a mass times a velocity squared, so 1 joule $= 1$ kg $\times \frac{\text{m}^2}{\text{s}^2}$.)

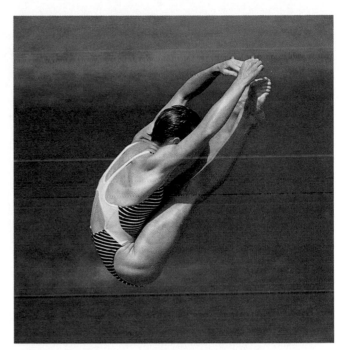

Figure 4.2 Understanding energy can help us understand both the graceful movements of a diver and the story of the universe.

The kinetic energy formula is easy to interpret. The m in the formula tells us that kinetic energy is proportional to mass: A 5-ton truck has 5 times the kinetic energy of a 1-ton car moving at the same speed. The v^2 tells us that kinetic energy increases with the *square* of the velocity: If you double your speed (e.g., from 30 km/hr to 60 km/hr), your kinetic energy becomes $2^2 = 4$ times greater.

Like potential energy, kinetic energy can appear in different forms. One of the most important is the way the motion of atoms and molecules creates temperature and heat.

Thermal Energy Suppose we want to know about the kinetic energy of the countless tiny particles (atoms and molecules) inside a rock or in the air or in a distant star. Each of these tiny particles has its own motion relative to surrounding particles, and these motions constantly change as the particles jostle one another. The result is that the particles inside a substance appear to move randomly: Any individual particle may be moving in any direction with any of a wide range of speeds.

Despite the seemingly random motion of particles within a substance, however, it's easy to measure the *average* kinetic energy of the particles—it's what we usually call **temperature**. A higher temperature simply means that, on average, the particles have more kinetic energy and hence are moving faster (Figure 4.3). (Kinetic energy depends on both mass and speed, but for a particular set of particles, greater kinetic energy means higher speeds.) The speeds of particles within a substance can be surprisingly fast. For example, the air molecules around you move at typical speeds of about 500 meters per second (about 1,000 miles per hour).

The energy contained *within* a substance as measured by its temperature is often called **thermal energy**. Thus, thermal energy represents the collective kinetic energy of the many individual particles moving within a substance.

Temperature and Heat The concepts of *temperature* and *heat* are not the same. To understand the difference, imagine the following experiment (but don't try it!). Suppose you heat your oven to 500°F. Then you open the oven door, quickly thrust your arm inside (without touching anything), and immediately remove it. What will happen to your arm? Not much. Now suppose you boil a pot of water. Although the temperature of boiling water is only 212°F, you would be badly burned if you put your arm in the pot, even if you remove it very quickly. Why does your arm burn so much more quickly in the boiling water than in the hotter oven? It happens because thermal energy content depends on both the temperature and the total number of particles, which is much larger in a pot of water (Figure 4.4).

Let's look at what is happening on the molecular level. If air or water is hotter than your body, molecules striking your skin transfer some of their thermal energy to your arm. The high temperature in a 500°F oven means that the air molecules strike your skin harder, on average, than the molecules in a 212°F pot of boiling water. However, because the *density* is so much higher in the pot of water, many more molecules strike your skin each second. Thus, while each individual molecular collision transfers a little less thermal energy in the boiling water than in the oven, the sheer number of collisions in the water transfers so much thermal energy that your skin burns rapidly.

THINK ABOUT IT

In air or water that is colder than your body temperature, thermal energy transfers from you to the surrounding cold air or water. Use this fact to explain why falling into a 32°F (0°C) lake is much more dangerous than standing naked outside on a 32°F day.

Longer arrows mean higher average speed.

Figure 4.3 Temperature is a measure of the average kinetic energy of the particles (atoms and molecules) in a substance. The particles in the box on the right have a higher temperature because their average speeds are higher (assuming that both boxes contain particles of the same mass).

Figure 4.4 Both boxes have the same temperature, but the box on the right contains more *thermal energy* because it contains more particles.

The environment in space provides another example of the difference between temperature and heat. Surprisingly, the temperature in low Earth orbit is several thousand degrees. However, astronauts working in Earth orbit (e.g., outside the Space Shuttle) can get very cold and sometimes need heated space suits and gloves. The astronauts can get cold despite the high temperature because the extremely low density of space means that relatively few particles are available to transfer thermal energy to them. (You may wonder how the astronauts become cold given that the low density also means the astronauts cannot transfer much of their own thermal energy to the particles in space. It turns out that they lose their body heat by emitting *thermal radiation*, which we will discuss in Section 6.4.)

Mathematical Insight **4.2** **Density**

The term *density* usually refers to *mass density*, which quantifies how much mass is packed into each unit of volume. The more tightly matter is packed, the higher its density. In science, the most common unit of density is grams per cubic centimeter (a cubic centimeter is about the size of a sugar cube):

$$\text{density} = \frac{\text{mass (in g)}}{\text{volume (in cm}^3)}$$

For example, if a 30-gram rock has a volume of 10 cubic centimeters, we calculate its density as follows:

$$\text{density of rock} = \frac{\text{mass of rock}}{\text{volume of rock}} = \frac{30\text{ g}}{10\text{ cm}^3} = 3\,\frac{\text{g}}{\text{cm}^3}$$

A useful guide for putting densities in perspective is the density of water, which is 1 gram per cubic centimeter. (This isn't a coincidence—it's how the gram was defined.) Rocks, like the one with a density of 3 grams per cubic centimeter, are denser than water and therefore sink in water. Wood is less dense than water and therefore floats.

The concept of density is sometimes applied to things other than mass. For example, an average of about 25,000 people live on each square kilometer of Manhattan, so we say that Manhattan has a *population density* of 25,000 people per square kilometer. As another example, 1 liter of oil releases about 12 million joules of energy when burned, so we say that the *chemical energy density* of oil is about 12 million joules per liter.

Potential Energy

Potential energy can be stored in many different forms and is not always easy to quantify. Fortunately, it is easy to describe two types of potential energy that we use frequently in astronomy.

Gravitational Potential Energy

Gravitational potential energy is extremely important in astronomy. The conversion of gravitational potential energy into kinetic (or thermal) energy helps explain everything from the speed at which an object falls to the ground to the formation processes of stars and planets. The mathematical formula for gravitational potential energy can take a variety of forms, but in words the idea is simple: *The amount of gravitational potential energy released as an object falls depends on its mass, the strength of gravity, and the distance it falls.*

This statement explains the obvious fact that falling from a 10-story building hurts more than falling out of a chair. Your gravitational potential energy is much greater on top of the 10-story building than in your chair because you can fall much farther. Because your gravitational potential energy will be converted to kinetic energy as you fall, you'll have a lot more kinetic energy by the time you hit the ground after falling from the building than after falling from the chair. The additional kinetic energy means you'll hit the ground with a much greater speed.

Gravitational potential energy also helps us understand how the Sun became hot enough to sustain nuclear fusion. Before the Sun formed, its matter was contained in a large, cold, diffuse cloud of gas. Most of the individual gas particles were far from the center of this large cloud and therefore had considerable amounts of gravitational potential energy. As the cloud contracted under its own gravity, the gravitational potential energy of these particles was converted to thermal energy, eventually making the center of the cloud hot enough to ignite nuclear fusion.

Mass-Energy

Although matter and energy seem very different in daily life, they are intimately connected. Einstein showed that mass itself is a form of potential energy, often called **mass-energy**. The mass-energy of any piece of matter is given by the formula

$$E = mc^2$$

where E is the amount of potential energy, m is the mass of the object, and c is the speed of light. If the mass is measured in kilograms and the speed of light in meters per second, the resulting mass-energy has units of joules. The speed of light is a large number ($c = 3 \times 10^8$ m/s), and the speed of light squared is much larger still ($c^2 = 9 \times 10^{16}$ m²/s²). Thus, Einstein's formula implies that a relatively small amount of mass represents a huge amount of mass-energy.

Mass-energy can be converted to other forms of energy, but noticeable amounts of mass become other forms of energy only under special but important circumstances. The process of nuclear fusion in the core of the Sun converts some of the Sun's mass into energy, ultimately generating the sunlight that sustains most life on Earth. On Earth, nuclear reactors and nuclear bombs also work in accord with Einstein's formula. In nuclear reactors, the splitting (fission) of elements such as uranium or plutonium converts some of the mass-energy of these materials into heat, which is then used to generate electrical power. In an H-bomb, nuclear fusion similar to that in the Sun uses a small amount of the mass-energy in hydrogen to devastating effect. Incredibly, a 1-megaton H-bomb that could destroy a major city requires the conversion of only about 0.1 kilogram of mass (about 3 ounces) into energy (Figure 4.5).

Just as the formula $E = mc^2$ tells us that mass can be converted into other forms of energy, it also tells us that energy can be transformed into mass. In *particle accelerators*, scientists accelerate subatomic particles to extremely

---

Mathematical Insight **4.3** **Mass-Energy**

It's easy to calculate mass-energies with Einstein's formula $E = mc^2$. Once we calculate an energy, we can compare it to other known energies.

Example: Suppose a 1-kilogram rock were completely converted to energy. How much energy would it release? Compare this to the energy released by burning 1 liter of oil.

Solution: The total mass-energy of the rock is given by $E = mc^2$, where m is the 1-kg mass and $c = 3 \times 10^8$ m/s:

$$E = mc^2 = 1 \text{ kg} \times \left(3 \times 10^8 \, \frac{\text{m}}{\text{s}}\right)^2$$

$$= 1 \text{ kg} \times \left(9 \times 10^{16} \, \frac{\text{m}^2}{\text{s}^2}\right)$$

$$= 9 \times 10^{16} \, \frac{\text{kg} \times \text{m}^2}{\text{s}^2}$$

$$= 9 \times 10^{16} \text{ joules}$$

Burning 1 liter of oil releases 12 million joules (see Table 4.1). Dividing the mass-energy of the rock by the energy released by burning 1 liter of oil, we find:

$$\frac{9 \times 10^{16} \text{ joules}}{1.2 \times 10^7 \text{ joules}} = 7.5 \times 10^9$$

That is, if the rock could be converted completely to energy, its mass would supply as much energy as 7.5 billion liters of oil—roughly the amount of oil used by *all* cars in the United States in a week. Unfortunately, no technology available now or in the foreseeable future can release all the mass-energy of a rock.

Figure 4.5 The energy from this H-bomb comes from converting only about 0.1 kg of mass into energy in accordance with the formula $E = mc^2$.

high speeds. When these particles collide with one another or with a barrier, some of the energy released in the collision spontaneously turns into mass, producing a shower of subatomic particles. These showers of particles allow scientists to test theories about how matter behaves at extremely high temperatures such as those that prevailed in the universe during the first fraction of a second after the Big Bang. Among the most powerful particle accelerators in the world are Fermilab in Illinois, the Stanford Linear Accelerator in California, and CERN in Switzerland.

THINK ABOUT IT

Einstein's formula $E = mc^2$ is probably the most famous physics equation of all time. Considering its role as described in the preceding paragraphs, do you think its fame is well deserved?

Conservation of Energy

A fundamental principle in science is that, regardless of how we change the *form* of energy, the total *quantity* of energy never changes. Energy cannot be created, and it cannot be destroyed. This principle is called the **law of conservation of energy**. It has been carefully tested in many experiments and is a pillar upon which modern theories of the universe are built. Because of this law, the story of the universe is a story of the interplay of energy and matter: All actions in the universe involve exchanges of energy or the conversion of energy from one form to another.

For example, imagine that you've thrown a baseball. It is moving, so it has kinetic energy. Where did this kinetic energy come from? The baseball got its kinetic energy from the motion of your arm as you threw it. That is, some of the kinetic energy of your moving arm was transferred to the baseball. Your arm, in turn, got its kinetic energy from the release of chemical potential energy stored in your muscle tissues. Your muscles got this energy from the chemical potential energy stored in the foods you ate. The energy stored in the foods came from sunlight, which plants convert into chemical potential energy through photosynthesis. The radiative energy of the Sun was generated through the process of nuclear fusion, which releases some of the mass-energy stored in the Sun's supply of hydrogen. Thus, the ultimate source of the energy of the moving baseball is the mass-energy stored in hydrogen—which was created in the Big Bang.

We have described where the baseball got its kinetic energy. Where will this energy go? As the baseball moves through the air, some of its energy is transferred to molecules in the air, generating heat or sound. If someone catches the baseball, its energy will cause his or her hand to recoil and will also generate some heat and sound. Ultimately, the energy of the moving baseball will be converted to a barely noticeable amount of heat (thermal energy) in the air, the ground, or a person's hand, making it extremely difficult to track. Nevertheless, the energy will never disappear. According to present understanding, the total energy content of the universe was determined in the Big Bang. It remains the same today and will stay the same forever into the future.

4.3 The Material World

Now that we have seen how energy animates the matter in the universe, it's time to consider matter itself in greater detail. You are familiar with two basic properties of matter on Earth from everyday experience. First, matter can exist in different **phases**: as a **solid**, such as ice or a rock; as a **liquid**, such as flowing water or oil; or as a **gas**, such as air. Second, even in a particular phase, matter exists as a wide variety of different substances.

What is matter, and why does it have so many different forms? Let's follow the lead of the ancient Greek philosopher Democritus (c. 470–380 B.C.), who wondered what would happen if we broke a piece of matter, such as a rock, into ever smaller pieces. Democritus believed that the rock would eventually break into particles so small that nothing smaller could be possible. He called these particles *atoms,* a Greek term meaning "indivisible." (By modern definition, atoms are *not* indivisible because they are composed of even smaller particles.)

Building upon the beliefs of earlier Greek scientists, Democritus thought that all materials were composed of just four basic *elements:* fire, water, earth, and air. He proposed that the different properties of the elements could be explained by the physical characteristics of their atoms. Democritus suggested that atoms of water were smooth and round, so water flowed and had no fixed shape, while burns were painful because atoms of fire were thorny.

He imagined atoms of earth to be rough and jagged, like pieces of a three-dimensional jigsaw puzzle, so that they could stick together to form a solid substance. He even explained the creation of the world with an idea that sounds uncannily modern, suggesting that the universe began as a chaotic mix of atoms that slowly clumped together to form Earth.

Although Democritus was wrong about there being only four types of atoms and about their specific properties, he was on the right track. Today, we know that all ordinary matter is composed of **atoms** and that each different type of atom corresponds to a different chemical **element**. Among the most familiar chemical elements are hydrogen, helium, carbon, oxygen, silicon, iron, gold, silver, lead, and uranium. (Appendix D gives the periodic table of all the elements.)

The number of different material substances is far greater than the number of chemical elements because atoms can combine to form **molecules**. Some molecules consist of two or more atoms of the same element. For example, we breathe O_2, oxygen molecules made of two oxygen atoms. Other molecules, such as water (H_2O) and sulfuric acid (H_2SO_4), are made up of atoms of two or more different elements; such molecules are called **compounds**.

The chemical properties of a molecule are different from those of its individual atoms. For example, water behaves very differently than pure hydrogen or pure oxygen, even though each water molecule is composed of two hydrogen atoms and one oxygen atom, as indicated by the familiar symbol H_2O.

Atomic Structure

Atoms are incredibly small: Millions could fit end to end across the period at the end of this sentence, and the number in a single drop of water (10^{22}–10^{23} atoms) may exceed the number of stars in the observable universe. Yet atoms are composed of even smaller particles: **protons**, **neutrons**, and **electrons**. (Protons and neutrons are, in turn, made of even

smaller particles called *quarks* [Section S4.2]). Protons and neutrons are found in the tiny **nucleus** at the center of the atom. The rest of the atom's volume contains the electrons that surround the nucleus (Figure 4.6). Although the nucleus is very small compared to the atom as a whole, it contains most of the atom's mass, because protons and neutrons are each about 2,000 times more massive than an electron.

The properties of an atom depend mainly on the amount of **electrical charge** in its nucleus. Electrical charge is a fundamental physical property that is always conserved, just as energy is always conserved. We define the electrical charge of a proton as the basic unit of positive charge, which we write as $+1$. The electron has an electrical charge that is precisely opposite that of a proton, so we say it has negative charge (-1). Neutrons are electrically neutral; that is, they have no charge.

Oppositely charged particles attract one another, and similarly charged particles repel one another. The attraction between the positively charged protons in the nucleus and the negatively charged electrons that surround it holds an atom together. Ordinary atoms have identical numbers of electrons and protons, making them electrically neutral overall. (You may wonder why electrical repulsion doesn't cause the positively charged protons in a nucleus to fly apart from one another. It tries, but it is overcome by an even stronger force that holds nuclei together, called the *strong force* [Section S4.2].)

Although electrons can be thought of as tiny particles, they are not quite like tiny grains of sand, and they don't really orbit the nucleus the way planets orbit the Sun. Instead, the electrons in an atom are "smeared out," forming a kind of cloud that surrounds the nucleus and gives the atom its apparent size. The electrons aren't really cloudy, but it is impossible to pinpoint their positions.

In Figure 4.6, you can see that the electrons give the atom a size far larger than its nucleus even though they represent only a tiny portion of the atom's mass. If we imagine an atom on a scale on which its nucleus is the size of your fist, its electron cloud would be many miles wide.

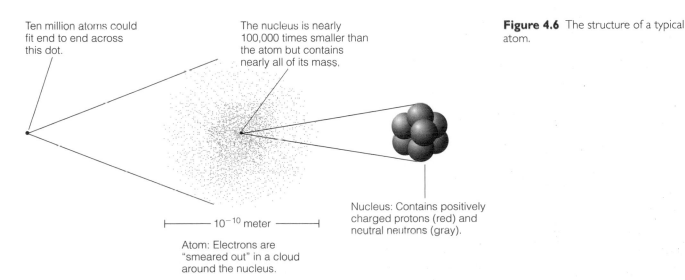

Ten million atoms could fit end to end across this dot.

The nucleus is nearly 100,000 times smaller than the atom but contains nearly all of its mass.

Figure 4.6 The structure of a typical atom.

├—————— 10^{-10} meter ——————┤

Atom: Electrons are "smeared out" in a cloud around the nucleus.

Nucleus: Contains positively charged protons (red) and neutral neutrons (gray).

The Illusion of Solidity

Bang your hand on a table. Although the table feels solid, it is made almost entirely of empty space! Nearly all the mass of the table is contained in the nuclei of its atoms. But the volume of an atom is more than a trillion times the volume of its nucleus, so relatively speaking the nuclei of adjacent atoms are nowhere near to touching one another. The solidity of the table comes about from a combination of electrical interactions between the charged particles in its atoms and the strange quantum laws governing the behavior of electrons. If we could somehow pack all the table's nuclei together, the table's mass would fit into a microscopic speck. Although *we* cannot pack matter together in this way, nature can and does—in *neutron stars,* which we will study in Chapter 18.

Each different chemical element contains a different number of protons in its nucleus. This number is its **atomic number**. For example, a hydrogen nucleus contains just one proton, so its atomic number is 1. A helium nucleus contains two protons, so its atomic number is 2.

The *combined* number of protons and neutrons in an atom is called its **atomic mass number**. The atomic mass number of ordinary hydrogen is 1 because its nucleus is just a single proton. Helium usually has two neutrons in addition to its two protons, giving it an atomic mass number of 4. Carbon usually has six protons and six neutrons, giving it an atomic mass number of 12.

While every atom of a given element contains exactly the same number of protons, the number of neutrons can vary. For example, all carbon atoms have six protons, but they may have six, seven, or eight neutrons. Versions of an element with different numbers of neutrons are called **isotopes** of the element (Figure 4.7).

To name the isotopes of an element, we use their atomic mass numbers. For example, the most common isotope of carbon, with six neutrons, has atomic mass number 6 protons + 6 neutrons = 12. We call it carbon-12. The other isotopes of carbon are carbon-13 (its six protons and seven neutrons give it atomic mass number 13) and carbon-14 (its six protons and eight neutrons give it atomic mass number 14). We can also write isotopes by writing the atomic mass number as a superscript to the left of the element symbol: ^{12}C, ^{13}C, ^{14}C. We read ^{12}C as "carbon-12." Figure 4.7 summarizes some of this basic atomic terminology.

The symbol ^{4}He represents helium with an atomic mass number of 4. ^{4}He is the most common form of helium, containing two protons and two neutrons. What does the symbol ^{3}He represent?

Phases of Matter

Everyday experience tells us that the same substance can exist in different phases depending on the temperature. The main difference between phases is how tightly neighboring particles are bound together. As a substance is heated, the average kinetic energy of its particles increases, enabling the particles to break the bonds holding them to their neigh-

Figure 4.7 Terminology of atoms.

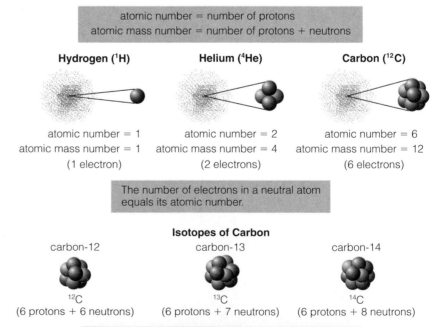

atomic number = number of protons
atomic mass number = number of protons + neutrons

Hydrogen (^{1}H)	Helium (^{4}He)	Carbon (^{12}C)
atomic number = 1	atomic number = 2	atomic number = 6
atomic mass number = 1	atomic mass number = 4	atomic mass number = 12
(1 electron)	(2 electrons)	(6 electrons)

The number of electrons in a neutral atom equals its atomic number.

Isotopes of Carbon

carbon-12	carbon-13	carbon-14
^{12}C	^{13}C	^{14}C
(6 protons + 6 neutrons)	(6 protons + 7 neutrons)	(6 protons + 8 neutrons)

Different isotopes of a given element contain the same number of protons but different numbers of neutrons.

bors. Each change in phase corresponds to the breaking of a different kind of bond. Phase changes occur in all substances. Let's consider what happens when we heat water, starting from its solid phase, ice (Figure 4.8):

● **Solid phase.** Below 0°C (32°F), water molecules have a relatively low average kinetic energy, and each molecule is bound tightly to its neighbors, making the *solid* structure of ice.

● **Liquid phase.** As the temperature increases (but remains below freezing), the rigid arrangement of the molecules in ice vibrates more and more. At 0°C, the molecules have enough energy to break the solid bonds of ice. The molecules can then move relatively freely among one another, allowing the water to flow as a *liquid*. Even in liquid water, a loose bond between adjacent molecules keeps them close together.

● **Gas phase.** When a water molecule breaks free of all bonds with its neighbors, we call it a molecule of water vapor, which is a *gas*. Molecules in the gas phase move independently of other molecules. Even at temperatures at which water is a solid or liquid, a few molecules will have enough energy to enter the gas phase. We call the process **evaporation** when molecules escape from a liquid and **sublimation** when they escape from a solid. As temperatures rise, the rates of sublimation and evaporation increase, eventually changing all the solid and liquid water into water vapor.

What happens if we continue to raise the temperature of water vapor? As the temperature rises, the molecules move faster, making collisions among them more violent. These collisions eventually split the water molecules into their component atoms of hydrogen and oxygen. The process by which the bonds that hold the atoms of a molecule together are broken is called **molecular dissociation**.

At still higher temperatures, collisions can break the bonds holding electrons around the nuclei of individual atoms, allowing the electrons to go free. The loss of one or more negatively charged electrons leaves a remaining atom with a net positive charge. Such charged atoms are called **ions**. The process of stripping electrons from atoms is called **ionization**.

Thus, at high temperatures, what once was water becomes a hot gas consisting of freely moving electrons and positively charged ions of hydrogen and oxygen. This type of hot gas, in which atoms have become ionized, is called a **plasma**, sometimes referred to as "the fourth phase of matter." Other chemical substances go through similar

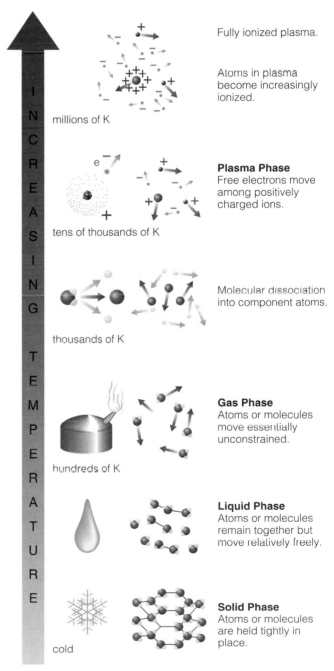

Fully ionized plasma.

Atoms in plasma become increasingly ionized.

millions of K

Plasma Phase
Free electrons move among positively charged ions.

tens of thousands of K

Molecular dissociation into component atoms.

thousands of K

Gas Phase
Atoms or molecules move essentially unconstrained.

hundreds of K

Liquid Phase
Atoms or molecules remain together but move relatively freely.

Solid Phase
Atoms or molecules are held tightly in place.

cold

Figure 4.8 The general progression of phase changes.

phase changes, but the temperatures at which phase changes occur depend on the types of atom or molecules involved.

The degree of ionization in a plasma depends on the temperature and the composition of the gas. A neutral hydrogen atom contains only one electron, which balances the single positive charge of the one proton in its nucleus. Thus, hydrogen can be ionized only once, and the remaining hydrogen ion, designated H$^+$, is simply a proton. Oxygen, with atomic number 8, has eight electrons when it is neutral, so it can be ionized multiple times. *Singly ionized* oxygen is missing one electron, so it has a charge of $+1$ and is designated O$^+$. *Doubly ionized* oxygen, or O^{++}, is missing two electrons; *triply ionized* oxygen, or O^{+3}, is missing three electrons; and so on. At extremely high temperatures, oxygen can be *fully ionized*, in which case all eight electrons are stripped away and the remaining ion has a charge of $+8$.

4.4 Energy in Atoms

So far, we've seen two different ways in which atoms have energy. First, by virtue of their mass, they possess mass-energy in the amount mc^2. Second, they possess kinetic energy by virtue of their motion. Atoms also contain energy in a third way: as *electrical potential energy* in the distribution of their electrons around their nuclei.

The simplest case is that of hydrogen, which has only one electron. Remember that an electron tends to be "smeared out" into a cloud around the nucleus. When the electron is "smeared out" to the minimum extent that nature allows, the atom contains its smallest possible amount of electrical potential energy, and we say that the atom is in its **ground state** (Figure 4.9). If the electron somehow gains energy, it becomes "smeared out" over a greater volume, and we say that the atom is in an **excited state**. If the electron gains enough energy, it can escape the atom completely, in which case the atom has been *ionized*.

Perhaps the most surprising aspect of atoms was discovered in the 1910s, when scientists realized that electrons in atoms can have only *particular* energies (which correspond to particular sizes and shapes of the electron cloud). As a simple analogy, suppose you're washing windows on

a building. If you use an adjustable platform to reach high windows, you can stop the platform at any height above the ground (Figure 4.10a). But if you use a ladder, you can stand only at *particular* heights—the heights of the rungs of the ladder—and not at any height in between (Figure 4.10b). The possible energies of electrons in atoms are like the possible heights on a ladder. Only a few particular energies are possible, and energies between these special few are not possible.

The possible energy levels of the electron in hydrogen are represented like the steps of a ladder in Figure 4.11. The ground state, or level 1, is the bottom rung of the ladder. Its energy is labeled zero because the atom has no excess electrical potential energy to lose. Each subsequent rung of the ladder represents a possible excited state for the electron. Each level is labeled with the electron's energy above the ground state in units of **electron-volts**, or **eV** (1 eV = 1.60×10^{-19} joule). For example, the energy of an electron in energy level 2 is 10.2 eV greater than that of an electron in the ground state. That is, an electron must gain 10.2 eV of energy to "jump" from level 1 to level 2. Similarly, jumping from level 1 to level 3 requires gaining 12.1 eV of energy.

Unlike a ladder built for climbing, the rungs on the electron's energy ladder are closer together near the top. The top itself represents the energy of ionization—if the electron gains this much energy, 13.6 eV above the ground state in the case of hydrogen, the electron breaks free from the atom. (Any excess energy beyond that needed for ionization becomes kinetic energy of the free-moving electron.)

Because energy is always conserved, an electron cannot jump to a higher energy level unless its atom gains the energy from somewhere else. Generally, the atom gains this energy either from the kinetic energy of another particle colliding with it or from the absorption of energy carried by light. Similarly, when an electron falls to a *lower* energy

ground state   excited state   ionization

Figure 4.9 In its ground state, an electron is "smeared out" to the minimum extent allowed by nature. Adding energy can raise the electron to an excited state that occupies a larger volume. Adding enough energy can ionize the atom.

Orbiting Electrons

Most people have been taught to think of electrons in atoms as "orbiting" the nucleus like tiny planets orbiting a tiny sun. But this representation is simply not true. Electrons and other subatomic particles do not behave at all like baseballs, rocks, or planets. In fact, the behavior of subatomic particles is so strange that the human mind may be incapable of visualizing it. Nevertheless, the *effects* of electrons are easy to observe, and one of the most important effects is that electrons give atoms their size. Thus, we say that electrons in atoms are "smeared out" into an "electron cloud" around the nucleus. This description is rather vague, but it is far more accurate than the misleading picture of electrons circling the nucleus like tiny planets.

a A window washer on an adjustable platform can be at any height.

b A window washer on a ladder can be at only the particular heights of the steps. Similarly, electrons in an atom can have only particular energy levels.

Figure 4.10 A window-washing analogy to energy levels of electrons.

ionization level	13.6 eV
level 4	12.8 eV
level 3	12.1 eV
level 2	10.2 eV
level 1 (ground state)	0 eV

Figure 4.11 Energy levels for the electron in a hydrogen atom. (There are many more closely spaced energy levels between level 4 and the ionization level.) The electron can jump between energy levels only if it gains or loses the precise amount of energy separating the levels. In Chapter 6, we'll see how this fact helps us learn the chemical composition of distant objects by seeing the "fingerprints" that electron jumps leave in light.

level, it either transfers its energy to another particle through a collision or emits light that carries the energy away. The key point is this: *Electron jumps can occur only with the particular amounts of energy representing differences between possible energy levels.*

The result is that electrons in atoms can absorb or emit only particular amounts of energy and not other amounts in between. For example, if you attempt to provide a hydrogen atom in the ground state with 11.1 eV of energy, the atom won't accept it because it is too high to boost the electron to level 2 but not high enough to boost it to level 3.

THINK ABOUT IT

We will see in Chapter 6 that light comes in "pieces" called *photons* that carry specific amounts of energy. Can a hydrogen atom absorb a photon with 11.1 eV of energy? Why or why not? Can it absorb a photon with 10.2 eV of energy? Explain.

If you think about it, the idea that electrons in atoms can jump only between particular energy levels is quite bizarre. It is as if you had a car that could go around a track only at particular speeds and not at speeds in between. How strange it would seem if your car suddenly jumped from 5 miles per hour to 20 miles per hour without first passing through a speed of 10 miles per hour! In

scientific terminology, the electron's energy levels in an atom are said to be *quantized*, and the study of the energy levels of electrons (and other particles) is called *quantum mechanics*.

Electrons have quantized energy levels in all atoms, not just in hydrogen. Moreover, the allowed energy levels differ from element to element and even from one ion of an element to another ion of the same element. This fact holds the key to the study of distant objects in the universe. As we will see in Chapter 6 when we study light, the different energy levels of different elements allow light to carry "fingerprints" that can tell us the chemical composition of distant objects.

THE BIG PICTURE

Putting Chapter 4 into Context

In this chapter, we discussed the concepts of energy and matter in some detail. Key "big picture" ideas to draw from this chapter include the following:

- Energy and matter are the two basic ingredients of the universe. Therefore, understanding energy and matter, and the interplay between them, is crucial to understanding the universe.

- Energy is always conserved, and we can understand many processes in the universe by following how energy changes from one form to another in its interactions with matter. Don't forget that mass itself is a form of potential energy, called mass-energy.

- The strange laws of quantum mechanics govern the interactions of matter and energy on the atomic level. Electrons in atoms can have only particular energies and not energies in between, and each element has a different set of allowed energy levels.

4.1 Matter and Energy in Everyday Life

- *How do we measure energy in science?* The standard unit of energy is called a joule. It is equivalent to $\frac{1}{4,184}$ of a food Calorie.

- *What are the three basic categories of energy?* Kinetic energy is energy of motion. Potential energy is stored energy that can be released later. Radiative energy is energy carried by light.

4.2 A Scientific View of Energy

- *How is temperature different from heat?* Temperature is a measure of the average kinetic energy of the many individual atoms or molecules in a substance. Heat depends on both temperature and density: At a particular temperature, a denser substance contains more thermal energy.

- *What is gravitational potential energy?* It is energy that can be released by an object falling under the force of gravity. The amount of an object's gravitational potential energy depends on its mass, the strength of gravity, and how far it could fall.

- *What is the meaning of $E = mc^2$?* This formula describes the potential energy of mass itself. *E* is the energy stored in a piece of matter of mass *m*, and *c* is the speed of light.

- *Why is the law of conservation of energy so important?* It tells us that energy can be neither created nor destroyed. Energy can only be exchanged between objects or transformed from one form to another.

4.3 The Material World

- *What is the basic structure of an atom?* An atom consists of a tiny nucleus made of protons and neutrons surrounded by a "smeared out" cloud of electrons that gives the atom its size.

- *What is the difference between atomic number and atomic mass number?* Atomic number is the number of protons in an atom's nucleus. Atomic mass number is the sum of the number of protons and neutrons.

- *How do phases of matter change with increasing temperature?* Most substances are solid at low temperature. As temperature rises, the substance may melt into liquid and then evaporate into gas. (Some material may *sublimate* directly from solid to gas.) As temperature rises further, molecules (if any) will dissociate, and atoms will be ionized to make a plasma.

4.4 Energy in Atoms

- *How is energy stored in atoms?* Atoms have electrical potential energy by virtue of the distribution of electrons around the nucleus. Electrons can have only particular amounts of electrical potential energy, not amounts in between. Electrons can jump between the allowed energy levels only by gaining or losing the precise amounts of energy separating levels.

- *How do energy levels differ from one chemical element to another?* Every chemical element has its own unique set of energy levels.

❓ Does It Make Sense?

Decide whether each statement makes sense and explain why it does or does not.

1. The sugar in my soda will provide my body with about a million joules of energy.

2. When I drive my car at 30 miles per hour, it has more kinetic energy than it does at 10 miles per hour.

3. If you put an ice cube outside the Space Station, it would take a very long time to melt, even though the temperature in Earth orbit is several thousand degrees (Celsius).

4. Someday soon, scientists are likely to build an engine that produces more energy than it consumes.

5. Two isotopes of the element rubidium differ not only in their number of neutrons, but also in their number of protons.

6. According to the laws of quantum mechanics, an electron's energy in a hydrogen atom can jump suddenly from 10.2 eV to 12.1 eV, without ever having any in-between energy such as 10.9 eV.

7. Two ions, each carrying a positive charge of $+1$, will attract each other electrically.

8. In particle accelerators, scientists can create particles where none existed previously by converting energy into mass.

Problems

(Quantitative problems are marked with an asterisk.)

9. *Types of Energy.* Briefly describe and differentiate between *kinetic energy, potential energy,* and *radiative energy.* For each type of energy, give at least two examples of objects that either have it or use it. Explain clearly.

10. *Energy Conservation in Astronomy.* What is the law of *conservation of energy?* How is it important in astronomy?

11. *The Nature of Atoms.* Briefly define *atom, element,* and *molecule.* How was Democritus's idea of atoms similar to the modern concept of atoms? How was it different?

12. *Atomic Structure and Size.* Briefly describe the structure of an atom. How big is an atom? How big is the *nucleus* in comparison to the entire atom?

13. *Evaporation and Sublimation.* Briefly explain why a few atoms (or molecules) are always in gas phase around any solid or liquid. Then explain how *sublimation* and *evaporation* are similar and how they are different.

14. *Energy in Atoms.* How are the possible energy levels of electrons in atoms similar to the possible gravitational potential energies of a person on a ladder? How are they different?

15. *Gravitational Potential Energy.*

    a. Why does a bowling ball perched on a cliff ledge have more gravitational potential energy than a baseball perched on the same ledge?

    b. Why does a diver on a 10-meter platform have more gravitational potential energy than a diver on a 3-meter diving board?

    c. Why does a 100-kg satellite orbiting Jupiter have more gravitational potential energy than a 100-kg satellite orbiting Earth, assuming both satellites orbit at the same distance from the planet centers?

16. *Einstein's Famous Formula.*

    a. What is the meaning of the formula $E = mc^2$? Be sure to define each variable.

    b. How does this formula explain the generation of energy by the Sun?

    c. How does this formula explain the destructive power of nuclear bombs?

17. *Atomic Terminology Practice.*

    a. The most common form of iron has 26 protons and 30 neutrons in its nucleus. State its atomic number, atomic mass number, and number of electrons if it is electrically neutral.

    b. Consider the following three atoms: Atom 1 has 7 protons and 8 neutrons; atom 2 has 8 protons and 7 neutrons; atom 3 has 8 protons and 8 neutrons. Which two are *isotopes* of the same element?

    c. Oxygen has atomic number 8. How many times must an oxygen atom be ionized to create an O^{+5} ion? How many electrons are in an O^{+5} ion?

    d. Consider fluorine atoms with 9 protons and 10 neutrons. What are the atomic number and atomic mass number of this fluorine? Suppose we could add a proton to this fluorine nucleus. Would the result still be fluorine? Explain. What if we added a neutron to the fluorine nucleus?

    e. The most common isotope of gold has atomic number 79 and atomic mass number 197. How many protons and neutrons does the gold nucleus contain? If it is electrically neutral, how many electrons does it have? If it is triply ionized, how many electrons does it have?

    f. The most common isotope of uranium is ^{238}U, but the form used in nuclear bombs and nuclear power plants is ^{235}U. Given that uranium has atomic number 92, how many neutrons are in each of these two isotopes of uranium?

18. *The Fourth Phase of Matter.*

    a. Explain why nearly all the matter in the Sun is in the plasma phase.

    b. Based on your answer to part (a), explain why plasma is the most common phase of matter in the universe.

    c. If plasma is the most common phase of matter in the universe, why is it so rare on Earth?

19. *Energy Level Transitions.* The labeled transitions below represent an electron moving between energy levels in hydrogen. Answer each of the following questions and explain your answers.

    a. Which transition could represent an electron that *gains* 10.2 eV of energy?

    b. Which transition represents an electron that *loses* 10.2 eV of energy?

    c. Which transition represents an electron that is breaking free of the atom?

    d. Which transition, as shown, is *not* possible?

    e. Describe the process taking place in transition A.

*20. *Energy Comparisons.* Use the data in Table 4.1 to answer each of the following questions.

    a. Compare the energy of a 1-megaton hydrogen bomb to the energy released by a major earthquake.

    b. If the United States obtained all its energy from oil, how much oil would be needed each year?

    c. Compare the Sun's annual energy output to the energy released by a supernova.

*21. *Moving Candy Bar.* Metabolizing a candy bar releases about 10^6 joules. How fast must the candy bar travel to have the same 10^6 joules in the form of kinetic energy? (Assume the candy bar's mass is 0.2 kg.) Is your answer faster or slower than you expected?

*22. *Calculating Densities.* Find the average density of the following objects in grams per cubic centimeter.

    a. A rock with volume 15 cm^3 and mass 50 g.

    b. Earth, with its mass of 6×10^{24} kg and radius of about 6,400 km. (*Hint:* The formula for the volume of a sphere is $\frac{4}{3} \times \pi \times$ radius^3.)

    c. The Sun, with its mass of 2×10^{30} kg and radius of about 700,000 km.

*23. *Spontaneous Human Combustion.* Suppose that, through a horrific act of an angry god (or a very powerful alien, if you prefer), all the mass in your body was suddenly converted into energy according to the formula $E = mc^2$. How much energy would be produced? Compare this to the energy released by a 1-megaton hydrogen bomb (see Table 4.1). What effect would your disappearance have on the surrounding region?

*24. *Fusion Power.* No one has yet succeeded in creating a commercially viable way to produce energy through nuclear fusion. However, suppose we could build fusion power plants using the hydrogen in water as a fuel. Based on the data in Table 4.1, how much water would we need each minute in order to meet U.S. energy needs? Could such a reactor power the entire United States with the water flowing from your kitchen sink? Explain. (*Hint:* Use the annual U.S. energy consumption to find the energy consumption per minute, and then divide by the energy yield from fusing 1 liter of water to figure out how many liters would be needed each minute.)

Discussion Questions

25. *Knowledge of Mass-Energy.* Einstein's discovery that energy and mass are equivalent has led to technological developments that are both beneficial and dangerous. Discuss some of these developments. Overall, do you think the human race would be better or worse off if we had never discovered that mass is a form of energy? Defend your opinion.

26. *Perpetual Motion Machines.* Every so often, someone claims to have built a machine that can generate energy perpetually from nothing. Why isn't this possible according to the known laws of nature? Why do you think claims of perpetual motion machines sometimes receive substantial media attention?

27. *Indoor Pollution.* Since sublimation and evaporation are very similar processes, why is sublimation generally much more difficult to notice? Discuss how sublimation, particularly from plastics and other human-made materials, can cause "indoor pollution."

28. *Democritus and the Path of History.* Besides his belief in atoms, Democritus held several other strikingly modern notions. For example, he maintained that the Moon was a world with mountains and valleys and that the Milky Way was composed of countless individual stars—ideas that weren't generally accepted until the time of Galileo, more than 2,000 years later. Unfortunately, we know of Democritus's work only secondhand because none of the 72 books he is said to have written survived the destruction of the Library of Alexandria. How do you think history might have been different if the work of Democritus had not been lost?

MEDIA EXPLORATIONS

Web Projects

Take advantage of the useful Web links on www.astronomyplace.com to assist you with the following projects.

1. *Energy Comparisons.* Using information from the Energy Information Administration Web site, choose some aspect of U.S. or world energy use that interests you. Write a short report on this issue.

2. *Nuclear Power.* There are two basic ways to generate energy from atomic nuclei: through nuclear fission (splitting nuclei) and through nuclear fusion (combining nuclei). All current nuclear reactors are based on fission, but fusion would have many advantages if we could develop the technology. Research some of the advantages of fusion and some of the obstacles to developing fusion power. Do you think fusion power will be a reality in your lifetime? Explain.

5 The Universal Laws of Motion

5.1 Describing Motion: Examples from Daily Life
- What is the difference between speed, velocity, and acceleration?
- What is the acceleration of gravity?
- How can you tell when a net force is acting on an object?
- Have you ever been weightless? Have you ever been massless?

5.2 Newton's Laws of Motion
- What are Newton's three laws of motion?
- Why does a spinning skater spin faster as she pulls in her arms?

5.3 The Force of Gravity
- What is the universal law of gravitation?
- What types of orbits are possible according to the law of gravitation?
- How can we determine the mass of a distant object?

5.4 Tides
- Why are there two high tides on Earth each day?
- Why are tides on Earth caused primarily by the Moon rather than by the Sun?
- Why is Earth's rotation gradually slowing down?
- Why does the Moon always show the same face to Earth?

5.5 Orbital Energy and Escape Velocity
- What is orbital energy?
- Will a spacecraft passing by a planet be "sucked in"?
- How can an object achieve escape velocity?

5.6 The Acceleration of Gravity
- How does the acceleration of gravity depend on the mass of a falling object?

If I have seen farther than others, it is because I have stood on the shoulders of giants.

Isaac Newton

Everything in the universe is in constant motion, from the random meandering of molecules in the air to the large-scale drifting of galaxies in superclusters. Despite the vast difference in the scale of these motions, just a few physical laws describe them all. The elucidation of these laws over the past several centuries is surely one of the greatest scientific triumphs of all time.

The Copernican revolution provided much of the impetus for the discovery of the laws of motion, and Galileo discovered some of those laws through his experiments. But the task of putting all the pieces together and discovering the precise mechanics of gravity fell to Sir Isaac Newton, one of the most influential human beings of all time. In this chapter, we'll discuss Newton's discoveries and why they are so important to modern astronomy.

As in Chapter 4, much of the subject matter of this chapter may already be familiar to you. Again, don't worry if this is not the case. The material is not difficult, and studying it carefully will greatly enhance your understanding of the astronomy in the rest of the book as well as of many everyday phenomena.

5.1 Describing Motion: Examples from Daily Life

Think about what happens when you throw a ball to a dog: The dog runs and catches it. Now think about the complexity of this trick. The ball leaves your hand traveling in some particular direction with some particular amount of kinetic energy. As the ball rises, gravity converts some of its kinetic energy into potential energy, slowing the ball's rise until it reaches the top of its trajectory. Then gravity transforms the potential energy of the ball back into kinetic energy, bringing it back toward the ground. Meanwhile, the ball may lose some of its kinetic energy to air resistance or may be pushed by gusts of wind. Despite this complexity, the dog still catches the ball.

We humans can perform an even better trick: We have learned how to figure out where the ball will land even before throwing it, and we can perform this trick with extraordinary precision. Understanding how we perform this trick and applying it to problems of motion throughout the universe require understanding the laws that govern motion.

We all have a great deal of experience with motion and natural intuition as to how motion works, so we begin our discussion with some familiar examples. You probably are familiar with all the terms defined in this section, but their scientific definitions may differ subtly from those you use in casual conversation.

Speed, Velocity, and Acceleration

The concepts we use to determine the trajectory of a ball, a rocket, or a planet are familiar to you from driving a car. The speedometer indicates your **speed**, usually in units of both miles per hour (mi/hr) and kilometers per hour (km/hr). For example, 100 km/hr is a speed. Your **velocity** is your speed in a certain direction: "100 km/hr going due north" describes a velocity. It is possible to change your velocity without changing your speed, for example, by maintaining a steady 60 km/hr as you drive around a curve. Because your direction is changing as you round the curve, your *velocity* is also changing—even though your *speed* is constant.

Whenever your velocity is changing, you are experiencing **acceleration**. You are undoubtedly familiar with the term *acceleration* as it applies to *increasing* speed, such as accelerating away from a stop sign while driving. In science, we also say that you are accelerating when you slow down or turn (Figure 5.1). Slowing occurs when acceleration is in a direction opposite to the motion. In this case, we say that your acceleration is negative, causing your velocity to decrease. Turning changes your direction, which means a change in velocity and thus involves acceleration even if your speed remains constant.

You don't feel anything when you are traveling at *constant velocity*, which is why you don't feel any sensation of motion when you're traveling in an airplane on a smooth flight. In contrast, you can often feel effects of acceleration: As you speed up in a car you feel yourself being pushed back into your seat, as you slow down you feel yourself being pulled forward from the seat, and as you drive around a curve you lean outward because of your acceleration.

The Acceleration of Gravity

One of the most important types of acceleration is that caused by gravity, which makes objects accelerate as they fall. In a famous (though probably apocryphal) experiment that involved dropping weights from the Leaning Tower of Pisa, Galileo demonstrated that gravity accelerates all objects by the same amount, regardless of their mass. This

We say that this car is accelerating because its velocity is increasing.

We say that this car is accelerating because its direction is changing as it turns, which means its velocity is changing even though its speed stays constant.

We say that this car is accelerating because its velocity is decreasing. Decreasing velocity is still acceleration, although it is a negative acceleration.

Figure 5.1 Speeding up, turning, and slowing down are all examples of acceleration.

fact may be surprising because it seems to contradict everyday experience: A feather floats gently to the ground, while a rock plummets. However, this difference is caused by air resistance. If you dropped a feather and a rock on the Moon, where there is no air, both would fall at exactly the same rate.

Find a piece of paper and a small rock. Hold both at the same height, one in each hand, and let them go at the same instant. The rock, of course, hits the ground first. Next crumple the paper into a small ball and repeat the experiment. What happens? Explain how this experiment suggests that gravity accelerates all objects by the same amount.

The acceleration of a falling object is called the **acceleration of gravity**, abbreviated *g*. On Earth, the acceleration of gravity causes falling objects to fall faster by 9.8 meters per second (m/s), or about 10 m/s, with each passing second. For example, suppose you drop a rock from a tall building. At the moment you let it go, its speed is 0 m/s. After 1 second, the rock will be falling downward at about 10 m/s. After 2 seconds, it will be falling at about 20 m/s. In the absence of air resistance, its speed will continue to increase by about 10 m/s each second until it hits the ground (Figure 5.2). We therefore say that the acceleration of gravity is about 10 *meters per second per second*, or 10 *meters per second squared*, which we write as 10 m/s^2. (More precisely, $g = 9.8$ m/s^2.)

Momentum and Force

Imagine that you're innocently stopped in your car at a red light when a bug flying at a velocity of 30 km/hr due south slams into your windshield. What will happen to your car? Not much, except perhaps a bit of a mess on your windshield. Next imagine that a 2-ton truck runs the red light

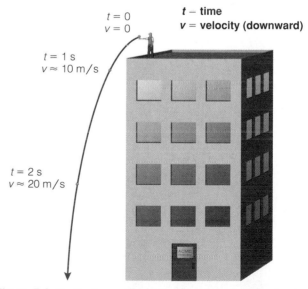

Figure 5.2 On Earth, gravity causes falling objects to accelerate downward at about 10 m/s^2. That is, a falling object's downward velocity increases by about 10 m/s with each passing second. (Gravity does not affect horizontal velocity.) More precisely, $g = 9.8$ m/s^2.

and hits you head-on with the same velocity as the bug. Clearly, the truck will cause far more damage.

Scientifically, we say that the truck imparts a much larger jolt than the bug because it transfers more **momentum** to you. Momentum describes a combination of mass and velocity. We can describe the momentum of the truck before the collision as "2 tons moving due south at 30 km/hr," while the momentum of the bug is perhaps "1 gram moving due south at 30 km/hr." Mathematically, momentum is defined as mass × velocity.

In transferring some of its momentum to your car, the truck (or bug) exerts a force on your car. More generally, a **force** is anything that can cause a change in momentum. You are familiar with many types of force besides collisional force. For example, if you shift into neutral while driving along a flat stretch of road, the forces of air resistance and road friction will continually sap your car's momentum (transferring it to molecules in the air and the pavement), slowing your velocity until you come to a stop.

The mere presence of a force does not always cause a change in momentum. For example, if the engine works hard enough, a car can maintain constant velocity—and hence constant momentum—despite air resistance and road friction. In this case, the force generated by the engine to turn the wheels precisely offsets the forces of air resistance and road friction that act to slow the car, and we say that no **net force** is acting on the car. More generally, forces of some kind are always present, such as the force of grav-

ity or the electromagnetic forces acting between atoms. The net force acting on an object represents the combined effect of all the individual forces put together. A change in momentum occurs only when the net force is not zero.

As long as an object is not shedding (or gaining) mass, a change in momentum means a change in velocity (because momentum = mass × velocity), which means an acceleration. Hence, any net force will cause acceleration, and all accelerations must be caused by a force. That is why you feel forces when you accelerate in your car.

Mass and Weight

Up until now, we've been glossing over one key term: *mass.* Your **mass** refers to the amount of matter in your body, which is different from your *weight*. Imagine standing on a scale in an elevator (Figure 5.3). When the elevator is stationary or moving at constant velocity, the scale reads your "normal" weight. When the elevator is accelerating upward, the floor exerts a greater force than it does when you are at rest. You feel heavier, and the scale verifies your greater apparent weight.* When the elevator accelerates downward, the floor and the scale are dropping away, so your weight

---

*Many physics texts distinguish between *true weight,* which is due only to the effects of gravity on mass, and the *apparent weight* that a scale reads when other forces (such as in an accelerating elevator) also act. In this book, "weight" refers to apparent weight, except when stated otherwise.

elevator stationary or moving at constant velocity

Normal weight

elevator accelerating upward

Heavier-than-normal weight

elevator accelerating downward

Lighter-than-normal weight

elevator in free-fall

Weightless

Figure 5.3 Mass is not the same as weight. The man's mass never changes, but his apparent weight does.

is reduced. Thus, your weight varies with the elevator's acceleration, but your mass remains the same. Be sure to recognize that your weight is greater than its "normal" value only during this *acceleration*, not while the elevator moves at constant velocity. (You can verify these facts by taking a small bathroom scale with you on an elevator.)

More precisely, your apparent **weight** describes the *force* that acts on your mass. It depends on the strength of gravity and other forces acting on you (such as the force due to the elevator's acceleration). Thus, while your mass is the same anywhere, your weight can vary. For example, your mass would be the same on the Moon as on Earth, but you would weigh less on the Moon because of its weaker gravity.

Free-Fall, Weightlessness, and Orbit

If the cable breaks so that the elevator is in **free-fall**, the floor drops away at the same rate that you fall. You lose contact with the scale, so your apparent weight is zero and you feel **weightless**. In fact, you are in free-fall whenever nothing is *preventing* you from falling. For example, you are in free-fall when you jump off a chair or spring from a diving board or trampoline. Surprising as it may seem, you have therefore experienced weightlessness many times in your life and can experience it right now simply by jumping off your chair. Of course, your weightlessness lasts for only the very short time until you hit the ground.

Astronauts are weightless for much longer periods because orbiting spacecraft are in a constant state of free-fall. To understand why, imagine a cannon that shoots a ball horizontally from a tall mountain. Once launched, the ball falls solely because of gravity and hence is in free-fall.

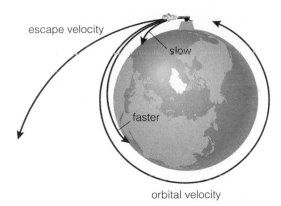

Figure 5.4 The faster the cannonball is shot, the farther it goes before hitting the ground. If it goes fast enough, it will continually "fall around," or orbit, Earth. With an even faster speed, it may escape Earth's gravity altogether.

The faster the cannon shoots the ball, the farther it goes (Figure 5.4). If the cannonball could go fast enough, its motion would keep it constantly "falling around" the Earth and it would never hit the ground (as long as we neglect air resistance). This is precisely what spacecraft such as the Space Shuttle and the Space Station do as they orbit Earth. Their constant state of free-fall makes these spacecraft and everything in them weightless.

THINK ABOUT IT

In the *Hitchhiker's Guide to the Galaxy* books, author Douglas Adams says that the trick to flying is to throw yourself at the ground and miss. Although this phrase does not really explain flying, which involves lift from air, it does describe an orbit fairly well. Explain.

Some orbits carry objects on paths that never return. If something is launched outward from Earth or some other world faster than a special speed called **escape velocity**, then gravity cannot stop its outward motion. Look again at Figure 5.4. If the cannon fires the ball at a speed too slow for orbit, it crashes to Earth (two such cases are shown by the two curves that hit the surface). If it fires the ball just fast enough, the ball ends up in low-Earth orbit (shown by the complete circular orbit). If the cannon fires the ball even faster, the ball will end up in a higher (and more elliptical) orbit. With sufficient firing speed, the ball will achieve escape velocity. It will fly outward from Earth, slowing along its way but never stopping its outward progress. The escape velocity from Earth's surface is about 40,000 km/hr (25,000 mi/hr).

5.2 Newton's Laws of Motion

The human "trick" of being able to figure out where a ball will land before it is thrown or to predict how other motions will unfold requires understanding precisely how forces affect objects in motion. The complexity of motion in daily life might lead you to guess that the laws governing how forces affect motion would also be complex. For example, if you watch a falling piece of paper waft lazily to the ground, you'll see it rock irregularly back and forth in a seemingly unpredictable pattern. However, the complexity of this motion arises because the paper is affected by a variety of forces, including gravity and the changing forces caused by air currents. If you could analyze the forces individually, you'd find that each force affects the paper's motion in a simple, predictable way.

Galileo was the first to glean hints of the remarkable simplicity of motion. As we have seen, his many discoveries included learning that all falling objects accelerate at the same rate and that moving objects will not come to rest unless some force acts to stop them [Section 3.4]. However, many fundamental questions about motion remained unanswered at the time of Galileo's death in January 1642. The answers would come soon, though, because Isaac Newton (1642–1727) was born later that same year.

Newton Unified Earth and the Heavens

Newton was born prematurely in Lincolnshire, England, on Christmas day in 1642. His father, a farmer who never learned to read or write, died 3 months before he was born. Newton had a difficult childhood and showed few signs of unusual talent. He attended Trinity College at Cambridge, where he earned his keep by performing menial labor, such as cleaning the boots and bathrooms of wealthier students and waiting on their tables.

Shortly after he graduated, the plague hit Cambridge, and Newton returned home. By his own account, he experienced a moment of inspiration in 1666 when he saw an apple fall to the ground and suddenly understood that gravity is universal. In that moment, Newton shattered the remaining vestiges of the Aristotelian view of the world, which for centuries in Europe had been taken as near-gospel truth (see Special Topic: Aristotle, p. 66).

Aristotle's beliefs included many ideas about the physics of motion, which he had used to support the idea of an Earth-centered cosmos. Aristotle had also maintained that the heavens were totally distinct from Earth, so that physical laws on Earth did not apply to heavenly motion. By the time Newton saw the apple fall, the Copernican revolution had displaced Earth from a central position, and Galileo's experiments had shown that the laws of physics were not what Aristotle had believed.

Newton's sudden insight delivered the final

Sir Isaac Newton (1642–1727)

blow to Aristotle's physics. He realized that the force that brought the apple to the ground and the force that held the Moon in orbit were the same. With that insight, Newton eliminated the distinction between the Earth and the heavens, bringing both together in one *universe*. Newton's insight also heralded the birth of the modern science of *astrophysics* (although the term wasn't coined until much later). Astrophysics applies physical laws discovered on Earth to phenomena throughout the cosmos.

Over the next 20 years, Newton's work completely revolutionized mathematics and science. He quantified the laws of motion and gravity, publishing them in 1687 in a book usually known as *Principia*, short for *Philosophiae Naturalis Principia Mathematica* ("Mathematical Principles of Natural Philosophy"). He also conducted crucial experiments regarding the nature of light, built the first reflecting telescopes, and invented the branch of mathematics called calculus. The compendium of Newton's discoveries is so tremendous that it would take a complete book just to describe them, and many more books to describe their influence on civilization. When Newton died in 1727, at age 84, English poet Alexander Pope composed the following epitaph:

> *Nature, and Nature's laws lay hid in the Night.*
> *God said,* Let Newton be! *and all was Light.*

Next, we'll discuss **Newton's three laws of motion,** which describe how forces affect motion. In the following section, we'll discuss Newton's discoveries about gravity.

Newton's First Law of Motion

Newton's first law of motion restates Galileo's discovery that objects will remain in motion unless a force acts to stop them. We call it "Newton's first law" because it was the first of the three laws of motion that he enumerated in his book *Principia*. It can be stated as follows:

> **In the absence of a net (overall) force acting upon it, an object moves with constant velocity.**

Thus, objects at rest (velocity = 0) tend to remain at rest, and objects in motion tend to remain in motion with no change in either their speed or their direction.

The idea that an object at rest should remain at rest is rather obvious: A car parked on a flat street won't suddenly start moving for no reason. But what if the car is traveling along a flat, straight road? Newton's first law says that the car should keep going forever *unless* a force acts on it. You know that the car eventually will come to a stop if you take your foot off the gas pedal, so we must conclude that one or more forces are stopping the car—in this case forces arising from friction and air resistance.* If the car were in space, and therefore unaffected by friction or air, it would keep moving forever (though gravity would eventually alter

---

*Why doesn't gravity help stop the car? A force can affect motion only if it is acting along the direction of motion. Because gravity acts downward, it cannot affect the motion of a car traveling along a flat (level) road.

its speed and direction). That is why interplanetary space-craft, once launched into space, need no fuel to keep going.

Newton's first law also explains why you don't feel any sensation of motion when you're traveling in an airplane on a smooth flight. As long as the plane is traveling at constant velocity, no net force is acting on it or on you. Therefore, you feel no different from how you would feel at rest. You can walk around the cabin, play catch with a person a few rows forward, or relax and go to sleep just as though you were "at rest" on the ground.

Newton's Second Law of Motion

Newton's second law of motion tells us what happens to an object when a net force *is* present. We have already said that a net force changes an object's momentum, accelerating it in the direction of the force. Newton's second law quantifies this relationship, which can be stated in two equivalent ways:

force = rate of change in momentum

force = mass × acceleration (or $F = ma$)

This law explains why you can throw a baseball farther than you can throw a shot-put. For both the baseball and the shot-put, the force delivered by your arm equals the product of mass and acceleration. Because the mass of the shot-put is greater than that of the baseball, the same force from your arm gives the shot-put a smaller acceleration. Due to its smaller acceleration, the shot-put leaves your hand with less speed than the baseball and thus travels a shorter distance before hitting the ground.

We can also use Newton's second law of motion to understand acceleration around curves. Suppose you swing a ball on a string around your head (Figure 5.5a). The ball is accelerating even if it has a steady speed, because it is constantly changing direction. What makes it accelerate? According to Newton's second law, the taut string must be applying a force to the ball. We can understand this force by considering what happens when the string breaks and the force disappears (Figure 5.5b). In that case, the ball simply flies off in a straight line. Thus, when the string is intact, the force must be pulling the ball *inward* to keep it from flying off. Because acceleration must be in the same direction as the force, we conclude that the ball has an inward acceleration as it moves around the circle.

The same idea helps us understand the force on a car moving around a curve or a planet orbiting around the Sun. In the case of a car, the force comes from friction between the tires and the road. The tighter the curve or the faster the car is going, the greater the force needed to keep the car moving around it. If the force is not great enough, the car skids outward. For a planet moving around the Sun, the force pulling inward is gravity, which we'll discuss shortly. Thus, an orbiting planet is always accelerating toward the Sun. Indeed, it was Newton's discovery of the precise nature of this acceleration that helped him deduce the law of gravity.

Newton's Third Law of Motion

Think for a moment about standing still on the ground. The force of gravity acts downward on you, so if this force were acting alone, Newton's second law would demand that you be accelerating downward. The fact that you are not falling means that the ground must be pushing back up on you with exactly the right amount of force to offset gravity. This fact is embodied in Newton's third law of motion:

For any force, there always is an equal and opposite reaction force.

According to this law, your body exerts a gravitational force on the Earth identical to the one the Earth exerts on you, except that it acts in the opposite direction. In this mutual pull, the ground just happens to be caught in the middle. Because other forces (between atoms and molecules in the Earth) keep the ground stationary with respect to the

---

Mathematical Insight **5.1**   **Units of Force, Mass, and Weight**

Newton's second law, $F = ma$, shows that the units of force are equal to a unit of mass multiplied by a unit of acceleration. For example, if a mass of 1 kg accelerates at 10 m/s^2, the magnitude of the responsible force is:

$$\text{force} = \text{mass} \times \text{acceleration}$$

$$= 1 \text{ kg} \times 10 \, \frac{\text{m}}{\text{s}^2} = 10 \, \frac{\text{kg} \times \text{m}}{\text{s}^2}$$

$$= 10 \text{ newtons}$$

The standard unit of force is the *kilogram-meter per second squared*, called the **newton** for short.

We can now further clarify the difference between mass and weight. When you stand on a scale, it records the downward force that you exert on it, which is equal and opposite to the upward force it exerts on you. If the scale were suddenly pulled out from under your feet, you would begin accelerating downward with the acceleration of gravity. Thus, when you are standing still, the scale must support you with a force equal to your mass times the acceleration of gravity. Your weight must also equal this force (but in an opposite direction):

$$\text{weight} = \text{mass} \times \text{acceleration of gravity}$$

Your apparent weight may differ from this value if forces besides gravity are affecting you.

Like any force, weight has units of mass times acceleration. Thus, although we commonly speak of weights in *kilograms*, this usage is not technically correct: Kilograms are a unit of mass, not of force. You may safely ignore this technicality as long as you are dealing with objects on Earth that are not accelerating. In space or on other planets, the distinction between mass and weight is important and cannot be ignored.

Figure 5.5 Newton's second law of motion tells us that an object going around a curve has an acceleration pointing toward the inside of the curve.

a When you swing a ball on a string, the ball moves in a circle because the string exerts a force that pulls the ball inward. The acceleration must also be inward.

b If the string breaks, the force disappears and the ball flies off in a straight line with velocity *v*.

What Makes a Rocket Launch?

If you've ever watched a rocket launch, it's easy to see why many people believe that the rocket "pushes off" the ground. In fact, the ground has nothing to do with the rocket launch. The rocket takes off because of momentum conservation. Rocket engines are designed to expel hot gas with an enormous amount of momentum. To balance the explosive force driving gas out the back of the rocket, an equal and opposite force must propel the rocket forward, keeping the total momentum—gas plus rocket—unchanged. Thus, rockets can be launched horizontally as well as vertically, and a rocket can be "launched" in space (e.g., from a space station) with no need for any nearby solid ground.

center of the Earth, the opposite, upward force is transmitted to you by the ground, holding you in place. Newton's third law also explains how rockets work: Engines generate an explosive force driving hot gas out the back, which creates an equal and opposite force propelling the rocket forward.

Figure 5.6 summarizes Newton's three laws of motion.

Conservation of Momentum

If you look more closely at Newton's laws, you can see that they all reflect aspects of a deeper principle: the *conservation of momentum*. Like the amount of energy, the total amount of momentum in the universe is conserved—that is, it does not change.

● Newton's first law says that an individual object's momentum will not change at all if the object is left alone.

● Newton's second law says that a force can change the object's momentum, but . . .

Figure 5.6 Newton's three laws of motion.

A baseball accelerates as the pitcher applies a force by moving his arm. (Once released, this force and acceleration cease, so the ball's path changes only due to gravity and effects of air resistance.)

A spaceship needs no fuel to keep moving in space.

a Newton's first law of motion: An object will remain in motion at constant velocity unless a force acts to change its speed or direction.

b Newton's second law of motion: force = mass × acceleration.

A rocket is propelled upward by a force equal and opposite to the force with which gas is expelled out its back.

c Newton's third law of motion: For any force, there is always an equal and opposite reaction force.

- Newton's third law says that another equal and opposite force simultaneously changes some other object's momentum by a precisely opposite amount.

Notice that total momentum always remains unchanged.

On a pool table, momentum conservation is rather obvious. When one ball hits another ball "dead on," the first ball stops and the second one takes off with the speed of the first (Figure 5.7). Sometimes, momentum conservation is less apparent. When you jump into the air, how do you get your upward momentum? As your legs propel you skyward, they are actually pushing Earth in the other direction, giving Earth's momentum an equal and opposite kick. However, Earth's huge mass renders its acceleration undetectable. During your brief flight, the gravitational force between you and Earth pulls you back down, transferring your momentum back to Earth. Again, the total momentum remains the same at all times.

Conservation of Angular Momentum

In astronomy, a special kind of momentum is particularly important. Consider an ice skater spinning in place. She certainly has some kind of momentum, but it is a little different from the momentum we've discussed previously because she's not actually going anywhere. Her "spinning momentum" is called **angular momentum**. (The term *angular* arises because each spin involves turning through an *angle* of 360°.)

Any object that is spinning or orbiting has angular momentum. We can write a simple formula for the angular momentum of an object moving in a circle:

$$\text{angular momentum} = m \times v \times r$$

where m is the object's mass, v is its speed around the circle, and r is the radius of the circle (Figure 5.8).

Just as momentum can be changed only by a force, the angular momentum of any object can be changed only by a twisting force, or **torque**. For example, opening a door requires rotating the door on its hinges. Making it rotate means giving it some angular momentum, which you can do by applying a torque (Figure 5.9). The amount of torque depends not only on how much force you use to push on the door, but also on *where* you push. The farther out from the hinges you push, the more torque you can apply, and the easier it is to open the door.

THINK ABOUT IT

Use the idea of torque to explain why it is easier to change a tire with a long wrench than with a short wrench.

When no net torque is present, a law very similar to Newton's first law applies—the **law of conservation of angular momentum**:

In the absence of net torque (twisting force), the total angular momentum of a system remains constant.

A spinning ice skater illustrates this law. Because there is so little friction on ice, the ice skater essentially keeps a con-

Figure 5.7 Momentum conservation as demonstrated on a pool table.

Figure 5.8 The angular momentum of an object moving in a circle is $m \times v \times r$.

stant angular momentum. When she pulls in her extended arms, she effectively decreases her radius. Therefore, for the product $m \times v \times r$ to remain unchanged, her velocity of rotation must increase (Figure 5.10).

The law of conservation of angular momentum arises frequently in astronomy, because many objects spin or orbit without being affected by any significant torque. For example, angular momentum is generally conserved for rotating planets, planets orbiting a star, and rotating galaxies. In fact, conservation of angular momentum explains why a planet doesn't need any fuel to keep it rotating: In the absence of any torque to slow its rotation, the planet would keep rotating at the same rate forever.

Figure 5.9 Opening a door requires applying a torque. Given the same amount of force, the torque on the door is greater if you push farther from the hinges (the door's rotation axis).

low torque

moderate torque

most torque

In the product $m \times v \times r$, extended arms mean larger radius and smaller velocity of rotation.

Bringing in her arms decreases her radius and therefore increases her rotational velocity.

Figure 5.10 A spinning skater conserves angular momentum.

 Orbits and Kepler's Laws Tutorial, Lessons 1–4

5.3 The Force of Gravity

Newton's three laws of motion apply generally to *all* types of motion. They tell us how motion depends on *any* force, regardless of the source of the force. Gravity is the major force at work on the astronomical scale, and Newton himself discovered how gravity works.

The Universal Law of Gravitation

Newton described the force of gravity mathematically in what we now call the **universal law of gravitation**. Three simple statements summarize this law:

- Every mass attracts every other mass through the force called *gravity*.

- The force of attraction between any two objects is *directly proportional* to the product of their masses. For example, doubling the mass of *one* object doubles the force of gravity between the two objects.

- The force of attraction between two objects decreases with the *square* of the distance between their centers. That is, the force follows an **inverse square law** with distance. For example, doubling the distance between two objects weakens the force of gravity by a factor of 2^2, or 4.

 THINK ABOUT IT

How does the gravitational force between two objects change if the distance between them triples? If the distance between them drops in half?

Mathematically, Newton's universal law of gravitation is written:

$$F_g = G \frac{M_1 M_2}{d^2}$$

where F_g is the force of gravitational attraction, M_1 and M_2 are the masses of the two objects, and d is the distance between their *centers* (Figure 5.11). The symbol G is a constant called the **gravitational constant**. Its numerical value was not known to Newton but has since been measured by experiments to be $G = 6.67 \times 10^{-11}$ m³/(kg × s²). Throughout the rest of this book, we will see many examples of the universal law of gravitation in action.

$$F_g = G \frac{M_1 M_2}{d^2}$$

Figure 5.11 The universal law of gravitation. It is an *inverse square law* because the force declines with the square of the distance *d*. For example, doubling the distance *d* weakens the force by a factor of $2^2 = 4$.

The "Why" of Kepler's Laws

For some 70 years after Kepler published his laws of planetary motion [Section 3.4], scientists debated *why* these laws hold true. Newton solved the mystery, showing that Kepler's laws are consequences of the laws of motion and the universal law of gravitation.

Let's begin with Kepler's first two laws of planetary motion (see Figures 3.19 and 3.20). Kepler's first law states that planetary orbits are ellipses with the Sun at one focus, and his second says that a planet moves faster when it is nearer the Sun and slower when it is farther from the Sun. Newton showed that these behaviors are mathematical consequences of his laws.

We can understand why by thinking about conservation of angular momentum. The angular momentum of an orbiting planet is $m \times v \times r$, where *m* is the mass of the planet, *v* is its orbital speed, and *r* is its distance ("radius") from the Sun. Conservation of angular momentum means that the product $m \times v \times r$ always stays the same. A planet on a circular orbit can therefore keep orbiting forever with no need for fuel. The distance from the Sun (*r*) is always the same on a circular orbit, as is the planet's mass (*m*). Thus, the orbital speed (*v*) must also stay constant to conserve angular momentum. This means the planet keeps orbiting as long as no other force disturbs it. Ellipses can work just as well as circles for conserving angular momentum, because the product $m \times v \times r$ will remain constant as long as the orbital speed (*v*) goes up when the distance from the Sun (*r*) goes down, and vice versa. That is exactly what Kepler's second law says. As long as no other force acts to change the planet's orbit, it will continue on its elliptical path forever.

We can understand Kepler's third law—that average orbital speed is higher for planets close to the Sun and lower for planets far from the Sun—by thinking about how gravity affects the speed of an orbiting object. Remember that an object can orbit Earth only if it is moving fast enough to continually "fall around" Earth (see Figure 5.4). In a similar way, an orbiting planet is continually "falling around" the Sun, with its orbital speed determined by the strength of the gravitational force trying to pull it into the Sun. Planets closer to the Sun feel a stronger gravitational pull and therefore must move at a faster average speed to maintain their "falling around" orbits. This is just what Kepler's third law tells us (see Figure 3.21).

Newton's explanation of Kepler's laws sealed the triumph of the Copernican revolution. Prior to Newton, it was still possible to see Kepler's model of planetary motion as "just" another model, albeit one that fit the observational data far better than any previous model. By explaining Kepler's laws in terms of basic laws of physics, Newton removed virtually all remaining doubt about the legitimacy of the Sun-centered solar system.

Generalizing Kepler's Laws

Newton's work did much more than explain the orbits of planets in our solar system. Newton showed that the physics underlying the Copernican system could be used to explain the structure of the cosmos. He did this by extending Kepler's laws in three crucial ways:

- *Newton generalized Kepler's first two laws of planetary motion to apply to all orbiting objects.* For example, the orbits of a satellite around Earth, of a moon around a planet, and of an asteroid around the Sun are all ellipses in which the orbiting object moves faster at the nearer points in its orbit and slower at the farther points. Moreover, Newton showed that two objects attracted by gravity actually both orbit a point between them. Although this is true of all pairs of orbiting objects, it is easiest to see with binary star systems: Each of the two stars orbits the other star on an elliptical path. As seen from afar, this means that two orbiting objects both move around their **center of mass**—the point at which the two objects would balance if they were somehow connected (Figure 5.12). We will see in Chapter 9 how this fact is being used to discover planets around other stars.

- *Newton found that ellipses are not the only possible orbital paths* (Figure 5.13). Kepler was right when he found that ellipses (which include circles) are the only possible shapes for **bound orbits**—orbits in which an object goes around another object over and over again. (The term *bound orbit* comes from the idea that gravity creates a *bond* that holds the objects together.) However, Newton discovered that objects can also follow **unbound orbits**—paths that bring an object close to another object just once. For example, many comets that enter the inner solar system follow unbound orbits. They come in from afar just once, loop around the Sun, and never return. More specifically, Newton showed that the allowed orbital paths are ellipses, parabolas, and hyperbolas—which together are known as the "conic sections," because they can be made by slicing through a cone at different angles.

- *Newton found that Kepler's third law could be generalized in a way that enables us to calculate the masses of distant objects.* The precise statement of Kepler's third law is $p^2 = a^3$, where *p* is a planet's orbital period in years and *a* is the planet's average distance from the Sun in AU. Newton found that this statement is actually a special case of a more general equation that we call **Newton's version of Kepler's third law** (see Mathematical Insight 5.2). This more general equation allows us to measure orbital period and distance in any units we wish (rather than only in years and AU, respectively) and also shows that the relation-ship between the orbital period and the average distance depends on the masses of the orbiting objects. Thus, measuring the orbital period and average distance of an orbiting object enables us to calculate the mass of the object it orbits.

Mathematical Insight   **5.2**   **Using Newton's Version of Kepler's Third Law**

In its original form, Kepler's third law reads $p^2 = a^3$, where p is a planet's orbital period in years and a is the planet's average distance from the Sun in AU. By working with his equations of motion and the universal law of gravitation, Newton found that Kepler's law is only a special case of a more general law. Newton's version of Kepler's third law reads as follows:

$$p^2 = \frac{4\pi^2}{G(M_1 + M_2)}a^3$$

The term $4\pi^2$ is simply a number ($4\pi^2 \approx 4 \times 3.14^2 = 39.44$), and G is the gravitational constant. As before, p and a are the orbital period and distance, respectively, of one object orbiting another (such as of a planet orbiting the Sun), with the distance measured from the center of one object to the center of the other. The terms M_1 and M_2 are the masses of the two objects.

For example, for a planet orbiting the Sun, M_1 and M_2 are the masses of the planet and the Sun, and for two stars orbiting each other in a binary star system, M_1 and M_2 are the masses of the two individual stars. Any units for p and a can be used in Newton's equation, as long as the units are consistent with the units used for the gravitational constant G.

Newton's version of Kepler's third law gives us the power to measure the masses of distant objects. Any time we measure the orbital period and distance of an orbiting object, we can use Newton's equation to calculate the sum $M_1 + M_2$ of the two objects involved in the orbit. If one object is much more massive than the other, we essentially learn the mass of the massive object. For example, in the case of a planet orbiting the Sun, the sum $M_{Sun} + M_{planet}$ is pretty much just M_{Sun} because the Sun is so much more massive than any of the planets. Thus, knowing the orbital period and distance from the Sun of any planet allows us to calculate the mass of the Sun.

The following examples show some of the remarkable power of Newton's version of Kepler's third law.

Example 1: Use the fact that Earth orbits the Sun in 1 year at an average distance of 150 million km (1 AU) to calculate the mass of the Sun.

Solution: For Earth orbiting the Sun, Newton's version of Kepler's third law takes the form:

$$(p_{earth})^2 = \frac{4\pi^2}{G(M_{Sun} \times M_{Earth})}(a_{Earth})^3$$

However, because the Sun is so much more massive than Earth, the sum of their masses is approximately the mass of the Sun alone: $M_{Sun} + M_{Earth} \approx M_{Sun}$. Using this approximation, the equation becomes:

$$(p_{earth})^2 \approx \frac{4\pi^2}{G \times M_{Sun}}(a_{Earth})^3$$

Because we already know Earth's orbital period (p_{Earth}) and distance (a_{Earth}), this equation contains only one unknown: the Sun's mass (M_{Sun}). We can therefore find the Sun's mass by solving the equation for this one unknown. We multiply both sides by M_{Sun} and divide both sides by $(p_{Earth})^2$:

$$M_{Sun} \approx \frac{4\pi^2}{G}\frac{(a_{Earth})^3}{(p_{Earth})^2}$$

We now plug in the known values. Earth's orbital period is $p_{Earth} = 1$ year, which is the same as 3.15×10^7 seconds. Its average distance from the Sun is $a_{Earth} \approx 150$ million km, or 1.5×10^{11} m. Using these values and the experimentally measured value $G = 6.67 \times 10^{-11}$ m^3/(kg $\times$ s^2), we find:

$$M_{Sun} \approx \frac{4\pi^2}{\left(6.67 \times 10^{-11} \dfrac{m^3}{kg \times s^2}\right)}\frac{(1.5 \times 10^{11} \ m)^3}{(3.15 \times 10^7 \ s)^2}$$

$$= 2 \times 10^{30} \ kg$$

The mass of the Sun is about 2×10^{30} kg. Simply by knowing Earth's orbital period and distance from the Sun and the gravitational constant G, we have used Newton's version of Kepler's third law to "weigh" the Sun.

Example 2: A *geosynchronous satellite* orbits Earth in the same amount of time that Earth rotates: 1 sidereal day. If a geosynchronous satellite is also in an equatorial orbit, it is said to be *geostationary* because it remains fixed in the sky (i.e., it maintains a constant altitude and direction) as seen from the ground (see problem 21). Calculate the orbital distance of a geosynchronous satellite.

Solution: A satellite is much less massive than Earth, so $M_{Earth} + M_{satellite} \approx M_{Earth}$ and we can use Newton's version of Kepler's third law in the form:

$$(p_{satellite})^2 \approx \frac{4\pi^2}{G + M_{Earth}}(a_{satellite})^3$$

Because we want to know the satellite's distance, we solve for $a_{satellite}$ by multiplying both sides of the equation by $(G \times M_{Earth})$, dividing both sides by $4\pi^2$, and then taking the cube root of both sides:

$$a_{satellite} \approx \sqrt[3]{\frac{G \times M_{Earth}}{4\pi^2}(p_{satellite})^2}$$

We know that $p_{satellite} = 1$ sidereal day $\approx 86,164$ seconds. You should confirm that substituting this value and the mass of Earth yields $a_{satellite} \approx 42,000$ km. Thus, a geosynchronous satellite orbits at a distance of 42,000 km above the *center* of the Earth, which is about 35,600 km above Earth's surface.

Two Stars of Equal Mass

center
of mass

star 1   star 2

a For two stars of the same mass, the center of mass lies halfway between them.

Star 1 Is More Massive Than Star 2

center
of mass

star 1   star 2

b For two stars with different masses, the center of mass lies closer to the more massive one than to the less massive one.

Sun Is Much More Massive Than Planet

center
of mass

Sun   planet

c For a planet orbiting the Sun, the center of mass of the Sun and planet may lie *inside* the Sun—which is why we generally don't notice the Sun's own orbital motion around this center of mass. However, the motion does exist, and this type of motion has allowed us to discover planets around other stars [Section 9.6].

Figure 5.12 The idea of *center of mass* for a pair of orbiting objects. If we imagine the two objects to be connected by an invisible rod (which has no mass of its own), the center of mass is the point at which a fulcrum could balance the rod.

For example, Newton's version of Kepler's third law allows us to calculate the mass of the Sun from Earth's orbital period (1 year) and its average distance from the Sun. Similarly, measuring the orbital period and distance of one of Jupiter's moons allows us to calculate Jupiter's mass, and measuring the orbital periods and distances of stars in a binary system can allow us to determine their masses. Thus, Newton's version of Kepler's third law provides the primary means by which we determine masses throughout the universe.

Newton's version of Kepler's third law also explains another important characteristic of orbital motion. It shows that the orbital period of a *small* object orbiting a much more massive object depends only on its orbital distance, not on its mass. That is why an astronaut does not need a tether to stay close to the Space Shuttle or the Space Station during a space walk (Figure 5.14). Even

Figure 5.14 ▶ Newton's version of Kepler's third law enables us to measure masses throughout the universe. In addition, it shows that when one object orbits a much more massive object, the orbital period depends only on its average orbital distance. Thus, the astronaut and the Space Shuttle share the same orbit despite their different masses—even as both orbit Earth at a speed of some 25,000 km/hr.

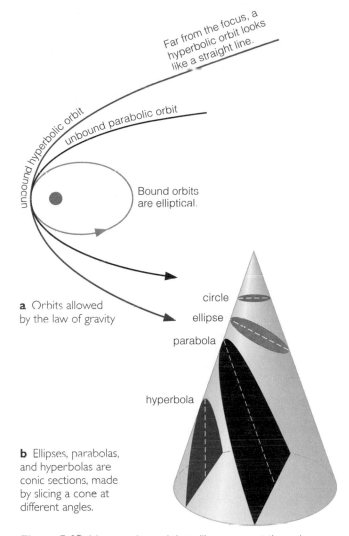

Far from the focus, a hyperbolic orbit looks like a straight line.

unbound hyperbolic orbit

unbound parabolic orbit

Bound orbits are elliptical.

a Orbits allowed by the law of gravity

circle

ellipse

parabola

hyperbola

b Ellipses, parabolas, and hyperbolas are conic sections, made by slicing a cone at different angles.

Figure 5.13 Newton showed that ellipses are not the only possible orbital paths. Orbits can also be unbound parabolas and hyperbolas.

though the spacecraft is much bigger than the astronaut, both are much smaller than Earth and thus stay together because they have the same orbital distance and hence the same orbital period.

Newton's universal law of gravitation has applications that go far beyond explaining Kepler's laws. In the rest of this chapter, we'll explore three important concepts that we can understand with the help of the universal law of gravitation: tides, orbital energy and escape velocity, and the acceleration of gravity.

5.4 Tides

If you've spent time near an ocean, you're probably aware of the rising and falling of the tide twice each day. What causes the tides, and why are there two each day?

We can understand the basic idea by examining the gravitational attraction between Earth and the Moon. Gravity attracts Earth and the Moon toward each other (with the Moon staying in orbit as it "falls around" Earth), but it affects different parts of Earth slightly differently: Because the strength of gravity declines with distance, the side of Earth facing the Moon feels a slightly stronger gravitational attraction than the side facing away from the Moon. As shown in Figure 5.15, this creates two tidal bulges, one facing the Moon and one opposite the Moon. Earth's rotation carries your location through each of

tidal bulge opposite Moon

tidal bulge toward Moon

Not to scale! The real tidal bulge raises the oceans by only about 2 meters.

Figure 5.15 Tidal bulges face toward and away from the Moon because of the difference in the strength of the gravitational attraction in parts of Earth at different distances from the Moon. (Arrows represent the strength and direction of the gravitational attraction toward the Moon.) There are two daily high tides as any location on Earth rotates through the two tidal bulges.

the two bulges each day, creating two high tides. Low tides occur when your location is at the points halfway between the two tidal bulges.

In a simple sense, the reason there are *two* daily high tides is that the oceans facing the Moon bulge because they are being pulled out from the Earth, while the oceans opposite the Moon bulge because the Earth is being pulled out from under them. However, a better way to look at tides is to recognize that the attraction toward the Moon gets progressively weaker with distance throughout the Earth. This varying attraction creates a "stretching force," or **tidal force**, that stretches the *entire* Earth—land and ocean—along the Earth–Moon line. The tidal bulges are more noticeable for the oceans than for the land only because liquid water flows more readily than solid rock.

The two "daily" high tides come slightly more than 12 hours apart. Because the Moon orbits Earth while Earth rotates, the Moon is at its highest point (i.e., on the meridian) at any location about every 24 hours 50 minutes, rather than every 24 hours. Thus, the tidal cycle of two high tides and two low tides actually takes about 24 hours 50 minutes, with each high tide occurring about 12 hours 25 minutes after the previous one.

The height and timing of tides can vary considerably from place to place around the Earth, depending on factors such as latitude, the orientation of the coastline (e.g., north-facing or west-facing), and the depth and shape of any channel through which the rising tide must flow. For example, while the tide rises gradually in most locations, the incoming tide near the famous abbey on Mont-Saint-Michel, France, moves much faster than a person can swim (Figure 5.16). In centuries past, the Mont was an island twice a day at high tide but was connected to the mainland at low tide. (Today, a man-made land bridge keeps the island connected to the mainland.) Many pilgrims drowned when they were caught unprepared by the inrushing tide. Another unusual tidal pattern occurs in coastal states along the northern shore of the Gulf of Mexico. There, topography

The Origin of Tides

Many people believe that tides arise because the Moon pulls Earth's oceans toward it. But if that were the whole story, there would be a bulge only on the side of Earth facing the Moon, and hence only one high tide each day. The correct explanation for tides must account for why Earth has *two* tidal bulges.

Only one explanation works: Earth must be stretching from its center in both directions (toward and away from the Moon). Once you see this, it becomes clear that tides must come from the *difference* between the force of gravity on one side of Earth and that on the other, since a difference makes Earth stretch. In fact, stretching due to tides affects many objects, not just Earth. Many moons are stretched into oblong shapes by tidal forces caused by their parent planets, and mutual tidal forces stretch close binary stars into teardrop shapes. In regions where gravity is extremely strong, such as near a black hole, tides could even stretch spaceships or people [Section 18.4].

Figure 5.16 Photographs of high and low tide at the abbey at Mont-Saint-Michel, France, one of the world's most popular tourist destinations. Here the tide rushes in much faster than a person can swim. Before a causeway was built (visible to the left, with cars on it), the Mont was accessible by land only at low tide. At high tide, it became an island.

and other factors combine to make only one noticeable high tide and low tide each day.

Spring and Neap Tides

The Sun also exerts a tidal force on Earth, causing Earth to stretch along the Sun–Earth line. You might at first guess that the Sun's tidal force would be more than the Moon's, since the Sun's mass is more than a million times that of the Moon. Indeed, the *gravitational* force between Earth and the Sun is much greater than that between Earth and the Moon, which is why Earth orbits the Sun. However, the much greater distance to the Sun (than to the Moon) means that the *difference* in the Sun's pull on the near and far sides of Earth is relatively small, and the overall tidal force caused by the Sun is only about one-third that caused by the Moon (Figure 5.17).

When the tidal forces of the Sun and the Moon work together, as is the case at both new moon and full moon, we get the especially pronounced *spring tides* (so named because the water tends to "spring up" from the Earth). When the tidal forces of the Sun and the Moon oppose each other, as is the case at first- and third-quarter moon, we get the relatively small tides known as *neap tides*.

THINK ABOUT IT

Explain why any tidal effects (on Earth) caused by the other planets would be extremely small (in fact, so small as to be unnoticeable).

neap tides

spring tides

a At new moon and full moon, the tidal forces from both the Moon and the Sun stretch Earth along the same line, leading to enhanced *spring tides*.

Figure 5.17 Tides on Earth also depend on tidal force from the Sun, which is about one-third as strong as that from the Moon. In these diagrams, yellow arrows represent tidal force due to the Sun, which causes Earth to stretch along the Sun–Earth line, and black arrows represent tidal force due to the Moon, which causes Earth to stretch along the Earth–Moon line.

b At first- and third-quarter moon, the Sun's tidal force stretches Earth along a line perpendicular to the Moon's tidal force. The tides still follow the Moon, because the Moon's tidal force is greater, but they are reduced in size, making *neap tides*.

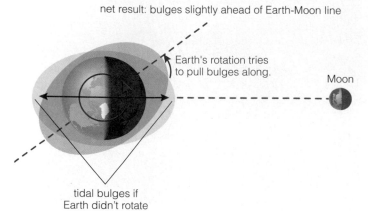

net result: bulges slightly ahead of Earth-Moon line

Earth's rotation tries to pull bulges along.

Moon

tidal bulges if Earth didn't rotate

Figure 5.18 If Earth always kept the same face to the Moon, the tidal bulges would stay fixed on the Earth–Moon line. But Earth's rotation tries to pull them along with it. The result is that the bulges stay nearly fixed relative to the Moon but in a position slightly ahead of the Earth–Moon line. (The effect is exaggerated here for clarity.)

Tidal Friction

So far, we have talked as if Earth slides smoothly through the tidal bulges as it rotates. But because tidal forces stretch the Earth itself, the process involves some friction, called **tidal friction**. In essence, the friction arises because the tidal bulges try to stay on the Earth–Moon line, while Earth's rotation tries to pull the bulges around with it. The resulting "compromise" puts the bulges just ahead of the Earth–Moon line at all times (Figure 5.18), ensuring that Earth feels continuous friction as it rotates through the tidal bulges.

This tidal friction has two important effects. First, it causes Earth's rotation to slow gradually, so that the length of a day gradually gets longer. Second, it makes the Moon move gradually farther from Earth: The slight excess mass in Earth's tidal bulge exerts a gravitational attraction that tends to pull the Moon slightly ahead in its orbit. This pulling ahead makes it harder for Earth's overall gravity to hold on to the Moon, and as a result the Moon moves slightly farther from Earth.

These two effects are barely noticeable on human time scales. For example, tidal friction increases the length of a day by only about 1 second every 50,000 years. (On shorter time scales, the length of the day fluctuates by up to a second or more per year due to slight changes in Earth's internal mass distribution, which is why "leap seconds" are occasionally added to or subtracted from the year.) But the effects add up over billions of years. Early in Earth's history, a day may have been only 5 or 6 hours long and the Moon may have been one-tenth or less its current distance from Earth.

These changes in Earth's rotation and the Moon's orbit provide a remarkable example of conservation of angular momentum. The amount of angular momentum Earth loses as its rotation slows is precisely the same as the amount the Moon gains through its growing orbit.

Synchronous Rotation

Tidal friction has had even more dramatic effects on the Moon. Recall that the Moon always shows (nearly) the same face to Earth [Section 2.5]. This trait is called **synchronous rotation**, because it means that the Moon's rotation period and orbital period are the same (see Figure 2.22). Synchronous rotation may seem like an extraordinary coincidence, but it is a natural consequence of tidal friction.

The Moon probably once rotated much faster than it does today. But just as the Moon exerts a tidal force on Earth, the Earth exerts a tidal force on the Moon. In fact, because of its larger mass, Earth exerts a greater tidal force on the Moon than vice versa. This tidal force stretches the Moon along the Earth–Moon line, creating two tidal bulges similar to those on Earth. As long as the Moon rotated through these bulges, the motion created tidal friction that slowed the Moon's rotation. But once the Moon's rotation slowed to the point at which the Moon and its bulges rotated at the same rate—that is, synchronously with the orbital period—there was no further source for tidal friction. The Moon has stayed in synchronous rotation ever since, with its two tidal bulges permanently fixed along the Earth–Moon line. (The Moon's diameter is not greater on this line, so we cannot see any outward sign of the Moon's tidal bulges, but the Moon does have mass concentrations along the line of the tidal bulges.)

Tidal friction has led to synchronous rotation in many other cases. Most of the moons of the jovian planets rotate

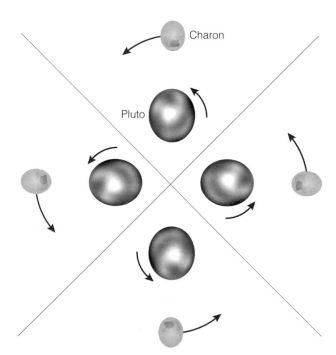

Charon

Pluto

Figure 5.19 Pluto and Charon rotate synchronously with each other, so that each always shows the same face to the other. If you stood on Pluto, Charon would remain stationary in your sky, always showing the same face (but going through phases like the phases of our Moon). Similarly, if you stood on Charon, Pluto would remain stationary in your sky, always showing the same face (and going through phases). Tidal bulges are exaggerated in this figure.

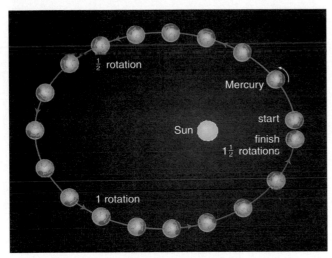

Figure 5.20 These "snapshots" of Mercury show that it rotates exactly one and a half times per orbit, or three times for every two orbits. The rotation pattern ensures that the tidal bulges are always aligned with the Sun at perihelion, when the tidal forces are strongest. The orbital eccentricity is also exaggerated.

synchronously. For example, Jupiter's four large moons (Io, Europa, Ganymede, and Callisto) keep nearly the same face toward Jupiter at all times. Pluto and its moon Charon *both* rotate synchronously: Like two dancers, they always keep the same face toward each other (Figure 5.19). If Earth and the Moon were to stay together, tidal friction would eventually (in a few hundred billion years) create a similar situation: Earth would always show the same face to the Moon.

Some moons and planets exhibit variations on synchronous rotation. For example, Mercury rotates exactly three times for every two orbits of the Sun (Figure 5.20). This pattern ensures that Mercury's tidal bulge always aligns with the Sun at perihelion, where the Sun exerts its strongest tidal force. As you study astronomy, you will encounter many more cases where tides and tidal friction play important roles.

5.5 Orbital Energy and Escape Velocity

Consider a satellite in an elliptical orbit around Earth. Its gravitational potential energy is greatest when it is farthest from Earth, and smallest when it is nearest Earth. Conversely, its kinetic energy is greatest when it is nearest Earth and moving fastest in its orbit, and smallest when it is farthest from Earth and moving slowest in its orbit. Throughout its orbit, its total **orbital energy**—the sum of its kinetic and gravitational potential energies—must be conserved.

Because any change in its orbit would mean a change in its total orbital energy, a satellite's orbit around Earth cannot change if it is left completely undisturbed. If the

satellite's orbit *does* change, it must somehow have gained or lost energy. For a satellite in low-Earth orbit, Earth's thin upper atmosphere exerts a bit of drag that can cause the satellite to lose energy and eventually plummet back to Earth. The satellite's lost orbital energy is converted to thermal energy in the atmosphere, which is why a falling satellite usually burns up. Raising a satellite to a higher orbit requires that it gain energy by firing one of its rockets. The chemical potential energy of the rocket fuel is converted to gravitational potential energy as the satellite moves to a higher orbit.

Generalizing from the satellite example shows that conservation of energy has a very important implication for motion throughout the cosmos: *Orbits cannot change spontaneously.* For example, an asteroid or a comet passing near a planet cannot spontaneously be "sucked in" to crash on the planet. It can hit the planet only if its current orbit already intersects the planet's surface or if it somehow gains or loses orbital energy so that its new orbit intersects the planet's surface. Of course, if the asteroid or comet *gains* energy, something else must *lose* exactly the same amount of energy, and vice versa.

One way two objects can exchange orbital energy is through a **gravitational encounter**, in which they pass near enough so that each can feel the effects of the other's gravity. For example, Figure 5.21 shows a gravitational encounter between Jupiter and a comet headed toward the Sun on an unbound orbit. The comet's close passage by Jupiter allows the comet and Jupiter to exchange energy: The comet loses orbital energy and changes to a bound, elliptical orbit, and Jupiter must gain the energy the comet loses. However, because Jupiter is so much more massive than the comet, the effect on Jupiter is unnoticeable.

More generally, when two objects exchange orbital energy, we expect one to lose energy and fall to a lower orbit while the other gains energy and is thrown to a higher orbit. If an object gains enough energy, it may end up in

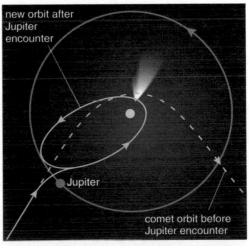

Figure 5.21 Depiction of a comet in an unbound orbit of the Sun that happens to pass near Jupiter. The comet loses orbital energy to Jupiter, thereby changing to a bound orbit around the Sun.

an unbound orbit that allows it to *escape* from the gravitational influence of the object it is orbiting. For example, if we want to send a space probe to Mars, we must use a large rocket that gives the probe enough energy to achieve an unbound orbit (relative to Earth) and ultimately escape Earth's gravitational influence.

Although it would probably make more sense to say that the probe achieves "escape energy," we instead say that it achieves *escape velocity* (see Figure 5.4). For example, the escape velocity from Earth's surface is about 40,000 km/hr, or 11 km/s, meaning that this is the minimum velocity required to escape Earth's gravity if you start near the surface. The escape velocity does not depend on the mass of the escaping object—*any* object must travel at a velocity of 11 km/s to escape from Earth, whether it is an individual atom or molecule escaping from the atmosphere, a spacecraft being launched into deep space, or a rock blasted into the sky by a large impact. Escape velocity *does* depend on whether you start from the surface or from someplace high above the surface. Because gravity weakens with distance, it takes less energy—and hence a lower escape velocity—to escape from a point high above Earth than from Earth's surface.

5.6 The Acceleration of Gravity

Throughout the remainder of the text, we will see many more applications of the universal law of gravitation. For now, let's look at just one more: Galileo's discovery that the acceleration of a falling object is independent of its mass.

If you drop a rock, the force acting on the rock is the force of gravity. The two masses involved are the mass of Earth and the mass of the rock, denoted M_{Earth} and M_{rock}, respectively. The distance between their *centers* is the distance from the *center of Earth* to the center of the rock. If the rock isn't too far above Earth's surface, this distance is approximately the radius of Earth, R_{Earth} (about 6,400 km). That is, $d \approx R_{Earth}$. Thus, the force of gravity acting on the rock is:

$$F_g = G\frac{M_{Earth}M_{rock}}{d^2} \approx G\frac{M_{Earth}M_{rock}}{(R_{Earth})^2}$$

According to Newton's second law of motion ($F = ma$), this force is equal to the product of the mass and the acceleration of the rock. That is:

$$G\frac{M_{Earth}\cancel{M_{Rock}}}{(R_{Earth})^2} = \cancel{M_{Rock}}a_{rock}$$

Note that M_{rock} "cancels" because it appears on both sides of the equation (as a multiplier), giving Galileo's result that the acceleration of the rock—or of any falling object—does not depend on the object's mass.

The fact that objects of different mass fall with the same acceleration struck Newton as an astounding coincidence, even though his own equations showed it to be so. For the next 240 years, this seemingly odd coincidence

---

Mathematical Insight **5.3** **Calculating the Escape Velocity**

A simple formula allows us to calculate the escape velocity from any planet, moon, or star:

$$v_{escape} = \sqrt{\frac{2 \times G \times M}{R}}$$

where *M* is the object's mass, *R* is the starting distance above the object's center, and *G* is the gravitational constant. If you use this formula to calculate the escape velocity from an object's surface, replace *R* with the object's radius.

Example 1: Calculate the escape velocity from the Moon. Compare it to that from Earth.

Solution: The mass and radius of the Moon are, respectively, $M = 7.4 \times 10^{22}$ kg and $R = 1.7 \times 10^6$ m. Plugging these numbers into the escape-velocity formula, we find:

$$v_{escape} = \sqrt{\frac{2 \times \left(6.67 \times 10^{-11}\ \frac{m^3}{kg \times s^2}\right) \times \left(7.4 \times 10^{22}\ kg\right)}{1.7 \times 10^6\ m}}$$

$$\approx 2{,}400\ m/s = 2.4\ km/s$$

The escape velocity from the Moon is 2.4 km/s, or less than one-fourth the 11-km/s escape velocity from Earth.

Example 2: Suppose a future space station orbits Earth in geosynchronous orbit, which is 42,000 km above the center of Earth (see Mathematical Insight 5.2). At what velocity must a spacecraft be launched from the station to escape Earth? Is there any advantage to launching from the space station instead of from Earth's surface?

Solution: We use the escape-velocity formula with the mass of Earth ($M_{Earth} = 6.0 \times 10^{24}$ kg) and the distance of the orbit above the center of Earth ($R = 42{,}000$ km $= 4.2 \times 10^7$ m):

$$v_{escape} = \sqrt{\frac{2 \times \left(6.67 \times 10^{-11}\ \frac{m^3}{kg \times s^2}\right) \times \left(6.0 \times 10^{24}\ kg\right)}{4.2 \times 10^7\ m}}$$

$$= 4{,}400\ m/s = 4.4\ km/s$$

The escape velocity from geosynchronous orbit is 4.4 km/s—considerably lower than the 11-km/s escape velocity from Earth's surface. Thus, it requires substantially less fuel to launch the spacecraft from the space station than from Earth. (In addition, the spacecraft would already have the orbital velocity of the space station.) Of course, this assumes that the space station is already in place and that the spacecraft is assembled at the space station.

---

remained just that—a coincidence—in the minds of scientists. However, in 1915 Einstein discovered that it is not a coincidence at all. Rather, it reveals something deeper about the nature of gravity and of the universe. The new insights were described by Einstein in his *general theory of relativity* (the topic of Chapter S3).

Putting Chapter 5 into Context

We've covered a lot of ground in this chapter, from the scientific terminology of motion to the story of how universal motion was understood by Newton. Be sure you understand the following "big picture" ideas:

- Understanding the universe requires understanding motion. Although the terminology of the laws of motion may be new to you, the *concepts* are familiar from everyday experience. Think about your own experiences of motion so that you'll better understand the less familiar astronomical applications of the laws of motion.

- Motion may seem very complex, but it can be understood simply through Newton's three laws of motion. By combining these laws with his universal law of gravitation, Newton was able to explain how gravity holds planets in their orbits and much more—including how satellites can reach and stay in orbit, the nature of tides, and why the Moon rotates synchronously with Earth.

- Perhaps even more important, Newton's discoveries showed that the same physical laws we observe on Earth apply throughout the universe. The universality of physics opens up the entire cosmos as a possible realm of human study.

Mathematical Insight · **5.4** · **The Acceleration of Gravity**

The text shows that the acceleration of a falling rock near the surface of Earth is:

$$a_{rock} = G \times \frac{M_{Earth}}{(R_{Earth})^2}$$

Because this formula applies to *any* falling object on Earth, it is the *acceleration of gravity, g*. Calculating *g* is easy. Simply look up Earth's mass (6.0×10^{24} kg) and radius (6.4×10^6 m), and then "plug in":

$$g = G \times \frac{M_{Earth}}{(R_{Earth})^2}$$
$$= \left(6.67 \times 10^{-11} \frac{m^3}{kg \times s^2}\right) \times \frac{6.0 \times 10^{24} \text{ kg}}{(6.4 \times 10^6 \text{ m})^2}$$
$$= 9.8 \frac{m}{s^2}$$

We can find the acceleration of gravity on the surface of any other world by using the same formula, using the other world's mass and radius instead of Earth's.

Example 1: What is the acceleration of gravity on the surface of the Moon?

Solution: We use the Moon's mass (7.4×10^{22} kg) and radius (1.7×10^6 m) to find:

$$g_{Moon} = G \times \frac{M_{Moon}}{(R_{Moon})^2}$$
$$= \left(6.67 \times 10^{-11} \frac{m^3}{kg \times s^2}\right) \times \frac{7.4 \times 10^{22} \text{ kg}}{(1.7 \times 10^6 \text{ m})^2}$$
$$= 1.7 \frac{m}{s^2}$$

The acceleration of gravity on the Moon is 1.7 m/s², or about one-sixth that on Earth. Thus, objects on the Moon weigh about one-sixth of what they would weigh on Earth. If you can lift a 50-kilogram barbell on Earth, you'll be able to lift a 300-kilogram barbell on the Moon.

Example 2: The Space Station orbits at an altitude roughly 300 kilometers above Earth's surface. What is the acceleration of gravity at this altitude?

Solution: Because the Space Station is significantly above Earth's surface, we cannot use the approximation $d \approx R_{Earth}$ that we used in the text. Instead, we must go back to Newton's second law, set the gravitational force on the Space Station equal to its mass times acceleration, and then solve for its acceleration:

$$G \times \frac{M_{Earth} M_{station}}{d^2} = M_{station} \times a_{station}$$
$$\Rightarrow a_{station} = G \times \frac{M_{Earth}}{d^2}$$

In this case, the distance *d* is the 6,400-km radius of Earth *plus* the 300-km altitude of the station, or $d = 6{,}700$ km $= 6.7 \times 10^6$ m. Thus, the gravitational acceleration of the Space Station when orbiting Earth is:

$$a_{station} = G \times \frac{M_{Earth}}{d^2}$$
$$= \left(6.67 \times 10^{-11} \frac{m^3}{kg \times s^2}\right) \times \frac{6.0 \times 10^{24} \text{ kg}}{(6.7 \times 10^6 \text{ m})^2}$$
$$= 8.9 \frac{m}{s^2}$$

The acceleration of gravity in low-Earth orbit is 8.9 m/s², or only slightly less than the 9.8 m/s² acceleration of gravity at Earth's surface.

5.1 Describing Motion: Examples from Daily Life

- *What is the difference between speed, velocity, and acceleration?* Speed is the rate at which an object is moving. Velocity is speed in a certain direction. Acceleration is a change in velocity, meaning a change in either speed or direction.

- *What is the acceleration of gravity?* It is the acceleration of an object falling to the ground because of gravity. On Earth's surface, it is 9.8 m/s^2.

- *How can you tell when a net force is acting on an object?* A net force must be acting whenever the object's momentum is changing. Momentum is the product of mass and velocity.

- *Have you ever been weightless? Have you ever been massless?* You are weightless every time you jump, because you are in free-fall while in the air. You have never been massless, because mass is a basic property of the matter in your body.

5.2 Newton's Laws of Motion

- *What are Newton's three laws of motion?* (1) In the absence of a net force acting upon it, an object moves with constant velocity. (2) Force = rate of change in momentum (or force = mass × acceleration). (3) For any force, there is always an equal and opposite reaction force.

- *Why does a spinning skater spin faster as she pulls in her arms?* On ice, there is little friction and therefore little torque, so the skater's angular momentum is conserved as she spins in accordance with the law of conservation of angular momentum. Pulling in her arms makes her "radius" smaller, so she must spin faster to keep the same angular momentum.

5.3 The Force of Gravity

- *What is the universal law of gravitation?* The force of gravity is directly proportional to the product of the objects' masses and declines with the square of the distance between their centers:

$$F_{\text{g}} = G \frac{M_1 \times M_2}{d^2}$$

- *What types of orbits are possible according to the law of gravitation?* Orbiting objects may follow bound orbits in the shape of ellipses (or circles) and unbound orbits in the shape of parabolas or hyperbolas.

- *How can we determine the mass of a distant object?* Newton's version of Kepler's third law allows us to calculate the mass of a distant object if another object orbits it and we can measure the orbital distance and period.

5.4 Tides

- *Why are there two high tides on Earth each day?* The Moon's gravity stretches Earth along the Earth–Moon line so that it bulges both toward and away from the Moon.

- *Why are tides on Earth caused primarily by the Moon rather than by the Sun?* Earth's gravitational attraction to the Sun is stronger than its gravitational attraction to the Moon, but tides are caused by the *difference* between the strength of the gravitational attraction across Earth's diameter. This difference is greater for the gravitational force due to the Moon because the Moon is so much closer than the Sun.

- *Why is Earth's rotation gradually slowing down?* Tidal friction, caused by the way the tidal bulges exert drag on Earth, causes a gradual slowing of Earth's rotation. A related consequence of tidal friction is the Moon's increasing distance from Earth.

- *Why does the Moon always show the same face to Earth?* The Moon's synchronous rotation is a result of tidal forces. The Moon may once have rotated much faster, but tidal friction slowed its rotation until it became synchronous with its orbit, at which point tidal friction could not slow the orbit any further.

5.5 Orbital Energy and Escape Velocity

- *What is orbital energy?* It is the combined kinetic and gravitational potential energy of an orbiting object.

- *Will a spacecraft passing by a planet be "sucked in"?* No. Energy must be conserved, so an object's orbital energy cannot change unless it gains or loses energy to something else.

- *How can an object achieve escape velocity?* For an object to escape the gravitational attraction of some world, it must travel away from that world fast enough so that gravity will never make it return. More technically, the object must gain enough energy (say, by firing a rocket) so that it ends up on an unbound orbit—an orbit on which it never returns to its starting point.

5.6 The Acceleration of Gravity

- *How does the acceleration of gravity depend on the mass of a falling object?* It doesn't. All falling objects fall with the same acceleration (on a particular planet).

❓ Does It Make Sense?

Decide whether each statement makes sense and explain why it does or does not.

1. If you could go shopping on the Moon to buy a pound of chocolate, you'd get a lot more chocolate than if you bought a pound on Earth.

2. Suppose you could enter a vacuum chamber (on Earth), that is, a chamber with no air in it. Inside this chamber, if you dropped a hammer and a feather from the same height at the same time, both would hit the bottom at the same time.

3. When an astronaut goes on a space walk outside the Space Station, she will quickly float away from the station unless she has a tether holding her to the station or constantly fires thrusters on her space suit.

4. Newton's version of Kepler's third law allows us to calculate the mass of Saturn from orbital characteristics of its moon Titan.

5. If we could magically replace the Sun with a giant rock that has precisely the same mass, Earth's orbit would not change.

6. The fact that the Moon rotates once in precisely the time it takes to orbit Earth once is such an astonishing coincidence that scientists probably never will be able to explain it.

7. Venus has no oceans, so it could not have tides even if it had a moon (which it doesn't).

8. If an asteroid passed by Earth at just the right distance, it would be captured by Earth's gravity and become our second moon.

Problems

(Quantitative problems are marked with an asterisk.)

9. *Speed and Velocity.* How does *speed* differ from *velocity?* Give an example in which you can be traveling at constant speed but not at constant velocity.

10. *Momentum and Force.* What is *momentum?* How can momentum be affected by a *force?* What do we mean when we say that momentum will be changed only by a *net force?*

11. *Free-Fall and Weightlessness.* What is *free-fall,* and why does it make you *weightless?* Briefly describe why astronauts are weightless in the Space Station.

12. *Orbiting Spaceship.* Why does a spaceship require a high speed to achieve orbit? What would happen if it were launched with a speed greater than Earth's *escape velocity?*

13. *Newton's Laws of Motion.* State each of *Newton's three laws of motion.* For each law, give an example of its application.

14. *Tidal Friction.* What is *tidal friction?* Briefly describe how tidal friction has affected Earth and led to the Moon's *synchronous rotation.*

15. *Orbital Change.* Explain why orbits cannot change spontaneously. How can atmospheric drag cause an orbit to change? How can a *gravitational encounter* cause an orbit to change?

16. *Understanding Acceleration.*

   a. Some schools have an annual ritual that involves dropping a watermelon from a tall building. Suppose it takes 6 seconds for the watermelon to fall to the ground (which would mean it's dropped from about a 60-story building). If there were no air resistance, so that the watermelon would fall with the acceleration of gravity, how fast would it be going when it hit the ground? Give your answer in m/s, km/hr, and mi/hr.

   b. As you sled down a steep, slick street, you accelerate at a rate of 4 m/s^2. How fast will you be going after 5 seconds? Give your answer in m/s, km/hr, and mi/hr.

   c. You are driving along the highway at a speed of 70 miles per hour when you slam on the brakes. If your acceleration is at an average rate of -20 miles per hour per second, how long will it take to come to a stop?

17. *Spinning Skater.* Suppose an ice skater wants to *start* spinning. Explain why she won't start spinning if she simply stomps her foot straight down on the ice. How should she push off on the ice to start spinning? Why? What should she do when she wants to stop spinning?

18. *The Gravitational Law.* Use the universal law of gravitation to answer each of the following questions.

   a. How does quadrupling the distance between two objects affect the gravitational force between them?

   b. Compare the gravitational force between Earth and the Sun to that between Jupiter and the Sun. Jupiter's mass is 318 times Earth's mass, and its distance from the Sun is 5.2 times Earth's distance.

   c. Suppose the Sun were magically replaced by a star with twice as much mass. What would happen to the gravitational force between Earth and the Sun?

19. *Head-to-Foot Tides.* You and Earth attract each other gravitationally, so you should also be subject to a tidal force resulting from the difference between the gravitational attraction felt by your feet and that felt by your head (at least when you are standing). Explain why you can't feel this tidal force.

20. *Eclipse Frequency in the Past.* Over billions of years, the Moon has gradually been moving farther from Earth. Thus, the Moon used to be substantially nearer to Earth than it is today.

   a. How would the Moon's past angular size in our sky compare to its present angular size? Why?

   b. How would the length of a lunar month in the past compare to the length of a lunar month today? Why? (*Hint:* Think about Kepler's third law as it would apply to the Moon orbiting Earth.)

   c. Based on your answers to parts (a) and (b), would eclipses (both solar and lunar) have been more or less common in the past? Why?

21. *Geostationary Orbit.* A satellite in geostationary orbit appears to remain stationary in the sky as seen from any particular location on Earth.

a. Briefly explain why a geostationary satellite must orbit Earth in 1 *sidereal* day, rather than 1 solar day.

b. Communications satellites, such as those used for television broadcasts, are often placed in geostationary orbit. The transmissions from such satellites are received with satellite dishes, such as those that can be purchased for home use. In one or two paragraphs, explain why geostationary orbit is a convenient orbit for communications satellites.

22. *Elevator to Orbit.* Suppose that someday we build a giant elevator from Earth's surface to geosynchronous orbit. The top of the elevator would then have the same orbital distance and period as any satellite in geosynchronous orbit.

a. Suppose you were to drop an object out of the elevator at its top. Explain why the object would appear to float right next to the elevator rather than falling.

b. Briefly explain why (not counting the huge costs for construction) the elevator would make it much cheaper and easier to put satellites in orbit or to launch spacecraft into deep space.

*23. *Understanding Kepler's Third Law.* Use Newton's version of Kepler's third law to answer the following questions. (*Hint:* The calculations for this problem are so simple that you will not need a calculator.)

a. Imagine another solar system, with a star of the same mass as the Sun. Suppose there is a planet in that solar system with a mass twice that of Earth orbiting at a distance of 1 AU from the star. What is the orbital period of this planet? Explain.

b. Suppose a solar system has a star that is four times as massive as our Sun. If that solar system has a planet the same size as Earth orbiting at a distance of 1 AU, what is the orbital period of the planet? Explain.

*24. *Gees.* Acceleration is sometimes measured in *gees*, or multiples of the acceleration of gravity: 1 gee ($1g$) means $1 \times g$, or 9.8 m/s^2; 2 gees ($2g$) means $2 \times g$, or 2×9.8 m/s^2 = 19.6 m/s^2; and so on. Suppose you experience 6 gees of acceleration in a rocket.

a. What is your acceleration in meters per second squared?

b. You will feel a compression force from the acceleration. How does this force compare to your normal weight?

c. Do you think you could survive this acceleration for long? Explain.

*25. *New Comet.* Imagine that a new comet is discovered and studies of its motion indicate that it orbits the Sun with a period of 1,000 years. What is the comet's average distance (semimajor axis) from the Sun? (*Hint:* Use Kepler's third law in its original form.)

*26. *Measuring Masses.* Use Newton's version of Kepler's third law to answer each of the following questions.

a. The Moon orbits Earth in an average time of 27.3 days at an average distance of 384,000 kilometers. Use these facts to determine the mass of Earth. You may neglect the mass of the Moon and assume $M_{Earth} + M_{Moon} \approx M_{Earth}$. (The Moon's mass is about $\frac{1}{80}$ of Earth's.)

b. Jupiter's moon Io orbits Jupiter every 42.5 hours at an average distance of 422,000 kilometers from the center of Jupiter. Calculate the mass of Jupiter. (Io's mass is very small compared to Jupiter's.)

c. Calculate the orbital period of the Space Shuttle in an orbit 300 kilometers above Earth's surface.

d. Pluto's moon Charon orbits Pluto every 6.4 days with a semimajor axis of 19,700 kilometers. Calculate the *combined* mass of Pluto and Charon. Compare this combined mass to the mass of Earth.

*27. Calculate the escape velocity from each of the following.

a. The surface of Mars (mass = $0.11M_{Earth}$, radius = $0.53R_{Earth}$).

b. The surface of Mars's moon Phobos (mass = 1.1×10^{16} kg, radius = 12 km).

c. The cloud tops of Jupiter (mass = $317.8M_{Earth}$, radius = $11.2R_{Earth}$).

d. Our solar system, starting from Earth's orbit. (*Hint:* Most of the mass of our solar system is in the Sun; $M_{Sun} = 2.0 \times 10^{30}$ kg.)

e. Our solar system, starting from Saturn's orbit.

*28. *Weights on Other Worlds.* Calculate the acceleration of gravity on the surface of each of the following worlds. How much would *you* weigh, in pounds, on each of these worlds?

a. Mars (mass = $0.11M_{Earth}$, radius = $0.53R_{Earth}$).

b. Venus (mass = $0.82M_{Earth}$, radius = $0.95R_{Earth}$).

c. Jupiter (mass = $317.8M_{Earth}$, radius = $11.2R_{Earth}$). Bonus: Given that Jupiter has no solid surface, how could you weigh yourself on Jupiter?

d. Jupiter's moon Europa (mass = $0.008M_{Earth}$, radius = $0.25R_{Earth}$).

e. Mars's moon Phobos (mass = 1.1×10^{16} kg, radius = 12 km).

Discussion Questions

29. *Aristotle and Modern English.* Aristotle believed that Earth was made from the four elements fire, water, earth, and air, while the heavens were made from *ether* (literally, "upper air"). The literal meaning of *quintessence* is "fifth element," and the literal meaning of *ethereal* is "made of ether." Look up these words in the dictionary. Discuss how their modern meanings are related to Aristotle's ancient beliefs.

30. *Tidal Complications.* The ocean tides on Earth are much more complicated than they might at first seem from the simple physics that underlies tides. Discuss some of the factors that make the real tides so complicated and how these factors affect the tides. Some factors to consider: the distribution of land and oceans; the Moon's varying distance from Earth in its orbit; the fact that the Moon's orbital plane is not perfectly aligned with the ecliptic and neither the Moon's orbit nor the ecliptic is aligned with Earth's equator.

MEDIA EXPLORATIONS

For a complete list of media resources available, go to www.astronomyplace.com and choose Chapter 5 from the pull-down menu.

 ## Astronomy Place Web Tutorials

Tutorial Review of Key Concepts

Use the interactive **Tutorial** at www.astronomyplace.com to review key concepts from this chapter.

Orbits and Kepler's Laws Tutorial

Lesson 1 Gravity and Orbits

Lesson 2 Kepler's First Law

Lesson 3 Kepler's Second Law

Lesson 4 Kepler's Third Law

Supplementary Tutorial Exercises

Use the interactive **Tutorial Lessons** to explore the following questions.

Orbits and Kepler's Laws Tutorial, Lesson 1

1. How fast should you fire the cannonball to get it to orbit Earth?

2. How fast should you fire the cannonball to get it to escape into space?

3. How did the Apollo astronauts demonstrate that mass does not affect the rate at which objects fall?

Orbits and Kepler's Laws Tutorial, Lesson 4

1. Use the tool in Lesson 4 to predict the orbital period of an asteroid that has an orbital radius of 4AU ($p =$ _____).

2. Use Kepler's third law, $p^2 = a^3$, to check your prediction ($a^3 =$ _____). Now take the square root of the result ($p =$ _____). Is your orbital period the same as the answer you calculated for question 1?

3. Use the tool to predict the orbital radius of an asteroid that has a period of two years. Does your answer agree with what you would expect from Kepler's third law?

4. Does the eccentricity of the orbit affect your answers to the last three questions?

 ## Exploring the Sky and Solar System

Of the many activities available on the **Voyager: SkyGazer CD-ROM** accompanying your book, use the following files to observe key phenomena covered in this chapter.

Go to the **File: Basics** folder for the following demonstrations.

1. Planet Paths

2. Planet Orrery

3. Follow a Planet

Go to the **File: Demo** folder for the following demonstrations.

1. Earth and Venus

2. Hyakutake at Perihelion

3. Pluto's Orbit

Go to the **Explore** menu for the following demonstrations.

1. Solar System

2. Paths of the Planets

Movies

Check out the following narrated and animated short documentary available on www.astronomyplace.com for a helpful review of key ideas covered in this chapter.

Orbits in the Solar System Movie

Web Projects

Take advantage of the useful Web links on www.astronomyplace.com to assist you with the following projects.

1. *Space Station.* Visit a NASA site with pictures from the Space Station. Choose two photos that illustrate some facet of Newton's laws of motion or gravity. Explain how what is going on is related to Newton's laws.

2. *Tide Tables.* Find a tide table or tide chart for a beach town that you'd like to visit. Briefly explain how to read the table or chart, and discuss any differences between the actual tidal pattern and the idealized tidal pattern described in this chapter.

3. *Space Elevator.* Read more about space elevators (see problem 22) and how they might make it easier and cheaper to get to Earth orbit or beyond. Write a short report about the feasibility of building a space elevator, and briefly discuss the pros and cons of such a project.

6 Light
The Cosmic Messenger

May the warp be the white light of
*   morning,*
May the weft be the red light of evening,
May the fringes be the falling rain,
May the border be the standing rainbow.
Thus weave for us a garment of
*   brightness.*

Song of the Sky Loom
(Native American)

Ancient observers could discern only the most basic features of the light that they saw—such as color and brightness. Over the past several hundred years, we have discovered that light carries far more information. For instance, analysis of light with special instruments can reveal to us the chemical composition of distant objects, their temperature, and how fast they rotate.

It is fortunate that light can convey so much information. Our present spacecraft can reach only objects within our solar system, and except for an occasional meteorite falling from the sky, the cosmos does not come to us. In contrast to the limited reach of spacecraft, light travels throughout the universe, carrying its treasury of information wherever it goes. Light, the cosmic messenger, brings the stories of distant objects to our home here on Earth.

 Light and Spectroscopy Tutorial, Lesson 1

6.1 Light in Everyday Life

Even without opening your eyes, it's clear that light is a form of energy—scientists call it *radiative energy* [Section 4.1]. Outside on a hot, sunny day, you can feel the radiative energy of sunlight being converted to thermal energy as it strikes your skin. On an economic level, electric companies charge you for the energy needed by your light bulbs (and other appliances). The rate at which a light bulb uses energy (converts electrical energy to light and heat) is usually printed on it—for example, "100 watts." A watt is a unit of **power**, which describes the *rate* of energy use. A power of 1 watt means that 1 joule of energy is being used each second:

$$1 \text{ watt} = 1 \text{ joule/s}$$

Thus, for every second that you leave a 100-watt light bulb turned on, you will have to pay the utility company for 100 joules of energy. Interestingly, the power requirement of an average human—about 10 million joules per day—is about the same as that of a 100-watt light bulb.

Another basic property of light is what our eyes perceive as *color*. You've probably seen a prism split light into the rainbow of light called a **spectrum** (Figure 6.1). You can also produce a spectrum with a **diffraction grating**, a piece of plastic or glass etched with many closely spaced lines. The colors in a spectrum are pure forms of the basic colors red, orange, yellow, green, blue, and violet. The wide variety of all possible colors comes from mixtures of these basic colors in varying proportions. *White* is what we see when the basic colors are mixed in roughly equal proportions. Your television takes advantage of this fact to simulate a huge range of colors by combining only three specific colors of red, green, and blue light.

If you have a magnifying glass handy, hold it close to your TV set to see the individual red, blue, and green dots. If you don't have a magnifying glass, try splashing a few droplets of water onto your TV screen (carefully!). What do you see? What are the drops of water doing?

Energy carried by light can interact with matter in four general ways:

- **Emission:** When you turn on a lamp, electric current flowing through the filament of the light bulb heats it to a point at which it *emits* visible light.

- **Absorption:** If you place your hand near a lit light bulb, your hand *absorbs* some of the light, and this absorbed energy makes your hand warmer.

- **Transmission:** Some forms of matter, such as glass or air, *transmit* light. That is, they allow light to pass through them.

- **Reflection:** A mirror *reflects* light in a very specific way, similar to the way a rubber ball bounces off a hard surface, so that the direction of a reflected beam of light

Figure 6.1 A prism reveals that white light contains a spectrum of colors from red to violet.

Figure 6.2 Reflection and scattering.

angle of incidence = angle of reflection

30°   30°

a A mirror reflects light along a path determined by the angle at which the light strikes the mirror.

screen   projection
                                                             booth

b A movie screen scatters light into an array of beams that reach every member of the audience. The pages in this book do the same thing, which is why you can read them from different angles and distances.

depends on the direction of the incident (incoming) beam of light (Figure 6.2a). Sometimes reflection is more random, so that an incident beam of light is **scattered** in many different directions. The screen in a movie theater, for example, scatters a narrow beam of light from the projector into an array of beams that reach every member of the audience (Figure 6.2b).

Materials that transmit light are said to be **transparent**, and materials that absorb light are called **opaque**. Many materials are neither perfectly transparent nor perfectly opaque. For example, dark sunglasses and clear eyeglasses are both at least partially transparent, but the dark glasses absorb more light and transmit less.

Particular materials can affect different colors of light differently. For example, red glass transmits red light but absorbs other colors. A green lawn reflects (scatters) green light but absorbs all other colors.

Now let's put all these ideas together and think about what happens when you walk into a room and turn on the light switch (Figure 6.3). The light bulb begins to emit white light, which is a mix of all the colors in the spectrum. Some of this light exits the room, transmitted through the windows. The rest of the light strikes the surfaces of objects inside the room, and each object's material properties determine the colors absorbed or reflected. The light coming from each object therefore carries an enormous amount of information about the object's location, shape and structure, and material makeup. You acquire this information when light enters your eyes, where it is absorbed by special cells (called *cones* and *rods*) that use the energy

of the absorbed light to send signals to your brain. Your brain interprets the messages carried by the light, recognizing materials and objects in the process we call *vision*.

Light carries much more information than your ordinary vision can recognize. Modern instruments can reveal otherwise hidden details in the spectrum of light. Learning to interpret these details is the key to unlocking the vast amount of information carried by light.

 Light and Spectroscopy Tutorial, Lesson 1

6.2 Properties of Light

Despite our familiarity with light, its nature remained a mystery for most of human history. The first real insights into the nature of light came with experiments performed by Isaac Newton in the 1660s. It was already well known that passing light through a prism produced a rainbow of colors, but the most common belief held that the colors were a property of the prism rather than of the light itself. Newton dispelled this belief by placing a second prism in front of the light of just one color, such as red, from the first prism. He found that the color did not change any further, thereby proving that the colors were not a property of the prism but must be part of the white light itself.

Newton guessed that light, with all its colors, is made up of countless tiny particles. However, later experiments by other scientists demonstrated that light behaves like waves. Thus began one of the most important debates in scientific history: Is light a wave or a particle?

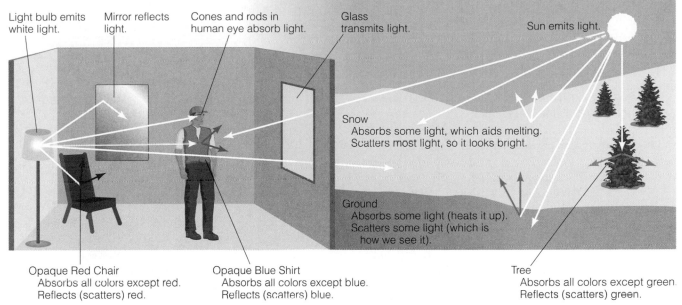

Light bulb emits white light.

Mirror reflects light.

Cones and rods in human eye absorb light.

Glass transmits light.

Sun emits light.

Snow
Absorbs some light, which aids melting.
Scatters most light, so it looks bright.

Ground
Absorbs some light (heats it up).
Scatters some light (which is how we see it).

Opaque Red Chair
Absorbs all colors except red.
Reflects (scatters) red.

Opaque Blue Shirt
Absorbs all colors except blue.
Reflects (scatters) blue.

Tree
Absorbs all colors except green.
Reflects (scatters) green.

Figure 6.3 When light strikes any piece of matter in the universe, that matter reacts in one or a combination of four ways: emission, absorption, transmission, and reflection.

Particles and Waves in Everyday Life

Marbles, baseballs, and individual atoms are all examples of *particles.* A particle of matter can sit still or it can move from one place to another. If you throw a baseball at a wall, it moves from your hand to the wall.

Now imagine tossing a pebble into a pond (Figure 6.4). The ripples moving out from the place where the pebble lands are *waves,* consisting of *peaks,* where the water is higher than average, and *troughs,* where the water is lower than average. If you watch as the waves pass by a floating leaf, you'll see the leaf rise up with the peak and drop down with the trough, but the leaf itself does *not* move across the pond's surface with the wave. This observation tells us that the water is moving up and down but not outward. That is, the wave carries *energy* outward from the place where the pebble landed but does not carry matter along with it. In a sense, a particle is a *thing,* while a wave is a *pattern* revealed by its interaction with particles.

THINK ABOUT IT

Hold a piece of rope with one end in each hand. Make waves move along the rope by shaking one end up and down. Watch the motion of the peaks and troughs. As a peak moves along the rope, does any material move with it? Explain.

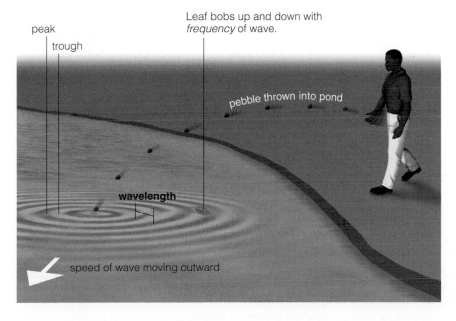

peak

trough

Leaf bobs up and down with *frequency* of wave.

pebble thrown into pond

wavelength

speed of wave moving outward

Figure 6.4 Tossing a pebble into a pond generates waves. The waves carry energy outward, but matter does *not* travel outward. Instead, matter—like a leaf or the water itself—moves up and down as the waves pass by.

Three basic properties characterize the waves moving outward through the pond. Their **wavelength** is the distance between adjacent peaks. Their **frequency** is the number of peaks passing by any point each second. If a passing wave causes the leaf to bob up and down twice per second, its frequency is 2 **cycles per second** (referring to the up and down "cycles" of the passing waves). Cycles per second often are called **hertz** (**Hz**), so we can also describe this frequency as 2 Hz. The third basic characteristic of the waves is the **speed** at which any peak travels across the pond. (A fourth characteristic of a wave is its amplitude, or height from trough to peak. Amplitude is related to the brightness of light, but we will not use it in our study of light in this book.)

The wavelength, frequency, and speed of a wave are related by a simple formula, which we can understand with the help of an example. Suppose a wave has a wavelength of 1 centimeter and a frequency of 2 hertz. The wavelength tells us that each time a peak passes by, the wave peak has traveled 1 centimeter. The frequency tells us that two peaks pass by each second. Thus, the speed of the wave must be 2 centimeters per second. If you try a few more similar examples, you'll find that the general rule is

$$\text{wavelength} \times \text{frequency} = \text{speed}$$

Photons and Electromagnetic Waves

In our everyday lives, waves and particles appear to be very different: No one would confuse the ripples on a pond with a baseball. However, experiments show that light behaves as *both* a wave and a particle. Like ripples on a pond, light can make (charged) particles bob up and down. Like particles, light comes in individual "pieces," called **photons**, that can hit a wall one at a time. We will discuss the implications of this "wave–particle duality" in Chapter S4. Here we need to discuss only how we measure the wave and particle properties of light.

Let's look first at the wave nature of light. Although waves don't carry material along with them, something must vibrate to transmit energy along a wave. For example, water waves are the up-and-down vibrations of the water surface, and sound waves are the back-and-forth vibrations of the air as it responds to changing pressure. In the case of light, it is electric and magnetic *fields* that vibrate.

The concept of a **field** is a bit abstract. Fields associated with forces, such as electric and magnetic fields, describe how these forces affect a particle placed at any point in space. For example, we say that Earth has a *gravitational field* because if you place an object above Earth's surface, gravity exerts a downward force that pulls the object toward the center of the Earth. That is, any object placed in a gravitational field feels the force of gravity. In a similar way, a charged particle (such as an electron) placed in an *electric field* or a *magnetic field* feels electric or magnetic forces.

Light involves both electric and magnetic fields. More specifically, light is an **electromagnetic wave**—a wave in

a If you could line up a row of electrons, they would wriggle up and down as light passes by, showing that light carries a vibrating electric field. Light also carries a magnetic field (not shown) that vibrates perpendicular to the direction of the electric field vibrations.

Wavelength is the distance between adjacent peaks of the electric field.

Frequency is the number of waves (cycles) passing any point each second.

All light travels with speed c = 300,000 km/s.

b Characteristics of light waves. Because all light travels at the same speed, light of longer wavelength must have lower frequency.

Figure 6.5 Light is an electromagnetic wave characterized by a wavelength and a frequency.

which these electric and magnetic fields vibrate. Like a leaf on a rippling pond, an electron will bob up and down when an electromagnetic wave passes by. If you could set up a row of electrons, they would wriggle like a snake (Figure 6.5a). The wavelength is the distance between adjacent peaks of the electric or magnetic field, and the frequency is the number of peaks that pass by any point each second (Figure 6.5b).

All light travels at the same speed (in a vacuum)—about 300,000 kilometers per second, or 3×10^8 m/s—regardless of its wavelength or frequency. Therefore, because speed = wavelength × frequency, light with a shorter wavelength must have a higher frequency, and vice versa.

Now let's look at the particle nature of light by considering light to be made up of photons. From this point of view, each photon is a distinct entity. Just as a moving baseball carries a specific amount of kinetic energy, each photon of light carries a specific amount of radiative energy. The shorter the wavelength of the light (or, equivalently, the higher its frequency), the higher the energy of the photons (Figure 6.5). For example, a photon with a wavelength of 100 nanometers (nm) has more energy than a photon with a 120-nm wavelength. (A nanometer [nm] is a billionth of a meter: 1 nm = 10^{-9} m. Many astronomers work with a unit called the Angstrom [Å]: 1 nm = 10Å.)

 Light and Spectroscopy Tutorial, Lesson 1

6.3 The Many Forms of Light

Because light consists of electromagnetic waves, light is often called *electromagnetic radiation* and the spectrum of light is called the **electromagnetic spectrum**. Photons of light can have *any* wavelength or frequency, so in principle the complete electromagnetic spectrum extends from a

Figure 6.6 The electromagnetic spectrum. The unit of frequency, hertz, is equivalent to waves (or cycles) per second. For example, 10^3 hertz means that $10^3 = 1{,}000$ wave peaks pass by a point each second.

wavelength of zero to infinity. We have names for different portions of the electromagnetic spectrum (Figure 6.6).

The **visible light** that we see with our eyes has wavelengths ranging from about 400 nm at the blue end of the rainbow to about 700 nm at the red end. Light with wavelengths somewhat longer than red light is called **infrared**,

because it lies beyond the red end of the rainbow. **Radio waves** are the longest-wavelength light. Thus, radio waves are a form of light, *not* a form of sound.

On the other side of the spectrum, light with wavelengths somewhat shorter than blue light is called **ultraviolet**, because it lies beyond the blue (or violet) end of

---

Mathematical Insight **6.1** **Wavelength, Frequency, and Energy**

As described in the text, the speed of any wave is the product of its wavelength and its frequency. Because all forms of light travel at the same speed (in a vacuum), $c = 3 \times 10^8$ m/s, we can write:

$$\lambda \times f = c$$

where λ (the Greek letter *lambda*) stands for wavelength and f stands for frequency. This formula is simple but revealing: Because the speed c is constant, the formula tells us that frequency must go up when wavelength goes down, and vice versa. Solving this formula allows us to find the wavelength of light if we know the frequency, or to find the frequency if we know the wavelength:

$$\lambda = \frac{c}{f} \quad \text{or} \quad f = \frac{c}{\lambda}$$

The formula for the radiative energy (E) carried by a photon of light is:

$$E = h \times f$$

where h is a number called *Planck's constant* ($h = 6.626 \times 10^{-34}$ joule $\times$ s). Thus, energy increases in proportion to the frequency of the photon. Because $f = c/\lambda$, we can also write this formula as:

$$E = \frac{hc}{\lambda}$$

showing that the energy is inversely proportional to the wavelength of the photon.

Example 1: The numbers on a radio dial for FM radio stations are their frequencies in megahertz (MHz), or millions of hertz. If your favorite radio station is "93.3 on your dial," it broadcasts radio waves with a frequency of 93.3 million cycles per second. What is the wavelength of these radio waves?

Solution: We know the speed of light and the frequency, so the wavelength is:

$$\lambda = \frac{c}{f} = \frac{3 \times 10^8 \, \frac{m}{s}}{93.3 \times 10^6 \, \frac{1}{s}} = 3.2 \text{ m}$$

Note that when we work with frequency in equations, the "cycles" do not show up as a unit. That is, the units of frequency are simply 1/s, or "per second."

Example 2: The average wavelength of visible light is about 550 nanometers (1 nm = 10^{-9} m). What is the frequency of this light?

Solution: This time we know the wavelength, so the frequency of the light is:

$$f = \frac{c}{\lambda} = \frac{3 \times 10^8 \, \frac{m}{s}}{550 \times 10^{-9} \, m} = 5.45 \times 10^{14} \, \frac{1}{s}$$

The frequency of visible light is about 5.5×10^{14} cycles per second, or about 550 trillion Hz.

Example 3: What is the energy of a visible-light photon with wavelength 550 nm?

Solution: We know the wavelength, so the energy is:

$$E = \frac{hc}{\lambda}$$

$$= \frac{(6.626 \times 10^{-34} \text{ joule} \times s) \times (3 \times 10^8 \, \frac{m}{s})}{550 \times 10^{-9} \, m}$$

$$= 3.6 \times 10^{-19} \text{ joule}$$

Note that this energy for a single photon is extremely small compared to, say, the energy of 100 joules used each second by a 100-watt light bulb.

the rainbow. Light with even shorter wavelengths is called **X rays**, and the shortest-wavelength light is called **gamma rays**. You can see that visible light is an extremely small part of the entire electromagnetic spectrum: The reddest red that our eyes can see has only about twice the wavelength of the bluest blue, but the radio waves from your favorite radio station are a billion times longer than the X rays used in a doctor's office.

Because wavelengths decrease as we move from the radio end toward the gamma-ray end of the spectrum, the frequencies and energies must increase. Visible-light photons

happen to have enough energy to activate the molecular receptors in our eyes. Ultraviolet photons, with a shorter wavelength than visible light, carry more energy—enough to harm our skin cells, causing sunburn or skin cancer. X-ray photons have enough energy to transmit easily through skin and muscle but not so easily through bones or teeth. That is why doctors and dentists can see our underlying bone structures on photographs taken with X-ray light.

Interactions between light and matter depend on the types of light and matter involved. A brick wall is opaque to visible light but transmits radio waves, and glass that is transparent to visible light can be opaque to ultraviolet light. In general, certain types of matter tend to interact more strongly with certain types of light, so each type of light carries different information about distant objects in the universe. Astronomers therefore seek to observe light of all wavelengths, using telescopes adapted to detecting each different form of light, from radio to gamma rays.

 Light and Spectroscopy Tutorial, Lessons 2–4

6.4 Light and Matter

Whenever matter and light interact, matter leaves its fingerprints. Examining the color of an object is a crude way of studying the clues left by the matter it contains. For example, a red shirt absorbs all visible photons except those in the red part of the spectrum, so we know that it must contain a dye with these special light-absorbing characteristics. If we take light and disperse it into a spectrum, we can see the spectral fingerprints in more detail.

Figure 6.7 shows a schematic spectrum of light from a celestial body such as a planet. The spectrum is a graph

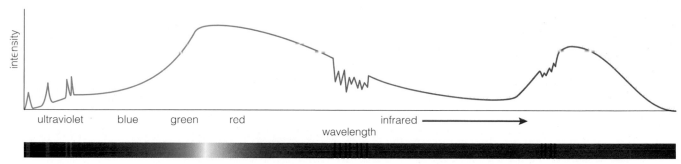

Figure 6.7 A schematic spectrum obtained from the light of a distant object. The "rainbow" at bottom shows how the light would appear when viewed through a prism or diffraction grating. The graph shows the corresponding intensity of the light at each wavelength. Note that the intensity is high where the rainbow is bright and low where it is dim (such as in places where the rainbow shows dark lines).

that shows the amount of radiation, or **intensity**, at different wavelengths. At wavelengths where a lot of light is coming from the celestial body, the intensity is high, while at wavelengths where there is little light, the intensity is low. Our goal is to see what the bumps and wiggles in this graph tell us about the celestial body in question. Let's begin by going through a short list of interactions between matter and light.

Figure 6.8 An atom emits or absorbs light only at specific wavelengths that correspond to changes in the atom's energy as an electron jumps between its allowed energy levels.

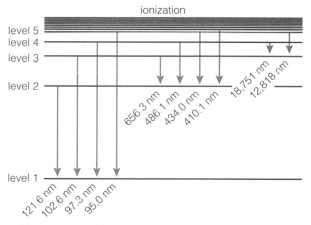

a Photons emitted by various energy-level transitions in hydrogen.

Absorption and Emission by Thin Gases

Thin or low-density gases can absorb light. For example, ozone in Earth's atmosphere absorbs ultraviolet light from space, preventing it from reaching the ground. Gases can also emit light, which is what makes neon lights and interstellar clouds glow so beautifully. But *how* do gases absorb or emit light?

Remember that the electrons in atoms can have only specific energies, somewhat like the specific heights of the rungs on a ladder [Section 4.4]. If an electron in an atom is bumped from a lower energy level to a higher one—by a collision with another atom, for example—it will eventually fall back to the lower level. The energy that the atom loses when the electron falls back down must go somewhere, and often it goes to *emitting* a photon of light. The emitted photon must have exactly the same amount of energy that the electron loses, which means that it has a specific wavelength (and frequency).

Figure 6.8a shows the allowed energy levels in hydrogen, along with the wavelengths of the photons emitted by various downward *transitions* of an electron from a higher energy level to a lower one. For example, the transition from level 2 to level 1 emits an ultraviolet photon of wavelength 121.6 nm, and the transition from level 3 to level 2 emits a red visible-light photon of wavelength

410.1 nm   434.0 nm   486.1 nm   656.3 nm

b The visible emission line spectrum from heated hydrogen gas. These lines come from transitions in which electrons fall from higher energy levels to level 2.

410.1 nm   434.0 nm   486.1 nm   656.3 nm

c If we pass white light through a cloud of cool hydrogen gas, we get this absorption line spectrum. These lines come from transitions in which electrons jump from energy level 2 to higher levels.

Figure 6.9 Visible-light emission line spectra for helium, sodium, and neon. The patterns and wavelengths of lines are different for each element, giving each a unique spectral fingerprint.

656.3 nm.* If you heat some hydrogen gas so that collisions are continually bumping electrons to higher energy levels, you'll get an **emission line spectrum** consisting of the photons emitted as each electron falls back to lower levels (Figure 6.8b).

> **THINK ABOUT IT**
>
> If nothing continues to heat the hydrogen gas, all the electrons eventually will end up in the lowest energy level (the ground state, or level 1). Use this fact to explain why we should *not* expect to see an emission line spectrum from a very cold cloud of hydrogen gas.

Photons of light can also be absorbed, causing electrons to jump *up* in energy—but only if an incoming photon happens to have precisely the right amount of energy. For example, just as an electron moving downward from level 2 to level 1 in hydrogen emits a photon of wavelength 121.6 nm, absorbing a photon with this wavelength will cause an electron in level 1 to jump up to level 2.

Suppose a lamp emitting white light illuminates a cloud of hydrogen gas from behind. The cloud will absorb photons with the precise energies needed to bump electrons in the hydrogen atoms from a low energy level to a higher one, while all other photons pass right through the cloud. The result is an **absorption line spectrum** that looks like a rainbow with light missing at particular wavelengths (Figure 6.8c). (The absorbed photons are generally re-emitted quickly but in all directions, so there is absorption

---

*Astronomers call transitions between level 1 and other levels the *Lyman* series of transitions. The transition between level 1 and level 2 is Lyman α, between level 1 and level 3 Lyman β, and so on. Similarly, transitions between level 2 and higher levels are called *Balmer* transitions. Other sets of transitions also have names.

along the line-of-sight to the source of illumination behind the cloud.)

If you compare the bright emission lines in Figure 6.8b to the dark absorption lines in Figure 6.8c, you will see that the lines occur at the same wavelengths regardless of whether the hydrogen is absorbing or emitting light. Absorption lines simply correspond to upward jumps of the electrons between energy levels, while emission lines correspond to downward jumps.

The energy levels of electrons in each chemical element are unique [Section 4.4]. As a result, each element produces its own distinct set of spectral lines, giving it a unique "spectral fingerprint." For example, Figure 6.9 shows emission line spectra for helium, sodium, and neon.

> **THINK ABOUT IT**
>
> Three common examples of objects with emission line spectra are storefront neon signs, fluorescent light bulbs, and the yellow sodium lights used in many cities at night. When you look at objects of different colors under such lights, do you see their normal colors? Why or why not?

The unique spectral fingerprint of each chemical element makes spectral analysis extremely useful. When you see the fingerprint of a particular element, you immediately know that the gas producing the spectrum contains this element. For example, Figure 6.10 shows the spectral fingerprints of hydrogen, helium, oxygen, and neon in an emission line spectrum from the Orion Nebula.

Not only does each chemical element produce a unique spectral fingerprint, but *ions* of a particular element (atoms that are missing one or more electrons) produce fingerprints different from those of neutral atoms [Section 4.3]. For example, the spectrum of doubly ionized neon (Ne^{++}) is different from that of singly ionized

Figure 6.10 The emission line spectrum of the Orion Nebula in a portion of the ultraviolet (about 350–400 nm). The lines are identified with the chemical elements or ions that produce them (He = helium; O = oxygen; Ne = neon). The many hydrogen lines are all transitions from high levels to level 2 (see Figure 6.8a).

neon (Ne^+), which in turn is different from that of neutral neon (Ne). These differences can help us determine the temperature of a hot gas or plasma. At higher temperatures, more highly charged ions will be present, so we can estimate the temperature by identifying the ions that are creating spectral lines.

Just as atoms and ions can absorb or emit light at particular wavelengths, so can *molecules*. Like electrons in atoms, the electrons in molecules can have only particular energies, and therefore molecules produce spectral lines when electrons in them change energy levels. However, because molecules are made of two or more atoms bound together, they can also have energy due to vibration or rotation (Figure 6.11a). It turns out that, just as its electrons can be in only specific energy levels, a molecule can rotate or vibrate only with particular amounts of energy. Thus, a molecule can absorb or emit a photon when it changes its rate of vibration or rotation.

Because molecules can change energy in three different ways, their spectra look very different from the spectra of individual atoms. Molecules produce a spectrum with many sets of tightly bunched lines, called **molecular bands** (Figure 6.11b). The energy jumps in molecules are usually smaller than those in atoms and therefore produce lower-energy photons. Thus, most molecular bands lie in the infrared rather than in the visible or ultraviolet. That is one reason why infrared telescopes and instruments are so important to astronomers.

Thermal Radiation: Every Body Does It

In a low-density gas, individual atoms or molecules are essentially independent of one another [Section 4.3]. That is why thin, low-density clouds of gas produce relatively simple emission or absorption spectra, with lines (or bands) in locations determined by the energy levels of their constituent atoms (or molecules). But what happens in an opaque object, such as a star, a planet, or you? (Although saying that a star is opaque may sound strange, it is true because we cannot see *through* a star.)

Photons of light emitted inside an opaque object cannot easily escape to the outside. Instead, they are quickly absorbed by an atom or molecule, which quickly reemits the photon—but often with a slightly different wavelength and in a different direction. In effect, the emitted photons bounce randomly around inside the object, constantly exchanging energy with its atoms or molecules.

We saw in Section 4.2 that the "bouncing around" of atoms and molecules tends to randomize their kinetic energies, giving them an average kinetic energy that we characterize as the object's *temperature*. In a similar way, the "bouncing around" of the photons inside an opaque object randomizes their radiative energies, and the resulting photon energies also depend only on the object's temperature. The photons emitted by an opaque object therefore produce a characteristic spectrum that depends only on the object's temperature. Because the spectrum depends only

Figure 6.11 Like atoms and ions, molecules also emit or absorb light at specific wavelengths.

rotation   vibration

a We can think of a two-atom molecule as two balls connected by a spring. Although this model is overly simplistic, it illustrates how molecules can rotate and vibrate. The rotations and vibrations can have only particular amounts of energy and therefore produce unique spectral fingerprints.

b This spectrum of molecular hydrogen (H_2) shows that molecular spectra consist of lines bunched into broad *molecular bands*.

on temperature, we call it a **thermal radiation** spectrum (sometimes called a *blackbody* spectrum).

No real object is "perfectly" opaque (absorbing and reemitting *all* radiation that strikes it), but many objects make close approximations. For example, almost all familiar objects—including the Sun, the planets, and even you—glow with light that approximates thermal radiation.

Two simple rules describe how a thermal radiation spectrum depends on the temperature of the emitting object:

- Rule 1: *Hotter objects emit more total radiation per unit surface area.* The radiated energy is proportional to the *fourth* power of the temperature expressed in Kelvin (*not* in Celsius or Fahrenheit). For example, a 600 K object has twice the temperature of a 300 K object and therefore radiates $2^4 = 16$ times as much total energy per unit surface area. (*Per unit surface area* is important. For example, if the 300 K object has 16 times as much surface area as the 600 K object, the *total* power emitted by both objects will be the same.)

- Rule 2: *Hotter objects emit photons with a higher average energy,* which means a shorter average wavelength (and higher average frequency).

You can see the first rule in action by playing with a light that has a dimmer switch: When you turn the switch up, the filament in the light bulb gets hotter and the light brightens. When you turn it down, the filament gets cooler and the light dims. (You can verify the changing temperature by placing your hand near the bulb.) You can see the second rule in action by observing a fireplace poker (Figure 6.12). When the poker is relatively cool, it emits only infrared radiation, which we cannot see. As it gets hot, it begins to glow red ("red hot"). If the poker continues to heat

Figure 6.12 A fireplace poker gets brighter as it is heated, demonstrating rule 1 for thermal radiation (hotter objects emit more total radiation per unit surface area). In addition, its "color" moves from infrared to red to white as it is heated, demonstrating rule 2 (hotter objects emit photons with higher average energy).

up, the average wavelength of the emitted photons gets shorter, moving toward the blue end of the visible spectrum. By the time it gets very hot, the mix of colors emitted by the poker looks white ("white hot").

Figure 6.13 shows several idealized thermal radiation spectra. Let's interpret the spectra based on the two rules. The spectra of hotter objects show bigger "humps" because they emit more total radiation per unit area (rule 1). Hotter objects also have the peaks of their humps at shorter wavelengths because of the higher average energy of their photons (rule 2).

You can see that hotter objects emit more light at *all* wavelengths, but the biggest difference appears at the shortest wavelengths. An object with a temperature of 310 K, which is about human body temperature, emits mostly in the infrared and emits no visible light at all—which explains why we don't glow in the dark! A relatively cool star, with a 3,000 K surface temperature, emits mostly red light, which is why some bright stars in our sky appear reddish,

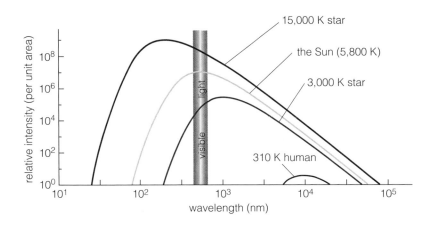

Figure 6.13 Graphs of idealized thermal radiation spectra. Note that hotter objects emit more radiation per unit surface area (intensity) at every wavelength, demonstrating rule 1 for thermal radiation. The peaks of the spectra occur at shorter wavelengths (higher energies) for hotter objects, demonstrating rule 2 for thermal radiation. Notice that the graph uses power-of-10 scales on both axes, which makes it possible to see all the curves even though the differences between them are quite large.

Mathematical Insight **6.2** **Laws of Thermal Radiation**

The two rules of thermal radiation have simple mathematical formulas. Rule 1, called the *Stefan–Boltzmann law* (named after its discoverers), is expressed as:

$$\text{emitted power per square meter} = \sigma T^4$$

where σ (Greek letter *sigma*) is a constant, $\sigma = 5.7 \times 10^{-8}$ watt/($m^2 \times$ Kelvin^4). Note that this equation is for power *per unit area*. Finding the total power radiated by an object requires multiplying the power per unit area by the object's surface area.

Rule 2, called *Wien's law*, is expressed approximately as:

$$\lambda_{max} \approx \frac{2,900,000}{T \,(\text{Kelvin})} \text{ nm}$$

where λ_{max} *(lambda-max)* is the wavelength (in nanometers) of maximum intensity, which is the peak of the hump in a thermal radiation spectrum.

Example: Consider a perfectly opaque object with a temperature of 15,000 K. How much power does it emit per square meter? What is its wavelength of peak intensity?

Solution: The emitted power per square meter from a 15,000 K object is:

$$\sigma T^4 = 5.7 \times 10^{-8} \frac{\text{watt}}{m^2 \times K^4} \times (15,000 \text{ K})^4$$

$$= 2.9 \times 10^9 \frac{\text{watt}}{m^2}$$

Its wavelength of maximum intensity is:

$$\lambda_{max} \approx \frac{2,900,000}{15,000 \,(\text{Kelvin})} \text{ nm} \approx 190 \text{ nm}$$

This wavelength is in the ultraviolet portion of the electromagnetic spectrum.

such as Betelgeuse (in Orion) and Antares (in Scorpius). The Sun's 5,800 K surface emits most strongly in green light (around 500 nm), but the Sun looks yellow or white to our eyes because it also emits other colors throughout the visible spectrum. Hotter stars emit mostly in the ultraviolet, but because our eyes cannot see ultraviolet they appear blue-white in color. If an object were heated to a temperature of millions of degrees, it would radiate mostly X rays. Some astronomical objects are indeed hot enough to emit X rays, such as disks of gas encircling exotic objects like neutron stars and black holes (see Chapter 18).

Summary of Spectral Formation

We can now summarize the circumstances under which objects produce thermal, absorption line, or emission line spectra. These rules are often called *Kirchhoff's laws*.

● Any opaque object produces thermal radiation over a broad range of wavelengths. If the object is hot enough

to produce visible light, as is the filament of a light bulb, we see a smooth, *continuous* rainbow when we disperse the light through a prism or a diffraction grating (Figure 6.14a). On a graph of intensity versus wavelength, the rainbow becomes the characteristic hump of a thermal radiation spectrum.

● When thermal radiation passes through a thin cloud of gas, the cloud leaves fingerprints that may be either absorption lines or emission lines, depending on its temperature.

   ● If the background source of thermal radiation is hotter than the cloud, the balance between emission and absorption in the cloud's spectral lines tips toward absorption. We then see absorption lines cutting into the thermal spectrum. This is the case when the light from the hot light bulb passes through a cool gas cloud (Figure 6.14b). On the graph of intensity versus wavelength, these absorption lines create dips in the thermal radiation spectrum. The

Figure 6.14 This diagram summarizes the "rules," often called *Kirchhoff's laws,* that determine whether we see thermal, absorption line, or emission line spectra. (**a**) An opaque object, such as a light bulb filament, produces a continuous spectrum of thermal radiation. (**b**) If thermal radiation passes through a thin gas that is cooler than the emitting object, dark absorption lines are superimposed on the continuous spectrum. (**c**) If the cloud of gas is viewed against a cold, dark background or is warmer than the background source of light, it produces an emission line spectrum.

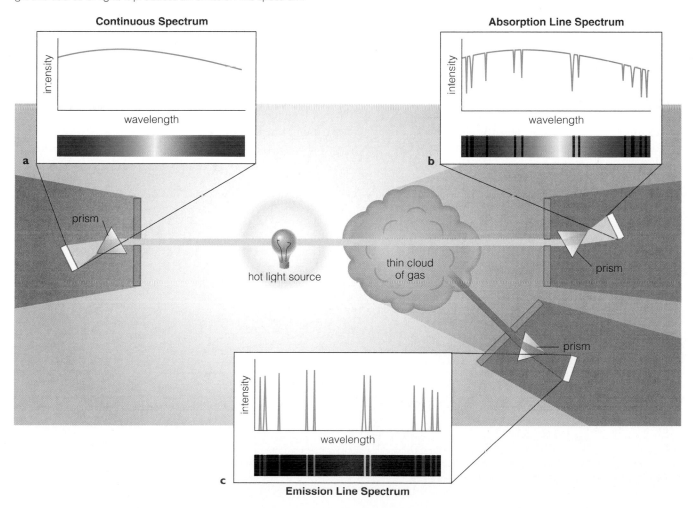

width and depth of each dip depend on how much light is absorbed by the atoms or molecules responsible for the line.

- If the background source (if any) is colder than the cloud and the cloud is warm enough to emit its own photons, the spectrum is dominated by bright emission lines produced by the cloud's atoms and molecules (Figure 6.14c). These lines create narrow peaks on a graph of intensity versus wavelength.

Our Sun is a great example of these rules in action. The chapter-opener photo (p. 152) shows the Sun's visible-light spectrum. It is an absorption line spectrum with hundreds of spectral lines (spread over many rows instead of one long row, so it fits on the page). What creates this spectrum? The interior of the hot Sun produces a continuous spectrum of thermal radiation, but this light must pass through the gas near the Sun's visible surface, or photosphere [Section 15.2], before it escapes to space. The gas in the photosphere is quite hot by earthly standards (some 6,000 K), but it is cooler than the gas in the underlying layers. As a result, the photosphere acts like a thin, cool cloud in front of a hot light source, thereby producing the Sun's absorption line spectrum.

Reflected Light

We've now covered enough material to understand spectra emitted by objects generating their own light, such as stars and clouds of interstellar gas. But most of our daily experience involves *reflected* (or *scattered*) light. The source of the light is thermal radiation from the Sun or a lamp. After this light strikes the ground, clouds, people, or other objects, we see only the wavelengths of light that are reflected. For example, a red sweatshirt absorbs blue light and re-flects red light, so its visible spectrum looks like the thermal radiation spectrum of its light source—the Sun—but with blue light missing.

In the same way that we distinguish lemons from limes, we can use color information in reflected light to learn about celestial objects. Different fruits, different rocks, and even different atmospheric gases reflect and absorb light at different wavelengths. Although the absorption features that show up in spectra of reflected light are not as distinct as the emission and absorption lines for thin gases, they still provide useful information. For example, the surface materials of a planet determine how much light of different colors is reflected or absorbed. The reflected light gives the planet its color, while the absorbed light heats the surface and helps determine its temperature.

Putting It All Together

Figure 6.15 again shows the complicated spectrum we began with in Figure 6.7, but this time with labels indicating the processes responsible for its various features. What can we say about this object from its spectrum? The hump of thermal emission peaking in the infrared shows that this object has a surface temperature of about 225 K, well below the freezing point of water. The absorption bands in the infrared come mainly from carbon dioxide, which tells us that the object has a carbon dioxide atmosphere. The emission lines in the ultraviolet come from hot gas in a high, thin layer of the object's atmosphere. The reflected light looks like the Sun's 5,800 K thermal radiation except that much of the blue light is missing, so the object must be reflecting sunlight and must look red in color. Perhaps by now you have guessed that this figure represents the spectrum of the planet Mars.

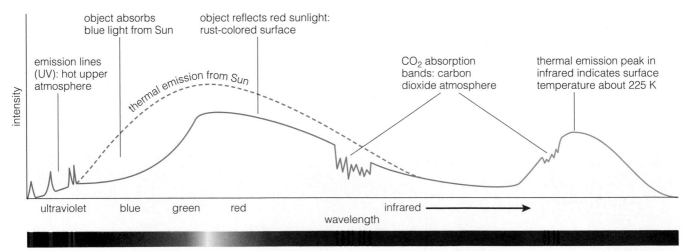

Figure 6.15 The spectrum of Figure 6.7, with interpretation. We can conclude that the object looks red in color because it absorbs more blue light than red light from the Sun. The absorption lines tell us that the object has a carbon dioxide atmosphere, and the emission lines tell us that its upper atmosphere is hot. The hump in the infrared (near the right of the diagram) tells us that the object has a surface temperature of about 225 K. It is a spectrum of the planet Mars.

6.5 The Doppler Shift

You're probably already amazed at the volume of information contained in light, but there is still more. In particular, light contains information about motion. We can determine the radial motion (toward or away from us) of a distant object from changes in its spectrum caused by the **Doppler effect**.

You've probably noticed the Doppler effect on the *sound* of a train whistle near train tracks (Figure 6.16). As the train approaches, its whistle is relatively high pitched. As the train recedes, the sound is relatively low pitched. Just as the train passes by, you hear the dramatic change from high to low pitch—a sort of "weeeeeeee–ooooooooooh" sound. To visualize the Doppler effect, imagine that the train's sound waves are bunched up ahead of it, resulting in shorter wavelengths and thus the high pitch you hear as the train approaches. Behind the train, the sound waves are stretched out to longer wavelengths, resulting in the low pitch you hear as the train recedes.

The Doppler effect causes similar shifts in the wavelengths of light. If an object is moving toward us, its entire spectrum is shifted to shorter wavelengths. Because shorter wavelengths are bluer when we are dealing with visible light, the Doppler shift of an object coming toward us is called a **blueshift**. If an object is moving away from us, its light is shifted to longer wavelengths. We call this a **redshift** because longer wavelengths are redder when we are dealing with visible light. Astronomers use *blueshift* and *redshift* even when not dealing with visible light.

Spectral lines provide the reference points we use to identify and measure Doppler shifts (Figure 6.17). For example, suppose we recognize the pattern of hydrogen lines in the spectrum of a distant object. We know the **rest wavelengths** of the hydrogen lines—that is, their wavelengths in stationary clouds of hydrogen gas—from laboratory experiments in which a tube of hydrogen gas is heated so the wavelengths of the spectral lines can be measured. If the hydrogen lines from the object appear at longer wavelengths, then we know that they are redshifted and the object is moving away from us. The larger the shift, the faster the object is moving. If the lines appear at shorter wavelengths, then we know that they are blueshifted and the object is moving toward us.

THINK ABOUT IT

Suppose the hydrogen emission line with a rest wavelength of 121.6 nm (the transition from level 2 to level 1) appears at a wavelength of 120.5 nm in the spectrum of a particular star. Given that these wavelengths are in the ultraviolet, is the shifted wavelength closer to or farther from blue visible light? Why, then, do we say that this spectral line is *blueshifted*?

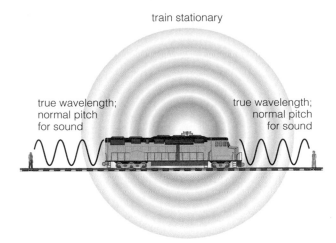

train stationary

true wavelength; normal pitch for sound

true wavelength; normal pitch for sound

a Each circle represents the crests of sound waves going in all directions from the train whistle. The circles represent wave crests coming from the train at different times, say, 1/10 second apart.

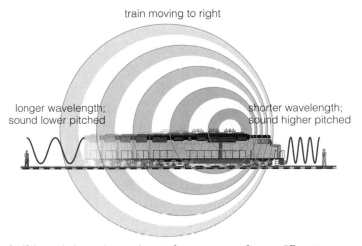

train moving to right

longer wavelength; sound lower pitched

shorter wavelength; sound higher pitched

b If the train is moving, each set of waves comes from a different location. Thus, the waves appear bunched up in the direction of motion and stretched out in the opposite direction.

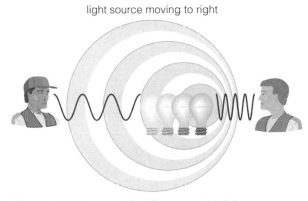

light source moving to right

c We get the same basic effect from a moving light source.

Figure 6.16 The Doppler effect.

Lines at rest wavelengths
(as measured in a laboratory).

Object 1
Lines redshifted: Object is moving
away from us.

Object 2
Greater redshift: Object is moving
away faster than Object 1.

Object 3
Lines blueshifted: Object is
moving toward us.

Object 4
Greater blueshift: Object is moving
toward us faster than Object 3.

Figure 6.17 Spectral lines provide the crucial reference points for measuring Doppler shifts.

In general, the Doppler shift tells us only part of an object's full motion—the part that is directed toward or away from us (the object's *radial* component of motion). Doppler shifts do not give us any information about how fast an object is moving across our line of sight (the object's *tangential* component of motion). For example, consider three stars all moving at the same speed, with one moving directly away from us, one moving across our line of sight, and one moving diagonally away from us

(Figure 6.18). The Doppler shift will tell us the full speed only of the first star. The Doppler shift will not measure any speed for the second star, because none of its motion is directed toward or away from us. For the third star, the Doppler shift will tell us only the part of the star's speed that is directed away from us. (To measure how fast an object is moving across our line of sight, we must observe it long enough to notice how its position gradually shifts across our sky.)

Mathematical Insight **6.3** **The Doppler Shift**

As long as an object's radial velocity is small compared to the speed of light (i.e., less than a few percent of c), we can use a simple formula to calculate the radial velocity (toward or away from us) of an object from its Doppler shift:

$$\frac{\text{radial velocity}}{\text{speed of light}} = \frac{\text{shifted wavelength} - \text{rest wavelength}}{\text{rest wavelength}}$$

If the result is positive, the object has a redshift and is moving away from us. A negative result means that the object has a blueshift and is moving toward us. The formula can also be written symbolically as:

$$\frac{v}{c} = \frac{\Delta\lambda}{\lambda_0} \quad \text{or} \quad v = \frac{\Delta\lambda}{\lambda_0} \times c$$

where v is the object's radial velocity, c is the speed of light, λ_0 is the rest wavelength of a particular spectral line, and $\Delta\lambda$ is its wavelength shift (positive for a redshift and negative for a blueshift).

Example: The rest wavelength of one of the visible lines of hydrogen is 656.285 nm. This line is easily identifiable in the spectrum of the bright star Vega, but it appears at a wavelength of 656.255 nm. What is the radial velocity of Vega?

Solution: The line's wavelength in Vega's spectrum is slightly shorter than its rest wavelength, so the line is blueshifted and Vega's radial motion is *toward* us. Vega's radial velocity is:

$$\frac{656.255 \text{ nm} - 656.285 \text{ nm}}{656.285 \text{ nm}} \times 300,000 \frac{\text{km}}{\text{s}}$$

$$= (-4.57 \times 10^{-5}) \times (3 \times 10^5) \frac{\text{km}}{\text{s}}$$

$$= -13.7 \frac{\text{km}}{\text{s}}$$

The negative answer confirms that Vega is moving *toward* us at 13.7 km/s.

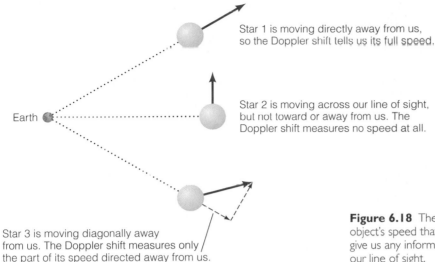

Star 1 is moving directly away from us, so the Doppler shift tells us its full speed.

Star 2 is moving across our line of sight, but not toward or away from us. The Doppler shift measures no speed at all.

Earth

Star 3 is moving diagonally away from us. The Doppler shift measures only the part of its speed directed away from us.

Figure 6.18 The Doppler shift measures only the portion of an object's speed that is directed toward or away from us. It does not give us any information about how fast an object is moving across our line of sight.

The Doppler effect not only tells us how fast a distant object is moving toward or away from us but also can reveal information about motion *within* the object. For example, suppose we look at spectral lines of a planet or star that happens to be rotating (Figure 6.19). As the object rotates, light from the part of the object rotating toward us will be blueshifted, light from the part rotating away from us will be redshifted, and light from the center of the object won't be shifted at all. The net effect, if we

look at the whole object at once, is to make each spectral line appear *wider* than it would if the object were not rotating. The faster the object is rotating, the broader in wavelength the spectral lines become. Thus, we can determine the rotation rate of distant objects by measuring the width of their spectral lines. We will see later that the Doppler effect on the spectra of celestial objects reveals even more information.

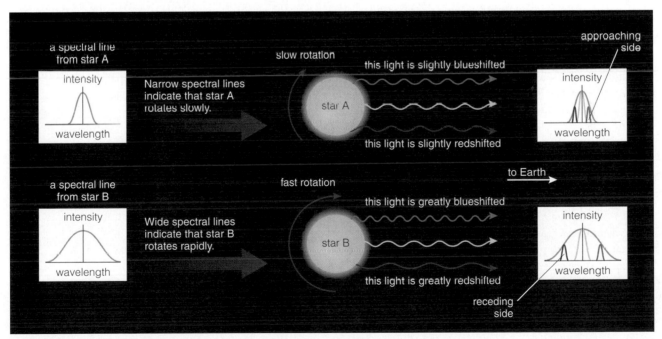

Figure 6.19 The Doppler effect broadens the widths of the spectral lines of rotating objects. One side of a rotating object is moving toward us, creating a blueshift, while the other side is rotating away from us, creating a redshift. The faster the rotation, the greater the spread in wavelength between the light from the two sides. When we look at a spectral line for the object as a whole, we see light from all parts of the object—that is, the parts with blueshifts as well as the parts with redshifts—at the same time. Thus, the greater Doppler shifts in a fast-rotating object make the overall spectral line wider because the light is spread over a greater range of wavelengths.

Putting Chapter 6 into Context

This chapter was devoted to one essential purpose: understanding how to read the messages contained in the spectra of distant objects. "Big picture" ideas that will help you keep your understanding in perspective include the following:

● There is far more to light than meets the eye. By dispersing light into a spectrum with a prism or a diffraction grating, we discover a wealth of information about the object from which the light has come. Most of what we know about the universe comes from information that we receive in the form of light.

● The visible light that our eyes can see is only a small portion of the complete electromagnetic spectrum. Different portions of the spectrum may contain different pieces of the story of a distant object, so it is important to study spectra at many wavelengths.

● The spectrum of any object is determined by interactions of light and matter. These interactions can produce *emission lines, absorption lines,* or *thermal radiation*. The spectra of most objects contain some degree of all three of these. Some spectra also include transmitted light (from a light source behind the object) and reflected light.

● By studying the spectra of a distant object, we can determine its composition, surface temperature, motion toward or away from us, rotation rate, and more.

SUMMARY OF KEY CONCEPTS

6.1 Light in Everyday Life

• *What is the difference between energy and power?* Power is the rate at which energy is used. The standard unit of power is 1 watt = 1 joule/s.

• *What are the four ways in which light and matter can interact?* Matter can emit light, absorb light, transmit light, or reflect (or scatter) light.

6.2 Properties of Light

• *In what way is light a wave?* Light is an electromagnetic wave—a wave of vibrating electric and magnetic fields. Light is characterized by a wavelength and a frequency. It travels at the speed of light, *c*, which is always the same.

• *In what way is light made of particles?* Light comes in individual photons, each with a specific energy that depends on its frequency (or wavelength).

• *How are wavelength, frequency, and energy related for photons of light?* Frequency increases when wavelength decreases and vice versa, because frequency × wavelength = speed of light. Energy is proportional to frequency.

6.3 The Many Forms of Light

• *What do we call the various forms of light that make up the electromagnetic spectrum?* In order of increasing frequency and energy (or decreasing wavelength), the forms of light are radio waves, infrared, visible light, ultraviolet, X rays, and gamma rays.

6.4 Light and Matter

• *How can we use emission or absorption lines to determine the composition of a distant object?* Emission or absorption lines occur only at specific wavelengths corresponding to particular energy-level transitions in atoms or molecules. Each chemical element has a unique spectral signature consisting of a particular set of emission or absorption lines.

• *Why do all material objects emit light?* All material objects are made of atoms that have thermal energy, and interactions among these atoms continually convert some of that thermal energy into radiative energy (light). Thus, all material objects emit a kind of light called thermal radiation. The spectrum of the emitted thermal radiation depends on the object's temperature and includes visible light only for fairly hot objects like light bulb filaments or stars. Objects with relatively low temperatures, such as planets and people, emit primarily infrared light and no visible light.

• *What are the two rules of thermal radiation?* Rule 1: Hotter objects emit more total radiation per unit area. Rule 2: Hotter objects emit photons with a higher average energy.

6.5 The Doppler Shift

• *What is a Doppler shift?* It is a shift in the wavelength of an object's light caused by its motion toward or away from us.

• *What do we learn from a redshift or blueshift?* The sizes of these shifts tell us how fast the object is moving away from us (redshift) or toward us (blueshift). The Doppler shift does not tell us about motion across our line of sight.

• *How does a star's rotation affect its spectral lines?* Because of Doppler shifts, stars that rotate faster have broader spectral lines.

❓ Does It Make Sense?

Decide whether each statement makes sense and explain why it does or does not.

1. If you could view a spectrum of light reflecting off a blue sweatshirt, you'd find the entire rainbow of color (looking the same as a spectrum of white light).

2. Because of their higher frequency, X rays must travel through space faster than radio waves.

3. If the Sun's surface became much hotter (while the Sun's size remained the same), the Sun would emit more ultra violet light but less visible light than it currently emits.

4. A black sheet of paper absorbs all the light that falls on it and emits no radiative energy at all.

5. If you could see infrared light, you would see the backs of your eyelids when you closed your eyes.

6. If you had X-ray vision, then you could read this entire book without turning any pages.

7. If you want to see an object that is too cold to emit visible light, you should try looking at it with an instrument that can detect ultraviolet light.

8. If a distant galaxy has a substantial redshift (as viewed from our galaxy), then anyone living in that galaxy would see a substantial redshift in a spectrum of the Milky Way Galaxy.

Problems

(Quantitative problems are marked with an asterisk.)

9. *Light Transmission.* What does it mean for a material to be *transparent?* To be *opaque?* Give at least two examples each of materials that are partially or fully transparent and of materials that are partially or fully opaque.

10. *Wave Definitions.* Define each of the following terms as it applies to waves: *wavelength, frequency, cycles per second, hertz, speed.*

11. *Photon Properties.* What is a *photon?* In what way is a photon like a particle? In what way is it like a wave?

12. *Types of Spectra.* Summarize the circumstances under which objects produce *thermal, emission line,* or *absorption line* spectra.

13. *Doppler Basics.* Describe the *Doppler effect* for light and what we can learn from it. What does it mean to say that radio waves are *blueshifted?*

14. *Spectral Summary.* Clearly explain how studying an object's spectrum can allow us to determine each of the following properties of the object.

    a. The object's surface chemical composition.

    b. The object's surface temperature.

    c. Whether the object is a thin cloud of gas or something more substantial.

    d. Whether the object has a hot upper atmosphere.

    e. The speed at which the object is moving toward or away from us.

    f. The object's rotation rate.

15. *Planetary Spectrum.* Suppose you take a spectrum of light coming from a planet that looks blue to the eye. Do you expect to see any visible light in the planet's spectrum? Is the visible light emitted by the planet, reflected by the planet, or both? Which (if any) portions of the visible spectrum do you expect to find "missing" in the planet's spectrum? Explain your answers clearly.

16. *Hotter Sun.* Suppose the surface temperature of the Sun were about 12,000 K, rather than 6,000 K.

    a. How much more thermal radiation would the Sun emit?

    b. How would the thermal radiation spectrum of the Sun be different?

    c. Do you think it would still be possible to have life on Earth? Explain.

17. *The Doppler Effect.* In hydrogen, the transition from level 2 to level 1 has a rest wavelength of 121.6 nm. Suppose you see this line at a wavelength of 120.5 nm in Star A, at 121.2 nm in Star B, at 121.9 nm in Star C, and at 122.9 nm in Star D. Which stars are coming toward us? Which are moving away? Which star is moving fastest relative to us (either toward or away from)? Explain your answers without doing any calculations.

18. *The Expanding Universe.* Recall from Chapter 1 that we know the universe is expanding because (1) all galaxies outside our Local Group are moving away from us and (2) more distant galaxies are moving faster. How do you think Doppler shift measurements allow us to know these two facts?

*19. *Human Wattage.* A typical adult uses about 2,500 Calories of energy each day.

    a. Using the fact that 1 Calorie is about 4,000 joules, convert the typical adult energy usage to units of joules per day.

    b. Use your answer from part (a) to calculate a typical adult's average *power* requirement, in watts. Compare this to that of a light bulb.

*20. *Wavelength, Frequency, and Energy.*

    a. What is the frequency of a visible light photon with wavelength 550 nm?

    b. What is the wavelength of a radio photon from an "AM" radio station that broadcasts at 1,120 kilohertz? What is its energy?

    c. What is the energy (in joules) of an ultraviolet photon with wavelength 120 nm? What is its frequency?

    d. What is the wavelength of an X-ray photon with energy 10 keV (10,000 eV)? What is its frequency? (*Hint:* Recall that 1 eV = 1.60×10^{-19} joule.)

*21. *How Many Photons?* Suppose that all the energy from a 100-watt light bulb came in the form of photons with wavelength 600 nm. (This is not quite realistic; see problem 24.)

   a. Calculate the energy of a *single* photon with wavelength 600 nm.

   b. How many 600-nm photons must be emitted each second to account for all the light from this 100-watt light bulb? Based on your answer, explain why we don't notice the particle nature of light in our everyday lives.

*22. *Taking the Sun's Temperature.* The Sun radiates a total power of about 4×10^{26} watts into space. The Sun's radius is about 7×10^8 meters.

   a. Calculate the average power radiated by each square meter of the Sun's surface. (*Hint:* The formula for the surface area of a sphere is $A = 4\pi r^2$.)

   b. Using your answer from part (a) and the Stefan–Boltzmann law (see Mathematical Insight 6.2), calculate the average surface temperature of the Sun. (*Note:* The temperature calculated this way is called the Sun's *effective temperature*.)

*23. *Doppler Calculations.* Calculate the speeds of each of the stars described in problem 17. Be sure to state whether each star is moving toward or away from us.

*24. *Understanding Light Bulbs.* A standard (incandescent) light bulb uses a hot tungsten coil to produce a thermal radiation spectrum. The temperature of this coil is typically about 3,000 K.

   a. What is the wavelength of maximum intensity for a standard light bulb? Compare this to the 500-nm wavelength of maximum intensity for the Sun. Also explain why standard light bulbs must emit a substantial portion of their radiation as invisible, infrared light.

   b. Overall, do you expect the light from a standard bulb to be the same as, redder than, or bluer than light from the Sun? Why? Use your answer to explain why professional photographers use a different type of film for indoor photography than for outdoor photography.

   c. *Fluorescent* light bulbs primarily produce emission line spectra rather than thermal radiation spectra. Explain why, if the emission lines are in the visible part of the spectrum, a fluorescent bulb can emit more visible light than a standard bulb of the same wattage.

   d. Today, *compact fluorescent* light bulbs are designed to produce so many emission lines in the visible part of the spectrum that their light looks very similar to the light of standard bulbs. However, they are much more energy efficient: A 15-watt compact fluorescent bulb typically emits as much visible light as a standard 75-watt bulb. Although compact fluorescent bulbs generally cost more than standard bulbs, is it possible that they could save you money? Besides initial cost and energy efficiency, what other factors must be considered?

Discussion Questions

25. *The Changing Limitations of Science.* In 1835, French philosopher Auguste Comte stated that the composition of stars could never be known by science. Although spectral lines had been seen in the Sun's spectrum by that time, not until the mid-1800s did scientists recognize that spectral lines give clear information about chemical composition (primarily through the work of Foucault and Kirchhoff). Why might our present knowledge have seemed unattainable in 1835? Discuss how new discoveries can change the apparent limitations of science. Today, other questions seem beyond the reach of science, such as the question of how life began on Earth. Do you think such questions will ever be answerable by science? Defend your opinion.

26. *Your Microwave Oven.* Microwaves is a name sometimes given to light near the long-wavelength end of the infrared portion of the spectrum. A *microwave oven* emits microwaves that have just the right wavelength needed to cause energy level jumps in water molecules. Use this fact to explain how a microwave oven cooks your food. Why doesn't a microwave oven make a plastic dish get hot?

For a complete list of media resources available, go to www.astronomyplace.com and choose Chapter 6 from the pull-down menu.

 Astronomy Place Web Tutorials

Tutorial Review of Key Concepts

Use the interactive **Tutorials** at www.astronomyplace.com to review key concepts from this chapter.

Light and Spectroscopy Tutorial

Lesson 1 Radiation, Light, and Waves

Lesson 2 Spectroscopy

Lesson 3 Atomic Spectra—Emission and Absorption Lines

Lesson 4 Thermal Radiation

Doppler Effect Tutorial

Lesson 1 Understanding the Doppler Shift

Lesson 2 Using Emission and Absorption Lines to Measure the Doppler Shift

Supplementary Tutorial Exercises

Use the interactive **Tutorial Lessons** to explore the following questions.

Light and Spectroscopy Tutorial, Lesson 3

1. Why do neon lights come in so many different colors?

2. How can we tell the difference between lithium and carbon in the spectrum of a distant object?

3. How will an interstellar cloud affect the spectrum of light that we see from a star that lies behind it?

Doppler Effect Tutorial, Lesson 1

1. As the source of a sound passes you, which two properties will you hear change? Which change is due to the Doppler effect?

2. Which property of a sound wave is responsible for the pitch that your ears hear? How does the Doppler effect change the sound you hear?

3. How can the Doppler effect for light be used to measure the speed of a car?

Doppler Effect Tutorial, Lesson 2

1. How can we use precise wavelengths of spectral lines to measure the speed of a distant object?

2. Does it matter whether lines are in emission or absorption for the Doppler effect? Explain.

Web Projects

Take advantage of the useful Web links on www.astronomyplace.com to assist you with the following projects.

1. *Kids and Light.* Visit one of the many Web sites designed to teach middle and high school students about light. Read the content and try the activities. If you were a teacher, would you find the site useful for your students? Why or why not? Write a one-page summary of your conclusions.

2. *Light Bulbs.* Learn about alternatives to standard incandescent light bulbs, such as fluorescent lights, compact fluorescent lights, and halogen lights. Write a short report summarizing some of the advantages and disadvantages of each technology.

3. *Medical Imaging.* Learn about CAT scans or other technologies for medical imaging of the human body. How do they work? How are such technologies similar to those employed by astronomers to learn about the universe? Write a short report summarizing your findings.

7 Telescopes
Portals of Discovery

We are in the midst of a great revolution in human understanding of the universe. Astonishing new discoveries about the early history of the universe, about the lives of galaxies and stars, and of planets around other stars are frequent features of the daily news.

The primary fuel for this astronomical revolution comes from recent and significant advances in telescope technology. A growing number of ever-larger telescopes are operating around the world, and new technologies have vastly improved the quality of data that can be obtained with telescopes on the ground. Meanwhile, telescopes lofted into space are offering views of the heavens unobstructed by Earth's atmosphere while also allowing us to study wavelengths of light that do not penetrate to the ground. By studying light from across the entire spectrum, we can learn far more about the universe than we can by studying visible light alone.

Because telescopes are the portals through which we study the universe, understanding them can help you understand both the triumphs and the limitations of modern astronomy. In this chapter, we will explore the basic principles by which telescopes work, along with some of the technological advances fueling the current revolution in astronomy.

7.1 Eyes and Cameras: Everyday Light Sensors

We observe the world around us with the five basic senses: touch, taste, smell, hearing, and sight. We learn about the world by using our brains to analyze and interpret the data recorded by our senses. The science of astronomy progresses similarly. We collect data about the universe, and then we analyze and interpret the data to develop theories about how the universe works. Within our solar system, we can analyze some matter directly. We can sample Earth's surface and study meteorites that fall to Earth. Occasionally, we send spacecraft to sample the surfaces or atmospheres of other worlds. Aside from these few samples collected in our own solar system, nearly all other data about the universe come to us in the form of light.

Astronomers collect light with telescopes and record light with photographic film or electronic detectors. You are already familiar with many of the basic principles because of your everyday experience with eyes and cameras.

The Eye

The eye is a remarkably complex organ, but its basic components are a *lens,* a *pupil,* and a *retina* (Figure 7.1). The retina contains light-sensitive cells (called *cones* and *rods*) that, when triggered by light, send signals to the brain via the optic nerve.

The lens of an eye acts much like a simple glass lens. Light rays that enter the lens farther from the center are bent more, and rays that pass directly through the center are not bent at all. In this way, parallel rays of light, such as the light from a distant star, converge to a point called the **focus** (Figure 7.2a). If you have perfect vision, the focus of your lens is on your retina. That is why distant stars appear as *points* of light to our eyes or on photographs.

Light rays that come from different directions, such as from different parts of an object, converge at different points to form an **image** of the object (Figure 7.2b). The place where the image appears in focus is called the **focal plane** of the lens. In an eye with perfect vision, the focal plane is on your retina. (The retina actually is curved, rather than a flat plane, but we will ignore this detail.) The image formed by a lens is upside down but is flipped right side up by your brain, where the true miracle of vision occurs.

The pupil controls the amount of light that enters the eye by adjusting the size of its opening. The pupil dilates (opens wider) in low light levels, allowing your eye to gather as many photons as possible. It constricts in bright light so that your eye is not overloaded with light.

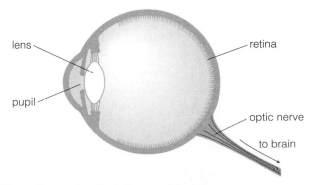

Figure 7.1 A simplified diagram of the human eye.

Cover one eye and look toward a bright light. Then look immediately into a mirror and compare the openings of your two pupils. Which one is wider, and why? Why do eye doctors dilate your pupils during eye exams?

One way to quantify the sharpness of your vision is to measure the smallest angle over which your eyes can tell that two dots—or two stars—are distinct. Imagine looking at a car heading toward you on a long, straight road at

Figure 7.2 These diagrams show how lenses and eyes focus light.

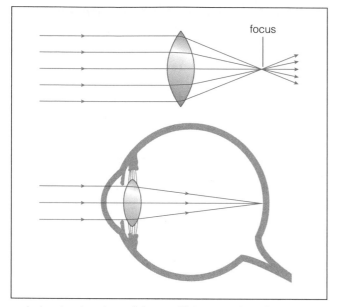

focus

a A glass lens bends parallel rays of light to a point called the focus of the lens. In an eye with perfect vision, the lens focuses light on the retina.

night. When the car is very far away, your eyes cannot distinguish the headlights individually, and they look like one light. As the car comes closer, the angular separation of the two headlights gets bigger, and you eventually see two distinct headlights (Figure 7.3). The smallest angular separation at which you can tell that the headlights are distinct is the **angular resolution** of your eyes.

At best, the human eye has an angular resolution of about 1 arcminute ($\frac{1}{60}°$), meaning that two stars will appear distinct if they lie farther than 1 arcminute apart in the sky. Two stars separated by *less* than 1 arcminute appear as a single point of light. Most stars are actually members of binary or multiple-star systems, but these systems appear to our eyes as single stars because the angular separation of the individual stars is far below the angular resolution of our eyes.

Cameras and Film

The basic operation of a camera is quite similar to that of an eye (Figure 7.4). The camera lens plays the role of the lens of the eye, and film plays the role of the retina. The camera is "in focus" when the film lies in the focal plane of the lens. Fancier cameras even have an adjustable circular opening, called the camera's *aperture,* that controls the amount of light entering the camera just as the pupil controls the amount of light entering the eye.

The chemicals in photographic film change in response to light, thereby recording an image. Recording the image of objects on film offers at least two important advantages over simply looking at them or drawing them. First, an image on film is much more reliable and detailed than a drawing. Second, the eye continually and automatically sends images to the brain, but we can use a camera's

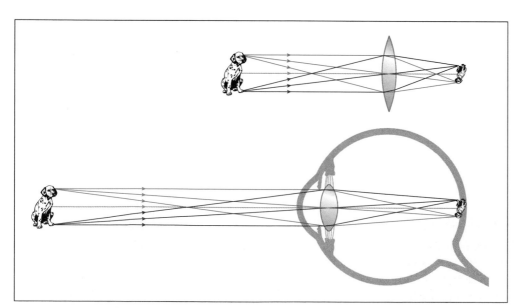

b Light from different parts of an object focuses at different points to make an image of the object. Note that the image formed on the retina is upside-down. The world does not look upside down to you because your brain flips the image as it processes the information it receives from the retina.

shutter to control the amount of time, called the **exposure time**, over which light collects on a single piece of film. A longer exposure time means that more photons reach the film, allowing more opportunities for the light-sensitive chemicals to change in response to light. As a result, long-exposure photographs can reveal images of objects far too faint for our eyes to see.

Electronic Detectors

Thanks to the advantages of film over the human eye, the advent of photography in the mid-1800s spurred a leap in astronomical data collection. However, photographic film is far from ideal for recording light. Most photons striking film have no effect at all: Fewer than 10% of the visible photons reaching the film cause a change in the light-sensitive chemicals.* Thus, your film may record nothing if too little light reaches it. At the other extreme, a long exposure time can *overexpose* (or *saturate*) your film,

---

*The probability that an individual photon will be detected is called the *quantum efficiency*. It is about 1% for the human eye, up to about 10% for film, and 90% or more for electronic detectors.

Figure 7.3 The angular separation of the headlights increases as the car comes closer.

affecting so large a fraction of the light-sensitive chemicals that details of the image are lost.

Today, digital cameras and camcorders come equipped to record images with electronic detectors called *charge-coupled devices*, or **CCDs**, that can accurately record

---

Mathematical Insight 7.1 Angular Separation

The angular separation of two points is related to their actual separation and distance. The figure shows two points with an actual separation s between them. The angle α is the angular separation between the two points when viewed from a distance d.

As long as d is much larger than s, we can think of s as a small portion of an imaginary circle with radius d. The circumference of a circle is $2 \times \pi \times$ radius, so the circumference of the dotted circle in the figure is $2\pi d$. Thus, the separation s represents a fraction $s/(2\pi d)$ of this circumference. We find the angle α by multiplying this fraction by the 360° in a full circle:

$$\alpha = \frac{s}{2\pi d} \times 360°$$

Example 1: Suppose the two headlights on a car are separated by 1.5 meters and you are looking at the car from a distance of 500 meters. What is the angular separation of the headlights? Can your eyes resolve the two headlights?

Solution: The separation of the headlights is $s = 1.5$ m, and their distance is $d = 500$ m. Thus, their angular separation is

$$\alpha = \frac{1.5 \text{ m}}{2\pi \times 500 \text{ m}} \times 360° = 0.17°$$

The angular resolution of the human eye is about $\frac{1}{60}° \approx 0.017°$, or 10 times smaller than the angular separation of the two headlights. Thus, your eyes can easily resolve the two headlights at a distance of 500 meters.

Example 2: The angular diameter of the Moon is about 0.5°, and the Moon is about 380,000 km away. Estimate the diameter of the Moon.

Solution: We are given $\alpha = 0.5°$ and $d = 380,000$ km. We are looking for s, the diameter of the Moon in this case. We first solve the angular separation equation for s:

$$\alpha = \frac{s}{2\pi d} \times 360° \quad \Rightarrow \quad s = \frac{2\pi d}{360°} \times \alpha$$

Now we substitute the given values:

$$s = \frac{2\pi \times 380,000 \text{ km}}{360°} \times 0.5° \approx 3,300 \text{ km}$$

This estimate is fairly close to the Moon's actual diameter (3,476 km), which we could find by using more precise values for the Moon's angular diameter and distance.

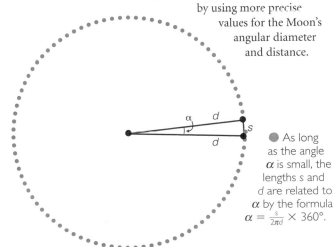

● As long as the angle α is small, the lengths s and d are related to α by the formula $\alpha = \frac{s}{2\pi d} \times 360°$.

film

shutter

lens

a Basic components of a camera.

Figure 7.4 A camera works much like an eye.

shutter

film or electronic detector

b When the shutter is open, light passes through the lens to form an image on the film or electronic detector.

CCD

Figure 7.5 Digital cameras capture light electronically with CCDs. A CCD is a silicon chip divided into a grid of squares called picture elements, or pixels.

individual pixel (typical digital cameras have 1 million–6 million pixels)

90% or more of the photons that strike them (Figure 7.5). A CCD is a chip of silicon carefully engineered to be extraordinarily sensitive to photons. The chip is physically divided into a grid of squares called *picture elements,* or **pixels** for short. When a photon strikes a pixel, it causes a bit of electric charge to accumulate. Each subsequent photon striking the same pixel adds to this accumulated electric charge. After an exposure is complete, a computer measures the total electric charge in each pixel, thus determining how many photons have struck each one. The overall image is stored in the computer as an array of numbers representing the results from each pixel. At this point, the image can be manipulated through techniques of *image processing* to bring out details that otherwise might be missed in analyzing the image.

7.2 Telescopes: Giant Eyes

Our naked eyes allow us to admire the beauty of the night sky, but for most astronomical purposes they are completely inadequate. Their small size limits their angular resolution and the amount of light they can collect. They are sensitive only to visible light. And they are attached to bodies that require a pleasantly warm atmosphere for survival—an atmosphere that distorts light and prevents much of the electromagnetic spectrum from reaching the ground.

Telescopes solve all these problems. We can design telescopes to compensate for the distorting effects of our atmosphere, or we can launch them into space. We can build telescopes and detectors that are sensitive to light our eyes cannot see, such as radio waves or X rays. Most important, a The basic design of a refracting telescope. The eyepiece contains a small lens that brings the collected light to a focus for an observer looking through it.

Figure 7.6 Refracting telescopes.

b The 1-meter refractor at the University of Chicago's Yerkes Observatory is the world's largest refracting telescope.

telescopes function as giant eyes, collecting far more light with far better angular resolution than our naked eyes.

Basic Telescope Design

Telescopes come in two basic designs: *refracting* and *reflecting*. A **refracting telescope** operates more like an eye, using transparent glass lenses to focus the light from distant objects (Figure 7.6). The earliest telescopes, including those built by Galileo, were refracting telescopes. The world's largest refracting telescope, completed in 1897, has a lens that is 1 meter (40 inches) in diameter and a telescope tube that is 19.5 meters (64 feet) long.

A **reflecting telescope** uses a precisely curved **primary mirror** to gather and focus light (Figure 7.7). This mirror focuses light at the **prime focus**, which lies *in front of* the mirror. A **secondary mirror**, placed just below the prime focus, reflects the light to a location that is more convenient for viewing or for attaching instruments. Figure 7.8 shows three common designs: The secondary mirror may reflect light through a hole in the primary mirror (a *Cassegrain focus*), to a hole in the side (a *Newtonian focus*, particularly common in smaller telescopes), or to a third mirror that deflects light toward instruments that are not attached to the telescope (a *coudé focus* or a *Nasmyth focus*). Although the secondary mirror blocks some of the light entering a reflecting telescope, this does not cause a serious problem because only a small fraction of the incoming light is blocked.

Today, reflecting telescopes are used for most astronomical research, mainly for two practical reasons. First, because light passes *through* the lens of a refracting telescope, lenses must be made from clear, high-quality glass with precisely shaped surfaces on both sides. In contrast, only the reflecting surface of a mirror must be precisely shaped, and the quality of the underlying glass is not a factor. Second, large glass lenses are extremely heavy and can be held in place only by their edges. Since the large lens is at the top of a refracting telescope, it is difficult to stabilize refracting telescopes and to prevent large lenses from deforming. The primary mirror of a reflecting telescope is mounted at the bottom, where its weight presents a far less serious problem. (A third problematic feature of lenses, called *chromatic aberration*, occurs because a lens brings different colors of light into focus at slightly different places. This problem can be minimized by using combinations of lenses.)

Fundamental Properties of Telescopes

The power of any telescope, whether refracting or reflecting, is characterized by two fundamental properties: its **light-collecting area**, which describes the telescope's size in terms of how much light it can collect, and its **angular resolution**, which tells us how much detail we can see in the telescope's images. Let's examine each property in a bit more detail.

a The basic design of a reflecting telescope. The primary mirror focuses light at the prime focus, which lies in front of it. A camera (or secondary mirror) placed at the prime focus will block some of the incoming light. This is not a serious problem, as long as it blocks only a small fraction of the incoming light.

Figure 7.7 Reflecting telescopes.

b Reflecting telescopes are used for most astronomical research. This photograph shows the Gemini North telescope, located on the summit of Mauna Kea, Hawaii. The primary mirror, visible at the bottom of the larger lattice tube, is 8 meters in diameter. The prime focus is located above the primary mirror in the smaller tube-shaped structure. A secondary mirror, located just below the prime focus, reflects light back down through the hole in the center of the primary mirror.

Cassegrain Focus

Newtonian Focus

Coudé Focus

Figure 7.8 Alternative designs for reflecting telescopes. Each of these designs incorporates a small, secondary mirror just below the prime focus. The secondary mirror reflects the light to a location that is convenient for viewing or for attaching instruments.

Light-Collecting Area The "size" of a telescope is usually described by the diameter of its primary mirror (or lens). For example, a "5-meter telescope" is a telescope with a primary mirror measuring 5 meters in diameter. The light-collecting area is proportional to the *square* of the mirror diameter. (For a round mirror surface, the mirror area is the area of a circle, or $\pi \times (radius)^2$.) Thus, a 2-meter telescope has $2^2 = 4$ times the light-collecting area of a 1-meter telescope.

For more than 40 years after its opening in 1947, the 5-meter Hale telescope on Mount Palomar outside San Diego remained the world's most powerful. (A Russian telescope built in 1976 was larger, but it suffered optical quality problems that made it far less useful.) Building larger telescopes posed a technological challenge, because large mirrors tend to sag under their own weight.

Technological innovations have made it possible to build very large, low-weight mirrors, fueling a boom in large telescope construction (Table 7.1). For example, each

Magnification and Telescopes

You are probably familiar with how binoculars, magnifying glasses, and telephoto lenses for cameras make objects appear larger in size—the phenomenon we call magnification. Many people assume that astronomical telescopes are also characterized by their magnification. In fact, professional astronomers are much more interested in a telescope's light-collecting area and angular resolution because they do not look through a telescope with their own eyes. A light-gathering detector records the image instead, and that image can later be magnified on a computer to any size desired. However, magnifying an image in this way cannot improve the sharpness of the image, which is determined by the angular resolution of the telescope that created it.

Table 7.1 Largest Optical (Visible-Light) Telescopes

Size	Telescope Name	Sponsor	Location	Operational Date	Special Features
10.4 m	Gran Telescopio Canarias	Spain, Mexico, U. Florida	Canary Islands	2004*	Segmented primary mirror based on mirrors for Keck telescopes
10 m	Keck I	Cal Tech, U. California, NASA	Mauna Kea, HI	1993	Primary mirror consists of 36 1.8-m hexagonal segments
10 m	Keck II	Cal Tech, U. California, NASA	Mauna Kea, HI	1996	Twin of Keck I; future plans for interferometry with Keck I
9.2 m	Hobby–Eberly	U. Texas, Penn State, Stanford, Germany	Mt. Locke, TX	1997	Consists of 91 1-m segments, for a total diameter of 11 m, but only 9.2 m can be used at a time; designed primarily for spectroscopy
9.2 m	South African Large Telescope	South Africa, Rutgers, UW—Madison, UNC—Chapel Hill, Dartmouth, Carnegie-Mellon, 5 others	South Africa	2004*	Based on design of Hobby–Eberly telescope
2 × 8.4 m	Large Binocular Telescope	U. Arizona, Ohio State U., Italy, Germany	Mt. Graham, AZ	2004*	Two 8.4-m mirrors on a common mount, giving light-collecting area of 11.8-m telescope
4 × 8.2 m	Very Large Telescope	European Southern Observatory	Cerro Paranal, Chile	2000	Four separate 8-m telescopes designed to work individually or together as the equivalent of a 16-m telescope
8.3 m	Subaru	Japan	Mauna Kea, HI	1999	Japan's first large telescope project
8 m	Gemini North and South	U.S., U.K., Canada, Chile, Brazil, Argentina	Mauna Kea, HI (North); Cerro Pachon, Chile (South)	1999	Twin telescopes, one in each hemisphere
6.5 m	Magellan I and II	Carnegie Institute, U. Arizona, Harvard, U. Michigan, MIT	Las Campanas, Chile	2000/2002	Twin 6.5-m telescopes, known respectively as the Walter Baade and Landon Clay telescopes
6.5 m	MMT	Smithsonian Institution, U. Arizona	Mt. Hopkins, AZ	2000	Replaced an older telescope in the same observatory

*Scheduled completion date.

a The two Keck telescopes sit atop Mauna Kea in Hawaii.

Figure 7.9 The Keck telescopes.

b The Keck telescopes use 36 hexagonal mirrors to function as a single 10-meter primary mirror. These mirrors make the honeycomb pattern that surrounds the hole enclosing the man in this photo.

of the twin 10-meter Keck telescopes in Hawaii has a primary mirror consisting of 36 smaller mirrors that function together (Figure 7.9). When you realize that each Keck telescope has more than a million times the light-collecting area of the human eye, you can understand why modern telescopes have so dramatically enhanced our ability to observe the universe.

THINK ABOUT IT

What is the population of your hometown? Suppose everyone in your hometown looked at the dark sky at the same time. How would the total amount of light entering everyone's eyes compare with the light collected by a 10-meter telescope? Explain.

Angular Resolution Large telescopes can have remarkable angular resolution. For example, the 2.4-meter Hubble Space Telescope has an angular resolution of about 0.05 arcsecond (for visible light), which would allow you to read this book from a distance of about 800 meters (a half mile).

In principle, larger telescopes have better angular resolution. That is, they can distinguish finer details. However, the angular resolution may be degraded if a telescope is poorly made, and effects caused by Earth's atmosphere can limit the angular resolution of ground-based telescopes (see Section 7.4).

The ultimate limit to a telescope's resolving power comes from the properties of light. Because light is an electromagnetic wave [Section 6.2], beams of light can interfere with one another like overlapping sets of ripples on a pond (Figure 7.10). This *interference* causes a blurring of images that limits a telescope's angular resolution even when all other conditions are perfect. That is why even a high-

quality telescope in space, such as the Hubble Space Telescope, cannot have perfect angular resolution* (Figure 7.11).

---

*In fact, the Hubble Space Telescope's primary mirror was made with the wrong shape, which further limited its angular resolution until corrective optics were installed in 1993. The corrective optics consist of small mirrors that correct for most of the blurring created by the primary mirror.

Figure 7.10 This computer-generated image shows how overlapping sets of ripples on a pond interfere with one another. The effects of the two sets of ripples add in some places, making the water rise extra high or fall extra low, and cancel in other places, making the water surface flat. Light waves also exhibit interference. (The colors in this image are for visual effect only.)

Figure 7.11 When examined in detail, a Hubble Space Telescope image of a star has rings (represented as green and purple in figure) resulting from the wave properties of light. With higher angular resolution, the rings would be smaller.

The angular resolution that a telescope could achieve if it were limited only by the interference of light waves is called its **diffraction limit**. (*Diffraction* is a technical term for the specific effects of interference that limit telescope resolution.) It depends on both the diameter of the primary mirror and the wavelength of the light being observed.

Larger telescopes have smaller diffraction limits for any particular wavelength of light. However, achieving a particular angular resolution requires a larger telescope for longer-wavelength light. That is why, for example, a radio telescope must be far larger than a visible-light telescope to achieve the same angular resolution.

7.3 Uses of Telescopes

Astronomers use many different kinds of instruments and detectors to extract the information contained in the light collected by a telescope. Every astronomical observation is unique, but most observations fall into one of three basic categories:*

- **Imaging** yields pictures of astronomical objects. At its most basic, an imaging instrument is simply a camera. Astronomers sometimes place *filters* in front of the camera that allow only particular colors or wavelengths of light to pass through. Most of the richly hued images in this and other astronomy books are made by combining images recorded through different filters (Figure 7.12).

- **Spectroscopy** involves spreading light into a spectrum. Instruments called *spectrographs* use diffraction

---

*Some astronomers include a fourth general category called *photometry*, which is the accurate measurement of light intensity from a particular object at a particular time. We do not list this as a separate category because today's detectors can do photometry at the same time that they are being used for imaging, spectroscopy, or timing.

Mathematical Insight **7.2** **The Diffraction Limit**

A simple formula gives the diffraction limit of a telescope in arcseconds:

$$\text{diffraction limit} \approx 2.5 \times 10^5 \times \left(\frac{\text{wavelength of light}}{\text{diameter of telescope}} \right)$$
$$\text{(arcseconds)}$$

The wavelength of light and the diameter of the telescope must be in the same units.

Example 1: What is the diffraction limit of the 2.4-m Hubble Space Telescope for visible light with a wavelength of 500 nm?

Solution: For light with a wavelength of 500 nm (500×10^{-9} m), the diffraction limit of the Hubble Space Telescope is approximately:

$$\text{diffraction limit} = 2.5 \times 10^5 \times \left(\frac{\text{wavelength}}{\text{telescope diameter}} \right)$$

$$= 2.5 \times 10^5 \times \frac{500 \times 10^{-9} \text{ m}}{2.4 \text{ m}}$$

$$= 0.05 \text{ arcsecond}$$

Example 2: Suppose you wanted to achieve a diffraction limit of 0.001 arcsecond for visible light of wavelength 500 nm. How large a telescope would you need?

Solution: First we solve the diffraction limit equation for telescope diameter:

$$\text{diffraction limit} \approx 2.5 \times 10^5 \times \left(\frac{\text{wavelength}}{\text{telescope diameter}} \right)$$

$$\Downarrow$$

$$\text{telescope diameter} \approx 2.5 \times 10^5 \times \left(\frac{\text{wavelength}}{\text{diffraction limit}} \right)$$

Now we substitute the given values for the wavelength and the diffraction limit:

$$\text{telescope diameter} = 2.5 \times 10^5 \times \frac{500 \times 10^{-9} \text{ m}}{0.001 \text{ arcsecond}}$$

$$= 125 \text{ m}$$

An optical telescope would need a mirror diameter (or separation between mirrors, for an interferometer) of 125 meters—longer than a football field—to achieve an angular resolution of 0.001 arcsecond.

actual light
collected

blue filter

green filter

red filter

Images are combined to show full color.

VIS

Figure 7.12 In astronomy, color images are usually constructed by combining several images taken through different filters.

focused starlight

primary mirror

3. Diffraction grating: Disperses reflected light into a spectrum.

1. Slit: Allows only the light from the object of interest to pass through.

5. CCD: Records an image of the spectrum.

4. Camera mirror: Focuses the spectrum onto the CCD.

2. Collimating mirror: Makes all the reflected rays parallel.

Figure 7.13 The basic design of a spectrograph. In this diagram, the spectrograph is attached to the bottom of a reflecting telescope, with light entering the spectrograph through a hole in the primary mirror. A narrow slit (or small hole) at the entrance to the spectrograph allows only light from the object of interest to pass through. This light bounces from the collimating mirror to the diffraction grating, which disperses the light into a spectrum. The dispersed light is then focused by the camera mirror and recorded by a CCD, giving an image of the spectrum.

gratings (or other devices) to disperse light into spectra, which are then recorded with a detector such as a CCD (Figure 7.13).

● **Timing** tracks how an object's brightness varies with time. For a slowly varying object, a timing experiment may be as simple as comparing a set of images or spectra obtained on different nights. For more rapidly varying sources, specially designed instruments essentially make rapid multiple exposures, in some cases recording the arrival time of every individual photon.

Major observatories typically have several different instruments capable of each of these tasks, and some instruments can perform all three basic tasks.

Images of Nonvisible Light

Many astronomical images show nonvisible light. For example, Figure 7.14 is an "X-ray image." What exactly does this mean, given that X rays are invisible to the eye? The easiest way to understand the idea is to think about X rays at a doctor's office [Section 6.3]. When the doctor makes an "X ray of your arm," he or she uses a machine that sends X rays through your arm. The X rays that pass through are recorded on a piece of X ray–sensitive film. We cannot see the X rays themselves, but we can see the image left behind on the film. Astronomical images work the same

Figure 7.14 X rays are invisible, but we can color-code the information recorded by an X-ray detector to make an image of the object as it would appear in X rays. This image, from NASA's Chandra X-Ray Observatory, shows the region (roughly 700 light-years on a side) in the center of the Andromeda Galaxy. The blue dot in the center marks what astronomers believe to be a supermassive black hole [Section 21.5]. The rest of the image is color-coded, with yellow indicating the most intense X rays. Most of the other bright X-ray sources are probably X-ray binaries [Section 18.3], star systems in which a neutron star or black hole orbits a normal star.

way. Whether a detector records visible or nonvisible light, it makes an image showing the intensity of the light.

The simplest way to display a nonvisible image is in black and white, with brighter regions of the image corresponding to brighter light coming from the object. (A doctor's X ray is actually a negative, making bones look bright because few X rays pass through them.) However, because it is much easier for the human eye to perceive color differences than shades of gray, nonvisible images are often color-coded. In Figure 7.14, for example, the colors (except for the blue dot) correspond to the intensity of the recorded X rays. The brightest regions are yellow, while the violet and black represent regions that emit few, if any, X rays.

All images recorded in nonvisible light must necessarily use some kind of color or gray-scale coding. Sometimes the colors correspond to different intensities of light, sometimes to different wavelengths, and sometimes to physical properties.

Color-coded images are common even outside astronomy. For example, medical images from CAT scans and MRIs are usually displayed in color, even though neither type of imaging uses visible light. What do you think the colors mean in CAT scans and MRIs? How are the colors useful to doctors?

Spectral Resolution

As we discussed in Chapter 6, a spectrum can reveal a wealth of information about an object, including its chemical composition, temperature, and rotation rate. However, just as the amount of information we can glean from an

image depends on the angular resolution, the information we can glean from a spectrum depends on the **spectral resolution**. The higher the spectral resolution, the more detail we can see (Figure 7.15).

In principle, astronomers would always like the highest possible spectral resolution. However, higher spectral resolution comes at a price. A telescope collects only so much light in a given amount of time, and the spectral resolution depends on how widely this light is spread out by the spectrograph. If the spectrograph spreads out the light too much, the spectrum may become so dim that nothing can be recorded without a very long exposure time. Thus, for the same telescope, making a spectrum of an object requires a longer exposure time than making an image, and high-resolution spectra take longer than low-resolution spectra.

 Light and Spectroscopy Tutorial, Lesson 1

7.4 Atmospheric Effects on Observations

A telescope on the ground does not look directly into space but rather looks through Earth's sometimes murky atmosphere. Earth's atmosphere creates several problems for astronomical observations. The most obvious problem is weather—a visible-light telescope is useless under cloudy skies.

Light Pollution

Another problem is that our atmosphere scatters the bright lights of cities, creating what astronomers call **light pollution**, which can obscure the view even for the best

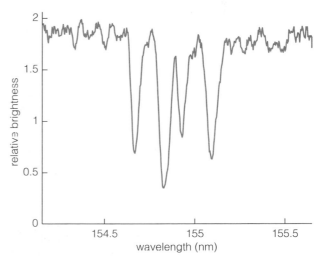

a Lower spectral resolution.

b Higher spectral resolution.

Figure 7.15 Higher spectral resolution means we can see more details in the spectrum. Both (**a**) and (**b**) show ultraviolet spectra of the same object for the same wavelength band, but the higher spectral resolution in (**b**) enables us to see individual spectral lines that appear merged together in (**a**). (The spectrum shows absorption lines created when interstellar gas absorbs light from a more distant star.)

telescopes. For example, the 2.5-meter telescope at Mount Wilson, the world's largest when it was built in 1917, would still be very useful today if it weren't located so close to the lights of what was once the small town of Los Angeles.

Similar but less serious light pollution hinders many other telescopes, including the Mount Palomar telescopes near San Diego and the telescopes of the National Optical Astronomy Observatory on Kitt Peak near Tucson. Fortunately, many communities are working to reduce light pollution. Placing reflective covers on the tops of streetlights directs more light toward the ground, rather than toward the sky. Using "low pressure" (sodium) lights also helps. Because these lights shine most brightly in just a few wavelengths of light, special filters can be used to absorb these wavelengths while transmitting most of the light from the astronomical objects under study. An extra benefit is that both reflective covers and low-pressure lights offer significant energy savings to communities that use them.

Atmospheric Distortion of Light

A somewhat less obvious problem is the distortion of light by the atmosphere. The ever-changing motion, or **turbulence**, of air in the atmosphere bends light in constantly shifting patterns. This turbulence causes the familiar twinkling of stars. For astronomers, this twinkling is a significant problem, because it blurs astronomical images.

If you look down a long, paved street on a hot day, you'll notice the images of distant cars and buildings rippling and distorting. How are these distortions similar to the twinkling of stars? Why do you think these distortions are more noticeable on hot days than on cooler days?

Mitigating Atmospheric Effects

To some extent, we can mitigate effects of weather, light pollution, and atmospheric distortion by choosing appropriate sites for observatories. The key criteria are that the sites be dark (limiting light pollution), dry (limiting rain and clouds), calm (limiting turbulence), and high (placing them above at least part of the atmosphere). Islands are often ideal, and the 4,300-meter (14,000-foot) summit of Mauna Kea on the Big Island of Hawaii is home to many of the world's best observatories (Figure 7.16).

Technology can help with some of the problems caused by the atmosphere. Putting a telescope on an airplane takes it above much of the atmosphere, allowing many infrared observations not possible from the ground. NASA's new airborne observatory, called SOFIA (Stratospheric Observatory for Infrared Astronomy), will have a 2.5-meter infrared telescope looking out through a large hole cut in the fuselage of a Boeing 747 airplane (Figure 7.17). Advanced technologies make it possible for the airplane to

Figure 7.16 Observatories on the summit of Mauna Kea in Hawaii.

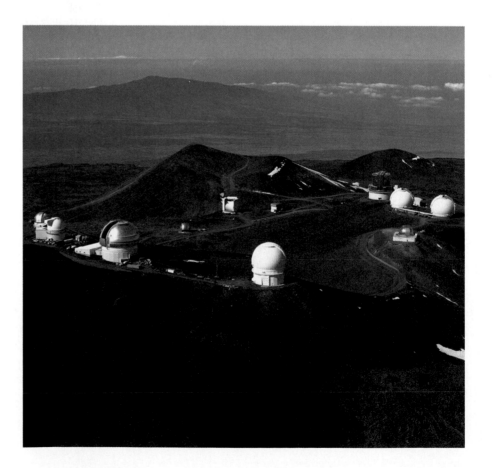

Twinkle, Twinkle Little Star

Twinkling, or apparent variation in the brightness and color of stars, is *not* intrinsic to the stars. Instead, just as light is bent by water in a swimming pool, starlight is bent by Earth's atmosphere. Air turbulence causes twinkling because it constantly changes how the starlight is bent. Hence, stars tend to twinkle more on windy nights and at times when they are near the horizon (and therefore are viewed through a thicker layer of atmosphere).

Planets also twinkle, but not nearly as much as stars. Because planets have a measurable angular size, the effects of turbulence on any one ray of light are compensated for by the effects of turbulence on others, reducing the twinkling seen by the naked eye (but making planets shimmer noticeably in telescopes).

While twinkling may be beautiful, it blurs telescopic images. Avoiding the effects of twinkling is one of the primary reasons for putting telescopes in space. There, above the atmosphere, stars do not twinkle and telescopes can record sharp astronomical images.

Figure 7.17 This photograph shows the airplane for NASA's airborne observatory, SOFIA, during a test flight. The 2.5-meter telescope will be located in the fuselage behind the painted black rectangle, which will open so the telescope can view the heavens. SOFIA is scheduled to begin observations in late 2004.

fly smoothly despite its large hole and to keep the telescope pointed accurately at observing targets during flight.

Perhaps the most amazing recent technology is **adaptive optics**, which can eliminate most atmospheric distortion. Atmospheric turbulence causes a stellar image to dance around in the focal plane of a telescope. Adaptive optics essentially make the telescope's mirrors do an opposite dance, canceling out the atmospheric distortions (Figure 7.18). The mirror shape (usually the secondary mirror) must change slightly many times each second to compensate for the rapidly changing atmospheric distortions. A computer calculates the necessary changes by monitoring the distortions of a bright star near the object under study.

In some cases, if there is no bright star near the object of interest, the observatory shines a laser into the sky to create an *artificial star* (a point of light in Earth's atmosphere) that can be monitored for distortions.

The ultimate solution to atmospheric distortion is to put telescopes in space, above the atmosphere. That is one reason why the Hubble Space Telescope (Figure 7.19) was built—and why it is so successful even though its 2.4-meter primary mirror is much smaller than the mirrors of many ground-based telescopes.

Atmospheric Absorption of Light

Earth's atmosphere poses one major problem that no Earth-bound technology can overcome: It prevents most

Figure 7.18 The technology of adaptive optics can enable a ground-based telescope to overcome most of the blurring caused by Earth's atmosphere. (Both of these images were taken in near-infrared light with the Canada-France-Hawaii telescope and are shown in false color)

a Atmospheric distortion makes this ground-based image of a double star look like a single star.

b When the same telescope is used with adaptive optics, the two stars are clearly distinguishable. The angular separation between the two stars is 0.38 arcsecond.

a This photograph shows astronauts working on the telescope in the Space Shuttle cargo bay during a servicing mission in 1997.

Figure 7.19 The Hubble Space Telescope orbits Earth. Although it orbits at a relatively low altitude, it is high enough to be above the distorting effects of Earth's atmosphere and to observe infrared and ultraviolet light.

b This diagram shows the basic components of the Hubble Space Telescope.

forms of light from reaching the ground at all. Figure 7.20 shows the depth to which different forms of light penetrate Earth's atmosphere. Only radio waves, visible light, parts of the infrared spectrum, and the longest wavelengths of ultraviolet light reach the ground.

If we studied only visible light, we'd be missing much of the picture. Planets are relatively cool and emit primarily infrared light. The hot upper layers of stars like the Sun emit ultraviolet and X-ray light. Some violent events even produce gamma rays that travel through space to Earth. Indeed, most objects emit light over a broad range of wavelengths. Observing these wavelengths requires placing

Figure 7.20 Diagram showing the approximate depths to which different wavelengths of light penetrate Earth's atmosphere. Note that most of the electromagnetic spectrum—except for visible light, a small portion of the infrared, and radio—can be observed only from very high altitudes or from space.

Closer to the Stars?

Many people mistakenly believe that space telescopes are advantageous because their locations above Earth make them closer to the stars. You can quickly realize the error of this belief by thinking about scale. On the scale of the Voyage model solar system that we discussed in Section 1.2, the Hubble Space Telescope is so close to the surface of the millimeter-diameter Earth that you would need a microscope to resolve its altitude, while the nearest stars are thousands of kilometers away. Thus, the distances to the stars are effectively the same whether a telescope is on the ground or in space. The real advantages of space telescopes all arise from their being above Earth's atmosphere and thus not subject to the many observational problems it presents.

7.5 Telescopes Across the Spectrum

Today, astronomers study light across the entire spectrum. Telescopes for nonvisible wavelengths often require very different designs than optical telescopes.

Infrared and Ultraviolet Telescopes

Light from much of the infrared and ultraviolet portions of the spectrum behaves enough like visible light to be recorded by an optical telescope, as long as it is equipped with appropriate detectors and mirror coatings. Of course, the telescope must be in space in order to receive significant ultraviolet light.

Near the extreme-wavelength ends of the infrared or ultraviolet, telescopes need special technology. Extreme ultraviolet light (the shortest wavelengths of ultraviolet) behaves like X rays, which we'll discuss next. Extreme infrared light (the longest wavelengths of infrared) poses observing difficulties because ordinary telescopes are warm enough to emit significant amounts of long-wavelength infrared light, and this telescope emission would interfere with any attempt to observe these wavelengths from the cosmos. One solution to this problem is to make the telescope so cold that it emits very little infrared radiation. NASA's Space Infrared Telescope Facility (SIRTF), launched

telescopes in space (or, in a few cases, very high in the atmosphere on airplanes or balloons).

The Hubble Space Telescope is the only major visible-light observatory in space (it is also used for infrared and ultraviolet observations), but many less famous observatories are in Earth orbit for the purpose of making observations in nonvisible wavelengths. Table 7.2 lists some of the most important existing and planned telescopes in space.

Table 7.2 Selected Present and Future Major Observatories in Space

Name	Launch Year	Lead Space Agency	Special Features
Hubble Space Telescope	1990	NASA	Optical, infrared, and ultraviolet observations
Far Ultraviolet Spectroscopic Explorer (FUSE)	1999	NASA	Ultraviolet spectroscopy
Chandra X-Ray Observatory	1999	NASA	X-ray imaging and spectroscopy
X-Ray Multi-Mirror Mission (XMM)	1999	ESA[*]	European-led mission for X-ray spectroscopy
High Energy Transient Explorer (HETE-2)	2000	NASA	Study of gamma-ray bursts
Wilkinson Microwave Anisotropy Probe (WMAP)	2001	NASA	Study of the cosmic microwave background
International Gamma-Ray Astrophysics Laboratory (INTEGRAL)	2002	ESA[*]	Gamma-ray imaging, spectroscopy, and timing
Space Infrared Telescope Facility (SIRTF)	2003[†]	NASA	Infrared observations of the cosmos
Swift	2003[†]	NASA	Study of gamma-ray bursts.
Kepler	2006[†]	NASA	Transit search for extrasolar Earth-like planets
Space Interferometry Mission (SIM)	2009[†]	NASA	First mission for optical interferometry in space
James Webb Space Telescope (JWST)	2010[†]	NASA	Follow-on to Hubble Space Telescope
Terrestrial Planet Finder (TPF)	2014[†]	NASA	Search for Earth-like planets around other stars

[*]European Space Agency.
[†]Scheduled launch year.

in 2003, is cooled with liquid helium to just a few degrees above absolute zero.

X-Ray and Gamma-Ray Telescopes

X rays have sufficient energy to penetrate many materials, including living tissue and ordinary mirrors. While this property makes X rays very useful to medical doctors, it gives astronomers headaches. Trying to focus X rays is somewhat like trying to focus a stream of bullets. If the bullets are fired directly at a metal sheet, they will puncture or damage the sheet. However, if the metal sheet is angled so that the bullets barely graze its surface, then it will slightly deflect the bullets. Specially designed mirrors can deflect X rays in much the same way. Such mirrors are called **grazing incidence** mirrors because X rays merely graze their surfaces as they are deflected toward the focal plane. X-ray telescopes, such as NASA's Chandra X-Ray Observatory, generally consist of several nested grazing incidence mirrors (Figure 7.21).

If you look straight down at your desktop, you probably cannot see your reflection. But if you glance along the desktop surface (or another smooth surface, such as that of a book), you should see reflections of objects in front of you. Explain how these reflections represent *grazing incidence* for visible light.

Gamma rays can penetrate even grazing incidence mirrors and therefore cannot be focused in the traditional sense. Capturing such high-energy light at all requires detectors so massive that the photons cannot simply pass through them. The largest gamma-ray observatory to date was the 17-ton Compton Gamma Ray Observatory (Figure 7.22), which was launched in 1991 and destroyed in a controlled crash to Earth in 2000.

Radio Telescopes and Interferometry

Radio telescopes use large metal dishes as "mirrors" to reflect radio waves. However, the long wavelengths of radio waves mean that very large telescopes are necessary to achieve

Figure 7.21 The mirrors used to focus X rays in X-ray telescopes must be designed and arranged differently than the mirrors used in visible-light telescopes.

a This diagram shows how grazing incidence mirrors can focus X rays. An X ray that enters the telescope on a path that grazes one of the mirrors in the first set gets deflected to a mirror in the second set, and from there to the focus.

Mirror elements are 0.8 m long and from 0.6 m to 1.2 m in diameter.

b Mirrors are usually cylindrical, rather than flat, as shown in this diagram of the mirror set for the Chandra X-Ray Observatory.

c This photograph shows actual Chandra mirrors during assembly; Chandra was launched into Earth orbit in 1999.

Figure 7.22 This photograph shows the Compton Gamma Ray Observatory being deployed from the Space Shuttle in 1991. Compton revolutionized our understanding of the gamma-ray universe before its demise (in a controlled reentry into the atmosphere).

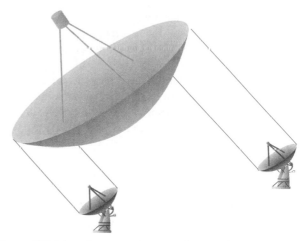

Figure 7.24 Interferometry gives two (or more) small radio dishes the angular resolution of a much larger dish. However, their total light-collecting area is only the sum of the light-collecting areas of the individual dishes.

reasonable angular resolution. The largest radio dish in the world, the Arecibo telescope, stretches 305 meters (1,000 feet) across a natural valley in Puerto Rico (Figure 7.23). Despite its large size, Arecibo's angular resolution is only about 1 arcminute at commonly observed radio wavelengths (e.g., 21 cm [Section 19.2])—a few hundred times worse than the visible-light resolution of the Hubble Space Telescope.

In the 1950s, radio astronomers developed an ingenious technique for improving the angular resolution of radio telescopes: They learned to link two or more individual telescopes to achieve the angular resolution of a much larger telescope (Figure 7.24). This technique is often called **interferometry** because it works by taking advantage of the wavelike properties of light that cause interference (see Figure 7.10). The procedure relies on precisely timing when radio waves reach each dish and on supercomputers to analyze the resulting interference patterns.

One famous example of radio interferometry, the Very Large Array (VLA) near Socorro, New Mexico, consists of 27 individual radio dishes that can be moved along railroad tracks laid down in the shape of a Y (Figure 7.25). The light-gathering capability of the VLA's 27 dishes is equal to their combined area, equivalent to that of a single telescope 130 meters across. The VLA's angular resolution, achieved by spacing the individual dishes as widely as possible, is

Figure 7.23 The Arecibo radio telescope stretches across a natural valley in Puerto Rico. At 305 meters across, it is the world's largest single radio dish.

Figure 7.25 The Very Large Array (VLA) in New Mexico consists of 27 telescopes that can be moved along train tracks. The telescopes work together through interferometry and can achieve an angular resolution equivalent to that of a single radio telescope almost 40 kilometers across.

equivalent to that of a single radio telescope with a diameter of almost 40 kilometers. Today, astronomers can achieve even higher angular resolution by linking radio telescopes around the world.

In principle, interferometry can improve angular resolution not only for radio waves but also for any other form of light. In practice, interferometry becomes increasingly difficult for light with shorter wavelengths. Nevertheless, astronomers have begun to succeed at infrared and visible interferometry and are testing technologies for X-ray interferometry. One reason why *two* Keck telescopes were built on Mauna Kea is so that they can be used for infrared and optical interferometry. The potential value of such interferometers is enormous. In the future, infrared interferometers may be able to obtain spectra from individual planets around other stars, allowing spectroscopy that could determine the compositions of their atmospheres and help determine whether they harbor life.

The Future of Astronomy in Space

It will always be cheaper and easier to do astronomy from ground-based observatories, but space is likely to play an ever-increasing role in astronomy. NASA is already at work on follow-ons to the Hubble Space Telescope, such as the James Webb Space Telescope (JWST), which may launch as early as 2010. For the more distant future, many astronomers dream of an observatory on the far side of the Moon

(Figure 7.26). Because the Moon has no atmosphere, it offers all the advantages of telescopes in space while also offering the ease of operating on a solid surface.

Putting Chapter 7 into Context

In this chapter, we've focused on the technological side of astronomy: the telescopes that we use to learn about the universe. Keep in mind the following "big picture" ideas as you continue to learn about astronomy:

● Technology drives astronomical discovery. Every time we build a bigger telescope, develop a more sensitive detector, or open up a new wavelength region to study, we learn more about the universe than was possible before.

● Telescopes work much like giant eyes, enabling us to see the universe in great detail. New technologies for making larger telescopes, along with advances in adaptive optics and interferometry, are making ground-based telescopes more powerful than ever.

● For the ultimate in observing the universe, space is the place! Telescopes in space allow us to detect light from across the entire spectrum while also avoiding the distortion caused by Earth's atmosphere.

Figure 7.26 An artist's conception of a possible future lunar observatory. The Moon has no atmosphere to create distortion and has a solid surface that makes it ideal as a home for astronomical observatories.

7.1 Eyes and Cameras: Everyday Light Sensors

- *How does a lens form an image?* A lens redirects parallel rays of light from the object being viewed so that they converge at a point on the focal plane to form an image.

- *Which is better, an angular resolution of 1° or of 2°?* The smaller angle means we can see finer details.

7.2 Telescopes: Giant Eyes

- *What's the difference between a refracting telescope and a reflecting telescope?* A refractor forms an image by bending light through a lens. A reflector forms an image by focusing light with mirrors.

- *What are the two most important properties of a telescope?* The two most important properties of a telescope are its light-collecting area, which determines how much light it gathers, and its angular resolution, which determines how much detail we can see in its images.

7.3 Uses of Telescopes

- *What are the three primary uses of telescopes?* Telescopes' three primary uses are imaging to create pictures of distant objects, spectroscopy to study the spectra of distant objects, and timing to study how a distant object's brightness changes with time.

- *How do we represent images of nonvisible light?* Detectors can record light that our eyes cannot see, and we can then represent the recorded light with some kind of color coding to reveal details that would otherwise be invisible to our eyes.

7.4 Atmospheric Effects on Observations

- *What is light pollution?* It is light from human activity that can interfere with astronomical observations.

- *Why do stars appear to twinkle?* Stars themselves do not really twinkle. Rather, they appear to twinkle in our sky because of the way their light is distorted as it passes through our atmosphere. Thus, astronauts and observatories above our atmosphere do not see any twinkling.

- *Which of the problems that our atmosphere poses for ground-based astronomy cannot be solved with technology?* Technology cannot do anything about the fact that our atmosphere absorbs most of the light in the electromagnetic spectrum. To see light that does not penetrate to the ground, telescopes must be put in space.

7.5 Telescopes Across the Spectrum

- *Why do we need different kinds of telescopes to collect different forms of light?* Photons of different energy behave differently and require different collection strategies.

- *What is interferometry used for?* It allows two or more small telescopes to achieve the angular resolution of a much larger telescope, enabling us to see more detail in astronomical images.

❓ Does It Make Sense?

Decide whether each statement makes sense and explain why it does or does not.

1. The image was blurry because the photographic film was not placed at the focal plane.

2. By using a CCD, I can photograph the Andromeda Galaxy with a shorter exposure time than I would need with photographic film.

3. Thanks to adaptive optics, the telescope on Mount Wilson can now make ultraviolet images of the cosmos.

4. New technologies will soon allow astronomers to use X-ray telescopes on Earth's surface.

5. Thanks to interferometry, a properly spaced set of 10-meter radio telescopes can achieve the angular resolution of a single, 100-kilometer radio telescope.

6. Thanks to interferometry, a properly spaced set of 10-meter radio telescopes can achieve the light-collecting area of a single, 100-kilometer radio telescope.

7. I have a reflecting telescope in which the secondary mirror is bigger than the primary mirror.

8. An observatory on the Moon's surface could have telescopes monitoring light from all regions of the electromagnetic spectrum.

Problems

(Quantitative problems are marked with an asterisk.)

9. *Angular Resolution.*

    a. Briefly describe why a smaller angle is better when it comes to angular resolution.

b. Suppose that two stars are separated in the sky by 0.1 arcsecond. What will you see if you look at them with a telescope that has an angular resolution of 0.01 arcsecond? What will you see if you look at them with a telescope that has an angular resolution of 0.5 arcsecond?

10. *Light-Collecting Area.*

   a. How much greater is the light-collecting area of one of the 10-meter Keck telescopes than that of the 5-meter Hale telescope?

   b. Suppose astronomers built a 100-meter telescope. How much greater would its light-collecting area be than that of the 10-meter Keck telescope?

11. *Diffraction Limit.* What is the *diffraction limit* of a telescope, and how does it depend on the telescope's size and the particular wavelength of light being observed?

12. *Telescopes in Space.* Briefly describe the advantages of putting telescopes in space.

13. *Telescopes Across the Spectrum.* Why is it useful to study light from across the spectrum? Briefly describe the characteristics of telescopes used to observe different portions of the spectrum.

14. *Project: Twinkling Stars.* Using a star chart, identify 5–10 bright stars that should be visible in the early evening. On a clear night, observe each of these stars for a few minutes. Note the date and time, and for each star record the following information: approximate altitude and direction in your sky, brightness compared to other stars, color, how much the star twinkles compared to other stars. Study your record. Can you draw any conclusions about how brightness and position in your sky affect twinkling? Explain.

*15. *Angular Separation Calculations.*

   a. Two light bulbs are separated by 0.2 meter and you look at them from a distance of 2 kilometers. What is the angular separation of the lights? Can your eyes resolve them? Explain.

   b. The diameter of a dime is about 1.8 centimeters. What is its angular diameter if you view it from across a 100-meter-long football field?

*16. *Calculating the Sun's Size.* The angular diameter of the Sun is about 0.5° (about the same as that of the Moon). Use this fact and the Sun's average distance of about 150 million km to estimate the diameter of the Sun. Compare your result to the Sun's actual diameter of 1.392 million km.

*17. *Viewing a Dime with the HST.* The Hubble Space Telescope (HST) has an angular resolution of about 0.05 arcsecond. How far away would you have to place a dime (diameter = 1.8 cm) for its angular diameter to be 0.05 arcsecond? (*Hint:* Start by converting an angular diameter of 0.05 arcsecond into degrees.)

*18. *Close Binary System.* Suppose that two stars in a binary star system are separated by a distance of 100 million km and are located at a distance of 100 light-years from Earth. What is the angular separation of the two stars? Give your answer in both degrees and arcseconds. Can the Hubble Space Telescope resolve the two stars?

*19. *Diffraction Limit of the Eye.* Calculate the diffraction limit of the human eye, assuming a lens size of 0.8 cm, for visible light of 500-nm wavelength. How does this compare to the diffraction limit of a 10-meter telescope?

*20. *The Size of Radio Telescopes.* What is the diffraction limit of a 100-meter radio telescope observing radio waves with a wavelength of 21 cm? Compare this to the diffraction limit of the 2.4-meter Hubble Space Telescope for visible light. Use your results to explain why radio telescopes must be much larger than optical telescopes to be useful.

*21. *Hubble's Field of View.* Large telescopes often have small fields of view. For example, the Hubble Space Telescope's (HST) new advanced camera has a field of view that is roughly square and about 0.06° on a side.

   a. Calculate the angular area of the HST's field of view in square degrees.

   b. The angular area of the entire sky is about 41,250 square degrees. How many pictures would the HST have to take with its camera to obtain a complete picture of the entire sky?

   c. Assuming that it requires an average of 1 hour to take each picture, how long would it take to acquire the number of pictures you calculated in part (b)? Use your answer to explain why astronomers would like to have more than one large telescope in space.

Discussion Questions

22. *Science and Technology Funding.* Technological innovation clearly drives scientific discovery in astronomy, but the reverse is also true. For example, Newton's discoveries were made in part to explain the motions of the planets, but they have had far-reaching effects on our civilization. Congress often must make decisions between funding programs with purely scientific purposes ("basic research") and programs designed to develop new technologies. If you were a member of Congress, how would you try to allocate spending between basic research and technology? Why?

23. *A Lunar Observatory.* Do the potential benefits of building an astronomical observatory on the Moon justify its costs at the present time? If it were up to you, would you recommend that Congress begin funding such an observatory? Defend your opinions.

For a complete list of media resources available, go to www.astronomyplace.com and choose Chapter 7 from the pull-down menu.

 ## Astronomy Place Web Tutorials

Tutorial Review of Key Concepts

Use the interactive **Tutorial** at www.astronomyplace.com to review key concepts from this chapter.

Light and Spectroscopy Tutorial

Lesson 2 Spectroscopy

Lesson 3 Atomic Spectra—Emission and Absorption Lines

Lesson 4 Thermal Radiation

Supplementary Tutorial Exercises

Use the interactive **Tutorial Lessons** to explore the following questions.

Light and Spectroscopy Tutorial, Lessons 2–4

1. How would an object's spectrum influence the choice of telescope used to observe it?

2. Consider the spectra of hot stars as compared to those of cooler stars. Which stars would be best observed with an ultraviolet telescope? Explain.

3. Are there spectral features that can be observed only with telescopes in space? Explain.

Web Projects

Take advantage of the useful Web links on **www.astronomyplace. com** to assist you with the following projects.

1. *Major Ground-Based Observatories.* Take a virtual tour of one of the world's major astronomical observatories. Write a short report on why the observatory is useful to astronomy.

2. *Space Observatory.* Visit the Web site of a major space observatory, either existing or under development. Write a short report about the observatory, including its purpose, its orbit, and how it operates.

3. *Really Big Telescopes.* Several studies are under way in hopes of building telescopes far larger than any now in operation. Learn about one or more of these projects (such as OWL, the Swedish 50-m Optical Telescope, or the California Extremely Large Telescope), and write a short report about the telescope's prospects and potential capabilities.

PART

III

LEARNING FROM OTHER WORLDS

8 Welcome to the Solar System

How vast those Orbs must be, and how inconsiderable this Earth, the Theatre upon which all our mighty Designs, all our Navigations, and all our Wars are transacted, is when compared to them. A very fit consideration, and matter of Reflection, for those Kings and Princes who sacrifice the Lives of so many People, only to flatter their Ambition in being Masters of some pitiful corner of this small Spot.

**Christiaan Huygens,
Dutch astronomer and scholar (c. 1690)**

Our ancestors long ago recognized the motions of the planets through the sky, but it has been only a few hundred years since we learned that we are part of a solar system centered on the Sun. Even then, we knew little about the other planets until the advent of large telescopes. More recently, our understanding of other worlds has exploded with the dawn of space exploration. We've lived in this solar system all along, but only now are we getting to know it.

In this chapter, we'll explore our solar system like newcomers to the neighborhood. We'll first look at the broad patterns we observe in the solar system and the general characteristics of the objects within it. We'll then take a brief tour of the individual worlds, starting from the Sun and moving outward through the planets. Finally, we'll discuss the use of spacecraft to explore the solar system, examining how we are coming to learn so much more about our neighbors.

 Formation of the Solar System Tutorial, Lesson 2

8.1 Comparative Planetology

Galileo's telescopic observations began a new era in astronomy in which the Sun, Moon, and planets could be studied for the first time as *worlds,* rather than as mere lights in the sky. Since that time, people have studied these worlds in different ways. Sometimes we study them individually—for example, when we seek to map the geography of Mars or to probe the atmospheric structure of Jupiter. Other times we compare the worlds to one another, seeking to understand their similarities and differences. This latter approach is called **comparative planetology**. Astronomers use the term *planetology* broadly to include moons, asteroids, and comets as well as planets.

The essence of comparative planetology lies in the idea that we can learn more about an individual world, including our own Earth, by studying it in the context of other objects in our solar system. It is much like learning more about a person by getting to know his or her family, friends, and culture. While we still can learn much by studying planets individually, the comparative planetology approach has demonstrated its value in at least three key ways.

First, it has given us deep new insights into the physical processes that have shaped Earth and other worlds—insights that can help us better understand and manage our own planet. Second, it has helped us learn about the origin and history of our solar system as a whole, giving us a better understanding of how we came to exist here on Earth. Third, it has allowed us to apply lessons from our solar system to the study of the many recently discovered planetary systems around other stars. These lessons help us understand both the general principles that govern star systems and the specific circumstances under which Earth-like planets—and possibly life—might exist elsewhere.

We will use the comparative planetology approach for most of our study of the solar system in this book. For example, rather than simply learning about the mountains and craters we have discovered on each individual world, we will focus on understanding how these features came to exist and why the features on one world are similar to or different from those on other worlds. Before we begin our comparative study, however, it's useful to have a general sense of the solar system as a whole.

 Orbits and Kepler's Laws Tutorial, Lessons 2–4

8.2 The Layout of the Solar System

Much of what we now know about the solar system has come from studying its general layout and organization. A good way to gain familiarity with this layout is to imagine having the perspective of an alien spacecraft making its first scientific survey of our solar system. What would we see?

Figure 8.1 shows the solar system as we approach it from outside Pluto's orbit. Orbits and orbital directions are shown, along with the rotation axis and direction of rotation for each planet. Remember that the Sun and planets are all quite small compared to the distances between them [Section 1.2]. On a scale at which the Sun is the size of a grapefruit, the planets are spread out over an area equivalent to that of about 300 football fields. They range in size from a dust speck for Pluto, to the tip of a ball point for Earth, to a marble for Jupiter (see Figures 1.7 and 1.8).

Of course, the Sun and planets are much more interesting than the vast empty spaces between them. A first scientific survey would focus on the widely spaced worlds—their orbits, sizes, compositions, densities, and any moons or ring systems. Table 8.1 summarizes these properties of the nine planets and the asteroids and comets. In mapping

Close-up of Inner Solar System

Mars

Earth

Mercury

Sun

Venus

The Solar System

Nepture

Pluto

Jupiter

Saturn

Uranus

Orbits to scale; planet sizes exaggerated about one million times. Sun not to scale.

Figure 8.1 The layout of our solar system as it would appear from beyond Pluto (if we could see planetary orbits). Notice that all the planets orbit the Sun in the same direction. Each planet's rotation axis is also shown, with a circling arrow to indicate the direction of rotation. (Planetary tilts in this diagram are aligned in the same plane for easier comparison.)

Table 8.1 Planetary Facts*

Photo	Planet	Relative Size	Average Distance from Sun (AU)	Min/Max Distance from Sun (AU)	Inclination of Orbit to Ecliptic	Orbital Period	Average Equatorial Radius (km)	Rotation Period	
	Mercury	·	0.387	0.308/ 0.466	7.00°	87.9 days	2,440	58.6 days	
	Venus	•	0.723	0.718/ 0.728	3.39°	225 days	6,051	243 days	
	Earth	•	1.00	0.983/ 1.017	0.00°	1.00 year	6,378	23.93 hours	
	Mars	·	1.52	1.38/ 1.66	1.85°	1.88 years	3,397	24.6 hours	
	Most asteroids	·	2–3	—	—	—	≤500	—	
	Jupiter	●	5.20	4.95/ 5.45	1.31°	11.9 years	71,492	9.93 hours	
	Saturn	●	9.54	9.01/ 10.05	2.48°	29.4 years	60,268	10.6 hours	
	Uranus	●	19.2	18.3/ 70.1	0.77°	83.8 years	25,559	17.2 hours	
	Neptune	●	30.1	29.8/ 30.4	1.77°	164 years	24,764	16.1 hours	
	Pluto	·	39.5	29.3/ 48.8	17.14°	248 years	1,160	6.39 days	
	Most comets	·	10–50,000	—	—	—	A few km?	—	

*Appendix E gives a more complete list of planetary properties.
†Surface temperatures for all objects except Jupiter, Saturn, Uranus, and Neptune, for which cloud-top temperatures are listed.
‡Several asteroids and a Kuiper belt comet have moons. It's not yet known if this is common.
§Includes water (H_2O), methane (CH_4), and ammonia (NH_3).
‖Comets passing close to the Sun warm considerably, especially their outer layers.

Axis Tilt	Mass (Earth = 1)	Average Density (g/cm^3)	Surface Gravity (Earth = 1)	Escape Velocity (km/s)	Average Surface (or Cloud Tops) Temperature[†]	Composition	Known Moons (2003)	Rings?
0.0°	0.06	5.43	0.38	4.43	700 K (day) 100 K (night)	Rocks, metals	0	No
177.3°	0.82	5.24	0.91	10.4	740 K	Rocks, metals	0	No
23.5°	1.00	5.52	1.00	11.2	290 K	Rocks, metals	1	No
25.2°	0.11	3.93	0.38	5.03	240 K	Rocks, metals	2	No
—	—	1.5–3	—	—	170 K	Rocks, metals	Note[‡]	No
3.1°	318	1.33	2.53	59.5	125 K	H, He, hydrogen compounds[§]	61	Yes
26.7°	95.2	0.70	1.07	35.5	95 K	H, He, hydrogen compounds[§]	31	Yes
97.9°	14.5	1.32	0.91	21.3	60 K	H, He, hydrogen compounds[§]	21	Yes
28.8°	17.1	1.64	1.14	23.6	60 K	H, He, hydrogen compounds[§]	11	Yes
119.6°	0.002	2.0	0.07	1.25	40 K	Ices, rock	1	No
—	—	<1?	—	—	A few K[‖]	Ices, dust	Note[‡]	No

the orbits, we would find that they are nearly circular. Figure 8.2 shows our view of these orbits from high above Earth's North Pole. Orbit directions for major moons are also shown.

Careful study of the figures and the table reveals several general features of our solar system. Let's investigate, grouping our observations into four major categories: the patterns of motion of the planets and the moons, the types of planets, the asteroids and comets, and the oddities that don't fit the patterns.

THINK ABOUT IT

Before reading any further, study Figures 8.1 and 8.2 and Table 8.1. Describe some of the patterns you recognize. For example, do you notice any pattern to the directions in which the planets orbit the Sun? Do the planets seem to fall into any natural groupings by size, location, or composition? Are the distinctions clear-cut? Explain.

Patterns of Motion

Figures 8.1 and 8.2 show that our solar system is quite organized. Planetary orbits follow noticeable patterns rather than being randomly strewn about. The most important patterns of motion in our solar system include:

- All planetary orbits are nearly circular and lie nearly in the same plane.

- All planets orbit the Sun in the same direction—counterclockwise as viewed from high above Earth's North Pole.

- Most planets rotate in the same direction in which they orbit (counterclockwise as viewed from above the North Pole), with fairly small axis tilts. The Sun also rotates in this same direction.

- Most of the solar system's large moons exhibit similar properties in their orbits around their planets—for example, orbiting in their planet's equatorial plane in the same direction that the planet rotates.

Figure 8.2 View of the solar system from high above Earth's North Pole. Notice that all orbits except Mercury and Pluto are nearly circular. The short, curved arrow for each planet indicates the planet's direction of rotation. The circles around each planet except Mercury and Venus indicate the orbital direction of the planet's major moons. (Mercury and Venus have no moons.) Also notice that most major moons orbit in the same direction that their planet rotates.

These orderly patterns are no accident. Rather, they are consequences of the fact that our entire solar system formed from the gravitational collapse of a single cloud of gas and dust. In Chapter 9, we will discuss how this collapse led to these orderly motions.

Terrestrial and Jovian Planets

Another fact that jumps out from Figures 8.1 and 8.2 is that the orbits of the four inner planets are grouped quite closely together compared to the orbits of the outer planets. Careful study of Table 8.1 shows that the differences go far beyond orbital spacing. In fact, most of the planets fall into one of two major groups: the rocky *terrestrial planets* and the gas-rich *jovian planets*.

The **terrestrial planets** are the four planets of the inner solar system—Mercury, Venus, Earth, and Mars. The word *terrestrial* means "Earth-like" (*terra* is Latin for "Earth"), so their designation stems from their similarities to Earth. The four terrestrial planets are relatively small, have solid, rocky surfaces, and have an abundance of metals deep in their interiors. They have few moons, if any, and none have rings. We often count our Moon as a fifth terrestrial world, because it shares these general characteristics, although it's not technically a planet.

The **jovian planets** are the four large planets of the outer solar system—Jupiter, Saturn, Uranus, and Neptune. The word *jovian* means "Jupiter-like," so these planets' designation comes from their similarities to Jupiter. The jovian planets have little in common with the terrestrial planets. They are much larger in size and lower in average density than the terrestrial planets, and each has rings and numerous moons. Most important, their composition is very different from that of the terrestrial worlds. The jovian planets are made mostly of hydrogen, helium, and **hydrogen compounds**—compounds containing hydrogen, such as water (H_2O), ammonia (NH_3), and methane (CH_4).

Because they are made mostly of substances that are gases under earthly conditions, the jovian planets are sometimes called "gas giants." They do not have solid surfaces and look like balls of gas from the outside. However, the intense pressures and temperatures of the jovian planet interiors transform these gases into phases unlike anything we ordinarily see on Earth. Table 8.2 contrasts the general traits of the terrestrial and jovian planets.

Notice that Pluto is left out in the cold, both literally and figuratively. On one hand, it is small and solid like the terrestrial planets. On the other hand, Pluto is far from the Sun, cold, and made of low-density ices. Pluto's orbit is also more elliptical and more inclined to the ecliptic plane (the plane of Earth's orbit) than the orbit of any other planet. For a long time, scientists considered Pluto to be a lone misfit. However, as we will discuss later, recent discoveries suggest that Pluto actually may be just one of many similar objects that roam the solar system beyond the orbit of Neptune.

Table 8.2 Comparison of Terrestrial and Jovian Planets

Terrestrial Planets	Jovian Planets
Smaller size and mass	Larger size and mass
Higher density	Lower density
Made mostly of rock and metal	Made mostly of hydrogen, helium, and hydrogen compounds
Solid surface	No solid surface
Few (if any) moons and no rings	Rings and many moons
Closer to the Sun (and closer together), with warmer surfaces	Farther from the Sun (and farther apart), with cool temperatures at cloud tops

Asteroids and Comets

A third major feature of the solar system is the existence of vast numbers of small objects orbiting the Sun. These objects fall into two major groups: asteroids and comets.

Asteroids are small, rocky bodies that orbit the Sun much like planets, but they are much smaller than planets. Even the largest of the asteroids have radii of only a few hundred kilometers, which means they are dwarfed by our Moon. Most asteroids are found within the relatively wide gap between the orbits of Mars and Jupiter that marks the **asteroid belt** (see Figure 8.1). They orbit the Sun in the same direction and nearly in the same plane as the planets. More than 10,000 asteroids have been identified and cataloged, but these are probably only the largest among a much greater number of small asteroids.

Figure 8.3 shows one of the few asteroids that have been photographed up close. Its appearance is probably typical of most asteroids. Like most small objects in the solar system, it is not spherical in shape.

Comets are also small objects that orbit the Sun, but they differ from asteroids in many ways. You are probably familiar with the occasional appearance of a comet in the inner solar system, where it may become visible to the naked eye with a long, beautiful tail (Figure 8.4). These visitors, which may delight sky watchers for a few weeks or months, are actually quite rare among comets. The vast majority of comets never visit the inner solar system, instead orbiting the Sun in the extreme outer reaches of the solar system. Indeed, based on the frequency with which we see comets in the inner solar system, astronomers calculate that something like a trillion (10^{12}) comets must inhabit the outskirts of our solar system. Thus, comets are both more numerous and more distant than asteroids. Comets also differ in composition from asteroids—they are made largely of ices (such as water ice, ammonia ice, and methane ice) rather than rock.

Studies of the orbits of the relatively few comets that enter the inner solar system have shown that comets come

Figure 8.3 The asteroid Eros (photographed from the NEAR spacecraft). Its appearance is probably typical of most asteroids. Eros is about 40 kilometers in length. Like other small objects in the solar system, it is not spherical.

from two distinct regions of space. The first region, called the **Kuiper belt** (pronounced *koy-per*), extends roughly from the orbit of Neptune to about three times Neptune's distance from the Sun (that is, from about 30 to 100 AU). Like the asteroids of the asteroid belt, the Kuiper belt comets orbit the Sun in the same direction as the planets, and their orbits generally lie close to the plane of planetary orbits. A few Kuiper belt comets are large enough for us to see through telescopes, though they are tiny compared to the terrestrial planets. In fact, some of the known Kuiper belt comets are

Figure 8.4 Comet Hale–Bopp, photographed over Boulder, Colorado, during its appearance in 1997.

nearly as large as Pluto, leading scientists to suspect that Pluto may simply be an unusually large member of this group.

The second and much larger region, called the **Oort cloud** (*Oort* rhymes with *court*), may extend more than one-fourth of the way to the nearest stars. Comets in this region have orbits around the Sun that are inclined at all angles to the plane of planetary orbits and go in every possible direction around the Sun. Thus, the Oort cloud would look roughly spherical in shape if we could see it. It is so vast, however, that even with a trillion comets each comet is typically separated from the next by more than a billion kilometers.

Figure 8.5 summarizes and contrasts the general characteristics of asteroids and the two major groups of comets.

THINK ABOUT IT

Recall the scale model of the solar system discussed in Chapter 1, in which our solar system fits on the National Mall in Washington, D.C. (see Figure 1.8b) but the nearest stars lie more than 4,000 kilometers away. Using this same scale, briefly describe where we would find the asteroid belt, the Kuiper belt, and the Oort cloud. What would we see in each of these regions?

Exceptions to the Rules

We have identified three general characteristics of our solar system: its patterns of motion, the division of its planets into two major categories, and the presence of numerous small objects known as asteroids and comets. The fourth key characteristic is that the general rules have a few notable exceptions. For example, while most of the planets rotate in the same direction that they orbit, Uranus and Pluto orbit of Oort cloud comet entering inner solar system

typical asteroid orbit

orbit of Jupiter

orbit of Neptune

typical Kuiper belt comet orbit

Oort cloud orbits have
random tilts and eccentricities.

Not to scale

Asteroids	Kuiper Belt Comets	Oort Cloud Comets
Made mostly of rock	Made mostly of ice	Made mostly of ice
Orbit in inner solar system (most in asteroid belt)	Orbit between Neptune and about 100 AU from the Sun	Orbit at great distance, as far as one-fourth of the way to nearby stars
Orbit in the same direction and nearly the same plane as the planets	Orbit in the same direction and nearly the same plane as the planets	Orbits inclined at every possible angle and going in all possible directions

Figure 8.5 Properties of asteroids and comets summarized both visually (above) and in words (left). No Oort cloud comet orbit can be considered typical, because the orbits are randomly oriented. The figure shows the orbit of one of the few Oort cloud comets that enters the inner solar system. Most spend their entire orbit within the distant Oort cloud. The diagram is not to scale. The inner solar system is enlarged to show more detail, the Kuiper belt extends to about three times the orbital distance of Neptune, and the Oort cloud is much farther away than can be shown here.

rotate nearly on their sides, and Venus rotates "backward"—clockwise, rather than counterclockwise, as viewed from high above Earth's North Pole. Similarly, while most large moons orbit their planets in the same direction that their planets rotate, Neptune has a large moon (called Triton) that goes in the opposite direction.

One of the most interesting exceptions concerns our own Moon. While the other terrestrial planets have either no moons (Mercury and Venus) or very tiny moons (Mars, with two moons), Earth has one of the largest moons in the solar system. Just as exceptional people make the world a more interesting place, the exceptions in our solar system make it a more interesting subject of study.

Figure 8.6 summarizes the four broad features of the solar system that we have discussed.

8.3 A Brief Tour of the Solar System

Now that we have looked at the overall features of our solar system, it is time to look more closely at its individual members. In this section, we will take a brief tour through our solar system, beginning at the Sun and stopping at each of the nine planets. We will highlight a few of the most interesting features of each world we visit, preparing ourselves for more detailed study in coming chapters. Along the side of each page, the planets are shown to scale. Their locations in the Voyage scale model solar system (discussed in Chapter 1) are shown at the bottom of each page.

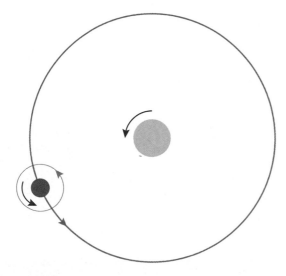

Large bodies in the solar system have orderly motions.
All planets and most satellites have nearly circular orbits going in the same direction in nearly the same plane. The Sun and most of the planets rotate in this same direction as well.

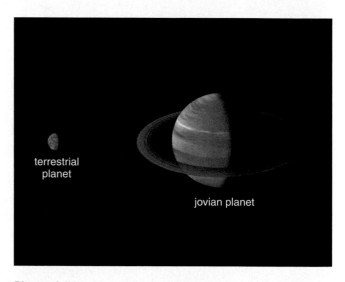

Planets fall into two main categories: small, rocky terrestrial planets near the Sun and large, hydrogen-rich jovian planets farther out. The jovian planets have many moons and rings made of rock and ice. Pluto does not fit in either category.

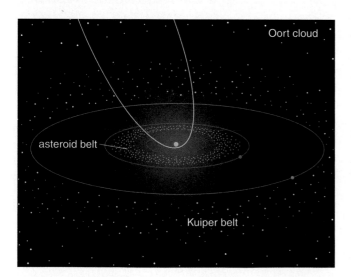

Swarms of asteroids and comets populate the solar system.
Asteroids are concentrated in the asteroid belt, and comets populate the regions known as the Kuiper belt and the Oort cloud.

Figure 8.6 Four major characteristics of the solar system.

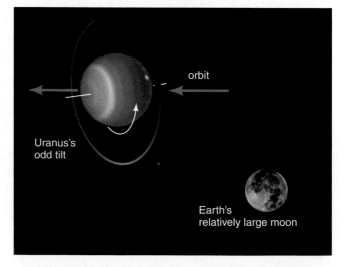

Several notable exceptions to these general trends stand out, such as planets with unusual axis tilts or surprisingly large moons, and moons with unusual orbits.

The Sun

- Radius: 695,000 km
- Mass (Earth = 1): 333,000

The Sun is by far the largest and brightest object in our solar system. It contains more than 99.9% of the solar system's total mass, making it more than a thousand times more massive than everything else in the solar system combined.

The Sun's surface looks solid in photographs (Figure 8.7), but it is actually a roiling sea of hot (about 6,000°C, or 10,000°F) hydrogen and helium gas. The surface is speckled with sunspots that appear dark in photographs only because they are slightly cooler than their surroundings. Solar storms sometimes send streamers of hot gas soaring far above the surface.

The Sun is gaseous throughout. If you could plunge into the Sun, you'd find ever-higher temperatures as you went deeper. The source of the Sun's energy lies deep in its core, where the temperatures and pressures are so high that the Sun becomes a nuclear fusion power plant. Each second, fusion transforms about 600 million tons of the Sun's hydrogen into 596 million tons of helium. The "missing" 4 million tons becomes energy in accord with Einstein's famous formula $E = mc^2$ [Section 4.2]. Despite losing 4 million tons of mass each second, the Sun contains so much hydrogen that it has already shone steadily for almost 5 billion years and will continue to shine for some 5 billion years more.

Earth shown for size comparison

a A visible-light photograph of the Sun's surface. The dark splotches are sunspots—each large enough to swallow several Earths.

b This ultraviolet photograph, from the *SOHO* spacecraft, shows a huge streamer of hot gas on the Sun. The image of Earth was added for size comparison.

Figure 8.7 The Sun contains more than 99.9% of the total mass in our solar system.

The Sun is certainly the most influential object in our solar system. Its gravity governs the orbits of the planets. The Sun's heat is the primary influence on planetary temperatures, and it is the source of virtually all the visible light in our solar system—the Moon and planets shine only by virtue of the sunlight they reflect. In addition, high-energy particles flowing outward from the Sun (the solar wind [Section 9.3]) help shape the magnetic fields of the planets and can influence planetary atmospheres. Nevertheless, we can understand almost all the present characteristics of the planets without knowing much more about the Sun than what we have just discussed. We'll therefore save further study of the Sun for Chapter 15, where we will study it as our prototype for understanding other stars.

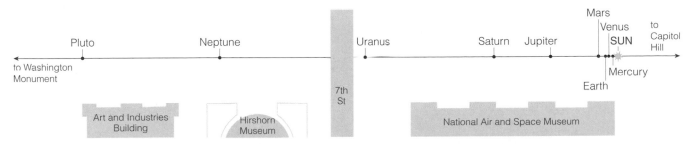

The Voyage scale model solar system represents sizes and distances in our solar system at one ten-billionth of their actual values (see Figure 1.7). The strip along the side of the page shows the sizes of the Sun and planets on this scale, and the map above shows their locations in the Voyage model on the National Mall in Washington, D.C. The Sun is about the size of a large grapefruit on this scale.

Mercury

● Average distance from the Sun: 0.39 AU

● Radius: 2,440 km

● Mass (Earth = 1): 0.055

Figure 8.8 This image shows what it would look like to be orbiting a few hundred kilometers above Mercury's surface with your back toward the Sun. Among the stars, you can see Earth and the Moon as the blue speck and its tiny companion. The view was created using imagery from NASA's *Mariner 10* spacecraft but with computer manipulation to provide color and the orbital viewpoint. The inset (left) is a composite photograph of the full disk of Mercury by *Mariner 10;* the blank strip at the upper right was not photographed. (Image above from the Voyage scale model solar system, developed by the Challenger Center for Space Science Education, the Smithsonian Institution, and NASA. Image created by ARC Science Simulations © 2001.)

Mercury, the innermost planet of our solar system, is the smallest planet except for Pluto (Figure 8.8). It is a desolate, cratered world with no active volcanoes, no earthquakes, no wind, no rain, and no life. Because there is virtually no air to scatter sunlight or color the sky, you could see stars even in the daytime if you stood on Mercury with your back toward the Sun. You wouldn't want to stay long, however, because the ground on Mercury's day side is nearly as hot as hot coals (about 425°C). Nighttime would not be much more comfortable. With no atmosphere to retain heat during the long nights (which last about 3 months), temperatures plummet below −150°C (about −240°F)—far colder than Antarctica in winter.

Mercury is the least studied of the terrestrial worlds. Its proximity to the Sun makes it difficult to observe through telescopes, and it has been visited by just one spacecraft. *Mariner 10* collected data during three rapid flybys of Mercury in 1974–1975, obtaining images of only one hemisphere (shown in Figure 8.8). Mercury has craters almost everywhere, ancient lava flows, and tall, steep cliffs that run hundreds of kilometers in length. As we'll discuss in Chapter 10, the cliffs may in fact be wrinkles from an episode of "planetary shrinking" early in Mercury's history.

The influence of Mercury's gravity on *Mariner 10*'s orbit allowed scientists to determine Mercury's mass and to make inferences about its interior composition and structure. Mercury appears to be made mostly of iron, making it the most metal-rich of the terrestrial planets. Scientists hope to learn much more about Mercury soon. A NASA spacecraft called *Messenger* is scheduled to begin Mercury observations in 2007, and a European mission called *Bepi-Colombo* has a Mercury visit planned for 2011.

Figure 8.9 The image above shows an artistic rendition of the surface of Venus. The surface topography is based on data from NASA's *Magellan* spacecraft, and the colors represent what scientists believe our eyes would see on Venus. The inset (left) shows the full disk of Venus as photographed by NASA's *Pioneer Venus Orbiter*. This photograph was taken with cameras sensitive to ultraviolet light. With visible light, cloud features cannot be distinguished from the general haze. (Image above from the Voyage scale model solar system, developed by the Challenger Center for Space Science Education, the Smithsonian Institution, and NASA. Image by David P. Anderson, Southern Methodist University © 2001.)

Venus

- Average distance from the Sun: 0.72 AU
- Radius: 6,051 km
- Mass (Earth = 1): 0.815

Venus, the second planet from the Sun, is nearly identical in size to Earth. Its surface is completely hidden from view by dense clouds, so we knew little about it until a few decades ago, when cloud-penetrating radar finally allowed us to study Venus in detail (Figure 8.9). Because we knew so little about it, science fiction writers used its Earth-like size, thick atmosphere, and closer distance to the Sun to speculate that it might be a lush, tropical paradise—a "sister planet" to Earth.

The reality is far different. We now know that an extreme *greenhouse effect* bakes its surface to an incredible 450°C (about 850°F), trapping heat so effectively that nighttime offers no relief. Day and night, Venus is hotter than a pizza oven. All the while, the thick atmosphere bears down on the surface with a pressure equivalent to that nearly a kilometer (0.6 mile) beneath the ocean's surface on Earth. Even if you could somehow survive these hazards, you'd find no oxygen to breathe and no water to drink. Far from being a beautiful sister planet to Earth, Venus resembles a traditional view of hell.

The surface of Venus has mountains, valleys, and craters and shows many signs of past (and possibly present) volcanic activity. Aside from these superficial similarities, however, the geology of Venus's surface appears to be quite different from that of Earth. We will devote much of Chapters 10 and 11 to understanding how and why Venus turned out so different from Earth, both in its surface geology and in its atmosphere.

a This image (left), computer generated from satellite data, shows the striking contrast between the daylight and nighttime hemispheres of Earth. The day side reveals little evidence of human presence, but at night our presence is revealed by the lights of human activity (mostly from cities as well as from agricultural, oil, and gas fires). (From the Voyage scale model solar system, developed by the Challenger Center for Space Science Education, the Smithsonian Institution, and NASA. Image created by ARC Science Simulations © 2001.)

b Earth and the Moon, shown to scale. The Moon is about one-fourth as large as Earth in diameter, while its mass is about $\frac{1}{80}$ Earth's mass. If you wanted to show the distance between Earth and the Moon on the same scale, you'd need to hold these two photographs about 6.1 meters (20 feet) apart.

Figure 8.10 Earth, our home planet.

Earth

● Average distance from the Sun: 1.00 AU

● Radius: 6,378 km

● Mass (Earth = 1): 1.00

Beyond Venus, we next encounter our home planet. Although Earth is a barely visible speck when viewed on the scale of our entire solar system, it is the only known oasis of life. Even if we someday discover microscopic life on other worlds in our solar system, Earth is the only world on which humans could survive without a protective enclosure. It is the only planet with oxygen for us to breathe and ozone to shield us from deadly solar radiation. It is the only planet with abundant surface water to nurture life. Temperatures are pleasant for us because Earth's atmosphere contains just enough carbon dioxide and water vapor to maintain a moderate greenhouse effect.

Despite its small size, Earth is striking in its beauty. Blue oceans cover nearly three-fourths of the surface, broken by the continental land masses and scattered islands. The polar caps are white with snow and ice, and white clouds are scattered above the surface. At night, the glow of artificial lights clearly reveals the presence of an intelligent civilization (Figure 8.10a).

Earth is the first planet on our tour with a moon. Although the Moon is much smaller than Earth, it is surprisingly large in relative size compared to most of the moons of other planets (Figure 8.10b). How Earth acquired such a large moon has long been one of the major mysteries of our solar system.

Mars

- Average distance from the Sun: 1.52 AU

- Radius: 3,397 km

- Mass (Earth = 1): 0.107

The next planet on our tour is Mars, the last of the terrestrial worlds in our solar system (Figure 8.11). Mars is larger than Mercury and the Moon but smaller than Venus and Earth. About half Earth's size in diameter, Mars has a mass about 10% that of Earth. Mars has two tiny moons, Phobos and Deimos, that look much like typical asteroids and may once have roamed freely in the asteroid belt. These moons are too small to influence Mars in any of the important ways that our Moon influences Earth (such as tides).

Mars is a world of wonders, with extinct volcanoes that dwarf the largest mountains on Earth, a great canyon that runs nearly one-fifth of the way around the planet, and polar caps made of frozen carbon dioxide ("dry ice") and water ice. Although Mars is frozen today, the presence of dried-up riverbeds and rock-strewn floodplains offers clear evidence that Mars was warm and wet sometime in the distant past. Thus, Mars may once have been hospitable for life, though its wet era probably ended at least 3 billion years ago. Mars's dramatic change in planetary fortune provides one of the most interesting stories in planetary science, and we will explore it in Chapter 11.

Mars looks almost Earth-like in photographs taken by spacecraft on its surface, but you wouldn't want to visit without a space suit. The air pressure is far less than that on top of Mount Everest, the temperature is usually well below freezing, the trace amounts of oxygen would not be nearly enough to breathe, and the lack of atmospheric ozone would leave you exposed to deadly ultraviolet radiation from the Sun.

Mars is the most studied planet besides Earth. More than a dozen spacecraft have flown past, orbited, or landed on Mars, and plans are in the works for many more missions to Mars. We may even send humans to Mars within our lifetime. Overturning rocks in ancient riverbeds or chipping away at ice in the polar caps, explorers will help us learn whether Mars has ever been home to life.

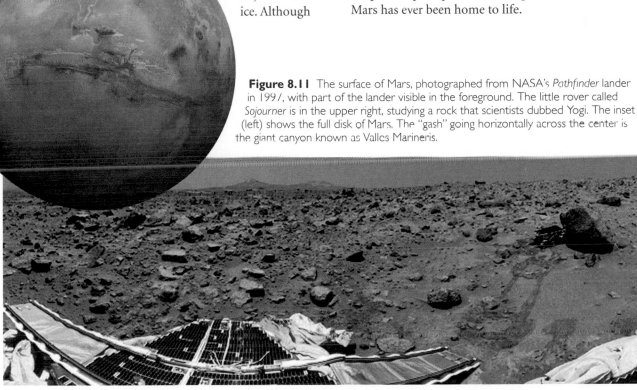

Figure 8.11 The surface of Mars, photographed from NASA's *Pathfinder* lander in 1997, with part of the lander visible in the foreground. The little rover called *Sojourner* is in the upper right, studying a rock that scientists dubbed Yogi. The inset (left) shows the full disk of Mars. The "gash" going horizontally across the center is the giant canyon known as Valles Marineris.

Jupiter

- Average distance from the Sun: 5.20 AU

- Radius: 71,492 km

- Mass (Earth = 1): 317.9

To reach the orbit of Jupiter from Mars, we must traverse a distance that is more than double the total distance from the Sun to Mars, passing through the asteroid belt along the way. Upon our arrival, we would find a planet much larger than any we have seen so far (Figure 8.12). Jupiter is the prototype for the group we refer to as the jovian planets.

Figure 8.12 This image shows what it would look like to be orbiting near Jupiter's moon Io as Jupiter comes into view. Notice the Great Red Spot to the left of Jupiter's center. The extraordinarily dark rings discovered in the Voyager missions are exaggerated to make them visible. This computer visualization was created using data from both NASA's Voyager and Galileo missions. (From the Voyage scale model solar system, developed by the Challenger Center for Space Science Education, the Smithsonian Institution, and NASA. Image created by ARC Science Simulations © 2001.)

Jupiter is so different from the planets of the inner solar system that we must adopt an entirely new mental image of the term *planet*. Its mass is more than 300 times that of Earth, and its volume is more than 1,000 times that of Earth. Its most famous feature—a long-lived storm called the Great Red Spot—is itself large enough to swallow two or three Earths. Like the Sun, Jupiter is made primarily of hydrogen and helium and has no solid surface. If you plunged deep into Jupiter, you would be crushed by the increasing gas pressure long before you ever reached its core.

Jupiter reigns over at least 60 moons and a thin set of rings (too faint to be seen in most photographs). The four largest moons—Io, Europa, Ganymede, and Callisto—are often called the *Galilean moons* because they were discovered by Galileo shortly after he first turned his telescope toward the heavens [Section 3.4]. Each is a fascinating world in its own right. Each of these four moons is larger than our Moon, and Ganymede is larger than the planet Mercury. Io is the most volcanically active place in the solar system. Europa has an icy crust that may hide a subsurface ocean of liquid water, making it a promising place to search for life. Ganymede and Callisto may also have subsurface oceans, and their surfaces have many features that remain mysterious. As we'll discuss further in Chapter 12, the moons of Jupiter and the other jovian planets are at least as interesting as the planets themselves.

Figure 8.13 This computer simulation, built upon imagery from the Voyager 1 mission, re-creates the striking view as the spacecraft passed by the planet. We see the shadow of the rings on Saturn's sunlit face, and the rings become lost in Saturn's shadow on the night side. (From the Voyage scale model solar system, developed by the Challenger Center for Space Science Education, the Smithsonian Institution, and NASA. Image created by ARC Science Simulations © 2001.)

Saturn

- Average distance from the Sun: 9.54 AU

- Radius: 60,268 km

- Mass (Earth = 1): 95.18

Saturn is the second largest planet in our solar system after Jupiter, and nearly twice as far from the Sun. Saturn is only slightly smaller than Jupiter in diameter, but it is considerably less massive (about one-third Jupiter's mass) because it is less dense. Like Jupiter, Saturn is made mostly of hydrogen and helium and has no solid surface.

Saturn is famous for its spectacular rings (Figure 8.13). Although all four of the giant outer planets have rings, only Saturn's rings can be seen easily through a small telescope. The rings may look solid from a distance, but this appearance is deceiving. If you could wander into the rings, you'd find yourself surrounded by countless individual particles of rock and ice, ranging in size from dust grains to city blocks. Each ring particle orbits Saturn like a tiny moon. The rings do not touch Saturn's surface—any ring particle that wandered into the atmosphere would quickly burn up.

At least 30 moons orbit Saturn. Most are the size of small asteroids or comets, but a few are much larger. Saturn's largest moon, Titan, is bigger than the planet Mercury. Titan is blanketed by a thick atmosphere. On Titan's surface, you'd find an atmospheric pressure even greater than on Earth, and you could inhale air with roughly the same nitrogen content as air on Earth. However, you'd need to bring your own oxygen, and you'd certainly want a warm space suit for protection against the frigid outside temperatures. NASA's *Cassini* spacecraft, designed to explore Saturn and its rings and moons, carries a probe that will be dropped to Titan's surface. *Cassini* is scheduled to arrive at Saturn in mid-2004.

THINK ABOUT IT

By the time you read this book, *Cassini* will be either nearing or already at Saturn. Find out the current location of the spacecraft (from news reports or the mission Web site). If it has not yet arrived, when should you expect to start hearing about it in the news? If it has already arrived, has it made any notable discoveries?

Uranus

- Average distance from the Sun: 19.19 AU

- Radius: 25,559 km

- Mass (Earth = 1): 14.54

Uranus (normally pronounced YUR-uh-nus) is much smaller than either Jupiter or Saturn but still much larger than Earth (Figure 8.14). It is made largely of hydrogen, helium, and hydrogen compounds such as methane (CH_4). Methane gas gives Uranus its pale blue-green color. Like the other giants of the outer solar system, Uranus lacks a solid surface. At least 21 moons orbit Uranus, along with a set of rings somewhat similar to those of the other jovian planets—though they are dark and difficult to see.

The entire Uranus system—planet, rings, and moon orbits—is tipped on its side compared to the rest of the planets. This unusual orientation may be the result of a cataclysmic collision suffered by Uranus as it was forming some 4.6 billion years ago. It also makes for the most extreme pattern of seasons on any planet. If you lived on a platform floating in Uranus's atmosphere near the north pole, you'd have continuous daylight for half of each orbit, or 42 years. Then, after a very gradual sunset, you'd enter into a 42-year-long night.

Only one spacecraft has visited Uranus. *Voyager 2* flew past all four of the jovian planets before heading out of the solar system. Much of our current understanding of Uranus comes from that mission, though powerful new telescopes are also capable of studying this planet. Scientists would love an opportunity to study Uranus and its interesting rings and moons in much greater detail, but no missions to Uranus are currently under development.

Figure 8.14 This image shows a view from a vantage point high above Uranus's moon Ariel. The ring system is shown, although it would actually be too dark to see from this vantage point. Computer simulation based upon data from NASA's Voyager 2 mission. (From the Voyage scale model solar system, developed by the Challenger Center for Space Science Education, the Smithsonian Institution, and NASA. Image created by ARC Science Simulations © 2001.)

Figure 8.15 This image shows what it would look like to be orbiting Neptune's moon Triton as Neptune itself comes into view. The dark rings are exaggerated to make them visible in this computer simulation using data from NASA's Voyager 2 mission. (From the Voyage scale model solar system, developed by the Challenger Center for Space Science Education, the Smithsonian Institution, and NASA. Image created by ARC Science Simulations © 2001.)

Neptune

- Average distance from the Sun: 30.06 AU

- Radius: 24,764 km

- Mass (Earth = 1): 17.13

Neptune looks nearly like a twin of Uranus, with very similar size and composition, although it is more strikingly blue (Figure 8.15). Like Uranus, Neptune has been visited only by the *Voyager 2* spacecraft. No further missions are currently planned.

Neptune has rings and at least eight moons. Its largest moon, Triton, is larger than the planet Pluto and is one of the most fascinating moons in the solar system. Its icy surface has features that appear to be somewhat like geysers but spew nitrogen gas into the sky. Even more surprisingly, Triton is the only large moon in the solar system that orbits its planet "backward"—that is, in a direction opposite to the direction in which Neptune rotates. Understanding how such a large moon ended up orbiting Neptune backward has given us new insights into the history of the outer solar system [Section 12.5].

Pluto

- Average distance from the Sun: 39.54 AU
- Radius: 1,160 km
- Mass (Earth = 1): 0.0022

At its average distance from the Sun, Pluto lies far beyond the orbit of Neptune—as far as the distance between the orbits of Uranus and Neptune. It is easy to imagine that Pluto must be cold and dark. From Pluto, the Sun would appear as little more than a bright light among the stars, offering little comfort or warmth.

A quick glance at the characteristics of Pluto in Table 8.1 shows it to be out of character with the rest of the planets. Pluto is neither large and gaseous like the jovian planets nor rocky like the terrestrial planets. Instead, it is very small—the smallest planet by

far—and is made mostly of ices. Pluto's orbit is also unusual, being both quite elliptical and substantially inclined to the plane of the other planets. Pluto actually comes closer to the Sun than Neptune for 20 years out of each 248-year orbit. The last such period ended in 1999, and the next won't begin until 2263. Its oddball nature gives it more in common with Kuiper belt objects than with the other eight planets, leading to debate over whether it should be called a "planet" at all [Section 13.5].

Pluto has a moon, Charon, that is about half Pluto's size in diameter and one-eighth Pluto's mass. Our best telescopic views of Pluto and Charon reveal little detail (Figure 8.16), and no spacecraft has yet visited these distant worlds. However, a Pluto mission called New Horizons is well into development. If all goes well, it will be launched in 2006 and will arrive at Pluto in 2015.

Figure 8.16 Pluto and Charon, as photographed by the Hubble Space Telescope. Because no spacecraft has been to Pluto, we do not yet have clear pictures of this tiny planet or its moon.

8.4 Exploring the Solar System

How have we learned so much about the solar system? Much of our knowledge comes from telescopic observations, using both ground-based telescopes and telescopes in Earth orbit such as the Hubble Space Telescope. In one case—our Moon—we have learned a lot by sending astronauts to explore the terrain and bring back rocks for laboratory study. The recent revolution in our understanding of the solar system has come largely through robotic spacecraft that actually visit the worlds we wish to study. In this section, we'll briefly investigate the types of robotic spacecraft used in exploring the solar system.

Spacecraft Basics

The spacecraft we send to explore the planets are robots suited for long space journeys. They carry specialized equipment for scientific study. All spacecraft have control computers, power sources such as solar cells, propulsion systems, and devices to point cameras and other instruments precisely at their targets. Robotic spacecraft operate primarily with preprogrammed instructions. They carry radios for communication, allowing them to receive additional instructions from Earth and to send home the data they collect. Most robotic spacecraft make one-way trips from Earth, never physically returning but sending their data back from space in the same way we send radio and television signals around the world.

Broadly speaking, the robotic missions we send to explore other worlds fall into one of four major categories:

SPECIAL TOPIC Discovering the Outermost Planets

The planets Mercury, Venus, Mars, Jupiter, and Saturn were all known to ancient people. Each can be seen with the naked eye and each clearly wanders among the fixed stars of the constellations. In contrast, the three planets beyond Saturn are all relatively recent discoveries in human history.

Uranus was the first "discovered" planet. Although faintly visible to the naked eye, Uranus is so faint and moves so slowly in its 84-year orbit of the Sun that ancient people did not recognize it as a planet. Uranus even appeared as a star on some sky charts prior to its "discovery" as a planet in 1781. English astronomer William Herschel discovered Uranus with the help of his sister, Caroline Herschel, who helped build his telescopes and assisted with much of the observing. William Herschel originally suggested naming the planet *Georgium Sidus*, Latin for *George's star*, in honor of his patron, King George III. Fortunately, the idea of "Planet George" never caught on. Instead, many eighteenth- and nineteenth-century astronomers referred to the new planet as *Herschel*. The modern name Uranus—after the mythological father of Saturn—was first suggested by one of Herschel's contemporaries, astronomer Johann Bode. It was generally accepted by the mid–nineteenth century.

Neptune's discovery followed next, and it represented an important triumph for Newton's law of universal gravitation and the young science of astrophysics. By the mid-1800s, careful observations of Uranus had shown its orbit to be slightly inconsistent with that predicted by Newton's law of gravity—at least if it was being influenced only by the Sun and the other known planets. In the early 1840s in England, a student named John Adams suggested that the inconsistency could be explained by a previously unseen "eighth planet" orbiting the Sun beyond Uranus. According to the official story, he used Newton's theory to predict the location of the planet but was unable to convince British astronomers to carry out a telescopic search. However, recently discovered documents suggest his prediction may not have been as precise as the official history suggests. Meanwhile, in the summer of 1846, French astronomer Urbain Leverrier made very precise calculations independently. He sent a letter to Johann Galle of the Berlin Observatory suggesting a search for the eighth planet. On the night of September 23, 1846, Galle pointed his telescope to the

position suggested by Leverrier. There, within 1° of its predicted position, he saw the planet Neptune. Hence, Neptune's discovery truly was made by mathematics and physics and was only confirmed with a telescope.

As a side note to this story, Leverrier had such faith in Newton's universal law of gravitation that he also suggested a second unseen planet, this one orbiting closer to the Sun than Mercury. He got this idea because other astronomers had identified slight discrepancies between Mercury's actual orbit and the orbit predicted by Newton's theory. He assumed that the planet, which he called Vulcan, had not yet been seen because it was so close to the Sun. Leverrier died in 1877, still believing that Vulcan would someday be discovered. In fact, Vulcan does not exist. Mercury's actual orbit does not match the orbit predicted by Newton's law of gravity because Newton's theory is *not* the whole story of gravity. About 40 years after Leverrier's death, Einstein showed that Newton's theory is only an approximation of a broader theory of gravity, known today as Einstein's general theory of relativity. Einstein's theory predicts an orbit for Mercury that matches its actual orbit. This match was one of the first key pieces of evidence in favor of Einstein's theory [Section S3.5].

Pluto was discovered in 1930 by American astronomer Clyde Tombaugh, culminating a search that began when astronomers analyzed the orbit of Neptune. The story of Pluto's discovery at first seemed much the same as that of Neptune's. Just as discrepancies between the predicted orbit and the actual orbit of Uranus led to the prediction that Neptune must exist, apparent discrepancies between the predicted orbit and the actual orbit of Neptune suggested the existence of an even more distant "ninth planet." Tombaugh found Pluto just 6° from the position in the sky where this ninth planet had been predicted to lie, so it seemed that the search had been successful. However, while initial estimates suggested that Pluto was much larger than Earth, we now know that Pluto has a radius of only 1,160 kilometers and a mass of just 0.002 Earth mass—making it far too small to affect the orbit of Neptune. In retrospect, the supposed orbital irregularities of Neptune appear to have been errors in measurement. No "ninth planet" is needed to explain the orbit of Neptune. The discovery of Pluto was just a happy coincidence.

- **Flyby**: A spacecraft on a flyby goes past a world just once and then continues on its way.

- **Orbiter**: An orbiter is a spacecraft that orbits the world it is studying, allowing longer-term study during its repeated orbits.

- **Lander** or **probe**: These spacecraft are designed to land on a planet's surface or to probe a planet's atmosphere by flying through it.

- **Sample return mission**: A sample return mission involves a spacecraft designed to return to Earth carrying a sample of the world it has studied.

The choice of spacecraft type depends on both scientific objectives and cost. In general, a flyby is the lowest-cost way to visit another planet, and some flybys gain more "bang for the buck" by visiting multiple planets. For example, *Voyager 2* flew past Jupiter, Saturn, Uranus, and Neptune before continuing on its way out of our solar system (Figure 8.17).

Flybys Flybys tend to be cheaper than other missions because they are generally less expensive to launch into space. Launch costs depend largely on weight, and onboard fuel is a significant part of the weight of a spacecraft heading to another planet. Once a spacecraft is on its way, the lack of friction or air drag in space means that it can maintain its orbital trajectory through the solar system without using any fuel at all. Fuel is needed only when the spacecraft needs to change from one trajectory (orbit) to another.

Moreover, with careful planning, some trajectory changes can be made by taking advantage of the gravity of other planets. Look closely at the *Voyager 2* trajectory in Figure 8.17. You'll see that it made significant trajectory changes as it passed by Jupiter and Saturn. In effect, it made these changes for free by using gravity to bend its path rather than by burning fuel. (This technique is known as a "gravitational slingshot.")

Study the *Voyager 2* trajectory in Figure 8.17. Given that Saturn orbits the Sun every 29 years, Uranus orbits the Sun every 84 years, and Neptune orbits the Sun every 165 years, would it be possible to send another flyby mission to all four jovian planets if we launched it now? Explain.

Although a flyby offers only a relatively short period of close-up study, it can provide valuable scientific information. Flybys generally carry small telescopes, cameras, and spectrographs. Because these instruments are brought within a few tens of thousands of kilometers of other worlds (or closer), they can obtain much higher resolution images and spectra than even the largest current telescopes viewing these worlds from Earth. In addition, flybys sometimes give us information that would be very difficult to obtain from Earth. For example, *Voyager 2* helped us discover Jupiter's rings and learn about the rings of Saturn, Uranus, and Neptune through views in which the rings were backlit by the Sun. Such views are possible only from beyond each planet's orbit.

Figure 8.17 The trajectory of *Voyager 2*, which made flybys of each of the four jovian planets in our solar system.

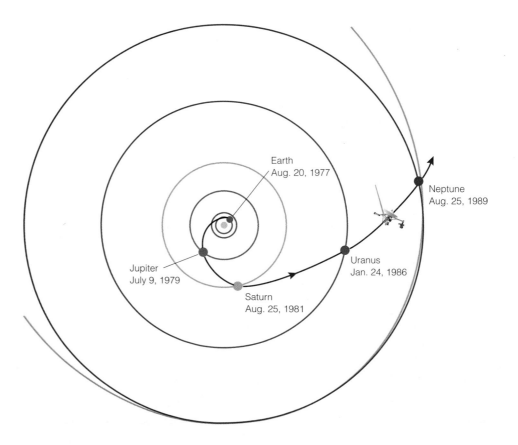

Flybys may also carry instruments to measure local magnetic field strength or to sample interplanetary dust. The gravitational effects of the planets and their moons on the spacecraft itself provide information about object masses and densities. Like the backlit views of the rings, these types of data cannot be gathered from Earth. Indeed, most of what we know about the masses and compositions of moons comes from data gathered by spacecraft that have flown past them.

Orbiters

An orbiter can study another world for a much longer period of time than a flyby. Like the spacecraft used for flybys, orbiters often carry cameras, spectrographs, and instruments for measuring the strength of magnetic fields. Some missions to Venus and Mars have used radar to make precise altitude measurements of surface features. In the case of Venus, radar observations give us our only clear look at the surface, because we cannot otherwise see through the planet's thick cloud cover (see Figure 8.9).

An orbiter is generally more expensive than a flyby for an equivalent weight of scientific instruments, primarily because it must carry added fuel to change from an interplanetary trajectory to a path that puts it in orbit around another world. Careful planning can minimize the added expense. Two recent Mars orbiters, *Mars Global Surveyor* and *Mars Odyssey,* saved on fuel costs by carrying only enough fuel to enter highly elliptical orbits around Mars. In each case, the spacecraft settled into the smaller, more circular orbit needed for scientific observations by skimming the Martian atmosphere at the low point of every elliptical orbit. Atmospheric drag slowed the spacecraft with each orbit and, over several months, circularized the spacecraft orbit. (This technique of using the atmosphere to slow the spacecraft and change its orbit is called *aerobraking.*)

Landers or Probes

The most "up close and personal" study of other worlds comes from spacecraft that send probes into the atmospheres or landers to the surfaces. For example, in 1995, the *Galileo* spacecraft dropped a probe into Jupiter's atmosphere [Section 12.3]. The probe collected temperature, pressure, composition, and radiation measurements for about an hour as it descended before being destroyed by Jupiter's high interior pressures and temperatures.

On planets with solid surfaces, a lander can offer close-up surface views, local weather monitoring, and the ability to carry out automated experiments. Some landers, such as the *Pathfinder* lander that arrived on Mars in 1997, carry robotic rovers able to venture across the surface (see Figure 8.11). Landers typically require fuel to slow their descent to a planetary surface, but clever techniques can reduce cost. For example, *Pathfinder* hit the surface of Mars at crash-landing speed but was protected by a cocoon of air bags deployed on the way down. These air bags allowed the lander to bounce along the surface of Mars for more than a kilometer before it finally came to rest (Figure 8.18).

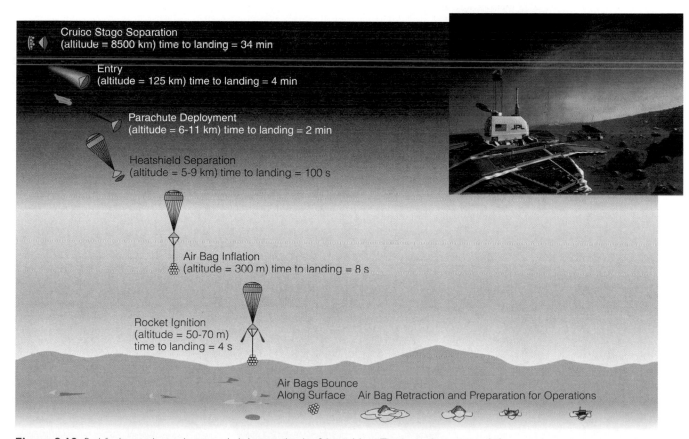

Figure 8.18 *Pathfinder* used parachutes and air bags to land safely on Mars. The inset shows an artist's conception of the lander on the surface.

Sample Return Missions While probes and landers can carry out experiments on surface rock or atmospheric samples, the experiments must be designed in advance and must fit inside the spacecraft. These limitations make scientists long for missions that will scoop up samples from other worlds and return them to Earth for more detailed study. To date, the only sample return missions have been to the Moon. Astronauts collected samples during the Apollo missions, and the Soviet Union sent robotic spacecraft to collect rocks from the Moon in the early 1970s. Many scientists are working toward a sample return mission to Mars. They hope to begin such a mission within the next decade or so. (A slight variation on the theme of a sample return mission is the Stardust mission, currently en route to collect a sample of comet dust and return it to Earth in 2006.)

Combination Spacecraft Many missions combine elements of more than one type of spacecraft. For example, the Galileo mission to Jupiter included an orbiter that studied Jupiter and its moons as well as the probe that entered Jupiter's atmosphere. The Cassini mission, which will arrive at Saturn in 2004, included flybys of Venus, Earth, and Jupiter. It involves both an orbiter to study Saturn and its moons and a probe, called the *Huygens probe,* that will descend through the atmosphere of Saturn's moon Titan (Figure 8.19).

Exploration Past, Present, and Future

Over the past several decades, studies using both telescopes on Earth and robotic spacecraft have allowed us to learn the general characteristics of all the major planets and moons in our solar system as well as the general characteristics of asteroids and comets. Telescopes will continue to play an important role in future observations, but for detailed study we will probably continue to depend on spacecraft.

Table 8.3 lists some significant robotic missions of the past and present. The next few years promise many new discoveries as missions arrive at their destinations. Over the longer term, all the world's major space agencies have high hopes of launching numerous and diverse missions to study our solar system. For example, nearly a dozen missions are on the drawing board for Mars alone. Of course, these hopes are subject to the vagaries of politics and budgets for scientific exploration. If these missions come to fruition, however, the next couple of decades will see spacecraft that answer many specific questions about the nature of our solar system and its many worlds.

Figure 8.19 The Cassini mission to Saturn.

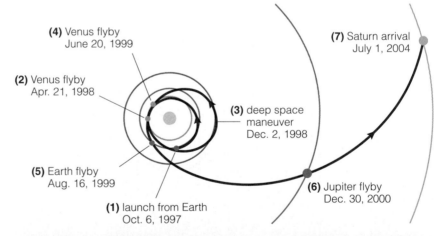

a The trajectory of *Cassini* from Earth to Saturn.

(4) Venus flyby
June 20, 1999

(7) Saturn arrival
July 1, 2004

(2) Venus flyby
Apr. 21, 1998

(3) deep space maneuver
Dec. 2, 1998

(5) Earth flyby
Aug. 16, 1999

(6) Jupiter flyby
Dec. 30, 2000

(1) launch from Earth
Oct. 6, 1997

b Artist's conception of the Cassini mission's orbiter, with the *Huygens* probe (the disk-shaped object to the lower left of the orbiter) descending toward Titan.

Table 8.3 Selected Robotic Missions to Other Worlds

Name	Type	Destination	Lead Space Agency	Arrival Year	Mission Highlights
Stardust	Flyby and sample return	Comet Wild 2	NASA	2004	To collect dust from comet and return it to Earth in 2006
Deep Impact	Flyby and "lander"	Comet Tempel 1	NASA	2005[†]	"Lander" to hit at 10 km/sec, flyby spacecraft to study the impact
Cassini	Orbiter	Saturn	NASA	2004[†]	Includes lander (called *Huygens*) for Titan
Mars Exploration Rovers	Lander	Mars	NASA	2004[†]	Rovers *Spirit* and *Opportunity* to study the surface in two locations
Mars Express	Orbiter and lander	Mars	ESA*	2004[†]	Climate and surface studies of Mars
Nozomi	Orbiter	Mars	Japan	2004[†]	Japanese-led mission to study Martian atmosphere
Mars Odyssey 2001	Orbiter	Mars	NASA	2001	Detailed study of Martian surface features and composition
Near Earth Asteroid Rendezvous (NEAR)	Orbiter	Eros (asteroid)	NASA	2000	First spacecraft dedicated to in-depth study of an asteroid
Mars Global Surveyor	Orbiter	Mars	NASA	1997	Detailed imaging of surface from orbit
Mars Pathfinder	Lander	Mars	NASA	1997	Carried *Sojourner*, the first robotic rover on Mars
Galileo	Orbiter	Jupiter	NASA	1995	Dropped probe into Jupiter; close-up study of moons
Magellan	Orbiter	Venus	NASA	1990	Detailed radar mapping of surface of Venus
Voyager 1	Flyby	Jupiter, Saturn	NASA	1979, 1980	Unprecedented views of Jupiter and Saturn; continuing out of solar system
Voyager 2	Flyby	Jupiter, Saturn, Uranus, Neptune	NASA	1979, 1981, 1986, 1989	Only mission to Uranus and Neptune; continuing out of solar system
Vikings 1 and 2	Orbiter and lander	Mars	NASA	1976	First landers on Mars

*European Space Agency.
[†]Scheduled arrival year.

THE BIG PICTURE

Putting Chapter 8 into Context

In this chapter, we've introduced the major characteristics of our solar system, briefly toured our planetary neighbors, and discussed how we study the solar system with robotic spacecraft. Keep in mind the following "big picture" ideas as you continue your study of the solar system:

● Our solar system is not a random collection of objects moving in random directions. Rather, it is highly organized, with clear patterns of motion and with most objects falling into simply defined categories.

● Each planet has its own unique and interesting features. Becoming familiar with the planets is an important first step in understanding the root causes of their similarities and differences.

● Much of what we now know about the solar system comes from spacecraft exploration. Choosing the type of mission to send to a planet involves many considerations, from the scientific to the purely political. Many missions are currently under way, offering us hope of learning much more in the near future.

8.1 Comparative Planetology

- *What can we learn by comparing the planets to one another?* Comparing the planets leads to deeper insights into the physical processes that shape Earth and the other planets than we can get by studying the planets individually. We also learn more about the origin and history of the solar system as a whole, and we can then apply ideas from our solar system to other planetary systems.

8.2 The Layout of the Solar System

- *What are the major patterns of motion in our solar system?* All planets orbit the Sun in the same direction and with nearly circular orbits in nearly the same plane. The Sun and most planets rotate in the same direction that they orbit. Most large moons orbit their planets in the same direction as well.

- *What are the two major types of planets?* The two major types of planets are the small, rocky terrestrial planets and the large, hydrogen-rich jovian planets.

- *Where do we find asteroids and comets in our solar system?* Most asteroids reside in the asteroid belt between Mars and Jupiter. Comets are found in two main regions: the Kuiper belt, which begins near the orbit of Neptune, and the much more distant Oort cloud.

- *What are a few important exceptions to these general rules?* Uranus and Pluto rotate sideways compared to their orbits. Venus rotates "backward." Triton orbits Neptune backward from what we'd expect. Earth has a surprisingly large Moon.

8.3 A Brief Tour of the Solar System

- *How does the Sun influence the planets?* Its gravity governs planetary orbits, its heat is the primary influence on planetary surface temperatures, it is the source of virtually all the visible light in our solar system, and high-energy particles from the Sun influence planetary atmospheres and magnetic fields.

- *What do you find particularly interesting about each planet?* Examples: Mercury has extreme days and nights, tall steep cliffs, and large iron content. Venus has an extreme greenhouse effect. Earth is an oasis of life. Mars shows evidence of a past wet era. Jupiter has a hydrogen and helium atmosphere and many moons. Saturn has rings and a moon, Titan, that is larger than Mercury. Uranus and its moons make up a system tipped on its side compared to other planets. Neptune's largest moon, Triton, has nitrogen "geysers" and a "backward" orbit. Pluto is a "misfit" among the planets.

8.4 Exploring the Solar System

- *What are four major categories of spacecraft missions?* Spacecraft missions may be flybys, orbiters, landers or probes, or sample return missions.

- *What are some of the most important missions to the planets in our solar system?* Important missions include the Voyager multiplanet flybys, missions to Mars, the Galileo mission to Jupiter, and the Cassini mission to Saturn.

True or False?

Decide whether each statement is true or false. Briefly explain why.

1. Pluto orbits the Sun in the opposite direction of all the other planets.

2. If we were to discover a Kuiper belt comet that is as large as the planet Mercury, we would classify it as a terrestrial planet.

3. Comets in the Kuiper belt and Oort cloud have long, beautiful tails that we can see when we look through telescopes.

4. Asteroids are made of essentially the same materials as the terrestrial planets.

5. The mass of the Sun compared to the mass of all the planets combined is like the mass of an elephant compared to the mass of a cat.

6. On average, Venus is the hottest planet in the solar system—even hotter than Mercury.

7. The weather conditions on Mars today are much different than they were in the distant past.

8. Moons cannot have atmospheres, active volcanoes, or liquid water.

9. Saturn is the only planet in the solar system with rings.

10. Several sample return missions are currently en route to the terrestrial planets.

Problems

11. *Solar System Trends.* Study the planetary data in Table 8.1 to answer each of the following.

   a. Notice the relationship between distance from the Sun and surface temperature. Describe the trend, explain why it exists, and explain any notable exceptions to the trend.

   b. The text says that planets can be classified as either terrestrial or jovian, with Pluto as a misfit. Describe in general how the columns for density, composition, and distance from the Sun support this classification.

   c. Which column would you use to find out which planet has the shortest days? Do you see any notable differences in the length of a day for the different types of planets? Explain.

   d. Describe the trend you see in orbital periods and explain the trend in terms of Kepler's third law.

   e. Which planets would you expect to have seasons? Why?

   f. Which column tells you how much a planet's orbit deviates from a perfect circle? Based on that column, are there any planets for which you would expect the surface temperature to vary significantly over the course of each orbit? Explain.

   g. By studying the table data, briefly describe how escape velocity is related to mass and radius. Is the trend what you expect based on what you learned about escape velocity in Chapter 5?

   h. Suppose you weigh 100 pounds. State how much you would weigh on each of the other planets in our solar system. (*Hint:* Recall from Chapter 5 that weight is mass times the acceleration of gravity. The surface gravity column tells you how the acceleration of gravity on other planets compares to that on Earth.)

12. *Left Out in the Cold.* In what ways does Pluto resemble a terrestrial planet? In what ways does it resemble a jovian planet? In what ways does it resemble neither?

13. *Comparing Leftovers.* Apart from their orbital properties, how are comets different from asteroids? Which are more numerous, comets or asteroids?

14. *Patterns of Motion.* In one or two paragraphs, explain why the existence of orderly patterns of motion in our solar system should suggest that the Sun and the planets all formed at one time from one cloud of gas, rather than as individual objects at different times.

15. *Two Classes of Planets.* In terms a friend or roommate would understand, write one or two paragraphs explaining why we say that the planets fall into two major categories and what those categories are.

16. *Planetary Tour.* Based on the brief planetary tour in this chapter, which planet besides Earth do you think is the most interesting, and why? Defend your opinion clearly in two or three paragraphs.

Discussion Questions

17. *Where Would You Go?* Suppose you could visit any one of the planets or moons in our solar system for one week. Which object would you choose to visit, and why?

18. *Planetary Priorities.* Suppose you were in charge of developing and prioritizing future planetary missions for NASA. What would you choose as your first priority for a new mission (that is, a mission that is not already being developed)? Describe in detail the type of mission, its goals, and how you would convince Congress to support its cost.

For a complete list of media resources available, go to www.astronomyplace.com and choose Chapter 8 from the pull-down menu.

Astronomy Place Web Tutorials

Tutorial Review of Key Concepts

Use the following interactive **Tutorials** at www.astronomyplace.com to review key concepts from this chapter.

Scale of the Universe Tutorial

Lesson 1 Distances of Scale: Our Solar System

Formation of the Solar System Tutorial

Lesson 1 Comparative Planetology

Supplementary Tutorial Exercises

Use the interactive **Tutorial Lessons** to explore the following questions.

Scale of the Universe Tutorial, Lesson 1

1. How big is an astronomical unit (AU) in comparison to more familiar units?

2. Why do we use AUs to describe distances in our solar system? Is the AU a good unit for describing distances to stars? Explain.

3. How big is the Sun relative to the Moon's orbit of Earth?

4. Explore the layout of the solar system, focusing on the scale of the various planetary orbits. Which of the key patterns discussed in this chapter are evident from the scaled view? Which require deeper study to uncover?

Formation of the Solar System Tutorial, Lesson 1

1. What features distinguish the terrestrial planets from the jovian planets?

2. What is the difference between axis tilt, orbital eccentricity, and orbital inclination? What patterns do we see in these properties among the planets?

3. Which solar system bodies constitute "exceptions to the rules," and in what ways?

Exploring the Sky and Solar System

Of the many activities available on the *Voyager: SkyGazer* **CD-ROM** accompanying your book, use the following files to observe key phenomena covered in this chapter.

Go to the **File: Basics** folder for the following demonstrations.

1. Saturn's Phases

2. Tracking Venus

3. Planet Panel

Go to the **File: Demo** folder for the following demonstrations.

1. Earth and Venus

2. Trailing Saturn

3. Triple Conjunction of 7 BC

Go to the **Explore** menu for the following demonstrations.

1. Solar System

2. Paths of Planets

Movies

Check out the following narrated and animated short documentaries available on www.astronomyplace.com for a helpful review of key ideas covered in this chapter.

Orbits in the Solar System Movie

History of the Solar System Movie

Web Projects

Take advantage of the useful Web links on www.astronomyplace.com to assist you with the following projects.

1. *Current Mission.* Visit the Web page for one of the current missions listed in Table 8.3. Write a one- to two-page summary of the mission's basic design, goals, and current status.

2. *Mars Missions.* Go to the home page for NASA's Mars Exploration Program. Write a one- to two-page summary of the plans for future exploration of Mars.

9 Formation of the Solar System

The evolution of the world may be compared to a display of fireworks that has just ended: some few red wisps, ashes and smoke. Standing on a cooled cinder, we see the slow fading of the suns, and we try to recall the vanished brilliance of the origin of the worlds.

**G. Lemaître (1894–1966),
astronomer and Catholic priest**

How did Earth come to be? How old is it? Is it unique? Our ancestors could do little more than guess at the answers to these questions, and they imagined Earth to be fundamentally different from objects in the heavens. Today we see Earth as just one of many worlds. Modern science tells us that Earth and the rest of our solar system formed from a great cloud of gas and dust about 4.6 billion years ago.

Our scientific ideas of how the solar system formed explain more than our cosmic origins. They are also the key to understanding the nature of planets. If the planets in our solar system all formed together, then their differences must be attributable to physical processes that occurred during the birth and subsequent evolution of the solar system. Thus, learning about the origin of the solar system forms the basis of our comparative study of the planets.

In this chapter, we'll explore the current scientific theory of how our solar system came to be. We'll focus most of our attention on how the theory explains the major characteristics of our solar system that we studied in Chapter 8. We'll also discuss how scientists have determined when the solar system was born, and we'll learn how recent discoveries of planets around other stars have forced us to reconsider a few details of the basic theory.

9.1 The Origin of the Solar System: The Nebular Theory

The development of any scientific theory is an interplay between observations and attempts to explain those observations [Section 3.5]. Hypotheses that seem to make sense at one time might later be dismissed because they fail to explain new data. For example, ancient Greek ideas about Earth's origins probably seemed quite reasonable when people assumed that Earth was the center of the universe, but they no longer made sense after Kepler and Galileo proved that Earth is a planet going around the Sun. By the end of the seventeenth century, the Copernican revolution and Newton's discovery of the universal law of gravitation [Section 5.3] had given us a basic understanding of the layout and motion of the planets and moons in our solar system. It was only natural that scientists would begin to speculate about how this system came to be.

We generally credit two eighteenth-century scientists with proposing the hypothesis that ultimately blossomed into our modern scientific theory of the origin of the solar system. Around 1755, German philosopher Immanuel Kant proposed that our solar system formed from the gravitational collapse of an interstellar cloud of gas. The same idea was put forth independently about 40 years later by French mathematician Pierre-Simon Laplace (who apparently was unaware of Kant's proposal). Because an interstellar cloud is usually called a *nebula* (Latin for "cloud"), their idea became known as the *nebular hypothesis.*

The nebular hypothesis remained popular throughout the nineteenth century. By the early twentieth century, however, scientists had found a few aspects of our solar system that the nebular hypothesis did not seem to explain very well—at least in its original form as described by Kant and Laplace. While some scientists sought to modify the nebular hypothesis, others looked for entirely different ideas about how the solar system might have formed.

During much of the first half of the twentieth century, the nebular hypothesis faced stiff competition from a hypothesis proposing that the planets represent debris from a near-collision between the Sun and another star. According to this *close encounter hypothesis,* the planets were formed from blobs of gas that had been gravitationally pulled out of the Sun during the near-collision.

Today, the close encounter hypothesis has been discarded (for reasons we'll discuss shortly). Meanwhile, so much evidence has accumulated in favor of the nebular hypothesis that it has achieved the status of a scientific theory—the **nebular theory** of our solar system's birth. The nebular theory has won out over competing hypotheses because it alone has proved capable of explaining virtually everything we have learned about the solar system.

Remember that a hypothesis can achieve the status of a scientific theory only if it offers a detailed, physical model that explains a broad range of observed facts. In Chapter 8, we discussed four major observed characteristics of our solar system (Figure 8.6). Any scientific theory designed to explain the formation of our solar system must account for all four characteristics:

1. **Patterns of motion**. The theory must successfully explain the orderly patterns of motion that we observe in the solar system. That is, it must explain why all planets orbit the Sun in the same direction and with nearly circular orbits in nearly the same plane, why the Sun and most planets rotate in the same direction that they orbit, and why most large moons also orbit their planets in the same direction.

2. **Two types of planets**. The theory must explain why planets fall into two main categories: the small, rocky terrestrial planets near the Sun and the large, hydrogen-rich jovian planets farther out.

3. **Asteroids and comets**. The theory must be able to explain the existence of huge numbers of asteroids and comets and why these small objects reside primarily in the regions we call the asteroid belt, the Kuiper belt, and the Oort cloud.

4. **Exceptions to the patterns**. Finally, the theory must explain all the general patterns while at the same time making allowances for the exceptions to the general rules, such as the odd axis tilt of Uranus and the existence of Earth's large Moon.

The close encounter hypothesis was discarded because calculations showed that it could not account for either the observed orbital motions of the planets or the neat division of the planets into two major categories (terrestrial and jovian). In addition, this model required a highly improbable event. Given the vast separation between star systems in our region of the galaxy, the chance that any two stars would pass close enough to cause a substantial gravitational disruption is extremely remote. While we can't rule out the possibility that our Sun might have been one of the rare stars to experience such a low-probability event, it's inconceivable that it could have happened often enough to account for the many other stars now known to have planets.

Meanwhile, the nebular theory has developed in a way that not only explains the major characteristics of our own solar system but also allows us to test it against observations of other star systems and with computer simulations of planetary formation. In the rest of this chapter, we will see how the modern nebular theory explains the four characteristics of our solar system and how it is consistent with what we know of star systems other than our own.

9.2 Orderly Motion in a Collapsing Nebula

The nebular theory begins with the idea that our solar system was born from a cloud of gas that collapsed under its own gravity. Strong observational evidence supports this idea, because stars that appear to be in the process of formation today are always found within interstellar clouds. A single star-forming cloud, such as the Orion Nebula, may give birth to thousands of individual stars over a period of many millions of years.

Figure 9.1 shows a small portion of the Orion Nebula, in which young stars—possibly accompanied by planets—are currently being born from small pieces of the giant interstellar cloud. Our own solar system presumably formed as gravity collapsed a similar small piece of a very large cloud. We refer to the collapsed piece of cloud that formed our own solar system as the **solar nebula**.

Collapse of the Solar Nebula

Before its collapse began, the gas that made up the solar nebula was probably spread out over a roughly spherical region a few light-years in diameter. This gas was extremely

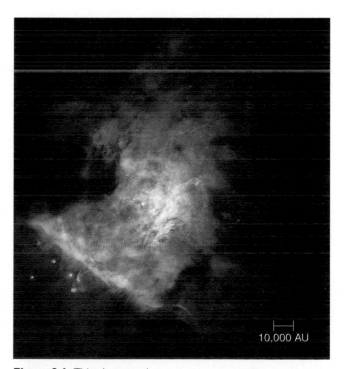

10,000 AU

Figure 9.1 This photograph shows the central region of the Orion Nebula, an interstellar cloud in which star systems—possibly including planets—are forming. The photo is a composite of more than a dozen separate images taken with the Hubble Space Telescope. (See Figure 1.4a for a complete view of the Orion Nebula.)

low in density and extremely cold. The start of the collapse may have been triggered by a cataclysmic event such as the impact of a shock wave from a nearby exploding star.

Once started, the collapse of the solar nebula continued because the force of gravity exerted on the cloud grew stronger as the cloud shrank in size. To understand why, remember that the strength of gravity follows an inverse square law with distance [Section 5.3]. Because the mass of the cloud remained the same as it shrank, the strength of gravity would have increased as the diameter of the cloud decreased. For instance, when the distance decreased by half, the force of gravity would have increased by a factor of four.

As the solar nebula shrank in size, its density, temperature, and shape all underwent dramatic change. These changes were the result of three important physical processes, summarized in Figure 9.2:

- *Heating.* The temperature of the solar nebula increased as it collapsed. Such heating represents energy conservation in action [Section 4.2]. As the cloud shrank, its gravitational potential energy was converted to the kinetic energy of individual gas particles falling inward. These particles crashed into one another, converting the kinetic energy of their inward fall to the random motions of thermal energy. Because the collapse caused mass to concentrate near the center, density and temperature also became higher near the center.

- *Spinning.* Like an ice skater pulling in her arms as she spins, the solar nebula rotated faster and faster as it shrank in radius. This increase in rotation rate represents conservation of angular momentum in action [Section 5.2]. The rotation of the cloud may have been imperceptibly slow before its collapse began, but the cloud's shrinkage made fast rotation inevitable. The rapid rotation helped ensure that not all the material in the solar nebula collapsed into the center: The greater the angular momentum of a rotating cloud, the more spread out it will be.

- *Flattening.* The solar nebula flattened into a disk. This flattening is a natural consequence of collisions between particles in a spinning cloud. A cloud may start with any size or shape, and different clumps of gas within the cloud may be moving in random directions at random speeds. When the cloud collapses, these different clumps collide and merge, giving the new clumps a velocity that is the average of their differing velocities. The result is that the random motions of the clumps in the original cloud become more orderly as the cloud collapses, changing the cloud's original lumpy shape into a rotating, flattened disk. Similarly, collisions between clumps of material in highly elliptical orbits reduce their eccentricities, making their orbits more circular.

These three processes—heating, spinning, and flattening—explain the tidy layout of our solar system. By the time the solar nebula shrank to a diameter of about 200 AU—

roughly twice the present-day diameter of Pluto's orbit—it had become a flattened, spinning disk. Its highest temperatures and densities were near its center, where the Sun formed. The planets were born in the spinning disk surrounding the forming Sun.

The three processes produced orderly motion. The direction in which all planets orbit the Sun today is the direction in which the disk was spinning. The Sun rotates in the same direction. Computer models show that the planets would also have tended to rotate in this same direction as they formed. The fact that collisions in the disk tended to make highly elliptical orbits more circular explains why most planets in our solar system have nearly circular orbits.

You can create a simple analogy to the organized motions of the solar system by sprinkling pepper into a bowl of water and stirring it quickly in random directions. Because the water molecules are always colliding with one another, the motion of the pepper grains will tend to settle down into a slow rotation representing the average of the original, random velocities. Try the experiment several times, stirring the water differently each time. Do the random motions ever exactly cancel out, resulting in no rotation at all? Describe what occurs, and explain how it is similar to what took place in the solar nebula.

Because these processes occur naturally, we expect them to be common throughout the universe. For example, flattening tends to occur anywhere that orbiting particles can collide. This explains why we find so many cases of flat disks: the disks of spiral galaxies like the Milky Way, the disks of planetary rings, and the *accretion disks* surrounding many neutron stars and black holes [Section 17.3]. In addition, because we expect all three processes to occur in any collapsing cloud of gas, other star systems should have undergone similar changes as they were born. Thus, the formation of a disk in which planets can form seems to be a natural part of the star formation process, and we expect such disks to be common.

Testing the Model

We have described a model that explains how our solar nebula may have formed from interstellar gas. The model may seem quite reasonable, but we cannot accept it without hard evidence. Fortunately, we have strong observational evidence to support the nebular theory. A collapsing nebula should emit thermal radiation [Section 6.4], primarily in the infrared. We've detected infrared radiation from many nebulae where star systems appear to be forming today. We've even seen structures around other stars that appear to be flattened, spinning disks (Figure 9.3).

Other support for this model comes from computer simulations of the formation process. A simulation begins with a set of data representing the conditions we observe

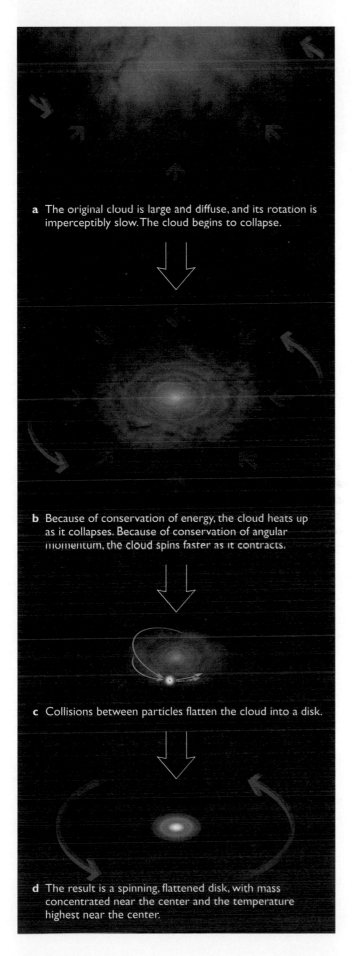

a The original cloud is large and diffuse, and its rotation is imperceptibly slow. The cloud begins to collapse.

b Because of conservation of energy, the cloud heats up as it collapses. Because of conservation of angular momentum, the cloud spins faster as it contracts.

c Collisions between particles flatten the cloud into a disk.

d The result is a spinning, flattened disk, with mass concentrated near the center and the temperature highest near the center.

in interstellar clouds. Then, with the aid of a computer, we apply the laws of physics to simulate the conditions in a real cloud as they change over time. These computer simulations successfully reproduce most of the general characteristics of motion in our solar system, suggesting that the nebular theory is on the right track. To date, however, the simulations do not fully explain the observed pattern of spacing between the planets, demonstrating that the theory is not yet complete. Nevertheless, the nebular theory does explain the generally orderly motions that we observe in our solar system.

 Formation of the Solar System Tutorial, Lesson 3

9.3 Two Types of Planets

Our next task is to explain the formation of two types of planets. To do that, we need to know the composition of the solar nebula, which means we need to know how the solar nebula itself came to be. Thus, we will look first at the really big picture—the universe—and learn about the galactic recycling process that produced the solar nebula. Then we can return to our solar system and discuss how planets formed in the disk swirling around the young Sun.

Galactic Recycling and the Composition of the Solar Nebula

According to the evidence as it is understood today, the universe was born in the Big Bang (see Figure 1.3). Hydrogen and helium (and a trace of lithium) were the only chemical elements present when the universe was young. All the heavier elements have been produced since that time by stars. When they die, massive stars explode, spewing much of their content back into interstellar space. This material can then be recycled into new generations of stars. Figure 9.4 summarizes the galactic recycling process. (We'll study this process in more detail in Chapter 19.)

Although this process of creating heavy elements in stars and recycling them within the galaxy has probably gone on for most of the 14-billion-year history of our universe, only a small fraction of the original hydrogen and helium has been converted into heavier elements. Study of our own solar system shows that it is made of roughly 98% hydrogen and helium and only 2% everything else (by mass). While 2% may sound small, it was more than enough to form the rocky terrestrial planets—and us. Thus, we are "star stuff" because we and Earth are made from elements forged in stars that lived and died long ago.

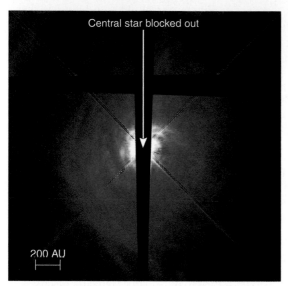

a This Hubble Space Telescope photo shows an edge-on view of a disk of dust surrounding the star Beta Pictoris. The light of the star itself is blacked out, allowing the disk to be seen.

b This Hubble Space Telescope photo shows a face-on view of a disk around the star AB Auriga. The black cross was used to block out the light of the central star.

Figure 9.3 Evidence of disks around other stars.

c Each disk-shaped "blob" in this photograph is a disk of material orbiting a star, perhaps much like the disk in which the planets of our own solar system formed.

d This Hubble Space Telescope photo shows a dust disk around the star HD141569A, part of a triple star system. The light from the star is blacked out. The wisps and clumps may show the influence of the neighboring stars and possibly unseen planets.

THINK ABOUT IT

Could a solar system like ours have formed with the first generation of stars after the Big Bang? Explain.

The churning and mixing of the gas in the solar nebula ensured that its composition was about the same throughout. How, then, did the planets end up being of two basic types with two very different compositions? To answer this question, we must investigate how the material in the spinning disk of the solar nebula came together to form the planets.

Condensation: Sowing the Seeds of Planets

In the center of the collapsing solar nebula, gravity drew together enough material to form the Sun. In the surrounding disk, however, the gaseous material was too spread out for gravity alone to clump it up. Instead, material had to begin clumping in some other way and to grow in size until gravity could start pulling it together into planets. In essence,

planet formation required the presence of "seeds"—solid chunks of matter around which gravity could ultimately build planets. Understanding how these seeds formed and grew into planets is the key to explaining the differences between the terrestrial and the jovian planets.

The basic process of seed formation was probably much like the formation of snowflakes in clouds on Earth: When the temperature is low enough, some atoms or molecules in a gas may bond and solidify. (Pressures in the solar nebula were generally too low to allow the condensation of liquid droplets.) The general process in which solid (or liquid) particles form in a gas is called **condensation**—we say that the solid particles *condense* out of the gas. Different materials condense at different temperatures, so fully understanding how the seeds of planets formed requires understanding the materials that were present in the solar nebula.

On the basis of their condensation properties, the ingredients of the solar nebula fell into four major categories, summarized in Table 9.1:

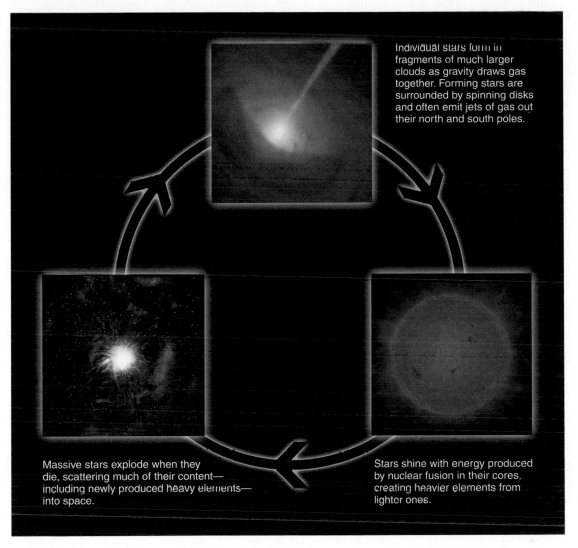

Individual stars form in fragments of much larger clouds as gravity draws gas together. Forming stars are surrounded by spinning disks and often emit jets of gas out their north and south poles.

Massive stars explode when they die, scattering much of their content—including newly produced heavy elements—into space.

Stars shine with energy produced by nuclear fusion in their cores, creating heavier elements from lighter ones.

Figure 9.4 The galactic recycling process.

- **Hydrogen and helium gas.** These gases made up about 98% of the solar nebula's mass. They never condense under the conditions present in a nebula. Thus, the vast majority of the material in the solar nebula remained gaseous.

- **Hydrogen compounds.** These molecules containing hydrogen, such as methane (CH_4), ammonia (NH_3), and water (H_2O), solidify into **ices** below about 150 K. (This is colder than the condensation temperatures for these compounds on Earth because of the much lower pressure in the solar nebula.) These compounds made up about 1.4% of the solar nebula's mass.

- **Rock.** Rock includes the familiar materials of rocks on Earth's surface, such as silicon-based minerals. (A *mineral* is a piece of rock with a particular chemical composition and structure.) Rock made up only about 0.4% of the nebula's mass. Our everyday experience tends to make us think of rock as being solid, but rocky material is gaseous at high temperatures. The precise temperature at which rocky material can condense depends on the type of rock, but it is typically in the range of

500–1,300 K. Thus, rocky material was gaseous where temperatures were above this range and formed solid particles where temperatures were below it.

- **Metals.** This category includes familiar metals on Earth, such as iron, nickel, and aluminum. Metals typically represent elements heavier than those in rocks and therefore were even rarer in the solar nebula, making up only about 0.2% of the nebula's mass. Metals also tend to have higher condensation temperatures than rocks—typically in the range of 1,000–1,600 K. Thus, metals were fully gaseous only where temperatures were higher than about 1,600 K and formed solid particles where temperatures were below this range.

THINK ABOUT IT

Consider a region of the solar nebula in which the temperature was about 1,300 K. What fraction of the material in this region was gaseous? What were the solid particles in this region made of? After you have answered these questions, do the same for a region with a temperature of 100 K. Would the 100 K region be closer to or farther from the Sun? Explain.

Table 9.1 Materials in the Solar Nebula A summary of the four types of materials present in the solar nebula, along with examples of each type and their typical condensation temperatures. The squares represent the relative proportions of each type (by mass).

	Metals	Rock	Hydrogen Compounds	Hydrogen and Helium Gas
Examples	iron, nickel, aluminum	various minerals	water (H_2O) methane (CH_4) ammonia (NH_3)	hydrogen, helium
Typical Condensation Temperature	1,000–1,600 K	500–1,300 K	<150 K	do not condense in nebula
Relative Abundance (by mass)	▪	▪	▪	■
	0.2%	0.4%	1.4%	98%

Because temperatures varied widely across the solar nebula, different types of solid seeds must have formed in different regions (Figure 9.5). In the innermost regions of the nebula near the forming Sun, where the temperature was above 1600 K, it was too hot for any material to condense. Everything remained gaseous, including all the metal and rock. Slightly farther out but still within the present orbit of Mercury, the temperature was slightly lower. Small particles of metal were able to condense. Near the distance of Mercury's orbit, bits of rock joined the mix. Moving outward past the orbits of Venus and Earth, more varieties of rock minerals condensed. In the region where the asteroid belt would eventually be located, temperatures were low enough to allow dark, carbon-rich minerals to condense, along with minerals containing small amounts of water. Where temperatures were low enough (<150 K),

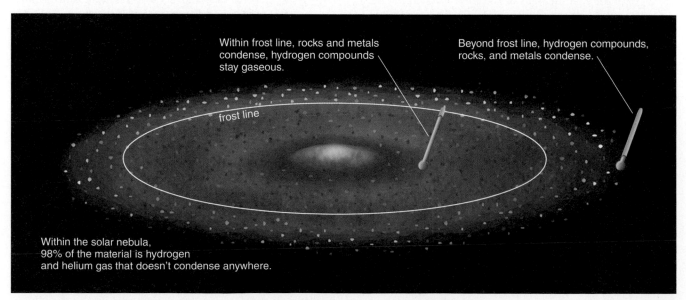

Within frost line, rocks and metals condense, hydrogen compounds stay gaseous.

Beyond frost line, hydrogen compounds, rocks, and metals condense.

frost line

Within the solar nebula, 98% of the material is hydrogen and helium gas that doesn't condense anywhere.

Figure 9.5 Temperature differences in the solar nebula led to different kinds of condensed materials, sowing the seeds for two different kinds of planets.

hydrogen compounds condensed into ices. Such low temperatures occurred only beyond the **frost line**, which lay between the present-day orbits of Mars and Jupiter.

The frost line marked an important dividing point in the solar nebula. Within it, temperatures were too high for hydrogen compounds to condense. The only solid particles were made of metal and rock—which together made up only 0.6% of the nebula's mass. Beyond the frost line, where hydrogen compounds could condense, the solid particles included ices as well as metal and rock. The presence of ices meant that the abundance of solid seeds became dramatically greater beyond the frost line, because hydrogen compounds were nearly three times as abundant in the nebula as metal and rock combined. Thus, the stage was set for the birth of two types of planets: planets born from seeds of metal and rock in the inner solar system and planets born from seeds of ice (as well as metal and rock) in the outer solar system.

Accretion: Assembling Planetesimals and the Terrestrial Planets

The first solid particles that condensed from the solar nebula were microscopic in size. They orbited the forming Sun with the same orderly, circular paths as the gas from which they condensed. Individual particles therefore moved at nearly the same speed as neighboring particles, so "collisions" were more like gentle touches. Although the particles were far too small to attract each other gravitationally at this point, they were able to stick together through electrostatic forces—the same "static electricity" that makes hair stick to a comb. Small particles thereby began to combine into larger ones. As the particles grew in mass, gravity began to aid the process of their sticking together, accelerating their growth. The process of growth by colliding and sticking is called **accretion**.

The growing objects formed by accretion are called **planetesimals**, which means "pieces of planets." Small planetesimals can have almost any random shape, much as small asteroids and comets have a variety of shapes.

Planetesimals that grow large—more than a few hundred kilometers across—inevitably become spherical. The reason is gravity: With enough mass, an object's own gravity becomes so strong that it can deform even solid rock. Because gravity acts to pull everything toward the center, it transforms massive objects into spheres.

The growth of planetesimals was rapid at first. As planetesimals grew larger, they had both more surface area and more gravity to attract other planetesimals. Some probably grew to hundreds of kilometers in only a few million years—a long time in human terms, but only about $\frac{1}{1,000}$ the present age of the solar system. However, once the planetesimals reached these relatively large sizes, further growth became more difficult.

Gravitational encounters between planetesimals tended to alter their orbits, particularly those of the smaller planetesimals [Section 5.5]. With different orbits crossing each other, collisions between planetesimals tended to occur at higher speeds and hence became more destructive. Such collisions tended to produce fragmentation more often than accretion. Only the largest planetesimals avoided being shattered and thus were able to grow into full-fledged planets. Figure 9.6 summarizes the growth of the planets through accretion.

The sizes and compositions of the planetesimals reflected the seeds from which they accreted. In the inner solar system, only rocky and metallic particles could condense and accrete into planetesimals. Because rock and metal made up a much smaller fraction of the solar nebula than the hydrogen compounds that condensed into ices beyond the frost line, the planetesimals of the inner solar system were relatively small and could build only small planets.

Theoretical evidence in support of this model comes from computer simulations of the accretion process. Observational evidence comes from meteorites, many of which appear to be surviving fragments from the early period of condensation in our solar system [Section 13.3]. Meteorites often contain metallic grains embedded in a variety of rocky minerals (Figure 9.7), just as we expect for the planetesimals of the inner solar system. Meteorites thought to come from the distance of the asteroid belt contain abundant carbon-rich materials, and some contain water—again as we expect for planetesimals that formed in that region.

In the outer solar system, where the much more abundant hydrogen compounds could condense, the planetesimals were made mostly of ices and grew much larger than planetesimals in the inner solar system. The solid objects that reside in the outer solar system today, such as comets and the moons of the jovian planets, still show this icy composition. We know that they contain a much higher proportion of ice and a lower proportion of rock and metal than the rocky terrestrial worlds, because their average densities are so much lower. However, the growth of icy planetesimals cannot be the whole story of jovian planet formation, because the jovian planets themselves are *not* made mostly of ice. Instead, they contain large amounts of hydrogen and helium. Let's explore why.

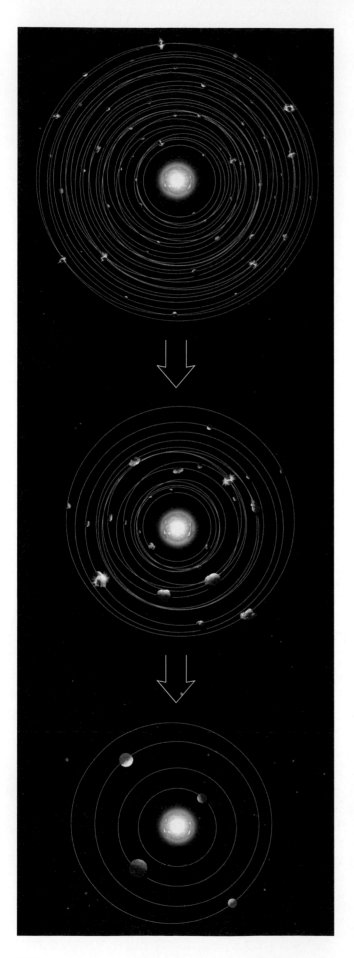

◄ Figure 9.6 These diagrams show how planetesimals gradually accrete into terrestrial planets. Early in the accretion process, there are many Moon-size planetesimals on crisscrossing orbits (top). As time passes, a few planetesimals grow larger by accreting smaller ones, while others shatter in collisions (center). Ultimately, only the largest planetesimals avoid shattering and grow into full-fledged planets (bottom). Diagram not to scale.

Capturing Nebular Gas: Making the Jovian Planets

The additional piece in the story of the formation of the jovian planets involves the way large, icy planetesimals gravitationally captured hydrogen and helium gas. Remember that 98% of the solar nebula was hydrogen and helium gas. Thus, the planetesimals represented just a tiny fraction of the total mass in any region of the nebula. In the inner solar system, the abundant hydrogen and helium gas did not play much of a role in planet development. The small sizes of the terrestrial planets kept their gravity relatively weak, and the high temperatures in the inner solar nebula meant that almost all of the hydrogen and helium gas could escape from these small worlds even when it was temporarily captured. In the outer solar system, the larger sizes of the planetesimals and the lower temperatures of the nebula allowed the story to play out quite differently.

Some of the icy planetesimals of the outer solar system grew to masses many times that of Earth. At these large masses, gravity was strong enough to capture and hold the abundant hydrogen and helium gas that surrounded the planetesimals. As they accumulated substantial amounts of gas, the gravity of these growing planets grew larger still, allowing them to capture even more gas.

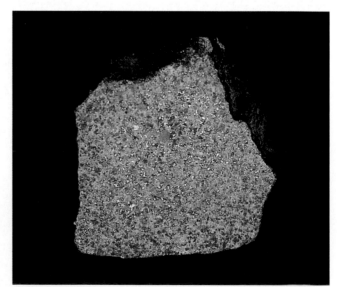

Figure 9.7 Shiny flakes of metal are clearly visible in this slice through a meteorite (a few centimeters across), mixed in among the rocky material. Such metallic flakes are just what we would expect to find if condensation really occurred in the solar nebula as described by the nebular theory.

By capturing more and more gas, the jovian planets grew so much that they ultimately bore little resemblance to the icy seeds from which they started. Instead, they ended up with large abundances of the hydrogen and helium that had always dominated the mass of the solar nebula. The fact that the jovian planets captured so much gas from the solar nebula explains their large sizes, low densities, and very different compositions from the terrestrial planets.

The process by which the jovian planets accumulated their gas also explains why these planets tend to have many moons. As gravity drew gas from the solar nebula toward the young jovian planets, the same processes that created the disk of gas in which the planets formed—heating, spinning, and flattening—created similar but smaller disks around these planets. In essence, each jovian planet formed at the center of its own "miniature solar nebula," surrounded by its own spinning disk of gas (Figure 9.8). Moons accreted from the icy planetesimals within these disks, which explains why jovian moons tend to contain large proportions of ice. Because they formed within these flattened, spinning disks, the moons tend to orbit in the same direction as their planet rotates, with nearly circular orbits lying close to the equatorial plane of their parent planet. (This model can also explain the presence of planetary rings around the jovian planets, which we'll discuss in Chapter 12.)

The only major characteristic of the jovian planets not clearly explained by our model is their wide spacing in the outer solar system. However, astronomers have come up with reasonable hypotheses that are consistent with the overall theory. For example, the strengthening gravity of the growing planets may have allowed them to gobble up smaller neighbors, leaving the jovian planets widely spaced.

The Solar Wind: Clearing the Nebula

The vast majority of the hydrogen and helium gas in the solar nebula never became part of any planet. What became of it? Apparently, it was swept into interstellar space by the **solar wind**, which consists of charged particles that are continually blown off the surface of the Sun in all directions. Although the solar wind is fairly weak today, observations of other stars show that winds tend to be much stronger in young stars. Thus, the young Sun should have had a strong solar wind—strong enough to have swept huge quantities of gas out of the solar system.

The clearing of the gas sealed the compositional fate of the planets. If the gas had not been cleared soon after the planets formed, it might have continued to cool until hydrogen compounds could have condensed into ices even in the inner solar system. In that case, the terrestrial planets might have accreted abundant ice, and perhaps hydrogen and helium gas as well, changing their basic nature. At the other extreme, if the gas had been blown out too early, the raw materials of the planets might have been swept away before the planets could fully form.

The young Sun's strong solar wind also helps explain what was once considered a surprising aspect of the Sun's rotation. According to the law of conservation of angular

Figure 9.8 Large icy planetesimals in the cold, outer regions of the solar nebula captured significant amounts of hydrogen and helium gas. This process created solar nebulae in miniature. In our solar system, four of these miniature solar nebulae formed, from which each of the jovian planets and many of their satellites formed. The inset painting (left) is located within the entire solar nebula as shown.

momentum, the spinning disk of the solar nebula should have spun fastest near its center, where most of the mass became concentrated. Thus, the young Sun should have been rotating very fast. Today the Sun rotates quite slowly, with each full rotation taking about a month. If the young Sun really did rotate fast, as our theory seems to demand, how did its rotation slow down?

Angular momentum cannot simply disappear, but it is possible to transfer angular momentum from one object to another—and then get rid of the second object. A spinning skater can slow her spin by grabbing her partner and then pushing him away. In the 1950s, scientists realized that the young Sun's rapid rotation would have generated a magnetic field far stronger than that of the Sun today. This strong magnetic field actually helped create the strong solar wind. More generally, the strong magnetic field would have made the young Sun much more active on its surface than it is today [Section 15.5]. For example, large "solar storms" and sunspots would have been much more common, causing the emission of much more ultraviolet and X-ray light. This high-energy radiation from the young Sun ionized gas in the solar nebula, creating many charged particles.

As we will discuss in more detail in Chapter 15, charged particles and magnetic fields tend to stick together. As the Sun rotated, its magnetic field dragged the charged particles along faster than the rest of the nebula, adding to their angular momentum. As the particles were gaining angular momentum, the Sun was losing it. The young Sun's strong solar wind then blew these particles into interstellar space, leaving the Sun with the greatly diminished angular momentum and much slower rotation that we see today (Figure 9.9).

Although we cannot prove that the young Sun really did lose angular momentum in this way, support for the idea comes from observations of other stars. When we look at young stars that have recently formed in interstellar clouds, we find that nearly all of them rotate rapidly and have strong magnetic fields and strong winds [Section 17.2]. Older stars, in contrast, almost invariably rotate slowly, like our Sun. This suggests that nearly all stars have their original rapid rotations slowed by transferring angular momentum to charged particles in their disks—particles that are later swept away—just as our theory suggests happened with the Sun.

 Formation of the Solar System Tutorial, Lesson 4

9.4 Explaining Leftovers and Exceptions to the Rules

So far, we have seen how the nebular theory explains our solar system's patterns of motion and its two types of planets. We still have two characteristics of our solar system left to explain: the existence and orbital properties of the myriad asteroids and comets, and the exceptions to the general rules. These two characteristics turn out to be closely related—both involve the planetesimals that remained among the forming planets in the early solar system.

The Origin of Asteroids and Comets

The strong wind from the young Sun cleared excess gas from the solar nebula, but many planetesimals remained scattered between the newly formed planets. These "leftovers" became asteroids and comets. Like the planetesimals

Figure 9.9 Charged particles in the solar nebula tend to move with the Sun's magnetic field (represented by the purple loops). As the magnetic field lines rotate with the Sun, these charged particles are dragged through the slower-moving disk. Friction between the charged particles and the rest of the disk slows the Sun's rotation (and also slightly speeds up the disk). The Sun's relatively slow rotation probably resulted from this process. (Particle sizes are highly exaggerated.)

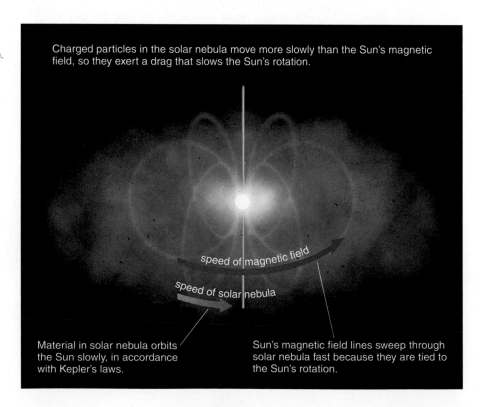

Charged particles in the solar nebula move more slowly than the Sun's magnetic field, so they exert a drag that slows the Sun's rotation.

speed of magnetic field

speed of solar nebula

Material in solar nebula orbits the Sun slowly, in accordance with Kepler's laws.

Sun's magnetic field lines sweep through solar nebula fast because they are tied to the Sun's rotation.

that formed the planets, those that became asteroids and comets were made by condensation and accretion in the solar nebula. Thus, their orbits and compositions followed the same patterns: nearly circular orbits in the same plane as that of the planets, with planetesimals of metal and rock in the inner solar system and icy planetesimals in the outer solar system.

Asteroids are the rocky leftover planetesimals of the inner solar system. The four inner planets must have swept up most of the planetesimals near their orbits so that relatively few leftovers remained between them. However, no planet formed between Mars and Jupiter (a fact attributable to effects from Jupiter's strong gravity [Section 12.2]), so this region—the asteroid belt—must have been dense with rocky leftover planetesimals. The asteroid belt probably once contained enough rocky planetesimals to form another terrestrial planet, but most of them ultimately crashed into the inner planets or were ejected from the solar system. Thus, the total mass of the asteroids in the asteroid belt today is only a tiny fraction of the mass of any terrestrial planet. Asteroid orbits tend to be somewhat more elliptical and inclined than planetary orbits due to gravitational "tugs" from Jupiter and gravitational encounters among the asteroids.

Comets are the icy leftover planetesimals of the outer solar system. A brief look at how these planetesimals must have interacted with the young jovian planets explains how they came to be concentrated in two distinct regions: the Kuiper belt, beyond the orbit of Neptune, and the Oort cloud, at a much greater distance.

The icy planetesimals that cruised the spaces between Jupiter, Saturn, Uranus, and Neptune couldn't grow to more than a few kilometers in size before suffering either a collision or a close gravitational encounter with one of the jovian planets. Recall that when a small object passes near a large planet, the planet is hardly affected but the small object may be flung off at high speed [Section 5.5]. Thus, the planetesimals that escaped being swallowed up by the jovian planets tended to be flung off at high speeds in all directions. Some may have been cast away at such high speeds that they completely escaped the solar system and now drift through interstellar space. Billions of others of these small, icy planetesimals ended up on orbits with very large average distances from the Sun. These became the comets of the Oort cloud. The random directions in which they were flung into the Oort cloud explain why this region is roughly spherical in shape. Thus, the comets of the Oort cloud seem to have originated closer to the Sun than the comets of the Kuiper belt, even though they now reside much farther away.

Beyond the orbit of Neptune, the icy planetesimals were much less likely to be destroyed by collisions or cast off by gravitational encounters. Instead, they remained on orbits going in the same directions as planetary orbits and stayed concentrated relatively near the plane of planetary orbits. These are the comets of the Kuiper belt. Because they stayed relatively close together, they were able to con-

tinue their accretion. They did not grow as large as the icy planetesimals that served as the seeds for the jovian planets, probably because the density of material was too low at their great distances from the Sun. Nevertheless, many of them probably grew to hundreds or even thousands of kilometers in diameter. The planet Pluto is probably one of these large Kuiper belt comets, which explains why Pluto is so different in character from the other eight planets of our solar system [Section 13.5].

Evidence that asteroids and comets really are leftover planetesimals comes from analysis of meteorites, spacecraft visits to comets and asteroids, and computer simulations of solar system formation. The nebular theory actually *predicts* the existence of both the Oort cloud and the Kuiper belt—a prediction first made in the 1950s. Thus, the discoveries, beginning in the 1990s, of numerous objects orbiting in the Kuiper belt represent a triumph for the nebular theory. Comets in the distant Oort cloud remain undetectable with present technology, but there is little doubt that they exist. Their presence is the only way to explain the comets that enter the inner solar system from afar.

The Heavy Bombardment

Now that we have explained the existence and properties of asteroids and comets, we are ready to explain the exceptions to the solar system's general rules. Most of the exceptions can be attributed to the effects of collisions or close encounters with leftover planetesimals.

The asteroids and comets that exist today probably represent only a small fraction of the vast numbers of leftover planetesimals that roamed the young solar system. The rest are now gone. While some of these "lost" planetesimals may have been ejected from the solar system or flung into the Oort cloud, many others must have collided with the planets. The vast majority of these collisions occurred in the first few hundred million years of our solar system's history, during the period we call the **heavy bombardment**.

Every world in our solar system must have been pelted by impacts during the heavy bombardment (Figure 9.10). However, we see **impact craters**—the scars left behind when an asteroid or a comet collides with a solid planet or moon—only on some worlds today. The jovian planets lack impact craters because they lack solid surfaces and thus simply "swallow" objects that collide with them. Earth has relatively few impact craters on its surface because the many craters of the heavy bombardment were erased long ago by erosion and other geological processes. Worlds on which we see many impact craters, such as the Moon and Mercury, must have undergone little change since the time of the heavy bombardment. Indeed, one way to estimate the age of a world's surface (the time since the surface last changed in a substantial way) is to count craters: A surface with many craters must still look much as it did when the heavy bombardment ended, about 4 billion years ago.

The many impacts that occurred as the planets were still forming explain an aspect of Earth that is critical to

Figure 9.10 Around 4 billion years ago, Earth, its Moon, and the other planets were heavily bombarded by leftover planetesimals. This painting shows the young Earth and Moon faintly glowing with the heat of accretion, and an impact in progress on Earth.

our existence. Remember that the terrestrial planets were built from planetesimals made of metal and rock. These planetesimals should have contained virtually no water or other hydrogen compounds at all, because it was too hot for these compounds to condense in our region of the solar nebula. How, then, did Earth come to have the water that makes up our oceans? More generally, how did any of the terrestrial planets end up with material that could make gaseous atmospheres, liquid water, or solid ice?

The answer is that water, along with other hydrogen compounds, must have been brought to the terrestrial planets by the impact of planetesimals that formed farther from the Sun. We don't yet know whether these planetesimals came primarily from the asteroid belt, where rocky planetesimals contained small amounts of water and other hydrogen compounds, or whether they were comets containing huge amounts of ice. In either case, the water we drink and the air we breathe probably once were part of planetesimals floating beyond the orbit of Mars.

Jupiter's gravity played a major role in flinging asteroids and comets into the inner solar system. Given this fact, what can you say about Jupiter's role in bringing water to Earth? How might Earth be different if Jupiter had never formed?

Captured Moons

We can easily explain the orbits of most large jovian planet moons by their formation in a disk that swirled around the forming planet. However, some moons have unusual orbits—orbits in the "wrong" direction (opposite the rotation of their planet) or with large inclinations to the planet's

equator. These unusual moons are probably leftover planetesimals that were captured into orbit around a planet.

It's not easy for a planet to capture a moon. An object cannot switch from an unbound orbit (for example, an asteroid whizzing by Jupiter) to a bound orbit (for example, a moon orbiting Jupiter) unless it somehow loses orbital energy [Section 5.5]. Captures probably occurred when the capturing planet had a very extended atmosphere or, in the case of the jovian planets, its own "miniature solar nebula." Passing planetesimals could have been slowed by friction with the gas, just as artificial satellites are slowed by drag in encounters with Earth's atmosphere. If friction reduced a passing planetesimal's orbital energy enough, it could have become an orbiting moon. Because of the random nature of the capture process, the captured moons would not necessarily orbit in the same direction as their planet or in its equatorial plane.

The two small moons of Mars—Phobos and Deimos—probably were asteroids captured by this process (Figure 9.11). They resemble asteroids seen in the asteroid belt and are much darker and lower in density than Mars itself.

Many of the small moons of the jovian planets probably also were captured planetesimals. For example, Jupiter has several groups of small moons that share elliptical and tilted orbits, with some groups orbiting in the "wrong" direction. Astronomers speculate that each group may be the result of a single captured asteroid that broke into pieces during the capture process.

At least one large moon in the solar system also may have been captured: Neptune's moon Triton [Section 12.5]. Triton's unusual orbit and its similarities to Pluto suggest that it may have been captured from the Kuiper belt.

The only exceptional moon that cannot be explained by the capture process is our own. Our Moon is much too large to have been captured by a small planet like Earth. We can also rule out the possibility that our Moon formed simultaneously with Earth. If the Moon and Earth had formed together, both would have accreted from planetesimals of the same type and would therefore have approximately the same composition and density. That is not the case. The Moon's density is considerably lower than Earth's, indicating that it has a very different average composition and could not have formed in the same way or at the same time as our planet. We must look for another way to explain the origin of our surprisingly large Moon.

Giant Impacts and the Formation of Our Moon

The largest leftover planetesimals may have been huge, perhaps the size of Mars. When one of these planet-size planetesimals collided with a planet, the spectacle would have been awesome. Such a **giant impact** could have significantly altered a planet's fate. Let's begin by considering the origin of our Moon, which is now believed to have been the result of a giant impact that occurred during the late stages of our planet's accretion.

a Phobos

b Deimos

Figure 9.11 The two moons of Mars, shown here in photos taken by the Viking spacecraft, are probably captured asteroids. Phobos is only about 13 km across and Deimos is only about 8 km across—making each of these two moons small enough to fit within the boundaries of a typical large city.

What would have happened if a Mars-size object had collided with the young Earth? Depending on exactly where and how fast the object struck Earth, the blow might have tilted Earth's axis, changed its rotation rate, or completely shattered our planet. The most interesting case arises when we consider what would have happened if the impact had blasted away rock from Earth's outer layers (mantle and crust) and sent this material into orbit around our planet. According to computer simulations, this orbiting material could have reaccreted to form our Moon (Figure 9.12).

Today, such a giant impact is the leading hypothesis for explaining the origin of our Moon. Strong support for this hypothesis comes from two features of the Moon's composition. First, the Moon's overall composition is quite similar to that of Earth's outer layers—just as we should expect if it was made from material blasted away from those layers. Second, the Moon has a much smaller proportion of easily vaporized ingredients (such as water) than Earth. This fact supports the hypothesis because these ingredients would have been vaporized by the heat of the impact. As gases, they would not have participated in the process of reaccretion that formed the Moon.

Giant impacts probably also explain many of the other exceptions to the general trends. For example, a giant impact may have contributed to the slow, backward rotation of Venus, which may have had a "normal" rotation before the impact occurred. Giant impacts may also have been responsible for tilting the axes of many planets (including Earth) and perhaps for tipping Uranus on its side. Pluto's moon Charon may have formed in a giant impact similar to the one that formed our Moon. Mercury's surprisingly high density may be the result of a giant impact that blasted away its rocky outer layers, leaving behind a planet made almost entirely of the metal that was once in its core.

Unfortunately, we can do little to test whether a particular giant impact really occurred billions of years ago. The difficulty in proving the giant impact hypothesis makes the idea controversial, and most planetary scientists didn't

Figure 9.12 Artist's conception of the impact of a Mars-size object with Earth, as may have occurred soon after Earth's formation. The ejected material comes mostly from the outer rocky layers and accretes to form the Moon, explaining why the Moon is poor in metal.

take the idea seriously when it was first proposed. But no other idea so effectively explains the formation of our Moon and the other "oddities" we've discussed. Moreover, giant impacts certainly should have occurred, given the number of large leftover planetesimals predicted by the nebular theory. Random giant impacts are therefore considered the most likely explanation for many of the exceptions to our general planetary rules.

The Nebular Theory: A Summary and a Question

The nebular theory, which we have now described in detail, has successfully explained all four major characteristics of our solar system. Of course, that does not mean that it explains everything about the formation of our solar system. As we'll discuss in Section 9.6, discoveries of other planetary systems have already forced us to revise parts of the theory. Nevertheless, because the basic theory is strongly supported by the available evidence, it is widely accepted today. Figure 9.13 summarizes our current understanding of how our solar system formed. The entire process of planet formation probably took no more than a few tens of millions of years, about 1% of the current age of the solar system.

Assuming the nebular theory is correct, was our solar nebula "destined" to form the solar system we see today? Probably not. The first stages of planet formation were orderly and inevitable according to the nebular theory. Nebular collapse, condensation, and the first stages of accretion were relatively gradual processes that probably would happen all over again if we turned back the clock. However, the final stages of accretion, and giant impacts in particular, are inherently random in nature and probably would not happen again in just the same way. A larger or smaller planet might form at Earth's location and might suffer from a larger giant impact or from no giant impact at all. We don't yet know whether these differences would fundamentally alter the solar system or simply change a few "minor" details—such as the possibility of life on Earth.

9.5 How Old Is the Solar System?

The nebular theory accounts for the major physical properties of our solar system, supporting the idea that all the planets formed at about the same time from the same cloud of gas. But *when* did it all happen, and how can we know? The answer is that the solar system began forming about 4.6 billion years ago, a fact we learn by determining the age of the oldest rocks in the solar system.

The concept of a rock's age is a bit subtle. A rock is made from atoms that were forged in stars and therefore are much older than Earth. Atoms are not stamped with any date of manufacture, and old atoms are indistinguishable from young ones. By the age of a rock we mean the time since its atoms became locked together in their pres-

ent arrangement, which in most cases means the time *since the rock last solidified.*

Radiometric Dating

The method by which we measure the age of a rock is known as **radiometric dating**. This method relies on careful measurement of the proportions of various atoms and isotopes in the rock.

Remember that each chemical element is uniquely characterized by the number of protons in its nucleus. Different *isotopes* of the same element differ only in their number of neutrons [Section 4.3]. A **radioactive isotope** is an isotope prone to spontaneous change. Many common elements have radioactive isotopes. For example, most carbon comes in the stable isotope carbon-12, which has six protons and six neutrons, but this carbon is generally found mixed with small amounts of the radioactive isotope carbon-14, which has six protons and eight neutrons (see Figure 4.7). Some elements, such as uranium, are never stable—every isotope of uranium is radioactive.

When the nucleus of a radioactive isotope undergoes spontaneous change, we say that **radioactive decay** has occurred. Radioactive decay can occur in a variety of ways but always involves a change in either the number of protons or the number of neutrons in the nucleus, or both. For example, the radioactive decay of carbon-14 changes it into nitrogen-14, while uranium-238 decays (in several steps) into lead-206. For any particular decay process, the original isotope before decay is called the *parent isotope,* and the resulting isotope after decay is called the *daughter isotope.* The daughter isotope remains trapped in the rock as long as the rock remains solid.

For any single nucleus, radioactive decay is an instantaneous event. However, in a large collection of atoms, individual nuclei will decay at different times. Thus, if we start with a collection of atoms of a particular parent isotope, we will find that it gradually transforms itself into a collection of atoms of the daughter isotope. The rate at which this transformation occurs is characterized by the parent isotope's **half-life**, the time it would take for half of the parent nuclei in the collection to decay. Every radioactive isotope decays with a precise half-life that we can measure in the laboratory. Some isotopes have half-lives as short as a fraction of a second, while others have half-lives of billions of years. (We can still measure these half-lives in the laboratory because we only need to observe the decay of a tiny fraction of the nuclei to calculate how long it would take for half of them to decay.)

The essence of radiometric dating is the fact that, as long as we know an isotope's half-life, we can determine the age of a rock from the precise ratio of parent and daughter atoms within it. As an example, consider the radioactive decay of potassium-40, which decays into argon-40 with a half-life of 1.25 billion years. Suppose a rock originally contained 1 microgram of potassium-40 and no argon-40. The half-life of 1.25 billion years means that half of the

Figure 9.13 A summary of the process by which our solar system formed.

Large, diffuse interstellar gas cloud (solar nebula) contracts under gravity.

As it contracts, the cloud heats, flattens, and spins faster, becoming a spinning disk of dust and gas.

Sun will be born in center.   Planets will form in disk.

Warm temperatures allow only metal/rock "seeds" to condense in inner solar system.

Hydrogen and helium remain gaseous, but other materials can condense into solid "seeds" for building planets.

Cold temperatures allow "seeds" to contain abundant ice in outer solar system.

Terrestrial planets are built from metal and rock.

Solid "seeds" collide and stick together. Larger ones attract others with their gravity, growing bigger still.

The seeds of jovian planets grow large enough to attract hydrogen and helium gas, making them into giant, mostly gaseous planets; moons form in disks of dust and gas that surround the planets.

Solar wind blows remaining gas into interstellar space.

Terrestrial planets remain in inner solar system.

Jovian planets remain in outer solar system.

"Leftovers" from the formation process become asteroids (metal/rock) and comets (mostly ice).

Not to scale

original potassium-40 will have decayed into argon-40 after 1.25 billion years, leaving the rock with $\frac{1}{2}$ microgram of potassium-40 and $\frac{1}{2}$ microgram of argon-40. Half again will have decayed by the end of the next 1.25 billion years, so after 2.5 billion years the rock will contain $\frac{1}{4}$ microgram of potassium-40 and $\frac{3}{4}$ microgram of argon-40. After three half-lives, or 3.75 billion years, only $\frac{1}{8}$ microgram of potassium-40 remains, while $\frac{7}{8}$ microgram has become argon-40. Figure 9.14 summarizes how the amount of potassium-40 gradually declines with time and the corresponding rise in the amount of argon-40. Because the rate of decay is so precise, we can use it to determine the age of a rock in which we measure the relative proportions of potassium-40 and argon-40.

Radiometric dating is simple in principle, but in practice it requires careful laboratory work and a good understanding of "rock chemistry." For example, suppose you find a rock that contains equal numbers of atoms of potassium-40 and argon-40. If you assume that all the argon came from potassium decay, then it must have taken precisely one half-life for the rock to end up with equal amounts of parent and daughter isotopes. You could therefore conclude that the rock is 1.25 billion years old (the half-life of potassium-40). However, this conclusion is correct only if you were right in assuming that the rock lacked argon-40 when it formed. In this case, knowing a bit of rock chemistry helps. Potassium-40 is a natural ingredient of many minerals in rocks, but argon-40 is a gas that never combines with other elements and did not condense in the

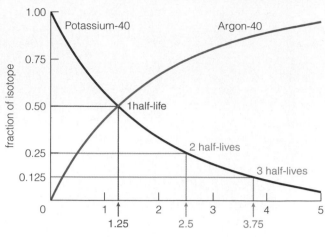

Figure 9.14 Potassium-40 is radioactive, decaying into argon-40 with a half-life of 1.25 billion years. The red line shows the decreasing amount of potassium-40, and the blue line shows the increasing amount of argon-40. The remaining amount of potassium-40 drops in half with each successive half-life.

solar nebula. Therefore, if you find argon-40 gas trapped inside minerals, you can be sure that it came from radioactive decay of potassium-40. Another assumption is that none of the argon-40 gas has escaped from the rock. Rock chemistry tells us that as long as the rock has not been significantly heated since the time it formed, even atoms of gases will remain trapped within it.

Mathematical Insight **9.1**　**The Mathematics of Radioactive Decay**

The amount of a radioactive substance decays by half with each half-life, so we can express the decay process with a simple formula relating the current amount of a radioactive substance in a rock to the original amount:

$$\frac{\text{current amount}}{\text{original amount}} = \left(\frac{1}{2}\right)^{t/t_{\text{half}}}$$

where t is the time since the rock formed and t_{half} is the half-life of the radioactive material. We can solve this equation for the age t by taking the logarithm of both sides and rearranging the terms. The resulting general equation for the age is:

$$t = t_{\text{half}} \times \frac{\log_{10}\left(\dfrac{\text{current amount}}{\text{original amount}}\right)}{\log_{10}\left(\dfrac{1}{2}\right)}$$

Example: You heat and chemically analyze a small sample of a meteorite. Potassium-40 and argon-40 are present in a ratio of approximately 0.85 unit of potassium-40 atoms to 9.15 units of gaseous argon-40 atoms. (The units are unimportant, because only the relative amounts of the parent and daughter materials matter.) How old is the meteorite?

Solution: Because no argon gas could have been present in the meteorite when it formed, the 9.15 units of argon must originally have been potassium. The sample must therefore have started with $0.85 + 9.15 = 10$ units of potassium-40, of which 0.85 unit remains. Thus, for calculating the age, the *current amount* of potassium-40 is 0.85 unit, and the *original amount* is 10 units. With these values and the above equation, the age of the meteorite is:

$$t = 1.25 \text{ billion yr} \times \frac{\log_{10}\left(\dfrac{0.85}{10}\right)}{\log_{10}\left(\dfrac{1}{2}\right)}$$

$$= 1.25 \text{ billion yr} \times \left(\frac{-1.07}{-0.301}\right)$$

$$= 4.45 \text{ billion yr}$$

This meteorite solidified about 4.45 billion years ago. (*Note:* This example captures the essence of radiometric dating, but real cases generally require more detailed analysis.)

For another example, consider rocks from the ancient lunar highlands [Section 10.3]. These rocks contain minerals with a very small amount of uranium-238, which ultimately decays to lead-206 with a half-life of almost 4.5 billion years. Lead and uranium have very different chemical behaviors, and some minerals start with virtually no lead. Laboratory analysis of such minerals in lunar rocks shows that they now contain almost equal proportions of atoms of uranium-238 and lead-206. We conclude that half the original uranium-238 has decayed, turning into the same number of lead-206 atoms. The lunar rock therefore must be about one half-life old, or almost 4.5 billion years old. More precise work shows the oldest lunar rocks to be about 4.4 billion years old.

THINK ABOUT IT

If future scientists examine the same lunar rocks 4.5 billion years from now, what proportions of uranium-238 and lead-206 will they find?

Earth Rocks, Moon Rocks, and Meteorites

How can we measure the age of the solar system? Rocks from Earth cannot give us a precise answer. Geological activity, such as plate tectonics and volcanoes, has ensured that virtually all rocks present on the early Earth have since melted and resolidified. Because radiometric dating tells us only how long it has been since a rock last solidified, we cannot use Earth rocks to determine the age of our planet, let alone the age of the solar system. Nevertheless, Earth rocks tell us that the solar system is old: The oldest Earth rocks solidified about 4 billion years ago, so Earth and the solar system must be older still.

THINK ABOUT IT

Suppose you could do radiometric dating on a chunk of lava recently spewed out of Kilauea, an active volcano on the island of Hawaii. What would you learn? Explain.

Studies of Moon rocks take us back somewhat farther in time. Astronauts have brought back numerous rocks from the Moon. As we've seen, the oldest of these are about 4.4 billion years old. This is presumably less than the age of the Moon itself, because even the oldest portions of the Moon's surface should have melted and resolidified during the heavy bombardment. However, it means that the solar system must be more than 4.4 billion years old. It also tells us that the giant impact thought to have created the Moon must have occurred more than 4.4 billion years ago.

To go all the way back to the origin of the solar system, we must find rocks that have not melted or vaporized since they first condensed in the solar nebula. Meteorites that have fallen to Earth are our source of such rocks. Many meteorites appear to have remained unchanged since they condensed and accreted in the early solar system. Careful analysis of radioactive isotopes in meteorites shows that

the oldest ones formed about 4.55 billion years ago, so this time must mark the beginning of accretion in the solar nebula. Rounding upward, we say that our solar system is about 4.6 billion years old. Because the planets apparently accreted within a few tens of millions of years, Earth and the other planets formed about 4.5 billion years ago. These ages are less than a third the 14-billion-year age of our universe. We therefore live in a solar system that is in early middle age in the context of the universe.

A Trigger for the Collapse?

What might have happened 4.6 billion years ago to start the collapse of our solar nebula? Radioactive elements in meteorites give us some clues. Radioactive elements are made only deep inside stars or in violent stellar explosions (supernovae). Thus, the presence of these elements in meteorites and on Earth underscores the fact that our solar system is made from the remnants of past generations of stars.

The discovery of daughter isotopes of short-lived radioactive isotopes tells us more. The rare isotope xenon-129 is found in some meteorites. Xenon is gaseous even at extremely low temperatures, so it could not have condensed and become trapped in planetesimals forming in the solar nebula. Instead, any xenon present in meteorites must be a product of radioactive decay. Xenon-129 is a decay product of iodine-129. Because iodine-129 has a half-life of just 17 million years, this iodine must have condensed within the solar nebula no more than a few tens of millions of years after it was produced by the explosion of another star. Otherwise, too much of the iodine-129 would have decayed for significant amounts to have condensed. This reasoning suggests that a nearby star exploded not long before the solar nebula began its collapse. The shock wave from the exploding star may even have triggered the collapse of the solar nebula by giving gravity a little extra push. Once gravity got started, the rest of the collapse was inevitable.

 Extrasolar Planetary Detection Tutorial, Lessons 1–3

9.6 Other Planetary Systems

Barely a decade ago, we had no conclusive proof that planets existed around any star besides our Sun. Today, we know of more than 100 planets orbiting other stars. The discovery of these planets represents a triumph of modern technology. It also has profound philosophical implications. Knowing that planets are common increases the chance that we might someday find life elsewhere, perhaps even intelligent life.

The discovery of other planetary systems also allows us to test our theory of the solar system's birth in new settings. If the nebular theory is correct, it should be able to explain the observed properties of planets that orbit other stars. In this section, we'll discuss how we have detected these planets and what we have learned about them to date.

We'll also discuss how our theory of solar system formation is holding up in light of the new discoveries.

Detecting Extrasolar Planets

You might think that the easiest way to discover **extrasolar planets**, or planets around other stars, would be to photograph them through powerful telescopes. Unfortunately, such direct detection of extrasolar planets remains beyond our capabilities. To understand the difficulty, imagine trying to see planets like Earth or Jupiter orbiting a nearby star. As we discussed in Chapter 1, seeing an Earth-like planet orbiting the nearest star besides the Sun would be like looking from San Francisco to see a pinhead orbiting just 15 meters from a grapefruit in Washington, D.C. Seeing a Jupiter-like planet would be like trying to see a marble about 80 meters from the grapefruit-size star.

The scale alone would make the task quite challenging, but it is further complicated by the fact that a Sun-like star would be a *billion times* brighter than the light reflected from any planets. Because even the best telescopes blur the light from stars at least a little, the small blips of planetary light would be overwhelmed by the glare of scattered starlight. Astronomers are working on technologies that may overcome this problem, such as special interferometers [Section 7.5] that can block out the bright starlight and allow us to see the dim light of planets. Nevertheless, all discoveries of extrasolar planets to date have been made on the basis of indirect evidence—acquired not by seeing the planets themselves but by studying the light of the stars they orbit.

To date, the most commonly employed strategy has involved searching for a planet by watching for the small gravitational tug it exerts on its star. In most cases, the easiest way to find this tug is by identifying small Doppler shifts in the star's spectrum [Section 6.5]. An orbiting planet causes its star to alternately move slightly toward and away from us, which makes the star's spectral lines shift alternately toward the blue and toward the red (Figure 9.15). Current techniques can measure a star's velocity to within 3 meters per second—jogging speed. In many cases, this is good enough to allow us to find gravitational tugs caused by planets the size of Jupiter or Saturn. For example, the data in Figure 9.15b reveal a planet with roughly 60% the mass of Jupiter orbiting the star 51 Pegasi. This planet lies so close to its star that its "year" lasts only four of our days, and its surface temperature is probably over 1,000 K.

Measuring the Properties of Extrasolar Planets

The Doppler technique for finding extrasolar planets has been used for the vast majority of planet discoveries to date. By carefully analyzing the Doppler shift data, scientists can determine both the average distance and the eccentricity of a newly discovered planet's orbit. In some cases, the data can even tell us whether the star has more than one planet. If two or more planets each exert a noticeable gravitational tug on the star, it is the combined effect of these tugs that causes the star's motion and shows up in its Doppler shift. The precise pattern of the Doppler shift allows scientists to determine the orbit of each planet responsible for the gravitational tugs.

The Doppler technique also tells us the planet's mass, though not always precisely. The amount of a Doppler shift depends both on the mass of the orbiting object and on the tilt of its orbit (Figure 9.16). Thus, for example, a lower-mass planet in an edge-on orbit and a higher-mass planet in a more tilted orbit can both produce the same Doppler shift.

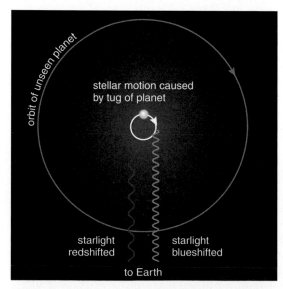

a Doppler shifts allow us to detect the slight motion of a star caused by an orbiting planet.

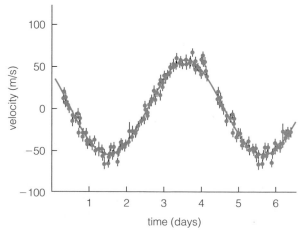

b A periodic Doppler shift in the spectrum of the star 51 Pegasi shows the presence of a large planet with an orbital period of about 4 days. Dots are actual data points; bars through dots represent measurement uncertainty.

Figure 9.15 The Doppler technique for discovering extrasolar planets.

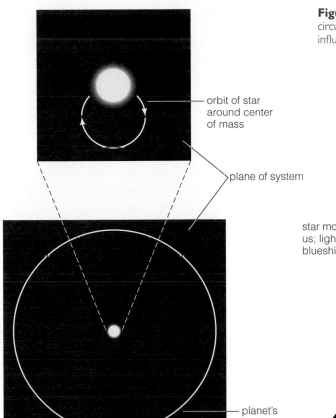

Figure 9.16 If we observe the Doppler shift of a star that moves in a circular orbit under the influence of a planet, the amount of shift will be influenced by the orbit's tilt, or inclination.

orbit of star around center of mass

plane of system

planet's orbit

a If the star–planet system is viewed pole-on, no to-and-fro velocity will be observed, and therefore no changing Doppler shift. Thus, we cannot detect the existence of this planet by the Doppler technique.

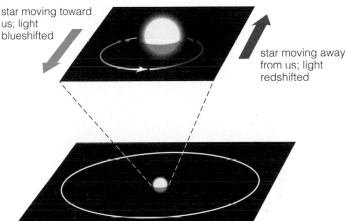

star moving toward us; light blueshifted

star moving away from us; light redshifted

b If the system appears edge on to us, as in this diagram, we see the maximum Doppler shift. If the orbital plane is tilted at some angle between edge-on and face-on, the Doppler shift tells us only the portion of the star's speed that is directed toward or away from us and thus gives an underestimate of the star's full orbital speed.

Because the Doppler shift comes only from the part of the orbital motion directed toward or away from us, the measured mass will be precise only for a planet in an edge-on orbit. In all other cases, the mass measured by the Doppler technique is actually the planet's minimum possible mass. In most cases, the planet's true mass will be no more than double this minimum mass, so the minimum mass is still a relatively good estimate. In one case (called Gliese 876), the star's changing velocity *and* varying position on the sky have been measured. This allowed the orbital tilt to be directly measured, so we know the planet's mass fairly precisely.

Figure 9.17 shows the average distances and minimum masses for the first 77 extrasolar planets discovered. The same information is given for our solar system for comparison. All the masses are much more similar to Jupiter's mass than to Earth's mass, telling us that we have not yet discovered any planets that might be like Earth (the smallest is more than 50 times the mass of Earth). This is not too surprising: A massive planet like Jupiter has a much greater gravitational effect on its star than a small planet like Earth. We do not yet know whether Earth-like planets are common or rare, because we have no data on which to base any conclusion.

While the Doppler technique tells us about planetary masses and orbits, it does not tell us anything about a planet's size (radius). However, there is a way to get such data. If a planet happens to have an edge-on orbit as viewed from Earth, the planet should pass directly in front of its star with each orbit, creating a sort of mini-eclipse that astronomers call a **transit**. (We see transits in our own solar system when Mercury or Venus passes in front of the Sun as viewed from Earth—see Figure S1.5.)

In 1999, astronomers doing follow-up studies on the extrasolar planets discovered earlier through the Doppler technique found that one of those planets transits its star every $3\frac{1}{2}$ days, causing a 1.7% drop in the star's brightness for about 2 hours (Figure 9.18). We can conclude that the planet must be just large enough to block 1.7% of the star's light, which allows us to make a fairly precise estimate of the planet's radius. In addition, because the transits tell us the planet is in an edge-on orbit, we know that the mass measured by the Doppler technique is precise. Knowing both the planet's mass and its size allows us to calculate its average density, which turns out to be similar to that of the jovian planets in our solar system. Although this transit technique may sound "high tech," the star involved is bright

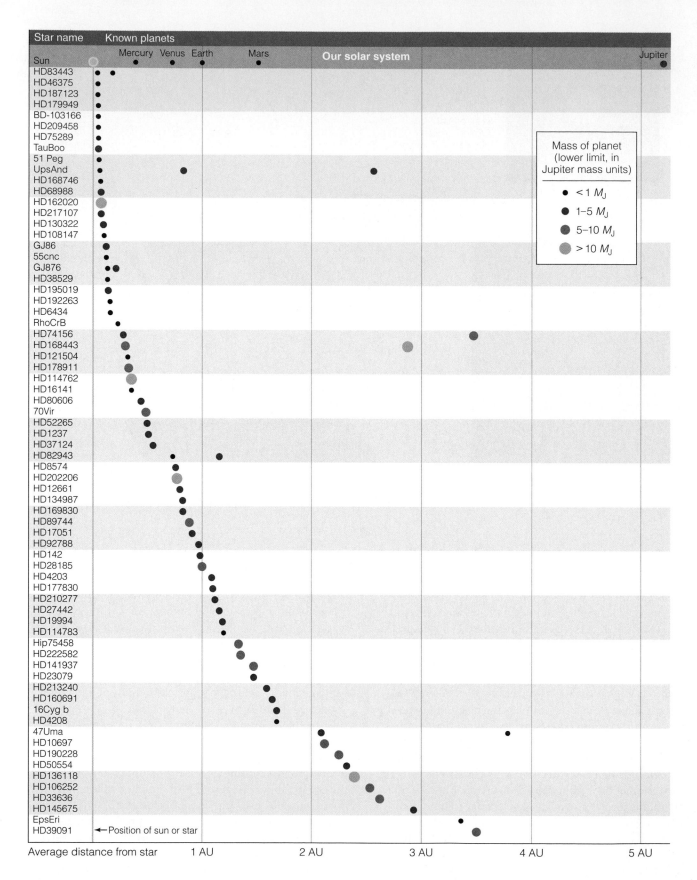

Figure 9.17 This diagram shows the orbital distances and approximate masses of the first 77 planets discovered around other stars. Most of the planets found so far are closer to their stars and more massive than the planets in our solar system. (Planet sizes are not to scale.)

a Artist's conception of the planet as it passes directly in front of its star as seen from Earth.

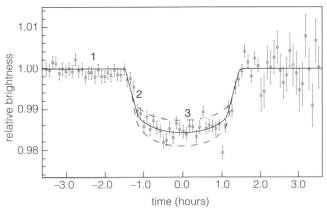

b These data show the 1.7% drop in the star's brightness that proved the planet is passing in front of the star as seen from Earth.

Figure 9.18 Careful measurements of the brightness of the star called HD209458 revealed that an orbiting planet passes directly in front of it as seen from Earth, which means that the planet's orbit must be edge-on as seen from Earth.

enough to be seen easily with a small telescope, and a 1.7% change in brightness is large enough to measure with an inexpensive "photometer." Thus, with relatively inexpensive equipment you can witness the transits of this planet for yourself, confirming in at least this one case that a planet really is orbiting another star.

Many astronomers are searching for new extrasolar planets using the transit method. So far, at least one planet has been detected by this way and then confirmed by the Doppler method. The transit method has the advantage that vast numbers of stars can be studied simultaneously, including stars at greater distances. The method is also sensitive to smaller planets than can be detected by the Doppler method. More discoveries by this method are expected.

Transiting planets must be relatively rare—the vast majority of planets do not have edge-on orbits as seen from Earth. However, they hold the best hope for near-term discoveries of Earth-size planets too small to be detected by the Doppler method. The dimming caused by the transit of an Earth-size planet would be only about 0.01% of a

star's normal brightness. However, the transits would keep repeating with each orbit of the planet, making it possible for us to recognize them if our technology allowed us to detect such small changes in stellar brightness. NASA is currently developing a mission called *Kepler,* scheduled for launch in 2007, which will look for transits caused by Earth-size (or even smaller) planets. Kepler will carefully monitor the brightness of 100,000 stars over a period of 4 years. If small planets are common, Kepler should find hundreds of them.

Lessons for Solar System Formation

The discovery of extrasolar planets presents us with an opportunity to test our theory of solar system formation. Can our existing theory explain other planetary systems, or will we have to go back to the drawing board? The discoveries have already presented at least one significant challenge.

The new challenge arises from the surprising orbits of many of the extrasolar planets. Based on their masses alone (see Figure 9.17), we'd expect the planets to be jovian in nature, because they are all much more massive than Earth. Looking at the average distances in Figure 9.17, however, shows that many of these planets orbit quite close to their star—a characteristic that seems more in line with the terrestrial planets in our solar system. Some orbit even closer to their star than Mercury orbits the Sun. In addition, many of the extrasolar planets have highly elliptical orbits, rather than the nearly circular orbits of the planets in our solar system. How could jovian planets orbit so close to their star in such elliptical orbits? The nebular theory clearly predicts that jovian planets should form only in the cold, outer regions of star systems, where icy materials can condense, and that planets should end up with nearly circular orbits.

As we learned in Chapter 3, if repeated observations conflict with a theory, then scientists must revise or reject the theory. Shortly after the first extrasolar planets were found, scientists did consider that something might be fundamentally wrong with the nebular theory of solar system formation. However, astronomers could not find any fundamental flaws in our basic understanding of how planets form, nor could they discover ways that jovian planets might form so close to their star. They also recognized that the Doppler technique might be discovering relatively rare exceptions rather than general rules.

The Doppler technique makes it much easier to find large planets with close-in orbits, because such planets exert greater gravitational tugs on their stars than do smaller or more distant planets. Thus, discovering the first known extrasolar planets may have been like glancing at animals in a rain forest: The jungle appears full of brightly colored parrots and frogs, but far more animals fail to catch our eye. Some recent planetary discoveries support this view. We now know of several planets that have orbits more consistent with what we expect of jovian planets.

We still must explain the surprising planets, even if they eventually prove to be relatively rare. Because scientists could not explain how jovian planets could *form* with such unusual orbits, they began to consider how the planets might *migrate* to these orbits after their formation.

How might such planetary migration occur? Calculations show that friction with the abundant gas and dust in the disks of newly formed solar systems can exert a drag on young planets, causing them to spiral slowly toward their sun. In our own solar system, this drag is not thought to have played a significant role because the solar wind cleared out the gas before it could have much effect. But the wind may kick in later in at least some other solar systems, allowing time for jovian planets to migrate substantially inward. In some cases, the wind might kick in so late that the planets end up crashing into their stars.

Another possible explanation for some of the observed extrasolar planets involves close encounters between young jovian planets in the outer regions of a disk. Such an encounter might send one planet out of the star system entirely while the other is sent inward in a highly elliptical orbit. Other possibilities include the idea that waves propagating through a gaseous disk could lead to inward migration, or that a jovian planet could migrate inward as a result of multiple close encounters with much smaller planetesimals. (We have some evidence that the jovian planets in our solar system have migrated by this latter mechanism.)

The bottom line is that discoveries of extrasolar planets have shown us that our nebular theory of solar system formation was incomplete. It explained the formation of planets and the simple layout of a solar system such as ours. However, it needed modification to explain the differing layouts of other solar systems. While we are not yet sure of the precise mechanism involved, planets clearly can move from their birthplace. A much wider range of solar system arrangements now seems possible than we had guessed before the discovery of extrasolar planets.

THINK ABOUT IT

Look back at the discussion of the nature of science in Chapter 3, especially the definition of a scientific theory. Does our theory of solar system formation qualify as a scientific theory even though we have recently learned it needs modification? Does this mean the theory was "wrong" before the modifications were made? Explain.

Implications for Terrestrial Worlds

Planetary migration may have important implications for the question of whether terrestrial planets are common or rare. The migration of a large planet is likely to cause significant disruption to the inner regions of a solar system. If the migration occurs before terrestrial planets have finished forming, the material that would have accreted onto the terrestrial worlds might instead be swallowed by the larger world. Even if the formation process is essentially complete, gravitational encounters between large planets and small ones nearly always send the smaller ones scattering. A large planet migrating inward would tend to drive less massive planets into its star. Thus, terrestrial planets would be unlikely to survive around stars where jovian planets have migrated substantially inward.

The question of whether terrestrial planets are common therefore may hinge on how common planetary migration proves to be. Based on the planets discovered to date, migration might seem to be very common, making the prospect of finding Earth-like worlds seem bleak. Keep in mind, however, that our current detection strategies may make it much easier to find planetary systems with migration than systems without it.

Of the several hundred Sun-like stars that we have examined so far, we have found evidence of planets around only about 1 in 10. This means we have not detected planets around the majority of Sun-like stars. This majority may include many stars with planetary systems much like our own—systems that our current detection techniques are unable to find.

We should begin to get more answers with the Kepler mission, which may be able to find Earth-size planets. More definitive results should follow within a couple of decades, when we hope to have high-resolution interferometers in space. These interferometers might not only obtain direct images of distant planets but also might allow us to obtain their spectra and determine whether they have life. Then, at last, we will know whether solar systems like ours—and planets like Earth—are rare or common.

THE BIG PICTURE

Putting Chapter 9 into Context

We've described the current scientific theory of solar system formation. We've seen how it explains the major characteristics we observe and how it can be extended to other planetary systems. As you continue your study of the solar system, keep in mind the following "big picture" ideas:

- The nebular theory of solar system formation gained wide acceptance because of its success in explaining the major characteristics of our solar system.

- Most of the general characteristics of the solar system were determined by processes that occurred very early in the solar system's history, which began some 4.6 billion years ago.

- Chance events may have played a large role in determining how individual planets turned out. No one knows how different our solar system might be if it started over.

- Planet-forming processes are apparently universal. The discovery of other planetary systems has inaugurated an exciting new era in planetary science.

9.1 The Origin of the Solar System: The Nebula Theory

- *What four characteristics of our solar system must be explained by any formation theory?* A successful theory must explain patterns of motion, the differences between terrestrial and jovian planets, the presence of asteroids and comets, and exceptions to the rules.

- *What is the basic idea behind the nebular theory?* Our solar system formed from a giant cloud of gas and dust, with the Sun forming at the center of the cloud and the planets forming in the spinning, flattened disk of material that orbited the young Sun.

9.2 Orderly Motion in a Collapsing Nebula

- *What was the solar nebula?* The solar nebula was the interstellar cloud from which our solar system was born.

- *How did gravitational collapse affect the solar nebula?* The nebula heated up, spun faster, and flattened into a disk.

- *What produced the orderly motion we observe in the solar system today?* The orderly motion of the solar system reflects the original motion of the spinning disk of the solar nebula. The plane of this disk became the plane in which the planets now orbit the Sun, and the direction of the disk's spin became the direction of the planetary orbits.

9.3 Two Types of Planets

- *What key fact explains why there are two types of planets?* Nearer to the Sun the frost line, temperatures were so high that only metal and rock could condense into solid particles. Beyond the frost line, cooler temperatures allowed hydrogen compounds to condense into solid particles of ice. Thus, the solid material beyond the frost line included abundant ice in addition to rock and metal.

- *How did the terrestrial planets form?* Terrestrial planets formed inside the frost line from condensation of solid grains of metal and rock that accreted into planetesimals which grew into planets.

- *How did the jovian planets form?* Jovian planets formed beyond the frost line from condensation of solid grains of metal, rock, and lots of ice that accreted into icy planetesimals. The capture of hydrogen and helium gas by the largest icy planetesimals made "miniature solar nebulae." The jovian planets formed at the centers of those nebulae, while moons accreted from ice in the spinning disks.

9.4 Explaining Leftovers and Exceptions to the Rules

- *How are asteroids and comets related to the planetesimals that formed the planets?* Asteroids are leftover planetesimals of the inner solar system, and comets are leftovers of the outer solar system.

- *What was the heavy bombardment?* It was the period early in our solar system's history during which the planets were bombarded by many leftover planetesimals.

- *How do we explain the exceptions to the rules?* Collisions or close encounters with leftover planetesimals can explain the exceptions.

- *How do we think our Moon formed?* A Mars-size leftover planetesimal slammed into Earth, blasting rock from Earth's outer layers into orbit, where it reaccreted to form the Moon.

9.5 How Old Is the Solar System?

- *How do we measure the age of a rock?* Radiometric dating involves looking for radioactive isotopes and their decay products within the rock. By comparing the present-day amounts of an isotope and its decay product, we can determine how much of the isotope must have been present when the rock solidified. We can then use the half-life of the isotope to determine how long it has been undergoing decay within the rock, which tells us the rock's age.

- *How old is the solar system, and how do we know?* The solar system is about 4.6 billion years old, an age we determine from radiometric dating of the oldest meteorites.

9.6 Other Planetary Systems

- *When did we first learn of planets beyond our solar system?* The first discoveries were made in the mid-1990s.

- *Have we ever actually photographed an extrasolar planet?* No. We have detected them only indirectly through their observable effects on the stars they orbit.

- *What new lessons have we learned by investigating other planetary systems?* Planetary systems exhibit a surprising range of layouts, suggesting that jovian planets sometimes migrate inward from where they are born. This lesson may have important implications for the likelihood of finding other Earth-like worlds.

❓ Surprising Discoveries?

Suppose we found a solar system with the property described. (These are *not* real discoveries.) In light of our theory of solar system formation, decide whether the discovery should be considered reasonable or surprising. Explain.

1. A solar system has five terrestrial planets in its inner solar system and three jovian planets in its outer solar system.

2. A solar system has four large jovian planets in its inner solar system and seven small planets made of rock and metal in its outer solar system.

3. A solar system has 10 planets that all orbit the star in approximately the same plane. However, 5 planets orbit in one direction (e.g., counterclockwise), while the other 5 orbit in the opposite direction (e.g., clockwise).

4. A solar system has 12 planets that all orbit the star in the same direction and in nearly the same plane. The 15 largest moons in this solar system orbit their planets in nearly the same direction and plane as well. However, several smaller moons have highly inclined orbits around their planets.

5. A solar system has six terrestrial planets and four jovian planets. Each of the six terrestrial planets has at least five moons, while the jovian planets have no moons at all.

6. A solar system has four Earth-size terrestrial planets. Each of the four planets has a single moon that is nearly identical in size to Earth's Moon.

7. A solar system has many rocky asteroids and many icy comets. However, most of the comets orbit the star in a belt much like the asteroid belt of our solar system, while the asteroids inhabit regions much like the Kuiper belt and Oort cloud of our solar system.

8. A solar system has several planets similar in composition to the jovian planets of our solar system but similar in mass to the terrestrial planets of our solar system.

Problems

(Quantitative problems are marked with an asterisk.)

9. *Ingredients of the Planets.* Describe the four categories of materials in the solar nebula by their condensation properties and abundance. Which ingredients are present in terrestrial planets? In jovian planets?

10. *Solar Wind.* What is the solar wind, and what roles did it play in the early solar system?

11. *The Beginning of Our Solar System?* What evidence suggests that a nearby stellar explosion may have triggered the collapse of the solar nebula?

12. *Finding Other Worlds.* Briefly summarize current techniques for detecting extrasolar planets.

13. *A Cold Solar Nebula.* Suppose the entire solar nebula had cooled to 50 K before the solar wind cleared it away. How would the composition and sizes of the planets of the inner solar system be different from what we see today? Explain your answer in a few sentences.

14. *No Gas Capture.* Suppose the solar wind had cleared away the solar nebula before the seeds of the jovian planets could gravitationally draw in hydrogen and helium gas. How would the planets of the outer solar system be different? Would they still have many moons? Explain your answer in a few sentences.

15. *Angular Momentum.* Suppose our solar nebula had begun with much more angular momentum than it did. Do you think planets could still have formed? Why or why not? What if the solar nebula had started with zero angular momentum? Explain your answers in one or two paragraphs.

*16. *Dating Lunar Rocks.* You are analyzing Moon rocks that contain small amounts of uranium-238, which decays into lead with a half-life of about 4.5 billion years.

   a. In a rock from the lunar highlands, you determine that 55% of the original uranium-238 remains, while the other 45% has decayed into lead. How old is the rock?

   b. In a rock from the lunar maria, you find that 63% of the original uranium-238 remains, while the other 37% has decayed into lead. Is this rock older or younger than the highlands rock? By how much?

*17. *Carbon-14 Dating.* The half-life of carbon-14 is about 5,700 years.

   a. You find a piece of cloth painted with organic dye. By analyzing the dye, you find that only 77% of the carbon-14 originally in the dye remains. When was the cloth painted?

   b. A well-preserved piece of wood found at an archaeological site has 6.2% of the carbon-14 it must have had when it was living. Estimate when the wood was cut.

   c. Is carbon-14 useful for establishing Earth's age? Why or why not?

*18. *Martian Meteorite.* Some unusual meteorites thought to be chips from Mars contain small amounts of radioactive thorium-232 and its decay product, lead-208. The half-life for this decay process is 14 billion years. Analysis of one such meteorite shows that 94% of the original thorium remains. How old is this meteorite?

*19. *51 Pegasi.* The star 51 Pegasi has about the same mass as our Sun. A planet discovered around it has an orbital period of 4.23 days. The mass of the planet is estimated to be 0.6 times the mass of Jupiter. Use Kepler's third law to find the planet's average distance (semimajor axis) from its star. Briefly explain why, according to our theory of solar system formation, it is surprising to find a planet this size orbiting at this distance. (*Hint:* Because the mass of 51 Pegasi is about the same as the mass of our Sun, you can use Kepler's third law in its original form, $p^2 = a^3$ [Section 3.4]. Be sure to convert the period into years before using this equation.)

*20. *Transit of HD209458.* The star HD209458, which has a transiting planet, is roughly the same size as our Sun, which has a radius of about 700,000 kilometers. The planetary transits block 1.7% of the star's light.

a. Calculate the radius of the transiting planet. (*Hint:* The brightness drop tells us that the planet blocks 1.7% of the *area* of the star's visible disk. The area of a circle is $\pi \times (\text{radius})^2$.)

b. The mass of the planet is approximately 0.6 times the mass of Jupiter, and Jupiter's mass is about 1.9×10^{27} kilograms. Calculate the average density of the planet. Give your answer in grams per cubic centimeter. Compare this density to the average densities of Saturn (0.7 g/cm^3) and Earth (5.5 g/cm^3). Is the planet terrestrial or jovian in nature? (*Hint:* To find the volume of the planet, use the formula for the volume of a sphere: $V = \frac{4}{3} \times \pi \times (\text{radius})^3$. Be careful with unit conversions.)

Discussion Questions

21. *Theory and Observation.* Discuss the interplay between theory and observation that has led to our modern theory of solar system formation. What role does technology play in allowing us to test this theory?

22. *Random Events in Solar System History.* According to our theory of solar system formation, numerous random events, such as giant impacts, had important consequences for the way our solar system turned out. Can you think of other random events that might have caused the planets to form very differently? If a different set of random events had occurred, what important properties of our solar system would have turned out the same and what properties might have been different?

23. *Lucky to Be Here?* Considering the overall process of solar system formation, do you think it was likely for a planet like Earth to have formed? Could random events in the early history of the solar system have prevented our being here today? What implications do your answers have for the possibility of Earth-like planets around other stars? Defend your opinions.

MEDIA EXPLORATIONS

For a complete list of media resources available, go to www.astronomyplace.com and choose Chapter 9 from the pull-down menu.

 Astronomy Place Web Tutorials

Tutorial Review of Key Concepts

Use the following interactive **Tutorials** at www.astronomyplace.com to review key concepts from this chapter.

Formation of the Solar System Tutorial

Lesson 1 Comparative Planetology

Lesson 2 Formation of the Protoplanetary Disk

Detecting Extrasolar Planets Tutorial

Lesson 1 Taking a Picture of a Planet

Lesson 2 Stars' Wobbles and Properties of Planets

Lesson 3 Planetary Transits

Supplementary Tutorial Exercises

Use the interactive **Tutorial Lessons** to explore the following questions.

Formation of the Solar System Tutorial, Lesson 2

1. Study the animations of the formation of the solar nebula. Why does the nebula flatten as it collapses?

2. Use the tool in the tutorial to make a graph of temperatures in the solar nebula. How does the temperature vary with the distance from the center?

3. Why didn't any planets form within 0.3 AU of the Sun?

Detecting Extrasolar Planets Tutorial, Lesson 2

1. Using the tool provided, explain how weekly measurements allow us to determine the orbital period of the extrasolar planet.

2. Use the tool to vary the mass of the planet. How does its mass affect the Doppler shifts in its star's light?

3. Use the tool to vary the orbital radius of the planet. How does the orbital radius affect the Doppler shifts in its star's light?

Detecting Extrasolar Planets Tutorial, Lesson 3

1. Under what conditions can we view a planetary transit of another star?

2. How does the change in brightness during a transit depend on the planet's properties?

Movies

Check out the following narrated and animated short documentaries available on www.astronomyplace.com for a helpful review of key ideas covered in this chapter.

 History of the Solar System Movie

 Orbits in the Solar System Movie

Web Projects

Take advantage of the useful Web links on www.astronomyplace.com to assist you with the following projects.

1. *New Planets.* Find the latest information on extrasolar planet discoveries. Create a personal "planet journal," complete with illustrations as needed, with a page for each of at least three recent discoveries of new planets. On each journal page, note the technique that was used to find the planet, give any information we have about the nature of the planet, and discuss how the planet does or does not fit in with our current understanding of planetary systems.

2. *The Kepler Mission.* The Kepler mission is the first funded mission designed expressly to look for Earth-size planets around other stars. Go to the Kepler Web site and learn more about the mission. Write a one- to two-page summary of the mission's goals and its current status.

3. *Planet-Hunting Interferometers.* Other future missions will use interferometry to learn about extrasolar planets. Go to the Web site for one future interferometry mission under consideration, such as SIM, TPF, or Darwin. For the mission you choose, write a one- to two-page summary of the mission's goals and its current status.

10 Planetary Geology
Earth and the Other Terrestrial Worlds

Nothing is rich but the inexhaustible wealth of nature. She shows us only surfaces, but she is a million fathoms deep.

Ralph Waldo Emerson

We have seen that the Sun and the planets of our solar system were all born together in the solar nebula. We have traced the basic properties of the planets—their sizes, compositions, orbits, and rotation—to events that occurred in the very early history of our solar system. These basic properties probably have not changed in the nearly 4.5 billion years since the planets formed, but other important planetary changes have occurred through time.

The rest of our study of the solar system will focus on understanding the histories of the planets. We will begin by exploring the geology of the terrestrial worlds in this chapter and then turn to the terrestrial atmospheres in Chapter 11. In Chapters 12 and 13, we will explore the jovian worlds and the many small bodies in the solar system. Finally, in Chapter 14, we'll take everything we've learned and apply it to the question of why Earth is different in a way that makes it possible for us to be here talking about it.

As you study the planets, moons, and other bodies in the solar system, keep in mind that despite 4.5 billion years of change, virtually all of the features we see today should be traceable to the properties these objects had at their birth. We will proceed in a way that is designed to help reveal these connections. In this chapter, we will first look inside the terrestrial worlds and then focus on geological processes that shape terrestrial world surfaces. We will tour our sister terrestrial worlds to understand their similarities and differences and conclude by tying the present-day surfaces of the terrestrial worlds to their birth properties.

Venus

Earth

Mercury

Mercury is heavily cratered but also has long, steep cliffs—one is visible here as the long curve that crosses through the crater near the center of the image.

The central structure is a tall, twin-peaked volcano on Venus. Both Venus images are based on radar data from the *Magellan* spacecraft, because Venus's thick clouds prevent us from seeing the surface in visible light.

Earth shows a variety of geological features visible in this photo from orbit.

Formation of the Solar System Tutorial, Lesson 1

10.1 Planetary Surfaces

Our planet's surface generally seems solid and steady, but once in a while it offers a reminder that nothing about it is permanent. If you live in Alaska or California, you've probably felt the ground shift beneath you in an earthquake. If you live in Washington State, you or your parents may have seen the awesome eruptions of Mount St. Helens. If you live in Hawaii, a visit to the still-active Kilauea Volcano will remind you that you live on mountains of volcanic rock protruding from the ocean floor—mountains that became islands only after growing larger through millions of years of eruptions.

Volcanoes and earthquakes are not the only processes acting to reshape Earth's surface. They are not even the most dramatic: Far greater change can occur on the rare occasions when an asteroid or a comet slams into Earth. Much more gradual processes can also have great effect if they continue over long periods of time. The Colorado River causes only small changes in the landscape from year to year, but its unrelenting flow over the past few million years carved the Grand Canyon. The Rocky Mountains were once twice as tall as they are today, having been cut down in size as the result of tens of millions of years of erosion by wind, rain, and ice. Entire continents even move slowly about, completely rearranging the map of Earth every few hundred million years. Together, gradual and sudden changes are constantly reshaping our planet. Earth's present surface bears little resemblance to the surface with which it was born some 4.5 billion years ago.

Earth is not alone in having undergone tremendous change since its birth. The surfaces of all five terrestrial worlds—Mercury, Venus, Earth, the Moon, and Mars—must have looked quite similar when they were young. All five were made of rocky material that had condensed in the solar nebula, and all five were subjected early on to the impacts of the heavy bombardment [Section 9.4]. The great differences in their present-day appearance must therefore be the result of changes that have occurred through time. Ultimately, these changes must be traceable to fundamental properties of the planets.

Figure 10.1 shows global views of the terrestrial surfaces to scale, along with sample close-up views from orbit. (The images of Venus are based on radar data, because photographs from space show only its thick cloud cover. See Figure 8.9.) Some of the profound differences between these worlds are immediately obvious. Mercury and the Moon still show the scars of their battering during the heavy bombardment—they are densely covered by craters except in areas that appear to be volcanic plains. Bizarre bulges and odd volcanoes dot the surface of Venus. Mars, despite its middling size, has the solar system's largest volcanoes and a huge canyon cutting across its surface. Mars also has numerous features that appear to have been shaped by running water. Earth has surface features similar to all those on the other terrestrial worlds, and more—including a unique layer of living organisms that covers almost the entire surface of the planet.

Our purpose in this chapter is to understand how these profound differences among the terrestrial planets came to be. We will be discussing what is often called **planetary geology**, the study of surface features and the processes that create them. *Geology* literally means "the study of

Mars

Earth's Moon

The Moon's surface is heavily cratered in most places.

Mars has impact craters like the one near the upper right and also has features that look much like dried-up riverbeds.

Figure 10.1 Global views to scale, along with sample close-ups viewed from orbit, of each of the five terrestrial worlds. All the photos were taken with visible light, except the Venus photos, which are based on radar data.

Earth" (the root *geo* means "Earth"), so planetary geology is an extension of this science to other worlds. Note that we use the term *planetary geology* for the study of any solid world, even if it is a moon rather than a planet.

● 10.2 Inside the Terrestrial Worlds

We can see only the surfaces of planets, but most of the processes that shape planetary surfaces are a result of what goes on inside them. This should not be too surprising, because the largest surface features are only tiny blips compared to a planet's overall size. For example, the tallest mountains on Earth rise only about 10 kilometers above sea level, less than $\frac{1}{600}$ of Earth's radius. These mountains could be represented by grains of sand on a typical globe. Thus, we cannot expect to understand what creates mountains and other surface features unless we first understand how planets work on the inside.

Find a globe of Earth that shows mountains in raised relief (that is, a globe that has bumps to represent mountains). Are the heights of the mountains correctly scaled relative to Earth's size? Explain.

Learning About Planetary Interiors

Even on Earth, our deepest drills have barely pricked the surface, penetrating only a few kilometers into the ground. Thus, we have never directly sampled Earth's deep interior, let alone the interior of any other world. How can we possibly know anything about what the planets' interiors look like?

Our most detailed information about Earth's interior comes from **seismic waves**, vibrations that travel both through Earth's interior and along its surface after an earthquake. In much the same way that shaking a present offers clues about what's inside, the vibrations that follow earthquakes can tell us about the inside of Earth (see box, p. 257). Careful analysis of seismic waves has allowed scientists to map Earth's interior structure in great detail. We also have some seismic data for the Moon thanks to monitoring stations left behind by the Apollo astronauts.

Learning about the interiors of other terrestrial worlds is more difficult, but several types of observation can help. One key piece of information comes from comparing a planet's overall average density to the density of its surface rock. We can determine a planet's average density from its mass and size. (Recall that density is mass divided by volume). We can find a planet's mass by applying Newton's version of Kepler's third law to the orbital properties of its moons or of spacecraft that have visited it [Section 5.3]. We find its size through careful study of telescopic or space-craft images. Through direct and indirect measurements, we know that surface rocks of all the terrestrial worlds are much less dense than the world as a whole. This tells us that much denser material must reside in their interiors than on their surfaces.

We can gain further information about a planet's interior from precise gravity measurements, studies of the planet's magnetic field, and observations of its surface rock. The strength of gravity tends to differ slightly from place to place on a planet's surface. Careful measurements made by spacecraft can detect these gravity variations, providing clues about the underlying interior structure of the planet. Magnetic fields are generated deep within planetary interiors, so the properties of a planet's magnetic field can also tell us about interior conditions. Finally, because some surface rock is lava that rose to the surface from deep within the interior, study of this rock can tell us about interior composition.

Interior Structure

Together, the types of observations we have described give us a fairly clear picture of the interiors of the five terrestrial worlds. In all five cases, the interiors have distinct layers that play an important role in geology. We can categorize the layers in two different ways: by density and by strength. Both categorizations are useful in geology.

Layering by Density You are probably familiar with the idea of categorizing interior layers by density. By density, each terrestrial planet has three basic layers:

- **Core:** The highest-density material, consisting primarily of metals such as nickel and iron, resides in the central core.

- **Mantle:** Rocky material, composed of a variety of minerals that contain silicon and oxygen (and other elements), forms the thick mantle that surrounds the core.

- **Crust:** The lowest-density rock, such as granite and basalt, forms the thin crust, essentially representing the world's outer skin.

Figure 10.2 shows these interior layers for each of the five terrestrial worlds. While the basic structure is the same in each case, the relative sizes of the layers vary. For example, Mercury's core is much larger in proportion to its overall size than that of any other planet. This large core is probably due in part to Mercury's closeness to the Sun, where it was too hot for rock to condense as readily as it did at greater distances in the solar nebula. Mercury therefore was built from more metal-rich planetesimals, and we would expect it to have a relatively large metal core. However, detailed models show that this idea alone cannot fully explain the huge size of Mercury's core in proportion to its mantle. That discrepancy is the primary reason why scientists suspect that Mercury suffered a giant impact early in its history [Section 9.4], an impact that blasted away much of its original mantle and left behind the surprising interior structure we find today.

Although not shown in the figure, Earth's metallic core actually consists of two distinct regions: a solid *inner core* and a molten (liquid) *outer core* (see the box "Seismic Waves"). Because temperature rises with depth in a planet, it may seem surprising that Earth's inner core is solid while the outer core is molten. The explanation is that the inner

core is kept solid by the higher pressure at this depth, even though the temperature is also higher. Venus may have a similar core structure, but without seismic data we cannot be sure.

The distinct layering of the terrestrial worlds may remind you of the way liquids separate by density. For example, in a mixture of oil and water, the less dense oil rises to the top while the denser water sinks to the bottom. The process in which gravity separates materials by density is

---

SPECIAL TOPIC   Seismic Waves

Just as stomping on the floor generates reverberations that can be felt many steps away from the place where the stomping occurs, an earthquake generates vibrations that can travel through the Earth. These vibrations are called *seismic waves* (*seismic* comes from the Greek word for "shake"). An earthquake generally does all its damage near the place where it occurs, but the seismic waves it generates reach nearly everywhere on Earth. Even far from the earthquake, the seismic waves shake the ground mildly when they arrive, and their precise pattern can be recorded by instruments known as seismographs. By comparing seismograph recordings made at many locations around the world, scientists can pinpoint the location of an earthquake and can also learn about the ground the seismic waves have traveled through.

Seismic waves come in two basic types that are analogous to two ways you can generate waves in a Slinky (see the figure). Pushing and pulling on one end of a Slinky (while someone holds the other end still) generates a wave in which the Slinky is bunched up in some places and stretched out in others. Waves like this in rock are called P waves. The *P* stands for *primary,* because these waves travel fastest and are the first to arrive after an earthquake, but it is easier to think of it as *pressure* or *pushing.* (Sound also travels as a pressure wave and thus is quite similar to P waves.) Shaking a Slinky slightly up and down generates an up-and-down motion

all along its length. Such up-and-down (or side-to-side) waves in rock are called S waves. The *S* stands for *secondary* but is easier to remember as *shear* or *side-to-side.*

The two types of waves can tell us about the phase (liquid or solid) of the material they encounter. P waves can travel through almost any material, because molecules can always push on their neighbors no matter how weakly they are bound together. S waves, however, generally travel only through solids. They cannot travel through liquids or gases because the bonds between neighboring molecules are too weak to transmit up-and-down or sideways forces. Studies of seismic waves show that only P waves reach the side of the world opposite an earthquake, telling us that the S waves have been stopped by a liquid layer. That is how we know that Earth must have a liquid layer in its outer core (see the figure).

Seismic waves can tell us much more. The precise speeds and directions of all seismic waves depend on the composition, density, pressure, temperature, and phase (solid or liquid) of the material they pass through. Careful analysis of seismic waves therefore offers a very detailed picture of Earth's interior. Geologists hope someday to have seismographs on all the terrestrial worlds so they can make similarly detailed maps of their interiors.

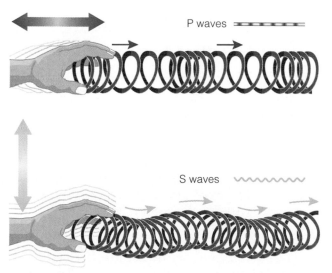

P waves

S waves

a Slinky examples demonstrating P and S waves.

● Seismic waves provide a probe of a planet's interior

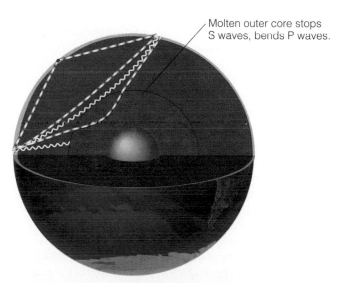

Molten outer core stops S waves, bends P waves.

b Because S waves do not reach the side of Earth opposite an earthquake, we infer that part of Earth's core is molten.

Figure 10.2 Interior structures of the terrestrial worlds, shown to scale and in order of decreasing size. The color-coded regions show the core-mantle-crust layering, which is based on density. Dashed lines indicate the lithosphere, which encompasses the crust and part of the mantle on each world and is defined by the strength of the rock rather than the density. (The thicknesses of the crust and lithosphere on Venus and Earth are exaggerated to make them visible in this figure.)

called **differentiation** (because it results in layers made of *different* materials). The layered interiors of the terrestrial worlds tell us that all the worlds underwent differentiation at some time in the past. They also tell us that the planets must have been essentially liquid at one time, to allow dense materials to sink and light materials to rise. Because liquefying metal and rock requires high temperatures, the interiors must once have been much hotter than they are today. We'll discuss how interiors became hot shortly. We'll also see that their subsequent cooling explains much of the difference we find in the geology of terrestrial planets today.

Layering by Strength: The Lithosphere The second way to categorize interior layers is by the strength of their rock. Although the activity of volcanoes may give the impression that Earth is molten inside, in most cases the lava actually rises upward from a fairly narrow layer of rock in the upper mantle that is only partially molten. The vast majority of Earth's interior is solid rock, with the exception of the outer core. However, the interior rock is not equally strong throughout, and differences in rock strength are often more important than differences in density for understanding many geological processes.

The idea that rock can vary in strength may sound strange, since we often think of rock as the very definition of strength. Remember that all atoms and molecules are mostly empty space and that electrical bonds between molecules are what make a rock feel solid [Section 4.3]. Although solid bonds are strong, they can still break and re-form, especially when subjected to sustained stress. Thus, over millions and billions of years, rocky material can slowly deform and flow. The long-term behavior of rock is much like that of the popular toy Silly Putty, which breaks like a brittle solid when you pull it sharply but deforms and stretches when you pull it slowly (Figure 10.3).

The idea that rock can deform and flow explains why small moons and asteroids are "potato-shaped" while larger worlds are spherical. The weak gravity of a small object is unable to overcome the rigidity of its rocky material, so the object retains the shape it had when it was born. For a larger world, gravity can overcome the strength of solid rock, slowly deforming and molding it into a spherical shape. Gravity will make any rocky object more than about 500 kilometers in diameter into a sphere within about 1 billion years. Larger worlds become spherical more quickly, especially if they are molten (or gaseous) at some point in their history.

The strength of a rock depends on three characteristics: its composition, its temperature, and the surrounding

Figure 10.3 Silly Putty stretches when pulled slowly but breaks cleanly when pulled rapidly. Rock behaves just the same, but on a longer time scale.

Internal Heat: The Driver of Geological Activity

The geology we see on planetary surfaces is shaped by processes that occur in planetary interiors. In particular, interior heat drives most geological activity on the terrestrial worlds. Volcanoes can erupt on Earth because Earth is quite hot inside—hot enough that, in at least some cases, rock melts into molten lava. In contrast, the Moon has no active volcanoes, because its interior is too cool to melt rock and push it to the surface. If we want to understand the profound differences among the terrestrial worlds, we must understand why they differ so much in their internal heat.

Internal heat has little to do with the Sun. It's often said that the Sun is the ultimate source of all our energy, but this statement applies only to energy received on a world's surface. On Earth's surface, for example, solar heating is some 20,000 times stronger than the heating from the interior. However, virtually none of this solar energy penetrates more than a few meters into the ground. Internal heat is a product of the planets themselves, not of the Sun.

How Interiors Get Hot A hot interior contains a lot of thermal energy, and the law of conservation of energy tells us that this energy must have come from somewhere [Section 4.2]. Three sources of energy explain nearly all the interior heat of the terrestrial worlds (Figure 10.4):

- **Heat of accretion:** heat generated at the time the planets accreted from planetesimals.

- **Heat from differentiation:** heat generated at the time the planets differentiated into their core-mantle-crust structures.

- **Heat from radioactive decay:** heat generated as radioactive nuclei in the planetary interior undergo natural decay.

pressure. Composition affects rock strength because different minerals are held together by molecular bonds of different strengths. Some rocks also contain traces of water, which can act as a lubricant and weaken rock. Temperature affects rock strength in a way you'd probably expect: Warm rocks are weaker and more deformable than cooler rocks of the same type (think of Silly Putty that has been warmed by your hands). High pressure can compress rocks slightly, and compression tends to make them stronger. The very high pressures found deep in planetary interiors can compress rocks so much that they stay solid even at temperatures that would melt them under lower pressure.

The interior layering of a planet by rock strength looks somewhat different from the core-mantle-crust layering by density. In terms of rock strength, the outer layer of relatively rigid rock is called the **lithosphere** (*lithos* is Greek for "stone"). The lithosphere generally encompasses the crust and part of the upper mantle, as shown by the dashed circles in Figure 10.2. You can see that Earth and Venus have very thin lithospheres, while the lithospheres of Mars, Mercury, and the Moon include most or all of their mantles. Beneath the lithosphere, the higher temperatures allow rock to deform and to flow much more easily. Thus, the lithosphere essentially is a layer of rigid rock that "floats" on the softer rock beneath.

COMMON MISCONCEPTIONS

Pressure and Temperature

You might think that Earth's interior is hot just because the pressures are high. After all, if we compress a gas from low pressure to high pressure, it heats up. The same is not necessarily true of rock. High pressure compresses rock only slightly, so the compression causes little increase in temperature. Thus, while high pressures and temperatures sometimes go together in planets, they don't have to. In fact, after all the radioactive elements decay (billions of years from now), Earth's deep interior will become quite cool even though the pressure will be the same as it is today. The temperatures inside Earth and the other planets can remain high only if there is an internal source of heat, such as accretion, differentiation, or radioactivity.

Accretion

Gravitational potential energy is converted to kinetic energy.

Kinetic energy is converted to thermal energy.

Differentiation

Light materials rise to the surface.

Dense materials fall to the core, converting gravitational potential energy to thermal energy.

Radioactive Decay

Mass-energy contained in nuclei is converted to thermal energy.

Figure 10.4 The three main sources of internal heat in terrestrial planets. Only radioactive decay is a major heat source today.

Accretion deposits energy brought in from afar by colliding planetesimals. An incoming planetesimal has a lot of gravitational potential energy when it is far away. As it approaches a forming planet, its gravitational potential energy is converted to kinetic energy, causing it to accelerate. Upon impact, much of the kinetic energy is converted to heat, adding to the thermal energy of the planet.

Differentiation also converts gravitational potential energy into thermal energy. In this case the conversion occurs as materials separate by density within the interior. Remember that differentiation occurs as denser materials sink and less dense materials rise in a molten interior. This process adds mass to the planet's core and reduces the mass of the outer layers. Thus, much of the planet's mass effectively moves inward, losing gravitational potential energy. The friction generated as the mass separates by density converts the lost gravitational potential energy to thermal energy, heating the interior. The same thing happens when you drop a brick into a pool: As the brick sinks to the bottom, friction with the surrounding water heats the pool—though the amount of heat from a single brick is too small to notice.

Radioactive decay affects the terrestrial worlds because the rock and metal planetesimals that built them contained radioactive isotopes of elements such as uranium, potassium, and thorium. When radioactive nuclei decay, subatomic particles fly off at high speeds, colliding with neighboring atoms and heating them. In essence, this transfers some of the mass-energy ($E = mc^2$) of the radioactive nucleus to the thermal energy of the planetary interior.

Accretion and differentiation deposited heat in planetary interiors billions of years ago, but radioactive decay continues to heat the terrestrial planets to this day. However, because the rate of radioactive decay declines with time, it was a more significant heat source when the planets were young.

THINK ABOUT IT

Most radioactive elements in the terrestrial worlds are chemically bound in rock, not metal. Based on this fact, should Mercury have had more or less heat from radioactive decay than the Moon? Explain.

The combination of the three heat sources explains how the terrestrial interiors became so hot that they melted, allowing them to differentiate into their core-mantle-crust structures. The many violent impacts that occurred during the latter stages of accretion deposited so much energy that the outer layers of the young planets began to melt. At that point, heat from radioactive decay and differentiation (which began as soon as dense materials started to sink through the outer layers) ensured that the interiors would melt throughout.

How Interiors Cool Off: The Role of Planetary Size

The interiors of all the terrestrial worlds started out quite hot, but they have been cooling ever since. Just as heat from the inside of a hot potato gradually leaks outward, heat gradually leaks outward from planetary interiors. Once the heat reaches the surface, the planet radiates it away to space. Remember that all objects emit *thermal radiation* characteristic of their temperatures [Section 6.4]. Because of their relatively low surface temperatures, planets radiate almost entirely in the infrared portion of the spectrum.

The single most important factor in how long it takes a planet to cool is its size: Larger planets stay hot longer. You can see why size is the critical factor by picturing a large planet as a smaller planet wrapped in extra layers of rock. The extra rock acts as insulation, so it takes much longer for interior heat to reach the surface. If you add the fact that a larger world contains more heat in the first place, it becomes clear that a large world will take much longer to cool than a small world. The relatively small size of the Moon and Mercury probably allowed their interiors to cool within a billion years or so after their formation, while Earth's active volcanoes prove that our planet's interior still remains quite hot.

THINK ABOUT IT

Would you expect the interior temperature of Venus to be more similar to that of the Moon or to that of Earth? What about the interior temperature of Mars? Explain.

Mantle Convection and Lithospheric Thickness

In planets that are still hot inside, such as Earth, the most important connection between internal heat and geological activity is the movement of rock within the mantle. Hot rock gradually rises within the mantle, slowly cooling as it rises. Cooler rock at the top of the mantle gradually falls. The process in which hot material expands and rises while cooler material contracts and falls is called **convection**. It can occur whenever there is strong heating from below. For example, you can see convection in a pot of soup on a hot burner. You may also be familiar with convection in weather—warm air near the ground tends to rise while cool air above tends to fall.

The ongoing process of convection creates individual **convection cells** within the mantle. Figure 10.5 shows a few mantle convection cells as small circles, with arrows indicating the direction of flow. Keep in mind that mantle convection primarily involves solid rock, not molten rock. Because solid rock flows quite slowly, mantle convection is a very slow process. At the typical rate of mantle convection on Earth, it would take about 100 million years for a piece of rock to be carried from the base of the mantle to the top.

Mantle convection carries hot rock upward, but not all the way to the surface. The convection stops at the base of the lithosphere, where the rock is too strong to flow as readily as in the lower mantle. In fact, mantle convection is closely tied to lithospheric thickness. As a planet's interior gradually cools, the rigid lithosphere grows thicker and convection occurs only deeper inside the planet (or stops entirely). You can see how this has affected the terrestrial worlds by looking again at Figure 10.2. Earth, which remains quite hot inside, has a thin lithosphere to go with its strong mantle convection. Venus is probably quite similar

---

Mathematical Insight **10.1** **The Surface Area–to–Volume Ratio**

We can see why large planets cool more slowly than smaller ones by thinking about their relative surface areas and volumes. Consider Earth and the Moon. Both started out hot inside but continually radiate heat away from their surfaces to space. As heat escapes from the surface, more heat flows upward from the interior to replace it. This process will continue until the interior is no hotter than the surface. Because all the heat escapes from the surface, the key factor in a world's heat-loss rate is its *surface area*.

The surface area of a sphere is given by the formula $4\pi \times (\text{radius})^2$, so the ratio of Earth's surface area to that of the Moon is:

$$\frac{\text{Earth surface area}}{\text{Moon surface area}} = \frac{4\pi \times (r_{\text{Earth}})^2}{4\pi \times (r_{\text{Moon}})^2}$$

$$= \left(\frac{r_{\text{Earth}}}{r_{\text{Moon}}}\right)^2 = \left(\frac{6{,}378 \text{ km}}{1{,}738 \text{ km}}\right)^2 \approx 13$$

Thus, Earth has about 13 times more area from which to radiate its heat away.

However, the total amount of heat deposited inside a world by radioactive decay roughly depends on its mass or, if we assume that we're dealing with worlds of similar density, its volume. The formula for the volume of a sphere is $\frac{4}{3}\pi \times (\text{radius})^3$, so the volume ratio for Earth and the Moon is:

$$\frac{\text{Earth volume}}{\text{Moon volume}} = \frac{\frac{4}{3}\pi \times (r_{\text{Earth}})^3}{\frac{4}{3}\pi \times (r_{\text{Moon}})^3}$$

$$= \left(\frac{r_{\text{Earth}}}{r_{\text{Moon}}}\right)^3 = \left(\frac{6{,}378 \text{ km}}{1{,}738 \text{ km}}\right)^3 \approx 50$$

Thus, Earth's interior contains some 50 times more heat than the Moon's interior. This larger amount of internal heat more than offsets the larger radiating area. That is why Earth is still hot inside while the Moon has mostly cooled off.

The general idea that larger objects cool more slowly than smaller ones is described by the mathematical idea of the *surface area–to–volume ratio*. As the Earth–Moon example shows, larger objects have relatively smaller surface areas compared to their volumes. Thus, they lose internal heat more slowly and also take longer to be heated from the outside. This principle explains many everyday phenomena. For example, crushing a cube of ice into smaller pieces increases the total amount of surface area, while the total volume of ice remains the same. Thus, the crushed ice will cool a drink more quickly than would the original ice cube.

1. Convection
Hot rock rises and cooler rock falls in a mantle convection cell.

2. Conduction
After convection brings heat to the base of the lithosphere, conduction carries heat through the rigid lithosphere to the surface.

3. Radiation
At the surface, energy is radiated into space.

Figure 10.5 In a planet that is still hot inside, the mantle can undergo convection. Individual convection cells are regions in which hot rock gradually rises upward while cool rock gradually falls. Arrows indicate the direction of flow in the convection cells. Note that convection stops at the base of the lithosphere. Heat moves through the lithosphere by conduction and is then radiated away into space.

to Earth in this respect, though it may have a thicker lithosphere (for reasons we will discuss later). With their thick lithospheres, Mercury and the Moon have no mantle convection. Mars, intermediate in size, has cooled significantly but probably retains some internal heat. Whether Mars has any remaining convection in its lower mantle remains an open question.

THINK ABOUT IT

In general, what kind of planet would you expect to have the *thickest* lithosphere: the largest planet, the smallest planet, the planet closest to the Sun, or the planet farthest from the Sun? Why?

From the base of the lithosphere to the surface, heat continues upward through a process called *conduction*. Conduction is the same process by which a hot potato transfers its heat to your cooler hand when you pick it up. It occurs through the microscopic jiggling of many individual atoms or molecules. The faster the jiggling, the more thermal energy [Section 4.2]. The jiggling molecules bump into one another, transferring thermal energy from the faster-moving molecules deeper inside the planet to the slower-moving ones nearer the surface. When the heat reaches the surface, the planet radiates the thermal energy away into space.

Planetary Cores and Magnetic Fields

Although we have discussed everything we need for our study of surface geology, there is another planetary feature affected by interior structure: the presence or absence of a **magnetic field**. You are probably familiar with the general pattern of the magnetic field created by an iron bar (Figure 10.6a). A planet's magnetic field is generated by a process more similar to that of an *electromagnet*, in which

the magnetic field arises as a battery forces charged particles (electrons) to move along a coiled wire (Figure 10.6b).

Planets do not have batteries, but they may have charged particles in motion in their metallic cores (Figure 10.6c). In the Earth, for example, molten metals rise and fall in convection cells in the liquid outer core. At the same time, the molten material spins each day with Earth's rotation. The result is that electrons in the molten metals move within Earth's outer core in much the same way they move in an electromagnet, generating Earth's magnetic field.

We can generalize these ideas to other worlds. There are three basic requirements for a global magnetic field:

1. An interior region of electrically conducting fluid (liquid or gas), such as molten metal.

2. Convection in that layer of fluid.

3. At least moderately rapid rotation.

Earth is unique among the terrestrial worlds in meeting all three requirements, which is why it is the only terrestrial world with a strong magnetic field. The Moon has no magnetic field, either because it lacks a metallic core altogether or because its core has long since solidified and ceased convecting. Probably because of similar core solidification, Mars also has virtually no magnetic field today—though it probably had one in the distant past [Section 11.6]. Venus probably has a molten metal layer much like that of Earth, but either its convection or its 243-day rotation period is too slow to generate a magnetic field. Mercury remains an enigma: It possesses a measurable magnetic field despite its small size and slow, 59-day rotation. The reason may be Mercury's huge metal core, which may still be partly molten and convecting.

The same three requirements for a magnetic field also apply to jovian planets and stars. For example, Jupiter's strong magnetic field comes from its rapid rotation and its layer of convecting, metallic hydrogen that conducts electricity [Section 12.4]. The Sun's magnetic field is generated by the combination of convection of ionized gas (plasma) in its interior and rotation.

THINK ABOUT IT

Recall that the Sun had a strong solar wind when it was young (which cleared out the remaining gas of the solar nebula [Section 9.3]) because its magnetic field was much stronger than it is today. What single factor explains why the young Sun had a much stronger magnetic field than the Sun today?

Magnetic fields can be very useful. On Earth, they help us navigate. Their presence or absence and their strength on any planet provide important information about the planetary interior. Aside from this utility, magnetic fields have virtually no effect on the structure of a planet itself. They can, however, form a protective *magnetosphere* surrounding a planet, with potentially profound effects on a planet's atmosphere and any inhabitants [Section 11.3].

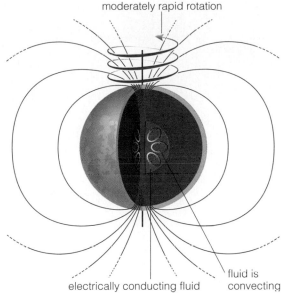

moderately rapid rotation

electrically conducting fluid

fluid is convecting

a This photo shows how a bar magnet influences iron filings (small black specks) around it. The *magnetic field lines* (red) represent this influence graphically.

b A similar magnetic field is created by an electromagnet, which is essentially a coiled wire attached to a battery. The field is created by the battery-forced motion of charged particles (electrons) along the wire.

c A planet's magnetic field also arises from the motion of charged particles. The charged particles in a terrestrial planet are in a molten metallic core, and their motion arises from the planet's rotation and interior convection.

Figure 10.6 Sources of magnetic fields.

Shaping Planetary Surfaces Tutorial, Lessons 1–3

10.3 Shaping Planetary Surfaces: The Four Basic Geological Processes

We have discussed how the terrestrial planets work on the inside. We are now ready to turn to their surface features. When we look around the Earth, we find an apparently endless variety of geological surface features. The diversity increases when we survey the other planets. But on closer examination, geologists have found that almost all the observed surface features result from just four major geological processes:

- **Impact cratering:** the blasting of bowl-shaped *impact craters* by asteroids or comets striking a planet's surface.

- **Volcanism:** the eruption of molten rock, or *lava*, from a planet's interior onto its surface.

- **Tectonics:** the disruption of a planet's surface by internal stresses.

- **Erosion:** the wearing down or building up of geological features by wind, water, ice, and other phenomena of planetary weather.

In this section, we'll examine each of these four processes and discover that they can be traced back to fundamental planetary properties. We'll begin our discussion of the geological processes by focusing on impact cratering and what it can tell us about a planetary surface.

Impact Cratering

The scarred faces of the Moon and Mercury (see Figure 10.1) attest to the battering that the terrestrial worlds have taken from leftover planetesimals, such as comets and asteroids. While the Moon and Mercury bear the most obvious scars, all the terrestrial worlds have suffered similarly.

The Impact Process Impacting objects, or *impactors*, typically hit a planet at speeds between 30,000 and 250,000 km/hr (10–70 km/s). At such tremendous speeds, impactors pack enough energy to vaporize solid rock and excavate a *crater* (the Greek word for "cup"). We have never witnessed a major impact on one of the terrestrial worlds (though we have witnessed one on Jupiter [Section 13.6]), but laboratory experiments have studied the impact process. Figure 10.7 illustrates what happens when an impactor hits a rocky planetary surface, and Figure 10.8 shows a prototypical impact crater. A large crater may have a central peak, which forms when the center rebounds after impact (in much the same way water rebounds after you drop a pebble into it).

Craters are generally circular because an impact blasts out material in all directions, no matter the direction of the incoming impactor. Craters are generally about 10 times wider than the impactors that created them and about 10–20% as deep as they are wide. For example, a 1-kilometer-wide impactor creates a crater about 10 kilometers wide and 1–2 kilometers deep. Debris from the blast shoots high above the surface and then rains down over a large area. If the impact is large enough, some of the ejected material can completely escape from the planet.

Figure 10.7 Artist's conception of the impact process.

Figure 10.8 This photo shows a crater, named Tycho, on the Moon. Note the classic shape and central peak. In large craters such as this one, the center can rebound just as water does after you drop a pebble into it.

Crater Crowding Reveals Surface Ages Photographs of the Moon immediately reveal two important features of impact cratering (Figure 10.9). First, small craters far outnumber large ones. This observation is simply explained: There are many more small asteroids and comets orbiting the Sun than large ones. Second, the degree of crowding among craters can vary greatly from place to place. In the regions on the Moon known as the **lunar highlands**, craters are so crowded that we see craters on top of other craters. In the regions known as the **lunar maria** (which were themselves once large craters), we see only a few craters on top of a generally smooth surface. Indeed, the maria got their name because they look much like oceans when seen from afar. *Maria* (singular, *mare*) is Latin for "seas."

Why are different parts of the Moon so different in their crater crowding? Might some parts of the Moon have

escaped heavy cratering? On the contrary, the whole Moon must have been densely covered with craters during the period of heavy bombardment, early in the solar system's history [Section 9.4]. Some process must have erased these craters from the maria after the heavy bombardment ended. As we'll discuss further in Section 10.4, the maria were some of the largest, deepest craters that formed during the heavy bombardment and must have been filled with many smaller craters from later impacts. However, lava later welled up and flooded the maria, covering all the craters within them.

Radiometric dating of rocks from the different regions confirms these ideas. Rocks from the heavily cratered lunar highlands date back as far as 4.4 billion years. Rocks from the maria are hundreds of millions of years younger, dating to between 3.0–3.9 billion years ago. Because the maria contain only about 3% as many craters as the highlands even though they are just a few hundred million years younger, the heavy bombardment must have subsided quite early in the solar system's history. Compared to the period of heavy bombardment, relatively few impacts have occurred in the last 4 billion years.

More precise study of lunar cratering has allowed scientists to reconstruct the rate of impacts during much of the Moon's history. Because impacts are essentially random events, the same changes in impact rate over time must apply to all the terrestrial worlds. The degree of crater crowding on any planetary surface therefore provides a good measure of the age of that surface. A surface that is crowded with craters must have remained essentially undisturbed since the end of the heavy bombardment. A surface with few craters must have had its ancient craters erased by more recent geological processes.

Lunar maria are huge impact basins that were flooded by lava. Only a few small craters appear on the maria.

Lunar highlands are ancient and heavily cratered.

b This flat map shows the entire surface of the Moon in the same way a flat map of Earth represents the entire globe. The lunar highlands are heavily cratered. Radiometric dating shows that these portions of the lunar surface are quite ancient, with some highland rocks dating to over 4.4 billion years ago. The darkly colored lunar maria are about a half billion or more years younger. Their relatively few craters tell us that the impact rate had dropped dramatically by the time they formed.

a This Apollo photograph of the Moon shows some areas to be much more heavily cratered than others. This view of the Moon is not the one we see from Earth.

Figure 10.9 These photos of the Moon show us that crater crowding is closely related to surface age.

VIS

For example, Earth must have been battered by at least as many impacts as the Moon, but relatively few impact craters remain to be seen because our planet's surface is constantly being reshaped. Thus, Earth's surface is quite young compared to the age of the solar system. In fact, Earth has probably been struck by more impacts and more violent impacts than the Moon, because Earth's stronger gravity would tend to draw in more impactors and accelerate them to higher speeds by the time they reach the surface.

THINK ABOUT IT

Mercury is heavily cratered, much like the lunar highlands. Venus has relatively few craters. Which planet has the older surface? Is there any difference in the ages of the two planets as a whole? Explain how the surfaces of Mercury and Venus can have different ages while the planets themselves are the same age.

Crater Shapes Reveal Surface Conditions and History
Although impact craters are almost always circular, details of their shapes can provide important information about geological conditions on a planetary surface. To illustrate how, Figure 10.10 contrasts three craters found on Mars. The crater in Figure 10.10a has a fairly simple bowl shape, suggesting that it formed just as we expect from the basic cratering process as the impactor slammed into the rocky surface of Mars. The crater in Figure 10.10b also has a fairly

"standard" shape, but the extra large bump in its center and the ripples that extend beyond its rim suggest that it was made in mud. The impact apparently vaporized underground water or ice, lubricating the flow of ejected material away from the crater. The shape of the crater in Figure 10.10c shows obvious signs of erosion. It lacks a sharp rim, and its floor no longer has the classic bowl shape. In this case, the erosion may be the result of ancient rainfall on Mars, and the sculpted crater bottom probably was once a lake. Studies of crater shapes on other worlds provide similar clues to their surface conditions and history.

Planetary Properties and Impact Craters Of the four geological processes, impact cratering is the most widespread among the terrestrial worlds and the least dependent on planetary properties. All five terrestrial worlds were densely cratered during the heavy bombardment, and planetary properties had little effect on the cratering rate (aside from the fact that the stronger gravity of larger worlds tended to pull in more impactors). The present-day abundance of impact craters, however, is determined by planetary properties. Worlds with many craters, like the Moon and Mercury, must have had little geological activity since the solar system was young. Worlds with few craters, like Venus and Earth, must have had their ancient craters erased by other geological processes. As we'll see shortly, these other processes depend heavily on a planet's fundamental properties.

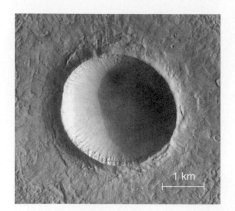

a Many craters are bowl-shaped.

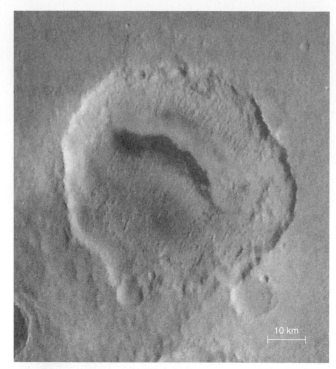

c Ancient Martian rains apparently eroded this crater. (See Figure 10.28b for a stunning close-up view of the floor of this crater.)

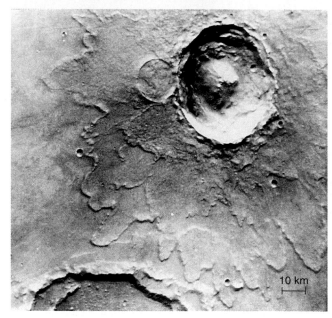

b Impacts into icy ground may result in the ejection of muddy debris.

Figure 10.10 Crater shapes on Mars tell us about Martian geology. These photos were taken from orbit by the Viking spacecraft. Crater shapes reveal the surface history and surface conditions of a planet.

Volcanism

The second major geological process is volcanism. On Earth, we generally associate volcanism with the mountains we call volcanoes. In the broader context of planetary geology, we use the term *volcanism* to refer to any eruption of molten lava onto the surface, whether it comes from a tall volcano or simply rises through a crack in a planet's lithosphere.

Causes of Eruptions Volcanism occurs when underground molten rock, technically called *magma,* finds a path through the lithosphere to the surface (Figure 10.11). Magma rises for two main reasons. First, molten rock is generally less dense than solid rock, so it has a natural tendency to rise. Second, a *magma chamber* may be squeezed by tectonic forces, driving the magma upward under pres-

sure. Any trapped gases expand as magma rises, sometimes leading to dramatic eruptions. The same molten rock that is called *magma* when it is underground is called *lava* once it erupts onto the surface.

Lava Viscosity and Volcanic Flows The result of an eruption depends on how easily the molten lava flows across the surface. The term **viscosity** describes the "thickness" of a liquid. For example, liquid water has a low viscosity that allows it to flow quickly and easily, while thick and slow-flowing honey and molasses have much higher viscosities.

Different types of molten rock have different viscosities. One of the most common types of rock in volcanic eruptions is **basalt**. A mixture of many different minerals that tend to melt at relatively low temperatures, basalt can become molten even while surrounding rock remains solid. Basalt is relatively high in density but relatively low in viscosity, primarily because it is made of relatively short molecular chains that don't tangle with each other and therefore can flow easily.

The viscosity of basalt (or any other type of molten rock) also depends on its temperature. The hotter the lava, the more easily the chains of molecules can jiggle and slide along one another and the lower the viscosity. Water and trapped gases also affect viscosity. While the molten rock is underground as magma, water and trapped gases can act as lubricants, decreasing the magma's viscosity. When the molten rock erupts as lava, however, vaporized water and other gases may form bubbles that *increase* the viscosity. You can see how bubbles increase viscosity by playing with bubmagma in upper mantle

a Hot magma erupts to the surface.

b Eruption of an active volcano on the flanks of Kilauea on the Big Island in Hawaii.

Figure 10.11 Volcanism is the process by which underground molten rock finds its way to a planetary surface. Volcanism requires substantial internal heat.

ble bath. When it is still bubbly, you can build a tall "mountain" with the foam, indicating its relatively high viscosity. When the bubbles are gone, the viscosity is more like that of plain water, and you cannot build much with it at all.

Types of Volcanic Flow Broadly speaking, lava of different viscosities can shape three different types of volcanic flow:

- The runniest basalt lavas flow far and flatten out before solidifying, creating vast **volcanic plains** such as the lunar maria (Figure 10.12a).

- Somewhat more viscous basalt lavas solidify before they can completely spread out, resulting in **shield volcanoes** (so named because of their shape). Shield volcanoes can be very tall but are not very steep. Examples include the mountains of the Hawaiian islands on Earth and Olympus Mons on Mars (Figure 10.12b).

- The most viscous lavas can't flow very far before solidifying and therefore build up tall, steep **stratovolcanoes**. Examples of stratovolcanoes on Earth include Mount Fuji in Japan, Mount Kilimanjaro in Tanzania, Mount Shasta in California, Mount Hood in Oregon, Mount Rainier and Mount St. Helens in Washington State, and Mount Etna in Sicily (Figure 10.12c).

Outgassing From our standpoint as living beings, the most important effect of volcanism is how it has made Earth's oceans and atmosphere possible. None of the terrestrial worlds were able to capture significant amounts of gas from the solar nebula. The young terrestrial worlds therefore contained water and gas only as a result of impacts. This water and gas was trapped in their interiors in much the same way the gas in a carbonated beverage is trapped in a pressurized bottle. Volcanism releases this gas. When molten rock erupts onto the surface as lava, the release of pressure expels the trapped gases in a process we call **outgassing**. Outgassing can be either quite dramatic, during a volcanic eruption (Figure 10.13a), or more gradual, when gas escapes from volcanic vents (Figure 10.13b). Virtually all the gas that made the atmospheres of Venus, Earth, and Mars—and the water vapor that rained down to form Earth's oceans—originally was released from the planetary interiors by outgassing.

Planetary Properties and Volcanism We can trace terrestrial volcanism directly to one major factor: the planet's size. Volcanism requires substantial interior heat. All the terrestrial worlds probably had some degree of volcanism when their interiors were young and hot. As the interior cools, however, volcanism subsides. The Moon and Mercury lack volcanism today because their small size allowed their interiors to cool long ago. Earth has active volcanism because it is large enough to still have a hot interior. Venus, nearly the same size as Earth, must also still be hot inside and probably also has active volcanism, though we lack conclusive evidence of eruptions in the past few tens of millions of years. Mars, if it still has any active volcanism at all, must have it at a much lower level than in the distant past.

THINK ABOUT IT

How does the connection between volcanism and outgassing explain the lack of a substantial atmosphere on Mercury and the Moon?

Lava plains (maria) on
the Moon

a Low-viscosity lava makes flat lava
plains. The long, winding channel
near the upper left in the photo was
made by a river of molten lava.

Olympus Mons (Mars)

b Medium-viscosity lava makes
shallow-sloped shield volcanoes.

Mount Hood (Earth)

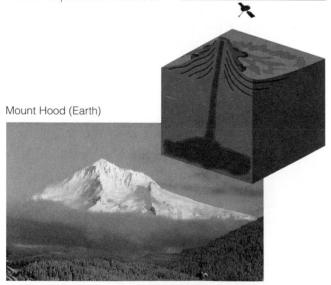

c High-viscosity lava makes steep-sloped stratovolcanoes.

Figure 10.12 Volcanoes produce different types of features
depending primarily on the viscosity of the lava erupted.

a This photo shows the eruption of Mount St. Helens (in Washington State) on May 18, 1980. Note the tremendous amount of
outgassing that accompanies the eruption.

b An eruption in Volcanoes National Park, Hawaii, exhibits more
gradual outgassing from a volcanic vent.

Figure 10.13 Examples of outgassing, which released the gases
that ultimately made the atmospheres of Venus, Earth, and Mars.

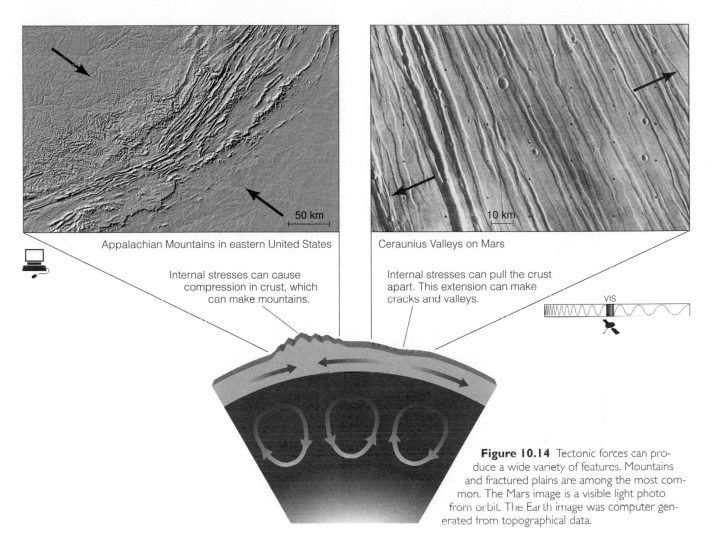

Appalachian Mountains in eastern United States

Internal stresses can cause compression in crust, which can make mountains.

Ceraunius Valleys on Mars

Internal stresses can pull the crust apart. This extension can make cracks and valleys.

Figure 10.14 Tectonic forces can produce a wide variety of features. Mountains and fractured plains are among the most common. The Mars image is a visible light photo from orbit. The Earth image was computer generated from topographical data.

Tectonics

We are now ready to discuss tectonics, the third major geological process. The root of the word *tectonics* comes from Greek legend, in which Tecton was a carpenter. In geology, tectonics refers to "carpentry" performed on planetary surfaces—that is, the action of internal forces and stresses on the lithosphere to create surface features.

Tectonics and Mantle Convection Ongoing tectonics can occur only on planets with mantle convection, such as Earth and Venus today and Mars in the past. Several types of internal stresses can drive tectonic activity (Figure 10.14). For example, the lithosphere can be compressed in places where adjacent convection cells are pushing the underlying mantle rock together. Such crustal compression helped create the Appalachian Mountains of the eastern United States. In other places, mantle convection may tend to pull apart the crust, making cracks and valleys. Examples of such cracks and valleys include the Guinevere Plains on Venus, the Ceraunius Valleys on Mars, and New Mexico's Rio Grande Valley. (The river named the Rio Grande came *after* the valley formed from tectonic processes.) Tectonic stresses can also occur on a more local scale. For example, the weight of a

newly formed volcano can bend or crack the lithosphere beneath it, and a plume of hot material rising from the deep interior can push up on the lithosphere to create a bulge.

On Earth, the ongoing stress of mantle convection ultimately fractured Earth's lithosphere into more than a dozen pieces, or *plates*. These plates move over, under, and around each other in a process we call **plate tectonics**. The movements of the plates explain nearly all of Earth's major geological features, including the arrangement of the continents, the nature of the seafloor, and the origin of earthquakes. Among the terrestrial worlds in our solar system, plate tectonics appears to be unique to Earth. We will therefore save further discussion of plate tectonics for Chapter 14, where we will study how our home planet is different from other terrestrial worlds.

Planetary Properties and Tectonics Tectonic activity usually goes hand in hand with volcanism, because both require internal heat and are ultimately traceable to planetary size. The most common tectonic features are tied to mantle convection, which can occur only on worlds that are large enough to remain hot inside. Thus, tectonic activity on both Venus and Earth occurs because of their relatively large sizes. We are not sure whether Mars has any ongoing mantle convection to support tectonics. On worlds that have cooled, such as the Moon and Mercury, the era of tectonic activity is long past.

The degree of tectonic activity depends on both the strength of mantle convection and the strength and thickness of the lithosphere. Stronger mantle convection tends to lead to greater tectonic activity, while a strong and thick lithosphere is better able to withstand internal stress and thus resists tectonic activity. As we will discuss in more detail later, differences in lithospheric thickness may explain the differences in tectonic activity on Venus and Earth.

Erosion

The last in our list of four major geological processes is erosion. *Erosion* is a blanket term for a variety of processes that break down or transport rock through the action of ice, liquid, or gas. The shaping of valleys by glaciers (ice), the carving of canyons by rivers (liquid), and the shifting of sand dunes by wind (gas) are all examples of erosion.

Although we often associate erosion with the breakdown of existing features, erosion also builds geological features. Sand dunes, river deltas, and lake bed deposits are all examples of features built by erosion. Indeed, much of the surface rock on Earth was built by erosion. Over long periods of time, erosion has piled sediments into layers on the floors of oceans and seas, forming what we call **sedimentary rock**. Thus, the gorgeous layered rock of the Grand Canyon and other river-carved valleys was itself built by erosion. Figure 10.15 shows several examples of erosion on Earth.

Planetary Properties and Erosion The presence or absence of erosion can also be traced back to fundamental planetary properties, although the linkage is a little more complex than that for volcanism and tectonics. Erosion can occur only on a planet with a significant atmosphere. Otherwise, there can be no surface liquids, no wind, and no climate change to cause glacial movements. An atmosphere is not the only requirement for erosion, however. A stationary atmosphere that simply bears down without stirring the surface would not cause any erosion. Erosion comes from processes such as wind and rain, which we generally refer to as "weather." Thus, the basic requirements for erosion are the existence of an atmosphere and significant weather affecting the surface.

Among the terrestrial worlds, we can trace the requirements for erosion to three fundamental planetary properties:

● *Planetary size.* Size is important because a terrestrial world can have an atmosphere only if it has had significant outgassing, and outgassing is part of volcanism. In addition, size determines the strength of gravity, and atmospheric gases are more easily lost to space on worlds with weaker gravity.

● *Distance from the Sun.* Distance affects atmospheric temperature, which in turn affects whether gas can

remain in an atmosphere [Section 11.5]. If all else is equal (such as planetary size), the higher temperatures on a world closer to the Sun will make it much easier for atmospheric gases to escape to space. Conversely, the colder temperatures on a world farther from the Sun may cause atmospheric gases to freeze out. Either extreme can make an atmosphere too thin to lead to much erosion.

● *Rotation rate.* The primary driver of winds and other weather on a planet with an atmosphere is its rotation rate: Faster rotation means stronger winds and storms [Section 11.4].

The Moon and Mercury lack significant atmospheres because they lack outgassing today, and any atmospheric gases they had in the distant past have been lost to space. Mars has a very thin atmosphere because much of the water vapor and carbon dioxide outgassed in its past lies frozen in its polar caps or beneath the surface. Venus and Earth probably had similar amounts of outgassing, but cooler temperatures on Earth led to condensation and the ultimate removal of most of this gas from Earth's atmosphere. In contrast, most of Venus's gas remained in its atmosphere, so its atmosphere is much thicker than Earth's. However, Venus has little erosion because its slow rotation gives it virtually no wind on its surface.

Putting It All Together

We've focused on the four geological processes individually, but we've also seen that they all interact with one another to shape planetary surfaces. For example, volcanism produces lava flows that can erase old impact craters, and outgassing can create atmospheres and lead to erosion.

More important, we've seen that all the geological processes ultimately are connected to fundamental planetary properties. Size is by far the most important property, because it determines whether a terrestrial world has active volcanism and tectonics. We've also seen that distance from the Sun and rotation rate play a role in erosion. Figure 10.16 summarizes the key connections that we have discussed between fundamental properties and geological processes.

THINK ABOUT IT

Imagine that Earth was only as large as Mercury but still had the same distance from the Sun and the same rotation rate. Would there be erosion on Earth today? Why or why not?

●

10.4 A Geological Tour: The Moon and Mercury

Now that we've examined the processes that shape the geology of the terrestrial worlds and their fundamental connections to planetary properties, we're ready to explain how

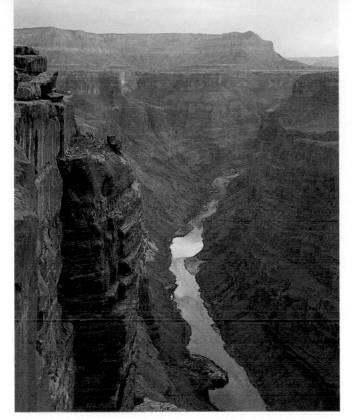

a The Colorado River continues to carve the Grand Canyon after millions of years.

b Glaciers created Yosemite Valley during ice ages.

d Erosional debris also creates geological features, as seen in this river delta.

c Wind erosion wears away rocks and builds up sand dunes, such as these in a California desert.

Figure 10.15 A few examples of water and wind erosion on Earth.

each world ended up as it did. In this and the next two sections, we'll take a brief geological "tour" of the terrestrial worlds. We could organize the tour by distance from the Sun, by density, or even by alphabetical order, but we'll choose the planetary property that has the strongest effect on geology: size, which controls a planet's internal heat. Proceeding from smallest to largest, we begin with the Moon and Mercury (Figure 10.17).

The Moon

On a clear night, you can see much of the Moon's global geological history with your naked eye. The Moon is unique in this respect. Other planets are too far away for us to see any surface details, and from the ground we can see only local features on Earth.

The Moon has long since lost most of its internal heat, so it has no active volcanism or tectonics. It has no atmosphere and therefore no erosion. Impacts have clearly played an important geological role, but the vast majority of the craters on the Moon formed billions of years ago during

The Role of Planetary Size

Small Terrestrial Planets

cold, solid interior

tectonic and volcanic activity end after a billion years or so

many ancient craters

little outgassing, atmosphere lost due to low gravity

no atmosphere, therefore no erosion

Large Terrestrial Planets

warm, convecting interior

ongoing tectonic and volcanic activity

most ancient craters erased

lots of outgassing

strong gravity retains atmosphere, erosion possible

The Role of Distance from the Sun

Planets Close to the Sun

warm surfaces

no oceans, rain, snow, or ice

less erosion

hot atmosphere leads to more atmospheric escape

Sun

Planets at Intermediate Distances from the Sun

surfaces at moderate temperatures

oceans, rain, snow, and ice

more erosion

Planets Far from the Sun

cool surfaces

ice and snow, but no rain or oceans

less erosion

The Role of Planetary Rotation

Slow Rotation

less wind

less weather

less erosion

Rapid Rotation

more wind

more weather

more erosion

Figure 10.16 A planet's fundamental properties of size, distance from the Sun, and rotation rate determine its geological history. This illustration shows the role of each key property separately. A planet's overall geological evolution depends on the combination of all the effects shown above. Note also that a planet's size and rotation affect the likelihood of a magnetic field (not shown in this diagram).

a The Moon

b Mercury

Figure 10.17 The Moon and Mercury, shown to scale. The Mercury photo is a composite of photos from the *Mariner 10* spacecraft [Section 8.3], which photographed only parts of the surface.

the heavy bombardment. However, the Moon apparently did have some volcanism and tectonics in the distant past, when its interior was still hot.

Lunar Volcanism and Tectonics The most noticeable features of volcanism are the lunar maria (see Figure 10.9). The relatively few tectonic features on the Moon are also found within the maria. These features, which look much like surface wrinkles, probably were created by small-scale tectonic stresses that arose when the lava that flooded the maria cooled and contracted (Figure 10.18).

To understand how the maria formed, we must look far back to the Moon's youth. The Moon's presumed formation in a giant impact [Section 9.4] makes its birth quite different from that of the other terrestrial worlds, but it still had the same three internal heat sources: accretion (from the material blasted away from Earth in the giant impact), differentiation, and radioactive decay. As the Moon's interior underwent differentiation, the lowest-density molten rock rose to the surface. There it formed what is sometimes called a *magma ocean*, because it would have looked like a

Figure 10.18 The features that look like wrinkles within maria like this one (Mare Imbrium) arose from tectonic stresses associated with the cooling and contracting of the lava that flooded the maria between 3 and 4 billion years ago.

glowing hot ocean of molten rock. The magma ocean cooled and solidified quite rapidly. This explains why the impact craters of the heavy bombardment still show over much of the Moon's surface.

The largest impacts of the heavy bombardment violently fractured the Moon's lithosphere beneath the huge craters they created—craters that would later become the maria. The maria did not flood with lava immediately, because the mantle contained no molten rock. The heat of accretion and differentiation had already leaked away. For a while, however, radioactive decay gradually reheated the interior. About 3–4 billion years ago, mantle material remelted, and molten rock welled up through the cracks in the lithosphere, flooding the maria with lava. The individual maria are fairly circular because they are essentially flooded craters (and craters are almost always round). Their dark color is from the dense, iron-rich rock that rose up from the lunar mantle as molten lava. Figure 10.19 summarizes how the lunar maria formed.

The lunar lavas that filled the maria must have been among the runniest (least viscous) in the solar system. The lava plains of the maria cover a large fraction of the Moon's surface, indicating that the lava spread easily and far. The low viscosity of the lunar lava may have been a result of the Moon's formation. The giant impact would have vaporized water and released other trapped gases, so these materials would not have become part of the Moon when it reaccreted from the giant impact debris. With no trapped gases, the Moon's lava would have lacked the gas bubbles that increase lava viscosity on Earth and other worlds. The lunar lava was so runny that in places it flowed like rivers of molten rock, carving out long, winding channels (see Figure 10.12a).

The Moon Today: Rare Impacts and Little More

The Moon's era of geological activity is long gone. Today, the Moon is a desolate and nearly unchanging place. Occasional, rare impacts may occur in the future, but we are unlikely ever to witness one. Little happens on the Moon, aside from the occasional visit of robotic spacecraft or astronauts from Earth (Figure 10.20a).

The only ongoing geological change on the Moon is a very slow "sandblasting" of the surface by **micrometeorites**, sand-size particles from space. These tiny particles burn up as meteors in the atmospheres of Earth, Venus, and Mars but rain directly onto the surface of the airless Moon. The micrometeorites gradually pulverize the surface rock, which explains why the lunar surface is covered by a thin layer of powdery "soil." The Apollo astronauts left their footprints in this powdery surface (Figure 10.20b). Pulverization by micrometeorites is a very slow process, and the astronauts' footprints will last millions of years before they are finally erased. Over millions and billions of years, the micrometeorite impacts can smooth out rough crater rims.

Aside from the micrometeorite sandblasting and rare larger impacts, the Moon has been "geologically dead" since the maria formed more than 3 billion years ago. Nevertheless, the Moon remains a prime target of future exploration. Further studies of lunar geology will help us understand the history of cratering in the solar system and the processes that shaped the worlds early in their histories. The Moon also offers an ideal location for an astronomical observatory [Section 7.5]. Someday, perhaps in the not too distant future, humans will build a lunar colony, with living quarters underground for protection against dangerous solar radiation.

a Just over 4 billion years ago, shortly after the Moon formed, it was completely covered in craters. The largest impacts left deep craters and fractured the lithosphere.

b Between 4 and 3 billion years ago, radioactive heating produced very "runny" lava which erupted through the fractured lithosphere to fill the deepest craters, forming the lunar maria.

c Since the interior cooled about 3 billion years ago, no more eruptions have occurred. Impact cratering (which still occurs everywhere at low levels) has affected the appearance of the maria, but the maria are still only lightly cratered.

Figure 10.19 Geological evolution of the Moon's surface.

b The Apollo astronauts left clear footprints, like this one, in the Moon's powdery "soil." The powder is the result of gradual pulverization of surface rock by micrometeorites. Micrometeorites will eventually erase the astronauts' footprints, but not for millions of years.

a Astronaut Gene Cernan takes the Lunar Roving Vehicle for a spin during the final Apollo mission to the Moon (Apollo 17, December 1972).

Figure 10.20 Little ever happens on the Moon today, aside from occasional visits by beings or robots from Earth.

Mercury

Mercury looks so much like the Moon that it's often difficult to tell which world you are looking at in surface photos. The similarities extend to all the geological processes, though there are also a few differences. Mercury's closeness to the Sun and slow rotation make it a world of extremes. The combination of its 88-day orbit and its 59-day rotation gives Mercury days and nights that last about 3 Earth months each [Section 8.1]. As the Sun rises above the horizon for 3 months of daylight, the equatorial surface temperature soars to 425°C. At night or in shadow, the temperature falls below −150°C.

Impact Cratering on Mercury Impact craters are visible almost everywhere on Mercury, indicating an ancient surface (see Figure 10.17). However, Mercury's craters are less crowded together than the craters of the lunar highlands, suggesting that many of the craters that formed on Mercury during the heavy bombardment were later covered up by molten lava (Figure 10.21a).

The largest single surface feature on Mercury is a huge impact crater called the *Caloris Basin* (Figure 10.21b). The Caloris Basin spans more than half of Mercury's radius, and its multiple rings bear witness to the violent impact that created it. The impact must have reverberated through-

out the planet—we see evidence of violent surface shaking on the precise opposite side of Mercury from the Caloris Basin (Figure 10.21c). The Caloris Basin has few craters within it, indicating that it must have formed at a time when the heavy bombardment was subsiding.

Volcanism and Tectonics on Mercury Mercury shows evidence of both volcanism and tectonics in its distant past. However, both processes apparently occurred rather differently on Mercury than on the Moon. Unlike lunar volcanism, which flooded the huge maria, volcanism on Mercury apparently created only small lava plains (see Figure 10.21a). We have found no evidence of lava plains as large as the lunar maria, but the smaller plains appear all over Mercury's surface. Early in Mercury's history, when it still retained some heat from its accretion and differentiation, heat from radioactive decay may have accumulated enough to remelt parts of the interior. This second phase of interior melting may have led to the volcanism that covered early craters and created the lava plains. Together, the less crowded cratering and the many lava plains suggest that Mercury had at least as much volcanism as the Moon.

Mercury exhibits evidence of past tectonics quite different from anything we have found on any other terrestrial world. The evidence comes in the form of tremendous

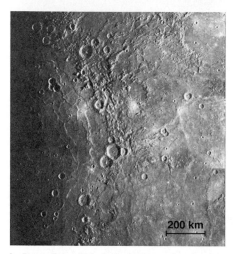

a Close-up view of Mercury's surface, showing impact craters and smooth regions where lava apparently covered up craters.

b Part of the Caloris Basin (note the circular rings near the left), a large impact crater on Mercury.

c The surface on the precise opposite side of Mercury from the Caloris Basin. The landscape shows evidence of violent shaking, presumably caused by the impact that made the Caloris Basin on the other side of the planet.

Figure 10.21 Features of impact cratering and volcanism on Mercury. (Photos from *Mariner 10*.)

cliffs that appear over much of the planet (Figure 10.22). These sheer cliffs have vertical faces up to 3 or more kilometers high and typically run for hundreds of kilometers across the surface. The cliffs probably formed when tectonic forces compressed the crust, causing the surface to crumple. Because crumpling would have shrunk the portions of the surface it affected, Mercury as a whole could not have stayed the same size unless other parts of the surface expanded. However, we find no evidence of "stretch marks" on Mercury. Can it be that the whole planet simply shrank?

Apparently so. Early in its history, Mercury gained more internal heat from accretion and differentiation than the Moon because of its larger size and greater iron content. This heat swelled the size of the large iron core. Later, as the core cooled, it contracted by perhaps as much as 20 kilometers in radius. The mantle and lithosphere must have contracted along with the core, generating the tectonic stresses that created the great cliffs. The contraction probably also closed off any remaining volcanic vents, ending Mercury's period of volcanism.

Mercury Summary Mercury is larger than the Moon but still much smaller than the other terrestrial planets. Thus, while it probably retained its internal heat a bit longer than the Moon did, its interior cooled relatively quickly. All the volcanic and tectonic features we see on Mercury probably formed within its first billion years. Like the Moon, Mercury has been geologically dead for most of its history.

10.5 A Geological Tour: Mars

Mars is much larger than the Moon and Mercury (see Figure 10.1), so we expect it to have a more interesting and varied geological history. However, it is much smaller than Earth: Its radius is about half that of Earth, and its mass is only about 10% that of Earth. Mars is also the most distant of the terrestrial planets from the Sun, orbiting about 50% farther from the Sun than does Earth. Mars's size and distance from the Sun have dictated much of its geological history.

Early telescopic observations of Mars revealed uncanny resemblances to Earth. A Martian day is just over 24 hours, and the Martian rotation axis is tilted about the same amount as Earth's. Mars has polar caps, which early observers assumed to be water ice. We now know that they are composed mostly of frozen carbon dioxide. Mars even shows seasonal variations over the course of its year (about 1.9 Earth years). Early observers attributed seasonal color changes to changes in vegetation. We now know that the changes arise from annual shifts in the winds that deposit dust on the Martian surface [Section 11.6].

In 1877, Italian astronomer Giovanni Schiaparelli reported seeing a network of 79 linear features on Mars

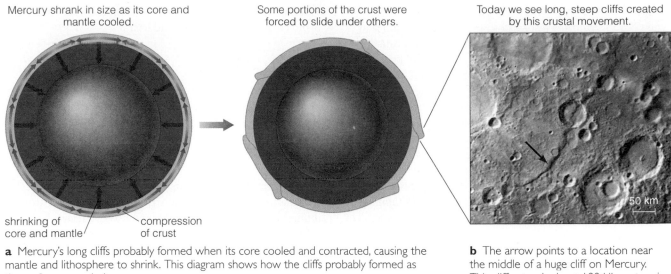

Mercury shrank in size as its core and mantle cooled.

Some portions of the crust were forced to slide under others.

Today we see long, steep cliffs created by this crustal movement.

50 km

shrinking of core and mantle

compression of crust

a Mercury's long cliffs probably formed when its core cooled and contracted, causing the mantle and lithosphere to shrink. This diagram shows how the cliffs probably formed as the surface crumpled.

Figure 10.22 Long cliffs on Mercury offer evidence that the entire planet shrank early in its history, perhaps by as much as 20 kilometers in radius.

VIS

b The arrow points to a location near the middle of a huge cliff on Mercury. This cliff extends about 100 kilometers in length, and its vertical face is as much as 2 kilometers tall. (Photo from *Mariner 10*.)

through his telescope. He named these features *canali*, by which he meant the Italian word for "channels," but it was frequently translated as "canals." Inspired by what sounded like evidence of intelligent life, American astronomer Percival Lowell commissioned the building of an observatory in Flagstaff, Arizona, for the study of Mars. The Lowell Observatory opened in 1894. Barely a year later, Lowell published detailed maps of the Martian canals and the first of three books in which he argued that they were the work of an advanced

civilization. He suggested that Mars was falling victim to unfavorable climate changes and the canals had been built to carry water from the poles to thirsty cities elsewhere.

Lowell's work drove rampant speculation about Martians and fueled science fiction fantasies such as H. G. Wells's *The War of the Worlds* (published in 1898). The public mania drowned out the skepticism of astronomers who saw no canals through their telescopes or in their photographs (Figure 10.23). The debate over the existence of Martian

Figure 10.23 The image at left is a telescopic photo of Mars. The image at right is a drawing of Mars made by Percival Lowell. As you can see, Lowell had a vivid imagination. Nevertheless, if you blur your eyes while looking at the photo, you might see how some of the features resemble what Lowell thought he saw.

VIS

canals and cities was not entirely put to rest until 1965, when NASA's *Mariner 4* spacecraft flew past Mars and sent back photos of a barren, cratered surface. Lowell's canals were nowhere to be seen, confirming that they had been the work of a vivid imagination looking at blurred telescopic images.

Several other spacecraft studied Mars in the 1960s and 1970s, culminating with the impressive Viking missions in 1976. Each of the two Viking spacecraft deployed an orbiter that mapped the Martian surface from above and landers that returned surface images and regular weather reports for almost 5 years. More than 20 years passed before the next successful missions to Mars, *Mars Pathfinder* and *Mars Global Surveyor* in 1997. *Mars Odyssey* began orbiting Mars in 2001.

Three missions are scheduled to reach Mars in late 2003 and early 2004: Japan's Nozomi, which will study Mars from orbit; Mars Express, a joint effort of NASA and European nations that includes both an orbiter and a lander; and NASA's *Mars Exploration Rovers,* a pair of landers that will rove about the surface. Future missions to Mars are scheduled roughly every 2 years to coincide with optimal orbital alignments of Earth and Mars.

Martian Geology

Past and present spacecraft have taught us most of what we know about Mars. Figure 10.24 shows the full surface of Mars, based on observations from the Mars Global Surveyor mission. Close examination of the figure reveals extensive evidence of all four geological processes on Mars.

Impacts on Mars Aside from the polar caps, the most striking feature is the dramatic difference in terrain around different parts of the planet. Much of the southern hemisphere has relatively high elevation and is scarred by numerous large impact craters, including the very large crater known as the Hellas Basin. In contrast, the northern plains show few impact craters and tend to be below the average Martian surface level. The differences in crater crowding clearly show that the southern highlands are a much older surface than the northern plains, which must have had

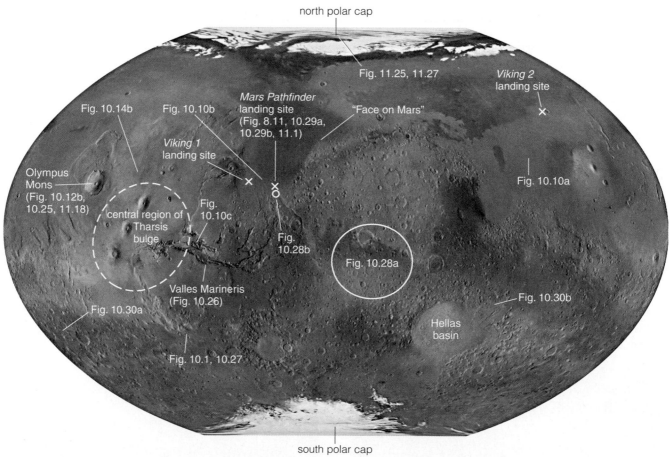

Figure 10.24 This image is a composite made by combining more than 1,000 images with more than 200 million altitude measurements from the Mars Global Surveyor mission. It shows the full surface of Mars. We see extensive evidence of all four geological processes on Mars. Several key geological features are indicated, and the locations of features shown in close-up photos elsewhere in this chapter are marked. The total surface area of Mars is about one-fourth that of Earth (roughly equal to the area of Earth's continents without the oceans). A map of Earth on the same scale would fill about two pages in this book.

Figure 10.25 Olympus Mons, the largest shield volcano in the solar system, covers an area the size of Arizona and rises higher than Mount Everest on Earth. Note the tall cliff around its rim and the central volcanic crater from which lava erupted.

their early craters erased by other geological processes. Further study suggests that volcanism was the most important of these processes. No one knows why volcanism affected the northern plains so much more than the southern highlands or why the two regions differ so much in elevation.

THINK ABOUT IT

Which fundamental planetary property (size, distance from the Sun, or rotation rate) explains why Earth does not have the same combination of heavily cratered and lightly cratered surfaces as Mars?

Volcanism on Mars More dramatic evidence of volcanism on Mars comes from several towering shield volcanoes. One of these, Olympus Mons, is the largest known volcano in the solar system (Figure 10.25; see also Figure 10.12b). Its base is some 600 kilometers across, large enough to cover an area the size of Arizona. Its peak stands about 26 kilometers above the average Martian surface level, or some three times as high as Mount Everest stands above sea level

on Earth. Much of Olympus Mons is rimmed by a cliff that in places is 6 kilometers high.

Olympus Mons and several other large volcanoes are concentrated on or near the continent-size *Tharsis Bulge*. Tharsis, as it is usually called, is some 4,000 kilometers across, and most of it rises several kilometers above the average Martian surface level. It was probably created by a long-lived plume of rising mantle material that bulged the surface upward and provided the magma for the eruptions that built the giant volcanoes.

Could volcanoes still be active on Mars? Until recently, it seemed unlikely. We expect Mars to be much less volcanically active than Earth, because its smaller size has allowed its interior to cool much more. In addition, Martian volcanoes show enough impact craters on their slopes to suggest that they have been inactive for the past billion years. However, analysis of meteorites that appear to have come from Mars (so-called Martian meteorites [Section 13.3]) offers a different perspective. Radiometric dating of these meteorites shows some of them to be made of volcanic rock that solidified from molten lava as recently as 180 million years ago—quite recent in the 4.5-billion-year history of the solar system. Given this evidence of geologically recent volcanic eruptions, it is likely that Martian volcanoes will erupt again someday, though not necessarily in our lifetimes. Nevertheless, the Martian interior is presumably cooling and its lithosphere thickening. At best, Mars will become as geologically dead as the Moon and Mercury within a few billion years.

Tectonics and Valles Marineris Mars shows no evidence of global tectonics like that on Venus and Earth, but some features appear to be tectonic in nature. One of the most prominent is a long, deep system of valleys called *Valles Marineris* (Figure 10.26). Named for the *Mariner 9* spacecraft that first imaged it, Valles Marineris is as long as the United States is wide and almost four times as deep as Earth's Grand Canyon.

No one knows exactly how Valles Marineris formed. Parts of the canyon are completely enclosed by high cliffs on all sides, so neither flowing lava nor water could have been responsible. Extensive cracks on its western end run up against the Tharsis Bulge (see Figure 10.24), suggesting a connection between the two features. Valles Marineris may have formed through tectonic stresses accompanying

Figure 10.26 Valles Marineris is a huge valley on Mars created in part by tectonic stresses.

the uplift of material that created the Tharsis Bulge, cracking the surface and leaving the tall cliff walls of the valleys.

Martian Water and Erosion

Mars is the only planet besides Earth that shows significant evidence of erosion. Figure 10.27 no doubt reminds you of dry riverbeds on Earth seen from above. These channels were almost certainly carved by running water, though no one knows whether the water came from runoff after rainfall, from erosion by water-rich debris flows, or from an underground source. Regardless of the specific mechanism,

the channels clearly seem to be features of erosion by liquid water. (Water is suspected, rather than other liquids, because it is the only substance in the solar system that could have been liquid under Martian conditions and that is sufficiently abundant to have created so many erosion features.)

If you were to visit Mars, however, the idea that parts of the surface were shaped by flowing water might seem quite strange. No liquid water exists anywhere on the surface of Mars today. We know this not only because we've studied most of the surface in reasonable detail, but also because the surface conditions would not allow liquid water to be present as lakes, rivers, or even puddles. In most places

---

SPECIAL TOPIC The Face on Mars

Among the tens of thousands of photographs snapped as part of the Viking missions to Mars in the late 1970s, one achieved special fame. The *Viking 1* orbiter snapped a picture showing a feature on Mars that looked remarkably like a human face (Figure 1). The "face on Mars" soon spawned a cottage industry, with proponents arguing that it must be the work of a lost Martian civilization or of alien beings visiting our solar system. It still appears regularly in supermarket tabloids.

The feature in the picture certainly resembles a face. However, it seemed a near-certainty that the human likeness was a coincidence of geology and camera angle. Because of the intense public interest surrounding the face on Mars, NASA made it a target of further observations with *Mars Global Surveyor*, which entered Martian orbit in 1997.

New photographs soon showed that the scientific analysis had been correct: The feature did not look like a face when it was viewed at higher resolution. Figure 2 compares the original Viking photo with a more recent photo from *Mars Global Surveyor*. Although a few die-hard proponents of the face have not given up, the scientific conclusion is clear.

For anyone heartbroken at the thought that Mars has never harbored a civilization, Mars offers other reasons to smile. In searching through hundreds of other photos from Mars missions,

NASA scientists came upon a crater that shows a "happy face" on Mars (Figure 3).

Figure 2 Two photos of the "face on Mars." The photo on the left is another image from *Viking 1* in 1976 (a blowup of the face in Figure 1). The much higher resolution photo on the right is from *Mars Global Surveyor*.

Figure 1 This *Viking 1* image shows the "face on Mars" (near top center) and the surrounding area. The face is on a tall mesa about 1 kilometer across. Some of the other features were claimed to represent a "fortress" and "pyramids." The black dots are instrument artifacts.

Figure 3 This Viking image shows a crater on Mars that appears to be showing a "happy face."

Figure 10.27 This photo, taken by the *Viking Orbiter,* shows what appear to be dried-up riverbeds. Toward the top of the image we see many individual tributaries, which merge into the larger "river" near the lower right. These channels are found in the heavily cratered southern highlands (see the location of this image on the map in Figure 10.24), indicating that they formed more than 3 billion years ago. You can even see a few small impact craters on top of the riverbeds. This area has almost as many craters as the lunar maria, which is what allows us to estimate the age of the channels.

and at most times, Mars is so cold that any liquid water would immediately freeze into ice. Even when the temperature rises above freezing, as it often does at midday near the equator, the air pressure is so low that liquid water would quickly evaporate. If you donned a spacesuit and took a cup of water outside your pressurized spaceship, the water would either freeze or boil away (or both) almost immediately.

Ancient Rains, Lakes, and Oceans

Given the absence of liquid water today, the Martian features of erosion must have been made at a time in the past when liquid water could flow. Because the channels seen in Figure 10.27 contain numerous impact craters, we conclude that they were created more than 3 billion years ago.

Other evidence also argues that Mars had rain and surface water in the distant past. In the ancient, heavily cratered terrain of the southern highlands, rainfall appears to have eroded the rims of large craters and erased smaller craters altogether (Figure 10.28a). Some craters, such as the one shown in Figure 10.28b, appear to have held lakes. Ancient rains may have filled the crater, allowing sediments to build up from material that settled to the bottom. The sculpted patterns in the crater bottom may have been created as erosion exposed layer upon layer of sedimentary rock, much as the action of the Colorado River exposed the layers visible in the walls of the Grand Canyon on Earth.

Perhaps the most intriguing idea is that a great ocean once may have filled the low-lying regions in the north (Figure 10.28c). Careful study shows the presence of features that resemble shorelines along the boundaries of this possible ocean, although the evidence is not particularly strong. Further evidence comes from valleys that appear to have formed as lakes overflowed their shores (Figure 10.28d).

Whether or not a full ocean existed, it seems that liquid water was plentiful on Mars during its first billion years.

Billion-Year-Old Floodplains

Craters in younger, less heavily cratered regions of the Martian surface show far less erosion than the ancient craters of the southern highlands, suggesting that little or no rain has fallen since 3 billion years ago. The warm, wet era of Mars must already have ended by that time, leaving the planet cold and dry.

However, we still find regions that appear to be vast floodplains, suggesting that catastrophic floods occasionally occurred even after the era of rainfall had ended. Such flooding could have occurred even if the temperature and pressure were as low as they are today. Although liquid water cannot last long on the surface, a large release of floodwaters could survive long enough to carve significant features before freezing or boiling away. Based on crater crowding, most of the floodplains appear to be between about 1 billion and 3 billion years old.

The *Mars Pathfinder* spacecraft landed at a site that may have witnessed one of the biggest floods in the history of the solar system. As seen from orbit, this region shows winding channels and streamlined islands that appear to have been formed by raging water (Figure 10.29a). We can trace the flow of the water upstream, but we find no surface features that look like a reservoir from which the water might have escaped. We therefore conclude that the water must have emerged from underground. Perhaps volcanic heat melted underground ice, unleashing the flood.

On the surface of the floodplain (Figure 10.29b), the *Mars Pathfinder* lander released the *Sojourner* rover (named after Sojourner Truth, an African-American heroine of the Civil War era who traveled the nation advocating equal rights for women and blacks). The six-wheeled rover, no larger than a microwave oven, carried cameras and instruments to measure the chemical composition of nearby rocks. Together the lander and rover confirmed the flood hypothesis: Rocks of many different types are jumbled together and stacked against each other, just as we find in the aftermath of floods on Earth.

When Did the Water Last Flow?

Prior to the arrival of *Mars Global Surveyor* in 1997, most scientists assumed that no water had flowed at the surface of Mars during the past several hundred million years or more. However, images from *Mars Global Surveyor* and the more recent *Mars Odyssey* offer tantalizing hints of water flows in much more modern times.

The strongest evidence for liquid water in recent times comes from photos of gullies on crater and channel walls. In Figure 10.30, for example, note the striking similarity to the gullies we see on almost any eroded slope on Earth. The gullies may have formed when underground water broke out in small flash floods, carrying boulders and soil down the slope. Alternatively, snow may accumulate on the crater walls in winter and melt away in spring. Because the gullies are relatively small (note the scale bar in Figure 10.30), they

a This photo shows a broad region of the southern highlands on Mars. The indistinct rims of the many large craters appear to have been eroded by ancient rains. Relatively few smaller craters are present, suggesting that these have been erased by erosion. A few craters show no signs of erosion, suggesting that they formed after the rains ended.

b This close-up view of a crater bottom (the crater shown in Figure 10.10c) offers evidence that the crater once held a lake. The sculpted patterns appear to be layers of sedimentary rock that were laid down at a time when the crater was filled with water.

c This map shows Mars color-coded by elevation. Blue areas are low-lying areas. Red, brown, and white areas are progressively higher. Some scientists speculate that the vast low-lying region in the north may once have held an ocean.

Figure 10.28 Evidence that Mars had rainfall and surface water more than 3 billion years ago.

d This computer-generated perspective view shows how a Martian valley forms a natural passage between two possible ancient lakes (shaded blue). Vertical relief is exaggerated 14 times to reveal the topography.

a This photo from orbit shows a region several hundred kilometers across, with winding valleys and sculpted islands that appear to have been made by raging floodwaters.

b *Mars Pathfinder* landed in the midst of this floodplain, taking surface photos that show the jumbled rocks left behind as the floodwaters departed. The *Sojourner* rover is visible near the large rock.

Figure 10.29 Evidence of catastrophic flooding on Mars. The era of rainfall appears to have ended more than 3 billion years ago, but floods have occurred more recently. They may be caused by the release of water melted from underground ice.

should be gradually covered over by blowing sand during Martian dust storms. Thus, gullies that are still clearly visible must be no more than a few million years old. Geologically speaking, this time is short enough to make it quite likely that water flows are still forming gullies today.

While we do not yet know whether Mars has near-surface water flows today, Mars clearly has substantial amounts of water in some form. The polar caps contain at least some water ice in addition to frozen carbon dioxide. We often find water vapor and ice crystals in the Martian atmosphere, sometimes forming clouds (see the clouds around Olympus Mons in Figure 10.12b). More significant, instruments aboard *Mars Odyssey* have discovered what

Figure 10.30 (**a**) This photograph from the *Mars Global Surveyor* shows gullies on a crater wall. Scientists suspect they were formed by water melting under the protective cover of snowpack that forms in the winter, though many alternative hypotheses have been proposed. The gullies are geologically young, but no one yet knows whether similar gullies are forming today. (**b**) This photo shows a closeup of another crater wall with an arrow pointing to a possible remnant of snow pack.

Figure 10.31 This map, made with data from *Mars Odyssey*, represents the hydrogen content of the Martian surface soil. The blue areas contain the most hydrogen, probably because they represent regions in which the top meter or so of surface soil contains water ice.

appears to be a large quantity of water ice frozen in the top meter or so of the surface soil (Figure 10.31). If water ice exists on Mars this close to the surface, even more water probably lies deeper underground. If there is still volcanic heat on Mars, this water may sometimes melt and flow.

Even if water does still flow on occasion, Mars clearly was much warmer and wetter at times in the past than it is today. Ironically, Percival Lowell's supposition that Mars was drying up has turned out to be basically correct, although in a very different way than he imagined. We'll discuss the reasons for Mars's dramatic climate change in Chapter 11.

Life on Mars?

Percival Lowell's visions of Martian cities and canals turned out to be pure fantasy. However, as we'll discuss further in Chapter 24, the discovery of water on Mars makes simpler life a legitimate possibility. Life on Earth probably arose before the solar system was a billion years old [Section 14.5], or during the same time that Mars apparently was warm and wet. Could life have arisen on Mars as well? If so, it probably never got larger than microbes [Section 24.2]. Hunting for fossils of early Martian life will require collecting Martian rocks and studying them under the microscope. (Chapter 24 will also discuss a controversial claim that a meteorite from Mars holds fossil evidence of life.) Moreover, if any pockets of liquid water have persisted underground for billions of years, microbes could possibly still live beneath the Martian surface.

The possibility of past or present life makes Mars an even more attractive place to study than its interesting geology alone. If political and budgetary considerations allow, we could send humans to Mars within a couple of decades. The search for microscopic fossils or living microbes will probably be long and hard. A good place to start might be in Valles Marineris. The canyon is so deep in places that some of its walls probably once were several kilometers under-

ground and hence may have been exposed to subsurface liquid water early in Mars's history. Because these walls are now accessible, they offer us the prospect of studying rocks from underground without having to drill deep into the surface. Within our lifetimes, we may witness astronauts making the perilous descent into the deepest canyon in the solar system and finding answers to the question of whether life has ever existed on Mars.

10.6 A Geological Tour: Venus

The surface of Venus is searing hot with brutal pressure [Section 8.3], making it seem quite unlike the "sister planet" to Earth it is sometimes called. However, beneath the surface, Venus must be quite similar to Earth. Venus is nearly the same size as Earth (see Figure 10.1), with a radius only about 5% smaller than Earth's. Venus and Earth are also quite close together relative to the overall size of the solar system, so both should have been built from the same kinds of planetesimals. The very similar densities of Venus and Earth (see Table 8.1) offer further evidence that both planets should have the same overall composition.

With the same basic size and the same basic composition, we expect the interiors of Venus and Earth to be quite similar in structure and to retain about the same level of internal heat today. Nevertheless, we see ample evidence of "skin-deep" differences in surface geology. In this section, we'll focus on the geology of Venus and its differences from the geology of Earth.

A Global View of Venus

Venus's thick cloud cover prevents us from seeing through to its surface, but we can study its geological features with radar since radio waves can pass through clouds. *Radar mapping* involves bouncing radio waves off the surface

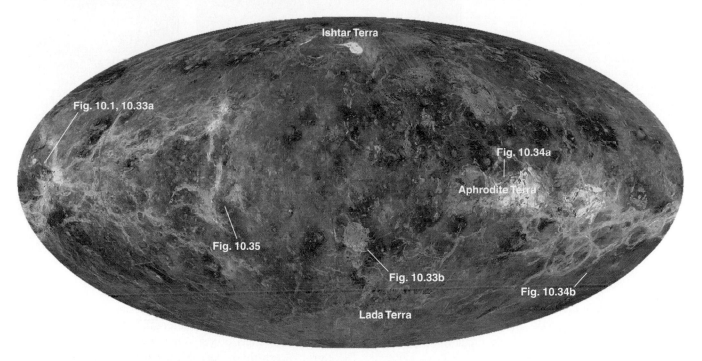

Ishtar Terra

Fig. 10.1, 10.33a

Fig. 10.34a

Aphrodite Terra

Fig. 10.35

Fig. 10.33b

Fig. 10.34b

Lada Terra

radio

Figure 10.32 This image shows the full surface of Venus as revealed by radar observations from the *Magellan* spacecraft. *Magellan* mapped 98% of the surface over a period of 24 months. Bright regions in this radar image represent rough areas or high altitudes, which often correspond to areas that have experienced greater geological activity. The structures show evidence of extensive tectonic activity, particularly around the planet's equatorial regions, but there is no evidence of plate tectonics as on Earth. Maxwell Montes, the planet's highest mountains, rise 12 km above the mean elevation.

from a spacecraft and using the reflections to create three-dimensional images of the surface. From 1990 to 1993, the *Magellan* spacecraft used radar to map the surface of Venus, discerning features as small as 100 meters across. Scientists have named almost all the geological features on Venus for goddesses and famous women.

Figure 10.32 shows a global map of Venus based on the *Magellan* radar observations. Venus has some impact craters, but far fewer than Mercury, the Moon, or Mars. Most of Venus is covered by relatively smooth, rolling plains. There are a few mountain ranges, but not nearly as many as on Earth. Three large, elevated "continents" (the three regions labeled "Terra" in Figure 10.32) are the biggest features on the surface of Venus.

Impacts, Volcanism, and Tectonics on Venus

The relatively few impact craters on Venus are distributed fairly uniformly over the entire planet, suggesting that the surface is about the same age everywhere. Precise crater counts suggest that the surface is about 1 billion years old. Apparently, the entire surface of Venus was somehow "repaved" about a billion years ago. This repaving erased all craters that had been made previously, including any remaining from the heavy bombardment.

We do not know how much of the repaving was due to tectonic processes and how much was due to volcanism,

but volcanism has clearly played an important role on Venus. The surface shows abundant evidence of lava flows and many volcanic mountains. Some are shield volcanoes (Figure 10.33a), indicating basaltic eruptions like those that formed the islands of Hawaii on Earth. Other volcanoes have steeper sides, indicating eruptions of a higher-viscosity lava (Figure 10.33b).

The most remarkable features on Venus are tectonic in origin. Its crust is quite contorted. In some regions, the surface appears to be fractured in a regular pattern (Figure 10.34a; see also Figure 10.14). Many of the tectonic features are associated with volcanic features, suggesting a strong linkage between volcanism and tectonics on Venus. One striking example of this linkage is a roughly circular feature called a *corona* (Latin for "crown"), which probably resulted from a hot, rising plume in the mantle (Figure 10.34b). The plume pushed up on the crust, forming concentric tectonic stretch marks on the surface. The plume also forced lava to the surface, dotting the area with volcanoes.

Does Venus, like Earth, have plate tectonics to move around pieces of its lithosphere? Convincing evidence for plate tectonics might include deep trenches and linear mountain ranges like those we see on Earth. *Magellan* found no such evidence on Venus. Some scientists have suggested that plate tectonics may have played a role in the surface repaving that occurred a billion years ago. This hypothesis is controversial, however. Even if it were true, it

Figure 10.33 Examples of volcanoes on Venus. Each image shows a region a few hundred kilometers across (see their locations in Figure 10.32). These images were made from radar data from the *Magellan* spacecraft. The dark and light areas correspond to how well radio waves are reflected by the surface, not visible light. Nonetheless, geological features stand out well. The yellow color is not real and is used only to help highlight the differences between light and dark in the radar images.

100 km

a The two peaks near the center are shield volcanoes, probably much like the volcanoes that created the Hawaiian islands on Earth.

10 km

b The round blobs are steeper stratovolcanoes, apparently built from a higher-viscosity lava.

would suggest that Venusian plate tectonics occurs in occasional episodes rather than as an ongoing process like that on Earth. In either case, it seems clear that Venus presently lacks Earth-style plate tectonics.

The apparent lack of plate tectonics is a major mystery. Earth's lithosphere fractured due to mantle convection, and Venus should have mantle convection similar to that on Earth. Venus's lithosphere must differ from Earth's in some way that has prevented Earth-style plate tectonics. Exactly how or why the two lithospheres may differ is not known, but one hypothesis ties the differences to Venus's hot and dry atmosphere. Venus is so hot that any water in its crust and mantle has probably been baked out over time and lost from the atmosphere [Section 11.6]. The atmosphere today contains virtually no water, so no water is available to re-enter the crust and replace the water baked out in the past.

Water tends to soften and lubricate rock, so its loss would have tended to thicken and strengthen Venus's lithosphere. A thicker, stronger lithosphere is less likely to fracture and thus may explain the lack of plate tectonics on Venus. We'll revisit this mystery when we study Earth's plate tectonics in Chapter 14.

Venusian Erosion

We might naively expect Venus's thick atmosphere to produce strong erosion, but the view both from orbit and on the surface suggests otherwise. The Soviet Union landed two probes on Venus's surface in 1975. Before they were destroyed by the intense surface heat, the probes returned images of a bleak, volcanic landscape with little evidence of erosion (Figure 10.35).

10 km

a Tectonic forces have fractured and twisted the crust in this region.

50 km

b A mantle plume probably created the round corona, which is surrounded by tectonic stress marks.

Figure 10.34 Examples of tectonic features on Venus. These images were made from radar data from the *Magellan* spacecraft.

radio

Figure 10.35 This photo from the Soviet Union's *Venera* lander shows the surface around the base of the spacecraft (visible at lower right). This region of the planet is covered in volcanic plains. Many volcanic rocks are visible, hardly affected by erosion despite their age of nearly a billion years.

Why is erosion so weak on Venus, despite the thick atmosphere? We can trace the answer to two root causes. First, Venus is so hot that no rain of any kind ever touches the surface. As we'll discuss in the next chapter, the high temperature is ultimately traceable to Venus's proximity to the Sun. Second, the surface on Venus has virtually no wind or weather because of Venus's slow rotation. Venus rotates so slowly—once every 243 days—that its atmosphere barely stirs the surface.

The lack of strong erosion on Venus leaves the tectonic contortions of its surface exposed to view, even though some are probably close to a billion years in age. Earth's terrain might look equally stark and rugged from space if not for the softening touches of wind, rain, and life.

THINK ABOUT IT

Suppose Venus rotated as fast as Earth. How would this change its relative levels of volcanism, tectonics, and erosion?

Present and Future Geology on Venus

In geological terms, the billion years since Venus's entire surface was repaved by volcanism and tectonics is not that long. It is, after all, within the most recent 25% of the solar system's history. Moreover, based on Venus's size, its interior should still be hot and should have mantle convection much like Earth's interior. Together, these facts suggest that Venus should remain geologically active today. Most scientists believe this is indeed the case, although we have no direct proof.

Because of both its internal similarities to Earth and its surface differences, we can probably learn much about our own planet by studying Venus further. Unfortunately, Venus is probably the most difficult planet to study of the terrestrial worlds. Its thick cloud cover always hides the surface from orbiting cameras, so only improved radar mapping will tell us more about its global geology. The harsh surface conditions make it difficult for landers to survive. Sending astronauts to Venus seems almost completely out of the question. The European Space Agency (ESA) is currently pursuing a fast-track development of a mission called

Venus Express, which will undertake an in-depth study of Venus's atmosphere.

 Shaping Planetary Surfaces Tutorial, Lessons 2, 3

10.7 Earth and Geological Destiny

We've toured our neighboring terrestrial worlds and their geological histories. In each case, we've found clear connections between a world's fundamental properties—such as size, distance from the Sun, and rotation rate—and its past and present geology. It seems that a planet's geological destiny is essentially determined in its early history. Thus, the same basic ideas should allow us to understand the geology of our own planet.

Figure 10.36 summarizes the key trends we've seen among the terrestrial worlds with respect to impact cratering, volcanism, and tectonics. All the worlds have been affected similarly by impact cratering. Volcanism and tectonics depend on interior heat, and the length of time a world can stay hot depends on its size. The smallest worlds cooled quickly, and have not had volcanism or tectonics for billions of years. The larger worlds have stayed hot longer, allowing volcanism and tectonics to stay active longer.

Earth's basic geology indeed fits the patterns observed on other worlds. Early on, the heavy bombardment must have scarred Earth as it did other worlds. Earth's large size allowed subsequent geological processes to erase most impact craters from the surface. Volcanism shaped many surface features and led to the outgassing that created Earth's early atmosphere. The continents and many other features were shaped by tectonics.

However, while Earth generally fits within the overall patterns, Earth's geology is unique in at least two ways: ongoing plate tectonics and rampant erosion caused by water. Both features are intimately connected to Earth's atmosphere and to the long-term climate stability that has made our existence possible. We will return to these topics in Chapter 14 when we study the unique features of our planet, including life.

period of heavy cratering

low levels of cratering continue

modern appearance

Earth

Venus ?

Mars ? · · · ·

Mercury

Moon

planets formed

time (billions of years)

today

amount of volcanic/tectonic activity

Figure 10.36 This diagram summarizes the geological histories of the terrestrial worlds. The brackets along the top indicate that impact cratering has affected all worlds similarly, with the vast majority of impacts occurring during the heavy bombardment early in the solar system's history. The red arrows represent volcanic and tectonic activity. A thicker and darker arrow means more volcanic/tectonic activity: Venus and Earth have been quite active, Mars is intermediate, and Mercury and the Moon have had very little volcanic/tectonic activity. The arrow lengths indicate how long the activity has persisted. We are certain of ongoing activity only on Earth. Venus is likely to remain active, and Mars may still have low-level volcanism. The planets are shown in order of size, because size is the major factor that determines volcanic/tectonic activity. Erosion is not shown, because it has played a significant role only on Earth (ongoing) and on Mars (where it was quite significant in the past and endures at low levels today).

THE BIG PICTURE

Putting Chapter 10 into Context

In this chapter, we have explored the geology of the terrestrial worlds. As you continue your study of the solar system, keep the following "big picture" ideas in mind:

● We can explain the wide variety of surface features on terrestrial worlds in terms of just four basic geological processes: impact cratering, volcanism, tectonics, and erosion.

● The extent to which different geological processes operate on the different worlds depends largely on their fundamental properties, especially their size.

● A planet's geology is largely destined from its birth—which means we will be able to predict the geology of as-yet-undiscovered planets once we know their fundamental properties.

● Earth has been affected by the same geological processes affecting the other terrestrial worlds. However, Earth also has some unique geological characteristics, such as plate tectonics and rampant erosion, that may be very important to our existence.

10.1 Planetary Surfaces

- *How do the surfaces of terrestrial planets differ from one another?* Mercury and the Moon are heavily cratered, with some volcanic plains. Venus has volcanoes and bizarre bulges. Mars shows varied geology, including volcanoes and evidence of running water. Earth shows features similar to those on all the other terrestrial worlds, and more.

10.2 Inside the Terrestrial Worlds

- *What is the basic interior structure of a terrestrial world?* In order of decreasing density and depth, the interior structure consists of core, mantle, and crust. The crust and part of the mantle together make up the rigid lithosphere.

- *How do interiors get hot?* Sources of heat as a planet forms are accretion and differentiation. Radioactive decay deposits heat over longer times, though it generated more heat early in the solar system's history.

- *Why is planetary size so important to internal heat and geology?* Size determines how fast a hot interior cools. Only large planets can retain significant heat and maintain mantle convection for billions of years.

- *Why is Earth the only terrestrial world with a strong magnetic field?* A planetary magnetic field requires three things: an interior layer of electrically conducting fluid, convection of that fluid, and rapid rotation. Among the terrestrial planets, only Earth has all three characteristics.

10.3 Shaping Planetary Surfaces: The Four Basic Geological Processes

- *What are the four basic geological processes?* The four basic geological processes are impact cratering, volcanism, tectonics, and erosion.

- *How are the geological processes connected to fundamental planetary properties?* Impacts affect all planets similarly, but larger planets are more likely to have their impact craters erased by other geological processes. Volcanism and tectonics require interior heat, which is retained only by larger planets. Erosion can occur only on planets with substantial atmospheres. This requires large size to allow outgassing to make an atmosphere, moderate distance from the Sun to prevent the gas from escaping or freezing, and wind or rain. Wind can occur only with moderately fast rotation.

10.4 A Geological Tour: The Moon and Mercury

- *What are the major geological features of the Moon and Mercury?* Both the Moon and Mercury are heavily cratered. The Moon's lava plains are large and localized in the maria, while Mercury's lava plains are small and globally distributed. Mercury has more tectonic features. Many large cliffs on Mercury appear to be tectonic in origin.

- *How did the lunar maria form?* Large impacts during the heavy bombardment fractured the lithosphere beneath the huge craters they created. A few hundred million years later, heat from radioactive decay melted mantle material, which welled up through the fractures and flooded the craters.

- *Why do we think Mercury shrank in size when it was young?* Long, high cliffs show that Mercury's surface was compressed, but there are no features to suggest surface expansion.

10.5 A Geological Tour: Mars

- *Why did many people once believe that intelligent beings lived on Mars?* Percival Lowell claimed to see canals and popularized the idea that they were made by intelligent beings. The canals do not exist.

- *What features of the four geological processes do we see on Mars?* Geological features on Mars include a dramatic difference in crater crowding on different parts of the surface; numerous tall volcanoes and the large Tharsis Bulge; the huge canyon of Valles Marineris, shaped at least in part by tectonics; and abundant evidence of water erosion.

- *What evidence suggests a past warm and wet period on Mars?* Surfaces older than 3 billion years appear to have been eroded by rainfall or running water.

- *What evidence suggests more recent water flows on Mars?* Some younger regions of Mars appear to have suffered catastrophic floods between 1 and 3 billion years ago. Small gullies suggest far more recent water flows at or near the surface.

10.6 A Geological Tour: Venus

- *How do we study the surface of Venus?* We study the surface of Venus through radar observations from spacecraft.

continued ▶

- *What happened to Venus about a billion years ago?* Its entire surface was apparently repaved by some combination of volcanism and tectonics.

- *Why isn't there much erosion on the surface of Venus?* Venus's rotation is too slow to produce weather, despite its thick atmosphere.

- *Is Venus still geologically active?* Venus is probably still geologically active, but we have no direct proof.

10.7 Earth and Geological Destiny

- *In what sense was the geology of each terrestrial world destined from birth?* Size is the key factor because it determines how long volcanism and tectonics continue. Size is also necessary for erosion, because terrestrial atmospheres come from volcanic outgassing.

- *In our solar system, what geological features are unique to Earth?* Earth is unique in having plate tectonics and rampant, ongoing erosion.

❓ Surprising Discoveries?

Suppose we were to make the following discoveries. (These are *not* real discoveries.) In light of your understanding of planetary geology, decide whether the discovery should be considered reasonable or surprising. Explain your answer, if possible tracing your logic back to the terrestrial world's formation properties.

1. The next mission to Mercury photographs part of the surface never seen before and detects vast fields of sand dunes.

2. New observations show that several of the volcanoes on Venus have erupted within the past few million years.

3. A Venus radar mapper discovers extensive regions of layered sedimentary rocks, similar to those found on Earth.

4. Radiometric dating of rocks brought back from one lunar crater shows that the crater was formed only a few tens of millions of years ago.

5. Seismographs placed on the surface of Mercury reveal frequent and violent earthquakes.

6. New orbital photographs of craters on Mars that have gullies also show pools of liquid water to be common on the crater bottoms.

7. Clear-cutting in the Amazon rain forest on Earth exposes vast regions of ancient terrain that is as heavily cratered as the lunar highlands.

8. Drilling into the Martian surface, a robotic spacecraft discovers liquid water a few meters beneath the slopes of a Martian volcano.

Problems

9. *Planetary Geology.* Define the term *planetary geology* in everyday terms. Why do we believe that differences between planetary surfaces can be traced to fundamental properties (size, distance from the Sun, rotation) instead of random occurrences that only affected individual planets?

10. *Seeing Inside.* Briefly explain how we learn about planetary interiors.

11. *Understanding Interiors.* What is differentiation, and how did it lead to the core-mantle-crust structures of the terrestrial worlds? What is a lithosphere, and why is it geologically important? Describe the conditions under which convection occurs. What is mantle convection, and how is it related to lithospheric thickness?

12. *Clues from Craters.* Describe what we can learn from impact craters. How is crater crowding related to surface age?

13. *Volcano Vocabulary.* What is viscosity, and how does it affect the shapes of volcanoes and the characteristics of lava flows? What is outgassing, and how is it important to our existence?

14. *Tectonics Terms.* What do we mean by *plate tectonics,* and how it is different from "regular" tectonics? Describe how mantle convection can make different kinds of features on a planet's surface.

15. *Earth vs. Moon.* Why is the Moon heavily cratered, but not Earth? Explain in one or two paragraphs, relating your answer back to their fundamental properties.

16. *Comparative Erosion.* Consider erosion on Mercury, Venus, the Moon, and Mars. Which of these four worlds has the greatest erosion? Why? For each world, write a paragraph explaining its level of erosion, tracing it back to fundamental properties.

17. *Miniature Mars.* Suppose Mars had turned out to be significantly smaller than its current size—say, the size of our Moon. How would this have affected the number of geological features due to each of the four major geological processes? Do you think Mars would still be a good candidate for harboring extraterrestrial life? Summarize your answers in two or three paragraphs.

18. *Predictive Geology.* Suppose another star system has a rocky terrestrial planet twice as large as Earth. In one or two paragraphs, describe the type of geology you would expect it to have.

19. *Mystery Planet.* It's the year 2098, and you are designing a robotic mission to a newly discovered planet around a nearby star that is nearly identical to our Sun. The planet is as large in radius as Venus, rotates with the same daily period as Mars, and lies 1.2 AU from its star. Your spacecraft will orbit but not land on the planet.

a. Some of your colleagues believe that the planet has no metallic core. How could you support or refute their hypothesis? (*Hint:* Remember that metal is dense and electrically conducting.)

b. Other colleagues suspect that the planet has no atmosphere, but the instruments designed to study the planet's atmosphere fail due to a software error. However, the spacecraft can still photograph geological features. How could you use the spacecraft's photos of geological features to determine whether a significant atmosphere is (or was) present on this planet?

20. *Dating Planetary Surfaces.* We have discussed two basic techniques for determining the age of a planetary surface: studying the abundance of impact craters and radiometric dating of surface rocks. Which technique seems more reliable? Which technique is more practical? Explain.

21. *Experiment: Geological Properties of Silly Putty.* Roll room-temperature Silly Putty into a ball and measure its diameter. Place the ball on a table and gently place one end of a heavy book on it. After 5 seconds, measure the height of the squashed ball. Repeat the experiment two more times, the first time warming the Silly Putty in hot water before you start and the second time cooling it in ice water before you start. How do the different temperatures affect the rate of "squashing"? How does the experiment relate to planetary geology? Explain.

22. *Experiment: Planetary Cooling in a Freezer.* To simulate the cooling of planetary bodies of different sizes, use a freezer and two small, *plastic* containers of similar shape but different size. Fill each container with cold water and put both in the freezer at the same time. Checking every hour or so, record the time and your estimate of the thickness of the "lithosphere" (the frozen layer) in the two tubs. How long does it take the water in each tub to freeze completely? Describe in a few sentences the relevance of your experiment to planetary geology. *Extra credit:* Plot your results on a graph with time on the *x*-axis and lithospheric thickness on the *y*-axis. What is the ratio of the two freezing times?

23. *Amateur Astronomy: Observing the Moon.* Any amateur telescope has resolution adequate to identify geological features on the Moon. The light highlands and dark maria should be evident, and shadowing is visible near the line between night and day. Try to observe the Moon near first- or third-quarter phase. Sketch or photograph the Moon at low magnification, and then zoom in on a region of interest. Again sketch or photograph your field of view, label its features, and identify the geological process that created them. Look for craters, volcanic plains, and tectonic features. Estimate the size of each feature by comparing it to the size of the whole Moon (radius = 1,738 km).

Discussion Questions

24. *What Is Predictable?* We've found that much of a planet's geological history is destined from its birth. Briefly explain why, and discuss the level of detail that is predictable. For example, was Mars's general level of volcanism predictable? Could we have predicted a mountain as tall as Olympus Mons or a canyon as long as Valles Marineris? Explain.

25. *Better Understanding of Earth.* Suppose you received funding for a single mission to one of the terrestrial worlds and your goal was to learn something that could help us better understand our own planet. Which terrestrial world would you choose as the mission target, and why? Be sure to consider all relevant factors, including what your mission would help us learn, the difficulty of the mission, and its cost.

26. *Worth the Effort?* Politicians often argue over whether planetary missions are worth the expense involved. Based on what we have learned by comparing the geologies of the terrestrial worlds, do you think the missions that have given us this knowledge have been worth their expense? Defend your opinion.

For a complete list of media resources available, go to www.astronomyplace.com and choose Chapter 10 from the pull-down menu.

Astronomy Place Web Tutorials

Tutorial Review of Key Concepts

Use the following interactive **Tutorials** at www.astronomyplace.com to review key concepts from this chapter.

Shaping Planetary Surfaces Tutorial

Lesson 1 The Four Geological Processes

Lesson 2 What do Geological Processes Depend On?

Lesson 3 Planet Surface Evolution

Supplementary Tutorial Exercises

Use the interactive **Tutorial Lessons** to explore the following questions.

Shaping Planetary Surfaces Tutorial, Lesson 1

1. Watch the impact cratering animation. How does the crater size depend on the size of the impactor?

2. Watch the tectonics animation. What happens in the mantle to cause compression and mountain building?

3. What happens in the mantle to cause tectonic stretching?

Shaping Planetary Surfaces Tutorial, Lesson 2

1. How has the rate of crater formation in the solar system changed over time?

2. Why do we see so many more impact craters on the Moon and Mercury than on Earth?

3. Why is erosion so much more dominant on Earth than on any of the other terrestrial worlds?

Exploring the Sky and Solar System

Of the many activities available on the *Voyager: SkyGazer* CD-ROM accompanying your book, use the following files to observe key phenomena covered in this chapter.

Go to the **File: Demo** folder for the following demonstrations.

1. Earth and Venus

2. Venus-Earth-Moon

Go to the **Explore** menu for the following demonstrations.

1. Solar System

2. Paths of the Planets

Web Projects

Take advantage of the useful Web links on www.astronomyplace. com to assist you with the following projcts.

1. *Planetary Geology.* Visit the Nine Planets Web site, which has summaries of features on each of the planets. Choose one of the terrestrial worlds, and learn about at least two geological features on it that we did not highlight in this chapter. Are the features consistent with what we expect based on the ideas of "geological destiny" discussed in this chapter? Explain.

2. *Water on Mars.* Go to the home page for NASA's Mars Exploration Program, and look for the latest evidence concerning recent water flows on Mars. Write a few paragraphs describing the new evidence and what it tells us.

3. *"Coolest" Surface Photo.* Visit the Astronomy Picture of the Day Web site, and search for past images of the terrestrial worlds. Look at many of them, and choose the one you think is the "coolest." Make a printout, write a short description of what it shows, and explain what you like about it.

11 Planetary Atmospheres

Earth and the Other Terrestrial Worlds

For the first time in my life, I saw the horizon as a curved line. It was accentuated by a thin seam of dark blue light—our atmosphere. Obviously this was not the ocean of air I had been told it was so many times in my life. I was terrified by its fragile appearance.

Ulf Merbold, astronaut (Germany)

Life as we know it would be impossible on Earth without our atmosphere. This layer of gas surrounding the planet supplies the oxygen we breathe, shields us from harmful ultraviolet and X-ray radiation from the Sun, protects us from continual bombardment by micrometeorites, generates rain-giving clouds, and traps just enough heat to keep Earth habitable. We couldn't have designed a better atmosphere if we'd tried.

Despite forming under similar conditions, our neighboring planets have ended up with atmospheres utterly inhospitable to us. No other planet in our solar system has air that we could breathe or temperature and pressure conditions in which we could survive without a space suit. Mars, only slightly farther from the Sun than Earth, has an atmosphere that leaves its surface cold, dry, and barren. Venus, only slightly closer to the Sun, has an atmosphere that makes its surface conditions resemble a classical view of hell.

How did Earth end up with such fortunate conditions? To fully appreciate what happened on Earth, we must first understand what happened on all the terrestrial worlds.

11.1 Atmospheric Basics

All the planets of our solar system have an atmosphere to at least some degree, as do several of the solar system's larger moons. The jovian planets are essentially atmosphere throughout, because they are made largely of gaseous material. In contrast, atmospheric gases around solid bodies—including the terrestrial worlds, jovian moons with atmospheres, and Pluto—never make up more than a miniscule fraction of a world's total mass. Nevertheless, just a few basic principles can explain nearly all atmospheric properties. In this chapter, we will concentrate on these basic principles as they apply to the atmospheres of the terrestrial worlds. Later we will see that similar principles apply to virtually any planetary atmosphere.

A Quick Tour of the Terrestrial Atmospheres

The terrestrial atmospheres are even more varied than the terrestrial geologies. Figure 11.1 shows global and surface views of each of the five terrestrial worlds. Table 11.1 lists general characteristics of their atmospheres. A quick scan of the figure and the table reveals the vast differences between the worlds.

The surface pressure column in Table 11.1 tells us that the Moon and Mercury have so little atmosphere that it's quite reasonable to call them "airless" worlds. These worlds therefore have no wind or weather of any kind. What little atmosphere they possess consists mostly of individual atoms blasted off the surface by micrometeorites, the solar wind, or energetic solar photons. Once released into the

Venus

Mercury

Figure 11.1 Views of the terrestrial worlds and their atmospheres from orbit and from the surface. The surface views for Mercury and Venus are artist's conceptions; the others are photos. The global views are visible-light photos taken from spacecraft. (Venus appears in gibbous phase as it was seen by the *Galileo* spacecraft during its Venus flyby en route to Jupiter.)

Table 11.1 Atmospheres of the Terrestrial Worlds

World	Composition	Surface Pressure*	Average Surface Temperature	Winds, Weather Patterns	Clouds, Hazes
Mercury	helium, sodium, oxygen	10^{-14} bar	day: 425°C (797°F); night: −175°C (−283°F)	none: too little atmosphere	none
Venus	96% carbon dioxide (CO_2) 3.5% nitrogen (N_2)	90 bars	470°C (878°F)	slow winds, no violent storms, acid rain	sulfuric acid clouds
Earth	77% nitrogen (N_2) 21% oxygen (O_2) 1% argon H_2O (variable)	1 bar	15°C (59°F)	winds, hurricanes	H_2O clouds, pollution
Moon	helium, sodium, argon	10^{-14} bar	day: 125°C (257°F); night: −175°C (−283°F)	none: too little atmosphere	none
Mars	95% carbon dioxide (CO_2) 2.7% nitrogen (N_2) 1.6% argon	0.007 bar	−50°C (−58°F)	winds, dust storms	H_2O and CO_2 clouds, dust

*1 bar ≈ the pressure at sea level on Earth.

"air," these atoms bounce around the surface like miniature rubber balls. They rarely collide with one another, and they eventually stick to the surface after bounding around for a few hours or a few days. The gas density of these atmospheres is far too low to scatter sunlight. Even in broad daylight, you would see a pitch-black sky surrounding the bright Sun. If you could condense the entire atmosphere of the Moon or Mercury into solid form, you would have so little material that you could almost store it in a dorm room.

At the other extreme, Venus is completely shrouded by a thick atmosphere. If you stood on the surface of Venus, you'd feel a searing temperature higher than that of a self-cleaning oven, and a crushing pressure 90 times greater than that on Earth. A deep-sea diver would have to go nearly 1 kilometer (0.6 mile) beneath the ocean surface on Earth to feel comparable pressure. Although you could not swim through the Venusian air, its density is about 10% that of water. Moving through this thick air would feel like a cross between swimming and flying. Surface views of Venus are perpetually overcast, with weak sunlight filtering through the thick clouds above.

Venus's atmosphere consists almost entirely of carbon dioxide (CO_2), and it has virtually no molecular oxygen (O_2). Thus, you could not breathe the air on Venus even if

Earth

Earth's Moon

Mars

Figure 11.2 Earth's atmosphere, visible in this photograph from the Space Shuttle, makes a very thin layer over Earth's surface. Most of the air is in the lowest 10 kilometers of the atmosphere, visible along the edge of the planet. Above an altitude of about 60 kilometers, the air is so thin that the sky is black even in the daytime.

you cooled it to a comfortable temperature. The slow rotation of the planet means there are no storms. High in the atmosphere, droplets of corrosive chemicals like sulfuric acid condense and rain down, but the raindrops evaporate in the warm lower atmosphere before ever hitting the ground.

The atmosphere of Mars is also made mostly of carbon dioxide, but there is much less of it than on Venus. The result is very thin air with a pressure so low that your body tissues would bulge painfully if you stood on the surface without wearing a full space suit. In fact, the pressure and temperature are both so low that liquid water would rapidly disappear, with some evaporating and some freezing into ice [Section 10.5]. Extensive evidence of flowing water in Mars's distant past tells us that the entire planet has undergone dramatic climate change. Mars has seasons like Earth because of its similar axis tilt, and it is occasionally engulfed in planet-wide dust storms.

Earth's atmosphere, composed mostly of nitrogen and oxygen, is the only atmosphere under which human life is possible outside of an artificial environment. The atmosphere provides our planet with just enough warmth and pressure to enable water to cycle between all three phases (solid ice, liquid water, and gaseous water vapor). It protects us from harmful solar radiation and produces the weather patterns that variously bring us days of sunshine, clouds, and rain or snow.

Our primary goal in this chapter is to understand why the terrestrial atmospheres are so different. Before we examine the details, however, we need to discuss a few basic properties of atmospheres. We'll begin with the most basic question: What is an atmosphere?

What Is an Atmosphere?

An *atmosphere* is a layer of gas that surrounds a world. In most cases, it is a surprisingly thin layer. On Earth, for example, about two-thirds of the air in the atmosphere lies within 10 kilometers of Earth's surface (Figure 11.2). You could represent this air on a standard globe with a layer only as thick as a dollar bill. Above an altitude of about 60 kilometers, the air is so thin that the sky is black even in the daytime, much like the sky on the Moon.

The air that makes up any atmosphere is a mixture of many different gases that may consist either of individual atoms or of molecules. In general, temperatures in the terrestrial atmospheres are low enough (even on Venus) for atoms to combine into molecules. For example, the air we breathe consists of *molecular* nitrogen (N_2) and oxygen (O_2), as opposed to individual atoms (N or O). Other common molecules in terrestrial atmospheres include water (H_2O) and carbon dioxide (CO_2).

Atmospheric Pressure Collisions of individual atoms or molecules in an atmosphere create *pressure*. On Earth, for example, the nitrogen and oxygen molecules in the air fly around at average speeds of about 500 meters per second [Section 4.2]—fast enough to cross your bedroom a hundred times in 1 second. Given that a single breath of air contains more than a billion trillion molecules, you can imagine how frequently molecules collide. On average, each molecule in the air around you will suffer a million collisions in the time it takes to read this paragraph.

A balloon offers a good example of how pressure works in a gas (Figure 11.3a). The air molecules inside a balloon exert pressure pushing outward as they constantly collide with the balloon walls. At the same time, air molecules outside the balloon collide with the walls from the other side, exerting pressure that by itself would make the balloon collapse. A balloon stays inflated when the inward pressure and the outward pressure are balanced. (We are neglecting the tension in the rubber of the balloon walls.) Imagine that you blow more air into the balloon. The extra molecules inside mean more collisions with the balloon wall, momentarily making the pressure inside greater than the pressure outside. The balloon therefore expands until the inward and outward pressures are again in balance.

If you heat the balloon, the gas molecules begin moving faster and therefore collide harder with the inside of the balloon, which also momentarily increases the inside pressure until the balloon expands. As it expands, the pressure

Pressure inside and outside balloon is equal.

Adding molecules to inside of balloon increases pressure inside balloon, causing balloon to expand until inside and outside pressures are again equal.

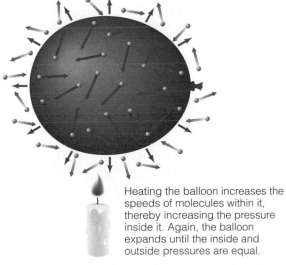

Heating the balloon increases the speeds of molecules within it, thereby increasing the pressure inside it. Again, the balloon expands until the inside and outside pressures are equal.

a

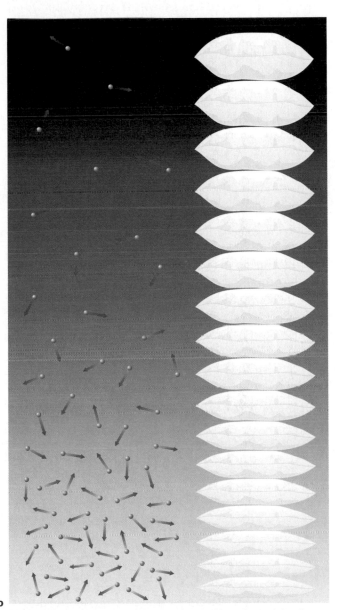

b

The pressure in each layer is enough to hold up the weight of layers above, counteracting the force of gravity.

Figure 11.3 Understanding an atmosphere requires understanding the meaning of gas pressure and thinking about how gravity holds down the gas in an atmosphere. (**a**) Gas pressure depends on the density of molecules and the speeds of these molecules. A balloon offers a good way to think about gas pressure. (Note that we have neglected the tension in the balloon itself, which also contributes to the pressure.) (**b**) Atmospheric density and pressure decrease with altitude for the same reason that a giant stack of pillows would be more compressed at the bottom than at the top.

inside it decreases and the balloon comes back into pressure balance. Conversely, cooling a balloon makes it contract, because the outside pressure momentarily exceeds the inside pressure. You can watch this contraction happen by putting a balloon in a refrigerator or freezer.

We can understand **atmospheric pressure** by applying similar principles. Gas in an atmosphere is held down by gravity. The atmosphere above any altitude therefore has some weight that presses downward, tending to compress the atmosphere beneath it. At the same time, the fast-moving molecules exert pressure in all directions, including upward, which tends to make the atmosphere expand. Planetary atmospheres exist in a perpetual balance between the downward weight of their gases and the upward push of their gas pressure (Figure 11.3b).

The higher you go in an atmosphere, the less the weight of the gas above you. Thus, the pressure must also become less as you go upward, which explains why the pressure decreases as you climb a mountain or ascend in an airplane. You can visualize what happens by imagining the atmosphere as a very big stack of pillows. The pillows at the bottom are highly compressed because of the weight of all the pillows above. Going upward, the pillows are less and less compressed because less weight lies on top of them.

The common unit of pressure is the **bar** (as in *barometer*), roughly equal to Earth's atmospheric pressure at sea level. If you gathered up all the air directly above any 1 square inch of Earth's surface, you would find that it weighs about 14.7 pounds (6.7 kg). We therefore say that 1 bar is a pressure equivalent to about 14.7 pounds per square inch. You don't feel the weight of the atmosphere above you for two reasons: because the pressure pushes in all directions and because the fluids in your body push outward with an equivalent pressure. Nevertheless, you certainly notice changes in pressure. Even slight changes can cause your ears to "pop." More extreme changes, such as those that affect deep-sea divers when they rise too rapidly, can be deadly. Atmospheric pressure varies considerably among the terrestrial worlds (see Table 11.1).

THINK ABOUT IT

Suppose Mars had exactly the same total amount of air above each square inch of surface as does Earth. Would the atmospheric pressure be higher or lower? Explain. (*Hint:* Does the strength of gravity on the two planets have an effect on pressure?)

Where Does an Atmosphere End? Atmospheres do not have clear upper boundaries but instead gradually fade away with increasing altitude. The higher you go in an atmosphere, the lower the pressure and density become. At some point, the density becomes so low that we can't really think of the gas as "air" anymore. Collisions between atoms or molecules are rare at these altitudes, and the gas is so thin that it would look and feel as if you had already entered space.

On Earth, this tenuous upper atmosphere extends for several hundred kilometers above Earth's surface. The Space

Shuttle, the Space Station, and many other satellites orbit Earth within these outer reaches of the atmosphere (Figure 11.4). The low-density gas may be barely noticeable under most conditions, but it still exerts drag on orbiting spacecraft. That is why satellites in low-Earth orbit slowly spiral downward, eventually burning up as they reenter the denser layers of the atmosphere [Section 5.5]. To prevent the Space Station from burning up in the same way, the Space Shuttle or other rockets provide it with periodic boosts.

What Do Atmospheres Do?

It may be fairly obvious that Earth's atmosphere is important to our existence, but the full range of atmospheric effects is more than most people realize. To provide context for everything else we will study in this chapter, we list here a summary of the key effects of planetary atmospheres. The following sections will explain and expand upon these effects and their consequences. For the moment, simply familiarize yourself with the effects. Later, you can refer back to this list to make sure you understand why each effect occurs.

- Atmospheres make planetary surfaces warmer than they would be otherwise, through what we call the *greenhouse effect*. The greenhouse effect can be quite dramatic, as it is on Venus. It also explains Earth's pleasant climate. Atmospheres also can distribute heat around a planet, reducing pole-to-equator and day-night temperature variations. [Section 11.2]

- Atmospheres absorb and scatter light. They can absorb high-energy radiation from the Sun, including solar ultraviolet and X rays. They can scatter visible light, making the daytime sky bright. Worlds without substantial atmospheres, such as the Moon and Mercury, have black skies both day and night. [Section 11.3]

- Atmospheres create pressure. If the pressure is great enough (and the temperature is not too high), water may exist in liquid form. Liquid water can in turn lead to substantial water erosion. It is also thought to be critical to the existence of life. [Section 11.4]

- Atmospheres can create wind and weather and play a major role in long-term climate change. [Section 11.4]

- Interactions between atmospheric gases and the solar wind can create a magnetosphere around planets with strong magnetic fields [Section 11.3]. The magnetosphere can protect the atmosphere from loss of gas to the solar wind and can lead to beautiful auroras.

 Surface Temperature of Terrestrial Planets Tutorial, Lessons 1–4

11.2 The Greenhouse Effect and Planetary Temperature

Perhaps the single most important effect of an atmosphere is its ability to make a planetary surface warmer than it

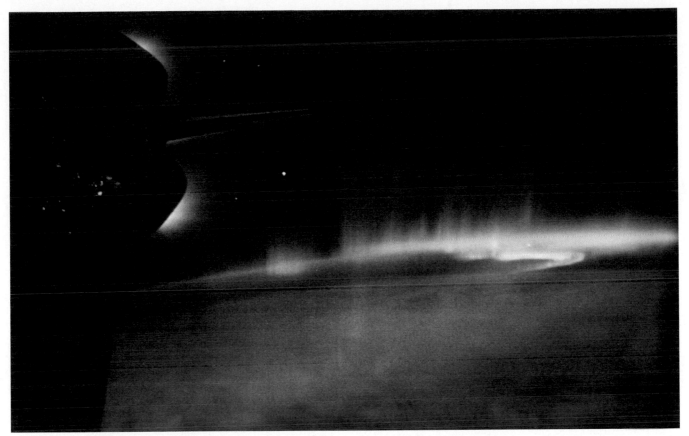

Figure 11.4 The atmosphere does not end suddenly. Some low-density gas is present even at altitudes of hundreds of kilometers, comprising what we call the exosphere. This photograph shows the visible glow of gas in the exosphere around the tail of an orbiting Space Shuttle. The aurora visible in the background is additional proof of the great vertical extent of the atmosphere. Charged particles from the magnetosphere interact with Earth's thin atmosphere above 100 kilometers.

would be otherwise. This planetary warming is caused by what we call the **greenhouse effect**.

You've probably heard of the greenhouse effect—it is an important part of the environmental problem known as global warming [Section 14.6]. You may be surprised to learn that the greenhouse effect is also critical to the existence of life on Earth. Without the greenhouse effect, Earth's surface would be too cold for liquid water to flow and for life to flourish. In this section, we'll explore how the greenhouse effect works and what it does to the temperatures of the terrestrial worlds.

How the Greenhouse Effect Works

The basic idea behind the greenhouse effect is quite simple (Figure 11.5). Sunlight consists mostly of visible light, which passes easily through most atmospheric gases to reach a planet's surface. Some of this visible light is absorbed by the ground (the rest is reflected). The ground returns the absorbed energy back toward space, but in the form of infrared rather than visible light. The reason why planets emit infrared light is that their temperatures are too low to emit visible light. Remember that all objects emit thermal radiation by virtue of their temperatures [Section 6.4]. Plan-

etary temperatures are in a range in which nearly all emitted energy is in the infrared.

The greenhouse effect works by "trapping" some of the infrared light emitted by the planet, slowing its return to space. Thus, the key to the greenhouse effect is gases that absorb infrared light effectively. Infrared light is best absorbed by molecules that can easily begin rotating and vibrating (see Figure 6.11). Molecules that are particularly good at absorbing infrared light—such as water vapor (H_2O), carbon dioxide (CO_2), and methane (CH_4)—are called **greenhouse gases**.

When a greenhouse gas molecule absorbs a photon of infrared light, it begins to rotate and vibrate and then reemits a photon of infrared light. This photon can be emitted up, down, or sideways and is usually absorbed by another greenhouse molecule, which does the same thing. The net result is that greenhouse gases tend to slow the escape of infrared radiation from the lower atmosphere, while their molecular motions heat the surrounding air. In this way, the greenhouse effect makes the surface and the lower atmosphere warmer than they would be from sunlight alone. The more greenhouse gases present, the greater the degree of surface warming.

Visible light passes through atmosphere.

Greenhouse gases absorb and re-emit infrared radiation, thereby heating the lower atmosphere.

Surface absorbs visible light and emits thermal radiation in infrared.

Figure 11.5 The greenhouse effect. The lower atmosphere becomes warmer than it would be if it had no greenhouse gases such as water vapor, carbon dioxide, and methane.

The name *greenhouse effect* is a bit of a misnomer. The name comes from botanical greenhouses, usually built mostly of glass, which keep the air inside them warmer than it would be otherwise. Greenhouses, however, trap heat simply by not letting hot air rise, very different from the way infrared absorption traps heat and leads to the "greenhouse effect" in planetary atmospheres.

THINK ABOUT IT

Clouds on Earth are made of water (H_2O), which acts as a very effective greenhouse gas even when the water vapor condenses into droplets in clouds. Use this fact to explain why clear nights tend to be colder than cloudy nights.

What If There Were No Greenhouse Effect?

Our overview of the greenhouse effect captures its essence, but we need to examine it in a little more depth to fully appreciate how it warms the terrestrial worlds. We can do so by asking what planetary temperatures would be in the absence of a greenhouse effect.

Planetary Energy Balance With or without a greenhouse effect, the amount of energy a planet receives from the Sun must be precisely balanced with the amount of energy it returns to space. Otherwise, the planet would rapidly heat up (if it received more energy than it returned) or cool down (if it returned more energy than it received). Incoming energy comes in the form of sunlight. The planet returns this energy to space through some combination of reflection of sunlight and emission of infrared radiation. Figure 11.6 summarizes this energy balance. (Remember that a planet's surface temperature is not significantly affected by internal heat sources like radioactivity.)

The greenhouse effect cannot change the amount of incoming sunlight and thus cannot change the amount of

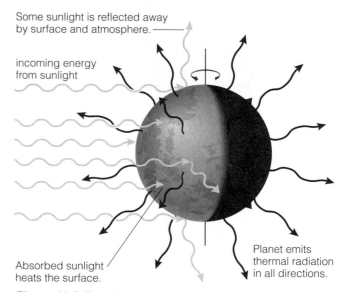

Some sunlight is reflected away by surface and atmosphere.

incoming energy from sunlight

Absorbed sunlight heats the surface.

Planet emits thermal radiation in all directions.

Figure 11.6 To maintain a steady temperature, a planet must emit precisely as much energy as it absorbs from sunlight.

energy the planet returns to space. So how can the greenhouse effect make a planet warmer while the overall energy balance remains unchanged? A good analogy is the way a blanket keeps you warm at night. The blanket itself does not warm your body but instead keeps you warm by slowing the escape of your body heat. In much the same way, the greenhouse effect warms a planet by slowing the escape of infrared radiation from its surface. In essence, the greenhouse effect increases the amount of energy in the lower atmosphere at any given time—making it warmer—while leaving the total amount of energy that escapes to space unchanged.

"No Greenhouse" Planetary Temperatures The ideas of energy balance make it easy to understand a planet's surface temperature in the absence of a greenhouse effect. In that case, a planet's average surface temperature would be affected by only two things:

- *The planet's distance from the Sun,* which determines the amount of energy incoming as sunlight. The closer a planet is to the Sun, the greater the intensity of the incoming sunlight.

- *The planet's overall reflectivity,* which determines the relative proportions of incoming sunlight that the planet reflects and absorbs. (You may sometimes see reflectivity described by the term *albedo.*) The higher the reflectivity, the less light absorbed and the cooler the planet.

A planet's reflectivity depends on its surface composition and color. The darker the surface, the more light it absorbs and the less it reflects. For example, clouds, snow, and ice reflect 70% or more of the light that hits them, while rocks typically reflect only about 20% of the light that hits them and absorb the other 80%.

Who gets hotter in the summer sun, you wearing a white shirt or a person in black next to you? Explain.

Both distance from the Sun and reflectivity have been measured for all the terrestrial worlds. With a little mathematics (see Mathematical Insight 11.1), it's possible to use these measurements to calculate the temperature any planet would have without greenhouse gases. For example, such calculations show that Earth's **global average temperature**—that is, the average temperature for the entire planet—would be a chilly $-17°C$ ($-1°F$) without the greenhouse effect. Because this "no greenhouse" temperature is well below the freezing point of water, we conclude that our planet would be frozen over in the absence of a greenhouse effect. Clearly, the greenhouse effect is very important to life on Earth.

Greenhouse Warming of the Terrestrial Worlds

Table 11.2 lists the distance and reflectivity for each terrestrial world and compares the world's calculated "no greenhouse" temperatures to actual surface temperatures. The "no greenhouse" temperatures for Mercury and the Moon lie between their actual day and night temperatures, just as we should expect on worlds with very little atmosphere and hence no greenhouse effect. For the other planets, however, the greenhouse effect plays an important role.

The greenhouse effect is fairly weak on Mars, where the actual average surface temperature is only 5°C higher than the "no greenhouse" temperature. Earth has a moderate greenhouse effect that makes the actual global average temperature about 15°C (59°F), which is 32°C warmer than the "no greenhouse" temperature. Venus has an extreme greenhouse effect that bakes the surface to a temperature that is more than 500°C hotter than it would be otherwise.

Mathematical Insight ● **11.1** **"No Greenhouse" Temperatures**

As discussed in the text, the "no greenhouse" temperature of a planet in our solar system depends only on the planet's distance from the Sun and its overall reflectivity. From those two factors and the Sun's brightness, we can derive the following formula for a planet's "no greenhouse" temperature:

$$T = 280 \text{ K} \times \sqrt[4]{\frac{(1 - \text{reflectivity})}{d^2}}$$

where d is the distance from the Sun in AU and the reflectivity is stated as a fraction between 0 and 1. For example, if a planet reflects 20% of the incoming sunlight, its reflectivity would be stated as 0.2. The symbol $\sqrt[4]{}$ means the fourth root, or $\frac{1}{4}$ power.

We will not derive the formula here, but you can see how it makes sense. The term $(1 - \text{reflectivity})$ is the proportion of sunlight that the planet absorbs, which is the light that heats its surface. This term is divided by d^2 because the amount of energy in sunlight (per unit area) declines with the square of distance from the Sun, just as gravity declines with the square of distance. Thus, the full term $(1 - \text{reflectivity})/d^2$ represents the total amount of energy that the planet absorbs from sunlight (per unit area) each second. This energy warms the surface, and the surface returns the energy to space by thermal radiation [Section 6.4]. The energy emitted by thermal radiation depends on temperature raised to the fourth power (see Mathematical Insight 6.2). Calculating the planet's temperature therefore requires taking the fourth root of the absorbed energy, or the fourth root of $(1 - \text{reflectivity})/d^2$. The 280 K in the formula happens to be the temperature of a perfectly black planet at 1 AU from the Sun.

Example: Calculate the "no greenhouse" temperature of Mercury.

Solution: Table 11.2 gives Mercury's distance from the Sun as $d = 0.387$ AU and its reflectivity as 11%, or 0.11 when expressed as a decimal fraction. Plugging these numbers into the formula gives Mercury's "no greenhouse" temperature:

$$T = 280 \text{ K} \times \sqrt[4]{\frac{(1 - \text{reflectivity})}{d^2}}$$

$$= 280 \text{ K} \times \sqrt[4]{\frac{(1 - 0.11)}{0.387^2}} = 437 \text{ K}$$

We can convert the Kelvin temperature to Celsius by subtracting 273 (see Mathematical Insight 4.1), giving us the "no greenhouse" temperature of 164°C shown for Mercury in Table 11.2.

Table 11.2. The Greenhouse Effect on the Terrestrial Worlds

World	Average Distance from Sun (AU)	Reflectivity	"No Greenhouse" Average Surface Temperature*	Actual Average Surface Temperature	Greenhouse Warming (actual temperature minus "no greenhouse" temperature)
Mercury	0.387	11%	164°C	425°C (day), −175°C (night)	—
Venus	0.723	72%	−43°C	470°C	513°C
Earth	1.00	36%	−17°C	15°C	32°C
Moon	1.00	7%	0°C	125°C (day), −175°C (night)	—
Mars	1.52	25%	−55°C	−50°C	5°C

*The "no greenhouse" temperature is calculated by assuming no change to the atmosphere other than lack of greenhouse warming. Thus, for example, Venus ends up with a lower "no greenhouse" temperature than Earth even though it is closer to the Sun, because the high reflectivity of its bright clouds means that it absorbs less sunlight than Earth.

If you think about it, this greenhouse warming is quite remarkable. Consider Earth. Our atmosphere is made almost entirely of nitrogen and oxygen, which have no effect on infrared light and do not contribute to the greenhouse effect. (Molecules with only two atoms, especially those with two of the same kind of atom, such as N_2 and O_2, are poor infrared absorbers because they have very few ways to vibrate and rotate.) Thus, greenhouse gases are only trace constituents of Earth's atmosphere, but they warm our planet enough to raise the global average temperature from well below freezing to well above it.

THINK ABOUT IT

Suppose nitrogen and oxygen were greenhouse gases. Assuming our atmosphere still had the same composition and density, how would Earth be different?

COMMON MISCONCEPTIONS

The Greenhouse Effect Is Bad

The greenhouse effect is often in the news, usually in discussions about environmental problems, but the greenhouse effect itself is not a bad thing. In fact, we could not exist without it. The "no greenhouse" temperature of Earth is well below freezing. Thus, the greenhouse effect is the only reason why our planet is not frozen over. Why, then, is the greenhouse effect discussed as an environmental problem [Section 14.4]? It is because human activity is adding more greenhouse gases to the atmosphere—which might change Earth's climate. While the greenhouse effect makes Earth livable, it is also responsible for the searing 470°C temperature of Venus—proving that it's possible to have too much of a good thing.

11.3 Atmospheric Structure

You may have heard different regions of Earth's atmosphere referred to by different names. For example, you may know that one relatively high region is called the *stratosphere* or that the lower region near the surface is called the *troposphere*. This layering of the atmosphere, often called the **atmospheric structure**, is important to understanding how atmospheres affect planets. In this section, we'll investigate how and why atmospheres have their structures. Along the way, we'll also answer some interesting questions, such as how ozone is important to life on Earth and why the sky is blue.

Interactions with Light and Atmospheric Structure

The key to understanding atmospheric structure lies in interactions between atmospheric gases and energy from the Sun. Essentially all of the energy in terrestrial atmospheres comes from sunlight. (Jovian planets get some of their atmospheric energy from internal heat.) In the previous section, we saw how the atmosphere's interaction with visible and infrared light can have a profound effect on a planet's surface temperature. Here, we'll see that other wavelengths of light also have important effects.

Most light coming from the Sun is visible light, but the Sun also emits significant amounts of ultraviolet light and X rays. The light emitted from a planet is almost entirely in the infrared. Atmospheric gases interact with each of these forms of light in different ways (summarized in Figure 11.7):

● X rays have enough energy to knock electrons free from almost any atom or molecule; that is, they ionize [Section 4.3] the atoms or molecules they strike. Thus,

X rays ionize (knock electrons off) almost any gas and dissociate (break apart) molecules when they are absorbed.

Ultraviolet photons dissociate molecules when they are absorbed.

Most visible-light photons are simply transmitted, though some are scattered.

Infrared photons are absorbed by molecules, causing them to vibrate and rotate.

Figure 11.7 The primary effects when light of different energies strikes common atmospheric gases.

X rays are easily absorbed by virtually all atmospheric gases, ionizing the gases in the process.

- Ultraviolet photons can break apart some molecules. For example, a molecule of water vapor (H_2O) can split apart if it absorbs an ultraviolet photon. Weakly bonded molecules, such as ozone (O_3), are particularly good at absorbing ultraviolet light, splitting apart in the process.

- Visible-light photons usually pass through atmospheric gases without being absorbed, but some are *scattered* so that their direction changes.

- As we discussed earlier, infrared light can be absorbed by greenhouse gases—molecules such as water vapor (H_2O), carbon dioxide (CO_2), and methane (CH_4)—that easily begin rotating and vibrating.

The absorption of any kind of light tends to warm gases. This warming leads directly to atmospheric structure.

Earth's Atmospheric Structure

Atmospheric structure varies among the terrestrial worlds. We'll start by examining the most familiar case. Figure 11.8 shows Earth's atmospheric structure. Both pressure and density drop rapidly with increasing altitude, so they have little effect on atmospheric layering. Temperature changes

with altitude in a more complex way. These varied temperatures help define the following four major layers of Earth's atmosphere (going upward from the ground):

- The **troposphere** is the lowest layer of the atmosphere. The temperature drops with altitude in the troposphere, something you've probably noticed if you've ever climbed a mountain.

- The **stratosphere** begins where the temperature stops dropping and instead begins to rise with altitude. High in the stratosphere, the temperature falls again. (Technically, the region where the temperature falls again is called the *mesosphere,* but we will not make this distinction in this book.)

- The **thermosphere** begins at a high altitude where the temperature again starts to rise.

- The **exosphere** is the uppermost region in which the atmosphere gradually fades away into space. The gas in the exosphere is so low in density that we usually don't think of it as "air" at all, but it can still exert drag on orbiting satellites (see Figure 11.4).

Figure 11.8 also shows where different forms of light are absorbed in the atmosphere. We'll next go layer by layer from the ground up to see how interactions between light and gas explain Earth's atmospheric structure.

Visible Light: Warming the Surface and Coloring the Sky For the most part, only visible light from the Sun can penetrate all the way through the atmosphere to the ground. Visible light reaches the ground because atmospheric gases are generally transparent to it. The ground warms as it absorbs visible light. (Some of the lowest-energy ultraviolet light also reaches the ground, where it can cause sunburns. On sunscreen labels, this low-energy ultraviolet light is called "UVA" and "UVB," with the latter having slightly higher energy.)

Although most visible-light photons pass through Earth's atmosphere without being disturbed, a small proportion is scattered randomly around the sky. This scattering is the reason our sky is bright rather than dark (which is why we cannot see stars in the daytime). Without scattering, our sky would look like the lunar sky does to an astronaut, with the Sun just a very bright circle set against a black, star-studded background. Scattering also prevents shadows on Earth from being pitch black. On the Moon, shadows receive no scattered sunlight and are extremely cold and dark.

The scattering of visible light also explains why our sky is blue. Visible light consists of all the colors of the rainbow, but not all the colors are scattered equally. Gas molecules scatter blue (higher-energy) light much more effectively than red (lower energy) light. The difference in scattering is so great that, for practical purposes, we can imagine that only the blue light gets scattered. When the Sun is overhead, this scattered blue light reaches our eyes from all directions and

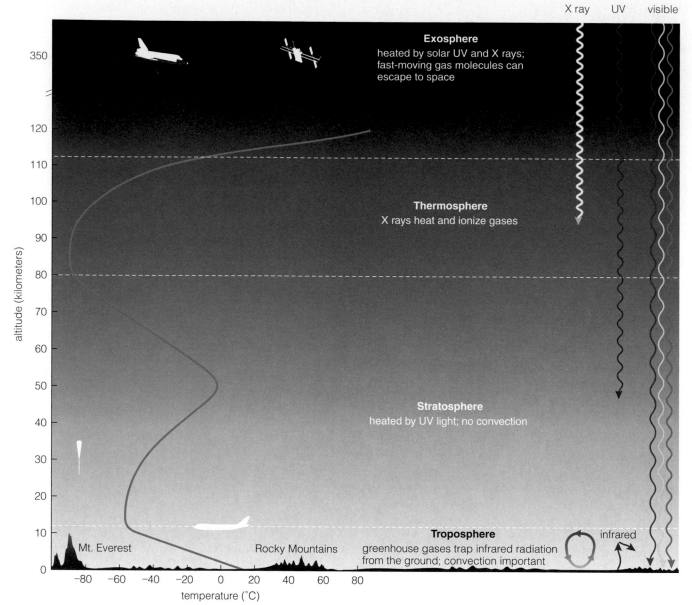

Figure 11.8 Earth's average atmospheric structure. The background color represents the air density and pressure, both of which decline rapidly with altitude. The curve shows average temperatures for each altitude. The squiggly arrows to the right represent photons and show where they are typically absorbed in the atmosphere.

the sky appears blue (Figure 11.9). At sunset or sunrise, the sunlight must pass through a greater amount of atmosphere on its way to us. Most of the blue light is scattered away, leaving only red light to color the sky.

Light scattering also helps explain the sky colors on other planets. The atmosphere of Mars is too thin to cause significant scattering, so Mars does not get the blue sky effect of Earth. Instead, the sky is a pale red from the presence in the atmosphere of reddish dust from the surface. Venus has so much atmosphere that its sky is similar to what we see at sunset. The thick atmosphere scatters nearly all the blue light away, leaving only reddish light to reach the ground. The sky on Venus therefore is dimly lit and reddish orange in color.

Infrared Light and the Troposphere The troposphere gets its structure because of the way atmospheric gases interact with infrared light radiated upward by the ground. That is, the troposphere is warmed by the greenhouse effect. Because the infrared light absorbed in the troposphere is coming from the surface, more is absorbed closer to the ground than at higher altitudes. That is why the temperature drops with altitude in the troposphere. (The relatively small amount of infrared light coming from the Sun is mostly absorbed by atmospheric molecules before it reaches the ground and thus has relatively little effect on the atmosphere.)

The drop in temperature with altitude, combined with the relatively high density of air in the troposphere, explains why the troposphere is the only layer of the atmosphere

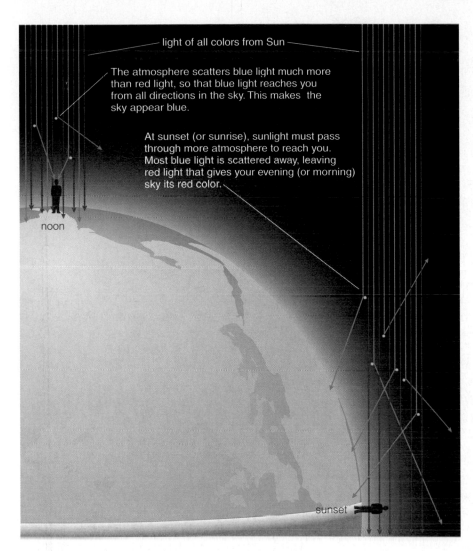

light of all colors from Sun

The atmosphere scatters blue light much more than red light, so that blue light reaches you from all directions in the sky. This makes the sky appear blue.

At sunset (or sunrise), sunlight must pass through more atmosphere to reach you. Most blue light is scattered away, leaving red light that gives your evening (or morning) sky its red color.

noon

sunset

Figure 11.9 This diagram summarizes why the sky is blue and sunsets (and sunrises) are red. Atmospheric gases scatter blue light more than they scatter red light. During most of the day, you therefore see blue photons coming from most directions in the sky, making the sky look blue. Only the red photons reach your eyes at sunrise or sunset, when the light must travel a longer path through the atmosphere to reach you.

with storms. The primary cause of storms is the churning of air by convection, in which warm air rises and cool air falls. Remember that convection occurs only when there is strong heating from below [Section 10.2]. In the troposphere, the heating from the ground can drive convection. Indeed, the troposphere gets its name from this convection, since *tropos* is Greek for "turning."

Ultraviolet Light and the Stratosphere

The primary source of heat in the troposphere—the absorption of infrared radiation from the ground—is not important in higher layers of the atmosphere. Once infrared radiation from the ground reaches the top of the troposphere, it travels essentially unhindered to space. Thus, we need to consider only the effects of sunlight in investigating atmospheric structure above the troposphere.

As we move upward through the stratosphere, the primary source of atmospheric heating is absorption of solar ultraviolet light by ozone. This heating tends to be stronger at higher altitudes, because almost all the ultraviolet light from the Sun gets absorbed before it reaches the bottom of the stratosphere. As a result, the lower stratosphere gets

Why Is the Sky Blue?

If you ask around, you'll find a wide variety of misconceptions about why the sky is blue. For example, some people think the sky is blue because of light reflecting from the oceans. Anyone who lives in Nebraska will immediately know that this must be wrong—Nebraska is too far from any ocean for its sky to be influenced by such reflection. Other people claim that "air is blue." It's not obvious what someone might mean by this statement, but it's clearly wrong. If air molecules emitted blue light, then air would glow blue even in the dark. If air molecules were blue because they reflected blue light and absorbed red light, then no red light could reach us at sunset. The real explanation for the blue sky is light scattering, as shown in Figure 11.9. This explanation not only makes sense for our blue sky but also explains our red sunsets.

Higher Altitudes Are Always Colder

Many people think that the low temperatures in the mountains are just the result of lower pressures, but Figure 11.8 shows that it's not that simple. The higher temperatures near sea level on Earth are a result of the greenhouse effect, which traps more heat at lower altitudes. If Earth had no greenhouse gases, mountaintops wouldn't be so cold—or, more accurately, sea level wouldn't be so warm.

warmer with increasing altitude—the opposite of the situation in the troposphere.

Convection cannot occur in the lower stratosphere because heat cannot rise if the air is even hotter higher up. The lack of convection makes the air relatively stagnant and *stratified* (or layered)—with layers of warm air overlying cooler air—which is how the stratosphere gets its name. The lack of convection also means that the stratosphere essentially has no weather and no rain. Pollutants that reach the stratosphere—such as ozone-destroying chlorofluorocarbons (CFCs) [Section 14.6]—remain there for decades.

A planet can have a stratosphere *only* if its atmosphere contains molecules that are particularly good at absorbing ultraviolet photons. Ozone plays this role on Earth, but the lack of oxygen in the atmospheres of the other terrestrial worlds means that they also lack ozone (O_3). As a result, Earth is the only terrestrial world with a stratosphere, at least in our solar system. (The jovian planets have stratospheres due to other ultraviolet-absorbing molecules [Section 12.3].)

X Rays and the Thermosphere
Because nearly all gases are good X-ray absorbers, X rays from the Sun are absorbed by the first gases they encounter as they enter the atmosphere. Because the density of gas in the exosphere is too low to absorb significant amounts of the incoming X rays, these gases are absorbed in the next layer, the thermosphere. Temperatures rise quite high in the thermosphere (*thermos* is Greek for "hot"), but you wouldn't feel hot there because the density and pressure are so low [Section 4.2]. Virtually no X rays penetrate beneath the thermosphere, which is why X-ray telescopes are useful only on very high flying balloons, rockets, and spacecraft.

In addition to heating the thermosphere, X rays ionize a small but important fraction of its gas. The portion of the thermosphere that contains most of the ionized gas is called the *ionosphere*. Thus, the ionosphere is part of the thermosphere, not a separate layer. The ionosphere is very important to radio communications on Earth. Most radio broadcasts are completely reflected back to Earth's surface by the ionosphere—almost as though Earth were wrapped in aluminum foil. Without reflection by the iono-

sphere, radio communication would work only between locations in sight of each other.

The Exosphere
The region above the thermosphere, where the atmosphere blends into space, is called the **exosphere** (*exo* means "outermost" or "outside"). The gas density in the exosphere is so low that collisions between atoms or molecules are very rare, although the high temperature means that gas particles move quite rapidly. Lightweight gas molecules sometimes reach escape velocity [Section 5.1] and fly off into space.

Comparative Structures of the Terrestrial Atmospheres

We've discussed everything we need to understand the structures of all the terrestrial atmospheres. The Moon and Mercury have so little gas that they essentially contain only an exosphere and have no structure to speak of. Figure 11.10 contrasts the structures of the lower atmospheres of Venus, Earth, and Mars. The following key ideas should make sense to you:

● All three planets have a warm troposphere at the bottom of the atmosphere, created by the greenhouse effect, and a warm thermosphere at the top, where solar X rays are absorbed.

● Only Earth has the extra "bump" of a stratosphere, because only Earth has a layer of ultraviolet-absorbing gas (ozone). Without the warming in our unique layer of stratospheric ozone, the middle altitudes of Earth's atmosphere would be almost as cold as those on Mars.

● All three planets are warmer at the surface than they would be without the greenhouse effect, although the amount of warming is far greater for Venus than for Earth or Mars.

THINK ABOUT IT

Would astronauts on the Moon need protection from solar X rays? What about astronauts on Mars? Explain.

Magnetospheres and the Solar Wind

We have yet to consider one more type of energy coming from the Sun: the low-density breeze of charged particles that we call the *solar wind* [Section 9.3]. The solar wind does not significantly affect atmospheric structure, but it can have other important effects. The effects depend largely on whether a planet has a strong enough magnetic field [Section 10.2] to divert the charged particles of the solar wind.

Worlds without a strong magnetic field are impacted directly by the subatomic solar wind particles. For example, solar wind particles impact the exospheres of Venus and Mars and the surfaces of Mercury and the Moon. In contrast, Earth's strong magnetic field creates a **magneto-**

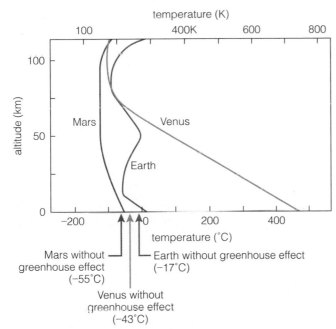

temperature (K)

Figure 11.10 Temperature profiles for Venus, Earth, and Mars. Note that Venus and Earth are considerably warmer than they would be without the greenhouse effect. (The thermospheres and exospheres, which extend above the top of the graph, are qualitatively alike.)

sphere that acts like a protective bubble surrounding the planet (Figure 11.11a). (Jovian planets also have magnetospheres [Section 12.4].)

Charged particles cannot easily pass through a magnetic field, so a magnetosphere diverts most of the solar wind around a planet. However, some solar wind particles infiltrate Earth's magnetosphere at its most vulnerable points, near the magnetic poles. Once they are inside the magnetosphere, the magnetic field forces charged particles to spiral along magnetic field lines. In Earth's magnetosphere, the ions and electrons accumulate to make **charged particle belts** encircling our planet. These belts are often called the *Van Allen belts* after their discoverer. (Similar charged particle belts are found in jovian planet magnetospheres.) The high energies of the particles in charged particle belts can be hazardous to spacecraft and astronauts passing through them.

Charged particles trapped in the magnetosphere create the beautiful spectacle of light we call the **aurora** (Figure 11.11b). Variations in the solar wind can buffet the magnetosphere and give energy to particles trapped there. If a trapped particle gains enough energy, it can follow the magnetic field all the way down to Earth's atmosphere, where it collides with atmospheric atoms and molecules. These collisions cause the atoms and molecules to radiate and produce the moving lights of the aurora. Because the charged particles follow the magnetic field, auroras are most common near the magnetic poles and are best viewed at high latitudes. In the Northern Hemisphere, the aurora is often called the *aurora borealis*, or northern lights. In the Southern Hemisphere, it is called the *aurora australis*, or south-

ern lights. The aurora can also be photographed from space, where the lights can look much like surf in the upper atmosphere (Figure 11.11c). We have also seen auroras on jovian planets.

The solar wind varies in strength at different times. For example, it tends to be strongest after the Sun ejects an unusually large amount of plasma into space. Explain why more people are likely to see the aurora after such events.

11.4 Weather and Climate

So far, we've talked mostly about *average* conditions in planetary atmospheres, such as the average temperature on the surface and at each altitude above the surface. From our experience on Earth, however, we know that surface and atmospheric conditions constantly change. These changes are generally described as *weather* or *climate*. Although all atmospheric layers can undergo such changes, we'll focus on the troposphere where the effects are greatest.

Weather and climate are closely related, but they are not quite the same thing. In any particular location, some days may be hotter or cooler, clearer or cloudier, or calmer or stormier than others. The ever-varying combination of winds, clouds, temperature, and pressure is what we call **weather**. The complexity of weather makes it difficult to predict, and at best the local weather can be predicted only a week or so in advance.

Climate is the long-term average of weather, which means it can change only on much longer time scales. For example, Antarctic deserts have a cold, dry climate, while tropical rain forests have a hot, wet climate. Neither type of climate changes naturally in periods of less than a few decades, and climates are often stable for hundreds or thousands of years.

Weather and climate can be hard to distinguish on a human time scale. For example, a drought lasting a few years may be the result of either random weather fluctuations or of a gradual change in which the climate is becoming drier. The difficulty in distinguishing between random weather and real climate trends is an important part of the debate over global warming. For now, we'll focus on understanding weather and climate generally.

Global Wind Patterns

If you turn on any television weather report, it will quickly become obvious that winds blow across our planet in rather distinctive patterns. Certainly, in any particular place the wind's direction and strength may change rapidly. On a global scale, however, certain wind patterns are always the same. For example, storms moving in from the Pacific hit the west coast of the United States first and then make their way eastward across the Rocky Mountains, the Great Plains,

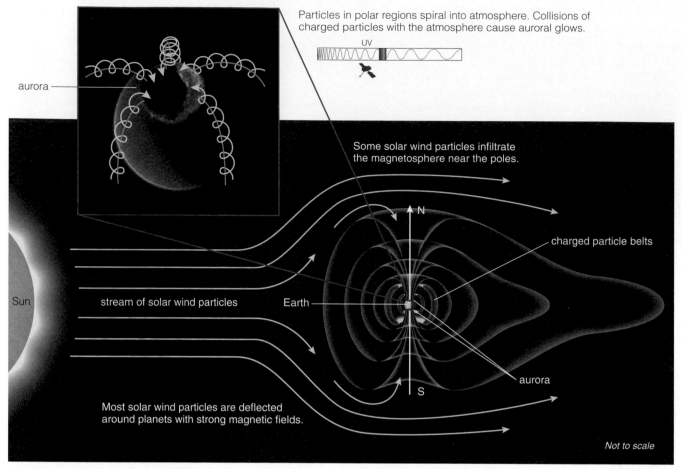

Particles in polar regions spiral into atmosphere. Collisions of charged particles with the atmosphere cause auroral glows.

Some solar wind particles infiltrate the magnetosphere near the poles.

charged particle belts

stream of solar wind particles

Earth

Sun

aurora

Most solar wind particles are deflected around planets with strong magnetic fields.

aurora

Not to scale

a Earth's magnetosphere and its interaction with the solar wind. The inset is a photo of a ring of auroras around the North Pole. The crescent is the day side of Earth.

b Aurora photographed in Mount Rainier National Park, Washington State.

VIS

c Earth's aurora as seen from the Space Shuttle.

VIS

Figure 11.11 A planet's magnetosphere acts like a protective bubble that shields the surface from charged particles coming from the solar wind. Among the terrestrial planets, only Earth has a strong enough magnetic field to create a magnetosphere. Earth's magnetosphere allows charged particles to strike the atmosphere only near the poles, thereby creating the phenomena of the aurora borealis and aurora australis.

and to the east coast. Winds that stay the same make up what we call Earth's **global wind patterns** (or *global circulation*).

Figure 11.12 shows Earth's global wind patterns. Note how the wind direction changes back and forth with latitude. What causes this strange pattern of winds? Two factors explain it: atmospheric heating and planetary rotation. We'll examine each factor in turn. Along the way, we'll see how we might expect global wind patterns to be different on other planets.

THINK ABOUT IT

Suppose you wanted to take a sailing trip from North America to Asia. Does it matter how far north or south you plot your course? Explain. (*Hint*: Study Figure 11.12.)

Atmospheric Heating and Circulation Cells

The first major factor affecting global wind patterns is atmospheric heating. In general, the equatorial regions of a planet receive more heat from the Sun than do polar regions. The excess heat makes the atmosphere expand above the equator so that it rises upward and moves toward the poles. Near the poles, cool air descends and flows toward the equator. If this process acted alone, the result would be two huge **circulation cells**, one in each hemisphere, transport-

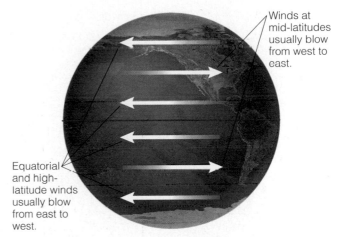

Winds at mid-latitudes usually blow from west to east.

Equatorial and high-latitude winds usually blow from east to west.

Figure 11.12 Schematic of Earth's global wind patterns. At mid-latitudes, such as over North America, the surface winds blow from the west.

ing air between the equator and the poles (Figure 11.13). (The circulation cells are often called *Hadley cells* after the man who first suggested their existence in 1735.) The circulation cells resemble convection cells in that they carry warm air upward and cool air downward, but they are much larger.

SPECIAL TOPIC Weather and Chaos

Scientists today have a very good understanding of the physical laws and mathematical equations that govern the behavior and motion of atoms in the air, oceans, and land. Why, then, do we have so much trouble predicting the weather? To understand why weather is so unpredictable, we must look at the nature of scientific prediction.

Suppose you want to predict the location of a car on a road 1 minute from now. You need two basic pieces of information: where the car is now, and how it is moving. If the car is now passing Smith Road and heading north at 1 mile per minute, it will be 1 mile north of Smith Road in 1 minute.

Now suppose you want to predict the weather. Again, you need two basic types of information: (1) the current weather and (2) how weather changes from one moment to the next.

You could attempt to predict the weather by creating a "model world." For example, you could overlay a globe of Earth with graph paper and then specify the current temperature, pressure, cloud cover, and wind within each square. These are your starting points, or initial conditions. Next, you could input all the initial conditions into a computer, along with a set of equations (physical laws) that describe the processes that can change weather from one moment to the next.

Suppose the initial conditions represent the weather around Earth at this very moment and you run your computer model to predict the weather for the next month in New York City. The model might tell you that tomorrow will be warm and sunny, with cooling during the next week and a major storm passing through a month from now. Now suppose you run the model again but

make one minor change in the initial conditions—say, a small change in the wind speed somewhere over Brazil. This slightly different initial condition will not change the weather prediction for tomorrow in New York City. For next week's weather, the new model may yield a slightly different prediction. For next month's weather, however, the two predictions may not agree at all!

The disagreement between the two predictions arises because the laws governing weather can cause very tiny changes in initial conditions to be greatly magnified over time. This extreme sensitivity to initial conditions is sometimes called the *butterfly effect*: If initial conditions change by as much as the flap of a butterfly's wings, the resulting prediction may be very different.

The butterfly effect is a hallmark of *chaotic systems*. Simple systems are described by linear equations in which, for example, increasing a cause produces a proportional increase in an effect. In contrast, chaotic systems are described by nonlinear equations, which allow for subtler and more intricate interactions. For example, the economy is nonlinear because a rise in interest rates does not automatically produce a corresponding change in consumer spending. Weather is nonlinear because a change in the wind speed in one location does not automatically produce a corresponding change in another location.

Despite their name, chaotic systems are not necessarily random. In fact, many chaotic systems have a kind of underlying order that explains the general features of their behavior even while details at any particular moment remain unpredictable. In a sense, many chaotic systems—like the weather—are "predictably unpredictable."

Figure 11.13 Circulation cells. Heat rises above the equator, setting up a flow of warm air toward the poles at high altitudes and a flow of cool air toward the equator near the surface. The planet's rotation is neglected for the moment.

Cool air descends over poles.

Warm air rises over equator.

The circulation cells transport heat both from lower to higher altitudes and from the equator to the poles. If the circulation is very efficient and the atmosphere is dense enough to carry a lot of thermal energy, planetary temperatures can be equalized from equator to pole. Such is the case on Venus, where temperatures near the poles are essentially the same as at the equator. The circulation is less efficient on Earth, but it still makes the poles much warmer than they would be in the absence of circulation. On Mars, the circulation cells transport very little heat because the atmosphere is so thin, so the poles remain much colder than the equator.

Rotation and the Coriolis Effect

The second major factor affecting global wind patterns is a planet's rotation. If the rotation is fast enough, it can split each of the two huge circulation cells into several smaller ones. The key to understanding how rotation splits circulation cells is the way air can be diverted by what is called the **Coriolis effect** (named for the French physicist who first explained it).

It's easy to understand the Coriolis effect on a spinning merry-go-round (Figure 11.14). The regions near the outer edge travel at faster speed than the regions near the center because they have a greater distance to travel around the axis with each rotation. If you sit near the edge and roll a ball toward the center, the ball begins with your relatively high speed around the axis. As it rolls inward, the ball's high speed around the axis makes it move ahead of the slower-moving inner regions. On a merry-go-round rotating counterclockwise, the ball therefore deviates *to the right* instead of heading straight inward. The deviation caused by the merry-go-round's rotation is an example of the Coriolis effect.

The Coriolis effect also makes the ball deviate to the right if you roll it outward from a position near the center.

In that case, the ball starts with your slower speed around the center and lags behind as it rolls outward to faster-moving regions. Again, it deviates to the right. If the merry-go-round rotates clockwise rather than counterclockwise, the deviations go to the left instead of the right.

The Coriolis effect alters the path of air on a rotating planet in much the same way. On a rotating planet, equatorial regions travel faster than polar regions around the rotation axis (see Figure 1.14). On Earth, air moving away from the equator deviates ahead of Earth's rotation to the east, and air moving toward the equator deviates behind Earth's rotation to the west. In either case, the deviation is to the *right* in the Northern Hemisphere and to the *left* in the Southern Hemisphere (Figure 11.15).

THINK ABOUT IT

The Coriolis effect also is important for long-range missiles. Suppose you are in the Northern Hemisphere on Earth and are trying to send a missile to hit a target that lies 5,000 kilometers due north of you. Should you aim directly at the target, somewhat to the left, or somewhat to the right? Explain.

One result is that air flowing toward low-pressure zones ends up circulating around them counterclockwise in the Northern Hemisphere and clockwise in the Southern Hemisphere. Uneven heating and cooling of the Earth's surface can create high- and low-pressure regions. Low-pressure regions ("L" on weather maps) are often associated with storms, including hurricanes. Because the circulating winds prevent air from flowing in to equalize the pressure, low-pressure regions tend to remain stable for many days. During this time, the global winds can carry them thou-

On a merry-go-round spinning counterclockwise, the Coriolis effect makes rolling balls deviate to the right.

Figure 11.14 The Coriolis effect on a merry-go-round rotating counterclockwise. A ball rolled inward starts with a speed around the center faster than that of regions nearer the center. This "extra" speed makes the ball deviate to the right. A ball rolled outward starts with a speed around the center slower than that of regions farther out. This "lag" also makes the ball deviate to the right. If the merry-go-round were rotating clockwise, both balls would veer to the left.

The Coriolis effect makes moving air deviate to its right in the Northern Hemisphere.

Low-pressure regions ("L") draw in air from surrounding areas.

The deviations make air flow counterclockwise around low-pressure regions in the Northern Hemisphere.

The deviations make air flow clockwise around low-pressure regions in the Southern Hemisphere.

The Coriolis effect makes moving air deviate to its left in the Southern Hemisphere.

a On Earth, the Coriolis effect causes air to deviate toward the right in the Northern Hemisphere and toward the left in the Southern Hemisphere. This makes storms around low-pressure regions ("L") circulate counterclockwise in the Northern Hemisphere and clockwise in the Southern Hemisphere.

b In this photograph of Earth, you can see the opposite directions of storm circulation in the two hemispheres. Contrast the counterclockwise swirling of the Northern Hemisphere storm near the upper left with the clockwise swirls in the Southern Hemisphere.

Figure 11.15 The Coriolis effect works on rotating planets much as it does on rotating merry-go-rounds, because regions near the equator move at faster speed around the axis than regions near the poles.

sands of kilometers across the planet. High-pressure regions ("H" on weather maps) can also last for days and travel great distances, but they usually are storm-free.

We can now see why the Coriolis effect splits the circulation cells. Consider air flowing southward along the surface from the North Pole. Without rotation, this air would head straight for the equator (see Figure 11.13). On a rotating planet, the Coriolis effect tends to divert air moving north or south into east-west winds. The result is that each single circulation cell must split into smaller ones. On Earth, each of the two huge circulation cells splits into three smaller cells (Figure 11.16). Note that surface winds in the cells near the equator and near the poles flow toward the equator. Hence, the Coriolis effect diverts these winds to the west in both hemispheres. That is why global winds near the equator and near the poles go westward (see Figure 11.12). In contrast, surface winds in the mid-latitude cells flow toward the poles and are diverted eastward by the Coriolis effect. That is why global winds at mid-latitudes generally blow west to east and why storms generally hit the west coast of North America first and progress eastward.

The strength of the Coriolis effect on other planets depends on their size and rotation rate: Both faster rotation and larger size increase its strength. Earth is the only terrestrial planet with a Coriolis force strong enough to split the two large circulation cells. Venus has a very weak Coriolis effect because of its slow rotation. Mars has a weak Coriolis effect because of its small size, even though it has nearly the same 24-hour rotation rate as Earth. The strongest Coriolis effect in our solar system occurs on Jupiter, whose combination of large size and rapid rotation splits the circulation cells into many more smaller cells than on Earth [Section 12.3].

Figure 11.16 On a rapidly rotating planet like Earth, the Coriolis effect causes each of the two large circulation cells (see Figure 11.13) to split into three cells.

Toilets in the Southern Hemisphere

A common myth holds that water circulates "backward" in the Southern Hemisphere—that toilets flush the opposite way, water spirals down sink drains the opposite way, and so on. If you visit an equatorial town, you may even find charlatans who, for a small fee, will demonstrate how to change the direction of water spiraling down a drain simply by stepping across the equator. This myth sounds similar to the Coriolis effect—which really does make hurricanes spiral in opposite directions in the two hemispheres—but it is completely untrue.

The Coriolis effect is noticeable only when material moves significantly closer to or farther from Earth's rotation axis, which means scales of hundreds of kilometers. Its effect is completely negligible on small scales, such as the scale of toilets or drains. Apart from the shape of the basin, the only thing that affects the direction in which the water in a toilet or sink swirls is the initial angular momentum, which you can affect by swirling the water in one direction or the other. Conservation of angular momentum [Section 5.2] then dictates that the water will swirl faster and faster as it gets closer to the drain. You can find water in toilets and sinks swirling in either direction, and even tornadoes twisting in either direction, in both hemispheres.

Clouds and Precipitation

Winds are an important part of planetary weather, but so are rain, hail, and snow, which together are called *precipitation* in weather reports. Precipitation requires clouds. We may think of clouds as imperfections on a sunny day, but

they have profound effects on a planet. Clouds fundamentally change the appearance of a planet, reflecting sunlight back to space and thereby reducing the amount of sunlight that warms the surface. Clouds are also important to planetary geology, because they are necessary for precipitation, a major cause of erosion. Despite their significant effects, clouds generally are made from minor ingredients of the atmosphere (see Table 11.1).

Clouds are made from tiny droplets of liquid or tiny particles of ice that condense from atmospheric gases. Such condensation is usually linked to convection (Figure 11.17). On Earth, evaporation of surface water (or sublimation of ice and snow) releases water vapor into the atmosphere. Convection carries the water vapor to high, cold regions of the troposphere, where it condenses into droplets of liquid water or flakes of ice. Thus, clouds on Earth consist of countless tiny water droplets or ice flakes. You can feel this if you walk through a cloud on a mountaintop. The droplets or flakes grow within the clouds until they become so large that the upward convection currents can no longer hold them aloft. They then begin to fall toward the surface as rain, snow, or hail.

Strong convection means more clouds and precipitation. Thunderstorms often form on summer afternoons when the sunlight-warmed surface drives stronger convection. Earth's equatorial regions also experience more convection and more rain. When you combine this fact with the global wind patterns (see Figure 11.12), you can understand why lush jungles lie at the equator and deserts lie at latitudes 20°–30° north and south of the equator. The equatorial rain depletes the moisture in the air as the winds carry it north or south, leaving little moisture to rain on the desert areas.

Clouds can also form when winds blow over mountains. As the air is pushed to high altitude over a mountain, droplets form. These droplets evaporate as the air descends

Figure 11.17 The cycle of water on Earth's surface and in the atmosphere.

Drops and flakes grow larger.

Water vapor condenses into droplets or flakes, forming clouds.

Rain and snow fall to the surface.

Convection carries air to higher altitudes and lower temperatures.

Evaporation (or sublimation) releases water vapor into lower atmosphere.

to the valley or plains below. The result is that droplets constantly form in the air over the mountain, creating a cloud that remains stationary even though strong winds are blowing. This process forms clouds over terrestrial mountain ranges on Earth such as the Rockies and also over tall Martian mountains such as Olympus Mons (Figure 11.18). The clouds over the Martian mountains consist of water-ice crystals, and water-ice fog sometimes fills Martian canyons. (However, the total amount of water vapor in the Martian atmosphere is so small that if spread planet-wide it would make a frost layer less than $\frac{1}{100}$ millimeter thick.)

Long-Term Climate Change

Long-term climate changes are more difficult to notice than day-to-day variations in the weather, but geological evidence proves that they occur. On Earth, we are well aware of ice ages and warm periods in our planet's past. Other planets have gone through even more dramatic change. For example, our geological tour of Mars showed that it must have been warm and wet in the distant past but now is very cold and dry [Section 10.5].

Four major factors have been implicated in long-term climate change on the terrestrial worlds:

● **Solar brightening**: The Sun has grown gradually brighter with time, increasing the amount of heat and light that reaches the planets.

● **Changes in axis tilt**: The tilt of Earth and the other planets can change over long periods of time.

● **Changes in planetary reflectivity**: Changes in reflectivity mean either more sunlight absorbed (tending to warm the planet) or less sunlight absorbed (tending to cool the planet).

● **Changes in greenhouse gas abundance**: More abundant greenhouse gases result in greater warming by the atmosphere.

The first factor, solar brightening, affects the entire solar system but becomes noticeable only over periods of tens of millions of years or longer. Both theoretical models of the Sun and observations of other Sun-like stars tell us that the Sun has gradually brightened with age [Section 15.3]. The Sun today is about 30% brighter than it was when our solar system was young. The Sun's gradual brightening tends to increase planetary surface temperatures unless it is counteracted by some other factor, such as a decrease in the concentration of greenhouse gases. Thus, all other things being equal, the planets should have been cooler in the distant past than they are today. (However, as we will see in the cases of Earth and Mars, all else was not equal, and both planets may have been warmer in the past.)

The other three factors involve changes that affect a single planet. A planet's axis tilt may slowly change over thousands or millions of years because of small gravitational

Figure 11.18 High-altitude clouds (the blue streaks in the upper right corner) over Olympus Mons, photographed by the *Mars Global Surveyor*.

tugs from moons, other planets, or the Sun. For example, Earth currently has an axis tilt of about $23\frac{1}{2}°$, but the tilt varies over time between about 22° and 25°, primarily because of gravitational tugs from Jupiter and the other planets (Figure 11.19). These small changes in Earth's axis tilt affect the climate. A greater tilt makes seasons more extreme, making summers warmer and winters colder. The summer warmth tends to prevent ice from building up, making the whole planet warmer on average. In contrast, a smaller tilt tends to keep polar regions colder and darker on average, allowing ice to build up. The periods of smaller axis tilt and colder climate correlate well with Earth's ice ages (especially when considered along with other small

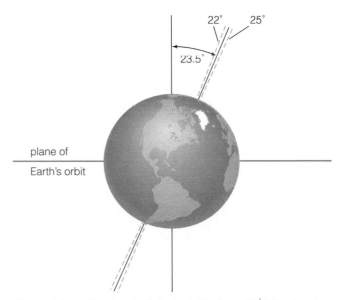

Figure 11.19 Earth's axis tilt is currently about $23\frac{1}{2}°$, but it varies over time between about 22° and 25°.

changes in Earth's rotation and orbit). Thus, they are thought to be a primary factor in climate changes on Earth. As we'll discuss shortly, Mars probably experiences much more extreme changes in axis tilt and climate than Earth, and similar climate cycles may occur on some of the icy moons of the outer solar system.

The third factor, changes in reflectivity, affects climate because it changes the proportion of sunlight absorbed and reflected. If a planet reflects more sunlight, it must absorb less. This can lead to planet-wide cooling. Microscopic dust particles (sometimes called *aerosols*) released by volcanic

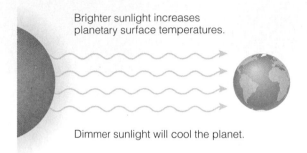

Brighter sunlight increases planetary surface temperatures.

Dimmer sunlight will cool the planet.

a Solar brightening.

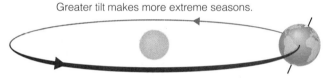

Greater tilt makes more extreme seasons.

Smaller tilt keeps polar regions colder and darker.

b Changes in axis tilt.

Higher reflectivity (from dust, smog in atmosphere, deforestation) leads to planetary cooling.

Lower reflectivity (e.g., due to paving) will warm the planet.

c Changes in planetary reflectivity.

An increase in greenhouse gases will warm the planet.

A decrease in greenhouse gases will cool the planet.

d Changes in greenhouse gas abundance.

Figure 11.20 Four major factors affecting long-term climate change.

eruptions can reflect sunlight and thus cool a planet. Small but measurable planet-wide cooling has been detected on Earth following a major volcanic eruption. The cooling can continue for years if the dust particles are blasted all the way into the stratosphere. Human activity is currently altering Earth's reflectivity, although we are not sure in which direction or by precisely how much. Smog particles can act like volcanic dust, reflecting sunlight before it reaches the ground. Deforestation also increases reflectivity because it removes sunlight-absorbing plants. On the other hand, paving a road with blacktop decreases reflectivity, because black roads absorb most of the sunlight that hits them. With different human activities altering Earth's reflectivity in different directions, the overall effects are not easy to gauge.

The fourth major cause of climate change is varying abundances of greenhouse gases. If the abundance of greenhouse gases increases, the planet generally will warm. If the planet warms enough, increased evaporation and sublimation may add substantial amounts of gas to the planet's atmosphere, leading to an increase in atmospheric pressure. Conversely, if the abundance of greenhouse gases decreases, the planet generally will cool, and atmospheric pressure may decrease as gases freeze.

Keep in mind that more than one process may be happening at any given time. Figure 11.20 summarizes these four processes involved in long-term climate change.

11.5 Sources and Losses of Atmospheric Gas

Of the factors that affect planetary climate, changes in greenhouse gas concentration appear to have had the greatest effect on the long-term climates of Venus, Earth, and Mars. Such changes generally occur not in isolation but rather as part of more general changes in the abundance of atmospheric gases. Thus, before we can understand the specific reasons for long-term climate changes on the terrestrial worlds, we must investigate how atmospheres gain and lose gas.

Each terrestrial world has gained and lost vast amounts of gas over time through a variety of processes. Figure 11.21 summarizes the various source and loss processes, which we will discuss in more detail in the rest of this section.

How Atmospheres Gain Gas

The jovian planets obtained their atmospheres by capturing gas from the solar nebula, but the terrestrial worlds were too small to capture significant amounts of gas before the solar wind cleared away the nebula [Section 9.3]. Any gas that the terrestrial worlds did capture from the solar nebula consisted primarily of hydrogen and helium (like the solar nebula as a whole). These light gases escaped easily from the terrestrial planets and are long gone. The atmospheres of the terrestrial planets therefore must have formed after the planets themselves.

Three Ways Atmospheres Gain Gas

a Three sources of atmospheric gas. Outgassing is the primary process for creating an atmosphere. Evaporation/sublimation occurs only after the atmosphere has been created. Bombardment is important only on nearly "airless" worlds like the Moon and Mercury, where it can create a very low-density atmosphere.

Five Ways Atmospheres Lose Gas

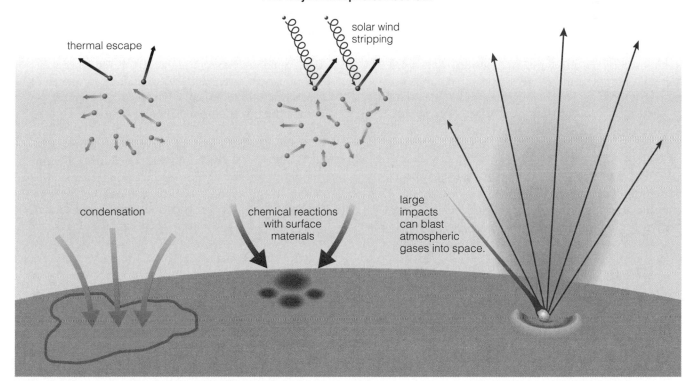

b Five processes for losing atmospheric gas once a planet has an atmosphere.

Figure 11.21 Atmosphere source and loss processes on terrestrial planets.

Where does the gas in terrestrial atmospheres come from? In general, three different processes can provide gas to terrestrial atmospheres:

● **Outgassing,** or the release by volcanism of gas trapped in interior rock [Section 10.3].

● **Evaporation/sublimation,** in which molecules break free from surface liquids or ices [Section 4.3] to become atmospheric gases.

● **Bombardment** by micrometeorites, solar wind particles, or high-energy solar photons, which if they reach the surface can blast individual atoms or molecules out of surface rock.

Outgassing by volcanic eruptions is by far the most important process. As we discussed in Chapter 10, virtually all of the gas that made the atmospheres of Venus, Earth, and Mars was released from their interiors by outgassing. The most common gases released are water (H_2O), carbon dioxide (CO_2), nitrogen (N_2), and sulfur-bearing gases (H_2S and SO_2). Among these materials, only water existed in modest quantities in the solar nebula. The others were created by chemical reactions that occurred *inside* the planets after gas became trapped. However, the modern-day atmospheres of Venus, Earth, and Mars do not have this same combination of atmospheric gases (see Table 11.1). Clearly, other source and loss processes can change atmospheric composition through time.

THINK ABOUT IT

Based on the fact that outgassing is the primary source of atmospheric gas on the terrestrial worlds, which fundamental property explains why Mercury and the Moon have virtually no atmosphere: planetary size, distance from the Sun, or rotation rate? Explain.

Evaporation (of liquid into gas) and sublimation (of solid into gas) can add substantial gas to an atmosphere, but only after the atmosphere is created initially by outgassing. Once outgassing creates an atmosphere, some of the gases may condense onto the surface. For example, Earth's oceans formed as outgassed water condensed and rained down to the surface [Section 14.3]. More generally, any surface liquids or ices will continually exchange gas with the atmosphere. If at any time the amount of gas being released by evaporation and/or sublimation is greater than the amount condensing (or being lost through other processes), the atmosphere will gain gas and surface pressure will increase. Because the gases released by evaporation or sublimation are usually greenhouse gases—such as water vapor on Earth and carbon dioxide on Mars—the gain of gas by evaporation/sublimation also tends to increase surface temperature.

Planets and moons on which outgassing and sublimation are negligible may still have very thin atmospheres created by bombardment. When micrometeorites, solar wind particles, or high-energy solar photons strike the surface, they can knock individual atoms or molecules free from surface rock. This bombardment can give the world a low-density atmosphere consisting of both vaporized surface materials and the particles bombarding the surface. As we'll discuss shortly, bombardment is the sole source of gas on "airless" worlds like the Moon and Mercury. Bombardment does not affect planets that have atmospheres from outgassing and evaporation/sublimation because their atmospheric gases prevent small particles and high-energy solar photons from reaching the surface.

How Atmospheres Lose Gas

Just as several processes can add gas to a planetary atmosphere, an atmosphere can also lose gas through several processes (see Figure 11.21):

- **Condensation** of atmospheric gas to make surface liquids or ices.
- **Chemical reactions** with surface rocks or liquids.
- **Solar wind stripping** of gases from the upper atmosphere.
- **Impacts** of asteroids or comets, which can blast atmospheric gas into space.
- **Thermal escape**, in which lightweight gas molecules are lost to space when they achieve escape velocity.

The first two processes, condensation and chemical reactions, return gas to the planet's surface or interior. This loss of gas may be permanent, or changing conditions on the planet may return the gas to the atmosphere at a later time. The other three processes always cause permanent change, since they involve loss of gas to space.

Loss by Condensation and Chemical Reactions Condensation generally goes hand in hand with the release of gas through evaporation/sublimation. On Earth, water vapor condenses into rain, snow, or hail. On Mars, it is cold enough for carbon dioxide to condense into solid ice, especially at the poles. That is why the Martian polar caps consist primarily of frozen carbon dioxide (also known as "dry ice").

Chemical reactions can remove gas from the atmosphere and incorporate it into rock. On Earth, for example, atmospheric carbon dioxide dissolves in the oceans, where it ultimately undergoes chemical reactions that make **carbonate rocks** (rocks rich in carbon and oxygen) such as limestone [Section 14.3]. These chemical reactions are the primary reason why our atmosphere, unlike that of Venus, is not full of carbon dioxide. Volcanoes have outgassed huge amounts of carbon dioxide into Earth's atmosphere over its history, but nearly all of this carbon dioxide has been removed from the atmosphere and converted into carbonate rocks by chemical reactions.

Loss by Solar Wind Stripping and Impacts Gases are lost forever if they are stripped away into space by the solar wind. Stripping occurs when solar wind particles sweep ionized gases from high in the atmosphere into space. Remember, however, that a magnetosphere can protect a planet from solar wind particles. Thus, planets with strong magnetic fields (like Earth) are less susceptible to this type of atmospheric loss.

Impacts can also cause permanent loss of gas. The same impacts that create large craters on the surface [Section 10.3] may blast away significant amounts of atmospheric gas. Loss of gas due to impacts was most common during the heavy bombardment, which ended some 4 billion years ago. Impacts should cause greater loss of gas on smaller worlds, where gravity's hold on the atmosphere is weaker. Mars probably lost a significant amount of atmospheric gas in this way early in its history.

Loss by Thermal Escape Thermal escape occurs when an atom or a molecule of atmospheric gas achieves escape

velocity and flies off into space, never to return. Whether gases are lost by thermal escape depends on three distinct factors:

1. The planet's escape velocity, or the speed at which a particle must travel to escape the pull of gravity. In general, more massive planets have higher escape velocities [Section 5.5].

Mathematical Insight **Thermal Escape from an Atmosphere**

Whether an atmosphere loses gas by thermal escape depends on how many of the gas particles (atoms or molecules) are moving faster than the escape velocity. In general, particles of a particular type in a gas move at a wide range of speeds that depend on the temperature.

For example, the figure shows the range of speeds of sodium atoms at the temperature of the Moon's extremely thin atmosphere. The peak in the figure represents the most common speed of the sodium atoms. This speed, called the *peak thermal velocity* of the sodium atoms, is given by the following formula:

$$v_{thermal} = \sqrt{\frac{2kT}{m}}$$

where m is the mass of a single atom, T is the temperature in Kelvin, and $k = 1.38 \times 10^{-23}$ joule/Kelvin is *Boltzmann's constant*. This formula shows that higher temperatures mean higher thermal velocities for the atoms or molecules in a gas. It also shows that, for a given temperature, particles of lighter gases (smaller m) move at faster speeds than particles of heavier gases.

If the peak thermal velocity of a particular type of gas is higher than the escape velocity, then most of the gas particles will be traveling fast enough to escape the atmosphere, and the planet will lose this atmospheric gas to space fairly quickly. However, even if the peak thermal velocity is lower than the escape velocity, the wide range of particle speeds in a gas means that some small fraction of the gas particles may be traveling faster than the escape

velocity. These fast-moving particles will escape as long as they're moving upward and don't hit another particle on their way out.

Collisions among the remaining gas particles will continually ensure that a few always attain escape velocity, and thus the atmosphere can slowly leak away into space. In general, most of the gas particles of a particular type will escape over the age of the solar system if the peak thermal velocity is 20% or more of the escape speed.

Example: The temperature on the dayside of the Moon is $T = 400$ K. Calculate the peak thermal velocities of both hydrogen atoms and sodium atoms. Then explain why the Moon (escape velocity ≈ 2.4 km/s) cannot retain a hydrogen atmosphere for very long but retains a thin exosphere of sodium. The mass of a hydrogen atom is 1.67×10^{-27} kg; the mass of a sodium atom is 23 times greater, or 3.84×10^{-26} kg.

Solution: From the formula, the peak thermal velocity of the hydrogen atoms on the dayside of the Moon is:

$$v_{thermal} = \sqrt{\frac{2 \times (1.38 \times 10^{-23} \frac{joule}{K}) \times (400 \text{ K})}{1.67 \times 10^{-27} \text{ kg}}}$$

$$\approx 2,600 \text{ m/s} = 2.6 \text{ km/s}$$

Similarly, the thermal velocity of the sodium atoms is:

$$v_{thermal} = \sqrt{\frac{2 \times (1.38 \times 10^{-23} \frac{joule}{K}) \times (400 \text{ K})}{3.84 \times 10^{-26} \text{ kg}}}$$

$$\approx 540 \text{ m/s} = 0.54 \text{ km/s}$$

The peak thermal velocity of the hydrogen atoms is slightly greater than the Moon's escape velocity of about 2.4 km/s. Thus, the Moon cannot retain a hydrogen atmosphere. However, the peak thermal velocity of sodium is only about one-fifth, or about 20%, of the escape velocity. Thus, the Moon cannot hold a sodium atmosphere indefinitely, but it can hold some of the sodium atoms that are continually entering its atmosphere through bombardment.

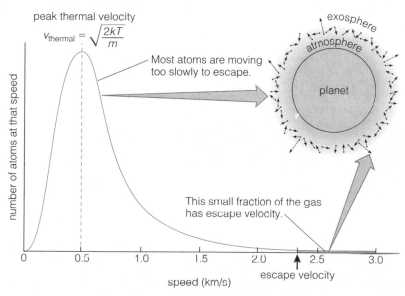

peak thermal velocity

$$v_{thermal} = \sqrt{\frac{2kT}{m}}$$

Most atoms are moving too slowly to escape.

number of atoms at that speed

This small fraction of the gas has escape velocity.

speed (km/s)

escape velocity

exosphere

atmosphere

planet

● In a gas at a given temperature, different atoms are always moving at different speeds. This plot shows the range of speeds of sodium atoms in the lunar atmosphere.

2. The temperature. Individual particles in a gas move at a wide range of random speeds [Section 4.2], but the higher the temperature, the faster their *average* speed and the more likely that some particles will reach escape velocity.

3. The mass of the gas particles. Because all gas particles have the same average kinetic energy at a particular temperature, lightweight gases such as hydrogen or helium move at faster average speeds than heavier ones such as oxygen or carbon dioxide (because kinetic energy depends on both mass and velocity [Section 4.2]). Lightweight gases are therefore more likely to achieve escape velocity.

All the terrestrial planets are small enough and warm enough for hydrogen and helium to escape. Thus, they were unable to hold any hydrogen or helium gas that they may have captured from the solar nebula when they were very young. Venus, Earth, and Mars are able to retain heavier gases in their atmospheres, but they can lose light gases—including hydrogen atoms released when molecules such as H_2O are broken apart by solar ultraviolet radiation.

The Bombardment Exospheres of the Moon and Mercury

On worlds with substantial atmospheres, such as Venus, Earth, and Mars, the many ways in which the gain and loss processes can interact with one another can make their atmospheric and climate histories quite complex. The Moon and Mercury, in contrast, offer very simple examples of the gain and loss processes at work. We'll therefore investigate these simple cases before we explore more complex cases in the next section.

We usually don't think of the Moon and Mercury as having atmospheres, but they are not totally devoid of gas. However, their gas is far too thin to absorb solar radiation or to have any kind of atmospheric structure such as a troposphere, stratosphere, or thermosphere. In essence, the Moon and Mercury have only the thin gas of an exosphere, in which the density is so low that collisions between atoms or molecules are rare. The gas particles therefore can rise as high as their speeds allow—sometimes even escaping to space. Their exospheres thus extend thousands of kilometers into space (Figure 11.22).

Source and Loss Processes on the Moon and Mercury

Bombardment is the only ongoing source of gas on the Moon and Mercury. Volcanic outgassing ceased long ago on these small worlds, and any gas it once released is long gone. Some may have been stripped away by the solar wind or lost to impacts, but it would have been lost to thermal escape anyway. Mercury cannot hold much of an atmosphere because its small size and high daytime temperature mean that nearly all gas particles eventually achieve escape velocity. Because it is smaller than Mercury, gases escape

a Mercury's atmosphere. Color-coded contours represent the gas density from lowest (black) to highest (red). Mercury's partially illuminated disk is shown schematically by the inset photograph. The Sun lies to the left.

b The Moon's atmosphere. This composite image represents the gas density from lowest (green) to highest (red). The Moon itself was blocked to make this image, resulting in the central black area of no data. The Moon's size is shown schematically by the inset photo.

Figure 11.22 The atmospheres of Mercury and the Moon—which are essentially exospheres only—viewed through instruments sensitive to emission lines from sodium atoms (the same emission lines that give sodium street lights their yellow color). Although these are extremely low-density atmospheres, they extend quite high.

about as easily from the Moon even though it is also somewhat cooler.

While bombardment continually releases small amounts of gas on the Moon and Mercury, the gas never accumulates because it is lost as quickly as it is gained. Some of the relatively few particles blasted upward by bombardment achieve escape velocity and escape directly to space. The rest bounce around like tiny rubber balls, arcing hundreds of kilometers into the sky before crashing back down to the surface. Each gas particle typically bounces a few dozen times before being absorbed into the surface.

Ice in Polar Craters Spacecraft observations made in the late 1990s strongly suggest that the Moon has water ice at the bottoms of craters near its poles. The bottoms of these polar craters lie in nearly perpetual shadow, keeping them so cold that the water remains perpetually frozen. The water probably came from condensation following impacts of ice-rich comets. The water in the comet vaporized upon impact. Water molecules that did not escape then bounced randomly around the surface. Some molecules would have settled into craters near the poles, where they condensed into ice. If humans ever establish large-scale colonies on the Moon, the ice in the polar craters offers a potential source of water. Mercury also has polar craters with their bottoms in perpetual shadow, so it could also have polar ice. Recent radar observations of Mercury suggest that this is indeed the case.

11.6 The Climate Histories of Mars, Venus, and Earth

We began this chapter with the goal of understanding how and why the terrestrial atmospheres came to differ so profoundly. The cases of the Moon and Mercury have proved to be quite easy to understand. However, because outgassing supplied substantial early atmospheres to the three larger worlds—Mars, Venus, and Earth—their very different present atmospheres must reflect atmospheric gains and losses that have occurred through time. In this final section, we'll explore the past and present climates of these three planets. As in our geology tour, we'll proceed in order of increasing planetary size, because size is the most impor-

tant factor in volcanism and the outgassing that made these planetary atmospheres.

Mars

Mars is only 40% larger in radius than Mercury, but its surface reveals a much more fascinating and complex atmospheric history. As we discussed in Chapter 10, photographs show clear evidence that water once flowed on the Martian surface. In the distant past, Mars must have been very different from its current "freeze-dried" state.

Martian Weather Today The present-day surface of Mars looks much like some deserts or volcanic plains on Earth (see Figures 10.29 and 11.1). However, its thin atmosphere makes it unlike any place on Earth. The atmosphere is so thin that it creates only a weak greenhouse effect (see Table 11.2) despite being made mostly of the greenhouse gas carbon dioxide. The temperature is usually well below freezing, with a global average of about $-53°C$ ($-63°F$), and the atmospheric pressure is less than 1% that on the surface of Earth. The lack of oxygen means that Mars lacks an ozone layer, so much of the Sun's damaging ultraviolet radiation passes unhindered to the surface.

Mars has seasons much like those on Earth because of its similar axis tilt, but they last about twice as long (because a Martian year is almost twice as long as an Earth year). On Mars, however, the shape of its orbit introduces a second important effect. Mars's more elliptical orbit puts it significantly closer to the Sun during the southern hemisphere summer and farther from the Sun during the southern hemisphere winter (Figure 11.23). Mars therefore has more extreme seasons in its southern hemisphere—that is, shorter, hotter summers and longer, colder winters—than in its northern hemisphere.

The seasons are the cause of a major feature of Mars's weather: winds blowing from the summer pole to the winter pole. Polar temperatures at the winter pole drop so low (about $-130°C$) that carbon dioxide condenses into "dry ice" at the polar cap. At the same time, frozen carbon dioxide at the summer pole sublimates into carbon dioxide gas (Figure 11.24). During the peak of summer, nearly all the carbon dioxide may sublime from the summer pole, leaving only a residual polar cap of water ice (Figure 11.25). The atmospheric pressure therefore increases at the summer

Seasons on Mars

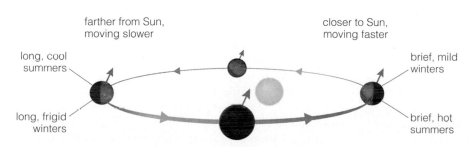

farther from Sun, moving slower

long, cool summers

long, frigid winters

closer to Sun, moving faster

brief, mild winters

brief, hot summers

Figure 11.23 The ellipticity of Mars's orbit makes seasons more extreme (hotter summers and colder winters) in the southern hemisphere than in the northern hemisphere.

October 1996   January 1997   March 1997

Figure 11.24 These images from the Hubble Space Telescope show Mars in polar projection so that the north pole is in the center of the images. The image at left shows the polar cap just as spring begins in the northern hemisphere, when the cap is near its maximum size. The middle image is in mid-spring, and the image at right shows the polar cap in early summer. The sublimation of carbon dioxide ice at the summer pole occurs at the same time that carbon dioxide is condensing at the winter pole. These changes drive winds blowing from the summer pole to the winter pole.

200 km

Figure 11.25 This image, from *Mars Global Surveyor*, shows the residual north polar cap during northern hemisphere summer. The white material is primarily water ice, with dust mixed in.

pole and decreases at the winter pole, driving strong pole-to-pole winds. Overall, as much as one-third of the total carbon dioxide of the Martian atmosphere moves seasonally between the north and south polar caps. The combination of circulation cells, the Coriolis effect, and pole-to-pole circulation gives Mars very dynamic weather.

The direction of the pole-to-pole winds on Mars changes with the alternating seasons. Sometimes these winds initiate huge dust storms, particularly when the more extreme summer approaches in the southern hemisphere (Figure 11.26). At times the surface is so shrouded by airborne dust that no surface markings can be distinguished and large dunes form and shift with the strong winds. As the dust settles out onto the surface, it can change the color or reflectivity over vast areas, creating the seasonal changes in appearance that fooled some astronomers in the late 1800s and early 1900s into thinking they saw changes in vegetation [Section 10.5]. The dust storms also leave Mars with a perpetually dusty, pale pink sky.

Mars Climate and Axis Tilt The weather on Mars does not change much from one year to the next. However, changes in its axis tilt probably cause Mars to undergo longer-term cycles of climate change. Theoretical calculations suggest that Mars's axis tilt varies far more than that of Earth—from as little as 0° to 60° or more on time scales of hundreds of thousands to millions of years. This extreme variation arises for two reasons. First, Jupiter's gravity has a greater effect on the axis of Mars than on Earth, because Mars's orbit is closer to Jupiter's orbit. Second, Earth's axis is stabilized by the gravity of our relatively large Moon. Mars's two tiny moons, Phobos and Deimos (see Figure 9.11), are far too small to offer any stabilizing influence on its axis.

As we discussed earlier, changes in axis tilt affect both the severity of the seasons and the global average temperature. When Mars's axis tilt is small, the poles may stay in a perpetual deep freeze for tens of thousands of years. With more carbon dioxide frozen at the poles, the atmosphere becomes thinner, lowering the pressure and weakening the greenhouse effect. When the axis is highly tilted, the summer pole becomes quite warm, allowing substantial amounts

September 4, 2001

Figure 11.26 These two Hubble Space Telescope photos contrast the appearance of Mars in the presence and absence of a global dust storm. If you look carefully at the first image, you can see localized dust storms near both polar caps (look toward the upper-right edge of the southern cap). The second image shows how, just over two months later, these storms had grown into a planet-wide dust storm.

of water ice to sublime, along with carbon dioxide, into the atmosphere. The pressure therefore increases, and Mars becomes warmer as the greenhouse effect strengthens—although probably not by enough to allow liquid water to become stable at the surface. The Martian polar regions show layering that probably reflects changes in climate due to the changing axis tilt (Figure 11.27).

Why Mars Froze Climate changes due to changing axis tilt may have important effects on Mars, but they do not explain the most dramatic climate change in Martian history. Remember that the surface shows strong evidence of a warm, wet period ending more than 3 billion years ago

Figure 11.27 This photo from *Mars Global Surveyor* shows a region about 40 kilometers wide at the edge of a Martian polar cap. Note the alternating layers of ice and dust, which are probably the result of climate changes caused by the changing axis tilt of Mars.

5 km

[Section 10.5]. We see evidence of erosion by rainfall, and perhaps even lakes and oceans in that ancient time. Mars apparently underwent a major and permanent climate change over 3 billion years ago, turning a wet and warm planet into a frozen world.*

The warm and wet period is explainable only if Mars once had a very different atmosphere than it has today. Mars must have had both a much stronger greenhouse effect to make it warm enough for water to flow and a much higher atmospheric pressure to keep water from boiling away. Computer simulations show that these requirements can be met by a carbon dioxide atmosphere about 400 times as dense as the current Martian atmosphere. Such an atmosphere would have made the surface pressure about three times that on Earth today. Additional greenhouse warming, perhaps by carbon dioxide ice clouds or atmospheric methane, may also have been necessary given the fact that the Sun was dimmer when the solar system was young.

The idea that Mars might have had such a dense atmosphere is reasonable. Mars has plenty of ancient volcanoes, and calculations show that these volcanoes could have supplied all the carbon dioxide needed to warm the planet. Moreover, if Martian volcanoes outgassed carbon dioxide and water in the same proportions as do volcanoes on Earth, Mars would have had enough water to fill oceans tens or even hundreds of meters deep.

The bigger question is not whether Mars once had a denser atmosphere but how all that atmospheric gas disappeared. In particular, Mars must somehow have lost most of its carbon dioxide gas. This loss would have weakened the greenhouse effect until the planet essentially froze over.

*Some recent research disputes the idea that Mars's climate was milder for much of its early history. Warmer temperatures and liquid water may have been present only after large impacts vaporized ground ice. This water vapor would have rained out and refrozen as ground ice in just a few years. This idea is controversial.

The fate of the lost gas is not completely clear, but we can think about it in terms of the five atmospheric loss processes. Some carbon dioxide is locked up in the polar caps (condensation), and some may have become chemically bound to surface rock. Some of the gas in the early Martian atmosphere may also have been blasted away by large impacts. Recent data suggest that the final blow to the Martian climate was probably stripping by solar wind particles. Early in Mars's history, the warm convecting interior produced a global magnetic field that created a protective magnetosphere around the planet. As the small planet cooled, the magnetic field weakened, making the atmosphere vulnerable to solar wind particles. Most of the carbon dioxide that remained in the atmosphere at the end of the warm and wet period was probably stripped away by these particles. Had Mars kept its magnetic field, it might have retained much of its atmosphere and might have a more moderate climate today.

Much of the water once present on Mars is also probably gone for good (apart from ice in Mars's polar soil; see Figure 10.31). Some may have been lost in the same manner as the carbon dioxide, but Mars also lost water in another way. Mars lacks an ultraviolet-absorbing stratosphere, which means that water molecules are easily broken apart by ultraviolet photons. Once water molecules were broken apart in the Martian atmosphere, the hydrogen atoms would have been lost rapidly to space through thermal escape. Some of the remaining oxygen was probably lost by solar wind stripping. The rest probably was drawn out of the atmosphere through chemical reactions with surface rock. This oxygen literally rusted the Martian rocks, giving the "red planet" its distinctive tint.

The history of the Martian atmosphere holds important lessons for us on Earth. Mars apparently was once a world with pleasant temperatures and with streams, rain, glaciers, lakes, and possibly oceans. It had all the necessities for life as we know it. This once hospitable planet turned into a frozen and barren desert at least 3 billion years ago, and it is unlikely that Mars will ever again be warm enough for its frozen water to flow. Any life that may have existed on Mars is either extinct or hidden away in a few choice locations, such as in underground water near not-quite-dormant volcanoes. As we consider the possibility of future climate change on Earth, Mars presents us with an ominous example of how drastically things can change.

Venus

Venus presents a stark contrast to Mars. Its thick carbon dioxide atmosphere creates an extreme greenhouse effect that bakes the surface to unbearable temperatures. Why Venus is so different from Mars is fairly easy to understand. Its larger size allowed it to retain more interior heat, leading to greater volcanism [Section 10.6]. The associated outgassing produced the vast quantities of carbon dioxide in Venus's atmosphere, and this carbon dioxide creates the strong greenhouse heating.

Venus becomes more mysterious when we compare it to Earth. Because Venus and Earth are so similar in size, we expect both planets to have had similar levels of volcanism and outgassing. In particular, both planets should have outgassed about the same total amounts of water vapor and carbon dioxide. As we discussed in Section 11.5, Earth's atmosphere has only relatively small amounts of these gases because they have been removed by condensation and chemical reactions. Water vapor condensed to make Earth's oceans, and the carbon dioxide was removed from the atmosphere by being dissolved in water and undergoing chemical reactions that incorporated it into carbonate rocks. In contrast, most or all of the carbon dioxide on Venus remains in its atmosphere, while water is almost totally absent from both the atmosphere and the surface. Let's take a deeper look at Venus to see if we can figure out why it turned out to be so different from Earth.

Weather on Venus Before we look at the climate history of Venus, let's start with a quick current weather update: The weather forecast for the surface of Venus today and every day is dull, dull, dull.

Venus's slow rotation (243 Earth days) means a very weak Coriolis effect. As a result, Venus has little wind on its surface and never has hurricane-like storms. The top wind speeds measured by the Soviet Union's Venera landers were only about 6 kilometers per hour. No rain falls, because any gases that condense in the cool upper atmosphere evaporate long before they reach the ground. Venus's thick atmosphere distributes heat planet-wide with its two large circulation cells, so the poles are no cooler than the equator and night is just as searingly hot as day. Venus has no seasons because it has virtually no axis tilt.* Thus, temperatures are the same year-round.

While Venus has very dull surface weather, more interesting things occur higher up in its atmosphere. Because its high surface temperature drives strong convection, Venus is covered with bright, reflective clouds (Figure 11.28). The clouds are made of droplets containing sulfuric acid (H_2SO_4). These droplets form high in the troposphere, where temperatures are 400°C cooler than on the surface. The droplets sometimes fall through the upper troposphere as sulfuric acid rain, but rising temperatures make them evaporate at least 30 kilometers above the surface. The upper atmosphere of Venus also has some fast winds, though we don't know precisely why.

The sulfuric acid clouds likely reveal something important about volcanism on Venus. Sulfuric acid is formed by chemical reactions involving sulfur dioxide (SO_2) and

---

*Numerically, Venus's axis tilt is usually written as 177.4°. This may sound large, but notice that it is nearly 180°—the same tilt as 0° but "upside down." Venus's tilt is written in this way because it rotates backward compared to its orbit, and backward rotation is the same thing as forward rotation that is upside down.

Figure 11.28 Clouds on Venus are all that can be seen in this ultraviolet image taken by the *Pioneer Venus Orbiter.*

water, both of which come from volcanic outgassing. Neither gas can remain in the atmosphere forever. We'll discuss the loss of water shortly. Sulfur dioxide is removed by chemical reactions with surface rocks. The presence of sulfuric acid clouds therefore suggests that volcanoes on Venus must still be active—at least on geological time scales (erupting within the past 100 million years). Without ongoing outgassing, the necessary gases would by now have disappeared.

Water on Venus Our understanding of planetary geology suggests that Venus should have outgassed about as much water vapor as Earth. But Venus today is incredibly dry, with a surface far too hot to have either liquid water or ice. It is even too hot for water to be chemically bound in surface rock. The only place where Venus could conceivably have much water is in the atmosphere, but the atmosphere has little water either. Overall, the total amount of water on Venus is about 10,000 times less than the total amount on Earth. If Venus has outgassed as much water vapor as Earth, where did it all go?

One possibility, of course, is that Venus was "born dry" and never had much water. However, this would require that the types of impacts that brought water to Earth somehow missed Venus. Unless such impacts were very rare and Earth was the lucky beneficiary of those few that occurred, Venus and Earth ought to have accumulated similar amounts of water. Most scientists therefore consider it more likely that Venus once had lots of water but lost it permanently.

As we'll discuss shortly, Venus's proximity to the Sun kept surface temperatures higher than those on Earth, ultimately leading to its strong greenhouse effect. The high surface temperatures prevented outgassed water vapor from condensing, so it remained in the atmosphere. Some

water molecules made their way into the upper atmosphere, where solar ultraviolet photons broke them apart. Thermal escape sent the freed hydrogen atoms into space, never to return, so the water molecules could not re-form. The remaining oxygen was lost to a combination of chemical reactions with surface rocks and stripping by the solar wind (made easy by Venus's lack of a magnetic field). Other water molecules then rose to the upper atmosphere, suffering the same fate. Acting over billions of years, this process can easily explain the loss of an ocean's worth of water from Venus. In fact, Venus may have lost so much water that its crust and mantle also dried out.

Proving that Venus really did lose so much water is difficult, but evidence comes from the gases that didn't escape. While most hydrogen nuclei contain just a single proton, a tiny fraction of all hydrogen atoms (about 1 in 50,000 on Earth) contain a neutron in addition to the proton, making the isotope of hydrogen that we call *deuterium* [Section 4.3]. Water molecules that contain an atom of deuterium (called "heavy water") behave chemically just like ordinary water and can be broken apart by ultraviolet light just as easily. However, a deuterium atom is twice as heavy as an ordinary hydrogen atom and thus does not escape to space as easily when the water molecule is broken apart. If Venus lost a huge amount of hydrogen (from water molecules) to space, the rare deuterium atoms would have been more likely to remain behind than the ordinary hydrogen atoms. Measurements show that this is the case. The fraction of deuterium among hydrogen atoms is a hundred times higher on Venus than on Earth (though both fractions are very small). Thus, a substantial amount of water must have been broken apart and its hydrogen lost to space. We cannot determine exactly how much water Venus has lost, but it certainly seems plausible that it really did outgas as much water as Earth and has lost virtually all of it.

The Runaway Greenhouse Effect The loss of water explains why Venus retains so much carbon dioxide in its atmosphere and hence why its surface is so hot. Remember that most of the outgassed carbon dioxide on Earth has been converted by chemical reactions into carbonate rocks. These reactions occur only *after* the carbon dioxide has dissolved in the oceans. Because Venus has no oceans in which to dissolve carbon dioxide, the gas remains in the atmosphere, resulting in an atmosphere thick with carbon dioxide and a strong greenhouse effect.

The real question in understanding the difference between Venus and Earth is why Earth kept its water while Venus did not. The answer involves the planets' distances from the Sun. Let's consider what would happen if we moved Earth to Venus's orbit. Sunlight is more intense at this closer distance to the Sun. The greater intensity of sunlight would almost immediately raise Earth's global average temperature by about 30°C, from its current 15°C to about 45°C (113°F). Although this is still well below the boiling point of water, the higher temperature would lead to increased evaporation of water from the oceans. The higher temperature would also allow the atmosphere to hold more water vapor before the vapor condensed to make rain.

The combination of more evaporation and greater atmospheric capacity for water vapor would substantially increase the total amount of water vapor in Earth's atmosphere. Because water vapor is a greenhouse gas, this added water vapor would strengthen the greenhouse effect, driving temperatures a little higher. The higher temperatures, in turn, would lead to even more ocean evaporation and more water vapor in the atmosphere—strengthening the greenhouse effect even further. In other words, we'd have a "positive feedback loop" in which each little bit of additional water vapor would mean higher temperature and even more water vapor (Figure 11.29). The process would rapidly spin out of control, resulting in a **runaway greenhouse effect**. The planet would heat up until the oceans were completely evaporated and the carbonate rocks had released all their carbon dioxide back into the atmosphere. By the time the runaway process was complete, temperatures on our "moved Earth" would be even higher than they are on Venus today, thanks to the combined greenhouse effects of carbon dioxide and so much water vapor in the atmosphere.

This scenario suggests a simple explanation for the difference between Earth and Venus. Even though the two planets' distances from the Sun are not that different, the difference was apparently critical. On Earth, the water rained down to make oceans, the oceans dissolved carbon dioxide, and chemical reactions locked away the carbon dioxide in carbonate rocks. The result is that our atmosphere contains just enough greenhouse gases to make our planet pleasantly warm. On Venus, the greater intensity of sunlight made it just enough warmer that oceans either never formed or soon evaporated. Without oceans to dissolve carbon dioxide and make carbonate rock, carbon dioxide accumulated in the atmosphere, leading to a runaway greenhouse effect.

Suppose we could magically replace Venus's actual atmosphere with an atmosphere identical to Earth's. Would you expect the surface temperature to allow for liquid water? Explain.

The next time you see Venus shining brightly as the morning or evening "star," consider the radically different path it has taken from that taken by Earth—and thank

If Earth moved to Venus's orbit. . .

Water vapor increases greenhouse effect, raising temperature further.

Runaway Greenhouse

More intense sunlight immediately raises Earth's surface temperature by about 30°C.

Greater warmth increases evaporation, and warmer air holds more water vapor.

As the oceans finish evaporating, carbonate rocks decompose, releasing CO_2. Earth becomes hotter than Venus.

Figure 11.29 This diagram shows how, if Earth were placed at Venus's distance from the Sun, the runaway greenhouse effect would cause the oceans to evaporate completely.

your lucky star. If Earth had formed a bit closer to the Sun or if the Sun had been slightly hotter, our planet might have suffered the same greenhouse-baked fate.

Did Venus Have a More Pleasant Past? Venus's ultimate fate may have been sealed by its closeness to the Sun, but it's possible that Venus might have been more moderate in its early history. Remember that the Sun has gradually brightened with age. Thus, sunlight was less intense on all the planets when they were young. Some 4 billion years ago, the intensity of sunlight at Venus was not much greater than it is at Earth today. Rain might well have fallen on the young planet Venus, and oceans could have formed. It's even conceivable that life could have arisen on the young Venus.

As the Sun gradually brightened, however, any liquid water or life on Venus was doomed. The runaway greenhouse effect raised the temperature so high that all the water evaporated. In the upper atmosphere, the water molecules were broken apart by ultraviolet light, and the hydrogen escaped to space. If Venus had oceans in its youth, the water is now gone forever.

We'll probably never know for sure whether Venus ever had oceans. The global "repaving" of Venus's surface by tectonics and volcanism [Section 10.6] would long ago have covered up any shorelines or other geological evidence of past oceans, and any gases that might once have been incorporated into surface rock would have been baked out by the high surface temperatures. If the climate once was pleasant, no evidence survives to tell the tale.

Earth's Surprising Climate

Venus and Mars are quite different today, but their histories share a common theme: dramatic climate change. The change went in the direction of freezing on Mars and in the direction of heating on Venus, but the lesson appears to be that major climate change is to be expected on terrestrial worlds with atmospheres. If so, Earth offers a striking exception to this rule. Geological evidence shows that Earth's climate has been remarkably (though not perfectly) stable throughout its history despite the Sun's gradual brightening. Moreover, this climate stability has been crucial to the long-term survival of life and our current existence.

How has Earth avoided the extreme fates of Venus and Mars? By now you have learned enough to realize that the answer ultimately must be traceable to Earth's basic properties, such as its size and distance from the Sun. However, the details involve some fairly subtle interactions between

Earth's geology and atmosphere—and perhaps life itself. The lesson of planetary exploration is that in order to understand Earth we must understand the solar system as a whole: its formation, geological processes, and atmospheres, as well as the jovian planets, asteroids, and comets. We are more than halfway through this list already. After two more chapters we will have come full circle, ready to explore and explain Earth's unique circumstances in Chapter 14.

Putting Chapter 11 into Context

With what we have learned in this chapter and the previous chapter, we now have a complete "big picture" view of how the terrestrial worlds started out so similar yet ended up so different. As you continue, keep in mind the following important ideas:

- Atmospheric properties differ widely among the terrestrial worlds, but we can trace these differences to root causes. For example, only the larger worlds have significant atmospheres. The smaller worlds lack the internal heat needed for volcanism and the associated outgassing that releases atmospheric gases, and they also lack the gravity necessary to retain atmospheric gases.

- Atmospheres affect planets in a variety of ways. For example, they absorb and scatter light, distribute heat, and create the weather that can lead to erosion. Perhaps most important, the greenhouse effect allows atmospheres to make a planet warmer than it would be otherwise.

- The climate histories of Venus and Mars suggest that major climate change should be the rule, not the exception. Mars was once warm and wet, but its small size and lack of a magnetic field caused it to lose gas and freeze over some 3 billion or more years ago. Venus may once have had oceans, but its proximity to the Sun doomed it to a runaway greenhouse effect.

- We are here to talk about these things today only because Earth somehow has managed to be the exception, a planet on which climate has remained relatively consistent. A full understanding of why our planet has been so fortunate—from our perspective—must await a deeper look at Earth's unique geology, atmosphere, and life.

11.1 Atmospheric Basics

- *What are the general atmospheric properties of the five terrestrial worlds?* The Moon and Mercury are essentially airless worlds with very little atmospheric gas. Venus has a thick carbon dioxide atmosphere and high surface temperature and pressure. Mars has a thin carbon dioxide atmosphere, with surface temperatures usually below freezing and pressure too low for liquid water. Earth has a nitrogen/oxygen atmosphere and pleasant surface temperature and pressure.

- *What is atmospheric pressure?* Atmospheric pressure is the result of countless collisions between atoms and molecules in a gas. It is measured in *bars,* where 1 bar is roughly equal to Earth's atmospheric pressure at sea level.

- *How do atmospheres affect planets?* Atmospheres absorb and scatter light, create pressure, warm a planet's surface and distribute heat, create weather, and interact with the solar wind.

11.2 The Greenhouse Effect and Planetary Temperature

- *What is the greenhouse effect?* The greenhouse effect is planetary warming caused by the absorption of infrared light from a planet's surface by greenhouse gases such as carbon dioxide, methane, and water vapor.

- *How would planets be different without the greenhouse effect?* They would be colder, with temperatures determined only by distance from the Sun and reflectivity.

- *How important is the greenhouse effect on Venus, Earth, and Mars?* All three planets are warmed by the greenhouse effect, but it is weak on Mars, moderate on Earth, and very strong on Venus. Global average temperatures on both Venus and Earth would be below freezing without the greenhouse effect.

11.3 Atmospheric Structure

- *What is the basic structure of Earth's atmosphere?* Pressure and density decrease rapidly with altitude. Temperature drops with altitude in the troposphere, rises with altitude in the lower part of the stratosphere, and rises again in the thermosphere and exosphere.

- *How do interactions with light explain atmospheric structure?* Solar X rays heat and ionize gas in the ther-mosphere. Solar ultraviolet is absorbed by molecules such as ozone, heating the stratosphere. Visible light warms the surface, which radiates infrared light that warms the troposphere.

- *How do the atmospheric structures of Venus and Mars differ from that of Earth?* Venus and Mars lack an ultraviolet-absorbing stratosphere.

- *What is a magnetosphere?* Created by a global magnetic field, a magnetosphere acts like a protective bubble surrounding a planet. It diverts charged particles from the solar wind, channeling some to the magnetic poles, where they can lead to auroras.

11.4 Weather and Climate

- *What is the difference between weather and climate?* Weather refers to short-term changes in wind, clouds, temperature, and pressure. Climate is the long-term average of weather.

- *What creates global wind patterns?* Atmospheric heating at the equator creates two huge equator-to-pole circulation cells. If the Coriolis effect is strong enough, these large cells may split into smaller cells. This split occurs on Earth but not on Venus (because of slow rotation) or on Mars (because of small size).

- *What causes rain or snow to fall?* Convection carries evaporated (or sublimated) water vapor to high, cold altitudes, where it condenses into droplets or ice flakes, forming clouds. When the droplets or ice flakes get large enough, convection can no longer hold them aloft, and they fall as rain, snow, or hail.

- *What four factors can cause long-term climate change?* Factors involved in long-term climate change are the gradual brightening of the Sun over the history of the solar system, changes in a planet's axis tilt, changes in a planet's reflectivity, and changes in a planet's abundance of greenhouse gases.

11.5 Sources and Losses of Atmospheric Gas

- *How can an atmosphere gain or lose gas?* Gains come from outgassing, evaporation/sublimation, or bombardment, the last only if there is very little atmosphere. Gas can be returned to the planet's surface or interior through condensation or chemical reactions with surface materials. Gas can be lost to space through solar wind stripping, by being blasted away by asteroid or comet impacts, or by thermal escape.

- *Why do the Moon and Mercury lack substantial atmospheres?* They have no current source for outgassing and are too small and warm to hold any atmosphere they may have had in the past. They have small amounts of gas above their surfaces only because of bombardment by solar wind particles.

11.6 The Climate Histories of Mars, Venus, and Earth

- *What are the major seasonal features of Martian weather today?* Seasonal changes in temperature cause carbon dioxide to alternately condense and sublime at the poles, driving pole-to-pole winds and sometimes creating huge dust storms.

- *Why did Mars's early warm and wet period come to an end?* Mars once had a thick carbon dioxide atmosphere and a strong greenhouse effect. Most of the carbon dioxide was eventually lost to space, probably because the cooling interior could no longer create a strong magnetic field to protect the atmosphere from the solar wind. As carbon dioxide was lost, the greenhouse effect weakened until the planet froze.

- *Why is Venus so hot?* At its distance from the Sun, any liquid water was destined to evaporate, ultimately driving a runaway greenhouse effect that dried up the planet and heated it to its extreme temperature.

- *Could Venus ever have had oceans?* Maybe. Venus probably outgassed plenty of water vapor, and deuterium measurements provide evidence of vast water loss over time. Early in the solar system's history, when the Sun was dimmer, this water vapor could have condensed to make rain and oceans, though we cannot be sure.

- *After studying Mars and Venus, why do we find Earth's climate surprising?* Mars and Venus both underwent dramatic and permanent climate change early in their histories. Earth somehow has maintained a relatively stable climate even as the Sun has warmed with time.

❓ True or False?

Decide whether each of the following statements is true or false. Explain your reasoning.

1. If Earth's atmosphere did not contain molecular nitrogen, X rays from the Sun would reach the surface.

2. When Mars had a thicker atmosphere in the past, it also had a stratosphere.

3. If Earth rotated faster, hurricanes would be more common and more severe.

4. Mars would still have seasons even if its orbit around the Sun were perfectly circular rather than elliptical.

5. If the solar wind were much stronger, Mercury might develop a carbon dioxide atmosphere.

6. Earth's oceans must have formed at a time when no greenhouse effect operated on Earth.

7. Mars once may have been warmer than it is today, but it could never have been warmer than Earth because it is farther from the Sun than is Earth.

8. If Earth had as much carbon dioxide in its atmosphere as Venus, our planet would be too hot for liquid water to exist on the surface.

Problems

9. *Blue Skies Everywhere?* Briefly explain why the sky is blue on Earth and why it is not blue on Venus or Mars.

10. *Atmospheric Layers.* Why is there convection in the troposphere but not in the stratosphere? Why is Earth the only terrestrial world with a stratosphere?

11. *The Thinnest Atmospheres.* What is the origin of the very thin atmospheres of the Moon and Mercury? How might it be possible for these worlds to have water ice in polar craters?

12. *Clouds of Venus.* Table 11.2 shows that Venus's temperature in the absence of the greenhouse effect is lower than Earth's, even though it is closer to the Sun.

    a. Explain this unexpected result in one or two sentences.

    b. Suppose Venus had neither clouds nor greenhouse gases. What do you think would happen to the temperature of Venus? Why?

    c. How are clouds and volcanoes linked on Venus? What change in volcanism might result in the disappearance of clouds? Explain.

13. *Atmospheric Structure.* Study the curve showing Earth's average atmospheric structure in Figure 11.8. Sketch similar curves, and explain each curve in words, to show how the

atmospheric structure would be different in each of the following cases.

a. Suppose Earth had no greenhouse gases.

b. Suppose the Sun emitted no ultraviolet light.

c. Suppose the Sun had a higher output of X rays.

14. *Magic Mercury.* Suppose we could magically give Mercury the same atmosphere as Earth. Assuming this magical intervention happened only once, would Mercury be able to keep its new atmosphere? Explain.

15. *A Swiftly Rotating Venus.* Suppose Venus had rotated as rapidly as Earth throughout its history. Briefly explain how and why you would expect it to be different in terms of each of the following: geological processes, atmospheric circulation, magnetic field, and climate history. Write a few sentences about each.

16. *Coastal Winds.* During the daytime, heat from the Sun tends to make the air temperature warmer over land near the coast than over the water offshore. At night, when land cools off faster than the sea, the temperatures tend to be cooler over land than over the sea. Use these facts to predict the directions in which winds generally blow during the day and at night in coastal regions. For example, do the winds blow out to sea or in toward land? (*Hint:* How do these conditions resemble those that cause planetary circulation cells?) Explain your reasoning in a few sentences. (Diagrams might help.)

17. *Sources and Losses.* Choose one process by which atmospheres can gain gas and one by which they can lose gas. For each process, write a few sentences that describe it and how it depends on each of the following fundamental planetary properties: size, distance from the Sun, and rotation rate.

18. *Project: Atmospheric Science in the Kitchen.*

a. Find an empty plastic bottle, such as a water bottle with a screwtop that makes a good seal. Warm the air inside by filling the bottle partway with hot water and then shaking and emptying the bottle. Seal the bottle and place it in the refrigerator or freezer. What happens after 15 minutes? Explain why this happens, imagining that you can see the individual air molecules.

b. You may have noticed that loose ice cubes in the freezer gradually shrink away to nothing or that frost sometimes builds up in old freezers. What technical terms used in this chapter describe these phenomena? On what terrestrial planet do these same processes play a major role in changing the atmospheric pressure?

*19. *Habitable Planet Around 51 Pegasi?* The star 51 Pegasi is approximately as bright as our Sun and has a planet that orbits at a distance of only 0.051 AU.

a. Suppose the planet reflects 15% of the incoming sunlight. Calculate its "no greenhouse" average temperature. How does this temperature compare to that of Earth?

b. Repeat part (a), but assume that the planet is covered in bright clouds that reflect 80% of the incoming sunlight.

c. Based on your answers to parts (a) and (b), do you think it is likely that the conditions on this planet are conducive to life? Explain.

*20. *Escape from Venus.*

a. Calculate the escape velocity from Venus's exosphere, which begins about 200 kilometers above the surface. (*Hint:* You'll need the radius of Venus and the formula in Mathematical Insight 5.3.)

b. Calculate and compare the thermal speeds of hydrogen and deuterium atoms at the exospheric temperature of 350 K. The mass of a hydrogen atom is 1.67×10^{-27} kg, and the mass of a deuterium atom is about twice the mass of a hydrogen atom.

c. In a few sentences, comment on the relevance of these calculations to the question of whether Venus has lost large quantities of water.

Discussion Questions

21. *Lucky Earth.* The climate histories of Venus and Mars make it clear that it's not "easy" to get a pleasant climate like that of Earth. How does this affect your opinion about whether Earth-like planets might exist around other stars? Explain.

22. *Terraforming Mars.* Some people have suggested that we might be able to carry out planet-wide engineering of Mars that would cause its climate to warm and its atmosphere to thicken. This type of planet engineering is called *terraforming*, because its objective is to make a planet more Earth-like and thus easier for humans to live on. Discuss possible ways to terraform Mars, at least in principle. Do any of these ideas seem practical? Defend your opinions.

23. *Terraforming Venus.* As a follow-up to the previous question, think of any ways in which it would be possible to terraform Venus. Again, discuss the possibilities in principle as well as their practicality.

For a complete list of media resources available, go to www.astronomyplace.com and choose Chapter 11 from the pull-down menu.

 Astronomy Place Web Tutorials

Tutorial Review of Key Concepts

Use the following interactive **Tutorial** at www.astronomyplace.com to review key concepts from this chapter.

Surface Temperature of Terrestrial Planets Tutorial

Lesson 1 Energy Balance

Lesson 2 Role of Planet's Distance from the Sun

Lesson 3 Role of Planet's Albedo

Lesson 4 Role of Planet's Atmosphere

Supplementary Tutorial Exercises

Use the interactive **Tutorial Lessons** to explore the following questions.

Surface Temperature of Terrestrial Planets Tutorial, Lesson 1

1. How does the rate at which a planet emits radiation depend on the amount of sunlight it receives?

2. Does a planet's size affect its temperature? Explain.

3. How does a planet's rotation rate affect its day and night temperatures? Does rotation rate affect the average temperature? Explain.

Surface Temperature of Terrestrial Planets Tutorial, Lesson 2

1. Use the tool in Lesson 2 to predict Mercury's surface temperature. How does this compare to Mercury's actual day and night temperatures? Explain.

2. Use the tool in Lesson 2 to predict Earth's surface temperature. Why isn't this Earth's actual surface temperature? Explain.

Movies

Check out the following narrated and animated short documentary available on www.astronomyplace.com for a helpful review of key ideas covered in this chapter.

History of the Solar System Movie

Web Projects

Take advantage of the useful Web links on www.astronomyplace.com to assist you with the following projects.

1. *Spacecraft Study of Atmospheres.* Learn about a current or planned mission to study the atmosphere of one of the terrestrial worlds (possibly including Earth). Write a one- to two-page essay describing the mission and what we hope to learn from it.

2. *Martian Weather.* Find the latest weather report for Mars from spacecraft and other satellites. What season is it in the northern hemisphere? When was the most recent dust storm? What surface temperature was most recently reported from Mars's surface, and where was the lander located? Summarize your findings by writing a 1-minute script for a television news update on Martian weather.

3. *Mars Colonization.* Visit the Web site of a group that advocates human colonization of Mars, such as the Mars Society. Learn about the challenges of human survival on Mars and about prospects for terraforming Mars. Overall, do you think colonization of Mars is a good idea? Do you think terraforming is a good idea? Write a short essay describing what you've learned and defending your opinions.

12 Jovian Planet Systems

LEARNING GOALS

12.1 The Jovian Worlds: A Different Kind of Planet
- What are the major features of the jovian planets?
- Why are jovian planets so different from terrestrial planets?

12.2 Jovian Planet Interiors
- What is Jupiter's interior structure?
- Why is Saturn almost as big as Jupiter in radius?
- How do the jovian planet interiors differ, and why?

12.3 Jovian Planet Atmospheres
- How is Jupiter's atmospheric structure similar to Earth's?
- Why does Jupiter have three distinct cloud layers?
- What is the Great Red Spot?
- How do other jovian atmospheres compare to Jupiter's?

12.4 Jovian Planet Magnetospheres
- How does Jupiter's magnetosphere compare with Earth's?
- How does Jupiter's magnetosphere interact with its moon Io?
- How do other jovian magnetospheres compare to Jupiter's?

12.5 A Wealth of Worlds: Satellites of Ice and Rock
- Why does active geology occur more readily on small icy worlds than on small rocky worlds?
- What makes Io so volcanically active?
- Why do we suspect a subsurface ocean on Europa?
- What are some key features of the large moons Ganymede, Callisto, Titan, and Triton?

12.6 Jovian Planet Rings
- What do Saturn's rings look like?
- How do other ring systems compare to Saturn's?
- What is the origin of planetary rings?

Do there exist many worlds, or is there but a single world? This is one of the most noble and exalted questions in the study of Nature.

St. Albertus Magnus (1206–1280)

n Roman mythology, the namesakes of the jovian planets are rulers among gods: Jupiter is the king of the gods, Saturn is Jupiter's father, Uranus is the lord of the sky, and Neptune rules the sea. However, our ancestors could not have forseen the true majesty of the four jovian planets. The smallest, Neptune, is large enough to contain the volume of more than 50 Earths. The largest, Jupiter, has a volume some 1,400 times that of Earth. These worlds are totally unlike the terrestrial planets. They are essentially giant balls of gas, with no solid surface on which to stand.

Why should we care about a set of worlds so different from our own? Apart from satisfying natural curiosity about the diversity of the solar system, the jovian planet systems serve as a testing ground for our general theories of comparative planetology. Are our theories good enough to predict the nature of these planets and their elaborate satellite systems, or are different processes at work? We'll find out in this chapter as we investigate the nature of the jovian planets, along with their intriguing moons and beautifully complex rings.

 Formation of the Solar System Tutorial, Lesson 1

12.1 The Jovian Worlds: A Different Kind of Planet

The jovian planets have long held an important position in many human cultures. Jupiter is the third brightest object that appears regularly in our night skies (after the Moon and Venus), and ancient astronomers carefully charted its 12-year trek through the constellations of the zodiac. Though Saturn is fainter and slower moving, people of many cultures also mapped its 29-year circuit of the zodiac.

Scientific study of the jovian worlds dates to Galileo in the early 1600s. However, scientists did not fully appreciate the differences between terrestrial and jovian planets until the 1760s, when they established the absolute scale of the solar system (that is, planetary distances in units such as kilometers instead of AU). The discoveries of Uranus

in 1781 and Neptune in 1846 gave astronomers two more jovian worlds to study.

Once distances were established, scientists could calculate the truly immense sizes of the jovian planets from their angular sizes [Section 7.1], which they measured through telescopes. Knowing the distance scale also allows calculating jovian planet masses by applying Newton's version of Kepler's third law to observed orbital characteristics of the jovian moons [Section 5.3]. Together, the measurements of size and mass revealed the low densities of the jovian planets, proving that these worlds are very different from Earth.

The slow pace of discovery of details about the jovian planets and their moons gave way to revolutionary advances with the era of spacecraft exploration. The first spacecraft to visit the outer planets, *Pioneer 10* and *Pioneer 11*, flew past Jupiter and Saturn in the early 1970s. The *Voyager* missions followed less than a decade later [Section 8.4]. The results were spectacular. Figure 12.1 shows a montage of the jovian planets compiled by the Voyager spacecraft. Exploration of the jovian planet systems continued with *Galileo*, which went into orbit around Jupiter in 1995. The exciting Cassini mission is scheduled to arrive at Saturn in 2004. Our interest in our own solar system's jovian planets has been heightened by the discovery of more than 100 presumably similar planets around other stars [Section 9.6].

General Properties of the Jovian Planets

Jupiter, Saturn, Uranus, and Neptune are so different from the terrestrial planets that we must create an entirely new mental image of the term *planet*. Table 12.1 summarizes the bulk properties of these giant worlds. The most obvious difference between the jovian and terrestrial planets is size. Earth's mass in comparison to that of Jupiter is like the mass of a squirrel compared to that of a full-grown person. By volume, Jupiter is some 1,400 times as large as Earth, which makes Earth like a large marble compared to a basketball-sized Jupiter. (Remember that volume is proportional to the cube of the radius. Jupiter's radius is 11.2 times Earth's radius, so its volume is about $11.2^3 = 1,405$ times Earth's volume.)

Jovian Planet Composition The overall compositions of the jovian planets—particularly Jupiter and Saturn—are more similar to that of the Sun than to any of the terrestrial worlds. Their primary chemical constituents are hydrogen and helium, although their atmospheres also contain many hydrogen compounds such as methane (CH_4), ammonia (NH_3), and water (H_2O). They have rocky material only deep in their cores, but even there the pressures and temperatures are so extreme that nothing resembles a solid "surface." In a sense, the general structures of the jovian planets are opposite those of the terrestrial planets. Rather than having thin atmospheres around relatively large rocky bodies, the jovian planets have relatively small, dense cores surrounded by massive layers of gas. A spacecraft descending into Jupiter, for example, would have to travel

Figure 12.1 Jupiter, Saturn, Uranus, and Neptune, shown to scale with Earth for comparison. These four jovian planets are very different in character from the terrestrial worlds.

VIS

some 60,000 kilometers before reaching the core. (For jovian planets, the core is defined to include rock, metal, and hydrogen compounds, which are all much denser than hydrogen and helium.)

The jovian planets may be Sun-like in composition, but they are not stars because they do not have nuclear fusion in their cores. Stars have nuclear fusion because

their large masses make gravity strong enough to compress their cores to extremely high temperatures and densities [Section 15.3]. Calculations show that gravity is strong enough to allow nuclear fusion only for objects with a mass at least 80 times that of Jupiter. Some people have called Jupiter a "failed star" for this reason, but it might better be thought of as a very successful planet.

Table 12.1. Comparison of Bulk Properties of the Jovian Planets

Planet		Average Distance from Sun (AU)	Mass (Earth masses)	Equatorial Radius (Earth radii)	Average Density (g/cm³) (Earth = 5.52 g/cm³)	Bulk Composition
Jupiter		5.20	318	11.2	1.33	Mostly H, He
Saturn		9.54	95	9.46	0.71	Mostly H, He
Uranus		19.2	14	3.98	1.24	Hydrogen compounds and rock, H and He
Neptune		30.1	17	3.81	1.67	Hydrogen compounds and rock, H and He

Jovian Planet Rotation The jovian planets rotate much more rapidly than any of the terrestrial worlds. However, precisely defining their rotation rates can be difficult because they lack solid surfaces. We can measure a terrestrial rotation period simply by watching the apparent movement of a mountain or crater as a planet rotates. On jovian planets, we can observe only the movements of clouds. Cloud movements can be deceptive, because their apparent speeds are a combination of winds and planetary rotation. In addition, observations of clouds at different latitudes suggest that the jovian planets rotate at different speeds near their equators than near their poles. (The Sun also exhibits such *differential rotation*.)

We can measure the rotation rates of the jovian interiors by tracking emissions from charged particles trapped in their magnetospheres [Section 11.3]. This technique tells us the rotation period of the magnetosphere, which should be the same as the rotation period deep in the interior, where the magnetic field is generated. Measurements show that the jovian "day" ranges from about 10 hours on Jupiter and Saturn to 16–17 hours on Uranus and Neptune.

THINK ABOUT IT

How would you measure the rotation period of Earth if you made telescopic observations from Mars? How might Earth's clouds mislead you if you focused on them?

Jovian Planet Shapes Even the shapes of the jovian planets are somewhat different from the shapes of the terrestrial planets. Gravity makes the jovian planets approximately spherical, but their rapid rotation rates make them slightly squashed (Figure 12.2). Material near the equator, where speeds around the rotation axis are highest, is flung outward in the same way you feel yourself flung outward when you ride on a merry-go-round. As a result, the jovian planets bulge noticeably around their equators. The size of the equatorial bulge depends on the balance between the strength of gravity pulling material inward and the rate of rotation pushing material outward. The balance tips most strongly toward flattening on Saturn, with its rapid 10-hour rotation period and relatively weak surface gravity—Saturn is about 10% wider at its equator than at its poles. In addition to altering a planet's appearance, the equatorial bulge exerts an extra gravitational pull that helps keep moons and rings aligned with the equator.

Moons and Rings

After size, perhaps the most noticeable difference between the jovian and the terrestrial planets involves moons and rings. The terrestrial planets are nearly isolated worlds, with only Earth (one moon) and Mars (two tiny moons) orbited by any moons at all. In contrast, each of the jovian planets is orbited by many moons and rings.

In total, more than 100 moons are now known to orbit the jovian planets. A few are as large as terrestrial worlds and exhibit fascinating geology, which we will discuss in Section 12.5. One large moon, Titan, even has a thick atmosphere. The vast majority of the moons are quite small and may be captured asteroids [Section 9.4]. Additional small moons will likely be discovered.

All four jovian planets have rings, although only Saturn's rings are easily visible from Earth (and in photos like Figure 12.1), for reasons we will discuss in Section 12.6. Rings are composed of countless small pieces of rock and ice, each orbiting its planet like a tiny moon. The rings look flat because the particles all orbit in essentially the same plane. They look somewhat solid because of the sheer number of small particles. Rings are located closer to the planets than any of their moderate- or large-size moons, but the inner edge of the rings is still well above the planets' cloudtops.

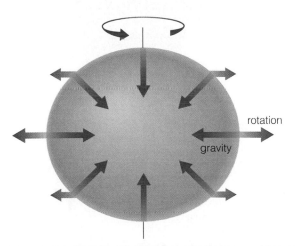

a Gravity by itself would make a planet spherical, but rapid rotation flattens out the spherical shape by flinging material near the equator outward.

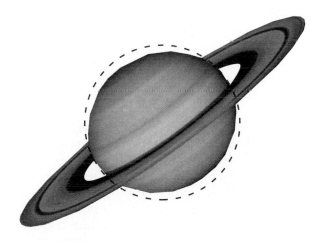

b Saturn is clearly not spherical. Compare its actual shape to the dashed circle.

Figure 12.2 The jovian planets are not quite spherical.

Why So Different?

Why are the jovian planets so different from the terrestrial planets? We can trace almost all the differences to the formation of the solar system.

Recall that the terrestrial planets accreted from planetesimals of rock and metal and ended up too small to capture significant amounts of the abundant hydrogen and helium gas in the solar nebula [Section 9.3]. The jovian planets formed far from the Sun, where condensation of ices meant that much more solid matter was available to accrete into planetesimals. As the massive ice/rock cores of the future jovian planets accreted, their strong gravity captured gas from the surrounding nebula, enabling the planets to grow to their large sizes.

Some of the captured gas formed flattened "miniature solar nebulae" around the jovian planets (look back at Figure 9.8). Just as solid grains condensed and accreted into planets in the full solar nebula, solid grains condensed and accreted into moons in these "miniature solar nebulae." The small, flattened nebulae probably also explain the origin of the material that now makes up the rings. However, as we'll discuss in Section 12.6, rings probably undergo dramatic change over millions and billions of years, so that the rings we see today are probably quite different in appearance from the rings of the distant past and future.

12.2 Jovian Planet Interiors

We are now ready to begin studying the jovian planets in more detail. As with the terrestrial planets, we'll begin by looking deep inside them. Probing jovian planet interiors is even more difficult than probing terrestrial interiors. Nevertheless, we can learn a great deal by combining careful observations, laboratory studies, and theoretical models.

Learning About Jovian Interiors

The first clue to interior structure comes from average density. As noted earlier, we learned about the low densities of the jovian planets from Earth-based measurements of their sizes and masses. Spacecraft have taught us much more. Just as with the terrestrial planets, detailed observations of the strength of a planet's gravity and of its magnetic field can help us put together a model of interior structure [Section 10.2]. For the jovian planets, the amount by which their shapes deviate from perfect spheres also helps us probe their interiors. For example, computer models using shape data tell us that Saturn's core makes up a larger fraction of its mass than does Jupiter's core. Laboratory studies tell us how the ingredients of the jovian planets (especially hydrogen and helium) act under the tremendous temperatures and pressures that exist deep below the cloudtops.

Nowadays, advanced computer models successfully match the observed sizes, densities, atmospheric composi-

tions, and shapes of the jovian planets. We therefore believe that we have a fairly clear understanding of their interiors. We'll begin our discussion of the jovian interiors by using Jupiter as a prototype.

Inside Jupiter

The jovian planets have no solid surfaces, but they still have fairly distinct interior layers. In Jupiter's case, the composition is probably nearly uniform throughout most of its interior—mostly hydrogen and helium. Its layers are defined by the phase (such as liquid or gas) of the material. The core contains rock, metal, and most of the hydrogen compounds. Figure 12.3 shows the basic interior layering of Jupiter, with Earth's interior compared to scale. Although Jupiter's core is only slightly larger in size than Earth, the tremendous weight of the overlying material compresses it to a much higher density. As a result, Jupiter's core is about 10 times as massive as the entire Earth.

Plunging into Jupiter To get a better sense of Jupiter's interior, imagine plunging head-on into Jupiter in a futuristic spacesuit that allows you to survive the ultra-extreme interior conditions. Near the cloudtops, you'll find the temperature to be a brisk 125 K (−148°C), the density to be a low 0.0002 g/cm^3, and the atmospheric pressure to be about 1 bar (the same as the pressure at sea level on Earth [Section 11.1]). The deeper you go, the higher the temperature, density, and pressure become.

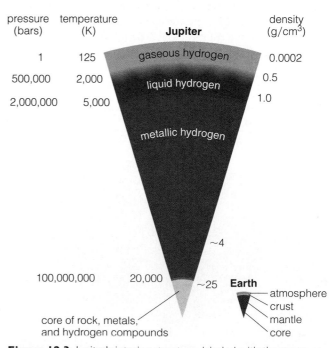

Figure 12.3 Jupiter's interior structure, labeled with the pressure, temperature, and density at various depths. Earth's interior structure is shown to scale for comparison. Note that Jupiter's core is only slightly larger than Earth but is about 10 times more massive.

By a depth of 7,000 km—about 10% of the way to the center—you'll find that the temperature has increased to a scorching 2,000 K, the density has reached 0.5 g/cm^3 (half that of water), and the pressure is about 500,000 bars. Under these conditions, hydrogen acts more like a liquid than a gas, which is why the layer that begins at this depth is labeled "liquid hydrogen" in Figure 12.3. The layer also contains some helium and hydrogen compounds.

At a depth of 14,000 km, the density has become about as high as that of water (1.0 g/cm^3). The temperature is near 5,000 K, almost as hot as the surface (but not the interior) of the Sun. The pressure has reached 2 million bars. This extreme pressure forces hydrogen into a compact form known as *metallic hydrogen.* This name is used because all the hydrogen atoms share electrons, like the atoms in more familiar metals on Earth. Like other metals, metallic hydrogen conducts electricity. This interior layer of metallic hydrogen plays an important role in generating Jupiter's strong magnetic field.

Continuing your descent, you'll reach Jupiter's core at a depth of 60,000 km, about 10,000 km from the center. The core temperature is some 20,000 K, and the pressure is about 100 million bars. The density of the core is about 25 g/cm^3, much denser than any material you'll find on Earth's surface. The core is a mix of hydrogen compounds, rock, and metal, but these materials bear little resemblance to familiar solids or liquids because of the combination of high temperature and extreme pressure. The core materials probably remain mixed together rather than separating into layers of different composition as in the terrestrial planets. As mentioned earlier, the total mass of the core is about 10 times that of Earth, so the core alone would make an impressive planet.

Jupiter's Internal Heat

Jupiter has a tremendous amount of internal heat, and like any hot object it must gradually lose this heat by emitting thermal radiation [Section 6.4]. In fact, Jupiter emits almost twice as much energy as it receives from the Sun. This heat contributes significant energy to Jupiter's upper atmosphere. (For comparison, Earth's internal heat contributes only 0.005% as much energy to the surface as does sunlight.)

What keeps Jupiter so hot inside? Jupiter's large size means it loses internal heat very slowly (Math Insight 10.1), but calculations show that the remaining heat of accretion and differentiation is not enough to explain Jupiter's present-day heat loss. Radioactive decay adds further heat, but still not enough to account for all the internal heat. The most likely explanation for Jupiter's excess heat is that the planet is still slowly contracting, as if it has not quite finished forming. Contraction converts gravitational potential energy to thermal energy, so ongoing contraction represents an ongoing source of internal heat. Although we have not measured any such contraction, theoretical considerations show that it is probably occurring. Moreover, calculations tell us that contraction could easily explain Jupiter's

internal heat, even if the contraction is so gradual that we have little hope of ever measuring it directly.

Regardless of the specific mechanism, Jupiter has undoubtedly lost substantial heat during the 4.5 billion years since its formation. Jupiter's interior must have been much warmer in the distant past, and this heat would have "puffed up" its atmosphere. Thus, Jupiter was larger and reflected more sunlight in the distant past, and it must have been even more prominent in Earth's sky than it is today.

Comparing Jovian Planet Interiors

We can build on what we've learned about Jupiter to investigate the other jovian planet interiors. Look back at Table 12.1. Although there are no obvious correlations among size, density, and composition, we can explain the jovian planets' properties by examining the behavior of gaseous materials in more detail.

Planetary Size and Density Building a planet of hydrogen and helium is a bit like making one out of very fluffy pillows (Figure 12.4a). Imagine assembling a planet pillow by pillow. As each new pillow is added, those on the bottom are compressed more by those above. As the lower layers are forced closer together, their mutual gravitational attraction increases, compressing them even further. At first the stack grows substantially with each pillow, but eventually the growth slows until adding pillows barely increases the height of the stack.

This analogy explains why Jupiter is only slightly larger than Saturn in radius even though it is more than three times as massive. The extra mass of Jupiter compresses its interior to a much greater extent, making Jupiter's average density almost twice Saturn's. More precise calculations show that Jupiter's radius is almost the maximum possible radius for a jovian planet. If much more gas were added to Jupiter, its weight would actually compress the interior enough to make the planet *smaller* rather than larger (Figure 12.4b). Thus, some extrasolar planets that are larger in mass than Jupiter [Section 9.6] are probably smaller than Jupiter in radius. In fact, the smallest stars are significantly smaller in radius than Jupiter, even though they are 80 times more massive. (Above 80 Jupiter masses, stars become larger with increasing mass due to the "puffing up" caused by increased fusion (Chapter 15).)

THINK ABOUT IT

Saturn's average density of 0.71 g/cm^3 is less than that of water. As a result, it has sometimes been said that Saturn could float on a giant ocean. Suppose there really were a gigantic planet with a gigantic ocean and we put Saturn on the ocean's surface. Would it float? If not, what would happen?

The pillow analogy suggests that a hydrogen/helium planet less massive than Saturn should have an even lower density. However, Uranus and Neptune have significantly

Figure 12.4
The relationship between mass and radius for jovian planets.

a Adding pillows to a stack may increase its height at first, but eventually it just compresses all the pillows in the stack. In a similar way, adding mass to a jovian planet would eventually just compress the planet to higher density without increasing the planet's radius.

b This graph shows how the radius of a hydrogen/helium planet depends on the planet's mass. Jupiter's radius is only slightly larger than Saturn's, although it is three times more massive. For a planet much more massive than Jupiter, gravitational compression would actually make it smaller in size.

higher densities than Saturn. We therefore conclude that Uranus and Neptune must have a significantly different composition than that of Saturn, with a much larger fraction of higher-density material such as hydrogen compounds and rock.

Interior Structure Computer models of the jovian interiors yield the somewhat surprising result that the cores of all four jovian planets are quite similar in composition. Each consists of a mixture of rock, metal, and hydrogen compounds, with a total mass about 10 times that of Earth. This suggests that all four jovian planets started from the same size "seeds" following accretion. Their differences must stem from having captured different amounts of additional hydrogen and helium gas from the solar nebula.

Jupiter captured so much gas that it ended up with 300 times Earth's mass. Saturn captured about one-fourth as much gas as Jupiter, while Uranus and Neptune captured

much smaller amounts of gas. This general pattern makes sense when you consider that icy planetesimals took longer to accrete in the outer solar system, where they were widely spread out. The more distant jovian planets didn't have time to capture as much gas from the solar nebula as Jupiter and Saturn did before the gas was blown into interstellar space by the solar wind [Section 9.3].

The similar cores of all four jovian planets also mean that their interior structures differ mainly in the hydrogen/helium layers that surround the cores (Figure 12.5). Saturn is large enough for interior densities, temperatures, and pressures to be high enough to force hydrogen into liquid and metallic forms just as in Jupiter. However, these conditions occur relatively deeper in Saturn than in Jupiter—because of its smaller size—resulting in less metallic hydrogen overall. Pressures within Uranus and Neptune are not high enough to form liquid or metallic hydrogen. However, their cores of rock, metal, and hydrogen compounds may

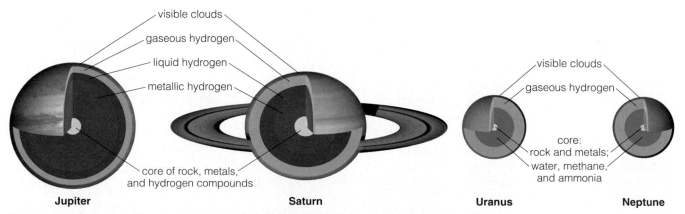

Figure 12.5 These diagrams compare the interior structures of the jovian planets (shown approximately to scale). All four planets have cores of rock, metal, and hydrogen compounds, with masses about 10 times the mass of Earth's core. They differ primarily in the hydrogen/helium layers that surround the cores. (The cores of Uranus and Neptune consist of both the rock/metal layers and the water/methane/ammonia layers.)

be liquid, making for very odd "oceans" buried deep inside them. The cores of Uranus and Neptune appear larger in radius than the cores of Jupiter and Saturn [Figure 12.5], even though they have about the same mass, because they are less compressed by the lower weight of overlying layers. The less extreme interior conditions also allow Uranus's and Neptune's cores to differentiate into separate layers of hydrogen compounds and rock/metal.

Internal Heat on Saturn, Uranus, and Neptune Like Jupiter, Saturn emits nearly twice as much energy as it receives from the Sun, suggesting that it must have some ongoing source of heat. However, Saturn's mass is too small for it to be generating all this heat by contracting like Jupiter. Instead, Saturn's pressure and its lower interior temperatures may allow helium to condense into liquid form at relatively high levels within the interior. The helium droplets slowly rain down to the deeper interior. This gradual helium rain represents a sort of ongoing differentiation [Section 10.2], because it means that higher-density material (the helium droplets) is still sinking inside the planet. Voyager observations confirmed that Saturn's atmosphere is somewhat depleted of helium (compared to Jupiter's atmosphere), just as we should expect if helium has been raining down into Saturn's interior for billions of years.

Neither Uranus nor Neptune has conditions that allow helium rain to form, and most of their original heat from accretion should have escaped long ago. This explains why Uranus emits virtually no excess internal energy. Neptune, however, is much more mysterious. Like Jupiter and Saturn, Neptune emits nearly twice as much energy as it receives from the Sun. The only reasonable explanation for Neptune's internal heat is that the planet is somehow still contracting, somewhat like Jupiter, thereby converting gravitational potential energy into thermal energy. However, we do not know why a planet of Neptune's size would still be contracting more than 4 billion years after its formation.

12.3 Jovian Planet Atmospheres

The jovian planets lack the geology of the terrestrial planets because they do not have solid surfaces. But whatever they lack in geology they more than make up for with their dynamic atmospheres. Fortunately, their atmospheric processes are quite similar to those we've already discussed for the terrestrial atmospheres, despite their much greater extents and very different compositions. Again, we'll begin with Jupiter.

Jupiter's Atmosphere

Jupiter's atmosphere is almost entirely hydrogen and helium—about 75% hydrogen and 24% helium (by mass). The rest consists of an assortment of hydrogen compounds.

Because oxygen, carbon, and nitrogen are the most common elements besides hydrogen and helium, the most common hydrogen compounds are methane (CH_4), ammonia (NH_3), and water (H_2O). Spectroscopy reveals the presence of more complex compounds, including acetylene (C_2H_2), ethane (C_2H_6), propane (C_3H_8), and larger molecules that can act like haze particles.

Although hydrogen compounds make up only a minuscule fraction of Jupiter's atmosphere, they are responsible for virtually all aspects of its appearance. Some of these compounds condense to form the clouds that are so prominent in telescope and spacecraft images. Others are responsible for Jupiter's great variety of colors. Without these compounds in its atmosphere, Jupiter would be a uniform, colorless ball of gas.

THINK ABOUT IT

Several gases in Jupiter's atmosphere—including methane, propane, and acetylene—are used as highly flammable fuels here on Earth. Jupiter also has plenty of lightning to provide sparks. Why don't these gases ignite in Jupiter's atmosphere? (*Hint:* What's missing from Jupiter's atmosphere that's necessary for ordinary fire?)

Jupiter is the only jovian planet that we've directly sampled. On December 7, 1995, following a 6-year trip from Earth, the *Galileo* spacecraft released a scientific probe the size of a large suitcase into Jupiter's atmosphere (Figure 12.6). The probe collected temperature, pressure, composition, and radiation measurements for about an hour as it descended,

Figure 12.6 Artist's conception of the Galileo probe entering Jupiter's atmosphere. This view, looking outward from beneath the clouds, shows the suitcase-size probe falling with its parachute extended above it.

until it was finally destroyed by the ever-increasing pressures and temperatures. The probe sent its data via radio signals back to *Galileo,* in orbit around Jupiter. The orbiting spacecraft relayed the data back to Earth.

Jupiter's Atmospheric Structure The Galileo probe helped confirm that the temperature structure of Jupiter's atmosphere is very similar to that of Earth (Figure 12.7). Let's follow its fiery descent through the atmosphere.

The probe plunged into Jupiter at a speed of over 200,000 km/hr, using a heat shield to prevent it from burning up immediately. Above the cloudtops, the probe found very low density gas heated to perhaps 1,000 K by solar X rays and by energetic particles from Jupiter's magnetosphere. This thin, hot gas makes up Jupiter's *thermosphere.*

Next, the probe encountered Jupiter's *stratosphere,* where solar ultraviolet photons are absorbed by a few minor ingredients in the atmosphere. This absorption gives the stratosphere a peak temperature of around 200 K. Chemical reactions driven by the solar ultraviolet photons also create a smoglike haze that masks the color and sharpness of the clouds below.

Below the stratosphere lies Jupiter's *troposphere,* where the increased density greatly slowed the probe's descent. The probe slowed further by jettisoning its heat shield and releasing a parachute. The temperature at the top of the

Figure 12.7 This graph shows the temperature structure of Jupiter's atmosphere. Jupiter has at least three distinct cloud layers because different atmospheric gases condense at different temperatures and hence at different altitudes.

troposphere is about 125 K. It then rises with depth because greenhouse gases trap both solar heat and Jupiter's own internal heat. The probe found tremendous winds and turbulence in the troposphere, where most of Jupiter's weather occurs. Strong convection drives this wind and weather. As on the terrestrial planets, convection occurs in the troposphere because warm air rises from the warmer regions below as cool air descends from higher altitudes. The Galileo probe stopped sending data at a depth of about 300 kilometers below the cloudtops, where it presumably was destroyed by the rising temperature and pressure.

The similarities between the atmospheric structures of Earth and Jupiter may seem surprising for two planets that are so different in other ways. However, the similarities confirm that atmospheres are governed primarily by interactions between sunlight and gases.

Jupiter's Cloud Layers The convection in Jupiter's troposphere is responsible for the thick clouds that enshroud Jupiter. As on Earth, convection sends gas upward through the troposphere until it reaches an altitude at which the temperature is low enough for condensation. Condensation of a gas into liquid droplets forms clouds. However, while only water vapor condenses to make clouds on Earth, Jupiter's atmosphere contains several gases that condense at different temperatures and hence at different altitudes.

Jupiter has three primary cloud layers, as shown in Figure 12.7. We can understand them by imagining that we could watch gas rising through the troposphere with convection. Deep in the troposphere, the gas includes three ingredients that will condense when temperatures are low enough: water (H_2O), ammonium hydrosulfide (NH_4SH), and ammonia (NH_3). The rising gas first encounters temperatures cool enough for water vapor to condense into liquid water but not cool enough for the other gases to condense. Thus, the lowest layer of clouds contains water droplets. As the remaining gas continues its rise, it next reaches an altitude at which ammonium hydrosulfide condenses to make the second cloud layer. Finally, after rising another 50 kilometers, the gas reaches an altitude at which ammonia (NH_3) condenses to make the upper cloud layer. The tops of the ammonia clouds are usually considered the "cloudtops" of Jupiter, which is why Figure 12.7 shows this altitude as zero altitude for Jupiter.

We can distinguish some of the cloud layers by their colors. Observations tell us that the high ammonia clouds are usually white, while the mid-level clouds of ammonium hydrosulfide are brown or red. These cloud colors in part explain Jupiter's striking appearance.

Jupiter's Global Circulation and Winds Jupiter has planet-wide circulation cells similar to those on Earth [Section 11.4]. As on Earth, solar heat causes equatorial air to expand and spill toward the poles, while cooler polar air flows toward the equator. Also as on Earth, the Coriolis effect splits the large equator-to-pole circulation cells into smaller cells. However, Jupiter's greater size and faster rotation make the Coriolis effect much stronger on Jupiter than on Earth. Instead of being split into just three smaller cells encircling each hemisphere, as on Earth, Jupiter's circulation cells split into many alternating bands of rising and falling air. These bands are easily visible as the "stripes" of alternating color in photographs of Jupiter (Figure 12.8a).

The bands of rising air (called *zones*) appear white because ammonia clouds form as the air rises to high, cool altitudes. Ammonia "snowflakes" rain down from the clouds, depleting the air of ammonia. The adjacent bands of falling air (called *belts*) therefore lack the ammonia needed to form white ammonia clouds. With no high-altitude clouds

We see whitish bands where rising air forms ammonia clouds.

We see reddish bands where air is depleted of ammonia, allowing us to see lower-altitude clouds of ammonium hydrosulfide.

The Coriolis effect diverts north-south winds into strong east-west winds.

Snow from ammonia cloud depletes air of ammonia.

reddish clouds at lower altitude

Rising air forms white ammonia cloud.

No cloud or snow in descending air on north and south sides of white ammonia cloud; reddish clouds below are visible.

These bands are warm, red, low-altitude clouds.

These bands are cool, white, high-altitude clouds.

a Jupiter's bands of color represent alternating regions of rising and falling air. The whitish bands look white because we see high-altitude ammonia clouds. The darker bands are reddish because these regions lack ammonia clouds, allowing us to see down to lower-altitude clouds.

VIS

b This infrared image was taken nearly simultaneously with the visible-light photo in (a). Brighter regions are higher in temperature. The white regions in (a) are dark here, because we are seeing cooler, higher-altitude clouds. The dark regions in (a) are brighter here, because we are seeing down to warmer, lower-altitude clouds. The Great Red Spot is an exception.

IR

Figure 12.8 This figure summarizes the reasons for Jupiter's banded appearance.

in the way, we can see down to the reddish clouds below, making these bands of falling air appear dark in color.

In essence, the distinction between Jupiter's bright and dark bands is analogous to the difference on Earth between cloudy, rainy equatorial regions (regions of generally rising air) and the clear, desert skies found at latitudes roughly 20°–30° north and south of the equator (regions of descending air). Infrared images confirm the temperature differences between Jupiter's bands of cool, high clouds and warm, low clouds (Figure 12.8b).

Jupiter's global wind patterns are shaped by these alternating bands of rising and falling air. As on Earth, the rising and falling air drives slow winds that are directed north or south, but Jupiter's strong Coriolis effect diverts these winds into fast east or west winds. The east-west winds have peak speeds above 400 km/hr, making hurricane winds on Earth seem mild by comparison. The winds are generally strongest at the equator and at the boundaries between bands of rising and falling air.

The Great Red Spot Jupiter's most famous feature is its **Great Red Spot**, which is visible in Jupiter's southern hemisphere (see Figure 12.1). The Great Red Spot is more than twice as wide as Earth. We know that it has been a consistent feature of Jupiter for at least three centuries, because it has been seen ever since telescopes became powerful enough to detect it. What is it?

The Great Red Spot is perhaps the most dramatic weather pattern in the solar system (Figure 12.9). It is sometimes likened to a hurricane, but the analogy is not quite correct. Hurricane winds circulate around low-pressure regions and therefore circulate *clockwise* in the southern hemisphere [Section 11.4]. The winds in the Great Red Spot circulate *counterclockwise,* indicating that it is a *high*-pressure storm, possibly kept spinning due to the influence of winds in the nearby horizontal bands.

Other, smaller storms are always brewing in Jupiter's atmosphere. They are small only in comparison to the Great Red Spot. Brown ovals are low-pressure storms with their cloudtops deeper in Jupiter's atmosphere, and white ovals are high-pressure storms topped with ammonia clouds. No one knows what drives Jupiter's storms, why Jupiter has only one Great Red Spot, or why the Great Red Spot has persisted so much longer than storms on Earth. Storms on Earth lose their strength when they pass over land, so perhaps Jupiter's biggest storms last for centuries simply because no solid surface is present to sap their energy.

Jupiter's Colors Jupiter's vibrant colors remain perplexing despite decades of intense study. The ammonia clouds found at high altitude are white, as expected. However, we'd also expect the lower clouds of ammonium hydrosulfide to be white, but we observe reds, tans, and browns. The colors probably come from small amounts of chemical compounds formed in reactions powered by Jupiter's internal heat or by lightning beneath the clouds. The identity of these chemical compounds is not known.

The bright colors of the Great Red Spot may be the most mysterious of all. We might expect its high-altitude clouds to be white like the high-altitude ammonia clouds elsewhere. Instead, of course, the Great Red Spot is red. The red colors may be the result of chemicals formed by interactions of the high-altitude gas and solar ultraviolet light, but we don't really know for sure.

Climate on Jupiter As far as we know, Jupiter's climate is steady and unchanging. Jupiter has no appreciable axis tilt and therefore has no seasons. In fact, Jupiter's polar temperatures are quite similar to its equatorial temperatures. Solar heating acting alone might leave the poles relatively cool, but heat from Jupiter's interior keeps the planet uniformly warm.

Comparing Jovian Planet Atmospheres

As with the jovian interiors, we can understand the jovian atmospheres by building upon what we have learned for Jupiter. We'll start by comparing their atmospheric structures.

Atmospheric Structure and Clouds The atmospheric structures of the four jovian planets are all quite similar (Figure 12.10). The primary difference among them is that the atmospheres get progressively cooler with increasing distance from the Sun, just as we would expect.

Each jovian planet has distinct cloud layers, with altitudes dictated by the atmospheric levels at which various gases can condense into liquid droplets or solid flakes. Jupiter and Saturn each have the same three layers of clouds: ammonia clouds at the level where the temperature is about 150 K, ammonium hydrosulfide clouds where the temperature is 200 K, and water clouds where the temperature is about 270 K. However, because Saturn's atmosphere is colder

Figure 12.9 This photograph shows Jupiter's Great Red Spot, a huge high-pressure storm that is large enough to swallow two or three Earths. The overlaid diagram shows a weather map of the region.

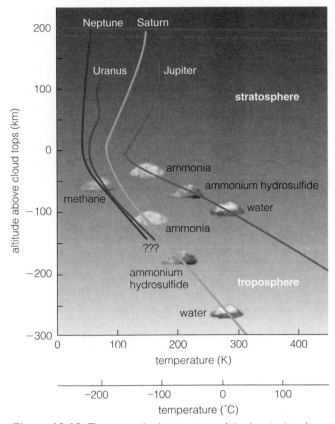

Figure 12.10 The atmospheric structures of the four jovian planets, showing the tropospheres and stratospheres only. All are quite similar, except that the atmospheres are colder on the planets farther from the Sun. Note that clouds of a particular composition always occur at about the same temperature. For example, the ammonia clouds on both Jupiter and Saturn form at the level at which the atmospheric temperature is about 150 K. The "???" at the bottom of the curves for Uranus and Neptune indicates that we do not have data beneath this level.

than Jupiter's at any particular altitude, the various cloud layers occur deeper within Saturn's atmosphere. In addition, Saturn's cloud layers are separated by greater vertical distances, because Saturn's weaker gravity causes less atmospheric compression than we find on Jupiter.

The clouds we see on Uranus and Neptune are made from flakes of methane snow. Methane can condense in the very cold upper tropospheres of Uranus and Neptune but not in the warmer tropospheres of Jupiter and Saturn. We do not know whether Uranus and Neptune have clouds of ammonia or other compounds at lower levels, because we have no data for their atmospheres at the depths where such clouds would occur.

Jovian Planet Weather Spacecraft visiting the jovian planets have monitored their cloud patterns and their weather close up (Figure 12.11). More distant weather observations have also been made by the Hubble Space Telescope and powerful ground-based telescopes. Although we have much more to learn, these observations have given us a general sense of the comparative weather of the four jovian planets.

Jupiter's weather phenomena (clouds, winds, storms) are by far the strongest and most active, but some weather is found on all four planets.

Like Jupiter, Saturn has "stripes" of alternating color and wind direction, though in more subdued colors. Saturn's atmosphere has small storms and an occasional large storm, but it lacks any feature as prominent as Jupiter's Great Red Spot. Surprisingly, Saturn's winds are even stronger than Jupiter's—the opposite of what we might expect from its weaker Coriolis effect. No one knows why the winds are so strong. We might expect seasons on Saturn, because it has an axis tilt similar to that of Earth. Some weather changes have been observed, but Saturn's internal heat keeps temperatures about the same year-round and planet-wide.

Neptune's atmosphere is also banded, and we have seen a high-pressure storm, called the Great Dark Spot, similar to Jupiter's Great Red Spot (though not as long-lasting). Like Saturn, Neptune has an axis tilt similar to Earth's but it has relatively little seasonal change because of its internal heat.

The greatest surprise in jovian weather comes from Uranus. Because Uranus is essentially tipped on its side, we would expect it to go through extreme seasonal changes during its 84-year orbit of the Sun. When *Voyager 2* flew past in 1986, Uranus's northern hemisphere was facing almost directly toward the Sun, and its southern hemisphere was in perpetual darkness. Photographs revealed virtually no clouds and no banded structure like those found on the other jovian planets. Scientists attributed the lack of weather to the lack of significant internal heat on Uranus. However, more recent observations from the Hubble Space Telescope show storms raging in Uranus's atmosphere (see Figure 12.11). The storms may be brewing because of the changing seasons as the southern hemisphere sees sunlight for the first time in decades.

Jovian Planet Colors The most striking difference among the jovian atmospheres is their colors. Starting with Jupiter's distinct reds, the jovian planets turn to Saturn's more subdued yellows, then to Uranus's faint blue-green tinge, and finally to Neptune's deep blue (see Figure 12.1).

Saturn's reds and tans almost certainly come from the same compounds that produce these colors on Jupiter—whatever those compounds may be. Just as on Jupiter, these compounds are probably created by chemical reactions beneath the cloud layers and carried upward by convection. However, because Saturn's cloud layers lie deeper in its atmosphere than Jupiter's, a thicker layer of tan "smog" overlies the clouds and washes out Saturn's colors. The greater vertical spread of Saturn's cloud layers also has an effect, because less light reaches the deeper cloud levels.

The blue colors of Uranus and Neptune come from methane gas, which is at least 20 times more abundant (by percentage) on these planets than on Jupiter or Saturn. Methane gas in the upper atmospheres of Uranus and Neptune absorbs red light, allowing only blue light to penetrate to the level at which the methane clouds condense. The

Jupiter

VIS

5,000 km

Uranus

IR

Figure 12.11 Close-ups of cloud patterns on the four jovian planets. The reddish spots in the Uranus image are the high clouds of newly brewing storms, and the Great Dark Spot is visible in the Neptune image to the left of center. The images for Jupiter, Saturn, and Neptune were taken in visible light. The image of Uranus was taken in infrared light. (You can compare Uranus's weather in this image to that a few years earlier, shown in Figure 12.40.)

20,000 km

Saturn

VIS

Neptune

VIS

methane clouds reflect this blue light upward, giving the planets their blue colors (Figure 12.12).

Neptune's deeper blue color must arise because more sunlight penetrates down to its clouds than reaches the clouds on Uranus. Because Neptune is farther from the Sun, we might expect the opposite. The explanation is probably that Uranus has much more smoglike haze than Neptune. This haze scatters sunlight before it reaches the level of the methane clouds. The extra haze on Uranus may be traceable to its long and extreme seasons. With one hemisphere remaining sunlit for decades, gases have plenty of time to interact with solar ultraviolet light and to make the chemical ingredients of the haze. Continuous sunlight may also explain why Uranus has a surprisingly hot thermosphere that extends thousands of kilometers above its cloudtops. As we'll discuss later, Uranus's thermosphere has an important influence on its rings.

light of all colors from Sun

Methane gas absorbs red light, transmits blue light.

Methane clouds reflect blue light to space.

methane gas mixed with hydrogen

methane clouds

Figure 12.12 Neptune and Uranus are blue because methane gas absorbs red light but transmits blue light. Clouds of methane-ice flakes reflect the transmitted blue light back to space.

12.4 Jovian Planet Magnetospheres

Each jovian planet is surrounded by a bubblelike magnetosphere consisting of the planet's magnetic field and the particles trapped within it. The jovian planets themselves dwarf the terrestrial planets, but their magnetospheres are even more impressive. The Voyager spacecraft carried instruments to measure magnetic fields and observe the charged particle belts encircling the planets. We learned not only about the magnetospheres themselves, but also about their interactions with the planets beneath them and the satellite systems around them.

Jupiter's Magnetosphere

Jupiter's awesome magnetic field is about 20,000 times stronger than Earth's [Section 11.3]. This strong magnetic field begins to deflect the solar wind some 3 million kilometers (about 40 Jupiter radii) before the wind even reaches Jupiter (Figure 12.13a). If our eyes could see this part of Jupiter's magnetosphere, it would be larger than the full moon in our sky. The solar wind sweeps around Jupiter's magnetosphere, stretching it out on the far side of Jupiter all the way to the orbit of Saturn.

Just as on Earth, the magnetosphere traps charged particles and makes them spiral along the magnetic field lines. The more energetic particles cause auroras as they follow the magnetic field into Jupiter's upper atmosphere, colliding with atoms and molecules and causing them to radiate (Figure 12.13b). Nearly all the charged particles in Earth's magnetosphere come from the solar wind. Jupiter has an additional source of even more charged particles: gas coming from its volcanically active moon, Io.

The magnetosphere, in turn, has important effects on Io and the other moons of Jupiter. The charged particles bombard the surfaces of Jupiter's icy moons, with each particle blasting away a few atoms or molecules. This process alters the surface materials and can even generate thin bombardment atmospheres, just as bombardment creates the thin atmospheres of Mercury and the Moon [Section 11.5]. On Io, the bombardment leads to the continuous escape of gases released by volcanic outgassing. As a result, Io loses atmospheric gas faster than any other object in the solar system. The escaping gases (sulfur, oxygen, and a hint of sodium) are ionized and feed a donut-shaped charged particle belt, called the *Io torus*, that approximately traces Io's orbit (Figure 12.14).

Comparing Jovian Planet Magnetospheres

The strength of each jovian planet's magnetic field depends primarily on the size of the electrically conducting layer buried in its interior (see Figure 12.5). Jupiter, with its thick layer of metallic hydrogen, has by far the strongest magnetic field. Saturn has a thinner layer of metallic hydrogen (due to its lower mass and lower pressures) and hence a correspondingly weaker magnetic field. Uranus and Neptune, smaller still, have no metallic hydrogen at all. Their relatively weak magnetic fields must be generated in their core "oceans" of hydrogen compounds, rock, and metal.

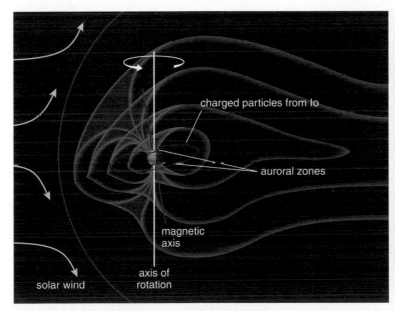

a Like Earth's magnetosphere (but much larger), Jupiter's magnetosphere has charged particle belts and auroras. In Jupiter's case most of the charged particles come from the volcanically active moon Io.

Figure 12.13 Jupiter's magnetosphere and auroras.

b Images of Jupiter's aurora in the ultraviolet are overlaid on a visible-wavelength image of Jupiter for reference (Hubble Space Telescope photos).

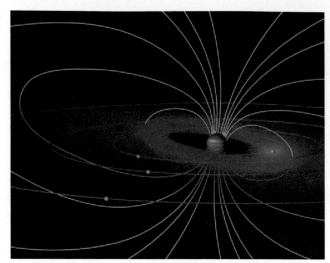

Figure 12.14 The bright red "donut" encircling Jupiter in this artist's conception is the charged particle belt known as the Io torus. It approximately traces Io's orbit (shown as the innermost of the red circles). (The other three red circles are the orbits of three other moons: Europa, Ganymede, and Callisto.) The Io torus is filled with ionized gases coming from volcanic outgassing on Io. The blue lines represent Jupiter's strong magnetic field.

The size of a planet's magnetosphere depends not only on the magnetic field strength, but also on the pressure of the solar wind against it. The pressure of the solar wind is weaker at greater distances from the Sun, so the magnetospheric "bubbles" surrounding more distant planets are larger than they would be if these planets were closer to the Sun. Thus, Uranus and Neptune have moderate-size magnetospheres despite their weak magnetic fields (Figure 12.15).

The jovian magnetic fields still pose some unsolved problems. Because magnetic fields are generated within the rotating interiors of planets, we expect them to be fairly closely aligned with planetary rotation. The magnetic fields of Jupiter and Saturn meet this expectation (10° tilt and 0° tilt, respectively), but those of Uranus and Neptune do not. Voyager observations showed that the magnetic field of Uranus is tipped by a whopping 60° relative to its rotation axis, and the magnetic field's center is also significantly offset from the planet's center. Neptune's magnetic field is inclined by 46° to its rotation axis. No one has yet explained these surprising magnetic-field tilts.

No other magnetosphere is as full of charged particles as Jupiter's, primarily because no other jovian planet has a

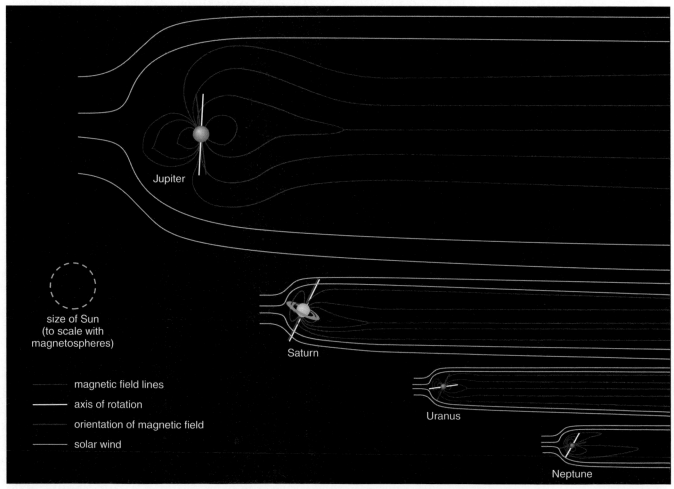

Figure 12.15 Comparison of jovian planet magnetospheres. The size of Jupiter's magnetosphere is particularly impressive in light of the greater pressure from the solar wind nearer the Sun. Note the significant tilts of the magnetic fields of Uranus and Neptune compared to their rotation axes. (Planets enlarged for clarity.)

satellite like Io. In addition, although all the magneto-spheres trap particles from the solar wind, Jupiter captures the most. Because trapped particles generate auroras, Jupiter has the brightest auroras, while those of the more distant jovian planets are progressively weaker.

 ## 12.5 A Wealth of Worlds: Satellites of Ice and Rock

The numerous moons orbiting the jovian planets exhibit an astonishing variety of features. Many are probably captured asteroids that have changed little in the past 4.5 billion years, but others show extensive signs of past geological activity. A few even have ongoing geological activity. After a brief general introduction to the moons, we will focus in this section on those moons that exhibit the most interesting geology.

Small, Medium, and Large Moons

Broadly speaking, the jovian moons can be divided into three groups: small moons less than about 300 km in diameter, medium-size moons ranging from about 300 to 1,500 km in diameter, and large moons more than 1,500 kilometers in diameter.

Figure 12.16 shows a montage of all the medium and large moons. These moons are planetlike in almost all ways. Each is spherical with a solid surface and its own unique geology. Some possess atmospheres, hot interiors, and even magnetic fields. The two largest—Jupiter's moon Ganymede and Saturn's moon Titan—are larger than the planet Mercury. Four others are larger than Pluto: Jupiter's moons Io, Europa, and Callisto, and Neptune's moon Triton.

Most of the medium and large moons probably also formed much like planets, except that they formed on orbits within the "miniature solar nebulae" surrounding the jovian planets rather than on orbits around the Sun. That explains why they generally have almost circular orbits that lie close to the equatorial plane of their parent planet and orbit in the same direction in which their planet rotates. (Triton is the one notable exception, as we will discuss shortly.) Nearly all of these moons also share an uncanny trait: They always keep the same face turned toward their planet, just as our Moon always shows the same face to Earth [Section 2.5]. This *synchronous rotation* arose from the strong tidal forces [Section 5.4] exerted by the jovian planets, which caused each moon to end up with equal periods of rotation and orbit regardless of how fast the moon rotated when it formed.

The medium and large moons have the most interesting geology, but they are far outnumbered by small moons. Many of the small moons are probably captured asteroids, which explains why so many of them have orbits with significant eccentricities, large orbital tilts, or even orbits that go backward relative to their planet's rotation. Dozens of

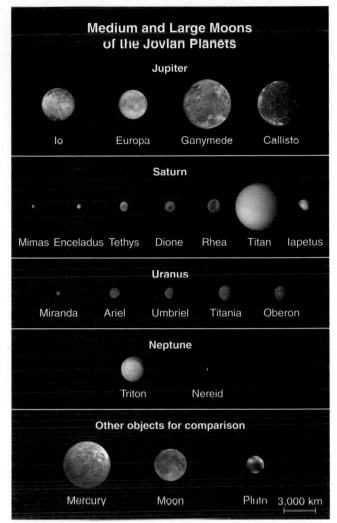

Figure 12.16 The medium-and large-size moons of the jovian planets, with sizes (but not distances) shown to scale. Mercury, the Moon, and Pluto are included for comparison.

the smallest moons have been discovered only within the past few years, and many more may yet be discovered. (See Table E.3 in Appendix E for a complete list of the known moons and their general properties.) The small moons' shapes generally resemble potatoes (Figure 12.17), because their gravities are too weak to force their rigid material into spheres.

Ice Geology

If you think about the role of planetary size in the geology of the terrestrial worlds, you might expect all of the jovian moons to be cold and geologically dead. After all, only two of the moons are even as large as Mercury, and Mercury has been geologically dead for more than 3 billion years. However, close-up photos from spacecraft have revealed several jovian moons with spectacular past and present geological activity.

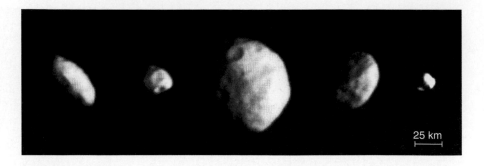

Figure 12.17 These photos from the Voyager spacecraft show five of Saturn's smaller moons. Note their irregular shapes. They are shown to scale. All are much smaller than the smallest moons shown in Figure 12.16. (These moons are all smaller than a few tens of kilometers in radius.)

25 km

How can these moons have so much more geological activity than similar-size (or larger) terrestrial worlds? Two factors play a role:

● In a few cases, most notably Jupiter's moons Io and Europa, the moons have an ongoing source of internal heat quite different from that found on any terrestrial world. We will discuss how this *tidal heating* works shortly.

● More generally, the jovian moons tend to be made from material with a high proportion of ice, and ice can undergo geological change at much lower temperatures than rock.

The icy composition of most jovian moons is a direct result of their formation far from the Sun. Remember that ice was able to condense into solid particles only beyond the frost line that lay between the orbits of Mars and Jupiter (see Figure 9.5). Thus, while the terrestrial worlds accreted only from rock and metal, the jovian moons accreted from material that contained a large proportion of ice. The different jovian planet systems themselves have distinct compositions: Jupiter's moons have the highest proportion of rock, and their ice is primarily water ice because it was too warm for ammonia or methane to condense at Jupiter's distance from the Sun. The moons of the more distant planets contain larger proportions of ice and contain methane and ammonia ice as well as water ice.

We can understand why icy worlds have more geological activity than similar-size rocky worlds by looking at three geologically important properties: strength, radioactive content, and melting point. In terms of strength, the differences for ice and rock might be smaller than you would expect. Ice at a temperature of 100 K (−173°C) is almost as rigid as rock at room temperature, so impact craters form similarly on icy and rocky worlds. Nevertheless, the difference in strength is great enough to cause ice mountains and ice cliffs on the jovian moons to sag and fade away more quickly than similar features of rock on the terrestrial planets.

Radioactive content affects geology because radioactive decay is an important source of internal heat. Most radioactive elements are found in rock, not ice. Thus, icy moons have relatively little internal heat generated by radioactive decay and hence lower internal temperatures than equivalent rocky moons. By itself, a lower internal temperature

would reduce geological activity. It is offset, however, by the much lower melting points of ices as compared to rocks. The heat of accretion alone was probably enough to melt the interiors and cause differentiation [Section 10.2] within most medium and large icy moons. As a result, many of these moons have internal structures similar to those of the terrestrial planets (see Figure 10.2), but with an additional thick layer of ice on the outside. The highly reflective ice may be mixed with varying amounts of less reflective "dirt," so not all the moons are bright in appearance.

All in all, the "ice geology" of the jovian moons bears many similarities to the rock geology of the terrestrial planets [Section 10.3]. *Impact cratering* during the heavy bombardment left all the moons with battered surfaces. *Volcanism* is certainly present on some jovian moons, at least when the definition is broadened to include the eruption of "molten ices" onto the surface, and *tectonics* seems likely as well. Both of these processes are driven by internal heat and can occur at much lower temperatures on icy moons than on rocky worlds. We expect erosion to be virtually nonexistent, because the moons (with the notable exception of Titan) lack the atmosphere needed for wind or rain.

Tour of the Moons: The Galilean Moons of Jupiter

We are now ready to embark on a brief tour of the medium- and large-size moons of the jovian planets. Our first stop is Jupiter, where four large moons exhibit some of the most interesting geology in the solar system. These moons were discovered by Galileo Galilei [Section 3.4] and therefore are known as the *Galilean moons*. Proceeding outward from Jupiter, we know them individually as Io, Europa, Ganymede, and Callisto (Figure 12.18)—names taken from the four mythological lovers of the Roman god Jupiter. Ganymede is the largest moon in the solar system. All the Galilean moons are large enough that they would count as planets if they orbited the Sun.

The densities of the Galilean moons decrease with distance from Jupiter, just as the densities of the planets tend to decrease with distance from the Sun. This fact suggests that similar condensation and accretion patterns occurred within the "miniature solar nebula" around Jupiter (with higher temperatures near the forming planet) as in the solar nebula as a whole. Io, the innermost of the four moons, is mostly rock and metal with little or no ice. Eu-

Figure 12.18 This set of photos, taken by the *Galileo* spacecraft orbiting Jupiter, shows global views of the four Galilean moons as we know them today (left to right): Io, Europa, Ganymede, and Callisto. Sizes are shown to scale.

ropa is mostly rock, but it formed with enough ice to give it an icy outer shell. Ganymede and Callisto are made from a mix of ice and rock.

Io: The Most Volcanically Active World in the Solar System

For anyone who thinks of moons as barren, geologically dead places like our own Moon, Io shatters the stereotype. When the Voyager spacecraft first photographed Io up close about two decades ago, we discovered a world with a surface so young that not a single impact crater has survived from past impacts. Moreover, Voyager cameras recorded volcanic eruptions in progress as the spacecraft passed by. We now know that Io is by far the most volcanically active world in our solar system. Its entire surface is pockmarked by large volcanoes, and eruptions have buried virtually every impact crater (Figure 12.19).

Many of Io's frequent eruptions are surprisingly similar to those of basalt volcanoes on Earth. Some of the taller

Figure 12.19 Io is the most volcanically active body in the solar system. Most of the black, brown, and red spots visible on the surface are recently active volcanic features. The white and yellow areas are sulfur and sulfur dioxide deposits from volcanic gases. Colors in this *Galileo* image are slightly enhanced.

jets of SO₂ gas　　SO₂ frost　　basalt lava flow

Figure 12.20
Many volcanoes on Io
are similar to basalt
volcanoes on Earth.

10 km

a When basaltic lava flows over sulfur-dioxide frost, the explosive sublimation creates huge plumes. The plume above this lava flow rises 80 km high.

VIS

200 km

b This enhanced-color photo shows fallout (dark patch) from a volcanic plume on Io. The fallout region covers an area the size of Arizona. (The orange ring is the fallout from another volcano.)

volcanoes may be shield volcanoes built from flows of basalt lava. Tall plumes of material sometimes rise from Io's frost-covered surface. These plumes are produced when hot lava causes surface frost (mostly sulfur dioxide) to flash-vaporize into gas and jet away (Figure 12.20a). The process is similar to what occurs when lava flows into the ocean on Earth, but Io's low gravity and very thin atmosphere let the plumes rise to altitudes of hundreds of kilometers. Fallout from the volcanic plumes can completely blanket an area the size of Arizona in a matter of months (Figure 12.20b).

Close-up photos of eruptions in progress show intensely hot lava glowing orange or red. Photos taken on the night side of Io show not only the hot volcanic vents, but also the thin sulfur dioxide atmosphere produced by the constant outgassing (Figure 12.21). This outgassing is the source of the ionized gas in the Io torus and Jupiter's magnetosphere (see Figure 12.14). Io's unusual red and orange colors come primarily from sulfur, which condenses on the surface after being outgassed by the volcanoes.

Tectonic processes are probably also active on Io, because tectonics and volcanism generally go hand in hand. However, volcanic eruptions are so frequent and cover the surface so thoroughly that any clear evidence of tectonic activity is likely to be buried.

Tidal Heating: A Different Kind of Internal Heating

Why is Io so geologically active? Its active volcanoes mean it must be quite hot inside, even though it is only about the size of our geologically dead Moon. Clearly, Io must have an additional internal heat source besides the combination of accretion, differentiation, and radioactive decay that heats the terrestrial interiors [Section 10.2]. This additional heat source is **tidal heating**. As its name implies, it arises from effects of tidal forces [Section 5.4].

Jupiter exerts a tidal force that keeps Io in synchronous rotation so that it keeps the same face toward Jupiter as it orbits. However, Io's orbit is slightly elliptical, so its orbital speed and distance from Jupiter vary. This variation means that the strength and direction of the tidal force change slightly as Io moves through each orbit, which in turn changes the size and orientation of Io's tidal bulges (Figure 12.22a). The result is that Io is continuously being flexed in different directions, which generates friction inside it. The flexing heats the interior in the same way that flexing warms Silly Putty. Tidal heating generates tremendous heat on Io—more than 200 times as much heat (per gram of mass) as the radioactive heat driving much of Earth's geology. That explains why Io is so volcanically active. (The energy for tidal heating ultimately comes from its orbital motion. Io is actually inching closer to Jupiter, but the effect is too small to be observed.)

But we are still left with a deeper question: Why is Io's orbit slightly elliptical, when almost all other large satellites' orbits are virtually circular? The answer lies in an inter-

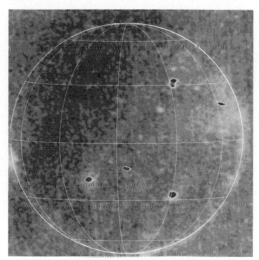

Figure 12.21 This false-color photo shows the glow of Io's volcanic vents (red) and atmosphere (green) when Io is in the darkness of Jupiter's shadow. The grid shows Io's outline, since the surface is not visible when shadowed by Jupiter.

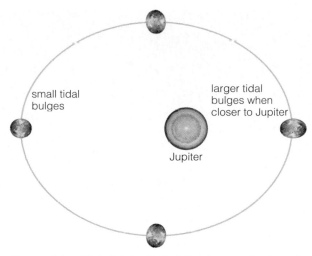

a Because Io's orbit is slightly elliptical, the size and direction of Io's tidal bulges change. These changes generate internal friction, which is the source of tidal heating. The bulges and orbital eccentricity are exaggerated in this diagram.

esting dance executed by Io and its neighboring moons. During the time Ganymede completes one orbit of Jupiter, Europa completes exactly two orbits and Io completes exactly four orbits (Figure 12.22b). The three moons therefore line up periodically, and the gravitational tugs they exert on one another add up over time.* Because the tugs are always in the same direction with each alignment, they tend to stretch out the orbits, making them slightly elliptical. The effect is much like that of pushing a child on a swing. If timed properly, a series of small pushes can add up to a *resonance* that causes the child to swing quite high. For the three moons, the resonance comes from the small gravitational tugs that repeat at each alignment.

The phenomenon of orbital periods falling into simple mathematical relationships is called an **orbital resonance**. As we will discuss later, resonances are quite common in the solar system. For example, they affect not only the Galilean moons but also planetary rings, the asteroid belt, and the Kuiper belt.

Because the orbital resonances make the orbits of all three moons (Io, Europa, and Ganymede) slightly elliptical, all three are subject to tidal heating. The heating is strongest on Io because it is closest to Jupiter and therefore feels the strongest tidal force from Jupiter, but tidal heating also occurs to a lesser extent on Europa and Ganymede.

Europa: The Water World? Europa's surface and crust are made almost entirely of water ice, and its bizarre, fractured appearance is proof enough that tidal heating has

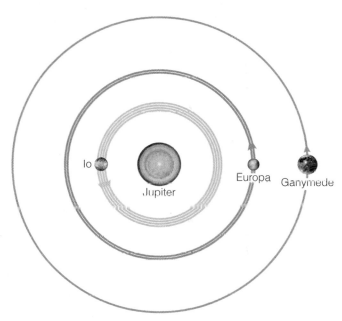

b But why is Io's orbit elliptical? About every seven Earth days (one Ganymede orbit, two Europa orbits, and four Io orbits), the three moons line up as shown. The small gravitational tugs repeat and make all three orbits slightly elliptical.

Figure 12.22 Tidal heating occurs when a moon's orbit is elliptical, so that its tidal bulges are constantly flexed in different directions. Tidal heating affects Io quite strongly and also affects Europa and Ganymede.

acted there (Figure 12.23). The icy surface is nearly devoid of impact craters and may be only a few million years old. Clearly, active geology covers up older craters on Europa. What could be doing the covering up?

Detailed measurements of the strength of gravity over different parts of Europa from the *Galileo* spacecraft, combined with theoretical modeling of its interior, suggest that Europa has a metallic core and a rocky mantle. Above the rocky mantle, Europa has enough water to make a layer of

---

*You may wonder why periodic alignments occur. Like synchronous rotation, they are not a coincidence but rather a consequence of feedback from the tides that the moons raise on Jupiter. The periodic tugs the moons exert on one another actually work to sustain the recurring alignments.

Figure 12.23 A global view of Europa's icy crust, as seen from the *Galileo* spacecraft. Colors are enhanced to bring out subtle details.

Figure 12.24 This diagram shows one model of Europa's interior structure. There is little doubt that the H₂O layer is real, but some question whether the material beneath the icy crust is actually liquid water or just relatively warm, convecting ice.

ice up to about 170 kilometers thick (Figure 12.24). Moreover, calculations suggest that tidal heating could melt much of the ice into liquid water.

If these ideas are all correct, then Europa might have a deep ocean of liquid water beneath a thin, icy shell. The liquid-water ocean would explain the lack of impact craters: Such an ocean might enhance tectonic activity that erases craters, or occasional upwelling of water might flood and cover up craters. If it really exists, the liquid ocean may be 100 kilometers or more deep, in which case it would con-tain more than twice as much liquid water as all of Earth's oceans combined.

Many scientists suspect that such an ocean exists, and numerous close-up photos of the surface of Europa support the idea. For example, some photos show what appear to be jumbled icebergs suspended in a region where liquid or slushy water has frozen (Figure 12.25a). Other evidence comes from double-ridged cracks on the surface (Figure 12.25b). These cracks may be created by tidal flexing that allows water to well up and build up ridges (Figure 12.25c). However, the photos alone do not allow us to rule out the possibility that the subsurface flows on Europa are of relatively warm, convecting ice rather than actual liquid water.

a Some regions show jumbled crust with icebergs, apparently frozen in slush.

b Close-up photos show that surface cracks have a double-ridged pattern.

Figure 12.25 Photographic evidence supporting the idea that Europa has a subsurface ocean.

Tidal flexing closes crack, grinds up ice.

Ridge builds up a little each time the crack opens and closes.

Tidal flexing opens crack. Debris in middle falls into crack.

c A possible mechanism for making the double-ridged surface cracks.

How can we distinguish between warm ice and liquid water beneath the crust? The case is by no means closed, but magnetic-field data from the *Galileo* spacecraft argue fairly strongly for liquid water. Moons seldom have magnetic fields, but Europa has one. Unlike Jupiter or Earth, which generate their own magnetic fields from within, Europa's magnetic field appears to be "induced" by Jupiter's strong magnetic field sweeping past as Jupiter rotates. The evidence for this idea is that Europa's magnetic field changes as Jupiter rotates—suggesting a direct response to Jupiter's magnetic field. This type of induced magnetic field requires a conducting liquid inside Europa. A salty ocean could be the needed electrical conductor, but ice could not. Most researchers consider this to be the best evidence yet that a liquid, salty ocean exists under Europa's icy surface. Moreover, the data require that the subsurface water be global in extent, not limited to just a few isolated liquid pockets. They also imply that this ocean is about as salty as Earth's

seas. Some of *Galileo*'s instruments found evidence for what appear to be salty compounds deposited on Europa's surface—possible seepage from a briny deep.

Taken together, the photographic and magnetic-field evidence makes a strong case for a deep ocean of liquid water on Europa. Perhaps as on Earth's seafloor, lava erupts from vents on Europa's seafloor, sometimes violently enough to jumble the icy crust above (Figure 12.26). Just possibly, tidal heating and the liquid ocean could make Europa a home to life [Section 24.2].

Ganymede: Largest Moon in the Solar System Photographs of Ganymede tell the story of a complex geological history (Figure 12.27). Like Europa, Ganymede has a surface of water ice. However, while Europa's surface appears relatively young everywhere, Ganymede has both very old and young regions. Dark regions of the surface are densely covered by impact craters, indicating that they are

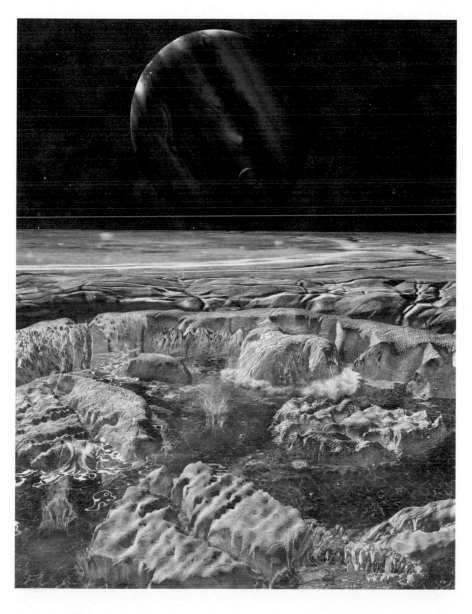

Figure 12.26 Tidal heating may give Europa a subsurface ocean beneath its icy crust. This artist's conception imagines a region where the crust has been disrupted by an undersea volcano.

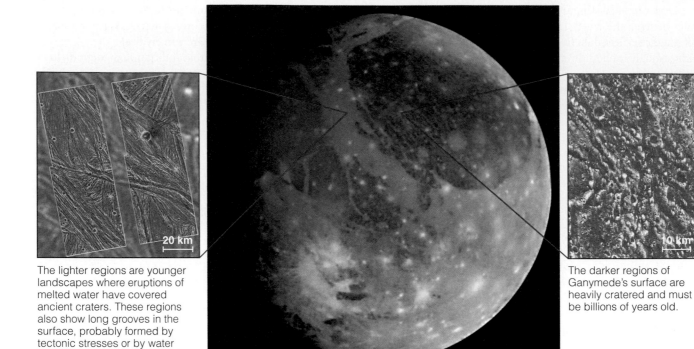

The lighter regions are younger landscapes where eruptions of melted water have covered ancient craters. These regions also show long grooves in the surface, probably formed by tectonic stresses or by water erupting along a surface crack and expanding as it refreezes.

The darker regions of Ganymede's surface are heavily cratered and must be billions of years old.

Figure 12.27 Ganymede, the largest moon in the solar system, has both old and young regions on its surface of water ice.

billions of years old. Light regions have long grooves and few craters and must be geologically young. In some cases, fairly sharp boundaries separate the two types of terrain (Figure 12.28).

The young terrain argues for occasional upwelling of liquid water or icy slush to the surface. This material would cover craters before refreezing, explaining why there are so few craters in this terrain. The long grooves may be made either by tectonic stresses or by the eruption of water or slush along a crack in the surface. As the water in the crack freezes, it expands and pushes outward, creating the groove.

If liquid water occasionally wells up to the surface, could it mean that Ganymede has a subsurface ocean like that thought to exist on Europa? The case for an ocean on Ganymede is less strong than the case for an ocean on Europa but is also bolstered by magnetic-field measurements. Like Europa, Ganymede has a magnetic field that varies with Jupiter's rotation, suggesting the presence of a salty ocean beneath the surface.

One difficulty with the idea of an ocean on Ganymede is figuring out where the heat to melt subsurface ice might come from. Tidal heating is much weaker on Ganymede than on Europa or Io and could not by itself supply enough heat. However, Ganymede's larger size means it should retain more heat from radioactive decay. Perhaps tidal heating and radioactive decay together provide enough heat

Figure 12.28 This close-up of Ganymede's surface shows a sharp boundary between heavily cratered ancient terrain on the left and younger "grooved terrain" on the right.

to make a liquid layer beneath the icy surface. If so, the lesser heating on Ganymede means its ocean probably lies at least 150 kilometers beneath the surface—much deeper than Europa's potential ocean.

Callisto: Last of the Galilean Moons

The outermost Galilean moon, Callisto, looks most like what scientists first expected for an outer solar system satellite: a heavily cratered iceball (Figure 12.29a). The bright patches on its surface are the result of impact cratering. Large impacts blasted out cleaner ice from deep within, spreading it out near the impact site. This cleaner ice reflects more light than the dirtier ice around it. The surface also holds some surprises. Close-up images show the surface to be covered by a dark, powdery substance concentrated in low-lying areas, leaving ridges and crests bright white (Figure 12.29b). The nature of this material and how it got there are unknown.

Despite its relatively large size (the third-largest moon in the solar system), Callisto lacks volcanic and tectonic features. Internal heating must be very weak. Callisto has no tidal heating because it shares no orbital resonances with other Galilean moons. In fact, gravity measurements by the *Galileo* spacecraft showed that Callisto's interior is so cool that it never underwent differentiation. Dense rock and lighter ice are thoroughly mixed throughout most of its interior.

The cool interior would seem to argue against any subsurface ocean on Callisto, but magnetic-field measurements argue to the contrary. Like Europa and Ganymede, Callisto has a magnetic field that suggests the presence of a salty ocean beneath a thick layer of ice. Perhaps Callisto's surface provides just enough insulation for heat from radioactive decay to melt water beneath the thick, icy crust.

Tour of the Moons: Titan and the Medium-Size Moons of Saturn

Leaving Jupiter behind, our moon tour takes us next to Saturn. In addition to its rings and numerous small moons, Saturn has six medium-size moons and one large moon—Titan, the second-largest moon in the solar system and the only moon with a thick atmosphere.

Titan: What Lies Beneath the Smog?

Titan's thick atmosphere hides its surface from view (Figure 12.30). Its reddish color comes from chemicals in its atmosphere much like those that make smog over cities on Earth. The atmosphere is about 90% nitrogen, not that different from the 77% nitrogen content of Earth's atmosphere. However, on Earth the rest of the atmosphere is mostly oxygen, while the rest of Titan's atmosphere consists of argon, methane (CH_4), ethane (C_2H_6), and other hydrogen compounds.

Methane and ethane are both greenhouse gases. They give Titan an appreciable greenhouse effect that makes it warmer than it would be otherwise. Still, because of its great distance from the Sun, its surface temperature is a frigid

a Callisto is heavily cratered, indicating an old surface that nonetheless may hide a deeply buried ocean. Colors are enhanced.

Figure 12.29 Callisto, the outermost of the four Galilean moons, shows no sign of volcanic or tectonic activity.

1 km

b Close-up photos show a dark powder overlying the low areas of the surface.

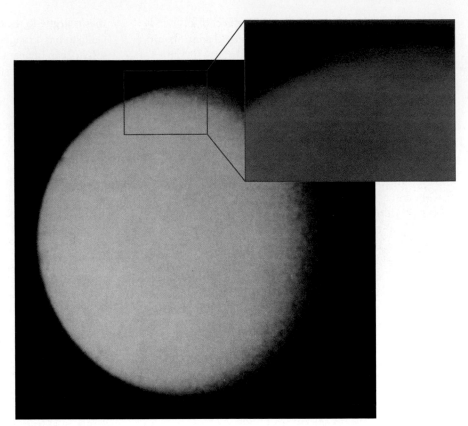

Figure 12.30 Titan, as photographed by *Voyager 2*, is enshrouded by a reddish smog. The inset shows just a portion of the moon. We can see the atmosphere clearly along its edge.

93 K ($-180°C$). The surface pressure on Titan is somewhat higher than on Earth—about 1.5 bars—which would be fairly comfortable if not for the lack of oxygen and the cold temperatures.

How did Titan end up with such an unusual atmosphere? At Saturn's distance from the Sun, temperatures in the solar nebula were cold enough for methane and ammonia to condense along with water into ice. Titan therefore is composed mostly of ices, including methane and ammonia ice. Some of this ice sublimed long ago to form an atmosphere on Titan. Over billions of years, solar ultraviolet light broke apart the ammonia molecules (NH_3) into nitrogen and hydrogen. The lighter hydrogen escaped from the atmosphere (through thermal escape [Section 11.5]), but the heavier nitrogen molecules remained and accumulated to give Titan its nitrogen atmosphere.

We encounter a particularly intriguing idea when we examine what happened to the methane gas that sublimed into Titan's atmosphere long ago. Methane is no longer present in large quantities, so some process must have removed it from the atmosphere. The most likely process is chemical reactions, triggered by solar ultraviolet light, that transform methane into ethane. If this idea is correct, atmospheric ethane may form clouds and rain on Titan, perhaps creating oceans of liquid ethane on the surface. Some methane may also be mixed in. Recent pictures of Titan taken with infrared light, which can penetrate through the clouds, show variations in surface brightness that lend at

least some support to the idea of ethane/methane oceans (Figure 12.31).

We should learn whether such frigid oceans really exist in late 2004, when the *Cassini* spacecraft will drop the *Huygens* probe toward Titan's surface (see Figure 8.19). The probe will collect data and take photographs on the way down and has enough battery power to survive for perhaps half an hour after it reaches the surface. Just in case, the probe is designed to float in liquid ethane. The probe should also help us determine whether volcanism or tectonics has played an important role in Titan's history.

THINK ABOUT IT

Would you expect Titan's surface to have many impact craters? Explain.

Saturn's Medium-Size Moons Saturn has six medium-size moons. In order of distance from the planet, they are: Mimas, Enceladus, Tethys, Dione, Rhea, and Iapetus (Figure 12.32). All are heavily cratered, indicating generally ancient surfaces, but each shows evidence of past geological activity.

All but Mimas have features suggesting past volcanism and/or tectonics. Light regions on the moons look like places where icy lava once flowed. Enceladus even has grooved terrain similar to that of Ganymede. Iapetus is particularly bizarre, with some regions distinctly bright and others distinctly dark. The dark regions appear to be covered by a

a This infrared image from the Keck telescope in Hawaii shows brightness variations. The dark areas could be oceans of liquid ethane and methane.

b Artist's conception of the surface of Titan, showing a possible ethane/methane sea.

Figure 12.31 Surface liquids on Titan?

thin veneer of dark material, but we don't know what it is or how it got there.

The geological activity may seem surprising when we consider the small size of these moons. All are considerably smaller than the Galilean moons. Enceladus, with its grooved terrain, is barely 500 kilometers across—small enough to fit inside the borders of Colorado. In fact, the geological activity is probably a demonstration of how effective low-temperature "ice geology" can be. As we discussed earlier, icy combinations of water, ammonia, and methane can melt, deform, and flow at remarkably low temperatures, producing a variety of volcanic and tectonic features.

Mimas, the smallest of the medium-size moons of Saturn, is essentially a heavily cratered iceball. One huge crater is sometimes called "Darth Crater," because of Mimas's resemblance to the Death Star in the *Star Wars* movies. (The crater's official name is Herschel.) The impact that created this crater probably came close to breaking Mimas apart.

Tour of the Moons:
The Medium-Size Moons of Uranus

Five medium-size moons orbit Uranus: Miranda, Ariel, Umbriel, Titania, and Oberon (see Figure 12.16), from innermost to outermost. Like other jovian moons, these moons are presumably made mostly of ice. Because of Uranus's great distance from the Sun, this ice includes a great deal of ammonia and methane as well as water.

These moons pose several puzzles. For example, Ariel and Umbriel are virtual twins in size, yet Ariel shows evi-

dence of volcanism and tectonics, while the heavily cratered surface of Umbriel suggests a lack of geological activity. Titania and Oberon also are twins in size, but Titania appears to have had much more geological activity than Oberon. No one knows why these two pairs of similar-size moons should vary so greatly in geological activity.

Miranda, the smallest of Uranus's medium-size moons, is the most surprising (Figure 12.33). We might expect Miranda to be a cratered iceball like Saturn's similar-size moon Mimas. Instead, Voyager images of Miranda show tremendous tectonic features and relatively few craters. Why should Miranda be so much more geologically active than Mimas? Our best guess is that Miranda had an episode of tidal heating billions of years ago, perhaps during a temporary orbital resonance with another moon of Uranus. Mimas apparently never had such an episode of tidal heating and thus lacks similar features.

Tour of the Moons:
Triton, the Backward Moon of Neptune

The last stop on our tour is Neptune, so distant that only two of its moons (Triton and Nereid) were known to exist before it was visited by the *Voyager 2* spacecraft. Nereid is Neptune's only medium-size moon, and Triton is its only large moon. Triton holds the distinction of being the coldest world in the solar system. It is even colder than Pluto because it reflects more of the weak sunlight that reaches the outskirts of the solar system.

Triton may appear to be a typical moon, but it is not. Its orbit is retrograde (it travels in a direction opposite that of Neptune's rotation) and highly inclined to Neptune's equator—telltale signs that Triton was captured from interplanetary space. However, rather than being small

Mimas

Enceladus

Tethys

Dione

Rhea

Iapetus

Figure 12.32 Saturn's six medium-size moons, which range in diameter from 390 km (Mimas) to 1,530 km (Rhea). All but Mimas show some evidence of volcanism and/or tectonics. (Photos not to scale.)

Figure 12.33 The surface of Miranda shows astonishing tectonic activity despite its small size. The cliff walls seen in the inset are higher than those of the Grand Canyon on Earth.

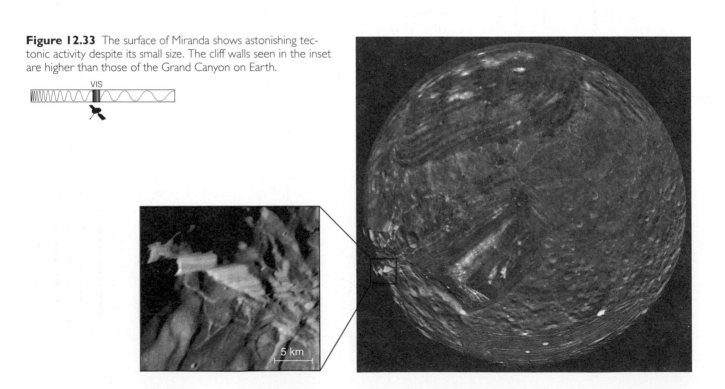

5 km

and potato-shaped like most captured asteroids, Triton is large, spherical, and icy. In fact, it is larger than the planet Pluto. As such, Triton presents many challenges to our understanding of satellite captures. Nonetheless, Triton almost certainly once orbited the Sun rather than Neptune.

Photos of Triton show a strange surface shaped by more than just impact cratering (Figure 12.34), although its many undisturbed craters suggest that major geological activity has subsided. Some regions show evidence of past icy volcanism, perhaps an icy equivalent to the volcanism that shaped the lunar maria. Other regions show wrinkly ridges (nicknamed "cantaloupe terrain") that appear tectonic in nature. Such geological activity may seem astonishing in light of Triton's extremely low (40 K) surface temperature, but again it is probably due to the "ice geology" of low-melting-point ices. The internal heating source for Triton's geological activity is not known, but it may have involved tidal heating. Triton probably had a very elliptical orbit and a more rapid rotation when it was first captured by Neptune. Tidal forces would have circularized its orbit, slowed it to synchronous rotation, and perhaps heated its interior enough to cause its past geological activity.

The sublimation of surface ices creates a thin atmosphere on Triton. Thin as it is, the atmosphere creates wind streaks on the surface, and unknown processes pump unusual plumes of gas and particles into the atmosphere. The combination of Triton's large orbital inclination and the substantial tilt of Neptune's rotation axis leads to extreme seasonal swings. The bright polar caps probably grow and shrink with seasons as gases sublimate and flow from pole to pole, much as occurs on Mars.

Assuming that Triton really is a captured moon, where did it come from? Could other, similar objects still be orbiting the Sun? Intriguingly, we know of at least one object that appears quite similar to Triton: the planet Pluto. But that's a story for the next chapter.

Satellite Summary: The Active Outer Solar System

Our brief tour took us to most of the medium and large moons in the solar system. The major lesson of the tour is that the geology of icy worlds is harder to predict than the geology of the terrestrial worlds. Size isn't everything when it comes to geological activity. Ices make "lava" at lower temperatures than do rocks, so an icy object can have more volcanic activity than a rocky object of the same size. Moons made of ices that include methane and ammonia may have even greater geological activity. In addition, tidal heating can supply added heat even when other internal heat sources are no longer important. This is certainly the case on Io and Europa and to lesser extents on Ganymede, Triton, and possibly Miranda. Figure 12.35 summarizes the differences between the geology of jovian moons and that of the terrestrial worlds.

b The flat regions in this close-up photo of Triton's icy surface look like the lava-filled impact basins of the lunar maria, though in Triton's case the "lava" would be molten ice or slush.

a Triton's southern hemisphere as seen by *Voyager 2*.

Figure 12.34 Neptune's moon Triton.

Terrestrial Planet Geology

- Radioactive decay heats the interior enough to cause volcanic and tectonic activity.
- Only large planets retain enough internal heat to stay geologically active today.
- Example: Mars (photo above) probably retains some internal heat. If it had been smaller, like Mercury, it would be geologically "dead" today. If it had been larger, like Earth, it would probably have much more active and ongoing tectonics and volcanism.

Jovian Moon Geology

- Tidal heating can cause tremendous geological activity on moons on elliptical orbits around massive planets.
- Even without tidal heating, icy materials can melt and deform at lower temperatures than rock, increasing the likelihood of geological activity.
- Together, these effects explain why icy moons are much more likely to have ongoing geological activity than rocky terrestrial worlds of the same size.
- Example: Ganymede (photo above) shows evidence of recent geological activity, even though it is similar in size to the geologically dead terrestrial planet Mercury.

Figure 12.35 Jovian planet moons can be much more geologically active than terrestrial worlds of similar size due to their icy compositions and a heating source (tidal heating) that is not important on the terrestrial worlds.

12.6 Jovian Planet Rings

We have completed our comparative study of the jovian planets and their major moons, but we have one more topic left to cover: their amazing rings. Saturn's rings have dazzled and puzzled astronomers since Galileo first saw them through his small telescope and suggested that they resembled "ears" on Saturn.

For a long time, Saturn's rings were thought to be unique in the solar system. We now know that all four jovian planets have rings. As we did for the planets themselves, we'll look at one ring system in detail and then examine the others for important similarities and differences. Saturn's rings are the clear choice as the standard for comparison.

Saturn's Rings

You can see Saturn's rings through a backyard telescope, but learning their nature requires higher resolution (Figure 12.36). Earth-based views make the rings appear to be continuous, concentric sheets of material separated by a large gap (called the *Cassini division*). Spacecraft images reveal these "sheets" to be made of many individual rings, each separated from the next by a narrow gap. Different regions of a ring can vary in brightness and transparency. But even these appearances are somewhat deceiving. If we could wander into the rings, we'd find that they are made

of countless individual particles ranging in size from large boulders to dust grains. All are far too small to be photographed even from spacecraft passing nearby.

Ring Particle Characteristics and Collisions Spectroscopy reveals that Saturn's ring particles are made of relatively reflective water ice. The rings look bright where they contain enough particles to intercept sunlight and scatter it back toward us. They appear more transparent where fewer particles are present.

Each individual ring particle orbits Saturn independently in accord with Kepler's laws. Thus, the rings are much like a myriad of tiny moons. However, the individual ring particles are so close together that they collide frequently. In the densest parts of the rings, each particle collides with another every few hours.

THINK ABOUT IT

Which ring particles travel faster: those closer to Saturn or those farther away? (*Hint:* Think about Kepler's third law.) Can you think of a way to confirm your answer with telescopic observations?

The frequent collisions explain why Saturn's rings are perhaps the thinnest known astronomical structure. They span over 270,000 kilometers in diameter but are only a few tens of *meters* thick. Indeed, the rings are so thin that

a Earth-based telescopic view of Saturn. The dark gap between the rings is called the *Cassini division*.

b Voyager image of Saturn's rings against the disk.

c Artist's conception of particles in a ring system. All the particles are moving slowly relative to one another and occasionally collide.

Figure 12.36 Zooming in on Saturn's rings.

they disappear from view when we see Saturn edge-on, as we do around the equinoxes of its 29.5-year orbit of the Sun.

To understand how collisions keep the rings thin, imagine what would happen to a ring particle on an orbit slightly inclined to the central ring plane. The particle would collide with other particles every time its orbit intersected the ring plane, and its orbital tilt would be reduced with

every collision. Before long, these collisions would force the particle to conform to the orbital pattern of the other particles. Similarly, a ring particle with an elliptical orbit would soon be forced into a circular orbit.

Rings and Gaps Close-up photographs show an astonishing number of rings, gaps, ripples, and other features—the total number of features may be as high as 100,000. Scientists are still struggling to explain all the features, but some general ideas are now clear.

Rings and gaps are caused by particles bunching up at some orbital distances and being forced out at others. This bunching happens when gravity nudges the orbits of ring particles in some particular way. One source of nudging is tiny moons—which we'll call *gap moons*—located within the rings themselves (Figure 12.37a). The gravity of a gap moon tugs gently on nearby particles, effectively nudging them farther away to other orbits.* This movement clears a gap in the rings near the moon's orbit. In some cases, two nearby gap moons can act like shepherds forcing particles between them into line and hence are referred to as *shepherd moons* (Figure 12.37b).

Ring particles also may be nudged by the gravity from larger, more distant moons. For example, a ring particle orbiting about 120,000 kilometers from Saturn's center will circle the planet in exactly half the time it takes the moon Mimas to orbit. Every time Mimas returns to a certain location, the ring particle will also be at its original location and therefore will experience the same gravitational nudge from Mimas. The nudges reinforce one another and eventually clear a gap in the rings. This Mimas 2:1 resonance is essentially the same type of orbital resonance that affects the orbits of Io, Europa, and Ganymede. It is responsible for the Cassini division—the large gap visible from Earth (see Figure 12.36a). Many similar resonances are responsible for other gaps. Resonances can even create vast numbers of beautiful ripples within the rings (Figure 12.38).

Although gap moons, shepherd moons, and orbital resonances are known to cause many of the features of Saturn's rings, we have not yet identified specific causes for the vast majority of ring features. Among the many remaining mysteries are dusty patches, called *spokes*, that can appear and undergo dramatic change in a matter of hours (Figure 12.39). They are probably particles of microscopic dust that have been levitated out of the ring plane by forces associated with Saturn's magnetic field.

Comparing Planetary Rings

The rings of Jupiter, Uranus, and Neptune are so much fainter than Saturn's that it took almost four centuries longer to discover them. Ring particles in these three systems are

---

* It may seem counterintuitive that a gap moon's gravitational attraction should nudge a ring particle *farther away,* but theoretical models show that this happens as a result of the combined gravitational interactions between the gap moon, the ring particles, and the planet.

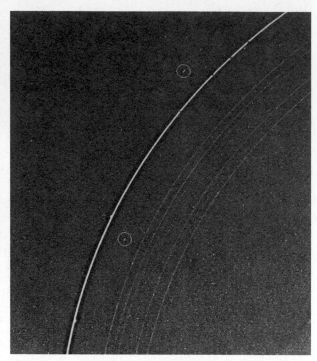

2000 km

b Voyager's best photograph of shepherd moons (circled) came during the *Voyager 2* Uranus flyby. Two satellites shepherd one of Uranus's rings. The satellite images are smeared out by their motion during the exposure. The inner moon is Cordelia, and the outer moon is Ophelia.

a Saturn's 20-km moon Pan (visible as little more than a dot in the inset) clears a gap in the rings.

Figure 12.37 The gravity of gap moons and shepherd moons affects ring structure.

20 km

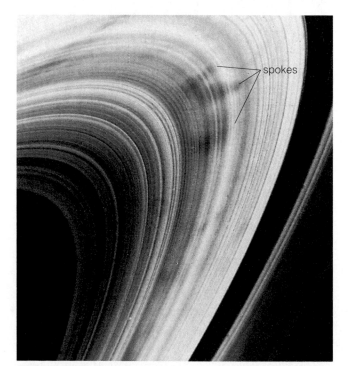

spokes

Figure 12.38 Orbital resonances between Saturn's larger moons and the ring particles create many features of the rings. The remarkable

ripples visible in this photograph are caused by a resonance with Saturn's moon Mimas. (The dark spots in the image are calibration marks for the camera.) An even stronger resonance with Mimas creates the large gap known as the Cassini division, shown in Figure 12.36a.

Figure 12.39 The dark patches in Saturn's rings are called spokes. They may be particles of dust lifted above the ring plane by forces from Saturn's magnetic field.

Jupiter — VIS

Saturn — VIS

Uranus — IR

Neptune — VIS

Figure 12.40 Four ring systems. The planets are not shown to scale. Uranus's rings were photographed by the Hubble Space Telescope, the others by Voyager. The Neptune frame is made of two images taken on either side of the bright planet.

far less numerous, generally smaller, and much darker. Despite these differences, a family portrait of the jovian ring systems shows many similarities (Figure 12.40). All rings lie in their planet's equatorial plane. Particle orbits are fairly circular, with small orbital tilts relative to the equator. Individual rings and gaps are probably shaped by gap moons and orbital resonances.

Uranus's rings were first discovered in 1977 during observations of a *stellar occultation*—a star passing behind Uranus as seen from Earth. During the occultation, the star "blinked" on and off nine times before it disappeared behind Uranus and nine more times as it emerged. Scientists concluded that these nine "blinks" were caused by nine thin rings encircling Uranus. Similar observations of stars passing behind Neptune yielded more confounding results: Rings appeared to be present at some times but not at others. Could Neptune's rings be incomplete or transient?

The Voyager spacecraft provided some answers. Voyager cameras first discovered thin rings around Jupiter in 1979. After next providing incredible images of Saturn's rings, *Voyager 2* photographed the rings of Uranus as it flew past in 1986. In 1989, *Voyager 2* passed by Neptune and found that it does, in fact, have partial rings—at least when seen from Earth. The space between the ring segments is filled with dust not detectable from Earth. In addition to the dust within the individual rings of Neptune, *Voyager 2* detected vast dust sheets between the widely separated rings of both Uranus and Neptune.

The differences among the ring systems present us with unsolved mysteries. The larger size, higher reflectivity, and much greater number of particles in Saturn's rings compel us to wonder whether different processes might be at work there. The reasons for the slight tilt and eccentricity of Uranus's thin rings is not yet understood, nor do we know why Neptune's rings contain dusty regions that make them appear as partial rings when viewed from Earth.

Origin of the Rings

Where do the rings come from? Did they form together with the planets out of the solar nebula, or are they a more recent phenomenon? An important clue is that the rings lie close to their planet in a region where tidal forces are very strong.

Within two to three planetary radii (of any planet), the tidal forces tugging apart an object become comparable to the gravitational forces holding it together. (This region is often called the *Roche zone*.) Only relatively small objects held together by nongravitational forces—such as the electromagnetic forces that hold solid rock, spacecraft, and human beings together—can avoid being ripped apart by the strong tidal forces in this region.

Because a large moon would be ripped apart by tidal forces in the region of the rings, scientists once suspected that Saturn's rings formed after a large moon came too close to the planet. However, moons don't simply "wander" away from their orbits, and a large moon could have been deflected toward Saturn only by a close gravitational encounter with an unusually large asteroid or comet that happened to pass nearby. Such encounters should be extremely rare. Back in the days when Saturn was the only planet known to have rings, it might have seemed possible that such a rare encounter was responsible for the rings. The discovery of rings around the other jovian planets all but ruled out this possibility, because it seems inconceivable that such an unlikely event could have occurred in all four cases.

If the rings did not originate with the destruction of a large moon, where did all the ring particles come from? Another idea that once seemed reasonable was that the ring particles are leftovers from accretion. In this scenario, the rings consist of particles that condensed in the miniature solar nebulae that surrounded each young jovian planet, but which were never able to accrete into a large moon because of the strong tidal forces close to the planet. However, scientists found that this idea is also fatally flawed. The same kinds of particles that produce meteors in Earth's atmosphere and micrometeorite impacts on the Moon [Section 10.4] must also bombard other objects throughout the solar system. Calculations show that such micrometeorite impacts will grind basketball-size ring particles into dust in a few million years. The dust particles are then slightly slowed in their orbits by the pressure of sunlight, causing them to spiral gradually into their planet. Ring particles may also be lost in other ways. For example, the thermosphere of Uranus extends into the ring region, exerting atmospheric drag that slowly causes ring particles to spiral into the planet. Between the effects of micrometeorite impacts and other processes, none of the abundant small particles that now occupy the jovian ring systems can have been there since the solar system was born more than 4 billion years ago.

We are therefore left with only one reasonable possibility for the origin of the rings: New particles are continually supplied to the rings, taking the place of older particles that have been ground to dust or have fallen into the planet. Scientists now believe that the source of these new particles lies with numerous small moons (such as the gap moons and shepherd moons) found within the ring systems. These moons are too small to be torn apart by the tidal forces in the rings but too large to have been ground to dust by micrometeorite impacts in 4.6 billion years. Moreover, we expect that many small moons should have accreted in the miniature solar nebulae that surrounded the jovian planets. Farther out from the planets, these small moons tended to continue their accretion, ultimately forming a few much larger moons. But within the ring regions, tidal forces prevented the small moons from growing into a large moon, explaining why so many small moons remain within the rings today.

Small moons provide an ongoing source of ring particles because tiny particles are continually released as these moons are pelted by micrometeorite impacts. In addition, rarer but larger impacts can shatter small moons completely, creating a supply of boulder-size ring particles. These boulders are then slowly ground into smaller ring particles by micrometeorites. Thus, the dust-to boulder-size particles in rings all ultimately come from the gradual dismantling of small moons that formed during the birth of the solar system.

The collisions that shatter small moons and generate a major source of ring particles must occur only occasionally and at essentially random times, which means that the number of particles in any particular ring system must change dramatically over millions and billions of years. Rings may be broad and bright when they are full of particles and almost invisible when particles are few. Thus, the brilliant spectacle of Saturn's rings may be a special treat of our epoch, one that could not have been seen a billion years ago and that may not last long on the time scale of our solar system.

THE BIG PICTURE

Putting Chapter 12 into Context

In this chapter, we saw that the jovian planets really are a "different kind of planet" and, indeed, a different kind of planetary system. The jovian planets dwarf the terrestrial planets. Even some of their moons are as large as terrestrial worlds. As you continue your study of the solar system, keep in mind the following "big picture" ideas:

● The jovian planets may lack solid surfaces on which geology can occur, but they are interesting and dynamic worlds with rapid winds, huge storms, strong magnetic fields, and interiors in which common materials behave in unfamiliar ways.

● Despite their relatively small sizes and frigid temperatures, many jovian moons are geologically active by virtue of their icy compositions. Ironically, it was the cold temperatures in the solar nebula that led to their icy compositions and hence their geological activity.

● Ring systems owe their existence to small moons formed from the "miniature solar nebulae" that produced the jovian planets billions of years ago. The rings we see today are composed of particles liberated from those moons surprisingly recently.

● Understanding the jovian planet systems forced us to modify many of our earlier ideas about the solar system, in particular by adding the concepts of ice geology, tidal heating, and orbital resonances. Each new set of circumstances we discover offers further opportunities to learn how our universe works.

12.1 The Jovian Worlds: A Different Kind of Planet

- *What are the major features of the jovian planets?* The jovian planets are largely composed of hydrogen, helium, and hydrogen compounds. They have no solid surfaces, a fast rotation, and slightly "squashed" shapes. They have many moons and ring systems.

- *Why are jovian planets so different from terrestrial planets?* The differences stem from differing formation processes in the solar nebula. The jovian planets formed in the cold, outer region of the solar system where ice could condense and planetesimals could grow much larger. These large, icy planetesimals drew in hydrogen and helium gas, forming "miniature solar nebulae." The large planets formed in the centers of these nebulae, and moons formed in the swirling disks.

12.2 Jovian Planet Interiors

- *What is Jupiter's interior structure?* Jupiter has a central core of hydrogen compounds, rocks, and metals. The next layer out contains metallic hydrogen, followed by a layer of liquid hydrogen, followed by a gaseous atmosphere. Pressure, density, and temperature are greatest in the center and decrease going outward.

- *Why is Saturn almost as big as Jupiter in radius?* Adding mass to a jovian planet does not necessarily increase its size, because the stronger gravity compresses the matter to greater density. Jupiter is near the maximum possible size for a jovian planet.

- *How do the jovian planet interiors differ, and why?* All have cores of about the same mass but have different amounts of surrounding hydrogen and helium. Accretion took longer in the more spread-out regions of the outer solar system, so the more distant planets captured less gas from the solar nebula before it was blown away by the solar wind.

12.3 Jovian Planet Atmospheres

- *How is Jupiter's atmospheric structure similar to Earth's?* It consists of a troposphere, stratosphere, and thermosphere created by similar interactions of gas and sunlight.

- *Why does Jupiter have three distinct cloud layers?* Different gases condense at different temperatures and therefore at different altitudes in Jupiter's atmosphere. Jupiter has three cloud layers, each at the altitude where a particular gas can condense.

- *What is the Great Red Spot?* The Great Red Spot is a giant, long-lived, high-pressure storm.

- *How do other jovian atmospheres compare to Jupiter's?* The atmospheric structures are all similar, but each is progressively colder with distance from the Sun. Saturn's atmosphere is the most similar to Jupiter's. Uranus and Neptune are cold enough to have a methane cloud layer, which leads to their blue colors.

12.4 Jovian Planet Magnetospheres

- *How does Jupiter's magnetosphere compare with Earth's?* Both have similar general properties, but Jupiter's magnetosphere is far larger, with a magnetic field 20,000 times stronger.

- *How does Jupiter's magnetosphere interact with its moon Io?* The magnetosphere contains charged particles coming from Io, which create the Io torus. The particles in turn bombard the surface of Io, leading to a very thin atmosphere.

- *How do other jovian magnetospheres compare to Jupiter's?* They are smaller, with weaker magnetic fields.

12.5 A Wealth of Worlds: Satellites of Ice and Rock

- *Why does active geology occur more readily on small icy worlds than on small rocky worlds?* Ices soften and melt at much lower temperatures than rock, allowing icy volcanism and tectonics at surprisingly low temperatures.

- *What makes Io so volcanically active?* Io is subjected to tidal heating caused by the tidal force of Jupiter as Io moves through its elliptical orbit. Io's orbit is elliptical because of orbital resonances with Europa and Ganymede.

- *Why do we suspect a subsurface ocean on Europa?* Photos show evidence of water flows on the surface. Magnetic-field measurements support the presence of a salty ocean, and enough tidal heating is present to melt a thick layer of ice beneath the surface.

- *What are some key features of the large moons Ganymede, Callisto, Titan, and Triton?* Ganymede is the largest moon in the solar system and might have a subsurface ocean. Callisto shows an ancient cratered surface but still could have a subsurface ocean. Titan is the only moon with a thick atmosphere. Triton was probably captured by Neptune.

continued ▶

12.6 Jovian Planet Rings

- *What do Saturn's rings look like?* From a distance the rings look fairly solid, but up close we see thousands of individual rings and gaps. Within them are countless individual particles.

- *How do other ring systems compare to Saturn's?* The other jovian rings contain fewer particles, are smaller in extent, and are darker in color.

- *What is the origin of planetary rings?* Ring particles are constantly destroyed through impacts or other processes. Thus, the rings must be replenished with new particles over time or disappear. Ring particles probably come from the dismantling of many small moons formed in the "miniature solar nebulae" that produced the jovian planets billions of years ago.

❓ Surprising Discoveries?

Suppose someone claimed to make the discoveries described below. (These are *not* real discoveries.) Decide whether each discovery should be considered reasonable or surprising. More than one right answer may be possible, so explain your answer clearly.

1. Saturn's core is pockmarked with impact craters and dotted with volcanoes erupting basaltic lava.

2. Neptune's deep blue color is not due to methane, as previously thought, but instead is due to its surface being covered with an ocean of liquid water.

3. A jovian planet in another star system has a moon as big as Mars.

4. An extrasolar planet is discovered that is made primarily of hydrogen and helium. It has approximately the same mass as Jupiter but is the same size as Neptune.

5. A new small moon is discovered to be orbiting Jupiter. It is smaller than any other of Jupiter's moons but has several large, active volcanoes.

6. A new moon is discovered to be orbiting Neptune. The moon orbits in Neptune's equatorial plane and in the same direction that Neptune rotates, but it is made almost entirely of metals such as iron and nickel.

7. An icy, medium-size moon is discovered to be orbiting a jovian planet in a star system that is only a few hundred million years old. The moon shows evidence of active tectonics.

8. A jovian planet is discovered in a star system that is much older than our solar system. The planet has no moons at all, but it has a system of rings as spectacular as the rings of Saturn.

Problems

9. *Jovian Planet Interiors.* Briefly summarize the techniques we use to study the interiors of the jovian planets. Briefly describe the internal heat sources of each of the four jovian planets.

10. *Comparing Jovian Planet Atmospheres.* Briefly describe Jupiter's global circulation and weather, and contrast them with those on the other jovian planets. Describe the possible origins of Jupiter's vibrant colors, and explain how and why the other jovian planets look different.

11. *Jovian Planet Moons.* Briefly describe how we categorize jovian moons by size. What is the origin of most of the medium and large moons? What is the origin of many of the small moons?

12. *Unusual Moons.* Describe the atmosphere of Titan. What evidence suggests that Titan may have oceans of liquid ethane? Describe Triton and explain why we think it is a captured moon.

13. *Sculpting Rings.* Briefly describe the effects of gap moons and orbital resonances on ring systems.

14. *The Importance of Rotation.* Suppose the material that formed Jupiter came together without any rotation so that no "jovian nebula" formed and the planet today wasn't spinning. How else would the jovian system be different? Think of as many effects as you can, and explain each in a sentence.

15. *The Great Red Spot.* Based on the infrared and visible images in Figure 12.8, is Jupiter's Great Red Spot warmer or cooler than nearby clouds? Does this mean it is higher or lower in altitude than the nearby clouds? Explain.

16. *Comparing Jovian Planets.* You can do comparative planetology armed only with telescopes and an understanding of gravity.

    a. The small moon Amalthea orbits Jupiter at about the same distance in kilometers at which Mimas orbits Saturn, yet Mimas takes almost twice as long to orbit. From this observation, what can you conclude about how Jupiter and Saturn differ? Explain.

    b. Jupiter and Saturn are not very different in radius. When you combine this information with your answer to part (a), what can you conclude? Explain.

17. *Minor Ingredients Matter.* Suppose the jovian planet atmospheres were composed 100% of hydrogen and helium rather than 98% of hydrogen and helium. How would the atmospheres be different in terms of color and weather? Explain.

18. *Observing Project: Jupiter's Moons.* Using binoculars or a small telescope, view the moons of Jupiter. Make a sketch of what you see, or take a photograph. Repeat your observations several times (nightly, if possible) over a period of a couple of weeks. Can you determine which moon is which? Can you measure the moons' orbital periods? Can you determine their approximate distances from Jupiter? Explain.

19. *Observing Project: Saturn's Rings.* Using binoculars or a small telescope, view the rings of Saturn. Make a sketch of what you see, or take a photograph. What season is it in Saturn's northern hemisphere? How far do the rings extend above Saturn's atmosphere? Can you identify any gaps in the rings? Describe any other features you notice.

*20. *Disappearing Moon.* Io loses about a ton (1,000 kg) of sulfur dioxide per second to Jupiter's magnetosphere.

   a. At this rate, what fraction of its mass would Io lose in 4.5 billion years?

   b. Suppose sulfur dioxide currently makes up 1% of Io's mass. When will Io run out of this gas at the current loss rate?

*21. *Ring Particle Collisions.* Each ring particle in the densest part of Saturn's rings collides with another about every 5 hours. If a ring particle survived for the age of the solar system, how many collisions would it undergo?

Discussion Questions

22. *A Miniature Solar System?* In what ways is the Jupiter system similar to a miniature solar system? In what ways is it different?

23. *Jovian Planet Mission.* We can study terrestrial planets up close by landing on them, but jovian planets have no surfaces to land on. Suppose that you are in charge of planning a long-term mission to "float" in the atmosphere of a jovian planet. Describe the technology you would use and how you would ensure survival for any people assigned to this mission.

24. *Pick a Moon.* Suppose you could choose any one moon to visit in the solar system. Which one would you pick, and why? What dangers would you face in your visit to this moon? What kinds of scientific instruments would you want to bring along for studies?

25. *Hot Jupiters.* Many of the newly discovered planets orbiting other stars are more massive than Jupiter but orbit much closer to their stars. Assuming that they would be Jupiter-like if they orbited at a greater distance from their star, how would you expect these new planets to differ from the jovian planets of our solar system? How would you expect their moons to differ? Explain.

MEDIA EXPLORATIONS

For a complete list of media resources available, go to www.astronomyplace.com and choose Chapter 12 from the pull-down menu.

 Astronomy Place Web Tutorials

Tutorial Review of Key Concepts

Use the following interactive **Tutorial** at www.astronomyplace.com to review key concepts from this chapter.

Formation of the Solar System Tutorial

Lesson 1 Comparative Planetology

Supplementary Tutorial Exercises

Use the interactive **Tutorial Lessons** to explore the following questions.

Formation of the Solar System Tutorial, Lesson 1

1. Contrast the general features of the jovian planets with those of the terrestrial planets.

2. Compare the properties of each of the four jovian planets. What properties do they all share? In what ways are they different?

 Exploring the Sky and Solar System

Of the many activities available on the *Voyager: SkyGazer* **CD-ROM** accompanying your book, use the following files to observe key phenomena covered in this chapter.

Go to the **File: Basics** folder for the following demonstrations.

1. Tracking Jupiter and Io

2. Saturn

Go to the **File: Demo** folder for the following demonstrations.

1. Backside of Jupiter

2. Locked on Dione

3. Three Moons on Jupiter

Web Projects

Take advantage of the useful Web links on www.astronomyplace.com to assist you with the following projects.

1. *The Galileo Mission to Jupiter.* Learn more about the Galileo mission and how it ended, and about ongoing scientific study of the Galileo data. Write a short summary of your findings.

2. *News from Cassini.* Find the latest news about the Cassini mission to Saturn. What is the current mission status? Write a short report about the mission's status and recent findings.

3. *Oceans of Europa.* The possibility of subsurface oceans on Europa holds great scientific interest and may even mean that life could exist there. Investigate plans for further study of Europa, either from Earth or with future spacecraft. Write a short summary of the plans and how they might help us learn whether an ocean really exists and, if so, what it might contain.

13 Remnants of Rock and Ice

Asteroids, Comets, and Pluto

13.1 Remnants from Birth
- What are the three major groups of small bodies found orbiting in the solar system?

13.2 Asteroids
- What are the general characteristics of asteroid sizes, shapes, and orbits?
- Why didn't a planet form in the region of the asteroid belt?
- How do we measure asteroid properties?

13.3 Meteorites
- How is a meteor different from a meteorite?
- How are meteorites categorized?
- Why do meteorites differ from one another?

13.4 Comets
- What are comets made of?
- What happens to a comet as it approaches the Sun?

- How do we know that vast numbers of comets reside in the Oort cloud and Kuiper belt?

13.5 Pluto: Lone Dog, or Part of a Pack?
- Why don't Pluto and Neptune collide?
- What is surprising about Pluto's density, and how might Pluto have come to be?
- Why do we think Pluto is a Kuiper belt comet?

13.6 Cosmic Collisions: Small Bodies Versus the Planets
- What happened to Jupiter in 1994?
- How often do small particles impact Earth?
- Why do we think the dinosaurs were driven extinct by an impact?
- Do future impacts pose a real threat to our civilization?
- What role do the jovian planets play in the impact threat?

As we look out into the Universe and identify the many accidents of physics and astronomy that have worked to our benefit, it almost seems as if the Universe must in some sense have known that we were coming.

Freeman Dyson

Asteroids and comets might at first seem insignificant in comparison to the planets and moons we've discussed so far, but there is strength in numbers. The trillions of small bodies orbiting our Sun are far more important than their small size might suggest.

The appearance of comets has more than once altered the course of human history when our ancestors acted upon superstitions related to comet sightings. More profoundly, asteroids or comets falling to Earth have scarred our planet with impact craters and have altered the course of biological evolution.

Asteroids and comets are also important because they are remnants from the birth of our solar system. As such, they provide an important testing ground for our theories of how our solar system came to be. In this chapter, we will explore asteroids, comets, and the rocks that fall to Earth as meteorites. We will also examine the smallest planet, Pluto, which may be a misfit among planets but is right at home among the smaller objects of our solar system. Finally, we will explore the dramatic effects of the occasional collisions between small bodies and large planets.

13.1 Remnants from Birth

The objects we've studied so far—terrestrial planets, jovian planets, and large moons—have changed dramatically since their formation. Most planets have been transformed in almost every way: Their interiors have differentiated, their surfaces have been reshaped by geological processes, and their atmospheres have changed as they have gained and lost gas. In contrast, many small bodies remain virtually unchanged since their formation some 4.5 billion years ago. Comets, asteroids, and meteorites carry the history of our solar system encoded in their compositions, locations, and numbers.

Using these small bodies to understand planetary formation is a bit like picking through the trash in a carpenter's shop to see how furniture is made. By studying the scraps, sawdust, paint chips, and glue, we can develop some understanding of how the carpenter builds furniture. Similarly, the "scraps" left over from the formation of our solar system give us an understanding of how the planets and larger moons came to exist. Indeed, much of our modern theory of solar system formation was developed from the study of asteroids, comets, and meteorites.

General Orbits of Small Bodies

We generally think of asteroids and comets together as "small bodies" of the solar system, because they are so much smaller than the terrestrial and jovian planets. As we discussed in Chapter 8, these small bodies fall into three fairly distinct groups (see Figure 8.5):

- **Asteroids** are rocky or metallic in composition. Most orbit the Sun between the orbits of Mars and Jupiter.

- **Kuiper belt comets** are made mostly of ice. They orbit the Sun in the same direction and nearly the same plane as the planets, but at distances ranging from about that of Neptune to about twice the distance of Pluto.

- **Oort cloud comets** are also made mostly of ice but orbit much farther from the Sun than Kuiper belt comets—in some cases, perhaps halfway to the nearest stars. In addition, their orbits are randomly inclined to the ecliptic plane.

The general difference between asteroids and comets is their composition, which reflects where they formed. Asteroids are made of rock and metal because they formed inside the frost line of the solar nebula [Section 9.3], while comets are icy because they formed beyond it.

A Note on Names

The terminology of small bodies has a long history that can sometimes be confusing. The word *comet* comes from the Greek word for "hair," and comets get their name from the long, hairlike tails they display on the rare occasions when they come close enough to the Sun to be visible in our sky. *Asteroid* means "starlike," but asteroids are starlike only in their appearance through a telescope, not in any more fundamental properties. The term *meteor*, literally "a thing in the air," refers to the flash of light emitted as a bit of interplanetary dust burns up in our atmosphere. Meteors are also sometimes called *shooting stars* or *falling stars* because some people once believed they really were stars falling from the sky. Larger chunks of rock that survive the plunge through the atmosphere and hit the ground are called *meteorites*, which means "associated with meteors." To keep potential confusion to a minimum, we will use only the following terms and definitions in this book:

- **Asteroid**: a rocky leftover planetesimal orbiting the Sun.

- **Comet**: an icy leftover planetesimal orbiting the Sun—regardless of its size, whether it has a tail, or where it resides or comes from (Kuiper belt or Oort cloud).

- **Meteor**: a flash of light in the sky caused by a particle entering the atmosphere, whether the particle comes from an asteroid or a comet.

- **Meteorite**: any piece of rock that has fallen to the ground from space, whether from an asteroid, a comet, or even another planet.

THINK ABOUT IT

Many everyday words have the same roots as the words used for the small bodies in the solar system. For example, what is the literal meaning of the word *disaster*, and how do you think this term arose? Why is the symbol * called an *asterisk*? What do we mean when we say a movie star has had a *meteoric career*? Why is a weather reporter called a *meteorologist*?

13.2 Asteroids

Asteroids are virtually undetectable by the naked eye and went unnoticed for almost two centuries after the invention of the telescope. The first asteroids were discovered about 200 years ago, at a time when astronomers thought they might discover a new planet in the relatively wide gap between the orbits of Mars and Jupiter. It took 50 years to discover the first 10 asteroids, which were all relatively large and bright. Today, advanced telescopes can discover far more than that every night. Asteroids are recognizable in telescopic images because they move noticeably relative to the stars over a short time period (Figure 13.1).

Newly discovered asteroids are first assigned a temporary identification based on the discovery year. An asteroid's orbit can be calculated from the law of gravity [Section 5.3], even after observation of only a fraction of a complete orbit. However, only about 10% of the asteroids identified in images have been followed well enough for their complete orbits to be calculated. The discoverer may then choose a name for the asteroid, subject to approval by the International Astronomical Union. The earliest discovered asteroids bear the names of mythological figures. More recent discoveries often carry names of scientists, cartoon heroes, pets, or rock stars.

General Characteristics of Asteroids

Asteroids, essentially chunks of rock, are thought to be planetesimals (or fragments of planetesimals) left over from the

Figure 13.1 Because asteroids orbit the Sun, they move relative to the stars just as planets do. In this long-exposure photograph, stars show up as white dots, and the motion of an asteroid makes it show up as a short streak.

early days of the solar system [Section 9.4]. The largest asteroid, Ceres, is about 1,000 kilometers in diameter—about half the diameter of Pluto. About a dozen others are large enough that we would call them medium-size moons if they orbited a planet.

Smaller asteroids are much more common. Although more than 150,000 asteroids have been cataloged, there may be a million or more asteroids with diameters of 1 kilometer or more. Despite their large numbers, asteroids don't add up to much in total mass. If we could put all the asteroids together and allow gravity to compress them into a sphere, they'd make an object less than 2,000 kilometers in diameter—far smaller than any terrestrial planet.

A handful of asteroids have been photographed up close by spacecraft (Figure 13.2). As we would expect for small objects, they are not spherical because gravity is not strong enough to compress their rocky material. The images also reveal numerous impact craters, telling us that asteroids, like planets and moons, have been battered by impacts. Indeed, many asteroids have odd shapes because they are fragments of larger asteroids that were shattered in collisions.

The vast majority of the asteroids that have been cataloged are located in the asteroid belt between the orbits of Mars and Jupiter (Figure 13.3). All asteroids orbit the Sun in the same direction as the planets, but their orbits tend to be more elliptical and more highly inclined to the ecliptic plane (often up to 20°–30°) than those of planets. Science fiction movies often show the asteroid belt as a crowded and hazardous place, but the asteroid belt is so large that even if it contains hundreds of thousands of asteroids they will all be quite far apart on average. The average distance between asteroids in the asteroid belt is millions of kilometers.

Not all asteroids are located in the asteroid belt. Two sets of asteroids, called *Trojan asteroids* (as in the Greek story of the Trojan War), share Jupiter's 12-year orbit around the Sun (see Figure 13.3). One clump of Trojan asteroids always stays 60° ahead of Jupiter in its orbit, while the other

a Gaspra (16 km long). Photographed by the *Galileo* spacecraft on its way to Jupiter.

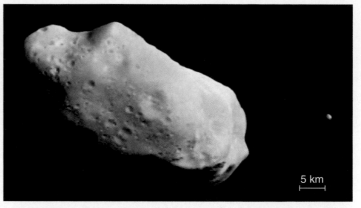

b Ida (53 km long), also photographed by the *Galileo* spacecraft en route to Jupiter. The small dot to the right is Dactyl, a tiny moon orbiting Ida.

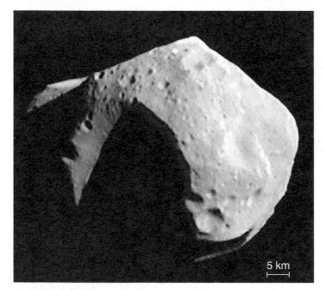

c Mathilde (59 km long), photographed by the Near-Earth Asteroid Rendezvous (NEAR) spacecraft on its way to Eros.

d Eros (40 km long), photographed by the NEAR spacecraft. NEAR orbited Eros for a year before its mission finally ended with a soft landing on the asteroid's surface.

Figure 13.2 Close-up views of asteroids studied by spacecraft.

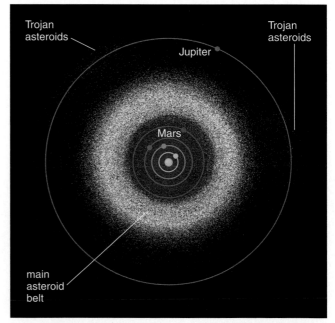

Figure 13.3 Calculated positions of 152,942 asteroids for midnight, 1 January 2004. The asteroids themselves are much smaller than the dots on this scale. The Trojan asteroids are found 60° ahead of and behind Jupiter in its orbit.

clump always stays 60° behind. The number of Trojan asteroids could be as large as the number in the asteroid belt, but the greater distance to the Trojan asteroids (and their darker surfaces) makes them more difficult to study or even count from Earth.

A relatively small number of asteroids have orbits that pass through the inner solar system—including the *near-Earth asteroids* that pass near Earth's orbit. These asteroids are probably "impacts waiting to happen." Many asteroids with similar orbits must have hit our planet in the past. In Section 13.6, we will discuss both past impacts and the future dangers posed by these asteroids.

Origin and Evolution of the Asteroid Belt

Why are asteroids concentrated in the asteroid belt, and why didn't a full-fledged planet form in this region? The answers probably lie with *orbital resonances* much like those that affect the Galilean moons of Jupiter and the rings of Saturn.

Remember that an orbital resonance occurs whenever two objects periodically line up with each other [Section 12.5]. Because gravity tugs at the objects in the same way at each alignment, the effects can build up over time, just as many small pushes can make a child swing quite high on a playground swing. Objects will periodically line up—and hence have an orbital resonance—whenever one object's orbital period is a simple ratio of another object's period, such as $\frac{1}{2}$, $\frac{1}{4}$, or $\frac{2}{5}$.

In the asteroid belt, orbital resonances occur between asteroids and Jupiter, the most massive planet by far. These resonances tend to clear gaps in the asteroid belt, much as resonances clear gaps in Saturn's rings [Section 12.6]. For example, any asteroid in an orbit that takes 6 years to circle the Sun—half of Jupiter's 12-year orbital period—would receive the same gravitational nudge from Jupiter every 12 years and thus would soon be pushed out of this orbit. The same is true for asteroids with orbital periods of 4 years ($\frac{1}{3}$ of Jupiter's period) and 3 years ($\frac{1}{4}$ of Jupiter's period). (Music offers an elegant analogy: When a vocalist sings into an open piano, the strings of notes "in resonance" with the voice—not just the note being sung but also notes with a half or a quarter of the note's frequency—are "nudged" and begin to vibrate.)

We can see the results of these orbital resonances on a graph of the number of asteroids with various orbital pe

riods (Figure 13.4). For example, notice the lack of asteroids with periods exactly $\frac{1}{2}$, $\frac{1}{3}$, or $\frac{1}{4}$ of Jupiter's—periods for which the gravitational tugs from Jupiter have cleared gaps

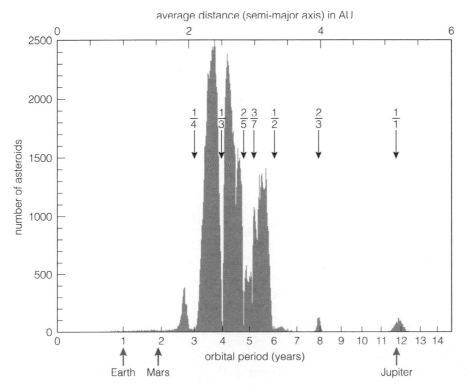

Figure 13.4 This graph shows the number of asteroids with different orbital periods. Average distance from the Sun is also labeled (along the top), because it is directly related to orbital period by Kepler's third law [Section 3.4]. Several gaps—places where few if any asteroids orbit the Sun—are labeled by the ratio of orbital period at that distance to Jupiter's orbital period. For example, the label $\frac{1}{3}$ means that the orbital period of objects with this average distance is $\frac{1}{3}$ of Jupiter's orbital period, or about 4 years. Most orbital resonances result in gaps, but some (such as $\frac{1}{1}$) happen to gather up asteroids. (Gaps are identified by *average* distance. Because asteroid orbits are elliptical, asteroids may pass through the gaps on parts of their orbits.)

in the asteroid belt. (The gaps are often called *Kirkwood gaps,* after their discoverer.)

The orbital resonances with Jupiter probably also explain why no planet ever formed between Mars and Jupiter. When the solar system was forming, this region of the solar nebula probably contained enough rocky material to form another planet as large as Earth or Mars. However, resonances with the young planet Jupiter disrupted the orbits of this region's planetesimals, preventing them from accreting into a full-fledged planet. Over the ensuing 4.5 billion years, ongoing orbital disruptions gradually kicked pieces of this "unformed planet" out of the asteroid belt altogether. Once booted from the asteroid belt, these objects either crashed into a planet or moon or were flung out of the solar system. Thus, the asteroid belt lost most of its original mass, which explains why the total mass of all its asteroids is now less than that of any terrestrial planet.

THINK ABOUT IT

Why are the gaps due to orbital resonances so easy to see in Figure 13.4 but so hard to see in Figure 13.3? (*Hint:* The top axis in Figure 13.4 is the *average* orbital distance.)

The overlapping elliptical orbits of the remaining asteroids still lead to a major collision somewhere in the asteroid belt every 100,000 years or so. Thus, over long periods of time, larger asteroids continue to be broken into smaller ones, with each collision also creating numerous dust-size particles. The asteroid belt has been grinding itself down in this way for over 4 billion years and will continue to do so for as long as the solar system exists.

Orbital resonances of a slightly different type explain why the Trojan asteroids are stable ahead of and behind Jupiter's orbit. In their case, any asteroid that wanders away from one of theses zones is nudged back into the zone by Jupiter's gravity.

Measuring Asteroid Properties

The general locations and orbits of asteroids tell us a lot about their origin, but we would like to know many other properties, such as their sizes, masses, densities, shapes, and compositions. Unfortunately, most asteroids appear as little more than points of light even when viewed through our largest Earth-based telescopes, and only a few have been visited by spacecraft. Measuring asteroid properties therefore requires clever techniques and careful study.

Asteroid Sizes We can directly measure the size of an asteroid only if we've seen it close up or in the rare cases when an asteroid is large enough to be resolved in a telescope. (We calculate physical size from the asteroid's angular size and distance [Section 7.1].) In many other cases, we can indirectly determine asteroid sizes by analyzing their light.

Asteroids shine with reflected sunlight, so an asteroid's brightness in our sky depends on its size, distance, and re-

flectivity. For example, if two asteroids at the same distance have the same reflectivity, the one that appears brighter must be larger in size. We can determine an asteroid's distance from its position in its orbit. Determining reflectivity is more difficult, but we can do it by comparing the asteroid's brightness in visible light to its brightness in infrared light. The visible light is reflected sunlight, while the infrared light is thermal radiation emitted by the asteroid itself. The infrared brightness therefore depends on the asteroid's temperature, which in turn depends on how much sunlight it absorbs. Thus, the comparison between visible and infrared brightness essentially tells us the proportions of incoming sunlight that an asteroid reflects and absorbs. Once we know these proportions, the asteroid's brightness and distance tell us its size. This technique has been used to make good size estimates for more than a thousand asteroids.

THINK ABOUT IT

Imagine that you are observing a newly discovered, bright asteroid. What additional observations would you need in order to determine whether it is very large, very close, or simply light-colored?

Asteroid Masses and Densities Determining an asteroid's density can offer valuable insights into its origin and makeup but is possible only if we first measure its mass and size. Mass measurements have been made in only a few cases, because they require observing an asteroid's gravitational effect on another object, such as a passing spacecraft or even a "moon" orbiting an asteroid (Figure 13.5). These few cases have already revealed that different asteroids can have very different densities.

For example, the asteroid Mathilde (see Figure 13.2c), visited by the NEAR spacecraft, has a density of 1.5 g/cm³. This density is so low that Mathilde cannot be a solid chunk of rock. Indeed, the impact that formed the huge crater on Mathilde's surface (the shadowed region in the photograph) would have disintegrated a solid object. Mathilde must be a loosely bound "rubble pile" that was able to absorb the shock of the impact without completely falling apart.

In contrast, the asteroid Eros (see Figure 13.2d) has a density of 2.4 g/cm³, which is close to the value expected for solid rock. Eros also has a subtle pattern of surface ridges and troughs that hint at a set of parallel faults and fractures in its interior.

Asteroid Shapes Determining asteroid shapes without close-up photographs is difficult but not impossible. For example, in some cases we can determine shapes by monitoring brightness variations as an asteroid rotates. A uniform spherical asteroid will not change its brightness as it rotates. A potato-shaped asteroid, however, will reflect more light when it presents its larger side toward the Sun and our telescopes (Figure 13.6a). In other cases, astronomers have bounced radar signals off asteroids that have passed close to Earth to determine their shapes (Figure 13.6b).

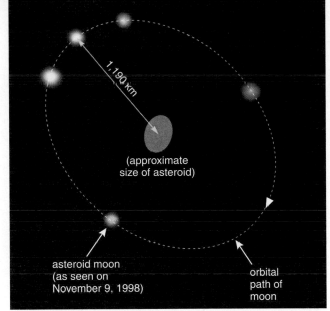

Figure 13.5 Asteroid masses can be measured only if we have observed the asteroid's gravitational effect on a passing spacecraft or, in some cases, if the asteroid has its own moon. One case of an asteroid with a moon is Ida, with its moon Dactyl (see Figure 13.2b). Another is shown here, in an image made by a telescope using adaptive optics [Section 7.4]. The image shows a small moon observed in five positions in its orbit around the asteroid Eugenia. The asteroid itself was blocked out during the observations and is shown as the gray oval in the image. Dozens of asteroid masses have been found by this method.

Shape measurements show that only the largest asteroid, Ceres (940-km diameter) is approximately spherical. The next two largest asteroids are somewhat oblong (Pallas, 540 km long, and Vesta, 510 km long). Smaller asteroids

have even odder shapes. The strength of gravity on these smaller asteroids is evidently less than the strength of their rock. In some cases, what appear to be single asteroids are probably two (or more) distinct objects held in contact by a weak gravitational attraction. As the case of Mathilde showed, some asteroids may be little more than weakly bound piles of rubble.

Asteroid Composition We can determine the surface compositions of asteroids through spectroscopy [Section 7.3]. Thousands of asteroids have been analyzed in this way. They fall into three main categories:

- Asteroids found in the outer regions of the asteroid belt are generally very dark and show absorption bands from carbon-rich materials.

- Asteroids found in the inner regions of the asteroid belt are generally more reflective and show absorption bands characteristic of rocky materials.

- A small number of asteroids have the spectral characteristics of metals such as iron.

Why do asteroids fall into these three different categories? Scientists have found answers in studies of meteorites.

13.3 Meteorites

Spectroscopy is a reasonably good method of determining the composition of an asteroid, but it would be far better if we could study an asteroid sample in a laboratory. No spacecraft has yet returned an asteroid sample. We have

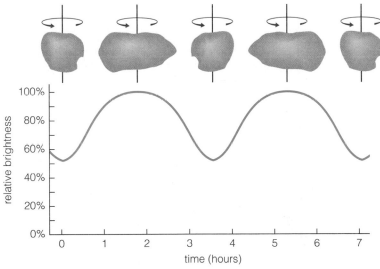

a Gaspra's irregular shape leads to large brightness variations as it rotates. The diagrams at the top show its appearance from Earth at different times during its rotation, and the graph shows the corresponding brightness at those times. Gaspra appears brighter when we are looking at a larger face because it reflects more sunlight.

b The asteroid Kleopatra is shaped like a bone. The image shows the shape based on calculations from the results of radar observations. The observations also suggest that Kleopatra is made almost entirely of metal.

Figure 13.6 In some cases, we can determine asteroid shapes from measurements made from Earth.

pieces of them nonetheless—the rocks called **meteorites** that fall from the sky.

Meteorite Falls

The reality of rocks falling from the sky wasn't always accepted. Stories of such events arose occasionally in human history, and sometimes even influenced history. For example, stories of "fallen stars" influenced the philosophy of the ancient Greek scientist Anaxagoras (see Figure 3.13), who concluded that planets and stars are flaming rocks in the heavens. Although he was not quite correct, his assumption that meteorites fell from the heavens made him the first person in history known to believe that the heavens and Earth are made of the same materials. Many later scientists regarded stories of fallen stars skeptically. Upon hearing of a meteorite fall in Connecticut, Thomas Jefferson (who took part in science as well as politics) reportedly said, "It is easier to believe that Yankee professors would lie than that stones would fall from heaven."

Today we know that rocks really do sometimes fall from the heavens. Meteorites are often blasted apart in their fiery descent through our atmosphere, scattering fragments over an area several kilometers across. A direct hit on the head by a meteorite would be fatal, but we have no reliable accounts of human deaths from meteorites. Meteorites have injured at least one human (by a ricochet, not a direct hit), damaged houses and cars, and killed animals (Figure 13.7). Of course, most meteorites fall into the ocean, which covers three-fourths of Earth's surface.

The Difference Between Meteors and Meteorites

It's easy to confuse meteors and meteorites, and in everyday language the two terms are often used interchangeably. Technically, however, a meteor is only a flash of light caused by a particle entering our atmosphere, not the particle itself. The vast majority of the particles that make meteors are no larger than peas and burn up completely before ever reaching the ground.

Only in rare cases are meteors caused by something large enough to survive the plunge through our atmosphere and leave a meteorite on the ground. Those cases make unusually bright meteors, called *fireballs*. Observers find a few meteorites each year by following the trajectories of fireballs. Meteorites found in this way can be especially valuable to science, for two reasons: We can plot their orbits to see exactly where they came from, and they can be collected before they have much time to be contaminated by terrestrial material.

Identifying Meteorites Most meteorites are difficult to distinguish from terrestrial rocks without detailed scientific analysis, but a few clues can help. Meteorites are usually covered with a dark, pitted crust resulting from their fiery

a The impact crater in this Chevrolet was created by a 12-kg meteorite. Analysis of videotapes of the meteor's fiery passage over several East Coast states confirmed its origin in the asteroid belt.

b No humans have been killed by meteorite falls, but . . . .

Figure 13.7 Two perspectives on the risks of being hit by a meteorite.

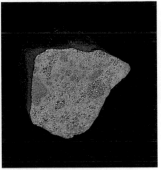

Stony primitive meteorite: metal flakes intermixed with rocky material.

Carbon-rich primitive meteorite: similar to the meteorite at left, but with carbon compounds added in.

Processed iron meteorite: similar to a planet's core.

Processed stony meteorite: similar to volcanic rocks or a planet's crust mantle.

a Primitive meteorites are remnants from the birth of the solar system.

b Processed meteorites are the shattered remains of larger objects that underwent differentiation.

Figure 13.8 There are two basic types of meteorites: primitive and processed. Each type has two sub-categories. (Each photo shows a flat face because the meteorite has been sliced open with a rock saw.) They are shown slightly larger than the actual size.

passage through the atmosphere. Some can be distinguished from terrestrial rocks by their metal content, which is high enough to attract a magnet hanging on a string. If you suspect that you have found a meteorite, many museums will analyze a small chip free of charge.

The ultimate judge of extraterrestrial origin is laboratory analysis. Meteorites can be identified by their very different isotope ratios and the presence of rare elements. Terrestrial rocks may differ from one another in composition, but all contain isotopes of particular elements in roughly the same ratios [Section 4.3]. The presence of certain rare elements, such as iridium, is a strong indicator of extraterrestrial origin. Nearly all of Earth's iridium sank to the core long ago and hence is absent from surface rocks.

More than 20,000 meteorites, found either by accident or by organized meteorite searches, have been identified and cataloged by scientists. Antarctica is one of the best places to hunt for meteorites, not because more fall there but because the icy surface makes them easier to find. Few terrestrial rocks end up on the ice, so a rock found on the ice has a good chance of being a meteorite. Moreover, the slow movement of Antarctic glaciers tends to carry meteorites along, concentrating them in small areas where the ice floes run into mountains. The majority of meteorites now in museums and laboratories were found in Antarctica.

Where Do Meteorites Come From?

The precise origin of meteorites was long a mystery, but in recent decades we've been able to determine where in our solar system they come from. The most direct evidence comes from the relatively few meteorite falls that have been observed or filmed. Such observations allow scientists to calculate the orbits that led the meteorites to crash to Earth,

telling us where the rocks originated. In every case so far, the results clearly show that the meteorites originated in the asteroid belt.

Detailed analysis of a meteorite's composition can also help us determine where it came from, because we can try to match its composition to the composition of known objects in the solar system. These studies confirm that nearly all meteorites originated in the asteroid belt, but they also hold a few surprises. Let's investigate what we've learned about meteorites and their origins.

Primitive and Processed Meteorites The most important clues about meteorite origin come from their compositions. The more than 20,000 known meteorites fall into two basic categories, each of which has two subcategories (Figure 13.8):

- **Primitive meteorites** appear to be remnants from the birth of our solar system. Radiometric dating [Section 9.5] shows them to be about 4.6 billion years old, meaning that they have remained essentially unchanged since they first accreted in the solar nebula. The vast majority of all meteorites are primitive. They can be divided into two groups:

  - Most primitive meteorites are "stony," composed of rocky minerals with a small but noticeable fraction of pure metallic flakes mixed in.

  - Other primitive meteorites also contain substantial amounts of carbon compounds, with some even including a small amount of water.

- **Processed meteorites** apparently once were part of a larger object that "processed" the original material of the solar nebula into another form. Radiometric dating shows that these meteorites are generally younger than the primitive meteorites. The processed meteorites can also be divided into two subcategories:

- Some processed meteorites resemble a terrestrial planet's core in composition: a high-density iron/nickel mixture with trace amounts of other metals.

- Other processed meteorites have much lower densities and are made of rocks more similar to a planet's crust or mantle. A few even have a composition remarkably close to that of the basalts erupted from terrestrial volcanoes [Section 10.3].

The Origin of Primitive Meteorites The structures of primitive meteorites confirm what their radiometric ages tell us: They are pieces of rock that accreted in the solar nebula and orbited the Sun for billions of years before finally falling to Earth. The individual flakes may represent the tiny particles that first condensed from the gas of the solar nebula. The small, roundish features visible in Figure 13.8a may be solidified droplets splashed out in the accretion process.*

Why do we find two types of primitive meteorites? The answer lies in where they formed in the solar nebula. Throughout the asteroid belt, temperatures were too warm for ices to condense, so all asteroids are made of rock and metal. However, beyond about 3 AU from the Sun, temperatures were low enough for carbon compounds to condense. Thus, the carbon-rich primitive meteorites must have accreted in what are now the outer regions of the asteroid belt, while the stony primitive meteorites formed in the inner regions. Laboratory studies of primitive meteorites confirm these origins. The composition of carbon-rich meteorites matches the composition of dark, carbon-rich asteroids in the outer part of the asteroid belt. The composition of stony meteorites matches the composition of asteroids in the inner part of the asteroid belt.

Curiously, most of the meteorites collected on Earth are of the stony variety even though most asteroids are of the carbon-rich variety. Evidently, orbital resonances are more effective at pitching asteroids our way from the inner part of the asteroid belt.

The Origin of Processed Meteorites The processed meteorites tell a more complex story. Their compositions look similar to the cores, mantles, or crusts of the terrestrial worlds. Thus, they must be fragments of larger asteroids that underwent *differentiation* [Section 10.2] in which their interiors melted so that metals sank to the center and rocks rose to the surface.

The processed meteorites with basaltic compositions must come from lava flows that occurred on the surfaces of large asteroids during a time when they had active volcanism. Perhaps as many as a dozen large asteroids were

---

*The roundish features are called *chondrules*, so primitive meteorites are technically known as *chondrites*. Processed meteorites are called *achondrites*, because they lack the chondrules (*a*- means "not").

geologically active shortly after the formation of the solar system. The old ages of processed meteorites tell us that this active period was short-lived—these asteroids must have cooled quickly. The basaltic meteorites simply may have been chipped off the surface of a large asteroid by smaller collisions. (By comparing meteorite and asteroid spectra, scientists have identified the asteroid Vesta as the likely source of some of these meteorites.)

The processed meteorites with core-like or mantle-like compositions must be fragments of larger asteroids that shattered in collisions. The relatively small number of metal-rich asteroids must share this origin. These shattered worlds not only lost a chance to grow into planets but sent many fragments onto collision courses with other asteroids and other planets, including Earth. Thus, processed meteorites essentially present us with an opportunity to study a "dissected planet."

Meteorites from the Moon and Mars If fragments such as the basaltic meteorites can be chipped off the surface of an asteroid, might they also have been chipped off a larger object such as a moon or planet? In fact, a few processed meteorites have been found that don't appear to match the compositions of asteroids. Instead, they appear to match the composition of either the Moon or Mars.

Careful analysis of meteorite compositions makes us very confident that we really do have a few meteorites that were once part of the Moon or Mars. Moderately large impacts can blast surface material from terrestrial worlds into interplanetary space. Once they are blasted into space, the rocks orbit the Sun until they come crashing down on another world. Calculations show that it is not surprising that we should have found a few meteorites chipped off the Moon and Mars in this way.

Study of these *lunar meteorites* and *Martian meteorites* is providing new insights into conditions on the Moon and Mars. In at least one case, a Martian meteorite may be offering clues about whether life ever existed on Mars [Section 24.2].

● 13.4 Comets

Humans have watched comets in awe for millennia. The occasional presence of a bright comet in the night sky was hard to miss before the advent of electric lights. In some cultures, these rare intrusions into the unchanging heavens foretold bad or good luck, and in most cultures little attempt was made to interpret the event in astronomical terms. (Comets were generally thought to be within Earth's atmosphere until proved otherwise by Tycho Brahe [Section 3.4].)

Nowadays, comets carry a different form of good luck. Thousands of amateur astronomers eagerly scan the skies with telescopes and binoculars in hopes of discovering a new comet. The first discoverers (up to three) who report their findings to the International Astronomical Union

a Comet Hyakutake

b Comet Hale–Bopp

Figure 13.9 Brilliant comets can appear at almost any time, as demonstrated by the back-to-back appearances of (**a**) comet Hyakutake in 1996 and (**b**) comet Hale–Bopp in 1997, photographed at Mono Lake in California.

have the comet named after them. Several comets are discovered every year, and most have never before been seen by human eyes. In 1996 and 1997, we were treated to back-to-back brilliant comets: comet Hyakutake and comet Hale–Bopp (Figure 13.9).

The surprise champion of comet discovery is the SOHO spacecraft, an orbiting solar observatory. SOHO has detected more than 500 "Sun-grazing" comets, most on their last pass by the Sun (Figure 13.10). More generally, comets are good luck for astronomers seeking to understand the solar system. These leftover planetesimals teach us about the history of the outer solar system just as asteroids teach us about the history of the inner solar system.

Figure 13.10 Comet SOHO-6's final blaze of glory. This "Sun-grazing" comet was observed by the SOHO spacecraft a few hours before it passed just 50,000 km above the Sun's surface. The comet did not survive its passage, due to the intense solar heating and tidal forces. In this image, the Sun is blocked out behind the large, orange disk; the Sun's size is indicated by the white circle.

Comet Basics

Comets are icy planetesimals that formed in the outer regions of the solar system and that became congregated in two distinct regions: the Kuiper belt and the Oort cloud [Section 9.4]. The vast majority of these comets remain perpetually frozen in the outer reaches of the solar system, well beyond the reach of our telescopes. However, a few venture inward on occasion, giving us an opportunity to study them as they warm up and grow long tails on their passage through the inner solar system. In the rest of this section, we'll see how careful study of these relatively few comets has allowed us to piece together a fairly good understanding of comet history.

Comet Orbits and Halley's Comet Once a comet is discovered, scientists can calculate its orbit from the law of gravity. Comets that visit the inner solar system have extremely eccentric orbits. In some cases, their orbits are unbound [Section 5.3], so they visit the inner solar system only once before being ejected into interstellar space. Many others travel so far from the Sun on the distant parts of their orbits that thousands of years pass between their returns to the inner solar system. These orbits prove that comets are visitors from the distant reaches of the outer solar system.

A few comets return more frequently to the inner solar system. The most famous of these frequent visitors is Halley's comet, named for English scientist Edmund Halley (1656–1742). Halley did not "discover" his comet but rather gained fame for recognizing that a comet seen in 1682 was the same one seen on a number of previous occasions. He used Newton's law of gravitation to calculate the comet's 76-year orbit. In a book published in 1705, Halley predicted that the comet would return in 1758. The comet was given his name when it reappeared as predicted, 16 years after his death. Halley's comet has returned on schedule ever since, including twice in the twentieth century—spectacularly in 1910 when it passed near Earth, and unimpressively in 1986 when it passed at a much greater distance. Its next visit will be in 2061.

Comets as "Dirty Snowballs"

Comets have been described as "dirty snowballs" because they are made of ices mixed with rocky dust. This composition is just what we expect for objects that formed in the cold outer solar system.

Spectra of comets show emission features from water molecules and other hydrogen compounds that condensed in the outer regions of the solar nebula. They also show emission from carbon dioxide and carbon monoxide, gases that condensed only in the very coldest and most distant regions of the solar nebula. More complex molecules, including some organic molecules, have been found in comet spectra. Indeed, some scientists speculate that much of the organic material that made life possible on Earth was brought here by comets.

We have not yet had the opportunity to study comet material in depth. Although lots of comet dust enters the atmosphere, these small, rocky particles burn up before reaching the ground. Any icy debris also vaporizes in the atmosphere. Comet samples should be earthbound soon, however. NASA's *Stardust* spacecraft is scheduled to pass near a comet (named Wild 2) in January 2004. *Stardust* will collect dust from the comet, returning it to Earth in 2006. Studies of the comet dust should help us understand comet origin in more detail. The comet dust could possibly contain some grains of interstellar dust that predate the birth of our solar system. If so, the Stardust mission will give us our first direct sample of the material from which stars are born.

The Flashy Lives of Comets

Comets are completely frozen when they are far from the Sun, and most are just a few kilometers across. A comet takes on an entirely different appearance when it comes close to the Sun. Solar heat warms the comet, giving it the following anatomy (Figure 13.11):

- The comet **nucleus** is the "dirty snowball" that is essentially the entire comet when it is far from the Sun and frozen.

- The **coma** is a large, dusty "atmosphere" surrounding the nucleus, made up of sublimated gas mixed with dust from the nucleus.

- The **tail** consists of gas and dust that can extend hundreds of millions of kilometers away from the coma. Most comets actually have two tails: a **plasma tail** made of ionized gas, and a **dust tail** made of small solid particles. Comet tails point away from the Sun.

Comet Nuclei

Comet nuclei are rarely observable with telescopes because they are quite small and are shrouded by the dusty coma when in the inner solar system. The European Space Agency's *Giotto* spacecraft provided our best view of a comet nucleus when it passed within a few hundred kilometers of the nucleus of Halley's comet in 1986. (The spacecraft was named for the medieval Italian painter Giotto, who was apparently inspired by a passage of Halley's comet to include a comet in a famous religious painting.)

The *Giotto* flyby revealed Halley's comet to have a dark, lumpy, potato-shaped nucleus about 16 kilometers long. Despite its icy composition, the nucleus is darker than charcoal, reflecting only 3% of the light that falls on it. (It doesn't take much rocky soil or carbon-rich material to

Figure 13.11 Anatomy of a comet. The inset photo is the nucleus of Halley's comet photographed by the *Giotto* spacecraft; the coma and tails shown are those of comet Hale–Bopp from a ground-based photo.

VIS

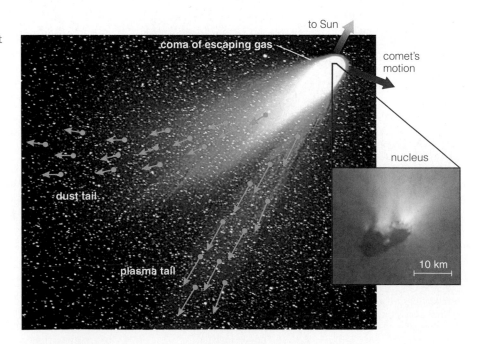

coma of escaping gas

to Sun

comet's motion

nucleus

dust tail

plasma tail

10 km

darken a comet.) The ice is not packed very solidly. Some estimates of the density of Halley's nucleus are considerably less than the density of water (1 g/cm^3), suggesting that the nucleus is part ice and part empty space.

Sublimation and the Coma

How can such a small nucleus put on such a spectacular show? Comets spend most of their lives in the frigid outer limits of our solar system. As a comet accelerates toward the Sun, its surface temperature increases, and ices begin to sublimate into gaseous form. Figure 13.12 summarizes the changes that occur during the comet's passage.

By the time the comet comes within about 5 AU of the Sun, sublimation begins to form a noticeable atmosphere that easily escapes the comet's weak gravity. The coma forms as the escaping atmosphere drags away dust particles that have been mixed with the sublimating ice. Sublimation rates increase as the comet approaches the Sun. Gases may jet away from patches on the nucleus at speeds of hundreds of meters per second. The jets can make faint pinwheel patterns within the coma due to the slow rotation of comet nuclei. The rapid

sublimation carries away so much dust that a spacecraft flyby is dangerous. The *Giotto* spacecraft, which approached Halley's nucleus at a relative speed of nearly 250,000 km/hr (70 km/sec), was sent reeling by dust impacts just minutes before closest approach, breaking radio contact with Earth.

Comet Tails

Comet tails are extensions of the coma. Because they look much like the wakes behind boats, many people mistakenly guess that tails extend behind comets as they travel through their orbits. In fact, comet tails generally point away from the Sun, regardless of the direction in which the comet is traveling. Comets have two tails because escaping gas and dust are influenced by the Sun in slightly different ways.

The escaping gases are ionized by ultraviolet photons from the Sun. Once they are ionized, the solar wind carries them straight outward away from the Sun at speeds of hundreds of meters per second. These ionized gases form the long, straight *plasma tail*.

Dust-size particles that escape from the comet experience a much weaker push from the Sun caused by the

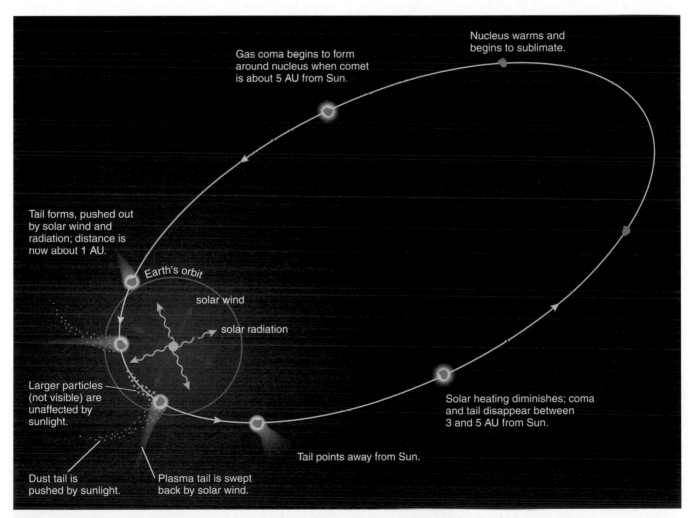

Figure 13.12 Comets exist as bare nuclei over most of their orbits and grow a coma and tails only when they approach the Sun. (Diagram not to scale.)

pressure of sunlight itself (called *radiation pressure* [Section 17.4]), rather than by the charged particles of the solar wind. Thus, while the *dust tail* also points generally away from the Sun, it has a slight curve back in the direction the comet came from.

Comets also eject some sand-to-pebble-size particles that are not affected by the solar wind or sunlight, due to their larger size. These particles form a third, invisible "tail" extending along the comet's orbit. These are the particles responsible for meteor showers, which we will discuss in more detail in Section 13.6.

After the Passage: Back to the Deep Freeze After the comet loops around the Sun and begins to head outward, sublimation declines, the coma dissipates, and the dust and plasma tails disappear. Nothing happens until the comet again comes sunward—in a century, a millennium, a million years, or perhaps never.

Active comets cannot last forever. A comet probably loses about 0.1% of its ice on every pass around the Sun, so it could not make more than a few hundred passages before losing most of its original ice. Changes in composition may end the comet's "life" even faster. In a close pass by the Sun, a comet may shed a layer of material a meter thick. Dust that is too heavy to escape accumulates on the surface. This thick, dusty deposit helps make comets dark and may eventually block the escape of interior gas, pre-

venting the comet from growing a coma or tails on future passes by the Sun.

No one is certain what happens to a comet after its ices can no longer sublime into gas and escape. Either the remaining dust disguises the dead comet as an asteroid, or the comet comes "unglued" without the binding effects of ices. In the latter case, the comet may simply disintegrate along its orbit.

The Origin of Comets

Comets are a rarity in the inner solar system, with only a handful coming within the orbit of Jupiter at any one time. However, because comets cannot last for very many passes by the Sun, comets must come from enormous reservoirs at great distances: the Kuiper belt and the Oort cloud (Figure 13.13).

The existence of these reservoirs was inferred by analyzing the orbits of comets that pass close to the Sun. Most comet orbits fit no pattern. They do not orbit the Sun in the same direction as the planets, and their elliptical orbits can be pointed in any direction. Comets Hyakutake and Hale–Bopp both fall into this class. Such comets are believed to be visitors from the Oort cloud.

Other comets have a pattern to their orbits. They travel around the Sun in the same plane and direction as the planets, though on very elliptical orbits that take them beyond

Figure 13.13 The Kuiper belt and the Oort cloud. Arrows indicate representative orbital motions of objects in the two regions. (Figure not to scale.)

Oort cloud:
• Extends out to about 50,000 AU
• Contains a trillion comets
• Comets formed near jovian planets but were flung into large, random orbits by gravitational encounters

Neptune's orbit

Kuiper belt:
• About 30–100 AU
• 100,000 comets more than 100 km across
• Comets orbit in the same plane and direction as planets
• Comets still in the region in which they formed
• Comets covered with dark carbon-rich compounds
• Many comets in orbital resonances with Neptune
• Pluto largest member of the group?

the orbit of Pluto. They also return much more frequently—typically every one to two centuries. Such comets are thought to be visitors from the Kuiper belt.

Comets from either the Oort cloud or the Kuiper belt may occasionally pass near enough to a planet to have their orbits significantly altered, sometimes becoming trapped in the inner regions of the solar system [Section 5.5]. Halley's comet must have passed near a jovian planet, causing it to end up on its current 76-year orbit of the Sun. Gravitational encounters with planets can also eject comets from the solar system altogether or send them crashing into the planets or into the Sun.

Oort Cloud Comets

Based on the number of comets with random orbits that fall into the inner solar system, we conclude that the Oort cloud must contain about a trillion (10^{12}) comets. However, Oort cloud comets normally orbit the Sun at such great distances—up to about 50,000 AU—that we have so far seen only those rare visitors to the inner solar system. Oort cloud comets are so far from the Sun that the gravity of neighboring stars can alter their orbits, preventing some comets from ever returning near the planets and sending others plummeting toward the Sun.

Our theory of solar system formation says that the comets of the Oort cloud formed in the vicinity of the jovian planets and were flung onto their large, random orbits by gravitational encounters with these planets. However, we need further studies to confirm that Oort cloud comets really formed in closer, warmer, and denser regions of the solar nebula than did their Kuiper belt counterparts. Spacecraft missions to the Oort cloud are prohibitive. We must be content to wait for comets to come to us, watching or visiting them as the Sun dissects them meter by meter.

Kuiper Belt Comets

The comets of the Kuiper belt are probably still in the same general region in which they formed, which explains why those that enter the inner solar system have orbits that share the plane and direction of planetary orbits. In the 1990s, astronomers began using new, large telescopes to search for Kuiper belt comets in their native environment of the outer solar system. The first discovery, given the name 1992QB1, was found orbiting beyond Pluto with an orbital period of 296 years. It is approximately 280 kilometers across, far larger than any comet known to visit the inner solar system.

Subsequent searches have identified many more objects with orbital properties suggesting that they are Kuiper belt comets. By 2003, more than 650 Kuiper belt comets had been discovered. Based on these discoveries, scientists infer that at least 100,000 comets more than 100 kilometers across must populate the Kuiper belt. Because smaller objects are more common, it's estimated that there must be another *billion* comets more than about 10 kilometers across—the typical size of comets that enter the inner solar system. Thus, the total mass of the Kuiper belt is far greater than that of the asteroid belt. Spectroscopy has provided some preliminary information on the composition of

Kuiper belt comets. Although they are undoubtedly rich in ices, the Kuiper belt comets are covered with dark, carbon-rich compounds.

How big can Kuiper belt comets be? The record for the largest known comet has fallen several times in the past decade. As of 2003, the record was held by an object orbiting a billion kilometers beyond Pluto in the Kuiper belt. The object, given the name Quaoar (pronounced *kwa-whar*) after a Native American god, is 1,300 kilometers across—more than half the diameter of Pluto. If a comet can be 1,300 kilometers across, could one be larger still? It's time to consider the odd planet Pluto.

 Formation of the Solar System Tutorial, Lesson 3

13.5 Pluto: Lone Dog, or Part of a Pack?

Pluto has seemed a misfit among the planets almost since the day it was discovered in 1930 (see the box "Discovering the Outermost Planets," p. 217). Its 248-year orbit is more elliptical and more inclined to the ecliptic plane than that of any other planet (Figure 13.14). At its nearest, it is actually closer to the Sun than Neptune, as was the case between 1979 and 1999. At its farthest, it is 50 AU from the Sun, putting it far beyond the realm of the jovian planets. Its size is also out of character with the other planets. At only 1,160 kilometers in radius, it is much smaller than any terrestrial planet and smaller than several moons of the jovian planets.

Despite the fact that Pluto sometimes comes closer to the Sun than Neptune, there is no danger of the two planets colliding. For every two Pluto orbits, Neptune circles the Sun three times. This stable orbital resonance means that Neptune is always a safe distance away whenever Pluto approaches its orbit. The two planets will probably continue their dance of avoidance until the end of the solar system.

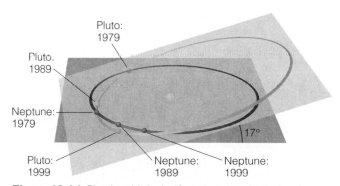

Figure 13.14 Pluto's orbit is significantly more elliptical and more tilted relative to the ecliptic than that of any other planet. It comes closer to the Sun than Neptune for 20 years in each 248-year orbit, as was the case between 1979 and 1999. There's no danger of a collision between Pluto and Neptune, because they share an orbital resonance in which Neptune completes three orbits for every two orbits by Pluto.

Figure 13.15 Pluto (left) and its moon Charon (right), photographed by the Hubble Space Telescope.

Pluto and Its Moon

Pluto's great distance and small size make it difficult to study, but the task became much easier with the 1978 discovery of Pluto's moon *Charon* (Figure 13.15). Charon orbits Pluto every 6.4 days. Combining this orbital period with Charon's distance from Pluto enabled astronomers to make an accurate measurement of Pluto's mass (by applying Newton's version of Kepler's third law [Section 5.3]). Astronomers also learned that Pluto rotates "backward" and with its rotation axis tipped almost on its side (see Figure 8.1).

Pluto's moon was discovered just in time—astronomers soon learned that Charon's orbit was about to go edge-on as seen from Earth, something that happens only every 124 years. From 1985 to 1990, astronomers monitored the combined brightness of Pluto and Charon as they alternately eclipsed each other every few days. Detailed analysis of brightness variations allowed the calculation of accurate sizes, masses, and densities for both Pluto and Charon, as well as the compilation of rough maps of their surface markings (Figure 13.16). We do not yet know why the surfaces have bright and dark patches.

Charon's diameter is more than half Pluto's, and its mass is about one-eighth the mass of Pluto. Furthermore, Charon orbits only 20,000 kilometers from Pluto. (For comparison, our Moon has a mass $\frac{1}{80}$ of Earth's and orbits 400,000 kilometers away.) Some astronomers argue that Pluto and Charon qualify as a "double planet." Others argue that neither is large enough to qualify as a planet at all.

Curiously, Pluto and Charon have slightly different densities and presumably slightly different compositions. Pluto's density, about 2 g/cm^3, is a bit high for the expected ice/rock mix in the outer solar system. Charon's density, 1.6 g/cm^3, is a bit low. The leading hypothesis is that Charon was created by a giant impact similar to that thought to have formed our Moon [Section 9.4]. A large, icy object impacting Pluto may have blasted away its low-density outer layers, which then formed a ring around Pluto and eventually reaccreted into the low-density moon Charon.

Pluto currently has a thin atmosphere of nitrogen and other gases formed by sublimation of surface ices. However, the atmosphere is steadily thinning because

Pluto's elliptical orbit is now carrying it farther from the Sun. (Pluto will not reach aphelion until 2113.) As Pluto recedes, its atmospheric gases are refreezing onto its surface.

Despite the cold, the view from Pluto would be stunning. Charon would dominate the sky, appearing almost 10 times larger in angular size than our Moon appears from Earth. Pluto and Charon's mutual tidal pulls long ago made them rotate synchronously with each other (see Figure 5.19). Thus, Charon is visible from only one side of Pluto and always shows the same face to Pluto. Moreover, the synchronous rotation means that Pluto's "day" is the same

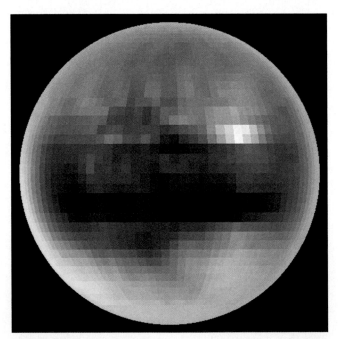

Figure 13.16 The surface of Pluto in approximate true color, as derived with computer processing from brightness measurements made during mutual eclipses between Pluto and Charon. The brown color probably comes from methane ice, but the origin of the dark equatorial band is unknown. The many squares in the image reflect the best picture that could be made with the available data. The squares indicate the size of the smallest area used in the calculation.

length as Charon's "month" (orbital period) of 6.4 Earth days. Charon therefore neither rises nor sets in Pluto's skies but instead hangs motionless as it cycles through its phases every 6.4 days. The Sun would appear more than a thousand times fainter than it appears here on Earth and would be no larger in angular size than Jupiter in our skies.

Pluto: Planet or Kuiper Belt Comet?

Pluto is an object in search of a category. A misfit among the planets, it seems somewhat less out of place when we consider it in the context of the Kuiper belt. After all, it orbits in the same vicinity as Kuiper belt comets, it has a comet-like composition of icy and rocky material, and it has a comet-like atmosphere that grows as it comes nearer the Sun in its orbit and fades as it recedes. Could Pluto be simply the largest known member of the Kuiper belt?

A closer look at the orbits of Kuiper belt comets uncovers uncanny resemblances to Pluto. Like Pluto, many Kuiper belt objects have stable orbital resonances with Neptune. In fact, more than a dozen Kuiper belt comets have the same orbital period and average distance from the Sun as Pluto itself (and are nicknamed "Plutinos"). Several Kuiper belt comets are even known to possess moons. Aside from Pluto's large size compared to the sizes of known comets, the most obvious difference between Pluto and the Kuiper belt comets is in surface brightness. Comet nuclei generally have surfaces that are quite dark from carbon compounds, while Pluto's high reflectivity indicates an icy surface.

This difference is easily explained. Whenever either Pluto or a smaller Kuiper belt comet approaches the Sun, surface ices sublimate into an atmosphere. These atmospheric gases completely escape from the smaller Kuiper belt comets, but Pluto's gravity holds them until they refreeze onto the surface. Over time, the smaller Kuiper belt comets become depleted in the reflective ices, leaving relatively more of the dark, carbon-rich compounds on their surfaces. Pluto has retained its reflective ices, giving it a bright, icy surface.

All things considered, Pluto stands out from other Kuiper belt comets only in its size. But this may not have been so unusual in the early solar system. Recall that Neptune's moon Triton is quite similar to (and larger than) Pluto and that its orbit indicates it must be a captured object. Thus, Triton was once a "planet" orbiting the Sun. Giant impacts of other large, icy objects may have created Charon and might even have given Uranus its large tilt. Many Pluto-size objects probably roamed the Kuiper belt in its early days. Those that did not become trapped in stable resonances may have fallen inward to be accreted by the jovian planets or been thrown outward and ejected from the solar system. Pluto may be the largest surviving member of the Kuiper belt. It remains possible that we will someday discover other objects similar in size to Pluto, though probably farther away or considerably darker. As of 2003, only a small percentage of the sky has been thoroughly searched for large Kuiper belt objects.

THINK ABOUT IT

What is *your* definition of a planet? How many planets are there, according to your criteria? Does your list include Pluto? Charon or Triton? Ceres or other asteroids? How big would a newly discovered Kuiper belt object have to be to meet your definition? What value does your definition serve? Explain.

Planet or not, Pluto remains the largest known object in the solar system never visited by spacecraft. Close-up observations of its surface, atmosphere, and unusual moon are sure to teach us as much about icy bodies as past missions did about rocky bodies. If all goes well, the first mission to Pluto will be launched in 2006, arriving there a decade later. Pluto's atmosphere will be one area of intense observation. Although it was expected to completely refreeze onto its surface by about 2020 as Pluto recedes from the Sun, recent observations have surprised astronomers by showing that the atmospheric density is actually increasing.

13.6 Cosmic Collisions: Small Bodies Versus the Planets

The hordes of small bodies orbiting the solar system are slowly shrinking in number through collisions with the planets and ejection from the solar system. Many more must have roamed the solar system in the days of the heavy bombardment, when most impact craters were formed [Section 9.4]. Plenty of small bodies still remain, however, and cosmic collisions still occur on occasion. These collisions have important ramifications for Earth as well as for the other planets.

Shoemaker–Levy 9 Impacts on Jupiter

We usually think about impacts in the context of solid bodies, because such impacts leave the long-lasting scars of impact craters. However, impacts must be even more common on the jovian planets than on the terrestrial planets, because of their larger size and stronger gravity. Astronomers have estimated that a major impact occurs on Jupiter about once every 1,000 years. We were privileged to witness one in 1994.

The dramatic event was predicted more than a year before impact, thanks to the discovery of the comet by the husband-and-wife team of Gene and Carolyn Shoemaker and their colleague David Levy. Because it was the trio's ninth joint discovery, the comet was designated *Shoemaker–Levy 9*, or *SL9* for short. SL9 was soon found to be a very unusual comet. Rather than having a single nucleus, it consisted of a string of comet nuclei lined up in a row (Figure 13.17a).

The reason for SL9's odd appearance became clear when astronomers calculated its orbit backward in time. The

10,000 km

a This photo shows comet Shoemaker–Levy 9 after it was broken apart by tidal forces after passing close to Jupiter. The kilometer-sized remains were shrouded in clouds of dust and gas.

crater chain

50 km

b This chain of craters on Callisto probably formed long ago from impacts by fragments of a comet that, like comet Shoemaker–Levy 9, was broken apart by tidal forces while passing close to Jupiter. These impacts probably occurred just hours after the comet broke apart.

VIS

Figure 13.17 Comets can be fragmented by passing close to Jupiter.

calculations showed that it had passed very close to Jupiter only a few months earlier (in July 1992). When it passed near Jupiter, it apparently was ripped apart by tidal forces. Prior to that event, SL9 probably was a single comet nucleus orbiting Jupiter, unnoticed by human observers. Crater chains on Jupiter's moons are evidence that similar breakups of comets near Jupiter have occurred in the past. For example, Figure 13.17b shows a chain of craters on Callisto, presumably formed when a string of comet nuclei crashed into its icy surface.

The orbital calculations also showed that, due to gravitational nudges from the Sun, SL9 was on a collision course with Jupiter. Astronomers had more than a year to plan for observations of the impacts, which were due in July 1994. The impacts of the 22 identified nuclei would take almost a week. Virtually every telescope on Earth and in space would be ready and watching. The only uncertainty was

whether anything dramatic would happen. Even the best Hubble Space Telescope images did not show whether the comet nuclei were massive, kilometer-size chunks of ice or merely insubstantial puffs of dust.

The answer came within minutes of the first impact. Infrared cameras recorded an intense fireball of hot gas rising thousands of kilometers above the impact site, which lay just barely on Jupiter's far side (Figure 13.18). As Jupiter's rapid rotation carried the scene into view, the collapsing plume smashed down on Jupiter's atmosphere, leading to another episode of heating and more infrared glow. The Hubble Space Telescope caught impact plumes in action as they rose into sunlight and collapsed back down, leaving dust clouds high in Jupiter's stratosphere. The impact "scars" lingered for months, but Jupiter has recovered—for now.

The impacts provided us with a unique opportunity to study material splashed out from well beneath those layers of Jupiter that are ordinarily visible. More important, we learned much about the impact process itself. Many scientists had correctly predicted the behavior of the rising gas plumes created by the impacts but had not considered the effects of the plumes themselves as they splashed back down onto Jupiter. These reimpacts by plume material turned out to have even more significant global effects than the original impacts. As the largest plumes crashed back down, each heated an area 10,000 kilometers across and created dark, dusty clouds that encircled the planet.

The SL9 impacts on Jupiter also provided two important sociological lessons. First, "Comet Crash Week" proved to be one of the best examples of international collaboration in history. With the aid of the Internet, scientists quickly and effectively shared data from observatories around the world. Second, extensive media coverage helped the event capture the public imagination, leading to awareness that impacts are not merely relegated to ancient geological history. Each individual fragment of SL9 crashed

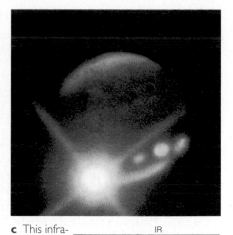

a The *Galileo* spacecraft, on its way to Jupiter at the time, got a direct view of the impacts on Jupiter's night side. The bright dot on the left shows one impact site.

b Less than 20 minutes after impact, the Hubble Space Telescope observed an impact plume rise thousands of kilometers above Jupiter's clouds and then collapse back down.

c This infrared photo shows the brilliant glow of a rising fireball from one of the impacts. Most of Jupiter appears dark because methane gas in its atmosphere absorbs infrared light.

d High clouds created by the impacts ring the southern hemisphere. These clouds reflect infrared light from the Sun because they lie high above the methane gas that absorbs infrared light.

e In this photo from the Hubble Space Telescope, the high, dusty clouds left by the impacts appear as dark "scars" on Jupiter.

f This Hubble Space Telescope photo shows how the impact scars became smeared due to atmospheric winds. The scars disappeared completely within a few months.

g Artist's conception of the impacts viewed from the surface of Io.

Figure 13.18 The impacts of comet Shoemaker–Levy 9 with Jupiter allowed astronomers their most direct view ever of cosmic collisions.

into Jupiter with an energy equivalent to that of a million hydrogen bombs. If such violent impacts can happen on other planets in our lifetime, could they also happen on Earth?

Meteors and Meteor Showers

Far smaller impacts happen on Earth all the time, lighting up the sky as meteors. If you watch the sky on a clear night, you'll typically see a few meteors each hour. Most meteors are created by single pieces of comet dust, sand-size to pea-size, that enter our atmosphere at speeds of up to 250,000 km/hr (70 km/s). (A small fraction of meteors may be dust from asteroids rather than comets, and the brightest meteors come from larger particles.)

The high speeds of these small impacts heat the particles and the surrounding atmosphere so much that we see a brief but brilliant flash. The particles are vaporized by the heat, disappearing into our atmosphere at altitudes of 50–100 kilometers. An estimated 25 million meteors enter the atmosphere worldwide every day, adding hundreds of tons of comet dust to Earth daily.

Even more comet dust enters the atmosphere when Earth crosses the orbit of a comet. Comet orbits tend to be filled with small particles ejected as the comets pass near the Sun. These particles rain down on our planet as Earth passes through a comet's orbit, producing a **meteor shower**. During a meteor shower on a clear night, you may see dozens of meteors per hour. Although the meteors in a shower all strike Earth on parallel tracks, they generally appear to radiate from a particular direction in the sky. You may have experienced a similar illusion while driving into a blizzard at night (Figure 13.19). Because more meteors impact in front of Earth than behind it (just as more snow hits the front windshield of a moving car), meteor showers are best observed in the predawn sky.

Table 13.1 Major Annual Meteor Showers

Shower Name	Approximate Date
Quadrantids	January 3
Lyrids	April 22
Eta Aquarids	May 5
Delta Aquarids	July 28
Perseids	August 12
Orionids	October 22
Taurids	November 3
Leonids	November 17
Geminids	December 14
Ursids	December 23

Meteor showers recur at about the same time each year because the orbiting Earth passes through a particular comet's orbit at the same time each year. For example, the meteor shower known as the *Perseids* occurs about August 12 each year—the time when Earth passes through the orbit of comet Swift–Tuttle. The Perseids get their name because the meteors appear to radiate from the constellation Perseus. Other annual meteor showers are similarly named for constellations and are listed in Table 13.1. Meteor showers may vary in intensity from year to year depending on how comet debris is distributed through the comet's orbit. You should watch news reports to find out whether a particular meteor shower is likely to be worth watching, but keep in mind that meteor forecasting is somewhat imprecise.

THINK ABOUT IT

The associated "comet" for the Geminid meteor shower is an object called Phaeton, which is classified as an asteroid because we've never seen a coma or tail associated with it. How is it possible that Phaeton looks like an asteroid today but once shed the particles that create the annual Geminid meteor shower?

Impacts and Mass Extinctions on Earth

Meteorites and impact craters bear witness to the fact that much larger impacts occasionally occur on Earth (Figure 13.20). Meteor Crater in Arizona formed about 50,000 years ago when a metallic asteroid roughly 50 meters across crashed to Earth with the explosive power of a 20-megaton hydrogen bomb. Although the crater is only a bit more than 1 kilometer across, the blast and ejecta probably battered an area covering hundreds of square kilometers. Meteor Crater is relatively small and recent. Despite the fact that erosion and other geological processes have erased most of Earth's impact craters, geologists have identified more than 100 impact craters on our planet. Some impacts apparently have had catastrophic consequences for life on Earth.

In 1978, while analyzing geological samples collected in Italy, the father-son team of Luis and Walter Alvarez and their colleagues made a startling discovery. They found that a thin layer of dark sediments deposited about 65 million years ago—about the time the dinosaurs went extinct—was unusually rich in the element iridium. Iridium is a metal that is rare on Earth's surface but common in meteorites. Subsequent studies found the same iridium-rich layer in 65-million-year-old sediments around the world (Figure 13.21).* The Alvarez team suggested a stunning

---

*The layer marks what geologists call the *K–T boundary*, because it separates sediments deposited in the Cretaceous and Tertiary periods (the "K" comes from the German word for Cretaceous, *Kreide*). The mass extinction that occurred 65 million years ago therefore is often called the *K–T event*.

hypothesis: The extinction of the dinosaurs was caused by the impact of an asteroid or comet.

In fact, the death of the dinosaurs was only a small part of the biological devastation that seems to have occurred 65 million years ago. The fossil record suggests that up to 99% of all living organisms died around that time and that up to 75% of all existing *species* were driven to extinction. This makes the event a clear example of what we call a **mass extinction**—the rapid extinction of a large fraction of all living species. Could it really have been caused by an impact?

Evidence for the Impact The Alvarezes' hypothesis was not immediately accepted and still generates some controversy, but it now seems clear that at least one major impact closely coincided with the death of the dinosaurs. Key evidence comes from further analysis of the sediment layer. Besides being unusually rich in iridium, this layer contains four other features: (1) unusually high abundances of several other metals, including osmium, gold, and platinum; (2) grains of "shocked quartz," quartz crystals with a distinctive structure that indicates they were formed under the high temperature and pressure conditions of an impact;

a When driving into a blizzard, the car is moving through falling snow, but the snowflakes appear to come from a single direction in front of the car.

b A meteor shower occurs as Earth moves through a swarm of comet debris, making it look as if all the meteors come from a single direction in front of Earth. Most particles strike the "front" face of Earth, making from midnight to dawn the best time to see meteors.

c Meteors are the streaks of light in this image of the 2001 Leonid meteor shower over Australia. Note how the meteors appear to come from a single point in the sky (in the constellation Leo). The large rock is Uluru, also known as Ayers Rock. The image is a digital composite of 22 separate photos, including one taken at sunset to illuminate the rock.

Figure 13.19 A meteor shower appears to radiate from a particular point in the sky.

The Manicouagan Lakes Crater in Canada was created by an impact that occurred about 200 million years ago. The crater has been heavily eroded by glaciers. It is about 100 kilometers across.

10 km

Meteor Crater in Arizona was created about 50,000 years ago by an asteroid impact. The crater is more than a kilometer across and almost 200 meters deep, but the asteroid that made it was only about 50 meters across.

100 m

Figure 13.20 This map shows the locations of known impact craters on Earth. Two representative craters are shown in the photos.

(3) spherical rock "droplets" of a type known to form when drops of molten rock cool and solidify in the air; and (4) soot (at some sites) that appears to have been produced by widespread forest fires.

All of these features point to an impact. The metal abundances look much like what we commonly find in meteorites rather than what we find elsewhere on Earth's surface. Shocked quartz is also found at other impact sites, such as Meteor Crater in Arizona. The rock "droplets" presumably were made from molten rock splashed into the air by the force and heat of the impact. The soot probably came from vast forest fires ignited by impact debris. Some debris would have been blasted so high that it rose above the atmosphere, spreading worldwide before falling back to Earth. On their downward plunge, the debris particles would have been heated by friction until they became a hot, glowing rain of rock. Like falling hot coals, the particles ignited fires as they reached the ground.

The clearest proof that an impact occurred 65 million years ago came with the 1991 discovery of a large impact crater that appears to match the age of the sediment layer. The crater, about 200 kilometers across, is located on the coast of Mexico's Yucatán peninsula, about half on land and half underwater (Figure 13.22). Its size indicates that it was created by the impact of an asteroid or a comet measuring about 10 kilometers across. (It is named the *Chicxulub crater,* after a nearby fishing village.)

The Mass Extinction How could the impact have caused a mass extinction? Figure 13.23 illustrates the impact and its immediate aftermath. The asteroid or comet slammed into Earth with the force of a hundred million hydrogen bombs. It apparently hit at a slight angle, sending a shower of red-hot debris across the continent of North America. A huge tidal wave sloshed more than 1,000 kilometers inland. Much of North American life may have been wiped out almost immediately. Not long after, the hot debris raining around

Figure 13.21 The arrow points to a layer of sediment laid down by the impact of 65 million years ago, which is linked to the extinction of the dinosaurs. At the time, the rock layers above the arrow did not exist. Dust from the impact and soot from global wildfires settled down through the atmosphere onto the seafloor that once occupied this location in Colorado. The sediment is about 2 centimeters thick at this location.

the rest of the world ignited fires that killed many other living organisms.

The longer-term effects were even more severe. Dust and smoke remained in the atmosphere for weeks or months, blocking sunlight and causing temperatures to fall as if Earth were experiencing a global and extremely harsh winter. The reduced sunlight would have stopped photosynthesis for up to a year, killing large numbers of species throughout the food chain. This period of cold may have been followed by a period of unusual warmth. Some evidence suggests that the impact site was rich in carbonate rocks, so the impact may have released large amounts of carbon dioxide into the atmosphere. The added carbon dioxide would have strengthened the greenhouse effect, and the months of global winter may have been followed by decades or longer of global summer.

The impact probably also caused chemical reactions in the atmosphere that produced large quantities of harmful compounds, such as nitrous oxides. These compounds dissolved in the oceans, where they probably were responsible for killing vast numbers of marine organisms. Acid rain

Figure 13.22 This false-color image, made with the aid of precision measurements of the local strength of gravity, shows an impact crater with its center near the northwest corner of the Yucatán. The red box on the inset map shows the region covered by the image. The white lines on the image correspond to the coastline and the borders of Mexican states.

Figure 13.23 Instant geology and biology. This sequence of 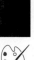 paintings shows an artist's conception of the impact thought to have caused a mass extinction—and the death of the dinosaurs—some 65 million years ago.

may have been another by-product, killing vegetation and acidifying lakes around the world.

Perhaps the most astonishing fact is not that 75% of all species died but that 25% survived. Among the survivors were a few small, rodentlike mammals. These mammals may have survived in part because they lived in underground burrows and managed to store enough food to outlast the global winter that followed the impact.

The evolutionary impact of the extinctions was profound. For 180 million years, dinosaurs had diversified into a great many species large and small, while mammals (which had arisen at almost the same time as the dinosaurs) had remained small and rodentlike. With the dinosaurs gone, mammals became the new kings of the planet. Over the next 65 million years, the small mammals rapidly evolved into a large assortment of much larger mammals—ultimately including us.

Controversies and Other Mass Extinctions There seems little doubt that a major impact coincided with the mass extinction of 65 million years ago, but does it tell the entire story? The jury is still out. The fossil record can be difficult to read when we are trying to understand events that happened in just a few years rather than a few million years. Some scientists suspect that the dinosaurs were already in decline and the impact was only the last straw. Others suggest that major volcanic eruptions also may have played a role.

Most recently, geologists have found that at least two other large craters—one in the North Sea and one in the Ukraine—date to approximately the same time as the crater in the Yucatán. Perhaps a swarm of impacts, rather than a single impact, caused the mass extinction. Unfortunately, the crater ages are not known well enough to allow us to determine whether the swarm came all at once, like the SL9 impacts on Jupiter, or whether individual impacts may have been separated by a few hundred thousand years.

Was the dinosaur extinction a unique event? Measuring precise extinction rates becomes more difficult as we look to older fossils, but there appear to have been at least four other mass extinctions during the past 500 million years. None of the other mass extinctions is as closely tied to an impact as the dinosaur extinction, but impacts almost certainly played a role. Sediments from the times of other mass extinctions reveal evidence similar to that found in the iridium-rich layer tied to the death of the dinosaurs. In some cases, impact craters have been found that date to about the right times. Much more research is needed, but impacts appear to have played a major role in shaping the history of life on Earth.

The Asteroid Threat: Real Danger or Media Hype?

On June 14, 2002, the 100-meter-wide asteroid 2002MN passed within 120,000 kilometers of Earth (less than a third of the Earth–Moon distance). There was no advance warning, and in fact the asteroid wasn't discovered until after it passed by. Although an asteroid of this size would not cause global devastation, it might kill thousands or millions of people in the unlikely event that it struck a large city. How concerned should we be about the possibility of objects like this striking Earth—or, worse yet, even larger objects such as those that caused past mass extinctions? We can analyze the threat by examining both the frequency of past impacts and the number of objects in space that pose a potential future threat.

No large impacts have left craters in modern times, but we know of at least one smaller impact that devastated a fairly large area. In 1908, a tremendous explosion occurred over Tunguska, Siberia (Figure 13.24). Entire forests were

Figure 13.24 Damage from the 1908 impact over Tunguska, Siberia.

flattened and set on fire, and air blasts knocked over people, tents, and furniture up to 200 kilometers away. Seismic disturbances were recorded at distances of up to 1,000 kilometers, and atmospheric pressure fluctuations were detected almost 4,000 kilometers away. The explosion, now estimated to have released energy equivalent to that of several atomic bombs, is thought to have been caused by a small asteroid no more than about 40 meters across. Atmospheric friction caused it to explode completely before it hit the ground, so it left no impact crater.

Objects of similar size to that of the Tunguska event probably strike our planet every century or so. They have gone unnoticed, presumably because they have always hit remote areas, with most probably hitting the ocean. Nevertheless, the death toll would be enormous if such an object struck a densely populated area.

Another way to gauge the threat is to look at asteroids that might strike Earth. The largest known near-Earth asteroid, Eros, is about 40 kilometers long (see Figure 13.2d). More than 600 smaller near-Earth asteroids larger than 1 kilometer in diameter have been detected by dedicated searches. Orbital calculations show that none of these known objects will impact Earth in the foreseeable future. However, the vast majority of near-Earth asteroids probably have not yet been detected. Astronomers estimate that hundreds more undiscovered near-Earth asteroids are larger than 1 kilometer across and that another 50,000 are larger than 100 meters across. NASA has set a goal of detecting 90% of all large near-Earth asteroids by 2008.

We do not have a similar way of studying the threat from comets, because they reside so far from the Sun. By the time we saw a comet plunging toward us from the outer solar system, we'd have at best a few years to prepare for the impact.

Figure 13.25 shows how often, on average, we expect Earth to be hit by objects of different sizes. Objects a few meters across probably impact Earth's atmosphere every few days, each liberating energy equivalent to that of an atomic bomb as it burns up. Larger impacts are obviously more devastating but thankfully more rare. Hundred-meter impacts that forge craters similar to Meteor Crater probably strike only about every 10,000 years or so. Impacts large enough to cause mass extinctions come tens of millions of years apart.

While it seems a virtual certainty that Earth will be battered by many more large impacts, the chance of a major impact happening in our lifetimes is quite small. Nevertheless, the chance is not zero, which is one reason why NASA is working to predict potential asteroid threats.

If we were to find an asteroid or a comet on a collision course with Earth, could we do anything about it? Many people have proposed schemes to save Earth by using nuclear weapons or other means to demolish or divert an incoming asteroid, but no one knows whether current technology is really up to the task. We can only hope that the threat doesn't become a reality before we're ready.

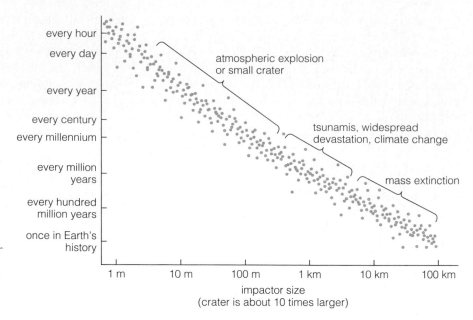

Figure 13.25 This graph shows that larger impactors (asteroids or comets) hit the Earth less frequently than smaller ones. The labels describe the effects of impacts of different sizes.

THINK ABOUT IT

Study Figure 13.25. About how often should we expect an impact large enough to wipe out our civilization? About how often should we expect an impact large enough to destroy a major city? Do you think the threat is serious enough to warrant action, or should we focus resources on other threats first? Defend your opinion.

While some people see danger in near-Earth asteroids, others see opportunity. Near-Earth asteroids bring valuable resources tantalizingly close to Earth. Iron-rich asteroids are particularly enticing, because they probably contain many precious metals that have mostly sunk to the core on Earth. In the not-too-distant future, it may prove technically feasible and financially profitable to mine metals from asteroids and bring these resources to Earth. It may also be possible to gather fuel and water from asteroids for use in missions to the outer solar system.

Small Bodies: The Jovian Connection

Throughout this chapter, as we've described the small bodies, we have noted the influence of Jupiter and the other jovian planets again and again. Jupiter disturbed the orbits of rocky planetesimals outside Mars's orbit, preventing a planet from forming and instead creating the asteroid belt there. The jovian planets ejected icy planetesimals from their range of orbits altogether, forming the distant Oort cloud of comets. Neptune continues to nudge the orbits of icy objects in the Kuiper belt.

In summary, the jovian planets were responsible for the creation of the Oort cloud and still exert their far-reaching control over the asteroid belt and the Kuiper belt through the subtler influence of orbital resonances. More-over, because impacts of asteroids and comets have played such an important role in the history of our planet, we find a deep connection between the jovian planets and the survival of life on Earth. Figure 13.26 summarizes the ways the jovian planets have controlled the motions of asteroids and comets, which in turn have profound effects on Earth—both past and future.

THE BIG PICTURE

Putting Chapter 13 into Context

In this chapter we focused on the solar system's smallest objects and found that they can have big consequences. Keep in mind the following "big picture" ideas as you continue your studies:

- Asteroids, comets, and meteorites may be small compared to planets, but they provide much of the evidence that has helped us understand how the solar system formed.

- Pluto is called the ninth planet, but it is much more similar to the thousands of Kuiper belt comets than to the other eight planets.

- The small bodies are subject to the gravitational whims of the largest. The jovian planets shaped the asteroid belt, the Kuiper belt, and the Oort cloud, and they continue to nudge objects onto collision courses with the planets.

- Collisions not only bring meteorites and leave impact craters but have profoundly affected life on Earth. The dinosaurs were probably wiped out by an impact, and future impacts pose a threat that we cannot ignore.

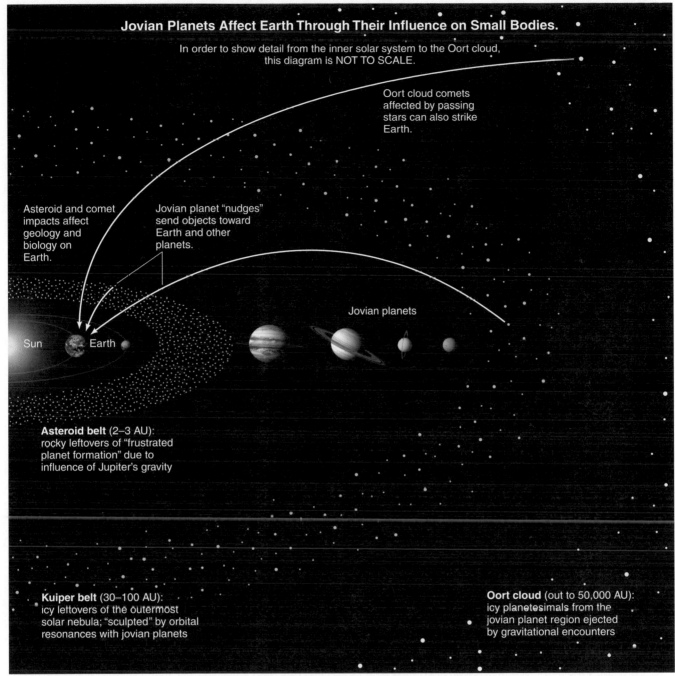

Jovian Planets Affect Earth Through Their Influence on Small Bodies.

In order to show detail from the inner solar system to the Oort cloud, this diagram is NOT TO SCALE.

Oort cloud comets affected by passing stars can also strike Earth.

Asteroid and comet impacts affect geology and biology on Earth.

Jovian planet "nudges" send objects toward Earth and other planets.

Jovian planets

Sun   Earth

Asteroid belt (2–3 AU): rocky leftovers of "frustrated planet formation" due to influence of Jupiter's gravity

Kuiper belt (30–100 AU): icy leftovers of the outermost solar nebula; "sculpted" by orbital resonances with jovian planets

Oort cloud (out to 50,000 AU): icy planetesimals from the jovian planet region ejected by gravitational encounters

Figure 13.26 The connections between the jovian planets, small bodies, and Earth. The gravity of the jovian planets helped shape both the asteroid belt and the Kuiper belt, and the Oort cloud consists of comets ejected from the jovian planet region by gravitational encounters with these large planets. Ongoing gravitational influences sometimes send asteroids or comets heading toward Earth.

13.1 Remnants from Birth

- *What are the three major groups of small bodies found orbiting in the solar system?* The three groups, distinguished by their orbits, are the asteroids, the comets of the Kuiper belt, and the comets of the Oort cloud.

13.2 Asteroids

- *What are the general characteristics of asteroid sizes, shapes, and orbits?* Most asteroids are small and potato shaped and orbit in the asteroid belt. Trojan asteroids share Jupiter's orbit. Near-Earth asteroids have orbits that pass near Earth's orbit.

- *Why didn't a planet form in the region of the asteroid belt?* Orbital resonances with Jupiter disrupted the orbits of planetesimals, preventing them from accreting into a planet. Resonances also cause the gaps in the asteroid belt today.

- *How do we measure asteroid properties?* We measure orbits from observations of asteroid motion and application of the law of gravity. We measure sizes by comparing infrared and visible brightness. We determine masses from gravitational effects on moons or passing spacecraft, and densities from size and mass. We can determine shapes from brightness changes with rotation or radar, and composition from spectroscopy.

13.3 Meteorites

- *How is a meteor different from a meteorite?* A meteor is a flash of light caused by a small particle entering our atmosphere. A meteorite is a rock that survives the plunge from space to reach the ground.

- *How are meteorites categorized?* Meteorites fall into two major categories: Primitive meteorites are remnants from the solar nebula, and processed meteorites are fragments of larger objects that underwent differentiation. Primitive meteorites may be either stony or carbon-rich. Processed meteorites may be either metallic like a planet's core or rocky like a planet's crust or mantle.

- *Why do meteorites differ from one another?* Whether a primitive meteorite is stony or carbon-rich depends on where it formed. Carbon-rich meteorites formed farther from the Sun (beyond about 3 AU), where it was cool enough for carbon compounds to condense. Processed meteorites are pieces of shattered larger objects. Metallic processed meteorites come from the cores of these objects, and rocky processed meteorites come from the mantles or crusts of these objects.

13.4 Comets

- *What are comets made of?* Comets are made of ice mixed with rocky dust, giving them a "dirty snowball" composition.

- *What happens to a comet as it approaches the Sun?* The nucleus—all the comet consists of when it is far away and frozen—heats up, and gases begin to sublimate from its surface. Escaping gases carry along some dust. The gas and dust form a coma and tails: a plasma tail of ionized gas and a dust tail.

- *How do we know that vast numbers of comets reside in the Oort cloud and Kuiper belt?* By calculating orbits of comets that we can see in the inner solar system, we learn that they come from the Kuiper belt or the Oort cloud. Based on the number of comets we have seen in the relatively short time we have been observing, we conclude that the Oort cloud and the Kuiper belt must contain enormous numbers of comets.

13.5 Pluto: Lone Dog, or Part of a Pack?

- *Why don't Pluto and Neptune collide?* A stable orbital resonance ensures that Pluto always remains a safe distance from Neptune, even though it is sometimes closer to the Sun than is Neptune.

- *What is surprising about Pluto's density, and how might Pluto have come to be?* Pluto's density is slightly higher than expected for material that condensed in the outer solar system. It may be the result of a giant impact that blasted away Pluto's low-density outer layers and led to the formation of its moon, Charon.

- *Why do we think Pluto is a Kuiper belt comet?* Pluto's composition and orbit are more similar to those of Kuiper belt comets than of other planets. Its size is not much bigger than that of other known Kuiper belt comets, and it is smaller than one object that almost certainly once roamed the Kuiper belt—Neptune's moon Triton.

13.6 Cosmic Collisions: Small Bodies Versus the Planets

- *What happened to Jupiter in 1994?* It was struck by a string of fragments of comet Shoemaker–Levy 9. Such impacts probably occur on Jupiter about once every 1,000 years, on average.

- *How often do small particles impact Earth?* An estimated 25 million small particles create meteors each day. Even more hit Earth during meteor showers.

- *Why do we think the dinosaurs were driven extinct by an impact?* Careful analysis of sediments from the time of the dinosaurs' demise shows that a major impact occurred at that time. An impact crater dating from that time has been found, and we have a plausible scenario to describe how the impact might have led to the mass extinction.

- *Do future impacts pose a real threat to our civilization?* The probability of a major impact in our lifetimes is very low, but not zero. The threat is still being assessed.

- *What role do the jovian planets play in the impact threat?* Gravitational influences of the jovian planets have shaped the asteroid belt, Kuiper belt, and Oort cloud. Thus, impacts of asteroids and comets are always linked in at least some way to past influences of the jovian planets.

? Surprising Discoveries?

Suppose we made the discoveries described below. (These are *not* real discoveries.) Decide whether each discovery should be considered reasonable or surprising. Explain.

1. A small asteroid that orbits within the asteroid belt has an active volcano.

2. Scientists discover a meteorite that, based on radiometric dating, is 7.9 billion years old.

3. An object that resembles a comet in size and composition is discovered to be orbiting in the inner solar system.

4. Studies of a large object in the Kuiper belt reveal that it is made almost entirely of rocky (as opposed to icy) material.

5. Astronomers discover a previously unknown comet that will produce a spectacular display in Earth's skies about 2 years from now.

6. A mission to Pluto finds that it has lakes of liquid water on its surface.

7. Geologists discover a crater from a 5-km object that impacted Earth more than 100 million years ago.

8. Archaeologists learn that the fall of ancient Rome was caused in large part by an asteroid impact in Asia.

Problems

9. *Clues from the Leftovers.* Why are comets, asteroids, and meteorites so useful to our understanding of the history of the solar system?

10. *Space Rock or Earth Rock?* How can we distinguish a meteorite from a terrestrial rock? Why do we find so many meteorites in Antarctica?

11. *Meteor Showers.* Explain how meteor showers are linked to comets. When is the best time of night to see meteor showers in general, and why?

12. *Know the Odds.* How often should we expect impacts of various sizes on Earth? How serious a threat do we face from impacts?

13. *Orbital Resonances.* Suppose orbital resonances had never been important anywhere in our solar system—for example, suppose Jupiter and the asteroids continued to orbit the Sun but did not affect each other through resonances. Describe at least five ways in which our solar system would be different. Which differences do you consider superficial, and which differences are more profound? Consider the effects of orbital resonances discussed both in this chapter and in Chapter 12.

14. *Life Story of an Iron Atom.* Imagine that you are an iron atom in a processed meteorite made mostly of iron that has recently fallen to Earth. Tell the story of how you got here, beginning from the time you were part of the gas in the solar nebula 4.6 billion years ago. Include as much detail as possible. Your story should be scientifically accurate but also creative and interesting.

15. *Asteroid Discovery.* You have discovered two new asteroids and named them Barkley and Jordan. Both lie at the same distance from Earth and have the same brightness when you look at them through your telescope, but Barkley is twice as bright as Jordan at infrared wavelengths. What can you deduce about the relative reflectivities and sizes of the two asteroids? Which would make a better target for a mission to mine metal? Which would make a better target for a mission to obtain a sample of a carbon-rich planetesimal? Explain.

*16. *Impact Energies.* A relatively small impact crater 20 km in diameter could be made by a comet 2 km in diameter traveling at 30 km/s (30,000 m/s).

   a. Assume that the comet has a total mass of 4.2×10^{12} kg. What is its total kinetic energy? (*Hint:* Use the formula for kinetic energy from Chapter 4; if you use mass in kg and velocity in m/s, the answer for kinetic energy will have units of joules.)

   b. Convert your answer from part (a) to an equivalent in megatons of TNT, the unit used for nuclear bombs. Comment on the degree of devastation the impact of such a comet could cause if it struck a populated region on Earth. (*Hint:* One megaton of TNT releases 4.2×10^{15} joules of energy.)

*17. *The "Near Miss" of Toutatis.* The 5-km asteroid Toutatis passed a mere 3 million km from Earth in 1992. Suppose Toutatis was destined to pass *somewhere* within 3 million km of Earth. Calculate the probability that this "somewhere" would have meant that it slammed into Earth. Based on your result, do you think it is fair to call the 1992 passage a "near miss"? Explain. (*Hint:* You can calculate the probability by considering an imaginary dartboard of radius 3 million km in which the bull's-eye has Earth's radius, 6,378 km.)

*18. *Comet Temperatures.* Find the "no greenhouse" temperatures for a comet at distances from the Sun of 50,000 AU (in the Oort cloud), 3 AU, and 1 AU (see Mathematical Insight 11.1). Assume that the comet reflects 3% of the incoming sunlight. At which location will the temperature be high enough for water ice to sublime (about 150 K)? How do your results explain comet anatomy? Explain.

19. *Project: Tracking a Meteor Shower.* Armed with an expendable star chart and a flashlight covered in red plastic, set yourself up comfortably to watch one of the meteor showers listed in Table 13.1. Each time you see a meteor, record its path on your star chart. Record at least a few dozen meteors, and try to determine the *radiant* of the shower—that is, the point in the sky from which the meteors appear to radiate. Does the meteor shower live up to its name?

20. *Project: Dirty Snowballs.* If there is snow where you live or study, make a dirty snowball. (The ice chunks that form behind tires work well.) How much dirt does it take to darken snow? Find out by allowing your dirty snowball to melt and measuring the approximate proportions of water and dirt afterward.

Discussion Questions

21. *Rise of the Mammals.* Suppose the impact 65 million years ago had not occurred. How do you think our planet would be different? For example, do you think that mammals still would eventually have come to dominate Earth? Would we be here? Defend your opinions.

22. *Could an Impact Cause a Nuclear War?* Imagine that an impact like the 1908 Tunguska impact occurred over a nuclear-armed country, such as Russia, India, Pakistan, or the United States. At first, the impact would look much like the result of a nuclear attack. Some military experts worry that in the panic of the moment a leader might order a "retaliatory" nuclear attack before learning that the explosion was caused by an impact rather than a bomb. Do you think an impact could start a nuclear war? What could we do to prevent this danger?

23. *Quantifying Impact Risk.* We face many hazards, and mitigating any of them can be expensive. One way to make decisions about risk management is to quantify risks, but impacts pose a difficulty because they are so rare. For example, the probability of a major impact that could kill half the world's population, or some 3 billion people, is estimated to be about 0.0000001 (10^{-7}) in any given year. If we multiply this low probability by the 3 billion people who would be killed, we find that the "average" risk from an impact is $10^{-7} \times (3 \times 10^9)$ or 300 people killed per year—about the same as the average number of people killed in commercial airline crashes. Do you think this is a valid way to assess the impact threat? Should we spend as much money to alleviate the impact threat as we spend on airline safety? Defend your opinions.

For a complete list of media resources available, go to www.astronomyplace.com and choose Chapter 13 from the pull-down menu.

 ## Astronomy Place Web Tutorials

Tutorial Review of Key Concepts

Use the following interactive **Tutorial** at www.astronomyplace.com to review key concepts from this chapter.

Formation of the Solar System Tutorial

Lesson 3 Formation of Planets

Supplementary Tutorial Exercises

Use the interactive **Tutorial Lesson** to explore the following questions.

Formation of the Solar System Tutorial, Lesson 3

1. Where were icy planetesimals able to form in the solar nebula?

2. Pluto is the smallest planet. Why isn't it made of metal and rock like the small terrestrial planets?

3. Pluto is in the outer solar system. Why didn't it capture hydrogen and helium gas like the jovian planets?

4. Suppose the solar wind had not cleared the solar nebula until much later in the history of the solar system. Do you think Pluto would have been different? Explain.

 ### Exploring the Sky and Solar System

Of the many activities available on the *Voyager: SkyGazer* CD-ROM accompanying your book, use the following files to observe key phenomena covered in this chapter.

Go to the **File: Basics** folder for the following demonstrations.

1. Orbit of Hale–Bopp

2. Pluto's Orbit

Go to the **File: Demo** folder for the following demonstrations:

1. Hale–Bopp Path

2. Hyakutake at Perihelion

3. Hyakutake Nears Earth

4. Pluto's Orbit

Movies

Check out the following narrated and animated short documentaries available on www.astronomyplace.com for a helpful review of key ideas covered in this chapter.

 History of the Solar System Movie

 Orbits in the Solar System Movie

Web Projects

Take advantage of the useful Web links on www.astronomyplace.com to assist you with the following projects.

1. *The NEAR Mission.* Learn about the NEAR mission to the asteroid Eros. How did the mission reach Eros? How did the mission end? What did it accomplish? Write a one to two-page summary of your findings.

2. *Stardust.* Learn about NASA's Stardust mission (launched in 1999) to return comet material to Earth. Write a short report about the mission status and its science.

3. *Asteroid and Comet Missions.* Learn about another proposed space mission to study asteroids or comets. For the mission you chose, write a short report about the mission plans, goals, and prospects for success.

4. *The New Horizons Mission to Pluto.* The New Horizons mission to visit Pluto is scheduled for launch in 2006. Find out the current status of the mission. What are its science goals? Summarize your findings in a few paragraphs.

14 Planet Earth
Seen in a New Light

Looking outward to the blackness of space, sprinkled with the glory of a universe of lights, I saw majesty—but no welcome. Below was a welcoming planet. There, contained in the thin, moving, incredibly fragile shell of the biosphere is everything that is dear to you, all the human drama and comedy. That's where life is; that's where all the good stuff is.

Loren Acton, U.S. astronaut

Perhaps you've heard Earth described as the "third rock from the Sun." At first glance, the perspective of comparative planetology may seem to support this dispassionate view of Earth. However, a deeper comparison between Earth and its neighbors reveals a far more remarkable planet.

Earth is the only planet with surface oceans of liquid water and substantial amounts of oxygen in its atmosphere. Its geological activity is more diverse than that of any other terrestrial world, and it is the only planet on which plate tectonics continually re-shapes the surface. Most important, Earth is the only world known to harbor life.

In this chapter, we will build upon what we've learned about other worlds to see our planet with new insight and understanding. We will explore how and why Earth is unique in our solar system. We will learn why Earth became suitable for life and how the origin and evolution of life have affected our planet's surface and atmosphere. Finally, we will consider a few potential threats to our planet's continued habitability for human life. We will find that our very survival may depend on how well we understand and respond to the lessons we've learned from our neighboring worlds.

14.1 How Is Earth Different?

Take a look at Figure 14.1. The abundant white clouds overlying an expanse of ocean show a planet unlike any other in our solar system. It is, of course, our own planet Earth. We are now ready to study it using the tools of comparative planetology we've developed in the preceding six chapters.

Figure 14.1 Dawn over the Atlantic Ocean, photographed from the Space Shuttle.

We have already discussed many features of Earth's geology and atmosphere, finding many similarities between our planet and others. For example, Earth is shaped by the same four geological processes as other planets: impact cratering, volcanism, tectonics, and erosion [Section 10.3]. Impact cratering has affected Earth at least as much as other worlds, though most evidence of past cratering has been erased. Still, more than 100 large impact craters have been identified on Earth (see Figure 13.20). Volcanism affects Earth regularly, with major volcanic eruptions every year. Erosion is rampant on Earth because of abundant rainfall, winds, and glaciation. Tectonics, which generally goes hand in hand with volcanism, probably has shaped Earth on a global scale more than any other single process. However, while Earth has many tectonic features similar to those seen on other worlds, it also has plate tectonics—a type of tectonics we have not found elsewhere.

Earth's atmosphere also has similarities with the atmospheres of other worlds, especially Venus and Mars. Like those atmospheres, ours originated from volcanic outgassing. Our atmosphere's basic structure is also similar to the atmospheres of other worlds, with a troposphere warmed by the greenhouse effect and a thermosphere heated by solar X rays (see Figure 11.8). However, Earth is the only planet with significant atmospheric oxygen and the only terrestrial world with an ultraviolet-absorbing stratosphere.

Earth is the only world in our solar system with temperature and pressure conditions that allow surface oceans of liquid water. Moreover, Earth's climate has remained stable enough for the oceans to persist for billions of years. In this respect, Earth is very different from Venus and Mars. Mars almost certainly had flowing water in the distant past [Section 10.5], and Venus could have had oceans early in its history [Section 11.6], but both planets underwent dramatic global climate change.

Water plays such an important role on Earth that some scientists treat it as a distinct planetary layer (called the *hydrosphere*) between the lithosphere and the atmosphere. Oceans cover nearly three-fourths of Earth's surface, with an average depth of about 3 kilometers (1.8 miles). On land, water flows through streams and rivers, fills lakes and underground water tables, and sometimes lies frozen in glaciers. Frozen water also creates the polar ice caps. The northern ice cap sits atop the Arctic Ocean and covers the large island of Greenland. The southern ice cap covers nearly the entire continent of Antarctica, with an average thickness of 2 kilometers (1.2 miles). The Antarctic ice sheet contains so much water that melting it would cause sea level to rise by more than 70 meters (220 feet).

THINK ABOUT IT

The vast majority of Earth's land lies well above sea level and is not at risk of flooding, even if all polar ice melted. Nevertheless, rising sea level could be of great concern to humans. Why? (*Hint:* Locate the largest cities of the world.)

Perhaps the most unique feature of Earth is its life. Even if we someday find life on Mars, Europa, or some other world in our solar system, Earth will still be unique in the abundance and diversity of its life. We find life nearly everywhere on Earth's surface, throughout the oceans, and even underground. (This layer of life on Earth is sometimes called the *biosphere*.) As we will see shortly, life helps shape many of Earth's physical characteristics and has radically altered the content of Earth's atmosphere.

In summary, we have identified five major ways in which Earth differs from other worlds in our solar system:

- **Plate tectonics**: Earth is the only planet with a surface shaped largely by this distinctive type of tectonics.

- **Atmospheric oxygen**: Earth is the only planet with significant oxygen in its atmosphere.

- **Surface liquid water**: Earth is the only planet on which temperature and pressure conditions allow surface water to be stable as a liquid.

- **Climate stability**: Earth differs from the other terrestrial worlds with significant atmospheres (Venus and Mars) in having a climate that has remained relatively stable throughout its history.

- **Life**: Earth is the only world known to have life, and it certainly has the most abundant and diverse life in our solar system.

We will devote most of the rest of this chapter to understanding these unique features of our planet. We will find that they do not exist in isolation but rather are closely tied to one another. For example, we'll see that atmospheric oxygen is a product of life, that life has thrived because of liquid water and climate stability, and that climate stability has been made possible in large part by the action of plate tectonics. Let's begin by examining Earth's unique geology of plate tectonics.

 Shaping Planetary Surfaces Tutorial, Lessons 1–3

14.2 Our Unique Geology

Most aspects of Earth's geology are unsurprising in light of what we know about other terrestrial worlds. Earth's size—largest of the five terrestrial worlds—explains why our planet remains geologically active. Radioactive decay has kept the interior hot for billions of years, supplying heat for ongoing volcanism and tectonics. The Moon and Mercury lack such activity because their small size has allowed them to cool. Mars, if it has any remaining active volcanism or tectonics, is clearly past its geological prime. Venus, similar in size to Earth, shows evidence of widespread geological activity. None of the other terrestrial worlds show evidence of plate tectonics, however.

In this section, we'll explore the nature of plate tectonics on Earth. This background will lay the groundwork for later sections, in which we'll explore the role of plate tectonics in Earth's climate stability and consider whether plate tectonics should be rare or common on planets around other stars.

The Theory of Plate Tectonics

The term *plate tectonics* refers to the scientific theory that explains much of Earth's surface geology as a result of the slow motion of *plates*—fractured pieces of the lithosphere. According to the theory, the lithosphere fractured because of stresses generated by underlying mantle convection [Section 10.2]. The plates essentially "float" over the mantle, gradually moving over, under, and around each other as convection moves Earth's interior rock. (Because the term *plate tectonics* refers to the theory, it is generally considered to be singular rather than plural, despite the *s* on *tectonics*.)

Earth's lithosphere is broken into more than a dozen plates, with known boundaries shown in Figure 14.2. On average, the plates move slowly and gradually. By monitoring positions on either side of a plate boundary with the

Figure 14.2 This relief map shows plate boundaries (solid lines), with arrows to represent directions of plate motion. Color represents elevation, progressing from blue (lowest) to red (highest). Labels identify some of the geological features discussed in the text.

global positioning system (GPS) [Section S1.6], geologists can now measure the motion of plates directly. Typical speeds are only a few centimeters per year— about the same speed at which your fingernails grow.

Development of the Theory of Plate Tectonics The theory of plate tectonics developed slowly during the twentieth century. We usually trace its beginning to an older idea proposed in 1912 by German meteorologist and geologist Alfred Wegener: *continental drift*, the idea that continents gradually drift across the surface of the Earth. Wegener got his idea in part from the puzzlelike fit of continents such as South America and Africa (Figure 14.3). In addition, he noted that similar types of distinctive rocks and rare fossils were found in eastern South America and western Africa, suggesting that these two regions once had been close together.

Despite these strong hints, no one at the time knew of a mechanism that could allow the continents to push their way through the solid rock beneath them. Wegener suggested that Earth's gravity and tidal forces from the Sun and Moon were responsible, but other scientists quickly showed that these forces were too weak to move continents around. As a result, Wegener's idea of continental drift was widely rejected by geologists for decades after he proposed it, even though his evidence of a "continental fit" for Africa and South America ultimately proved correct.

In the mid-1950s, scientists began to observe geological features that suggested a mechanism for continental motion. In particular, scientists discovered *mid-ocean ridges* along

Figure 14.3 The puzzlelike fit of South America and Africa.

which mantle material erupts onto the ocean floor, pushing apart the existing seafloor on either side. This **seafloor spreading** helped explain how the continents could move apart with time. Today, we understand that continents move in concert with underlying mantle convection, driven by the heat released from Earth's interior. Because this idea is quite different from Wegener's original notion of continents plowing through the solid rock beneath them, geologists no longer use the term "continental drift" and instead consider continental motion within the context of plate tectonics.

Seafloor Crust and Continental Crust In addition to the evidence of past continental motion and ongoing seafloor spreading, a third line of evidence shows the workings of plate tectonics: Earth's seafloors and continents are made up of two very distinct types of crust (Figure 14.4).

Seafloor crust is made of relatively high density *basalt*, which commonly erupts from volcanoes like those along mid-ocean ridges and in Hawaii (and probably from volcanoes on all the terrestrial worlds). Seafloor crust is typically only 5–10 kilometers thick, and radiometric dating shows that it is quite young—usually less than 200 million years old. Further evidence of the young age of seafloors comes from studies of impact craters. Large impacts should occur more or less uniformly over Earth's surface, and the oceans won't stop a large asteroid or comet from making a huge crater. However, we find far fewer large craters on the seafloor than on the continents, which means that seafloor craters must have been erased more recently.

Continental crust is made of lower-density rock (such as granite), and most of it is much older than the rock found in seafloor crust. Some rocks in the continental crust date to as old as 4.0 billion years. Continental crust is much thicker than seafloor crust—typically between 20 and 70 kilometers thick—but it sticks up only slightly higher because its sheer weight presses it down farther onto the mantle below. Unlike the high-density basalt of seafloor

seafloor crust (5–10 km thick, dense, young)

upper mantle

continental crust (20–70 km thick, lower-density, old)

Figure 14.4 Earth has two distinct kinds of crust.

crust, the low-density rock of continental crust appears to be uncommon on the other terrestrial worlds.

The two types of crust make it clear that Earth's surface undergoes continual change. New seafloor crust must continually emerge at sites of seafloor spreading, while continental crust gradually builds up over time.

The Conveyor-like Action of Plate Tectonics How did Earth come to have two types of crust? The answer lies in the way surface rock is recycled by plate tectonics. Over millions of years, the movements involved in plate tectonics act like a giant conveyor belt for Earth's lithosphere (Figure 14.5).

Mid-ocean ridges occur at places where mantle material rises upward, creating new seafloor crust and pushing plates apart. The newly formed basaltic crust cools and

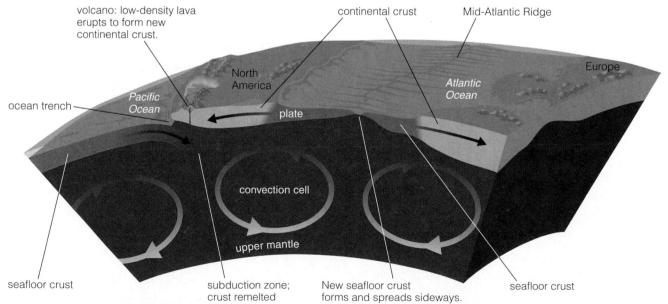

volcano: low-density lava erupts to form new continental crust.

continental crust

Mid-Atlantic Ridge

Europe

North America

Pacific Ocean

Atlantic Ocean

ocean trench

plate

convection cell

upper mantle

seafloor crust

subduction zone; crust remelted

New seafloor crust forms and spreads sideways.

seafloor crust

Figure 14.5 Plate tectonics acts like a giant conveyor belt that produces and recycles seafloor crust. New seafloor crust erupts from the mantle at mid-ocean ridges, where plates spread apart. At ocean trenches, the seafloor crust pushes under the less dense continental crust, returning the seafloor crust to the mantle. The subducting seafloor crust may partially melt, leading to volcanic eruptions that produce new continental crust.

contracts as it spreads sideways from the central ridge, giving seafloor spreading regions their characteristic ridged shape (see Figure 14.2). Worldwide along the mid-ocean ridges, new crust covers an area of about 2 square kilometers every year, enough to replace the entire seafloor within a geologically short time of about 200 million years.

Meanwhile, as new seafloor crust spreads over the surface, older crust is recycled into the mantle. This recycling occurs at deep ocean trenches, where the ocean depth can reach 8 kilometers (5 miles) or more. In these trenches, the dense seafloor crust of one plate pushes under the less dense continental crust of another plate in a process called **subduction**. Thus, seafloor crust remains on the surface only as long as it takes to move across the seafloor from a mid-ocean ridge to its subduction in a trench. This explains why seafloor crust is always geologically young.

Continental crust is produced by volcanoes near subduction zones. As the seafloor crust descends into the mantle, it is heated and may partially melt. The lowest-density material tends to melt first and then erupt from volcanoes, producing the lower-density rock that becomes new continental crust. Thus, volcanoes tend to be found along plate boundaries near subduction zones, and continental crust is less dense than seafloor crust. Moreover, because continental crust does not get recycled and is continually being produced by volcanoes, the continents are slowly growing

over time. Billions of years ago, Earth was even more of an ocean planet than it is today.

The conveyor-like process of plate tectonics undoubtedly is driven by the heat flow from mantle convection, although the precise relationship between the convection cells and the plates remains an active topic of research. For example, it is not clear whether mantle convection simply pushes plates apart at the sites of seafloor spreading or whether the plates are denser than the underlying material and pull themselves down at subduction zones.

Geological Features of Plate Tectonics

Mid-ocean ridges and deep-sea trenches are only the beginning of the geological features attributable to plate tectonics on Earth. Nearly all of Earth's major geological features have been shaped at least in part by plate tectonics.

Building Up Continents The present-day continents have been built up over billions of years by the ongoing production of continental crust. The continental crust does not simply pile up, however. Instead, the continents are continually reshaped by the volcanism and stresses associated with plate tectonics, as well as by erosion.

Figure 14.6 shows some of the complex geological history of North America. One of the youngest features

volcanic and tectonic mountains made over subduction zones

sediments deposited in ancient seas, later turned to rock

ancient continental crust, heavily eroded

Figure 14.6 The major geological features of North America record the complex history of plate tectonics. Only the basic processes behind the largest features are shown here.

former Pacific islands scraped onto continent during subduction

deep sedimentary layers formed by erosion of continent

mountains formed by repeated collisions with other continents is a string of islands that is not even part of the continent. Alaska's Aleutian islands are located over a region where one seafloor plate is subducting under another (see the plate boundaries in Figure 14.2). The subduction generates tremendous volcanic activity, and the Aleutians are a string of volcanoes that have grown tall enough to rise above sea level. As the process continues in the future, the islands will grow and merge. The same process has shaped Japan and the Philippines, each of which once contained many small islands that merged into the fewer islands we see today. As these islands continue to grow and merge, they may eventually create a new continent or merge with an existing one.

The west coast of North America was shaped by islands that merged into the existing continent. Alaska, British Columbia, Washington, Oregon, and most of California began as numerous volcanic islands in the Pacific. As the Pacific seafloor plate subducted under the continental plate, the islands essentially were scraped off the seafloor and stuck onto the continent. The subducting seafloor itself descended into the mantle. The islands remained "floating" above the mantle because they are made of lower-density continental crust.

THINK ABOUT IT

Find the islands of Indonesia on a map, and compare their location to the plate boundaries shown in Figure 14.2. What do you think is happening at the plate boundary near Indonesia? What do you think Indonesia looked like in the past, and what do you expect to happen to it in the future?

Other portions of North America have been shaped by interactions between plate tectonics and erosion. The Great Plains and the Midwest once were ancient seas filled with sediment that later turned to rock. The deep south formed from the buildup of sediments eroded from other parts of the continent and carried to these regions. Northeastern Canada features some of the oldest continental crust on Earth, worn down over time by erosion.

Tectonic Stresses on Continents Stresses associated with plate tectonics also play a major role in shaping continents. Partially molten granite may push upward on the continental rock over subduction zones. The rock gradually bulges upward, forming a mountain range. The Sierra Nevada range in California was pushed up in this way. Later, the upper layers of sedimentary rock [Section 14.5] eroded away to leave the granite exposed, which is why the Sierra Nevada is made largely of granite.

Mountain ranges may also be created when two continent-bearing plates collide. Because both colliding plates are made of low-density continental crust, neither plate can subduct under the other. The two plates therefore push against each other, and the resulting pressure can create tremendous mountain ranges. This process is currently

Figure 14.7 This satellite photo shows the Himalayas, which are still slowly growing as the plate carrying India pushes into the Eurasian plate. The arrow indicates the direction of the plate motion.

at work in building the Himalayas, located where a plate carrying the Indian subcontinent rammed into the plate carrying Eurasia (Figure 14.7). The Himalayas, already the tallest mountains on Earth, are still growing.

The Appalachian range in the eastern United States formed by the same process of colliding continental plates, repeated several times. The geological evidence shows that over a period of a few hundred million years North America collided twice with South America and then with western Africa. The Appalachians probably were once as tall as the Himalayas are now. Erosion has gradually transformed them into the fairly modest mountain range we see today. Similar processes contributed to the formation of the Rocky Mountains in the United States and Canada.

In places where continental plates are pulling apart, the crust thins and can create a large *rift valley*. The East African rift zone is an example (see Figure 14.2). This rift is slowly growing and will eventually tear the African continent apart. At that point, rock rising upward with mantle convection will begin to erupt from the valley floor, creating a new zone of seafloor spreading. A similar process tore

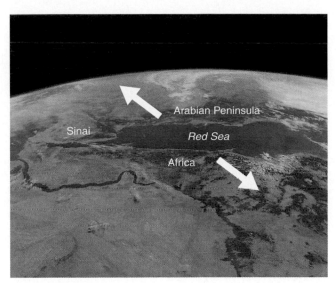

Figure 14.8 When continental plates pull apart, the crust thins and deep rift valleys form. This process tore the Arabian peninsula from Africa, forming the Red Sea. The arrows indicate the directions of the plate motions.

Figure 14.9 Along California's San Andreas fault (and many other places in the world), plates are sliding sideways past each other. Asterisks indicate recent earthquakes.

the Arabian Peninsula from Africa, creating the Red Sea (Figure 14.8).

At other places on Earth, plates slip sideways relative to each other along a **fault**—a fracture in the lithosphere. The San Andreas fault in California marks a line where the Pacific plate is moving northward relative to the continental plate of North America (Figure 14.9). In about 20 million years, this motion will bring Los Angeles and San Francisco together. The two plates do not slip smoothly against each other. Instead, their rough surfaces catch against each other. Tension builds up until it is so great it forces a rapid and violent shift, causing an earthquake. This friction between plates explains why earthquakes are so common along the San Andreas fault and other faults around the world.

Study the plate boundaries in Figure 14.2. Based on this diagram, explain why the west coast states of California, Oregon, and Washington are prone to more earthquakes and volcanoes than other parts of the United States. Find the locations of recent earthquakes and volcanic eruptions in the news. Do the locations fit the pattern depicted here?

The motions associated with earthquakes can raise mountains, level cities, and set the whole planet vibrating with seismic waves (Figure 14.10). In contrast to the usual motion of plates, which proceeds at a few centimeters per year, an earthquake can move plates by several *meters* in a few seconds. Although most earthquake faults lie along plate boundaries, some more ancient faults are found elsewhere. As a result, devastating earthquakes occasionally occur in regions that we tend to think of as being at low risk

from seismic events. For example, three of the largest earthquakes in U.S. history occurred in Missouri in 1811–1812.

Hot Spots

Not all volcanoes occur near plate boundaries. Sometimes, a plume of hot mantle material may rise in what we call a **hot spot**. The Hawaiian islands are the result of a hot spot that has been erupting basaltic lava for tens of millions of years, forming broad shield volcanoes [Section 10.3]. Plate tectonics gradually carries the Pacific plate over the hot spot, forming a chain of islands as different parts of the plate lie directly above the hot spot at different times (Figure 14.11). If plate tectonics were not moving the plate relative to the hot spot, a single huge volcano would have formed—perhaps looking somewhat like Olympus Mons on Mars (see Figure 10.25).

Today, most of the lava erupts on the "Big Island" of Hawaii, giving much of this island a young, rocky surface. About a million years ago, the Pacific plate lay farther to the southeast (relative to its current location), and the hot spot built the island of Maui. Before that, the hot spot created other islands, including Oahu (3 million years ago), Kauai (5 million years ago), and Midway (27 million years ago). The older islands are more heavily eroded. Midway has eroded so much that it barely rises above sea level.

The movement of the plate over the hot spot continues today, building underwater volcanoes that eventually will rise above sea level to become new Hawaiian islands. The growth of a future island, named Loihi, is already well under way—prime beach real estate should be available there in about a million years or so.

a This photo shows a place along the San Andreas fault where a road allows us to see how far the two sides of the fault moved in an earthquake. Note the break in the white lines running horizontally near the center, which were lined up prior to the earthquake.

b This photo shows damage from the 1995 earthquake in Kobe, Japan.

Figure 14.10 Plate tectonics on a human level. Earthquakes occur when plates slip violently along a fault.

Aleutian Trench

Kauai: heavily eroded valleys

Hawaii: recent lava flows

Midway: island eroded down to sea level

Loihi: future Hawaiian island (in about a million years)

motion of Pacific plate relative to mantle plume

Hawaiian islands

Figure 14.11 The Hawaiian islands are just the most recent of a very long string of volcanic islands made by a mantle hot spot. The black-and-white image of Loihi (lower right) was obtained by sonar, as it is still entirely underwater.

Figure 14.12 The geysers and hot springs of Yellowstone National Park get their heat from a hot spot underlying the location. This photo shows the Castle Geyser spraying liquid on the Upper Geyser Basin. The vent is about 2 meters across.

Hot spots can also occur beneath continental crust. For example, the geysers and hot springs of Yellowstone National Park are produced by a hot spot (Figure 14.12). The continental plate is moving in a southwestern direction over this hot spot.

THINK ABOUT IT

Find your hometown or current home on the map in Figure 14.2 or Figure 14.6 (or both). Based on what you have learned in this section, describe the changes your home has undergone over millions (or billions) of years.

Plate Tectonics Through Time

We can use the current motions of the plates to project the arrangement of continents millions of years into the past or the future. For example, at a speed of 2 centimeters per year, a plate will travel 2,000 kilometers in 100 million years. Figure 14.13 shows several past arrangements of the continents, along with one future arrangement. Note that about 200 million years ago the present-day continents were together in a single "supercontinent," sometimes called *Pangaea* (which means "all lands").

Mapping the sizes and locations of continents at even earlier times is more difficult. However, studies of magnetized rocks (which can record the orientation of ancient magnetic fields) and comparisons of fossils found in different places around the world have allowed geologists to map the movement of the continents much farther into the past. It seems that, over the past billion years or more, the continents have slammed together, pulled apart, spun around, and changed places on the globe. Central Africa once lay at Earth's South Pole, and Antarctica once was near the equator. The continents continue to move, and their current arrangement is no more permanent than any past arrangement.

 Surface Temperature of Terrestrial Planets Tutorial

14.3 Our Unique Atmosphere and Oceans

In Chapter 11, we learned that the major physical differences in the atmospheres of the terrestrial planets are easy to understand. For example, the Moon and Mercury lack substantial atmospheres because of their small size, and a very strong greenhouse effect causes the high surface temperature on Venus. We also learned how solar heating, seasonal changes, and the Coriolis effect that comes from rotation lead to differing global circulation patterns on different worlds.

Explaining differing atmospheric compositions is more challenging, especially when we consider the cases of Venus, Earth, and Mars. In all three cases, outgassing released the same gases—primarily water, carbon dioxide, and nitrogen. How, then, did Earth's atmosphere end up so different? We can break down this general question into four separate questions:

1. Why did Earth retain most of its outgassed water— enough to form vast oceans—while Venus and Mars lost theirs?

2. Why does Earth have so little carbon dioxide (CO_2) in its atmosphere, when Earth should have outgassed about as much of it as Venus?

3. Why does Earth have so much more oxygen (O_2) than Venus or Mars?

4. Why does Earth have an ultraviolet-absorbing stratosphere, while Venus and Mars do not?

To answer these questions, we must look at the history of Earth's atmosphere. The answers to all four questions turn out to be closely connected.

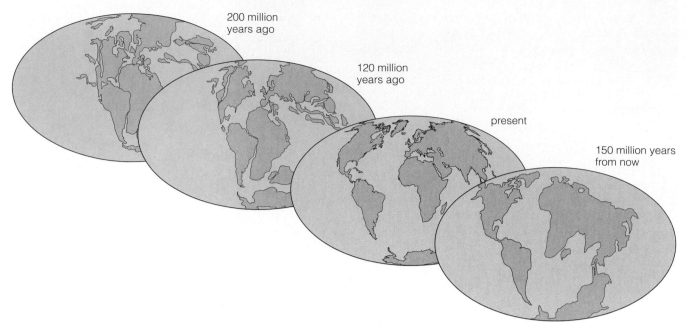

200 million
years ago

120 million
years ago

present

150 million years
from now

The Mediterranean Sea will become mountains, Australia will merge
with Antarctica, and California will slide northward to Alaska.

Figure 14.13 Past, present, and future arrangements of Earth's continents. The present continents were
all combined into a single "supercontinent" about 200 million years ago.

Why Earth Has Oceans

Venus, Earth, and Mars probably all outgassed plenty of
water vapor in their early histories. Four billion years ago,
all three planets may have had rainfall and surface water.
Evidence from tiny mineral grains (known as zircons)
suggests that Earth had oceans as early as 4.3–4.4 billion
years ago. Mars shows plenty of evidence of flowing water
before about 3 billion years ago. We may never know whether
Venus once had oceans. Early oceans on Venus certainly
seem plausible given that the young Sun was some 30%
fainter than the Sun is today and Venus may have been much
cooler at that time [Section 11.6].

Earth's oceans have persisted since their formation,
while Venus and Mars have lost whatever surface water they
once had. As we discussed in Chapter 11, Venus lost its
water as the runaway greenhouse effect set in. Water vapor
rose high in the atmosphere, where solar ultraviolet photons split apart the molecules. The hydrogen then escaped
to space, and the water molecules could not re-form. Venus
may well have lost an "ocean-full" of water in this way. Mars
probably lost some of its water in a similar way. The rest
is frozen at the polar caps and under the surface. For Mars,
the problem was the opposite of that for Venus. The greenhouse effect, instead of strengthening, weakened until the
planet froze.

In essence, Venus lost its water because the greenhouse
effect grew too strong, Mars lost its water because the greenhouse effect grew too weak, and Earth retained its water
because the greenhouse effect stayed "just right." How did
the greenhouse effect remain so moderate on Earth? We
will address that question next.

Where Is All the CO_2?

The simple answer to how the greenhouse effect remains so
moderate on Earth is that our atmosphere contains just the
right, small amount of the greenhouse gas carbon dioxide.
However, the low level of carbon dioxide is itself a surprise,
given that Earth should have outgassed about as much carbon dioxide throughout its history as Venus. Careful study
shows that the carbon dioxide is not "missing" at all but
rather is bound up in the oceans and rocks.

About 60 times as much carbon dioxide is dissolved in
the oceans as is present in the atmosphere, and far more is
trapped in rocks. These rocks can form only by chemical
reactions involving carbon dioxide dissolved in liquid water.
The process continuously pulls carbon dioxide out of the
atmosphere into the oceans and ultimately into rocks on
the seafloor. The total amount of carbon dioxide trapped in
the oceans and in rocks is some 170,000 times the amount
in our atmosphere today. If Venus's carbon dioxide could
be similarly pulled into rock, Venus would be left with a
nitrogen-rich atmosphere much like Earth's.

Thus, the existence of oceans and the low level of carbon dioxide in our atmosphere are intimately linked. Earth
apparently did outgas about as much carbon dioxide as

Venus, but the long-term presence of oceans allowed the carbon dioxide to be removed from the atmosphere and turned into rock. The result is an atmosphere with just enough greenhouse warming to ensure that the oceans have neither frozen completely nor boiled away.

Oxygen, Ozone, and the Stratosphere

Unlike water and carbon dioxide, molecular oxygen (O_2) is not a product of outgassing. In fact, no geological process can explain the great abundance of oxygen (about 21%) in Earth's atmosphere. Moreover, oxygen is a highly reactive chemical that would disappear from the atmosphere in just a few million years if it were not continuously resupplied.

Fire, rust, and the discoloration of freshly cut fruits and vegetables are everyday examples of chemical reactions that remove oxygen from the atmosphere (often called *oxidation reactions*). Similar reactions between oxygen and surface materials (especially iron-bearing minerals) give rise to the reddish appearance of much of Earth's rock and clay, including the beautiful reds of Arizona's Grand Canyon. Thus, we must explain not only how oxygen got into Earth's atmosphere in the first place, but also how the amount of oxygen remains relatively steady even while chemical reactions can remove it rapidly from the atmosphere.

The answer to the oxygen mystery is *life*. Plants and many microorganisms release oxygen through photosynthesis. Photosynthesis takes in CO_2 and releases O_2. The carbon becomes incorporated into a variety of components of living organisms. Today, plants and single-celled photosynthetic organisms return oxygen to the atmosphere in approximate balance with the rate at which animals and chemical reactions consume oxygen. Thus, the oxygen content of the atmosphere stays relatively steady. Earth originally developed its oxygen atmosphere at a time when photosynthesis added oxygen at a greater rate than the rate at which these processes could remove it from the atmosphere.

Suppose that, somehow, all photosynthetic life (e.g., plants) died out. What would happen to the oxygen in our atmosphere? Could animals, including us, still survive?

Life and oxygen also explain the presence of Earth's ultraviolet-absorbing stratosphere. In the upper atmosphere, chemical reactions involving solar ultraviolet light transform some of the O_2 into molecules of O_3, or *ozone*. The O_3 molecule is more weakly bound than the O_2 molecule, which allows it to absorb solar ultraviolet energy even better. The absorption of solar energy by ozone heats the upper atmosphere, creating the layer we call the stratosphere. This ozone layer prevents harmful ultraviolet ra-

diation from reaching the surface [Section 11.3]. Mars and Venus lack photosynthetic life and therefore have too little O_2 and consequently too little ozone to form a stratosphere.

Maintaining Balance

We have answered all four basic questions about Earth's atmosphere. Venus, Earth, and Mars all began on similar paths, releasing similar gases into their atmospheres by outgassing. Only Earth, however, has had conditions "just right" to maintain liquid oceans throughout almost all of its history. The oceans helped remove carbon dioxide from the atmosphere, keeping the greenhouse effect moderate and the climate pleasantly warm. Meanwhile, the rise of photosynthetic life added oxygen to Earth's atmosphere, and life now keeps the oxygen content steady despite the highly reactive nature of this gas. In the upper atmosphere, solar ultraviolet light helps convert the oxygen to ozone, explaining why Earth has a stratosphere (Figure 14.14).

A deeper look at what we have learned still leaves us with two major mysteries. First, while we've generally explained what happened to Earth's carbon dioxide, we have not yet explained *why* the amount in the atmosphere stays "just right." For example, why didn't the chemical reactions either remove all the carbon dioxide or leave so much of it that the oceans would boil away? Second, we've identified life as the source of oxygen, but we have not explained how life came to exist and develop in a way that maintains a steady oxygen level. We will address these questions in the next two sections.

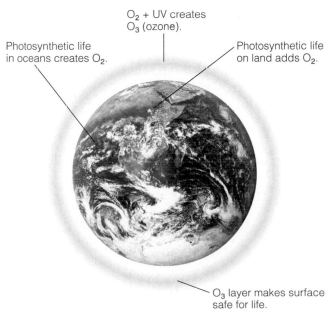

Figure 14.14 The origin of oxygen and ozone in Earth's atmosphere.

14.4 Climate Regulation and the Carbon Dioxide Cycle

Life and a stable climate are intimately connected. Nearly every planet, moon, asteroid, and comet in the solar system contains the basic chemical ingredients of life. However, abundant life is found only on Earth, and scientists believe the crucial reason is the presence of liquid water on its surface [Section 24.2]. Thus, the key to life's success on Earth has been the climate stability that has kept the oceans liquid for most of Earth's geological history. Earth has undergone significant temperature variations, but the return to moderate conditions after warming or cooling separates Earth from its neighbors whose climate change was one-way.

Why has Earth enjoyed relative climate stability while Venus overheated and Mars froze? The answer lies in an interplay among plate tectonics, the oceans, and the atmosphere that acts as a natural thermostat for our planet.

The Carbon Dioxide Cycle

The mechanism by which Earth maintains a fairly steady level of atmospheric carbon dioxide is called the **carbon dioxide cycle**, or the **CO_2 cycle** for short. (It is also sometimes called the *carbonate–silicate cycle* or simply the *carbon cycle*.) The CO_2 cycle continually moves carbon dioxide from the atmosphere to the ocean to rock and back to the atmosphere. Because it is a cycle, it has no starting or ending point. If we view the process as starting with atmospheric carbon dioxide, it proceeds as follows (Figure 14.15):

- Atmospheric carbon dioxide dissolves in the oceans.

- At the same time, rainfall erodes silicate rocks (rocks rich in silicon and oxygen) on Earth's continents and carries the eroded silicate minerals to the oceans.

- In the oceans, the silicate minerals react with dissolved carbon dioxide to form carbonate minerals. These minerals fall to the ocean floor, building up layers of carbonate rock such as limestone. (During the past half billion years or so, the carbonate minerals have been made by shell-forming sea animals, falling to the bottom in the seashells left after the animals die. Without the presence of animals, chemical reactions would do the same thing—and apparently did for most of Earth's history.)

- Over millions of years, the conveyor belt of plate tectonics carries the carbonate rocks to subduction zones, and subduction carries them down into the mantle.

- As they are pushed deeper into the mantle, some of the subducted carbonate rock melts and releases its carbon dioxide, which then erupts back into the atmosphere through volcanoes.

The CO_2 cycle maintains a balance between the supply of carbon dioxide released by volcanoes and the loss of carbon dioxide to the oceans and then to carbonate rocks. This balance can adjust itself slightly, which explains how Earth maintains a fairly stable climate.

Feedback Processes The regulatory role of the carbon dioxide cycle involves what we generally call **feedback processes**—processes in which a change in one property amplifies (*positive feedback*) or counteracts (*negative feedback*) the behavior of the rest of the system. You are probably familiar with audio feedback. If you bring a microphone too close to a loudspeaker, it picks up and amplifies small sounds from the speaker. These amplified sounds are again picked up by the microphone and further amplified, causing a loud screech. Audio feedback is an example of positive feedback. The screech usually leads to a form of negative feedback: The embarrassed person holding the microphone moves away from the loudspeaker, thereby stopping the positive audio feedback.

THINK ABOUT IT

Try to think of at least one everyday example of positive feedback and negative feedback.

The CO_2 Cycle as a Thermostat The CO_2 cycle acts as a thermostat for Earth, because the rate at which carbonate minerals form in the ocean is very sensitive to temperature. Carbonate minerals form faster at higher temperatures. Changes in the rate of carbonate mineral formation set up negative feedback processes that push Earth's temperature back toward "normal" whenever the planet heats or cools a bit (Figure 14.16):

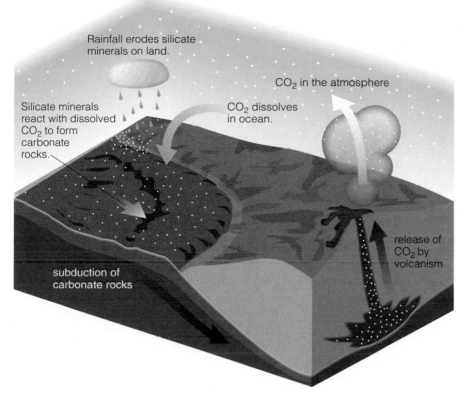

Rainfall erodes silicate minerals on land.

Silicate minerals react with dissolved CO_2 to form carbonate rocks.

CO_2 dissolves in ocean.

CO_2 in the atmosphere

release of CO_2 by volcanism

subduction of carbonate rocks

Figure 14.15 The CO_2 cycle continually moves carbon dioxide from the atmosphere to the ocean to rock and back to the atmosphere.

- If Earth warms up a bit, carbonate minerals form in the oceans at a more rapid rate. The rate at which the oceans dissolve CO_2 gas increases, thereby pulling CO_2 out of the atmosphere. The reduced atmospheric CO_2 concentration leads to a weakened greenhouse effect that counteracts the initial warming and cools the planet back down.

- If Earth cools a bit, carbonate minerals form more slowly in the oceans. The rate at which the oceans dissolve CO_2 gas decreases, allowing the CO_2 released by volcanism to build back up in the atmosphere. The increased CO_2 concentration strengthens the greenhouse effect and warms the planet back up.

too cool

too warm

Oceans dissolve less CO_2 from the atmosphere.

Fewer carbonate minerals form.

More CO_2 builds up in the atmosphere.

Earth's Thermostat: CO_2 Cycle

Strengthened greenhouse effect will warm Earth up.

Weakened greenhouse effect will cool Earth down.

Oceans dissolve more CO_2 from the atmosphere.

More carbonate minerals form.

Less CO_2 builds up in the atmosphere.

Figure 14.16 The carbon dioxide cycle acts as a thermostat for Earth through negative feedback processes. Cool temperatures cause atmospheric CO_2 to increase, and warm temperatures cause atmospheric CO_2 to decline.

Over long periods of time, these feedback mechanisms have kept Earth's climate fairly stable despite changes in the Sun's brightness, changes in the rate of volcanism (which change the output of CO_2 into the atmosphere), and other climate effects. For example, the young Earth was warm enough for liquid-water oceans despite a 30% dimmer Sun, because the atmosphere contained enough extra carbon dioxide to make the greenhouse effect strong enough to allow water to condense as rain. In fact, the greenhouse effect may have been so strong at early times that the young Earth may have been considerably warmer than Earth today, with a global average temperature perhaps as high as 85°C (185°F). Once the oceans formed and plate tectonics began, the feedback mechanisms of the CO_2 cycle acted to keep the temperature in a moderate range for billions of years.

Long-Term Climate Change

While Earth's climate has remained stable enough for the oceans to remain at least partly liquid throughout history, the climate has not been perfectly steady. Numerous warmer periods and numerous ice ages have occurred. Such variations are possible because the CO_2 cycle does not act instantly. When something begins to change the climate, it takes time for the feedback mechanisms of the CO_2 cycle to come into play because of their dependence on the gradual action of mineral formation in the oceans and of plate tectonics.

Calculations show that the time to stabilize atmospheric CO_2 through the CO_2 cycle is about 400,000 years. That is, if the amount of CO_2 in the atmosphere were to rise due to, say, increased volcanism, it would take some 400,000 years for the CO_2 cycle to restore temperatures to their current values. (This time scale applies to ocean/atmosphere equilibrium only. The time scale for crust recycling is much longer, while shorter-term climate variations in atmospheric CO_2 concentration can occur through factors besides the inorganic CO_2 cycle, such as cycling of carbon dioxide by life.)

THINK ABOUT IT

Today, the primary factor leading to increased carbon dioxide in the atmosphere is the release of carbon dioxide by human activity (such as the burning of fossil fuels and deforestation). Should we expect the carbon dioxide cycle to solve the problem by restoring the atmospheric concentration of CO_2 to its preindustrial level? Explain.

Ice Ages Let's briefly investigate Earth's climate variability. **Ice ages** occur when the global average temperature drops by a few degrees. The slightly lower temperatures lead to increased snowfall that may cover continents with ice down to fairly low latitudes. For example, the northern United States was entirely covered with glaciers during the peak of the most recent ice age, which ended only about 10,000 years ago.

The causes of ice ages are complex and are not fully understood. Over periods of tens or hundreds of millions of years, the climate has been influenced at least in part by the Sun's gradual brightening and the changing arrangement of the continents around the globe. During the past few million years—a period too short for solar changes or continental motion to have a significant effect—the ice ages appear to have been strongly influenced by small changes in Earth's axis tilt and other characteristics of Earth's rotation and orbit [Section 11.4].

Snowball Earth Geologists have recently discovered evidence of several particularly long and deep ice ages between 750 and 580 million years ago. During these periods, glaciers appear to have advanced all the way to the equator. If so, the global temperature must have dropped low enough for the oceans, which cover most of Earth's surface, to begin to freeze worldwide. This ocean freezing would have set up a positive feedback process that would have cooled Earth even further, because ice reflects much more sunlight than water. (About 90% of incoming sunlight is reflected by ice, compared to about 5% by water.) With less sunlight being absorbed, the surface would cool further, which in turn would allow the oceans to freeze further. Geologists suspect that in this way our planet may have entered the periods we now call **snowball Earth** (Figure 14.17). We do not know precisely how extreme the cold became. Some models suggest that at the peak of a snowball Earth period, the global average temperature may have been as low as −50°C (−58°F) and the oceans may have been frozen to a depth of 1 kilometer or more.

How did Earth recover from a snowball phase? Even if Earth's surface got cold enough for the ocean surface to freeze completely, the interior still remained hot. As a result, volcanism would have continued to add CO_2 to the atmosphere. Oceans covered by ice would have been unable to absorb this CO_2 gas, and the CO_2 content of the atmosphere would have gradually built up and strengthened the greenhouse effect. As long as the ocean surfaces remained frozen, the CO_2 buildup would have continued. Scientists suspect that, over a period of 10 million years or so, the CO_2 content of our atmosphere may have increased 1,000-fold. Eventually, the strengthening greenhouse effect would have warmed Earth enough to start melting the ocean surface ice. The feedback processes that started the snowball Earth episode now moved quickly in reverse.

As the ocean surface melted, more sunlight would be absorbed (again, because liquid water absorbs more and reflects less sunlight than ice), warming the planet further. Moreover, because the CO_2 concentration was so high, the warming would have continued well past current temperatures—perhaps taking the global average temperature to higher than 50°C (122°F). Thus, in just a few centuries, Earth would have emerged from a snowball phase into a hothouse phase. Geological evidence supports the occurrence of dramatic increases in temperature at the end of each snowball Earth episode. Earth then slowly recovered

▶ **Figure 14.17** The CO_2 cycle rescues Earth from a "snowball" phase.

from the hothouse phase as the CO_2 cycle removed carbon dioxide from the atmosphere. This recovery would have taken the 400,000 years required for stabilization of atmospheric CO_2.

THINK ABOUT IT

Suppose Earth did not have plate tectonics. Could the planet ever recover from a "snowball" phase? Explain.

The snowball Earth episodes must have had severe consequences for any life on Earth at the time. Indeed, the end of the snowball Earth episodes roughly coincides with a dramatic increase in the diversity of life on Earth (called the Cambrian explosion, which we'll discuss shortly). Some scientists suspect that the environmental pressures caused by the snowball Earth periods may have led to a burst of evolution. If so, we might not be here today if not for the dramatic climate changes of the snowball Earth episodes.

Earth's Long-Term Future Climate Despite going through periods of ice ages, snowball Earth, and unusual warmth, Earth's climate has remained suitable for life for some 4 billion years. However, the continuing brightening of the Sun ultimately will overheat our planet.

The numerous feedback processes that affect climate make it difficult to make precise predictions about future climate. According to some climate models, the warming Sun could cause Earth to begin losing its water as soon as a billion or so years from now. If so, then life on Earth has already completed about 75% of its history on this planet. However, uncertainties in the models make it quite possible that they are wrong. The CO_2 cycle may well keep the climate steady for much more than the next billion years. If so, water and life could persist for as long as 3–4 billion years to come.

About 3–4 billion years from now, the Sun will have grown so warm that sunlight on Earth will be as intense as it is on Venus today. The effect will be the same as if we moved Earth to Venus's orbit today (see Figure 11.29)—a runaway greenhouse effect. The rising temperature on Earth will cause increased ocean evaporation (if the oceans still exist), and the water vapor in the atmosphere will further increase the greenhouse effect. The positive feedback won't stop until the oceans have evaporated away and all the carbon dioxide has been released from carbonate rocks. Our planet will become a Venus-like hothouse, with temperatures far too high for liquid water to exist.

Thus, Earth's habitability for life will come to an end between 1 and 4 billion years from now. Although this may seem depressing, remember that a billion years is a very long time—equivalent to some 10 million human lifetimes

An extended cold spell causes oceans to start freezing.

— growing polar caps

— volcanic outgassing

Lowered reflectivity causes further cooling, ending in "snowball Earth."

Frozen oceans stop CO_2 cycle so CO_2 outgassed by ongoing volcanism builds up in atmosphere.

Strong greenhouse effect melts "snowball Earth," results in "hothouse Earth."

CO_2 cycle restarts, pulling CO_2 into oceans, reducing greenhouse effect to normal.

and far longer than humans have existed so far. If you want to lose sleep worrying about the future, there are far more immediate threats, some of which we will discuss in Section 14.6.

Are We "Lucky" to Have a Stable Climate?

One of the goals of comparative planetology is to learn what we might expect on other planets in other star systems. In particular, we'd like to know the likelihood of finding other terrestrial planets with Earth-like conditions, perhaps including life. We have found that the key difference between Earth, Venus, and Mars—at least in terms of making conditions suitable for life—has been our planet's long-term climate stability. This climate stability has in turn been made possible by the carbon dioxide cycle, which requires plate tectonics. Thus, plate tectonics seems to be a crucial requirement for any planet on which life can thrive and evolve over billions of years.

Because Earth has plate tectonics while similar-size Venus does not, some people suspect that plate tectonics may occur only under very rare circumstances. In that case, Earth may be very "lucky" to have plate tectonics and climate regulation, and Earth-like planets may be extremely rare. On the other hand, Venus's lack of plate tectonics may be traceable to its nearer distance to the Sun, and any planet with a size and orbit similar to those of Earth might be expected to have plate tectonics and a carbon dioxide cycle.

How might Venus's distance from the Sun account for its lack of plate tectonics? One scenario traces the explanation to the drying out of Venus caused by the runaway greenhouse effect. On Earth, volcanoes release water to the atmosphere, but plate tectonics eventually returns this water to the crust and mantle. As a result, Earth's interior never dries out. Volcanoes also would have released water to the atmosphere on Venus. However, because Venus lost this water to space, it could not be returned to Venus's interior. Therefore, we expect Venus's crust and mantle to be very dry. Because the presence of water tends to soften rock, a dry crust and mantle would tend to have thickened and strengthened Venus's lithosphere. In that case, the lack of plate tectonics may simply reflect the fact that a strong lithosphere resisted fracturing into plates. This hypothesis is controversial, but it underscores the surprising possibility that a planet's atmosphere can affect its interior, not just the other way around.

In summary, we do not yet understand the interplay of geology and atmospheres well enough to know whether plate tectonics should be rare or common on other terrestrial worlds. Until we do, we will not know whether Earth is "lucky" or just one of many planets in the Milky Way Galaxy that have had a climate stable enough to allow for abundant life.

14.5 Life on Earth

We are now ready to turn to Earth's most extraordinary feature: life. We've already seen that life plays a major role in shaping our planet's atmosphere by producing its oxygen. Careful study of fossils, the petrified remains of living organisms, has enabled scientists to piece together much of the story of life on Earth, including the interplay between life and Earth's atmosphere and climate.

Studying Past Life

Our theories of Earth's past are based on observations of the planet today, combined with the laws of physics that control the behavior of everything from planets to atoms to galaxies. As we turn our attention to biology, can we apply the scientific approach [Section 3.5] in the same way? Absolutely. We have observations of past life in the form of fossils, and we have come to understand the chemical behavior of life's building blocks through laboratory experiments. For most of Earth's history, the fossil record reveals the astonishing story of life on our planet, a story that must be matched with a coherent theory. However, for the earliest stages of life, no identifiable fossils of any kind have yet been discovered, and we must rely instead on an understanding of the chemistry of life's building blocks. In recent years, laboratory experiments have yielded fundamental advances that give us a clearer idea of how chemistry may have given way to biology early in Earth's history.

How Fossils Are Made For the past 3.5 billion years of Earth's history, fossils provide "concrete" evidence of how life has changed. Most fossils formed when dead organisms fell to the bottom of a sea (or other body of water) and were gradually buried by layers of sediment. The sediments were produced by erosion on land and carried by rivers to the sea. Over millions of years, sediments pile up on the seafloor, and the weight of the upper layers compresses underlying layers into rock. Erosion or tectonic activity later exposes the fossils (Figure 14.18).

In some places, such as the Grand Canyon, the sedimentary layers record billions of years of Earth's history (Figure 14.19). The distinctive layers visible in the walls of the Grand Canyon reflect different conditions at the time each layer was deposited. For example, layers will look different depending on the precise composition of the sediments settling to the bottom or the types of organisms leaving fossils.

What Fossils Tell Us The key to reconstructing the history of life is to determine the dates at which fossil organisms lived. The *relative* ages of fossils found in different layers are easy to determine: Deeper layers formed earlier and contain more ancient fossils. Radiometric dating [Section 9.5] confirms these relative ages and gives us fairly precise absolute ages for fossils.

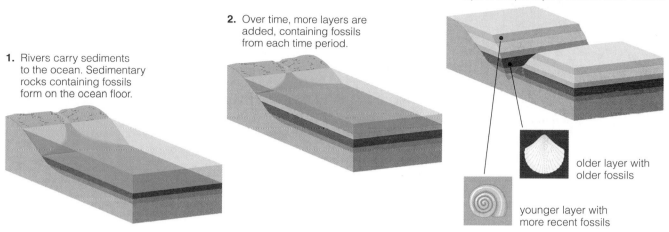

1. Rivers carry sediments to the ocean. Sedimentary rocks containing fossils form on the ocean floor.

2. Over time, more layers are added, containing fossils from each time period.

3. Tectonic stresses and sea level changes push the seafloor upward, exposing sedimentary rocks. Erosion by rivers reveals layers; deeper layers contain older fossils.

older layer with older fossils

younger layer with more recent fossils

Figure 14.18 Formation of sedimentary rock. Each layer represents a particular time and place in Earth's history and is characterized by fossils of organisms that lived in that time and place.

Figure 14.19 The rock layers of the Grand Canyon record 2 billion years of Earth's history.

Figure 14.20 Dinosaur fossils.

Some fossils are remarkably well preserved (Figure 14.20), though the vast majority of dead organisms decay completely and leave no fossils behind. Fossils become more difficult to find and study as we look to early epochs in Earth's history, for several reasons. For example, ancient organisms that lacked skeletons left fewer fossils, erosion erases much old fossil evidence, and subduction destroys fossils carried deep beneath Earth's surface. Nevertheless, scientists have found and studied thousands of fossils not only of the large animals that lived in the past couple hundred million years, but also of smaller animals, plants, and even microscopic, single-celled organisms.

The Origin of Life

How did life arise? We don't know the precise answer to this question, and it may well be impossible to know exactly what occurred to produce the first life on Earth. Nevertheless, scientists today can construct plausible scenarios that explain how chemistry on the early Earth may ultimately have led to the origin of life. Let's briefly investigate how the transition from chemistry to biology might have occurred.

Evidence of a Common Ancestor
One key piece of evidence concerning the origin of life comes from careful study of living organisms today: It seems quite clear that every living organism shares a common ancestor. Whether life arose once or multiple times, one particular type of organism came to dominate the entire Earth.

Several lines of evidence support the idea that all life evolved from a common ancestor. The first line of evidence comes from the fact that all known life-forms share uncanny chemical resemblances to one another. These similarities are not expected on simple chemical grounds and are too improbable and too numerous to be considered coincidences. For example, all known organisms build proteins from essentially the same set of 20 amino acids, even though more than 70 amino acids are found in nature. Similarly, all living organisms use the same basic molecule (called ATP) to store energy within cells, and all use molecules of DNA to transmit their genes from one generation to the next.

The second line of evidence supporting the idea of a common ancestor comes from the fact that all living organisms share nearly the same *genetic code* to read the instructions chemically encoded in DNA. You are probably familiar with the structure of DNA, which consists of two long strands (somewhat like the interlocking strands of a zipper) wound together in the spiral shape known as a double helix (Figure 14.21). Living organisms reproduce by copying DNA molecules and passing them on to their descendants. Each letter shown along the DNA strands represents one of four *chemical bases,* denoted by A, G, T, and C (for the first letters of their chemical names). The instructions for assembling the cell are written in the precise arrangement of these four chemical bases, with each set of three bases making a particular genetic "word." For example, the "word" CAG tells the cell to add one particular amino acid when building a protein, and TCA tells it to add a completely different amino acid.

We know of no reason why living organisms should follow a particular genetic code. Biochemists believe that DNA could use different sequences of these bases—or possibly even different bases altogether—to encode the same genetic information. Thus, the fact that all living organisms use the same genetic code implies a common ancestry.

The third line of evidence for a common ancestor comes from detailed comparisons of DNA among different organisms. The DNA instructions for a particular cell function, such as the instructions for building a single protein, make up what we call a *gene.* Very different organisms often have quite similar genes. For example, some bacterial genes have instructions for cellular molecules quite similar to those found in humans—itself a suggestion that people and bacteria share a common ancient ancestry.

By mapping similarities and differences between DNA sequences in particular genes, biologists can determine how closely two organisms are related to each other. For example, two organisms whose DNA sequences differ in five places for a particular gene are probably more distantly related than two organisms whose gene sequences differ in only one place. Many such comparisons have shown all living organisms to be related in a way depicted schematically by the "tree of life" in Figure 14.22. Although details in the structure of this tree remain uncertain, it clearly shows that living organisms share a common ancestry. As it is currently understood, the tree of life shows that life on Earth is divided into three major groupings (known as the domains Bacteria, Archaea, and Eukarya). Plants and ani-

Figure 14.21 DNA is the genetic material of all life on Earth. Left: A model of a small piece of a DNA molecule. Right: This diagram shows that a DNA molecule is made from two intertwined strands, with "ladder steps" connecting pairs of chemical bases. Note that A always connects to T, and G always connects to C. The genetic code is a language in which triplets (along a single strand) of the chemical bases make "words" that represent particular amino acids.

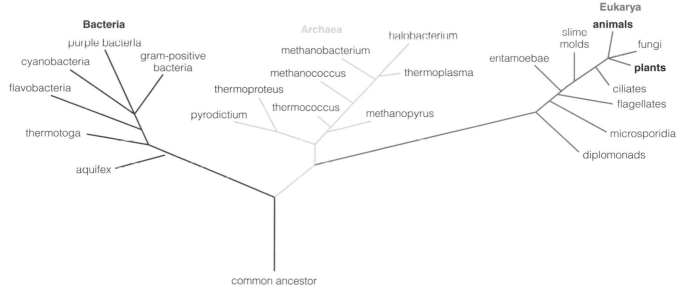

Figure 14.22 The tree of life, showing evolutionary relationships determined by comparison of DNA sequences in different organisms. Just two small branches represent *all* plant and animal species.

mals represent only two tiny branches of the great diversity of life on Earth.

The Nature of Early Life Comparison of DNA sequences also allows at least reasonable guesses as to which organisms found on Earth today most resemble the common ancestor. The answer appears to be organisms living in the deep oceans around seafloor volcanic vents called *black smokers* (after the dark, mineral-rich water that flows out of them) and in hot springs in places such as Yellowstone (Figure 14.23). These organisms thrive in temperatures as high as 110°C (230°F). (The high pressures at the seafloor prevent the water from boiling despite the high temperature.) Unlike most life at the Earth's surface, which depends on sunlight, the ultimate energy source for these organisms is chemical reactions in water heated volcanically by the internal heat of Earth itself.

That early organisms might have lived in such "extreme" conditions may at first seem surprising, but such conditions probably were quite common on the early Earth.

a Life around a black smoker deep beneath the ocean surface.

b This aerial photo shows a hot spring in Yellowstone National Park that is filled with colorful bacterial life. The path in the lower left gives an indication of scale.

Figure 14.23 Life in extreme conditions. Such conditions may have been common on the early Earth, and DNA studies suggest that life may have first arisen in extreme environments.

Moreover, these organisms do not depend on sunlight, a trait that was almost certainly a requirement for early life. Getting energy from sunlight involves photosynthesis, a complex molecular process that probably evolved well after the first organisms began living off Earth's internal heat.

THINK ABOUT IT

What other worlds in our solar system could have structures similar to black smokers or hot springs? What does the evidence that life originated in such locations mean for the possibility of finding life on those worlds?

The Transition from Chemistry to Biology Knowing that all life shares a common ancestor does not tell us how that common ancestor first arose. The step from simple chemical building blocks to our common DNA-bearing ancestor (from nonlife to life) is huge, to say the least. No fossils record the transition, and it has never been duplicated in the laboratory.

Nevertheless, laboratory experiments offer plausible scenarios for the origin of life. Experiments using chemical ingredients that we know were present on the early Earth (from the geological record) show that many complex organic molecules could have arisen naturally and spontaneously. For example, experiments that mix early Earth ingredients and "spark" the chemicals with electricity to simulate lightning or other energy sources have produced virtually all the major molecules of life, including all the amino acids and DNA bases. Many of these same molecules are found today in meteorites, suggesting that impacts may also have been an important source of organic molecules on the early Earth.

Laboratory experiments also show that mixing a warm, dilute solution of such organic molecules with naturally occurring sand or clay allows them to self-assemble into much more complex molecules. Strands of RNA (which look much like a single strand of DNA) up to nearly 100 bases in length have been produced in this way in the laboratory. Because some RNA is capable of self-replication, much like DNA, many biologists now presume that RNA was the original genetic material of life on Earth.

Other experiments have shown that microscopic, enclosed membranes also self-assemble under conditions expected on the early Earth. Self-replicating RNA molecules may have become enclosed in such membranes. In essence, this would have allowed the "pre-cells" to compete with one another. Those in which RNA replicated faster and more accurately were more likely to spread, leading to a type of positive feedback that would have encouraged even faster and more accurate replication. Given millions of years of chemical reactions occurring all over the Earth, RNA might eventually have evolved into DNA, making true living organisms.

Figure 14.24 summarizes the steps by which chemistry might have become biology on Earth. We may never know whether life really arose in this way, but it certainly seems possible.

THINK ABOUT IT

Questions about the origin of life inevitably raise religious issues. The scenario in Figure 14.24 appears to involve only random chemistry and requires no divine intervention. Nevertheless, polls show that many evolutionary biologists believe in God, and religious leaders, including the Pope of the Catholic Church, have concluded that evolution is not at odds with their religions. Regardless of your own personal beliefs, explain how someone could reconcile belief in God with the scenario we have discussed for the origin of life.

Could Life Have Migrated to Earth? Our scenario suggests that life could have arisen naturally here on Earth. However, an alternative possibility is that life arose somewhere else first—perhaps on Venus or Mars—and then migrated to Earth on meteorites. Remember that we have collected meteorites that have been blasted by impacts from the surfaces of the Moon and Mars [Section 13.3]. Calculations suggest that Venus, Earth, and Mars all should have exchanged many tons of rock, especially in the early days of the solar system when impacts were more common.

The idea that life could travel through space to land on Earth once seemed outlandish. After all, it's hard to imagine a more forbidding environment than that of space, with no air, no water, and constant bombardment by dangerous radiation from the Sun and stars. However, the presence of organic molecules in meteorites and comets tells us that the building blocks of life can indeed survive in the space

1. Synthesis of organic precursor molecules 2. Origin of self-replicating RNA 3. Origin of membrane-enclosed precells 4. Origin of true cells with RNA genome 5. Evolution of modern cells with DNA genome

Figure 14.24 A summary of the steps by which chemistry on the early Earth may have led to the origin of life.

a These large mats photographed at Shark Bay, Western Australia, are colonies of microbes known as "living stromatolites."

Figure 14.25 Rocks called stromatolites offer evidence of photosynthetic life as early as 3.5 billion years ago.

b The bands visible in this section of a modern-day mat are formed by layers of sediment adhering to different types of microbes.

c This section of a 3.5-billion-year-old stromatolite shows a structure nearly identical to that of a living mat. Thus, it offers strong evidence of having been made by microbes, including some photosynthetic ones, that lived 3.5 billion years ago.

environment, and tests have shown that some microbes can survive for years in space.

In a sense, Earth, Venus, and Mars have been "sneezing" on each other for billions of years. Life could conceivably have originated on any of these three planets and been transported to the others. It's an intriguing thought, but it does not change our basic scenario for the origin of life—it simply moves it from one planet to another.

When Did Life Begin on Earth? (More Than 3.5 Billion Years Ago)

Fossil evidence suggests that life was already thriving on Earth by 3.5 billion years ago, and possibly several hundred million years before that. One strong line of evidence comes from fossils of bacterial "colonies" called *stromatolites.* These fossils reveal structures nearly identical to those produced by some living bacteria today (Figure 14.25). Radiometric dating shows that some stromatolites are 3.5 billion years old.

Other evidence comes from what appear to be fossils of single-celled organisms. Such fossils have been found in rocks dating to 3.2–3.5 billion years ago, though some controversy remains over whether the oldest of these "fossils" are really the remains of organisms or simply unusual mineral structures (Figure 14.26).

Finding evidence of life before 3.5 billion years ago is much more challenging. Few rocks survive intact from such ancient times, making it unlikely that we'll find fossils even if organisms made them at the time. Nevertheless, the fact that organisms were already advanced enough to build

stromatolites by 3.5 billion years ago suggests that more primitive organisms must have lived even earlier. Some evidence supports this idea. More ancient rocks that have undergone too much change to leave fossils intact may still hold carbon that was once part of living organisms.

Carbon has two stable isotopes: carbon-12, with six protons and six neutrons in its nucleus, and carbon-13, which has one extra neutron. Living organisms incorporate carbon-12 slightly more easily than carbon-13. As a result, the fraction of carbon-13 is always a bit lower in fossils than in rock samples that lack fossils. All life and all fossils tested to date show the same characteristic ratio of the two carbon isotopes. Dating very ancient rocks can be difficult, but some rocks that are more than 3.85 billion years old show the same ratio of carbon-12 to carbon-13, suggesting that these rocks contain remnants of life.

We do not yet have any fossil evidence of life before 3.85 billion years ago. Given the overall rarity of such ancient rocks—and the fact that even those few that we find have been transformed substantially by geological processes—we may never know for sure whether life existed before that time. However, mineral evidence suggests that oceans of liquid water were present on Earth as early as 4.4 billion

Figure 14.26 Microfossils of ancient living cells? This microscopic photograph shows structures that some researchers believe to be ancient fossil cells dating to 3.5 billion years ago. Other researchers argue that the structures were formed by nonbiological processes.

years ago, so it's conceivable that life could have been present as well.

If life did originate much before 3.85 billion years ago, it may have been in for a rough time. The transition from intolerable conditions during planetary formation to more hospitable conditions for life was gradual. The heavy bombardment that continued after the end of accretion gradually died away over a period of a few hundred million years. Large impacts capable of sterilizing the planet by temporarily boiling all the oceans may have occurred as late as 4.0–3.8 billion years ago. Indeed, some biologists speculate that life may have arisen several times in Earth's early history, only to be extinguished by the hostile conditions. If so, life might have turned out very different if one less or one more sterilizing impact had occurred.

You may have noticed that, while we've been talking about life, we haven't actually defined the term. In fact, it's surprisingly difficult to draw a clear boundary between life and nonlife. How would *you* define life? Explain your reasoning.

Early Evolution in the Oceans (>3.5–2.0 Billion Years Ago)

At least 3.5 billion years ago, life thrived in the oceans in the form of single-celled organisms. Individual organisms that survived and reproduced passed on copies of their DNA to the next generation. However, the transmission of DNA from one generation to the next is not always perfect.

An organism's DNA may be altered by occasional copying errors or by external influences—such as ultraviolet light from the Sun or exposure to toxic or radioactive chemicals. Any change in the base sequence of an organism's DNA is called a **mutation**. Many mutations are lethal, killing the cell in which the mutation occurs. Some, however, may improve a cell's ability to survive and reproduce. The cell then passes on this improvement to its offspring.

The process by which mutations that make an organism better able to survive get passed on to future generations is called **natural selection** (because nature "selects" the more advantageous form of the gene). The idea of natural selection was proposed by Charles Darwin (1809–1882) as a way to explain evolution, which simply means "change with time." Other scientists had already accumulated substantial evidence of gradual change in animal species, but no one before Darwin saw how that change might occur. Darwin compiled a tremendous body of evidence supporting the idea that natural selection is the primary mechanism by which evolution proceeds. Over time, natural selection might help individuals of a species become better able to compete for scarce resources. It might also lead to the development of entirely new species from old ones. In essence, the fossil record provides strong evidence that evolution *has* occurred, while natural selection explains *how* it occurs.

More than 100 years of ongoing research has only given further support to Darwin's idea, which is why we call it the *theory* of evolution. Perhaps the strongest support for the theory has come from the discovery of DNA and of the way mutations underlie natural selection. Today, biologists routinely witness evolution occurring before their eyes among laboratory microorganisms, or over periods of just a few decades among plants and animals subjected to some kind of environmental stress.

Evolution rapidly diversified life on Earth. Processes as complex as photosynthesis evolved over a period of a few hundred million years or less. From a human perspective, however, evolution may seem to have made a slow start. For at least a billion years after life first arose, the most complex life-forms were still single-celled. Some 2 billion years ago, the land was still inhospitable because of the lack of a protective ozone layer. Continents much like those today were surrounded by oceans teeming with life, but, despite pleasant temperatures and plentiful rainfall, the land itself was probably as barren as Mars is today.

The Rise of Oxygen (Beginning About 2 Billion Years Ago)

The process of photosynthesis appears to have developed quite early in the history of life. Single-celled organisms known as cyanobacteria may have been producing oxygen through photosynthesis as early as 3.5 billion years ago. However, the oxygen did not immediately begin to accumulate in the atmosphere. For over a billion years, chemical reactions with surface rocks pulled oxygen back out of the atmosphere as fast as the cyanobacteria could produce it. But these tiny organisms were abundant and persistent, and eventually the surface rock was so saturated by oxygen that the rate of oxygen removal slowed down. At that point, some 2 billion years ago, oxygen began to accumulate in the atmosphere.

We do not know how rapidly oxygen accumulated or precisely when it reached its present concentration in our atmosphere. Some fossil studies suggest that the oxygen abundance was less than about 10% of its current level as recently as 550 million years ago. Clear evidence of an oxygen level near the current value does not appear in the fossil record until about 200 million years ago. That is when we first find charcoal in the fossil record, imply-ing that our atmosphere contained enough oxygen for fires to burn.

The buildup of oxygen had two far-reaching effects for life on Earth. First, it made possible the development of oxygen-dependent *animals* (Figure 14.27). No one knows precisely how or when the first oxygen-dependent organism

---

SPECIAL TOPIC The Theory of Evolution

If you live in the United States, you're undoubtedly familiar with the public debate about evolution—in particular, about whether the theory of evolution should be taught in schools. As is often the case with inflamed debates, much of the controversy comes from misunderstandings on both sides. On the antievolution side, there's a general lack of understanding of why scientists are near-unanimous in their acceptance of the theory of evolution. On the scientists' side, there's often a lack of respect for people's individual religious beliefs.

This is a science textbook. We will not go into the religious side of the issue other than to say that every person is entitled to his or her beliefs, whether or not these beliefs agree with prevailing scientific thought. However, the theory of evolution is a crucial part of the foundation of the modern scientific view of our world. Thus, if you are taking a science course, you need to understand why scientists are nearly unanimous in their acceptance of Darwin's basic theory. Having this understanding may or may not affect your personal beliefs about our human origins, but it will at least make it possible for you to continue your study of science.

To begin with, we need to distinguish between the *occurrence* of evolution and the *theory* of evolution. Evidence that evolution has occurred—meaning gradual change in the nature of life on Earth—comes primarily from the fossil record but also from studies in laboratories and localized environments. Scientists had found evidence for the occurrence of evolution long before Darwin. The first claim that life evolves from simple to more complex forms is attributed to Anaximander, who lived more than 2,500 years ago (see Figure 3.13).

Darwin is a central figure because he documented more evidence for the occurrence of evolution than anyone before him, and more important because he also laid out a clear model to explain *how* evolution occurs. His model of how evolution occurs through natural selection is what we call the *theory* of evolution. Darwin first published his evidence and his theory in his 1859 book *The Origin of Species.*

Perhaps because Darwin filled his book with so much detailed and carefully documented evidence, many people are surprised to learn that the basic case for the theory of evolution comes down to very simple logic. As described by biologist Stephen Jay Gould (1941–2002), Darwin built his case with "two undeniable facts and an inescapable conclusion." The logic goes like this:

- *Fact 1: overproduction and struggle for survival.* Any localized population of a species has the potential to produce far more offspring than the local environment can possibly support with resources such as food and shelter. This overproduction leads to a struggle for survival among the individuals of the population.

- *Fact 2: individual variation.* Individuals in a population of any species vary in many heritable traits (that is, traits passed from parents to offspring). No two individuals are exactly alike, and some individuals possess traits that make them better able to compete for food and other vital resources.

- *The inescapable conclusion: unequal reproductive success.* In the struggle for survival, those individuals whose traits best enable them to survive and reproduce will, on average, leave the largest number of offspring that in turn survive to reproduce. Therefore, in any local environment, heritable traits that enhance survival and successful reproduction will become progressively more common in succeeding generations. It is this unequal reproductive success that Darwin called *natural selection.* That is, over time, advantageous genetic traits will naturally win out (be "selected") over less advantageous traits because they are more likely to be passed down through many generations.

Darwin backed up this logic with so much specific evidence that the vast majority of biologists accepted his theory within a decade of the publication of his book. As noted in the main text, the evidence supporting Darwin's theory has only become stronger since that time.

Today, our detailed understanding of how mutations explain evolution on a molecular level leave little room for doubt that evolution really does occur through natural selection. Thus, the theory of evolution is a true scientific theory according to the definition we discussed in Section 3.5. Like other major scientific theories, such as the theory of gravity or the theory of the atom, it is supported by strong and wide-ranging evidence. This does not mean the theory will never be modified—indeed, scientists are constantly learning more about how evolution works. But it means that the evidence is far too strong to ignore and that any future changes to the theory will still have to explain all the evidence already gathered in its support.

The development of the theory of evolution is a clear example of how science works—with the requirement that theory match observations regardless of personal opinions about the theory. The fact that some religious texts and authorities are opposed to the theory of evolution has no bearing on its *scientific* validity.

Where does this discussion leave us with regard to the question of teaching evolution in schools? Let's start with what it does *not* tell us. It does not tell us that everyone needs to accept the theory of evolution—whether you choose to believe the theory is up to you. Nor does it tell us whether divine intervention may have a role in the progression of evolution—the science does not say anything one way or the other about the existence or role of any God. What it *does* tell us is that the theory of evolution is a clear and crucial part of the study of science. For students to learn how science works and how we've achieved the advances of modern technology—especially modern biology and medicine—the theory of evolution needs to be in the science curriculum.

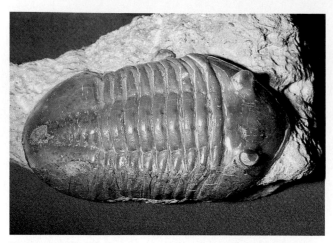

Figure 14.27 Thanks to the buildup of oxygen in the atmosphere (and small amounts dissolved in the oceans), animals like these trilobites became possible. Trilobites lived in the oceans and were among the most complex animals a few hundred million years ago. The largest specimens reached 75 cm (30 in.) in length.

Figure 14.28 This graph shows human population over the past 12,000 years. Note the tremendous population growth that has occurred in just the past few centuries.

appeared in the oceans, but for that animal the world was filled with food. You can imagine how quickly these creatures must have spread around the world. The fact that other organisms could now be eaten changed the "rules of the game" for evolution.

The second major effect of the oxygen buildup was the formation of the ozone layer, which made it safe for life to move onto the land. Plants appeared on land about 475 million years ago. Animals, taking advantage of this new food source, followed soon after.

An Explosion of Diversity (Beginning About 540 Million Years Ago)

As recently as 540 million years ago—some 3 billion or more years after life first appeared on Earth—most life-forms were still single-celled and tiny. The fossil record reveals a dramatic diversification of life beginning at about that time, and over the next 40 million years animals diversified into all 33 basic body plans (phyla) that we find on Earth today. This remarkable flowering of animal diversity occurred in such a short time relative to the history of Earth that it is often called the *Cambrian explosion.* (*Cambrian* is the name geologists give to the period from about 540 million to 500 million years ago.)

Some species have been more successful and more adaptable than others. Dinosaurs dominated the landscape of Earth for more than 100 million years. Their sudden demise 65 million years ago paved the way for the evolution of large mammals—including humans. The earliest humans appeared on the scene only a few million years ago, or after 99.9% of Earth's history to date had already gone by. Our few centuries of industry and technology have come after 99.99999% of Earth's history.

Despite our recent arrival (in geological terms), modern humans are by some measures the most successful

species ever to inhabit Earth. Humans survive and prosper in virtually every type of land environment. With proper equipment, we can survive underwater, and a few people have lived under the sea for extended periods of time in experimental habitats. We are even developing the technology to survive away from our planet in the inhospitable environment of space. Moreover, our population has been growing exponentially for the past few centuries, roughly doubling in just the past 40 years (Figure 14.28).

The tremendous growth of the human population is a testament to our success as a species, but it also suggests a chilling analogy: Exponential growth is a hallmark of a cancerous tumor, which may seem very successful at surviving while it is growing but ultimately dies when it kills its host. Could our tremendous success be killing our host, the Earth? We can gain some perspective on this question by looking at a few lessons the solar system teaches us about our relationship with Earth.

THINK ABOUT IT

The pace of change in human existence has accelerated since the advent of modern humans a few million years ago: Civilization arose about 10 thousand years ago, the industrial revolution began just a couple of centuries ago, and the computer era started just a few decades ago. Do you think the pace of change can continue to accelerate? What do you think the next stage will be?

14.6 Our Future Survival: Lessons from Other Worlds

The United States and other nations have carried out the exploration of the solar system for many reasons, among them scientific curiosity, national pride, and technological advancement. The most important return on this investment, however, may be something that was not originally planned: an improved understanding of our own planet and an enhanced ability to grasp some important global issues. In this section, we will explore three particularly important issues about which we have gained important insights through comparative planetology: global warming, ozone depletion, and mass extinctions.

Global Warming

Venus stands as a searing example of the effects of large amounts of atmospheric carbon dioxide. Earth remains habitable only because natural processes have locked up most of its carbon dioxide on the seafloor. However, humans are now tinkering with this distinction between the two planets by adding carbon dioxide and other greenhouse gases to Earth's atmosphere.

The primary way we release carbon dioxide is through the burning of fossil fuels (coal, natural gas, and oil), the carbon-rich remains of plants that died millions of years ago. Other human actions, such as deforestation and the subsequent burning of trees, also affect the carbon dioxide balance of our atmosphere. The expected result is **global warming**—an increase in Earth's global average temperature caused by human input of greenhouse gases into the atmosphere.

Is Global Warming Real?

The potential threat of global warming is a hot political issue because alleviating the threat would require finding new energy sources and making other changes that would dramatically affect the world's economy. However, a major research effort has gradually added to our understanding of the potential threat, particularly in the past decade. Although some controversy still exists about how serious the problem of global warming might be, at least three facts are now clear.

First, measurements show that Earth has indeed warmed up over the past 50 years, by about 0.5°C. The warming trend has continued and gotten stronger in recent years. Of course, warming can occur naturally, so by itself this fact does not necessarily implicate humans as the cause of the warming.

Second, human activity is clearly increasing the amounts of greenhouse gases in the atmosphere. Figure 14.29 shows the rise in CO_2 concentration over the past half-century (which is as long as these data have been collected). Data from ice cores show that the total rise in CO_2 concentration since the dawn of the industrial age has been much greater still. If current trends continue, by 2050 the CO_2 concentration may be more than double what it was in the early 1800s. Other greenhouse gases (such as methane) are also being added to the atmosphere by human activity. Thus, the observed warming over the past few decades has coincided with a rise in greenhouse gas concentration caused by human activity. Again, this fact does not prove that human activity is the cause of the warming, but it makes it a distinct possibility.

Third, the basic mechanism of the greenhouse effect [Section 11.2] is so simple and so well understood that no doubt remains that a continually rising concentration of greenhouse gases would eventually make our planet warm up. Geological evidence and evidence from other planets back up this fact. We have seen how added greenhouse gases warmed our planet in the past (such as in the hothouse phases following snowball Earth episodes) and how the greenhouse effect makes Venus unbearably hot. Thus, while we cannot be certain that human activity has caused the global warming observed to date, we will eventually cause our planet to warm up if we continue to add greenhouse gases to the atmosphere.

In an attempt to better understand the connection between human activity and rising global temperatures, scientists build sophisticated computer models of the climate. Creating these models is quite challenging. While the long-term consequences of increased greenhouse gas concentration are clear, the short-term consequences are much more difficult to predict. On time scales of years, decades, and even centuries, numerous feedback mechanisms might counter or enhance greenhouse warming.

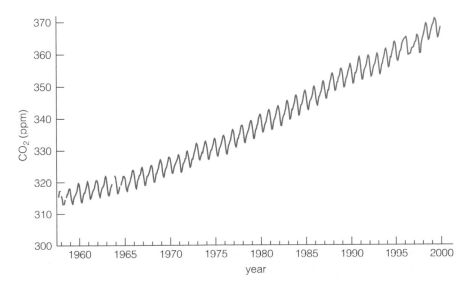

Figure 14.29 These data, collected for many years on Mauna Loa, Hawaii, show the increase in atmospheric carbon dioxide concentration. Yearly wiggles represent seasonal variations. The concentration is measured in parts per million (ppm). For example, the current concentration of about 370 ppm means there are about 370 CO_2 molecules among every 1 million air molecules (370 ppm is equivalent to $370 \div 1,000,000 = 0.00037 = 0.037\%$).

One set of feedback mechanisms involves clouds, which potentially could play a role in both positive and negative feedback. Increased greenhouse gases can lead to more evaporation, which makes more clouds. Increased cloudiness, in turn, tends to block sunlight and cool the planet, providing negative feedback to counter the warming effects of the extra greenhouse gases in our atmosphere. (This type of negative feedback could not operate indefinitely, because the warming would resume once Earth became completely cloud-covered.) On the other hand, clouds can also trap heat. Under some circumstances, therefore, the clouds could give positive feedback, accelerating the warming of our planet.

Another and more frightening feedback mechanism involves other human activity. Coal-burning industries giving off CO_2 also release sulfate particles into the atmosphere. These particles tend to reflect sunlight, so they increase Earth's reflectivity, which tends to cool down our planet. At first, this might seem to suggest that the pollutants from coal burning have the beneficial effect of limiting global warming from the increased greenhouse gas concentration. In reality, it can only postpone and worsen the problem. Sulfate particles fall out of the atmosphere in a matter of years, while the carbon dioxide will remain in the atmosphere for millennia.

The scientific uncertainties about such feedback mechanisms lead to corresponding uncertainties in predicting just how much warming will occur in the short term. Nevertheless, observations indicate that temperatures are, in fact, increasing globally in a manner consistent with the rise predicted by computer models of the climate (Figure 14.30). This agreement between models and observations, combined with the three facts discussed above, explains why widespread consensus now exists among scientists that global warming is real and is the result of human activity.

Consequences of Global Warming

Clearly, global warming would mean a higher global average surface temperature for Earth. Beyond that simple statement, however, the issue of consequences becomes much more complicated and the uncertainties involved in prediction even greater.

Changing weather patterns would ensure that different regions of Earth experience different degrees of warming. Some regions would even become colder. Other regions might experience more rainfall or might become deserts. The greater overall warmth of the atmosphere would tend to mean more evaporation from the oceans, leading to more numerous and more intense storms. Coastal regions would be hit by more hurricanes. Severe thunderstorms and associated tornadoes would strike more frequently. Winter blizzards would be more severe and more damaging.

Another serious threat comes from rising sea level. Water expands very slightly as it warms—so slightly that we don't notice the change in a glass of water, but enough that warming oceans could increase sea level by up to a meter or more during this century. Such increases in sea level would threaten many coastal regions with more severe

Figure 14.30 This graph compares the predictions of climate models with observed temperature changes (red line) since about 1860. The thickness of the green swath for the models represents the range of variation between several different models. These models take into account both natural factors (such as slight changes in the Sun's brightness and the effects of volcanic eruptions) and emission of greenhouse gases by humans. The agreement is not perfect—telling us we still have much to learn—but it is good enough to give us great confidence that greenhouse gases are, indeed, producing global warming.

flooding. Sea level has already risen some 20 centimeters in the last century. The tiny island nation of Tuvalu, composed of coral atolls that barely rise above sea level, is currently planning for the evacuation of its entire population.

An even greater increase in sea level may occur if polar ice and glaciers melt. Such melting appears to be occurring to some extent already. For example, the famous "snows" (glaciers) of Mount Kilimanjaro are already in retreat and may be gone within the next decade or so. Even small amounts of melting of the Greenland or Antarctic ice sheets could raise sea level enough to cause severe coastal flooding worldwide and might ultimately leave many villages and cities underwater. Moreover, some climate models suggest that polar melting could alter important ocean currents, including the Gulf Stream that flows northeast through the Atlantic off the east coast of the United States. Because the Gulf Stream plays a major role in regulating the climate of the eastern United States, Canada, and Europe, any change in its flow would surely cause dramatic climate changes for these regions.

Secondary effects, such as those arising from ecological changes, pose an even more intractable problem. For example, it is difficult to know how forests and other ecosystems will respond to any climate changes induced by global warming. As a result, we cannot easily predict the impact of such climate changes on food production, fresh water availability, or other issues critical to the well-being of human populations.

Given the current uncertainties, no one can predict the precise impact of global warming over the next century. However, the lesson from the solar system is clear. Our studies of the planets show that surface temperatures depend on many factors and that positive and negative feedback processes can alter climates in surprising ways. Dramatic and deadly change can occur unexpectedly. We should be very careful about tampering with the finely balanced mechanisms that control Earth's climate.

THINK ABOUT IT

If you were a political leader, how would *you* deal with the uncertain threat of global warming?

Ozone Depletion

For 2 billion years, Earth's ozone layer has shielded the surface from hazardous ultraviolet radiation. In the past few decades, humans have begun to destroy this shield. Satellite measurements suggest that ozone levels have fallen by as much as a few percent worldwide in recent decades. Declining ozone levels mean that more solar ultraviolet radiation reaches the surface. Humans face a greater risk from skin cancer, while plants and animals suffer more genetic damage from mutations—with unknown consequences for life on Earth as a whole.

The idea that human activity might damage the ozone layer was first suggested in the early 1970s, but the magnitude of the problem was not recognized until the discovery of an **ozone hole** over Antarctica in the mid-1980s (Figure 14.31). The "hole" is a place where the concentration of ozone in the stratosphere is dramatically lower than it would be naturally. The ozone hole appears for a few months during each Antarctic spring and has gradually worsened over the years, though there is some sign that the situation is beginning to improve.

The apparent cause of the ozone depletion is human-made chemicals known as CFCs (chlorofluorocarbons). CFCs were invented in the 1930s and have been widely used in air conditioners and refrigerators, in the manufacture of packaging foam, as propellants in spray cans, and in industrial solvents used in the computer industry. Indeed, CFCs once seemed like an almost ideal chemical. They are useful in many industries, cheap and easy to produce, and chemically inert, which means they do not burn, break down, or react with anything on Earth's surface.

Ironically, CFCs' inertness is also their downfall. Because CFCs are gases and are not destroyed by chemical reactions in the lower atmosphere, they eventually rise intact into the stratosphere. There they are broken down by the Sun's ultraviolet light. One by-product is chlorine, a very reactive gas that catalytically destroys ozone without being consumed in the process. On average, a single chlorine atom in the stratosphere can destroy 100,000 ozone molecules before it is itself consumed in some other chemical reaction. Chlorine's ability to attack ozone is enhanced at low temperatures, which explains why ozone depletion was first observed over Antarctica.

Mars and Venus both played an important role in our understanding of ozone and chlorine. Mars lacks an ozone layer, so ultraviolet radiation reaches the surface. As a result, the surface is sterile: Terrestrial life could not survive on the surface, and the building blocks of life would be torn apart by exposure to the Sun. Thus, Mars teaches us about the consequences of a lack of ozone. The study of Venus first alerted scientists to the specific danger of ozone depletion on Earth. Chlorine is a minor but important ingredient of Venus's atmosphere, and models of its chemical cycles showed that chlorine could rapidly destroy weak molecules such as ozone. When the models for Venus were altered to apply to Earth's atmosphere, they showed the connection between CFCs and ozone depletion.

This tale of three planets may have a happy ending, thanks to the awareness it brought us of the dangers posed by ozone destruction. Today, international treaties ban the

Figure 14.31
The extent of the ozone hole over Antarctica in 1979 (left) and 1998 (right). The ozone hole grew substantially in those 20 years. Red indicates normal ozone concentration, and blue indicates severe depletion.

ozone abundance

production of CFCs. (Previously existing CFCs may still be used and recycled). If these treaties are adhered to, ozone depletion probably will not be a serious problem in the future. However, it may take as long as 50 years for the stratosphere to fully recover. Most of the CFCs ever produced are still in intact air conditioners, refrigerators, or other products, and most of these will eventually escape into the atmosphere. Moreover, because the stratosphere lacks the weather of the troposphere [Section 11.3], CFCs (and their breakdown products) remain in the stratosphere for decades after they arrive.

The greater future danger lies in the possibility of treaty failures. Some replacements for CFCs may be only marginally less damaging to ozone than CFCs themselves. In addition, the replacements are generally much more expensive to produce than CFCs, and a large global black market now exists for CFCs produced in violation of the treaties. Like most global issues, the final outcome of the ozone problem remains in doubt.

Mass Extinctions

When geology was a young science, its practitioners thought that entire mountain ranges and huge valleys were formed suddenly by cataclysmic events. Later geologists denounced this *catastrophism* and replaced it with *uniformitarianism*, which holds that geological change occurs gradually through processes acting over very long periods. Uniformitarianism certainly holds for three of our four geological processes. Impact cratering, however, is more like the "instant geology" of catastrophism. Many geologists initially rejected the importance of impact cratering as a geological process. They were eventually convinced of its importance by the unambiguous evidence of impacts on the Moon, by astronomical studies of comets and asteroids, and by careful analysis of the few impact craters still visible on Earth.

Scientific ideas about evolution are undergoing a similar transformation. Biologists once assumed that evolution always proceeded gradually, with new species arising slowly and others occasionally going extinct. It now appears that the vast majority of all species in Earth's history died out in sudden *mass extinctions,* for which impacts are a prime suspect [Section 13.6]. The mass extinction that killed off the dinosaurs is only the most famous of these events and is not even the most severe.

While some species could have died out from direct effects of impacts (such as the blast wave or global fires caused by fallout), most probably died due to environmental effects and the disappearance from their ecosystem of

species on which they depended. For example, the deaths of certain plants would have limited the food sources for animals that depended on them. Indeed, while the survival or extinction of species appears to be largely random during mass extinctions, species at the top of the food chain—like the dinosaurs—seem especially vulnerable. In a sense, mass extinctions clear the slate, allowing evolution to get a fresh start with a new set of dominant species.

The story of mass extinctions teaches us about the danger of large impacts, but it also teaches a potentially more important lesson. Even in the absence of an impact, Earth may be undergoing a mass extinction right now. In the normal course of evolution, the rate of species extinction is fairly low, perhaps one species lost per century. Today, human activity is causing extinction at a far greater rate. The best-known cases involve relatively large and wide-ranging animals, such as the extinct passenger pigeon (Figure 14.32) or the nearly extinct Siberian tiger. However, most of the estimated 10 million or more species on our planet live in localized habitats, and most of these species have not even been cataloged. The destruction of just a few square kilometers of forest may mean the extinction of species that live only in that area.

According to some estimates, human activity is driving species to extinction so rapidly that up to half of today's species may be gone by the end of this century. On the scale of geological time, the disappearance of half the world's species over just a few hundred years would certainly qualify as another of Earth's mass extinctions. Are we unwittingly clearing the way for a new set of dominant species?

Figure 14.32 In the early 1800s, billions of passenger pigeons lived in the eastern United States. Migrating flocks sometimes darkened the skies for days. Farmers considered them pests and began to kill them in huge numbers. The last living passenger pigeon died in the Cincinnati (Ohio) Zoo in 1914. (This lifelike photo actually shows a stuffed bird in a museum diorama.) Today, human activity is driving many species, both large and small, extinct. The rate of extinction is so rapid that on geological time scales it may soon resemble one of Earth's past mass extinctions.

Putting Chapter 14 into Context

Humans have observed the Sun, the Moon, and the planets for thousands of years, but only recently did we learn that these other worlds have much to teach us about Earth. Through our study of solar system formation and comparative planetology, we have learned to look at Earth in a new light. Keep in mind the following "big picture" ideas:

● Earth has been shaped by the same geological and atmospheric processes that shaped the other terrestrial worlds. Earth is not a special case from a planetary point of view but rather a place where natural processes led to conditions conducive to life.

● Most of Earth's unique features can be traced to the fact that abundant water has remained liquid throughout our planet's history—thanks to the climate regulation of the carbon dioxide cycle and plate tectonics, which probably is a result of our distance from the Sun and the size of our planet.

● Life arose early on Earth and played a crucial role in shaping our planet's history. The abundant oxygen in our atmosphere is just one of many phenomena that demonstrate how life can transform a planet.

● Humans are ideally adapted to Earth today, but we have no guarantee that Earth will remain as hospitable in the future. The study of our solar system teaches us how planets can change.

SUMMARY OF KEY CONCEPTS

14.1 How Is Earth Different?

- *What are the five major ways in which Earth differs from other worlds in our solar system?* Earth differs in having plate tectonics, atmospheric oxygen, surface liquid water, climate stability, and life.

14.2 Our Unique Geology

- *How does plate tectonics act like a giant conveyor belt for Earth's lithosphere?* New crust emerges at mid-ocean ridges, leading to seafloor spreading. Subduction at trenches sends seafloor crust plunging back into the mantle.

- *Why does Earth have two types of crust?* Dense seafloor crust emerges at mid-ocean ridges and remains on the surface only until it subducts under continental crust. Near subduction zones, the seafloor crust begins to melt, releasing a lower-density lava that erupts from volcanoes to make new continental crust.

- *How does plate tectonics explain mountain ranges, earthquakes, and island chains such as Hawaii?* Mountain ranges are built up by volcanic eruptions near subduction zones or by rocks bulging up where continental plates collide. Earthquakes occur along plate boundaries, including boundaries where two plates are sliding past each other, sometimes getting stuck and allowing pressure to build until it releases violently in an earthquake. The Hawaiian islands are a chain of volcanic islands still forming as a plate moves over a mantle hot spot.

- *How has plate tectonics changed Earth's appearance through time?* Plate tectonics has moved continents around for much if not all of Earth's history. About 200 million years ago, the continents were all together in one large "supercontinent." The total continental area increases over time as new continental crust is made.

14.3 Our Unique Atmosphere and Oceans

- *Where did all the water in our oceans come from?* Water vapor trapped in Earth's interior after formation was outgassed by volcanoes and condensed to make rain, accumulating into the oceans.

- *Where is all of Earth's outgassed carbon dioxide?* The vast majority of it is locked up in carbonate rock, formed by chemical reactions after carbon dioxide dissolves in the oceans.

- *Why does Earth's atmosphere have oxygen and a stratosphere?* Life releases oxygen from carbon dioxide through photosynthesis. Atmospheric oxygen enables ozone to form and absorb ultraviolet light in the stratosphere.

14.4 Climate Regulation and the Carbon Dioxide Cycle

- *What is the carbon dioxide cycle?* Carbon dioxide dissolves in the oceans and reacts with silicate minerals to make carbonate rock. Plate tectonics carries the carbonate rock to subduction zones, where it plunges into the mantle. The carbonate rock then melts, releasing its carbon dioxide into the atmosphere through volcanic eruptions, and the cycle continues.

- *How does the carbon dioxide cycle regulate Earth's climate?* Feedback processes in the carbon dioxide cycle tend to counteract any warming by removing carbon dioxide from the atmosphere and to counteract cooling by adding carbon dioxide to the atmosphere.

continued ▶

- *Is Earth's climate perfectly stable?* No. Ice ages and perhaps even snowball Earth episodes have occurred in the past. In the future, the warming Sun will eventually lead to a runaway greenhouse effect.

14.5 Life on Earth

- *When did life arise on Earth?* Life arose at least 3.5 billion years ago, and perhaps much earlier than that.

- *How might chemicals on the early Earth have made living organisms?* Chemicals known to have been present on the early Earth react naturally to make complex organic molecules, including chains of RNA. Some of the RNA may have become self-replicating and ultimately could have led to DNA and full, self-replicating organisms.

- *How did life diversify?* Once living organisms with DNA existed, they evolved as mutations changed their DNA. Natural selection then allowed organisms that were better adapted to their environment to survive and pass their adaptations to future generations.

- *How did life create our oxygen atmosphere?* Cyanobacteria began releasing oxygen through photosynthesis, but for at least a billion years chemical reactions with surface rock removed the oxygen as fast as it was produced. Oxygen began to build up in the atmosphere only after the surface rock was saturated with it.

- *What was the Cambrian explosion?* It was an explosion in the diversity of life, especially animal life, that took place over about 40 million years, beginning about 540 million years ago.

14.6 Our Future Survival: Lessons from Other Worlds

- *Why is global warming a potential threat to our future?* Warming can change local climates, raise sea level, alter ocean currents, and alter ecosystems in ways that we cannot fully anticipate.

- *What is ozone depletion, and why is it dangerous?* Ozone depletion is a reduction in stratospheric ozone caused by human-made chemicals known as CFCs. It allows more ultraviolet light to reach the surface, increasing mutation rates among living organisms.

- *In what sense are we now causing a mass extinction?* At current rates, in this century we may drive so many species extinct that it will be one of the great mass extinctions of geological history.

❓ Surprising Discoveries?

Suppose we made the discoveries described below. (These are *not* real discoveries.) Decide whether each discovery should be considered reasonable or surprising. Explain. (In some cases, both views can be defended.)

1. A fossil of an organism that died more than 300 million years ago, found in the crust near a mid-ocean ridge.

2. Evidence that fish once swam in a region that is now high on a mountaintop.

3. A "lost continent" on which humans had a great city just a few thousand years ago but that now resides deep underground near a subduction zone.

4. A planet in another solar system that has an Earth-like atmosphere with plentiful oxygen but no life of any kind.

5. A planet in another solar system that has an ozone layer but no ordinary oxygen (O_2) in its atmosphere.

6. Evidence that the early Earth had more carbon dioxide in its atmosphere than Earth does today.

7. The discovery of life on Mars that also uses DNA as its genetic molecule and that uses a genetic code very similar to that used by life on Earth.

8. Evidence that plate tectonics completely turned off during snowball Earth episodes in the past, so that volcanism decreased until the snowball Earth episodes were over.

Problems

9. *Building Continents.* How are continents built up? Describe a few of the processes that have shaped North America.

10. *Two Paths Diverged.* Briefly explain why Earth has oceans and very little atmospheric carbon dioxide, while similar-size Venus has a thick carbon dioxide atmosphere.

11. *How Do We Know?* What evidence tells us that all life today shares a common ancestor? What is the earliest evidence for life, and what should we infer from the lack of earlier evidence?

12. *Change in Formation Properties.* Consider Earth's size, distance, composition, and rotation rate. Choose one property and suppose it had been different (e.g., smaller size or greater distance). Describe how this change might have affected Earth's subsequent history and the possibility of life on Earth.

13. *Growing Population.* Since about 1950, human population has grown with a doubling time of about 40 years; that is, the population doubles every 40 years. The current population (in 2002) is about 6 billion. If population continues to grow with a doubling time of 40 years, what will it be in 40 years? In 80 years? Do you think this will actually happen? Why or why not?

14. *Feedback Processes in the Atmosphere.* As the Sun gradually brightens in the future, how can the CO_2 cycle respond to reduce the warming effect? Which parts of the cycle will be affected? Is this an example of positive or negative feedback?

15. *Ozone Signature.* Suppose a powerful future telescope is able to take a spectrum of a terrestrial planet around another star. The spectrum reveals the presence of significant amounts of ozone. Why would this discovery strongly suggest the presence of life on this planet? Would it tell us whether the life is microbial or more complex? Summarize your answers in one or two paragraphs.

16. *Explaining Ourselves to the Aliens.* Imagine that someday we make contact with intelligent aliens. Many facets of life we take for granted may be utterly incomprehensible to them: music (perhaps they have no sense of hearing), money (perhaps there has never been a need), love (perhaps they don't feel this emotion), meals (perhaps they photosynthesize). Write a one-page description attempting to explain one of these concepts, or another of your choosing. Remember that even the words in your explanation require definition. Start at the lowest possible level.

Discussion Questions

17. *Evidence of Our Civilization.* Imagine a future archaeologist, say 10,000 years from now, trying to piece together a picture of human civilization at our time. What types of evidence of our civilization are most likely to survive for 10,000 years? For example, consider buildings, infrastructure such as highways and water pipes, and information such as books and computer data. What geological processes are likely to destroy evidence over the next 10,000 years? Next, discuss the evidence that will remain, and why, for an archaeologist living 100 million years from now.

18. *Defining Life.* Try to come up with a concise definition of life. Then evaluate whether each of the following three cases meets your definition: (i) A self-replicating molecule that is produced in a laboratory. (ii) A virus, which is essentially a packet of DNA (or RNA) encased in a microscopic shell of protein. Viruses cannot replicate themselves, but rather reproduce by infecting living cells and "hijacking" the cells' reproduction machinery to make copies of the viral DNA and proteins. (iii) Imagine that humans someday travel to other stars and discover a planet populated by what appear to be robots programmed to mine metal, refine it, and assemble copies of themselves. They reproduce themselves, learn, and even evolve into more sophisticated forms over time.

19. *Global Warming.* What, if anything, should we be doing to alleviate the threat of global warming that we are not doing already? Defend your opinion.

20. *Mass Extinction.* The fossil record suggests that the dominant animal species are nearly always victims in a mass extinction. If we are causing a mass extinction, do you think we will be victims of it as well, or will we be able to adapt to the changes so that we survive even while so many other species go extinct? Defend your opinion.

21. *Cancer of the Earth?* In the text, we discussed how the spread of humans over the Earth resembles, at least in some ways, the spread of cancer in a human body. Cancers end up killing themselves because of the damage they do to their hosts. Do you think we are in danger of killing ourselves through our actions on Earth? If so, what should we do to alleviate this danger? Overall, do you think the cancer analogy is valid or invalid? Defend your opinions.

For a complete list of media resources available, go to www.astronomyplace.com and choose Chapter 14 from the pull-down menu.

 Astronomy Place Web Tutorials

Tutorial Review of Key Concepts

Use the following interactive **Tutorials** at www.astronomyplace.com to review key concepts from this chapter.

Shaping Planetary Surfaces Tutorial

Lesson 1 The Four Geological Processes

Lesson 2 What Do Geological Processes Depend On?

Lesson 3 Planet Surface Evolution

Surface Temperature of Terrestrial Planets Tutorial

Lesson 1 Energy Balance

Lesson 2 Role of Planet's Distance from the Sun

Lesson 3 Role of Panet's Albedo

Lesson 4 Role of Planet's Atmosphere

Supplementary Tutorial Exercises

Use the interactive **Tutorial Lessons** to explore the following questions.

Shaping Planetary Surfaces Tutorial, Lesson 3

1. How has Earth's geology differed from that of the other terrestrial planets and why?

2. Suppose Earth had been born at half its current size. How would you expect it to have been different? Would life still be possible? Explain.

3. Suppose Earth had been born 10% larger than its current size. How would you expect it to have been different? Would life still be possible? Explain.

Surface Temperature of Terrestrial Planets Tutorial, Lesson 4

1. How would Earth's temperature be different if the carbon dioxide concentration doubled but all other gas concentrations remained the same?

2. A small warming could lead to increased evaporation of water. What would happen to the temperature if, along with the doubling of carbon dioxide as in question 1, the atmospheric water vapor concentration increased by 10% (to 1.1 times its current value)?

3. How might increased cloud cover offset the warming due to the greenhouse effect?

4. Increased cloud cover also means increased atmospheric water vapor. Use the tools to explore the balance between warming due to more water vapor and cooling due to more cloud cover. If we continued to add greenhouse gases to the atmosphere for centuries, what would ultimately happen to our planet's temperature? Explain.

Movies

Check out the following narrated and animated short documentary available on www.astronomyplace.com for a helpful review of key ideas covered in this chapter.

 Search for Extraterrestrial Life Movie

Web Projects

Take advantage of the useful Web links on www.astronomyplace.com to assist you with the following projects.

1. *Human Threats to Earth.* Write an in-depth research report, three to five pages in length, about current understanding and controversy regarding one of the following issues: global warming, ozone depletion, or the loss of species on Earth due to human activity. Be sure to address both the latest knowledge about the issue and proposals for alleviating any dangers associated with it. End your report by making your own recommendations about what, if anything, needs to be done to prevent damage to Earth.

2. *The Origin of Life.* Learn about one recent scientific discovery or controversy that concerns the origin of life on Earth. Write a one-page report about the discovery and how it fits in with other ideas about the origin of life on Earth.

3. *Volcanoes and Earthquakes.* Learn about one major earthquake or volcanic eruption that occurred during the past decade. Report on the geological conditions that led to the event, as well as on its geological and biological consequences.

4. *Local Geology.* Write a one- to two-page report about the geology of an area you know well, perhaps the location of your campus or your hometown. Describe the major geological features of the area and how they were formed.

A DEEPER LOOK AT NATURE

S2 Space and Time

Henceforth space by itself, and time by itself, are doomed to fade away into mere shadows, and only a kind of union of the two will preserve an independent reality.

Hermann Minkowski, 1908

The universe consists of matter and energy moving through *space* with the passage of *time*. Up to this point in the book, we have discussed the concepts of space and time as though they are absolute and distinct—just as they appear in everyday life. But what if this appearance is deceiving?

About a century ago, Albert Einstein discovered that space and time are not what they appear to be. Instead, space and time are intertwined in a remarkable manner described by Einstein's *theory of relativity*. Because space and time are such fundamental concepts, understanding relativity is important to understanding the universe.

The theory of relativity is *not* difficult to understand, despite popular myths to the contrary. It does, however, require us to think in new ways. That is our task in this chapter and the next. By the time we are finished, you will see that Einstein brought about a revolution in human thinking with many important ramifications for understanding our place in the universe.

S2.1 Einstein's Revolution

Imagine that, with the aid of a long tape measure, you carefully measure the distance you walk from home to work to be 5.0 kilometers. You wouldn't expect any argument about this distance. For example, if a friend drives her car along the same route and measures the distance with her car's odometer, she ought to get the same measurement of 5.0 kilometers—as long as her odometer is working properly.

Likewise, you would expect agreement about how much time it takes you to walk from home to work. Suppose your friend continues driving and you call her on a cellular phone just as you leave your house at 8:00 A.M. and again just as you arrive at work at 8:45 A.M. You'd certainly be surprised if she argued that your walk took an amount of time other than 45 minutes.

Distances and times appear absolute and distinct in our daily lives. We expect everyone to agree on the distance between two points, such as the locations of home and work. We also expect agreement about the time between two events, such as leaving home and arriving at work. Thus, it came as a huge surprise to everyone when in 1905 Albert Einstein showed that these expectations are not strictly correct.

With extremely precise measurements, the distance you measure between home and work will be *different* from the distance measured by a friend in a car, and you and your friend will also disagree about the time it takes you to walk to work. At ordinary speeds, the differences will be so small as to be unnoticeable. But if your friend could drive at a speed close to the speed of light, the differences would be substantial.

Disagreements about distances and times are only the beginning of the astonishing ideas contained in Einstein's **theory of relativity**. Einstein developed this theory in two parts. His **special theory of relativity**, published in 1905, showed how space and time are intertwined but did not deal with the effects of gravity. His **general theory of relativity**, published in 1915, offered a surprising new view of gravity—a view that we will use to help us understand topics such as the expansion and fate of the universe and the strange objects known as *black holes*.

In this chapter, we will focus on the new view of space and time in Einstein's special theory of relativity. In particular, we will see how this theory supports each of the following ideas:

- Nothing can travel faster than the speed of light (in a vacuum), and no material object can even reach the speed of light.

- If you carefully observe anyone or anything moving by you at a speed close to the speed of light, time will run more slowly for the moving object. That is, a person moving by you ages more slowly than you, a clock moving by you ticks more slowly than your clock, a computer moving by you runs more slowly than your similar computer, and so on.

- If you observe two events to occur simultaneously, such as flashes of light in two different places at the same time, a person moving by you at a speed close to the speed of light will not agree that the two events were simultaneous.

- If you carefully measure the size of something moving by you at a speed close to the speed of light, you will find that its length (in the direction of its motion) is shorter than it would be if the object were not moving.

- If you could measure the mass of something moving by you at a speed close to the speed of light, you would find its mass to be greater than the mass it would have if it were stationary. Einstein's famous equation, $E = mc^2$, follows from this fact.

Although the consequences of special relativity may sound like science fiction or fantasy, their reality is supported by a vast body of observational and experimental evidence. The consequences also follow logically from a few simple ideas. If you keep an open mind and think deeply as you read, you'll soon understand the basic ideas of relativity.

What Is Relativity?

Suppose a supersonic airplane is flying at a speed of 1,650 km/hr from Nairobi, Kenya, to Quito, Ecuador. How fast is the plane going? At first, this question sounds trivial—we have just said that the plane is going 1,650 km/hr.

But wait. Nairobi and Quito are both nearly on Earth's equator, and the equatorial speed of Earth's rotation is the same 1,650 km/hr speed at which the plane is flying [Section 1.3]. Moreover, the east-to-west flight from Nairobi to Quito is opposite the direction of Earth's rotation (Figure S2.1). Thus, if you could observe the plane from far off in space, it would appear to stay put *while Earth rotated beneath it.* When the flight began, you would see the plane lift straight off the ground in Nairobi. The plane would then remain stationary while Earth's rotation carried Nairobi away from it and Quito toward it. When Quito finally reached the plane's position, the plane would drop straight down to the ground.

We have two alternative viewpoints about the plane's flight. People on Earth say that the plane is traveling westward across the surface of the Earth. Observers in space say that the plane is stationary while Earth rotates eastward beneath it. Both viewpoints are equally valid.

In fact, there are many other equally valid viewpoints about the plane's flight. Observers looking at the solar system as a whole would see the plane moving at a speed

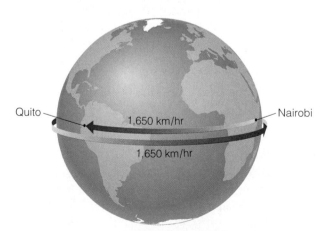

Figure S2.1 A plane flying at 1,650 km/hr from Nairobi to Quito (westward) travels precisely opposite Earth's eastward rotation. Thus, viewed from afar, the plane remains stationary while Earth rotates underneath it.

of more than 100,000 km/hr—Earth's speed in its orbit around the Sun. Observers living in a distant galaxy would see the plane moving away from them at a very high speed, carried along with Earth and the Milky Way Galaxy by the expansion of the universe. The only thing all these observers would agree on is that the plane is traveling at 1,650 km/hr *relative to* the surface of the Earth.

This example shows that questions like "Who is really moving?" and "How fast are you going?" have no absolute answers. Einstein's *theory of relativity* gets its name from the fact that it tells us that measurements of motion, as well as measurements of time and space, make sense only when we describe whom or what they are being measured relative to.

THINK ABOUT IT

Suppose you are running on a treadmill and the readout says you are going 8 miles per hour. What is the 8 miles per hour measured relative to? How fast are you going relative to the ground? How fast would an observer on the Moon see you going? Describe a few other possible viewpoints on your speed.

Note that the theory of relativity does *not* say that *everything* is relative. In particular, the theory claims that two things in the universe are absolute:

1. The laws of nature are the same for everyone.

2. The speed of light is the same for everyone.

As we'll see shortly, all the astounding consequences of relativity follow directly from these two seemingly innocuous statements.

Making Sense of Relativity

One reason why relativity has a reputation for being difficult to grasp, despite its underlying simplicity, is that most of its ideas and consequences are not obvious in everyday life. They become obvious only when we deal with speeds close to the speed of light or with gravitational fields far stronger than that of Earth. Because we don't commonly experience such extreme conditions, we have no *common* sense about them. Thus, to say that relativity violates common sense is not really accurate. The theory of relativity is perfectly consistent with everything we have come to expect in daily life.

Making sense of relativity requires only that you learn to view your everyday experiences from a new, broader perspective. Fortunately, you have learned to broaden your perspective in a similar way before. At a very young age, you learned "common sense" meanings for *up* and *down*: *Up* is above your head, *down* is toward your feet, and things tend to fall down. One day, however, you learned that Earth is round. When you looked at a globe with the Northern Hemisphere on the top, you were immediately confronted with a **paradox**—a situation that *seems* to violate common

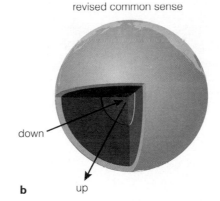

early-childhood common sense

up

down

a   Australia: They should "fall off."

revised common sense

down

b   up

Figure S2.2 Learning that Earth is round helps children revise their "common sense" understanding of *up* and *down*.

sense or to contradict itself. Your common sense told you that Australians should fall off the Earth (Figure S2.2a), but they don't. To resolve this paradox, you were forced to accept that your "common sense" about *up* and *down* was incorrect. You therefore revised your common sense to accept that *up* and *down* are determined relative to the center of the Earth (Figure S2.2b).

THINK ABOUT IT

Why do you suppose most maps and globes show the Northern Hemisphere on the top and the Southern Hemisphere on the bottom? If you hung your map upside down and rewrote the words so they read right-side up, would the map be equally valid?

As a teenager, Einstein wondered what the world would look like if he could travel at or beyond the speed of light. He inevitably encountered paradoxes when he thought about this question. Ultimately, he resolved the paradoxes only when he recognized that our commonsense ideas about space and time must change if we are to extend them to the realm of very high speeds or very strong gravitational fields. Just as we all once learned a new common sense about up and down, we now must learn a new common sense about space and time.

S2.2 Relative Motion

We will study relativity with the aid of *thought experiments* in which we create imaginary situations and logically follow them through to their conclusions. A starting point for our thought experiments will be the assumption that the two absolutes of relativity are true.

The first of the two absolutes, that the laws of nature are the same for everyone, is probably not surprising. If you're on an airplane with the shades drawn during a very smooth flight, you won't feel any sensation of motion. Thus, you should expect to get the same results from any experiments you perform on the airplane that someone else would

get performing the same experiments on the ground. These equivalent results prove that the laws of nature are the same in the airplane and on the ground.

In the language of relativity, the ground and the airplane represent different **reference frames** (or *frames of reference*) because they are moving relative to each other. The idea that the laws of nature are the same for everyone means that they do not depend on your frame of reference.

The second absolute of relativity, that the speed of light is the same for everyone, is far more surprising. In general, we expect people in different reference frames to give different answers for the speed of the same moving object. For example, suppose you roll a ball down the aisle of an airplane. The ball rolls slowly in your airplane reference frame, but it moves fast—because it also has the airplane's speed—in the reference frame of a person on the ground (Figure S2.3a). However, if you turn on a flashlight and measure the speed of the emitted light, a person on the ground will find exactly the same speed for the light beam (Figure S2.3b). That is, people in different reference frames can disagree about the speeds of material objects, but everyone always agrees on the speed of light—regardless of where the light comes from.

How can we know that everyone always measures the same speed of light? Einstein reached this conclusion because it was the only way he could resolve the paradoxes he encountered when he thought about traveling at the speed of light. Observations and experiments are the ultimate judge of any scientific theory. Many observations and experiments, some of which we'll study later in the chapter, have verified that the speed of light really is an absolute in nature. *The absoluteness of the speed of light is an experimentally verified fact.*

This statement may not seem important, but it will force you to let go of many of your intuitive beliefs about how the universe works. We will investigate why this statement forces such a change by constructing a series of thought experiments about relative motion viewed from different reference frames. If you follow the logic of the thought experiments carefully, you will come to understand the strange predictions of relativity.

a Observers in different reference frames measure different speeds for material objects.

b Everyone measures the same speed of light.

Figure S2.3 Unlike the case with the speed of material objects, the speed of light is the same in all reference frames.

Later, we will investigate the experimental evidence that shows these predictions to be accurate. Because special relativity does not deal with the effects of gravity, the thought experiments are easiest to visualize by imagining that we are in spaceships in deep space, far from any gravitational fields, and floating freely without engine power. Because everything in and around these spaceships is weightless and floats freely, we call the reference frames of these spaceships **free-float frames** (more traditionally called *inertial reference frames*).

Thought Experiments at Ordinary Speeds

First, to make sure you're comfortable with the ideas of relative motion in everyday life, let's analyze a few thought experiments involving ordinary speeds.

Thought Experiment 1 (Figure S2.4) Imagine that you are floating freely in a spaceship. Because you feel no sensation of motion, you perceive yourself to be at rest, or traveling at zero speed. As you look out your window, you see your friend Jackie in her own spaceship, moving away at a constant speed of 90 km/hr. How does the situation appear to Jackie?

We can answer the question by logically analyzing the experimental situation. Jackie has no reason to think she is moving. Like you, she is in a free-float frame, floating freely in a spaceship with the engines off. Jackie will therefore say that *she* is at rest and that *you* are moving away from her at 90 km/hr.

Both points of view—yours and Jackie's—are equally valid. You both would find the same results for any experiments performed in your own spaceship, and your research would lead you to exactly the same laws of nature. You could argue endlessly about who is really moving, but your argument would be pointless because all motion is relative.

Thought Experiment 2 (Figure S2.5) We begin with the same situation as in Thought Experiment 1, but this time you put on your space suit and strap yourself to the outside of your spaceship. You happen to have a baseball, which you throw in Jackie's direction at a speed of 100 km/hr. How is the ball moving relative to Jackie?

From your point of view, Jackie and the ball are both going in the same direction. Jackie is going 90 km/hr and the ball is going 100 km/hr, so the ball is going 10 km/hr faster than Jackie. The ball will therefore overtake and pass her.

From Jackie's point of view, *she* is stationary and *you* are moving away from her at 90 km/hr. Thus, she sees the ball moving toward her at 10 km/hr—the ball's speed of 100 km/hr relative to you *minus* the 90 km/hr at which you are moving relative to her. Note that both you and Jackie agree that the ball will pass her at a *relative* speed of 10 km/hr.

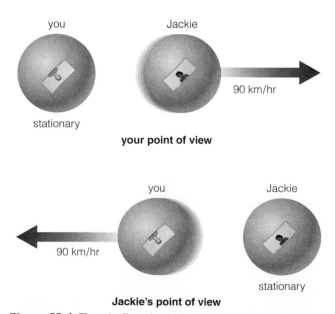

Figure S2.4 Thought Experiment 1.

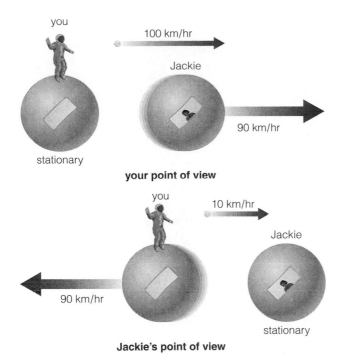

Figure S2.5 Thought Experiment 2.

Thought Experiment 3 (Figure S2.6) This time you throw a baseball in Jackie's direction at 90 km/hr. From your point of view, the ball is traveling at exactly the same speed as Jackie. Therefore, you'll see the ball forever chasing her through space, neither catching up nor falling behind.

From Jackie's point of view, the 90 km/hr at which you threw the ball exactly matches your 90 km/hr speed away from her. Therefore, the ball is *stationary* in her reference

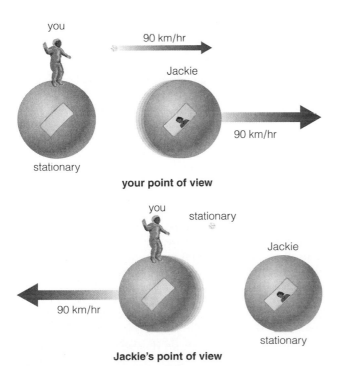

Figure S2.6 Thought Experiment 3.

frame. Think about this for a moment: Before you throw the baseball, Jackie sees it moving away from her at 90 km/hr because it is in your hand. At the moment you release the baseball, it suddenly becomes stationary in Jackie's reference frame, floating in space at a fixed distance from her spaceship. Many hours later, after you have traveled far away, Jackie will still see the ball floating in the same place. If she wishes, she can put on her spacesuit and go out to retrieve it, or she can just leave it there. From her point of view, it's not going anywhere, and neither is she.

In Thought Experiment 3, suppose instead that you throw the ball in Jackie's direction at 80 km/hr. In that case, what will Jackie see the ball doing? Next, suppose that Jackie is moving *toward* you rather than *away* from you. What would she see the ball doing in that case?

Thought Experiments at High Speeds

The absoluteness of the speed of light did not come into play in our first three thought experiments because the speeds were small compared to the speed of light. For example, 100 km/hr is less than *1 ten-millionth* of the speed of light. Now let's raise the speeds much higher and explore the strange consequences of the absoluteness of the speed of light.

Thought Experiment 4 (Figure S2.7) Imagine that Jackie is moving away from you at 90% of the speed of light, or 0.9*c*. (Recall that *c* is the symbol for the speed of light, which is about 300,000 km/s.) How does the situation appear to Jackie?

Other than the much higher speed, this situation is just like that in Thought Experiment 1. Jackie perceives *herself* to be at rest and sees *you* moving away from her at 0.9*c*.

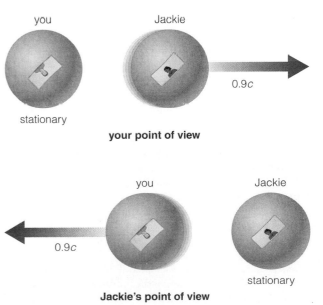

Figure S2.7 Thought Experiment 4.

Thought Experiment 5 (Figure S2.8) Now, instead of throwing a baseball, you climb out of your spaceship and point a flashlight in Jackie's direction. How is the beam of light moving relative to Jackie?

From your point of view, Jackie and the flashlight beam are both going in the same direction. Jackie is going at 90% of the speed of light, or 0.9*c*, and the light beam is going at the full speed of light, or *c*. Thus, you see the light beam going 0.1*c* faster than Jackie.* Nothing should be surprising so far.

From Jackie's point of view, *she* is stationary and *you* are moving away from her at 0.9*c*. Following the "old common sense" used in our earlier thought experiments, we would expect Jackie to see the light moving toward her at 0.1*c*—the light's speed of *c* minus your speed of 0.9*c*. *But this answer is wrong!* Relativity tells us that the speed of light is always the same for everyone. Therefore, Jackie must see the beam of light coming toward her at *c, not* at 0.1*c*.

In this case, you and Jackie no longer agree about her speed *relative* to the speed of the light beam: She'll see the light beam pass by her at the speed of light, *c*, but you'll see it going only 0.1*c* faster than she is going. By our old common sense, this result sounds preposterous. However, we found it by using simple logic, starting with the assumption that the speed of light is the same for everyone. As long as this assumption is true—and, remember, the absoluteness

---

*You can't actually *see* a light beam moving forward. When we say that you see the beam moving at the speed of light, we really mean that you would find it to be moving at this speed if you made a careful measurement with instruments at rest in your reference frame.

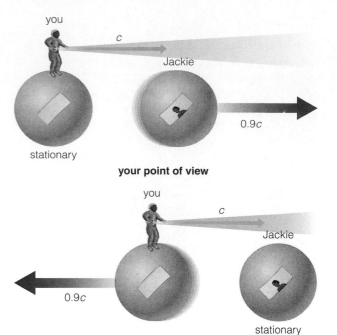

your point of view

Jackie's point of view

Figure S2.8 Thought Experiment 5.

of the speed of light is an experimentally verified fact—our conclusions follow logically.

Suppose Jackie is moving away from you at a speed of 0.99999*c*, which is short of the speed of light by only 0.00001*c*, or 3 km/s. How much faster than Jackie will you see the light going? How fast will Jackie see the light going as it passes her?

SPECIAL TOPIC   What If Light Can't Catch You?

If you're like most students learning about relativity for the first time, you're probably already looking for loopholes in the logic of our thought experiments. For example, confronted with Thought Experiment 6, you might be tempted to ask, "What happens if you're traveling away from some planet faster than light, so the light from the planet can't catch you?" While it's surely true that light couldn't catch you if you were going faster than the speed of light, it also makes the question somewhat moot: If you can't see light from the planet, there is no way for you to know that the planet is there.

In fact, what relativity really tells us about the speed of light is that it is a limit on the speed at which *information* can be transmitted. There are numerous circumstances in the universe in which, on philosophical grounds, an object may *seem* to be exceeding the speed of light. However, these circumstances do not provide any means of sending information or objects at speeds faster than the speed of light.

As an example, consider the implications of the fact that the universe is expanding [Section 1.3]. Recall that the more distant a galaxy, the faster the expansion of the universe is carrying it away from us. In principle, somewhere far away, there could be a point

beyond which the expansion is carrying galaxies away from us faster than the speed of light. However, we cannot observe such galaxies because their light cannot reach us. Scientifically, we say that such galaxies are beyond the bounds of our observable universe, and the observable universe is the only "universe" that we can study.

Other examples of things that *seem* to move faster than the speed of light arise frequently in the strange world of quantum mechanics (see Chapter S4). According to quantum principles, measuring a particle in one place can (in certain specific circumstances) affect a particle in another place *instantaneously*—even if the particle is many light-years away. In fact, this process has been observed in laboratories over short distances. This instantaneous effect of one particle on another may at first seem to violate relativity, but it does not. The built-in randomness of quantum mechanics prevents this technique from being used to transmit useful information to a distant point. Moreover, if we wish to confirm that the second particle really was affected, we will have to receive a signal carrying information about the particle—and that signal can travel no faster than the speed of light.

You Can't Reach the Speed of Light

You might be wondering what Jackie would see if she were moving away from you at the speed of light or faster—which is essentially the same question Einstein asked when he wondered how the world would look if he could travel at or beyond the speed of light. However, once we accept the absoluteness of the speed of light, it follows that neither she nor you nor any other material object can ever reach the speed of light, let alone exceed it.

Thought Experiment 6 You have just built the most incredible rocket imaginable, and you are taking it on a test ride. Soon you are going faster than anyone had ever imagined possible—and then you put the rocket into second gear! You keep going faster and faster and faster. Here is the key question: Are you ever traveling faster than the speed of light?

Before we answer this question, the fact that all motion is relative forces us to answer another question: In what reference frame is your speed being measured? Let's begin with *your* reference frame. Imagine that you turn on your rocket's headlights. Because the speed of light is the same for everyone, you must see the headlight beams traveling at the speed of light—which means they are racing ahead of your rocket at a speed you'll measure to be 300,000 km/s. The fact that you'll see your headlight beams racing away is true no matter how long you have been firing your rocket engines. Thus, you cannot possibly keep up with your own headlight beams.

The fact that you cannot outrace your headlight beams, combined with the fact that the speed of light is the same for everyone, means that no observer can ever see you reach or exceed the speed of light. Observers in different reference frames will measure your speed differently, but *all* observers will agree on two key points: (1) Your headlight beams are moving out ahead of you, and (2) these light beams are traveling at $c = 300,000$ km/s. Clearly, if you are being outraced by your headlights and if the headlights are traveling at the speed of light, you must be traveling *slower* than the speed of light. It does not matter who is measuring your speed. It can be you, someone on Earth, or anyone else in any other reference frame. No one can ever observe you to be traveling as fast as a light beam.

In case you are still not convinced, let's turn the situation around. Imagine that, as you race by some planet, a person on the planet turns on a light beam. Because the speed of light is absolute, you will see the light beam race past you at $c = 300,000$ km/s. The person on the planet will also see the light traveling at $c = 300,000$ km/s and will see the light outrace you. Thus, again, everyone will agree that you are traveling slower than the speed of light.

The same argument applies to any moving object. As long as the speed of light is absolute, no material object can reach or exceed it. Building a spaceship to travel at the speed of light is not a mere technological challenge—it simply cannot be done.

S2.3 The Reality of Space and Time

In the beginning of this chapter, we listed five key predictions of the theory of relativity. So far, we have shown only how the first one—that no material object can reach or exceed the speed of light—follows from the two absolutes of relativity. Let's continue on and see why our old conceptions of time and space must be revised.

Time Differs in Different Reference Frames

In preparation for our next thought experiment, imagine that you are on a moving train tossing a ball straight up so that it bounces straight back down from the ceiling. How do the path and speed of the ball appear to an observer along the tracks outside the train?

As shown in Figure S2.9, the outside observer sees the ball going forward with the train at the same time that it is going up and down, so the ball's path always slants forward. Because the outside observer sees the ball moving with this forward speed *in addition to* its up-and-down speed while you see the ball going only up and down, he

Inside the train, the ball goes up and down.

Outside the train, the ball appears to be going faster: It has the same up-and-down speed, plus the forward speed of the train.

The faster the train is moving, the faster the ball appears to be going to the outside observer.

Figure S2.9 A ball tossed straight up and down on a moving train follows a slanted path according to an outside observer.

would measure a *faster* overall speed for the ball than you would. If the train is moving slowly, the outside observer sees the ball's path slant forward only slightly and would say that the ball's overall speed is only slightly faster than you would report. If the train is moving rapidly, the ball's path leans much farther forward, and the ball appears to be going considerably faster to the outside observer than to you.

Thought Experiment 7 (Figure S2.10) Inside her spaceship, Jackie has a laser on her floor that is pointed up to a mirror on her ceiling. She momentarily flashes the laser light and uses a very accurate clock to time how long it takes the light to travel from the floor to the ceiling and back. As Jackie zips by you at a speed close to the speed of light, you observe her experiment through a window in her spaceship. Using your own very accurate clock, you also time the laser light's trip from Jackie's floor to her ceiling and back.

Because Jackie, the laser, and the mirror are all moving from your point of view, the light's path looks slanted as it goes from floor to ceiling and back. Thus, from your point of view, the light travels a *longer* path in going from the floor to the ceiling than it does from Jackie's point of view, just as the ball tossed in a train took a longer path from the outside observer's point of view.

By our old common sense, this would be no big deal. You and Jackie would agree on how long it takes the light to go from the floor to the ceiling and back, just as you and an outside observer would agree on how long it takes you to toss a ball up and down on a train. You would explain the light's longer path by saying that the light is moving faster relative to you than it is relative to Jackie, just as the outside observer sees the tossed ball moving faster because of the forward motion of the train. However, because we are dealing with *light*, you and Jackie must both see its speed to be the *same*—300,000 km/s—even though you see its path slanted forward with the movement of the spaceship.

Jackie's point of view

your point of view

Figure S2.10 Thought Experiment 7.

So what? Both you and Jackie see the light traveling at the same speed, but you see it traveling a longer distance—and, at a given speed, it takes more time to travel a longer distance. Thus, *your clock will record more time* than Jackie's as the light travels from floor to ceiling and back. That is, if you watch Jackie's clock, you will see it running slower than your own. It doesn't matter what you and Jackie use to measure the time for the light trip. *Anything* that can measure time will be going slower in Jackie's reference frame than in yours, including mechanical clocks, electrical clocks, heartbeats, and biochemical reactions. Our astonishing conclusion: From your point of view, *time itself* is running slower for Jackie.

How much slower is time running for Jackie? It depends on her speed relative to you. If she is moving slowly compared to the speed of light, you will scarcely be able to detect the slant, and your clock and Jackie's clock will tick at nearly the same rate. The faster she is moving, the more slanted the light path you see and the greater the difference between the rate of her clock and that of yours. Generalizing, we reach the following conclusion:

> From your point of view, time runs slower in the reference frame of anyone moving relative to you. The faster the other reference frame is moving, the slower time passes within it.

This effect is called **time dilation** because it tells us that time is *dilated,* or expanded, in a moving reference frame.

The Relativity of Simultaneity

Our old common sense tells us that everyone must agree on whether two events happen at the same time or one event happens before another. If you see two apples—one red and one green—fall from two different trees and hit the ground at the same time, you expect everyone else to agree that they landed at the same time (assuming you've accounted for any difference in the light travel times from the two trees). If you see the green apple land before the red apple, you are very surprised if someone else says that the red apple hit the ground first. Prepare yourself to be surprised, because our next thought experiment will show that observers in different reference frames will not necessarily agree about the order or simultaneity of events that occur in different places.

Before we go on, note that observers in different reference frames *must* agree about the order of events that occur in the *same* place. For example, suppose you grab a cookie and eat it. In your reference frame, both events (picking up the cookie and eating it) occur in the same place. Thus, it could not possibly be the case that someone else would see you eat the cookie before you pick it up.

Thought Experiment 8 Jackie is coming toward you in a brand new, extra-long spaceship. She is in the center of her spaceship, which is totally dark. At the instant that Jackie happens to pass you, two lights flash at either end of her spaceship. After a brief moment, the two flashes reach you at *exactly the same time*: a flash of green light from the front of her spaceship, and a flash of red light from the rear of her spaceship (Figure S2.11a). However, during the very short time that the light flashes are traveling toward Jackie, her speed carries her toward the point where you saw the green flash occur and away from the point where you saw the red flash occur (Figure S2.11b). Therefore, the green flash will reach her before the red flash: You'll see her illumi-

nated first by green light and then by red light. Jackie must also see herself first turn green and then turn red, agreeing with you that the green light reached her before the red light.

So far, nothing is surprising. But remember that Jackie considers herself to be in the center of a *stationary* space-ship. From her point of view, the flashes from the front and the rear both travel toward her at the *same speed* (the speed of light), and both have the same distance to travel. There-fore, because the green flash from the front of the spaceship

---

Mathematical Insight **S2.1** The Time Dilation Formula

Look at the different paths that you and Jackie see the light take from the floor to the ceiling in Figure S2.10. We can form a right triangle from these paths, which we can use to find an exact formula for time dilation. Recall that *distance = speed × time*, and note how the sides of the triangle are constructed in Figure 1.

● *You* see the light take a slanted path from Jackie's floor to ceil-ing because of her motion relative to you. Let's use t to repre-sent the time that *you* measure as the light travels this path. Because the light travels at the speed of light, c, the length of this path is $c \times t$.

● The distance that *you* see Jackie travel while the light goes from her floor to her ceiling is $v \times t$, where v is her speed relative to you.

● We know that Jackie measures time differently than you, so we'll use t' (read "t-prime") to represent the time *she* measures as the light travels from her floor to her ceiling. The distance she sees the light travel therefore is $c \times t'$.

Because we have a right triangle, we can solve for the time on Jackie's clock (t') in terms of the time on your clock (t) by using the Pythagorean theorem and a bit of algebra. (Remember that the Pythagorean theorem states that $a^2 + b^2 = c^2$ for a right triangle with side lengths a, b, and c, where c is the hypotenuse.)

Start with the Pythagorean theorem:

$$(ct')^2 + (vt)^2 = (ct)^2$$

Expand the squares:

$$c^2t'^2 + v^2t^2 = c^2t^2$$

Subtract v^2t^2 from both sides:

$$c^2t'^2 = c^2t^2 - v^2t^2$$

Divide both sides by c^2t^2:

$$\frac{t'^2}{t^2} = \frac{c^2 - v^2}{c^2}$$

Simplify:

$$\frac{t'^2}{t^2} = \frac{c^2}{c^2} - \frac{v^2}{c^2} = 1 - \left(\frac{v}{c}\right)^2$$

Take the square root of both sides:

$$\frac{t'}{t} = \sqrt{1 - \left(\frac{v}{c}\right)^2} \quad \text{or} \quad t' = t\sqrt{1 - \left(\frac{v}{c}\right)^2}$$

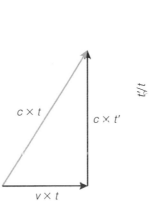

Figure 1 The setup for finding the time dilation formula.

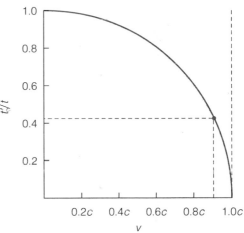

Figure 2 The time dilation factor graphed against speed.

The final result, called the *time dilation formula*, tells us the *ratio* of time in a moving reference frame to time in a reference frame at rest. This ratio is graphed against speed in Figure 2. Note that $t'/t \approx 1$ at speeds that are small compared to the speed of light, meaning that clocks in both reference frames tick at about the same rate. As v approaches c, the amount of time passing in the moving reference frame gets smaller and smaller compared to the time passing in the reference frame at rest.

Example: Suppose Jackie is moving past you at a speed of $0.9c$. While 1 hour passes for you, how much time passes for Jackie?

Solution: Jackie's speed of $v = 0.9c$ means that $v/c = 0.9$. The variable t represents *your* time, so $t = 1$ hour. Substituting into the time dilation formula yields:

$$t' = t\sqrt{1 - \left(\frac{v}{c}\right)^2}$$
$$= (1 \text{ hr}) \sqrt{1 - (0.9)^2}$$
$$= (1 \text{ hr}) \sqrt{1 - 0.81}$$
$$= (1 \text{ hr}) \sqrt{0.19} \approx 0.44 \text{ hr}$$

or about 26 minutes. Thus, while 1 hour passes for you, only 26 minutes pass for Jackie. You can also find this answer from the graph in Figure 2. At a speed of $v = 0.9c$, the graph shows that the ratio of Jackie's time to your time is 0.44.

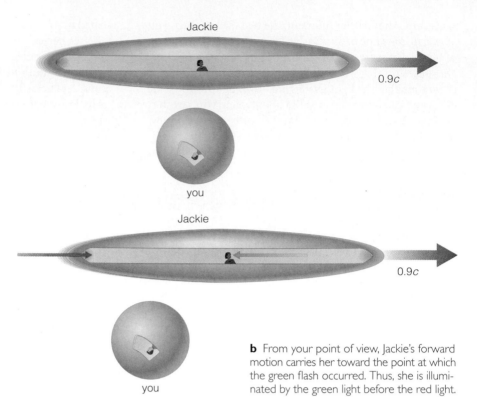

a In your reference frame, the green and red flashes occur simultaneously at the instant when Jackie is immediately adjacent to you.

Figure S2.11 Thought Experiment 8. Green and red flashes of light come from the two ends of Jackie's spaceship. Both you and Jackie will agree that the green light reaches her before the red light. You'll say that both flashes occurred simultaneously and that the green light reached her first because of her forward motion. However, because Jackie thinks she is stationary, she can explain the fact that the green light reaches her first only by concluding that it flashed before the red light.

b From your point of view, Jackie's forward motion carries her toward the point at which the green flash occurred. Thus, she is illuminated by the green light before the red light.

reaches her *before* the red flash from the rear, she must conclude that the green flash happened first and the red flash happened some time later. Thus, while you say that the two flashes were simultaneous, Jackie disagrees: She says that the green flash occurred before the red flash. Both of you are equally correct, because the order or simultaneity of events depends on your frame of reference.

Effects on Length and Mass

The fact that time is different in different reference frames means that lengths (or distances) and masses must also be affected, although the explanations are a bit subtler. The following two thought experiments use the idea of time dilation to help us understand the effects on length and mass.

Thought Experiment 9 Jackie is back in her original spaceship, coming toward you at high speed. As usual, both you and she agree on your *relative* speed. You disagree only about who is stationary and who is moving. Now imagine that Jackie tries to measure the length of your spaceship as she passes by you.

In your reference frame, you'll see Jackie's time running slower. Because her clocks record less time than yours as she passes from one end of your spaceship to the other, she must measure the length of your spaceship to be shorter than you measure it to be. (Recall that distance is *speed* × *time*.) Similarly, you will measure Jackie's spaceship as having a shorter length than it would have if it were at rest in your reference frame, an effect called **length contraction**. Figure S2.12 shows that lengths are affected only in the direc-

tion of motion. Her spaceship is shorter from your point of view, but its height and depth are unaffected. Generalizing, we reach the following conclusion:

> From your point of view, lengths of objects moving by you (or distances between objects moving by you) are shorter in their direction of motion than they would be if the objects were at rest. The faster the objects are moving, the shorter the lengths.

Thought Experiment 10 Imagine that Jackie has an identical twin sister with an identical spaceship, and suppose that her sister is at rest in your reference frame while Jackie is moving by you at high speed. At the instant Jackie passes by, you give both Jackie and her sister identical pushes (Figure S2.13). If Jackie and her sister are truly identical, the force of your push should have the same effect on each of them. For example, it might cause each of them to gain 1 km/s of speed relative to you. However, because you'll see Jackie's time running slower than yours and her sister's, you will conclude that Jackie feels the force of your push for a *shorter* time than her sister feels your push. (For example,

Figure S2.12 Thought Experiment 9. People and objects moving relative to you are contracted in their direction of motion.

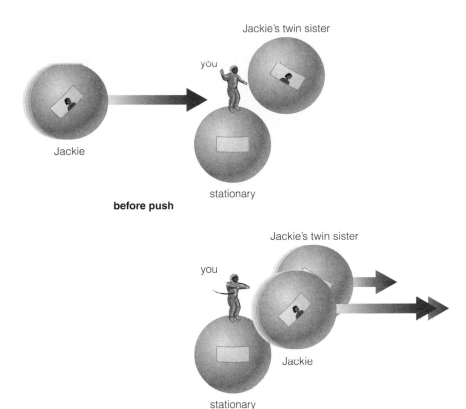

Figure S2.13
Thought Experiment 10.

before push

after push

if you and Jackie's sister measure the duration of the push to be 1 microsecond, Jackie's clock will show the duration to be less than 1 microsecond.)

Because Jackie feels the force of your push for a shorter time than her sister, the push must have a *smaller* effect on Jackie's velocity. In other words, you'll find that your push has less effect on Jackie than on her sister, despite the fact that you gave them identical pushes. According to Newton's laws of motion [Section 5.2], the only way the same push can have a smaller effect on Jackie's velocity is if her mass is *greater* than her sister's mass. This effect is sometimes called **mass increase:**

> **From your point of view, objects moving by you have greater mass than they have at rest. The faster an object is moving, the greater the increase in its mass.**

Mass increase provides another way of understanding why no material object can reach the speed of light. The faster an object is moving relative to you, the greater mass you'll find it to have. Thus, at higher speeds, the same force will have less effect on an object's velocity. In fact, as an object approaches the speed of light, you will find its mass to be heading toward infinity. No force can accelerate an infinite mass, so the object can never gain that last little bit of speed needed to push it to the speed of light.

Velocity Addition

We have just one more important effect to discuss: As the next thought experiment shows, you and Jackie will dis-

agree about the speed of a material object moving relative to both of you. The only speed you will agree on is the speed of light.

Thought Experiment 11 (Figure S2.14) Jackie is moving toward you at 0.9c. Your friend Bob jumps into his spaceship and starts heading in Jackie's direction at 0.8c (from your point of view). How fast will Jackie see him approaching her?

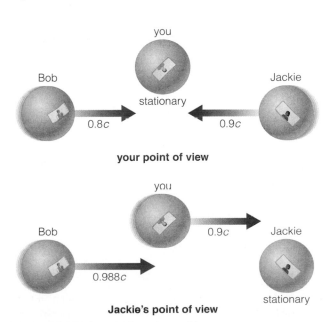

your point of view

Jackie's point of view

Figure S2.14 Thought Experiment 11.

By our old common sense, Jackie should see Bob coming toward her at $0.9c + 0.8c = 1.7c$. We know this answer is wrong, however, because $1.7c$ is faster than the speed of light. Jackie must see Bob coming toward her at a speed less than c, but he will be coming faster than the speed of $0.9c$ at which she sees you coming. Thus, she'll conclude that Bob's speed is somewhere between $0.9c$ and c. (In this case, it turns out that she'll see him coming at $0.988c$. See Mathematical Insight S2.2.)

S2.4 Is It True?

We've shown how all the major predictions of special relativity follow directly from the absoluteness of the speed of light and from the fact that the laws of nature are the same for everyone. Despite the clear logic of our thought experiments, our conclusions remain tentative until they pass observational or experimental tests. Are the bizarre predictions of relativity true?

The Absoluteness of the Speed of Light

The first thing we might wish to test is the surprising premise of relativity: the absoluteness of the speed of light. In principle, we can test this premise by measuring the speed of light coming from many different objects and going in many different directions and verifying that the speed is always the same.

The speed of light was first measured in the late 1600s, but experimental evidence for the *absoluteness* of the speed of light did not come until 1887, when A. A. Michelson and E. W. Morley performed their now-famous *Michelson–Morley experiment*. This experiment showed that the speed of light is not affected by the motion of Earth around the Sun. Countless subsequent experiments verified and ex-

---

Mathematical Insight **S2.2** **Formulas of Special Relativity**

We found a formula for time dilation in Mathematical Insight S2.1. Although we will not go through the derivations, it is possible to find similar formulas for length contraction and mass increase. In summary, the three formulas are:

$$\underset{\text{(moving frame)}}{\text{time}} = \underset{\text{(rest frame)}}{\text{time}} \times \sqrt{1 - \left(\frac{v}{c}\right)^2}$$

$$\underset{\text{(moving frame)}}{\text{length}} = (\text{rest length}) \times \sqrt{1 - \left(\frac{v}{c}\right)^2}$$

$$\text{moving mass} = \frac{(\text{rest mass})}{\sqrt{1 - \left(\frac{v}{c}\right)^2}}$$

There's also a simple formula for velocity addition. Suppose you see Jackie moving at speed v_1 and Jackie sees a second object moving relative to her at speed v_2. By our old common sense, you would see the second object moving at speed $v_1 + v_2$. However, the speed you actually see is:

$$\text{speed of second object} = \frac{v_1 + v_2}{1 + \left(\frac{v_1}{c} \times \frac{v_2}{c}\right)}$$

Example 1: Length Contraction. Suppose Jackie is moving by you at $0.99c$. Because her spaceship is the same model as yours, you know that it is 100 meters long when it is at rest. How long is her spaceship as it moves by you?

Solution: We use the length contraction formula to calculate the length of the moving spaceship:

$$\underset{\text{(moving frame)}}{\text{length}} = (\text{rest length}) \times \sqrt{1 - \left(\frac{v}{c}\right)^2}$$
$$= (100 \text{ m}) \times \sqrt{1 - (0.99)^2} = 14 \text{ m}$$

You would measure her spaceship as only 14 meters long, instead of its rest length of 100 meters.

Example 2: Mass Increase. A fly has a mass of 1 gram at rest. It is an unusual fly, however, in that it can travel at $0.9999c$. What is the mass of the fly at that speed?

Solution: We use the mass increase formula to calculate the mass of the moving fly:

$$\text{moving mass} = \frac{(\text{rest mass})}{\sqrt{1 - \left(\frac{v}{c}\right)^2}}$$
$$= \frac{1 \text{ g}}{\sqrt{1 - (0.9999)^2}} = 70.7 \text{ g}$$

At $0.9999c$, the mass of the fly is almost 71 grams, or more than 70 times its rest mass.

Example 3: Velocity Addition. Jackie is moving toward you at $0.9c$. Your friend Bob jumps into his spaceship and starts heading in Jackie's direction at $0.8c$ (from your point of view). How fast will Jackie see him approaching her? (See Figure S2.14.)

Solution: According to Jackie, your speed is $v_1 = 0.9c$. Bob's speed *relative to you* is $v_2 = 0.8c$ in the same direction. Thus, she sees him coming at:

$$\underset{\text{(relative to Jackie)}}{\text{Bob's speed}} = \frac{v_1 + v_2}{1 + \left(\frac{v_1}{c} \times \frac{v_2}{c}\right)}$$
$$= \frac{0.9c + 0.8c}{1 + (0.9 \times 0.8)}$$
$$= \frac{1.7c}{1.72} = 0.988c$$

Jackie sees Bob moving at almost 99% of the speed of light.

tended the results of the Michelson–Morley experiment: The speed of light is always the same.

Experimental Tests of Special Relativity

Although *we* cannot yet travel at speeds at which the effects of relativity should be obvious, tiny subatomic particles can reach such speeds. In machines called *particle accelerators*, physicists accelerate subatomic particles to speeds near the speed of light and study what happens when the particles collide. The collisions involve large amounts of kinetic energy, some of which is converted into mass-energy that emerges as a shower of newly produced particles [Section 4.2]. Many of these particles have very short lifetimes, at the end of which they decay (change) into other particles.

For example, a particle called the π^+ ("pi plus") meson has a lifetime of about 18 nanoseconds (billionths of a second) when produced at rest. But π^+ mesons produced at speeds close to the speed of light in particle accelerators last much longer than 18 nanoseconds—just as predicted by the time dilation formula.

Particle accelerators also offer experimental evidence that nothing can reach the speed of light. It is relatively easy to get particles traveling at 99% of the speed of light in particle accelerators. However, no matter how much more energy is put into the accelerators, the particle speeds get only fractionally closer to the speed of light. Some particles have been accelerated to speeds within 0.00001% of the speed of light, but none have ever reached the speed of light.

Although the effects of relativity are obvious only at very high speeds, modern techniques of measuring time are so precise that effects can be measured even at ordinary speeds. For example, a 1975 experiment compared the amount of time that passed on an airplane flying in circles to the time that passed on the ground. Over 15 hours, the airborne clocks lost a bit under 6 nanoseconds to the ground clocks, matching the result expected from relativity.

A Great Conspiracy?

Perhaps you're thinking "I still don't believe it." After all, how can you know that the scientists who report the experimental evidence are telling the truth? Perhaps physicists are making up the whole thing as part of a great conspiracy designed to confuse everyone else so they can take over the world!

What you need is evidence that you can see for yourself. What about nuclear energy? Einstein's famous formula $E = mc^2$, which explains the energy release in nuclear reactions, is a direct consequence of the special theory of relativity—and one that you can derive for yourself with a bit of algebra (see Mathematical Insight S2.3). Every time you see film of an atomic bomb, use electrical power from a nuclear power plant, or feel the energy of sunlight that was generated in the Sun by nuclear fusion, you are gaining experimental evidence of relativity.

Another test you can do yourself is to look through a telescope at a binary star system. If the speed of light were *not* absolute, the speed at which light from each star comes toward Earth would depend on its velocity toward us in the binary orbit (Figure S2.15). Imagine, for example, that one star is currently moving directly away from us in its orbit.

SPECIAL TOPIC Measuring the Speed of Light

The only reason it is difficult to measure the speed of light is that light travels so fast. If you stand a short distance from a mirror and turn on a light, the reflection seems to appear instantaneously. Such observations led Aristotle (384–322 B.C.) to conclude that light travels at infinite speed, a view that was still held by many scientists as recently as the late 1600s.

One way to make the measurement easier is to place a mirror at increasingly great distances. If the speed of light truly were infinite, the reflection would always appear instantaneously. However, if it takes time for the light to travel to and from the mirror, you should eventually find a delay between the time you turn on the light and the time you see the reflection. Galileo tried a version of this experiment using the distance between two tall hills, but he was not able to detect any delay (instead of using a mirror, he stationed an assistant on the distant hill to signal back when he saw the light). He concluded that the speed of light, if not infinite, was too fast to be measured between hills on Earth with the technology of his day.

A delay in seeing reflected light was first detected in 1675 by the Danish astronomer Olaus Roemer, who used the four largest moons of Jupiter as his "mirrors." By that time, the orbital periods of the moons were well known, so it was possible to predict the precise moments at which each moon would be eclipsed by Jupiter.

To his surprise, Roemer found that the eclipses occurred progressively earlier than expected during those times of year when Earth was moving toward Jupiter in its orbit and progressively later when Earth was moving away from Jupiter. He realized that the eclipses actually were occurring at the predicted times, but the light was taking longer to reach us when we were farther from Jupiter.

Roemer's observations proved that the speed of light is finite. Using estimates of the Sun–Earth and Sun–Jupiter distances that were available in 1675, Roemer calculated the speed of light to be 227,000 km/s. Redoing his calculations using the presently known values of these distances yields 300,000 km/s.

As technology advanced, it became possible to measure light travel time between mirrors at much closer distances. In 1849 and 1850, the French physicists Fizeau and Foucault (also famous for the Foucault pendulum) performed a series of experiments using rotating mirrors to measure the speed of light much more precisely. Modern devices for measuring the speed of light take advantage of the wave properties of light, particularly the fact that light waves can interfere with one another. Such devices, called *interferometers*, were refined by A. A. Michelson in the early 1880s and used in the Michelson–Morley experiment. Details of how these experiments work can be found in many physics texts.

Light from the star at this point would approach us at speed $c - v$. Some time later, when the same star is moving toward us in its orbit, its light would approach us at speed $c + v$. This light would therefore tend to catch up with the light that the star emitted from the other side of its orbit. If the orbital speed and distance were just right, we might see the same star on both sides of the orbit at once! More generally, because the light from each star would come toward us at a different speed from each point in its orbit, we would see each star in multiple positions in its orbit simultaneously. Thus, each star would appear as a short line of light rather than as a point—if the speed of light were not absolute. The fact that we always see distinct stars in binary systems demonstrates that the speed of light *is* absolute.*

---

*This conclusion would not follow if light waves were carried by a medium in the same way that sound waves are carried by air. Scientists in the 1800s believed that such a medium, which they called the *ether*, permeated all of space. The Michelson–Morley experiment ruled out the existence of such a medium, so we are left with the conclusion that c is absolute.

Finally, you can explore the paradoxes that would occur if the speed of light were not absolute. For example, imagine that two cars, both traveling at about 100 km/hr,

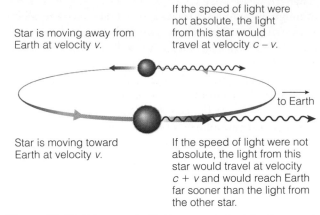

Figure S2.15 If the speed of light were *not* absolute, the speed at which light from each star in a binary system comes toward Earth would depend on its velocity toward us in the binary orbit.

Mathematical Insight **S2.3** **Deriving $E = mc^2$**

We can derive the formula $E = mc^2$ from the mass increase formula, calling the moving mass m and the rest mass m_0:

$$\text{moving mass} = \frac{(\text{rest mass})}{\sqrt{1 - \left(\dfrac{v}{c}\right)^2}}$$

or

$$m = m_0 \left(1 - \frac{v^2}{c^2}\right)^{-\frac{1}{2}}$$

We also need a special mathematical approximation: If x is small compared to 1, then:

$$(1 + x)^{-\frac{1}{2}} \approx 1 - \frac{1}{2}x \quad \text{(for } x \text{ small compared to 1)}$$

You can verify this approximation by using your calculator to check that it holds for a few small values of x, such as $x = 0.05$ or $x = 0.001$.

Note what happens if we substitute $x = -v^2/c^2$ in this approximation:

Start with the approximation:

$$(1 + x)^{-\frac{1}{2}} \approx 1 - \frac{1}{2}x$$

Substitute $x = -v^2/c^2$:

$$\left(1 + -\frac{v^2}{c^2}\right)^{-\frac{1}{2}} \approx 1 - \frac{1}{2}\left(-\frac{v^2}{c^2}\right)$$

Simplify:

$$\left(1 - \frac{v^2}{c^2}\right)^{-\frac{1}{2}} \approx 1 + \frac{1}{2}\frac{v^2}{c^2}$$

We can use this new form of the approximation to rewrite the mass increase formula. The condition that x must be small compared to 1 now means that $-v^2/c^2$ must be small. That is, the speed v must be small compared to the speed of light, c.

We start with the mass increase formula:

$$m = m_0 \left(1 - \frac{v^2}{c^2}\right)^{-\frac{1}{2}}$$

Notice that the term in parentheses on the right is the term for which we found an approximation (bottom of left column). We replace this term with its approximation, so that the mass increase formula now reads:

$$m \approx m_0 \left(1 + \frac{1}{2}\frac{v^2}{c^2}\right)$$

We expand the right side to find:

$$m \approx m_0 + \frac{1}{2}\frac{m_0 v^2}{c^2}$$

Finally, we multiply both sides by c^2, so the formula becomes:

$$mc^2 \approx m_0 c^2 + \frac{1}{2}m_0 v^2$$

You may recognize the last term on the right as the *kinetic energy* [Section 4.2] of an object with mass m_0. Because the other two terms also have units of mass multiplied by speed squared, they also must represent some kind of energy. Einstein recognized that the term on the left represents the *total* energy of a moving object. He then noticed that, even if the speed is *zero* ($v = 0$) so that there is *no* kinetic energy, the equation states that the total energy is *not* zero but instead is $m_0 c^2$. In other words, when an object is not moving at all, it still contains energy by virtue of its mass. Thus, $E = mc^2$ is a direct consequence of Einstein's theory of relativity.

collide at an intersection (Figure S2.16). You witness the collision from far down one street. If the speed of light were *not* absolute, the light from the car that was coming toward you would have a speed of *c* + 100 km/hr, while the light from the other car would approach you only at a speed of *c*. You therefore would see the car coming toward you reach the intersection slightly *before* the other car and thus would see events unfold differently than the passengers in the car or eyewitnesses in other locations. This difference would be scarcely noticeable, because 100 km/hr is only about *one-millionth* the speed of light, but the difference would increase if you could watch from *very* far away. For example, if you had a super telescope and watched such a collision on a planet in a galaxy 1 million light-years from Earth, you'd see the first car reach the intersection a *year* before the second car. This is a serious paradox: From the viewpoint of the passengers in the cars, they have collided, yet you saw one car reach the collision point long before the other car had even started its journey!

If *c* were not absolute, you would see car A reach the collision point before car B.

Figure S2.16 If the speed of light were *not* absolute, the light from a car coming toward you would approach you faster than the light from a car going across your line of sight.

THINK ABOUT IT

The preceding paradox presents us with a choice. If the speed of light is *not* absolute, different people can witness the same events in very different ways. If the speed of light *is* absolute, measurements of time and space are relative. Einstein preferred the latter choice. Do you? Explain.

Of course, like any scientific theory, the theory of relativity can never be *proved* beyond all doubt. However, it is supported by a tremendous body of evidence, some of which you can see for yourself. This evidence is real and cannot be made to disappear. If anyone ever comes up with an alternative theory, the new theory will still have to explain the many experimental results that seem to support relativity so well.

S2.5 Toward a New Common Sense

We've used thought experiments to show that our old common sense doesn't work, and we've discussed how actual experiments verify the ideas of our thought experiments. But we haven't yet figured out what new common sense should replace the old. Surprisingly, another thought experiment that may at first make everything seem even more bizarre will help us understand what is really going on.

Thought Experiment 12 Suppose Jackie is moving by you at a speed close to the speed of light. From our earlier thought experiments, we know that you'll measure her time as running slow, her length as having contracted, and her mass as having increased. But what would *she* say?

From Jackie's point of view, she's not going anywhere—you are moving by *her* at high speed. Because the laws of nature are the same for everyone, she must reach exactly the same conclusions from her point of view that you reach from your point of view. That is, she'll say that *your* time is running slow, *your* length is contracted, and *your* mass is increased!

Now we have a severe argument on our hands. Imagine that you are looking into Jackie's spaceship with a super telescope. You can clearly see that her time is running slow because everything she does is in slow motion. You send her a radio message saying, "Hi, Jackie! Why are you doing everything in slow motion?" Because the radio message travels at the absolute speed of light, Jackie has no trouble receiving your message, and she responds with her own radio message back to you.

As you listen to her response, you'll hear it in slow motion—"Hheeeelllllooooo tthhheeerrrreee"—thus verifying that her time is running slow. However, if you record the entire message and use your computer to speed up her voice so that it sounds normal, you'll hear Jackie say, "I'm not moving in slow motion, *you are!*"

You can argue back and forth all you want, but it will get neither of you anywhere. Then you come up with a brilliant idea. You hook up a video camera to your telescope and record a film showing that Jackie's clock is moving slower than yours. You put your videotape in a very fast rocket and shoot it off toward her. When the video arrives and she watches it, she'll see that you are right and that *she* is in slow motion.

Before you declare victory in the argument, one slight problem remains. Jackie had the same brilliant idea, and a rocket from Jackie arrives with a videotape that she made. You put it into the videotape player and watch what appears to be clear proof that Jackie is right—her tape shows that *you* are in slow motion!

How can it be that you see Jackie's time running slow while she sees your time running slow? Think back to our earlier discussion of up and down, and imagine that an American child and an Australian child are talking on the telephone. The Australian says, "Isn't the Moon beautiful up in the sky right now?" The American replies, "What are

Figure S2.17 The Moon is up for the Australian observer but down for the American observer. (Not to scale.)

you talking about? The Moon isn't up right now!" According to childhood common sense about up and down, the two children appear to be contradicting each other and could argue endlessly. However, once they realize that up and down are measured *relative* to the center of the Earth, they realize that the argument stems only from incorrect definitions of *up* and *down* (Figure S2.17).

Similarly, the argument between you and Jackie arises because you are using the old common sense in which we think of space and time as absolutes and expect the speed of light to be relative. The theory of relativity tells us that we have it backward. The speed of light is the absolute, and time and space are relative.

By the new common sense, the fact that you and Jackie disagree about whose time is running slow is no more surprising than the fact that the two children disagree about whether the Moon is up or down. The disagreement is meaningless because it involves inadequate definitions of time and space. What counts are *results*, and every experiment

that you perform will agree with every experiment that Jackie performs. You are both experiencing the same laws of nature, albeit in ways different from those our old common sense would have suggested.

S2.6 Ticket to the Stars

The fact that we cannot exceed the speed of light might at first make distant stars seem forever out of reach, but time dilation and length contraction actually offer a ticket to the stars—if we can ever build spaceships capable of traveling at speeds close to the speed of light.

Suppose you want to take a trip to the star Vega, about 25 light-years away. Further, suppose you have a ship that can travel at very close to the speed of light—say, at $0.999c$. The trip will take you about 25 years from our point of view on Earth (since you are going at nearly the speed of light over a distance of 25 light-years), and the return trip will take another 25 years. If you leave in the year 2025, you will arrive at Vega in the year 2050 and return to Earth in 2075.

However, from your point of view, you remain stationary while Earth rushes away from you and Vega rushes toward you at $0.999c$. You'll therefore find the length from Earth to Vega contracted from its rest length of 25 light-years—with the length contraction formula, the contracted distance turns out to be just over 1 light-year. Because Vega is coming toward you at $0.999c$ and has only 1 light-year to travel from your point of view, you'll be at Vega in only about 1 year. Your return trip to Earth will also take about 1 year, so the round-trip time is only about 2 years from your point of view (Figure S2.18). If you leave at age 40, you'll return as a 42-year-old.

Figure S2.18 At high speed, a traveler to a distant star will age less than people back home on Earth.

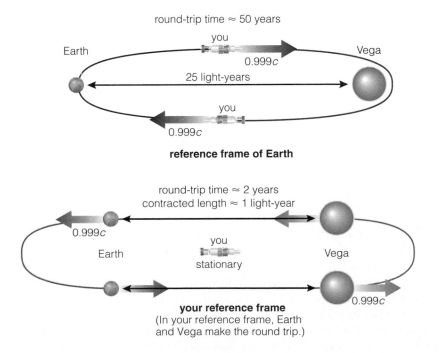

Although it sounds contradictory by our old common sense, both points of view are correct.* If you leave in 2025 at age 40, you'll return to Earth at age 42—but in the year 2075. That is, while you will have aged only 2 years, all your surviving friends and family will be 50 years older than when you left.

You could make even longer trips within your lifetime with a sufficiently fast spaceship. For example, the Andromeda Galaxy is about 2.5 million light-years away, so the round-trip to any star in the Andromeda Galaxy would take at least 5 million years from the point of view of observers on Earth. However, if you could travel at a speed within 50 parts in 1 trillion of the speed of light (i.e., $c - 5 \times 10^{-11}c$), the trip would take only about 50 years from your point of view. You could leave Earth at age 30 and return at age 80—but you would return to an Earth on which your friends, your family, and everything you knew had been gone for 5 million years.

In terms of time, relativity offers only a one-way ticket to the stars. You can go a long distance and return to the *place* that you left, but you cannot return to the epoch from which you left.

---

*Given that you see the Earth–Vega distance contracted, shouldn't you also claim that it is *Earth's* time running slow, rather than yours? This question underlies the so-called *twin paradox,* in which you make the trip while your twin sister stays home on Earth (see Chapter S3). The resolution comes from the fact that, because you must turn around at Vega, you effectively change reference frames relative to Earth at least once during your trip. A careful analysis of the changing reference frames is beyond the scope of this book, but it turns out that you do indeed measure less total time than your stay-at-home twin.

Putting Chapter S2 into Context

In this chapter, we have studied Einstein's special theory of relativity, learning that space and time are intertwined in remarkable ways that are quite different from what we might expect from everyday experience. As you look back on our new viewpoint about space and time, keep in mind the following "big picture" ideas:

● The ideas of relativity all derive from two simple facts: The laws of nature are the same for everyone, and the speed of light is absolute. The thought experiments in this chapter simply derived the consequences of these facts.

● Thought experiments are useful, but the ultimate judge of any theory is observation and experiment. The theory of relativity has been extensively tested and verified.

● Although the ideas of relativity may sound strange at first, you can understand them simply and logically if you allow yourself to develop a "new common sense" that incorporates them.

● Space and time are properties of the universe itself, and the new understanding of them gained through the theory of relativity will enable you to better appreciate how the universe works.

SUMMARY OF KEY CONCEPTS

S2.1 Einstein's Revolution

- *What is "relative" about the theory of relativity?* The theory is based on the idea that all *motion* is relative. That is, there is no correct answer to the question of who or what is really moving in the universe, so motion can be described only for one object relative to another.

- *What is absolute according to the theory of relativity?* (1) The laws of nature are the same for everyone, and (2) the speed of light is the same for everyone.

- *How are paradoxes useful for understanding relativity?* Because paradoxes *seem* to violate common sense or to be self-contradictory, many of the ideas of relativity can best be understood by confronting the paradoxes and finding their underlying resolutions.

S2.2 Relative Motion

- *What do we mean by a reference frame in relativity?* Two (or more) objects share the same reference frame if they are *not* moving relative to each other. In that case, the objects will experience the passage of time and measurements of distance and mass in the same way. Time, distance, and mass will be different for objects in different reference frames.

- *Why can't you reach the speed of light?* Light always travels at the same speed, so your own light (light that you emit or reflect) is always moving ahead of you at the speed of light. All other observers will also see your light moving at the speed of light—and because it is moving ahead of you, the observers will always conclude that you are moving slower than the speed of light.

continued ▶

S2.3 The Reality of Space and Time

- *How are time, space, and mass different for a moving object than for an object at rest?* If you observe an object moving by you at high speed, you'll find that its time is running slower than yours, its length is shorter than its length when at rest, and its mass is greater than its mass when at rest.

- *Will observers in different reference frames agree that two events happened at the same time?* They will not agree unless both events also occur in the same place. In general, when an observer in one reference frame sees two events happen simultaneously, observers in other reference frames will claim that one event preceded the other.

S2.4 Is It True?

- *How have experiments and observations verified that the speed of light is always the same?* The Michelson–Morley experiment showed that the speed of light is not affected by the motion of Earth around the Sun. Observations of binary star systems also confirm that the speed of light is unaffected by motion, because otherwise we would not see the two stars in binary systems as distinct points of light.

- *How have experiments verified other predictions of the special theory of relativity?* Experiments in particle accelerators can measure how time and mass are affected for subatomic particles moving at speeds very close to the speed of light. The results are in perfect agreement with the predictions of the theory. The predictions have also been verified at relatively low speeds in aircraft and spacecraft. Nuclear power plants and nuclear bombs release energy in accord with the formula $E = mc^2$, which is also a prediction of special relativity.

S2.5 Toward a New Common Sense

- *If you observe time running slow in a spaceship moving by you at high speed, how do passengers in the spaceship view your time?* They view your time as running slow. Because you both are experiencing the same laws of nature and your motion is relative, you must each see the same effects on the other.

S2.6 Ticket to the Stars

- *How does special relativity offer us a ticket to the stars?* Although the theory tells us that journeys to the stars will always take many years from the point of view of Earth, it also tells us that time for the passengers will be much shorter if they travel at speeds close enough to the speed of light. Thus, the passengers may be able to make very distant journeys within their lifetimes, even though their friends back on Earth will not be there to greet them when they return.

❓ True or False?

Decide whether each of the following statements is true or false, and clearly explain why.

1. Einstein proved that everything is relative.

2. An object moving by you at very high speed will appear to have a higher density than it has at rest. (*Hint:* Think about the effects on both length and mass.)

3. Suppose you and a friend are standing at opposite sides of a room, and you each pop a peanut into your mouth at precisely the same instant. According to the theory of relativity, it is possible for a person moving past you at high speed to observe that you ate your peanut before your friend ate hers.

4. Suppose you and a friend are standing at opposite sides of a room, and you each pop a peanut into your mouth at precisely the same instant. According to the theory of relativity, it is possible for a person moving past you at high speed to observe that you ate cashews rather than peanuts.

5. Because we can't build spaceships that travel at nearly the speed of light, we have never been able to test whether time dilation really occurs.

6. The detonation of a nuclear bomb is a test of the special theory of relativity.

7. If a person is moving past you at a speed close to the speed of light, you will see that person's time running slow, while he or she will see your time running fast.

8. If you had a sufficiently fast spaceship, you could leave today, make a round-trip to a star 500 light-years away, and return home to Earth in the year 2020.

Problems

9. *Stationary Bike.* Suppose you are riding on a stationary bike and the speedometer says you are going 30 km/hr. What does this number mean? What does it tell you about the idea of relative motion?

10. *Relative Motion Practice.* In all of the following, assume that you and your friends are in free-float reference frames.

    a. Bob is coming toward you at a speed of 75 km/hr. You throw a baseball in his direction at 75 km/hr. What does he see the ball doing?

b. Shawn is traveling away from you at a speed of 120 km/hr. He throws a baseball that, according to him, is going 100 km/hr in your direction. What do you see the ball doing?

c. Carol is going away from you at 75 km/hr, and Sam is going away from you in the opposite direction at 90 km/hr. According to Carol, how fast is Sam going?

d. Consider again the situation in part (c). Suppose you throw a baseball in Sam's direction at a speed of 120 km/hr. What does Sam see the ball doing? What does Carol see the ball doing?

11. *Moving Spaceship.* Suppose you are watching a spaceship go past you at a speed close to the speed of light.

a. How do clocks on the spaceship run, compared to your own clocks?

b. If you could measure the length, width, and height of the spaceship as it passed by, how would these measurements compare to your measurements of the spaceship's size if it were stationary?

c. If you could measure the mass of the spaceship, how would it compare to its rest mass?

d. How would a passenger on the spaceship view your time, size, and mass?

12. *Relativity of Simultaneity.* Consider the situation in Thought Experiment 8, about the green and red flashes of light at opposite ends of Jackie's spaceship. Suppose your friend Bob is traveling in a spaceship in the opposite direction from Jackie. Further imagine that he is also precisely aligned with you and Jackie at the instant the two flashes of light occur (in your reference frame).

a. According to Bob, is Jackie illuminated first by the green flash or the red flash? Explain.

b. According to you, which flash illuminates Bob first? Why?

c. According to Bob, which flash occurred first? Explain. How does Bob's view of the order of the flashes compare to your view and to Jackie's view?

*13. *Time Dilation 1.* A clever student, after learning about the theory of relativity, decides to apply his knowledge in order to prolong his life. He decides to spend the rest of his life in a car, traveling around the freeways at 55 miles per hour (89 km/hr). Suppose he drives for a period of time during which 70 years pass in his house. How much time will pass in the car? (*Hint:* If you are unable to find a difference, be sure to explain why.)

*14. *Time Dilation 2.* An even more clever student, upon realizing the folly of the student in the previous problem, decides on a better approach for prolonging her life. She decides to spend time cruising around the local solar neighborhood at a speed of $0.95c$ (95% of the speed of light). How much time will pass on her spacecraft during a period in which 70 years pass on Earth?

*15. *Time Dilation with Subatomic Particles.* Recall that a π^+ meson produced at rest has a lifetime of 18 nanoseconds

(1.8×10^{-8} s). Thus, in its own reference frame, a π^+ meson will always "think" it is at rest and therefore will decay after 18 nanoseconds. Suppose a π^+ meson is produced in a particle accelerator at a speed of $0.998c$.

a. Use the time dilation formula to calculate how long scientists will see the particle last before it decays.

b. Briefly explain how an experiment like this helps verify the special theory of relativity.

*16. *Travel to the Stars.* Suppose you stay home on Earth while your twin sister takes a trip to a distant star and back in a spaceship that travels at 99% of the speed of light. If both of you are 25 years old when she leaves and you are 45 years old when she returns, how old is your sister when she gets back?

*17. *The Betelgeuse Red Stars.* Like the fans in Boston, the fans of interstellar baseball at Betelgeuse (in the constellation Orion) have been deprived of a championship team for a long time. In fact, the Betelgeuse Red Stars have not won the Universe Series for nearly 200,000 years. (They did, however, come very close to winning 75,000 years ago. Their defeat was sealed only when a routine ground ball went through the legs of the infamous Zargon Buckner.) Realizing that time may be running out for their team—Betelgeuse is expected to explode as a supernova within the next 100,000 years—the Red Stars management has decided to break some league rules (hopefully without getting caught) in hopes of winning the series.

The team has extended a lucrative offer to Hideo Nomo, of planet Earth, if he will leave his home and join the Betelgeuse Red Stars as their new pitcher. Although he was reluctant to leave friends and family behind, Nomo finally was swayed to join the Red Stars. Interestingly, in an interview with the *Intergalactic Press,* Nomo said it was the travel opportunity, rather than the money, that lured him to Betelgeuse. Nomo was given a ticket to travel to Betelgeuse on an express spaceship at 95% of the speed of light. During the trip, he decided to try some pitching in the ship's on-board stadium. He found that, with the replacement body parts provided by the Red Stars management, his fastball was considerably improved: He was now able to throw a pitch at 80% of the speed of light. Assuming that he throws a pitch in the same direction the spacecraft is traveling, use the formula for velocity addition to calculate how fast we would see the ball moving from Earth.

*18. *Racing a Light Beam, Part 1.* A long time ago, in a galaxy far away, there was a civilization that hosted an Olympic competition every 4 of their years. One year, during an Olympiad held in the city of Sole in the nation of Kira, a sprinter by the name of Jo shattered their world record for the 100-meter dash. Alas, Jo was disqualified for having ingested illegal substances, and his record was eliminated from the books. Rather than sulk, however, Jo decided that human competition was too easy. He announced that he would instead race a beam of light.

Sponsors lined up, crowds gathered, and the event was sold to a pay-per-view audience. Everything was set. The starting gun was fired. Jo raced out of the starting blocks and shattered the old world record, running the 100 meters in 8.7 seconds. The light beam, represented by a flashlight turned on at precisely the right moment, of course emerged from its "blocks" at the speed of light. How long did it take the light beam to cover the 100-meter distance? What can you say about the outcome of this race?

*19. *Racing a Light Beam, Part 2.* Following his humiliation in the race against the light beam, Jo went into hiding for the next 2 years. By that time, most people had forgotten about both him and the money they had wasted on the pay-per-view event. However, Jo was secretly in training during his hiding. He worked out hard and tested new performance-enhancing substances. One day, he emerged from hiding and called a press conference. "I'm ready for a rematch," he announced.

Sponsors were few this time and spectators scarce in the huge Olympic stadium where Jo and the flashlight lined up at the starting line. Those who were there will never forget what they saw, although it all happened very quickly. Jo blasted out of the starting blocks at 99.9% of the speed of light. The light beam, emitted from the flashlight, took off at the speed of light. The light beam won again—but barely! After the race, TV commentators searched for Jo, but he seemed to be hiding again. Finally, they found him in a corner of the locker room, sulking under a towel. "What's wrong? You did great!" said the commentators. Jo looked back sadly, saying, "Two years of training and experiments, for nothing!" Let's investigate what happened.

a. As seen by spectators in the grandstand, how much faster than Jo is the light beam?

b. As seen by Jo, how much faster is the light beam than he is? Explain your answer clearly.

c. Using your results from parts (a) and (b), explain why Jo can say that he was beaten just as badly as before, while the spectators can think he gave the light beam a good race.

d. Although Jo was disappointed by his performance against the light beam, he did experience one pleasant surprise: The 100-meter course seemed short to him. In Jo's reference frame during the race, how long was the 100-meter course?

Discussion Questions

20. *Common Sense.* Discuss the meaning of the term *common sense.* How do we develop common sense? Can you think of other examples, besides the example of the meaning of *up* and *down,* in which you've had to change your common sense? Do you think that the theory of relativity contradicts common sense? Why or why not?

21. *Photon Philosophy.* Extend the ideas of time dilation and length contraction to think about how the universe would look if you were a photon traveling at the speed of light. Do you think there's any point to thinking about how a photon "perceives" the universe? If so, discuss any resulting philosophical implications. If not, explain why not.

22. *Ticket to the Stars.* Suppose that we someday acquire the technology to travel among the stars at speeds near the speed of light. Imagine that many people make journeys to many places. Discuss some of the complications that would arise from people aging at different rates depending on their travels.

MEDIA EXPLORATIONS

Web Projects

Take advantage of the useful Web links on www.astronomyplace.com to assist you with the following projects.

1. *Relativity Simulations.* Explore some of the simulations of the effects of special relativity available on the Web. Write a short report on what you learn.

2. *Einstein's Life.* Learn more about Einstein's life and work and how he has influenced the modern world. Write a short essay describing some aspect of his life or work.

3. *The Michelson–Morley Experiment.* Find details about the famous Michelson–Morley experiment. Write a one-page description of the experiment and its results, including a diagram of the experimental setup.

S3 Spacetime and Gravity

Nature conceals her secrets because she is sublime, not because she is a trickster.

Albert Einstein

What is gravity? Newton considered gravity to be a mysterious force that somehow reached across vast distances of space to hold the Moon in orbit around Earth and the planets in orbit around the Sun. His law of gravity explained the actions and consequences of this mysterious force but said nothing about how the force was transmitted through space.

Einstein removed the mystery of how gravity acts at a distance. As he extended his theory of relativity, Einstein found that he could explain gravity in terms of the structure of space and time. In Einstein's view, the orbits of the Moon and the planets are as natural as motion in a straight line.

As we investigate Einstein's revolutionary view of gravity, we will see that the consequences of his discoveries abound in astronomy and explain phenomena ranging from the peculiar orbit of Mercury to the expansion of the universe. We will also see how space and time merge into a four-dimensional *spacetime* and that our universe is a strange world of curved space and altered time containing *black holes* through which material can leave the universe, never to return.

S3.1 Einstein's Second Revolution

Imagine that you and everyone around you believe the Earth to be flat. As a wealthy patron of the sciences, you decide to sponsor an expedition to the far reaches of the world. You select two fearless explorers and give them careful instructions. Each is to journey along a perfectly straight path, but they are to travel in opposite directions. You provide each with a caravan for land-based travel and boats for water crossings, and you tell each to turn back only after discovering "something extraordinary."

Some time later, the two explorers return. You ask, "Did you discover something extraordinary?" To your surprise, they answer in unison, "Yes, but we both discovered the same thing: We ran into each other, despite having traveled in opposite directions along perfectly straight paths."

Although this outcome would be extraordinarily surprising if you truly believed the Earth to be flat, we are not surprised because we know that the Earth is round (Figure S3.1). In a sense, the explorers followed the *straightest possible paths*, but these "straight" lines follow the curved surface of the Earth.

Now let's consider a somewhat more modern scenario. You are floating freely in a spaceship somewhere out in space. Hoping to learn more about space in your vicinity, you launch two small probes along straight paths in opposite directions. Each probe is equipped with a camera that transmits pictures back to your spaceship. Imagine that, to your astonishment, the probes one day transmit pictures of each other! That is, although you launched them in opposite directions and neither has ever fired its engines, the probes have somehow met. In fact, this situation arises quite naturally with any orbiting objects in space. If you launch two probes in opposite directions from a space station, they will meet as they orbit Earth.

Since the time of Newton, we've generally explained the curved paths of the two probes as an effect caused by the force of gravity. However, by analogy with the explorers journeying in opposite directions on the Earth, might we instead conclude that the probes meet because *space* is somehow curved? The idea that space could be curved certainly sounds strange at first. While it's easy to visualize a surface curving *through* space, our mind cannot visualize three-dimensional space as being curved. The idea that space can be curved lies at the heart of Einstein's second revolution—a revolutionary view of gravity contained in his *general theory of relativity,* published in 1915.

From the special theory of relativity, we already know that space and time are inextricably linked. In fact, the three dimensions of space and the one dimension of time together form an inseparable, *four*-dimensional combination called **spacetime**. General relativity tells us that matter shapes the "fabric" of spacetime in a manner analogous to

Figure S3.1 Travelers going in opposite directions along paths that are as straight as possible will meet as they go around the Earth.

Figure S3.2 A rubber-sheet analogy to spacetime: Matter distorts the "fabric" of four-dimensional spacetime in a manner analogous to the way heavy weights distort a taut, two-dimensional rubber sheet. The greater the mass, the greater the distortion of spacetime.

the way heavy weights distort a taut rubber sheet or trampoline (Figure S3.2). Of course, we cannot place weights "upon" spacetime because all matter exists *within* spacetime, and we cannot visualize distortions of spacetime. However, using a rubber sheet as an analogy, we can begin to appreciate the principles of general relativity that govern the structure of space and time.

It is difficult to overstate the significance of general relativity to our understanding of the universe. For example, the following ideas all come directly from Einstein's general theory of relativity:

● Gravity arises from distortions of spacetime. It is *not* a mysterious force that acts at a distance. The presence of mass causes the distortions, and the resulting distortions determine how other objects move through spacetime.

● Time runs slow in gravitational fields. The stronger the gravity, the slower time runs.

● *Black holes* can exist in spacetime, and falling into a black hole means leaving the observable universe.

● The universe has no boundaries and no center, yet it might still have a finite volume.

● Large masses that undergo rapid changes in motion or structure emit *gravitational waves* that travel at the speed of light.

●
S3.2 The Equivalence Principle

Special relativity shows that there is no single, absolute answer to the question "Who is moving?" when two people pass each other at a constant velocity. Each individual can claim to be at rest, and each claim is equally valid. How-

ever, the situation seems quite different when accelerations are involved.

Imagine that you and your friend Jackie from Chapter S2 are both floating freely in space when your rocket engine fires (Figure S3.3). Jackie sees you accelerating away, with your speed growing ever faster, so she sends you a radio message saying, "Good-bye, have a nice trip!"

Suppose you start from rest in Jackie's reference frame and she sees you accelerate at 1g ($\approx$ 10 m/s^2). Approximately how fast will Jackie see you going after 1 second? After 10 seconds? After a minute? (*Hint:* See Section 5.1.)

From your perspective, it is Jackie who is receding into the distance at ever-faster speeds, so you reply: "Thanks, but I'm not going anywhere. You're the one accelerating into the distance." On closer examination, however, your reply seems to have a flaw. If you had been moving at constant velocity relative to Jackie, you both would have been floating freely, making it impossible to determine who was "really" moving. However, your acceleration makes you feel a force. Instead of floating weightlessly like Jackie, you are held to the floor of your spaceship. If you happen to be accelerating at 1g, or 9.8 m/s^2, you'll feel just as you feel when you are stationary on Earth, with your normal Earth weight. Thus, Jackie can respond: "Oh yeah? If you're not going anywhere, why are you stuck to the floor of your spaceship, and why do you have your engines turned on? Furthermore, if I'm accelerating, why don't I feel any forces?"

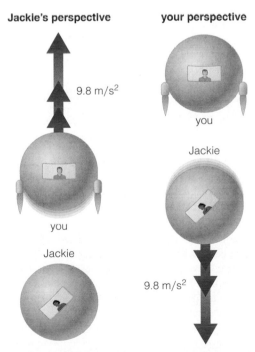

Figure S3.3 Jackie is floating freely in her spaceship. Your engines are firing, and you feel a force allowing you to stand on the floor of your spaceship.

a closed room on Earth

a closed room accelerating
through space at 1*g*

9.8 m/s^2

Figure S3.4 The equivalence principle states that the effects of gravity are exactly equivalent to the effects of acceleration. Thus, according to the equivalence principle, you cannot tell the difference between being in a closed room on Earth and being in a closed room accelerating through space at 1*g*.

You must admit that Jackie is asking very good questions. It certainly *looks* as if you really are the one who is moving and that you *cannot* legitimately claim to be stationary. Einstein didn't like the appearance of situations like this one, because he believed that *all* motion should be relative.

In 1907, Einstein hit upon what he later called "the happiest thought of my life." His revelation consisted of the idea that, whenever you feel weight (as opposed to weightlessness), you can equally well attribute it to effects of either acceleration or gravity. This idea is called the **equivalence principle**. Stated more precisely, it says:

> **The effects of gravity are exactly equivalent to the effects of acceleration.**

To clarify the meaning of the equivalence principle,* imagine that you are sitting inside with doors closed and window shades pulled down when your room is magically removed from Earth and sent hurtling through space with an acceleration of 1*g* (Figure S3.4). According to the equivalence principle, you have no way of knowing that you've left Earth. Any experiment you performed, such as dropping balls of different weights, would yield the same results you'd get on Earth.

---

*Technically, this equivalence holds only within small regions of space. Over larger regions, we can detect *tidal forces* that can arise from gravity but not from acceleration.

SPECIAL TOPIC Einstein's Leap

Given that the similarities in the effects of gravity and acceleration were well known to scientists as far back as the time of Newton, you may be wondering why the equivalence principle is so surprising. The answer is that the similarities were generally attributed to coincidence—although a very puzzling coincidence. It is as if other scientists imagined that nature was showing them two boxes, one labeled "effects of gravity" and the other labeled "effects of acceleration." They shook, weighed, and kicked the boxes but could never find any obvious differences between them. They concluded: "What a strange coincidence! The boxes seem the same from the outside even though they contain different things." Einstein's revelation was, in essence, to look at the boxes and say that it is not a coincidence at all. The boxes appear the same from the outside because they contain the same thing.

In many ways, Einstein's assertion of the equivalence principle represented a leap of faith, although it was a faith he would willingly test through scientific experiment. He made the assertion because he thought the universe would make more sense if it were true, not because of any compelling observational or experimental evidence for it at the time. This leap of faith sent him on a path far ahead of his scientific colleagues.

From a historical viewpoint, special relativity was a "theory waiting to happen" because it was needed to explain two significant problems left over from the nineteenth century: the perplexing constancy of the speed of light, demonstrated in the Michelson–Morley experiment [Section S2.4], and some seeming peculiarities of the laws of electromagnetism. Indeed, several other scientists were very close to discovering the ideas of special relativity when it was published by Einstein in 1905, and *someone* was bound to come up with special relativity around that time.

General relativity, in contrast, was a tour de force by Einstein. He recognized that unsolved problems remained after completing the theory of special relativity, and he alone took the leap of faith required to accept the equivalence principle. Without Einstein, general relativity probably would have remained undiscovered for several decades beyond 1915, the year he completed and published the theory.

you

Jackie

9.8 m/s²

Figure S3.5 If you hover over a cliff while Jackie falls, you feel weight, you need your engines to keep you from falling, and Jackie is weightless because she is in free-fall.

Now let's go back to Jackie's questions. She claims that *you* must be accelerating because you feel weight as you stand in your spaceship. The equivalence principle tells us that, with equal validity, you can claim to feel weight because of gravity. From this perspective, space is filled with a gravitational field pointing "downward" toward the back of your spaceship. You are stationary only because your rocket engine prevents you from falling, and you feel weight just as a person hovering in a helicopter over the Earth does. Jackie, because she is not using her engines, is falling through the gravitational field. As you may recall, anyone in *free-fall* feels weightless [Section 5.1]. In summary, you can claim that the situation is much as it would be if you were hovering over a cliff while Jackie had fallen over the edge (Figure S3.5). Thus, you can respond: "Sorry, Jackie, but I still say that you have it backward. I'm using my engines to prevent my spaceship from falling, and I feel weight because of *gravity*. You're weightless because you're in free-fall. I hope you won't be hurt by hitting whatever lies at the bottom of this gravitational field!"

The equivalence principle allows us to claim that *all* motion is relative. Thus, it is the starting point for general

relativity. Just as we derived the strange consequences of special relativity from the idea that the speed of light is absolute, the astounding predictions of general relativity follow from the equivalence principle. And, just as special relativity led us to recognize some underlying truths about nature—such as that space and time are different for observers in different reference frames—general relativity also leads us to a new and deeper understanding of the universe.

S3.3 Understanding Spacetime

It's easy to say that you can equally well attribute your weight to effects of gravity or acceleration, but the two effects tend to *look* very different. A person standing on the surface of the Earth appears to be motionless, while an astronaut accelerating through space continually gains speed. How can gravity and acceleration produce such similar effects when they look so different? The theory of general relativity answers that we're not seeing the whole picture. Instead of looking just at the three dimensions of space, we must learn to "look" at the *four* dimensions of spacetime.

Four Dimensions

The concept of **dimension** describes the number of independent directions in which movement is possible. A **point** has zero dimensions. If you were a geometric prisoner confined to a point, you'd have no place to go. Sweeping a point back and forth along one direction generates a **line** (Figure S3.6). The line is one-dimensional because only one direction of motion is possible (going backward is considered the same as going forward by a negative distance). Sweeping a line back and forth generates a two-dimensional **plane.** The two directions of possible motion are, say, lengthwise and widthwise. Any other direction is just a combination of these two. If we sweep a plane up and down, it fills **three-dimensional space**, with the three independent directions of length, width, and depth.

We live in three-dimensional space and thus cannot visualize any direction that is distinct from length, width, and depth (and combinations thereof). However, just because we cannot *see* "other" directions doesn't mean they don't exist. Thus, we can imagine sweeping space back and forth in some "other" direction to generate a **four-dimensional space**.

Although we have no hope of visualizing a four-dimensional space, we can easily describe it mathematically.

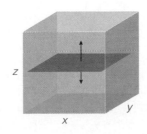

Figure S3.6 We can generate one-, two-, and three-dimensional spaces by sweeping a point and a line back and forth and a plane up and down, respectively.

In algebra, we do one-dimensional problems with the single variable *x*, two-dimensional problems with the variables *x* and *y*, and three-dimensional problems with the variables *x*, *y*, and *z*. A four-dimensional problem simply requires adding a fourth variable, as in *x*, *y*, *z*, and *w*. We could continue to five dimensions, six dimensions, and many more. Any space with more than three dimensions is called a **hyperspace**, which means "beyond space."

Spacetime

Spacetime is a four-dimensional space in which the four directions of possible motion are length, width, depth, and *time*. Note that time is not "the" fourth dimension. It is simply one of the four. (However, time differs in an important way from the other three dimensions. See Mathematical Insight S3.1.)

We cannot picture all four dimensions of spacetime at once, but we can imagine what things would look like if we could. In addition to the three spatial dimensions of spacetime that we ordinarily see, every object would be stretched out through time. Objects that we see as three-dimensional in our ordinary lives would appear as "solid" four-dimensional objects in spacetime. If we could see in four dimensions, we could look through time just as easily as we look to our left or right. If we looked at a person, we could see every event in the person's life. If we wondered what really happened during some historical event, we'd simply look to find the answer.

Try to imagine how you would look in four dimensions. How would your body, stretched through time, appear? Imagine that you bumped into someone on the bus yesterday. What would this event look like in spacetime?

This spacetime view of objects provides a new way of understanding why different observers can disagree about measurements of time and distance. Because we can't visualize four dimensions, we'll use a three-dimensional analogy. Suppose you give the same book to many different people and ask each person to measure the book's dimensions. Everyone will get the same results, agreeing on the three-dimensional structure of the book (Figure S3.7a). Now suppose instead that you show each person only a two-dimensional picture of the book rather than the book itself. The pictures can look very different, even though they show the same book in all cases (Figure S3.7b). If the people believed that the two-dimensional pictures reflected reality, they might argue endlessly about what the pictured object really looks like.

In our ordinary lives, we perceive only three dimensions, and we assume that this perception reflects reality. But spacetime is actually four-dimensional. Just as different people can see different two-dimensional pictures of the same three-dimensional book, different observers can see

a A book has an unambiguous three-dimensional shape.

b Two-dimensional pictures of the book can look very different.

Figure S3.7 Two-dimensional views of a three-dimensional object (like a book) can appear different, even though the object has only a single real shape. In a similar way, observers in different reference frames may measure space and time differently (because they perceive only three dimensions at once) even though they are all observing the same four-dimensional spacetime reality.

different three-dimensional "pictures" of the same spacetime reality. These different "pictures" are the differing perceptions of time and space of observers in different reference frames. Thus, different observers will get different results when they measure time, length, or mass, even though they are all actually looking at the same spacetime reality. In the words of a famous textbook on relativity:

> *Space is different for different observers.*
> *Time is different for different observers.*
> *Spacetime is the same for everyone.*[*]

Spacetime Diagrams

Suppose you drive your car along a straight road from home to work as shown in Figure S3.8a. At 8:00 A.M., you leave your house and accelerate to 60 km/hr. You maintain this speed until you come to a red light, where you decelerate to a stop. After the light turns green, you accelerate again to 60 km/hr, which you maintain until you slow to a stop when you reach work at 8:10. What does your trip look like in spacetime?

If we could see all four dimensions of spacetime, we'd see all three dimensions of your car and your trip stretched out through the 10 minutes of time taken for your trip. We

---

[*]From E. F. Taylor and J. A. Wheeler, *Spacetime Physics*, 2d ed., Freeman, 1992.

can't visualize all four dimensions at once, but in this case we have a special situation: Your trip progressed along only one dimension of space because you took a straight road. Therefore, we can represent your trip in spacetime by drawing a graph showing your path through one dimension of space on the horizontal axis and your path through time on the vertical axis (Figure S3.8b). This type of graph is called a **spacetime diagram**.

The car's path through four-dimensional spacetime is called its **worldline**. Any particular point along a worldline represents a particular **event**. That is, an event is a specific place and time. For example, the lowest point on the worldline in Figure S3.8b represents the event of your leaving your house: The place is 0 km from home, and the time is 8:00 A.M. You can see three very important properties of any worldline in Figure S3.8b:

1. The worldline of an object at rest is vertical (i.e., parallel to the *time* axis). The object is going nowhere in space, but it still moves through time.

2. The worldline of an object moving at constant velocity is straight but slanted. The more slanted the worldline, the faster the object is moving.

3. The worldline of an accelerating object is curved. If the object's speed is increasing, its worldline curves

toward the horizontal. If its speed is decreasing, its worldline gradually becomes more vertical.

In Figure S3.8b, we used units of minutes for time and kilometers for distance. In relativity it is easiest to work with spacetime diagrams in which we use units related to the speed of light, such as *seconds* for time and *light-seconds* for distance. In this case, light follows 45° lines on the spacetime diagram because it travels 1 light-second of distance with each second of time. For example, suppose you are sitting still in your chair, so that your worldline is vertical (Figure S3.9a). If at some particular time you flash a laser beam pointed to your right, the worldline of the light goes diagonally to the right. If you flash the laser to the left a few seconds later, its worldline goes diagonally to the left. Worldlines for several other objects are shown in Figure S3.9b.

THINK ABOUT IT

Explain why, in Figure S3.9, the worldlines of objects we see in our everyday life would be nearly vertical.

We can use spacetime diagrams to clarify the relativity of time and space. Suppose you see Jackie moving past you in a spaceship at 0.9c. Figure S3.10a shows the spacetime

a A trip from home to work on a straight road.

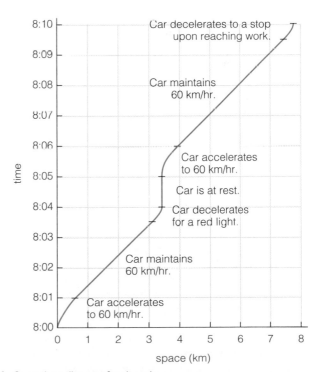

b Spacetime diagram for the trip.

Figure S3.8 The process of creating a spacetime diagram for a simple car trip.

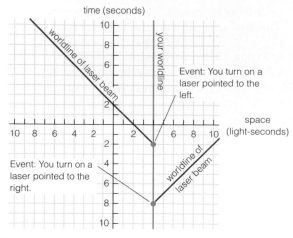

a Light follows 45° lines on a spacetime diagram that uses units of seconds for time and light-seconds for space.

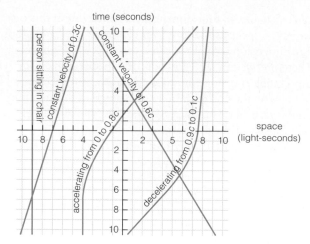

b This spacetime diagram shows several sample worldlines. Note that objects at rest have vertical worldlines, objects moving at constant velocity have straight but slanted worldlines, and accelerating objects have curved worldlines.

Figure S3.9 Spacetime diagrams marked with units of seconds for time and light-seconds for space.

diagram from your point of view: You are at rest and therefore have a vertical worldline, while Jackie is moving and has a slanted worldline. Of course, Jackie claims that *you* are moving by her and therefore would draw the spacetime

diagram shown in Figure S3.10b, in which her worldline is vertical and yours is slanted. Special relativity tells us that you would see Jackie's time running slow, while she sees *your* time running slow (along with effects on length and

---

Mathematical Insight **S3.1** **Spacetime Geometry**

Working with spacetime geometry is mathematically complex because the four dimensions are not all equivalent. In particular, time enters the equations of spacetime geometry differently than do the three dimensions of space.

To gain insight into the nature of spacetime geometry, consider two points in a plane separated by amounts $x = 3$ along the horizontal axis and $y = 4$ along the vertical axis (see the left side of the figure). You may recall from geometry that the *distance* between the two points is given by $\sqrt{x^2 + y^2}$. Now consider the same two points viewed from a coordinate system that happens to be rotated so that both points lie along the x-axis, as in the right side of the figure. In this coordinate system, the horizontal separation is $x = 5$ and the vertical separation is $y = 0$. However, the distance $\sqrt{x^2 + y^2}$ between the points is still the same.

This fact should not be surprising: *Distance* is a real, physical quantity, while the x and y separations are artifacts of a chosen coordinate system. The same idea holds if we add a z-axis, perpendicular to both x and y (you can represent the z-axis with a pencil that sticks straight up out of the page), to make a three-dimensional coordinate system. Different stationary observers using different coordinate systems can disagree about the x, y, and z separations, but they will always agree on the distance $\sqrt{x^2 + y^2 + z^2}$.

We can think of spacetime as having a fourth axis, which we will call the t-axis, for time. We might expect that, just as different observers always agree on the three-dimensional distance between two points, they will also agree on some kind of four-dimensional "distance" that has the formula $\sqrt{x^2 + y^2 + z^2 + t^2}$. However, it turns out that different observers will instead agree on the value of the quantity $\sqrt{x^2 + y^2 + z^2 - t^2}$. This value is called the *inter-*

val. (Technically, the interval formula should use ct rather than t so that all the terms have units of length.) That is, different observers can disagree about the values of x, y, z, and t separating two events, but all will agree on the interval between the two events.

The *minus sign* that goes with the time dimension in the interval formula is what makes the geometry of spacetime surprisingly complex. For example, the three-dimensional distance between two points can be zero only if the two points are in the same place, but the interval between two events can be zero even if they are in different places in spacetime, as long as $x^2 + y^2 + z^2 = t^2$. (For example, the interval is zero between any two events connected by a light path on a spacetime diagram.) If you study general relativity further, you will see many more examples of how this strange geometry comes into play.

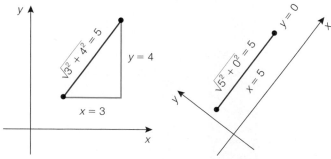

● The distance between two points in a plane is the same regardless of how we set up a coordinate system.

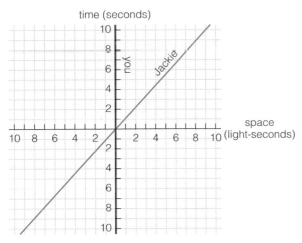

a The spacetime diagram from your point of view.

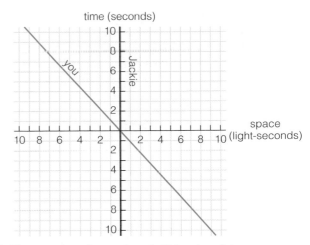

b The spacetime diagram from Jackie's point of view.

Figure S3.10 Spacetime diagrams for the situation in which Jackie is moving by you at 0.9c.

mass). We know there is no contradiction here, but simply a problem with our old common sense about space and time.

From a four-dimensional perspective, the problem is that the large angle between your worldline and Jackie's means that neither of you is looking at the other "straight-on" in spacetime. Thus, you and Jackie are both looking at the same four-dimensional reality but from different three-dimensional perspectives. It's not surprising that, like two people looking at each other cross-eyed, if you see Jackie's time running slow she sees the same thing when she looks at you.

Spacetime Curvature

So far, we've been viewing spacetime diagrams drawn on the flat pages of this book. However, as we discussed in the beginning of this chapter, spacetime can be curved. Unfortunately, while it's easy to visualize the curvature of a two-dimensional surface, we have no hope of visualizing the curvature of space or spacetime. What, then, do we mean when we talk about space or spacetime being curved?

The answer lies in the rules of geometry, which are easiest to study on two-dimensional surfaces. Consider Earth's curved surface (Figure S3.11a). There really is no such thing as a "straight" line on the Earth. Instead, the shortest and *straightest possible* path between two points on Earth's surface is a piece of a **great circle**—a circle whose center is at the center of Earth. For example, the equator is a great circle, and any "line" of longitude is part of a great circle. However, circles of latitude (besides the equator) are *not* great circles because their centers are *not* at the center of Earth.

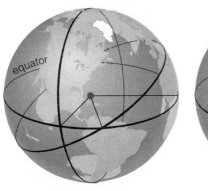

The three circles shown are great circles because their centers are at the center of the Earth.

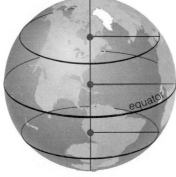

Except for the equator, lines of latitude are not great circles.

a A great circle is any circle on the surface of Earth that has its center at the center of Earth.

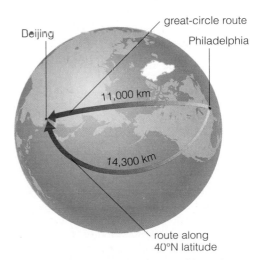

b The shortest and straightest possible path between two points on Earth is always a piece of a great circle.

Figure S3.11 The straightest possible path between any two points on a sphere must be a segment of a great circle.

If you are seeking the shortest and straightest route between two cities, you must follow a *great-circle route*. For example, Philadelphia and Beijing are both at about 40°N latitude, but the shortest route between them does *not* follow the circle of 40°N latitude. Instead, it follows a great-circle route that extends far to the north (Figure S3.11b).

THINK ABOUT IT

Find a globe and locate New Orleans and Katmandu (Nepal). Explain why the shortest route between these two cities goes almost directly over the North Pole. Why do you think airplanes generally follow great-circle routes?

Now let's contrast the rules of geometry on the surface of Earth or any other sphere with the rules of geometry in a flat plane. You are probably familiar with the rules of geometry illustrated in Figure S3.12a, such as that two parallel lines never meet and that the circumference of a circle is $2\pi r$. However, as shown in Figure S3.12b, these rules do *not* hold if we draw lines and shapes on the surface of a sphere. For example, if we draw two great circles that look parallel in one place, they do *not* obey the "flat plane" rule that parallel lines never meet. Instead, they eventually converge. Similarly, the circumference of a circle on the surface of the sphere is *not* $2\pi r$. Instead, it is *less* than $2\pi r$.

Generalizing these ideas to more than two dimensions, we say that space, or spacetime, has a **flat geometry** if the rules of geometry for a flat plane hold. For example, if the circumference of a circle in space really *is* $2\pi r$, then space has a flat geometry. (Flat geometry is also known as *Euclidean geometry*, after the Greek mathematician Euclid [c. 325–270 B.C.].) However, if the circumference of a circle in space turns out to be less than $2\pi r$, we say that space has a **spherical geometry** because the rules are those that hold on the surface of a sphere. Flat and spherical geometries are two of three general types of geometry. The third general type of geometry is called **saddle-shaped geometry** (also called *hyperbolic geometry*) because its rules are most easily visualized on a two-dimensional surface shaped like a saddle (Figure S3.12c). In this case, the circumference of a circle is *greater* than $2\pi r$.

The actual geometry of spacetime turns out to be a mixture of all three general types. That is, just as a two-dimensional surface can look flat in some places, like a piece of a sphere in others, and like a piece of a saddle in still others, different regions of spacetime obey different sets of geometrical rules.

"Straight" Lines in Curved Spacetime

One of the keys to understanding spacetime is being able to tell whether an object is following the *straightest possible path* between two points in spacetime. However, given that we can visualize neither the time part of spacetime nor the curvature of spacetime, how do we know whether an object is traveling on the straightest possible path through spacetime?

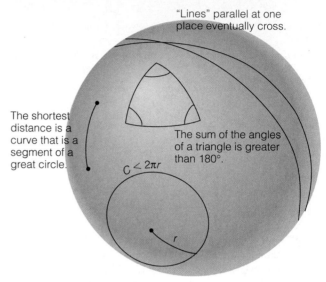

a Rules of flat geometry.

b Rules of spherical geometry.

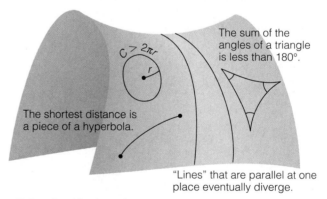

c Rules of saddle-shaped geometry.

Figure S3.12 These diagrams contrast three basic types of geometry.

Einstein used the equivalence principle to provide the answer. According to the equivalence principle, we can attribute a feeling of weight either to experiencing a force generated by acceleration or to being in a gravitational field. Similarly, any time we feel weightless, we may attribute it either to being in free-fall or to traveling at constant velocity far from any gravitational fields. Because traveling at constant velocity means traveling in a straight line, Einstein reasoned that objects experiencing weightlessness for *any* reason must be traveling in a "straight" line—that is, along a line that is the straightest possible path between two points in spacetime.

If you are floating freely, then your worldline is following the straightest possible path through spacetime. If you feel weight, then you are not on the straightest possible path.

This conclusion provides us with a remarkable way to examine the geometry of spacetime. Recall that any orbit is a free-fall trajectory. The Space Station is always free-falling toward Earth, but its forward velocity always moves it ahead just enough to "miss" hitting the ground. Earth is constantly free-falling toward the Sun, but its orbital speed keeps us going around and around instead of ever hitting the Sun. According to the equivalence principle, all orbits represent paths of objects that are following the straightest possible path through spacetime. Thus, the shapes and speeds of orbits reveal the geometry of spacetime, which leads us to an entirely new view of gravity.

THINK ABOUT IT

Suppose you are standing on a scale in your bathroom. Is your worldline following the straightest possible path through spacetime? Explain.

S3.4 A New View of Gravity

Newton's law of gravity claims that every mass exerts a gravitational attraction on every other mass, no matter how far away it is. However, on close examination, this idea of "action at a distance" is rather mysterious. For example, how does Earth feel the Sun's attraction and know to orbit it? Newton himself was troubled by this idea. A few years after publishing his law of gravity in 1687, Newton wrote:

That one body may act upon another at a distance through a vacuum, . . . and force may be conveyed from one to another, is to me so great an absurdity, that I believe no man, who has . . . a competent faculty in thinking, can ever fall into it.*

Einstein's general theory of relativity removes the idea of "action at a distance" by stating that Earth feels *no* forces—it simply follows the straightest possible path through spacetime. Thus, the fact that Earth goes around the Sun tells us that spacetime itself is curved. In other words:

What we perceive as gravity arises from the curvature of spacetime.

We cannot picture curvature of four-dimensional spacetime, but a two-dimensional analogy illustrates many of the basic ideas. We use a stretched rubber sheet to repre-

sent spacetime and assume that there is no friction on the rubber sheet, just as there is no friction in space. We represent the Sun by placing a heavy mass on the rubber sheet, which causes the sheet to curve and form a bowl-like depression (Figure S3.13).

Freely moving objects follow paths that are as straight as possible given the curvature of the rubber sheet. The central mass is not grabbing them, communicating with them, or doing anything else to influence their motion. The path of an object coming from far away bends as it passes by the central mass. This path is the type we called an *unbound orbit* in Chapter 5. Other objects are "trapped" in the bowl and therefore follow circular or elliptical orbits around the central mass. By analogy, Einstein tells us that planets orbit the Sun because the Sun's mass curves spacetime in such a way that no other motion could be more natural. In summary:

Mass causes spacetime to curve, and the curvature of spacetime determines the paths of freely moving masses.

The rubber-sheet analogy to spacetime is useful, but we should keep three very important subtleties in mind:

● The rubber sheet is supposed to represent the universe, and it makes no sense to think of placing a mass "upon" the universe. Instead, we should think of the masses as being *within* the rubber sheet.

● The rubber sheet allows us to picture orbits in only two dimensions. For example, it allows us to show that different planets orbit at different distances from the Sun and that some have more highly elliptical orbits than others, but it does not allow us to represent the fact that not all planets orbit the Sun in the same plane.

● The rubber-sheet analogy does not show the *time* part of spacetime at all. Bound orbits on the sheet or in space appear to return to the same point with each circuit of the Sun. However, objects cannot return to the same point in spacetime because they always move forward through time. For example, with each orbit of the Sun, Earth returns to the same place in space (relative to the Sun) but to a time that is a year later (Figure S3.14).

e elliptical orbit   c circular orbit   u unbound orbit

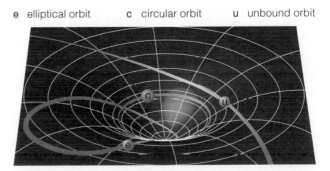

Figure S3.13 On a rubber sheet, marbles "orbit" a central mass by following the straightest paths that are possible.

---

*Letter from Newton, 1692–1693, as quoted in J. A. Wheeler, *A Journey into Gravity and Spacetime*, Scientific American Library, 1990, p. 2.

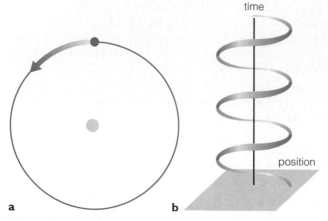

time

position

a   **b**

Figure S3.14 (**a**) If we ignore time, Earth appears to return to the same point with each orbit of the Sun. (**b**) If we include a time axis, we see that Earth never returns to the same point in space-time because it always moves forward in time.

The Strength of Gravity

The more spacetime curves, the stronger gravity becomes. The rubber-sheet analogy shows that there are two basic ways to increase the curvature of spacetime. First, a larger mass results in greater curvature at particular distances away from it. For example, the Sun curves spacetime more than any planet, and Earth curves spacetime more than the Moon. Second, for an object of a given mass, spacetime

curvature is greater near the object's surface if the object is denser (and hence smaller in size). For example, suppose we could compress the Sun into a type of "dead" star called a *white dwarf* [Section 18.2]. Because its total mass is still the same, there is no effect on the curvature of spacetime far from the Sun, but spacetime is much more curved near the compressed Sun's surface (Figure S3.15a). Thus, gravity feels much stronger on a white dwarf star than on the Sun.

If we continued to compress the Sun to smaller and smaller size, spacetime would curve more and more at the Sun's surface. Eventually, we could create a bottomless pit in spacetime, or what we call a **black hole** (Figure S3.15b). Nothing can escape from within a black hole, and we can never again detect or observe an object that falls into a black hole. The boundary that marks the "point of no return" is called the **event horizon**, because events that occur within this boundary can have no influence on our observable universe. Thus, a black hole is truly a hole in the observable universe. (We'll discuss black holes further in Chapter 18.)

Gravitational Time Dilation

Given that gravity arises from curvature of spacetime, you should not be surprised to learn that gravity affects time as well as space. We can learn about the effects of gravity on time by considering the effects of accelerated motion and then invoking the equivalence principle.

spacetime around the Sun today

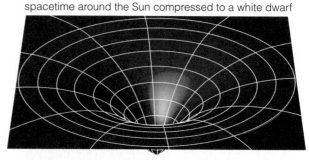

spacetime around the Sun compressed to a white dwarf

a The left diagram represents the curvature of space around the Sun today, with the yellow circle in the center representing the Sun itself. The right diagram represents the curvature that would result if the Sun were compressed to the size of a white dwarf (about the size of Earth). There is no change beyond the old boundaries of the Sun, but the curvature becomes much greater within the region formerly occupied by the Sun.

b If we kept compressing the Sun, the curvature within the region where the Sun used to be would be greater and greater, eventually creating a *black hole* in the universe.

spacetime around the Sun compressed to a black hole

Figure S3.15 These rubber-sheet diagrams represent the curvature of space around an object with the mass of our Sun. Remember that these diagrams have limitations, as described in the three bullets on page 463.

Imagine that you and Jackie are floating weightlessly at opposite ends of a spaceship. You both have watches that flash brightly each second and that you have synchronized beforehand. Because you are both floating freely with no relative motion between you, you are both in the same reference frame. Therefore, you will see each other's watches flashing at the same rate.

Now suppose you fire the spaceship engines so that the spaceship begins to accelerate, with you at the front and Jackie at the back. When the ship begins accelerating, you and Jackie will no longer be weightless. The acceleration introduces an even more important change into the situation, which we can understand by imagining the view of someone floating weightlessly outside the spaceship. Remember that observers moving at different relative speeds are in different reference frames. When the spaceship is accelerating, its speed is constantly increasing relative to the outside observer, which means that both you and Jackie are constantly changing reference frames. Moreover, the flashes from your watches take a bit of time to travel the length of the spaceship. Thus, by the time a particular flash from Jackie's watch reaches you (or a flash from your watch reaches Jackie), both of your reference frames are different from what they were at the time the flash was emitted.

Because you are in the *front* of the accelerating spaceship, your changing reference frames are always carrying you *away* from the point at which each of Jackie's flashes is emitted. Thus, the light from each of her flashes will take a little *longer* to reach you than it would if the ship were not

accelerating. As a result, instead of seeing Jackie's flashes 1 second apart, you'll see them coming a little *more* than 1 second apart. That is, you'll see Jackie's watch flashing slower than yours (Figure S3.16a). You will therefore conclude that time is running slow at the back end of the spaceship.

From Jackie's point of view at the *back* of the accelerating spaceship, her changing reference frames are always carrying her *toward* the point at which each of your flashes is emitted. Thus, the light from each of your flashes will take a little *less* time to reach her than it would if the ship were not accelerating, so she'll see them coming a little *less* than 1 second apart. She will see your watch flashing *faster* than hers and conclude that time is running *fast* at the front end of the spaceship. Note that you and Jackie agree: Time is running slower at the back end of the spaceship and faster at the front end. The greater the acceleration of the spaceship, the greater the difference in the rate at which time passes at the two ends of the spaceship.

Now we apply the equivalence principle, which tells us that we should get the same results for a spaceship at rest in a gravitational field as we do for a spaceship accelerating through space. Thus, if the spaceship were at rest on a planet, time would also have to be running slower at the bottom of the spaceship than at the top (Figure S3.16b). That is, time must run slower at lower altitudes than at higher altitudes in a gravitational field. This effect is known as **gravitational time dilation**.

The stronger the gravity—and hence the greater the curvature of spacetime—the larger the factor by which

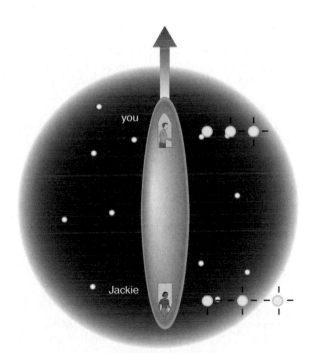

a A thought experiment with flashing watches shows that time runs slower at the back of an accelerating spaceship. The yellow dots represent the flashes from the watches, and the spacing between the dots represents the time between the flashes.

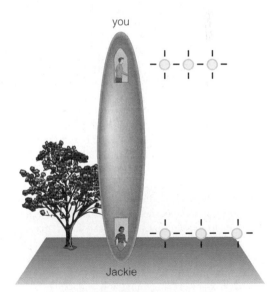

b By the equivalence principle, time must also run slower at lower altitudes in a gravitational field.

Figure S3.16 A thought experiment demonstrating why time must run slower when gravity is stronger.

time runs slow. Time runs slower on the surface of the Sun than on Earth, and slower on the surface of a white dwarf star than on the Sun. Perhaps you've already guessed that the extreme case is a black hole: As seen by anyone watching from a distance, time comes to a stop at the event horizon. If you could observe clocks placed at varying distances from the black hole, you'd see that clocks nearer the event horizon run slower and clocks *at* the event horizon show time to be frozen.

THINK ABOUT IT

Where would you age more slowly, on Earth or on the Moon? Explain.

The Geometry of the Universe

On a small scale, the surface of Earth appears flat in some places and curved by hills and valleys in others. However, when we expand our view to the entire Earth, it's clear that the overall geometry of Earth's surface is like the surface of a sphere. In a similar way, but in two more dimensions, the four-dimensional spacetime of our universe presumably has some overall geometry determined by the masses within it. This overall geometry must be one of the three general types of geometry discussed earlier: flat, spherical, or saddle-shaped (see Figure S3.12).

In geometry, a plane is infinite in extent. An idealized, saddle-shaped (hyperbolic) surface is also infinite in extent. In the same way, spacetime would be infinite in extent if the universe were flat or saddle-shaped overall, and therefore the universe would have no center and no edges. If the overall geometry of the universe were spherical, spacetime would be finite like the surface of Earth, but it would still have no center and no edges. Just as you can sail or fly around Earth's surface endlessly, you could fly through the universe forever and never encounter an edge. And just as the *surface* of Earth has no center—New York is no more "central" than Beijing or any other place on Earth's surface—there is no center to the universe. (The three-dimensional Earth *does* have a center, but this center is not part of the two-dimensional surface of Earth. Similarly, if four-dimensional spacetime is spherical, any "center" will be visible only by looking in at least five dimensions. Such a center has no

---

SPECIAL TOPIC The Twin Paradox

Imagine two twins, one of whom stays on Earth while the other takes a high-speed trip to a distant star and back. In Chapter S2, we said that the twin who takes the trip will age less than the twin who stays home on Earth. Shouldn't the traveling twin be allowed to claim that she stayed stationary while Earth made a trip away from her and back? And in that case, shouldn't the twin on Earth be the one who ages less? This question underlies the so-called *twin paradox*. It can be analyzed in several different ways. We will take an approach that offers some insights into the nature of spacetime.

Suppose you and Jackie are floating weightlessly next to each other with synchronized watches. While you remain weightless, Jackie uses her engines to accelerate a short distance away from you, decelerate to a stop a bit farther away, and then turn around and return. From your point of view, Jackie's motion means that you'll see her watch ticking slower than yours. Thus, upon her return, you expect to find that less time has passed for her than for you. But how does Jackie view the situation?

The two of you can argue endlessly about who is really moving, but one fact is obvious to both of you: During the trip, you remained weightless while Jackie felt *weight* holding her to the floor of her spaceship. Jackie can account for her weight in either of two ways. First, she can agree with you that she was the one who accelerated. However, because we know that time runs slow in an accelerating spaceship, she'll therefore agree that her watch ran slower than yours. Alternatively, she can claim that she felt weight because her engines counteracted a gravitational field in which she was stationary while you fell freely, but we also know that time runs slow in gravitational fields. Therefore, she'll still agree that her watch ran slower than yours. Thus, no matter how you or Jackie looks at it, the result is the same: Less time passes for Jackie.

The left side of Figure 1 shows a spacetime diagram for this experiment. You and Jackie both moved between the same two events in spacetime: the start and end points of Jackie's trip. However, your path between the two events is shorter than Jackie's. Because we have already concluded that less time passes for Jackie, we are led to a remarkable insight about the passage of time:

> Between any two events in spacetime, more time passes on the shorter (and hence straighter) path.

The maximum amount of time you can record between two events in spacetime occurs if you follow the straightest possible path—that is, the path on which you are weightless.

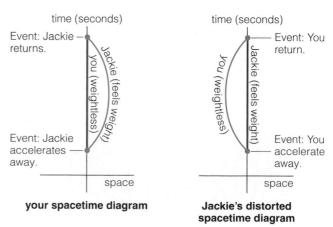

Figure 1 A person floating weightlessly must be following the straightest possible path through spacetime (left). Because this is not the case in Jackie's diagram (right), her diagram must be distorted.

meaning, because it is not part of the four spacetime dimensions of our universe.)

S3.5 Is It True?

Starting from the principle of equivalence, we've used logic and analogies to develop the ideas of general relativity. However, as always, we should not accept these theoretical conclusions unless they withstand observational and experimental tests. Like the predictions of special relativity, those of general relativity have faced many tests and have passed with flying colors.

Mercury's Peculiar Orbit

The first observational test passed by the theory of general relativity involved observations of the orbit of Mercury. Newton's law of gravity predicts that Mercury's orbit should precess slowly around the Sun because of the gravitational influences of other planets (Figure S3.17). Care-

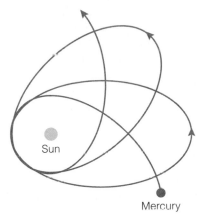

Figure S3.17 Mercury's orbit slowly precesses around the Sun. The amount of precession with each orbit is highly exaggerated in this picture.

ful observations of Mercury's orbit during the 1800s showed that it does indeed precess, with each precession cycle taking more than 20,000 years. However, careful calculations made with Newton's law of gravity could not completely

The subtlety arises if Jackie chooses to claim that she is at rest and attributes her weight to gravity. In that case, she might be tempted to draw the spacetime diagram on the right in Figure 1, on which she appears to have the shorter path through spacetime. The rule that more time passes on shorter paths would then seem to imply that *your* watch should have recorded less time than Jackie's, contradicting our earlier claims. The contradiction is an illusion. If Jackie wishes to assert that she felt gravity, she must also claim that the gravity she felt implies that spacetime is curved in her vicinity. Therefore, she should not have drawn a spacetime diagram on a flat piece of paper.

Jackie's problem is analogous to that of a pilot who plans a trip from Philadelphia to Beijing on a flat map of Earth (Figure 2). On the flat map, it appears that the pilot has plotted the straightest possible path. However, this appearance is an illusion: The shortest and straightest path really is the great-circle route that appears curved on the flat map of Earth. A flat map of Earth distorts reality because the actual geometry of Earth's surface is spherical.

Just as the distortions in a map of the world do not change the actual distances between cities, the way we choose to draw a spacetime diagram does not alter the reality of spacetime. The solution to the twin paradox is that the two twins do not share identical situations. The twin who turns around at the distant star must have a more strongly curved worldline than the stay-at-home twin. Thus, more time must pass for the stay-at-home twin, and the traveler does indeed age less during the journey.

Figure 2 On a flat map of Earth, what looks like a straight line is not really as straight as possible.

Figure S3.18 Light from Star A passes through a more highly curved region of spacetime than light from Star B, making the angular separation of the two stars appear smaller than their true angular separation.

true position of Star A

apparent position of Star A

true and apparent position of Star B

account for the observed precession. Although the discrepancy was small, further observations verified that it was real.

Einstein was aware of this discrepancy and had hoped that he would be able to explain it from the time he first thought of the equivalence principle in 1907. When he finally succeeded in November 1915, he was so excited that he was unable to work for the next three days. He later called the moment of this success the high point of his scientific life.

In essence, Einstein showed that the discrepancy arose because Newton's law of gravity assumes that time is absolute and space is flat. In reality, time runs slower and space is more curved on the part of Mercury's orbit that is nearer the Sun. The equations of general relativity take this distortion of spacetime into account, providing a predicted orbit for Mercury that precisely matches its observed orbit.

THINK ABOUT IT

Suppose the perihelion of Mercury's orbit was even closer to the Sun than it actually is. Would you expect the discrepancy between the actual orbit and the orbit predicted by Newton's laws to be greater than or less than it actually is? Explain.

Gravitational Lensing

We can also test Einstein's claim that space is curved by observing the trajectories of light rays moving through the universe. Because light always travels at the same speed, never accelerating or decelerating, light must always follow the straightest possible path. If space itself is curved, then light paths will appear curved as well.

Suppose we could carefully measure the angular separation between two stars during the daytime just when the light from one of the stars passes near the Sun. The curvature of space near the Sun forces the light beam passing closer to the Sun to curve more than the light beam from the other star (Figure S3.18). Therefore, the angular separation of the two stars will appear smaller than their true angular separation (which we would know from nighttime measurements). This effect was first observed during a total eclipse in 1919. This second spectacular success of general relativity brought Einstein worldwide fame.

Even more dramatic effects occur when a distant star or galaxy, as seen from Earth, lies directly behind another object with a strong gravitational field (Figure S3.19). The

mass of the intervening object curves spacetime in its vicinity, altering the trajectories of light beams passing nearby. Different light paths can curve so much that they end up converging at Earth, grossly distorting the appearance of the star or galaxy. Depending on the precise four-dimensional geometry of spacetime between us and the observed star or galaxy, the image we see may be magnified or distorted into arcs, rings, or multiple images of the same object (Figure S3.20). This type of distortion is called **gravitational lensing**, analogous to the bending of light by a glass lens.

Effects on Time

The prediction of gravitational time dilation can be tested by comparing clocks located in places with different gravitational field strengths. Even in Earth's weak gravity, experiments have demonstrated that clocks at low altitude tick more slowly than identical clocks at higher altitude. Although the effect would add up to only a few billionths of a second over a human lifetime, the differences agree precisely with the predictions of general relativity.

Surprisingly, it's even easier to compare the passage of time on Earth with the passage of time on the surface of the Sun and other stars. Because stellar gases emit and absorb *spectral lines* with particular frequencies [Section 6.4], they serve as natural atomic clocks. Suppose that, in a laboratory on Earth, we find that a particular type of gas emits a spectral line with a frequency of 500 trillion cycles per second. If this same gas is present on the Sun, it will also

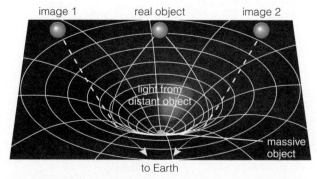

Figure S3.19 Gravitational lensing can create distorted or multiple images of a distant object whose light passes by a massive object on its way to Earth.

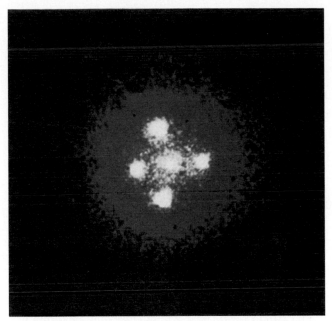

a In this case of gravitational lensing, called Einstein's Cross, the gravity of a foreground galaxy (center) bends light from a single bright background object so that it reaches us along four different paths—creating four distinct images of a single object.

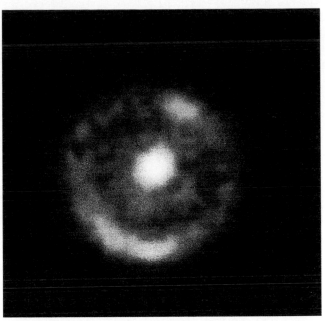

b When one galaxy lies directly behind another, the foreground galaxy can bend light on all sides so that the light converges on Earth, forming an Einstein Ring like that pictured here.

Figure S3.20 Examples of gravitational lensing.

emit a spectral line with a frequency of 500 trillion cycles per second.

General relativity claims that time should be running very slightly slower on the Sun than on Earth. That is, 1 second on the Sun lasts *longer* than 1 second on Earth, or, equivalently, a second on Earth is shorter than a second on the Sun. Thus, during 1 second on Earth, we will see *fewer* than 500 trillion cycles from the gas on the Sun. Because lower frequency means longer or redder wavelength, the spectral lines from the Sun ought to be *redshifted*. This redshift has nothing to do with the Doppler shifts that we see from moving objects [Section 6.5]. Instead, it is a **gravitational redshift**, caused by the fact that time runs slow in gravitational fields. Gravitational redshifts have been measured for spectral lines from the Sun and from many other stars. The results agree with the predictions of general relativity.

The Search for Gravitational Waves

If the curvature of space suddenly changes somewhere, the change can have effects on distant parts of the universe. For example, the effect of a star suddenly imploding or exploding is rather like the effect of dropping a rock into a pond. It generates ripples of curvature that propagate outward through space. Similarly, two massive stars orbiting each other closely and rapidly generate ripples of curvature in space rather like those of a blade turning in water. Einstein called these ripples **gravitational waves**. Similar in charac-

ter to light waves but far weaker, gravitational waves have no mass and travel at the speed of light.

The distortions of space carried by gravitational waves should compress and expand objects they pass by. In principle, we could detect gravitational waves by looking for such waves of compression and expansion, but these effects are expected to be extremely weak. No one has yet succeeded in detecting gravitational waves. However, a new observatory dedicated to the search for gravitational waves has recently begun operations. Called the Laser Interferometer Gravitational Wave Observatory, or LIGO, it consists of two large detectors—one in Louisiana and one in Washington State—that search in tandem for telltale signs of gravitational waves.

Despite the lack of direct detection, we are quite certain that gravitational waves exist because of a special set of observations carried out over the past quarter-century. In 1974, astronomers Russell Hulse and Joseph Taylor discovered an unusual binary star system in which both stars are highly compressed (they are neutron stars [Section 18.3]), allowing them to orbit each other extremely closely and rapidly. General relativity predicts that this system should be emitting a substantial amount of energy in gravitational waves. If the system is losing energy to these waves, the orbits of the two stars should be steadily decaying. Observations show that the rate at which the orbital period is decreasing matches the prediction of general relativity, a strong suggestion that the system really is losing energy by emitting gravitational waves (Figure S3.21).

Figure S3.21 The decrease in the orbital period of the Hulse–Taylor binary star system matches what we expect if the system is emitting gravitational waves.

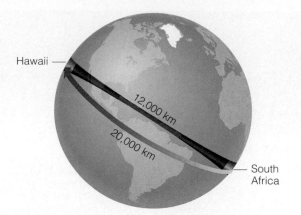

Figure S3.22 If you could take a shortcut *through* the Earth, the trip from Hawaii to South Africa would be shorter than is possible on the *surface* of the Earth.

S3.6 Hyperspace, Wormholes, and Warp Drive

If you're a fan of *Star Trek, Star Wars,* or other science fiction, you've seen spaceships bounding about the galaxy with seemingly little regard for Einstein's prohibition on traveling faster than the speed of light. In fact, these stories do not necessarily have to violate the precepts of relativity as long as they exploit potential "loopholes" in the known laws of nature.

Let's begin with an analogy. Suppose you want to take a trip between Hawaii and South Africa, which happen to lie diametrically opposite on the Earth (Figure S3.22). Ordinarily, we are restricted to traveling along Earth's surface by car, boat, or plane, and the most direct trip would cover about 20,000 kilometers. Suppose you could somehow drill a hole through the center of the Earth and fly through the hole from Hawaii to South Africa. In that case, the trip would be only about 12,000 kilometers.

Now consider a trip from Earth to the star Vega, about 25 light-years away. From the point of view of someone who stays home on Earth, this trip must take at least 25 years in each direction. However, suppose space happens to be curved in such a way that Earth and Vega are much closer together as viewed from a multidimensional *hyperspace,* just as Hawaii and South Africa are closer together if we can go through the Earth than if we must stay on its surface. Further, suppose there is a tunnel through hyperspace, often called a **wormhole**, through which we can travel (Figure S3.23). If the tunnel is short—perhaps only a few

kilometers in length—then a spaceship would need to travel only a few kilometers through the wormhole to go from Earth to Vega. The trip might take only a few minutes in each direction! Relativity is not violated because the spaceship has not exceeded the speed of light. It has simply taken a shortcut through hyperspace.

If no wormhole is available, perhaps we might discover a way to "jump" through hyperspace and return to the universe anywhere we please. Such hyperspace jumps are the fictional devices used for space travel in the *Star Wars* movies. Alternatively, we might discover a way to warp spacetime to our own specifications, thereby allowing us to make widely separated points in space momentarily touch in hyperspace. This fictional device is the basic premise behind *warp drive* in the *Star Trek* series.

Do wormholes really exist? If so, could we really travel through them? Is it possible that we might someday discover a way to jump into hyperspace or create a warp drive? Our current understanding of physics is insufficient to answer these questions definitively. For the time being, the known laws of physics do not prohibit any of these exotic forms of travel. These loopholes are therefore ideal for science fiction writers, because they might allow rapid travel among distant parts of the universe without violating the established laws of relativity.

However, many scientists believe we will eventually find that these exotic forms of travel are *not* possible. Their primary objection is that wormholes seem to make time travel possible. If you could jump through hyperspace to another place in our universe, couldn't you also jump back to another *time*? If you used a trip through hyperspace to travel into the past, could you prevent your parents from ever meeting?

The paradoxes we encounter when we think about time travel are severe and seem to have no resolution. Most scientists therefore believe that time travel will prove to be impossible, even though we don't yet know any laws of physics that prohibit it. In the words of physicist Stephen Hawking, time travel should be prohibited "to keep the world safe for historians."

Figure S3.23 The curved sheet represents our universe, in which a trip from Earth to Vega covers a distance of 25 light-years. This trip could be much shorter if a wormhole existed that created a shortcut through hyperspace.

If time travel is not possible, it is much more difficult to see how shortcuts through hyperspace could be allowed. Nevertheless, neither time travel nor travel through hyperspace can yet be categorically ruled out in the same way that we can rule out the possibility of exceeding the speed of light. Until we learn otherwise, the world remains safe for science fiction writers who choose their fictional space travel techniques with care, avoiding any conflicts with relativity and other known laws of nature.

S3.7 The Last Word

We now know that space and time are intertwined in ways that would have been difficult to imagine before Einstein's work. For the last word in our study of relativity, we turn to Einstein himself. Here is what he said about a month before his death on April 18, 1955:

> *"Death signifies nothing. . . . the distinction between past, present, and future is only a stubbornly persistent illusion."**

---

*This quotation was found with the aid of Alice Calaprice, author of *The Quotable Einstein*, Princeton University Press, 1996.

THE BIG PICTURE

Putting Chapter S3 into Context

Just as the Copernican revolution overthrew the ancient belief in an Earth-centered universe, Einstein's revolution overthrew the ancient belief that space and time are distinct and absolute. We have explored Einstein's revolution in some detail in the past two chapters. Keep in mind the following "big picture" ideas:

- We live in four-dimensional spacetime. Disagreements among different observers about measurements of time and space occur because the different observers are looking at this four-dimensional reality from different three-dimensional perspectives.

- Gravity arises from curvature of spacetime. Once we recognize this fact, the orbits of planets, moons, and all other objects are perfectly natural consequences of the curvature, rather than results of a mysterious "force" acting over great distances.

- Although the predictions of relativity may seem quite bizarre, they have been verified many times by observations and experiments.

- Some questions remain well beyond our current understanding. In particular, we do not yet know whether travel through hyperspace might be possible, allowing some of the imaginative ideas of science fiction to become reality.

S3.1 Einstein's Second Revolution

- *What is the primary topic of the general theory of relativity?* The general theory of relativity is primarily a theory of gravity, stating that the force of gravity arises from distortions of spacetime.

- *What is spacetime?* Spacetime is the four-dimensional combination of space and time that forms the "fabric" of our universe.

S3.2 The Equivalence Principle

- *What is the equivalence principle?* The effects of gravity are exactly equivalent to the effects of acceleration.

- *How does general relativity allow for all motion to be relative?* Special relativity shows that motion at constant velocity is always relative, but the fact that an accelerating object feels a force makes it seem as if accelerated motion can be distinguished from nonmotion. By telling us that the effects of gravity and acceleration are equivalent, general relativity makes all motion relative: An object that is feeling a force can be explained equally well as a moving object undergoing acceleration or as a stationary object feeling effects of gravity.

S3.3 Understanding Spacetime

- *What do we mean by dimensions?* Each dimension represents an independent direction of possible motion. In three-dimensional space, the three dimensions of length, width, and height are perpendicular to one another. A four-dimensional space has a fourth dimension perpendicular to all three of the others. We cannot visualize this, but it can still exist.

- *Does spacetime differ for different observers?* No. Both time and space differ for observers in different reference frames, but the combination of spacetime is the same for everyone.

- *How are spacetime diagrams useful?* Spacetime diagrams can clarify the relativity of time and space. Objects with vertical worldlines are stationary. Objects with slanted worldlines are moving at constant velocity (relative to the person making the spacetime diagram). Objects with curved worldlines are accelerating.

- *What are the three possible geometries of spacetime?* The three possible geometries are a flat geometry, in which the ordinary laws of flat (Euclidean) geometry apply; a spherical geometry, in which lines that start out parallel tend to converge; and a saddle-shaped geometry, in which lines that start out parallel tend to diverge. Spacetime may have different geometries in different places.

- *Are there straight lines in curved spacetime?* If spacetime is curved, then no line can be perfectly straight. However, any worldline that follows the *straightest possible* path is a worldline on which an object or person would feel weightless.

S3.4 A New View of Gravity

- *How does mass affect spacetime?* Mass causes spacetime to curve, and the curvature of spacetime determines the paths of freely moving masses.

- *How would an ordinary star, a white dwarf, and a black hole of the same mass differ in spacetime?* Because all three objects have the same mass, far from their surfaces they would affect spacetime in precisely the same way. Up close, however, the white dwarf would distort spacetime much more than the ordinary star, and the black hole would distort spacetime so much that it essentially would form a bottomless pit—a true hole in the universe.

- *According to general relativity, how does gravity affect time?* Time runs slower in places where gravity is stronger.

S3.5 Is It True?

- *How have experiments and observations verified the predictions of the general theory of relativity?* Observations of the precession of Mercury's orbit match the precession predicted by Einstein's theory. Observations of stars during eclipses and photos of gravitational lensing provide spectacular confirmation of the idea that light can travel curved paths through space. Gravitational redshifts observed in the light of objects with strong gravity confirm the slowing of time predicted by general relativity.

- *What are gravitational waves, and do they really exist?* General relativity predicts that accelerating masses produce gravitational waves that travel at the speed of light. Observations of binary neutron stars provide solid indirect evidence that gravitational waves really exist.

S3.6 Hyperspace, Wormholes, and Warp Drive

- *What is a wormhole?* A wormhole is a theoretical construct that connects two remote parts of the universe with a "tunnel" through spacetime.

- *Is it really possible to travel through hyperspace or wormholes or to use warp drive to circumvent the limitation on speeds greater than the speed of light?* No one knows. No known physical laws prevent these possibilities from being reality. However, if any one of them proves to be real, then cause and effect might not be absolute, a proposition troubling to many scientists.

❓ True or False?

Decide whether each of the following statements is true or false, and clearly explain why.

1. Imagine that you are sitting in an enclosed room with no windows. According to the principle of equivalence, there are no experiments you could perform to tell you whether you were still on Earth or in a spaceship accelerating through space at 1*g*.

2. According to the principle of equivalence, if you feel weight, you must be standing on a planetary surface.

3. A person moving by you at high speed will measure time and space differently than you, but you will both agree that there is just a single spacetime reality.

4. The term *black hole* is a misnomer, because a black hole is neither black nor a hole of any kind.

5. Time runs slightly slower on the surface of the Sun than it does here on Earth.

6. Telescopes sometimes see multiple images of a single object, just as we should expect from the general theory of relativity.

7. The general theory of relativity offers nice explanations for a number of paradoxes that Einstein thought about, but no real evidence supports this theory.

8. According to the general theory of relativity, it is impossible to travel through hyperspace or to use anything like *Star Trek*'s "warp drive."

Problems

9. *Alternative Geometries.* Briefly describe how the rules of geometry are different depending on whether the geometry is *flat, spherical,* or *saddle-shaped.*

10. *Worldlines at Low Speed.* Make a spacetime diagram and draw worldlines for each of the following situations. Explain your drawings.
    a. A person sitting still in a chair.
    b. A person driving by at a constant velocity of 50 km/hr.
    c. A person driving by at a constant velocity of 100 km/hr.
    d. A person accelerating from a stop sign to a speed of 50 km/hr.
    e. A person decelerating from 50 km/hr to a stop.

11. *Worldlines at High Speed.* Make a spacetime diagram on which the time axis is marked in seconds and the space axis is marked in light-seconds. Assume you are floating weightlessly and therefore consider yourself at rest. You see Sebastian moving to your right at 0.5*c* and Michaela moving to your left at 0.7*c*. Sebastian passed your location 2 seconds ago, and Michaela passed your location 4 seconds ago. Draw worldlines for Sebastian, Michaela, and yourself. Explain your diagrams.

12. *Galileo and the Equivalence Principle.* Galileo demonstrated that all objects near Earth's surface should fall with the same acceleration, regardless of their mass. According to general relativity, why shouldn't the mass of a falling object affect its rate of fall? Explain in one or two paragraphs.

13. *Long Trips at Constant Acceleration.* On a realistic trip to the stars, we could not suddenly jump to a speed near the speed of light without being killed by the forces associated with the sudden acceleration. Thus, a more realistic trip would have us accelerate at a comfortable rate, such as 1*g*, until we are halfway to our destination and then decelerate at the same rate until we reach our destination. Explain why we would be comfortable with this acceleration. By our own reckoning, would we notice anything unusual about lengths, masses, or the passage of time on our spaceship? Why or why not?

14. *Long Trips at Constant Acceleration: Earth Time.* Suppose you stay on Earth and watch a spaceship leave on a long trip at a constant acceleration of 1*g*.
    a. At an acceleration of 1*g*, approximately how long will it take before you see the spaceship traveling away from Earth at *half* the speed of light? Explain. (Use $g = 9.8$ m/s².)
    b. Describe how you will see its speed change as it continues to accelerate. Will it keep gaining speed at a rate of 9.8 m/s each second? Why or why not?
    c. Suppose the ship travels to a star that is 500 light-years away. According to you back on Earth, *approximately* how long will this trip take? Explain.

15. *Movie Science Fiction.* Choose a popular science fiction movie that involves interstellar travel and study it closely as you watch it. What aspects of the movie are consistent with relativity and other laws of physics? What aspects of the movie violate the laws of relativity or other laws of physics? Write a two- to three-page summary of your findings.

16. *Research: The Eötvös Experiment.* Galileo's result that all objects fall to Earth with the same acceleration (neglecting air resistance) is very important to general relativity—if it were not true, general relativity would be in serious trouble. Describe the experiments of Baron Roland von Eötvös, who tested Galileo's conclusions in the late 1800s. How did the results of these experiments influence Einstein as he worked on general relativity? Write a one- to two-page summary of your findings.

17. *Research: Wormholes.* Some scientists have thought seriously about wormholes and their consequences. Find and read a popular article or book about wormholes. Write a short summary of the article, and discuss your opinion of the implications of wormholes raised in the article.

*18. *Long Trips at Constant Acceleration: Spaceship Time.* Consider again the spaceship on a long trip with a constant acceleration of 1*g*. Although the derivation is beyond the scope of this book, it is possible to show that, as long as the ship is gone from Earth for many years, the amount of time that passes on the spaceship during the trip is approximately:

$$T_{\text{ship}} = \frac{2c}{g} \ln\left(\frac{g \times D}{c^2}\right)$$

In this formula, *D* is the distance to the destination and ln stands for the natural logarithm. (Your calculator probably has a key for taking natural logarithms [usually labeled "ln"], so you can use this formula even if you are not familiar with them.) If D is in meters, $g = 9.8$ m/s^2, and $c = 3 \times 10^8$ m/s, the answer will be in units of *seconds*.

a. Suppose the ship travels to a star that is 500 light-years away. How much time will pass on the ship? Compare this to the amount of time that passes on Earth (see problem 14c). (*Hint:* Be sure you convert the distance from light-years to meters and convert your answer from seconds to years.)

b. Suppose the ship travels to the center of the Milky Way Galaxy, about 28,000 light-years away. How much time will pass on the ship? Compare this to the amount of time that passes on Earth.

c. The Andromeda Galaxy is about 2.5 million light-years away. Suppose you had a spaceship that could constantly accelerate at 1g. Could you go to the Andromeda Galaxy and back within your lifetime? Explain. What would you find when you returned to Earth?

Discussion Questions

19. *Relativity and Fate.* In principle, if we could see all four dimensions of spacetime, we could see future events as well as past events. In his novel *Slaughterhouse Five,* writer Kurt Vonnegut used this idea to argue that our futures are predetermined and that there is no such thing as free will. Do you agree with this argument? Why or why not?

20. *Philosophical Implications of Relativity.* According to our description of spacetime, you exist in spacetime as a "solid" object stretching through time. In that sense, you cannot erase anything you've ever said or done from spacetime. If we could see in four dimensions, we would be able to see everything you've ever said or done. Do you think these ideas have any important philosophical implications? Discuss.

21. *Wormholes and Causality.* Suppose it turns out that travel through wormholes *is* possible and that it is possible to travel into the past. Discuss some of the paradoxes that would occur. In light of these paradoxes, do you believe that travel through wormholes will turn out to be prohibited? Why or why not?

For a complete list of media resources available, go to www.astronomyplace.com and choose Chapter S3 from the pull-down menu.

 Astronomy Place Web Tutorials

Tutorial Review of Key Concepts

Use the following interactive **Tutorial** at www.astronomyplace.com to review key concepts from this chapter.

Black Holes Tutorial

Lesson 1 What Are Black Holes?

Lesson 2 The Search for Black Holes

Supplementary Tutorial Exercises

Use the interactive **Tutorial Lessons** to explore the following questions.

Black Holes Tutorial, Lesson 1

1. Define a black hole in terms of escape velocity. How does this definition compare to the definition in terms of curvature of spacetime?

2. What is the Schwarzschild radius? How is it related to the event horizon?

3. How does the Schwarzschild radius depend on the mass of the object that forms the black hole?

4. Are there limits to the mass of a black hole (on either the low or high ends)? Explain.

Black Holes Tutorial, Lesson 2

1. What evidence makes us think that black holes really exist?

2. How do we think black holes may form?

3. What do we mean by supermassive black holes? Where do we think they exist?

Web Projects

Take advantage of the useful Web links on www.astronomyplace.com to assist you with the following projects.

1. *Person of the Century.* Read the article "A Brief History of Relativity" by Stephen Hawking (on the *Time* magazine Web site) to learn why *Time* chose Einstein as their "Person of the Century." Write a short essay explaining the reasons and whether you agree with the choice. Defend your opinion.

2. *Gravity Probe-B.* A NASA satellite called *Gravity Probe-B* is designed to test some of the predictions of general relativity with great precision. Learn more about this satellite and its mission, and write a short report about your findings.

3. *LIGO.* Learn more about the Laser Interferometer Gravitational Wave Observatory (LIGO) and how it is attempting to detect gravitational waves. Write a short report on your findings.

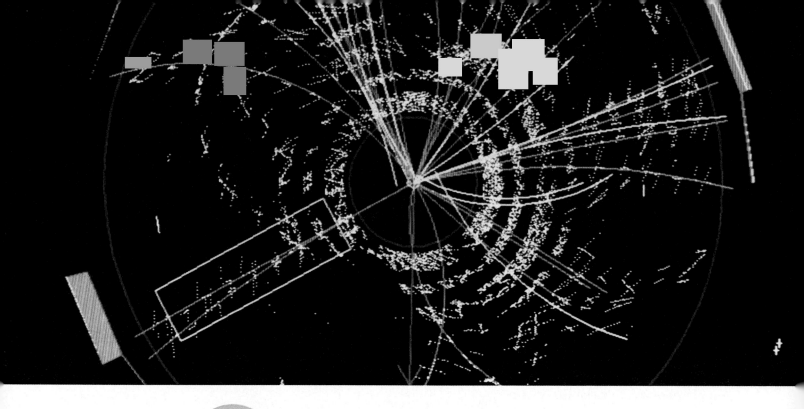

S4 Building Blocks of the Universe

*There is a theory which states that
if ever anyone discovers exactly what
the Universe is for and why it is here, it
will instantly disappear and be replaced
by something even more bizarre and
inexplicable.*

*There is another which states that
this has already happened.*

**Douglas Adams, from
The Restaurant at the End of the Universe**

The microscopic realm of atoms and nuclei seems far removed from the vast realm of planets, stars, and galaxies. Nevertheless, much of what we know about the cosmos today would have remained mysterious without a thorough understanding of these tiny particles. They are the building blocks from which all else is made, and the behavior of very large objects frequently depends on the laws that govern their tiniest pieces.

We've already seen that matter and energy behave in some strange ways when we break them down into small units. For example, we know that electrons in atoms can have only specific energies and that photons act sometimes like particles and sometimes like waves.

In this chapter, we will examine the laws of nature that underlie the structure of matter. We will look more deeply into the building blocks of nature, investigating current knowledge of the fundamental particles and forces that make up the universe. We will see that the strange laws of the microscopic world play a fundamental role in diverse processes such as nuclear fusion in the Sun and the collapse of a star into a black hole.

S4.1 The Quantum Revolution

Around the same time that Einstein was discovering the principles of relativity, he and others were also investigating the behavior of matter and energy. Their discoveries in this area were no less astonishing. In 1905, the same year he published his special theory of relativity, Einstein showed that light behaves like particles (photons) in addition to behaving like waves [Section 6.2]. In 1911, British physicist Ernest Rutherford (1871–1937) discovered that atoms

consist mostly of empty space, begging the question of how matter can ever feel solid. In 1913, Danish physicist Niels Bohr (1885–1962) suggested that electrons in atoms can have only particular energies—that is, electron energies are *quantized* [Section 4.4]. Thus, the realm of the very small is often called the *quantum realm,* and the science of the quantum realm is called **quantum mechanics.**

Other scientists soon built upon the work of Einstein, Rutherford, and Bohr. By the mid-1920s, our ideas about the structure and nature of atoms and subatomic particles were undergoing a total revolution. The repercussions of this *quantum revolution* continue to reverberate today. They have forced us to reexamine our "common sense" about the fundamental nature of matter and energy. They have also driven a technological revolution, because the laws of quantum mechanics make modern electronics possible. Most important, at least from an astronomical point of view, the combination of new ideas and new technology is enabling us to look ever deeper into the heart of matter and energy—the ultimate building blocks of the universe.

This chapter discusses key ideas of the quantum revolution that are important to the study of astronomy. The following are among these surprising new ideas:

- The protons, neutrons, and electrons that we usually consider the building blocks of atoms are not truly fundamental. Instead, the building blocks of ordinary matter are *quarks* and *leptons,* which, in turn, belong to a category called *fermions.* Photons belong to an entirely distinct category of particles called *bosons.*

- *Antimatter* is real and is readily produced in the laboratory. When a particle and its antiparticle meet, the result is mutual annihilation and release of energy.

- Just four forces govern all interactions between particles: gravity, electromagnetism, and the strong and weak nuclear forces. In fact, these four forces are themselves manifestations of a smaller number of truly fundamental forces. Perhaps a single unified force rules all of nature.

- Our everyday common sense tells us that particles and waves are different, but the quantum laws show that *all* tiny particles exhibit the same *wave–particle duality* that Einstein demonstrated for photons.

- The quantum laws have important astronomical consequences. For example, a strange quantum effect called *degeneracy pressure* can prevent the core of a dying star from collapsing, *quantum tunneling* helps make nuclear fusion possible in the Sun, and phantom-like *virtual particles* may be important to the ultimate fate of black holes and the universe itself.

No matter how bizarre the quantum laws may seem, they lead to concrete predictions that can be tested experimentally and observationally. Some of these tests require sophisticated technological equipment found only in advanced physics laboratories. Others require billion-dollar particle accelerators. But some are performed every day, right before your eyes. Every time you see a ray of sunlight, you are seeing the product of nuclear reactions that are made possible by the quantum laws. Every time you turn on a computer, a calculator, a television, or any other "high-tech" electronic device, the strange laws of quantum mechanics are being put to work for your benefit.

S4.2 Fundamental Particles and Forces

More than 2,400 years ago, Democritus proposed that matter is made from building blocks that he called *atoms* [Section 4.3]. He believed that atoms were **fundamental particles**, the most basic units of matter, impossible to divide any further. By this definition, the particles we now call atoms are not truly fundamental. By the 1930s, we had learned that atoms themselves are made of protons, neutrons, and electrons. Following this realization, scientists briefly hoped that these three particles were the true fundamental building blocks of the universe. However, under more extreme conditions, matter starts to display greater variety, and strange new particles begin to appear.

Physicists can generate many unusual particles with the aid of **particle accelerators** (sometimes called *atom smashers*), such as Fermilab near Chicago and CERN on the border between Switzerland and France (Figure S4.1). The large magnets inside particle accelerators accelerate familiar particles such as electrons or protons to very high speeds—often extremely close to the speed of light. When these particles collide with one another or with a stationary target, they release a substantial amount of energy within a very small space. Some of this energy spontaneously turns

into mass, producing a shower of particles. (Recall that $E = mc^2$ tells us not only that mass can turn into energy, but also that energy can turn into mass.) Scientists recognize particles of different types by their differing behavior. Whenever scientists observe a particle that behaves in previously unseen ways, it is cataloged as a new type of particle and given a name.

Hundreds of different particles had been discovered by the early 1960s, and scientists began to wonder whether any of them were truly fundamental. At about that time, physicist Murray Gell-Mann proposed a scheme in which all these particles could be built from just a few fundamental components. Gell-Mann's scheme has since blossomed into what physicists call the *standard model* for the structure of matter. The standard model has proved very successful and has even predicted the existence of new particles later discovered in particle accelerators. In this section, we briefly describe the fundamental particles according to the standard model, along with their special relationships with the forces of nature.

Properties of Particles

Each particular type of subatomic particle, such as an electron, a proton, or a neutron, has its own peculiar behavior that is determined by just a few basic properties. The most important of these basic properties are *mass, charge,* and *spin*. Mass is already very familiar to you, and the effects of charge, such as static electricity or lightning, are also part of your everyday experience. Spin, on the other hand, is a property evident only in the quantum realm of the very small.

The word *spin* indicates that this property is related to angular momentum. Recall that a spinning ice skater or a spinning baseball has rotational angular momentum [Section 5.2]. By analogy, we say that a subatomic particle, such as an electron, has **spin angular momentum**—or **spin**, for short—as it "spins" on its axis. However, because an electron is not a particle in the same sense as a baseball, it does not actually spin in the same sense as a baseball. *Spin* is simply a term used to describe the angular momentum that belongs to the electron. Just as all electrons have exactly the same mass and electric charge, all electrons have exactly the same amount of spin. Similar considerations hold for all other types of subatomic particle: Every particle of a particular type has a particular amount of mass, charge, and spin.

Spin is a particularly important property for subatomic particles. All particles fall into one of two broad classes based on their spin: the **fermions**, named for Enrico Fermi (1901–1954), and the **bosons**, named for Satyendra Bose (1894–1974).[*] The most familiar fermions are elec-

Figure S4.1 Aerial photograph of the Fermilab particle accelerator in Illinois.

---

[*]Physicists measure the angular momentum of subatomic particles in units of Planck's constant divided by 2π. In these special units, *fermions* have half-integer spins (e.g., $\frac{1}{2}, \frac{3}{2}, \frac{5}{2}$), and *bosons* have integer spins (e.g., 0, 1, 2).

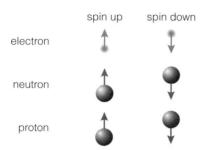

spin up   spin down

electron

neutron

proton

Figure S4.2 The two possible states of an electron's, neutron's, or proton's spin are spin up and spin down, represented by arrows.

trons, neutrons, and protons. The most familiar bosons are photons.

One important aspect of spin is its orientation: For fermions, spin can be oriented in two ways, usually called *spin up* and *spin down*. These two orientations, which correspond to the two opposite senses of rotation (clockwise and counterclockwise), are often represented by arrows (Figure S4.2). Keep in mind that denoting a particle's spin by a small dot with an arrow is a representation of convenience and that subatomic particles are *not* tiny spinning balls.

Quarks and Leptons: The Building Blocks of Matter

While electrons appear to be truly fundamental, protons and neutrons are made from even smaller particles, called **quarks**. Protons and neutrons each contain two different types of quarks: the **up quark**, which has an electric charge of $+\frac{2}{3}$, and the **down quark**, which has an electric charge of $-\frac{1}{3}$. (Charge values are given relative to the -1 charge of electrons.) Two up quarks and one down quark form a proton, giving it an overall charge of $+\frac{2}{3} + \frac{2}{3} - \frac{1}{3} = +1$. One up quark and two down quarks form a neutron, making it neutral: $+\frac{2}{3} - \frac{1}{3} - \frac{1}{3} = 0$ (Figure S4.3).

To sum up so far, the fundamental building blocks of atoms are the up quark, the down quark, and the electron. But what about the hundreds of other particles discovered by scientists? The standard model organizes all these particles into a relatively simple hierarchy. First, as we discussed earlier, particles are classified as either fermions or bosons, depending on their *spin*. Quarks, electrons, protons, and neutrons are all fermions. The fermions in turn fall into two groups: those made from quarks and those not

proton   neutron

two up quarks   one up quark
one down quark   two down quarks

Figure S4.3 The quark composition of protons and neutrons.

made from quarks. Fermions not made from quarks, such as electrons, are called **leptons**.*

THINK ABOUT IT

In most high school science classes, students are told that the fundamental building blocks of atoms are protons, neutrons, and electrons. However, based on the preceding discussion, explain why our current understanding holds that the fundamental building blocks of atoms are *quarks* and *leptons*.

Particles made from quarks can consist of either two or three quarks. Interestingly, experiments in particle accelerators indicate that a single quark cannot exist in isolation but rather must always live either in a pair or in a threesome with other quarks.

We can determine what types of quarks a particle is made of by analyzing the particle's behavior. In the standard model, six different types (or "flavors") of quarks are needed to explain the characteristics of all particles observed to date. We have already met the up and down quarks that make up protons and neutrons. The other four types of quarks have the rather exotic names *strange, charmed, top,* and *bottom.*

Six different types of leptons go with the six types of quarks: the electron and the *electron neutrino,* the *muon* and the *mu neutrino,* and the *tauon* and the *tau neutrino.* All the quarks and leptons are listed in Table S4.1. All six quarks and all six leptons have been detected in experiments. The detection of the top quark was a particularly impressive success of the standard model. It was first predicted to exist in the 1970s and was verified experimentally in 1995.

At this point, you may be wondering what all these bizarrely named particles have to do with astronomy. In part, the answer is simply that these particles are the fundamental building blocks of everything, from atoms to people, planets, stars, and galaxies. In addition, in Chapter 23 we'll see that the events that unfolded during the first fraction of a second after the Big Bang depended critically on

---

*There is also a name for the particles made from quarks. They are called *hadrons.* However, in this book we refer to them simply as "particles made from quarks."

Table S4.1 Fundamental Fermions

The Quarks	The Leptons
Up	Electron
Down	Electron neutrino
Strange	Muon
Charmed	Mu neutrino
Top	Tauon
Bottom	Tau neutrino

the various types of fundamental particles. We'll also see that *neutrinos*, in particular, are important in several astronomical processes, including nuclear fusion, the explosion of stars, and the fate of the universe.

Incidentally, neutrinos get their name, which means "little neutral ones," from the fact that they are electrically neutral and extremely lightweight—far less massive than electrons. No one has yet succeeded in measuring the precise masses of neutrinos, but they turn out to be extremely common—neutrinos outnumber protons, neutrons, and electrons combined by a factor of roughly a billion. As a result, their total mass could be significant, though it is unlikely to be more than a tiny fraction of the total mass of the universe.

Antimatter

Anyone who has watched *Star Trek* has heard of **antimatter**. Although *Star Trek* is science fiction, antimatter is not. It really exists. In fact, every quark and every lepton has a corresponding *antiquark* and *antilepton*. The antiparticle is like an exact opposite of its corresponding ordinary particle. For example, an *antielectron* (also called a *positron*) is identical to an ordinary electron except that it has a positive charge instead of a negative charge.

When a particle and its corresponding antiparticle meet, the result is mutual **annihilation** (Figure S4.4a). The combined mass of the particle and antiparticle turns completely into energy in accord with $E = mc^2$. Because our universe is made predominantly of ordinary matter, antimatter generally does not last very long. Whenever an antiparticle is produced, it quickly meets an ordinary particle, and the two annihilate each other to make energy.

This process also works in reverse. When conditions are right, pure energy can turn into a particle–antiparticle pair. For example, whenever an electron "pops" into existence, an antielectron also pops into existence with it (Figure S4.4b). This process of **pair production** happens routinely in particle accelerators here on Earth and on a much grander scale in outer space. In fact, during the first few moments after the Big Bang, the universe's energy fields

were so intense that particle–antiparticle pairs popped rapidly in and out of existence at virtually every point in space.

When we include antiparticles, the total number of types of quarks and leptons really is twice as high as shown in Table S4.1. There are really 12 quarks: the six quarks listed and their six corresponding antiquarks (e.g., up quark and anti–up quark). Similarly, there are 12 leptons: the six listed and their six corresponding antileptons. The net total of 24 different fundamental particles is quite complex.* As we'll see shortly, this complexity leads many scientists to believe that additional simplifying principles of particle physics still await discovery.

Forces

Without forces, the universe would be infinitely boring, a uniform sea of fundamental particles drifting aimlessly about. Forces supply the means through which particles communicate and exchange momentum, attracting or repelling one another depending on their properties. For example, particles with mass interact with other massive particles via the force of *gravity*, and particles with charge interact with other charged particles via the force of *electromagnetism*.

Electromagnetism rules the processes of chemistry and biology, grouping electrons and protons into atoms, atoms into molecules, and molecules into living cells. Gravity drives the action on larger scales, holding people on planets, planets in solar systems, and stars in galaxies.

Besides gravity and electromagnetism, we know of two other fundamental forces in the universe, called the **strong force** and the **weak force**. Both act only on extremely short distance scales—so short that these forces can be felt only *within* atomic nuclei. You can see why the strong force must exist by remembering that the nuclei of all elements except hydrogen (which has just a single proton) contain more than one proton. Protons are positively charged, so the electromagnetic force pushes them apart. If the electromagnetic force were unopposed, atomic nuclei would fly apart. The strong force, so named because it is strong enough to overcome electromagnetic repulsion, is what holds nuclei together. The weak force, also very important in nuclear reactions, is a bit more subtle. All particles made from quarks respond to the strong force, but neutrinos, for example, feel only the weak force.

According to the standard model, each force is transmitted by an **exchange particle** that transfers momentum between two interacting particles. Photons are the exchange particle for the electromagnetic force. That is, photons carry electromagnetic force through the universe. Motions of electrons in a star create photons. The photons then

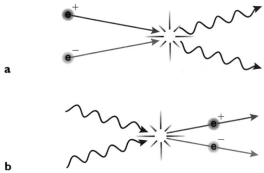

Figure S4.4 (**a**) In this representation, an electron and a positron (antielectron) annihilate each other to make energy in the form of a pair of photons. (**b**) Here, a concentration of energy leads to pair production of an electron and a positron.

---

*Moreover, each of the six quarks and six antiquarks is believed to come in three distinct varieties, called *colors*. (The term *color* is not meant to be literal. Rather, it describes a property of quarks that we cannot visualize.) For example, the up quark comes in three colors, often called red, green, and blue.

cross light-years of space to our eyes, where they generate an electromagnetic disturbance that we see as starlight.

Similarly, *gravitons* carry the gravitational force through the universe.* *Gluons,* which get their name because they act like glue to bind nuclei together, carry the strong force. The weak force is carried by particles sometimes called *weak bosons.* (All of the exchange particles are bosons.) The four forces and their exchange particles are listed in Table S4.2.

Table S4.2 also has columns describing the relative strengths of the four forces. For example, within an atomic nucleus the strong force is about 100 times stronger than the electromagnetic force, enabling the strong force to hold nuclei together. Outside atomic nuclei, the strong and weak forces vanish, leaving only gravity and electromagnetism to be felt in our daily lives.

Given the weakness of gravity in comparison to the other forces, you might wonder why it is important at all. The answer is that the other forces are out of the picture when we deal with large masses. The strong and weak forces vanish beyond atomic nuclei. The electromagnetic force cancels itself out in large objects, because such objects contain virtually equal numbers of protons and electrons, making them electrically neutral and unresponsive to the electromagnetic force. In contrast, all the matter we know about has positive mass, so gravity never cancels itself out. Thus, gravity is the only force left to act between massive enough objects. That is why gravity dominates the universe on large scales.

The Quest for Simplicity

The standard model involves four forces that mediate interactions between particles built from six types of quarks and six types of leptons, plus their corresponding antiparticles. It explains many experimental and observational results quite successfully. However, some scientists think that this model is still too complicated. They seek an even more basic theory of matter that reduces the number of forces and fundamental particles.

Gravitons is the name for the particles that correspond to gravitational waves in general relativity, analogous to the correspondence between photons and electromagnetic waves. Gravitons have not yet been detected.

Theoretical work in the 1970s, verified experimentally in the 1980s, showed that the electromagnetic and weak forces are really just two different aspects of a single force, called the *electroweak force.* Many scientists hope that future discoveries will show three or even all four forces to be simply different aspects of a single, unified force governing *all* interactions in nature. As we will see in Chapter 23, these *unified theories* might be necessary to understanding the goings-on during the first fraction of a second after the Big Bang.

S4.3 The Uncertainty Principle

So far, we've used the word *particle* as if we were talking about tiny balls of matter. However, subatomic particles frequently act like waves, too. Unfortunately, our intuition is not accustomed to this dual wave–particle nature of subatomic particles. Nevertheless, physicists managed to invent the science of quantum mechanics, which enables us to determine with great accuracy the properties of matter.

Two fundamental laws that lie at the heart of quantum mechanics lead to most of its bizarre predictions. The first of these two laws is the **uncertainty principle**, discovered by Werner Heisenberg (1901–1976) in 1927. Here is one way of stating it:

> **The more we know about where a particle is located, the less we can know about its momentum, and the more we know about its momentum, the less we can know about its position.**

We can illuminate the meaning of the uncertainty principle by considering how we might measure the trajectories of a baseball and an electron. In the case of a baseball, we could photograph it with a blinking strobe light. The resulting photograph shows us both where the ball was and where it was going at each moment in time (Figure S4.5). In scientific terms, knowing the path of the baseball means that we are measuring both its *location* and its *velocity* at each instant or, equivalently, its *location* and its *momentum.* (Recall that momentum is mass times velocity.)

Table S4.2 The Four Forces

Force	Relative Strength Within Nucleus*	Relative Strength Beyond Nucleus	Exchange Particles	Major Role
Strong	100	0	Gluons	Holding nuclei together
Electromagnetic	1	1	Photons	Chemistry and biology
Weak	10^{-5}	0	Weak bosons	Nuclear reactions
Gravity	10^{-43}	10^{-43}	Gravitons	Large-scale structure

*The force laws for the strong and weak forces are more complex than the inverse square laws for the electromagnetic force and gravity, so the numbers given for the strong and weak forces are rough approximations.

Figure S4.5 A photograph of a ball taken with a blinking strobe light allows us to determine both where the ball was and where it was going at each moment in time.

Now imagine trying to observe an electron in the same way. We will detect the electron only if it manages to scatter some of the photons streaming by it. However, whereas the photons are tiny particles of light compared to a baseball, they are quite large compared to the minuscule electron. The precision with which we can pinpoint the electron's location depends on the wavelengths of the photons. If we use visible light with a wavelength of 500 nanometers, we can measure the electron's location only to within 500 nanometers—about *5,000 times* the size of a typical

atom. That is, if we see a flash from a row of 5,000 atoms, we do not even know which atom contains the electron that caused the flash!

To locate the electron more precisely, we must use shorter-wavelength light, such as ultraviolet or X rays. Now we encounter our next problem. To determine the electron's path, we must observe the flashes from its interactions with one photon after another. Yet each photon's energy delivers a "kick" that disturbs the electron and thereby changes the momentum we are trying to measure. The higher the energy

---

SPECIAL TOPIC Does God Play Dice?

Suppose we knew all the laws of nature and, at some particular moment in time, what every single particle in the universe was doing. Could we then predict the future of the universe for all time?

Until the twentieth century, nearly every philosopher would have answered yes. The idea was so pervasive that many philosophers concluded that God was like a watchmaker: God simply started up the universe, and the future was forever after determined. This idea that everything in the universe is predictable from its initial state is called *determinism,* and a universe that runs predictably, like a watch, is called a *deterministic universe.*

The discovery of the uncertainty principle shattered the idea of a deterministic universe by telling us that, at best, we can only make statements about the *probability* of the precise future location of a subatomic particle. Because everything is made of subatomic particles, the uncertainty principle implies a degree of built-in randomness to the universe.

The idea that nature is governed by probability rather than certainty unsettled many people, including Einstein. Although he was well aware that the theories of quantum physics had survived many experimental tests, Einstein maintained a belief that the theories were incomplete. He believed that scientists would one day discover a deeper level of nature at which uncertainty would

be removed. To summarize his philosophical objections to uncertainty, Einstein said, "God does not play dice."

Einstein did more than simply object on philosophical grounds. He also proposed a number of thought experiments in which he showed that the uncertainty principle implied paradoxical results. Claiming that such paradoxes made no sense, he argued that the uncertainty principle must not be correct. In the years since Einstein's death in 1955, advances in technology have made performing some of Einstein's thought experiments possible. The results have proved to be just as paradoxical as Einstein claimed they would be. That is, the experiments have confirmed the uncertainty principle at the same time that they have posed apparent logical paradoxes.

What can we make of an idea, such as the uncertainty principle, that seems to violate common sense at the same time that it survives every experimental test? Under the tenets of science, experiment is the ultimate judge of theory, and we must accept the results despite philosophical objections. In a sense, Einstein's objection that "God does not play dice" reflected his beliefs about how the universe *should* behave. Niels Bohr argued instead that nature need not fit our preconceptions with his famous reply to Einstein: "Stop telling God what to do."

---

of the photon—which means the shorter its wavelength—the more it alters the electron's momentum.

It is almost as if nature is playing a perverse trick on us. Locating the electron precisely requires hitting it with a short-wavelength photon, but the high energy of this photon prevents us from determining the electron's momentum. Conversely, measuring the electron's momentum requires hitting it with a low-energy photon that will not disturb it much. Because low-energy light has long wavelengths, we'll no longer have a very good idea of where the electron is located.

THINK ABOUT IT

Colloquially, we often express the uncertainty principle by saying that we can't know both where a particle is and where it is going. How does this statement relate to the more precise statement that we can't know both the particle's location and its momentum? In what ways is the colloquial statement accurate, and in what ways is it an oversimplification?

The uncertainty principle applies to *all* particles, not just to electrons. In fact, it applies even to large "particles" such as baseballs, but it is unnoticeable at this level. Consider what happens when we look at a baseball with visible light of 500-nanometer wavelength. Just as with the electron, we can locate any part of the baseball only to within 500 nanometers. However, an uncertainty of 500 nanometers (about 0.00002 inch) is negligible in comparison to the size of the baseball. Moreover, the energy of visible light is so small compared to the energy of the baseball (including its mass-energy) that it has no noticeable effect on the baseball's momentum.

That is why Newton's laws work perfectly well when we deal with the motion of baseballs, cars, planets, or other objects in the macroscopic world. (The prefix *macro* comes from the Greek word for "large," and the term *macroscopic* is used to contrast the large world of objects visible to the naked eye with the microscopic world visible only through microscopes.) Newton's laws fail us in the microscopic quantum realm, where we must deal with the implications of the uncertainty principle.

Wave–Particle Duality

Our thought experiment suggested that a particle such as an electron somehow "hides" its precise path from us. However, the uncertainty principle runs deeper than this—it implies that the electron *does not even have* a precise path. From this point of view, the concepts of location and

---

Mathematical Insight **S4.1** **Electron Waves in Atoms**

In the text, we've said that an electron in an atom is "smeared out" over some volume of space. In fact, the physics is much more precise than this vague statement implies. If we choose to view the electron as a wave, an electron in an atom can be regarded as a *standing wave.*

You are probably familiar with standing waves on a string that is anchored in place at its two ends, such as a violin string. Such waves are called standing waves because each point on the string vibrates up and down but the wave does not appear to move along the length of the string. Moreover, because the string is anchored at both ends, only wave patterns with a half-integer (e.g., $0, \frac{1}{2}, 1, \frac{3}{2}, 2$) number of wavelengths along the string are possible. Other patterns, such as having three-fourths of a wavelength along the string, are not possible without breaking one end of the string away from its anchor point (see the figure).

An electron viewed as a standing wave is anchored by the electromagnetic force holding it in the atom. Like the waves on a string, only particular wave patterns are possible. However, these patterns are more complex than waves on a string because they are three-dimensional wave patterns. The allowed wave patterns for an electron in an atom can be calculated with the famous *Schrödinger equation,* developed by Erwin Schrödinger in 1926. These allowed wave patterns correspond directly to the allowed energies of the electron in the atom. Thus, the Schrödinger equation enables scientists to predict what energies should be allowed in different atoms. The fact that the Schrödinger equation successfully predicts the energy levels that are measured in the laboratory (by analysis of spectral lines) is one of the great triumphs of quantum mechanics.

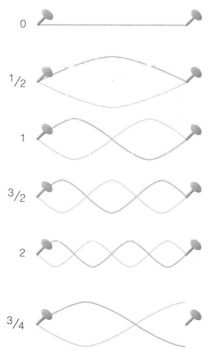

● Standing waves on a string must have a half-integer number of wavelengths. Any other number of wavelengths, such as $\frac{3}{4}$, is not possible without breaking the string away from one of its anchors.

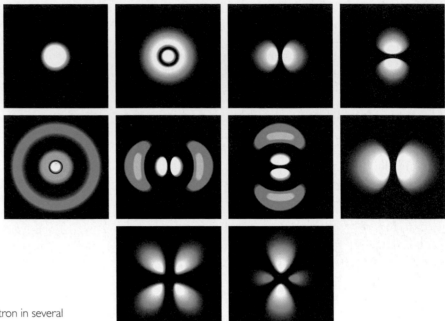

Figure S4.6 Probability patterns for the electron in several energy levels of hydrogen.

momentum do not exist independently for electrons in the way they appear to exist for objects in everyday life. Instead of imagining the electron as following some complex but hidden path, we need to think of the electron as being "smeared out" over some volume of space [Section 4.3]. This "smearing out" of electrons and other particles holds the essence of the idea of wave–particle duality.

If we choose to regard the electron as a particle, we are imagining that we can locate it precisely by hitting it with a short-wavelength photon. In that case, we know the precise location of the electron at each instant but can never predict where it will be at the next instant. We see the electron in one place at one moment and in another place at another moment, but we have no idea how it passed through the regions in between. A mathematical description of quantum mechanics allows us to calculate the *probability* that we'll find the electron in any particular place at any particular time. For example, Figure S4.6 shows the probability patterns for the electron in several energy levels of hydrogen. The brighter the region on the diagram, the higher the probability of finding the electron at any particular instant. From the point of view of the electron as a particle, the "smeared out" electron cloud around the hydrogen nucleus represents those places where we are most likely to find the electron at any instant.

Alternatively, we can choose to regard the electron as a wave. We can measure the momentum of a wave, such as that of a ripple moving across a pond, but we cannot say that a wave has a single, precise location (Figure S4.7). Instead, the wave is spread out over some region of the water in the same way that an electron is "smeared out" over some volume of space.

The fact that electrons exhibit both particle and wave properties demonstrates that our common sense from the

Figure S4.7 A wave has a well-defined momentum (represented by the arrows) but not a single, precise location.

macroscopic world does not translate well to the quantum world. We say that electrons, like photons of light, exhibit wave–particle duality. In fact, all subatomic "particles" exhibit wave–particle duality. We call them *particles* only for convenience. Like a photon of light, each particle has a wavelength.* When the wavelength of the particle is small, we can locate the particle fairly precisely, but the particle's momentum is highly uncertain. When the wavelength of the particle is large, its momentum becomes well defined, but its location grows fuzzy.

Quantifying the Uncertainty Principle

We can quantify the uncertainty principle with a simple mathematical statement:

$$\begin{array}{c} \text{uncertainty} \\ \text{in location} \end{array} \times \begin{array}{c} \text{uncertainty} \\ \text{in momentum} \end{array} \approx \text{Planck's constant}$$

---

*Electrons usually have very small wavelengths and therefore can be used to locate other particles with high precision. This is the principle behind *electron microscopes*, in which short-wavelength electrons are used to study microscopic objects. Whereas the resolution of visible-light microscopes is limited to the roughly 500-nanometer wavelength of visible light, electron microscopes can achieve resolutions of less than 0.1 nanometer.

Planck's constant is a fundamental constant in nature rather like the gravitational constant (*G*) in Newton's law of gravity and the speed of light. Its numerical value is 6.626×10^{-34} joule × s. (Planck's constant also appears in the formula for the energy of a photon [Section 6.2].)

This formula quantifies what we have already learned: Because the product of the uncertainties is roughly constant, when one uncertainty (either location or momentum) goes down, the other must go up. For example, if we determine the location of an electron with a particular amount of uncertainty, such as to within 500 nanometers, we can use the formula to calculate the amount of uncertainty in the electron's momentum. The numerical value of Planck's constant is quite small, which explains why uncertainties are scarcely noticeable for macroscopic objects.

A second way of writing the uncertainty principle is mathematically equivalent but leads to additional insights. Instead of expressing the uncertainty principle in terms of location and momentum, this alternative version expresses it in terms of the amount of *energy* that a particle has and *when* it has this energy. This version reads:

$$ \begin{array}{c} \text{uncertainty} \\ \text{in energy} \end{array} \times \begin{array}{c} \text{uncertainty} \\ \text{in time} \end{array} \approx \text{Planck's constant} $$

An amusing way to gain further appreciation for the uncertainty principle in both its forms is to imagine a game of "quantum baseball." Suppose a quantum pitcher is pitching an electron that you try to hit with your quantum bat. You'll find it extremely difficult! With the first version of the uncertainty principle, the problem is that you'll never know both where the electron is and where it is headed next. You might see it right in front of you, but because you don't know which direction it's going, you can't know whether to swing straight, up, down, or sideways. With the second version of the uncertainty principle, your problem is that you might know the electron's energy, which tells you how hard you need to swing, but you'll never know *when* to swing. Either way, your chances of hitting the electron are completely random, governed by probabilities that can be calculated with quantum mechanics.

THINK ABOUT IT

Explain why real baseball players don't have the same problems that would arise with quantum baseball. (*Hint:* How do the quantum uncertainties compare to the size of a real baseball and a real bat?)

S4.4 The Exclusion Principle

The uncertainty principle is the first of the two fundamental laws that lie at the heart of quantum mechanics. The second fundamental law, called the **exclusion principle**, was discovered by Wolfgang Pauli (1900–1958) in 1925. In its simplest sense, the exclusion principle says that two particles cannot be in the same place at the same time. (It

applies only to fermions, not to bosons.) A more complete understanding of the exclusion principle requires investigating the properties of particles a little more deeply.

The Quantum State of a Particle

Scientists use the term *state* to describe the current conditions of an object. For example, if you are relaxing in a chair, a scientist might say that you are in a "state of rest." A more precise description of your state in the chair might be something like "Your current state is a velocity of zero (at rest), a heart rate of 65 beats per minute, a breathing rate of 12 breaths per minute, a body temperature of 37°C, a metabolic rate of 200 Calories per hour," and so on.

THINK ABOUT IT

Suppose you have the following information about the current state of a friend: velocity of 5 km/hr, heart rate of 160 beats per minute, metabolic rate of 1,200 Calories per hour. Which of the following is your friend most likely doing: walking slowly as she reads a book, driving in her car, riding a bicycle down a hill, or swimming at a hard pace? How do you know?

Completely describing a person's state is quite complicated. Fortunately, describing the state of a subatomic particle such as an electron, proton, or neutron is much easier. In general, a complete description of the state of a subatomic particle—called its **quantum state**—specifies its location, momentum, orbital angular momentum, and spin to the extent allowed by the uncertainty principle. Like the energy of an electron in an atom, each property that describes a particle's quantum state is *quantized*, meaning that it can take on only particular values and not other values in between.

Statement and Meaning of the Exclusion Principle

Above, we described the exclusion principle in simple terms by saying that two particles cannot be in the same place at the same time. Now let us state the exclusion principle more precisely:

> **Two fermions of the same type cannot occupy the same quantum state at the same time.**

This principle has many important implications. One of the most important is in chemistry, where it dictates how electrons occupy their various states in atoms. For example, an electron occupying the lowest energy level in an atom necessarily has a particular amount of orbital angular momentum and a restricted range of locations. The electron's energy level fully determines its quantum state, except for its spin. Because electrons have only two possible spin states, up or down, only two electrons can occupy the lowest energy level (Figure S4.8). If you tried to put a third electron into the lowest energy level, it would have the

This energy level is "full."

spin up

spin down

A third electron must go to a higher energy level.

Figure S4.8 Only two electrons, one with spin up and the other with spin down, can share a particular energy level in an atom.

same spin—and hence the same quantum state—as one of the two electrons already there. The exclusion principle won't allow that, so the third electron must go into a higher energy level.

If you take a course in chemistry, you'll learn how a similar analysis of higher energy levels explains the chemical properties of all the elements, including their arrangement in the periodic table of the elements (see Appendix D).

The uncertainty principle and the exclusion principle together determine the sizes of atoms and of everything made of atoms, including your own body. The uncertainty principle ensures that electrons cannot be packed into infinitesimally tiny spaces. If you tried to confine an electron in too small a space in an atom, its momentum would become so large that the electromagnetic attraction of the nucleus could no longer retain it. The exclusion principle ensures that each electron has to have its own space. Thus, despite the fact that atoms are almost entirely empty, the two fundamental laws of quantum mechanics explain the solidity of matter. In the words of physicist Richard Feynman (1918–1988), "It is the fact that electrons cannot all get on top of each other that makes tables and everything else solid."

The exclusion and uncertainty principles also govern the behavior of tightly grouped protons, neutrons, and all other kinds of fermions. Just as these principles give atoms their size, they also determine the size of nuclei, because they limit how closely protons and neutrons can pack together. As we will see shortly, these quantum effects can even influence the lives of stars.

Before looking at some astronomical implications of the quantum laws, you should be aware that photons and other bosons do *not* obey the exclusion principle. For example, laser beams are so intense because many photons *can* be in the same state at the same time. In addition, it is possible under special conditions for two or more fermions to act together like a single boson. Such conditions lead to some amazing behavior, including *superconductivity,* in which electricity flows without any resistance, and *superfluidity,* in which extremely cold liquids can "creep" up the walls of a container. Many popular books on quantum laws explain these phenomena in more detail.

S4.5 Key Quantum Effects in Astronomy

The uncertainty principle and the exclusion principle lead to some truly strange consequences in the subatomic realm. Amazingly, this odd microscopic behavior produces important effects on much larger scales. In fact, we cannot fully understand how stars are born, shine brightly throughout their lives, and die unless we understand the implications of the quantum laws. In this section, we investigate three quantum effects of great importance in astronomy: degeneracy pressure, quantum tunneling, and virtual particles.

Degeneracy Pressure

Under ordinary conditions in gases, pressure and temperature are closely related. For example, suppose we inflate a balloon, filling it with air molecules. The individual molecules zip around inside the balloon, continually bouncing off its walls (Figure S4.9). The force of these molecules striking the walls of the balloon creates **thermal pressure**, which keeps the balloon inflated. If we cool the balloon by, say, putting it in a freezer, the molecules slow down. Slowing the molecules reduces the force with which they strike the walls, reducing the thermal pressure and shrinking the balloon. Heating the balloon speeds up the molecules, raising the thermal pressure and causing the balloon to expand. (Thermal pressure gets its name because it depends on temperature.)

Thermal pressure is the dominant type of pressure at low to moderate densities. However, quantum effects produce an entirely different type of pressure under conditions of extremely high density, one that does not depend on temperature at all. Consider what happens when we compress a *plasma,* a mixture of positively charged ions and free electrons [Section 4.3]. At first, the energy we expend in crushing the plasma makes the electrons and protons move faster and faster, increasing the pressure and the temperature. Now suppose we let the plasma cool off for a while. As the plasma cools, its pressure drops, enabling us to compress it further.

Continuing this process of cooling and compression, we can squeeze the plasma down to a very dense state. However, we cannot continue this process indefinitely. According to the exclusion principle, no two electrons can have exactly the same position, momentum, and spin. Just as in an atom, all the electrons can't get on top of one another at once. The compression must stop at some point, no matter how cold the plasma. This resistance to compression, stemming from the exclusion principle, is what we call **degeneracy pressure**.

The following analogy might help you visualize how degeneracy pressure works. Imagine a small number of people in an auditorium filled with folding chairs. Each person can move freely about and sit in any empty chair. The chairs represent available quantum states, the people represent electrons darting from place to place, and their motions represent thermal pressure as the electrons move from one quantum state to the next. The exclusion principle corresponds to the rule that two people cannot sit in the same chair at the same time. As long as the chairs greatly outnumber the people, two people will rarely fight over the same chair (Figure S4.10a).

Suppose ushers begin removing chairs from the front of the auditorium (compression), gradually forcing people to move to the back. Soon everyone has to crowd toward the back of the auditorium, where the number of remaining chairs is just slightly larger than the number of people. Now when a person moves to a particular chair, there's a good chance that it's already occupied. Ultimately, when the number of people equals the number of chairs, people can still move from place to place, but only if they swap seats with somebody else (Figure S4.10b). The ushers can't take away any more chairs, and the compression must stop. In other words, all the available states are filled.

The uncertainty principle also influences degeneracy pressure, though in a way that does not perfectly fit this analogy. In a highly compressed plasma, the available space for each electron is very small, so the electrons' positions must be precisely defined. According to the uncertainty principle, their momentum must then be extremely uncertain, which means that the electrons must be moving very fast on average. This requirement that highly compressed electrons have to move quickly holds *even if the object is cold.**

---

*In this sense, an object is *cold* if there's no way to get heat from it. Consider a plasma in which all the available momentum states are filled up to a certain level. To extract heat from the plasma, you'd have to slow down some of its particles. That means moving them to lower-momentum states, which are already taken. The exclusion principle thus prevents any energy from escaping, so the plasma is cold even though the electrons may be moving at high speeds.

This quantum-mechanical trade-off between position and momentum is at the root of degeneracy pressure. If you want to compress lots of electrons into a tiny space, you need to exert an enormous force to rein in their momentum.

Degeneracy pressure caused by the crowding of electrons, or **electron degeneracy pressure**, affects the lives of stars in several different ways. In some cases, it can prevent a collapsing cloud of gas from becoming a star in the first place, creating what is called a *brown dwarf* [Section 17.2]. In stars like the Sun, it determines how they begin burning helium near the end of their lives [Section 17.3]. When stars die, most leave behind an extremely dense stellar corpse called a *white dwarf*, which is also supported by electron degeneracy pressure [Section 18.2].

Not all stars meet this fate, because electron degeneracy pressure cannot grow infinitely strong. Under extreme compression, the average speed of the electrons begins to approach the speed of light. Nothing can go faster than the speed of light, so we eventually reach a limit to how much degeneracy pressure the electrons can exert. Once a dying star reaches that limit, electron degeneracy pressure cannot prevent it from shrinking further. The star then collapses until it becomes a ball of neutrons, called a *neutron star* [Section 18.3]. Neutron stars support themselves through **neutron degeneracy pressure**, which is just like electron degeneracy pressure except that it is caused by neutrons and occurs at much higher densities. Neutron degeneracy pressure comes into play only at much higher densities, because neutrons have much greater mass than electrons. A neutron near the speed of light possesses over 1,800 times more momentum than an electron at the same speed. Thus, the positions of neutrons can be over 1,800 times more precise, enabling them to occupy a much smaller volume of space.

Neutron degeneracy pressure cannot grow infinitely strong either. It begins to fail when the speed of the neutrons approaches the speed of light. According to our pres-

a When there are many more available quantum states (chairs) than electrons (people), an electron is unlikely to try to enter the same state as another electron. The only pressure comes from the temperature-related motion of the electrons, which is the thermal pressure.

b When the number of electrons (people) approaches the number of available quantum states (chairs), finding an available state requires that the electrons move faster than they would otherwise. This extra motion creates degeneracy pressure.

Figure S4.10 The auditorium analogy to degeneracy pressure. Chairs represent available quantum states, and people moving from chair to chair represent electrons.

ent understanding, nothing can stop the collapse of an object once its gravity overcomes neutron degeneracy pressure. Such an object collapses indefinitely, becoming the mysterious type of structure in spacetime that we call a *black hole.*

Quantum Tunneling

The next quantum effect we'll investigate arises from the uncertainty principle and has important implications not only in astronomy, but also in modern technology. Imagine that, as an unfortunate result of a case of mistaken identity, you're sitting on a bench in a locked jail cell (Figure S4.11a). Another bench is on the other side of the cell wall. If you could magically transport yourself from the bench on the inside to the bench on the outside, you'd be free. Alas, no such magic ever occurs for humans.

What if you were an electron? In that case, the uncertainty principle would prevent us from predicting your precise location. At best, we could state only the probability of your being in various locations. While the probability that you remain in your cell might be greatest, there is always some small probability that you will be found outside your cell. With a bit of luck, you might suddenly find yourself free, thanks to the uncertainty principle (Figure S4.11b). The process in which an electron or any other subatomic particle "magically" goes through a wall-like barrier is called **quantum tunneling**.

Figure S4.11 (**a**) A person is confined to the bench inside the jail cell, even though it would take no more energy to sit on the outside bench. (**b**) If you were an electron, you could "magically" become free through the process of quantum tunneling.

We can gain a deeper understanding of quantum tunneling by thinking in terms of the *energy* needed to cross a barrier. If you are sitting in a jail cell, the barrier is the wall of the cell. The reason you cannot escape the cell is that you don't have enough energy to crash through the wall. Just as the cell wall keeps you imprisoned, a barrier of electromagnetic repulsion can imprison an electron that does not have enough energy to crash through it. However, recall that we can write the uncertainty principle in the following form:

$$\begin{array}{c}\text{uncertainty} \\ \text{in energy}\end{array} \times \begin{array}{c}\text{uncertainty} \\ \text{in time}\end{array} \approx \text{Planck's constant}$$

Because of the uncertainty inherent in energy, there is always *some* chance that the electron will have more energy than we think at a particular moment, allowing it to cross the barrier anyway. From this point of view, quantum tunneling comes about because of uncertainty in energy rather than uncertainty in location.

Both points of view on quantum tunneling are equivalent, but the latter viewpoint illustrates a rather remarkable "loophole" in the law of conservation of energy. To cross the barrier, the particle must briefly gain some excess energy. Thanks to the uncertainty principle, this "stolen" energy need not come from anywhere as long as the particle returns it within a time period *shorter* than the uncertainty in time. In that case, we cannot be certain that any energy was ever missing! It's like stealing a dollar and putting it right back before anyone notices, so that no harm is done—except that the particle uses the stolen energy to cross the barrier before returning it.

This tale of phantom quantum energy thefts may at first sound utterly ridiculous, but the process of quantum tunneling is readily observed, and it is extremely important. In fact, the "microchips" used in all modern computers and many other modern electronic devices work because of quantum tunneling by electrons. We can control the rate of quantum tunneling, and hence the electric current, by adjusting the "height" of the energy barrier. The higher the energy barrier, the less likely it is that particles will tunnel through it.

Even more amazingly, our universe would look much different were it not for quantum tunneling. The nuclear fusion reactions that power stars occur when atomic nuclei smash together so hard that they stick. However, nuclei tend to repel each other, because they are positively charged (they contain only positive protons and neutral neutrons) and like charges repel. This repulsion creates an electromagnetic barrier that prevents nuclear fusion under most conditions. Inside the electromagnetic barrier, the attraction of the strong force takes over. But, even at the high temperatures inside stars, atomic nuclei don't have enough energy to crash through the electromagnetic barrier. Instead, they rely on quantum tunneling to sneak through to the region where the strong force dominates. In other words, quantum tunneling is what makes nuclear fusion possible in stars like our Sun. (Quantum tunneling is not as crucial to fusion in stars with core temperatures much higher than the Sun's.)

Virtual Particles

Now we will investigate the third quantum effect of importance to astronomy. The same "loophole" in the law of conservation of energy that allows particles to tunnel through otherwise impenetrable barriers permits even bigger crimes. Entire particles can "pop" into existence from nowhere—their mass made from stolen energy—as long as they "pop" back out of existence before anyone can verify that they ever existed. A somewhat fanciful analogy will help us understand this bizarre concept.

Imagine that the law of conservation of energy is enforced by a "great cosmic accountant." (In reality, of course, the law is enforced naturally.) Further, imagine that the cosmic accountant keeps a storehouse of energy in a large bank vault and ensures that any time something borrows some energy it returns the energy in a precisely equal amount. A particle that pops into existence is like a bank robber who steals some energy from the vault. If the particle is caught by the cosmic accountant, someone will have to pay for the stolen energy. However, the particle won't be caught as long as it pops back out of existence quickly enough, returning the stolen energy. The length of time the particle is allowed to exist is so short that the uncertainty principle prevents anyone from knowing that energy is missing.

Particles that pop in and out of existence before anyone can possibly detect them are called **virtual particles**. Although these particles are undetectable themselves, quantum theories predict that virtual particles should exert real, measurable effects on those particles we can detect. Remarkably, many of these predictions have been verified experimentally. In fact, modern theories of the universe propose that empty space—what we call a *vacuum*—actually "bubbles" with virtual particles that pop rapidly in and out of existence.

Imagine that you write a check for $100, but you have no money in your checking account. Your check is not necessarily doomed to bounce—as long as you deposit the needed $100 before the check clears. Explain how this $100 of "virtual money" is similar in concept to virtual particles.

Astronomically speaking, the most important process involving virtual particles concerns black holes. We can understand the process by extending our analogy to the case of a virtual electron popping into existence near a black hole. The virtual electron cannot pop into existence alone but rather must be accompanied by a virtual positron (an antielectron) so that electric charge is conserved. The virtual electron–positron pair must return the stolen energy before being caught by our imaginary "great cosmic accountant," and they do so by annihilating each other (Figure S4.12a). Suppose a virtual electron–positron pair appears very close to, but outside of, the event horizon of a black hole. The

a Pairs of virtual electrons and positrons can pop into existence, as long as they annihilate each other before they can be detected.

event horizon

b Near the event horizon of a black hole, one member of a virtual pair may cross the event horizon. The other particle is then left with no partner and cannot annihilate itself.

Figure S4.12 Virtual particles may become real near a black hole.

electron and positron are supposed to annihilate each other quickly, but a terrible problem comes up: One of the particles crosses the event horizon during its brief, virtual existence (Figure S4.12b). From the perspective of our cosmic accountant, this virtual particle was never accounted for in the first place, so there's no problem with the fact that it will never be seen again. But the other particle is suddenly caught like a deer in the headlights. Without its virtual mate, it has no way to annihilate itself and is caught red-handed by the great cosmic accountant.

If we strip away the fanciful imagery of a cosmic accountant, the end result is the creation of *real* particles, not virtual ones, just outside the event horizon of a black hole. Nothing is escaping from inside the black hole. Rather, these real particles are created from the gravitational potential energy of the black hole. Around the black hole, these real electrons and positrons annihilate each other, producing real photons that are radiated into space. To an outside observer, the black hole would appear to be radiating, even though nothing ever escapes from inside it. This effect was first predicted by Stephen Hawking in the 1970s and is therefore called **Hawking radiation**. As we've just seen, the ultimate source of Hawking radiation is the gravitational potential energy of the black hole. The continual emission of Hawking radiation causes the black hole to shrink slowly in mass, or *evaporate*, over very long periods of time.

The idea that black holes can evaporate remains untested, but if it is true it may have profound implications for both the origin and the fate of the universe. Some scientists speculate that black holes of all sizes might have been created during the Big Bang. If so, some of the smaller ones should be evaporating today. The fact that we have not yet detected any such evaporation sets limits on the number of small black holes that might have formed in the Big Bang.

At the other end of time, if the universe continues to expand forever, black holes may be the last large masses left after all the stars have died. In that case, the slow evaporation caused by Hawking radiation will mean that even black holes cannot last forever, and the universe eventually will contain nothing but a fog of photons and subatomic particles, separated from one another by incredible distances as the universe continues to grow in size.

THE BIG PICTURE

Putting Chapter S4 into Context

In this chapter, we have studied the quantum revolution and its astronomical consequences. As you look back, keep in mind the following "big picture" ideas:

- The quantum revolution can be considered the third great revolution in our understanding of the universe. The first was the Copernican revolution, which demolished the ancient belief in an Earth-centered universe. The second was relativity, which radically revised our ideas about space, time, and gravity. The quantum revolution has changed our ideas about the fundamental nature of matter and energy.

- Strange as the laws of quantum mechanics may seem, they can be readily tested and confirmed through observation and experiment. The quantum laws, like relativity, now stand on very solid ground. So far, they have passed every experimental test devised for them.

- The tiny quantum realm may seem remote from the large scales we are accustomed to in astronomy, but it is exceedingly important. The laws of quantum mechanics are necessary for understanding many astronomical processes, including nuclear fusion in the Sun, the degeneracy pressure that supports stellar corpses, and the possible evaporation of black holes.

S4.1 The Quantum Revolution

- *How did the advent of quantum mechanics change our world?* Quantum mechanics has revolutionized our understanding of particles and forces and made possible the development of modern electronic devices such as computers.

S4.2 Fundamental Particles and Forces

- *How are particles classified by spin?* All particles fall into one of two classes by spin: fermions and bosons. Fermions include all the particles that make up atoms, including electrons, neutrons, and protons. Bosons include photons and other particles that transmit forces, including gravitons, gluons, and weak bosons.

- *What are the fundamental building blocks of matter?* Quarks and leptons are the fundamental building blocks of matter. There are six known types of each. Two of the six known types of quarks make up protons and neutrons, while electrons are one of the six known types of leptons. Quarks and leptons are all fermions.

- *Does antimatter really exist?* Yes. Every particle has a corresponding antiparticle. In fact, reactions in particle accelerators always produce particles in particle–antiparticle pairs.

- *What are the four fundamental forces in nature?* The four fundamental forces are gravity, the electromagnetic force, the weak force, and the strong force.

S4.3 The Uncertainty Principle

- *What is the uncertainty principle?* The uncertainty principle tells us that we cannot simultaneously know the precise value of an object's position and momentum—or, equivalently, its energy and the precise time during which it has this energy.

- *Does the uncertainty principle apply to objects we use in everyday life?* No. For everyday objects, the uncertainty is so small that it has no noticeable effect. Nevertheless, the uncertainty is quantifiable and large enough to be crucial when we work with subatomic particles.

- *Are electrons waves or particles?* Under some circumstances electrons act like particles, while under other circumstances they behave like waves. In fact, according to the uncertainty principle, all subatomic particles exhibit "wave–particle duality" much like that of photons.

S4.4 The Exclusion Principle

- *What is the exclusion principle?* Two fermions of the same type cannot occupy the same quantum state at the same time. (This principle does not apply to bosons.)

- *How is the exclusion principle important to our existence?* The exclusion principle explains the different energy levels in atoms, which make all of chemistry possible. It also explains why electrons cannot all be on top of one another in atoms, a fact that gives atoms their physical size.

S4.5 Key Quantum Effects in Astronomy

- *What is degeneracy pressure, and how is it important in astronomy?* Degeneracy pressure is a type of pressure that can occur even in the absence of heat. It arises from the combination of the exclusion principle and the uncertainty principle. It is the dominant form of pressure in the astronomical objects known as brown dwarfs, white dwarfs, and neutron stars.

- *Do black holes last forever?* No. According to current theory, isolated black holes can gradually evaporate through quantum tunneling, emitting Hawking radiation in the process. Although the theoretical basis of this idea seems solid, it has not yet been observed in nature.

❓ True or False?

Decide whether each of the following statements is true or false, and clearly explain why.

1. Although there are six known types of quarks, ordinary atoms contain only two of these types.

2. Ordinary atoms contain two of the six known types of leptons.

3. There's no such thing as antimatter, except in science fiction.

4. According to the uncertainty principle, we can never be certain whether one theory is really better than another.

5. The exclusion principle applies to protons and neutrons as well as to electrons.

6. No known astronomical objects exhibit any type of degeneracy pressure.

7. Although we speak of four fundamental forces—gravity, electromagnetic, strong, and weak—it is likely that these forces are different manifestations of a smaller number of truly fundamental forces.

8. As bizarre as the effects of quantum mechanics may seem, their reality is supported by a huge amount of observational and experimental evidence.

Problems

9. *The Strong Force.* The strong nuclear force is the force that holds the protons and neutrons in the nucleus together. Based on the fact that atomic nuclei can be stable, briefly explain how you can conclude that the strong force must be even stronger than the electromagnetic force, at least over very short distances.

10. *Chemistry and Biology Are Electromagnetic.* All chemical and biological reactions involve the creation and breaking of chemical bonds, which are bonds between the electrons of one atom and the electrons of others. Given this fact, explain why the electromagnetic force governs all chemical and biological reactions. Also explain why the strong force, the weak force, and gravity play no role in these reactions.

11. *Why Does Gravity Dominate on Large Scales?* As shown in Table S4.2, the electromagnetic force between two charged particles is much greater than the strength of gravity between them, no matter how far apart they are. Nevertheless, it is *gravity*, rather than the electromagnetic force, that dominates the universe on large scales. Briefly explain why.

12. *Quantum Tunneling and Life.* In one or two paragraphs, explain the role of quantum tunneling in creating the elements from which we are made. (*Hint:* Recall that we are *star stuff* in the sense that the elements of our bodies were produced by nuclear fusion inside stars.)

*13. *Comparing Gravity and the EM Force.* In this problem, we compare the strength of gravity to the strength of the electromagnetic (EM) force for two interacting electrons. Because both electrons are negatively charged, they will want to *repel* each other according to the EM force. Because electrons have mass, they will want to *attract* each other according to gravity. Let's see which effect will dominate. You will need the following information for this problem:

   * The force law for gravitation is:

   $$F_g = G \frac{M_1 M_2}{d^2} \quad \left(G = 6.67 \times 10^{-11} \frac{N \times m^2}{kg^2} \right)$$

   where M_1 and M_2 are the masses of the two objects, d is the distance between them, and G is the gravitational constant. ("N" is the abbreviation for *newton,* the metric unit of force.)

   * The force law for electromagnetism is:

   $$F_{EM} = k \frac{q_1 q_2}{d^2} \quad \left(k = 9.0 \times 10^9 \frac{N \times m^2}{Coul^2} \right)$$

   where q_1 and q_2 are the *charges* of the two objects (in *Coulombs,* the standard unit of charge), d is the distance

between them, and k is a constant. ("Coul" is an abbreviation for *Coulomb.*)

   * The *mass* of an electron is 9.10×10^{-31} kg.

   * The *charge* of an electron is -1.6×10^{-19} Coul.

   a. Calculate the gravitational force, in *newtons,* that attracts the two electrons if they are separated by a distance of 10^{-10} meter (about the size of an atom).

   b. Calculate the electromagnetic force, in *newtons,* that repels the two electrons.

   c. How many times stronger is the electromagnetic repulsion than the gravitational attraction for the two electrons?

*14. *Evaporation of Black Holes.* The time it takes for a black hole to evaporate through the process of Hawking radiation can be calculated using the following formula, in which M is the mass of the black hole in kilograms and t is the lifetime of the black hole in seconds:

   $$t = 10{,}240 \, \pi^2 \, \frac{G^2 M^2}{hc^4} \quad \left(h = 6.63 \times 10^{-34} \frac{kg \times m^2}{s}; \right.$$
   $$\left. G = 6.67 \times 10^{-11} \frac{m^3}{kg \times s^2} \right)$$

   a. Without doing any calculations, explain how this formula implies that lower-mass black holes have much shorter lifetimes than more massive ones and that the evaporation process accelerates as a black hole loses mass.

   b. What is the lifetime of a black hole with the mass of the Sun ($M_{Sun} = 2.0 \times 10^{30}$ kg)? How does this compare to the current age of the universe?

   c. In Chapter 23, we will see that some scientists speculate that the universe will eventually consist only of gigantic black holes and scattered subatomic particles. The largest black holes that conceivably might form would have a mass of about a trillion (10^{12}) Suns. Calculate the lifetime of such a giant black hole. (*Hint:* Your calculator probably will be unable to handle the large numbers involved in this problem, so you will need to rearrange the numbers so that you can calculate the powers of 10 without your calculator.)

*15. *Mini–Black Holes.* Some scientists speculate that black holes of many different masses might have been formed during the early moments of the Big Bang. Some of these black holes might be mini–black holes, much smaller in mass than those that can be formed by the crush of gravity in today's universe.

   a. Calculate the lifetime of a mini–black hole with the mass of Earth (about 6×10^{24} kg). Compare this to the current age of the universe.

   b. Calculate the mass of a black hole that might have formed in the Big Bang and be completing the evaporation today. Compare this mass to the mass of Earth. For this calculation, assume that the universe is 12 billion years old.

Discussion Questions

16. *Common Sense Versus Experiment.* Even the most highly trained physicists find the results of quantum mechanics to be strange and counter to their everyday common sense, yet the predictions of quantum mechanics have passed every experimental test yet posed for them. Discuss whether it is important to reconcile our common sense with experimental results and, if so, how we can do so.

17. *The Meaning of the Uncertainty Principle.* When first hearing about it, many people assume that the uncertainty principle means that we cannot *measure* the position and momentum of a particle precisely. According to current understanding, it really tells us that the particle *does not have* a precise position and momentum in the sense that we would expect from everyday life. How do these two viewpoints differ? Discuss the different philosophical consequences of these two viewpoints.

18. *Antimatter Engines.* In the *Star Trek* series, starships are powered by matter–antimatter annihilation. Explain why matter–antimatter annihilation is the most efficient possible source of power. What practical problems would we face in developing matter–antimatter engines?

Web Projects

Take advantage of the useful Web links on www.astronomyplace.com to assist you with the following projects.

1. *Fermilab and CERN.* Visit the Web site for either the Fermilab or CERN particle accelerator. Take a virtual tour, and learn about the work at the laboratory. Write a short report about what you learn.

2. *The Higgs Particle.* Learn about how the so-called Higgs particle fits into the standard model and how physicists are searching for it. Write a short summary of your findings, including your opinion as to when we are likely to acquire evidence in support of (or against) the existence of the Higgs particle.

3. *Quantum Computing.* Learn how computer scientists hope to harness quantum effects to build computers much more powerful than any existing today. Briefly summarize the ideas, and write an essay stating your opinion concerning the benefits and drawbacks of developing this technology.

STELLAR
ALCHEMY

15 Our Star

I say Live, Live, because of the Sun,
The dream, the excitable gift.

Anne Sexton (1928–1974)

Astronomy today involves the study of the entire universe, but the root of the word *astronomy* comes from the Greek word for "star." Although we have learned a lot about the universe up to this point in the book, only now do we turn our attention to the study of the stars, the namesakes of astronomy.

When we think of stars, we usually think of the beautiful points of light visible on a clear night. The nearest and most easily studied star is visible only in the daytime—our Sun. Of course, the Sun is important to us in many more ways than as an object for astronomical study. The Sun is the source of virtually all light, heat, and energy reaching Earth, and life on Earth's surface could not survive without it.

In this chapter, we will study the Sun in some depth. We will learn how the Sun makes life possible on Earth. Equally important, we will study our Sun as a star so that in subsequent chapters we can more easily understand stars throughout the universe.

15.1 Why Does the Sun Shine?

Ancient peoples recognized the vital role of the Sun in their lives. Some worshiped the Sun as a god, and others created elaborate mythologies to explain its daily rise and set. Only recently, however, have we learned how the Sun provides us with light and heat.

Most ancient thinkers viewed the Sun as some type of fire, perhaps a lump of burning coal or wood. The Greek philosopher Anaxagoras (c. 500–428 B.C.) imagined the Sun to be a very hot, glowing rock about the size of the Greek peninsula of Peloponnesus (comparable in size to Massachusetts). Thus, he was one of the first people in history to believe that the heavens and Earth are made from the same types of materials.

By the mid-1800s, the size and distance of the Sun were reasonably well known, and scientists seriously began to address the question of how the Sun shines. Two early ideas held either that the Sun was a cooling ember that had once been much hotter or that the Sun generated energy from some type of chemical burning similar to the burning of coal or wood. Simple calculations showed that a cooling or chemically burning Sun could shine for a few thousand years—an age that squared well with biblically based estimates of Earth's age that were popular at the time. However, these ideas suffered from fatal flaws. If the Sun were a cooling ember, it would have been much hotter just a few hundred years earlier, making it too hot for civilization to have existed. Chemical burning was ruled out because it cannot generate enough energy to account for the rate of radiation observed from the Sun's surface.

A more plausible hypothesis of the late 1800s suggested that the Sun generates energy by contracting in size, a process called **gravitational contraction**. If the Sun were shrinking, it would constantly be converting gravitational potential energy into thermal energy, thereby keeping the Sun hot. Because of its large mass, the Sun would need to contract only very slightly each year to maintain its temperature—so slightly that the contraction would be unnoticeable. Calculations showed that the Sun could shine for up to about 25 million years generating energy by gravitational contraction. However, geologists of the late 1800s had already established the age of Earth to be far older than 25 million years, leaving astronomers in an embarrassing position.

Only after Einstein published his special theory of relativity, which included his discovery of $E = mc^2$, did the true energy-generation mechanism of the Sun become clear. We now know that the Sun generates energy by *nuclear fusion*, a source so efficient that the Sun can shine for about 10 billion years. Because the Sun is only 4.6 billion years old today [Section 9.5], we expect it to keep shining for some 5 billion more years.

According to our current model of solar-energy generation by nuclear fusion, the Sun maintains its size through a balance between two competing forces: gravity pulling inward and pressure pushing outward. This balance is called **gravitational equilibrium** (or *hydrostatic equilibrium*). It means that, at any point within the Sun, the weight of overlying material is supported by the underlying pressure. A stack of acrobats provides a simple example of this balance (Figure 15.1). The bottom person supports the weight of everybody above him, so the pressure on his body is very great. At each higher level, the overlying weight is less, so the pressure decreases. Gravitational equilibrium in the Sun means that the pressure increases with depth, making the Sun extremely hot and dense in its central core (Figure 15.2).

<hr>

THINK ABOUT IT

Earth's atmosphere is also in gravitational equilibrium, with the weight of upper layers supported by the pressure in lower layers. Use this idea to explain why the air gets thinner at higher altitudes.

Figure 15.1 An acrobat stack is in gravitational equilibrium: The lowest person supports the most weight and feels the greatest pressure, and the overlying weight and underlying pressure decrease for those higher up.

Figure 15.2 Gravitational equilibrium in the Sun: At each point inside, the pressure pushing outward balances the weight of the overlying layers.

Although the Sun today maintains its gravitational equilibrium with energy generated by nuclear fusion, the energy-generation mechanism of gravitational contraction was important in the distant past and will be important again in the distant future. Our Sun was born from a collapsing cloud of interstellar gas. The contraction of the cloud released gravitational potential energy, raising the interior temperature higher and higher—but not high enough to halt the contraction. The cloud continued to shrink because thermal radiation from the cloud's surface carried away much of the energy released by contraction, even while the interior temperature was rising. When the central temperature and density eventually reached the values necessary to sustain nuclear fusion, energy generation in the Sun's interior matched the energy lost from the surface in the form of radiation. With the onset of fusion, the Sun entered a long-lasting state of gravitational equilibrium that has persisted for the last 4.6 billion years.

About 5 billion years from now, when the Sun finally exhausts its nuclear fuel, the internal pressure will drop, and gravitational contraction will begin once again. As we will see later, some of the most important and spectacular processes in astronomy hinge on this ongoing "battle" between the crush of gravity and a star's internal sources of pressure.

In summary, the answer to the question "Why does the Sun shine?" is that about 4.6 billion years ago *gravitational*

contraction made the Sun hot enough to sustain nuclear fusion in its core. Ever since, energy liberated by fusion has maintained the Sun's *gravitational equilibrium* and kept the Sun shining steadily, supplying the light and heat that sustain life on Earth.

15.2 Plunging to the Center of the Sun: An Imaginary Journey

In the rest of this chapter, we will discuss in detail how the Sun produces energy and how that energy travels to Earth. First, to get a "big picture" view of the Sun, let's imagine you have a spaceship that can somehow withstand the immense heat and pressure of the solar interior and take an imaginary journey from Earth to the center of the Sun.

Approaching the Surface

As you begin your voyage from Earth, the Sun appears as a whitish ball of glowing gas. With spectroscopy [Section 7.3], you verify that the Sun's mass is 70% hydrogen and 28% helium. Heavier elements make up the remaining 2%.

The total power output of the Sun, called its **luminosity**, is an incredible 3.8×10^{26} watts. That is, every second, the Sun radiates a total of 3.8×10^{26} joules of energy into space (recall that 1 watt = 1 joule/s). If we could somehow capture and store just 1 second's worth of the Sun's lumi-

Figure 15.3 This photo of the visible surface of the Sun shows several dark sunspots.

Table 15.1 Basic Properties of the Sun

Radius (R_{Sun})	696,000 km (about 109 times the radius of Earth)
Mass (M_{Sun})	2×10^{30} kg (about 300,000 times the mass of Earth)
Luminosity (L_{Sun})	3.8×10^{26} watts
Composition (by percentage of mass)	70% hydrogen, 28% helium, 2% heavier elements
Rotation rate	25 days (equator) to 30 days (poles)
Surface temperature	5,800 K (average); 4,000 K (sunspots)
Core temperature	15 million K

nosity, it would be enough to meet current human energy demands for roughly the next 500,000 years!

Of course, only a tiny fraction of the Sun's total energy output reaches Earth, with the rest dispersing in all directions into space. Most of this energy is radiated in the form of visible light, but once you leave the protective blanket of Earth's atmosphere you'll encounter significant amounts of other types of solar radiation, including dangerous ultraviolet and X rays. Your spaceship will require substantial shielding to protect you from serious radiation burns caused by these high-energy forms of light.

Through a telescope, you can see that the Sun seethes with churning gases. At most times you'll detect at least a few **sunspots** blotching its surface (Figure 15.3). If you focus your telescope solely on a sunspot, you'll find that it is blindingly bright. Sunspots appear dark only in contrast to the even brighter solar surface that surrounds them. A typical sunspot is large enough to swallow the entire Earth, dramatically illustrating that the Sun is immense by any earthly standard. The Sun's radius is nearly 700,000 kilometers, and its mass is 2×10^{30} kilograms—about 300,000 times more massive than Earth.

Sunspots appear to move from day to day along with the Sun's rotation. If you watch very carefully, you may notice that sunspots near the solar equator circle the Sun faster than those at higher solar latitudes. This observation reveals that, unlike a spinning ball, the entire Sun does *not* rotate at the same rate. Instead, the solar equator completes one rotation in about 25 days, and the rotation period increases with latitude to about 30 days near the solar poles. Table 15.1 summarizes some of the basic properties of the Sun.

THINK ABOUT IT

As a brief review, describe how we measure the mass of the Sun using Newton's version of Kepler's third law. (*Hint:* Look back at Chapter 5.)

As you and your spaceship continue to fall toward the Sun, you notice an increasingly powerful headwind exerting a bit of drag on your descent. This headwind, called the **solar wind**, is created by ions and subatomic particles flowing outward from the solar surface. The solar wind helps shape the magnetospheres of planets [Sections 11.3, 12.4] and blows back the material that forms the tails of comets [Section 13.4].

A few million kilometers above the solar surface, you enter the solar **corona**, the tenuous uppermost layer of the Sun's atmosphere (Figure 15.4). Here you find the temperature to be astonishingly high—about 1 million Kelvin. This region emits most of the Sun's X rays. However, the density here is so low that your spaceship feels relatively little heat despite the million-degree temperature [Section 4.2].

Nearer the surface, the temperature suddenly drops to about 10,000 K in the **chromosphere**, the primary source of the Sun's ultraviolet radiation. At last you plunge through the visible surface of the Sun, called the **photosphere**, where the temperature averages just under 6,000 K. Although the photosphere looks like a well-defined surface from Earth, it consists of gas far less dense than Earth's atmosphere.

Throughout the solar atmosphere, you notice that the Sun has its own version of weather, in which conditions at a particular altitude differ from one region to another. Some regions of the chromosphere and corona are particularly hot and bright, while other regions are cooler and less dense. In the photosphere, sunspots are cooler than the surrounding surface, though they are still quite hot and bright by earthly standards. In addition, your compass goes crazy as you descend through the solar atmosphere, indicating that solar weather is shaped by intense magnetic fields. Occasionally, huge magnetic storms occur, shooting hot gases far into space.

Into the Sun

Up to this point in your journey, you may have seen Earth and the stars when you looked back, but as you slip beneath the photosphere, blazing light engulfs you. You are

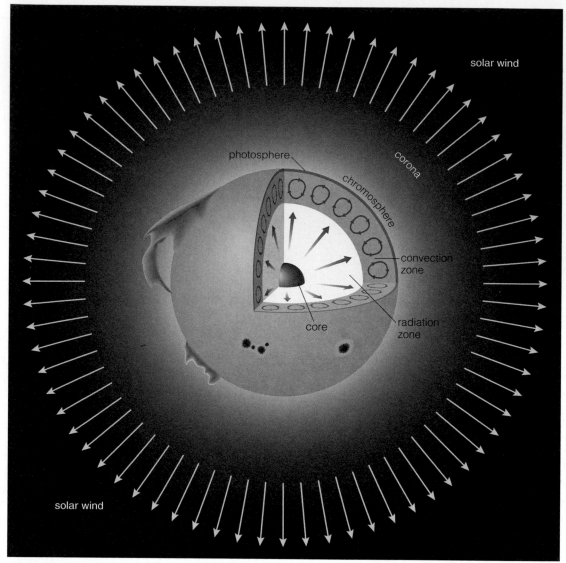

Figure 15.4 The basic structure of the Sun. Nuclear fusion in the solar *core* generates the Sun's energy. Photons of light carry that energy through the *radiation zone* to the bottom of the *convection zone*. Rising plumes of hot gas then transport the energy through the convection zone to the *photosphere*, where it is radiated into space. The photosphere, at a temperature of roughly 6,000 K, is relatively cool compared to the layers that lie above it. The temperature of the *chromosphere*, which is directly above the photosphere, exceeds 10,000 K. The temperature of the *corona*, extending outward from the chromosphere, can reach 1 million degrees. Because the coronal gas is so hot, some of it escapes the Sun's gravity, forming a *solar wind* that blows past Earth and out beyond Pluto.

inside the Sun, and your spacecraft is tossed about by incredible turbulence. If you can hold steady long enough to see what is going on around you, you'll notice spouts of hot gas rising upward, surrounded by cooler gas cascading down from above. You are in the **convection zone**, where energy generated in the solar core travels upward, transported by the rising of hot gas and falling of cool gas called *convection* [Section 10.2]. With some quick thinking, you may realize that the photosphere above you is the top of the convection zone and that convection is the cause of the Sun's seething, churning appearance.

As you descend through the convection zone, the surrounding density and pressure increase substantially, along with the temperature. Soon you reach depths at which the Sun is far denser than water. Nevertheless, it is still a *gas* (more specifically, a *plasma* of positively charged ions and free electrons) because each particle moves independently of its neighbors [Section 4.3].

About a third of the way down to the center, the turbulence of the convection zone gives way to the calmer plasma of the **radiation zone**, where energy is carried outward primarily by photons of light. The temperature rises to almost 10 million K, and your spacecraft is bathed in X rays trillions of times more intense than the visible light at the solar surface.

The Sun Is Not on Fire

We are accustomed to saying that the Sun is "burning," a way of speaking that conjures up images of a giant bonfire in the sky. However, the Sun does not burn in the same sense as a fire burns on Earth. Fires on Earth generate light through chemical changes that consume oxygen and produce a flame. The glow of the Sun has more in common with the glowing embers left over after the flames have burned out. Much like hot embers, the Sun's surface shines with the visible thermal radiation produced by any object that is sufficiently hot [Section 6.4].

However, hot embers quickly stop glowing as they cool, while the Sun keeps shining because its surface is kept hot by the energy rising from the Sun's core. Because this energy is generated by nuclear fusion, we sometimes say that it is the result of "nuclear burning"—a term that suggests nuclear changes in much the same way that "chemical burning" suggests chemical changes. Nevertheless, while it is reasonable to say that the Sun undergoes nuclear burning in its core, it is not accurate to speak of any kind of burning on the Sun's surface, where light is produced primarily by thermal radiation.

No real spacecraft could survive, but your imaginary one keeps plunging straight down to the solar **core**. There you finally find the source of the Sun's energy: nuclear fusion transforming hydrogen into helium. At the Sun's center, the temperature is about 15 million K, the density is more than 100 times that of water, and the pressure is 200 billion times that on the surface of Earth. The energy produced in the core today will take about a million years to reach the surface.

With your journey complete, it's time to turn around and head back home. We'll continue this chapter by studying fusion in the solar core and then tracing the flow of the energy generated by fusion as it moves outward through the Sun.

15.3 The Cosmic Crucible

The prospect of turning common metals like lead into gold enthralled those who pursued the medieval practice of alchemy. Sometimes they tried primitive scientific approaches, such as melting various ores together in a vessel called a crucible. Other times they tried magic. Their get-rich-quick schemes never managed to work. Today we know that there is no easy way to turn other elements into gold, but it *is* possible to transmute one element or isotope into another.

If a nucleus gains or loses protons, its atomic number changes and it becomes a different element. If it gains or

Figure 15.5 Nuclear fission splits a nucleus into smaller nuclei (not usually of equal size), while nuclear fusion combines smaller nuclei into a larger nucleus.

loses neutrons, its atomic mass changes and it becomes a different isotope [Section 4.3]. The process of splitting a nucleus into two smaller nuclei is called **nuclear fission**. The process of combining nuclei to make a nucleus with a greater number of protons or neutrons is called **nuclear fusion** (Figure 15.5). Human-built nuclear power plants rely on nuclear fission of uranium or plutonium. The nuclear power plant at the center of the Sun relies on nuclear fusion, turning hydrogen into helium.

Nuclear Fusion

The 15 million K plasma in the solar core is like a "soup" of hot gas, with bare, positively charged atomic nuclei (and negatively charged electrons) whizzing about at extremely high speeds. At any one time, some of these nuclei are on high-speed collision courses with each other. In most cases, electromagnetic forces deflect the nuclei, preventing actual collisions, because positive charges repel one another. If nuclei collide with sufficient energy, however, they can stick together to form a heavier nucleus (Figure 15.6).

Sticking positively charged nuclei together is not easy. The **strong force**, which binds protons and neutrons together in atomic nuclei, is the only force in nature that can

At low speeds, electromagnetic repulsion prevents the collision of nuclei.

At high speeds, nuclei come close enough for the strong force to bind them together.

Figure 15.6 Positively charged nuclei can fuse only if a high-speed collision brings them close enough for the strong force to come into play.

Figure 15.7 Hydrogen fuses into helium in the Sun by way of the proton–proton chain. In step 1, two protons fuse to create a deuterium nucleus consisting of a proton and a neutron. In step 2, the deuterium nucleus and a proton fuse to form helium-3, a rare form of helium. In step 3, two helium-3 nuclei fuse to form helium-4, the common form of helium.

overcome the electromagnetic repulsion between two positively charged nuclei [Section S4.2]. In contrast to gravitational and electromagnetic forces, which drop off gradually as the distances between particles increase (by an inverse square law [Section 5.3]), the strong force is more like glue or Velcro: It overpowers the electromagnetic force over very small distances but is insignificant when the distances between particles exceed the typical sizes of atomic nuclei. The trick to nuclear fusion, therefore, is to push the positively charged nuclei close enough together for the strong force to outmuscle electromagnetic repulsion.

The high pressures and temperatures in the solar core are just right for fusion of hydrogen nuclei into helium nuclei. The high temperature is important because the nuclei must collide at very high speeds if they are to come close enough together to fuse. (Quantum tunneling is also important to this process [Section S4.5].) The higher the temperature, the harder the collisions, making fusion reactions more likely at higher temperatures. The high pressure of the overlying layers is necessary because without it the hot plasma of the solar core would simply explode into space, shutting off the nuclear reactions. In the Sun, the pressure is high and steady, allowing some 600 million tons of hydrogen to fuse into helium every second.

Hydrogen Fusion in the Sun: The Proton–Proton Chain

Recall that hydrogen nuclei are nothing more than individual protons, while the most common form of helium consists of two protons and two neutrons. Thus, the overall hydrogen fusion reaction in the Sun is:

$$4\ {}^1H \longrightarrow 1\ {}^4He + \text{energy}$$

However, collisions between two nuclei are far more common than three- or four-way collisions, so this overall reaction proceeds through steps that involve just two nuclei at a time. The sequence of steps that occurs in the Sun is called the **proton–proton chain** because it begins with collisions between individual protons (hydrogen nuclei). Figure 15.7 illustrates the steps in the proton–proton chain:

Step 1. Two protons fuse to form a nucleus consisting of one proton and one neutron, which is the isotope of hydrogen known as *deuterium*. Note that this step converts a proton into a neutron, reducing the total nuclear charge from +2 for the two fusing protons to +1 for the resulting deuterium nucleus. The lost positive charge is carried off by a *positron*, the antimatter version of an electron with a positive rather than negative charge [Section S4.2]. A *neutrino*—a subatomic particle with a very tiny mass—is also produced in this step.* The positron won't last long, because it soon meets up with an ordinary electron, resulting

---

*Producing a neutrino is necessary because of a law called *conservation of lepton number:* The number of leptons (e.g., electrons or neutrinos [Chapter S4]) must be the same before and after the reaction. The lepton number is zero before the reaction because there are no leptons. Among the reaction products, the positron (antielectron) has lepton number −1 because it is antimatter, and the neutrino has lepton number +1. Thus, the total lepton number remains zero.

in the creation of two gamma-ray photons through matter–antimatter annihilation.

Step 2. A fair number of deuterium nuclei are always present along with the protons and other nuclei in the solar core, since step 1 occurs so frequently in the Sun (about 10^{38} times per second). Step 2 occurs when one of these deuterium nuclei collides and fuses with a proton. The result is a nucleus of helium-3, a rare form of helium with two protons and one neutron. This reaction also produces a gamma-ray photon.

Step 3. The third and final step of the proton–proton chain requires the addition of another neutron to the helium-3, thereby making normal helium-4. This final step can proceed in several different ways, but the most common route involves a collision of two helium-3 nuclei. Each of these helium-3 nuclei resulted from a prior, separate occurrence of step 2 somewhere in the solar core. The final result is a normal helium-4 nucleus and two protons.

Total reaction. Somewhere in the solar core, steps 1 and 2 must each occur twice to make step 3 possible. Six protons go into each complete cycle of the proton–proton chain, but two come back out. Thus, the overall proton–proton chain converts four protons (hydrogen nuclei) into a helium-4 nucleus, two positrons, two neutrinos, and two gamma rays.

Each resulting helium-4 nucleus has a mass that is slightly less (by about 0.7%) than the combined mass of the four protons that created it. Overall, fusion in the Sun converts about 600 million tons of hydrogen into 596 million tons of helium every second. The "missing" 4 million tons of matter becomes energy in accord with Einstein's formula $E = mc^2$. About 98% of the energy emerges as kinetic energy of the resulting helium nuclei and radiative energy of the gamma rays. As we will see, this energy slowly percolates to the solar surface, eventually emerging as the sunlight that bathes Earth. About 2% of the energy is carried off by the neutrinos. Neutrinos rarely interact with matter (because they respond only to the weak force [Section S4.2]), so most of the neutrinos created by the proton–proton chain pass straight from the solar core through the solar surface and out into space.

The Solar Thermostat

The rate of nuclear fusion in the solar core, which determines the energy output of the Sun, is very sensitive to temperature. A slight increase in temperature would mean a much higher fusion rate, and a slight decrease in temperature would mean a much lower fusion rate. If the Sun's rate of fusion varied erratically, the effects on Earth might be devastating. Fortunately, the Sun's central temperature is steady thanks to gravitational equilibrium—the balance between the pull of gravity and the push of internal pressure.

Outside the solar core, the energy produced by fusion travels toward the Sun's surface at a slow but steady rate. In this steady state, the amount of energy leaving the top of each gas layer within the Sun precisely balances the energy entering from the bottom (Figure 15.8). Suppose the core temperature of the Sun rose very slightly. The rate of nuclear fusion would soar, generating lots of extra energy. Because energy moves so slowly through the Sun, this extra energy would be bottled up in the core, causing an increase in the core pressure. The push of this pressure would temporarily exceed the pull of gravity, causing the core to expand and cool. With cooling, the fusion rate would drop back down. The expansion and cooling would continue until gravitational equilibrium was restored, at which point the fusion rate would return to its original value.

An opposite process would restore the normal fusion rate if the core temperature dropped. A decrease in core temperature would lead to decreased nuclear burning, a drop in the central pressure, and contraction of the core. As the core shrank, its temperature would rise until the burning rate returned to normal.

The response of the core pressure to changes in the nuclear fusion rate is essentially a *thermostat* that keeps the Sun's central temperature steady. Any change in the core temperature is automatically corrected by the change in the fusion rate and the accompanying change in pressure.

While the processes involved in gravitational equilibrium prevent erratic changes in the fusion rate, they also ensure that the fusion rate gradually rises over billions of years. Because each fusion reaction converts *four* hydrogen nuclei into *one* helium nucleus, the total number of *independent particles* in the solar core is gradually falling. This gradual reduction in the number of particles causes the solar core to shrink.

The slow shrinking of the solar core means that it must generate energy more rapidly to counteract the stronger compression of gravity, so the solar core gradually gets hotter as it shrinks. Theoretical models indicate that the Sun's core temperature should have increased enough to raise its fusion rate and the solar luminosity by about 30% since the Sun was born 4.6 billion years ago.

How did the gradual increase in solar luminosity affect Earth? Geological evidence shows that Earth's surface temperature has remained fairly steady since Earth finished forming more than 4 billion years ago, despite this 30% increase in the Sun's energy output, because Earth has its own thermostat. This "Earth thermostat" is the carbon dioxide cycle. By maintaining a fairly steady level of atmospheric carbon dioxide, the carbon dioxide cycle regulates the greenhouse effect that maintains Earth's surface temperature [Section 14.4].

"Observing" the Solar Interior

We cannot see inside the Sun, so you may be wondering how we can know so much about what goes on underneath its surface. Astronomers can study the Sun's interior in three different ways: through mathematical models of the

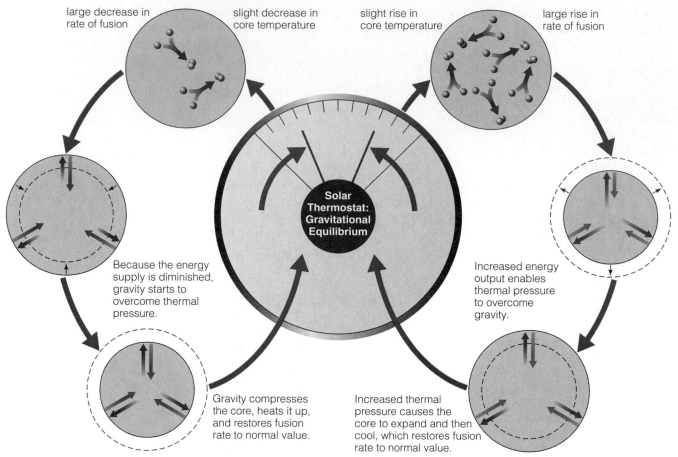

large decrease in
rate of fusion

slight decrease in
core temperature

slight rise in
core temperature

large rise in
rate of fusion

Solar
Thermostat:
Gravitational
Equilibrium

Because the energy
supply is diminished,
gravity starts to
overcome thermal
pressure.

Increased energy
output enables
thermal pressure
to overcome
gravity.

Gravity compresses
the core, heats it up,
and restores fusion
rate to normal value.

Increased thermal
pressure causes the
core to expand and then
cool, which restores fusion
rate to normal value.

Figure 15.8 The solar thermostat. Gravitational equilibrium regulates the Sun's core temperature. Everything is in balance if the amount of energy leaving the core equals the amount of energy produced by fusion. A rise in core temperature triggers a chain of events that causes the core to expand, lowering its temperature to its normal value. A decrease in core temperature triggers the opposite chain of events, also restoring the normal core temperature.

Sun, observations of "sun quakes," and observations of solar neutrinos.

Mathematical Models The primary way we learn about the interior of the Sun and other stars is by creating *mathematical models* that use the laws of physics to predict the internal conditions. A basic model uses the Sun's observed composition and mass as inputs to equations that describe gravitational equilibrium, the solar thermostat, and the rate at which solar energy moves from the core to the photosphere. With the aid of a computer, we can use the model to calculate the Sun's temperature, pressure, and density at any depth. We can then predict the rate of nuclear fusion in the solar core by combining these calculations with knowledge about nuclear fusion gathered in laboratories here on Earth.

Remarkably, such models correctly "predict" the radius, surface temperature, luminosity, age, and many other properties of the Sun. However, current models do not predict *everything* about the Sun correctly. Scientists are constantly working to discover what is missing from them. Successful prediction of so many observed characteristics

of the Sun gives us confidence that the models are on the right track and that we really do understand what is going on inside the Sun.

Sun Quakes A second way to learn about the inside of the Sun is to observe "sun quakes"—vibrations of the Sun that are similar to the vibrations of the Earth caused by earthquakes, although they are generated very differently. Earthquakes occur when Earth's crust suddenly shifts, generating *seismic waves* that propagate through Earth's interior [Section 10.2]. We can learn about Earth's interior by recording seismic waves on Earth's surface with seismographs.

Sun quakes result from waves of pressure (sound waves) that propagate deep within the Sun at all times. These waves cause the solar surface to vibrate when they reach it. Although we cannot set up seismographs on the Sun, we can detect the vibrations of the surface by measuring Doppler shifts [Section 6.5]. Light from portions of the surface that are rising toward us is slightly blueshifted, while light from portions that are falling away from us is slightly redshifted. The vibrations are relatively small but measurable (Figure 15.9).

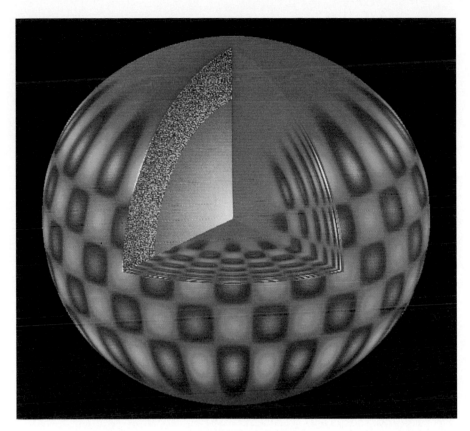

Figure 15.9 Vibrations on the surface of the Sun can be detected by Doppler shifts. In this schematic representation, red indicates falling gas, and blue indicates rising gas. The speckled region indicates the convection zone. The vibration pattern illustrated here is just one of many possible patterns. The overall vibration pattern of the Sun is a complex combination of patterns similar to this one.

In principle, we can deduce a great deal about the solar interior by carefully analyzing these vibrations. (By analogy to seismology on Earth, this type of study of the Sun is called *helioseismology*—*helios* means "sun.") Re-sults to date confirm that our mathematical models of the solar interior are on the right track (Figure 15.10). At the same time, they provide data that can be used to improve the models further.

Mathematical Insight **15.1** **Mass-Energy Conversion in the Sun**

We can calculate how much mass the Sun loses through nuclear fusion by comparing the input and output masses of the proton–proton chain. A single proton has a mass of 1.6726×10^{-27} kg, so four protons have a mass of 6.690×10^{-27} kg.

A helium-4 nucleus has a mass of only 6.643×10^{-27} kg, slightly less than the mass of the four protons. The difference is:

$$6.690 \times 10^{-27} \text{ kg} - 6.643 \times 10^{-27} \text{ kg} = 4.7 \times 10^{-29} \text{ kg}$$

which is 0.7%, or 0.007, of the original mass. Thus, for example, when 1 kilogram of hydrogen fuses, the resulting helium weighs only 993 grams, while 7 grams of mass turns into energy.

To calculate the *total* amount of mass converted to energy in the Sun each second, we use Einstein's equation $E = mc^2$. The total energy produced by the Sun each second is 3.8×10^{26} joules, so we can solve for the total mass converted to energy each second:

$$E = mc^2 \quad \Rightarrow \quad m = \frac{E}{c^2}$$

$$= \frac{3.8 \times 10^{26} \text{ joules}}{\left(3.0 \times 10^8 \, \frac{\text{m}}{\text{s}}\right)^2} = 4.2 \times 10^9 \text{ kg}$$

The Sun loses about 4 billion kilograms of mass every second, which is roughly equivalent to the combined mass of nearly 100 million people.

Example: How much hydrogen is converted to helium each second in the Sun?

Solution: We have already calculated that the Sun loses 4.2×10^9 kg of mass each second and that this is only 0.7% of the mass of hydrogen that is fused:

$$4.2 \times 10^9 \text{ kg} = 0.007 \times \text{mass of hydrogen fused}$$

We now solve for the mass of hydrogen fused:

$$\frac{\text{mass of}}{\text{hydrogen fused}} = \frac{4.2 \times 10^9 \text{ kg}}{0.007}$$

$$= 6.0 \times 10^{11} \text{ kg} \times \frac{1 \text{ metric ton}}{10^3 \text{ kg}}$$

$$= 6.0 \times 10^8 \text{ metric tons}$$

The Sun fuses 600 million metric tons of hydrogen each second, of which about 4 million tons becomes energy. The remaining 596 million tons becomes helium.

a Temperature at different radii within the Sun.

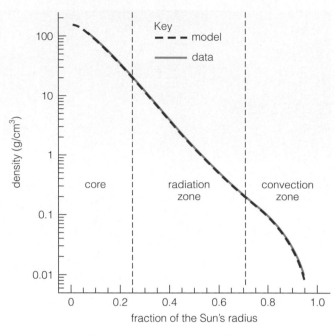

b Density at different radii within the Sun. (The density of water is 1 g/cm³.)

Figure 15.10 Agreement between mathematical models of solar structure and actual measurements of solar structure derived from "sun quakes." The red lines show predictions of mathematical models of the Sun. The blue lines show the interior structure of the Sun as indicated by vibrations of the Sun's surface. These vibrations tell us about conditions deep within the Sun because they are produced by sound waves that propagate through the Sun's interior layers.

Solar Neutrinos Another way to study the solar interior is to observe the neutrinos coming from fusion reactions in the core. Don't panic, but as you read this sentence about a thousand trillion solar neutrinos will zip through your body. Fortunately, they won't do any damage, because neutrinos rarely interact with anything. Neutrinos created by fusion in the solar core fly quickly through the Sun as if passing through empty space. In fact, while an inch of lead will stop an X ray, stopping an average neutrino would require a slab of lead more than 1 light-year thick! Clearly, counting neutrinos is dauntingly difficult, because virtually all of them stream right through any detector built to capture them.

THINK ABOUT IT

Is the number of solar neutrinos zipping through our bodies significantly lower at night? (*Hint:* How does the thickness of Earth compare with the thickness of a slab of lead needed to stop an average neutrino?)

Nevertheless, neutrinos *do* occasionally interact with matter, and it is possible to capture a few solar neutrinos with a large enough detector. Neutrino detectors are usually placed deep inside mines so that the overlying layers of rock block all other kinds of particles coming from outer space except neutrinos, which pass through rock easily. The

first major solar neutrino detector, built in the 1960s, was located 1,500 meters underground in the Homestake gold mine in South Dakota (Figure 15.11).

The detector for this "Homestake experiment" consisted of a 400,000-liter vat of chlorine-containing dry-cleaning fluid. It turns out that, on very rare occasions, a chlorine nucleus can capture a neutrino and change into a nucleus of radioactive argon. By looking for radioactive argon in the tank of cleaning fluid, experimenters could count the number of neutrinos captured in the detector.

From the many trillions of solar neutrinos that passed through the tank of cleaning fluid each second, experimenters expected to capture an average of just one neutrino per day. This predicted capture rate was based on measured properties of chlorine nuclei and models of nuclear fusion in the Sun. However, over a period of more than two decades, neutrinos were captured only about once every 3 days on average. That is, the Homestake experiment detected only about one-third of the predicted number of neutrinos. This disagreement between model predictions and actual observations came to be called the **solar neutrino problem**.

The shortfall of neutrinos found with the Homestake experiment led to many more recent attempts to detect solar neutrinos using more sophisticated detectors (Figure 15.12). The chlorine nuclei in the Homestake experiment could

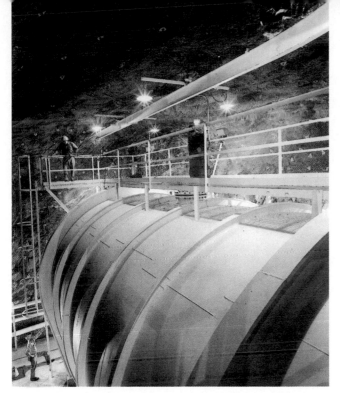

Figure 15.11 This tank of dry-cleaning fluid (visible underneath the catwalk), located deep within South Dakota's Homestake mine, was a solar neutrino detector. The chlorine nuclei in the cleaning fluid turned into argon nuclei when they captured neutrinos from the Sun.

a Scientists inspecting individual detectors within Super-Kamiokande.

capture only high-energy neutrinos that are produced by one of the rare pathways of step 3 in the proton–proton chain (not shown in Figure 15.7). More recent experiments can detect lower-energy neutrinos, including those produced by step 1 of the proton–proton chain, and therefore offer a better probe of fusion in the Sun. To date, all these experiments have found fewer neutrinos than current models of the Sun predict. This discrepancy between model and experiment probably means one of two things: Either something is wrong with our models of the Sun, or something is missing in our understanding of how neutrinos behave.

THINK ABOUT IT

Although the observed number of neutrinos falls short of theoretical predictions, experiments like Homestake have shown that at least some neutrinos are coming from the Sun. Explain why this provides direct evidence that nuclear fusion really is taking place in the Sun right now. (*Hint:* See Figure 15.7.)

For the moment, many physicists and astronomers are betting that we understand the Sun just fine and that the discrepancy has to do with the neutrinos themselves. One intriguing idea arises from the fact that neutrinos come in three types: electron neutrinos, muon neutrinos, and tau neutrinos [Section S4.2].

Fusion reactions in the Sun produce only electron neutrinos, and most solar neutrino detectors can detect only

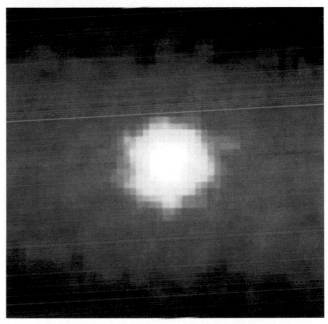

b An image of the Sun created by tracing the paths of neutrinos detected by Super-Kamiokande back to the Sun.

Figure 15.12 The Super-Kamiokande experiment in Japan is one of the world's premier neutrino detectors.

electron neutrinos. However, recent experiments have shown that some of the electron neutrinos might change into muon and tau neutrinos as they fly out through the solar plasma. In that case, our detectors would count fewer than the expected number of electron neutrinos. Early results from the Sudbury Neutrino Observatory in Canada, a new detector designed to search for all types of neutrinos, suggest that neutrinos changing type is indeed the solution to the solar neutrino problem. The observations are ongoing, and it will probably be several more years before this solution can be definitively confirmed.

Because of their roles in detecting solar neutrinos and identifying the solar neutrino problem, Raymond Davis, leader of the Homestake experiment, and Masatoshi Koshiba, leader of Super-Kamiokande, shared in the 2002 Nobel Prize for physics.

15.4 From Core to Corona

Energy liberated by nuclear fusion in the Sun's core must eventually reach the solar surface, where it can be radiated into space. The path that the energy takes to the surface is long and complex. In this section, we follow that long path.

The Path Through the Solar Interior

In Chapter 6, we discussed how atoms can absorb or emit photons. In fact, photons can also interact with any charged particle, and a photon that "collides" with an electron can be deflected into a completely new direction.

Deep in the solar interior, the plasma is so dense that the gamma-ray photons resulting from fusion travel only a fraction of a millimeter before colliding with an electron. Because each collision sends the photon in a random new direction, the photon bounces around the core in a haphazard way, sometimes called a *random walk*. With each random bounce, the photon drifts farther and farther, on average, from its point of origin. As a result, photons from the solar core gradually work their way outward (Figure 15.13). The technical term for this slow, outward migration of photons is **radiative diffusion** (to *diffuse* means to "spread out" and *radiative* refers to the photons of light or radiation).

Along the way, the photons exchange energy with their surroundings. Because the surrounding temperature declines as the photons move outward through the Sun, they are gradually transformed from gamma rays to photons of lower energy. (Because energy must be conserved, each gamma-ray photon becomes many lower-energy photons.) By the time the energy of fusion reaches the surface, the photons are primarily visible light. On average, the energy released in a fusion reaction takes about a million years to reach the solar surface.

Figure 15.13 A photon in the solar interior bounces randomly among electrons, slowly working its way outward in a process called radiative diffusion.

THINK ABOUT IT

Radiative diffusion is just one type of diffusion. Another is the diffusion of dye through a glass of water. If you place a concentrated spot of dye at one point in the water, each individual dye molecule begins a random walk as it bounces among the water molecules. The result is that the dye gradually spreads through the entire glass. Can you think of any other examples of diffusion in the world around you?

Radiative diffusion is the primary way by which energy moves outward through the *radiation zone,* which stretches from the core to about 70% of the Sun's radius (see Figure 15.4). Above this point, where the temperature has dropped to about 2 million K, the solar plasma absorbs photons more readily (rather than just bouncing them around). This point is the beginning of the solar *convection zone,* where the buildup of heat resulting from photon absorption causes bubbles of hot plasma to rise upward in the process known as **convection** [Section 10.2]. Convection occurs because hot gas is less dense than cool gas. Like a hot-air balloon, a hot bubble of solar plasma rises upward through the cooler plasma above it. Meanwhile, cooler plasma from above slides around the rising bubble and sinks to lower layers, where it is heated. The rising of hot plasma and sinking of cool plasma form a cycle that transports energy outward from the top of the radiation zone to the solar surface (Figure 15.14a).

The Solar Surface

Earth has a solid crust, so its surface is well defined. In contrast, the Sun is made entirely of gaseous plasma. Defining where the surface of the Sun begins is therefore something like defining the surface of a cloud: From a distance it looks quite distinct, but up close the surface is fuzzy, not sharp. We generally define the solar surface as the layer that appears distinct from a distance. This is the layer we identified as the *photosphere* when we took our imaginary journey into the Sun. More technically, the photosphere is the layer of the Sun from which photons finally escape into space after the million-year journey of solar energy outward from the core.

a This schematic diagram shows how hot gas (white arrows) rises while cooler gas (orange/black arrows) descends around it. Bright spots appear on the solar surface in places where hot gas is rising from below, creating the granulated appearance of the solar photosphere.

b Granulation is evident in this photo of the Sun's surface. Each bright granule is the top of a rising column of gas. At the darker lines between the granules, cooler gas is descending below the photosphere. Each granule is about 1,000 kilometers across.

Figure 15.14 Convection transports energy outward in the Sun's convection zone.

Most of the energy produced by fusion in the solar core ultimately leaves the photosphere as thermal radiation [Section 6.4]. The average temperature of the photosphere is about 5,800 K, corresponding to a thermal radiation spectrum that peaks in the green portion of the visible spectrum, with substantial energy coming out in all colors of visible light. The Sun appears whitish when seen from space, but in our sky the Sun appears somewhat more yellow—and even red at sunset—because Earth's atmosphere scatters blue light. It is this scattered light from the Sun that makes our skies blue [Section 11.3].

Although the average temperature of the photosphere is 5,800 K, actual temperatures vary significantly from place to place. The photosphere is marked throughout by the bubbling pattern of **granulation** produced by the underlying convection (Figure 15.14b). Each *granule* appears bright in the center, where hot gas bubbles upward, and dark around the edges, where cool gas descends. If we made a movie of the granulation, we'd see it bubbling rather like a pot of boiling water. Just as bubbles in a pot of boiling water burst on the surface and are replaced by new bubbles, each granule lasts only a few minutes before being replaced by other granules bubbling upward.

Sunspots and Magnetic Fields

Sunspots are the most striking features on the solar surface (Figure 15.15a). The temperature of the plasma in sunspots is about 4,000 K, significantly cooler than the 5,800 K plasma of the surrounding photosphere. If you think about this for a moment, you may wonder how sunspots can be so much cooler than their surroundings. Why doesn't the surrounding hot plasma heat the sunspots? Something must be

preventing hot plasma from entering the sunspots, and that "something" turns out to be magnetic fields.

Detailed observations of the Sun's spectral lines reveal sunspots to be regions with strong magnetic fields. These magnetic fields can alter the energy levels in atoms and ions and therefore can alter the spectral lines they produce. More specifically, magnetic fields cause some spectral lines to split into two or more closely spaced lines (Figure 15.15b). This effect (called the *Zeeman effect*) enables scientists to map magnetic fields on the Sun by studying the spectral lines in light from different parts of the solar surface.

Magnetic fields are invisible, but in principle we could visualize a magnetic field by laying out many compasses. Each compass needle would point to local magnetic north. We can represent the magnetic field by drawing a series of lines, called **magnetic field lines**, connecting the needles of these imaginary compasses (Figure 15.16a). The strength of the magnetic field is indicated by the spacing of the lines: Closer lines mean a stronger field (Figure 15.16b). Because these imaginary field lines are so much easier to visualize than the magnetic field itself, we usually discuss magnetic fields by talking about how the field lines would appear. Charged particles, such as the ions and electrons in the solar plasma, follow paths that spiral along the magnetic field lines (Figure 15.16c). Thus, the solar plasma can move freely *along* magnetic field lines but cannot easily move perpendicular to them.

The magnetic field lines act somewhat like elastic bands, being twisted into contortions and knots by turbulent motions in the solar atmosphere. Sunspots occur where the most taut and tightly wound magnetic fields poke nearly straight out from the solar interior. Sunspots tend to occur in pairs connected by a loop of magnetic field lines. These

a This close-up view of the Sun's surface (right) shows two large sunspots and several smaller ones. Both of the big sunspots are roughly as large as Earth.

Figure 15.15 Sunspots are regions of intense magnetic activity.

b Spectra of sunspots can be used to measure the strength of their magnetic fields. This image shows the spectrum of a sunspot and its surroundings. The sunspot region shows up as dark horizontal bands because it is darker than the rest of the solar surface in its vicinity. The vertical bands are absorption lines that are present both inside and outside the sunspots. The influence of strong magnetic fields within the sunspot region splits a single absorption line into three parts. Measuring the separation between these lines tells us the strength of the magnetic field within the sunspot.

tight magnetic field lines suppress convection within each sunspot and prevent surrounding plasma from sliding sideways into the sunspot. With hot plasma unable to enter the region, the sunspot plasma becomes cooler than that of the rest of the photosphere (Figure 15.17a).

The magnetic field lines connecting two sunspots often soar high above the photosphere, through the chromosphere, and into the corona (Figure 15.17b). These vaulted loops of magnetic field sometimes appear as **solar prominences**, in which the field traps gas that may glow for days or even weeks. Some prominences rise to heights of more than 100,000 kilometers above the Sun's surface (Figure 15.18).

The most dramatic events on the solar surface are **solar flares**, which emit bursts of X rays and fast-moving charged particles into space (Figure 15.19). Flares generally occur in the vicinity of sunspots, leading us to suspect that they occur when the magnetic field lines become so twisted and knotted that they can no longer bear the tension. The magnetic field lines suddenly snap like tangled elastic bands twisted beyond their limits, releasing a huge amount of energy. This energy heats the nearby plasma to 100 million K over the next few minutes to few hours, generating X rays and accelerating some of the charged particles to nearly the speed of light.

The Chromosphere and Corona

The high temperatures of the chromosphere and corona perplexed scientists for decades. After all, temperatures gradually decline as we move outward from the core to the top of the photosphere. Why should this decline suddenly reverse? Some aspects of this atmospheric heating remain a mystery today, but we have at least a general explanation: The Sun's strong magnetic fields carry energy upward from the churning solar surface to the chromosphere and corona.

More specifically, the rising and falling of gas in the convection zone probably shakes magnetic field lines beneath the solar surface. This shaking generates waves along the magnetic field lines that carry energy upward to the solar

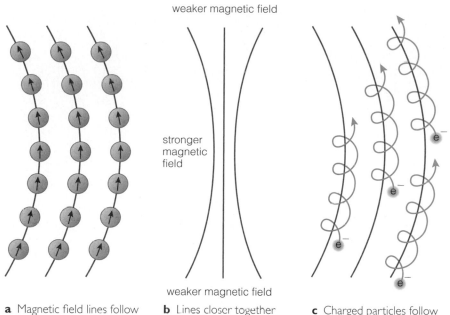

weaker magnetic field

stronger magnetic field

weaker magnetic field

a Magnetic field lines follow the directions that compass needles would point.

b Lines closer together indicate a stronger field.

c Charged particles follow paths that spiral along magnetic field lines.

Figure 15.16 We draw magnetic field lines to represent invisible magnetic fields.

atmosphere. Precisely how the waves deposit their energy in the chromosphere and corona is not known, but the waves agitate the low-density plasma of these layers, somehow heating them to high temperatures. Much of this heating appears to happen near where the magnetic field lines emerge from the Sun's surface.

According to this model of solar heating, the same magnetic fields that keep sunspots cool make the overlying plasma of the chromosphere and corona hot. We can test this idea observationally. The gas of the chromosphere and corona is so tenuous that we cannot see it with our eyes except during a total eclipse, when we can see the faint visible light scattered by electrons in the corona [Section 2.5]. However, the roughly 10,000 K plasma of the chromosphere emits strongly in the ultraviolet, and the million K plasma of the corona is the source of virtually all X rays coming

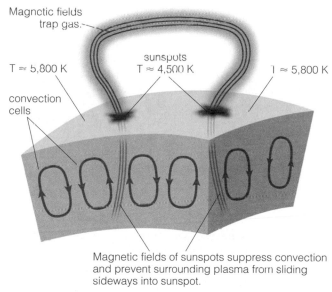

Magnetic fields trap gas.

$T \approx 5,800$ K

sunspots $T \approx 4,500$ K

$T \approx 5,800$ K

convection cells

Magnetic fields of sunspots suppress convection and prevent surrounding plasma from sliding sideways into sunspot.

a Pairs of sunspots are connected by tightly wound magnetic field lines.

Figure 15.17 Loops of magnetic field lines can arch high above the solar surface, reaching heights many times larger than Earth's diameter.

b This X-ray photo (from NASA's TRACE mission) shows gas trapped within looped magnetic field lines.

X-ray

Figure 15.18 A gigantic solar prominence erupts from the solar surface at the upper right of this ultraviolet-light photo (from the SOHO mission). The gas within this prominence, which is over 20 times the size of Earth, is quite hot but still cooler than the million-degree gas of the surrounding corona.

Figure 15.19 This photo (from TRACE) of ultraviolet light emitted by hydrogen atoms shows a solar flare erupting from the Sun's surface.

from the Sun. Figure 15.20 shows an X-ray image of the Sun. As the solar heating model predicts, the brightest regions of the corona tend to be directly above sunspot groups.

Some regions of the corona, called **coronal holes**, barely show up in X-ray images. More detailed analyses show that the magnetic field lines in coronal holes project out into space like broken rubber bands, allowing particles spiraling along them to escape the Sun altogether. These particles streaming outward from the corona constitute the *solar wind,* which blows through the solar system at an average speed of about 500 kilometers per second and has important effects on planetary surfaces, atmospheres, and magnetospheres. Well beyond the planets, the pressure of interstellar gas must eventually halt the solar wind. The Pioneer and Voyager spacecraft that visited the outer planets in the 1970s and 1980s are still traveling outward from our solar system and may soon encounter this "boundary" (called the *heliopause*) of the realm of the Sun.

The solar wind also gives us something tangible to study. In the same way that meteorites provide us with samples of asteroids we've never visited, solar wind particles captured by satellites provide us with a sample of material from the Sun. Analysis of these solar particles has reassuringly verified that the Sun is made mostly of hydrogen, just as we conclude from studying the Sun's spectrum.

Figure 15.20 An X-ray image of the Sun reveals the million-degree gas of the corona. Brighter regions of this image (yellow) correspond to regions of stronger X-ray emission. The darker regions are the coronal holes from which the solar wind escapes. (From the Yohkoh space observatory.)

15.5 Solar Weather and Climate

Individual sunspots, prominences, and flares are short-lived phenomena, somewhat like storms on Earth. They constitute what we call *solar weather* or **solar activity**. You know from personal experience that the Earth's weather is notoriously unpredictable. The same is true for the Sun: We cannot predict precisely when or where a particular sunspot or flare will appear. Earth's *climate*, on the other hand, is quite regular from season to season. So it is with the Sun, where despite day-to-day variations the general nature and intensity of solar activity follow a predictable cycle.

The Sunspot Cycle

Long before we realized that sunspots were magnetic disturbances, astronomers had recognized patterns in sunspot activity. The most notable pattern is the number of sunspots visible on the Sun at any particular time. Thanks to telescopic observations of the Sun recorded by astronomers since the 1600s, we know that the number of sunspots gradually rises and falls in a **sunspot cycle** with an average period of about 11 years (Figure 15.21a). At the time of **solar maximum**, when sunspots are most numerous, we may see dozens of sunspots on the Sun at one time. In contrast, we see few if any sunspots at the time of **solar minimum**. The frequency of prominences and flares also follows the sunspot cycle, with these events being most common at solar maximum and least common at solar minimum.

Although we'll call it an "11-year" cycle, the interval between solar maxima is sometimes as long as 15 years or as short as 7 years. The number of sunspots also varies dramatically (Figure 15.21a). In fact, sunspot activity virtually ceased between the years 1645 and 1715, a period sometimes called the *Maunder minimum* (after E. W. Maunder, who identified it in historical sunspot records).

Another feature of the sunspot cycle is a gradual change in the solar latitudes at which individual sunspots form and dissolve (Figure 15.21b). As a cycle begins at solar minimum, sunspots form primarily at mid-latitudes (30° to 40°) on the Sun. The sunspots tend to form at lower latitudes as the cycle progresses, appearing very close to the solar equator as the next solar minimum approaches.

A less obvious feature of the sunspot cycle is that something peculiar happens to the Sun's magnetic field at each solar minimum. The field lines connecting all pairs of sunspots (see Figure 15.17) tend to point in the same direction throughout an 11-year solar cycle (within each hemisphere). For example, all compass needles might point from the easternmost sunspot to the westernmost sunspot in a pair. However, as the cycle ends at solar minimum, the magnetic field reverses: In the subsequent solar cycle, the field lines connecting pairs of sunspots point in the opposite direction. Apparently, the entire magnetic field of the Sun flip-flops every 11 years.

The magnetic reversals hint that the sunspot cycle is related to the generation of magnetic fields on the Sun. They also tell us that the *complete* magnetic cycle of the Sun, called the *solar cycle,* really averages 22 years, since it takes two 11-year cycles before the magnetic field is back the way it started.

a This graph shows how the number of sunspots on the Sun changes with time. The vertical axis shows the percentage of the Sun's surface covered by sunspots. The cycle has a period of approximately 11 years.

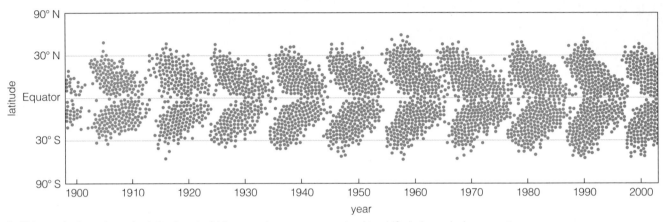

b This graph shows how the latitudes at which sunspot groups appear tend to shift during a single sunspot cycle. Each dot represents a group of sunspots and indicates the year (horizontal axis) and latitude (vertical axis) at which the group appeared.

Figure 15.21 Sunspot cycle during the past century.

What Causes the Sunspot Cycle?

The causes of the Sun's magnetic fields and the sunspot cycle are not well understood, but we believe we know the general nature of the processes involved. Convection is thought to dredge up weak magnetic fields generated in the solar interior, amplifying them as they rise. The Sun's rotation—faster at its equator than near its poles—then stretches and shapes these fields.

Imagine what happens to a magnetic field line that originally runs along the Sun's surface directly from the north pole to the south pole. At the equator, the field line circles the Sun in 25 days, but at higher latitudes the field line lags behind. Gradually, this rotation pattern winds the field line more and more tightly around the Sun (Figure 15.22). This process, operating at all times over the entire Sun, produces the contorted field lines that generate sunspots and other solar activity.

Investigating how the Sun's magnetic field develops and changes in time requires sophisticated computer models. Scientists are working hard on such models, but the behavior of these fields is so complex that approximations are necessary even with the best supercomputers. Using these computer models, scientists have successfully replicated some features of the sunspot cycle, such as changes in the number and latitude of sunspots and the magnetic field reversals that occur about every 11 years. However, much still remains mysterious, including why the period of the sunspot cycle varies and why solar activity is different from one cycle to the next.

Solar Activity and Earth

During solar maximum, solar flares and other forms of solar activity send large numbers of highly energetic charged particles (protons and electrons) toward Earth. Sometimes these particles travel in the smooth flow known as the solar wind. Other times they come in the form of huge magnetic bubbles called *coronal mass ejections*. Do these forms of solar weather affect Earth? In at least some ways, the answer is a definitive yes.

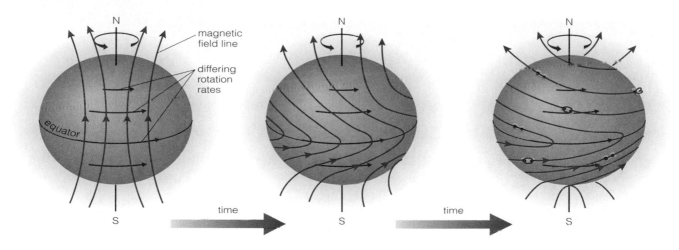

Figure 15.22 The Sun rotates more quickly at its equator than it does near its poles. Because gas circles the Sun faster at the equator, it drags the Sun's north-south magnetic field lines into a more twisted configuration. The magnetic field lines linking pairs of sunspots, depicted here as green and black blobs, trace out the directions of these stretched and distorted field lines.

The magnetic field associated with the solar wind constantly interacts with Earth's magnetic field. Occasionally these fields interconnect. When that happens, large amounts of energy are released from the magnetic field into the charged particles near the interconnection zone. Many of these energized particles then flow down Earth's magnetic field lines toward the poles (Figure 15.23a). Collisions between the charged particles and atoms in Earth's upper atmosphere cause electrons in the atoms to jump to higher energy levels [Section 4.4]. These excited atoms subsequently emit visible-light photons as they drop to lower energy levels,

creating the shimmering light of *auroras* (Figure 15.23b). Because coronal mass ejections are particularly energetic, the auroras they stimulate can be especially spectacular.

Particles streaming from the Sun after the occurrence of solar flares, coronal mass ejections, or other major solar storms can also have practical impacts on society. For example, these particles can hamper radio communications, disrupt electrical power delivery, and damage the electronic components in orbiting satellites. During a particularly powerful magnetic storm on the Sun in March 1989, the U.S. Air Force temporarily lost track of over 2,000 satellites,

SPECIAL TOPIC Long-Term Change in Solar Activity

Figure 15.21 shows that the sunspot cycle varies in length and intensity, and it sometimes seems to disappear altogether. With these facts as background, many scientists are searching for longer-term patterns in solar activity. Unfortunately, the search for longer-term variations is difficult because telescopic observations of sunspots cover a period of only about 400 years. Some naked-eye observations of sunspots recorded by Chinese astronomers go back almost 2,000 years, but these records are sparse, and naked-eye observations may not be very reliable. We can also guess at past solar activity from descriptions of solar eclipses recorded around the world: When the Sun is more active, the corona tends to have longer and brighter "streamers" visible to the naked eye.

Another way to gauge past solar activity is to study the amount of carbon-14 in tree rings. High-energy *cosmic rays* [Section 19.2] coming from beyond our own solar system produce radioactive carbon-14 in Earth's atmosphere. During periods of high solar activity, the solar wind tends to grow stronger, shielding Earth from some of these cosmic rays. Thus, production of carbon-14 drops when the Sun is more active. All the while, trees steadily

breathe in atmospheric carbon, in the form of carbon dioxide, and incorporate it year by year into each ring. We can therefore estimate the level of solar activity in any given year by measuring the level of carbon-14 in the corresponding ring. No clear evidence has yet been found of longer-term cycles of solar activity, but the search goes on.

Theoretical models predict a very long term trend of lessening solar activity. According to our theory of solar system formation, the Sun must have rotated much faster when it was young [Section 9.3]. Because a combination of convection and rotation generates solar activity, a faster rotation rate should have meant much more activity. Observations of other stars that are similar to the Sun but rotate faster confirm that these stars are much more active. We find evidence for many more "starspots" on these stars than on the Sun, and their relatively bright ultraviolet and X-ray emissions suggest that they have brighter chromospheres and coronas—just as we would expect if they are more active than the Sun.

a Interactions between Earth's magnetic field and the magnetic field of the solar wind can send energetic charged particles toward Earth's poles.

Figure 15.23 Particles from the Sun cause auroras on Earth.

b Aurora along the coast of Norway.

and powerful currents induced in the ground circuits of the Quebec hydroelectric system caused it to collapse for more than 8 hours. The combined cost of the loss of power in the United States and Canada exceeded $100 million. In January 1997, AT&T lost contact with a $200-million communications satellite, probably because of damage caused by particles coming from another powerful solar storm.

Satellites in low-Earth orbit are particularly vulnerable during solar maximum, when the increase in solar X rays and energetic particles heats Earth's upper atmosphere, causing it to expand. The density of the gas surrounding low-flying satellites therefore rises, exerting drag that saps their energy and angular momentum. If this drag proceeds unchecked, the satellites ultimately plummet back to Earth. Satellites in low orbits, including the Hubble Space Telescope and the Space Station, require occasional boosts to prevent them from falling out of the sky.

Connections between solar activity and Earth's climate are much less clear. The period from 1645 to 1715, when solar activity seems to have virtually ceased, was a time of exceptionally low temperatures in Europe and North America known as the *Little Ice Age.* Did the low solar activity cause these low temperatures, or was their occurrence

a coincidence? No one knows for sure. Some researchers have claimed that certain weather phenomena, such as drought cycles or frequencies of storms, are correlated with the 11- or 22-year cycle of solar activity. However, the data supporting these correlations are weak in many cases, and even real correlations may be coincidental.

Part of the difficulty in linking solar activity with climate is that no one understands how the linkage might work. Although emissions of ultraviolet light, X rays, and high-energy particles increase substantially from solar minimum to solar maximum, the total luminosity of the Sun barely changes at all. (The Sun becomes only about 0.1% brighter during solar maximum.) Thus, if solar activity really is affecting Earth's climate, it must be through some very subtle mechanism. For example, perhaps the expansion of Earth's upper atmosphere that occurs with solar maximum somehow causes changes in weather.

The question of how solar activity is linked to Earth's climate is very important, because we need to know whether global warming is affected by solar activity in addition to human activity. Unfortunately, for the time being at least, we can say little about this question.

Putting Chapter 15 into Context

In this chapter, we have examined our Sun, the nearest star. When you look back at this chapter, make sure you understand these "big picture" ideas:

● The ancient riddle of why the Sun shines is now solved. The Sun shines with energy generated by fusion of hydrogen into helium in the Sun's core. After a million-year journey through the solar interior and an 8-minute journey through space, a small fraction of this energy reaches Earth and supplies sunlight and heat.

● Gravitational equilibrium, the balance between pressure and gravity, determines the Sun's interior structure and maintains its steady nuclear burning rate.

The Sun achieved its long-lasting state of gravitational equilibrium when energy generation by fusion in the core came into balance with the energy lost through thermal radiation from the surface. If the Sun were not relatively steady, life on Earth might not have been possible.

● The Sun's atmosphere displays its own version of weather and climate, governed by solar magnetic fields. Solar weather has important influences on Earth.

● The Sun is important not only as our source of light and heat, but also because it is the only star near enough for us to study in great detail. In the coming chapters, we will use what we've learned about the Sun to help us understand other stars.

SUMMARY OF KEY CONCEPTS

15.1 Why Does the Sun Shine?

- *What process creates energy in the Sun?* Fusion of hydrogen into helium in the Sun's core generates the Sun's energy.

- *Why does the Sun's size remain stable?* The Sun's size remains stable because it is in gravitational equilibrium. The outward pressure of hot gas balances the inward force of gravity at every point within the Sun.

- *How did the Sun become hot enough for fusion in the first place?* As the Sun was forming, it grew hotter as it shrank in size because gravitational contraction converted gravitational potential energy into thermal energy. Gravitational contraction continued to shrink the Sun and raise its central temperature until the core became hot and dense enough for nuclear fusion.

15.2 Plunging to the Center of the Sun: An Imaginary Journey

- *What are the major layers of the Sun, from the center out?* The layers of the Sun are core, radiation zone, convection zone, photosphere, chromosphere, and corona.

- *What do we mean by the "surface" of the Sun?* We consider the photosphere to be the surface of the Sun because light can pass through the photosphere but cannot escape from deeper inside the Sun. Thus, photographs of visible light from the Sun show us what the photosphere looks like.

- *What is the Sun made of?* It is made almost entirely of hydrogen and helium (98% of the Sun's mass).

15.3 The Cosmic Crucible

- *Why does fusion occur in the Sun's core?* The core temperature and pressure are so high that colliding nuclei can come close enough together for the strong force to overcome electromagnetic repulsion and bind them together.

- *Why is energy produced in the Sun at such a steady rate?* The fusion rate is self-regulating like a thermostat. If the fusion rate increases for some reason, the added energy production puffs up and cools the core, bringing the rate back down. Similarly, a decrease in the fusion rate allows the core to shrink and heat, bringing the fusion rate back up.

- *Why was the Sun dimmer in the distant past?* Although the fusion rate is steady on short time scales, it gradually increases over billions of years, increasing the Sun's luminosity. The increase occurs because fusion gradually reduces the number of individual nuclei in the solar core. Four hydrogen nuclei are fused to make just one helium nucleus, causing the core to shrink and become hotter.

- *How do we know what is happening inside the Sun?* We can construct theoretical models of the solar interior using known laws of physics and then check

continued ▶

the models against observations of the Sun's size, surface temperature, and energy output as well as studies of "sun quakes" and solar neutrinos.

- *What is the solar neutrino problem? Is it solved?* Neutrino detectors capture fewer neutrinos coming from the Sun than models of fusion in the core predict. This discrepancy is called the solar neutrino problem. The problem now appears to be solved. Apparently, neutrinos can transform themselves among three different types as they travel from the solar core to Earth, while most detectors can capture only one type. Thus, the detectors capture fewer than the expected number of neutrinos.

15.4 From Core to Corona

- *How long ago did fusion generate the energy we now receive as sunlight?* Fusion created the energy we receive today about a million years ago. It takes about a million years for photons and then convection to transport energy through the solar interior to the photosphere. Once sunlight emerges from the photosphere, it takes only about 8 minutes to reach Earth.

- *How are sunspots, prominences, and flares related to magnetic fields?* Sunspots occur where strong magnetic fields trap and isolate gas from the surrounding plasma of the photosphere. The trapped gas cools, so the sunspots become cooler and darker than the rest of the photosphere. Sunspots tend to occur in pairs connected by a loop of magnetic field, which may rise high above the surface as a solar promi-

nence. The magnetic fields are twisted and contorted by the Sun's rotation, and solar flares may occur when the field lines suddenly snap and release their energy.

- *What is surprising about the temperature of the chromosphere and corona, and how do we explain it?* Temperature gradually decreases from the core to the photosphere but then rises again in the chromosphere and corona. These high layers of the Sun are probably heated by energy carried upward along the magnetic field lines by waves that are generated as turbulent motions in the convection zone shake the magnetic field lines.

15.5 Solar Weather and Climate

- *What is the sunspot cycle?* The sunspot cycle, or the variation in the number of sunspots on the Sun's surface, has an average period of 11 years. The magnetic field flip-flops every 11 years or so, resulting in a 22-year magnetic cycle. Sunspots first appear at mid-latitudes at solar minimum, then become increasingly more common near the Sun's equator as the next minimum approaches.

- *What effect does solar activity have on Earth and its inhabitants?* Particles ejected from the Sun by solar flares and other types of solar activity can affect communications, electrical power delivery, and the electronic circuits in space vehicles. The connections between solar activity and Earth's climate are not clear.

? Sensible Statements?

Decide whether each of the following statements is sensible and explain why it is or is not.

1. Before Einstein, gravitational contraction appeared to be a perfectly plausible mechanism for solar energy generation.

2. A sudden temperature rise in the Sun's core is nothing to worry about, because conditions in the core will soon return to normal.

3. If fusion in the solar core ceased today, worldwide panic would break out tomorrow as the Sun began to grow dimmer.

4. Astronomers have recently photographed magnetic fields churning deep beneath the solar photosphere.

5. Neutrinos probably can't harm me, but just to be safe I think I'll wear a lead vest.

6. If you want to see lots of sunspots, just wait for solar maximum!

7. News of a major solar flare today caused concern among professionals in the fields of communications and electrical power generation.

8. By observing solar neutrinos, we can learn about nuclear fusion deep in the Sun's core.

Problems

9. *Gravitational Contraction.* Briefly describe how gravitational contraction generates energy and when it was important in the Sun's history.

10. *Solar Characteristics.* Briefly describe the Sun's luminosity, mass, radius, and average surface temperature.

11. *Sunspots.* What are sunspots? Why do they appear dark in pictures of the Sun?

12. *Solar Fusion.* What is the overall nuclear fusion reaction in the Sun? Briefly describe the proton–proton chain.

13. *Models of the Sun.* Explain how mathematical models allow us to predict conditions inside the Sun. How can we be confident that the models are on the right track?

14. *Sun Quakes.* How are "sun quakes" similar to earthquakes? How are they different? Describe how we can observe them and how they help us learn about the solar interior.

15. *Energy Transport.* Why does the energy produced by fusion in the solar core take so long to reach the solar surface? Describe the processes of radiative diffusion and convection in the solar interior.

16. *The Photosphere.* Describe the appearance and temperature of the Sun's photosphere. What is granulation? How would granulation appear in a movie?

17. *Observing the Sun's Atmosphere.* Why is the chromosphere best viewed with ultraviolet telescopes? Why is the corona best viewed with X-ray telescopes?

18. *An Angry Sun.* A *Time* magazine cover once suggested that an "angry Sun" was becoming more active as human activity changed Earth's climate through global warming. It's certainly possible for the Sun to become more active at the same time that humans are affecting Earth, but is it possible that the Sun could be responding to human activity? Can humans affect the Sun in any significant way? Explain.

*19. *Number of Fusion Reactions in the Sun.* Use the fact that each cycle of the proton–proton chain converts 4.7×10^{-29} kg of mass into energy (see Mathematical Insight 15.1), along with the fact that the Sun loses a total of about 4.2×10^9 kg of mass each second, to calculate the total number of times the proton–proton chain occurs each second in the Sun.

*20. *The Lifetime of the Sun.* The total mass of the Sun is about 2×10^{30} kg, of which about 75% was hydrogen when the Sun formed. However, only about 13% of this hydrogen ever becomes available for fusion in the core. The rest remains in layers of the Sun where the temperature is too low for fusion.

  a. Based on the given information, calculate the total mass of hydrogen available for fusion over the lifetime of the Sun.

  b. Combine your results from part (a) and the fact that the Sun fuses about 600 billion kg of hydrogen each second to calculate how long the Sun's initial supply of hydrogen can last. Give your answer in both seconds and years.

  c. Given that our solar system is now about 4.6 billion years old, when will we need to worry about the Sun running out of hydrogen for fusion?

*21. *Solar Power Collectors.* This problem leads you through the calculation and discussion of how much solar power can be collected by solar cells on Earth.

  a. Imagine a giant sphere surrounding the Sun with a radius of 1 AU. What is the surface area of this sphere, in square meters? (*Hint:* The formula for the surface area of a sphere is $4\pi r^2$.)

  b. Because this imaginary giant sphere surrounds the Sun, the Sun's entire luminosity of 3.8×10^{26} watts must pass through it. Calculate the power passing through each square meter of this imaginary sphere in *watts per square meter*. Explain why this number represents the maximum power per square meter that can be collected by a solar collector in Earth orbit.

  c. List several reasons why the average power per square meter collected by a solar collector on the ground will always be less than what you found in part (b).

  d. Suppose you want to put a solar collector on your roof. If you want to optimize the amount of power you can collect, how should you orient the collector? (*Hint:* The optimum orientation depends on both your latitude and the time of year and day.)

*22. *Solar Power for the United States.* The total annual U.S. energy consumption is about 2×10^{20} joules.

  a. What is the average *power* requirement for the United States, in watts? (*Hint:* 1 watt = 1 joule/s.)

  b. With current technologies and solar collectors on the ground, the best we can hope is that solar cells will generate an average (day and night) power of about 200 watts/m^2. (You might compare this to the maximum power per square meter you found in problem 22b.) What total area would we need to cover with solar cells to supply all the power needed for the United States? Give your answer in both square meters and square kilometers.

  c. The total surface area of the United States is about 2×10^7 km^2. What fraction of the U.S. area would have to be covered by solar collectors to generate all of the U.S. power needs? In one page or less, describe potential environmental impacts of covering so much area with solar collectors. Also discuss whether you think these environmental impacts would be greater or less than the impacts of using current energy sources such as coal, oil, nuclear power, and hydroelectric power.

*23. *The Color of the Sun.* The Sun's average surface temperature is about 5,800 K. Use Wien's law (see Mathematical Insight 6.2) to calculate the wavelength of peak thermal emission from the Sun. What color does this wavelength correspond to in the visible-light spectrum? In light of your answer, why do you think the Sun appears white or yellow to our eyes?

Discussion Questions

24. *The Role of the Sun.* Briefly discuss how the Sun affects us here on Earth. Be sure to consider not only factors such as its light and warmth, but also how the study of the Sun has led us to new understanding in science and to technological developments. Overall, how important has solar research been to our lives?

25. *The Solar Neutrino Problem.* Discuss the solar neutrino problem and its potential solutions. How serious do you consider this problem? Do you think current theoretical models of the Sun could be wrong in any fundamental way? Why or why not?

26. *The Sun and Global Warming.* One of the most pressing environmental issues on Earth concerns the extent to which human emissions of greenhouse gases are warming our planet. Some people claim that part or all of the observed warming over the past century may be due to changes on the Sun, rather than to anything humans have done. Discuss how a better understanding of the Sun might help us understand the threat posed by greenhouse gas emissions. Why is it so difficult to develop a clear understanding of how the Sun affects Earth's climate?

MEDIA EXPLORATIONS

Web Projects

Take advantage of the useful Web links on www.astronomyplace.com to assist you with the following projects.

1. *Current Solar Activity.* Daily information about solar activity is available at NASA's Web site sunspotcycle.com. Where are we in the sunspot cycle right now? When is the next solar maximum or minimum expected? Have there been any major solar storms in the past few months? If so, did they have any significant effects on Earth? Summarize your findings in a one- to two-page report.

2. *Solar Observatories in Space.* Visit NASA's Web site for the Sun–Earth connection and explore some of the current and planned space missions designed to observe the Sun. Choose one mission to study in greater depth, and write a one- to two-page report on the mission status and goals and what it has taught or will teach us about the Sun.

3. *Sudbury Neutrino Observatory.* Visit the Web site for the Sudbury Neutrino Observatory (SNO) and learn how it has helped to solve the solar neutrino problem. Write a one- to two-page report describing the observatory, any recent results, and what we can expect from it in the future.

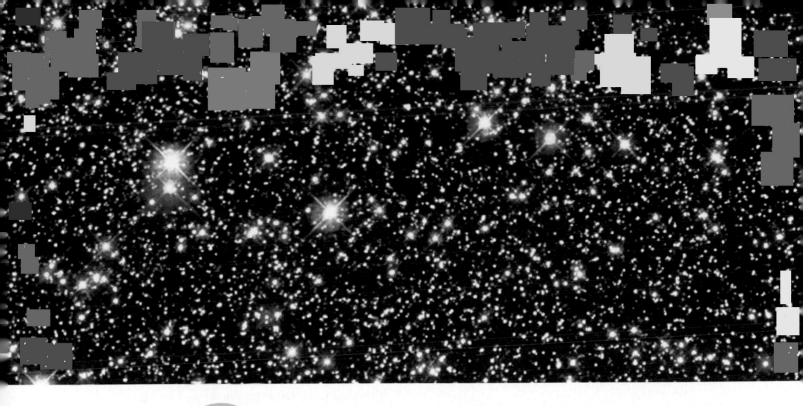

16 Properties of Stars

"All men have the stars," he answered, "but they are not the same things for different people. For some, who are travelers, the stars are guides. For others they are no more than little lights in the sky. For others, who are scholars, they are problems. For my businessman they were wealth. But all these stars are silent. You—you alone—will have the stars as no one else has them."

Antoine de Saint-Exupéry, from The Little Prince

On a clear, dark night, a few thousand stars are visible to the naked eye. Many more become visible through binoculars, and with a powerful telescope we can see so many stars that we could never hope to count them. Like individual people, each individual star is unique. Like the human family, all stars share much in common.

Today, we know that stars are born from clouds of interstellar gas, shine brilliantly by nuclear fusion for millions or billions of years, and then die, sometimes in dramatic ways. This chapter outlines how we study and categorize stars and how we have come to realize that stars, like people, change over their lifetime.

16.1 Snapshot of the Heavens

Imagine that an alien spaceship flies by Earth on a simple but short mission: The visitors have just 1 minute to learn everything they can about the human race. In 60 seconds, they will see next to nothing of each individual person's life. Instead, they will obtain a collective "snapshot" of humanity that shows people from all stages of life engaged in their daily activities. From this snapshot alone, they must piece together their entire understanding of human beings and their lives, from birth to death.

We face a similar problem when we look at the stars. Compared with stellar lifetimes of millions or billions of years, the few hundred years humans have spent studying stars with telescopes is rather like the aliens' 1-minute glimpse of humanity. We see only a brief moment in any star's life, and our collective snapshot of the heavens consists of such frozen moments for billions of stars. From this snapshot, we try to reconstruct the life cycles of stars while also analyzing what makes one star different from another.

Thanks to the efforts of hundreds of astronomers studying this snapshot of the heavens, stars are no longer

mysterious points of light in the sky. We now know that all stars form in great clouds of gas and dust. Each star begins its life with roughly the same chemical composition: About three-quarters of the star's mass at birth is hydrogen, and about one-quarter is helium, with no more than about 2% consisting of elements heavier than helium. During most of any star's life, the rate at which it generates energy depends on the same type of balance between the inward pull of gravity and the outward push of internal pressure that governs the rate of fusion in our Sun.

Despite these similarities, stars appear different from one another for two primary reasons: They differ in mass, and we see different stars at different stages of their lives.

The key that finally unlocked these secrets of stars was an appropriate classification system. Before the twentieth century, humans classified stars primarily by their brightness and location in our sky. The names of the brightest stars within each constellation still bear Greek letters designating their order of brightness. For example, the brightest star in the constellation Centaurus is Alpha Centauri, the second brightest is Beta Centauri, the third brightest is Gamma Centauri, and so on. However, a star's brightness and membership in a constellation tell us little about its true nature. A star that appears bright could be either extremely luminous or unusually nearby, and two stars that appear right next to each other in our sky might not be true neighbors if they lie at significantly different distances from Earth.

Today, astronomers classify a star primarily according to its *luminosity* and *surface temperature*. Our task in this chapter is to learn how this extraordinarily effective classification system reveals the true natures of stars and their life cycles. We begin by investigating how to determine a star's luminosity, surface temperature, and mass.

 Measuring Cosmic Distances Tutorial, Lesson 2

16.2 Stellar Luminosity

A star's **luminosity** is the total amount of power it radiates into space, which can be stated in *watts*. For example, the Sun's luminosity is 3.8×10^{26} watts [Section 15.2]. We cannot measure a star's luminosity directly, because its brightness in our sky depends on its distance as well as its true luminosity. For example, our Sun and Alpha Centauri A (the brightest of three stars in the Alpha Centauri system) are similar in luminosity, but Alpha Centauri A is a feeble point of light in the night sky, while our Sun provides enough light and heat to sustain life on Earth. The difference in brightness arises because Alpha Centauri A is about 270,000 times farther from Earth than is the Sun.

More precisely, we define the **apparent brightness** of any star in our sky as the amount of light reaching us *per unit area* (Figure 16.1). (A more technical term for apparent brightness is *flux*.) The apparent brightness of any light source obeys an *inverse square law* with distance, similar to the inverse square law that describes the force of gravity [Section 5.3]. If we viewed the Sun from twice Earth's

Photos of Stars

Photographs of stars, star clusters, and galaxies convey a great deal of information, but they also contain a few artifacts that are not real. For example, different stars seem to have different sizes in photographs, but stars are so far away that they should all appear as mere points of light. Stellar sizes in photographs are an artifact of how our instruments record light. Because of the problem of overexposure, brighter stars tend to appear larger than dimmer stars.

Overexposure can be a particular problem for photographs of globular clusters of stars and photographs of galaxies. These objects are so much brighter near their centers than in their outskirts that the centers are almost always overexposed in photographs that show the outskirts. That is why globular clusters and galaxies often look in photographs as if their central regions contain a single bright blob, when in fact the centers contain many individual stars separated by vast amounts of space.

Spikes around bright stars in photographs, often making the pattern of a cross with a star at the center, are another such artifact. These spikes are not real but rather are created by the interaction of starlight with the supports holding the secondary mirror in the telescope [Section 7.2]. The spikes generally occur only with point sources of light like stars, and not with larger objects like galaxies. When you look at a photograph showing many galaxies (for example, Figure 20.1), you can tell which objects are stars by looking for the spikes.

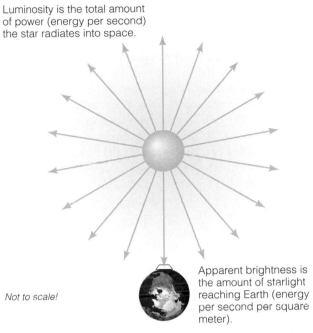

Luminosity is the total amount of power (energy per second) the star radiates into space.

Not to scale!

Apparent brightness is the amount of starlight reaching Earth (energy per second per square meter).

Figure 16.1 Luminosity is a measure of power, and apparent brightness is a measure of power per unit area.

1 AU
2 AU
3 AU

Figure 16.2 The inverse square law for light. At greater distances from a star, the same amount of light passes through an area that gets larger with the square of the distance. The amount of light per unit area therefore declines with the square of the distance.

distance, it would appear dimmer by a factor of $2^2 = 4$. If we viewed it from 10 times Earth's distance, it would appear $10^2 = 100$ times dimmer. From 270,000 times Earth's distance, it would look like Alpha Centauri A—dimmer by a factor of $270,000^2$, or about 70 billion.

Figure 16.2 shows why apparent brightness follows an inverse square law. The same total amount of light must pass through each imaginary sphere surrounding the star. If we focus our attention on the light passing through a small square on the sphere located at 1 AU, we see that the same amount of light must pass through *four* squares of the same size on the sphere located at 2 AU. Thus, each square on the sphere at 2 AU receives only $\frac{1}{2^2} = \frac{1}{4}$ as much light as the square on the sphere at 1 AU. Similarly, the same amount of light passes through *nine* squares of the same size on the sphere located at 3 AU. Thus, each of these squares receives only $\frac{1}{3^2} = \frac{1}{9}$ as much light as the square on the sphere at 1 AU. Generalizing, we see that the amount of light received per unit area decreases with increasing distance by the square of the distance—an inverse square law.

This inverse square law leads to a very simple and important formula relating the apparent brightness, luminosity, and distance of any light source. We will call it the **luminosity–distance formula**:

$$\text{apparent brightness} = \frac{\text{luminosity}}{4\pi \times (\text{distance})^2}$$

Because the standard units of luminosity are watts, the units of apparent brightness are *watts per square meter*. Because we can always measure the apparent brightness of a star, this formula provides a way to calculate a star's luminosity if we can first measure its distance or to calculate a star's distance if we somehow know its luminosity. (The luminosity–distance formula is strictly correct only if interstellar dust does not absorb or scatter the starlight along its path to Earth.)

Although watts are the standard units for luminosity, it's often more meaningful to describe stellar luminosities in comparison to the Sun by using units of **solar luminosity**: $L_{Sun} = 3.8 \times 10^{26}$ watts. For example, Proxima Centauri, the nearest of the three stars in the Alpha Centauri system and hence the nearest star besides our Sun, is only about 0.0006 times as luminous as the Sun, or $0.0006L_{Sun}$. Betelgeuse, the bright left-shoulder star of Orion, has a luminosity of $38,000L_{Sun}$, meaning that it is 38,000 times more luminous than the Sun.

Measuring Apparent Brightness

We can measure a star's apparent brightness by using a detector, such as a CCD, that records how much energy strikes its light-sensitive surface each second. For example, such a detector would record an apparent brightness of 2.7×10^{-8} watt per square meter from Alpha Centauri A. The only difficulties involved in measuring apparent brightness are making sure the detector is properly calibrated and, for ground-based telescopes, taking into account the absorption of light by Earth's atmosphere.

No detector can record light of all wavelengths, so we necessarily measure apparent brightness in only some small range of the complete spectrum. For example, the human eye is sensitive to visible light but does not respond to ultraviolet or infrared photons. Thus, when we perceive a star's brightness, our eyes are measuring the apparent brightness only in the visible region of the spectrum.

When we measure the apparent brightness in visible light, we can calculate only the star's *visible-light luminosity*. Similarly, when we observe a star with a spaceborne X-ray telescope, we measure only the apparent brightness in X rays and can calculate only the star's *X-ray luminosity*. We will use the terms **total luminosity** and **total apparent brightness** to describe the luminosity and apparent brightness we would measure *if* we could detect photons across the entire electromagnetic spectrum. (Astronomers refer to the total luminosity as the *bolometric* luminosity.)

Measuring Distance Through Stellar Parallax

Once we have measured a star's apparent brightness, the next step in determining its luminosity is to measure its distance. The most direct way to measure the distances to stars is with *stellar parallax*, the small annual shifts in a star's apparent position caused by Earth's motion around the Sun [Section 2.6].

Recall that you can observe parallax of your finger by holding it at arm's length and looking at it alternately with first one eye closed and then the other. Astronomers measure stellar parallax by comparing observations of a nearby star made 6 months apart (Figure 16.3). The nearby star appears to shift against the background of more distant stars because we are observing it from two opposite points of Earth's orbit. The star's **parallax angle** is defined as *half* the star's annual back-and-forth shift.

Measuring stellar parallax is difficult because stars are so far away, making their parallax angles very small. Even the nearest star, Proxima Centauri, has a parallax angle of only 0.77 arcsecond. For increasingly distant stars, the parallax angles quickly become too small to measure even with our highest-resolution telescopes. Current technology

---

Mathematical Insight **16.1**   **The Luminosity–Distance Formula**

We can derive the luminosity–distance formula by extending the idea illustrated in Figure 16.2. Suppose we are located a distance d from a star with luminosity L. The apparent brightness of the star is the power per unit area that we receive at our distance d. We can find this apparent brightness by imagining that we are part of a giant sphere with radius d, similar to any one of the three spheres in Figure 16.2. The surface area of this giant sphere is $4\pi \times d^2$, and the star's entire luminosity L must pass through this surface area. (The surface area of any sphere is $4\pi \times$ radius^2.) Thus, the apparent brightness at distance d is the power per unit area passing through the sphere:

$$\text{apparent brightness} = \frac{\text{star's luminosity}}{\text{surface area of imaginary sphere}}$$

$$= \frac{L}{4\pi \times d^2}$$

This is our luminosity–distance formula.

Example: What is the Sun's apparent brightness as seen from Earth?

Solution: The Sun's luminosity is $L_{Sun} = 3.8 \times 10^{26}$ watts, and Earth's distance from the Sun is $d = 1.5 \times 10^{11}$ meters. Thus, the Sun's apparent brightness is:

$$\frac{L}{4\pi \times d^2} = \frac{3.8 \times 10^{26} \text{ watts}}{4\pi \times (1.5 \times 10^{11} \text{ m})^2}$$

$$= 1.3 \times 10^3 \text{ watts/m}^2$$

The Sun's apparent brightness is about 1,300 watts per square meter at Earth's distance. It is the maximum power per unit area that could be collected by a detector on Earth that directly faces the Sun, such as a solar power (or *photovoltaic*) cell. In reality, solar collectors usually collect less power because Earth's atmosphere absorbs some sunlight, particularly when it is cloudy.

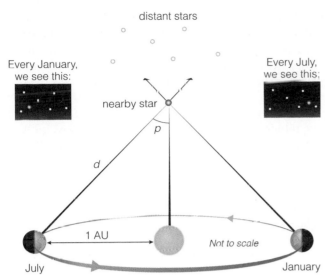

distant stars

Every January, we see this:

nearby star

p

d

1 AU

Not to scale

July

January

Every July, we see this:

Figure 16.3 Parallax makes the apparent position of a nearby star shift back and forth with respect to distant stars over the course of each year. If we measure the parallax angle p in arcseconds, the distance d to the star in parsecs is $\frac{1}{p}$. The angle in this figure is greatly exaggerated: All stars have parallax angles of less than 1 arcsecond.

allows us to measure parallax only for stars within a few hundred light-years—not much farther than what we call our *local solar neighborhood* in the vast, 100,000-light-year-diameter Milky Way Galaxy.

By definition, the distance to an object with a parallax angle of 1 arcsecond is 1 **parsec**, abbreviated **pc**. (The word *parsec* comes from the words *parallax* and *arcsecond*.) With a little geometry and Figure 16.3 (see Mathematical Insight 16.2), it is possible to show that:

$$1 \text{ pc} = 3.26 \text{ light-years} = 3.09 \times 10^{13} \text{ km}$$

If we use units of arcseconds for the parallax angle, a simple formula allows us to calculate distances in parsecs:

$$d \text{ (in parsecs)} = \frac{1}{p \text{ (in arcseconds)}}$$

For example, the distance to a star with a parallax angle of $\frac{1}{2}$ arcsecond is 2 parsecs, the distance to a star with a parallax angle of $\frac{1}{10}$ arcsecond is 10 parsecs, and the distance to a star with a parallax angle of $\frac{1}{100}$ arcsecond is 100 parsecs. Astronomers often express distances in parsecs or light-years interchangeably. You can convert quickly between them by remembering that 1 pc = 3.26 light-years. Thus, 10 parsecs is about 32.6 light-years; 1,000 parsecs, or 1 kiloparsec (1 kpc), is about 3,260 light-years; and 1 million parsecs, or 1 megaparsec (1 Mpc), is about 3.26 million light-years.

Enough stars have measurable parallax to give us a fairly good sample of the many different types of stars. For example, we know of more than 300 stars within about 33 light-years (10 parsecs) of the Sun. About half are binary star systems consisting of two orbiting stars or

---

Mathematical Insight **16.2** **The Parallax Formula**

Here is one of several ways to derive the formula relating a star's distance and parallax angle. Figure 16.3 shows that the parallax angle p is part of a right triangle, the side opposite p is the Earth–Sun distance of 1 AU, and the hypotenuse is the distance d to the object. You may recall that the *sine* of an angle in a right triangle is the length of its opposite side divided by the length of the hypotenuse. In this case, we find:

$$\sin p = \frac{\text{length of opposite side}}{\text{length of hypotenuse}} = \frac{1 \text{ AU}}{d}$$

If we solve for d, the formula becomes:

$$d = \frac{1 \text{ AU}}{\sin p}$$

By definition, 1 parsec is the distance to an object with a parallax angle of 1 arcsecond (1″), or 1/3,600 degree (because that $1° = 60'$ and $1' = 60''$). Substituting these numbers into the parallax formula and using a calculator to find that $\sin 1'' = 4.84814 \times 10^{-6}$, we get:

$$1 \text{ pc} = \frac{1 \text{ AU}}{\sin 1''} = \frac{1 \text{ AU}}{4.84814 \times 10^{-6}} = 206,265 \text{ AU}$$

That is, 1 parsec = 206,265 AU, which is equivalent to 3.09×10^{13} km or 3.26 light-years. (Recall that 1 AU = 149.6 million km.)

We need one more fact from geometry to derive the parallax formula given in the text. As long as the parallax angle, p, is small, $\sin p$ is proportional to p. For example, $\sin 2''$ is twice as large as $\sin 1''$, and $\sin \frac{1}{2}''$ is half as large as $\sin 1''$. (You can verify these examples with your calculator.) Thus, if we use $\frac{1}{2}''$ instead of 1″ for the parallax angle in the formula above, we get a distance of 2 pc instead of 1 pc. Similarly, if we use a parallax angle of $\frac{1}{10}''$, we get a distance of 10 pc. Generalizing, we get the simple parallax formula given in the text:

$$d \text{ (in parsecs)} = \frac{1}{p \text{ (in arcseconds)}}$$

Example: Sirius, the brightest star in our night sky, has a measured parallax angle of 0.379″. How far away is Sirius in parsecs? In light-years?

Solution: From the formula, the distance to Sirius in parsecs is:

$$d \text{ (in pc)} = \frac{1}{0.379} = 2.64 \text{ pc}$$

Because 1 pc = 3.26 light-years, this distance is equivalent to:

$$2.64 \text{ pc} \times 3.26 \frac{\text{light-years}}{\text{pc}} = 8.60 \text{ light-years}$$

multiple star systems containing three or more stars. Most are tiny, dim red stars such as Proxima Centauri—so dim that we cannot see them with the naked eye, despite the fact that they are relatively close. A few nearby stars, such as Sirius (2.6 parsecs), Vega (8 parsecs), Altair (5 parsecs), and Fomalhaut (7 parsecs), are white in color and bright in our sky, but most of the brightest stars in the sky lie farther away. Because so many nearby stars appear dim while many more distant stars appear bright, their luminosities must span a wide range.

The Magnitude System

Many amateur and professional astronomers describe stellar brightness using the ancient **magnitude system** devised by the Greek astronomer Hipparchus (c. 190–120 B.C.). The magnitude system originally classified stars according to how bright they look to our eyes—the only instruments available in ancient times. The brightest stars received the designation "first magnitude," the next brightest "second magnitude," and so on. The faintest visible stars were magnitude 6. We call these descriptions **apparent magnitudes** because they compare how bright different stars *appear* in the sky. Star charts (such as those in Appendix J) often use dots of different sizes to represent the apparent magnitudes of stars.

In modern times, the magnitude system has been extended and more precisely defined (see Mathematical Insight 16.3). As a result, stars can have fractional apparent

magnitudes, and a few bright stars have apparent magnitudes *less than* 1—which means *brighter* than magnitude 1. For example, the brightest star in the night sky, Sirius, has an apparent magnitude of −1.46. Appendix F gives the apparent magnitudes and solar luminosities for nearby stars and the brightest stars.

The modern magnitude system also defines **absolute magnitudes** as a way of describing stellar luminosities. A star's absolute magnitude is the apparent magnitude it would have *if* it were at a distance of 10 parsecs from Earth. For example, the Sun's absolute magnitude is about 4.8, meaning that the Sun would have an apparent magnitude of 4.8 *if* it were 10 parsecs away from us—bright enough to be visible, but not conspicuous, on a dark night.

Understanding the magnitude system is worthwhile because it is still commonly used. However, for the calculations in this book, it's much easier to work with the luminosity–distance formula, so we will avoid using magnitude formulas in this book.

 The Hertzsprung–Russell Diagram Tutorial, Lessons 1–3

16.3 Stellar Surface Temperature

The second basic property of stars (besides luminosity) needed for modern stellar classification is surface temperature. Measuring a star's surface temperature is somewhat easier than measuring its luminosity because the measure-

---

Mathematical Insight **16.3** **The Modern Magnitude Scale**

The modern magnitude system is defined so that each difference of 5 magnitudes corresponds to a factor of exactly 100 in brightness. For example, a magnitude 1 star is 100 times brighter than a magnitude 6 star, and a magnitude 3 star is 100 times brighter than a magnitude 8 star. Because 5 magnitudes corresponds to a factor of 100 in brightness, a single magnitude corresponds to a factor of $(100)^{1/5} \approx 2.512$.

The following formula summarizes the relationship between stars of different magnitudes:

$$\frac{\text{apparent brightness of Star 1}}{\text{apparent brightness of Star 2}} = (100^{1/5})^{m_2 - m_1}$$

where m_1 and m_2 are the apparent magnitudes of Stars 1 and 2, respectively. If we replace the apparent magnitudes with absolute magnitudes (designated M instead of m), the same formula applies to stellar luminosities:

$$\frac{\text{luminosity of Star 1}}{\text{luminosity of Star 2}} = (100^{1/5})^{M_2 - M_1}$$

Example 1: On a clear night, stars dimmer than magnitude 5 are quite difficult to see. Today, sensitive instruments on large telescopes can detect objects as faint as magnitude 30. How much more sensitive are such telescopes than the human eye?

Solution: We imagine that our eye sees "Star 1" with magnitude 5 and the telescope detects "Star 2" with magnitude 30. Then we compare:

$$\frac{\text{apparent brightness of Star 1}}{\text{apparent brightness of Star 2}} = (100^{1/5})^{30-5} = (100^{1/5})^{25}$$

$$= 100^5 = 10^{10}$$

The magnitude 5 star is 10^{10}, or 10 billion, times brighter than the magnitude 30 star, so the telescope is 10 billion times more sensitive than the human eye.

Example 2: The Sun has an absolute magnitude of about 4.8. Polaris, the North Star, has an absolute magnitude of −3.6. How much more luminous is Polaris than the Sun?

Solution: We use Polaris as Star 1 and the Sun as Star 2:

$$\frac{\text{luminosity of Polaris}}{\text{luminosity of Sun}} = (100^{1/5})^{4.8-(-3.6)} = (100^{1/5})^{8.4}$$

$$= 100^{1.7} \approx 2,500$$

Polaris is about 2,500 times more luminous than the Sun.

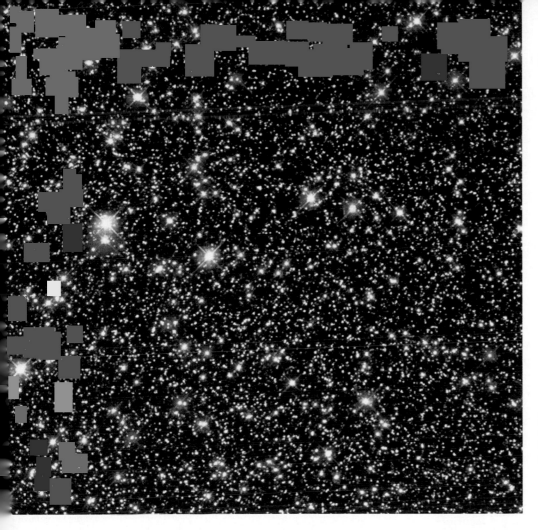

Figure 16.4 This Hubble Space Telescope view through the heart of our Milky Way Galaxy reveals that stars emit light of many different colors.

ment is not affected by the star's distance. Instead, we determine surface temperature directly from the star's color or spectrum. One note of caution: We can measure only a star's *surface temperature*, not its interior temperature. (Interior temperatures are calculated with theoretical models [Section 15.3].) When astronomers speak of the "temperature" of a star, they usually mean the surface temperature unless they say otherwise.

A star's surface temperature determines the color of light it emits [Section 6.4]. A red star is cooler than a yellow star, which in turn is cooler than a blue star. The naked eye can distinguish colors only for the brightest stars, but colors become more evident when we view stars through binoculars or a telescope (Figure 16.4).

Astronomers can determine the "color" of a star more precisely by comparing its apparent brightness as viewed through two different filters [Section 7.3]. For example, a cool star such as Betelgeuse, with a surface temperature of about 3,400 K, emits more red light than blue light and therefore looks much brighter when viewed through a red filter than when viewed through a blue filter. In contrast, a hotter star such as Sirius, with a surface temperature of about 9,400 K, emits more blue light than red light and looks brighter through a blue filter than through a red filter.

Spectral Type

The emission and absorption lines in a star's spectrum provide an independent and more accurate way to measure its surface temperature. Stars displaying spectral lines of highly ionized elements must be fairly hot, while stars displaying spectral lines of molecules must be relatively cool [Section 6.4]. Astronomers classify stars according to surface temperature by assigning a **spectral type** determined from the spectral lines present in a star's spectrum.

The hottest stars, with the bluest colors, are called spectral type O, followed in order of declining surface temperature by spectral types B, A, F, G, K, and M. The time-honored mnemonic for remembering this sequence, OBAFGKM, is "Oh Be A Fine Girl/Guy, Kiss Me!" Table 16.1 summarizes the characteristics of each spectral type.

Each spectral type is subdivided into numbered subcategories (e.g., B0, B1, . . . , B9). The larger the number, the cooler the star. For example, the Sun is designated spectral type G2, which means it is slightly hotter than a G3 star but cooler than a G1 star.

Invent your own mnemonic for the OBAFGKM sequence. To help get you thinking, here are two examples: (1) Only Bungling Astronomers Forget Generally Known Mnemonics; and (2) Only Business Acts For Good, Karl Marx.

History of the Spectral Sequence

You may wonder why the spectral types follow the peculiar order of OBAFGKM. The answer lies in the history of stellar spectroscopy.

Astronomical research never paid well, and many astronomers of the 1800s were able to do research only because of family wealth. One such astronomer was Henry Draper (1837–1882), an early pioneer of stellar spectroscopy. After Draper died in 1882, his widow made a series of large donations to Harvard College Observatory for the purpose of building upon his work. The observatory director, Edward Pickering (1846–1919), used the gifts to improve the facilities and to hire numerous assistants, whom he called "computers." Pickering added money of his own, as did other wealthy donors.

Most of Pickering's hired computers were women who had studied physics or astronomy at women's colleges such as Wellesley and Radcliffe. Women had few opportunities to advance in science at the time. Harvard, for example, did not allow women as either students or faculty. Pickering's project of studying and classifying stellar spectra provided plenty of work and opportunity for his computers, and many of the Harvard Observatory women ended up among the most prominent astronomers of the late 1800s and early 1900s.

One of the first computers was Williamina Fleming (1857–1911). Following Pickering's suggestion, Fleming classified stellar spectra according to the strength of their hydrogen lines: type A for the strongest hydrogen lines, type B for slightly weaker hydrogen lines, and so on to type O, for stars with the weakest hydrogen lines. Pickering published Fleming's classifications of more than 10,000 stars in 1890.

As more stellar spectra were obtained and the spectra were studied in greater detail, it became clear that the classification scheme based solely on hydrogen lines was inadequate. Ultimately, the task of finding a better classification scheme fell to Annie Jump Cannon (1863–1941), who joined Pickering's team in 1896 (Figure 16.5). Building on the work of Fleming and another of Pickering's computers, Antonia Maury (1866–1952), Cannon soon realized that the spectral classes fell into a natural order—but not the alphabetical order determined by hydrogen lines alone. Moreover, she found that some of the original classes overlapped others and could be eliminated. Cannon discovered that the natural sequence consisted of just a few of Pickering's original classes in the order OBAFGKM and also added the subdivisions by number.

Cannon became so adept that she could properly classify a stellar spectrum with little more than a momentary

Table 16.1 The Spectral Sequence

Spectral Type	Example(s)	Temperature Range
O	Stars of Orion's Belt	>30,000 K
B	Rigel	30,000 K–10,000 K
A	Sirius	10,000 K–7,500 K
F	Polaris	7,500 K–6,000 K
G	Sun, Alpha Centauri A	6,000 K–5,000 K
K	Arcturus	5,000 K–3,500 K
M	Betelgeuse, Proxima Centauri	<3,500 K

glance. During her lifetime, she personally classified over 400,000 stars. She became the first woman ever awarded an honorary degree by Oxford University, and in 1929 the League of Women Voters named her one of the 12 greatest living American women.

Figure 16.5 Women astronomers pose with Edward Pickering at Harvard College Observatory in 1913. Annie Jump Cannon is fifth from the left in the back row.

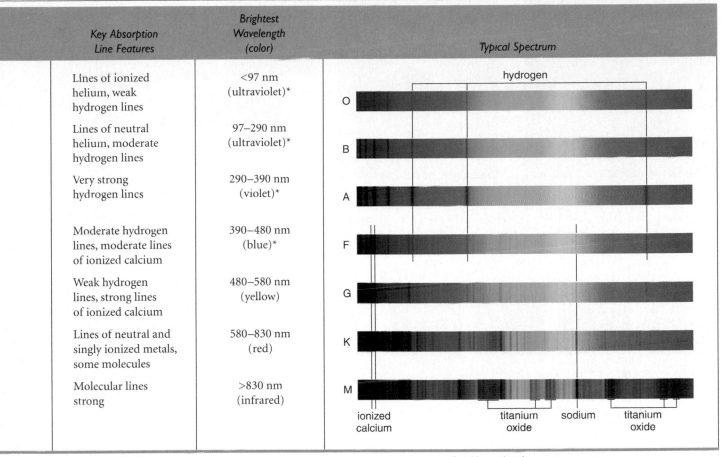

Key Absorption Line Features	Brightest Wavelength (color)	Typical Spectrum
Lines of ionized helium, weak hydrogen lines	<97 nm (ultraviolet)*	O
Lines of neutral helium, moderate hydrogen lines	97–290 nm (ultraviolet)*	B
Very strong hydrogen lines	290–390 nm (violet)*	A
Moderate hydrogen lines, moderate lines of ionized calcium	390–480 nm (blue)*	F
Weak hydrogen lines, strong lines of ionized calcium	480–580 nm (yellow)	G
Lines of neutral and singly ionized metals, some molecules	580–830 nm (red)	K
Molecular lines strong	>830 nm (infrared)	M

*All stars above 6,000 K look more or less white to the human eye because they emit plenty of radiation at all visible wavelengths.

The astronomical community adopted Cannon's system of stellar classification in 1910. However, no one at that time knew *why* spectra followed the OBAFGKM sequence. Many astronomers guessed, incorrectly, that the different sets of spectral lines reflected different compositions for the stars. The correct answer—that all stars are made primarily of hydrogen and helium and that a star's surface temperature determines the strength of its spectral lines—was discovered by Cecilia Payne-Gaposchkin (1900–1979), another woman working at Harvard Observatory.

Relying on insights from what was then the newly developing science of quantum mechanics, Payne-Gaposchkin showed that the differences in spectral lines from star to star merely reflected changes in the ionization level of the emitting atoms. For example, O stars have weak hydrogen lines because, at their high surface temperatures, nearly all their hydrogen is ionized. Without an electron to "jump" between energy

Cecilia Payne-Gaposchkin

levels, ionized hydrogen can neither emit nor absorb its usual specific wavelengths of light. At the other end of the spectral sequence, M stars are cool enough for some particularly stable molecules to form, explaining their strong molecular absorption lines. Payne-Gaposchkin described her work and her conclusions in a dissertation published in 1925. A later review of twentieth-century astronomy called her work "undoubtedly the most brilliant Ph.D. thesis ever written in astronomy."

 The Hertzsprung–Russell Diagram Tutorial, Lessons 1–3

16.4 Stellar Masses

The most important property of a star is its mass, but stellar masses are harder to measure than luminosities or surface temperatures. The most dependable method for "weighing" a star relies on Newton's version of Kepler's third law [Section 5.3]. This law can be applied only when we can measure both the orbital period and the average distance between the stars (semimajor axis) of the orbiting star system. Thus, in most cases we can measure stellar masses only in binary star systems in which we have determined the orbital properties of the two stars.

Figure 16.6 Mizar looks like one star to the naked eye but is actually a system of four stars. Through a telescope Mizar appears to be a visual binary made up of two stars, Mizar A and Mizar B, that gradually change positions, indicating that they orbit every few thousand years. However, each of these two "stars" is actually a spectroscopic binary, making a total of four stars. (The star Alcor appears very close to Mizar to the naked eye but does *not* orbit it.)

Figure 16.7 Each frame represents the relative positions of Sirius A and Sirius B at 10-year intervals from 1900 to 1970. The back-and-forth "wobble" of Sirius A allowed astronomers to infer the existence of Sirius B even before the two stars could be resolved in telescopic photos.

Types of Binary Star Systems

About half of all stars orbit a companion star of some kind. These star systems fall into three classes:

- A **visual binary** is a pair of stars that we can see distinctly (with a telescope) as the stars orbit each other. Mizar, the second star in the handle of the Big Dipper, is one example of a visual binary (Figure 16.6). Sometimes we observe a star slowly shifting position in the sky as if it were a member of a visual binary, but its companion is too dim to be seen. For example, slow shifts in the position of Sirius, the brightest star in the sky, revealed it to be a binary star long before its companion was discovered (Figure 16.7).

- An **eclipsing binary** is a pair of stars that orbit in the plane of our line of sight (Figure 16.8). When neither star is eclipsed, we see the combined light of both stars. When one star eclipses the other, the apparent bright-

Figure 16.8 The apparent brightness of an eclipsing binary system drops when either star eclipses the other.

ness of the system drops because some of the light is blocked from our view. A *light curve*, or graph of apparent brightness against time, reveals the pattern of the eclipses. The most famous example of an eclipsing binary is Algol, the "demon star" in the constellation Perseus (*algol* is Arabic for "the ghoul"). Algol becomes three times dimmer for a few hours about every 3 days as the brighter of its two stars is eclipsed by its dimmer companion.

● If a binary system is neither visual nor eclipsing, we may be able to detect its binary nature by observing Doppler shifts in its spectral lines [Section 6.5]. Such systems are called **spectroscopic binary** systems. If one star is orbiting another, it periodically moves toward us and away from us in its orbit. Its spectral lines show blueshifts and redshifts as a result of this motion (Figure 16.9). Sometimes we see two sets of lines shifting back and forth—one set from each of the two stars in the system (a *double-lined* spectroscopic binary). Other times we see a set of shifting lines from only one star because its companion is too dim to be detected (a *single-lined* spectroscopic binary). Each of the two stars in the visual binary Mizar is itself a spectroscopic binary (see Figure 16.6).

Star B spectrum at time 1: approaching, therefore blueshifted

1 approaching us

to Earth

Star B spectrum at time 2: receding, therefore redshifted

2 receding from us

Figure 16.9 The spectral lines of a star in a binary system are alternately blueshifted as it comes toward us in its orbit and redshifted as it moves away from us.

Measuring Masses in Binary Systems

Even for a binary system, we can apply Newton's version of Kepler's third law only if we can measure both the orbital period and the separation of the two stars. Measuring orbital period is fairly easy. In a visual binary, we simply observe how long each orbit takes (or extrapolate from part of an

---

Mathematical Insight

Orbital Separation and Newton's Version of Kepler's Third Law

Measurements of stellar masses rely on Newton's version of Kepler's third law [Section 5.3], for which we need to know the orbital period p and semimajor axis a. As described in the text, it's generally easy to measure p for binary star systems. We can rarely measure a directly, but we can calculate it in cases in which we can measure the orbital velocity of one star relative to the other.

If we assume that the first star traces a circle of radius a around its companion, the circumference of its orbit is $2\pi a$. Because the star makes one circuit of this circumference in one orbital period p, its velocity relative to its companion is:

$$v = \frac{\text{distance traveled in one orbit}}{\text{period of one orbit}} = \frac{2\pi a}{p}$$

Solving for a, we find:

$$a = \frac{pv}{2\pi}$$

Once we know both p and a, we can use Newton's version of Kepler's third law to calculate the *sum* of the masses of the two stars $(M_1 + M_2)$. We can then calculate the individual masses of the two stars by taking advantage of the fact that the relative velocities of the two stars around their common center of mass are inversely proportional to their relative masses.

Example: The spectral lines of two stars in a particular eclipsing binary system shift back and forth with a period of 2 years ($p = 6.2 \times 10^7$ seconds). The lines of one star (Star 1) shift twice as far as the lines of the other (Star 2). The amount of Doppler shift indicates an orbital speed of $v = 100,000$ m/s for Star 1 relative to

Star 2. What are the masses of the two stars? Assume that each of the two stars traces a circular orbit around their center of mass.

Solution: We will find the masses by using Newton's version of Kepler's third law, solved for the masses:

$$p^2 = \frac{4\pi^2}{G(M_1 + M_2)}a^3 \Rightarrow (M_1 + M_2) = \frac{4\pi^2}{G} \cdot \frac{a^3}{p^2}$$

We are given the orbital period $p = 6.2 \times 10^7$ s, and we find the semimajor axis a of the system from the given orbital velocity v:

$$a = \frac{pv}{2\pi} = \frac{(6.2 \times 10^7 \text{ s}) \times (100,000 \text{ m/s})}{2\pi}$$

$$= 9.9 \times 10^{11} \text{ m}$$

Now we calculate the sum of the stellar masses by substituting the values of p, a, and the gravitational constant G [Section 5.3] into the mass equation above:

$$(M_1 + M_2) = \frac{4\pi^2}{\left(6.67 \times 10^{-11} \dfrac{\text{m}^3}{\text{kg} \times \text{s}^2}\right)} \times \frac{(9.9 \times 10^{11} \text{ m})^3}{(6.2 \times 10^7 \text{ s})^2}$$

$$= 1.5 \times 10^{32} \text{ kg}$$

Because the lines of Star 1 shift twice as far as those of Star 2, we know that Star 1 moves twice as fast as Star 2, and hence that Star 1 is half as massive as Star 2. In other words, Star 2 is twice as massive as Star 1. Using this fact and their combined mass of 1.5×10^{32} kg, we conclude that the mass of Star 2 is 1.0×10^{32} kg and the mass of Star 1 is 0.5×10^{32} kg.

orbit). In an eclipsing binary, we measure the time between eclipses. In a spectroscopic binary, we measure the time it takes the spectral lines to shift back and forth.

Determining the average separation of the stars in a binary system is usually much more difficult. Except in rare cases in which we can measure the separation directly, we can calculate the separation only if we know the actual orbital speeds of the stars from their Doppler shifts. Unfortunately, a Doppler shift tells us only the portion of a star's velocity that is directly toward us or away from us [Section 6.5]. Because orbiting stars generally do not move directly along our line of sight, their actual velocities can be significantly greater than those we measure through the Doppler effect.

The exceptions are eclipsing binary stars. Because these stars orbit in the plane of our line of sight, their Doppler shifts can tell us their true orbital velocities.* Eclipsing binaries are therefore particularly important to the study of stellar masses. As an added bonus, eclipsing binaries allow us to measure stellar radii directly. Because we know how fast the stars are moving across our line of sight as one eclipses the other, we can determine their radii by timing how long each eclipse lasts.

THINK ABOUT IT

Suppose two orbiting stars are moving in a plane perpendicular to our line of sight. Would the spectral features of these stars appear shifted in any way? Explain.

 The Hertzsprung–Russell Diagram Tutorial, Lessons 1–3

16.5 The Hertzsprung–Russell Diagram

During the first decade of the twentieth century, a similar thought occurred independently to astronomers Ejnar Hertzsprung, working in Denmark, and Henry Norris Russell, working in the United States at Princeton University: Each decided to make a graph plotting stellar luminosities on one axis and spectral types on the other. Such graphs are now called **Hertzsprung–Russell (H–R) diagrams**. Soon after they began making their graphs, Hertzsprung and Russell uncovered some previously unsuspected patterns in the properties of stars. As we will see shortly, understanding these patterns and the H–R diagram is central to the study of stars.

A Basic H–R Diagram

Figure 16.10 displays an example of an H–R diagram.

- The horizontal axis represents stellar surface temperature, which, as we've discussed, corresponds to spectral type. Temperature increases *from right to left* because

---

*In other binaries, we can calculate an actual orbital velocity from the velocity obtained by the Doppler effect if we also know the system's orbital inclination. Astronomers have developed techniques for determining orbital inclination in a relatively small number of cases.

Hertzsprung and Russell based their diagrams on the spectral sequence OBAFGKM.

- The vertical axis represents stellar luminosity, in units of the Sun's luminosity (L_{Sun}). Stellar luminosities span a wide range, so we keep the graph compact by making each tick mark represent a luminosity 10 times larger than the prior tick mark.

Each location on the diagram represents a unique combination of spectral type and luminosity. For example, the dot representing the Sun in Figure 16.10 corresponds to the Sun's spectral type, G2, and its luminosity, $1 L_{Sun}$. Because luminosity increases upward on the diagram and surface temperature increases leftward, stars near the upper left are hot and luminous. Similarly, stars near the upper right are cool and luminous, stars near the lower right are cool and dim, and stars near the lower left are hot and dim.

THINK ABOUT IT

Explain how the colors of the stars in Figure 16.10 help indicate stellar surface temperature. Do these colors tell us anything about interior temperatures? Why or why not?

The H–R diagram also provides direct information about stellar radii, because a star's luminosity depends on both its surface temperature and its surface area or radius. Recall that surface temperature determines the amount of power emitted by the star *per unit area:* Higher temperature means greater power output per unit area [Section 6.4]. Thus, if two stars have the same surface temperature, one can be more luminous than the other only if it is larger in size. Stellar radii therefore must increase as we go from the high-temperature, low-luminosity corner on the lower left of the H–R diagram to the low-temperature, high-luminosity corner on the upper right.

Patterns in the H–R Diagram

Figure 16.10 also shows that stars do not fall randomly throughout the H–R diagram but instead fall into several distinct groups:

- Most stars fall somewhere along the **main sequence**, the prominent streak running from the upper left to the lower right on the H–R diagram. Our Sun is a main-sequence star.

- The stars along the top are called **supergiants** because they are very large in addition to being very bright.

- Just below the supergiants are the **giants**, which are somewhat smaller in radius and lower in luminosity (but still much larger and brighter than main-sequence stars of the same spectral type).

- The stars near the lower left are small in radius and appear white in color because of their high temperature. We call these stars **white dwarfs**.

When classifying a star, astronomers generally report both the star's spectral type and a **luminosity class** that describes the region of the H–R diagram in which the star falls. Table 16.2 summarizes the luminosity classes: Luminosity class I represents supergiants, luminosity class III represents giants, luminosity class V represents main-sequence stars, and luminosity classes II and IV are intermediate to the others. For example, the complete spectral classification of our Sun is G2 V. The G2 spectral type means it is yellow in color, and the luminosity class V means it is a main-sequence star. Betelgeuse is M2 I, making it a *red supergiant*. Proxima Centauri is M5 V—similar in color and

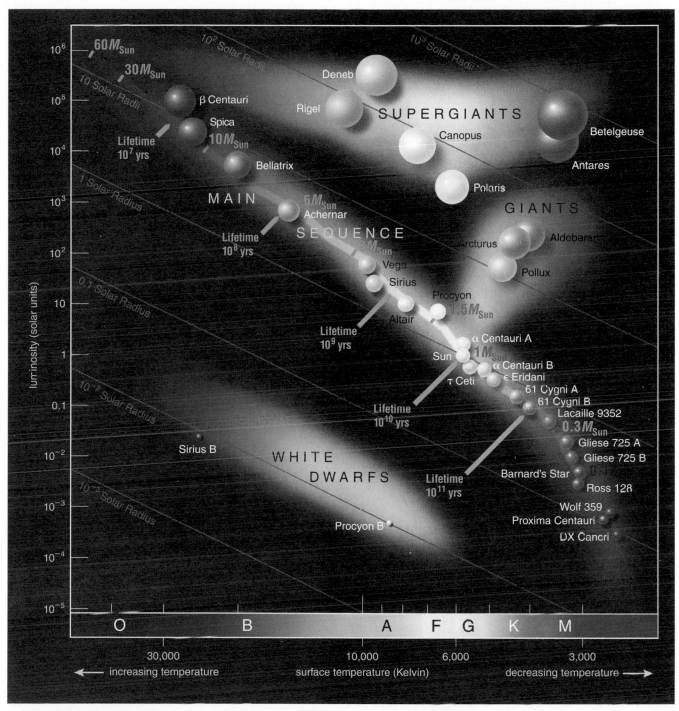

Figure 16.10 An H–R diagram, one of astronomy's most important tools, shows how the surface temperatures of stars (plotted along the horizontal axis) relate to their luminosities (plotted along the vertical axis). Several of the brightest stars in the sky are plotted here, along with a few of those closest to Earth. They are not drawn to scale—the diagonal lines, labeled in solar radii, indicate how large they are compared to the Sun. The lifetime and mass labels apply only to main-sequence stars (see Figure 16.11). (Star positions on this diagram are based on data from the Hipparcos satellite.)

surface temperature to Betelgeuse, but far dimmer because of its much smaller size. White dwarfs are usually designated with the letters *wd* rather than with a Roman numeral.

The Main Sequence

The common trait of main-sequence stars is that, like our Sun, they are fusing hydrogen into helium in their cores. Because stars spend the majority of their lives fusing hydrogen, most stars fall somewhere along the main sequence of the H–R diagram.

Why do main-sequence stars span such a wide range of luminosities and surface temperatures? By measuring the masses of stars in binary systems, astronomers have discovered that stellar masses decrease downward along the main sequence (Figure 16.11). At the upper end of the main sequence, the hot, luminous O stars can have masses as high as 100 times that of the Sun ($100M_{Sun}$). On the lower end, cool, dim M stars may have as little as 0.08 times the mass of the Sun ($0.08M_{Sun}$). Many more stars fall on the lower end of the main sequence than on the upper end, which tells us that low-mass stars are much more common than high-mass stars.

The orderly arrangement of stellar masses along the main sequence tells us that *mass* is the most important attribute of a hydrogen-burning star. Luminosity depends directly on mass because the weight of a star's outer layers determines the nuclear burning rate in its core. More weight means the star must sustain a higher nuclear burning rate in order to maintain gravitational equilibrium [Section 15.3].

Table 16.2 Stellar Luminosity Classes

Class	Description
I	Supergiants
II	Bright giants
III	Giants
IV	Subgiants
V	Main sequence

The nuclear burning rate, and hence the luminosity, is very sensitive to mass. For example, a $10M_{Sun}$ star on the main sequence is about 10,000 times more luminous than the Sun.

The relationship between mass and surface temperature is a little subtler. In general, a very luminous star must either be very large or have a very high surface temperature, or some combination of both. Stars on the upper end of the main sequence are thousands of times more luminous than the Sun but only about 10 times larger than the Sun in radius. Thus, their surfaces must be significantly hotter than the Sun's surface to account for their high luminosities. Main-sequence stars more massive than the Sun therefore have higher surface temperatures than the Sun, and those less massive than the Sun have lower surface temperatures. That is why the main sequence slices diagonally from the upper left to the lower right on the H–R diagram.

Mathematical Insight **16.5** **Calculating Stellar Radii**

Almost all stars are too distant for us to measure their radii directly. However, we can calculate a star's radius from its luminosity with the aid of the thermal radiation laws. As given in Mathematical Insight 6.2, the amount of thermal radiation emitted by a star of surface temperature *T* is:

$$\text{emitted power per unit area} = \sigma T^4$$

where the constant $\sigma = 5.7 \times 10^{-8}$ watt/(m^2 × Kelvin^4).

The luminosity *L* of a star is its power per unit area multiplied by its total surface area. If the star has radius *r*, its surface area is given by the formula $4\pi r^2$. Thus:

$$L = 4\pi r^2 \times \sigma T^4$$

With a bit of algebra, we can solve this formula for the star's radius *r*:

$$r = \sqrt{\frac{L}{4\pi\sigma T^4}}$$

Example: Betelgeuse has a luminosity of $38,000L_{Sun}$ and a surface temperature of about 3,400 K. What is its radius?

Solution: First, we must make our units consistent by converting the luminosity of Betelgeuse into watts. Remembering that $L_{Sun} = 3.8 \times 10^{26}$ watts, we find:

$$L_{Bet} = 38,000 \times L_{Sun} = 38,000 \times 3.8 \times 10^{26} \text{ watts}$$
$$= 1.4 \times 10^{31} \text{ watts}$$

Now we can use the formula derived above to calculate the radius of Betelgeuse:

$$r = \sqrt{\frac{L}{4\pi\sigma T^4}}$$

$$= \sqrt{\frac{1.4 \times 10^{31} \text{ watts}}{4\pi \times \left(5.7 \times 10^{-8} \dfrac{\text{watt}}{\text{m}^2 \times \text{K}^4}\right) \times (3,400 \text{ K})^4}}$$

$$= \sqrt{\frac{1.4 \times 10^{31} \text{ watts}}{9.6 \times 10^7 \dfrac{\text{watts}}{\text{m}^2}}} = 3.8 \times 10^{11} \text{ m}$$

The radius of Betelgeuse is about 380 billion meters or, equivalently, 380 million kilometers. Note that this is more than twice the Earth–Sun distance of 150 million kilometers.

Main-Sequence Lifetimes

A star has a limited supply of core hydrogen and therefore can remain as a hydrogen-fusing main-sequence star for only a limited time—the star's **main-sequence lifetime** (or *hydrogen burning lifetime*). Because stars spend the vast majority of their lives fusing hydrogen into helium, we sometimes refer to the main-sequence lifetime as simply the "lifetime." Like masses, stellar lifetimes vary in an orderly way as we move up the main sequence: Massive stars near the upper end of the main sequence have *shorter* lives than less massive stars near the lower end (see Figure 16.11).

Why do more massive stars live shorter lives? A star's lifetime depends on both its mass and its luminosity. Its mass determines how much hydrogen fuel the star initially contains in its core. Its luminosity determines how rapidly the star uses up its fuel. Massive stars live shorter lives because, even though they start their lives with a larger supply of hydrogen, they consume their hydrogen at a prodigious rate.

The main-sequence lifetime of our Sun is about 10 billion years [Section 15.1]. A 30-solar-mass star has 30 times more hydrogen than the Sun but burns it with a luminosity some 300,000 times greater. Consequently, its lifetime is roughly 30/300,000 = 1/10,000 as long as the Sun's—corresponding to a lifetime of only a few million years. Cosmically speaking, a few million years is a remarkably short time, which is one reason why massive stars are so rare: Most of the massive stars that have ever been born are long since dead. (A second reason is that lower-mass stars form in larger numbers than higher-mass stars [Section 17.2].)

The fact that massive stars exist at all at the present time tells us that stars must form continuously in our galaxy. The massive, bright O stars in our galaxy today formed only recently and will die long before they have a chance to complete even one orbit around the center of the galaxy.

THINK ABOUT IT

Would you expect to find life on planets orbiting massive O stars? Why or why not? (*Hint:* Compare the lifetime of an O star to the amount of time that passed from the formation of our solar system to the origin of life on Earth.)

On the other end of the scale, a 0.3-solar-mass star emits a luminosity just 0.01 times that of the Sun and consequently lives roughly 0.3/0.01 = 30 times longer than the Sun. In a universe that is now about 14 billion years old, even the most ancient of these small, dim M stars still survive and will continue to shine faintly for hundreds of billions of years to come.

Giants, Supergiants, and White Dwarfs

Giants and supergiants are stars nearing the ends of their lives because they have already exhausted their core hydrogen. Surprisingly, stars grow more luminous when they begin to run out of fuel. As we will discuss in the next chapter, a star generates energy furiously during the last stages of its life as it tries to stave off the inevitable crushing force of gravity. As ever-greater amounts of power well up from the core, the outer layers of the star expand, making it a giant or supergiant. The largest of these celestial behemoths have radii more than 1,000 times the radius of the Sun. If our Sun were this big, it would engulf the planets out to Jupiter.

Because they are so bright, we can see giants and supergiants even if they are not especially close to us. Many of the brightest stars visible to the naked eye are giants or supergiants. They are often identifiable by their reddish color. Nevertheless, giants and supergiants are rarer than main-sequence stars. In our snapshot of the heavens, we catch most stars in the act of hydrogen burning and relatively few in a later stage of life.

Giants and supergiants eventually run out of fuel entirely. A giant with a mass similar to that of our Sun ultimately ejects its outer layers, leaving behind a "dead" core in which all nuclear fusion has ceased. White dwarfs are these remaining embers of former giants. They are hot because they are essentially exposed stellar cores, but they are dim because they lack an energy source and radiate only their leftover heat into space. A typical white dwarf is no larger in size than Earth, although it may have a mass as great as that of our Sun. (Giants and supergiants with masses much larger than that of the Sun ultimately explode, leaving behind neutron stars or black holes as corpses [Section 17.4].)

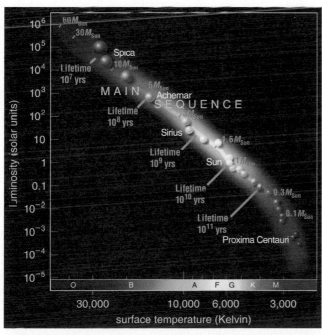

Figure 16.11 Along the main sequence, more massive stars are brighter and hotter but have shorter lifetimes. (Stellar masses are given in units of solar masses: $1 M_{Sun} = 2 \times 10^{30}$ kg.)

Figure 16.12 A typical light curve for a Cepheid variable star. Cepheids are giant, whitish stars whose luminosities regularly pulsate over periods of a few days to about a hundred days. The pulsation period of this Cepheid is about 50 days.

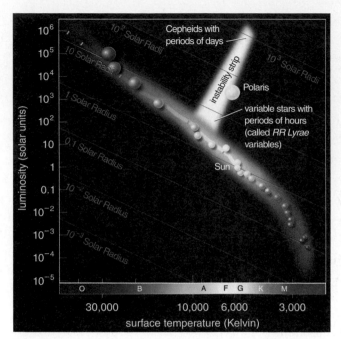

Figure 16.13 An H–R diagram with the instability strip highlighted.

Pulsating Variable Stars

Not all stars shine steadily like our Sun. Any star that significantly varies in brightness with time is called a *variable star*. A particularly important type of variable star has a peculiar problem with achieving the proper balance between the power welling up from its core and the power being radiated from its surface. Sometimes the upper layers of such a star are too opaque, so energy and pressure build up beneath the photosphere and the star expands in size. However, this expansion puffs the upper layers outward, making them too transparent. So much energy then escapes that the underlying pressure drops, and the star contracts again.

In a futile quest for a steady equilibrium, the atmosphere of such a **pulsating variable star** alternately expands and contracts, causing the star to rise and fall in luminosity. Figure 16.12 shows a typical light curve for a pulsating variable star, with the star's brightness graphed against time. Any pulsating variable star has its own particular period between peaks in luminosity, which we can discover easily from its light curve. These periods can range from as short as several hours to as long as several years.

Most pulsating variable stars inhabit a strip (called the *instability strip*) on the H–R diagram that lies between the main sequence and the red giants (Figure 16.13). A special category of very luminous pulsating variables lies in the upper portion of this strip: the **Cepheid variables**, or **Cepheids** (so named because the first identified star of this type was the star Delta Cephei).

Cepheids fluctuate in luminosity with periods of a few days to a few months. In 1912, another woman astronomer at Harvard, Henrietta Leavitt, discovered that the periods of these stars are very closely related to their luminosities: The longer the period, the more luminous the star. This **period–luminosity relation** holds because larger (and hence more luminous) Cepheids take longer to pulsate in and out in size.

Once we have measured the period of a Cepheid variable, we can use the period–luminosity relation to determine its luminosity. We can then calculate its distance with the luminosity–distance formula. In fact, as we'll discuss in Chapter 20, Cepheids provide our primary means of measuring distances to other galaxies and thus teach us the true scale of the cosmos. The next time you look at the North Star, Polaris, gaze upon it with renewed appreciation. Not only has it guided generations of navigators in the Northern Hemisphere, but it is also one of these special Cepheid variable stars.

 Stellar Evolution Tutorial, Lessons 1, 4

16.6 Star Clusters

All stars are born from giant clouds of gas. Because a single interstellar cloud can contain enough material to form many stars, stars almost inevitably form in groups. In our snapshot of the heavens, many stars still congregate in the groups in which they formed. These groups are of two basic types: modest-size **open clusters** and densely packed **globular clusters**.

Open clusters of stars are always found in the disk of the galaxy (see Figure 1.18). They can contain up to several thousand stars and typically span about 30 light-years (10 parsecs). The most famous open cluster is the *Pleiades*, a prominent clump of stars in the constellation Taurus (Figure 16.14). The Pleiades are often called the *Seven Sisters*, although only six of the cluster's several thousand stars are easily visible to the naked eye. Other cultures have other names for this beautiful group of stars. In Japanese it is called *Subaru*, which is why the logo for Subaru automobiles is a diagram of the Pleiades.

Globular clusters are found primarily in the halo of our galaxy, although some are in the disk. A globular clus-

Figure 16.14 A photo of the Pleiades, a nearby open cluster of stars. The most prominent stars in this open cluster are of spectral type B, indicating that the Pleiades are no more than 100 million years old, relatively young for a star cluster. The region shown here is about 11 light-years across.

ter can contain more than a million stars concentrated in a ball typically from 60 to 150 light-years across (20 to 50 parsecs). Its innermost part can have 10,000 stars packed within a region just a few light-years across (Figure 16.15). The view from a planet in a globular cluster would be marvelous, with thousands of stars lying closer than Alpha Centauri is to the Sun.

Because a globular cluster's stars nestle so closely, they engage in an intricate and complex dance choreographed by gravity. Some stars zoom from the cluster's core to its outskirts and back again at speeds approaching the escape velocity from the cluster, while others orbit the dense core more closely. When two stars pass especially close to each other, the gravitational pull between them deflects their trajectories, altering their speeds and sending them careening off in new directions. Occasionally, a close encounter boosts one star's velocity enough to eject it from the cluster. Through such ejections, globular clusters gradually lose stars and grow more compact.

Star clusters are extremely useful to astronomers for two key reasons:

1. All the stars in a cluster lie at about the same distance from Earth.

2. Cosmically speaking, all the stars in a cluster formed at about the same time (i.e., within a few million years of one another).

Astronomers can therefore use star clusters as laboratories for comparing the properties of stars, as yardsticks for measuring distances in the universe [Section 20.3], and as timepieces for measuring the age of our galaxy.

We can use clusters as timepieces because we can determine their ages from H–R diagrams of cluster stars.

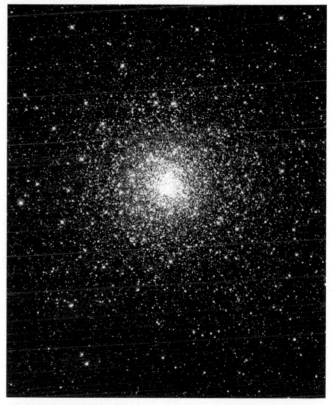

Figure 16.15 This globular cluster, known as M 80, is over 12 billion years old. The prominent reddish stars in this Hubble Space Telescope photo are red giant stars nearing the ends of their lives. The region pictured here is about 15 light-years across.

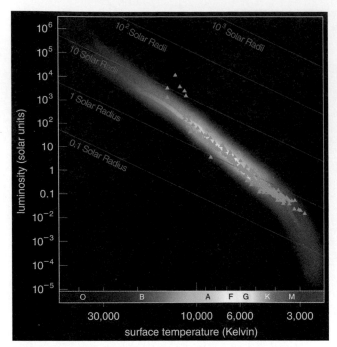

Figure 16.16 An H–R diagram for the stars of the Pleiades. Triangles represent individual stars. The Pleiades cluster is missing its upper main-sequence stars, indicating that these stars have already ended their hydrogen-burning lives. The main-sequence turnoff point at about spectral type B6 tells us that the Pleiades are about 100 million years old.

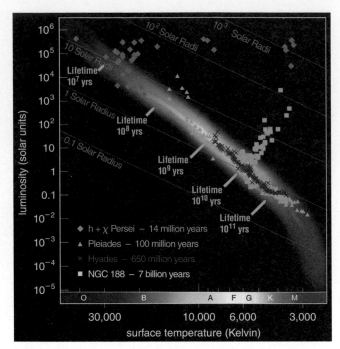

Figure 16.17 This H–R diagram shows stars from four clusters with very different ages. Each star cluster has a different main-sequence turnoff point. The youngest cluster, h + χ Persei, still contains main-sequence O and B stars (only the most massive stars are shown), indicating that it is only about 14 million years old. Stars at the main-sequence turnoff point for the oldest cluster, NGC 188, are only slightly more luminous and massive than the Sun, indicating an age of 7 billion years.

To understand how the process works, consider Figure 16.16, which shows an H–R diagram for the Pleiades. Most of the stars in the Pleiades fall along the standard main sequence, with one important exception: The Pleiades' stars trail away to the right of the main sequence at the upper end. That is, the hot, short-lived O stars are missing from the main sequence. Apparently, the Pleiades are old enough for its O stars to have already ended their hydrogen-burning lives. At the same time, they are young enough for some B stars to still survive on the main sequence.

The precise point on the H–R diagram at which the Pleiades' main sequence diverges from the standard main sequence is called the **main-sequence turnoff** point. In this cluster, it occurs around spectral type B6. The main-sequence lifetime of a B6 star is roughly 100 million years, so this must be the age of the Pleiades. Any star in the Pleiades that was born with a main-sequence spectral type hotter than B6 had a lifetime shorter than 100 million years and hence is no longer found on the main sequence. Stars with lifetimes longer than 100 million years are still fusing hydrogen and hence remain as main-sequence stars. Over the next few billion years, the B stars in the Pleiades will die out, followed by the A stars and the F stars. Thus, if we could make an H–R diagram for the Pleiades every few million years, we would find that the main sequence gradually grows shorter.

Comparing the H–R diagrams of other open clusters makes this effect more apparent (Figure 16.17). In each

case, we determine the cluster's age from the lifetimes of the stars at its main-sequence turnoff point:

$$\text{age of the cluster} = \text{lifetime of stars at main-}$$
$$\text{sequence turnoff point}$$

Stars in a particular cluster that once resided above the turnoff point on the main sequence have already exhausted their core supply of hydrogen, while stars below the turnoff point remain on the main sequence.

Suppose a star cluster is precisely 10 billion years old. On an H–R diagram, where would you expect to find its main-sequence turnoff point? Would you expect this cluster to have any main-sequence stars of spectral type A? Would you expect it to have main-sequence stars of spectral type K? Explain. (*Hint:* What is the lifetime of our Sun?)

The technique of identifying main-sequence turnoff points is our most powerful tool for evaluating the ages of star clusters. We've learned, for example, that most open clusters are relatively young and that very few are older than about 5 billion years. In contrast, the stars at the main-sequence turnoff points in globular clusters are usually less massive than our Sun (Figure 16.18). Because stars like our Sun have a lifetime of about 10 billion years and these

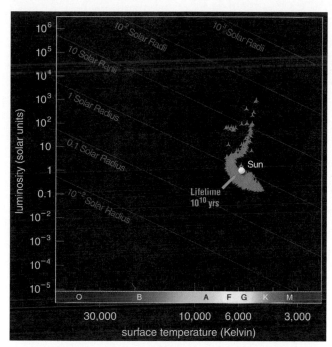

Figure 16.18 This H–R diagram shows stars from the globular cluster Palomar 3. The main-sequence turnoff point is in the vicinity of stars like our Sun, indicating an age for this cluster of around 10 billion years. A more technical analysis of this cluster places its age at around 12–14 billion years. (Stars in globular clusters tend to contain virtually no elements other than hydrogen and helium. Because of their different composition, these stars are somewhat bluer and more luminous than stars of the same mass and composition as our Sun.)

stars have already died in globular clusters, we conclude that globular-cluster stars are older than 10 billion years.

More precise studies of the turnoff points in globular clusters, coupled with theoretical calculations of stellar lifetimes, place the ages of these clusters at between 12 and

16 billion years, making them the oldest known objects in the galaxy. In fact, globular clusters place a constraint on the possible age of the universe: If stars in globular clusters are 12 billion years old, then the universe must be at least this old.

THE BIG PICTURE

Putting Chapter 16 into Context

We have classified the diverse families of stars visible in the night sky. Much of what we know about stars, galaxies, and the universe itself is based on the fundamental properties of stars introduced in this chapter. Make sure you understand the following "big picture" ideas:

● All stars are made primarily of hydrogen and helium, at least at the time they form. The differences between stars come about primarily because of differences in mass and age.

● Much of what we know about stars comes from studying the patterns that appear when we plot stellar surface temperatures and luminosities in an H–R diagram. Thus, the H–R diagram is one of the most important tools of astronomers.

● Stars spend most of their lives as main-sequence stars, fusing hydrogen into helium in their cores. The most massive stars live only a few million years, while the least massive stars will survive until the universe is many times its present age.

● Much of what we know about the universe comes from studies of star clusters. Here again, H–R diagrams play a vital role. For example, H–R diagrams of star clusters allow us to determine their ages.

SUMMARY OF KEY CONCEPTS

16.1 Snapshot of the Heavens

• *How can we learn about the lives of stars, which last millions to billions of years?* By taking observations of many stars, we can study stars in many phases of their life, just as we might study how humans age by observing all the humans living in a particular village at one time.

• *What are the two main elements in all stars?* All stars are made primarily of hydrogen and helium at birth.

• *What two basic physical properties do astronomers use to classify stars?* Stars are classified by their luminosity and surface temperature, which depend primarily on a star's mass and its stage of life.

16.2 Stellar Luminosity

• *What is luminosity, and how do we determine it?* A star's luminosity is the total power (energy per unit time) that it radiates into space. It can be calculated from a star's measured apparent brightness and distance, using the luminosity–distance formula:

$$\text{apparent brightness} = \frac{\text{luminosity}}{4\pi \times (\text{distance})^2}$$

• *How do we measure the distance to nearby stars?* The distance to nearby stars can be measured by parallax, the shift in the apparent position of a star with respect to more distant stars as Earth moves around the Sun.

continued ▶

- *How does the magnitude of a star relate to its apparent brightness?* The magnitude scale runs backward, so a star of magnitude 5 is brighter than a star of magnitude 18.

16.3 Stellar Surface Temperature

- *How are stars classified into spectral types?* From hottest to coolest, the major spectral types are O, B, A, F, G, K, and M. These types are futher subdivided into numbered categories. For example, the hottest A stars are type A0 and the coolest A stars are type A9, which is slightly hotter than F0.

- *What determines a star's spectral type?* The main factor in determining a star's spectral type is its surface temperature. Spectral type does not depend much on composition, because the compositions of stars—primarily hydrogen and helium—are nearly the same.

16.4 Stellar Masses

- *What is the most important property of a star?* A star's most important property is its mass, which determines its luminosity and spectral type at each stage of its life.

- *What are the three major classes of binary star systems?* A visual binary is a pair of orbiting stars that we can see distinctly through a telescope. An eclipsing binary reveals its binary nature because of periodic dimming that occurs when one star eclipses the other as viewed from Earth. A spectroscopic binary reveals its binary nature when we see the spectral lines of one or both stars shifting back and forth as the stars orbit each other.

- *How do we measure stellar masses?* We can directly measure stellar mass only in binary systems for which we are able to determine the period and separation of the two orbiting stars. We can then calculate the system's mass using Newton's version of Kepler's third law.

16.5 The Hertzsprung–Russell Diagram

- *What is the Hertzsprung–Russell (H–R) diagram?* The H–R diagram is the most important classification tool in stellar astronomy. Stars are located on the H–R diagram by their surface temperature (or spectral type) along the horizontal axis and their luminosity along the vertical axis. Surface temperature decreases from left to right on the H–R diagram.

- *What are the major features of the H–R diagram?* Most stars occupy the main sequence, which extends diagonally from lower right to upper left. The giants and supergiants inhabit the upper-right region of the diagram, above the main sequence. The white dwarfs are found near the lower left, below the main sequence.

- *How do stars differ along the main sequence?* All main-sequence stars are fusing hydrogen to helium in their cores. Stars near the lower right of the main sequence are lower in mass and have longer lifetimes than stars further up the main sequence. Lower-mass main-sequence stars are much more common than higher-mass stars.

- *What determines the length of time a star spends on the main sequence?* A star's mass determines how much hydrogen fuel it has and how fast it fuses that hydrogen into helium. The most massive stars have the shortest lifetimes because they fuse their hydrogen at a much faster rate than do lower-mass stars.

- *What are Cepheid variable stars, and why are they important to astronomers?* Cepheid variables are very luminous pulsating variable stars that follow a period–luminosity relation, which means we can calculate luminosity by measuring pulsation period. Once we know a Cepheid's luminosity, we can calculate its distance with the luminosity–distance formula. This technique enables us to measure distances to many other galaxies in which we have observed these variable stars.

16.6 Star Clusters

- *What are the two major types of star cluster?* Open clusters contain up to several thousand stars and are found in the disk of the galaxy. Globular clusters are much denser, containing hundreds of thousands of stars, and are found mainly in the halo of the galaxy. Globular-cluster stars are among the oldest stars known, with estimated ages of up to 12–14 billion years. Open clusters are generally much younger than globular clusters.

- *Why are star clusters useful for studying stellar evolution?* The stars in star clusters are all at roughly the same distance and, because they were born at about the same time, are all about the same age.

- *How do we measure the age of a star cluster?* The age of a cluster is equal to the main-sequence lifetime of the hottest, most luminous main-sequence stars remaining in the cluster. On an H–R diagram of the cluster, these stars sit farthest to the upper left and define the main-sequence turnoff point of the cluster.

Decide whether each of the following statements is true or false and clearly explain how you know.

1. Two stars that look very different must be made of different kinds of elements.

2. Sirius is the brightest star in the night sky, but if we moved it 10 times farther away it would look only one-tenth as bright.

3. Sirius looks brighter than Alpha Centauri, but we know that Alpha Centauri is closer because its apparent position in the sky shifts by a larger amount as Earth orbits the Sun.

4. Stars that look red-hot have hotter surfaces than stars that look blue.

5. Some of the stars on the main sequence of the H–R diagram are not converting hydrogen into helium.

6. The smallest, hottest stars are plotted in the lower left-hand portion of the H–R diagram.

7. Stars that begin their lives with the most mass live longer than less massive stars because it takes them a lot longer to use up their hydrogen fuel.

8. Star clusters with lots of bright, blue stars are generally younger than clusters that don't have any such stars.

9. All giants, supergiants, and white dwarfs were once main-sequence stars.

10. Most of the stars in the sky are more massive than the Sun.

Problems

11. *Similarities and Differences.* What basic composition are all stars born with? Why do stars differ from one another?

12. *Across the Spectrum.* Explain why we sometimes talk about wavelength-specific (e.g., visible-light or X-ray) luminosity or apparent brightness, rather than total luminosity and total apparent brightness.

13. *Determining Parallax.* Briefly explain how we calculate a star's distance in parsecs by measuring its parallax angle in arcseconds.

14. *Magnitudes.* What is the magnitude system? Briefly explain what we mean by the apparent magnitude and absolute magnitude of a star.

15. *Deciphering Stellar Spectra.* Briefly summarize the roles of Annie Jump Cannon and Cecilia Payne-Gaposchkin in discovering the spectral sequence and its meaning.

16. *Eclipsing Binaries.* Describe why eclipsing binaries are so important for measuring masses of stars.

17. *Basic H–R Diagram.* Draw a sketch of a basic Hertzsprung–Russell (H–R) diagram. Label the main sequence, giants, supergiants, and white dwarfs. Where on this diagram do we find stars that are cool and dim? Cool and luminous? Hot and dim? Hot and bright?

18. *Luminosity Classes.* What do we mean by a star's luminosity class? On your sketch of the H–R diagram from problem 17, identify the regions for luminosity classes I, III, and V.

19. *Pulsating Variables.* What are pulsating variable stars? Why do they vary periodically in brightness?

20. *H–R Diagrams of Star Clusters.* Explain why H–R diagrams look different for star clusters of different ages. How does the location of the main-sequence turnoff point tell us the age of the star cluster?

21. *Stellar Data.* Consider the following data table for several bright stars. M_v is absolute magnitude, and m_v is apparent magnitude.

Star	M_v	m_v	Spectral Type	Luminosity Class
Aldebaran	−0.2	+0.9	K5	III
Alpha Centauri A	+4.4	0.0	G2	V
Antares	−4.5	+0.9	M1	I
Canopus	−3.1	−0.7	F0	II
Fomalhaut	+2.0	+1.2	A3	V
Regulus	−0.6	+1.4	B7	V
Sirius	+1.4	−1.4	A1	V
Spica	−3.6	+0.9	B1	V

Answer each of the following questions, including a brief explanation with each answer.

a. Which star appears brightest in our sky?

b. Which star appears faintest in our sky?

c. Which star has the greatest luminosity?

d. Which star has the least luminosity?

e. Which star has the highest surface temperature?

f. Which star has the lowest surface temperature?

g. Which star is most similar to the Sun?

h. Which star is a red supergiant?

i. Which star has the largest radius?

j. Which stars have finished burning hydrogen in their cores?

k. Among the main-sequence stars listed, which one is the most massive?

l. Among the main-sequence stars listed, which one has the longest lifetime?

22. *Data Tables.* Study the spectral types listed in Appendix F for the 20 brightest stars and for the stars within 12 light-years. Why do you think the two lists are so different? Explain.

*23. *The Inverse Square Law for Light.* Earth is about 150 million km from the Sun, and the apparent brightness of the Sun in our sky is about 1,300 watts/m². Using these two facts and the inverse square law for light, determine the apparent brightness we would measure for the Sun *if* we were located at the following positions.

a. Half Earth's distance from the Sun.

b. Twice Earth's distance from the Sun.

c. Five times Earth's distance from the Sun.

*24. *The Luminosity of Alpha Centauri A.* Alpha Centauri A lies at a distance of 4.4 light-years and has an apparent brightness in our night sky of 2.7×10^{-8} watt/m². Recall that 1 light-year = 9.5×10^{12} km = 9.5×10^{15} m.

a. Use the luminosity–distance formula to calculate the luminosity of Alpha Centauri A.

b. Suppose you have a light bulb that emits 100 watts of visible light. (*Note:* This is *not* the case for a standard 100-watt light bulb, in which most of the 100 watts goes to heat and only about 10–15 watts is emitted as visible light.) How far away would you have to put the light bulb for it to have the same apparent brightness as Alpha Centauri A in our sky? (*Hint:* Use 100 watts as L in the luminosity–distance formula, and use the apparent brightness given above for Alpha Centauri A. Then solve for the distance.)

*25. *More Practice with the Luminosity–Distance Formula.* Use the luminosity–distance formula to answer each of the following questions.

a. Suppose a star has the same luminosity as our Sun (3.8×10^{26} watts) but is located at a distance of 10 light-years. What is its apparent brightness?

b. Suppose a star has the same apparent brightness as Alpha Centauri A (2.7×10^{-8} watt/m²) but is located at a distance of 200 light-years. What is its luminosity?

c. Suppose a star has a luminosity of 8×10^{26} watts and an apparent brightness of 3.5×10^{-12} watt/m². How far away is it? Give your answer in both kilometers and light-years.

d. Suppose a star has a luminosity of 5×10^{29} watts and an apparent brightness of 9×10^{-15} watt/m². How far away is it? Give your answer in both kilometers and light-years.

*26. *Parallax and Distance.* Use the parallax formula to calculate the distance to each of the following stars. Give your answers in both parsecs and light-years.

a. Alpha Centauri: parallax angle of 0.742″.

b. Procyon: parallax angle of 0.286″.

*27. *The Magnitude System.* Use the definitions of the magnitude system to answer each of the following questions.

a. Which is brighter in our sky, a star with apparent magnitude 2 or a star with apparent magnitude 7? By how much?

b. Which has a greater luminosity, a star with absolute magnitude +4 or a star with absolute magnitude −6? By how much?

*28. *Measuring Stellar Mass.* The spectral lines of two stars in a particular eclipsing binary system shift back and forth with a period of 6 months. The lines of both stars shift by equal amounts, and the amount of the Doppler shift indicates that each star has an orbital speed of 80,000 m/s. What are the masses of the two stars? Assume that each of the two stars traces a circular orbit around their center of mass. (*Hint:* See Mathematical Insight 16.4.)

*29. *Calculating Stellar Radii.* Sirius A has a luminosity of $26L_{Sun}$ and a surface temperature of about 9,400 K. What is its radius? (*Hint:* See Mathematical Insight 16.5.)

Discussion Question

30. *Classification.* Edward Pickering's team of female "computers" at Harvard University made many important contributions to astronomy, particularly in the area of systematic stellar classification. Why do you think rapid advances in our understanding of stars followed so quickly on the heels of their efforts? Can you think of other areas in science where huge advances in understanding followed directly from improved systems of classification?

For a complete list of media resources available, go to www.astronomyplace.com, and choose Chapter 16 from the pull-down menu.

 Astronomy Place Web Tutorials

Tutorial Review of Key Concepts

Use the interactive **Tutorials** at www.astronomyplace.com to review key concepts from this chapter.

Hertzsprung–Russell Diagram Tutorial

Lesson 1 The Hertzsprung–Russell (H–R) Diagram

Lesson 2 Determining Stellar Radii

Lesson 3 The Main Sequence

Measuring Cosmic Distances Tutorial

Lesson 2 Stellar Parallax

Stellar Evolution Tutorial

Lesson 1 Main-Sequence Lifetimes

Lesson 4 Cluster Dating

Supplementary Tutorial Exercises

Use the interactive **Tutorial Lessons** to explore the following questions.

Hertzsprung–Russell Diagram Tutorial, Lessons 1–3

1. If one star appears brighter than another, can you be sure that it is more luminous? Why or why not?

2. Answer each part of this question with either high, low, left, or right. On an H–R diagram, where will a star be if it is hot? Cool? Bright? Dim?

3. Why is there a relationship between stellar radii and locations on the H–R diagram?

4. Why is there a relationship between luminosity and mass for main-sequence stars on the H–R diagram?

Measuring Cosmic Distances Tutorial, Lesson 2

1. Explain how we measure distances with stellar parallax. Give an example.

2. Explain why we cannot use parallax to measure the distance to *all* stars.

Stellar Evolution Tutorial, Lesson 4

1. In the animation at the beginning of Lesson 4, list the order in which you saw the three differently colored stars in the cluster disappear, and explain why they disappeared in this order.

2. In the second animation in Lesson 4, in what order did you see stars on the main sequence disappear? Explain the reason for this.

3. How does the age of a dim star cluster of mostly small stars compare to a bright cluster with some giants in it? Explain your answer.

 Exploring the Sky and Solar System

Of the many activities available on the *Voyager: SkyGazer* CD-ROM accompanying your book, use the following files to observe key phenomena covered in this chapter.

Go to the **File: Basics** folder for the following demonstrations:

1. Large Stars

2. More Stars

3. Star Color and Size

Go to the **File: Demo** folder for the following demonstrations:

1. Circling the Hyades

2. Flying Around Pleiades

3. The Tail of Scorpius

Web Projects

Take advantage of the useful web links on www.astronomyplace.com to assist you with the following projects.

1. *Women in Astronomy.* Until fairly recently, men greatly outnumbered women in professional astronomy. Nevertheless, many women made crucial discoveries in astronomy throughout history. Do some research about the life and discoveries of a woman astronomer from any time period, and write a two- to three-page scientific biography.

2. *The Hipparcos Mission.* The European Space Agency's Hipparcos mission, which operated from 1989 to 1993, made precise parallax measurements for more than 40,000 stars. Learn about how Hipparcos allowed astronomers to measure smaller parallax angles than they could from the ground and how Hipparcos discoveries have affected our knowledge of the universe. Write a one- to two-page report on your findings.

17 Star Stuff

I can hear the sizzle of newborn stars,
and know anything of meaning, of the
fierce magic emerging here. I am witness
to flexible eternity, the evolving past,
and know I will live forever, as dust or
breath in the face of stars, in the
shifting pattern of winds.

Joy Harjo, Secrets from the Center of the World

We inhale oxygen with every breath. Iron-bearing hemoglobin carries this oxygen through the bloodstream. Chains of carbon and nitrogen form the backbone of the proteins, fats, and carbohydrates in our cells. Calcium strengthens our bones, while sodium and potassium ions moderate communications of the nervous system. What does all this biology have to do with astronomy? The profound answer, recognized only in the second half of the twentieth century, is that life is based on elements created by stars.

We've already discussed in general terms how the elements in our bodies came to exist. Hydrogen and helium were produced in the Big Bang, and heavier elements were created later by stars and scattered into space by stellar explosions. There, in the spaces between the stars, these elements mixed with interstellar gas and became incorporated into subsequent generations of stars.

In this chapter, we will discuss the origins of the elements in greater detail by delving into the lives of stars. As you read, keep in mind that no matter how far removed the stars may seem from our everyday lives, they actually are connected to us in the most intimate way possible: Without the births, lives, and deaths of stars, none of us would be here.

 Stellar Evolution Tutorial, Lesson 1

17.1 Lives in the Balance

The story of a star's life is in many ways the story of an extended battle between two opposing forces: gravity and pressure. The most common type of pressure in stars is **thermal pressure**—the familiar type of pressure that keeps a balloon inflated and that increases when the temperature or thermal energy increases.

A star can maintain its internal thermal pressure only if it continually generates new thermal energy to replace the energy it radiates into space. This energy can come from two sources: *nuclear fusion* of light elements into heavier ones and the process of *gravitational contraction,* which converts gravitational potential energy into thermal energy [Section 15.1].

These energy-production processes operate only temporarily, although in this case "temporarily" means millions or billions of years. In contrast, gravity acts eternally. Moreover, any time gravity succeeds in shrinking a star's core, the strength of gravity grows. (The force of gravity inside an object grows stronger if it either gains mass or shrinks in radius [Section 5.3].) Because a star cannot generate thermal energy forever, its ultimate fate depends on whether something other than thermal pressure manages to halt the unceasing crush of gravity.

The final outcome of a star's struggle between gravity and pressure depends almost entirely on its birth mass. All stars are born from spinning clumps of gas, but newborn stars can have masses ranging from less than 10% of the mass of our Sun to about 100 times that of our Sun. The most massive stars live fast and die young, proceeding from birth to explosive death in just a few million years. The lowest-mass stars, in contrast, consume hydrogen so slowly that they will continue to shine until the universe is many times older than it is today.

Because of the wide range of stellar masses, we can simplify our discussion of stellar lives by dividing stars into three basic groups:

- **Low-mass stars** are stars born with less than about two times the mass of our Sun, or less than 2 *solar masses* ($2M_{Sun}$) of material.

- **Intermediate-mass stars** have birth weights between about 2 and 8 solar masses.

- **High-mass stars** are those stars born with masses greater than about 8 solar masses.

Both low-mass and intermediate-mass stars swell into red giants near the ends of their lives and ultimately become white dwarfs. High-mass stars also become red and large in their latter days, but their lives end much more violently.

We will focus most of our discussion in this chapter on the dramatic differences between the lives of low- and high-mass stars. Because the life stages of intermediate-mass stars are quite similar to the corresponding stages of high-mass stars until the very ends of their lives, we include them in our discussion of high-mass stars.

Given the brevity of human history compared to the life of any star, you might wonder how we can know so much about stellar life cycles. As with any scientific inquiry, we study stellar lives by comparing theory and observation. On the theoretical side, we use mathematical models based on the known laws of physics to predict the life cycles of stars.

On the observational side, we study stars of different mass but the same age by looking in star clusters whose ages we have determined by main-sequence turnoff [Section 16.6]. Occasionally, we even catch a star in its death throes. Theoretical predictions of the life cycles of stars agree quite well with these observations.

In the remainder of this chapter, we will examine in detail our modern understanding of the life stories of stars and how they manufacture the variety of elements—the *star stuff*—that make our lives possible.

17.2 Star Birth

Stars are born from clouds of interstellar gas (Figure 17.1) and return much of that gas to interstellar space when they die. In Chapter 19, we will examine this star–gas–star cycle in more detail. Here we will focus on star formation

itself. The clouds that form stars tend to be quite cold, typically only 10–30 K. (Recall that 0 K is absolute zero, and temperatures on Earth are around 300 K.) They also tend to be quite dense compared to the rest of the gas between the stars, although they would qualify as a superb vacuum by earthly standards. Like the galaxy as a whole, star-forming clouds are made almost entirely of hydrogen and helium.

Star-forming clouds are sometimes called **molecular clouds**, because their low temperatures allow hydrogen atoms to pair up to form hydrogen molecules (H_2). The relatively rare atoms of elements heavier than helium can also form molecules, such as carbon monoxide or water, or tiny, solid grains of dust. More important, the cold temperatures and high densities allow gravity to overcome thermal pressure more readily in molecular clouds than elsewhere in interstellar space. If the thermal pressure in a molecular cloud is too weak to counteract the compressing force of gravity, then the cloud must undergo gravitational contraction. Because molecular clouds are generally lumpy, gravity

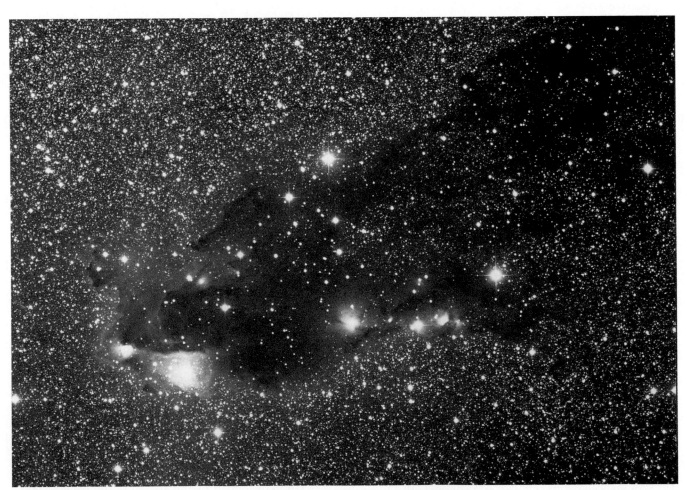

Figure 17.1 A star-forming cloud of molecular hydrogen gas in the constellation Scorpius extends from the upper-right corner of this photo through the center. The cloud appears dark because dust particles within it obscure the light radiated from more distant stars lying behind it. Blue-white blotches near the edges of the dark cloud are newly formed stars. They appear fuzzy because some of their light is reflecting off patchy gas in their vicinity. The region pictured here is about 50 light-years across.

Figure 17.2 Fragmentation of a molecular cloud. Gravity attracts matter to the densest regions of a molecular cloud. If gravity can overcome thermal pressure in these dense regions, they collapse to form even denser knots of gaseous matter known as *molecular cloud cores*. The cloud thus fragments into a number of pieces, each of which will form one or more new stars.

pulls the molecular gas toward the densest lumps, known as molecular cloud *cores*. A cloud thus fragments into numerous pieces, each of which will form one or more new stars (Figure 17.2).

From Cloud to Protostar

Gravitational contraction within each shrinking fragment of a molecular cloud releases thermal energy. Early in the process of star formation, the gas quickly radiates away much of this energy, preventing thermal pressure from building high enough to resist gravity. Because the temperature of the cloud remains below 100 K, it glows in long-wavelength infrared light (Figure 17.3).

The unopposed contraction initiated by gravity cannot continue indefinitely. As the cloud fragment contracts, the resulting increase in density makes the escape of radiation increasingly difficult. Eventually, its central regions grow completely opaque to infrared radiation, trapping the thermal energy produced by gravitational contraction. Because thermal energy can no longer escape easily, both the thermal pressure and the gas temperature at the center of the contracting region rise dramatically. This rising pressure begins to fight back against the crush of gravity, and the dense cloud fragment becomes a **protostar**— the clump of gas that will become a new star. Meanwhile, gaseous matter surrounding the protostar continues to rain down upon it, increasing its mass.

Disks and Jets

The rain of matter onto the protostar produces a **protostellar disk** similar to the spinning disk from which the planets of our solar system formed [Section 9.2]. A cloud

Betelgeuse

Orion
Nebula

Figure 17.3 This picture shows infrared radiation from star-forming regions in the constellation Orion. The colors correspond to the temperature of the emitting gas: Red is cooler, and white is hotter. Star formation is most intense in the yellow-white regions, which have been heated to 60–100 K (which is still quite cold). The Orion Nebula, home to many protostars, is the prominent yellow-white area near the bottom. Betelgeuse, a red supergiant, appears as a blue-white dot near the center. The region pictured here is about 80 light-years across at the distance of the Orion Nebula.

a A cloud fragment necessarily has some angular momentum.

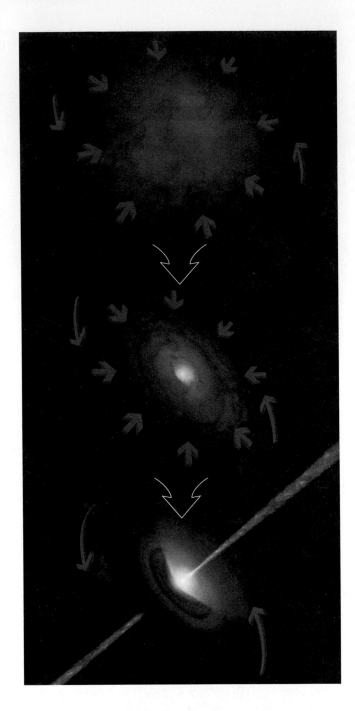

b This angular momentum causes it to spin faster and flatten into a protostellar disk as it collapses.

c In the late stages of collapse, the central protostar has a strong protostellar wind and may fire jets of high-speed gas outward along its rotation axis.

Figure 17.4 Artist's conception of star birth.

fragment begins its collapse with a certain amount of angular momentum (the sum total of the angular momentum of each gas particle within it), which might at first be unmeasurable. Contraction makes the angular momentum more obvious because conservation of angular momentum [Section 5.2] demands that the cloud fragment spin faster and faster as it shrinks. Thermal pressure can halt the infalling motion of the collapsing gas as it reaches the protostar but does nothing to stop its spinning motion. Much of the infalling matter therefore settles into a spinning protostellar disk orbiting the protostar (Figure 17.4). These disks sometimes coalesce into planetary systems like our solar system. We do not yet know how commonly this occurs.

THINK ABOUT IT

The term *protostellar disk* refers to any disk of material surrounding a protostar. Do you expect all protostellar disks to eventually give birth to planets? If so, why? If not, what do you think might prevent planets from forming in some protostellar disks?

The protostellar disk probably plays a large role in eventually slowing the rotation of the protostar. The protostar's rapid rotation generates a strong magnetic field. As the magnetic field lines sweep through the protostellar disk, they transfer some of the angular momentum to outlying material, slowing the disk's rotation [Section 9.3]. The strong magnetic field also helps generate a strong **protostellar wind**—an outward flow of particles similar to the *solar wind* [Section 15.2]. The protostellar wind may carry additional angular momentum from the protostar to interstellar space.

Rotation is probably also responsible for the formation of some binary star systems. Protostars that are unable to rid themselves of enough angular momentum spin too fast to become stable single stars and tend to split in two. Each of these two fragments can form a separate star. If the stars are particularly close together, the resulting pair is called a **close binary** system, in which two stars coexist in close proximity and rapidly orbit each other. Binary systems with wider separations can arise when neighboring protostars form close enough together for their mutual gravity to keep them from drifting apart.

Observations show that the late stages of a star's formation can be surprisingly violent. Besides the strong protostellar wind, many young stars also fire high-speed streams, or **jets**, of gas into interstellar space (Figure 17.5). No one knows exactly how protostars generate these jets, but two high-speed streams generally flow out along the rotation axis of the protostar, shooting in opposite directions. We also sometimes see glowing blobs of material along the jets (named *Herbig–Haro objects* after their discoverers). These blobs appear to be collections of gas swept up as the jet plows into the surrounding interstellar material. Together,

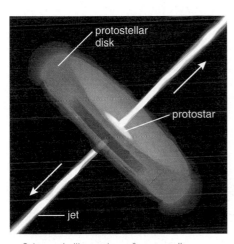

a Schematic illustration of protostellar disk–jet structure.

b Photograph of a protostellar disk and jet. We are seeing the disk edge-on, as in (**a**). The disk's top and bottom surfaces, illuminated by the protostar, are shown in green, and the jets emerging along the disk's axis are shown in red. The dark central layers of the disk block our view of the protostar itself.

c A wider-angle photograph of a jet emanating from a protostar (left) and ramming into surrounding interstellar gas (right).

Figure 17.5 Some protostars can be seen shooting jets of matter into interstellar space.

winds and jets play an important role in clearing away the cocoon of gas that surrounds a forming star, revealing the protostar within.

From Protostar to the Main Sequence

A protostar looks starlike, but its interior is not yet hot enough for fusion. The central temperature of a protostar may be only a million degrees or so when its wind and jets blow away the surrounding gas. To ignite fusion, the protostar needs to contract further to boost the central temperature.

Paradoxically, radiation of thermal energy from the surface of a protostar is what enables its central temperature to rise, because only half the thermal energy released by gravitational contraction is radiated away. The other half remains in the protostar's interior, raising its temperature. If the protostar did not lose thermal energy from its surface, it would not contract, and its central temperature would remain fixed. Early in this period of contraction, convection carries the protostar's thermal energy to the surface, as in the convection zone of the Sun [Section 15.4]. However, as the interior of the protostar heats up, photons can flow more easily, and radiative diffusion takes over from convection.

A protostar becomes a true star when its core temperature exceeds 10 million K, hot enough for hydrogen fusion to operate efficiently by the *proton–proton chain* [Section 15.3]. The ignition of fusion halts the protostar's gravitational contraction and marks what we consider the birth of a star. The new star's interior structure stabilizes because the energy produced in the center matches the amount radiated from its surface. The star is now a hydrogen-burning, *main-sequence* star [Section 16.5].

The length of time from the formation of a protostar to the birth of a main-sequence star depends on the star's mass. Massive stars do everything faster. The contraction of a high-mass protostar into a main-sequence star may take only a million years or less. A star like our Sun takes about 50 million years to go from the beginning of the protostellar stage to the main sequence. A very low-mass star of spectral type M may spend more than a hundred million years as a protostar. Thus, the most massive stars in a young star cluster may live and die before the smallest stars begin to fuse hydrogen in their cores and become main-sequence stars.

We can summarize the transitions that occur during star birth with a special type of H–R diagram. A standard H–R diagram shows luminosities and surface temperatures for many different stars [Section 16.5]. This special H–R diagram shows part of a **life track** (also called an *evolutionary track*) for a single star in relation to the standard main sequence. Each point along a star's life track represents its surface temperature and luminosity at some moment during its life.

Figure 17.6 shows a life track leading to the birth of a $1M_{Sun}$ star like our Sun. This prebirth period includes four distinct stages:

Stage 1. When the protostar first assembles from a collapsing cloud fragment, it is concealed within a shroud of dusty molecular gas. It becomes visible after the protostellar wind and jets disrupt this shroud. At this time, energy moves to the surface of the protostar primarily through convection. At the end of this stage the photosphere's temperature is about 3,000 K, placing it on the right side of the H–R diagram, and its surface is many times larger than that of the Sun, producing a luminosity somewhere between $10L_{Sun}$ and $100L_{Sun}$.

Figure 17.6 The life track of a $1M_{Sun}$ star from protostar to main-sequence star.

3. Surface temperature rises when radiation becomes the dominant mode of energy flow within the protostar.

2. The protostar shrinks and heats as gravitational potential energy is converted into thermal energy.

1. A protostar assembles from a collapsing cloud fragment. It is concealed beneath a shroud of dusty gas.

4. The fusion rate increases until it balances the energy radiated from the star's surface.

Stage 2. The protostar's surface temperature remains near 3,000 K as long as convection remains the dominant mechanism for transporting thermal energy to the surface. Gravitational contraction leads to a decrease in the protostar's luminosity, because its radius becomes smaller while its surface temperature stays nearly constant. Thus, the protostar's life track drops almost straight downward on the H–R diagram.

Stage 3. When energy transport within the protostar switches from convection to radiative diffusion [Section 15.4], the surface temperature begins to rise. This rise in surface temperature brings a slight rise in luminosity, even though the protostar continues to contract. The life track therefore bends toward higher surface temperature and slightly higher luminosity. During this stage, hydrogen nuclei begin to fuse into helium nuclei, but the energy released is small compared to the amount radiated away.

Stage 4. The core temperature and rate of fusion continue to increase gradually for a few tens of millions of years. Finally, the rate of fusion becomes high enough to balance the rate at which radiative energy escapes from the surface. At this point, fusion becomes self-sustaining. The star settles into its hydrogen-burning, main-sequence life.

Protostars of different masses go through similar stages as they approach the main sequence. Figure 17.7 illustrates life tracks for several protostars of different masses.

THINK ABOUT IT

Explain in your own words what we mean by a *life track* for a star. Why do we say that Figures 17.6 and 17.7 show only *pre-main-sequence* life tracks? In general terms, predict the appearance on Figure 17.7 of pre-main-sequence life tracks for a $25M_{Sun}$ star and for a $0.1M_{Sun}$ star.

Stellar Birth Weights

A single group of molecular clouds can contain thousands of solar masses of gas, which is why stars generally are born in clusters. We do not yet fully understand the processes that govern the clumping and fragmentation of these clouds into protostars with a wide variety of masses. However, we can observe the results.

In a newly formed star cluster, stars with low masses greatly outnumber stars with high masses. For every star with a mass between 10 and 100 solar masses, there are typically 10 stars with masses between 2 and 10 solar masses, 50 stars with masses between 0.5 and 2 solar masses, and a few hundred stars with masses below 0.5 solar mass. Thus, although the Sun lies toward the middle of the overall range of stellar masses, most stars in a new star cluster are less massive than the Sun. With the passing of time, the balance tilts even more in favor of the low-mass stars as the high-mass stars die away.

The masses of stars have limits. Theoretical models indicate that stars above about $100M_{Sun}$ generate power

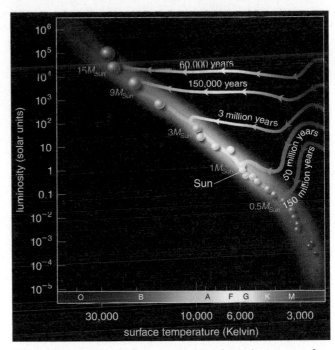

Figure 17.7 Life tracks from protostar to the main sequence for stars of different masses. A high-mass protostar becomes a main-sequence star much more quickly than a lower-mass protostar, and its luminosity does not decline as much as it collapses.

so furiously that gravity cannot contain their internal pressure. Such stars effectively blow themselves apart and drive their outer layers into space. Observations confirm the absence of stars much larger than $100M_{Sun}$—if any stars this massive existed nearby, they would be so luminous that we would easily detect them.

On the other end of the scale, calculations show that the central temperature of a protostar with less than $0.08M_{Sun}$ never climbs above the 10 million K threshold needed for efficient hydrogen fusion. Instead, a strange effect called **degeneracy pressure** halts the gravitational contraction of the core before hydrogen burning can begin.

The quantum mechanical origins of degeneracy pressure are discussed in Chapter S4. A simple way to visualize the origin of degeneracy pressure is with an analogy to an auditorium in which the chairs represent all possible places that subatomic particles (electrons, in this case) can be located and people represent the particles. Protostars with masses above $0.08M_{Sun}$ are like auditoriums with many more available chairs than people—the people (particles) can easily squeeze into a smaller section of the auditorium. However, the cores of protostars with masses below $0.08M_{Sun}$ are like auditoriums with so few chairs that the people (particles) fill nearly all of them. Because no extra chairs are available, the people (particles) cannot squeeze into a smaller section of the auditorium. This resistance to squeezing explains why degeneracy pressure halts gravitational contraction. Keep in mind that the degeneracy pressure arises *only* because particles have no place else to go. Thus, unlike thermal pressure, degeneracy pressure has nothing to do with temperature.

Because degeneracy pressure halts the collapse of a protostellar core with less than $0.08 M_{Sun}$ before fusion becomes self-sustaining, the result is a "failed star" that slowly radiates away its internal thermal energy, gradually cooling with time. Such objects, called **brown dwarfs**, occupy a fuzzy gap between what we call a planet and what we call a star. (Note that $0.08 M_{Sun}$ is about 80 times the mass of Jupiter.) Because degeneracy pressure does *not* rise and fall with temperature, the gradual cooling of a brown dwarf's interior does not weaken its degeneracy pressure. In the constant battle of any "star" to resist the crush of gravity, brown dwarfs are winners, albeit dim ones. Their degeneracy pressure will not diminish with time, so gravity will never gain the upper hand.

Brown dwarfs radiate primarily in the infrared and actually look deep red or magenta in color rather than brown. They are far dimmer than normal stars and therefore are extremely difficult to detect, even if they are quite nearby. The first brown dwarf was discovered in 1995— a $0.05 M_{Sun}$ object (called Gliese 229B) in orbit around a much brighter star (Gliese 229A). Many more brown dwarfs are now known. If the trend that makes small stars far more common than massive stars continues to masses below $0.08 M_{Sun}$, brown dwarfs might outnumber ordinary stars by a huge margin.

 Stellar Evolution Tutorial, Lesson 2

17.3 Life as a Low-Mass Star

In the grand hierarchy of stars, our Sun ranks as rather mediocre. We should be thankful for this mediocrity. If the Sun had been a high-mass star, it would have lasted only a few million years, dying before life could have arisen on Earth. Instead, the Sun has shone steadily for nearly 5 billion years and will continue to do so for about 5 billion more. In this section, we investigate the lives of low-mass stars like our Sun.

Slow and Steady

Low-mass stars spend their main-sequence lives fusing hydrogen into helium in their cores slowly and steadily via the *proton–proton chain* [Section 15.3]. As in the Sun, the energy released by nuclear fusion in the core may take a million years to reach the surface, where it finally escapes into space as the star's luminosity. The energy moves outward from the core through a combination of *radiative diffusion* and *convection* [Section 15.4]. Radiative diffusion transports energy through the random bounces of photons from one electron to another, and convection transports energy by the rising of hot plasma and the sinking of cool plasma.

Radiative diffusion is more effective at transporting energy outward in the deeper, hotter plasma near a star's core. In higher layers, where the temperature is cooler, some of the ions in the plasma retain electrons. These ions can absorb photons and thereby tend to prevent photons from continuing outward by radiative diffusion. When the energy welling up from fusion in the core reaches the point at which radiative diffusion is inhibited, convection must take over as the means of transporting energy outward. This point represents the beginning of a star's *convection zone* (Figure 17.8).

In the Sun, the temperature is cool enough to allow convection in the outer one-third of its interior. More massive stars have hotter interiors and hence shallower

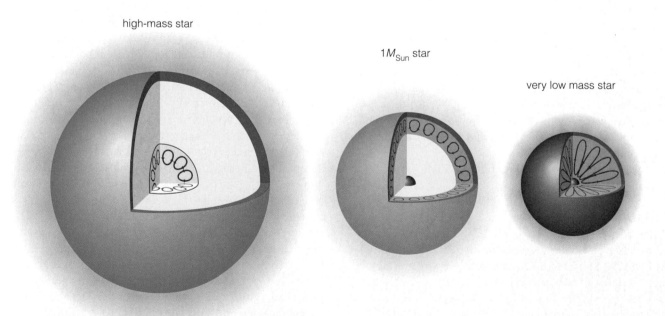

high-mass star

$1 M_{Sun}$ star

very low mass star

Figure 17.8 Among main-sequence stars, convection zones extend deeper in lower-mass stars. High-mass stars have convective cores but no convection zones near their surfaces.

convection zones. Lower-mass stars have cooler interiors and deeper convection zones. In very low mass stars, the convection zone extends all the way down to the core. The highest-mass stars have no convection zone at all near their surface, but they have a convective core because they produce energy so furiously.

The depth of the convection zone plays a major role in determining whether a star has activity similar to that of the sunspot cycle on our Sun [Section 15.5]. The Sun's activity arises from the twisting and stretching of its magnetic fields by convection and rotation. The most dramatically active stars are low-mass M stars that happen to have fast rotation rates in addition to their deep convection zones. The churning interiors of these stars are in a constant state of turmoil, twisting and knotting their magnetic field lines. When these field lines suddenly snap and reconfigure themselves, releasing energy from the magnetic field, the result can be a spectacular flare. For a few minutes or hours, the flare can produce more radiation in X rays than the total amount of light coming from the star in infrared and visible light. Life on a planet near one of these **flare stars** might be quite difficult.

A low-mass star gradually consumes its core hydrogen, converting it into helium over a period of billions of years. In the process, the declining number of independent particles in the core (four independent protons fuse into just one independent helium nucleus) causes the core to shrink and heat very gradually, pushing the luminosity of the main-sequence star slowly upward as it ages [Section 15.3]. The most dramatic changes occur when nuclear fusion exhausts the hydrogen in the star's core.

Red Giant Stage

The energy released by hydrogen fusion during a star's main-sequence life maintains the thermal pressure that holds gravity at bay. When the core hydrogen is finally depleted, nuclear fusion ceases in the star's core. With no fusion to supply thermal energy, the core pressure can no longer resist the crush of gravity, and the core begins to shrink more rapidly.

Surprisingly, the star's outer layers expand outward while the core is shrinking. On an H–R diagram, the star's life track moves almost horizontally to the right as the star grows in size to become a **subgiant** (Figure 17.9). As the expansion of the outer layers continues, the star's luminosity begins to increase substantially, and the star slowly becomes a **red giant**. For a $1M_{Sun}$ star, this process takes about a billion years, during which the star's radius increases about 100-fold and its luminosity grows by an even greater factor. (Like all phases of stellar lives, the process occurs faster for more massive stars and slower for less massive stars.) This process may at first seem paradoxical: Why does the star grow bigger and more luminous at the same time that its core is shrinking?

We can find the answer by considering the interior structure of the star. The core is now made of helium—the

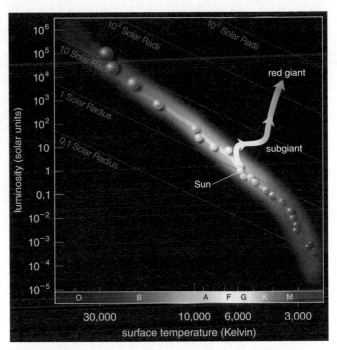

Figure 17.9 The life track of a $1M_{Sun}$ star on an H–R diagram from the end of its main-sequence life until it becomes a red giant.

"ash" left behind by hydrogen fusion—but the surrounding layers still contain plenty of fresh hydrogen. Gravity shrinks both the *inert* (nonburning) helium core and the surrounding *shell* of hydrogen, and the shell soon becomes hot enough to sustain hydrogen fusion (Figure 17.10). The shell becomes so hot that this **hydrogen shell burning** proceeds at a higher rate than core hydrogen fusion did during the star's main-sequence life.

The result is that the star becomes more luminous than ever before. Energy transport within the star is too slow to keep pace with this larger energy-generation rate. Because much of this new thermal energy is trapped within the star, thermal pressure builds up, pushing the surface of the star outward. What was once a fairly dim main-sequence star balloons into a luminous red giant. While the red giant is large on the outside, most of its mass is buried deep in a shrunken stellar core.

The situation grows more extreme as long as the helium core remains inert. Recall that, in a main-sequence star like the Sun, a rise in the fusion rate causes the core to inflate and cool until the fusion rate drops back down in a self-correcting process called the *solar thermostat* [Section 15.3]. Thermal energy generated in the hydrogen-burning shell of a red giant, however, cannot do anything to inflate the inert core that lies underneath. Instead, newly produced helium keeps adding to the mass of the helium core, amplifying its gravitational pull and shrinking it further. The hydrogen-burning shell shrinks along with the core, growing hotter and denser. The fusion rate in the shell consequently rises, feeding even more helium ash to the core. The star is caught in a vicious circle.

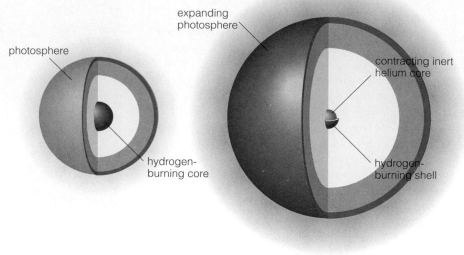

photosphere

contracting inert
helium core

hydrogen-
burning core

hydrogen-
burning shell

Figure 17.10 After a star ends its main-sequence life, its inert helium core contracts while hydrogen shell burning begins. The high rate of fusion in the hydrogen shell forces the star's upper layers to expand outward.

main-sequence star

expanding subgiant

The core and shell continue to shrink in size and the shell grows in luminosity, while thermal pressure continues to push the star's upper layers outward. This cycle breaks down only when the inert helium core reaches a temperature of about 100 million K, at which point helium nuclei can fuse together. (In a very low mass star, the inert helium core may never become hot enough to fuse helium. The core collapse will instead be halted by degeneracy pressure, ultimately leaving the star a *helium white dwarf*.) Throughout the expansion phase, a *stellar wind* carries away much more matter than the solar wind does for our Sun, but at much slower speeds.

Before you read on, briefly summarize why a star grows larger and brighter after it exhausts its core hydrogen. When does the growth of a red giant finally halt, and why? How would a star's red giant stage be different if the temperature required for helium fusion were around 200 million K, rather than 100 million K? Why?

Helium Burning

Recall that fusion occurs only when two nuclei come close enough together for the attractive *strong force* to overcome electromagnetic repulsion [Section 15.3]. Helium nuclei have two protons (and two neutrons) and hence a greater positive charge than the single proton of a hydrogen nucleus. The greater charge means that helium nuclei repel one another more strongly than hydrogen nuclei. **Helium fusion** therefore occurs only when nuclei slam into one another at much higher speeds than those needed for hydrogen fusion. Therefore, helium fusion requires much higher temperatures.

The helium fusion process (often called the "triple-alpha" reaction because helium nuclei are sometimes called

"alpha particles") converts three helium nuclei into one carbon nucleus:

$3\ ^4\mathrm{He}$ $1\ ^{12}\mathrm{C}$ energy

Energy is released because the carbon-12 nucleus has a slightly lower mass than the three helium-4 nuclei, and the lost mass becomes energy in accord with $E = mc^2$.

The ignition of helium burning in low-mass stars has one subtlety. Theoretical models show that, in the inert helium core of a low-mass star, the thermal pressure is too low to counteract gravity. Instead, the models show that the pressure fighting against gravity is *degeneracy pressure*—the same strange type of pressure that supports brown dwarfs. Because degeneracy pressure does *not* increase with temperature, the onset of helium fusion heats the core rapidly without causing it to inflate. The rising temperature causes the helium fusion rate to rocket upward in what is called a **helium flash**.

The helium flash dumps enormous amounts of new thermal energy into the core. In a matter of seconds, the rapidly rising thermal pressure becomes the dominant pressure pushing back against gravity. The core is no longer degenerate and begins to expand. This core expansion pushes the hydrogen-burning shell outward, lowering its temperature and its burning rate. The result is that, even though the star now has core helium fusion and hydrogen shell burning taking place simultaneously (Figure 17.11), the total energy production falls from its peak during the red giant phase. The reduced total energy output of the star reduces its luminosity, allowing its outer layers to contract from their peak size during the red giant phase. As the outer

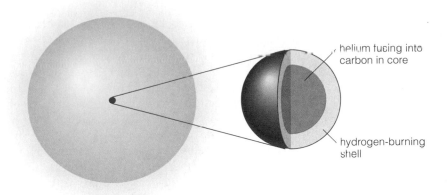

Figure 17.11 Core structure of a helium-burning star. Helium fusion causes the core and hydrogen-burning shell to expand and slightly cool, thereby reducing the overall energy generation rate in comparison to the rate during the red giant stage. The outer layers shrink back, so a helium-burning star is smaller than a red giant of the same mass.

layers contract, the star's surface temperature also increases somewhat.

Because the helium-burning star is now smaller and hotter than it was as a red giant, its life track on the H–R diagram drops downward and to the left (Figure 17.12a). The helium cores of all low-mass stars fuse helium into carbon at about the same rate, so these stars all have about the same luminosity. However, the outer layers of these stars can have different masses depending on how much mass they lost through their stellar winds. Stars that lost more mass end up with smaller radii and higher surface temperatures and hence are farther to the left on the H–R diagram.

In a cluster of stars, those stars that are currently in their helium-burning phase all have about the same luminosity but differ in surface temperature. Thus, on an H–R diagram for a cluster of stars, the helium-burning stars are arranged along a **horizontal branch** (Figure 17.12b). This cluster H–R diagram clearly shows the main-sequence turnoff point. Stars that have recently left the main sequence are subgiants on their way to becoming red giants. In the upper-right corner of the red giant region are stars that are almost ready for the helium flash. The stars that have already become helium-burning, horizontal-branch stars are slightly dimmer and hotter.

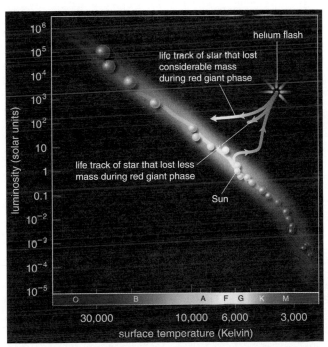

a After the helium flash, a star's surface shrinks and heats, so the star's life track moves downward and to the left on the H–R diagram.

Figure 17.12 The onset of helium fusion.

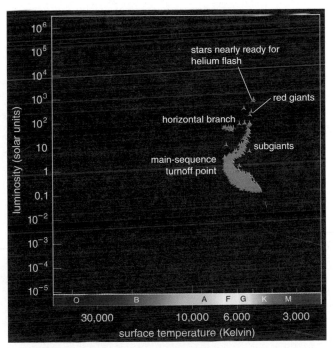

b This H–R diagram plots the luminosity and surface temperature of individual stars in a cluster (i.e., it does *not* show life tracks). Helium core–burning stars occupy the horizontal branch of the diagram. A star's position along the horizontal branch depends on how much mass it lost during the red giant phase.

Last Gasps

It is only a matter of time until a horizontal-branch star fuses all its core helium into carbon. The core helium in a low-mass star will run out in about a hundred million years. When the core helium is exhausted, fusion turns off, and the core begins to shrink once again under the crush of gravity.

The basic processes that changed the star from a main-sequence star into a red giant now resume, but this time helium fusion ignites in a shell around an inert carbon core. Meanwhile, the hydrogen shell still burns atop the helium layer. Both shells contract along with the inert core, driving their temperatures and fusion rates much higher. The luminosity of this double-shell star grows greater than ever, and its outer layers swell to an even huger size. On the H–R diagram, the star's life track once again turns upward (Figure 17.13). Theoretical models show that helium burning inside such a star never reaches equilibrium but instead proceeds in a series of **thermal pulses** during which the fusion rate spikes upward every few thousand years.

The furious burning in the helium and hydrogen shells cannot last long—maybe a few million years or less for a $1M_{Sun}$ star. The star's only hope of extending its life lies with the carbon core, but this is a false hope in the case of low-mass stars. Carbon fusion is possible only at temperatures above about 600 million K. Before the core of a low-mass star ever reaches such a lofty temperature, degeneracy pressure halts its gravitational collapse.

For a low-mass star with a carbon core, the end is near. The huge size of the dying star means that it has a very weak grip on its outer layers. As the star's luminosity and radius keep rising, matter flows from its surface at increasingly high rates. Meanwhile, during each thermal pulse, strong convection dredges up carbon from the core, enriching the surface of the star with carbon. Red giants whose photospheres become especially carbon-rich in this way are called **carbon stars**.

Carbon stars have cool, low-speed stellar winds, and the temperature of the gas in these winds drops with distance from the stellar surface. At the point at which the temperature has dropped to 1,000–2,000 K, some of the gas atoms in these slow-moving winds begin to stick together in microscopic clusters, forming small, solid particles of dust. These dust particles continue their slow drift with the stellar wind into interstellar space, where they become **interstellar dust grains**. The process of particulate formation is very similar to the formation of smoke particles in a fire. Thus, in a sense, carbon stars are the most voluminous polluters in the universe. However, this "carbon smog" is essential to life: Most of the carbon in your body (and in all life on Earth) was manufactured in carbon stars and blown into space by their stellar winds.

THINK ABOUT IT

Suppose the universe contained only low-mass stars. Would elements heavier than carbon exist? Why or why not?

Before a low-mass star dies, it treats us to one last spectacle. Through winds and other processes, the star ejects its outer layers into space. The result is a huge shell of gas expanding away from the inert, degenerate carbon core. The exposed core is still very hot and therefore emits intense ultraviolet radiation that ionizes the gas in the expanding shell. It now glows brightly as what we call a **planetary nebula**. Despite this name, planetary nebulae have nothing to do with planets. The name comes from the fact that nearby planetary nebulae look much like planets through small telescopes, appearing as simple disks. Through a larger telescope, more detail is visible. The famous Ring Nebula

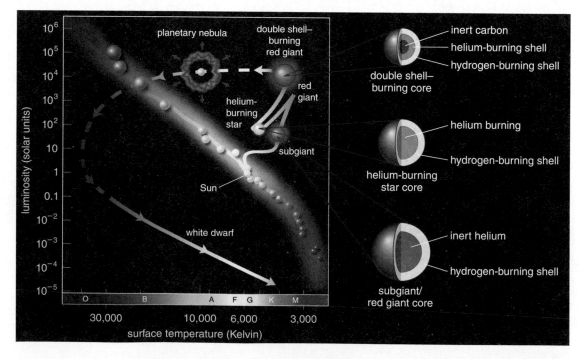

Figure 17.13 The life track of a $1M_{Sun}$ star from main-sequence star to white dwarf. Core structure is shown at key stages.

in the constellation Lyra is a planetary nebula. Many other planetary nebulae are also quite beautiful (Figure 17.14).

The glow of the planetary nebula fades as the exposed core cools and the ejected gas disperses into space. The nebula will disappear within a million years, leaving behind the cooling carbon core. On the H–R diagram, the life track now represents this "dead" core (see Figure 17.13). At first the life track heads to the left, because the core is initially quite hot and luminous. However, the life track soon veers downward and to the right as the remaining ember cools and fades. You already know these naked, inert cores by the name *white dwarfs* [Section 16.5].

In the ongoing battle between gravity and a star's internal pressure, white dwarfs are a sort of stalemate. As long as no mass is added to the white dwarf from some other source (such as a companion star in a binary system), neither the strength of gravity nor the strength of the degeneracy pressure that holds gravity at bay will ever change.

a Ring Nebula

b Eskimo Nebula

c Spirograph Nebula

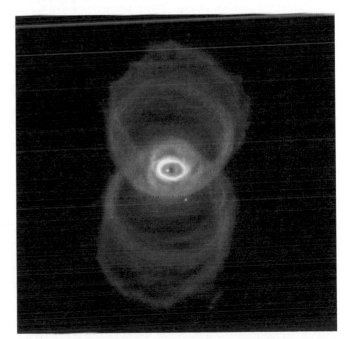

d Hourglass Nebula

Figure 17.14 Planetary nebulae occur when low-mass stars in their final death throes cast off their outer layers of gas, as seen in these photos from the Hubble Space Telescope. The hot core that remains ionizes and energizes the richly complex envelope of gas surrounding it.

Thus, a white dwarf is little more than a decaying corpse that will cool for the indefinite future, eventually disappearing from view as a *black dwarf*.

The Fate of Life on Earth

The evolutionary stages of low-mass stars are immensely important to Earth, because we orbit a low-mass star—the Sun. The Sun will gradually brighten during its remaining time as a main-sequence star, just as it has been brightening since its birth more than 4 billion years ago [Section 15.3]. The Sun's past brightening has not threatened the long-term survival of life on Earth, because Earth's climate self-regulates by adjusting the strength of the greenhouse effect (through the carbon dioxide cycle [Section 14.4]). However, this climate regulation will eventually break down as the Sun warms.

We still do not understand Earth's climate regulation well enough to be certain of when the warming Sun will begin to overheat Earth. Some climate models predict that the oceans will begin to evaporate about a billion years from now, while other models suggest that our planet's climate may remain stable much longer. All models agree that about 3–4 billion years from now, the Sun will have brightened enough to doom Earth to a runaway greenhouse effect like that on Venus [Section 10.6], causing the oceans to boil away. Temperatures on Earth will rise even more dramatically when the Sun finally exhausts its core supply of hydrogen, somewhere around the year A.D. 5,000,000,000.

Things will only get worse as the Sun grows into a red giant over the next several hundred million years. Just before helium flash, the Sun will be more than 100 times larger in radius and over 1,000 times more luminous than it is today (Figure 17.15). Earth's surface temperature will exceed 1,000 K. Clearly, any surviving humans will need to have found a new home. Saturn's moon Titan [Section 12.5] might not be a bad choice. Its surface temperature will have risen from well below freezing today to about the present temperature of Earth. The Sun will shrink somewhat after helium burning begins, providing a temporary lull in incineration while the Sun spends 100 million years as a helium-burning star.

Anyone who survives the Sun's helium-burning phase will need to prepare for one final disaster. After exhausting its core helium, the Sun will expand again during its last million years. Its luminosity will soar to thousands of times what it is today, and it will grow so large that solar prominences might lap at Earth's surface. Then it will eject its outer layers as a planetary nebula that will engulf Jupiter and Saturn and drift on past Pluto into interstellar space. If Earth is not destroyed, its charred surface will be cold and dark in the faint, fading light of the white dwarf that the Sun has become. From then on, Earth will be little more than a chunk of rock circling the corpse of our once-brilliant star.

 Stellar Evolution Tutorial, Lesson 3

17.4 Life as a High-Mass Star

Human life would be impossible without both low-mass stars and high-mass stars. The long lives of low-mass stars allow evolution to proceed for billions of years, but only high-mass stars produce the full array of elements on which life depends. Fusion of elements heavier than helium to produce elements heavier than carbon requires extremely high temperatures in order to overcome the larger electro-

---

SPECIAL TOPIC *Five Billion Years*

The Sun's demise in about 5 billion years might at first seem worrisome, but 5 billion years is a very long time. It is longer than Earth has yet existed, and human time scales pale by comparison. A single human lifetime, if we take it to be about 100 years, is only 2×10^{-8}, or two hundred-millionths, of 5 billion years. Because 2×10^{-8} of a human lifetime is about 1 minute, we can say that a human lifetime compared to the life expectancy of the Sun is roughly the same as 60 heart beats compared to a human lifetime.

What about human creations? The Egyptian pyramids have often been described as "eternal," but they are slowly eroding due to wind, rain, air pollution, and the impact of tourists. All traces of them will have vanished within a few hundred thousand years. While a few hundred thousand years may seem like a long time, the Sun's remaining lifetime is more than 1,000 times longer.

On a more somber note, we can gain perspective on 5 billion years by considering evolutionary time scales. During the past century, our species has acquired sufficient technology and power to destroy human life totally, if we so choose. However, even if

we make that unfortunate choice, some species (including many insects) are likely to survive.

Would another intelligent species ever emerge on Earth? There is no way to know, but we can look to the past for guidance. Many species of dinosaurs were biologically quite advanced, if not truly intelligent, when they were suddenly wiped out about 65 million years ago. Some small rodentlike mammals survived, and here we are 65 million years later. We therefore might guess that another intelligent species could evolve some 65 million years after a human extinction. If these beings also destroyed themselves, another species could evolve 65 million years after that, and so on.

Even at 65 million years per shot, Earth would have *nearly 80* more chances for an intelligent species to evolve in 5 billion years (5 billion ÷ 65 million ≈ 77). Perhaps one of those species will not destroy itself, and future generations might move on to other star systems by the time the Sun finally dies. Perhaps this species will be our own.

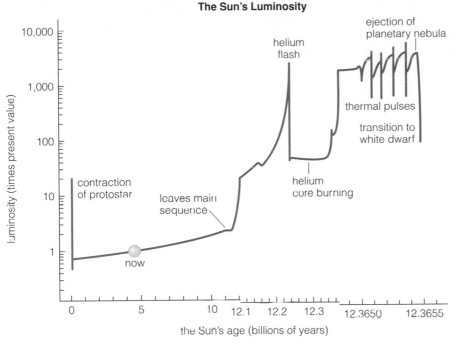

The Sun's Luminosity

luminosity (times present value)

10,000

1,000

100

10

1

contraction
of protostar

now

leaves main
sequence

helium
flash

ejection of
planetary nebula

thermal pulses

transition to
white dwarf

helium
core burning

0   5   10   12.1   12.2   12.3   12.3650   12.3655

the Sun's age (billions of years)

a Solar luminosity. The Sun's luminosity has gradually been rising over the past 4.6 billion years and will continue to rise slowly for another 5–6 billion years. Hydrogen shell burning will then drive a rapid increase in luminosity ending in a helium flash when the Sun's luminosity reaches about $1,000L_{Sun}$. The Sun will next enter a helium core–burning stage lasting about 100 million years. After the core helium runs out, the luminosity will rise once again, peaking at a few thousand L_{Sun} with a series of thermal pulses. Note that the time scale is expanded on the far right in order to show details of the final stages.

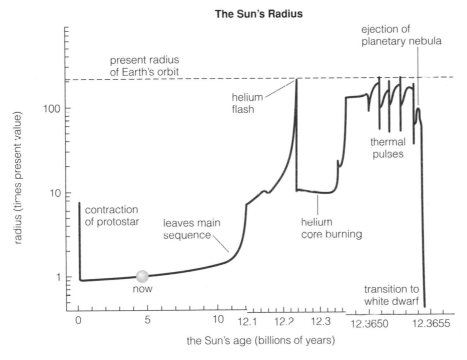

The Sun's Radius

radius (times present value)

present radius
of Earth's orbit

100

10

1

contraction
of protostar

now

leaves main
sequence

helium
flash

ejection of
planetary nebula

thermal
pulses

helium
core burning

transition to
white dwarf

0   5   10   12.1   12.2   12.3   12.3650   12.3655

the Sun's age (billions of years)

b Solar radius. The Sun's radius increases with each rise in solar luminosity and declines with each fall. During the Sun's first red giant stage, it will reach a size roughly equal to Earth's current orbit. (However, because the Sun's mass and gravitational pull will be reduced by its strong wind during this stage, Earth's orbit will move to a slightly larger distance.)

Figure 17.15 Evolution of the Sun. These graphs show calculations from models in which the main-sequence lifetime of the Sun is 11 billion years, slightly greater than the more commonly quoted 10-billion-year lifetime.

magnetic repulsion of more highly charged nuclei. Reaching such temperatures requires an extremely strong crush of gravity on a star's core—a crush that occurs only under the immense weight of the overlying gas in high-mass stars.

The early stages of a high-mass star's life are very similar to the early life stages of a low-mass star. However, in the final stages of their lives, the highest-mass stars proceed to fuse increasingly heavy elements until they have exhausted all possible fusion sources. When fusion ceases, gravity drives the core to implode suddenly, which, as we

will soon see, causes the star to self-destruct in the titanic explosion we call a *supernova*. The fast-paced life and cataclysmic death of a high-mass star—surely among the great dramas of the universe—are the topics of this section.

Brief but Brilliant

During the main-sequence life of a high-mass star, its strong gravity compresses its hydrogen core to higher temperatures than we find in lower-mass stars. You already know that the rate of fusion via the proton–proton chain increases substantially at higher temperatures. The even hotter core temperatures of high-mass stars enable protons to slam into carbon, oxygen, or nitrogen nuclei as well as into other protons. Although carbon, nitrogen, and oxygen make up less than 2% of the material from which stars form in interstellar space, this 2% is more than enough to be useful in a stellar core. The carbon, nitrogen, and oxygen act as catalysts for hydrogen fusion, making it proceed at a far higher rate than would be possible by the proton–proton chain alone. (A *catalyst* is something that aids the progress of a reaction without being consumed in the reaction.) The lives of high-mass stars are truly brief but brilliant.

The chain of reactions that leads to hydrogen fusion in high-mass stars is called the **CNO cycle**; the letters *CNO* stand for carbon, nitrogen, and oxygen, respectively. The six steps of the CNO cycle are shown in Figure 17.16. Just as in the proton–proton chain [Section 15.3], four hydrogen nuclei go in while one helium-4 nucleus comes out. The amount of energy generated in each reaction cycle therefore is the same as in the proton–proton chain: It is equal to the difference in mass between the four hydrogen nuclei and the one helium nucleus multiplied by c^2. The CNO cycle is simply another, faster way to accomplish hydrogen fusion.

THINK ABOUT IT

Did the very first high-mass stars in the history of the universe produce energy through the CNO cycle? Explain.

The escalated fusion rates in high-mass stars generate remarkable amounts of power. Many more photons stream from the photospheres of high-mass stars than from the Sun, and many more photons are bouncing around inside. Although photons have no mass, they act like particles and carry *momentum* [Section 5.1], which they can transfer to anything they hit, imparting a very slight jolt. The combined jolts from the huge number of photons streaming outward through a high-mass star apply a type of pressure called **radiation pressure**.

Radiation pressure can have dramatic effects on high-mass stars. In the most massive stars, radiation pressure is even more important than thermal pressure in keeping gravity at bay. Near the photosphere, the radiation pressure can drive strong, fast-moving stellar winds. The wind from a very massive star can expel as much as 10^{-5} solar mass of gas per year at speeds greater than 1,000 km/s. This wind would cross the United States in about 5 seconds and would send a mass equivalent to that of our Sun hurtling into space in only 100,000 years. Such a wind cannot last long because it would blow away all the mass of even a very massive star in just a few million years.

Advanced Nuclear Burning

The exhaustion of core hydrogen in a high-mass star sets in motion the same processes that turn a low-mass star into a red giant, but the transformation proceeds much more quickly. The star develops a hydrogen-burning shell, and its outer layers begin to expand outward. At the same

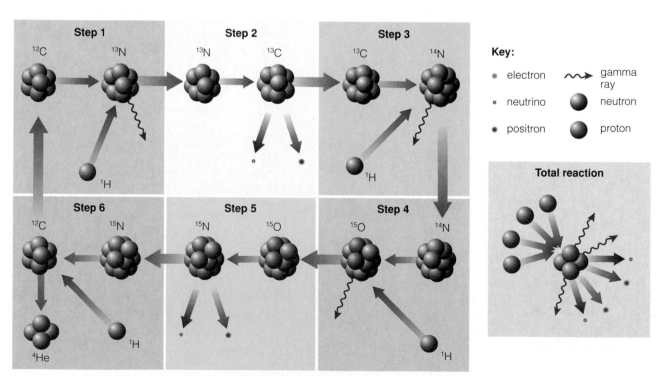

Figure 17.16 This diagram illustrates the six steps of the CNO cycle by which massive stars fuse hydrogen into helium. Note that the overall result is the same as that of the proton–proton chain: Four hydrogen nuclei fuse to make one helium nucleus. The carbon, nitrogen, and oxygen nuclei help the cycle proceed, but overall these nuclei are neither consumed nor created in the cycle.

time, the core contracts, and this gravitational contraction releases energy that raises the core temperature until it becomes hot enough to fuse helium into carbon. However, there is no helium flash in stars of more than 2 solar masses. The high core temperatures induced by core contraction keep the thermal pressure high, preventing degeneracy pressure from being a factor. Helium burning therefore ignites gradually, just as hydrogen burning did at the beginning of the star's main-sequence life.

A high-mass star fuses helium into carbon so rapidly that it is left with an inert carbon core after just a few hundred thousand years or less. Once again, the absence of fusion leaves the core without a thermal energy source to fight off the crush of gravity. The inert carbon core shrinks, the crush of gravity intensifies, and the core pressure, temperature, and density all rise. Meanwhile, a helium-burning shell forms between the inert core and the hydrogen-burning shell. The star's outer layers swell again.

Up to this point, the life stories of intermediate-mass stars ($2-8M_{Sun}$) and high-mass stars ($>8M_{Sun}$) are very similar, except that all stages proceed more rapidly in higher-mass stars. However, degeneracy pressure prevents the cores of intermediate-mass stars from reaching the temperatures required to burn carbon or oxygen into anything much heavier. These stars eventually blow away their upper layers and finish their lives as white dwarfs. The rest of a high-mass star's life, on the other hand, is unlike anything that a low- or intermediate-mass star ever experiences.

The crush of gravity in a high-mass star is so overwhelming that degeneracy pressure never comes into play in the collapsing carbon core. The gravitational contraction of the core continues, and the core temperature soon reaches the 600 million K required to fuse carbon into heavier elements. Carbon fusion provides the core with a new source of energy that restores the balance versus gravity, but only temporarily. In the highest-mass stars, carbon burning may last only a few hundred years. When the core carbon is depleted, the core again begins to collapse, shrinking and heating until it can fuse a still-heavier element. The star is engaged in the final phases of a desperate battle against the ever-strengthening crush of gravity. Each successive stage of core nuclear burning proceeds more rapidly than prior stages.

The nuclear reactions in the star's final stages of life become quite complex, and many different reactions may take place simultaneously (Figure 17.17). The simplest sequence of fusion stages involves **helium capture**—the fusing of helium nuclei into progressively heavier elements. (Some helium nuclei still remain in the core, but not enough to continue helium fusion efficiently.) Helium capture can fuse carbon into oxygen, oxygen into neon, neon into magnesium, and so on.

At high enough temperatures, a star's core plasma can fuse heavy nuclei to one another. For example, fusing carbon to oxygen creates silicon, fusing two oxygen nuclei creates sulfur, and fusing two silicon nuclei generates iron. Some of these heavy-element reactions release free neutrons,

Helium-capture reactions

Other reactions

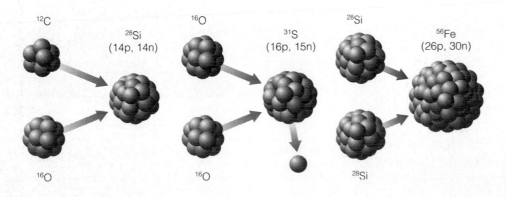

Figure 17.17 A few of the many nuclear reactions that occur in the final stages of a high-mass star's life. Fusion of two silicon nuclei and some other processes that lead to iron actually first produce nickel-56 (28 protons and 28 neutrons), but this decays rapidly to cobalt-56 (27 protons and 29 neutrons) and then to iron-56 (26 protons and 30 neutrons).

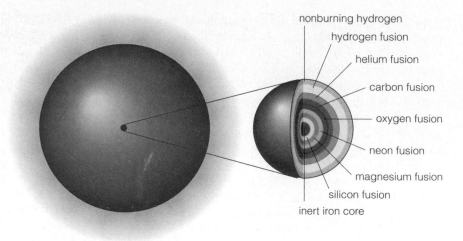

Figure 17.18 The multiple layers of nuclear burning in the core of a high-mass star during the final days of its life.

which may fuse with heavy nuclei to make still rarer elements. The star is forging the variety of elements that, in our solar system at least, became the stuff of life.

Each time the core depletes the elements it is fusing, it shrinks and heats until it becomes hot enough for other fusion reactions. Meanwhile, a new type of shell burning ignites between the core and the overlying shells of fusion. Near the end, the star's central region resembles the inside of an onion, with layer upon layer of shells burning different elements (Figure 17.18). During the star's final few days, iron begins to pile up in the silicon-burning core.

Despite the dramatic events taking place in its interior, the high-mass star's outer appearance changes only slowly. As each stage of core fusion ceases, the surrounding shell burning intensifies and further inflates the star's outer layers. Each time the core flares up again, the outer layers may contract a bit. The result is that the star's life track zigzags across the top of the H–R diagram (Figure 17.19). In very massive stars, the core changes happen so quickly that the outer layers don't have time to respond, and the star progresses steadily toward becoming a red supergiant.

Betelgeuse, the upper-left shoulder star of Orion, is the best-known red supergiant star. Its radius is over 500 solar radii, or more than twice the distance from the Sun to Earth. We have no way of knowing what stage of nuclear burning is now taking place in Betelgeuse's core. Betelgeuse may have a few thousand years of nuclear burning still ahead, or we may be seeing it as iron piles up in its core. If the latter is the case, then sometime in the next few days we will witness one of the most dramatic events that ever occurs in the universe.

Iron: Bad News for the Stellar Core

As a high-mass star develops an inert core of iron, the core continues shrinking and heating while iron continues to pile up from nuclear burning in the surrounding shells. If iron were like the other elements in prior stages of nuclear burning, this core contraction would stop when iron fusion ignited. However, iron is unique among the elements

in a very important way: It is the one element from which it is *not* possible to generate any kind of nuclear energy.

To understand why iron is unique, remember that only two basic processes can release nuclear energy: *fusion* of light elements into heavier ones, and *fission* of very heavy elements into not-so-heavy ones. Recall that hydrogen fusion converts four protons (hydrogen nuclei) into a helium nucleus that consists of two protons and two neutrons. Thus, the total number of *nuclear particles* (protons and neutrons combined) does not change. However, this fusion reaction generates energy (in accord with $E = mc^2$) because the *mass* of the helium nucleus is less than the combined mass of the

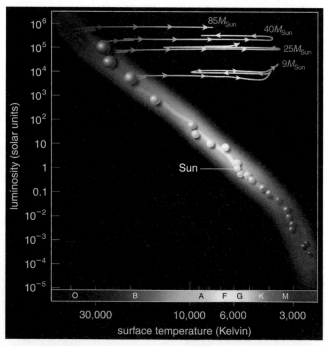

Figure 17.19 Life tracks on the H–R diagram from main-sequence star to red supergiant for selected high-mass stars. Labels on the tracks give the star's mass at the beginning of its main-sequence life. Because of the strong wind from such a star, its mass can be considerably smaller when it leaves the main sequence. (Based on models from A. Maeder and G. Meynet.)

four hydrogen nuclei that fused to create it—despite the fact that the *number* of nuclear particles is unchanged.

In other words, fusing hydrogen into helium generates energy because helium has a lower *mass per nuclear particle* than hydrogen. Similarly, fusing three helium-4 nuclei into one carbon-12 nucleus generates energy because carbon has a lower mass per nuclear particle than helium—which means that some mass disappears and becomes energy in this fusion reaction. In fact, the decrease in mass per nuclear particle from hydrogen to helium to carbon is part of a general trend shown in Figure 17.20.

The mass per nuclear particle tends to decrease as we go from light elements to iron, which means that fusion of light nuclei into heavier nuclei generates energy. This trend reverses beyond iron: The mass per nuclear particle tends to *increase* as we look to still heavier elements. As a result, elements heavier than iron can generate nuclear energy only through fission into lighter elements. For example, uranium has a greater mass per nuclear particle than lead, so uranium fission (which ultimately leaves lead as a by-product) must convert some mass into energy.

Iron has the lowest mass per nuclear particle of all nuclei and therefore cannot release energy by either fusion or fission. Thus, once the matter in a stellar core turns to iron, it can generate no further thermal energy or pressure. The iron core's only hope of resisting the crush of gravity lies with degeneracy pressure, but iron keeps piling up in the core until degeneracy pressure can no longer support it either. What ensues is the ultimate nuclear-waste catastrophe.

How would the universe be different if hydrogen, rather than iron, had the lowest mass per nuclear particle? Why?

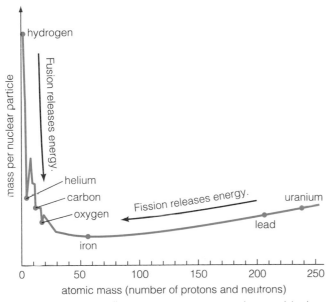

Figure 17.20 Overall, the average mass per nuclear particle declines from hydrogen to iron and then increases. Selected nuclei are labeled to provide reference points. (This graph shows the most general trends only. A more detailed graph would show numerous up-and-down bumps superimposed on the general trends.)

Supernova

The degeneracy pressure that briefly supports the inert iron core arises because the laws of quantum mechanics prohibit electrons from getting too close together [Section S4.5]. Once gravity pushes the electrons past the quantum mechanical limit, however, they can no longer exist freely. The electrons disappear by combining with protons to form neutrons, releasing neutrinos in the process (Figure 17.21). The electron degeneracy pressure suddenly vanishes, and gravity has free rein.

In a fraction of a second, an iron core with a mass comparable to that of our Sun and a size larger than that of Earth collapses into a ball of neutrons just a few kilometers across. The collapse halts only because the neutrons have a degeneracy pressure of their own. The entire core then resembles a giant atomic nucleus. If you recall that ordinary atoms are made almost entirely of empty space [Section 4.3] and that almost all their mass is in their nuclei, you'll realize that a giant atomic nucleus must have an astoundingly high density.

The gravitational collapse of the core releases an enormous amount of energy—more than a hundred times what the Sun will radiate over its entire 10-billion-year lifetime! Where does this energy go? It drives the outer layers off into space in a titanic explosion—a **supernova**. The ball of neutrons left behind is called a **neutron star**. In some cases, the remaining mass may be so large that gravity also overcomes neutron degeneracy pressure, and the core continues to collapse until it becomes a *black hole* [Section 18.4].

Theoretical models of supernovae successfully reproduce the observed energy outputs of real supernovae, but the precise mechanism of the explosion is not yet clear. Two general processes could contribute to the explosion. In the first process, neutron degeneracy pressure halts the gravitational collapse, causing the core to rebound slightly and ram into overlying material that is still falling inward. Until recently, most astronomers thought that this *core-bounce* process ejected the star's outer layers. Current models of supernovae, however, suggest that the more important process involves the neutrinos formed when electrons and protons combine to make neutrons. Although these ghostly particles rarely interact with anything [Section 15.3], so many are produced when the core implodes that they drive a shock wave that propels the star's upper layers into space.

The shock wave sends the star's former surface zooming outward at a speed of 10,000 km/s, heating it so that

Figure 17.21 During the final, catastrophic collapse of a high-mass stellar core, electrons and protons combine to form neutrons, accompanied by the release of neutrinos.

it shines with dazzling brilliance. For about a week, a supernova blazes as powerfully as 10 billion Suns, rivaling the luminosity of a moderate-size galaxy. The ejected gases slowly cool and fade in brightness over the next several months, but they continue to expand outward until they eventually mix with other gas in interstellar space. The scattered debris from the supernova carries with it the variety of elements produced in the star's nuclear furnace, as well as additional elements created when some of the neutrons produced during the core collapse slam into other nuclei. Millions or billions of years later, this debris may be incorporated into a new generation of stars.

The Origin of Elements

Before we leave the subject of massive-star life cycles, it's useful to consider the evidence that indicates we actually understand the origin of the elements. We cannot see inside stars, so we cannot directly observe elements being created in the ways we've discussed. However, the signature of nuclear reactions in massive stars is written in the patterns of elemental abundances across the universe.

For example, if massive stars really produce heavy elements (that is, elements heavier than hydrogen and helium) and scatter these elements into space when they die, the total amount of these heavy elements in interstellar gas should gradually increase with time (because additional massive stars have died). We should expect stars born recently to contain a greater proportion of heavy elements than stars born in the distant past because they formed from interstellar gas that contained more heavy elements.

Stellar spectra confirm this prediction: Older stars do indeed contain smaller amounts of heavy elements than younger stars. For very old stars in globular clusters, elements besides hydrogen and helium typically make up as little as 0.1% of the total mass. In contrast, about 2–3% of the mass of young stars that formed in the recent past is in the form of heavy elements.

We gain even more confidence in our model of elemental creation when we compare the abundances of different elements. For example, because helium-capture reactions add two protons (and two neutrons) at a time, we expect nuclei with even numbers of protons to outnumber those with odd numbers of protons that fall between them. Sure enough, even-numbered nuclei such as carbon, oxygen, and neon are relatively abundant (Figure 17.22). Similarly, because elements heavier than iron are made only by rare fusion reactions shortly before and during a supernova, we expect these elements to be extremely rare. Again, this prediction made by our model of nuclear creation is verified by observations.

Supernova Observations

The study of supernovae owes a great debt to astronomers of many different epochs and cultures. Careful scrutiny of the night skies allowed the ancients to identify several supernovae whose remains still adorn the heavens. The most

Figure 17.22 This graph shows the observed relative abundances of elements in the galaxy in comparison to the abundance of hydrogen. For example, the abundance of nitrogen is about 10^{-4}, which means that there are about $10^{-4} = 0.0001$ times as many nitrogen atoms in the galaxy as hydrogen atoms.

famous example concerns the Crab Nebula in the constellation Taurus. The Crab Nebula is a **supernova remnant**—an expanding cloud of debris from a supernova explosion (Figure 17.23).

A spinning neutron star lies at the center of the Crab Nebula, providing evidence that supernovae really do create neutron stars. Photographs taken years apart show that the nebula is growing larger at a rate of several thousand kilometers per second. Calculating backward from its present size, we can trace the nebula's birth to somewhere near A.D. 1100. Thanks to observations made by ancient astronomers, we can be even more precise.

Historical Observations The official history of the Sung Dynasty in China contains a record of a remarkable celestial event:

> In the first year of the period Chih-ho, the fifth moon, the day chi-ch'ou, a guest star appeared approximately several [degrees] southeast of Thien-kuan. After more than a year it gradually became invisible.

This description of the sudden appearance and gradual dimming of a "guest star" matches what we expect for a supernova, and the location "southeast of Thien-kuan" corresponds to the Crab Nebula's location in Taurus. Moreover, the Chinese date described in the excerpt corresponds to July 4, 1054, telling us precisely when the Crab supernova became visible on Earth. Descriptions of this particular supernova also appear in Japanese astronomical writings, in an Arabic medical textbook, and possibly in Native American paintings in the southwestern United

Figure 17.23 The Crab Nebula is the remnant of the supernova observed in A.D. 1054. This photograph was taken with the Very Large Telescope at the European Southern Observatory in Chile.

VIS

Before. The arrow points to the star observed to explode in 1987.

After. The supernova actually appeared as a bright point of light. It appears larger than a point in this photograph only because of overexposure.

Figure 17.24 Before and-after photos of the location of Supernova 1987A.

VIS

States. Curiously, European records do not mention this supernova, even though it would have been clearly visible.

Historical records of supernovae allow us to age-date the remnants we see today to determine the kinds of supernovae that produced them and to assess how frequently stars explode in our region of the Milky Way Galaxy. At least four supernovae have been observed during the past thousand years, appearing as brilliant new stars for a few months in the years 1006, 1054, 1572, and 1604.

The supernova of 1006, the brightest of these four, could be seen during the daytime and cast shadows at night. Supernovae may even have influenced human history. The Chinese were meticulous in recording their observations because they believed that celestial events foretold the future, and they may have acted in accord with such fortunetelling. The 1572 supernova was witnessed by Tycho Brahe and helped convince him and others that the heavens were not as perfect and unchanging as Aristotle had imagined [Section 3.4]. Kepler saw the 1604 supernova at a time when he was struggling to make planetary orbits fit perfect circles. Perhaps this "imperfection" of the heavens helped push him to test elliptical orbits instead.

Modern Observations: Supernova 1987A

No supernova has been seen in our own galaxy since 1604, but today astronomers routinely discover supernovae in other galaxies. The nearest of these extragalactic supernovae, and the only one near enough to be visible to the naked eye, burst into view in 1987. Because it was the first supernova detected that year, it was given the name **Supernova 1987A**.

Supernova 1987A was the explosion of a star in the *Large Magellanic Cloud,* a small galaxy that orbits the Milky Way and is visible only from southern latitudes. The Large Magellanic Cloud is about 150,000 light-years away, so the star really exploded some 150,000 years ago.

As the nearest supernova witnessed in four centuries, Supernova 1987A provided a unique opportunity to study a supernova and its debris in detail. Astronomers from all over the planet traveled to the Southern Hemisphere to observe it, and several orbiting spacecraft added observations in many different wavelengths of light.

Older photographs of the Large Magellanic Cloud allowed astronomers to determine precisely which star had exploded (Figure 17.24). It turned out to be a blue star, not the red supergiant expected when core fusion has ceased. The most likely explanation is that the star's outer layers were unusually thin and warm near the end of its life, changing its appearance from that of a red supergiant to a blue one. The surprising color of the preexplosion star demonstrates that we still have much to learn about supernovae. Reassuringly, most other theoretical predictions of stellar life cycles were well matched by observations of Supernova 1987A.

One of the most remarkable findings from Supernova 1987A was a burst of neutrinos, recorded by neutrino detectors in Japan and Ohio. The neutrino data confirmed that the explosion released most of its energy in the form of neutrinos, suggesting that we are correct in believing that the stellar core undergoes sudden collapse to a ball of neutrons. The capture of neutrinos from Supernova 1987A has spurred scientific interest in building more purposeful "neutrino telescopes." Perhaps these neutrino telescopes will open new fields of astronomical research in the coming decades.

THINK ABOUT IT

When Betelgeuse explodes as a supernova, it will be more than 10 times brighter than the full moon in our sky. If Betelgeuse had exploded a few hundred or a few thousand years ago, do you think it could have had any effect on human history? How do you think our modern society would react if we saw Betelgeuse explode tomorrow?

 Stellar Evolution Tutorial, Lessons 1–3

17.5 The Lives of Close Binary Stars

For the most part, stars in binary systems proceed from birth to death as if they were isolated and alone. The exceptions are close binary stars. Algol, the "demon star" in the constellation Perseus, consists of two stars that orbit each other closely: a $3.7M_{Sun}$ main-sequence star and a $0.8M_{Sun}$ subgiant.

A moment's thought reveals that something quite strange is going on. The stars of a binary system are born at the same time and therefore must both be the same age. We know that more massive stars live shorter lives, and therefore the more massive star must exhaust its core hydrogen and become a subgiant before the less massive star does. How, then, can Algol's less massive star be a subgiant while the more massive star is still burning hydrogen as a main-sequence star?

This so-called *Algol paradox* reveals some of the complications in ordinary stellar life cycles that can arise in close binary systems. The two stars in close binaries are near enough to exert significant tidal forces on each other [Section 5.4]. The gravity of each star attracts the near side of the other star more strongly than it attracts the far side. The stars therefore stretch into football-like shapes rather than remaining spherical. In addition, the stars become *tidally locked* so that they always show the same face to each other, much as the Moon always shows the same face to Earth.

During the time that both stars are main-sequence stars, the tidal forces have little effect on their lives. However, when the more massive star (which exhausts its core hydrogen sooner) begins to expand into a red giant, gas from its outer layers can spill over onto its companion. This **mass exchange** occurs when the giant grows so large that its tidally distorted outer layers succumb to the gravitational attraction of the smaller companion star. The companion then begins to gain mass at the expense of the giant.

The solution to the Algol paradox should now be clear (Figure 17.25). The $0.8M_{Sun}$ subgiant *used to be* much more massive. As the more massive star, it was the first to begin expanding into a red giant. As it expanded, however, so much of its matter spilled over onto its companion that it is now the less massive star.

The future may hold even more interesting events for Algol. The $3.7M_{Sun}$ star is still gaining mass from its subgiant companion. Thus, its life cycle is actually accelerating as its increasing gravity raises its core hydrogen fusion rate. Millions of years from now, it will exhaust its hydrogen and begin to expand into a red giant itself. At that point, it can begin to transfer mass *back* to its companion. Even stranger things can happen in other mass-exchange systems, particularly when one of the stars is a white dwarf or a neutron star. That is a topic for the next chapter.

Summary of Stellar Lives

We have seen that the primary factor determining how a star lives its life is its mass. Low-mass stars live long lives and die in planetary nebulae, leaving behind white dwarfs. High-mass stars live short lives and die in supernovae, leaving behind neutron stars and black holes. Both types of stars are crucial to life. Near the ends of their lives, low-mass stars can become carbon stars, which are the source of most of the carbon in our bodies. High-mass stars produce the vast array of other chemical elements on which life depends. Mass exchange between close binary stars can complicate these basic patterns. Figure 17.26 summarizes the life cycles of stars of different masses.

Algol shortly after its birth. The higher-mass main-sequence star (left) evolved more quickly than its lower-mass companion (right).

Algol at onset of mass transfer. When the more massive star expanded into a red giant, it began losing some of its mass to its main-sequence companion.

Algol today. As a result of the mass transfer, the red giant has shrunk to a subgiant, and the main-sequence star on the right is now the more massive of the two stars.

Figure 17.25 Artist's conception of the development of the Algol close binary system.

THE BIG PICTURE

Putting Chapter 17 Into Context

In this chapter, we answered the question of the origin of elements that we first discussed in Chapter 1. As you look back over this chapter, remember these "big picture" ideas:

- Virtually all elements besides hydrogen and helium were forged in the nuclear furnaces of stars. Carbon can be released from low-mass stars near the ends of their lives (carbon stars), and many other elements are released into space by massive stars in supernova explosions.

- The tug-of-war between gravity and pressure determines how stars behave from the time of their birth in a cloud of molecular gas to their sometimes violent death.

- Low-mass stars like our Sun live long lives and die with the ejection of a planetary nebula, leaving behind a white dwarf.

- High-mass stars live fast and die young, exploding dramatically as supernovae.

- Close binary stars can exchange mass, altering the usual course of stellar evolution.

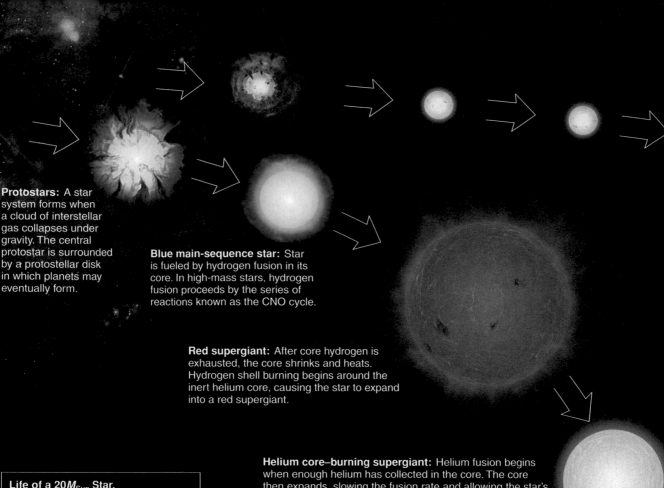

Protostars: A star system forms when a cloud of interstellar gas collapses under gravity. The central protostar is surrounded by a protostellar disk in which planets may eventually form.

Blue main-sequence star: Star is fueled by hydrogen fusion in its core. In high-mass stars, hydrogen fusion proceeds by the series of reactions known as the CNO cycle.

Red supergiant: After core hydrogen is exhausted, the core shrinks and heats. Hydrogen shell burning begins around the inert helium core, causing the star to expand into a red supergiant.

Life of a 20M_{Sun} Star.
Main-sequence lifetime: 8 million years
Duration of later stages: 1 million years

Helium core–burning supergiant: Helium fusion begins when enough helium has collected in the core. The core then expands, slowing the fusion rate and allowing the star's outer layers to shrink somewhat. Hydrogen shell burning continues at a reduced rate.

Multiple shell–burning supergiant: After core helium is exhausted, the core shrinks until carbon fusion begins, while helium and hydrogen continue to burn in shells surrounding the core. Late in its life, the star fuses heavier elements like carbon and oxygen in shells while iron collects in the inert core.

Neutron star: During the core collapse of the supernova, electrons combine with protons to make neutrons. The leftover core is therefore made almost entirely of neutrons.

Supernova: Iron cannot provide fusion energy, so it accumulates in the core until degeneracy pressure can no longer support it. Then the core collapses, leading to the catastrophic explosion of the star.

Yellow main-sequence star: Star is fueled by hydrogen fusion in its core, which converts four hydrogen nuclei into one helium nucleus. In low-mass stars, hydrogen fusion proceeds by the series of reactions known as the proton–proton chain.

Figure 17.26 Summary of stellar lives. The life stages of a high-mass star (on the left) and a low-mass star (on the right) are depicted in clockwise sequences beginning with the proto-stellar stage in the upper left corner. (Stars not drawn to scale.)

Red giant star: After core hydrogen is exhausted, the core shrinks and heats. Hydrogen shell burning begins around the inert helium core, causing the star to expand into a red giant.

Helium core–burning star: Helium fusion, in which three helium nuclei fuse to form a single carbon nucleus, begins when enough helium has collected in the core. The core then expands, slowing the fusion rate and allowing the star's outer layers to shrink somewhat. Hydrogen shell burning continues at a reduced rate.

Life of a 1M_{Sun} Star.
Main-sequence lifetime: 10 billion years
Duration of later stages: 1 billion years

Double shell–burning red giant: After core helium is exhausted, the core again shrinks and heats. Helium shell burning begins around the inert carbon core and the star enters its second red giant phase. Hydrogen shell burning continues.

Planetary nebula: The dying star expels its outer layers in a planetary nebula, leaving behind the exposed inert core.

White dwarf: The remaining white dwarf is made primarily of carbon and oxygen because the core never grew hot enough to fuse these elements into anything heavier.

17.1 Lives in the Balance

- *What kind of pressure opposes the inward pull of gravity during most of a star's life?* Thermal pressure, due to heat produced either by fusion or by gravitational contraction, opposes gravity during most of a star's life.

- *What basic stellar property determines how a star will live and die? Why?* A star's mass determines its fate, because it determines the star's luminosity, its spectral type, and the kind of remnant it leaves behind.

- *How do we categorize stars by mass?* Low-mass stars are those born with mass less than about $2M_{Sun}$. Intermediate-mass stars are those born with mass between about 2 and $8M_{Sun}$. High-mass stars are those born with mass greater than about $8M_{Sun}$.

17.2 Star Birth

- *Where are stars born?* Stars are born in cold, relatively dense molecular clouds—so named because they are cold enough for molecular hydrogen (H_2) to form.

- *What is a protostar?* A protostar is a compact clump of gas formed by gravitational contraction of a molecular cloud fragment. A protostar in the early stages of becoming a star is usually enshrouded in gas and dust. Because angular momentum must be conserved, a contracting protostar is often surrounded by a protostellar disk circling its equator. Outflowing matter from a protostar, in either a protostellar wind or two oppositely directed jets, eventually clears away the shroud of gas and dust.

- *What are the "prebirth" stages of a star's life?* (1) A protostar assembles from a cloud fragment and is bright in infrared light because gravitational contraction rapidly transforms potential energy into thermal energy. (2) Luminosity decreases as gravitational contraction shrinks the protostar's size. (3) Surface temperature rises and luminosity levels off when energy transport switches from convection to radiative diffusion. (4) Core temperature and rate of fusion gradually rise until energy production through fusion balances the rate at which the protostar radiates energy into space. At this point, the forming star becomes a main-sequence star.

- *What is a brown dwarf?* A brown dwarf is a "star" that never grows massive enough for efficient nuclear fusion in its core. Degeneracy pressure halts its gravitational contraction before the core gets hot enough for steady fusion.

17.3 Life as a Low-Mass Star

- *What are the major phases in the life of a low-mass star?* A low-mass star spends most of its life as a main-sequence star, generating energy by fusing hydrogen in its core. Then it becomes a red giant, with a hydrogen shell burning around an inert helium core. Next comes helium core burning, followed by double shell burning of hydrogen and helium shells around an inert carbon core. When the star dies, it ejects a planetary nebula, leaving behind a white dwarf.

- *How did past red giant stars contribute to the existence of life on Earth?* Red giants created and released much of the carbon that exists in the universe, including the carbon that is the basis of organic molecules on Earth.

- *What prevents carbon from fusing to heavier elements in low-mass stars?* Electron degeneracy pressure counteracts the crush of gravity, preventing the core of a low-mass star from ever getting hot enough for carbon fusion.

17.4 Life as a High-Mass Star

- *In what ways do high-mass stars differ from low-mass stars?* High-mass stars live much shorter lives than low-mass stars. High-mass stars fuse hydrogen via the CNO cycle, while low-mass stars fuse hydrogen via the proton–proton chain. High-mass stars die in supernovae, while low-mass stars die in planetary nebulae. Only high-mass stars can fuse elements heavier than carbon. A high-mass star may leave behind a neutron star or a black hole, while a low-mass star leaves behind a white dwarf. High-mass stars are far less common than low-mass stars.

- *How do high-mass stars produce elements heavier than carbon?* Late in their lives, high-mass stars undergo successive episodes of fusion of ever-heavier elements, producing elements as heavy as iron. Elements heavier than iron are produced by these stars when they die in supernovae.

- *What causes a supernova?* As a high-mass star ages, carbon and heavier elements can fuse via helium capture and other processes to form ever-heavier elements. Shells of increasingly heavy element fusion are created in the star's core. However, because fusion of iron uses up energy instead of releasing energy, an iron core cannot support the weight of the outer layers. The collapse of this core—which occurs in a fraction of a second—results in a supernova that nearly obliterates the star (perhaps leaving a black hole or a neutron star).

- *Do supernovae explode near Earth?* At least four supernovae have been observed in the Milky Way Galaxy during the past thousand years: In 1006, 1054, 1572, and 1604. Another supernova called Supernova 1987A was observed to explode in the Large Magellanic Cloud, a companion galaxy to the Milky Way, in 1987.

17.5 The Lives of Close Binary Stars

- *Why are the life stories of close binary stars different from those of single, isolated stars?* The transfer of mass from one star to its companion affects the life history (evolution) of both stars.

- *What is the Algol paradox?* The star Algol is a binary star in which the lower-mass star is in a more advanced stage of life than the higher-mass star. This is a paradox because both stars must have been born at the same time and lower-mass stars should live longer, not shorter, lives. The explanation is that the lower-mass star was once the higher-mass star, but as it grew into a giant it transferred much of its mass to its companion.

❓ Sensible Statements?

Decide whether each of the following statements is sensible and explain why it is or is not.

1. The iron in my blood came from a star that blew up over 4 billion years ago.

2. A protostellar cloud spins faster as it contracts, even though its angular momentum stays the same.

3. When helium fusion begins in the core of a low-mass star, the extra energy generated causes the star's luminosity to rise.

4. Humanity will eventually have to find another planet to live on, because one day the Sun will blow up as a supernova.

5. I sure am glad hydrogen has a higher mass per nuclear particle than many other elements. If it had the lowest mass per nuclear particle, none of us would be here.

6. I just discovered a $3.5M_{Sun}$ main-sequence star orbiting a $2.5M_{Sun}$ red giant. I'll bet that red giant was more massive than $3M_{Sun}$ when it was a main-sequence star.

7. If the Sun had been born as a high-mass star some 4.6 billion years ago, rather than as a low-mass star, the planet Jupiter would probably have Earth-like conditions today, while Earth would be hot like Venus.

8. If you could look inside the Sun today, you'd find that its core contains a much higher proportion of helium and a lower proportion of hydrogen than it did when the Sun was born.

Problems

Homes to Civilization? We do not yet know how many stars have Earth-like planets, nor do we know the likelihood that such planets might harbor advanced civilizations like our own. However, some stars can probably be ruled out as candidates for advanced civilizations. For example, given that it took a few billion years for humans to evolve on Earth, it seems unlikely that advanced life would have had time to evolve around a star that is only a few million years old. For each of the following stars, decide whether you think it is possible that it could harbor an advanced civilization. Explain your reasoning in one or two paragraphs.

9. A $10M_{Sun}$ main-sequence star.

10. A flare star.

11. A carbon star.

12. A $1.5M_{Sun}$ red giant.

13. A $1M_{Sun}$ horizontal branch star.

14. A red supergiant.

15. *Molecular Clouds.* What is a molecular cloud? Briefly describe the process by which a protostar and protostellar disk form from gas in a molecular cloud.

16. *Birth of a Close Binary.* Under what conditions does a close binary form?

17. *Protostellar Winds and Jets.* Describe some of the activity seen in protostars, such as strong protostellar winds and jets.

18. *Life Tracks.* What do we mean by a star's life track on an H–R diagram? How does an H–R diagram that shows life tracks differ from a standard H–R diagram?

19. *Degeneracy Pressure.* What is degeneracy pressure? How does it differ from thermal pressure? Explain why degeneracy pressure can support a stellar core against gravity even when the core becomes very cold.

20. *Hydrogen Shell Burning.* What happens to the core of a star when it exhausts its hydrogen supply? Why does hydrogen shell burning begin around the inert core?

21. *Helium Fusion.* Why does helium fusion require much higher temperatures than hydrogen fusion? Briefly describe the overall reaction by which helium fuses into carbon.

22. *Planetary Nebulae.* What is a planetary nebula? What happens to the core of a star after a planetary nebula occurs?

23. *Fate of the Sun.* Briefly describe how the Sun will change, and how Earth will be affected by these changes, over the next several billion years.

24. *Advanced Nuclear Burning.* Describe some of the nuclear reactions that can occur in high-mass stars after they exhaust their core helium. Why does this continued nuclear burning occur in high-mass stars but not in low-mass stars?

25. *Formation of the Elements.* Summarize some of the observational evidence supporting our ideas about how the elements formed and showing that supernovae really occur.

26. *Rare Elements.* Lithium, beryllium, and boron are elements with atomic numbers 3, 4, and 5, respectively. Despite their being three of the five simplest elements, Figure 17.22 shows that they are rare compared to many heavier elements. Suggest a reason for their rarity. (*Hint:* Consider the process by which helium fuses into carbon.)

27. *Future Skies.* As a red giant, the Sun's angular size in Earth's sky will be about 30°. What will sunset and sunrise be like? About how long will they take? Do you think the color of the sky will be different from what it is today? Explain.

28. *Research: Historical Supernovae.* As discussed in the text, historical accounts exist for supernovae in the years 1006, 1054, 1572, and 1604. Choose one of these supernovae and learn more about historical records of the event. Did the supernova influence human history in any way? Write a two- to three-page summary of your research findings.

Discussion Questions

29. *Connections to the Stars.* In ancient times, many people believed that our lives were somehow influenced by the patterns of the stars in the sky, a belief that survives to this day in astrology. Modern science has not found any evidence to support this belief but instead has found that we have a connection to the stars on a much deeper level: In the words of Carl Sagan, we are "star stuff." Discuss in some detail our real connections to the stars as established by modern astronomy. Do you think these connections have any philosophical implications in terms of how we view our lives and our civilization? Explain.

30. *Humanity in A.D. 5,000,000,000.* Do you think it is likely that humanity will survive until the Sun begins to expand into a red giant 5 billion years from now? Why or why not? If the human race does survive, how do you think people in A.D. 5,000,000,000 will differ from people today? What do you think they will do when faced with the impending death of the Sun? Debate these questions, and see if you and your friends can come to any agreement on possible answers.

For a complete list of media resources available, go to www.astronomyplace.com, and choose Chapter 17 from the pull-down menu.

 Astronomy Place Web Tutorials

Tutorial Review of Key Concepts

Use the interactive **Tutorial** at www.astronomyplace.com to review key concepts from this chapter.

Stellar Evolution Tutorial

Lesson 1 Main-Sequence Lifetimes

Lesson 2 Evolution of a Low-Mass Star

Lesson 3 Late Stages of a High-Mass Star

Supplementary Tutorial Exercises

Use the interactive **Tutorial Lessons** to explore the following questions.

Stellar Evolution Tutorial, Lesson 1

1. Use the tool for calculating stellar lifetimes to estimate the lifetimes of ten stars of different mass. Record the mass and lifetime for each of your ten stars.

2. Make a graph of your results from question 1, plotting mass on the *x*-axis and lifetime on the *y*-axis.

3. Based on your graph from question 2, briefly describe in words how lifetime depends on mass for a main-sequence star.

Stellar Evolution Tutorial, Lessons 2, 3

1. Study the animations for both the low-mass (Lesson 2) and high-mass (Lesson 3) stars. How are the lives of low-mass and high-mass stars similar? How are they different?

2. How do low-mass stars "move" on the H–R diagram as they go through their various stages of life?

3. How do high-mass stars "move" on the H–R diagram as they go through their various stages of life?

Exploring the Sky and Solar System

Of the many activities available on the *Voyager: SkyGazer* **CD-ROM** accompanying your book, use the following files to observe key phenomena covered in this chapter.

Go to the **File: Demo** folder for the following demonstrations:

1. Crab from Finland

Movies

Check out the following narrated and animated short documentaries available on www.astronomyplace.com for a helpful review of key ideas covered in this chapter.

Lives of Stars Movie

Double Stars Movie

Web Projects

Take advantage of the useful web links on www.astronomyplace.com to assist you with the following projects.

1. *Coming Fireworks in Supernova 1987A.* Astronomers believe that the show from Supernova 1987A is not yet over. In particular, sometime between now and about 2010, the expanding cloud of gas from the supernova is expected to ram into surrounding material, and the heat generated by the impact is expected to create a new light show. Learn more about how Supernova 1987A is changing and what we might expect to see from it in the future. Summarize your findings in a one- to two-page report.

2. *Picturing Star Birth and Death.* Photographs of stellar birthplaces (i.e., molecular clouds) and death places (e.g., planetary nebulae and supernova remnants) can be strikingly beautiful, but only a few such photographs are included in this chapter. Search the Web for additional photographs of these types. Look not only for photos taken in visible light, but also for photographs made from observations in other wavelengths of light. Put each photograph you find into a personal on-line journal, along with a one-paragraph description of what the photograph shows. Try to compile a journal of at least 20 such photographs.

18 The Bizarre Stellar Graveyard

Now, my suspicion is that the universe is not only queerer than we suppose, but queerer than we can suppose.

J. B. S. Haldane, Possible Worlds, 1927

Welcome to the afterworld of stars, the fascinating domain of white dwarfs, neutron stars, and black holes. To scientists, these dead stars are ideal laboratories for testing the most extreme predictions of general relativity and quantum theory. To most other people, the eccentric behavior of stellar corpses demonstrates that the universe is stranger than they had ever imagined.

Dead stars behave in unusual and unexpected ways that challenge our minds and stretch the boundaries of what we believe is possible. Stars that have finished nuclear burning have only one hope of staving off the crushing power of gravity: the strange quantum mechanical effect of degeneracy pressure. Even this strange pressure cannot save the most massive stellar cores, which collapse into oblivion as black holes. Prepare to be amazed by the eerie inhabitants of the stellar graveyard!

 Stellar Evolution Tutorial, Lesson 1

18.1 A Star's Final Battle

In the previous chapter, we saw that an ongoing "battle" between the inward crush of gravity and the outward push of pressure governs a star's life from the time when it first begins to form in an interstellar cloud to the time when it finally exhausts its nuclear fuel. Throughout most of this time, the pressure that holds gravity at bay is *thermal pressure,* which results from the heat produced as the star fuses light elements into heavier ones in its core. The nuclear fuel eventually runs out. In the end, after the star dies in a planetary nebula or supernova, the fate of the stellar corpse lies in the outcome of a final battle between gravity and pressure—but this time the source of the pressure is the quantum mechanical effect called *degeneracy pressure.*

We have already discussed how degeneracy pressure successfully resists the crush of gravity in the stellar corpses known as white dwarfs and neutron stars, as well as in several other cases (e.g., brown dwarfs and inert stellar cores). Because these objects are supported by degeneracy pressure, they are known collectively as *degenerate objects,* and the matter within them is called *degenerate matter.*

Degeneracy pressure arises when subatomic particles are packed as closely as the laws of quantum mechanics allow [Section S4.5]. More specifically, white dwarfs are supported against the crush of gravity by **electron degeneracy pressure**, in which the pressure arises from densely packed electrons. In neutron stars, *neutrons* are packed tightly together, thereby generating **neutron degeneracy pressure**.

White dwarfs and neutron stars would be strange enough if the story ended here, but it does not. Sometimes, gravity wins the battle with degeneracy pressure. When this happens, the stellar corpse collapses without end, crushing itself out of existence and forming a *black hole*. A black hole is truly a hole in the universe. If you enter a black hole, you leave our observable universe and can never return.

In this chapter, we will study the bizarre properties and occasional catastrophes of the stellar corpses known as white dwarfs, neutron stars, and black holes.

 Stellar Evolution Tutorial, Lesson 2

18.2 White Dwarfs

A **white dwarf** is the inert core left over after a star has ceased nuclear burning (Figure 18.1), so its composition reflects the products of the star's final nuclear-burning stage. The white dwarf left behind by a $1 M_{Sun}$ star like our Sun will be made mostly of carbon, the product of the star's final helium-burning stage. The cores of very low mass stars never become hot enough to fuse helium and thus end up as helium white dwarfs. Some intermediate-mass stars progress to carbon burning but do not create any iron (and hence do not explode as supernovae). These stars leave behind white dwarfs containing large amounts of oxygen or even heavier elements.

Despite the ordinary-sounding compositions of white dwarfs, their degenerate matter is unlike anything ever seen on Earth. Gravity and electron degeneracy pressure have battled to a draw in white dwarfs. Electron degeneracy pressure finally halts the collapse of a $1 M_{Sun}$ stellar corpse when it shrinks to about the size of Earth. If you recall that Earth is smaller than a typical sunspot, it should be clear that packing the entire mass of the Sun into the volume of Earth is no small feat. The density of such a white dwarf is so high that a pair of standard dice made from its material would weigh about 5 tons.

More massive white dwarfs are actually smaller in size than less massive ones. For example, a $1.3 M_{Sun}$ white dwarf is half the diameter of a $1.0 M_{Sun}$ white dwarf (Figure 18.2). The more massive white dwarf is smaller—even though it contains more matter—because its greater gravity can compress its matter to a much greater density. As a white dwarf becomes denser, its electrons must move faster, and degeneracy pressure becomes stronger [Section S4.5]. The most massive white dwarfs are therefore the smallest. They must be extremely dense for degeneracy pressure to balance their greater gravity.

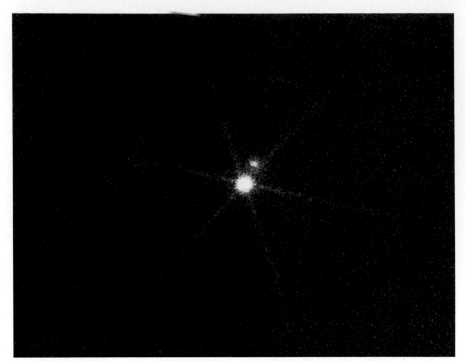

Figure 18.1 The binary star system Sirius as seen in X-ray light by the Chandra Telescope. Sirius A, the brightest star in the night sky to human eyes, is actually the dimmer of the two stars in this picture. Sirius B, its white dwarf companion, is much hotter and therefore much brighter in X-ray light. (The spikes emanating from Sirius B are not real. They are artifacts created by the telescope's optics.)

1.0M_{Sun} white dwarf   1.3M_{Sun} white dwarf

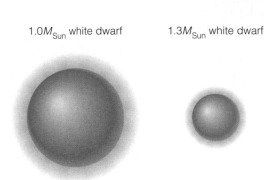

Figure 18.2 Contrary to what you might expect, more massive white dwarfs are actually *smaller* (and thus denser) than less massive white dwarfs.

The fact that adding mass to a white dwarf makes it shrink also explains why red giants become more luminous as they age [Section 17.3]. Degeneracy pressure supports the inert helium core of a low-mass red giant, so this core is essentially a white dwarf buried inside a star. As the hydrogen-burning shell deposits more helium ash onto the degenerate core, the mass of the core continually increases. Therefore, the core continually contracts, and the surrounding shell of hydrogen shrinks along with it. Because the shell gets hotter as it shrinks, its hydrogen fusion rate increases, steadily raising the luminosity of the red giant.

The White Dwarf Limit

Theoretical calculations show that the mass of a white dwarf cannot exceed a **white dwarf limit** of about 1.4M_{Sun}, commonly called the *Chandrasekhar limit* after its discoverer. The limit comes about because the electron speeds

are higher in more massive white dwarfs, approaching the speed of light in white dwarfs with masses near 1.4M_{Sun}. Neither electrons nor anything else can travel faster than the speed of light. Thus, electrons can do nothing to halt the crush of gravity in a stellar corpse with a mass above the white dwarf limit. Such an object must inevitably collapse into a compact ball of neutrons, at which point *neutron degeneracy pressure* can stop the crush of gravity.

Strong observational evidence supports this theoretical limit on the mass of a white dwarf. Many known white dwarfs are members of binary systems, and hence we can measure their masses [Section 16.4]. In every observed case, the white dwarfs have masses below 1.4M_{Sun}, just as we would expect from theory.

White Dwarfs in Close Binary Systems

Left to itself, a white dwarf will never succumb to the crush of gravity because its electron degeneracy pressure does not lessen even as the white dwarf cools into a cold black dwarf. However, white dwarfs in close binary systems do not necessarily rest in peace.

A white dwarf in a close binary system can gain substantial quantities of mass if its companion is a main-sequence or giant star. When a clump of mass first spills over from the companion to the white dwarf, it has some small orbital velocity. The law of conservation of angular momentum dictates that the clump must orbit faster and faster as it falls toward the white dwarf's surface. The infalling matter therefore forms a whirlpool-like disk around the white dwarf (Figure 18.3). Because the process in which material falls onto another body is called *accretion* [Section 9.3], this rapidly rotating disk is called an **accretion disk**.

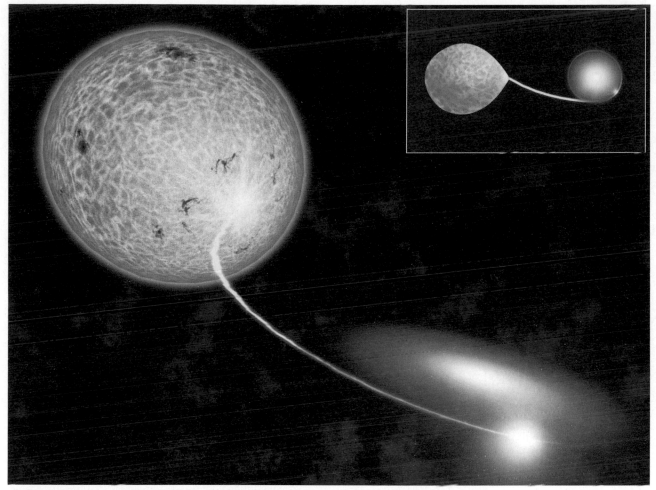

Figure 18.3 This artist's conception shows how mass spilling from a companion star (left) toward a white dwarf (right) forms an accretion disk around the white dwarf. The white dwarf itself is in the center of the accretion disk—too small to be seen on this scale. Matter streaming onto the disk creates a hot spot at the point of impact. The inset shows how the system looks from above the pole, which you can relate to the perspective shown in the painting.

THINK ABOUT IT

Explain why the infalling matter forms a *disk* around the white dwarf (as opposed to, say, a spherical distribution of matter). (*Hint:* See Sections 9.2 and 17.2.) How is an accretion disk similar to a disk in which planets form? How is it different?

Accreting gas gradually spirals inward through the accretion disk and eventually falls onto the surface of the white dwarf. This happens because the individual gas particles in the accretion disk obey Kepler's laws, just like anything else that orbits a massive body [Section 3.4]. Gas in the inner parts of the accretion disk moves faster than gas in the outer parts. Because of these differences in orbital speed, gas in any particular part of the accretion disk "rubs" against slower-moving gas just outside of it. This "rubbing" generates friction and heat in the same way that rubbing your palms together makes them warm. The friction slowly causes the orbits of individual gas particles to decay, making the orbits smaller and smaller until they fall onto the white dwarf surface. Meanwhile, additional gas spilling over from the companion constantly replenishes the accretion disk.

Accretion can provide the "dead" white dwarf with a new energy source. Theory predicts that the heat generated by friction should make the accretion disk hot enough to radiate optical and ultraviolet light, and sometimes even X rays. Thus, although accretion disks are far too small to be seen directly, we should be able to detect their intense ultraviolet or X-ray radiation.

Searches for this radiation have turned up strong evidence for accretion disks around many white dwarfs. In some cases, the brightness of these systems is highly variable—sudden increases in brightness by a factor of 10 or more may persist for a few days and then fade away, only to repeat a few weeks or months later. Such brightening probably arises when instabilities in the accretion disk cause some of the matter to fall suddenly onto the white dwarf surface, with an accompanying release of gravitational potential energy. (This type of brightening is sometimes called a *dwarf nova.*)

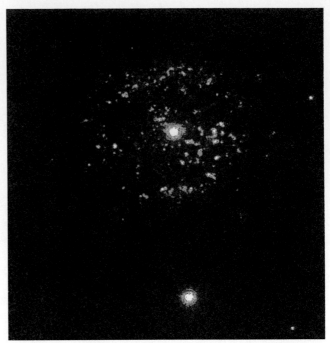

white dwarf   companion star

Hydrogen-rich gas spills into an accretion disk and forms a shell of hydrogen on the white dwarf.

Nova occurs when the shell becomes hot enough for a burst of hydrogen fusion.

a Diagram of the nova process.

b Hubble Space Telescope image showing blobs of gas ejected from the nova T Pyxidis. The bright spot at the center of the blobs is the binary star system that generated the nova.

VIS

Figure 18.4 A nova occurs when hydrogen fusion ignites on the surface of a white dwarf in a binary star system.

Accreting white dwarfs occasionally flare up even more dramatically. Remember that a white dwarf is "dead" because it has no hydrogen left to fuse. This situation changes for an accreting white dwarf. The gas spilling onto an accreting white dwarf comes from the upper layers of its companion star and thus is composed mostly of hydrogen. A thin shell of fresh hydrogen builds up on the surface of the white dwarf as more and more material rains down from the accretion disk. The pressure and temperature at the bottom of this shell increase as the shell grows, and hydrogen fusion ignites when the temperature exceeds 10 million K.

The white dwarf suddenly blazes back to life as its hydrogen shell burns. This thermonuclear flash causes the binary system to shine for a few glorious weeks as a **nova** (Figure 18.4a). A nova can radiate as brightly as 100,000 Suns. The nova generates heat and pressure, ejecting most of the material that has accreted onto the white dwarf. This material expands outward, creating a *nova remnant* that sometimes remains visible years after the nova explosion (Figure 18.4b).

Accretion resumes after a nova explosion subsides, so the entire process can repeat itself. The time between successive novae in a particular system depends on the rate at which hydrogen accretes on the white dwarf surface and on how highly compressed this hydrogen becomes. The compression of hydrogen is greatest for the most massive white

dwarfs, which have the strongest surface gravities. In some cases, novae have been observed to repeat after just a few decades. More commonly, accreting white dwarfs may have 10,000 years between nova outbursts.

Thus, according to our modern definitions, a nova and a supernova are quite different events: A nova is a relatively minor detonation of hydrogen fusion on the surface of a white dwarf in a close binary, while a supernova is the total explosion of a star. However, the word *nova* simply means "new." Historically a nova referred to any star that appeared to the naked eye where none was visible before. Because supernovae generate far more power than novae—the light of 10 billion Suns in a supernova versus that of 100,000 Suns in a nova—a very distant supernova can appear as bright in our sky as a nova that is relatively close. Thus, people could not distinguish between novae and supernovae prior to modern times.

White Dwarf Supernovae

Each time a nova occurs, the white dwarf ejects some of its mass. Each time a nova subsides, the white dwarf begins to accrete matter again. Theoretical models cannot yet tell us whether the net result is a gradual increase or decrease in the white dwarf's mass. Nevertheless, in at least some cases, accreting white dwarfs in binary systems continue to gain mass as time passes. If such a white dwarf gains enough mass, it can one day approach the $1.4M_{Sun}$ white dwarf limit. This day is the white dwarf's last.

As the white dwarf's mass approaches $1.4M_{Sun}$, its interior temperature rises high enough for carbon fusion to ignite. Because the white dwarf material is degenerate, carbon fusion ignites almost instantly throughout the star. This "carbon bomb" detonation is similar to the helium flash that occurs in low-mass red giants, but it releases far more energy [Section 17.3]. The white dwarf explodes completely in what we will call a **white dwarf supernova**.

THINK ABOUT IT

Why do we expect that most collapsing white dwarfs undergo *carbon* fusion, as opposed to fusion of hydrogen, helium, or some other element?

A white dwarf supernova shines as brilliantly as a supernova that occurs at the end of a high-mass star's life [Section 17.4]. To distinguish the two types, we'll refer to the latter as a **massive star supernova**. Both white dwarf supernovae and massive star supernovae result in the destruction of a star. A massive star supernova is thought inevitably to leave behind either a neutron star or a black hole. Nothing remains after the "carbon bomb" detonation of a white dwarf supernova.

Astronomers can distinguish between white dwarf and massive star supernovae by studying their light.* Because white dwarfs contain almost no hydrogen, the spectra of white dwarf supernovae show no spectral lines of hydrogen. In contrast, massive stars usually have plenty of hydrogen in their outer layers at the time they explode, so hydrogen lines are prominent in the spectra of most massive star supernovae.

A second way to distinguish the two types of supernovae is to plot *light curves* that show how their luminosity fades with time. Figure 18.5 contrasts typical light curves for white dwarf and massive star supernovae. Both types of supernovae reach a peak luminosity of about 10 billion Suns ($10^{10}L_{Sun}$), but white dwarf supernovae fade steadily while massive star supernovae fade in two distinct stages.

Not only are white dwarf supernovae dramatic, but they also provide one of the primary means by which we measure large distances in the universe. Because white dwarf supernovae always occur in white dwarfs that have just reached the $1.4M_{Sun}$ limit, their light curves all look amazingly similar, and their maximum luminosities are nearly identical. (The same is not true of massive star supernovae,

---

*Observationally, astronomers classify supernovae as *Type II* if their spectra show hydrogen lines, and *Type I* otherwise. All Type II supernovae are assumed to be massive star supernovae. However, a Type I supernova can be either a white dwarf supernova or a massive star supernova in which the star blew away all its hydrogen before exploding. Type I supernovae appear in three classes whose light curves differ, called *Type Ia*, *Type Ib*, and *Type Ic*. Only Type Ia supernovae are thought to be white dwarf supernovae.

Figure 18.5 Typical light curves for white dwarf and massive star supernovae.

because they involve stars of many different masses.) The fact that all white dwarf supernovae are nearly identical is extremely useful. Once we know the true luminosity of one white dwarf supernova, we essentially know the luminosities of them all. Thus, whenever we discover a white dwarf supernova in a distant galaxy, we can determine the galaxy's distance by using the luminosity–distance formula [Section 16.2].

 Stellar Evolution Tutorial, Lesson 3

18.3 Neutron Stars

White dwarfs with densities of 5 tons per teaspoon may seem incredible, but neutron stars are stranger still. A **neutron star** is the ball of neutrons created by the collapse of the iron core in a massive star supernova (Figure 18.6). Typically just 10 kilometers across yet more massive than the Sun, neutron stars are essentially giant atomic nuclei, with two important differences: (1) They are made almost entirely of neutrons, and (2) gravity, not the strong force, is what binds them together.

The force of gravity at the surface of a neutron star is truly awe-inspiring. Escape velocity is about half the speed of light. The strong gravity causes photons to emerge with a *gravitational redshift* that increases their wavelengths by about 30%, just as predicted by Einstein's general theory of relativity [Section S3.5]. If you foolishly chose to visit a neutron star's surface, your body would be squashed immediately into a microscopically thin puddle of subatomic particles.

Things would be only slightly less troubling if a bit of neutron star came to visit you. A paper clip made from neutron star material would outweigh Mount Everest. If such a paper clip magically appeared in your hand, you could not prevent it from falling. Down it would plunge, passing through the Earth like a rock falling through air. It would gain speed until it reached the center of the Earth, and its momentum would carry it onward until it slowed to a stop on the other side of our planet. Then it would fall back

Figure 18.6 This X-ray image from the Chandra X-Ray Observatory shows the supernova remnant G11.2-03, the remains of a supernova observed by Chinese astronomers in A.D. 386. The white dot at the center is the neutron star left behind by the supernova. The different colors correspond to emission of X rays in different wavelength bands. The region pictured is about 23 light-years across.

down again. If it came in from space, each plunge of the neutron star material would drill a different hole through the rotating Earth. In the words of Carl Sagan, the inside of the Earth would "look briefly like Swiss cheese" (until the melted rock flowed to fill in the holes) by the time friction finally brought the piece of neutron star to rest at the center of the Earth.

In the unfortunate event that an *entire* neutron star came to visit you, it would not fall at all. Because it would be only about 10 kilometers across, the neutron star would probably fit in your hometown. Remember, however, that it would be 300,000 times more massive than Earth. As a result, the neutron star's immense surface gravity would quickly destroy your hometown and the rest of civilization. By the time the dust settled, the former Earth would be a shell no thicker than your thumb on the surface of the neutron star.

Pulsars

Theorists first speculated about neutron stars in the 1930s. However, until observational proof of their existence arrived in 1967, most astronomers assumed that neutron stars were too strange to exist. The proof came largely through the efforts of Jocelyn Bell, then a 24-year-old graduate student at Cambridge University.

Bell had helped her adviser, Anthony Hewish, build a radio telescope ideal for discovering fluctuating sources of radio waves. She was busily trying to interpret the flood of data pouring out of this instrument in October 1967 when she noticed a peculiar signal. After ruling out other possibilities, she concluded that *pulses* of radio waves were arriving from somewhere near the direction of the con-

stellation Cygnus at precise 1.337301-second intervals (Figure 18.7).

No known astronomical object pulsated so regularly. In fact, the pulsations came at such precise intervals that they were nearly as reliable for measuring time as the most precise human-made clocks. For a while, the mysterious source of the radio waves was dubbed "LGM" for Little Green Men—only half-jokingly. Today we refer to such rapidly pulsing radio sources as **pulsars**.

The mystery of pulsars was soon resolved. By the end of 1968, astronomers had found two smoking guns: Pulsars sat at the centers of both the Crab Nebula and the Vela Nebula, the gaseous remnants of supernovae (Figure 18.8). The pulsars are neutron stars left behind by the supernova explosions.

The pulsations arise because the neutron star is spinning rapidly as a result of the conservation of angular momentum: As an iron core collapses into a neutron star, its rotation rate must increase as it shrinks in size. The collapse also bunches the magnetic field lines running through the core far more tightly, greatly amplifying the strength of the magnetic field. Shortly after the supernova event,

Figure 18.7 About 20 seconds of data from the first pulsar discovered by Jocelyn Bell in 1967. Arrows mark the pulses, which come precisely 1.337301 seconds apart.

Figure 18.8 This time-lapse image of the pulsar at the center of the Crab Nebula shows its main pulse recurring every 0.033 second. The fainter pulses are thought to come from the pulsar's other lighthouselike beam. (Photo from the Very Large Telescope of the European Southern Observatory.)

the magnetic field of the remaining neutron star is a trillion times stronger than Earth's.

These intense magnetic fields somehow direct beams of radiation out along the magnetic poles, although we do not yet know exactly how the radiation is generated. If a neutron star's magnetic poles do not align with its rotation axis (just as Earth's magnetic poles do not coincide with its geographic poles), the beams of radiation sweep round and round (Figure 18.9). Like lighthouses, these neutron stars actually emit a fairly steady beam of light, which we see as a pulse of light each time the beam sweeps past Earth.

Pulsars are not quite perfect clocks, because each revolution of a pulsar takes slightly longer. This gradual slowing

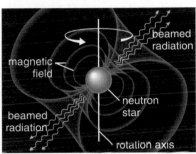

a A pulsar is a rapidly rotating neutron star that beams radiation along its magnetic axis.

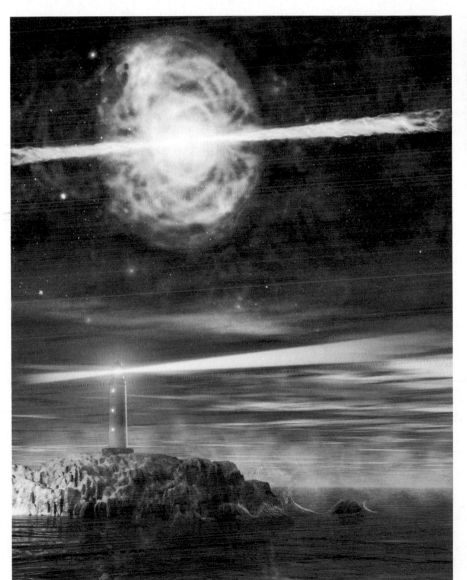

b This artwork likens a pulsar (top) to a lighthouse (bottom). If a pulsar's radiation beams are not aligned with its rotation axis, they will sweep through space. Each time one of these beams sweeps across Earth, we see a pulse of radiation.

Figure 18.9 Radiation from a neutron star can appear to pulse if the neutron star is rotating.

of the neutron star's spin occurs because the continual twirling of the magnetic field generates electromagnetic radiation that carries away energy and angular momentum. The pulsar in the Crab Nebula, for example, currently spins about 30 times per second. Two thousand years from now, it will spin less than half as fast. Eventually, a pulsar's spin slows so much and its magnetic field becomes so weak that we can no longer detect it. In addition, some spinning neutron stars may be oriented so that their beams do not sweep past our location. Thus, we have the following rule: All pulsars are neutron stars, but not all neutron stars are pulsars.

Suppose we do *not* see pulses from a particular neutron star and hence do not call it a pulsar. Is it possible that a civilization living in some other star system would see this neutron star as a pulsar? Explain.

We know that pulsars must be neutron stars because no other massive object could spin so fast. A white dwarf, for example, can spin no faster than about once per second. An increase in spin would tear it apart because its surface would be rotating faster than the escape velocity. Pulsars have been discovered that rotate as fast as 625 times per second. Only an object as extremely small and dense as a neutron star could spin so fast without breaking apart.

Interestingly, the first planets discovered outside our solar system orbit a pulsar. In 1992, the pulsar known as PSR B1257+12 was found to have a pulsation period that periodically becomes very slightly shorter or longer. A detailed analysis of the pulsation periods suggests that they can be explained by gravitational tugs of planetlike bodies on the pulsar. In fact, at least two planets appear to be orbiting PSR B1257+12, with orbital periods of about 67 and 98 days, respectively.

This discovery came as an immense surprise to astronomers, because these planets are orbiting the remains of a star that exploded. Why weren't the planets destroyed during their star's red supergiant stage? Why didn't they fly off into space when the supernova ejected most of the star's mass? Why didn't the supernova explosion kick the pulsar away from its planets? The answer is that the planets probably formed *after* the explosion. It's likely that the pulsar once had a stellar companion that came too close. Tidal forces would have ripped the companion star apart, with the debris forming a protoplanetary disk that then coalesced into planets.

Neutron Stars in Close Binary Systems

Like their white dwarf brethren, neutron stars in close binary systems can brilliantly burst back to life. As is the case with white dwarfs, gas overflowing from a companion star can create a hot, swirling accretion disk around the neutron star. However, in the neutron star's mighty gravitational field, infalling matter releases an amazing amount of gravitational potential energy. Dropping a brick onto a neutron star would liberate as much energy as an atomic bomb.

Because the gravitational energies of accretion disks around neutron stars are so tremendous, they are much hotter and much more luminous than those around white dwarfs. The high temperatures in the inner regions of the accretion disk make it radiate powerfully in X rays. Some close binaries with neutron stars emit 100,000 times more energy in X rays than our Sun emits in all wavelengths of light combined. Due to this intense X-ray emission, close binaries that contain accreting neutron stars are often called **X-ray binaries**. Their existence confirms many of the strange properties of neutron stars.

Today we know of hundreds of X-ray binaries in the Milky Way. These neutron star systems are concentrated in the disk, just like most of our galaxy's stars, gas, and dust. The emission from most X-ray binaries pulsates rapidly, confirming the identification of the X-ray source as a neutron star. In contrast to radio pulsars, whose spins slow with time, the pulsation rates of X-ray binaries tend to accelerate. Matter accreting onto the neutron star adds angular momentum, speeding up the neutron star's rotation (Figure 18.10). By the time the companion star finally stops overflowing, the neutron star may be rotating hundreds of times per second. Because such pulsars spin every few thousandths of a second, they are sometimes called **millisecond pulsars**.

One of the strangest millisecond pulsars is known as the "black widow" because it appears to be destroying its companion. The companion star must once have had a reasonable mass, but almost all of it has spilled over onto the black widow pulsar. The result is that the pulsar now

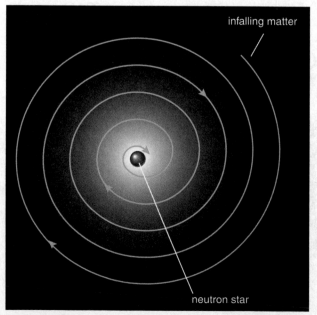

Figure 18.10 Matter accreting onto a neutron star adds angular momentum, increasing the neutron star's rate of spin.

spins every 1.6 milliseconds, while the companion orbits it every 9 hours. The companion also eclipses the pulsar for about an hour of each orbit. Such a long eclipse means that the companion must still have an extended atmosphere despite the fact that its mass is a mere $0.02M_{Sun}$. Evidently, the pulsar's energy output gradually evaporates matter from the surface of its companion. In a few million years, the companion star will have vanished entirely, leaving behind a solitary millisecond pulsar.

Like accreting white dwarfs that occasionally erupt into novae, accreting neutron stars sporadically erupt with enormous luminosities. Hydrogen-rich material from the companion star builds up on the surface of the neutron star. Pressures at the bottom of this accreted hydrogen shell, only a meter thick, maintain steady fusion. Helium accumulates beneath the hydrogen-burning shell, and helium fusion suddenly ignites when the temperature builds to 100 million K. The helium burns rapidly to carbon and heavier elements, generating a burst of energy that flows from the neutron star in the form of X rays. These **X-ray bursters** typically flare every few hours to every few days. Each burst lasts only a few seconds, but during that short time the system radiates power equivalent to 100,000 Suns, all in X rays. Within a minute after a burst, the X-ray burster cools back down and resumes accreting.

 Stellar Evolution Tutorial, Lessons 1, 2

18.4 Black Holes: Gravity's Ultimate Victory

We know that white dwarfs cannot exceed $1.4M_{Sun}$ because gravity overcomes electron degeneracy pressure above that mass. The mass of a neutron star has a similar limit. The precise *neutron star limit* is not known, but it certainly lies below $3M_{Sun}$. A collapsing stellar core that weighs more than $3M_{Sun}$ faces the ultimate oblivion: becoming a **black hole**.

A black hole originates in the collapse of the iron core that forms just prior to the supernova of a very high mass star. Any star born with more than about $8M_{Sun}$ dies in a supernova, but most of the star's mass is blown into space by the explosion. As a result, the collapsed cores left behind by most supernovae are neutron stars. However, theoretical models show that the most massive stars might not succeed in blowing away all their upper layers. If enough matter falls back onto the core, raising its mass above the neutron star limit, then neutron degeneracy pressure will not be able to fend off gravity.

A core whose mass exceeds the neutron star limit will continue to collapse catastrophically. Then, when the core has only a fraction of a second left, gravity plays its cruelest trick. Usually, the gravitational potential energy released as a star collapses boosts its temperature and pressure enough to fight off gravity. However, in a star destined to become a black hole, the enhanced temperature and pressure just make gravity stronger.

According to Einstein's theory of relativity, energy is equivalent to mass ($E = mc^2$) and thus must exert some gravitational attraction. The gravity of pure energy usually is negligible, but not in a stellar core collapsing beyond the neutron star limit. Here the energy associated with the temperature and pressure concentrated in the tiny core acts like additional mass, hastening the collapse. To the best of our current understanding, *nothing* can halt the crush of gravity. The core collapses without end, forming a black hole. Gravity has achieved its ultimate triumph.

A Hole in the Universe

The idea of a black hole was first suggested in the late 1700s by British philosopher John Mitchell and French physicist Pierre Laplace. It was already known from Newton's laws that the escape velocity from any object depends only on its mass and size. Making an object of a particular mass more compact raises its escape velocity [Section 5.5]. Mitchell and Laplace speculated about objects so compact that their escape velocity exceeded the speed of light. Because they worked in a time before it was known that light always travels at the same speed, they assumed that light emitted from such an object would behave like a rock thrown upward, slowing to a stop and falling back down.

Einstein's work showed that black holes are considerably more bizarre. He found that space and time are not distinct, as we usually think of them, but instead are bound up together as four-dimensional **spacetime** [Section S3.3]. Moreover, in his general theory of relativity, Einstein showed that what we perceive as gravity arises from *curvature of spacetime*.

The concept "curvature of spacetime" is quite strange, because we cannot even visualize four dimensions, let alone visualize their curvature. However, as discussed in more detail in Chapter S3, we can draw an analogy to spacetime curvature with a "rubber sheet" diagram in which we show how different masses would affect a stretched rubber sheet. In this analogy, a black hole is a region in which spacetime is stretched so far that it becomes a bottomless pit (Figure 18.11). Keep in mind that a black hole is not really shaped like a funnel. The illustration is only a two-dimensional analogy. Black holes are actually spherical, and they are black because not even light can escape from them.

A black hole really is a *hole* in the observable universe. The boundary between the inside of the black hole and the universe outside is called the **event horizon**. Within the event horizon, the escape velocity exceeds the speed of light, so nothing—not even light—can ever get out. The event horizon gets its name because information can never reach us from events that occur within it. According to general relativity, light always follows the *straightest possible path* through spacetime. If space happens to be curved, as it is near a black hole (or any other massive object), then the path of a light beam will also be curved, and no such path exits the event horizon.

a A two-dimensional representation of "flat" spacetime. The circumference of each circle is 2π times its radius.

Figure 18.11 Spacetime is strongly curved near a black hole.

b A two-dimensional representation of the "curved" spacetime around a black hole. The black hole's mass distorts spacetime, making the radial distance between two circles larger than it would be in "flat" spacetime.

As the rubber-sheet analogy in Figure 18.11b shows, it is possible to draw a series of concentric circles around a black hole. However, it is not really possible to measure a *radius* for these circles. Because of the extreme curvature of spacetime near a black hole, their centers are within the event horizon and hence are not part of our observable universe. Thus, we define the radius of a circle around a black hole as the radius it *would* have if geometry were flat (Euclidean), as shown in Figure 18.11a. (That is, radius = circumference ÷ 2π.)

The radius of the event horizon is known as the **Schwarzschild radius**, named for Karl Schwarzschild (1873–1916), who computed it from Einstein's general theory of relativity in 1916. The Schwarzschild radius of a black hole depends only on its mass: The larger the mass, the larger

the Schwarzschild radius. Schwarzschild computed his famous radius only a month after Einstein published his theory. He did the work while serving in the German Army on the Russian front during World War I. Sadly, he died less than a year later of an illness contracted during the war.

There is no way to tell what has fallen into a black hole in the past. A black hole that forms from the collapse of a stellar core has the mass of that core, but no recognizable material remains. Because stellar cores normally rotate, conservation of angular momentum dictates that black holes should be rotating rapidly when they form. Aside from mass and angular momentum, the only other measurable property of a black hole is its electrical charge. Most black holes are probably electrically neutral. Any other informa-

---

Mathematical Insight **18.1** **The Schwarzschild Radius**

The Schwarzschild radius (R_S) of a black hole is given by a simple formula:

$$\text{Schwarzschild radius} = R_S = \frac{2GM}{c^2}$$

where M is the black hole's mass, $G = 6.67 \times 10^{-11}$ m³/(kg × s²) is the gravitational constant, and $c = 3 \times 10^8$ m/s is the speed of light. With a bit of calculation, this formula can also be written as:

$$\text{Schwarzschild radius} = R_S = 3.0 \times \frac{M}{M_{\text{Sun}}} \text{ km}$$

Example 1: What is the Schwarzschild radius of a black hole with a mass of $10M_{\text{Sun}}$?

Solution: The latter version of the formula is easier to use in this case. We set $M = 10M_{\text{Sun}}$ to find:

$$R_S = 3 \times \frac{10M_{\text{Sun}}}{M_{\text{Sun}}} = 30 \text{ km}$$

The Schwarzschild radius of a $10M_{\text{Sun}}$ black hole is about 30 km.

Example 2: As far as we know, black holes in the present-day universe form only when an object's mass exceeds the roughly $3M_{\text{Sun}}$ neutron star limit. However, Stephen Hawking and others have speculated that much less massive *mini–black holes* might have formed during the Big Bang. Suppose a mini–black hole has the mass of Earth (about 6×10^{24} kg). What is its Schwarzschild radius?

Solution: In this case, the first version of the formula is more convenient:

$$R_S = \frac{2 \times \left(6.67 \times 10^{-11} \dfrac{\text{m}^3}{\text{kg} \times \text{s}^2}\right) \times (6 \times 10^{24} \text{ kg})}{\left(3 \times 10^8 \dfrac{\text{m}}{\text{s}}\right)^2}$$

$$\approx 0.009 \text{ m}$$

The mini–black hole would have a Schwarzschild radius of only 9 millimeters, making it small enough to fit on the tip of your finger. Don't try to hold it—it would weigh as much as the entire Earth!

---

tion carried by objects that plunge into a black hole is irrevocably lost from the universe.

Voyage to a Black Hole

Imagine that you are a pioneer of the future, making the first visit to a black hole. You've selected a black hole with a mass of $10 M_{Sun}$ and a Schwarzschild radius of 30 km. As your spaceship approaches the black hole, you fire its engines to put the ship on a circular orbit a few thousand kilometers above the event horizon. This orbit will be perfectly stable—there is no need to worry about getting "sucked in."

Your first task is to test Einstein's general theory of relativity. As discussed in Chapter S3, this theory predicts that time should run more slowly as the force of gravity grows stronger. It also predicts that light coming out of a strong gravitational field should show a redshift, called a *gravitational redshift*, that is due to gravity rather than to the Doppler effect. You test these predictions with the aid of two identical clocks whose numerals glow with blue light. You keep one clock aboard the ship and push the other one, with a small rocket attached, directly toward the black hole (Figure 18.12). The small rocket automatically fires its engines just enough so that the clock falls only gradually toward the event horizon. Sure enough, the clock on the rocket ticks slower as it heads toward the black hole, and its light becomes increasingly redshifted. When the clock reaches a distance of about 10 km above the event horizon, you see it ticking only half as fast as the clock on your spaceship, and its numerals are red instead of blue.

The rocket has to expend fuel rapidly to keep the clock hovering in the strong gravitational field, and it soon runs out of fuel. The clock plunges toward the black hole. From your safe vantage point inside the spaceship, you see the clock ticking slower and slower as it falls. However, you soon need a radio telescope to "see" it, as the light from the clock face shifts from the red part of the visible spectrum,

Black Holes Don't "Suck"

What would happen if our Sun suddenly became a black hole? For some reason, it has become part of our popular culture for most people to believe that Earth and the other planets would inevitably be "sucked in" by the black hole. This is not true. Although the sudden disappearance of light and heat from the Sun would be bad news for life, Earth's orbit would not change.

Newton's law of gravity tells us that the allowed orbits in a gravitational field are ellipses, hyperbolas, and parabolas [Section 5.3]. Note that "sucking" is not on the list! A spaceship would get into trouble only if it came so close to a black hole—within about three times its Schwarzschild radius—that gravity would deviate significantly from Newton's law. Otherwise, a spaceship passing near a black hole would simply swing around it on an ordinary orbit (ellipse, parabola, or hyperbola). In fact, because most black holes are so small—typical Schwarzschild radii are smaller than any star or planet, and smaller even than most asteroids—a black hole is actually one of the most difficult things in the universe to fall into by accident.

through the infrared, and on into the radio. Finally, its light is so far redshifted that no conceivable telescope could detect it. Just as the clock vanishes from view, you see that the time on its face has frozen to a stop.

Curiosity overwhelms the better judgment of one of your colleagues. He hurriedly climbs into a spacesuit, grabs the other clock, resets it, and jumps out of the airlock on a trajectory aimed straight for the black hole. Down he falls, clock in hand. He watches the clock, but because he and the clock are traveling together, its time seems to run normally and its numerals stay blue. From his point of view, time seems to neither speed up nor slow down. In fact, he'd say

Figure 18.12 Time runs slower on the clock nearer to the black hole, and gravitational redshift makes its glowing blue numerals appear red from your spaceship.

that *you* were the one with the strange time, as he would see your time running increasingly fast and your light becoming increasingly blueshifted. When the clock reads, say, 00:30, he and his clock pass through the event horizon. There is no barrier, no wall, no hard surface. The event horizon is a mathematical boundary, not a physical one. From his point of view, the clock keeps ticking. He is inside the event horizon, the first human being ever to leave our universe.

Back on the spaceship, you watch in horror as your overly curious friend plunges to his death. Yet, from your point of view, he will *never* cross the event horizon. You'll see time come to a stop for him and his clock just as he vanishes from view due to the huge gravitational redshift of light. When you return home, you can play a videotape for the judges at your trial, proving that your friend is still a part of our universe. Strange as it may seem, all of this is true according to Einstein's theory. From your point of view, your friend takes *forever* to cross the event horizon. From his point of view, it is but a moment's plunge before he leaves the universe.

The truly sad part of this story is that your friend did not live to experience the crossing of the event horizon. The force of gravity he felt as he approached the black hole grew so quickly that it actually pulled much harder on his feet than on his head, simultaneously stretching him lengthwise and squeezing him from side to side (Figure 18.13). In essence, your friend was stretched in the same way the oceans are stretched by the tides, except that the *tidal force* near the black hole is trillions of times stronger than the tidal force of the Moon on Earth [Section 5.4]. No human could survive it.

Figure 18.13 Tidal forces would be lethal near a black hole formed by the collapse of a star. The black hole would pull more strongly on the astronaut's feet than on his head, stretching him lengthwise and squeezing him from side to side.

If he had thought ahead, your friend might have waited to make his jump until you visited a much larger black hole, like one of the *supermassive black holes* thought to reside in the centers of many galaxies [Section 21.5]. A 1 billion M_{Sun} black hole has a Schwarzschild radius of 3 billion kilometers—about the distance from our Sun to Uranus. Although the gravitational forces at the event horizon of all black holes are equally great, the larger size of the supermassive black hole makes its tidal forces much weaker and hence nonlethal. Your friend could safely plunge through the event horizon.

Again, from your point of view, the crossing would take forever, and you would see time come to a stop for him just as he vanished from sight because of the gravitational redshift. Again, he would experience time running normally and would see time in the outside universe running increasingly fast as he approached the event horizon. Unfortunately, anything he saw would do him little good as he plunged to oblivion inside the black hole.

Singularity and the Limits to Knowledge

The center of a black hole is thought to be a place where gravity crushes all matter to an infinitely tiny and infinitely dense point called a **singularity**. This singularity is the point at which all the mass that created the black hole resides.

We can never know what really happens inside a black hole, because no information can ever emerge from within the event horizon. Nevertheless, we can use Einstein's theory of relativity to predict conditions inside the black hole, as long as we don't try to describe conditions too close to the singularity. The singularity itself is more puzzling, because the equations of modern physics yield conflicting predictions when we try to apply them to an infinitely compressed mass. General relativity predicts that spacetime should grow infinitely curved as it enters the pointlike singularity. Quantum physics predicts that, as a consequence of the uncertainty principle [Section S4.3], spacetime should fluctuate chaotically in regions smaller than 10^{-35} meter across. No current theory adequately accommodates these divergent claims.

We will not fully understand how a singularity behaves until we have developed a quantum theory of gravity that encompasses both general relativity and quantum mechanics. This uncertainty in our current knowledge is a gold mine for science fiction writers, who speculate about using black holes for exotic forms of travel through spacetime [Section S3.6].

Evidence for Black Holes

Have astronomers discovered any black holes yet? We think so, but the evidence is indirect. Black holes emit no light. Because they are impossible to see, we must look for their effects on surrounding matter. Black holes in close binaries should be among the easiest to identify. Gas overflowing a black hole's stellar companion will form a hot, X ray–

emitting accretion disk similar to the disks that circle accreting neutron stars. The X rays can escape because the disk emits them from well outside the event horizon.

We strongly suspect that a few X-ray binaries contain black holes rather than neutron stars. The trick is to tell the difference, because the accretion disk is likely to be just as hot and to emit about as many X rays whether it circles a neutron star or a black hole. However, we can distinguish between the two possibilities if we can measure the mass of the accreting object.

One of the most promising *black hole candidates* is in an X-ray binary called Cygnus X-1 (Figure 18.14). This system contains an extremely bright star with an estimated mass of $18 M_{Sun}$. Based on Doppler shifts of its spectral lines, this star orbits an unseen companion with a mass of about $10 M_{Sun}$. Although there is some uncertainty in these mass estimates, the mass of the invisible accreting object clearly exceeds the $3 M_{Sun}$ neutron star limit. Moreover, careful studies of variations in the system's X-ray emission indicate that this massive accreting object must be very small in size—far too small to be an ordinary star [Section 21.5]. Thus, based on our current knowledge, the accreting object in Cygnus X-1 cannot be anything other than a black hole.

Cygnus

Cygnus X-1

a The location of Cygnus X-1 in the sky.

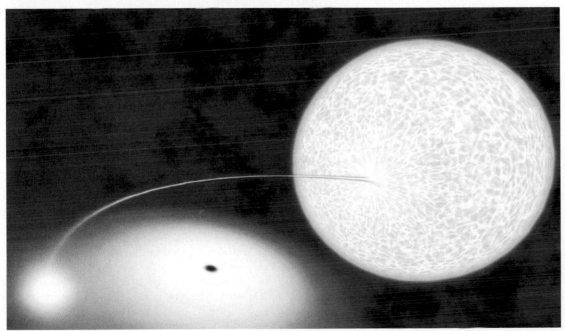

b Artist's conception of the Cygnus X-1 system. The X rays come from the high-temperature gas in the accretion disk surrounding the black hole. (The perspective in this painting is similar to that in Figure 18.3. See the inset in that figure.)

Figure 18.14 Cygnus X-1, a binary system containing a black hole candidate.

As we discussed earlier, some X-ray binaries that contain neutron stars emit frequent X-ray bursts and are called X-ray bursters. Could an X-ray binary that contains a black hole exhibit the same type of X-ray bursts? Why or why not? (*Hint:* What is the source of the X-ray bursts from an X-ray binary with a neutron star, and where is it located? Does a similar location exist for a system containing a black hole?)

Of course, confirming that black holes are real with 100% certainty is very difficult. However, our current theories successfully explain neutron stars, and the general theory of relativity that leads to the idea of black holes is also on solid ground. Unless something is dramatically wrong in our current theories about the mass limit of neutron stars or some other, unknown type of compact object can have a huge mass, black holes must be real.

18.5 The Mystery of Gamma-Ray Bursts

In the early 1960s, the United States began launching a series of top-secret satellites designed to look for gamma rays emitted by nuclear bomb tests. The satellites soon began detecting occasional bursts of gamma rays, typically lasting

---

SPECIAL TOPIC   Too Strange to Be True?

Theoretical calculations predicted the existence of neutron stars and black holes long before their observational discovery, but many astronomers considered these theoretical results too strange to be true.

The story begins with Subrahmanyan Chandrasekhar, an astrophysicist from India. Chandrasekhar was only 19 when he completed the calculations showing that there is a white dwarf limit of $1.4M_{Sun}$, and he boldly predicted that a more massive white dwarf would collapse under the force of gravity. He did this work in 1931 while traveling by ship to England, where

Subrahmanyan Chandrasekhar

he hoped to impress the eminent British astrophysicist Sir Arthur Stanley Eddington. However, Eddington ridiculed Chandrasekhar for believing that white dwarfs could collapse. Neutrons had not yet been discovered, little was known about fusion, and no one had any idea what supernovae were. The idea of gravity achieving an ultimate victory seemed nonsensical to Eddington, who speculated that some type of force must prevent gravity from crushing any object.

Sir Arthur Stanley Eddington

A few more radical thinkers took collapsing stars more seriously. A Russian physicist, Lev Davidovich Landau, independently computed the white dwarf limit in 1932. Neutrons were discovered just a few months later, and Landau speculated that stellar corpses above the white dwarf limit might collapse until neutron degeneracy pressure halted the crush of gravity. While most astronomers found the idea of neutron stars to be unacceptably weird, two European scientists who had emigrated to California, Fritz Zwicky and Walter Baade, were not so skeptical. Without knowing

Lev Davidovich Landau

of Landau's ideas, they also concluded that neutron stars were possible. In 1934, they suggested that a supernova might result when a stellar core collapses and forms a neutron star—an extraordinarily insightful guess. By 1938, physicist Robert Oppenheimer, working at Berkeley, was contemplating whether neutron stars had a limiting mass of their own. He and his coworkers concluded that the answer was yes and that neutron degeneracy pressure could not resist the crush of gravity when the mass rose above a few solar masses. Because no known force could keep such a star from collapsing indefinitely, Oppenheimer speculated that gravity would achieve ultimate victory, crushing the star into a black hole.

Robert Oppenheimer

Astronomers gradually came to accept Chandrasekhar's $1.4M_{Sun}$ white dwarf limit, because observations found no white dwarfs more massive than this. However, most astronomers held to a belief that high-mass stars would inevitably shed enough mass late in life to prevent the formation of a more massive collapsed object. Jocelyn Bell's 1967 discovery of pulsars shattered this belief.

Within a few months, Thomas Gold of Cornell University correctly suggested that the pulsars were spinning neutron stars. These discoveries forced astronomers to admit that nature was far stranger than they had expected. The verification that neutron stars really did exist made the prospect of the still-stranger black holes much less difficult to accept.

Chandrasekhar, who had long since moved to the University of Chicago, was awarded a Nobel Prize in 1984 for his lifelong contributions to astronomy. Landau won a Nobel Prize in 1962 for his work on condensed states of matter. Oppenheimer went on to lead the *Manhattan Project* that developed the atomic bomb in 1945. Eddington died in 1944, still convinced that white dwarf stars could not collapse.

Jocelyn Bell

Figure 18.15 A typical gamma-ray burst light curve, showing dramatic fluctuations in gamma-ray intensity over a period of 2 minutes.

a few seconds (Figure 18.15). It took several years for military scientists to become convinced that these **gamma-ray bursts** were coming from space, not from some sinister human activity. They publicized the discovery in 1973.

An increasingly large armada of satellites launched over the next two decades detected hundreds of other gamma-ray bursts. However, not only was their origin unknown, but it was very difficult to tell what direction the bursts came from. The problem is that gamma rays are very difficult to focus. A detector can record that it has been hit by a gamma ray but can provide little information on the direction of the gamma ray. With no specific proof, most astronomers assumed that gamma-ray bursts, like X-ray bursts, came from explosive events associated with neutron stars in X-ray binaries.

In 1991, NASA launched the *Compton Gamma Ray Observatory*, or *Compton* for short. It carried an array of eight detectors designed expressly to study gamma-ray bursts. By comparing the data recorded by all eight detectors, scientists could determine the direction of a gamma-ray burst within about 1°. The results were stunning. Compton detected gamma-ray bursts at a rate of about one per day and compiled a catalog of more than a thousand gamma-ray bursts within a few years. However, these bursts were *not* concentrated in the disk of the Milky Way like X-ray binaries and thus must not be associated with neutron stars in the disk of our galaxy.

So where do gamma-ray bursts come from? The very even distribution across the sky of gamma-ray bursts discovered by Compton ruled out the possibility that they come from anywhere in the Milky Way Galaxy. If the bursts came from objects distributed spherically about the Milky Way Galaxy, we would see a concentration of them in the direction of the galactic center. (Remember that we are located more than halfway out from the center of our galaxy.) Gamma-ray bursts therefore must originate far outside our own galaxy.

Additional evidence for this conclusion came in 1997, when astronomers first observed the afterglow of a gamma-ray burst in other wavelengths. The higher resolution possible in these other wavelengths allowed astronomers to pinpoint the burst's origin in a distant galaxy. Since then, several other bursts have also been traced to explosions in distant galaxies (Figure 18.16).

We now know that gamma-ray bursts come from very distant explosions, but we still face a great mystery: How can something so distant be so bright? The afterglows of some gamma-ray bursts can be seen with binoculars, even though they are coming from galaxies billions of light-years away—making them by far the most powerful bursts of energy that ever occur in the universe. If these bursts shine their light equally in all directions, like a light bulb, then the total luminosity of a burst can briefly exceed the combined luminosity of a million galaxies like our Milky Way! Because such a high luminosity is very difficult to explain, some scientists speculate that gamma-ray bursts channel their energy into narrow searchlight beams, like pulsars. A burst whose beam is pointed directly at Earth would look unusually bright. Even in this arrangement, the burst's luminosity would still surpass that of many thousands of galaxies like our Milky Way.

Astronomers still do not know what causes such massive outbursts of energy. One hypothesis suggests that the bursts come from the collision of *two* neutron stars in a binary system. As discussed in Chapter S3, the neutron stars in such a system gradually spiral in toward each other (because they lose energy to *gravitational waves* [Section S3.5]) and must eventually be destroyed in some type of catastrophic collision resulting in a black hole. However, why a neutron star collision would produce the peculiar spectra

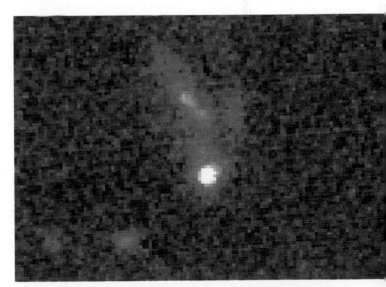

Figure 18.16 The bright dot near the center of this image is the visible-light afterglow of a gamma-ray burst, as seen by the Hubble Space Telescope. The elongated blob extending above the dot is the distant galaxy in which the burst occurred.

VIS

and light curves of gamma-ray bursts is not clear, and many scientists have raised theoretical objections to this model.

Another, more promising, hypothesis is that gamma-ray bursts come from unusually powerful supernovae. An ordinary supernova that forms a neutron star does not release enough energy to power the luminosity of the brightest gamma-ray bursts. However, a supernova that forms a black hole crushes even more matter into an even smaller radius, releasing many times more gravitational potential energy than one that forms a neutron star. This kind of event, sometimes called a *hypernova,* might be powerful enough to explain the most extreme gamma-ray bursts.

Some observational evidence supports this idea. The extremely massive stars that lead to hypernovae are very short-lived and should be found only in places where stars are actively forming. Indeed, all the distant galaxies known to have produced gamma-ray bursts also appear to be forming new stars. In addition, several relatively nearby gamma-ray bursts have been linked to supernova explosions. These observations have inspired many theoretical models that attempt to explain how hypernovae might generate gamma-ray bursts, but none has met with much success. Thus, we are left with one of the greatest mysteries in all of science: Gamma-ray bursts are the most powerful events in the universe, but we don't know how they are generated.

THE BIG PICTURE

Putting Chapter 18 into Context

We have now seen what happens to stars after they die. What a mind-bending experience! Nevertheless, try to keep these "big picture" ideas straight in your head:

- Despite the strange nature of stellar corpses, clear evidence exists for white dwarfs and neutron stars, and the case for black holes is very strong.

- White dwarfs, neutron stars, and black holes can all have close stellar companions from which they accrete matter. These binary systems produce some of the most spectacular events in the universe, including novae, white dwarf supernovae, and X-ray bursters.

- Black holes are truly holes in the observable universe that strongly warp space and time around them. The nature of black hole singularities remains beyond the frontier of current scientific understanding.

- Gamma-ray bursts were once thought to be related to neutron stars in our galaxy, but recent evidence indicates that this idea is wrong. At least some of them come from supernova explosions on distant galaxies.

SUMMARY OF KEY CONCEPTS

18.1 A Star's Final Battle

- *What determines the fate of a stellar core that has exhausted all its nuclear fuel?* A star's final state depends on whether degeneracy pressure can halt the crush of gravity. A white dwarf is supported by electron degeneracy pressure. A neutron star is supported by neutron degeneracy pressure. If neutron degeneracy pressure cannot halt the collapse, the core becomes a black hole.

18.2 White Dwarfs

- *What is a white dwarf?* A white dwarf is the inert core left over from a low-mass star, supported by electron degeneracy pressure.

- *Why can't white dwarfs weigh more than 1.4 times the mass of the Sun?* At masses greater than $1.4M_{Sun}$, the white dwarf can no longer support its own weight with electron degeneracy pressure. The electrons would have to "move" faster than the speed of light, which is physically impossible. Thus, white dwarfs that become more massive than $1.4M_{Sun}$ must collapse.

- *What is a nova?* A white dwarf in a binary system can acquire hydrogen from its companion that swirls toward the surface in an accretion disk. If enough hydrogen rains down on the white dwarf, the surface hydrogen layer will become hot enough to ignite nuclear fusion, essentially making a thermonuclear flash in which the star shines as brightly as 100,000 Suns for a few weeks as a nova.

- *What are white dwarf supernovae, and why are they helpful for measuring gigantic distances?* A white dwarf supernova arises from the explosion of a white dwarf triggered by carbon fusion when it gains enough mass to approach the $1.4M_{Sun}$ limit. The peak luminosity of the explosion exceeds 10 billion times that of the Sun. Because all these supernovae have nearly identical light curves and are so bright that they can be seen across the universe, we measure their distances from their apparent brightness in our sky.

18.3 Neutron Stars

- *What is a neutron star?* A neutron star is the ball of neutrons created by the collapse of the iron core in a massive star supernova. It resembles a giant atomic nucleus 10 kilometers across but is more massive than the Sun.

- *What is a pulsar?* Pulsars are rotating neutron stars with magnetic fields. Hot spots at the magnetic poles of these neutron stars send out beams of radiation. If the magnetic poles do not align with the poles of the rotation, the beamed radiation from the hot spots sweeps through space the way a lighthouse beam swings through the night. If this beam crosses Earth, we see it periodically appear and disappear in pulses.

- *Why do X-ray binaries emit so much X-ray radiation?* Neutron stars in binary systems can accrete hydrogen from their companions, forming dense, hot accretion disks. Because the inner regions of these disks are so hot, they radiate X rays.

- *What is the difference between X-ray bursters and novae?* Novae occur when hydrogen fusion suddenly ignites on the surface of a white dwarf in a binary system. In contrast, hydrogen fusion is steady on the surface of a neutron star in a binary system. However, the steady hydrogen burning builds up a layer of helium beneath the surface shell of hydrogen. Every few days, enough helium can build up for it to ignite suddenly in a burst of helium fusion, which causes an X-ray burst. For about a minute during an X-ray burst, the star produces power equivalent to that of 100,000 Suns.

18.4 Black Holes: Gravity's Ultimate Victory

- *What is the maximum mass of a neutron star?* The maximum mass of a neutron star is about $3M_{Sun}$. It is determined by the maximum amount of mass that neutron degeneracy pressure can support.

- *What is a black hole?* A black hole is a place where gravity has crushed matter into oblivion, creating a true hole in the universe from which nothing can ever escape, not even light.

- *Do black holes suck in objects?* Black holes don't suck in objects at large distances. If our Sun were instantly replaced by a $1M_{Sun}$ black hole, Earth would not be sucked into the black hole. The planets would continue to orbit normally.

- *What property of a black hole determines its "size"?* A black hole's "size" depends on its mass, because the mass determines the size of the black hole's event horizon, the boundary of the region from which not even light can escape.

- *What would you see if you watched someone falling into a black hole?* You'd see time slow down for them as they approached the black hole, and their light would be increasingly redshifted. They would never quite reach the event horizon, though they would soon disappear from view as their light became so redshifted that no instrument could detect it.

- *What observational evidence do we have for the existence of black holes?* We cannot see black holes directly, but we can infer their presence by their influence on their surroundings. For example, the study of X-ray binaries shows that some of these binary systems may have black holes rather than neutron stars. The X-ray behavior of black hole systems differs from that of neutron stars because a neutron star has a surface and a black hole does not. The most definitive evidence comes from measuring the orbit of the companion, from which we can infer the mass of the object with the accretion disk. If that object is more massive than $3M_{Sun}$, it is probably a black hole.

18.5 The Mystery of Gamma-Ray Bursts

- *How are gamma-ray bursts detected?* Because gamma rays from space do not penetrate the atmosphere, they can only be detected from high altitude or space. Gamma ray telescopes aboard satellites have detected gamma-ray bursts coming from every direction in the sky.

- *What might produce gamma-ray bursts?* They occur in distant galaxies, and they must be extremely large explosions—the most powerful bursts of energy since the Big Bang. We do not know their precise cause, although at least some appear to come from unusually powerful supernovae.

? Sensible Statements?

Decide whether each of the following statements is sensible and explain why it is or is not.

1. Most white dwarf stars have masses close to that of our Sun, but a few white dwarf stars are up to three times more massive than the Sun.

2. The radii of white dwarf stars in close binary systems gradually increase as they accrete matter.

3. White dwarf supernovae are useful distance indicators.

4. Before pulsars were discovered, no one knew for sure whether neutron stars existed.

5. If you want to find a pulsar, you might want to look near the remnant of a supernova described by ancient Chinese astronomers.

6. If a black hole 10 times more massive than our Sun were lurking just beyond Pluto's orbit, we'd have no way of knowing it was there.

7. If the Sun suddenly became a $1\,M_{Sun}$ black hole, the orbits of the nine planets would not change at all.

8. We can detect black holes with X-ray telescopes because matter falling into a black hole emits X rays after it smashes into the event horizon.

Problems

Life Stories of Stars. Write a one- to two-page life story for the following scenarios in problems 9 and 10. Each story should be detailed and scientifically correct but also creative. That is, it should be entertaining while at the same time proving that you understand stellar evolution. Be sure to state whether "you" are a member of a binary system.

9. You are a white dwarf of $0.8M_{Sun}$.

10. You are a neutron star of $1.5M_{Sun}$.

11. *Electrons and Neutrons.* What is the difference between electron degeneracy pressure and neutron degeneracy pressure? Which type supports a white dwarf? Why are neutron stars so much smaller in size than white dwarfs?

12. *Sizes of White Dwarfs.* Explain why more massive white dwarfs are smaller in size. How does this idea explain why red giants become more luminous as they age?

13. *Accretion Disks.* What is an accretion disk? Under what conditions does an accretion disk form? Explain how the accretion disk provides a white dwarf with a new source of energy that we can detect from Earth.

14. *Types of Supernovae.* Contrast the process of a white dwarf supernova with that of a massive star supernova. Observationally, how can we distinguish between these two types of supernovae?

15. *Birth of a Black Hole.* Briefly explain the process by which the core of a very high mass star can collapse to form a black hole.

16. *Event Horizon.* What is the event horizon of a black hole? How does it get its name? How is it related to the Schwarzschild radius?

17. *Into a Black Hole.* Suppose you are falling into a black hole. How will you perceive the passage of your own time? How will you perceive the passage of time in the universe around you? Briefly explain why your trip will be lethal if the black hole is relatively small in mass, and why you may survive crossing the event horizon of a supermassive black hole.

18. *Singularity.* What do we mean by the singularity of a black hole? How do we know that our current theories are inadequate to explain what happens at the singularity?

*19. *Neutron Star Density.* A typical neutron star has a mass of about $1.5M_{Sun}$ and a radius of 10 km.

   a. Calculate the average density of a neutron star, in *kilograms per cubic centimeter*.

   b. Compare the mass of 1 cm^3 of neutron star material to the mass of Mount Everest ($\approx 5 \times 10^{10}$ kg).

*20. *A Black Hole?* You've just discovered a new X-ray binary, which we will call *Hyp-X1* ("Hyp" for hypothetical). The system Hyp-X1 contains a bright, B2 main-sequence star orbiting an unseen companion. The separation of the stars is estimated to be 20 million km, and the orbital period of the visible star is 4 days.

   a. Use Newton's version of Kepler's third law to calculate the sum of the masses of the two stars in the system. (*Hint:* See Mathematical Insight 16.4.) Give your answer in both kilograms and solar masses ($M_{Sun} = 2.0 \times 10^{30}$ kg).

   b. Determine the mass of the unseen companion. Is it a neutron star or a black hole? Explain. (*Hint:* A main-sequence star with spectral type B2 has a mass of about $10M_{Sun}$.)

*21. *Schwarzschild Radii.* Calculate the Schwarzschild radius (in km) for each of the following.

   a. A $10^8 M_{Sun}$ black hole in the center of a quasar.

   b. A $5M_{Sun}$ black hole that formed in the supernova of a massive star.

   c. A mini–black hole with the mass of the Moon.

   d. A mini–black hole formed when a superadvanced civilization decides to punish you (unfairly) by squeezing you until you become so small that you disappear inside your own event horizon.

*22. *Challenge Problem: A Neutron Star Comes to Town.* Suppose a neutron star were suddenly to appear in your hometown. How thick a layer would Earth form as it wraps around the neutron star's surface? To make the problem easier, you may assume that the layer formed by Earth has the same average density as the neutron star. (*Hint:* Consider the mass of Earth to be distributed in a spherical *shell* over the surface of the neutron star and then calculate the thickness of such a shell with the same mass as Earth. The volume of a spherical shell is approximately its surface area times its thickness: $V_{shell} \approx 4\pi r^2 \times$ thickness. Because the shell will be thin, you can assume that its radius is the radius of the neutron star.)

Discussion Questions

23. *Too Strange to Be True?* Despite strong theoretical arguments for the existence of neutron stars and black holes, many scientists rejected the possibility that such objects could really exist until they were confronted with very strong observational evidence. Some people claim that this type of scientific skepticism demonstrates an unwillingness on the part of scientists to give up their deeply held scientific beliefs. Others claim that this type of skepticism is necessary for scientific advancement. What do you think? Defend your opinion.

24. *Black Holes in Popular Culture.* Phrases such as "it disappeared into a black hole" are now common in popular culture. Give a few examples in which the term *black hole* is used in popular culture but is not meant to be taken literally. In what ways are these uses correct in their analogies to real black holes? In what ways are they incorrect? Why do you think such an esoteric scientific idea as that of a black hole has so captured the public imagination?

For a complete list of media resources available, go to www.astronomyplace.com, and choose Chapter 18 from the pull-down menu.

Astronomy Place Web Tutorials

Tutorial Review of Key Concepts

Use the interactive **Tutorials** at www.astronomyplace.com to review key concepts from this chapter.

Black Holes Tutorial

Lesson 1 What Are Black Holes?

Lesson 2 The Search for Black Holes

Stellar Evolution Tutorial

Lesson 1 Main-Sequence Lifetimes

Lesson 2 Evolution of a Low-Mass Star

Lesson 3 Late Stages of a High-Mass Star

Supplementary Tutorial Exercises

Use the interactive **Tutorial Lessons** to explore the following questions.

Black Holes Tutorial, Lesson 1

1 What would have to happen to Earth to turn it into a black hole? Could this really happen?

2. Use the tool in the simulation to determine the Schwarzschild radii for ten black holes of different masses. Record the mass and Schwarzschild radius for each black hole.

3. Make a graph of your results from question 2, plotting mass on the *x*-axis and Schwarzschild radius on the *y*-axis.

4. Based on your graph from question 3, briefly describe in words how the Schwarzschild radius depends on mass for a black hole.

Black Holes Tutorial, Lesson 2

1. Briefly explain how scientists can detect a black hole.

2. Why does the search for black holes generally begin with a search for X-ray sources in the sky?

Stellar Evolution Tutorial, Lessons 2, 3

1. What kind of "corpse" is left behind by a low-mass star? Why?

2. What kind of "corpse" is left behind by a high-mass star? Why?

Web Projects

Take advantage of the useful web links on www.astronomyplace.com to assist you with the following projects.

1. *Gamma-Ray Bursts.* Go to the Web site for a mission studying gamma-ray bursts (such as HETE, INTEGRAL, or Swift) and find the latest information about gamma-ray bursts. Write a one- to two-page essay on recent discoveries and how they may shed light on the mystery of gamma-ray bursts.

2. *White Dwarf Supernovae.* Learn more about how astronomers are using white dwarf supernovae to determine distances for distant galaxies. What have these observations taught us about the universe? What have they taught us about white dwarf supernovae?

PART VI

GALAXIES AND BEYOND

19 Our Galaxy

The infinitude of creation is great enough to make a world, or a Milky Way of worlds, look in comparison with it what a flower or an insect does in comparison with the Earth.

Immanuel Kant

In previous chapters, we saw how stars forge new elements and expel them into space. We also studied how interstellar gas clouds enriched with these stellar by-products form new stars and planetary systems. These processes do not occur in isolation. Instead, they are part of a dynamic system that acts throughout our Milky Way Galaxy.

You are probably familiar with the idea that all living species on Earth interact with one another and with the land, water, and air to form a large, interconnected ecosystem. In a similar way, but on a much larger scale, our galaxy is a nearly self-contained system that cycles matter from stars into interstellar space and back into stars again. The birth of our solar system and the evolution of life on Earth would not have been possible without this "galactic ecosystem."

In this chapter, we will study our Milky Way Galaxy. We will investigate the galactic processes that maintain an ongoing cycle of stellar life and death, examine the structure and motion of the galaxy, and explore the mysteries of the galactic center. Through it all, we will see that we are not only "star stuff" but "galaxy stuff"—the product of eons of complex recycling and reprocessing of matter and energy in the Milky Way Galaxy.

19.1 The Milky Way Revealed

On a dark night, you can see a faint band of light slicing across the sky through several constellations, including Sagittarius, Cygnus, Perseus, and Orion. This band of light looked like a flowing ribbon of milk to the ancient Greeks, and we now call it the *Milky Way*. In the early 1600s, Galileo used his telescope to prove that the light of the Milky Way comes from a myriad of individual stars. Together these stars make up the kind of stellar system we call a *galaxy*, echoing the Greek word for "milk," *galactos*.

Today we know that our Milky Way Galaxy holds over 100 billion stars and is just one among tens of billions of galaxies in the observable universe. If we could stand outside our galaxy, we would see it as a flat **disk** of stars with a bright central **bulge**, spectacular **spiral arms**, and a dimmer, rounder **halo** surrounding everything (Figure 19.1). A few hundred **globular clusters** of stars [Section 16.6] circle our galaxy's center in orbits extending tens of thousands of light-years into the halo.

The Milky Way is a relatively large galaxy, so its gravity strongly influences smaller galaxies in its vicinity. Two small galaxies, known as the Large and Small Magellanic Clouds, orbit the Milky Way at distances of some 150,000 light-years (50 kpc).* (A third small galaxy, called the Sagittarius dwarf elliptical, lies even closer but is obscured from view by the Milky Way's galactic plane [Section 21.3].) Although the Magellanic Clouds are relatively small for galaxies (a few billion stars each), they are far larger than globular clusters (which typically contain a few hundred thousand stars). Both Magellanic Clouds are visible to the naked eye from the Southern Hemisphere.

Our knowledge of the Milky Way's true size and shape was long in coming. Clouds of interstellar gas and dust known collectively as the **interstellar medium** fill the galactic disk, obscuring our view when we try to peer directly through it. The dusty, smoglike nature of the interstellar medium hides most of our galaxy from us and long fooled astronomers into believing that we lived near our galaxy's center.

Astronomer Harlow Shapley finally proved otherwise in the 1920s, when he demonstrated that the Milky Way's globular clusters are centered on a point tens of thousands of light-years from our Sun. He concluded that this point, not our Sun, must be the center of the galaxy. We now know that our Sun lies in the outer part of the galactic disk, about 28,000 light-years (8.5 kpc) from its center.

Our own galaxy can be difficult to study not only because it is dusty, but also because we see it from the inside. The Milky Way's dustiness is no longer such a hindrance, because technologies developed in the past few decades allow us to observe the Milky Way's radio and infrared light. These wavelengths penetrate the enshrouding interstellar medium, enabling us to see into regions of the galaxy previously obscured from view. Still, determining our galaxy's structure from our location within the galactic disk is somewhat like trying to draw a picture of your house without ever leaving your bedroom. Just as it is easier to draw pictures of other houses that you can see out the window, it is easier to measure the sizes and shapes of other galaxies than it is to measure our own. Nevertheless, the Milky Way is the only galaxy whose inner workings we can examine up close.

---

*Recall that 1 parsec (pc) ≈ 3.26 light-years, so 1 kiloparsec (kpc) ≈ 3,260 light-years and 1 megaparsec (Mpc) ≈ 3.26 million light-years [Section 16.2].

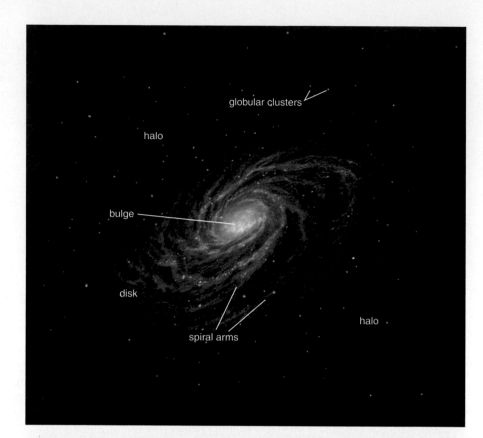

a Artist's conception of the Milky Way viewed from the outside.

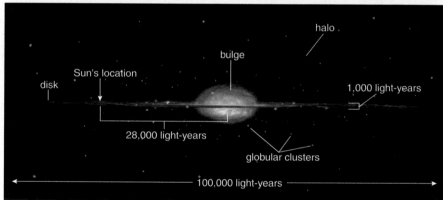

b Edge-on schematic view of the Milky Way.

Figure 19.1 The Milky Way Galaxy.

We once thought of the Milky Way as a band of light, and later we saw it as a collection of stars. Now we see it revealed as a dynamic and complex system. In the rest of this chapter, we will see that our galaxy in many ways is like a forest of stars. Its ecology is shaped by the cycle of stellar life and death, and its structure is determined by gravity and the majestic rotation of the galactic disk.

19.2 The Star–Gas–Star Cycle

Stars have formed, fused atomic nuclei, and exploded throughout the history of our galaxy. Generations of stars continually recycle the same galactic matter through their cores, gradually raising the overall abundance of elements made by fusion. Elements heavier than helium, usually called

*heavy elements** by astronomers, now constitute about 2% of the galaxy's gaseous content. The overall composition of the galaxy is about 70% hydrogen, 28% helium, and 2% heavy elements by mass.

The process of adding to the abundance of heavy elements, called **chemical enrichment**, is an inevitable by-product of continual star formation. However, the synthesis of elements is only one part of the galactic ecocycle. If the galaxy did not reincorporate these elements into new stars and their accompanying planetary systems, the mineral riches fused in the bellies of stars would be wasted.

---

*Astronomers sometimes refer to elements heavier than helium as *metals*—a very different use of the term than that used in daily life, where *metal* refers to substances like copper, silver, and gold that shine when polished and often are good electrical conductors.

Based on the idea of chemical enrichment, which types of stars must contain a higher proportion of heavy elements: stars in globular clusters or stars in open clusters? (*Hint:* Recall from Chapter 16 that stars in globular clusters are all very old, while stars in open clusters are relatively young.)

Holding onto the newly made elements released by stars is not an easy task. When a star explodes as a supernova, the ejected matter flies out at speeds of several thousand kilometers per second—far exceeding the escape velocity from the galaxy. Were it not for the interstellar medium, the new heavy elements released in the supernova would fly straight out of the Milky Way into intergalactic space. Instead, the blobs of matter expelled from the supernova collide with the interstellar medium, slow down, and eventually stop.

Supernovae of one star after another stir and heat the interstellar medium while feeding it new heavy elements. These elements eventually blend with the older, less chemically enriched hydrogen gas in the vicinity. Before new stars can form, this gas must cool and form clouds. The hot gas cools first into clouds of atomic hydrogen (that is, neutral hydrogen atoms as opposed to ionized hydrogen or hydrogen molecules) and then into clouds of molecular hydrogen (H_2). The cooling of interstellar gas into clouds of molecular hydrogen takes millions of years. These clouds subsequently give birth to new stars more highly enriched in heavy elements, thus completing the **star–gas–star cycle**

(Figure 19.2). Let's look at the stages of this cycle more closely.

Gas from Stars

All stars return much of their original mass to interstellar space in two basic ways: through stellar winds that blow throughout their lives, and through "death events" of planetary nebulae (for low-mass stars) or supernovae (for high-mass stars). Low-mass stars generally have weak stellar winds while they are on the main sequence. Their winds grow stronger and carry more material into space when they become red giants. By the time a low-mass star like the Sun ends its life with the ejection of a planetary nebula [Section 17.3], it has returned almost half its original mass to the interstellar medium (Figure 19.3).

High-mass stars lose mass much more dynamically and explosively. The powerful winds from supergiants and massive O and B stars recycle large amounts of matter into the galaxy. At the ends of their lives, these stars explode as supernovae. The high-speed gas ejected into space by these winds and supernovae sweeps up surrounding interstellar material, excavating a **bubble** of hot, ionized gas around the exploding star. Although the bubble in Figure 19.4 looks much like a soap bubble, it is actually an expanding shell of hot gas. The glowing surface is the edge of the expanding bubble, where gas piles up as the bubble sweeps outward through the interstellar medium.

SPECIAL TOPIC Discovering the Milky Way

The river of light in our sky that we call the Milky Way appears indistinct to our eyes. Galileo, looking through his telescope in 1610, was the first to realize that the Milky Way's light comes from innumerable faint stars, but the question of the size and shape of the Milky Way remained unanswered.

In the late 1700s, British astronomers William and Caroline Herschel (brother and sister) tried to determine the shape of the Milky Way more accurately by counting how many stars lay in each direction. Their approach suggested that the Milky Way's width was five times its thickness. More than a century later, in the early 1900s, Dutch astronomer Jacobus Kapteyn and his colleagues used a more sophisticated star-counting method to gauge the size and shape of the Milky Way. Their results seemed to confirm the general picture found by the Herschels and suggested that the Sun lay very near the center of the galaxy.

Kapteyn's results made astronomers with a sense of history slightly nervous. Only four centuries earlier, before Copernicus challenged the Ptolemaic system, astronomers had believed Earth was the center of the universe. Kapteyn's placement of the Sun near the Milky Way's center seemed to be giving Earth a central place again. Kapteyn knew that obscuring material could deceive us by hiding the rest of the galaxy like some kind of interstellar fog, but he found no evidence for such a fog.

While Kapteyn was counting stars, American astronomer Harlow Shapley was studying globular clusters. He found that these

clusters appeared to be centered around a point tens of thousands of light-years from the Sun. (His original estimate was that the center of the galaxy was 45,000 light-years from the Sun. Today's accepted distance is 28,000 light-years.) Shapley concluded that this point marked the true center of our galaxy and that Kapteyn must be wrong.

Today we know that Shapley was right. The Milky Way's interstellar medium is the "fog" that misled Kapteyn. Robert Trumpler, working at California's Lick Observatory in the 1920s, established the existence of this dusty gas by studying open clusters of stars. By assuming that all open clusters had about the same diameter, he estimated their distances from their apparent sizes in the sky, much as you might estimate the distances of cars at night from the separation of their headlights. He found that stars in distant clusters appeared too dim, just as a car's headlights might appear in foggy weather. That is, the distant stars seemed to have lower luminosities than similar stars nearby.

Trumpler concluded that light-absorbing material fills the spaces between the stars, partially obscuring the distant clusters and making them appear fainter than they would appear otherwise. Thus, we learned how interstellar material had been deceiving earlier astronomers, and that the stars visible in the night sky occupy a minuscule portion of the observable universe.

atomic-hydrogen clouds

molecular clouds

hot bubbles

Star–Gas–Star Cycle

star formation

supernovae and stellar winds

stellar burning/ heavy-element formation

Figure 19.2 A pictorial representation of the star–gas–star cycle. (For descriptions of individual photos, proceeding counterclockwise from lower right, see Figures 19.4, 19.7a, 19.13a, 19.10, 19.12, and 19.14.)

Figure 19.3 A dying low-mass star, like this one photographed by the Hubble Space Telescope, returns gas to the interstellar medium in a planetary nebula. This particular planetary nebula is known as the Retina Nebula. It is about 1 light-year across in the longer direction.

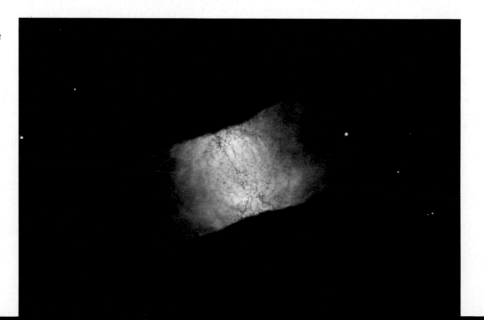

These hot, tenuous bubbles are quite common, filling roughly 20–50% of the Milky Way's disk. However, they are not always easy to detect. While some are hot enough to emit profuse amounts of X rays and others emit strongly in visible light, many others are evident only through radio emission from the shells of atomic hydrogen gas that surround them.

Shock Waves and Supernova Remnants

Supernovae generate shock waves—waves of pressure that move faster than the speed of sound. A shock wave sweeps up surrounding gas as it travels, creating a "wall" of fast-moving gas on its leading edge. When we observe a *supernova remnant*, we are seeing the aftermath of its shock wave, which compresses, heats, and ionizes all the interstellar gas it encounters [Section 17.4].

Figure 19.5 shows a young supernova remnant whose shocked gas is hot enough to emit X rays. In contrast, older supernova remnants are cooler because their shock waves have swept up more material and must share their energy among more particles (Figure 19.6). Eventually, the shocked gas radiates away most of its original energy, and the expanding wall of gas slows to subsonic speeds. As the energy dissipates and the gas cools, the supernova's cargo of new elements merges with the surrounding interstellar medium.

In addition to their role as the movers and shakers of the interstellar medium, shock waves from supernovae

Figure 19.4 This photo shows a bubble of hot, ionized gas blown by a wind from the hot star near its center. The bubble measures about 10 light-years from side to side.

Figure 19.5 This image from the Chandra X-Ray Observatory shows X-ray emission from hot gas in a young supernova remnant—the remnant from the supernova observed by Tycho Brahe in 1572. The most energetic X rays, represented in blue, are coming from 20-million-degree gas just behind the expanding shock wave of the remnant, which is about 20 light-years across. Less energetic X rays, represented in green and red, are coming from the 10-million-degree debris ejected by the exploded star. (The straight edge at the lower right appears because part of the image fell outside the view of the X-ray detector.)

a This large-scale view shows the entire Cygnus Loop supernova remnant glowing in visible light. The angular size of this remnant in our sky is six times that of the Moon, and it is about 130 light-years across.

b The close-up view, from the Hubble Space Telescope, displays the fine filamentary structure in a small piece of the remnant. Blue represents emission from singly ionized oxygen, green represents emission from atomic hydrogen, and red represents emission from singly ionized sulfur.

c A visible-light spectrum from the Cygnus Loop shows the strong emission lines that account for the distinct colors in the Hubble Space Telescope photo.

Figure 19.6 Emission of visible light from a supernova remnant, the Cygnus Loop.

can act as subatomic particle accelerators. Some of the electrons in supernova remnants accelerate to nearly the speed of light as they interact with the shock wave. These fast electrons emit radio waves as they spiral around magnetic field lines threading the supernova remnant (Figure 19.7). (This radio emission is sometimes called *synchrotron radiation*.)

Supernova remnants may also generate the **cosmic rays** that permeate the interstellar medium and bombard Earth's atmosphere. Cosmic rays are made of electrons, protons, and atomic nuclei that zip through interstellar space at close to the speed of light. Some cosmic rays penetrate Earth's atmosphere and reach Earth's surface. On average, about one cosmic-ray particle strikes your body each second. At the altitudes at which jet planes fly, high above most of Earth's protective atmosphere, the cosmic-ray bombardment rate is 100 times higher. Even more cosmic rays fun-

nel along magnetic field lines to Earth's magnetic poles, so for safety reasons some airlines restrict how often their flight crews cross the Earth's polar regions.

Superbubbles and Fountains

The bubble associated with a single supernova remnant can grow to a size of about a hundred light-years (30 pc). In some areas of the Milky Way we see cavities of hot gas over a thousand light-years (300 pc) wide. These huge cavities arise because stars tend to form in clusters. The hottest, most massive stars in a cluster can end their lives and explode within a few hundred thousand years of one another. The shock waves from the individual supernovae soon overlap, combining their energy into one very powerful shock wave. This extra-large shock wave forms an enormous **superbubble** in the interstellar medium. Subsequent supernovae

a Radio emission caused by electrons spiraling around magnetic field lines in the young supernova remnant Cassiopeia A. This remnant is about 10 light-years in diameter.

radio

b Photograph of X-ray emission from the hot gas of Cassiopeia A, as seen by the Chandra X-Ray Observatory. Red indicates the lowest-energy X rays and blue indicates the highest-energy X rays.

X-ray

Figure 19.7 Supernova remnant Cassiopeia A.

from the cluster explode inside the superbubble, adding even more energy.

In many places in our galactic disk, we see what appear to be elongated bubbles extending from young clusters of stars to distances of 3,000 light-years (about 1 kpc) or more above the disk. These probably are places where superbubbles have grown so large that they cannot be contained within the disk of the Milky Way (Figure 19.8). Once the superbubble breaks out of the disk, where nearly all of the Milky Way's gas resides, nothing remains to slow its expansion except gravity. Such a *blowout* is in some ways similar

to a volcanic eruption, but on a galactic scale: Hot plasma erupts from the disk and shoots high into the galactic halo (Figure 19.9).

Mounting evidence suggests that processes like blowouts continually cycle gas between the Milky Way's disk and the halo, an idea summarized in a model known as the **galactic fountain**. According to this model, fountains of hot, ionized gas rise from the disk into the halo through the elongated bubbles carved by blowouts. The gravity of the galactic disk slows the rise of the gas, eventually pulling it back down. Near the top of its trajectory, the ejected gas starts to cool and form clouds of atomic hydrogen. These clouds cool further as they plunge back down, ultimately rejoining the layer of atomic hydrogen gas in the disk (Figure 19.8d).

The galactic fountain model is plausible but difficult to verify. We do indeed see some hot gas high above the galaxy's disk. We also see cooler clouds that appear to be raining down from the halo. However, we can see this "rain" only directly above and below us, making it difficult to demonstrate beyond a doubt that galactic fountains circulate the products of supernovae throughout the Milky Way.

Atomic Hydrogen Gas

The hot, ionized gas in bubbles, superbubbles, and fountains is dynamic and widespread, but it is a relatively small fraction of the gas in the Milky Way. Atomic hydrogen gas is much more common. We can map its distribution in the Milky Way with radio observations. Atomic hydrogen emits a spectral line with a wavelength of 21 cm in the radio portion of the electromagnetic spectrum. We see the radio

---

MOVIE MADNESS

The Sound of Space

In many science fiction movies, a thunderous sound accompanies the demolition of a spaceship. If the moviemakers wanted to be more realistic, they would silence the explosion. On Earth, we perceive sound when sound waves—which are waves of alternately rising and falling pressure—cause trillions of gas atoms to push our eardrums back and forth. Although sound waves can and do travel through interstellar gas, the extremely low density of this gas means that only a handful of atoms per second would collide with something the size of a human eardrum. As a result, it would be impossible for a human ear (or a similar-size microphone) to register any sound. Despite the presence of sound waves and shock waves in space, the sound of space is silence.

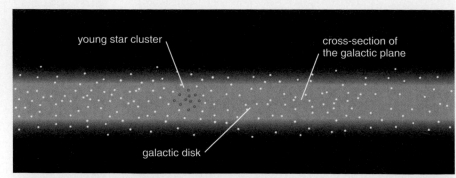

a A young star cluster forms deep within the gas layer of our galaxy's disk.

b Multiple supernova explosions within the star cluster create a superbubble filled with hot gas.

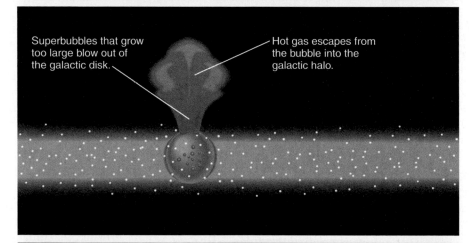

c Repeated supernova explosions can enlarge the superbubble until the galactic disk can no longer contain it. When that happens, the hot gas within the bubble blows out into the galaxy's halo.

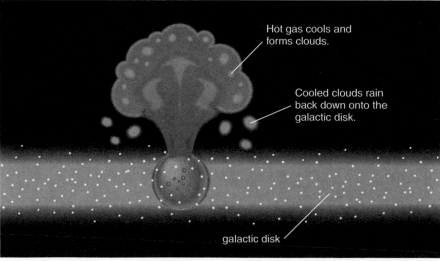

d According to the galactic fountain model for our interstellar medium, the hot gas blowing out into the galactic halo eventually cools into gas clouds that rain back down on the disk. This process may be an important part of the galaxy-wide recycling system that incorporates the products of supernova explosions into new generations of stars and planets.

Figure 19.8 This four-step schematic drawing illustrates how a superbubble can erupt out of the galactic disk, blowing hot gas enriched with newly made heavy elements into the galactic halo.

Figure 19.9 Supercomputers are sometimes used to simulate the growth and eruption of a superbubble into the galactic halo. This image from a supercomputer simulation shows a superbubble of hot gas (red) blowing out of the cooler gas (blue) in the galactic disk. Such blowouts can be quite turbulent, as is evident from the complex structure in this image.

emission from this **21-cm line** coming from all directions, telling us that atomic hydrogen gas is distributed throughout the galactic disk. Based on the overall strength of the 21-cm emission, the amount of atomic hydrogen gas in our galaxy must be about 5 billion solar masses, which is a few percent of the galaxy's total mass.

Atomic hydrogen gas tends to be found in two distinct forms: large, tenuous clouds of warm (10,000 K) atomic hydrogen and smaller, denser clouds of cool (100 K) atomic hydrogen. If you were to take an interstellar voyage across the Milky Way, you would spend the majority of your time cruising through regions of warm atomic hydrogen interspersed with bubbles of hot, ionized gas. Every thousand light-years or so, you would encounter a cooler, denser cloud of atomic hydrogen.

In the warm regions, you would detect about one atom per cubic centimeter and a weak magnetic field (weaker than Earth's magnetic field by a factor of about 100,000). In the cooler clouds, you would find a density of about 100 atoms per cubic centimeter and a stronger magnetic field.

Matter remains in the warm atomic hydrogen stage of the star–gas–star cycle for millions of years. Gravity slowly draws blobs of this gas together into tighter clumps, which radiate energy more efficiently as they grow denser. The blobs therefore cool and contract, forming the smaller clouds of cool atomic hydrogen. This process of shrinkage and coagulation takes a much longer time than the other steps in the journey from star death to star birth. The slowness of the transition from warm gas to cool clouds accounts

for the large amount of gaseous matter in the atomic hydrogen stage of the star–gas–star cycle.

Although we speak of clouds of *hydrogen*, all interstellar material actually has a composition of about 70% hydrogen, 28% helium, and 2% heavy elements by mass. Some of the heavy elements in regions of atomic hydrogen are in the form of tiny, solid **dust grains**: flecks of carbon and silicon minerals that resemble particles of smoke and form in the winds of red giant stars [Section 17.3]. Once formed, dust grains remain in the interstellar medium unless they are heated and destroyed by a passing shock wave or incorporated into a protostar. Although dust grains make up only about 1% of the mass of the atomic hydrogen clouds, they are responsible for the absorption of visible light that prevents us from seeing through the disk of the galaxy.

Molecular Clouds

As the temperature drops further in the center of a cool cloud of atomic hydrogen, hydrogen atoms combine into molecules, making a **molecular cloud**. Molecular clouds are the coldest, densest collections of gas in the interstellar medium. They often congregate into *giant molecular clouds* that hold up to a million solar masses of gas. The total mass of molecular clouds in the Milky Way is somewhat uncertain, but it is probably about the same as the total mass of atomic hydrogen gas—about 5 billion solar masses. Throughout much of this molecular gas, temperatures hover only a few degrees above absolute zero, and gas densities are a few hundred molecules per cubic centimeter.

Molecular hydrogen (H_2) is by far the most abundant molecule in molecular clouds, but it is difficult to detect because temperatures are usually too cold for the gas to produce H_2 emission lines. As a result, most of what we know about molecular clouds comes from observing spectral lines of molecules that make up only a tiny fraction of a cloud's mass. The most abundant of these molecules is carbon monoxide (CO, also a common ingredient in car exhaust). Carbon monoxide produces strong emission lines in the radio portion of the spectrum at the 10–30 K temperatures of molecular clouds (Figure 19.10). Many other molecules also produce radio emission lines, and astronomers have used these lines to identify more than 120 different kinds of molecules in molecular clouds. Among the more familiar are water (H_2O), ammonia (NH_3), and ethyl alcohol (C_2H_5OH).

Molecular clouds are heavy and dense compared to the rest of the interstellar gas and therefore tend to settle toward the central layers of the Milky Way's disk. This tendency creates a phenomenon you can see with your own eyes: the dark fissures running through the luminous band of light in our sky that we call the Milky Way [Section 2.1].

Gravitational forces in molecular clouds gather molecules into the compact *cores* that eventually become protostars [Section 17.2]. The final stages of star formation can be quite disruptive. Recall that protostars often have violent jets of gas spurting outward [Section 17.2]. The turbulence of these jets stirs up nearby regions of the molecular cloud, probably preventing other stars from forming in the vicinity of the growing protostar, at least for a while (Figure 19.11).

Completing the Cycle

Once a few stars form in a cluster, their radiation begins to erode the surrounding gas in the molecular cloud. Ultraviolet photons from high-mass stars heat and ionize the gas, and winds and radiation pressure push the ionized gas away. This kind of feedback prevents much of the gas in a molecular cloud from turning into stars.

The process of molecular cloud erosion is sometimes spectacular. Figure 19.12 shows the *Eagle Nebula,* a complex of clouds where new stars are currently forming. The dark, lumpy columns are molecular clouds. Off to the upper right (outside the picture), newly formed massive stars glow with ultraviolet radiation. This radiation sears the surface of the molecular clouds, destroying molecules and stripping electrons from atoms. As a result, matter "evaporates" from the molecular clouds and joins the hotter ionized gas encircling them. Only the densest knots of gas resist evaporation. Stars are forming in some of these dense knots, which remain compact while the rest of the cloud erodes. These star-forming knots are the tips of the dark protrusions on the columns of molecular gas in Figure 19.12.

We have arrived back where we started in the star–gas–star cycle. The most massive stars now forming in the Eagle Nebula will explode within a few million years, filling the region with hot gas and newly formed heavy elements. Farther in the future, this gas will once again cool and coalesce into molecular clouds, forming new stars, new planets, and maybe even new civilizations.

Despite the recycling of matter from one generation of stars to the next, the star–gas–star cycle cannot go on forever. With each new generation, some of the galaxy's gas becomes permanently locked away in brown dwarfs that never return material to space and in stellar corpses left behind when stars die (white dwarfs, neutron stars, and black holes). The interstellar medium therefore is slowly running out of gas, and the rate of star formation will gradually

Figure 19.10 Image of a molecular cloud in the constellation Orion, showing its complex structure. The picture was made by measuring Doppler shifts of emission lines from carbon monoxide molecules in different locations. The colors indicate gas motions: Relative to the cloud as a whole, bluer parts are moving toward us and redder parts are moving away from us. This enormous cloud is about 1,600 light-years distant and several hundred light-years across.

Figure 19.11 Jets from protostars disrupt nearby regions of molecular clouds. This photo shows a jet emanating from the protostar on the left. All along the jet in the center and on the far right of the photo, gas shocked by the jet is glowing in optical light.

Figure 19.12 A portion of the Eagle Nebula, as seen by the Hubble Space Telescope. The dark columns of gas are molecular clouds, and stars are currently forming in the densest parts of these clouds. Meanwhile, ultraviolet radiation from newly formed massive stars just off the upper right of the picture is searing the surfaces of these clouds, eroding their gas and causing them to glow. The dark protrusions on some of these clouds are knots of gas dense enough to resist this erosion. Arrows indicate two of the locations where dense knots are giving birth to stars. The whole region pictured here is about 5 light-years across.

taper off over the next 50 billion years or so. Eventually, star formation will cease.

Putting It All Together: The Distribution of Gas in the Milky Way

As we look at different regions of the galaxy, we see various stages of the star–gas–star cycle playing themselves out. Because the cycle proceeds over such a long period of time compared to a human lifetime, each stage appears to us as a snapshot. We therefore see the interstellar medium in a wide variety of manifestations, ranging from the tenuous million-degree gas of bubbles to the cold, dense gas of molecular clouds. Table 19.1 summarizes the different states in which we see interstellar gas in the galactic disk.

Figure 19.13 shows seven views of the disk of the Milky Way Galaxy. Each view represents a panorama made by photographing the Milky Way's disk in every direction

Table 19.1 Typical States of Gas in the Interstellar Medium

State of Gas	Primary Constituent	Approximate Temperature	Approximate Density (atoms per cm³)	Description
Hot bubbles	Ionized hydrogen	1,000,000 K	0.01	Pockets of gas heated by supernova shock waves
Warm atomic gas	Atomic hydrogen	10,000 K	1	Fills much of galactic disk
Cool atomic clouds	Atomic hydrogen	100 K	100	Intermediate stage of star–gas–star cycle
Molecular clouds	Molecular hydrogen	30 K	300	Regions of star formation
Molecular cloud cores	Molecular hydrogen	60 K	10,000	Star-forming clouds

from Earth. You can visualize how one of these views corresponds to the sky by imagining cutting it out, bringing its ends together to form a circular band, and then lining up the band with the Milky Way on a model of the celestial sphere. Each view shows the Milky Way as it appears in a different set of wavelengths of light and thus reveals different features of the galactic disk.

● Figure 19.13a shows variations in the intensity of radio emission from the 21-cm line of atomic hydrogen. Thus, it maps the distribution of atomic hydrogen gas, demonstrating that this gas fills much of the galactic disk.

● Figure 19.13b shows variations in the intensity of radio emission lines from carbon monoxide (CO) and therefore maps the distribution of molecular clouds.

a 21-cm radio emission from atomic hydrogen gas.

b Radio emission from carbon monoxide reveals molecular clouds.

c Infrared (60–100 μm) emission from interstellar dust.

d Infrared (1–4 μm) emission from stars that penetrates most interstellar material.

e Visible light emitted by stars is scattered and absorbed by dust.

f X-ray emission from hot gas bubbles (diffuse blobs) and X-ray binaries (pointlike sources).

g Gamma-ray emission from collisions of cosmic rays with atomic nuclei in interstellar clouds.

Figure 19.13 Panoramic views of the Milky Way in different bands of the spectrum. The center of the galaxy, which lies in the direction of the constellation Sagittarius, is in the center of each strip. The rest of each strip shows all other directions in the Milky Way disk as seen from Earth. (Imagine attaching the left and right ends of each strip to form a circular band that corresponds to the 360° band of the Milky Way in our sky.)

These cold, dense clouds are concentrated in a narrow layer near the midplane of the galactic disk.

- Figure 19.13c shows variations in the intensity of infrared emission (at wavelengths of 60–100 micrometers) from interstellar dust grains. The regions of strongest emission from dust correspond to the locations of molecular clouds in Figure 19.13b.

- Figure 19.13d shows infrared light from stars at wavelengths that penetrate clouds of gas and dust (1–4 micrometers). Thus, this image shows how our galaxy would look if there were no dust blocking our view. The galactic bulge is clearly evident at the center.

- Figure 19.13e shows the galactic disk in visible light, just as it appears in the night sky. (Of course, only part of the Milky Way is above the horizon at any one time.) Because visible light cannot penetrate interstellar dust, the dark blotches correspond closely to the bright patches of molecular radio emission and infrared dust emission in Figure 19.13b and c.

- Figure 19.13f shows the distribution of X-ray light from the galactic disk. The pointlike blotches in this view are mostly X-ray binaries [Section 18.3]. The rest of the X-ray emission comes primarily from hot gas bubbles. Because hot gas tends to rise into the halo, it is less concentrated toward the midplane than the atomic and molecular gas. (The prominent yellow blob on the lower right is the Vela supernova remnant.)

- Figure 19.13g shows gamma-ray emission from the Milky Way. Most of the gamma-ray emission is produced by collisions between cosmic-ray particles and atomic nuclei in interstellar clouds. Such collisions happen most frequently where gas densities are highest, so the gamma-ray emission corresponds closely to the locations of molecular and atomic gas. (Gamma rays from the pulsar at the center of the Vela supernova remnant are prominent on the lower right.)

THINK ABOUT IT

Carefully compare and contrast the different views of the Milky Way's disk in Figure 19.13. Why do regions that appear dark in some views appear bright in others? What kinds of general patterns do you notice?

19.3 Galactic Environments

The star–gas–star cycle has operated continuously since the Milky Way's birth, yet new stars are not spread evenly across the galaxy. Some regions seem much more fertile than others. Galactic environments rich in molecular clouds tend to spawn new stars easily, while gas-poor environments do not. A quick tour of some characteristic galactic environments will help you spot where the action is.

Out in the Halo

A census of stars in the Milky Way's *halo* would turn up many senior citizens and very few newborns. Most of the halo stars are old, red, and dim and much smaller in mass than our Sun. Halo stars also contain far fewer heavy elements than our Sun, sometimes having heavy-element proportions as low as 0.02% (in contrast to about 2% in the Sun). The relative lack of chemical enrichment in the halo indicates that its stars formed early in the galaxy's history—before many supernovae had exploded, adding heavy elements to star-forming clouds.

The halo's gas content corroborates this view. The halo is virtually gas-free compared to the disk, with very few detectable molecular clouds. Apparently, the bulk of the Milky Way's gas settled into the disk long ago. The halo environment is a place where lack of gas caused star formation to cease early in our galaxy's life. Now only very old stars still survive, and new stars are rarely born.

THINK ABOUT IT

How does the halo of our galaxy resemble the distant future fate of the galactic disk? Why?

Our Neighborhood

Our own stellar neighborhood is more active than the halo and typifies much of the galactic disk. Within about 33 light-years (10 pc), we know of over 300 stars. Most are dim, red, spectral type M stars. A few, including Sirius, Vega, Altair, and Fomalhaut, are bright, white stars younger than our Sun. While stars of many different ages and different proportions of heavy elements are in our neighborhood, no very massive, short-lived stars (i.e., spectral type O or B) are present. We therefore infer that stars form periodically in our neighborhood but that no star clusters have formed here recently.

Our quiet suburb of the galaxy has not always been as calm as it is today. X-ray telescopes in space reveal hot, X-ray–emitting gas coming from nearby in every direction. Surrounding this hot gas, at distances ranging up to a few hundred light-years (100 pc), lies a region of much cooler gas. Apparently, we and all our stellar neighbors live inside a hot bubble. The existence of this *Local Bubble* means that a number of supernovae must have detonated within our stellar neighborhood over the past several million years.

Hot-Star Hangouts

The hot spots in our galaxy are the neighborhoods of high-mass stars. Because hot, massive stars live fast and die young, they never get a chance to move very far from their birthmates. Thus, we find them in star clusters close to the molecular clouds from which they formed. These environments are highly active and extraordinarily picturesque.

We find colorful, wispy blobs of glowing gas known as **ionization nebulae** throughout the galactic disk, particularly

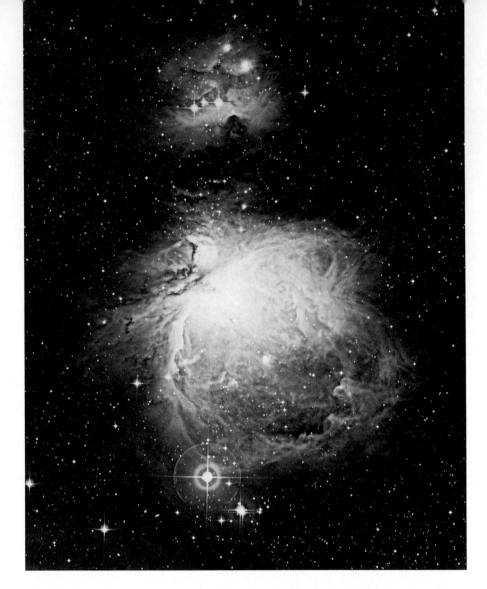

Figure 19.14 A photo of the Orion Nebula, an ionization nebula energized by ultraviolet photons from hot stars.

in the spiral arms. (Ionization nebulae are sometimes called *emission nebulae* or *H II regions*. H II is an abbreviation astronomers use for ionized hydrogen.) The energy that powers ionization nebulae comes from neighboring hot stars that irradiate them with ultraviolet photons. These photons ionize and excite the atoms in the nebulae, causing them to emit light. The Orion Nebula, about 1,500 light-years away in the "sword" of the constellation Orion, is among the most famous. Few astronomical objects can match its spectacular beauty (Figure 19.14).

Most of the striking colors in an ionization nebula come from particular spectral lines produced by particular atomic transitions. For example, the transition in which an electron falls from energy level 3 to energy level 2 in a hydrogen atom generates a red photon with a wavelength of 656 nanometers [Section 6.4]. Ionization nebulae appear predominantly red in photographs because of all the red photons released by this particular transition. (Jumps from level 2 to level 1 are even more common, but they produce ultraviolet photons that can be studied only with ultraviolet telescopes in space.) Transitions in other elements produce other spectral lines of different colors (Figure 19.15).

What Is a Nebula?

The term *nebula* means "cloud," but in astronomy it can refer to many different kinds of objects—a state of affairs that sometimes leads to misconceptions. Many astronomical objects look "cloudy" through small telescopes, and in past centuries astronomers called any such object a nebula as long as they were sure it wasn't a comet. For example, galaxies were called nebulae because they looked like either fuzzy round blobs or fuzzy spiral blobs.

Using the term *nebula* to refer to a galaxy now sounds somewhat dated, given the enormous differences between these distant star systems and the much smaller clouds of gas that populate the interstellar medium. Nevertheless, some people still refer to spiral galaxies as "spiral nebulae." Today, we generally use the term *nebula* to refer to true interstellar clouds, but be aware that the term is still sometimes used in other ways.

atomic hydrogen

doubly ionized oxygen

singly ionized oxygen

doubly ionized neon

atomic hydrogen

atomic hydrogen

atomic helium

atomic hydrogen

apparent brightness

wavelength (nanometers)

350   400   450   500

Figure 19.15 Spectrum of an ionization nebula in blue and green light. The prominent emission lines in the spectrum reveal the atoms and ions that emit most of the light. Through careful modeling of these lines, we can determine the nebula's chemical composition.

The blue and black tints in some nebulae have a different origin. Starlight reflected from dust grains produces the blue colors, because interstellar dust grains scatter blue light much more readily than red light (Figure 19.16). These so-called *reflection nebulae* are always bluer in color than the stars supplying the light. (The effect is similar to the scattering of sunlight in our atmosphere that makes the sky blue [Section 11.3].) The black regions of nebulae are dark,

dusty gas clouds that block our view of the stars beyond them. Figure 19.17 shows a multicolored nebula characteristic of a hot-star neighborhood.

THINK ABOUT IT

In Figure 19.17, identify the red ionized regions, the blue reflecting regions, and the dark obscuring regions. Briefly explain the origin of the colors in each region.

Figure 19.16 The blue tints in this nebula in the constellation Orion are produced by reflected light.

Figure 19.17 A photo of the Horsehead Nebula and its surroundings. (The region pictured is about 150 light-years in size.)

Detecting Dark Matter in a Spiral Galaxy Tutorial, Lessons 1, 2

19.4 The Milky Way in Motion

The internal motions of the Milky Way range from the chaos of stars soaring on randomly oriented orbits through the halo to the stately rotation of the galactic disk. These motions, combined with the law of gravity, can tell us how mass is distributed within the Milky Way. When we decipher these movements, we find that the matter binding stars and gas to our galaxy extends far beyond the stars of the galactic disk and that the galaxy's spiral arms are propagating waves of new star formation. Let's investigate the motion of the Milky Way in more detail.

Orbits in the Disk and Halo

If you could stand outside the Milky Way and watch it for a few billion years, the disk would resemble a huge merry-go-round. All the stars in the disk, including our Sun, orbit the center in the same direction. Moreover, like horses on a merry-go-round, individual stars bob up and down through

the disk as they orbit. The general orbit of a star around the galaxy arises from its gravitational attraction toward the galactic center, while the bobbing arises from the localized pull of gravity within the disk itself (Figure 19.18). A star that is "too far" above the disk is pulled back into the disk by gravity. Because the density of interstellar gas is too low to slow the star, it flies through the disk until it is "too far" *below* the disk on the other side. Gravity then pulls it back in the other direction. This ongoing process produces the bobbing of the stars.

The up-and-down motions of the disk stars spread the disk stars over a thickness of about 1,000 light-years—quite thin in comparison to the 100,000-light-year diameter of the disk. Each orbit takes over 200 million years in the vicinity of our Sun, and each up-and-down "bob" takes a few tens of millions of years.

The orbits of stars in the halo and bulge are much less organized than the orbits of stars in the disk (see Figure 19.18). Individual bulge and halo stars travel around the galactic center on more or less elliptical paths, but the orientations of these paths are relatively random. Neighboring halo stars can circle the galactic center in opposite directions. They swoop from high above the disk to far

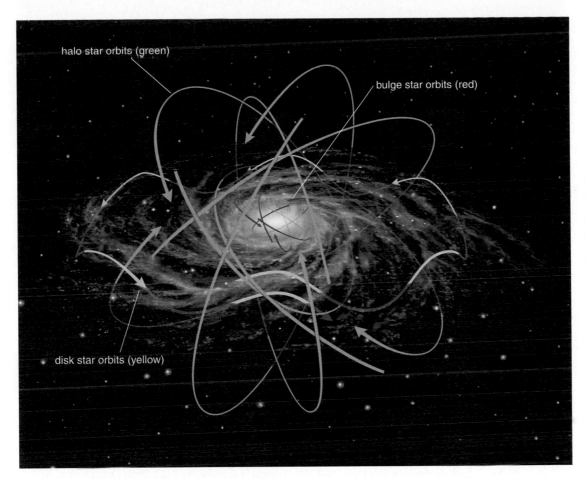

halo star orbits (green)

bulge star orbits (red)

disk star orbits (yellow)

Figure 19.18
Characteristic orbits of disk stars, bulge stars, and halo stars around the galactic center. The up-and-down motion of the disk star orbits is exaggerated compared to the radius of the orbit. (Orbits of stars in the Milky Way generally are not perfectly circular or elliptical because the Milky Way's mass is not concentrated at the center.)

below it and back again, plunging through the disk at velocities so high that the disk's gravity hardly alters their trajectories.

Near the Sun, we see several fast-moving halo stars in the course of hasty excursions through the local region of the disk. One such star is Arcturus, the fourth-brightest star in the night sky.

THINK ABOUT IT

Is there much danger that the Sun or Earth will someday be hit by a halo star swooping through the disk of the galaxy? Why or why not? (*Hint:* Think about the typical distances between stars, as illustrated by use of the 1-to-10-billion scale in Chapter 1.)

The Sun's orbital path around the galaxy is called the **solar circle**. Its radius is our 28,000-light-year distance from the galactic center. By measuring the speeds of globular clusters relative to the Sun, we've determined that the Sun and its neighbors orbit the center of the Milky Way at a speed of about 220 km/s (about 800,000 km/hr [Section 1.3]). Even at this speed, it takes the Sun about 230 million years to complete one orbit around the galactic center. Early dinosaurs ruled Earth when our Sun last visited this side of the galaxy.

Orbits and Galactic Mass

Recall that Newton's law of gravity determines how quickly objects orbit one another. This fact, embodied in Newton's version of Kepler's third law, allows us to determine the mass of a relatively large object when we know the period and average distance of a much smaller object in orbit around it [Section 5.3]. A closely related law, which we will call the *orbital velocity law,* allows us to "weigh" the galaxy using the Sun's orbital velocity and its distance from the galactic center. If we call the Sun's orbital velocity v and its distance (radius) from the galactic center r, the orbital velocity law tells us that the mass of the galaxy within the Sun's orbit (M_r) is:

$$M_r = \frac{r \times v^2}{G}$$

The orbital velocity law tells us only how much mass lies *within* the Sun's orbit. It does not tell us the mass of the entire galaxy, because matter lying outside the solar circle (the Sun's orbit) has very little effect on the Sun's orbital velocity. Every part of the galaxy exerts gravitational forces on the Sun as it orbits, but the net force from matter outside the solar circle is relatively small because the pulls from opposite sides of the galaxy virtually cancel one another.

In contrast, the net gravitational forces from mass within the solar circle all pull the Sun in the same direction—toward the galactic center. Thus, the Sun's orbital velocity responds almost exclusively to the gravitational pull of matter inside its orbit. Substituting the Sun's 28,000-light-year distance and 220-km/s orbital velocity into the orbital velocity law, we find that the total amount of mass within the solar circle is about 2×10^{41} kilograms, or about 100 billion solar masses.

The Distribution of Mass in the Milky Way

Just as we can use the orbit of the Sun to determine the mass of the galaxy within the solar circle, we can use the orbital motion of any other star to measure the mass of the Milky Way within the star's own orbital circle. In prin-

ciple, we could determine the complete distribution of mass in the Milky Way by applying the orbital velocity law to the orbits of stars at every different distance from the galactic center.

In practice, interstellar dust obscures our view of disk stars beyond a few thousand light-years, making it very difficult to measure stellar velocities. However, radio waves penetrate this dust, so we can see the 21-cm line from atomic hydrogen gas and measure its Doppler shift no matter where the gas is located in the galaxy. Such studies allow us to build a map of atomic hydrogen clouds. These maps have revealed our galaxy's large-scale spiral structure and also its overall pattern of rotation.

We can get a sense of the distribution of mass in the Milky Way by making a diagram called a **rotation curve**, which plots *rotational velocity* against *distance from the*

| Mathematical Insight | **19.1** | **The Orbital Velocity Law** |

As discussed in the text, the orbital velocity law allows us to use the orbital characteristics of a star or gas cloud to calculate the mass of the galaxy *within* the object's orbital circle. We can derive it by starting from Newton's version of Kepler's third law, using the version in which one object is much more massive than the other (see Mathematical Insight 5.2):

$$p^2 = \frac{4\pi^2}{G \times M} \times a^3$$

where M is the mass of the massive object and p and a are the orbital period and semimajor axis of a smaller orbiting object, respectively. Solving for M, we find:

$$M = \frac{4\pi^2 \times a^3}{G \times p^2}$$

In Mathematical Insight 16.4, we found that the orbital speed of an object with a circular orbit is:

$$v = \frac{\text{distance traveled in one orbit}}{\text{period of one orbit}} = \frac{2\pi a}{p}$$

Solving for p, we find:

$$p = \frac{2\pi a}{v}$$

Substituting this expression for p into Kepler's third law (solved for the mass M), we find:

$$M = \frac{4\pi^2 \times a^3}{G \times p^2}$$

$$= \frac{4\pi^2 \times a^3}{G \times \left(\frac{2\pi a}{v}\right)^2}$$

$$= \frac{4\pi^2 \times a^3}{G \times \frac{4\pi^2 a^2}{v^2}} = \frac{a \times v^2}{G}$$

Because we are dealing with a circular orbit, we can replace a with the radius r and write this equation as:

$$M_r = \frac{r \times v^2}{G}$$

The subscript r of M_r reminds us that we have calculated the mass only within a distance r of the galactic center.

Example: Calculate the mass of the Milky Way Galaxy within the solar circle.

Solution: The orbital velocity of the Sun around the center of the galaxy is $v = 220$ km/s $= 2.2 \times 10^5$ m/s. The radius of the solar circle is the Sun's 28,000-light-year distance from the galactic center. Using the fact that a light-year is about 9.46×10^{15} meters, we find this radius to be equivalent to $r = 2.6 \times 10^{20}$ m. We can now find the mass within the solar circle by substituting these values for v and r into the orbital velocity law:

$$M_r = \frac{r \times v^2}{G}$$

$$= \frac{(2.6 \times 10^{20} \text{ m}) \times \left(2.2 \times 10^5 \frac{\text{m}}{\text{s}}\right)^2}{6.67 \times 10^{-11} \frac{\text{m}^3}{\text{kg} \times \text{s}^2}}$$

$$= 1.9 \times 10^{41} \text{ kg}$$

The mass of the Milky Way Galaxy within the solar circle is about 2×10^{41} kg. The mass of the Sun is about 2×10^{30} kg. Thus, the mass within the solar circle is equivalent to about 10^{11}, or 100 billion, solar masses.

center. As a simple example of the concept, let's construct a rotation curve for a merry-go-round. Every object on a merry-go-round goes around the center with the same rotational period, but objects farther from the center move in larger circles. Thus, objects farther from the center move at faster speeds, and the rotation curve for a merry-go-round is a straight line that rises steadily outward (Figure 19.19a).

In contrast, the rotation curve for our solar system drops off with distance from the Sun because inner planets orbit at faster speeds than outer planets (Figure 19.19b). This drop-off in speed with distance occurs because virtually all the mass of the solar system is concentrated in the Sun. The gravitational force holding a planet in its orbit decreases with distance from the Sun, and a smaller force means a lower orbital speed. The rotation curve of any astronomical system whose mass is concentrated toward the center therefore drops steeply.

Figure 19.19c shows the rotation curve for the Milky Way Galaxy. Each individual dot represents the distance from the galactic center and the orbital speed of a particu-

SPECIAL TOPIC Determining Stellar Orbits

Astronomers learned how stars orbit the Milky Way by measuring the motions of many different stars relative to the Sun. Although these measurements are easy in principle, they can be difficult in practice.

Determining a star's precise motion relative to the Sun requires knowing its true velocity through space. However, our primary means of measuring speeds in the universe—the Doppler effect—can tell us only a star's *radial velocity*, the component of its velocity directed toward or away from us. If we want to know the star's true velocity, we must also measure its *tangential velocity*, the component of its velocity directed across our line of sight.

Tangential velocity is difficult to measure because of the vast distances to stars. Over tens of thousands of years, the tangential velocities of stars cause their apparent positions in our sky to change, changing the shapes of the constellations. These changes are far too small for human eyes to notice. However, we can measure tangential velocities for many stars by comparing telescopic photographs taken years or decades apart. For example, if photographs taken 10 years apart show that a star has moved across our sky by an angle of 1 arcsecond, we know the star is moving at an angular rate of 0.1 arcsecond per year. We can convert this angular rate of motion (often called the star's *proper motion*) to a tangential velocity if we also know the star's distance. For a given angular rate, the tangential velocity is greater for more distant stars (see Mathematical Insight 7.1).

Because the earliest telescopic photographs date only to the late nineteenth century, we can measure tangential velocities only for objects that have moved noticeably since that time. In general, this limits us to stars within a few hundred light-years of Earth. Thus, we can determine precise stellar orbits only for relatively nearby stars. For more distant stars (and galaxies), we can usually measure only radial velocities. This is one reason why we have only limited knowledge of large-scale motion in the universe.

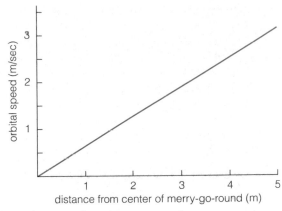

a A rotation curve for a merry-go-round.

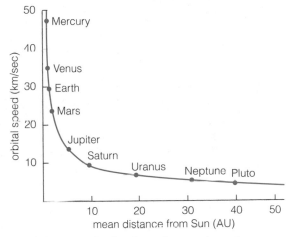

b The rotation curve for the planets in our solar system.

c The rotation curve for the Milky Way Galaxy.

Figure 19.19 Rotation curves show how the orbital speed of a system depends on distance from its center. The solar system's rotation curve declines with radius because its mass is concentrated at the center. The Milky Way's rotation curve is flat, indicating that the Milky Way's mass extends well beyond the Sun's orbit.

lar star or cloud of atomic hydrogen. The curve running through the dots represents a "best fit" to the data. You can see that, beyond the inner few thousand light-years, the orbital velocities remain approximately constant, leading to a flat rotation curve. This behavior contrasts sharply with the steeply declining rotation curve of the solar system. Thus, unlike the solar system, most of the mass of the Milky Way must *not* be concentrated at its center. Instead, the orbits of progressively more distant hydrogen clouds must encircle more and more mass. The solar circle encompasses about 100 billion solar masses, but a circle twice as large surrounds twice as much mass, and a larger circle surrounds even more mass. Because of the difficulty involved in finding clouds to measure on the outskirts of the galaxy, we have not yet found the "edge" of this mass distribution.

The flatness of the Milky Way's rotation curve came as an immense surprise to astronomers because it implies that most of our galaxy's mass must lie well beyond our Sun, tens of thousands of light-years from the galactic center. In fact, a more detailed analysis suggests that most of this mass is distributed in the galactic halo and that the halo might outweigh all the disk stars *combined* by a factor of 10. However, aside from the light of the relatively small number of halo stars, we have detected very little radiation coming from this enormous amount of mass.

Because we see so little light coming from outside the Sun's orbit, most of the halo's huge mass cannot be in the form of orbiting stars. The nature of this mass remains unknown—which means that we do not yet know the nature of the vast majority of matter in our own galaxy. We call this mysterious mass **dark matter** because it does not emit any light that we have yet detected. The nature of this dark matter is one of the greatest mysteries in astronomy today. We will investigate it in much more depth in Chapter 22.

Figure 19.20 Galaxy M51, a spiral galaxy with two prominent spiral arms as photographed by the Hubble Space Telescope. Note how blue the spiral arms are compared with the yellower tones of the bulge and the regions of the disk between the arms. Notice also that the blue arms contain many reddish blotches, which are ionization nebulae. Because blue stars live only for a few million years, the relative blueness of these spiral arms tells us that stars must be forming more actively here than elsewhere in the galaxy. (This portion of M51 is roughly 30,000 light-years across.)

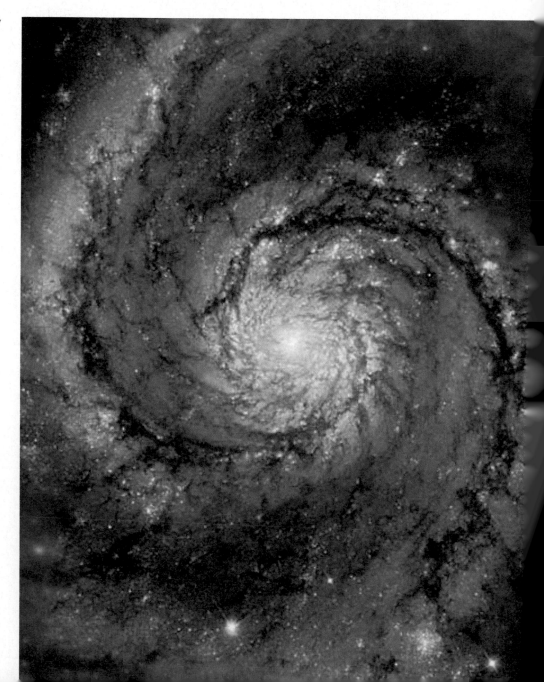

Spiral Arms

The disks of spiral galaxies like the Milky Way display sweeping spiral arms that can stretch tens of thousands of light-years from their bulges. At first glance, spiral arms look as if they ought to move with the stars, like the fins of a giant pinwheel in space. However, we know that spiral arms cannot be fixed patterns of stars that rotate along with the galaxy. The reason is that a spiral galaxy rotates like a whirlpool. That is, stars near the center of the galaxy complete an orbit in much less time than stars far from the center. If the spiral arms simply moved along with the stars, the central parts of the arms would complete several orbits around the galaxy as the outer parts orbited just once. This difference in orbital periods would eventually wind up the spiral arms into a tight coil. Because we generally don't see such tightly wound spiral arms in galaxies, we conclude that spiral arms are more like swirling ripples in a whirlpool than like the fins of a giant pinwheel. In fact, spiral arms are enormous waves of star formation that propagate through the gaseous disk of a spiral galaxy.

We know that spiral arms are actively forming new stars because they bear all the hallmarks of star formation. Detailed images reveal that these arms are home to numerous clusters of young, bright, blue stars surrounded by ionization nebulae (Figure 19.20). The stars between the arms are generally redder and older. We also see enhanced amounts of molecular and atomic gas in the spiral arms, and streaks of interstellar dust often obscure the inner sides of the arms themselves (Figure 19.21). Thus, spiral arms contain both young stars and the material necessary to make new stars.

The forces that drive these organized patterns of star formation are less obvious. Theoretical models suggest that disturbances called **spiral density waves** underlie the spiral arms. According to these models, spiral arms are places in a galaxy's disk where stars and gas clouds are more densely packed than elsewhere in the disk. Packing the stars closer together has little effect on the stars themselves—they are still much too widely separated to collide with each other. However, the large gas clouds do collide, and packing the clouds closer together enhances the force of gravity within them, triggering the formation of many new stars.

To visualize how spiral density waves propagate through a galaxy's disk, consider how traffic backs up behind a slow-moving tractor on a rural highway. Cars approaching the tractor slow down and bunch together. After cars pass the tractor, they speed up and spread out again. Thus, a pack of cars is always bunched up behind the tractor, even though the cars themselves are constantly flowing past it.

In a spiral density wave, gravity plays the role of the tractor, while stars and gas clouds play the role of the cars. The stars and gas clouds of a galaxy's disk are con-

Dust lane arises on inner edge of spiral arm where gas clouds crowd together.

Young blue stars are found on outer edge of spiral arm.

Ionization nebulae arise where newly forming blue stars are ionizing gas clouds.

Figure 19.21 The relationship of dust, gas, and new star clusters in a spiral arm.

stantly flowing through its spiral arms, but the extra density of matter caused by the spiral density wave underlying the spiral arm alters that flow. The extra matter exerts a gravitational force that pulls stars and gas clouds into the arm and tries to halt their escape as they move out the other side (Figure 19.22). This gravitational pull is not strong enough to trap the stars and gas clouds. However, like the tractor, it does temporarily slow them down, and this temporary slowdown produces a long-lasting pattern.

We call this kind of propagating disturbance a *wave* because, like a wave in water, it moves through matter without carrying that matter along with it. Also, just as with water waves, some sort of disturbance is needed to generate a density wave within a galaxy's disk. Gravitational tugs from other galaxies passing nearby are one possible source of disturbance. The whirlpool-like rotation of the disk then stretches the wave initiated by such a tug into a spiral shape. Once a gravitational tug sets a spiral density wave in motion, the wave will continue to move through the galaxy's disk, perhaps for billions of years.

To sum up, spiral arms are sites of prolific star formation. Stars are created more readily in spiral arms because gravity bunches interstellar gas clouds more tightly in these arms than elsewhere in a galaxy's disk. The underlying spiral density pattern that triggers this star formation does not move with the stars but instead propagates through the disk like a wave. Massive blue stars that form as gas clouds pass through a spiral arm die out quickly. These luminous hot stars therefore are found close to the spiral arms in which they formed, making the arms look bluer than the rest of the galaxy. Yellow and red stars live long enough to pass through many spiral arms and therefore are distributed more evenly throughout the galactic disk.

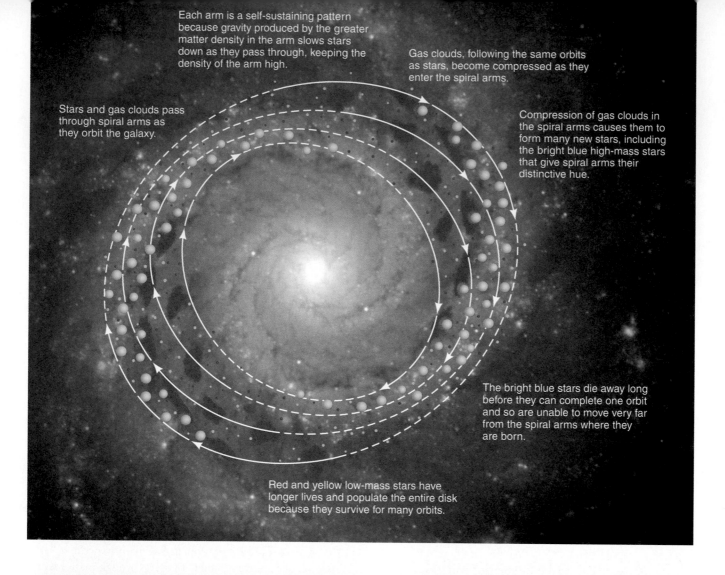

Each arm is a self-sustaining pattern because gravity produced by the greater matter density in the arm slows stars down as they pass through, keeping the density of the arm high.

Gas clouds, following the same orbits as stars, become compressed as they enter the spiral arms.

Stars and gas clouds pass through spiral arms as they orbit the galaxy.

Compression of gas clouds in the spiral arms causes them to form many new stars, including the bright blue high-mass stars that give spiral arms their distinctive hue.

The bright blue stars die away long before they can complete one orbit and so are unable to move very far from the spiral arms where they are born.

Red and yellow low-mass stars have longer lives and populate the entire disk because they survive for many orbits.

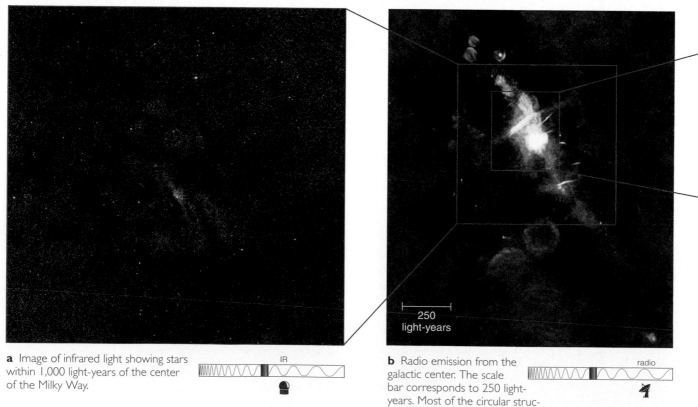

a Image of infrared light showing stars within 1,000 light-years of the center of the Milky Way.

IR

b Radio emission from the galactic center. The scale bar corresponds to 250 light-years. Most of the circular structures are supernova remnants.

250 light-years

radio

◀ **Figure 19.22** Schematic diagram of spiral density waves in a two-armed spiral galaxy. Brown blobs represent interstellar gas clouds, red and yellow dots represent low-mass stars, and blue dots represent high-mass stars. The white lines show how stars and gas clouds orbit within the galactic disk. These orbits pack gas clouds closer together in the spiral arms, causing them to produce new stars. Notice how the blue stars in the schematic diagram track the spiral arms in the underlying photograph and how the brown clouds track the dusty regions on the inner edges of those arms.

 Black Holes Tutorial, Lessons 1, 2

19.5 The Mysterious Galactic Center

The center of the Milky Way Galaxy lies in the direction of the constellation Sagittarius. This region of the sky does not look particularly special to our unaided eyes. However, if we could remove the interstellar dust that obscures our view, the elliptical shape of the galaxy's central bulge would be one of the night sky's most spectacular sights.

Because the Milky Way is relatively transparent to long-wavelength radiation, we can use radio and infrared telescopes to peer into its heart (Figure 19.23). Deep in the galactic center we find swirling clouds of gas and a cluster of several million stars. Bright radio emission traces out the magnetic fields that thread this turbulent region and outlines the shells of numerous supernova remnants. At its core sits a source of radio emission named Sagittarius A*

(pronounced "Sagittarius A-star"), or Sgr A* for short, that is quite unlike any other radio source in our galaxy.

The motions of stars and gas in Sgr A* indicate that it contains a few million solar masses within a very small region of space. Because the star cluster in Sgr A* cannot account for all that mass, astronomers have long suspected that Sgr A* contains a black hole at least 2 million times more massive than the Sun. These suspicions received a big boost in 2002 when astronomers monitoring infrared light from the galactic center observed a star swooping within 17 light-hours of this massive object (Figure 19.24). Applying Newton's version of Kepler's third law to the star's orbit led to a measured mass of around 3 to 4 million solar masses, all packed into a region just a little larger than our solar system. An object that massive within such a small space is almost certainly a black hole.

However, the behavior of this suspected black hole is puzzling. Most other suspected black holes are thought to accumulate matter through accretion disks that radiate brightly in X rays. These include black holes in binary star systems like Cygnus X-1 [Section 18.4] and some giant black holes at the centers of other galaxies that we will discuss in Chapter 21. If the black hole at the center of our galaxy had an accretion disk like these others, its X-ray light would easily penetrate the dusty gas of our galaxy, making it simple to detect. Yet the X-ray emission from Sgr A* has usually been relatively faint, leading some astronomers to doubt whether Sgr A* really contains a black hole.

Recent observations of Sgr A* made with the Chandra X-Ray Observatory have helped quell those doubts. In

Figure 19.23 The galactic center at infrared and radio wavelengths.

c Radio image showing vast threads of emission tracing magnetic field lines.

d Radio image of gas swirling around the suspected black hole known as Sgr A*, which is located at the position of the white dot.

e Infrared image showing stars in the immediate vicinity of Sgr A*. The scale bar indicates a distance of 1 light-year. The two arrows mark Sgr A*.

1 light-year

Figure 19.24 Motions of stars around the Milky Way's central black hole enable us to measure its mass.

a Numerous stars in this infrared image orbit around the suspected black hole known as Sgr A*. The star labeled S2 is one of the closest to this massive object.

b Very high resolution infrared observations were used to measure the location of the star S2 during the period from 1992 to 2002. The dates of the observations are given in fractional years (for example, 2001.50 corresponds to July 1, 2001). The star's motion sped up dramatically during 2002, as it passed within 17 light-hours of Sgr A*. Estimating the mass of Sgr A* from this star's orbit alone gives a value of around 3 million solar masses.

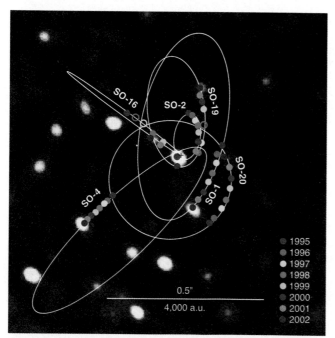

c Tracking the orbits of other stars around Sgr A* is providing even more information about its mass. (The star labeled SO-2 in this figure is the same as the one labeled S2 in part **a**.) Combining the orbital information for all these stars indicates that the mass of Sgr A* may be as large as 4 million solar masses. Continued monitoring of stellar orbits at the galactic center will help us refine this mass measurement.

Figure 19.25 This image of X-ray light from the Chandra X-Ray Observatory shows the central 60 light-years of our galaxy. The circled white dot in the middle of the image is an X-ray flare from the suspected black hole at the Milky Way's center. The flare probably came from a comet-size lump of matter that was torn apart by tidal forces just before it disappeared into the black hole.

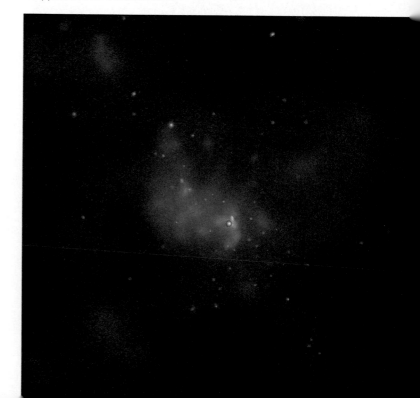

October 2000, an enormous X-ray flare lasting 3 hours was observed coming from the location of the suspected black hole (Figure 19.25). This erratic change in X-ray brightness probably came from a comet-size lump of matter that was torn apart by tidal forces just before it disappeared into the black hole. If we continue to observe similar X-ray flares from Sgr A*, then the explanation for its generally low X-ray brightness may be that matter falls into it in big chunks instead of in the smooth, swirling flow of an accretion disk. Until we better understand Sgr A*, it is sure to remain a favorite target for X-ray telescopes, infrared telescopes, and radio telescopes alike.

THE BIG PICTURE

Putting Chapter 19 into Context

In this chapter, we have explored our galaxy, focusing on the recycling of gas that continually produces new stars and on the motions of the Milky Way that reveal its mass. When you review this chapter, pay attention to these "big picture" ideas:

- The inability of visible light to penetrate deeply through interstellar gas and dust concealed the true nature of our galaxy until recent times. Modern astronomical instruments reveal the Milky Way Galaxy to be a dynamic system of stars and gas that continually gives birth to new stars and planetary systems.

- Stellar winds and explosions make interstellar space a violent place. Hot gas tears through the pervasive atomic hydrogen gas in the galactic disk, leaving expanding pockets of hot plasma and fast-moving clouds in its wake. All this violence might seem quite dangerous, but it performs the great service of mixing essential new elements throughout the Milky Way.

- We can use the orbital speeds of stars and gas clouds around the center of the galaxy to determine the distribution of mass in the Milky Way. To our great surprise, we've discovered that most of the mass of the galaxy lies in the halo, not in the galactic disk where most of the stars are located. The nature of this *dark matter* is one of the greatest mysteries in astronomy today.

- Although the elements from which we are made were forged in stars, we could not exist if stars were not organized into galaxies. The Milky Way Galaxy acts as a giant recycling plant, converting gas expelled from each generation of stars into the next and allowing some of the heavy elements to solidify into planets like our own.

SUMMARY OF KEY CONCEPTS

19.1 The Milky Way Revealed

- *What is the general structure of the Milky Way Galaxy?* The Milky Way Galaxy consists of a thin disk about 100,000 light-years in diameter with a central bulge and a spherical region called the halo that surrounds the entire disk.

- *Where is the Sun located within our galaxy?* The Sun is located in the disk about 28,000 light-years from the galactic center.

- *Can we see through our galaxy's interstellar medium?* Gas and dust absorb visible light, but some other wavelengths, notably infrared and radio, can pass through the gas and dust, allowing us to study regions of the galaxy whose visible light is blocked.

19.2 The Star–Gas–Star Cycle

- *What is the galaxy's star–gas–star cycle?* Stars are born from the gravitational collapse of gas clumps in molecular clouds. Near the ends of their lives, stars more massive than our Sun create elements heavier than hydrogen and helium and expel them into space via supernovae and stellar winds. The supernovae and winds create hot bubbles in the interstellar medium, but the gas moving outward with these bubbles gradually slows and cools. Eventually, this gas cools enough to condense into clouds of atomic hydrogen. Further cooling allows atoms of hydrogen and other elements to collect into molecules, producing molecular clouds. These molecular clouds then form stars, completing the star–gas–star cycle.

- *What would happen to the heavy elements made by massive stars if there were no interstellar medium?* Supernovae would still blow the heavy elements into space, but without an interstellar medium to slow them down these heavy elements would simply fly out of the galaxy into intergalactic space. Thus, no star–gas–star cycle would recycle these elements into subsequent generations of stars.

- *How does observing in different wavelengths allow us to study different parts of the interstellar medium?* Although all the Milky Way's interstellar gas has roughly the same composition—about 70% hydrogen, 28% helium, and 2% heavy elements (by mass)—

continued ▶

gas in different stages of the star–gas–star cycle produces different kinds of radiation. Molecules in molecular clouds and hydrogen atoms in atomic gas emit radiation at radio wavelengths. Interstellar dust absorbs visible light and converts the absorbed energy into infrared light. Hot gas in bubbles and superbubbles emits X rays. Gamma rays are emitted when cosmic rays collide with atomic nuclei.

19.3 Galactic Environments

- *How do halo stars differ from disk stars?* The halo generally contains only old, low-mass stars with a much smaller proportion of heavy elements than stars in the disk.

- *What does the environment around hot stars look like?* High-mass, hot stars energize spectacular ionization and reflection nebulae in the gas and dust around them. High-mass stars tend to be clustered together, so their neighborhoods are likely to have gas bubbles and shock waves from the winds and supernovae produced by nearby stars.

19.4 The Milky Way in Motion

- *How do stellar orbits in the disk differ from those in the halo?* Stars in the disk all orbit the galactic center in about the same plane and in the same direction. Halo stars also orbit the center of the galaxy, but their orbits are randomly inclined to the disk of the galaxy.

- *How do we determine galactic mass from stellar orbits?* By using a star's orbital speed and distance from the galactic center in the orbital velocity law, $M_r = (r \times v^2)/G$, we can calculate the mass of the galaxy that lies within the region enclosed by the star's orbit.

- *What is the significance of a rotation curve that is flat at large distances from the galactic center?* The Milky Way's flat rotation curve implies that the matter associated with our galaxy extends to great distances from the center. Thus, the Milky Way's mass is not concentrated toward the center but instead extends far into the halo. Because we do not detect light from all this mass in the halo, we call it *dark matter*.

- *Why are spiral arms bright?* Spiral arms are probably caused by disturbances known as spiral density waves that move through the galaxy's disk. The enhanced gravity in these waves compresses gas clouds that pass through them, causing the clouds to form many new stars. Bright but short-lived massive stars are therefore more common in spiral arms than elsewhere in the disk.

19.5 The Mysterious Galactic Center

- *What lies in the center of our galaxy?* Motions of stars near the center of our galaxy suggest that it contains a black hole about 3 to 4 million times more massive than the Sun. The black hole appears to be powering a bright source of radio emission known as Sgr A*. However, the gas thought to be disappearing into this black hole does not emit as much X-ray light as expected.

❓ Sensible Statements?

Decide whether each of these statements is sensible and explain why it is or is not.

1. We did not understand the true size and shape of our galaxy until NASA satellites were launched into the galactic halo, enabling us to see what the Milky Way looks like from the outside.

2. Planets like Earth probably didn't form around the very first stars because there were so few heavy elements back then.

3. If I could see infrared light, the galactic center would look much more impressive.

4. Many spectacular ionization nebulae are seen throughout the Milky Way's halo.

5. The carbon in my diamond ring was once part of an interstellar dust grain.

6. The Sun's velocity around the Milky Way tells us that most of our galaxy's dark matter lies within the solar circle.

7. We know that a black hole lies at our galaxy's center because numerous stars near it have vanished over the past several years, telling us that they've been sucked in.

8. If we could watch a time-lapse movie of a spiral galaxy over millions of years, we'd see many stars being born and dying within the spiral arms.

Problems

9. *Sketch the Galaxy.* Draw simple sketches of our galaxy as it would appear face-on and edge-on, identifying the disk, bulge, halo, and spiral arms.

10. *Heavy Elements.* What do we mean by heavy elements? How much of the Milky Way's gas is in the form of heavy elements? Where are heavy elements made?

11. *Chemical Enrichment.* Use the idea of chemical enrichment to explain why stars that formed early in the history of the galaxy contain a smaller proportion of heavy elements than stars that formed more recently.

12. *Interstellar Hydrogen.* Distinguish between the following forms of gas: ionized hydrogen, atomic hydrogen, and molecular hydrogen. What is the most common form of gas in the interstellar medium? What is the wavelength of the radio emission line characteristic of this gas?

13. *Superbubbles.* What is a superbubble? How is it made? Describe what happens in a blowout in which a superbubble breaks out of the galactic disk.

14. *Interstellar Dust.* What are dust grains? Where do they form?

15. *Molecular Clouds.* How do molecular clouds form? How do we observe them? Why do molecular clouds tend to settle toward the central layers of the Milky Way's disk?

16. *Cycling of Gas.* Will the star–gas–star cycle continue forever? Why or why not?

17. *Solar Circle.* At what speed does the Sun travel around the solar circle? About how long does one orbit take? Briefly explain how we can use the characteristics of the Sun's orbit to determine the mass of the Milky Way contained within the solar circle. What is the result?

18. *Pinwheels in Space?* How do we know that spiral arms are not simple "pinwheels" as they appear at first glance?

19. *Unenriched Stars.* Suppose you discovered a star made purely of hydrogen and helium. How old do you think it would be? Explain your reasoning.

20. *Enrichment of Star Clusters.* The gravitational pull of an isolated globular cluster is rather weak—a single supernova explosion can blow all the interstellar gas out of a globular cluster. How might this fact be related to observations indicating that stars ceased to form in globular clusters long ago? How might it be related to the fact that globular clusters are deficient in elements heavier than hydrogen and helium? Summarize your answers in one or two paragraphs.

21. *High-Velocity Star.* The average speed of stars relative to the Sun in the solar neighborhood is about 20 km/s (i.e., the speed at which we see stars moving toward or away from the Sun— *not* their orbital speed around the galaxy). Suppose you discover a star in the solar neighborhood that is moving relative to the Sun at a much higher speed, say 200 km/s. What kind of orbit does this star probably have around the Milky Way? In what part of the galaxy does it spend most of its time? Explain.

*22. *Mass from Rotation Curve.* Using velocities shown on the Milky Way's rotation curve in Figure 19.19c, along with the orbital velocity law (see Mathematical Insight 19.1), calculate the mass of the Milky Way Galaxy within each of the following distances. Give your answers in both kilograms and solar masses.

    a. 10,000 light-years.

    b. 30,000 light-years.

    c. 50,000 light-years.

23. *Research: Discovering the Milky Way.* Humans have been looking at the Milky Way since long before recorded history, but only in the past century did we verify the true shape of the galaxy and our location within it. Learn more about how conceptions of the Milky Way developed through history. What names did different cultures give the band of light they saw? What stories did they tell about it? How have ideas about the galaxy changed in the past few centuries? Try to locate diagrams that illustrate these changes. Write a two- to three-page summary of your findings.

Discussion Questions

24. *Galactic Ecosystem.* The introduction to this chapter likened the star–gas–star cycle in our Milky Way to the ecosystem that sustains life on Earth. Here on our planet, water molecules cycle from the sea to the sky to the ground and back to the sea. Our bodies convert atmospheric oxygen molecules into carbon dioxide, and plants convert the carbon dioxide back into oxygen molecules. How are the cycles of matter on Earth similar to the cycles of matter in the galaxy? How do they differ? Do you think the term *ecosystem* is appropriate to discussions of the galaxy?

25. *Galaxy Stuff.* In the chapters on stars, we learned why we are "star stuff." Based on what you've learned in this chapter, explain why we are also "galaxy stuff." Does the fact that the entire galaxy was involved in bringing forth life on Earth change your perspective on Earth or on life in any way? If so, how? If not, why not?

MEDIA EXPLORATIONS

For a complete list of media resources available, go to www.astronomyplace.com and choose Chapter 19 from the pull-down menu.

 Astronomy Place Web Tutorials

Tutorial Review of Key Concepts

Use the interactive **Tutorial** at www.astronomyplace.com to review key concepts from this chapter.

Detecting Dark Matter in a Spiral Galaxy Tutorial

Lesson 1 Introduction to Rotation Curves

Lesson 2 Determining the Mass Distribution

Supplementary Tutorial Exercises

Use the interactive **Tutorial Lessons** to explore the following questions.

Detecting Dark Matter in a Spiral Galaxy Tutorial, Lesson 1

1. Use the tool in the simulation to determine how the rotation curve for a merry-go-round changes as the merry-go-round spins faster. Describe the change in words.

2. Use the tool in the simulation to determine how the rotation curve for a star system changes as the mass of the central star changes. Describe the change in words.

3. Explain why the rotation curves for the merry-go-round and the solar system are different.

4. Compare the rotation curve of a galaxy with the rotation curve of the solar system.

Detecting Dark Matter in a Spiral Galaxy Tutorial, Lesson 2

1. How does orbital speed depend on the mass of the central object?

2. How does orbital speed depend on the distance from the central object?

3. What can we infer about the distribution of mass in our galaxy by observing the orbital speed of stars as they get farther from the center of our galaxy? Explain your answer.

Exploring the Sky and Solar System

Of the many activities available on the **Voyager: SkyGazer CD-ROM** accompanying your book, use the following files to observe key phenomena covered in this chapter.

Go to the **File: Basics** folder for the following demonstrations.

1. Milky Way

2. Wide Field Milky Way

3. Winter Milky Way

4. Lagoon Nebulae

Go to the **Explore** menu for the following demonstration.

1. Solar Neighborhood

Movies

Check out the following narrated and animated short documentary available on www.astronomyplace.com for a helpful review of key ideas covered in this chapter.

The Milky Way Galaxy Movie

Web Projects

Take advantage of the useful Web links on www.astronomyplace.com to assist you with the following projects.

1. *Images of the Star–Gas–Star Cycle.* Explore the Web to find pictures of nebulae and other forms of interstellar gas in different stages of the star–gas–star cycle. Assemble the pictures into a sequence that tells the story of interstellar recycling, with a one-paragraph explanation accompanying each image.

2. *The Galactic Center.* Search the Web for recent images of the galactic center, along with information about whether the center hides a massive black hole. Present a two- to three-page report, with pictures, giving an update on current knowledge about the center of the Milky Way Galaxy.

20 Galaxies

From Here to the Horizon

Thus the explorations of space end on a note of uncertainty.... Eventually, we reach the dim boundary—the utmost limits of our telescopes. There we measure shadows, and we search among ghostly errors of measurement for landmarks that are scarcely more substantial.

Edwin Hubble

Far beyond the Milky Way, we see many other galaxies—some similar to our own and some very different—scattered throughout space to the very limits of the observable universe. This fabulous sight inspires some fundamental questions: How far away are all these galaxies? How old is the universe? How big is it? Such questions might have seemed ridiculously speculative a century ago. Today, we believe we know the answers to all three questions with respectable accuracy.

Edwin Hubble, the man for whom the Hubble Space Telescope is named, provided the key discovery when he proved conclusively that galaxies exist beyond the Milky Way. The distances he measured revealed an astonishing fact: The more distant a galaxy is, the faster it moves away from us. Hubble's discovery dealt a mortal blow to the traditional belief in a static, eternal, and unchanging universe. The motions of the galaxies away from one another imply instead that the entire universe is expanding and that its age is finite.

This chapter follows the steps leading to this profound revelation as we acquaint ourselves with the different types of galaxies in the universe and then learn how to measure their distances. From the distances to other galaxies, we can infer the age of the universe and determine how large the observable universe really is.

20.1 Islands of Stars

Figure 20.1 shows an amazing image of a tiny patch of the sky taken by the Hubble Space Telescope. The telescope pointed in a single direction in the sky and collected all the light it could for 10 days. If you held a grain of sand at arm's length, the angular size of the grain would match the angular size of everything in this picture. This patch of the sky is jam-packed with galaxies of many sizes, colors, and shapes. Some look large, some small. Some are reddish, some whitish. Some appear round, and some appear flat. Counting the galaxies in this patch of the sky and multiplying by the number of such patches it would take to make a montage of the entire sky, we find that the observable universe contains over 80 billion galaxies. Small galaxies greatly outnumber large ones, yet the large ones produce most of the light in the universe.

Each of these blobs of light is an island of stars bound together by gravity. Like our own Milky Way, each is a dynamic system that has cycled hydrogen gas through stars for billions of years, producing new elements for future generations of stars. Yet reconstructing the histories of these galaxies in more detail poses a formidable challenge. Just as with stars, our observations capture only the briefest instant in any galaxy's life, leaving us to piece together the life story of a typical galaxy from pictures of different galaxies at various life stages. However, getting pictures of all the life stages of galaxies is not easy because young galaxies are found only at very great distances from Earth.

Our understanding of galactic lives is less complete than our understanding of stellar lives because much of the action happened so long ago. Nevertheless, we can proceed similarly in putting the story together. In the rest of this chapter we will study basic characteristics of galaxies, much as we began with basic characteristics of stars in Chapter 16. We will outline how astronomers classify galaxies and investigate how we measure their distances and estimate their ages. Along the way, we will learn how the first measurements of galaxy distances revolutionized the way we look at the universe. We now see the universe as an expanding realm within which the story of creation still unfolds. With this understanding, we'll study the lives of galaxies in the next chapter.

20.2 Galaxy Types

Astronomers classify galaxies into three major categories. **Spiral galaxies** look like flat white disks with yellowish bulges at their centers. The disks are filled with cool gas and dust, interspersed with hotter ionized gas as in the Milky Way, and usually display beautiful spiral arms. **Elliptical galaxies** are redder, more rounded, and often longer in one direction than in the other, like a football. Compared with spiral galaxies, elliptical galaxies contain very little cool gas and dust, though they often contain very hot, ionized gas. **Irregular galaxies** appear neither disklike nor rounded.

The sizes of all three types of galaxies span a wide range, from *dwarf galaxies* containing as few as 100 million (10^8)

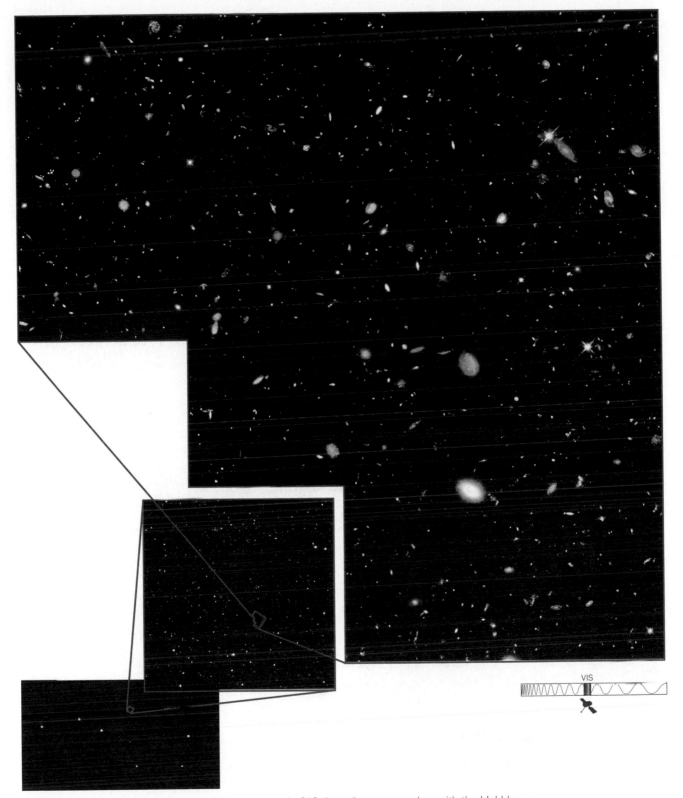

VIS

Figure 20.1 The Hubble Deep Field, an image composed of 10 days of exposures taken with the Hubble Space Telescope. Some of the galaxies pictured are three-quarters of the way across the observable universe. The field itself is located in the Big Dipper.

stars to *giant galaxies* with more than 1 trillion (10^{12}) stars. The differing colors of galaxies arise from the different kinds of stars that populate them: Spiral and irregular galaxies look white because they contain stars of all different colors and ages, while elliptical galaxies look redder because old, reddish stars produce most of their light.

a NGC 6744, a spiral galaxy thought to be very similar to our Milky Way.

b NGC 4414, a spiral galaxy whose disk is somewhat tilted to our line of sight.

c NGC 891, a spiral galaxy seen nearly edge-on. Note the central streak of dust associated with the disk.

Figure 20.2 This three-photo sequence shows spiral galaxies from three different perspectives ranging from nearly face-on to nearly edge-on. In each case, the region pictured is about 100,000 light-years across. (NGC stands for the *New General Catalog*, a listing of more than 7,000 objects published in 1888.)

THINK ABOUT IT

Take a moment and try to classify the larger galaxies in Figure 20.1. How many appear spiral? Elliptical? Irregular? Do the colors of galaxies seem related to their shapes?

Spiral Galaxies

Like the Milky Way, other spiral galaxies also have a thin *disk* extending outward from a central *bulge* (Figure 20.2).

Figure 20.3 NGC 4594, also known as the Sombrero Galaxy, an almost edge-on spiral with a large bulge and a dusty disk. A large but nearly invisible halo surrounds the entire galaxy. The bulge and halo together make up the spheroidal component of the galaxy.

The bulge itself merges smoothly into a *halo* that can extend to a radius of over 100,000 light-years. Because stars are more sparsely distributed in the halo, it is often hard to see in photographs.

Together, the bulge and halo of a spiral galaxy make up its **spheroidal component**, so named because of its rounded shape. Although no clear boundary divides the pieces of the spheroidal component, astronomers usually consider stars within 10,000 light-years of the center to be members of the bulge and those outside this radius to be members of the halo. Figure 20.3 shows a spiral galaxy with a substantial spheroidal component.

The **disk component** of a spiral galaxy slices directly through the halo and bulge. The disk of a large spiral galaxy like the Milky Way can extend 50,000 light-years or more from the center. The disks of all spiral galaxies contain an *interstellar medium* of gas and dust, but the amounts and proportions of the interstellar medium in molecular, atomic, and ionized forms differ from one spiral galaxy to the next. Spiral galaxies with large bulges generally have less interstellar gas and dust than those with small bulges.

Not all galaxies with disks are standard spiral galaxies. Some spiral galaxies appear to have a straight bar of stars cutting across the center, with spiral arms curling away from the ends of the bar. Such galaxies are known as *barred spiral galaxies* (Figure 20.4). In fact, astronomers suspect that the Milky Way itself is a barred spiral galaxy because the shape of its bulge appears to be somewhat elongated.

Other galaxies have disks but appear to lack spiral arms (Figure 20.5). These are called *lenticular galaxies*, because they look lens-shaped when seen edge-on (*lenticular* means "lens-shaped"). Although they look like spiral galaxies without arms, lenticular galaxies might more appropriately be

Figure 20.4 NGC 1300, a barred spiral galaxy about 150,000 light-years in diameter.

Figure 20.5 The central part of NGC 2787, a lenticular galaxy. A few streaks of dusty gas can be seen in this galaxy's disk, but it does not contain any noticeable spiral arms. The region pictured is about 4,400 light-years across.

considered an intermediate class between spirals and ellipticals, because they tend to have less cool gas than normal spirals but more than ellipticals.

Among large galaxies in the universe, most (75–85%) are spiral or lenticular. (Spiral and lenticular galaxies are much rarer among small galaxies.) Spiral galaxies are often found in loose collections of several galaxies, called **groups**, that extend over a few million light-years (Figure 20.6). Our Local Group is one example, with two large spirals: the Milky Way and the Great Galaxy in Andromeda (M 31). Lenticular galaxies are particularly common in **clusters** of galaxies, which can contain hundreds and sometimes thousands of galaxies extending over more than 10 million light-years (Figure 20.7).

Figure 20.6 Hickson Compact Group 87, a small group of galaxies consisting of a large edge-on spiral galaxy at the bottom of this photo, two smaller spiral galaxies at the center and upper left, and an elliptical galaxy to the right. The whole group is about 170,000 light-years in diameter. (The other objects in this photograph are foreground stars in our own galaxy.)

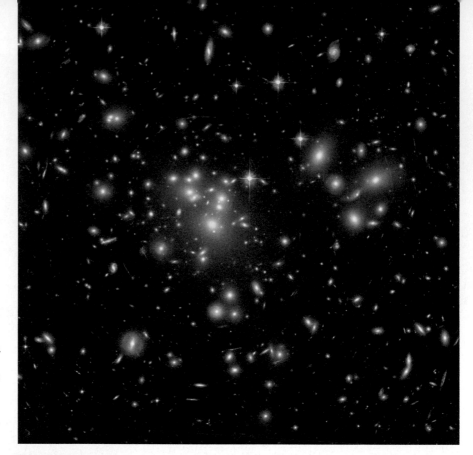

Figure 20.7 Central part of the galaxy cluster Abell 1689. The region pictured is about 2 million light-years across. Almost every object in this photograph is a galaxy belonging to the cluster. Yellowish elliptical galaxies outnumber the whiter spiral galaxies. A few stars from our own galaxy appear in the foreground, looking like white dots with four spikes in the form of a cross.

Elliptical Galaxies

The major difference between elliptical and spiral galaxies is that ellipticals lack a significant disk component (Figure 20.8). Thus, an elliptical galaxy has only a spheroidal component and looks much like the bulge and halo of a spiral galaxy. (In fact, elliptical galaxies are sometimes called *spheroidal galaxies*.)

Most of the interstellar medium in large elliptical galaxies consists of low-density, hot, X ray–emitting gas, making it much like the gas in bubbles and superbubbles in the Milky Way [Section 19.2]. Elliptical galaxies usually

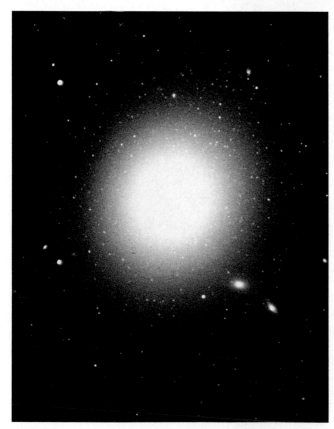

a M 87, a giant elliptical galaxy in the Virgo Cluster. The entire region shown is over 120,000 light-years across.

Figure 20.8 Elliptical galaxies.

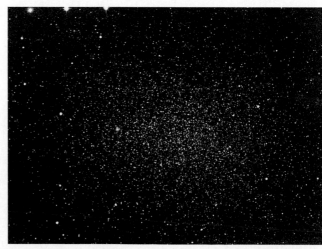

b Leo I, a dwarf elliptical galaxy in the Local Group. It is only about 2,500 light-years across.

a The Large Magellanic Cloud, a small companion to the Milky Way. It is about 30,000 light-years across.

b The Small Magellanic Cloud, a smaller companion to the Milky Way. It is about 18,000 light-years across.

Figure 20.9 Irregular galaxies.

c NGC 1313, an irregular galaxy with scattered patches of star formation. The region pictured is about 50,000 light-years across.

contain very little dust or cool gas, although they are not completely devoid of either. Some have relatively small and cold gaseous disks rotating at their centers. These disks might be the remnants of a collision with a spiral galaxy [Section 21.3].

Elliptical galaxies appear to be more social than spiral galaxies: They are much more common in clusters of galaxies than outside clusters. Elliptical galaxies make up about half the large galaxies in the central regions of clusters, while they represent only a small minority (about 15%) of the large galaxies found outside clusters. However, ellipticals are more common among small galaxies. Particularly small elliptical galaxies with less than a billion stars, called **dwarf elliptical galaxies**, are often found near larger spiral galaxies. At least 10 dwarf elliptical galaxies belong to the Local Group.

Irregular Galaxies

A small percentage of the large galaxies we see nearby fall into neither of the two major categories. This *irregular* class of galaxies is a miscellaneous class, encompassing small galaxies such as the Magellanic Clouds and "peculiar" galaxies that appear to be in disarray (Figure 20.9). These blobby star systems are usually white and dusty, like the disks of spirals. Telescopic observations probing deep into the universe show that distant galaxies are more likely to be irregular in shape than nearby galaxies. Because the light of more distant galaxies was emitted longer ago in the past, these observations tell us that irregular galaxies were more common when the universe was younger.

Hubble's Galaxy Classes

Edwin Hubble invented a system for classifying galaxies that remains widely used (Figure 20.10). It assigns the letter *E*, followed by a number, to each elliptical galaxy. The larger the number, the flatter the galaxy. An E0 galaxy is round, and an E7 galaxy is highly elongated. A spiral galaxy is assigned an uppercase *S*, or *SB* if it has a bar, and a lowercase *a*, *b*, or *c*. The lowercase letter indicates the size of the bulge and the dustiness of the disk. The bulge size decreases from *a* to *c*, while the amount of dusty gas increases.

For example, an Sa galaxy has a large bulge and a modest amount of dusty gas, and an SBc galaxy has a bar,

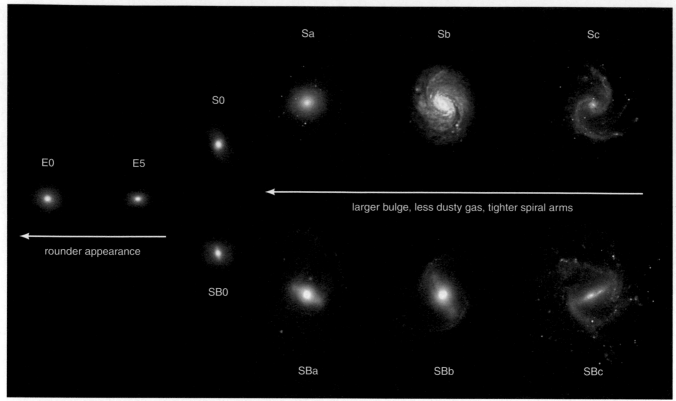

Figure 20.10 Hubble "tuning fork" diagram illustrating Hubble's galaxy classes. An elliptical galaxy is classified with the letter *E* and a number indicating how round it is (a lower number means a rounder galaxy). A spiral galaxy is classified with an *S*, or *SB* for a barred spiral, and a lowercase letter indicating the prominence of its bulge, the tightness of its spiral arms, and the amount of dusty gas it contains. Lenticular galaxies, an intermediate class between spiral and elliptical galaxies, are classified *S0* or *SB0*, depending on whether their bulge is bar-shaped.

a small bulge, and lots of dusty gas. The lenticular galaxies are designated S0, signifying their intermediate spot between spirals and ellipticals. Irregular galaxies are designated Irr.

Astronomers had once hoped that the classification of galaxies might yield deep insights, just as the classification of stars did in the early twentieth century. The Hubble classification scheme itself was suspected for a time to be an evolutionary sequence in which galaxies flattened and spread out as they aged. Unfortunately for astronomers, galaxies turn out to be far more complex than stars, and classification schemes like this one have not led to easy answers about their nature.

 Measuring Cosmic Distances Tutorial, Lessons 1–3

20.3 Measuring Cosmic Distances

To know more about galaxies than just their shape and color, we need to know how far away they are. However, measuring cosmic distances is one of the most challenging tasks we face when trying to understand galaxies and the universe as a whole. Our determinations of astronomical distances depend on a chain of methods that begins with knowing distances in our own solar system. At each link in this chain, we establish standards that help us forge the next link. Because each link depends on preceding links, any inaccuracies accumulate as we move up the chain. Nevertheless, we now can measure the distances to the farthest galaxies to an accuracy of within 15%.

The distance chain begins with radar measurements of the solar system's size. The second link involves parallax measurements of the distances to nearby stars [Section 16.2]. We will now follow the rest of this chain, link by link, to the outermost reaches of the observable universe.

Standard Candles

We can usually measure an object's apparent brightness with a detector, so we can use the luminosity–distance formula [Section 16.2] to calculate the object's luminosity if we know its distance or to calculate its distance if we know its luminosity. For example, suppose we see a distant street lamp and know that all street lamps of its type put out 1,000 watts of light. Then, once we measure the apparent brightness of the street lamp (in units of watts/m²), we can calculate its distance with the luminosity–distance formula.

An object such as a street lamp, for which we are likely to know the true luminosity, represents what astronomers call a **standard candle**. The term *standard candle* is meant to suggest a light source of a known, standard luminosity.

Unlike light bulbs, astronomical objects do not come marked with wattage. Thus, an astronomical object can serve as a standard candle only if we have some way of knowing its true luminosity without first measuring its apparent brightness and distance. Fortunately, many astronomical objects meet this requirement. For example, any star that is a twin of our Sun—that is, a main-sequence star with spectral type G2—should have about the same luminosity as the Sun. Thus, if we measure the apparent brightness of a Sun-like star, we can use the luminosity–distance formula to estimate its distance by assuming that it has the Sun's luminosity (3.8×10^{26} watts).

Beyond the few hundred light-years for which we can measure distances by parallax, we use standard candles for most cosmic distance measurements. These distance measurements always have some uncertainty, because no astronomical object is a perfect standard candle. The challenge of measuring astronomical distances comes down to the challenge of finding the objects that make the best standard candles. The more confidently we know an object's true luminosity, the more certain the distance we calculate with the luminosity–distance formula.

Main-Sequence Fitting

We can use Sun-like stars as standard candles because we know that they are similar to the Sun and because we can measure the Sun's luminosity quite easily. However, Sun-like stars are relatively dim, and we cannot detect them at great distances. To measure distances beyond 1,000 light-years or so, we need brighter standard candles.

An obvious first choice is to use brighter main-sequence stars. However, before we can use any main-sequence star as a standard candle, we must first have some way of knowing its true luminosity. Thus, we must follow two steps to use bright main-sequence stars as standard candles:

1. We identify a star cluster that is close enough for us to determine its distance by parallax and plot its H–R diagram. Because we know the distances to the cluster stars, we can use the luminosity–distance formula to establish their true luminosities from their apparent brightnesses.

2. We can look at stars in other clusters that are too far away for parallax measurements and measure their apparent brightnesses. If we assume that main-sequence stars in other clusters have the same true luminosities as their counterparts in the nearby cluster, we can calculate their distances with the luminosity–distance formula.

Twentieth-century astronomers laid the groundwork for this technique by calibrating the luminosities on a "standard" H–R diagram. This calibration relied largely on a single, nearby star cluster—the *Hyades Cluster* in the constellation Taurus. The Hyades distance is now known from its parallax, so the true luminosities of its stars can be calculated with the luminosity–distance formula.

Astronomers find the distances to other star clusters by comparing the apparent brightnesses of their main-sequence stars with those in the Hyades Cluster and assuming that all main-sequence stars of the same color have the

---

Mathematical Insight **20.1** **The Luminosity–Distance Relation Revisited**

As we saw in Mathematical Insight 16.1, the luminosity–distance formula arises from the fact that apparent brightness decreases with the square of the distance to a light source. That is why we usually write it in the following form:

$$\text{apparent brightness} = \frac{\text{luminosity}}{4\pi \times (\text{distance})^2}$$

We can usually measure apparent brightness. Thus, we generally use this formula either to calculate luminosity when we know distance or to calculate distance when we know luminosity. We can solve the formula to find luminosity by multiplying both sides by $4\pi \times (\text{distance})^2$:

$$\text{luminosity} = 4\pi \times \text{apparent brightness} \times (\text{distance})^2$$

We can solve for the distance by dividing both sides of this version of the formula by ($4\pi \times$ apparent brightness) and then taking the square root of both sides:

$$\text{distance} = \sqrt{\frac{\text{luminosity}}{4\pi \times \text{apparent brightness}}}$$

Example 1: You measure the apparent brightness of a particular star to be 2.5×10^{-10} watt/m^2. A parallax measurement shows

the star's distance to be 42 light-years, or about 4×10^{17} meters. What is the luminosity of the star?

Solution: We simply substitute into the formula for the luminosity:

$$\text{luminosity} = 4\pi \times \left(2.5 \times 10^{-10}\ \frac{\text{watt}}{\text{m}^2}\right) \times (4 \times 10^{17}\ \text{m})^2$$

$$= 5 \times 10^{26}\ \text{watts}$$

The star's luminosity is 5×10^{26} watts, or about 25% greater than the Sun's luminosity of 3.8×10^{26} watts.

Example 2: Find the distance to a Sun-like star ($L = 3.8 \times 10^{26}$ watts) whose apparent brightness at Earth is 1.0×10^{-10} watt/m^2.

Solution: We substitute into the formula for the distance:

$$\text{distance} = \sqrt{\frac{3.8 \times 10^{26}\ \text{watts}}{4\pi \times \left(1.0 \times 10^{-10}\ \frac{\text{watt}}{\text{m}^2}\right)}} = 5.5 \times 10^{17}\ \text{m}$$

The distance to the star is 5.5×10^{17} meters, or about 59 light-years.

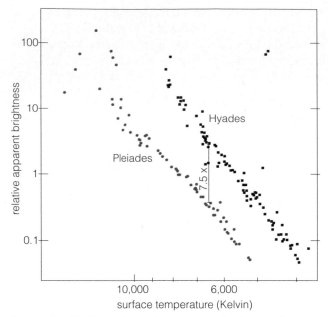

Figure 20.11 Comparison of the apparent brightness of stars in the Hyades Cluster with those in the Pleiades Cluster shows that the Pleiades are 2.75 times farther away because they are $2.75^2 = 7.5$ times dimmer.

.same luminosity (Figure 20.11). This technique of determining distances by comparing main sequences in different star clusters is called **main-sequence fitting**.

The Hyades Cluster does not contain stars of every spectral type. Building a complete, standard H–R diagram therefore required that astronomers use main-sequence fitting to find the distances to many nearby star clusters until every spectral type was represented. Today, with the standard H–R diagram well understood, we can use main-sequence fitting to measure distances to any star cluster near enough for us to identify individual main-sequence stars.

In principle, we need to measure the apparent brightness of only a single star in a cluster to determine the cluster's distance. Once we have determined the star's spectral type, we can use its expected true luminosity and the luminosity–distance formula to calculate its distance. This method is often used for measuring distances to isolated stars that are not members of clusters. The advantage of main-sequence fitting over using a single star in a cluster is that it reduces the uncertainty in the distance calculation. By comparing entire main sequences, we are essentially comparing many different luminosities at once, and thus we achieve much more precise distance measurements.

Cepheid Variables

Main-sequence fitting works well for measuring distances to star clusters throughout the Milky Way, but not for measuring distances to other galaxies. Most main-sequence stars are too faint to be seen in other galaxies, even with our largest telescopes. Instead, we need very bright stars to serve as standard candles for distance measurements beyond the

Milky Way. The most useful bright stars are the *Cepheid variables* [Section 16.5].

Recall that Cepheids are *pulsating variable stars* that follow a simple *period–luminosity relation:* The longer the time period between peaks in brightness, the greater the luminosity of the Cepheid variable star (Figure 20.12). All Cepheids of a particular period have very nearly the same luminosity (within about 10%), so Cepheids effectively scream out their true luminosities when we measure their periods. "Hey, everybody, my luminosity is 10,000 solar luminosities!" is how we would translate the message of a Cepheid variable whose brightness peaks every 30 days. Thus, Cepheid variables are the primary standard candles used to determine distances to nearby galaxies.

Cepheids also played a key role in the discovery that the Milky Way is only one of billions of galaxies in the universe. The only galaxies easily visible to the naked eye are the Andromeda Galaxy and the Large and Small Magellanic Clouds (which appear as little more than fuzzy, glowing blobs). By the mid-1700s, telescopes revealed that the Andromeda Galaxy was really a much larger spiral-shaped structure and that many similar spiral-shaped structures littered the sky. Many astronomers, who thought these spirals were spinning gas clouds inside our own galaxy, called them "spiral nebulae." Around the same time, some creative thinkers guessed the truth about "spiral nebulae"— that they are galaxies like the Milky Way.

Here is a 1755 quote from German philosopher Immanuel Kant, as translated by Edwin Hubble in his book *The Realm of the Nebulae:*

> *We see that scattered through space out to infinite distances, there exist similar systems of stars [i.e., galaxies], and that creation, in the whole extent of its infinite grandeur, is everywhere organized into systems whose members are in relation with one another. . . . A vast*

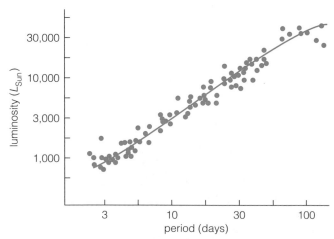

Figure 20.12 Cepheid period–luminosity relation. The data show that all Cepheids of a particular period have very nearly the same luminosity. (Cepheids actually come in two types with two different period–luminosity relations. The relation here is for Cepheids with heavy-element content similar to that of our Sun, or "Type I Cepheids.")

a Edwin Hubble at the Mount Wilson Observatory.

b The Hubble Space Telescope in orbit around Earth.

Figure 20.13 Edwin Hubble and his namesake.

*field lies open to discoveries and observation alone
will give the key.*

Following Kant, some later writers referred to galaxies as
"island universes," because they appeared to be vast, self-
contained communities of stars, isolated from one another
by chasms of empty space. However, these views could not

be conclusively established with the evidence available at
the time. As recently as the 1920s, many astronomers still
held that the "spiral nebulae" were merely clouds of gas
within the Milky Way—and therefore that the Milky Way
represented the entire universe.

Edwin Hubble put this debate to rest in 1924 (Figure 20.13). Using the new, 100-inch telescope atop southern California's Mount Wilson (Figure 20.14)—the largest
telescope in the world at the time—he discovered Cepheid
variables in the Andromeda Galaxy by comparing photographs of the galaxy taken days apart. He used the period–
luminosity relation to determine the luminosities of these
stars and then used these luminosities in the luminosity–
distance formula to compute their distances. (At the time,
Hubble was unaware that there actually are two different

Figure 20.14 The 100-inch telescope on
Mount Wilson, outside Los Angeles.

types of Cepheids, depending on their heavy-element content, with each type obeying a different period–luminosity relation. As a result, he underestimated the true distance of the Cepheid variable stars in Andromeda by about half.)

Hubble's distance measurements proved that the Andromeda Galaxy sat far beyond the outer reaches of stars in the Milky Way, demonstrating that it is a separate galaxy. This single stroke of scientific discovery dramatically changed our view of the universe. Rather than inhabiting a universe that ended with the Milky Way, we suddenly knew that we live in just one among billions of galaxies.

 Hubble's Law Tutorial, Lesson 1

Hubble's Law

Hubble's determination of the distance to the Andromeda Galaxy assured him a permanent place in the history of astronomy, but he didn't stop there. Hubble proceeded to estimate the distances to many more galaxies. Within just a few years, Hubble made one of the most astonishing discoveries in the history of science: that the universe is expanding.

Astronomers had known since the 1910s that the spectra of most "spiral nebulae" tended to be *redshifted*. In other words, each of the emission and absorption lines in their spectra appeared at longer (redder) wavelengths than expected (Figure 20.15). Recall that redshifts occur when the object emitting the radiation is moving away from us [Section 6.5]. Because Hubble had not yet proved that the "spiral nebulae" were distant galaxies, no one understood the true significance of their motions.

Figure 20.15 Redshifted galaxy spectrum. The gray line shows the spectrum of light originally emitted by the galaxy. The blue line shows the spectrum we observe, which is shifted by 5% to longer (redder) wavelengths, indicating that this galaxy is moving away from ours at 5% of the speed of light.

Following his discovery of Cepheids in Andromeda in 1924, Hubble and his coworkers spent the next few years busily measuring the redshifts of galaxies and estimating their distances. Because even Cepheids were too dim to be seen in most of these galaxies, Hubble needed brighter standard candles for his distance estimates. One of his favorite techniques was to assume that the brightest object in each galaxy—which he assumed to be a bright star—had about

---

Mathematical Insight **20.2** **Redshift**

The *redshift* of an object is the difference between the observed wavelength ($\lambda_{observed}$) of a line in the object's spectrum and the wavelength the line would have if the object were standing still (λ_{rest}). (The Greek letter λ is usually used to stand for wavelength.) Redshifts are best expressed as fractional differences:

$$\text{redshift} = z = \frac{\lambda_{observed} - \lambda_{rest}}{\lambda_{rest}}$$

We use the letter z to represent redshift. The value of z is the same for every line in a moving object's spectrum.

For a relatively nearby galaxy, with a redshift z that is much less than 1, we can find its velocity away from us with the following simple formula:

$$v = c \times z$$

where $c = 3.0 \times 10^5$ km/s is the speed of light. (A more complex formula can be used when the redshift is *not* much less than 1, but as we will discuss later in this chapter, the idea of speed becomes hard to define for such distant galaxies.)

Example: Two emission lines from hydrogen in the visible spectrum have rest wavelengths of 656.3 nm and 486.1 nm. Suppose

you see this pair of hydrogen emission lines at 662.9 nm and 491.0 nm in the spectrum of a distant galaxy. What is the redshift of the galaxy? How fast is it moving away from us?

Solution: For the two different lines, we find a redshift z of:

$$656.3\text{-nm line:} \quad z = \frac{662.9 \text{ nm} - 656.3 \text{ nm}}{656.3 \text{ nm}} = 0.010$$

$$486.1\text{-nm line:} \quad z = \frac{491.0 \text{ nm} - 486.1 \text{ nm}}{486.1 \text{ nm}} = 0.010$$

The redshift of the galaxy is $z = 0.010$. As we should expect, we find the same redshift from either line. (Astronomers generally try to measure the redshifts of at least two lines in any particular spectrum to make sure that they aren't mistaking a misidentified line for a shifted line.)

Because this redshift is much less than 1, we can use the simple formula to find the galaxy's speed:

$$v = c \times z = (3.0 \times 10^5 \text{ km/s}) \times 0.01 = 3,000 \text{ km/s}$$

The galaxy is receding from us at 3,000 km/s, equivalent to 1% of the speed of light.

Figure 20.16 Hubble's original velocity–distance diagram. Although Hubble underestimated the true galactic distances, he still discovered the general trend (represented by the straight lines) in which more distant galaxies move away from us at higher speeds. He determined the galaxy speeds by measuring redshifts in the galaxy spectra.

the same luminosity, so he could use these objects as standard candles.

In 1929, Hubble announced his conclusion: The more distant a galaxy, the greater its redshift and hence the faster it moves away from us (Figure 20.16). Hubble's original assertion was based on an amazingly small sample of galaxies. Even more incredibly, he had grossly underestimated the luminosities of his standard candles. The "brightest stars" he had been using as standard candles were really entire *clusters* of bright stars. Fortunately, Hubble was both bold and lucky. Subsequent studies of much larger samples of galaxies showed that they are indeed receding from us, but they are even farther away than Hubble thought.

We can express the idea that more distant galaxies move away from us faster with a very simple formula, now known as **Hubble's law**:

$$v = H_0 \times d$$

where *v* stands for velocity (sometimes called a *recession velocity*), *d* stands for distance, and H_0 (pronounced "H-naught") is a number called **Hubble's constant**. Astronomers generally quote the value of Hubble's constant in *kilometers per second per megaparsec* (km/s/Mpc). Dividing both sides of Hubble's law by H_0 puts it into a form in which we can use a galaxy's velocity to determine its distance:

$$d = \frac{v}{H_0}$$

In principle, applying this law is one of the best ways to determine distances to galaxies. First we find the velocity of a distant galaxy from its redshift. Then we divide this velocity by H_0 to find its distance. However, we encounter two major practical difficulties in finding distances in this way:

1. Galaxies do not obey Hubble's law perfectly because they can have velocities relative to one another caused by gravitational tugs that Hubble's law does not take into account.

2. Even when galaxies do obey Hubble's law well, the distances we find with it are only as accurate as our best measurement of Hubble's constant.

The first problem makes Hubble's law difficult to use for nearby galaxies. Within the Local Group, for example, Hubble's law does not work at all: The galaxies in the Local Group are gravitationally bound together with the Milky Way and therefore are *not* moving away from us in accord with Hubble's law. However, Hubble's law works fairly well for more distant galaxies. The recession speeds of galaxies at large distances are so great that any motions caused by the gravitational tugs of neighboring galaxies are tiny in comparison.

The second problem means that, even for distant galaxies, we can know only *relative* distances until we pin down the true value of H_0. For example, Hubble's law tells us that a galaxy moving away from us at 20,000 km/s is twice as far away as one moving at 10,000 km/s, but we can determine the actual distances of the two galaxies only if we know H_0. The quest to measure H_0 accurately has been one of the main missions of the Hubble Space Telescope.

THINK ABOUT IT

Suppose a galaxy is moving away from us at 10,000 km/s. Use Hubble's law in the form $d = v/H_0$ to calculate its distance (in Mpc) if $H_0 = 70$ km/s/Mpc. What is its distance if $H_0 = 60$ km/s/Mpc?

Seeking Distant Standards

The process of measuring H_0 is rather like the process of calibrating a scale. We can calibrate a scale by, for example, making sure that it reads 10 pounds when we place a 10-pound weight on it, 20 pounds when we place a 20-pound weight on it, and so on. Once it is properly calibrated, we can use it to weigh objects whose weights are not known.

In a similar way, astronomers calibrate Hubble's law by making sure that it gives correct distances for galaxies whose distances we already know from applying some other method. However, until recently, telescopic technology was insufficient for measuring distances accurately by other methods. We simply were unable to detect good standard candles, such as Cepheids, in galaxies much beyond the Local Group.

The superior resolving power of the Hubble Space Telescope now enables us to measure the periods and apparent brightnesses of Cepheid variable stars in galaxies as distant as the Virgo Cluster of galaxies (Figure 20.17), about 55 million light-years away. This distance may sound quite far, but it is still near enough that gravitational tugs cause noticeable deviations from Hubble's law. Nevertheless, these Cepheids can help us by enabling us to calibrate even brighter standard candles, such as supernovae.

White Dwarf Supernovae Recall that white dwarf supernovae are thought to represent exploding white dwarf stars that have reached the 1.4-solar-mass limit and should all have nearly the same luminosity [Section 18.2]. Thus, white dwarf supernovae should be good standard candles—once we know their true luminosities. Although only a few supernovae have been detected during the past century in galaxies within about 50 million light-years of the Milky Way, astronomers have kept careful records of these events. Today, we can look back at the light curves for these events to determine which of these past supernovae were white dwarf supernovae (as opposed to massive star supernovae).

Once we have identified a white dwarf supernova in the historical records, we can look for Cepheids in the same galaxy. We then use the Cepheids as standard candles to determine the distance to the galaxy in which the supernova occurred. Once we know the galaxy's distance, we can use

---

Mathematical Insight **20.3** **Understanding Hubble's Law**

Hubble's law is one of the most important tools of modern astronomy. Let's briefly investigate the units of Hubble's constant, H_0, and how it is used by astronomers.

Hubble's Constant

Astronomers commonly quote Hubble's constant in units of *kilometers per second per megaparsec*, abbreviated km/s/Mpc. What do these strange-sounding units mean? They tell us that if we measure the distance to a distant galaxy in megaparsecs, then Hubble's law will give us the galaxy's velocity in kilometers per second. The idea should become clear with some simple examples.

Example 1: Current measurements of the rate of universal expansion indicate that the value of Hubble's constant is $H_0 \approx 70$ km/s/Mpc. According to Hubble's law, what velocity should we expect for a galaxy located 10 megaparsecs (32.6 million light-years) away?

Solution: We are given the galaxy's distance, $d = 10$ Mpc. We plug this distance and the given value of Hubble's constant into Hubble's law:

$$v = H_0 \times d = 70 \, \frac{\text{km/s}}{\text{Mpc}} \times 10 \, \text{Mpc} = 700 \text{ km/s}$$

According to Hubble's law, this galaxy should be moving away from us at a speed of 700 km/s—assuming its speed is unaffected by other factors, such as the gravitational pull of other galaxies.

Example 2: Using the same value for Hubble's constant as in Example 1, what velocity would we expect for a galaxy located 11 megaparsecs away?

Solution: We could plug in the values in the same way we did for Example 1. However, there's an even easier method. We already know that a galaxy located at 10 megaparsecs would be moving away at a speed of 700 km/s. Hubble's constant of 70 km/s/Mpc tells us that for each megaparsec of distance, a galaxy should have an additional 70 km/s of speed. Thus, a galaxy at a distance of 11 Mpc should be moving away from us 70 km/s faster than a galaxy at a distance of 10 Mpc, which means it should be moving at 770 km/s. Keep in mind that the galaxy would have this precise velocity only if the galaxy's speed were determined by the expan-

sion of the universe alone and not by any other gravitational pulls—which is rarely the case.

Example 3: Hubble's constant is sometimes quoted in units of kilometers per second per million light-years. Assuming Hubble's constant is $H_0 = 70$ km/s/Mpc, what is its value in units of kilometers per second per million light-years?

Solution: One megaparsec is equivalent to 3.26 million light-years. The value of 70 km/s/Mpc tells us that a galaxy should have an additional 70 km/s of speed for each megaparsec of distance. Because 1 million light-years is smaller than a megaparsec by a factor of 3.26, the additional velocity per million light-years should be (70 km/s)/3.26, or about 21.5 km/s. We can see exactly how the units work out by doing the division:

$$H_0 = 70 \, \frac{\text{km/s}}{\text{Mpc}} \times \frac{1 \, \text{Mpc}}{3.26 \text{ million ly}} = 21.5 \text{ km/s/million ly}$$

A value for Hubble's constant of 70 kilometers per second per megaparsec is equivalent to a value of 21.5 kilometers per second per million light-years.

Using Hubble's Law to Estimate Distances

In reality, we rarely use Hubble's law to calculate velocities, because we can determine a galaxy's velocity directly by measuring its redshift. Instead, we use Hubble's law to estimate distances to far-off galaxies whose distances we are unable to measure with other techniques.

Example 4: Estimate the distance to a galaxy whose redshift indicates that it is moving away from us at a speed of 14,000 km/s. Assume that the value of Hubble's constant is $H_0 = 70$ km/s/Mpc.

Solution: We solve Hubble's law for distance and then plug in the galaxy's velocity and our value for Hubble's constant:

$$d = \frac{v}{H_0} = \frac{14,000 \text{ km/s}}{70 \text{ km/s/Mpc}} = 200 \text{ Mpc}$$

The galaxy is located approximately 200 megaparsecs away. Because 1 Mpc = 3.26 million light-years, this is equivalent to 652 million light-years.

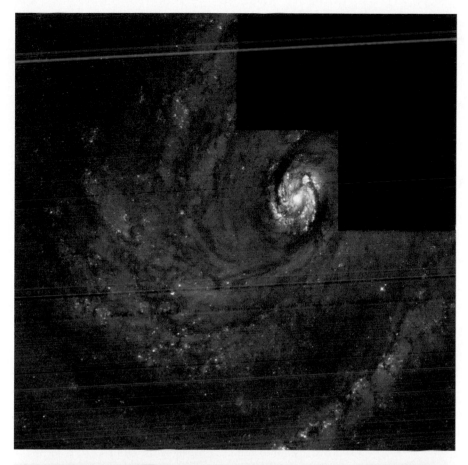

a Galaxy M 100 in the Virgo Cluster. The region pictured is about 40,000 light-years across.

b One of the Cepheid variable stars in M 100 varying in brightness over several weeks. Measurements of Cepheid variables place M 100 at about 56 million light years from Earth. The variable star is at the center of each picture.

Figure 20.17 Images from the Hubble Space Telescope project to measure Hubble's constant.

the luminosity–distance formula to determine the supernova's true luminosity.

This technique has allowed researchers using the Hubble Space Telescope to determine the true luminosities of several white dwarf supernovae. As expected, they are all about the same, and we can now use them as reliable standard candles (Figure 20.18). Because white dwarf supernovae are so bright—about 10 billion solar luminosities at

their peak—we can use them to measure distances to galaxies billions of light-years away (Figure 20.19).

Tully–Fisher Relation Although white dwarf supernovae are excellent standard candles, most galaxies have not hosted supernovae during the time that humans have been watching them. Thus, it would be useful if we could use galaxies themselves as standard candles.

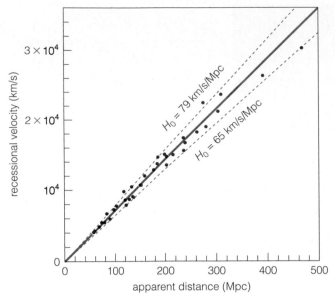

Figure 20.18 White dwarf supernovae can be used as standard candles to establish Hubble's law out to very large distances. The points on this figure show the apparent distances of white dwarf supernovae and the recessional velocities of the galaxies in which they exploded. The fact that these points all fall close to a straight line demonstrates that these supernovae are good standard candles. Calibrating the luminosities of selected white dwarf supernovae using Cepheid variables indicates that Hubble's constant lies between 65 and 79 km/s/Mpc.

Astronomers have discovered a close relationship between the total luminosities of spiral galaxies and the rotation speeds of their disks: The faster a spiral galaxy's rotation speed, the more luminous it is (Figure 20.20). This relationship, called the **Tully–Fisher relation** (after its discoverers), holds because both luminosity and rotation speed depend on the galaxy's mass. A galaxy's luminosity depends on the number of stars it contains, which is related to the total amount of matter within it, and the total amount of matter determines a galaxy's rotation speed [Section 19.4].

Figure 20.19 Photo of a white dwarf supernova in a galaxy 7 billion light-years away, halfway across the observable universe, recorded in March 1996 by the Hubble Space Telescope.

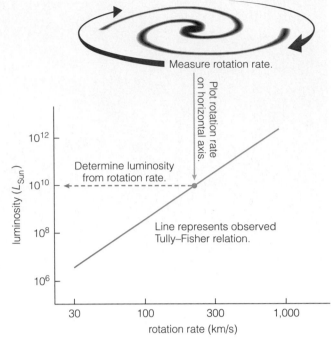

Figure 20.20 A schematic diagram of the Tully–Fisher relation. (The precise relation differs among different subclasses of spiral galaxies.)

We can measure the rotation speed of a spiral galaxy's disk by, for example, comparing the Doppler shifts of the portion of the disk rotating toward us and the portion rotating away from us. Once we have measured the rotation speed of a spiral galaxy, the Tully–Fisher relation tells us its true luminosity, which makes the galaxy itself a standard candle that we can use to determine its distance.

Despite the new calibrations of Hubble's law made possible by the Hubble Space Telescope, some uncertainties remain. These uncertainties should decrease as we measure the distances to more and more galaxies. As of 2003, the true value of H_0 appears to be somewhere between 65 and 79 km/s/Mpc.

Summary: The Distance Chain

Figure 20.21 summarizes the chain of measurements that allows us to determine ever-greater distances. With each link in the distance chain, however, uncertainties become somewhat greater. Thus, although we know the Earth–Sun distance at the base of the chain extremely accurately, distances to the farthest reaches of the observable universe remain uncertain by about 20%.

● **Radar ranging:** We measure distances within the solar system by bouncing radio waves off planets. Then, with the aid of some geometry, we use these radar measurements to determine the Earth–Sun distance.

● **Parallax:** We measure the distances to nearby stars by observing how their positions change, relative to the background stars, as Earth moves around the Sun. These

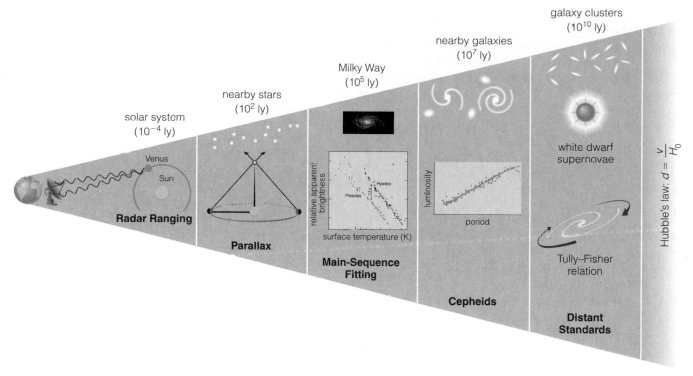

Figure 20.21 Measurement of cosmic distances relies on a chain of interlocking techniques. The chain begins with radar ranging to determine distances within our solar system and proceeds through parallax and standard candle techniques to Hubble's law, which provides distances to galaxies across the observable universe.

distances thus rely on our knowledge of the Earth–Sun distance.

- **Main-sequence fitting:** We know the distance to the Hyades Cluster in our Milky Way Galaxy through parallax. Comparing the observed intensities of its main-sequence stars to the intensities of those in other clusters gives us the distances to these other star clusters in our galaxy.

- **Cepheid variables:** By studying Cepheids in star clusters with distances measured by main-sequence fitting, we learn the precise period–luminosity relation for Cepheids. When we find a Cepheid in a more distant star cluster or galaxy, we can determine its true luminosity by measuring the period between its peaks in brightness and then use this true luminosity to determine the distance.

- **Distant standards:** By measuring distances to relatively nearby galaxies with Cepheids, we learn the true luminosities of white dwarf supernovae and the Tully–Fisher relation between the luminosities and rotation speeds of spiral galaxies. Thus, supernovae and entire galaxies become useful standard candles for measuring great distances in the universe.

- **Hubble's law:** Distances measured to galaxies with white dwarf supernovae and the Tully–Fisher relation allow us to calibrate Hubble's law and measure Hubble's constant, H_0. Once we know H_0, we can determine a galaxy's distance directly from its redshift.

 Hubble's Law Tutorial, Lessons 2, 3

20.4 Measuring Cosmic Ages

The ages of the galaxies we see in deep images of the universe, such as Figure 20.1, are directly related to their distances. Remember that light travels at a finite speed. Light from the most distant galaxies began its journey to Earth when the universe was a small fraction of its current age. We are therefore viewing these galaxies as they were when they were young, and the light we are detecting is ancient.

Because a deep image records galaxies at many different distances, it contains information about galaxies of many different ages. For example, we see galaxies 1 billion light-years away as they were 1 billion years ago and galaxies 2 billion light-years away as they were 2 billion years ago. The history of galaxies across the universe is therefore encoded in such images, but to decipher the clues they hold we need to know how the universe's history is related to its expansion. The intimate relationship between the universe's expansion rate and its age helps us age-date galaxies throughout the observable universe.

Universal Expansion

Galaxies all across the universe are moving away from one another in accordance with Hubble's law. This fact implies that galaxies must have been closer together in the past. Tracing this convergence back in time, we reason that all the matter in the observable universe started very close together and that the entire universe came into being at a single moment. Ever since that moment the universe has been expanding, and all the while the gravitational pull of each galaxy on every other galaxy has been working to slow the expansion rate. In the densest regions of the universe, gravity has won out over expansion. The Local Group no longer expands, nor do other clusters of galaxies, but the expansion continues on all larger scales.

Visualizing this expansion requires a little thought. Upon first hearing about the expansion of the universe, most people are tempted to think of the universe as a ball of galaxies expanding into a void. This impression is mistaken. The universe expands, but as far as we can tell it is not expanding into empty space. Each point moves away from every other point, and no matter where you are located the distribution of galaxies around you will look more or less the same. (This idea, called the *Cosmological Principle,* states that matter in the universe is evenly distributed on scales much larger than superclusters of galaxies. It is impossible to prove that the Cosmological Principle holds outside the observable regions of the universe, but within the observable universe matter does appear to be distributed more evenly as we look on increasingly larger scales [Section 22.5].)

How can the universe be expanding if it's not expanding *into* anything? Back in Chapter 1, we likened the expanding universe to a raisin cake baking [Section 1.3], but a cake has a center and edges that grow into empty space as it bakes. A better analogy is something that can expand but that has no center and no edges. The surface of a balloon can fit the bill, as can an infinite surface such as a flat sheet of rubber that extends to infinity in all directions. Because it's hard to visualize infinity, let's use a balloon as our model of the expanding universe.

In a balloon model of the universe, we interpret the balloon's *surface* as representing all three dimensions of space. The spaces inside and outside the balloon have no meaning in this model. Aside from this mental leap, the balloon model works well because its spherical surface has no center and no edges, just as no city is the center of Earth's surface and no edges exist where you could walk or sail off the Earth. We can represent clusters and groups of galaxies with plastic polka dots attached to the balloon, and we can make our model universe expand by inflating the balloon (Figure 20.22).

As the real universe expands, the distances between clusters of galaxies grow. However, the clusters themselves do not grow because gravity binds each cluster's galaxies together. In a similar way, as the surface of the balloon expands, the polka dots move apart, but the dots themselves don't grow. Suppose that, 1 second after you begin blowing into the balloon, dots A, B, C, and D are spaced 1 cm apart along a line. After 2 seconds, the distances between neighboring dots grow to 2 cm, and after 3 seconds to 3 cm. If you recorded the distances of the other dots from dot B once each second, your observations would look like this:

Measuring Distances from Dot B

Time	Dot A Distance	Dot B Distance	Dot C Distance	Dot D Distance
1 s	1 cm	0 cm	1 cm	2 cm
2 s	2 cm	0 cm	2 cm	4 cm
3 s	3 cm	0 cm	3 cm	6 cm

Miniature scientists on dot B might be tempted to think that they are at the center of some explosion. After all, they see every other dot moving away from them, and the farthest dots move fastest. However, miniature scientists on dot C would record a very similar set of observations:

Measuring Distances from Dot C

Time	Dot A Distance	Dot B Distance	Dot C Distance	Dot D Distance
1 s	2 cm	1 cm	0 cm	1 cm
2 s	4 cm	2 cm	0 cm	2 cm
3 s	6 cm	3 cm	0 cm	3 cm

The miniature scientists on dot C would also be tempted to think that they are at the center of some explosion. Eventually, the miniature scientists would realize that the balloon itself is expanding. Just as we cannot point to a single spot on a balloon's surface and say "the balloon is expanding from here," we cannot identify a single point in space from which the universe is expanding. Galaxies surround each point in the universe on all sides and move away from that point in exactly the same way.

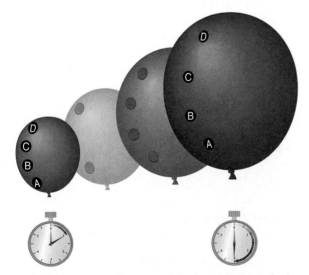

Figure 20.22 As the balloon expands, dots move apart in the same way that galaxies move apart in our expanding universe.

Hubble's Constant and Age

To see how the universe's expansion rate is related to its age, let's return to the miniature scientists living on dot B. Three seconds after the balloon began to expand, they would measure the following:

Dot A is 3 cm away and moving at 1 cm/s.

Dot C is 3 cm away and moving at 1 cm/s.

Dot D is 6 cm away and moving at 2 cm/s.

They could summarize these observations as follows: *Every dot is moving away from our home with a speed that is 1 cm/s for each 3 cm of distance.* Because the expansion of the balloon is uniform, scientists living on any other dot would come to the same conclusion. Each scientist living on the balloon would determine that the following formula relates the distances and velocities of other dots on the balloon:

$$v = \left(\frac{1 \text{ cm/s}}{3 \text{ cm}}\right) \times d \quad \text{or} \quad v = \left(\frac{1}{3 \text{ s}}\right) \times d$$

where v and d are the velocity and distance of any dot, respectively.

THINK ABOUT IT

Confirm that this formula gives the correct values for the speeds of dots C and D, as seen from dot B, 3 seconds after the balloon began expanding. How fast would a dot located 9 cm from dot B move, according to the scientists on dot B?

If the miniature scientists think of their balloon as a bubble, they might call the number relating distance to velocity—the term $\frac{1}{3 \text{ s}}$ in the above formula—the "bubble constant." An especially insightful miniature scientist might flip over the "bubble constant" and find that it is exactly equal to the time since the balloon started expanding. That

is, the "bubble constant" $\frac{1}{3 \text{ s}}$ tells them that the balloon has been expanding for 3 seconds. Perhaps you see where we are heading.

Just as the inverse of the "bubble constant" tells the miniature scientists that their balloon has been expanding for 3 seconds, the inverse of the Hubble constant, or $1/H_0$, tells us something about how long our universe has been expanding. The "bubble constant" for the balloon depends on when it is measured, but it is always equal to 1/(time since balloon started expanding). Similarly, the Hubble constant actually changes with time but stays roughly equal to 1/(age of the universe). We call it a constant because it is the same at all locations in the universe. Moreover, its value does not change noticeably on the time scale of human civilization.

Current estimates based on the value of Hubble's constant put the age of the universe between about 12 and 15 billion years. To derive a more precise value for the universe's age, we need to know whether the expansion has been speeding up or slowing down over time, a question we will examine more closely in Chapter 22. If the gravitational pull of each galaxy on every other galaxy has significantly slowed the expansion rate, then the universe's age is somewhat less than $1/H_0$. If some mysterious force has accelerated the expansion rate, then the universe's age is somewhat more than $1/H_0$.

Intriguingly, the oldest stars in globular clusters appear to be somewhere between 12 and 14 billion years old [Section 16.6]. These stars could not possibly have existed before the birth of the universe, so our age estimates based on Hubble's constant probably aren't far off the mark. The best available evidence as of 2003 suggests that the universe is indeed somewhere about 14 billion years old. If you are reading this book at a much later date, the age of the universe might be known to even greater precision.

Mathematical Insight **20.4** **Age from Hubble's Constant**

The reciprocal of Hubble's constant, or $1/H_0$, is directly related to the age of the universe. It tells us what the age of the universe would be *if* the expansion rate has remained constant since the beginning. We can calculate this age from the measured value of H_0. The best current estimates put H_0 between 65 and 79 km/s/Mpc, so let's take the average and assume that $H_0 = 72$ km/s/Mpc.

The easiest way to find the reciprocal of H_0 is to first rewrite it in more convenient units. A megaparsec is about 3.26 million light-years, so we can write our value of Hubble's constant as:

$$H_0 = 72 \frac{\text{km/s}}{\text{Mpc}} = \frac{72 \text{ km/s}}{(3.26 \times 10^6 \text{ ly})}$$

Because a light-year is the speed of light ($c = 300,000$ km/s) times 1 year, we can further rewrite this value of Hubble's constant as:

$$H_0 = \frac{72 \text{ km/s}}{(3.26 \times 10^6) \times (3 \times 10^5 \text{ km/s}) \times 1 \text{ yr}}$$

Taking the reciprocal, we find:

$$\frac{1}{H_0} = \frac{(3.26 \times 10^6) \times (3 \times 10^5) \times 1 \text{ yr}}{72}$$

$$= \frac{9.78 \times 10^{11} \times 1 \text{ yr}}{72}$$

$$\approx 1.36 \times 10^{10} \text{ yr}$$

The reciprocal is 1.36×10^{10} yr, or 13.6 billion years. In other words, if the expansion rate has never changed, a value of Hubble's constant of $H_0 = 72$ km/s/Mpc implies that the universe is 13.6 billion years old.

Lookback Time

All the galaxies we observe must be younger than the universe itself. Pictures of the farthest galaxies show us how galaxies looked when they were only a few billion years old, while pictures of the closest show galaxies nearly as old as the universe. However, when we try to specify exactly how the distances of galaxies relate to their ages, we run into complications.

Imagine photons of light from a supernova in a distant galaxy traveling toward us at the speed of light. If the supernova occurred in a galaxy 400 million light-years away, we are seeing a supernova that occurred 400 million years ago. But what do we mean by the distance of 400 million light-years? Because the universe is expanding, the distance between Earth and the supernova is greater today than it was at the time of the supernova event. Are we talking about the distance from the supernova to Earth when the supernova exploded, or are we referring to the distance between the supernova and Earth when the light arrived here? Do we mean that photons traveled through 400 million light-years of space?

When we start trying to measure large distances through the universe, our everyday notions of space-only or time-only become inadequate. The path light takes from the supernova to us is really a trajectory through *both* space *and* time, so we must think in terms of *spacetime.*

Figure 20.23 shows a *spacetime diagram* for the supernova and our observation of it [Section S3.3]. The horizontal axis shows distance (through space) from the Milky Way, and the vertical axis shows time. Photons of light from the

supernova travel toward us at the speed of light, starting at the place and time of the supernova explosion. By the time these photons reach us, the galaxy in which the supernova occurred has moved farther away from us.

Looking along the time axis, we see that it has taken 400 million years for the photons of light to reach us. We call this the **lookback time** to the supernova. In other words, a distant object's lookback time is the difference between the current age of the universe and the age of the universe when the light left the object. This quantity, which essentially combines both space and time, is much more meaningful than an actual distance when we discuss objects that are moving away from us with the expansion of the universe.

THINK ABOUT IT

Explain why there is no question about meaning when we talk about distances to objects within the Local Group, but distances become difficult to define when we talk about objects much farther away.

An object's lookback time is directly related to its redshift. We have seen that the redshifts of galaxies tell us how quickly they are moving away from us. In the context of an expanding universe, redshifts have an additional, more fundamental interpretation. Let's return one final time to the universe on the balloon. Suppose you draw wavy lines on the balloon's surface to represent light waves. As the balloon inflates, these wavy lines stretch out, and their wavelengths increase (Figure 20.24). This stretching closely resembles what happens to photons in an expanding universe. The expansion of the universe stretches out all the photons within it, shifting them to longer, redder wavelengths. We call this effect a **cosmological redshift**.

In a sense, we have a choice when we interpret the redshift of a distant galaxy: We can think of the redshift either as being caused by the Doppler effect as the galaxy moves away from us or as being caused by a photon-stretching, cosmological redshift. However, as we look to very distant galaxies, the ambiguity in the meaning of

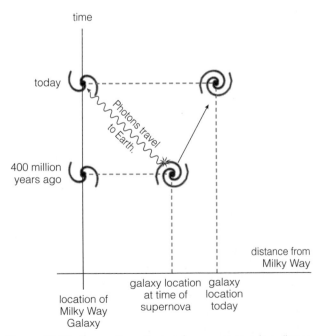

time

today

400 million
years ago

Photons travel
to Earth.

distance from
Milky Way

location of
Milky Way
Galaxy

galaxy location
at time of
supernova

galaxy
location
today

Figure 20.23 A spacetime diagram for a supernova in a distant galaxy. The lookback time to this supernova is 400 million years because that is how long its light took to reach us.

Figure 20.24 As the universe expands, photon wavelengths stretch like the wavy lines on this expanding balloon.

distance also makes it difficult to specify precisely what we mean by a galaxy's *speed.* Thus, it becomes preferable to interpret the redshift as being due to photon stretching in an expanding universe.

From this perspective, it is better to think of space itself as expanding, carrying the galaxies along for the ride, than to think of the galaxies as projectiles flying through a static universe. The cosmological redshift of a galaxy thus tells us how much space has expanded during the lookback time to that galaxy.

The Horizon of the Universe

What lies beyond the youngest, most distant galaxies? When we began our discussion of the expanding universe, we stressed that the universe as a whole does not seem to have an edge. Yet the universe does have a horizon, a place beyond which we cannot see. The **cosmological horizon** that marks the limits of the observable universe is a boundary in time, not in space. It exists because we cannot see back to a time before the universe began. For example, if the universe is 14 billion years old, then no object can have a lookback time greater than 14 billion years. Thus, the lookback time to the cosmological horizon is equal to the age of the universe.

For reasons we'll discuss in Chapter 23, we cannot see matter at the cosmological horizon, but if we could we would see it as it looked at the very beginning of time. One year from now, we would see this same matter as it looked 1 year after the beginning—which means the horizon will have moved somewhat farther from us. In fact, our horizon is continually expanding outward at the speed of light.

Beyond the Horizon

Perhaps you're thinking there must be something beyond the cosmological horizon. After all, the horizon encircles more and more matter with each passing second. This new matter had to come from somewhere, didn't it? The problem with this reasoning is that the cosmological horizon, unlike a horizon on Earth, is a boundary in *time,* not in space.

At any one moment, in whatever direction we look, the cosmological horizon lies at the beginning of time and encompasses a certain volume of the universe. At the next moment, the cosmological horizon still lies at the beginning of time, but it encompasses a slightly larger volume. However, that does not mean that the matter at the outskirts of the observable universe used to lie "beyond the horizon." When we peer into the distant universe, we are looking back in both space and time. We cannot look "past the horizon" because we cannot look back to a time before the universe began.

Our observable universe therefore grows larger in radius by 1 light-year with each passing year.*

Our quest to study galaxies and measure their distances has brought us to the very limits of the observable universe. The galaxies in Figure 20.1 at the beginning of this chapter extend from relatively nearby almost all the way to the horizon. Now it is time to examine what these galaxies have been doing for the past 10 billion or more years—the topic of our next chapter.

Putting Chapter 20 into Context

The picture could hardly get any bigger than it has in this chapter. We have reached the limits of the observable universe in both space and time. If your head hasn't already exploded, try to make sure that these "big picture" ideas fit inside:

- The universe is filled with galaxies that come in a variety of shapes and sizes. The most fundamental distinctions in galaxies are between *disk components,* with stars of many ages and abundant gas for new star formation, and *spheroidal components,* which contain old stars and little gas. Both components are present in spiral galaxies, while elliptical galaxies generally lack significant disks.

- Our measurements of the distances to faraway galaxies rely on a chain of distance measurements that begins with the Earth–Sun distance measured with the aid of radar ranging and continues with parallax and various standard-candle techniques. The distance chain allowed Hubble to measure distances to galaxies and thus to discover what we know as Hubble's law and the expansion of the universe.

- When we say that the universe is expanding, we mean that *space itself* is expanding. The universe is not expanding "into" anything.

- As the universe expands, it carries the galaxies within it along for the ride. From the current expansion rate and from estimates of the overall strength of gravity in the universe and the ages of the oldest stars, we infer that the universe is about 14 billion years old. The galaxies themselves must be somewhat younger than this.

- When we look out into the universe, we are looking back in time. Just short of the cosmological horizon, we are seeing the universe shortly after the moment of creation. We can see no farther.

---

*This definition of the cosmological horizon is strictly true only if the expansion has never accelerated. We will discuss possible accelerations of the expansion in Chapters 22 and 23.

20.1 Islands of Stars

- *How do we study the lives of galaxies?* We piece together observations of different galaxies at different stages of their lives to reconstruct the overall life cycles of galaxies.

20.2 Galaxy Types

- *What are the three major types of galaxies?* (1) Spiral galaxies have prominent disks and spiral arms. (2) Elliptical galaxies are rounder and redder than spiral galaxies and contain less cool gas and dust. (3) Irregular galaxies are neither disklike nor rounded in appearance.

- *What is the primary difference between spiral and elliptical galaxies?* A spiral galaxy has both a disk and a spheroidal component that includes the bulge and the halo. An elliptical galaxy has a similar spheroidal component but lacks a prominent disk component.

- *How are galaxy types different within clusters of galaxies?* Outside clusters, most large galaxies are spirals. Within clusters, elliptical galaxies are much more common, representing up to about half the large galaxies in the central regions of clusters.

20.3 Measuring Cosmic Distances

- *What is a standard candle, and how is it used to determine distance?* A standard candle is an object whose luminosity we can determine without knowledge of its distance. We can determine the distance to a standard candle by measuring its apparent brightness and applying the luminosity–distance formula.

- *Why do Cepheid variable stars make good standard candles?* Cepheids are very bright variable stars that follow a well-known period–luminosity relation. Measuring their period of variability tells us their luminosity.

- *What is Hubble's law, and where is it applicable?* Hubble's law tells us that more distant galaxies are moving away faster: $v = H_0 \times d$, where H_0 is Hubble's constant. It allows us to determine a galaxy's distance from the speed at which it is moving away from us, which we can measure from its Doppler shift. Hubble's law applies only to galaxies far enough away so that any motion due to gravitational pulls from other galaxies is small compared with their overall motion.

- *What are the links in the chain through which we build up measurements of distance throughout the universe?* Radar ranging gives us the Earth–Sun distance (1 AU), which allows us to calculate distances to nearby stars from their parallax. Main-sequence fitting allows us to measure the distances to numerous Cepheids, allowing us to establish their period–luminosity relation. This relationship then allows us to measure distances to nearby galaxies in which we can observe Cepheids. From the brightnesses of white dwarf supernovae and the Tully–Fisher relation between galaxy luminosity and rotation speed, we can determine the value of H_0 in Hubble's law, which we can use to measure vast distances in the universe.

20.4 Measuring Cosmic Ages

- *In what sense does the universe expand?* We say that the universe is expanding because its galaxies are continually moving farther apart. However, the universe is not a ball of galaxies expanding into empty space. As far as we can tell, the universe has no center and no edges.

- *How is Hubble's constant related to the age of the universe?* Hubble's constant tells us the rate at which galaxies are moving apart, so we can use it to determine how long it has been since they started moving apart. In particular, the inverse of Hubble's constant tells us how long it would have taken the universe to reach its present size *if* the expansion rate had never changed. It is only a rough estimate of the true age of the universe, because the expansion rate almost certainly has changed through time.

- *Why is lookback time a better way to describe the locations of distant objects than distances in light-years?* When we state that a galaxy's lookback time is, say, 400 million years, it means that its light has traveled for 400 million years to reach us at Earth. There is no ambiguity in this meaning. In contrast, a "distance" of 400 million light-years could mean its distance today or its distance at the time light left the galaxy—which would have been smaller than the distance today because galaxies were closer together in the past.

- *What defines the size of the observable universe?* The cosmological horizon—the place at which the lookback time is equal to the age of the universe—defines the size of the observable universe. We cannot see beyond the cosmological horizon because that would mean looking back to a time before the universe existed.

? Sensible Statements?

Decide whether each of the following statements is sensible and explain why it is or is not.

1. If you want to find elliptical galaxies, you'll have better luck looking in clusters of galaxies than elsewhere in the universe.

2. Cepheid variables make good standard candles because they all have exactly the same luminosity.

3. If the standard candles you are using are less luminous than you think they are, then the distances you determine from them will be too small.

4. Galaxy A is moving away from me twice as fast as Galaxy B. That probably means it's twice as far away.

5. The lookback time to the Andromeda Galaxy is about 2.5 million years.

6. I'd love to live in one of the galaxies at the very edge of the universe, because I want to see the black void into which the universe is expanding.

7. We can't see galaxies beyond the cosmological horizon because they are moving away from us at speeds faster than the speed of light.

8. We see distant galaxies as they were when the universe was younger, and any people living in those galaxies today would see the Milky Way as it was when the universe was younger.

Problems

9. *Components of Galaxies.* Distinguish between the disk component and the spheroidal component of a spiral galaxy. Which component includes the galaxy's spiral arms? Which includes its bulge? Which includes its halo?

10. *Not Quite Normal.* How does a barred spiral galaxy differ from a normal spiral galaxy? How does a lenticular galaxy differ from a normal spiral galaxy?

11. *Galaxies and Clusters.* How does a group of galaxies differ from a cluster of galaxies?

12. *Large Distances.* Describe how we can use Hubble's law to determine the distance to a distant galaxy. What practical difficulties limit the use of Hubble's law for measuring distances?

13. *White Dwarf Supernovae.* What makes white dwarf supernovae good standard candles? Briefly describe how we have learned the true luminosities of white dwarf supernovae.

14. *Universe on a Balloon.* In what ways is the surface of an expanding balloon a good analogy to the universe? In what ways is this analogy limited? Explain why a miniature scientist living in a polka dot on the balloon would observe all other dots to be moving away, with more distant dots moving away faster.

15. *Cosmological Redshift.* What do we mean by a cosmological redshift? How does our interpretation of a distant galaxy's redshift differ if we think of it as a cosmological redshift rather than as a Doppler shift?

16. *Distance Measurements.* The techniques astronomers use to measure distances are not so different from the ones you use every day. Describe how parallax measurements are similar to your visual depth perception. Describe how standard-candle measurements are similar to the way you estimate the distance to an oncoming car at night.

17. *Cepheids as Standard Candles.* Suppose you are observing Cepheids in a nearby galaxy. You observe one Cepheid with a period of 8 days between peaks in brightness, and another with a period of 35 days. Estimate the luminosity of each star. Explain how you arrived at your estimate. (*Hint:* See Figure 20.12.)

*18. *Counting Galaxies.* Estimate how many galaxies are pictured in Figure 20.1. Explain the method you used to arrive at this estimate. This picture shows about 1/30,000,000 of the sky, so multiply your estimate by 30,000,000 to obtain an estimate of how many galaxies like these fill the entire sky.

*19. *Cepheids in M 100.* Scientists using the Hubble Space Telescope have observed Cepheids in the galaxy M 100. Here are actual data for three Cepheids in M 100:

- Cepheid 1: luminosity = 3.9×10^{30} watts, brightness = 9.3×10^{-19} watt/m^2.

- Cepheid 2: luminosity = 1.2×10^{30} watts, brightness = 3.8×10^{-19} watt/m^2.

- Cepheid 3: luminosity = 2.5×10^{30} watts, brightness = 8.7×10^{-19} watt/m^2.

Compute the distance to M 100 with data from each of the three Cepheids. Do all three distance computations agree? Based on your results, estimate the uncertainty in the distance you have found.

*20. *Redshift and Hubble's Law.* Imagine that you have obtained spectra for several galaxies and have measured the observed wavelength of a hydrogen emission line that has a rest wavelength of 656.3 nm. Here are your results:

- Galaxy 1: Observed wavelength of hydrogen line is 659.6 nm.

- Galaxy 2: Observed wavelength of hydrogen line is 664.7 nm.

- Galaxy 3: Observed wavelength of hydrogen line is 679.2 nm.

a. Calculate the redshift, z, for each of the three galaxies.

b. From their redshifts, calculate the speed at which each of the galaxies is moving away from us. Give your answers both in km/s and as a fraction of the speed of light.

c. Estimate the distance to each galaxy from Hubble's law. Assume that $H_0 = 72$ km/s/Mpc.

Discussion Questions

21. *Cosmology and Philosophy.* One hundred years ago, many scientists believed that the universe was infinite and eternal, with no beginning and no end. When Einstein first developed his general theory of relativity, he found it predicted

that the universe should be either expanding or contracting. He believed so strongly in an eternal and unchanging universe that he modified the theory, a modification he would later call his "greatest blunder." Why do you think Einstein and others assumed that the universe had no beginning? Do you think that a universe with a definite beginning in time, some 14 billion or so years ago, has any important philosophical implications? Explain.

22. *Tired Light.* Hubble's law relates the redshifts of galaxies directly to their distances. The overwhelming majority of astronomers believe that these redshifts arise from the Doppler effect, an easily measurable and well-understood phenomenon that can be measured in the laboratory. However, a few astronomers argue that there might be some other explanation for galaxy redshifts based on physical effects that occur only on vast scales. For example, perhaps photons somehow get "tired" and lose energy as they cross immense intergalactic distances. Which explanation for redshifts do you prefer? Why is considering alternative explanations important in science?

For a complete list of media resources available, go to www.astronomyplace.com and choose Chapter 20 from the pull-down menu.

 Astronomy Place Web Tutorials

Tutorial Review of Key Concepts

Use the interactive **Tutorials** at www.astronomyplace.com to review key concepts from this chapter.

Measuring Cosmic Distances Tutorial

Lesson 1 Radar

Lesson 2 Stellar Parallax

Lesson 3 Standard Candles: Main-Sequence Stars and Cepheid Variables

Lesson 4 Standard Candles: White Dwarf Supernovae and Spiral Galaxies

Hubble's Law Tutorial

Lesson 1 Hubble's Law

Lesson 2 The Expansion of the Universe

Lesson 3 The Age of the Universe

Supplementary Tutorial Exercises

Use the interactive **Tutorial Lessons** to explore the following questions.

Measuring Cosmic Distances Tutorial, Lesson 1

1. How is radar used to determine distances in the solar system? Explain.

2. Why can't we use radar to measure distances to stars (besides the Sun)?

Measuring Cosmic Distances Tutorial, Lessons 3, 4

1. What makes it possible for a streetlight to appear much brighter than a star?

2. How does the apparent brightness of an object vary with the distance from it?

3. How can we measure the distance to a faraway galaxy in which we observe a white dwarf supernova?

4. How can we measure the distance to a faraway spiral galaxy in which we do *not* observe a supernova?

Hubble's Law Tutorial, Lesson 1

1. Are all distant galaxies (those not in our own local cluster) moving away from us or toward us? How do we know?

2. How can we use Hubble's law to determine the distance to a faraway galaxy?

 Exploring the Sky and Solar System

Of the many activities available on the *Voyager: SkyGazer* **CD-ROM** accompanying your book, use the following files to observe key phenomena covered in this chapter.

Go to the **File: Basics** folder for the following demonstration.

1. Galaxies in Coma

Movies

Check out the following narrated and animated short documentary available on www.astronomyplace.com for a helpful review of key ideas covered in this chapter.

From the Big Bang to the Galaxies Movie

Web Projects

Take advantage of the useful Web links on www.astronomyplace. com to assist you with the following projects.

1. *Galaxy Gallery.* Many fine images of galaxies are available on the Web. Collect several images of each major type and build a galaxy gallery of your own. Supply a descriptive paragraph about each galaxy.

2. *Hubble's Constant.* Measurements of Hubble's constant are rapidly growing more reliable. Find at least two measurements of H_0 determined in the past 5 years. Describe how these values were measured. What is the quoted uncertainty of each? By how much do they differ?

3. *Greatest Lookback Time.* Look for recent discoveries of objects with the largest lookback times (or redshifts). What is the current record for the most distant known object? What kind of object is it?

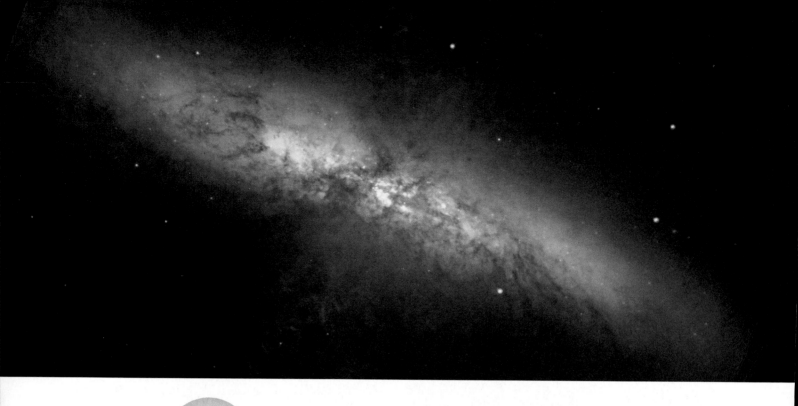

21 Galaxy Evolution

Reality provides us with facts so romantic that imagination itself could add nothing to them.

Jules Verne

The spectacle of galaxies strewn like beautiful islands across the universe invites us to ponder their origins. Some galaxies have majestic spiral arms, some are elliptical, and some are irregular. What processes sculpted them into their present shapes?

If we look closely, we see even more spectacular sights. A few nearby galaxies are engaged in titanic collisions with other galaxies. Many of these colliding galaxies are full of star-forming clouds in which new stars are born and massive stars explode 100 times more frequently than in the Milky Way. Other galaxies appear to harbor enormous black holes surrounded by gigantic accretion disks that generate extraordinary luminosities. Narrow streams of matter jet from a few of these galaxies into intergalactic space at nearly the speed of light.

The origins of galaxies and the incredible phenomena they sometimes display puzzled astronomers for much of the twentieth century, but today we are assembling a rudimentary understanding of how galaxies evolve. In this chapter, we will sift through the fascinating clues that hint at how galaxies formed and developed, pausing now and again to admire the fantastic spectacles that this search for our origins has uncovered.

21.1 Looking Back Through Time

In Chapter 20, we saw that galaxies come in a wide variety of shapes, colors, and sizes. This variety is interesting and beautiful to behold, but its significance goes much deeper. Remember that we would not be here if not for the galaxy-wide recycling processes that gradually transform the Milky Way's primordial gases into stars and planets [Section 19.2]. The same basic star–gas–star cycle that operates in our own galaxy governs the development of all galaxies, even though the outward appearances of galaxies can greatly differ. Thus, if we want to understand our cosmic origins, we must learn not only how the galaxies we see today formed from the

hydrogen and helium gas present in the early universe. We must also figure out how the process of galaxy formation produces galaxies that look so different. Because so many important questions about galaxies remain unanswered, the study of the formation and development of galaxies, or **galaxy evolution**, is one of the most active research areas in astronomy.

Our understanding of galaxy evolution is rapidly advancing largely because modern telescopes enable us to look back through time. When we look at a galaxy a billion light-years away, we are seeing it at a time when the universe was a billion years younger than it is today. When we look at a galaxy halfway to the cosmological horizon [Section 20.4], we see it as it was when the universe was only half its present age. Even closer to the cosmological horizon, we see galaxies as they were within the first few billion years after the Big Bang.

The linkage between a galaxy's distance and its age gives us a remarkable ability: Simply by photographing galaxies at different distances, we can assemble "family albums" showing galaxies in different stages of development. Figure 21.1 shows partial family albums for elliptical, spiral, and irregular galaxies. Each individual photograph shows a single galaxy at a single stage in its life. Grouping these photographs by galaxy type allows us to see how galaxies of a particular type have changed through time. Pictures of the farthest galaxies show galaxies in their childhood, and pictures of the nearest show mature galaxies as they are today.

In the last sentence above, we use the term *today* in a very broad sense. For example, if we look at a relatively nearby galaxy—one located, say, 20 million light-years away—we see it as it was 20 million years ago. In what sense is this "today"? (*Hint:* How does 20 million years compare to the age of the universe? To the lifetime of a massive star?)

We would like to tell the life stories of these galaxies from beginning to end as completely as we told the life stories of stars, but too many aspects of galaxy evolution remain mysterious. In broad terms, the story we seek should tell us how the hydrogen and helium gas in the early universe turned into the arrangements of stars we see in galaxies today. As we'll discuss, even this general story still presents some puzzles, such as why some galaxies turned out flattened and gas-rich while others did not.

We also need to explain some of the stranger aspects of galaxies. For example, observations show that some galaxies, called *starburst galaxies,* are currently forming stars at a tremendous rate. Starbursts must be only a temporary phase in the lives of these galaxies, or they would have turned all of the galaxies' gas into stars long ago. We have also found that some galaxies have extremely bright centers, known as *active galactic nuclei,* that sometimes outshine all the stars in the galaxy combined. *Quasars,* the most luminous active galactic nuclei, can radiate more light than a thousand galaxies like our Milky Way. Because quasars are much more common in very distant galaxies than in nearby

Young Galaxies

Ellipticals

Spirals

Irregulars

0 billion   3 billion   6 billion   9 billion   12 billion

approximate age of universe in years

Figure 21.1 Family albums for elliptical, spiral, and irregular galaxies at different ages along with some very young, distant galaxies. All of these galaxy photos have been drawn from the Hubble Space Telescope image in Figure 20.1. Light from the most distant galaxies in the photo shows us what these galaxies looked like when the universe was relatively young (up to 3 billion years old). Light from closer galaxies shows us what these galaxies looked like when the universe was somewhat older. By arranging galaxies in this way, we can see how galaxies of different types have evolved as the universe has aged. (Younger galaxies appear smaller because they are more distant.)

ones, we can conclude that they were much more common in the past than they are today. Perhaps all galaxies once harbored an active nucleus that shined brightly for a few hundred million years early in the universe's history. We simply don't yet know.

In this chapter, we'll discuss what is currently known about the lives of galaxies. We will first discuss how galaxies form and next examine why some galaxies end up as spirals while others end up as ellipticals. Then we will look in detail at the phenomena of starburst galaxies and active galactic nuclei and attempt to fit these dramatic stages into the overall puzzle of galaxy evolution. Finally, we will explore how very distant quasars are helping astronomers study galaxy evolution in a completely different way.

The light from these distant quasars traverses billions of light-years on its way to Earth, and their spectra bear the signatures of many gaseous clouds that lie between the quasars and Earth. Some of these clouds probably represent galaxies in their most youthful state. Thus, the study of quasar spectra not only may teach us about some early phases in the evolution of galaxies, but also may soon help us understand the complete history of galaxy evolution.

21.2 Galaxy Formation

The photographs in Figure 21.1, taken by the Hubble Space Telescope, show galaxies extending back to ages of just 1 or 2 billion years, but the first stars and galaxies formed even earlier than this. Observing these first stars and galaxies is a challenge not even Hubble can meet. Detecting them will require extremely large telescopes that are extraordinarily sensitive to infrared light. We have seen that the expansion of the universe shifts light from distant galaxies toward the red end of the spectrum [Section 20.3]. The redshifts of the most distant galaxies are so extreme that the visible and ultraviolet light they emitted comes to us in the form of infrared light. Ground-based telescopes can detect some of this infrared light, but Earth's atmosphere greatly complicates such observations [Section 7.4].

Within a decade or so, NASA hopes to launch a much larger, infrared-sensitive successor to the Hubble Space Telescope (called the James Webb Space Telescope), but for now we have little direct information about galaxy birth. Instead, we must try to weave several lines of less direct evidence into a single coherent story. In this section, we will discuss what astronomers have learned about galaxy formation from theoretical modeling, the properties of present-day galaxies, and the detailed structure of the Milky Way.

Modeling Galaxy Birth

Because our telescopes cannot yet see back to the time when galaxies formed their first stars, we must use theoretical modeling to study the earliest stages in galaxy evolution. The most successful models for galaxy formation assume the following:

- Hydrogen and helium gas filled all of space more or less uniformly when the universe was very young—say, in the first million years after its birth.

- This uniformity was not quite perfect, and certain regions of the universe were ever so slightly denser than others.

Beginning from these assumptions, which are supported by a mounting body of evidence (discussed in Chapter 23), we can model galaxy formation using well-established laws of physics to trace how the denser regions in the early universe grew into galaxies. The models show that the regions of enhanced density originally expanded along with the rest of the universe. However, the slightly greater pull of gravity in these regions gradually slowed their expansion. Within about a billion years, the expansion of these denser regions halted and reversed, and the material within them began to contract into **protogalactic clouds**, the clouds of matter that eventually formed galaxies.

According to the models, protogalactic clouds initially cooled as they contracted, radiating away their thermal energy, and the first generation of stars grew from the densest, coldest clumps of gas. The most massive of these stars lived and died within just a few million years, a short time compared to the time required for the collapse of a protogalactic cloud into a mature galaxy. The supernovae of these massive stars generated shock waves that heated the surrounding interstellar gas, slowing the collapse of young galaxies and the rate at which new stars formed within them. Some of these first-generation stars may still exist, but we have not yet found any of them.

Old Spheroids, Young Disks

The colors of galaxies provide a crucial clue to what happened next (Figure 21.2). Recall that the disks of spiral galaxies appear whitish with flecks of blue [Section 20.2]. Thus, the stars in the disks of spiral galaxies, sometimes referred to as the *disk population* (or *Population I*), include hot, blue stars as well as cool, red ones. Because only short-lived, massive stars look blue, we conclude that star formation

Figure 21.2 The colors of this spiral galaxy, known as NGC 1232, provide crucial clues about its development. Its disk and spiral arms look white with flecks of blue, indicating that the disk component contains stars of all colors, including both short-lived blue stars and long-lived red stars. The disk of a spiral galaxy must therefore be forming stars continually. In contrast, the reddish yellow tint of the bulge indicates that the galaxy's spheroidal component contains few short-lived blue stars. Thus, star formation in the spheroidal component must nearly have ceased.

must be an ongoing process in galactic disks—just as it is in the disk of the Milky Way.

In contrast, the stars in elliptical galaxies and in the bulges and halos of spiral galaxies, called the *spheroidal population* (or *Population II*), look reddish in color. The absence of blue stars among the spheroidal population indicates that most of these stars are old and that their star formation must have ceased long ago—just as is the case in the halo of the Milky Way [Section 19.3].

The characteristic motions of stars in disks and spheroids provide another important clue to their nature (Figure 21.3). Disk stars generally orbit in orderly circles in the same direction, and all galactic disks rotate much like the Milky Way's disk [Section 19.4]. Stars of spheroidal populations tend to orbit in random directions, much like stars in the bulge and halo of the Milky Way. Because the motions of stars in the two populations are so different, disks and spheroids must have formed in very different ways.

According to the most basic model for galaxy formation from a protogalactic cloud, the stars of the spheroidal population formed first. Early on, the gravity associated with a protogalactic cloud drew in matter from all directions, creating a cloud that was blobby in shape and had little or no measurable rotation. The orbits of stars forming within such a cloud could have had any orientation, accounting for the randomly oriented orbits of spheroidal population stars.

Later, at least in spiral galaxies, conservation of angular momentum caused the remaining gas to flatten into a spinning disk as it contracted under the force of gravity (Figure 21.4). This process was much like the process that leads to protoplanetary disks around young stars [Section 9.2], but on a much larger scale. Collisions among gas particles tended to average out their random motions, leading them

to acquire orbits in the same direction and in the same plane (see Figure 9.2). Stars that formed within this spinning disk were born on orbits moving at the same speed and in the same direction as their neighbors and thus became the disk population stars.

Something different must have happened during the development of elliptical galaxies to leave them without prominent disks. In the next section, we'll explore some possible reasons for this key difference. Nevertheless, the early stages of an elliptical galaxy's life story are probably similar to those of a spiral galaxy (Figure 21.4a,b).

Clues from the Milky Way

Because most spiral galaxies in today's universe look similar to our own Milky Way Galaxy, we can test our theories about the formation of spiral galaxies by studying the "fossil record" written within the stars of the Milky Way. The clues found to date support the basic picture but suggest that the full story of galaxy formation may be somewhat more complex.

All available evidence confirms that the stars in the Milky Way's halo are indeed old. The main-sequence turn-off points in H–R diagrams of globular clusters show that their stars were born at least 12 billion years ago [Section 16.6]. Individual halo stars (i.e., those not in globular clusters) and some of the bulge stars appear similarly old. Furthermore, the proportions of heavy elements in halo stars are much lower than in the Sun, indicating that they formed before many generations of supernovae had a chance to enrich the Milky Way's interstellar medium [Section 19.2].

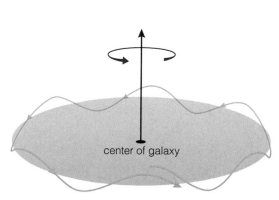

a Stellar orbits in disk component.

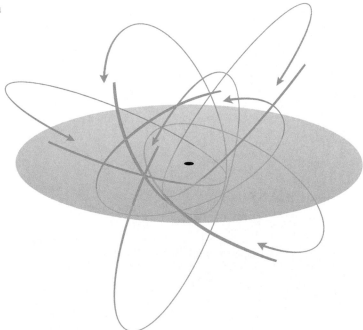

b Stellar orbits in spheroidal component.

Figure 21.3 Characteristic stellar orbits. Stars in the disk tend to orbit in the same direction around the center of the galaxy, with a bit of up-and-down motion. Stars in the bulge and halo also orbit the center of the galaxy, but the orientations of their orbits are more random.

a A protogalactic cloud contains only hydrogen and helium gas.

b Halo stars begin to form as the protogalactic cloud collapses.

c Conservation of angular momentum ensures that the remaining gas flattens into a spinning disk.

d Billions of years later, the star–gas–star cycle supports ongoing star formation within the disk. The lack of gas in the halo precludes further star formation outside the disk.

Figure 21.4 This four-picture sequence illustrates a simple schematic model of galaxy formation, showing how a spiral galaxy might develop from a protogalactic cloud of hydrogen and helium gas

However, careful study of heavy-element proportions suggests that our galaxy formed from a few different gas clouds. If the Milky Way had formed from a single protogalactic cloud, it would have steadily accumulated heavy elements during its inward collapse as stars formed and exploded within it. In that case, the outermost stars in the halo would be the oldest and the most deficient in heavy elements. Stars belonging to different globular clusters in the Milky Way's halo do indeed differ in age and heavy-element content, but these variations do not seem to depend on the stars' distance from the galactic center. The easiest way to account for the variations is to suppose that the Milky Way's earliest stars formed in relatively small protogalactic clouds, each with a few globular clusters, and that these clouds later collided and combined to create the full protogalactic cloud that became the Milky Way (Figure 21.5).

THINK ABOUT IT

If the preceding scenario is true, then the Milky Way suffered several collisions early in its history. Explain why we should not be surprised that galaxy collisions (or collisions between protogalactic clouds) were rather common in the distant past. (*Hint:* How did the average separations of galaxies in the past compare to their average separations today?)

Figure 21.5 Illustration of how the Milky Way's halo may have formed. The characteristics of stars in the Milky Way's halo suggest that several smaller gas clouds, already bearing some stars and globular clusters, may have merged to form the Milky Way's protogalactic cloud. These stars and star clusters remained in the halo while the gas settled into the Milky Way's disk.

Once the full protogalactic cloud was in place, its collapse and heavy-element enrichment should have continued in a more orderly fashion than previously. Support for this scenario comes from a layer of stars intermediate between the disk and the halo. The heavy-element content of stars and globular clusters in this intermediate layer does indeed depend on their distances from the galactic center. These stars are nearly as old as halo stars but formed before the spinning protogalactic cloud finished flattening into a disk. Their proportions of heavy elements suggest that, at the time the disk formed, the Milky Way contained about 10% as much material in the form of heavy elements as it does today.

After the disk formed, generations of stars lived and died in the star–gas–star cycle that gradually increased the abundance of heavy elements [Section 19.2]. Thus, the ages of stars in the Milky Way's disk range from newly born to 10 billion or more years old. New stars will continue to be born as long as enough gas remains in the disk.

Loose Ends

We seem to be well on our way to understanding how and why protogalactic clouds evolve into galaxies. Nevertheless, a few puzzles remain. For example, the oldest halo stars in our Milky Way have very low fractions of heavy elements, but these fractions are not zero. Where did the heavy elements in these stars come from? They probably came from an even earlier generation of stars, but we have not yet discovered any examples of first-generation stars with no heavy elements.

An even more fundamental question relates to the formation of the protogalactic clouds themselves. Our models assume that these clouds formed in regions of slightly enhanced density in the early universe, but where did these density enhancements come from? The question of the na-

ture and origin of density enhancements in the early universe is one of the major puzzles in astronomy. We'll revisit it in Chapters 22 and 23.

21.3 Why Do Galaxies Differ?

The idea that galaxies grow from contracting protogalactic clouds and form their stars over billions of years is consistent with what we know from studying the Milky Way and from our limited observations of very young galaxies. However, this simple model does not explain why galaxies come in such a great variety of shapes and sizes. We now believe that both the initial conditions of the protogalactic clouds and later interactions with other galaxies affected the development of all types of galaxies, leading to the great variety we see today. To help differentiate the role of initial conditions and that of subsequent interactions, let us now consider a question about galaxy evolution that is still unanswered: Why do spiral galaxies have gas-rich disks, while elliptical galaxies do not?

Destined from Birth?

Two plausible explanations for the differences between spiral galaxies and elliptical galaxies trace a galaxy's appearance back to the protogalactic cloud from which it formed:

● *Protogalactic spin.* A galaxy's present-day appearance might be determined by the spin of the protogalactic cloud from which it formed. If the original cloud has a significant amount of angular momentum, it will rotate quickly as it collapses. The galaxy it produces will therefore tend to form a disk, and the resulting galaxy would be a spiral galaxy. If the protogalactic cloud has little or no angular momentum, its gas might not form a disk at all, and the resulting galaxy would be elliptical (Figure 21.6a).

- *Protogalactic cooling.* Elliptical galaxies may arise from denser protogalactic clouds than spiral galaxies. Remember that large elliptical galaxies are most common in clusters of galaxies [Section 20.2]. Because galaxy clusters are unusually dense regions of the universe, we suspect that the gas in the clouds that formed these elliptical galaxies was also unusually dense. The rela-

tively high gas density in these clouds would enable the gas to radiate energy more effectively and to cool more quickly. This rapid cooling might allow gravity to collapse clumps of the protogalactic cloud's gas into stars before the gas can settle into a disk. The resulting galaxy would therefore lack a disk, making it an elliptical galaxy. In contrast, the lower-density clouds that became spiral galaxies would form stars more slowly, leaving plenty of gas to form a disk as the cloud collapsed (Figure 21.6b).

protogalactic cloud with less angular momentum —————————————→ elliptical galaxy

protogalactic cloud with more angular momentum ————————————→ spiral galaxy

a Protogalactic clouds with substantial angular momentum might become spiral galaxies because their spin forces them to become disklike as they contract. Protogalactic clouds with little angular momentum would contract into elliptical galaxies.

high-density protogalactic cloud cools faster —————————————→ elliptical galaxy

lower-density protogalactic cloud cools slower ————————————→ spiral galaxy

b Unusually dense protogalactic clouds cool fast and might become elliptical galaxies because their gas forms stars before it can settle into a disk. Less dense clouds that cool more slowly might allow more gas to settle into a disk, producing a spiral galaxy.

Figure 21.6 Two plausible explanations for the difference between spiral and elliptical galaxies.

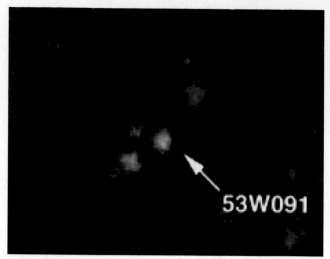

Figure 21.7 The light we are observing from this distant elliptical galaxy, LBDS53W091, left when the universe was about 4 billion years old. The redness of its light at this age indicates that all its stars formed during the first billion years or so after the birth of the universe. Its two red companions to the left are probably also elliptical galaxies with similar star-formation histories.

Some evidence for the latter scenario comes from a few giant elliptical galaxies at very great distances (Figure 21.7). These galaxies look very red even after we have accounted for their large redshifts. They apparently have no blue or white stars at all, indicating that new stars no longer form within these galaxies—even though we are seeing them as they were when the universe was only a few billion years old. This finding supports the idea that all the stars in these galaxies formed almost simultaneously, leaving no time for a disk to develop.

Shaped by Interactions?

The two previous scenarios, in which the formation of a gas-rich disk depends on the angular momentum or density of the protogalactic cloud, probably describe important parts of the overall story. However, they ignore one key fact: Galaxies rarely evolve in perfect isolation.

Think back to our scale-model solar system in Chapter 1—we made the Sun the size of a grapefruit. On this scale, the nearest star was like another grapefruit a few thousand kilometers away. Because the average distances between stars are so huge compared to the sizes of stars, direct star–star collisions are extremely rare. However, if we rescale the universe so that our *galaxy* is the size of a grapefruit, the Andromeda Galaxy is like another grapefruit only about 3 meters away, and a few smaller galaxies lie considerably closer. Thus, the average distances between galaxies are not tremendously larger than the sizes of galaxies, and collisions between galaxies are inevitable. Even the Milky Way Galaxy is not immune. About 80,000 light-years away, directly behind the galactic bulge, a small elliptical galaxy

(called the Sagittarius dwarf elliptical) is currently crashing through the Milky Way's disk.

Galactic Collisions Collisions between galaxies are spectacular events that unfold over hundreds of millions of years (Figure 21.8). In our short lifetimes, we can at best see a snapshot of a collision in progress, as evidenced by the distorted shape of the colliding galaxies. Galactic collisions must have been even more frequent when the universe was smaller and galaxies were even closer together. Photographs of galaxies at a variety of distances confirm that distorted-looking galaxies—probably galaxy collisions in progress—were more common in the early universe than they are today (Figure 21.9).

We can learn much more about galactic collisions with the aid of computer simulations that allow us to "watch" collisions that in nature take hundreds of millions of years to unfold. These computer models show that a collision between two spiral galaxies can create an elliptical galaxy (Figure 21.10). Tremendous tidal forces between the colliding galaxies tear apart the two disks, randomizing the orbits of their stars. Meanwhile, a large fraction of their gas sinks to the center of the collision and rapidly forms new stars. Supernovae and stellar winds eventually blow away the rest of the gas. When the cataclysm finally settles down, the merger of the two spirals has produced a single elliptical galaxy. Little gas is left for a disk, and the orbits of the stars have random orientations.

Evidence for Mergers Observations seem to confirm the idea that at least some elliptical galaxies result from collisions and subsequent mergers. Elliptical galaxies dominate the galaxy populations at the cores of dense clusters of galaxies, where collisions should be most frequent. This fact may mean that any spirals once present became ellipticals through collisions.

Stronger evidence comes from structural details of elliptical galaxies, which often attest to a violent past. Some elliptical galaxies have stars and gas clouds in their cores that rotate in a different direction from the rest of the galaxy, suggesting that they are leftover pieces of galaxies that merged in a past collision. Other elliptical galaxies are surrounded by shells of stars, which probably formed when stars were stripped out of smaller galaxies that once strayed too close and were destroyed (Figure 21.11).

The most decisive evidence that collisions affect the evolution of elliptical galaxies comes from observations of the **central dominant galaxies** found at the centers of many dense clusters. Central dominant galaxies are gigantic elliptical galaxies that apparently grew to a huge size by consuming other galaxies through collisions. Central dominant galaxies frequently contain several tightly bound clumps of stars that probably were the centers of individual galaxies before being swallowed by the giant (Figure 21.12). This process of *galactic cannibalism* can create central dominant galaxies more than 10 times as massive as the Milky Way, making them the largest galaxies in the universe.

a The Antennae, also known as NGC 4038/4039, are a pair of colliding spiral galaxies. The wide-field image taken from the ground (left) reveals their vast tidal tails, and the close-up from the Hubble Space Telescope shows the fine details of the burst of star formation at the center of the collision.

b NGC 6240, a galaxy collision farther along than the Antennae, is still highly disturbed, but the two galaxies have almost merged into one.

Figure 21.8 Galaxy collisions.

Age of Universe: 2–4 billion years

Age of Universe: 5–7 billion years

Figure 21.9 Hubble Space Telescope photographs of distorted young galaxies showing how they looked when the age of the universe was between 2 and 7 billion years. These close-up views of some of the most distant galaxies in Figure 20.1 correspond to the young galaxies at the left side of the family album in Figure 21.1. The strange shapes of these galaxies are not at all like the more familiar spiral and elliptical galaxies of the present-day universe, suggesting that they were undergoing collisions when their light began journeying to Earth. Because galaxies this distant often look distorted, we infer that collisions between galaxies were much more common early in the universe's history.

Figure 21.10 Several stages in a supercomputer simulation of a collision between two spiral galaxies that results in an elliptical galaxy. At least some of the elliptical galaxies in the present-day universe formed in this way. The whole sequence spans about 1.5 billion years.

Two simulated spiral galaxies approach each other on a collision course.

The first encounter begins to disrupt the two galaxies and sends them into orbit around each other.

As the collision continues, much of the gas in the disk of each galaxy collapses toward the center.

Gravitational forces between the two galaxies tear out long streamers of stars called tidal tails.

The centers of the two galaxies approach each other and begin to merge.

The single galaxy resulting from the collision and merger is an elliptical galaxy surrounded by debris.

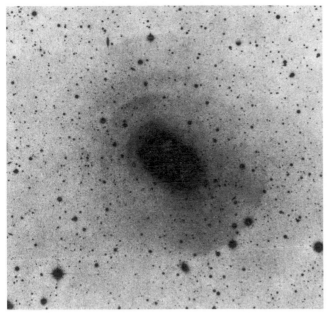

Figure 21.11 Shells of stars around the elliptical galaxy NGC 3923. This photograph has been specially processed to bring out faint details. It is a negative image in which starlight is black and dark space is white. The shells of stars surrounding the main part of the galaxy are thought to be the remains of one or more galaxies that collided with NGC 3923 sometime in the past. The orbits of these stars resemble the motion of a pendulum: They plunge back and forth through the main part of the galaxy, swinging from one side of the galaxy to the other. The shells represent the extreme ends of these plunging orbits, where the stars spend most of their time.

Stripping of Interstellar Gas?

Clusters of galaxies might also turn spiral galaxies into ellipticals simply by pulling out their interstellar gas. Hot gas fills the centers of galaxy clusters [Section 22.3]. When a spiral galaxy cruises through the center of such a cluster, the intracluster gas exerts drag forces that slow the galaxy's gas but not its stars. Thus, the galaxy's stars continue to move freely along their way, but the gas is left behind. If the disk has not yet formed many stars, then the galaxy will evolve to look more like an elliptical galaxy as its massive stars die away. Its disk will fade, while its bulge and halo will remain prominent. In contrast, if the disk has already formed a large number of stars when its gas is stripped, the remaining galaxy is more likely to look *lenticular* in shape [Section 20.2].

No Verdict Yet

We have identified several scenarios that may determine whether a galaxy becomes an elliptical or a spiral galaxy. Two of these scenarios depend on initial conditions: One suggests that elliptical galaxies form when protogalactic clouds have low angular momentum, and the other suggests that elliptical galaxies form as a result of rapid star formation in protogalactic clouds with relatively high density. But we have also found that interactions, such as galactic collisions

Figure 21.12 The central dominant galaxy of the cluster Abell 3827 contains multiple clumps of stars that probably once were the centers of individual galaxies. The image has been specially processed so that different colors represent different levels of brightness. This procedure makes the brightest clumps (which appear yellow) easier to see. This giant elliptical galaxy has apparently grown partly by consuming smaller galaxies that have collided with it.

or drag created by hot gas in galaxy clusters, can drastically influence the lives of galaxies and very likely cause some spiral galaxies to become elliptical.

Each of these mechanisms probably plays a role in galaxy evolution. We do not yet know whether one is more important than the others. Irregular galaxies, which make up only a few percent of large galaxies, probably arise from a variety of different and unusual circumstances. However, they probably are affected by the same processes that affect spirals and ellipticals.

21.4 Starburst Galaxies

We have discussed current ideas about galaxy formation and why galaxies come in different types. We now turn our attention to the incredible phenomena we see in some of our snapshots of galactic lives. We do not yet know how these phenomena fit into the overall story of galaxy evolution, but they are clearly important. We begin in this section by studying starburst galaxies. In the next section, we will look at quasars and other active galactic nuclei.

Most of the galaxies in the present-day universe appear comfortably settled, steadily forming new stars at modest rates. The Milky Way Galaxy produces an average of about one new star per year. At this rate, the Milky Way won't exhaust the interstellar gas in its disk until long after the Sun has died.

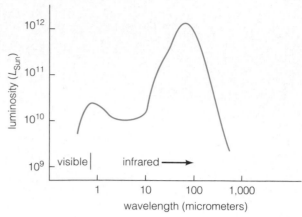

b The spectrum of Arp 220 shows that it emits most of its radiation as infrared light rather than visible light. Because dusty gas enshrouds the starburst, dust absorbs most of the visible starlight. It then reemits that energy as infrared radiation, which can more easily escape the dusty shroud. The total infrared luminosity, $10^{12}L_{Sun}$, is about a hundred times greater than that of our Milky Way, making Arp 220 one of the most luminous galaxies known.

Figure 21.13 Infrared observations of Arp 220, a large starburst galaxy.

a Infrared image of starburst galaxy Arp 220. The distorted appearance of this galaxy suggests that it is a merging system—the two brightest spots appear to be the bulges of the two galaxies that collided. The image shows a region about 10,000 light-years across.

Starburst galaxies are not so thrifty. Some form new stars at rates exceeding 100 stars per year, more than 100 times the star-formation rate in the Milky Way. These galaxies cannot possibly sustain such a torrid pace of star formation for long. At their current rates of star formation, they would consume *all* their interstellar gas in just a few hundred million years—a relatively short time compared to the 10 billion or more years since galaxies first formed. Starbursts are thus an important but temporary phase in the evolution of at least some galaxies.

Starburst galaxies are relatively new subjects of study. These voracious consumers of interstellar gas look peculiar at visible wavelengths because they are filled with star-forming molecular clouds, but these clouds conceal much of the action. Dust grains in the molecular clouds absorb most of the visible and ultraviolet radiation streaming from a starburst galaxy's many young stars. This radiation heats the dust grains to much higher temperatures than normal, and they ultimately reemit all the absorbed visible and ultraviolet energy as infrared light, with a peak wavelength of around 60,000 nanometers (60 micrometers).

Because photons at these wavelengths cannot penetrate Earth's atmosphere, the intensity of star-forming activity in starburst galaxies did not become fully apparent until the 1983 launching of the *Infrared Astronomy Satellite* (*IRAS*). Figure 21.13 shows an especially luminous starburst galaxy, along with its spectrum from radio through visible wavelengths. Note that the visible output of this galaxy is about 10 billion solar luminosities ($10^{10}L_{Sun}$), not very different from the total luminosity of the Milky Way. However, its infrared output is a trillion times that of our Sun ($10^{12}L_{Sun}$), making it 100 times brighter in infrared light than in visible light.

Galactic Winds

A star-formation rate 100 times that of the Milky Way also means an occurrence of supernovae at 100 times the Milky Way's rate. Just as in the Milky Way, each supernova in a starburst galaxy generates a shock wave that creates a *bubble* of hot gas [Section 19.2]. The shock waves from several nearby supernovae quickly overlap and blend into a much larger *superbubble*. In the Milky Way, the story usually ends here, but in a starburst galaxy the drama is just beginning. Supernovae continue to explode inside the superbubble, adding to its thermal and kinetic energy. When the superbubble starts to break through the disrupted gaseous disk, it expands even faster. Hot gas erupts into intergalactic space, creating a **galactic wind**.

Galactic winds consist of low-density but extremely hot gas, typically with temperatures of 10–100 million Kelvin (Figure 21.14). They do not emit much visible light, but they do generate X rays. X-ray telescopes in orbit have detected pockets of X-ray emission surrounding the disks of some starburst galaxies, presumably coming from the outflowing galactic wind. Sometimes we also see the glowing remnants of a punctured superbubble extending out into space.

Supernova-driven galactic winds have an even more dramatic impact when they occur in small starburst galaxies (Figure 21.15). The winds can blow out of small galaxies on all sides, driving away much of their gas. As a result, star formation in these galaxies may shut down for billions of years. Some of the small elliptical galaxies in the Local Group apparently have burst like this at least twice during their lifetimes. In one example, about half the stars in the small galaxy were born more than about 12 billion years

a This visible-light photograph of M 82 from the Subaru telescope shows violently disturbed gas (red) poking out from above and below the disk.

Figure 21.14 Two views of starburst galaxy M 82 showing how its wind is blowing out the top and bottom of its disk. The area shown in both pictures is about 16,000 light-years across.

b This X-ray photograph from the Chandra X-Ray Observatory shows the same region as the visible-light photograph in (a). The amber-colored region, roughly coinciding with the red area in the optical photograph, maps X-ray emission from hot gas blowing out of the disk. The bright dots probably correspond to accretion disks around black holes produced by some of the supernovae in the starburst.

Figure 21.15 This X-ray/visible composite photo of dwarf starburst galaxy NGC 1569 shows hot gas blowing out in several different directions. X rays from escaping gas at a temperature of several million degrees are shown in green, and optical light from disturbed hydrogen gas is shown in red.

ago, and the rest only 5–8 billion years ago. Presumably, each of these bursts of star formation ejected nearly all of the galaxy's gas. Star formation was put on hold for several billion years until enough gas could reaccumulate within the galaxy for a new starburst to ignite.

Dwarf galaxies that have undergone bursts of star formation tend to have fewer heavy elements than large galaxies. Why do you think that is? (*Hint:* What happens to the heavy elements produced by the burst of star formation?)

Causes of Starbursts

Many of the most luminous starburst galaxies look violently disturbed (see Figure 21.8). They are neither flat disks with symmetric spiral arms nor smoothly rounded balls of stars, like elliptical galaxies. Streamers of stars are strewn everywhere, often including two tidal tails streaming in opposite directions. Such galaxies are filled with dusty molecular clouds, and deep inside them we see two distinct clumps of stars that look like two different galactic centers. Their appearance suggests that such starbursts result from a collision between two gas-rich spiral galaxies. The collision apparently compresses the gas inside the colliding galaxies, leading to the burst of star formation.

The causes of smaller-scale starbursts are less clear. Some may result from direct collisions, while others may be caused by close encounters with other galaxies. The Large Magellanic Cloud is currently undergoing a period of rapid star formation, perhaps because of the tidal influence of the Milky Way. No matter what their exact causes turn out to be, starburst galaxies clearly represent an important piece in the overall puzzle of galaxy evolution.

 Black Holes Tutorial, Lessons 1, 2

21.5 Quasars and Other Active Galactic Nuclei

Starbursts may be spectacular, but some galaxies display even more incredible phenomena: extreme amounts of radiation, and sometimes powerful jets of material, emanating from deep in their centers (Figure 21.16). We generally refer to these unusually bright galactic centers as **active galactic nuclei** and reserve the term **quasar** for the very brightest of them. (Galaxies with active galactic nuclei are sometimes called *active galaxies*.) The brightest quasars shine more powerfully than 1,000 galaxies the size of the Milky Way.

The glory days of quasars are long past. We find quasars primarily at great distances, telling us that these blazingly luminous objects were most common billions

Figure 21.16 The active galactic nucleus in the elliptical galaxy M 87. The bright yellow spot is the active nucleus, and the blue streak is a jet of particles shooting outward from the nucleus at nearly the speed of light.

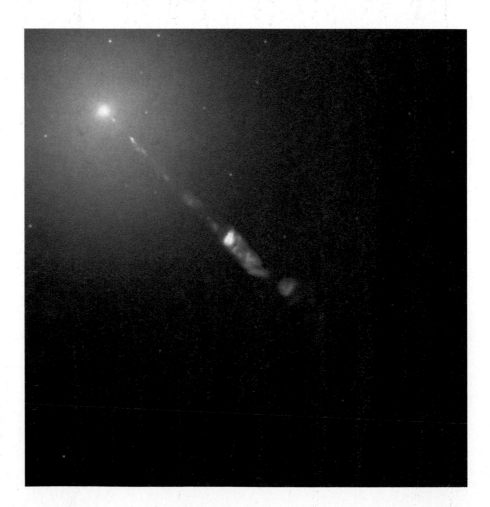

of years ago, when galaxies were in their youth. Because we find no nearby quasars and relatively few nearby galaxies with any type of active galactic nucleus, we conclude that the objects that shine as quasars in young galaxies must become dormant as the galaxies age. Thus, many nearby galaxies that now look quite normal—we don't yet know which ones—must have centers that once shone brilliantly as quasars.

What could possibly drive the incredible luminosities of quasars, and why did quasars fade away? A growing body of evidence points to a single answer: The energy output of quasars comes from gigantic accretion disks surrounding **supermassive black holes**—black holes with masses millions to billions of times that of our Sun. Before we study how these incredible powerhouses work, let's investigate the evidence that points to their existence.

The Discovery of Quasars

In the early 1960s, a young professor at the California Institute of Technology named Maarten Schmidt was busy identifying cosmic sources of radio-wave emission. Radio astronomers would tell him the coordinates of newly discovered radio sources, and he would try to match them with objects seen through visible-light telescopes. Usually the radio sources turned out to be normal-looking galaxies, but one day he discovered a major mystery: A radio source called 3C 273 looked like a blue star through a telescope but had strong emission lines at wavelengths that did not appear to correspond to any known chemical element. (The designation 3C 273 stands for 3rd Cambridge Radio Catalogue, object 273. A few other, similarly enigmatic radio sources were known at the time, but Schmidt's breakthrough came with 3C 273.)

After months of puzzlement, Schmidt suddenly realized that the emission lines were not coming from an unfamiliar element at all. Instead, they were emission lines of hydrogen hugely redshifted from their normal wavelengths (Figure 21.17). Schmidt calculated that the expansion of the universe was carrying 3C 273 away from us at 17% of the speed of light.

Schmidt computed the distance to 3C 273 using Hubble's law. Then he plugged this distance into the luminosity–distance formula [Section 16.2]. What he found was astonishing: 3C 273 has a luminosity of about 10^{39} watts, or well over a trillion (10^{12}) times that of our Sun—making it hundreds of times more powerful than the entire Milky Way Galaxy. Discoveries of similar but even more distant objects soon followed. Because the first few of these objects were strong sources of radio emission that looked like stars through visible-light telescopes, they were named "quasi-stellar radio sources," or *quasars* for short. Later, astronomers learned that most quasars are not such powerful radio emitters, but the name has stuck.

For many years, a debate raged among astronomers over whether Hubble's law could really be used to determine quasar distances. Some argued that quasars might have high redshifts for other reasons and therefore might be much nearer to us than Hubble's law would suggest. The vast majority of astronomers now consider this debate settled. Improved images show that quasars are indeed the centers of extremely distant galaxies and often are members of very distant galaxy clusters.

Most quasars lie more than halfway to the cosmological horizon. The lines in typical quasar spectra are shifted to more than three times their rest wavelengths, which tells us that the light from these quasars emerged when the universe was less than a third of its present age. The farthest known quasar as of early 2003 has spectral lines shifted to 7.4 times their rest wavelengths. The light we see from this distant quasar began its journey when the universe was only about 6% of its present age.

The extraordinary energy output of quasars emerges across an unusually wide swath of the electromagnetic spectrum. Quasars radiate approximately equal amounts of power from infrared wavelengths all the way through to gamma rays (Figure 21.18). They also produce strong emission lines. By comparison, most stars and galaxies emit primarily visible light. The wide spread of photon energies coming from quasars implies that they contain matter with a wide range of temperatures. How does this matter manage to radiate such a large luminosity?

Evidence from Nearby Active Galactic Nuclei

Quasars are difficult to study in detail because they are so far away. Luckily, some quasarlike objects are much closer to home. About 1% of present-day galaxies—that is, galaxies we see nearby—have active galactic nuclei that look very much like quasars, except that they are less powerful. (These galaxies are often called *Seyfert galaxies* after astronomer Carl Seyfert, who in 1943 grouped galaxies with active galactic nuclei into a special class.)

The spectra of these active nuclei range from infrared to gamma rays, just like those of quasars. The only real distinction between nearby active galactic nuclei and quasars is their luminosity. Generally, the power outputs of quasars swamp those of the galaxies that contain them, making the

Figure 21.17 Spectrum of the quasar 3C 273. The lines labeled Hβ, Hγ, and Hδ are hydrogen emission lines. Note their significant redshift in the quasar spectrum relative to the "comparison spectrum" that shows them at their rest wavelengths.

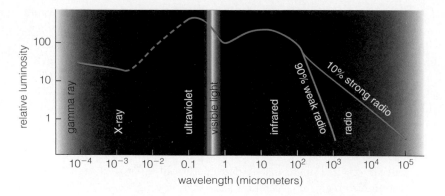

Figure 21.18 This schematic spectrum, representing the average of many quasar spectra, shows that all quasars emit strongly across most of the electromagnetic spectrum. However, only about 10% of quasars show strong radio emission. Quasar spectra also have strong emission lines, not shown in this diagram. (The dashed line shows a region of the extreme ultraviolet spectrum that is almost impossible to observe because radiation at these wavelengths cannot penetrate the Milky Way's interstellar gas.)

surrounding galaxies hard to detect (Figure 21.19). That is why quasars look "quasi-stellar." The galaxies surrounding dimmer active galactic nuclei are much easier to see because their centers are not so overwhelmingly bright.

The light-emitting regions of active galactic nuclei are so small that even the sharpest images do not resolve them. Our best visible-light images show only that active galactic nuclei must be smaller than 100 light-years across. Radio-wave images made with the aid of *interferometry* [Section 7.5] show that these nuclei are even smaller: less than 3 light-years across. Rapid changes in the luminosities of some active galactic nuclei point to an even smaller size.

To understand how variations in luminosity give us clues about an object's size, imagine that you are a master

of the universe and you want to signal one of your fellow masters a billion light-years away. An active galactic nucleus would make an excellent signal beacon, because it is so bright. However, suppose the smallest nucleus you can find is 1 light-year across. Each time you flash it on, the photons from the front end of the source reach your fellow master a full year before the photons from the back end. Thus, if you flash it on and off more than once a year, your signal will be smeared out. Similarly, if you find a source that is 1 light-day across, you can transmit signals that flash on and off no more than once a day. If you want to send signals just a few hours apart, you need a source no more than a few light-hours across.

Occasionally, the luminosity of an active galactic nucleus doubles in a matter of hours. The fact that we see a clear signal indicates that the source must be less than a few light-hours across. In other words, the incredible luminosities of active galactic nuclei and quasars are apparently being generated in a volume of space not much bigger than our solar system.

Radio Galaxies and Jets

In the early 1950s, a decade before the discovery of quasars, radio astronomers noticed that certain galaxies emit unusually large quantities of radio waves. Today we believe that these **radio galaxies** are another class of celestial powerhouse closely related to quasars. Upon close inspection, we find that much of the radio emission comes not from the galaxies themselves but rather from pairs of huge *radio lobes*, one on either side of the galaxy (Figure 21.20).

The radio waves from the lobes are produced by electrons and protons spiraling around magnetic field lines at nearly the speed of light. Astronomers of the 1950s guessed that this might be the source of the radio emission, but the implied amounts of energy in the lobes seemed implausibly large until the discovery of quasars. British theoretician Martin Rees led the way in showing that radio galaxies contain active galactic nuclei. He argued in 1971 that radio galaxies must have powerful **jets** spurting from their nuclei that transport energy to their lobes. That is, although the lobes lie far outside the visible galaxy, he asserted that the

Figure 21.19 The bright quasar in the nucleus of the galaxy just left of center outshines the rest of the galaxy, which appears disturbed, as if it has recently undergone a collision. Pictures like this one prove that quasars are extremely bright active galactic nuclei. (The entire region pictured is about 500,000 light-years across.)

VIS

Figure 21.20 Radio image of the radio galaxy Cygnus A recorded by the Very Large Array in New Mexico. The active galactic nucleus powering the radio emission sits at the location of the core. Two jets of particles shoot out of the core at nearly the speed of light. Hot spots occur where these jets ram into surrounding intergalactic gas. After these particles pass through a hot spot, they are deflected into the lobes like the spray of water from a firehose after it hits a wall. The distance between the lobes is about 400,000 light-years, several times larger than the extent of the galaxy in visible light.

ultimate source of their energy would be found in the galaxy's tiny nucleus. The nucleus must therefore be as powerful as a quasar. A few years later, detailed radio images began to support his claims.

Today, radio telescopes resolve the structure of radio galaxies in vivid detail. At the center of a typical radio galaxy sits a tiny *radio core*—the active galactic nucleus of the radio galaxy—less than a few light-years across. Two jets of plasma shoot out of the core in opposite directions. Frequently, only the jet tilted in our direction is visible. Using time-lapse radio images taken several years apart, we can track the motions of various plasma blobs in the jets. Some of these blobs move at close to the speed of light. The lobes lie at the ends of the jets, sometimes as much as a million light-years from the core. The relative prominence of these three elements—cores, jets, and lobes—varies from radio galaxy to radio galaxy. Thus, a gallery of typical radio galaxies exhibits a wide variety of sizes and shapes (Figure 21.21).

Quasars and radio galaxies may be much more similar than they appear. Indeed, many quasars, including 3C 273, have core–jet–lobe radio structures reminiscent of radio galaxies. For example, Figure 21.22 shows a series of pictures of a jet in which a blob of plasma is moving outward from a quasar at close to the speed of light.

Moreover, the active galactic nuclei of many radio galaxies seem to be concealed beneath donut-shaped rings of dark molecular clouds (Figure 21.23). Such structures may look like quasars when they are oriented so that we can see the active galactic nucleus at the center and look like the cores of radio galaxies when the ring of dusty gas dims our view of the central object. (A subset of active

galactic nuclei called *BL Lac objects* probably represents the centers of radio galaxies whose jets happen to point directly at us.)

Supermassive Black Holes

Astronomers have worked hard to envision physical processes that might explain how radio galaxies, quasars, and other active galactic nuclei release so much energy within such small central volumes. Only one explanation seems to fit: The energy comes from matter falling into a supermassive black hole (Figure 21.24). Gravity converts the potential energy of the infalling matter into kinetic energy. Collisions between infalling particles convert the kinetic energy into thermal energy, and photons carry this thermal energy away. As in X-ray binaries, we expect that the infalling matter swirls through an accretion disk before it disappears beneath the event horizon of the black hole [Section 18.4].

This method of energy generation can be awesomely efficient. The gravitational potential energy lost by a chunk of matter falling into a black hole is equivalent to its mass-energy, $E = mc^2$. As much as 10–40% of this energy can emerge as radiation before the matter crosses the event horizon. Thus, accretion by black holes is far more efficient at producing light than is nuclear fusion, which converts less than 1% of mass-energy into photons. As with accretion into the black holes formed in supernovae, the light is coming not from the black hole itself but rather from the hot gas surrounding it.

A variety of mechanisms explain why quasars and other active galactic nuclei radiate energy across the electromagnetic spectrum. The hot gas in and above the

a Radio galaxy 3C 353. The distance between the ends of the lobes is about 500,000 light-years.

b Radio galaxy 3C 219. The distance between the lobes is about 2.7 million light-years.

c Radio galaxy 3C 31. The lobes extend 1 million light-years to either side of the core.

Figure 21.21 Radio galaxy gallery. Note the double-lobe structure in all cases, but with very different shapes and sizes.

d Radio galaxy NGC 1265. The lobes of this radio galaxy are swept back because the galaxy is moving with respect to the surrounding intergalactic gas. The distance between the ends of the lobes is about 100,000 light-years.

accretion disk produces copious amounts of ultraviolet and X-ray photons. This radiation ionizes surrounding interstellar gas, energizing intense ionization nebulae that emit visible light. (The emission lines produced by these nebulae are the same ones Maarten Schmidt used to measure the first quasar redshifts.) The infrared light may come from rings of molecular clouds that encircle the active galactic nucleus (see Figure 21.23). As in starburst galaxies, dust grains in these molecular clouds absorb the high-energy light and reemit it as infrared light. This ring of

Figure 21.22 These close-up images of the jet in the quasar 3C 345, taken over a period of several years, show a blob of plasma (right) moving away from the core (left) at close to the speed of light. (From April 1981 to January 1984, the apparent distance from the blob to the core increased by about 10 light-years. The blob therefore seems to move faster than the speed of light, in violation of Einstein's theory of relativity. In fact, the blob's huge apparent speed is an illusion caused by the blob's motion at nearly the speed of light in a direction close to our line of sight.)

dense clouds could well be the reservoir for the accretion disk, supplying the matter that the giant black hole eventually consumes. Finally, the radio emission from active galactic nuclei comes from the fast-moving electrons that we sometimes see jetting from these nuclei at nearly the speed of light.

Although it's easy to understand why matter swirls inward through the accretion disk toward the black hole, explaining the powerful jets emerging from active galactic nuclei is more challenging. One plausible model for jet production relies on the twisted magnetic fields thought to accompany accretion disks (Figure 21.25).

As an accretion disk spins, it pulls the magnetic field lines that thread it around in circles. Charged particles start to fly outward along the field lines like beads on a twirling string. The particles careening along the field lines accelerate to speeds near the speed of light, forming a jet that shoots out into space. Some jets blast all the way through the galaxy's interstellar medium and penetrate into the much less dense intergalactic gas. The *hot spots* at the ends of the lobes in radio galaxies (see Figure 21.20) are the places where the jets are currently ramming into the intergalactic gas. When particles traveling down a jet hit the hot spot, they are deflected into the surrounding radio lobe like water from a firehose hitting a wall. The particles

Figure 21.23 Artist's conception of the central region of a radio galaxy. The active galactic nucleus, obscured by a ring of dusty molecular clouds, lies at the point from which the jets emerge. If viewed along a direction closer to the jet axis, the active nucleus would not be obscured and would look more like a quasar.

Figure 21.24 Artist's conception of an accretion disk surrounding a supermassive black hole. This picture represents only the very center of an object like that shown in Figure 21.23.

then fill the lobe with energy, generating powerful radio emission.

The supermassive black hole theory explains many of the observed features of quasars and other active galactic nuclei, but it is incomplete. We do not yet know what would create such giant black holes, nor do we know why quasars eventually run out of gas to accrete and stop shining. The preponderance of quasars during the first several billion years of the universe suggests that the formation of super-massive black holes is somehow linked to galaxy formation. This suggestion is supported by recent observations showing that the mass of a galaxy's central black hole is closely related to the mass of its spheroidal component. Some scientists have suggested that clusters of neutron stars resulting from extremely dense starbursts at the centers of galaxies might somehow coalesce to form an enormous black hole, but these speculations are still unverified. The origins of supermassive black holes remain mysterious.

Mathematical Insight **21.1** **Feeding a Black Hole**

As discussed in the text, 10–40% of the mass-energy of matter falling into a black hole can be radiated away as energy. (The precise value for a particular black hole depends on its rotation rate: Faster rotation allows more energy to be released.) Suppose 10% of the mass-energy is radiated away. Then the amount of energy radiated by mass m falling into a black hole is $E = \frac{1}{10}mc^2$. Equivalently, if we know the energy E radiated into space, then the amount of mass the black hole must accrete to radiate this energy is:

$$\text{accreted mass} = m = 10 \times \frac{E}{c^2}$$

Example: The most powerful quasars have luminosities of about 10^{40} watts. How much mass must the central black hole consume each second for its accretion disk to produce this luminosity? How many solar masses of material must it consume each year? Assume that 10% of the mass-energy of consumed mass is radiated away.

Solution: Recall that 1 watt = 1 joule/s, so the black hole must accrete enough mass to radiate 10^{40} joules each second. Thus, the amount of mass required to feed the black hole for 1 second is about:

$$m = 10 \times \frac{E}{c^2}$$

$$= 10 \times \frac{10^{40} \, \dfrac{\text{kg} \times \text{m}^2}{\text{s}^2}}{\left(3 \times 10^8 \, \dfrac{\text{m}}{\text{s}}\right)^2}$$

$$= 1.1 \times 10^{24} \, \text{kg}$$

Note that we replaced 1 joule with its equivalent units:

$$1 \, \text{joule} = \frac{\text{kg} \times \text{m}^2}{\text{s}^2}$$

The black hole must accrete 1.1×10^{24} kilograms of mass per second. We can convert this rate to solar masses per year as follows:

$$\frac{1.1 \times 10^{24} \, \text{kg}}{1 \, \text{s}} \times \frac{1 \, \text{solar mass}}{2.0 \times 10^{30} \, \text{kg}} \times \frac{3.1 \times 10^7 \, \text{s}}{1 \, \text{year}} = \frac{17 \, \text{solar masses}}{1 \, \text{year}}$$

Thus, the central black hole in a luminous quasar must consume the equivalent of about 17 Suns every year!

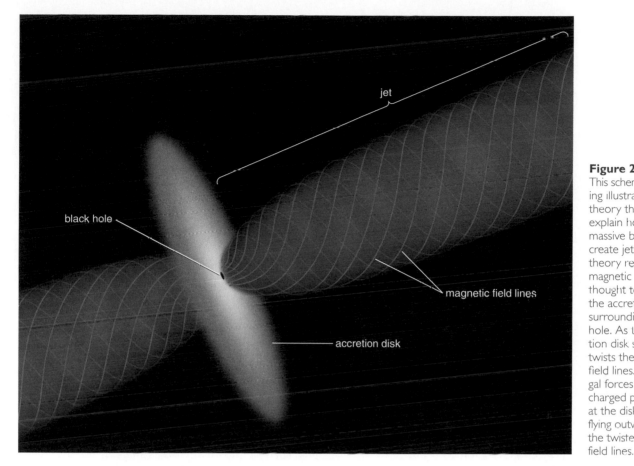

Figure 21.25
This schematic drawing illustrates one theory that might explain how supermassive black holes create jets. The theory relies on the magnetic field lines thought to thread the accretion disk surrounding a black hole. As the accretion disk spins, it twists the magnetic field lines. Centrifugal forces then send charged particles at the disk's surface flying outward along the twisted magnetic field lines.

Hunting for Monsters

Do supermassive black holes really drive the tremendous activity of radio galaxies, quasars, and other active galactic nuclei? The idea that such monsters even exist has been hard for some astronomers to swallow. Proving that we have found a black hole is tricky. Black holes themselves do not emit any light, so we need to infer their existence from the ways in which they alter their surroundings. In the vicinity of a black hole, matter should be orbiting at high speed around something invisible. We have already examined the evidence for a black hole at the center of our Milky Way [Section 19.5], but what about other galaxies?

The relatively nearby galaxy M 87 features a bright nucleus and a jet that emits both radio and visible light. Thus, it was already a prime black-hole suspect when astronomers pointed the Hubble Space Telescope at its core in 1994 (Figure 21.26). The spectra they gathered showed blueshifted emission lines on one side of the nucleus and redshifted emission lines on the other. This pattern of Doppler shifts is the characteristic signature of orbiting gas: On one side of the orbit the gas is coming toward us and hence is blueshifted, while on the other side it is moving away from us and is redshifted. The magnitude of these Doppler shifts shows that the gas, located up to 60 light-years from the center, is orbiting something invisible at a speed of hundreds of kilometers per second. This high-speed orbital motion indicates that the central object has a mass some 2–3 billion times that of our Sun.

Observations of NGC 4258, another galaxy with a visible jet, delivered even more persuasive evidence just 1 year later. A ring of molecular clouds orbits the nucleus of this galaxy in a circle less than 1 light-year in radius. We can pinpoint these clouds because they amplify the microwave emission lines of water molecules, generating beams of microwaves very similar to laser beams. (The word *laser* stands for "*l*ight *a*mplification by *s*timulated *e*mission of *r*adiation." These clouds contain *water masers.* The word *maser* stands for "*m*icrowave *a*mplification by *s*timulated *e*mission of *r*adiation.") The Doppler shifts of these emission lines allow us to determine the orbits of the clouds very precisely. Their orbital motion tells us that the clouds are circling a single, invisible object with a mass of 36 million solar masses. A supermassive black hole is the only thing we know of that could be so massive while remaining unseen.

We may never be 100% certain that we have discovered black holes in other galaxies. The best we can do is rule out all other possibilities. However, the hypothesis that gigantic black holes lie at the cores of quasars, nearby active galactic nuclei, and radio galaxies is so far withstanding the tests of time and thousands of observations.

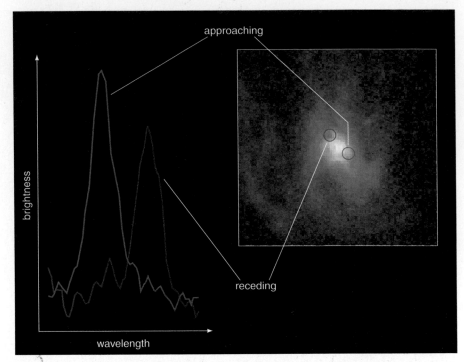

Figure 21.26 Doppler shifts of the emission lines from the gas disk in the elliptical galaxy M 87 indicate that it orbits a black hole 2–3 billion times more massive than the Sun. The inset photo is a Hubble Space Telescope photograph showing the gas disk. The red line shows an emission line from the side of the disk that is rotating away from us (receding). The blue line shows the same emission line from the side of the disk that is rotating toward us (approaching). Two circles, each about 60 light-years from the disk's center, show where these emission lines were measured. The Doppler shifts of these lines tell us how fast the disk is spinning—about 800 km/s. Applying Newton's laws gives us the mass of the central black hole.

 Hubble's Law Tutorial, Lesson 1

21.6 Shedding Light on Protogalactic Clouds

The most mysterious part of galaxy evolution is the part we've never observed: the formation and development of protogalactic clouds. In recent years, the study of quasars has begun to shed light on even this very early stage of galaxy evolution. The most distant quasars inhabit the outskirts of the observable universe. Photons from some of

these quasars began journeying to Earth when the universe was only 1 or 2 billion years old. Along the way, these photons have passed through numerous intergalactic hydrogen clouds. The vast majority of these clouds are too diffuse and wispy ever to become galaxies, but a few have as much hydrogen as the disk of the Milky Way. The thickest of these clouds may well be protogalactic clouds in the process of becoming galaxies.

Quasar spectra therefore contain valuable information about the properties of hydrogen clouds in the early universe. Because atoms tend to absorb light at very specific wavelengths, every time a light beam from a quasar

---

Mathematical Insight **21.2** **Weighing Supermassive Black Holes**

We weigh supermassive black holes the same way we weigh almost everything else in the universe: by measuring the velocity v and orbital radius r of the matter circling the central black hole. Given these measurements, we can apply the orbital velocity law from Mathematical Insight 19.1 to find the mass M_r within a distance r of the galactic center:

$$M_r = \frac{r \times v^2}{G}$$

Example: Doppler shifts show that ionized gas in the nucleus of the active galaxy M 87 orbits at a speed of about 800 km/s at a radius of 60 light-years, or approximately 5.6×10^{17} meters. Use these values to calculate the mass within 60 light-years of the galactic center.

Solution: Substituting the given values into the orbital velocity law, we find:

$$M_r = \frac{(5.6 \times 10^{17} \text{ m}) \times \left(8.0 \times 10^5 \frac{\text{m}}{\text{s}}\right)^2}{6.67 \times 10^{-11} \frac{\text{m}^3}{\text{kg} \times \text{s}^2}}$$

$$= 5.3 \times 10^{39} \text{ kg}$$

Converting kilograms to solar masses, we find:

$$M_r = (5.3 \times 10^{39} \text{ kg}) \times \frac{1 \text{ solar mass}}{2.0 \times 10^{30} \text{ kg}}$$

$$= 2.7 \times 10^9 \text{ solar masses}$$

The ionized gas is orbiting around a mass of about 2.7 billion solar masses. Presumably, nearly all of this mass is in the central black hole.

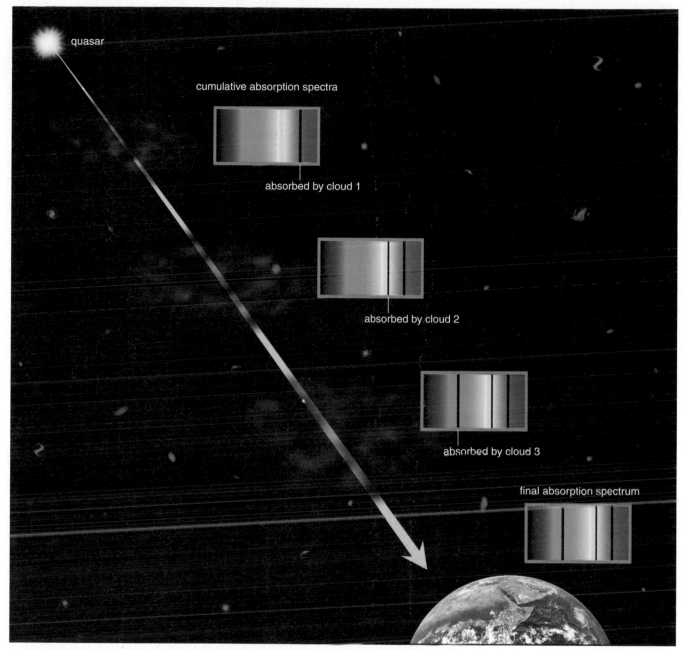

Figure 21.27 This schematic illustration shows how interstellar hydrogen clouds leave their mark on the spectra of quasars. As light travels from a distant quasar to Earth, it passes through many different clouds of hydrogen gas. Hydrogen atoms within each cloud absorb light of a very specific wavelength. Because these clouds are at different distances and therefore have different redshifts, we observe the wavelength corresponding to hydrogen atoms in a different place for each cloud. Thus, each line in the final spectrum tells us about a unique hydrogen cloud along the line of sight. As we study progressively redder lines, we learn about progressively younger clouds.

passes through an intergalactic or protogalactic cloud some of the atoms in the cloud absorb photons from the beam, creating an absorption line (Figure 21.27). Studies of these absorption lines in quasar spectra can tell us what happened in protogalactic clouds during the epoch of galaxy formation and thus provide clues about how galaxies evolve.

We are only beginning to learn how to read the clues that hydrogen absorption lines have etched into the spectra

of quasars, but the evidence gathered so far supports our general picture of spiral-galaxy evolution. The most prominent hydrogen absorption lines, thought to be associated with newly forming galaxies, are typically produced by the most distant clouds, indicating that the youngest galaxies are made mostly of gas. In fact, they contain about the same amount of mass in the form of hydrogen gas as older galaxies contain in the form of stars. The hydrogen lines from nearer clouds, which arise in more mature galaxies, are not nearly as strong. Presumably, a greater fraction of the gas in these galaxies has already collected into stars.

Absorption lines from elements other than hydrogen corroborate this picture. The lines from these heavy elements are more prominent in the mature galaxies than in the youngest galaxies, implying that the mature galaxies have experienced more supernovae, which have added heavy elements to their interstellar gas. This overall pattern of gradual heavy-element enrichment accompanied by the gradual diminishing of interstellar hydrogen agrees well with what we know about the Milky Way. All across the universe, stars in the gaseous disks of galaxies appear to have been forming steadily for over 10 billion years.

With every new study of a quasar spectrum, we learn more about the galaxies and intergalactic clouds that have left their mark on the light we observe from quasars. Perhaps someday soon these observations will help us fit together all the pieces of the galaxy evolution puzzle, and we at last will understand the whole glorious history of galaxy evolution in our universe. Until that time, we will keep peering deep into space and back into time, searching for the clues that will unlock the mysteries of cosmic evolution.

THE BIG PICTURE

Putting Chapter 21 into Context

We have not yet solved the whole puzzle of galaxy evolution, but in this chapter we have described some of its crucial pieces. As you look back, keep sight of these "big picture" ideas:

- Although we do not yet know the complete story of galaxy evolution, we are rapidly learning more. We know that galaxies grow from protogalactic clouds of gas, but collisions with neighboring galaxies have probably affected many galaxies.

- Galaxies in the present-day universe do not always evolve peacefully—many undergo temporary starbursts that can eject large fractions of their interstellar gas.

- The tremendous energy outputs of quasars and other active galactic nuclei, including those of radio galaxies, are probably powered by gas accreting onto supermassive black holes. The centers of many present-day galaxies must still contain the supermassive black holes that once enabled them to shine as quasars.

- Quasars are brilliant beacons whose light has crossed billions of light-years of space to reach us, passing through numerous intergalactic clouds and perhaps some newly forming galaxies on the way. These clouds and galaxies leave their mark on quasar spectra, revealing how protogalactic clouds were behaving before they formed most of their stars.

SUMMARY OF KEY CONCEPTS

21.1 Looking Back Through Time

- *What do we mean by galaxy evolution, and how do we study it?* Galaxy evolution is the study of how galaxies form and change with time. Because we are looking farther back into time as we look deeper into space, we can create "family albums" showing galaxies as they appeared at different times in the history of the universe.

21.2 Galaxy Formation

- *What two starting assumptions do we make in most models of galaxy formation?* (1) Hydrogen and helium gas filled all of space when the universe was young. (2) The distribution of matter in the universe was nearly but not quite uniform, so some regions of the universe were slightly denser than others.

- *How are galaxies thought to have formed?* Gravity slowed the expansion of matter in regions of the universe where the density was slightly greater than average. Within about a billion years after the birth of the universe, gravity stopped the expansion of these regions and began to pull matter together into protogalactic clouds. Halo stars began to form as protogalactic clouds collapsed into young galaxies. In galaxies that had enough remaining gas after this initial star formation, conservation of angular momentum ensured that the gas flattened into a spinning disk.

- *What does the "fossil record" of our Milky Way Galaxy tell us about galaxy formation?* The Milky Way's halo stars are very old and their orbits have random orientations, suggesting that they formed before the protogalactic cloud collapsed into a disk. The low abundances of heavy elements in halo stars tell us that they were born before the star–gas–star cycle significantly enriched the interstellar medium with heavy elements. The relationship between heavy-element abundance and distance from the galactic center suggests that our Milky Way formed not from a single protogalactic cloud but rather from the merger of several small protogalactic clouds.

21.3 Why Do Galaxies Differ?

- *How might a galaxy's birth properties have determined whether it ended up spiral or elliptical?* There are two basic possibilities: (1) Angular momentum tends to shape a collapsing gas cloud into a spinning disk. Thus, ellipticals may have formed from protogalactic clouds with relatively small amounts of angular momentum, and spiral galaxies may have formed from clouds with greater angular momentum. (2) Dense clouds tend to cool and form stars more rapidly. Thus, ellipticals may have formed from protogalactic clouds that started out with greater density, leading to a high rate of star formation that left little or no gas to collapse into a disk. Spirals may have started from lower-density protogalactic clouds in which a lower rate of star formation left enough gas to form a disk.

- *How might interactions between galaxies cause spiral galaxies to become elliptical?* Computer models show that a collision of spiral galaxies randomizes the orbits of their stars, while their combined gas sinks to the center and is quickly used up in a burst of rapid star formation.

- *What do observations of galaxy clusters tell us about the role of galaxy interactions?* They support the idea that at least some galaxies are shaped by collisions. Elliptical galaxies are more common in the centers of clusters—where collisions also are more common—suggesting that they may have formed from collisions of spiral galaxies. The central dominant galaxies found in cluster centers also appear to be the result of collisions, both because of their large size and because they sometimes contain multiple clumps of stars that probably once were the centers of individual galaxies.

21.4 Starburst Galaxies

- *What is a starburst galaxy?* A starburst galaxy is a galaxy that is forming new stars at a very high rate—sometimes more than 100 times the star formation rate of the Milky Way. This high rate of star formation can lead to a supernova-driven galactic wind.

- *How do we know that a starburst must be only a temporary phase in a galaxy's life?* The rate of star formation is so high that the galaxy would use up all its interstellar gas in just a few hundred million years—far shorter than the age of the universe.

- *What can cause starbursts?* Many starbursts apparently result from collisions between galaxies. Some starbursts may occur as a result of close encounters with other galaxies rather than from direct collisions.

21.5 Quasars and Other Active Galactic Nuclei

- *What are active galactic nuclei and quasars?* Active galactic nuclei are the unusually bright centers found in some galaxies. The brightest active galactic nuclei are called *quasars*. Active galactic nuclei (including quasars) generally radiate energy across much of the electromagnetic spectrum. In some cases, we see spectacular jets of material shooting out of these objects, sometimes forming huge lobes (revealed by radio observations) at great distances from the center of the galaxy.

- *What do we think is the source of power for active galactic nuclei?* We suspect that active galactic nuclei are powered by supermassive black holes that can exceed 1 billion solar masses. As matter falls into one of these supermassive black holes, it releases tremendous amounts of energy. A black hole is the only mechanism we know of that can account for the prodigious energy output of active galactic nuclei.

- *What place do active galactic nuclei have in the story of galaxy evolution?* Because quasars were much more common in the past, many galaxies probably once had very bright nuclei that have now gone dormant. If so, then many galaxies that now look quite normal have supermassive black holes at their centers. In fact, the masses of supermassive black holes in nearby galaxies appear closely related to the mass of the spheroidal component.

21.6 Shedding Light on Protogalactic Clouds

- *How do quasars let us study gas between the galaxies?* Each cloud of gas through which the quasar's light passes on its long journey to Earth produces a hydrogen absorption line in the quasar spectrum. Study of these absorption lines in quasar spectra allows us to study matter—including protogalactic clouds—that we cannot otherwise detect.

Decide whether each of the following statements is true and explain why it is or is not.

1. Galaxies that are more than 10 billion years old are too far away to see even with our most powerful telescopes.

2. Heavy elements ought to be much more common near the Milky Way's center than at its outskirts.

3. If the Andromeda Galaxy someday collides and merges with the Milky Way, then the resulting galaxy will probably be elliptical.

4. Starburst galaxies have been forming stars at the same furious pace ever since the universe was about a billion years old.

5. We know that some galaxies contain supermassive black holes because their centers are completely dark.

6. The black hole at the center of our own galaxy may once have powered an active galactic nucleus.

7. Radio galaxies emit only radio waves and no visible light.

8. Analyses of quasar light can tell us about intergalactic clouds that might otherwise remain invisible.

Problems

9. *Modeling Galaxy Birth.* How do theoretical models help us study galaxy birth?

10. *Stellar Populations.* How do disk population stars differ from spheroidal population stars? Briefly explain how the collapse of a protogalactic cloud is thought to lead to these two distinct populations of stars. How does this explain why the halo of our galaxy looks so much like an elliptical galaxy?

11. *Galaxy Collisions.* Briefly explain why we expect that collisions between galaxies should be relatively common, while collisions between stars are extremely rare. Why should galaxy collisions have been more common in the past than they are today?

12. *Infrared Starbursts.* Briefly explain why starburst galaxies often appear ordinary when they are observed in visible light but extraordinary when they are observed in infrared light.

13. *Galactic Winds.* What is a galactic wind? What causes it? How is it similar to a superbubble in the Milky Way, and how is it different?

14. *Quasars.* Briefly describe the discovery of quasars. What evidence convinced astronomers that the high redshifts of quasars really do imply great distances? Why can we learn more about quasars by studying nearby active galactic nuclei?

15. *Varying Luminosity.* Briefly explain how we can use variations in luminosity to set limits on the size of an object's emitting region. For example, if an object doubles its luminosity in 1 hour, how big can it be?

16. *Radio Galaxies.* What is a radio galaxy? Describe jets and radio lobes. Why do we think that the ultimate energy sources of radio galaxies lie in quasarlike galactic nuclei?

17. *Life Story of a Spiral.* Imagine that you are a spiral galaxy. Describe your life history from birth to the present day. Your story should be detailed and scientifically consistent, but also creative. That is, it should be entertaining while at the same time incorporating current scientific ideas about the formation of spiral galaxies.

18. *Life Story of an Elliptical.* Imagine that you are an elliptical galaxy. Describe your life history from birth to the present. There are several possible scenarios for the formation of elliptical galaxies, so choose one and stick to it. Be creative while also incorporating scientific ideas that demonstrate your understanding.

*19. *Your Last Hurrah.* Suppose you fell into an accretion disk that swept you into a supermassive black hole. Assume that, on your way down, the disk will radiate 10% of your mass energy, $E = mc^2$. (*Hint:* See Mathematical Insight 21.1.)

   a. What is your mass in kilograms? (*Hint:* Use the conversion 1 kg = 2.2 pounds.)

   b. Calculate how much radiative energy will be produced by the accretion disk as a result of your fall into the black hole.

   c. Calculate approximately how long a 100-watt light bulb would have to burn to radiate this same amount of energy.

*20. *The Black Hole in NGC 4258.* The molecular clouds circling the center of the active galaxy NGC 4258 orbit at a speed of about 1,000 km/s, with an orbital radius of 0.49 light-year = 4.8×10^{15} meters. Use the orbital velocity law (see Mathematical Insight 21.2) to calculate the mass of the central black hole. Give your answer both in kilograms and in solar masses ($1 M_{Sun} = 2.0 \times 10^{30}$ kg).

Discussion Questions

21. *The Case for Supermassive Black Holes.* The evidence for supermassive black holes at the center of galaxies is strong. However, it is very difficult to prove absolutely that they exist because the black holes themselves emit no light. We can only infer their existence from their powerful gravitational influences on surrounding matter. How compelling do you find the evidence? Do you think astronomers have proved the case for black holes beyond a reasonable doubt? Defend your opinion.

22. *Life in Colliding Galaxies.* Suppose the Milky Way were currently undergoing a collision with another large spiral galaxy. Do you think this collision would affect life on Earth? Why or why not? How different would the night sky look if our galaxy was in the midst of such a collision?

For a complete list of media resources available, go to www.astronomyplace.com and choose Chapter 21 from the pull-down menu.

 Astronomy Place Web Tutorials

Tutorial Review of Key Concepts

Use the following interactive **Tutorial** at www.astronomyplace.com to review key concepts from this chapter.

Black Holes Tutorial

Lesson 1 What Are Black Holes?

Lesson 2 The Search for Black Holes

Supplementary Tutorial Exercises

Use the interactive **Tutorial Lessons** to explore the following questions.

Black Holes Tutorial, Lessons 1, 2

1. What is the Schwarzschild radius for a black hole of 1 million solar masses? For black holes of 10 million and 100 million solar masses?

2. Describe how scientists determine if the center of a galaxy contains a supermassive black hole.

Movies

Check out the following narrated and animated short documentary available on www.astronomyplace.com for a helpful review of key ideas covered in this chapter.

From the Big Bang to the Galaxies Movie

Web Projects

Take advantage of the useful Web links on www.astronomyplace.com to assist you with the following projects.

1. *Future Missions.* The subject of galaxy evolution is a very active area of research. Look for information on current and future NASA missions involved in investigating galaxy evolution (such as the James Webb Space Telescope). How big are the planned telescopes? What wavelengths will they look at? When will they be launched? Write a short summary of a proposed mission.

2. *Greatest Redshift.* As of early 2003, the most distant quasar known has a redshift $z = 6.4$, meaning that the wavelengths of its light are $1 + 6.4 = 7.4$ longer than normal. Find the current record holder for the largest redshift. Write a one-page report describing the object and its discovery.

3. *The Quasar Controversy.* For many years, some astronomers argued that quasars were not really as distant as Hubble's law indicates. Research the history of the discovery of quasars and the debates that followed. What evidence led some astronomers to think quasars might be nearer than Hubble's law suggested? Why did most astronomers eventually conclude that quasars really are far away? Write a one- to two-page report summarizing your findings.

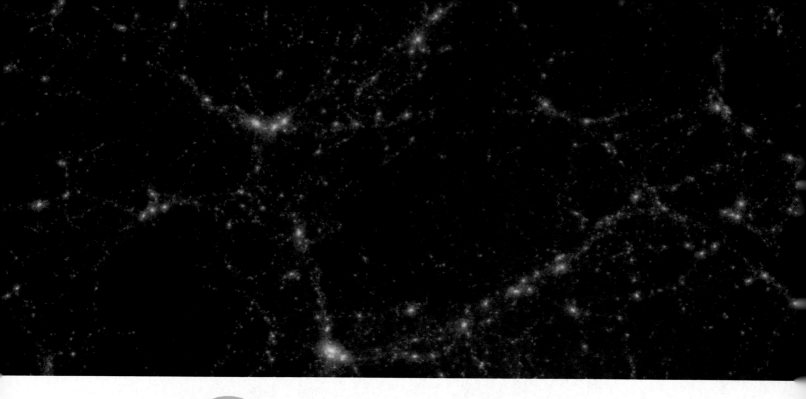

22 Dark Matter and the Fate of the Universe

The majority of the matter in the universe, the stuff that binds galaxies together with the force of its own gravity, is too dark to see. We know that this matter exists because we can detect its gravitational influence on other things, but what is this so-called *dark matter?* Do armies of Jupiter-size bodies, too dim for us to see at a distance, populate the voids between the stars? Could black holes be responsible? Or is dark matter an entirely new form of matter, still undiscovered here on Earth? We don't yet know the answer. Incredibly, we still haven't identified the most common form of matter in the universe, making dark matter one of the greatest cosmic mysteries.

In this chapter, we will investigate why most astronomers believe that dark matter exists. We'll investigate current hypotheses about the nature of dark matter. We'll see how it affects structures the size of galaxies and larger. We'll see how the fate of the universe relates to the question of just how much dark matter there is. Finally, we'll see that dark matter may not be the only unseen influence on the fate of the universe. Recent discoveries suggest that a mysterious "dark energy" may be causing the expansion of the universe to accelerate.

 Orbits and Kepler's Laws, Lessons 1–4

22.1 The Mystery of Dark Matter

We encountered dark matter when we studied the Milky Way's rotation in Chapter 19. We saw that atomic hydrogen clouds lying farther from the galactic center than our Sun orbit the galaxy at unexpectedly high speeds. We concluded that much of our galaxy's mass must lie beyond the distance of the Sun's orbit around the galactic center, distrib-

uted throughout the galaxy's spherical halo [Section 19.4]. Yet most of the Milky Way's *light* comes from stars lying closer to the galaxy's center than does our Sun. Together, these facts lead us to conclude that the halo contains large amounts of matter but this matter emits so little light that we cannot see it. Thus, we call it *dark matter.*

Does dark matter really exist? To claim that the galaxy is filled with a form of matter that we cannot yet identify may seem strange, but there are only two possible explanations for the high orbital velocities of the atomic hydrogen clouds: Either their velocities are caused by the gravitational attraction of unseen matter, or we are doing something wrong when we apply the law of gravity to understand their orbits. The latter possibility would imply that we do not understand how gravity operates on galaxy-size scales. However, our theories of gravity successfully account for many other cosmic phenomena. Thus, the vast majority of astronomers believe that we correctly understand gravity and that dark matter must really exist.

The evidence that the Milky Way contains dark matter qualifies as an interesting surprise on its own, but it is also a glimpse into a more profound mystery. If our suspicions about dark matter are correct, then the luminous part of the Milky Way's disk must be rather like the tip of an iceberg, marking only the center of a much larger clump of mass (Figure 22.1). A more detailed analysis involving the motions of nearby dwarf galaxies suggests that the total mass of dark matter in the Milky Way might be 10 times greater than the mass of visible stars.

Other galaxies are similar to our own in this respect, meaning that we probably cannot see most of the matter in

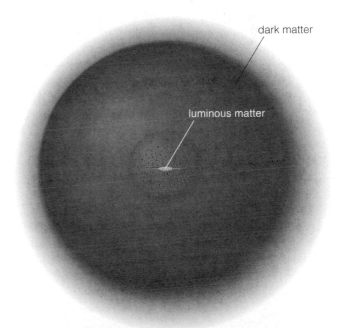

Figure 22.1 The dark matter associated with a spiral galaxy occupies a much larger volume than the galaxy's luminous matter. The radius of this dark-matter halo could be as much as 10 times larger than the galaxy's halo of stars.

the universe. This matter's darkness makes it challenging to study, but its overwhelming dominance of the universe makes it extremely important. Dark matter, by virtue of its mighty gravitational pull, seems to be the glue that holds galaxies and clusters of galaxies together.

Dark matter has even more profound implications for the fate of the universe. The universe seems destined to proceed along one of two paths. It might continue to expand forever, or it might someday stop expanding and begin to collapse. We are not yet certain which fate awaits our universe partly because the answer depends on the overall density of matter in the universe—and we cannot determine the overall density until we first determine how much dark matter is out there. Thus, the very fate of the universe hinges on the total amount of dark matter.

In this chapter, we will presume that we understand gravity correctly and consequently that dark matter pervades the universe. If our understanding of gravity is correct, then we already know quite a lot about dark matter, despite the fact that its composition remains a mystery. Measuring the total amount of dark matter in the universe is difficult, but we can more easily measure the amounts in individual galaxies and clusters of galaxies from its gravitational effects. We can even make some reasonable guesses as to what it will turn out to be. We'll begin by discussing how we "discover" dark matter and conclude by discussing what our current knowledge predicts about the ultimate fate of the universe.

 Detecting Dark Matter in a Spiral Galaxy Tutorial, Lessons 1–3

22.2 Dark Matter in Galaxies

The claim that dark matter far outweighs the visible matter in the universe might seem farfetched, but it rests on fundamental physical laws. Newton's laws of motion and gravity are among the most trustworthy tools in science. We have used them time and again to measure masses of celestial objects.

We found the masses of Earth and the Sun by applying Newton's version of Kepler's third law [Section 5.3]. We used this same law to calculate the masses of stars in binary star systems, revealing the general relationships between the masses of stars and their outward appearances. Newton's laws have also told us the masses of things we can't see directly, such as the masses of neutron stars in X-ray binaries and of black holes in active galactic nuclei. These laws have proved extremely reliable in many different applications. Thus, when Newton's laws tell us that dark matter exists, we are not inclined to ignore them.

We can determine the amount of dark matter in a galaxy if we know the galaxy's mass and luminosity. We can infer the total mass in stars from the galaxy's luminosity, so whatever additional mass remains unaccounted for must be dark. Measuring a galaxy's luminosity is relatively easy. We simply point a telescope at the galaxy in question, mea-

sure its apparent brightness, and calculate its luminosity from the luminosity–distance formula [Section 16.2]. Measuring the galaxy's total mass is more complicated. We would like to apply Newton's laws in the form of the orbital velocity law from Chapter 19 to matter orbiting as far from the galaxy's center as possible. However, the matter farthest from the center of a galaxy is extremely dim, so we need to look at more than just the galaxy's stars.

Weighing Spiral Galaxies

We can weigh a spiral galaxy by measuring the gravitational effects of the galaxy's mass on the orbits of objects in its disk. Even beyond the point in the disk at which starlight fades into the blackness of intergalactic space, we can still see radio waves from atomic hydrogen gas. We can therefore use Doppler shifts of the 21-cm emission line of atomic hydrogen [Section 19.2] to determine how quickly this gas moves toward us or away from us (Figure 22.2).

Galaxies beyond the Local Group have cosmological redshifts affecting all their spectral lines. However, on one side of a spiral galaxy the gas is rotating away from us, so its 21-cm line is redshifted a little more than the redshift of the galaxy as a whole. On the other side, the 21-cm line is blueshifted relative to the redshift of the galaxy as a whole because the gas is rotating toward us. From the Doppler shifts of these clouds, we can construct a *rotation curve*— a plot showing orbital velocities of gas clouds and stars— just as we did for the Milky Way (Figure 22.3) [Section 19.4]. A rotation curve contains all the information we need to measure the mass contained within the orbits of the outermost gas clouds. (Because the Doppler effect tells us only about the velocity of material directly toward or away from us, we must also take into account the tilt of the galaxy before we construct the rotation curve.)

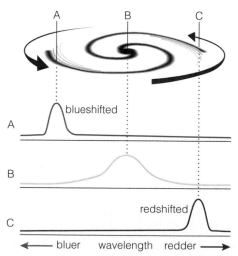

Figure 22.2 Measuring the rotation of a spiral galaxy with the 21-cm line of atomic hydrogen. Blueshifted lines on the left side of the disk show how fast that side is rotating toward us. Redshifted lines on the right side show how fast that side is rotating away from us.

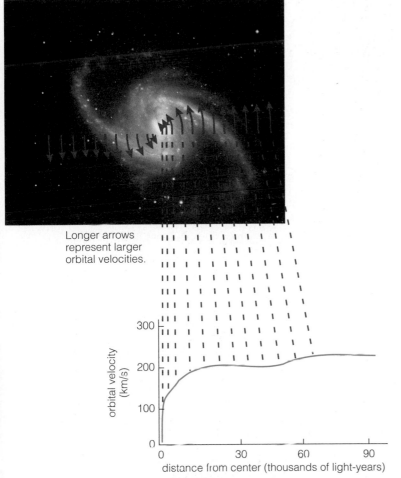

Longer arrows represent larger orbital velocities.

Figure 22.3 A rotation curve shows the orbital velocities of stars or gas clouds at different distances from a galaxy's center.

THINK ABOUT IT

As a brief review, draw a rotation curve for our solar system. How does the solar system rotation curve differ from the rotation curve of the Milky Way? What does that tell us about the distribution of matter in the Milky Way? Why? (*Hint:* See Chapter 19 for review.)

The rotation curves of most spiral galaxies turn out to be remarkably flat as far out as we can see, which means that the orbital speeds of gas clouds remain roughly constant with increasing distance from the galactic center (Figure 22.4). We saw in Chapter 19 that our Milky Way's

rotation curve remains more or less flat from well within the radius of the Sun's orbit (28,000 light-years) to beyond twice that radius. The rotation curves of some other spiral galaxies are flat to well beyond 150,000 light-years.

Because the orbital speeds of gas clouds tell us the amount of mass contained within their orbital paths, the flat rotation curves imply that a great deal of matter lies far from the galactic center [Section 19.4]. In particular, if we imagine drawing bigger and bigger circles around a galaxy, the flat rotation curves imply that we must keep encircling more and more matter. The mass of a spiral galaxy with gas clouds orbiting at 200 km/s at a distance of 150,000 light-years from its center must be at least 500 billion (5×10^{11}) solar masses.

Weighing Elliptical Galaxies

We must use a different technique to weigh elliptical galaxies, because most of them contain very little atomic hydrogen gas and hence do not produce detectable 21-cm radiation. Instead, we generally weigh the inner parts of elliptical galaxies by observing the motions of the stars themselves.

The motions of stars in an elliptical galaxy are disorganized, so we cannot assemble their velocities into a sensible rotation curve. Nevertheless, the velocity of each individual star still responds to the mass inside the star's orbit. At any particular distance from an elliptical galaxy's center, some stars are moving toward us and some are moving away from us. Thus, the spectral lines from the galaxy as a whole tend to be smeared out. Instead of a nice sharp line at a particular wavelength, we see a *broadened* line spanning a range of wavelengths reflecting the various Doppler shifts of the individual stars (Figure 22.5). The greater the broadening of the spectral line, the faster the stars must be moving.

When we compare spectral lines from different regions of an elliptical galaxy, we find that the speeds of the stars remain fairly constant as we look farther from the galactic center. Thus, just as in spirals, most of the matter in elliptical galaxies must lie beyond the distance where the light trails off and hence must be dark matter. However, we cannot determine the *total* amount of dark matter in elliptical galaxies as well as we can in spirals, because we cannot measure their masses as far from their centers. In ellipticals we can study only the motions of stars, while

Figure 22.4 Actual rotation curves of four spiral galaxies. They are all nearly flat over a wide range of distances from the center, indicating that dark matter is common in spiral galaxies.

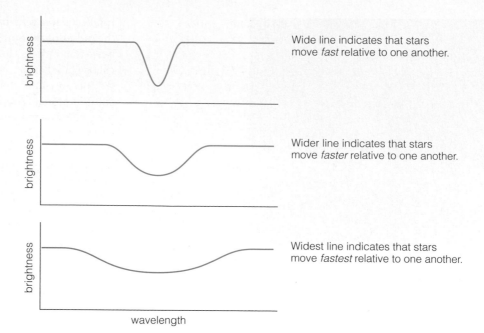

Wide line indicates that stars move *fast* relative to one another.

Wider line indicates that stars move *faster* relative to one another.

Widest line indicates that stars move *fastest* relative to one another.

wavelength

Figure 22.5 The broadening of absorption lines in an elliptical galaxy's spectrum tells us how fast its stars move relative to one another.

in spirals we are able to detect the 21-cm radiation from gas clouds well beyond the radii at which we no longer see individual stars.

Mass-to-Light Ratio

Now that we can measure a galaxy's approximate mass, we can determine how much dark matter the galaxy contains by comparing the galaxy's measured mass to its luminosity. For example, the Milky Way contains about 90 billion (9×10^{10}) solar masses of material within the Sun's orbit. However, the total luminosity of stars within this region is only about 15 billion (1.5×10^{10}) solar luminosities. Thus, on average, it takes 6 solar masses of matter to produce 1 solar luminosity of light in this region of the galaxy

(15 billion $\times$ 6 = 90 billion). We therefore say that the Milky Way's **mass-to-light ratio** within the Sun's orbit is about 6 solar masses per solar luminosity. This fact tells us that most matter is dimmer than our Sun, which is not surprising because we know that most stars are smaller and dimmer than our Sun. We find similar mass-to-light ratios for the inner regions of most other spiral galaxies.

THINK ABOUT IT

Because the Sun's mass is $1M_{Sun}$ and its luminosity is $1L_{Sun}$, its mass-to-light ratio is $1M_{Sun}/1L_{Sun}$ = 1 solar mass per solar luminosity. What is the mass-to-light ratio of a $1M_{Sun}$ red giant with luminosity $100L_{Sun}$? What is the mass-to-light ratio of a $1M_{Sun}$ white dwarf with luminosity $0.001L_{Sun}$?

Mathematical Insight **22.1** **Mass-to-Light Ratio**

Mathematically, an object's mass-to-light ratio is defined as its total mass in units of solar masses divided by its total luminosity in units of solar luminosities. Thus, by definition, the mass-to-light ratio of the Sun is:

$$\frac{1 \text{ solar mass}}{1 \text{ solar luminosity}} = \frac{1 \text{ solar mass}}{\text{per solar luminosity}}$$

A galaxy made entirely of stars just like the Sun would have a mass-to-light ratio identical to that of the Sun.

Example 1: What is the mass-to-light ratio of the matter inside the solar circle of the Milky Way?

Solution: To answer this question, we divide the 90 billion solar masses inside the solar circle by the 15 billion solar luminosities of the stars in the same region:

$$\frac{9 \times 10^{10} \text{ solar masses}}{1.5 \times 10^{10} \text{ solar luminosities}} = \frac{6 \text{ solar masses}}{\text{per solar luminosity}}$$

Because this mass-to-light ratio is larger than the Sun's ratio of 1, the average luminosity of objects in this region of the Milky Way must be less than that of the Sun.

Example 2: Suppose a galaxy contains 5×10^{11} solar masses within a radius of 150,000 light-years of its center, but its total luminosity is only 1.5×10^{10} solar luminosities. What is its mass-to-light ratio? What does this imply?

Solution: Again we divide the total mass by the total luminosity. The mass-to-light ratio within 150,000 light-years of this galaxy's center is:

$$\frac{5 \times 10^{11} \text{ solar masses}}{1.5 \times 10^{10} \text{ solar luminosities}} = \frac{33 \text{ solar masses}}{\text{per solar luminosity}}$$

Taken as a whole, the mass in this galaxy is far darker, on average, than the Sun or than the objects in the inner region of the Milky Way.

Elliptical galaxies contain virtually no luminous high-mass stars, so their stars are less bright on average than the stars in a spiral galaxy. In the inner regions of elliptical galaxies, the orbits of stars indicate a mass-to-light ratio of about 10 solar masses per solar luminosity—nearly double the mass-to-light ratio in the central region of the Milky Way. This is not surprising given that elliptical galaxies have dimmer stars.

The surprise in mass-to-light ratios—and one of the key pieces of evidence for the existence of dark matter—comes when we look to the outer reaches of galaxies. As we look farther from a galaxy's center, we find a lot more mass but not much more light. The mass-to-light ratios for entire spiral galaxies can be as high as 50 solar masses per solar luminosity, and the overall ratios in some dwarf galaxies can be even higher. We are forced to conclude that stars alone cannot account for the amount of mass present in galaxies.

For example, if a galaxy's overall mass-to-light ratio is 60 solar masses per solar luminosity and its stars account for only 6 solar masses per solar luminosity, then the remaining 90% of the galaxy's mass must be dark. This proportion of dark matter is typical of galaxies whose motions we can measure at large distances from the center. Thus, we conclude that the vast majority of the mass in galaxies is not in the form of stars.

Figure 22.6 Fritz Zwicky, discoverer of dark matter in clusters of galaxies. Zwicky had an eccentric personality, but some of his ideas that seemed strange in the 1930s proved correct many decades later.

22.3 Dark Matter in Clusters

The problem of dark matter in astronomy is not particularly new. In the 1930s, astronomer Fritz Zwicky was already arguing that clusters of galaxies held enormous amounts of this mysterious stuff (Figure 22.6). Few of his colleagues paid attention, but later observations supported Zwicky's claims.

We can now weigh clusters of galaxies in three different ways: by measuring the speeds of galaxies orbiting the center of the cluster, by studying the X-ray emission from hot gas between the cluster galaxies, and by observing how the clusters bend light as *gravitational lenses* [Section S3.5]. All three techniques indicate that clusters contain huge amounts of dark matter. Let's investigate each of these techniques more closely.

Orbiting Galaxies

Zwicky was one of the first astronomers to think of galaxy clusters as huge swarms of galaxies bound together by gravity. It seemed natural to him that galaxies clumped closely in space should all be orbiting one another, just like the stars in a star cluster. He therefore assumed that he could measure cluster masses by observing galaxy motions and applying the orbital velocity law.

Armed with a spectrograph, Zwicky measured the redshifts of the galaxies in a particular cluster and used these redshifts to calculate the speeds at which the individual galaxies are moving away from us. He determined the velocity of the cluster as a whole by averaging the velocities of its individual galaxies. He then estimated the speed of a galaxy around the cluster—that is, its orbital speed—by subtracting this average velocity from the individual galaxy's velocity. Finally, he plugged these orbital speeds into an appropriate form of the orbital velocity law [Section 19.4] to estimate the cluster's mass and compared this mass to the luminosity of the cluster.

To his surprise, Zwicky found that clusters of galaxies have huge mass-to-light ratios. They contain hundreds of solar masses for each solar luminosity of radiation they emit. He concluded that most of the matter within these clusters must be almost entirely dark. Many astronomers disregarded Zwicky's result, believing that he must have done something wrong to arrive at such a strange answer.

Today, far more sophisticated measurements of galaxy orbits in clusters confirm Zwicky's original finding. We cannot measure the actual orbital velocity and position of a single galaxy but can measure only its *radial velocity* along our line of sight (Figure 22.7). However, if we measure the radial velocities of many galaxies in a cluster, we can estimate the average orbital velocity of all the cluster's galaxies. Cluster masses found in this way imply lots of dark matter. The visible portions of the galaxies in a cluster represent less than 10% of the cluster's mass.

Figure 22.7 A galaxy's true velocity can be broken into a radial component along our line of sight and a tangential component across our line of sight. Doppler shifts tell us only the radial component. We cannot measure the tangential component for distant galaxies. Measurements of the radial velocities of galaxies in clusters indicate that clusters contain lots of dark matter. We can estimate the total amount of dark matter by assuming the galaxies' tangential velocities are similar to their radial velocity within the cluster.

to Earth

→ radial velocity (along our line of sight)

→ tangential velocity (across our line of sight)

→ true (total) velocity

X Rays from the Intracluster Medium

A second method for weighing a cluster of galaxies relies on X-ray observations of the hot gas that fills the space between the galaxies in the cluster (Figure 22.8). This gas, also known as the **intracluster medium** (*intra* means "within"), is so hot that it emits primarily X rays and therefore went undetected until the 1960s, when X-ray telescopes were finally launched above Earth's atmosphere. The temperature of this gas is tens of millions of degrees in many clusters and can exceed 100 million degrees in the largest clusters.

Even though the intracluster medium is invisible in optical light, its mass often exceeds the mass of all the visible stars in all the cluster's galaxies combined. The largest clusters of galaxies contain more than five times more matter in the form of hot gas than in the form of stars.

The temperature of the hot intracluster gas depends on the mass of the cluster itself, enabling us to use X-ray telescopes to measure the masses of galaxy clusters. The intracluster medium in most clusters is nearly in a state of *gravitational equilibrium*—that is, the outward gas pressure balances gravity's inward pull [Section 15.1]. In this state of balance, the average kinetic energies of the gas particles are determined primarily by the strength of gravity and hence by the amount of mass within the cluster. Because the temperature of a gas reflects the average kinetic energies of its particles, the gas temperatures we measure with X-ray telescopes tell us the average speeds of the X ray–emitting particles. We can then use these speeds and the orbital velocity law to weigh the cluster.

The results obtained with this method agree well with the results found by studying the orbital motions of the cluster's galaxies. Again, we find that mass-to-light ratios in clusters of galaxies generally exceed 100 solar masses per solar luminosity. Even after accounting for the hot intracluster gas, it is clear that clusters must contain huge quantities of dark matter binding all the galaxies together.

Scientists always take a risk when they publish what they think are ground-breaking results. If their results turn out to be in error, their reputations may suffer. In the case of dark matter, the pioneers in its discovery risked their entire careers. A case in point is Fritz Zwicky, with his proclamations in the 1930s about dark matter in clusters of galaxies. Most of his colleagues considered him an eccentric who leapt to premature conclusions.

Another pioneer in the discovery of dark matter was Vera Rubin, an astronomer at the Carnegie Institution. Working in the 1960s, she became the first woman to observe under her own name at California's Palomar Observatory, then the largest telescope in the world. (Another woman, Margaret Burbidge, was permitted to observe at Palomar earlier but was required to apply for time under the name of her husband, also an astronomer.) Rubin first saw the gravitational signature of dark matter in spectra she recorded of stars in the Andromeda Galaxy. She noticed that stars in the outskirts of Andromeda moved at surprisingly high speeds, suggesting a stronger gravitational attraction than could be explained by the mass of the galaxy's stars alone. In other words, she found that the rotation curve for Andromeda is relatively flat to great distances from the center, just as we now know is also the case for the Milky Way.

Working with a colleague, Kent Ford, Rubin constructed rotation curves for the hydrogen gas in many other spiral galaxies (by studying Doppler shifts in the spectra of hydrogen gas) and discovered that flat rotation curves are common. Although Rubin and Ford did not immediately recognize the significance of the results, they were soon arguing that the universe must contain substantial quantities of dark matter.

For a while, other astronomers had trouble believing the results. Some astronomers suspected that the bright galaxies studied by Rubin and Ford were unusual for some reason. So Rubin and Ford went back to work, obtaining rotation curves for fainter galaxies. They found flat rotation curves—a signature of dark matter—even in these galaxies. By the 1980s, the evidence compiled by Rubin, Ford, and other astronomers measuring rotation curves was so overwhelming that even the critics came around. Either the theory of gravity was wrong, or they had discovered dark matter in spiral galaxies.

a Virtually every object pictured in this visible-light photo is a galaxy in the Coma Cluster. Measuring the motions of these galaxies reveals that the Coma Cluster contains about 10^{15} solar masses of matter. Because the total luminosity of all the cluster's galaxies is less than $10^{13} L_{Sun}$, the mass-to-light ratio of the cluster exceeds 100 solar masses per solar luminosity. We therefore conclude that the cluster contains far more dark matter than luminous matter. (The picture shows the central 3 million light-years of the cluster.)

Figure 22.8 The Coma Cluster of galaxies in both visible light and X-ray light.

b This false-color image shows X-ray emission from the extremely hot gas (the intracluster medium) that fills the Coma Cluster. The temperature of this gas—almost 100 million degrees—also indicates that the Coma Cluster contains about 10^{15} solar masses of matter. Thus, the X-ray observations confirm the amount of dark matter estimated from the visible-light observations of galaxy motions. (The whole X-ray map shows a region about 14 million light-years across. The dark red and orange regions are bright in X-ray light and mark the center of the cluster. The green and blue regions are less bright.)

Gravitational Lensing

We have so far relied exclusively on methods derived from Newton's laws, such as the orbital velocity law, to measure galaxy and cluster masses. These laws keep telling us that the universe holds far more matter than we can see. Can we trust these laws? Today, we have another tool for measuring masses: *gravitational lensing*.

Gravitational lensing occurs because masses distort spacetime—the "fabric" of the universe [Section S3.3]. Massive objects can therefore act as **gravitational lenses** that bend light beams passing nearby. This prediction of Einstein's theory of general relativity was first verified in 1919 during an eclipse of the Sun [Section S3.5]. Because the light-bending angle of a gravitational lens depends on the mass of the object doing the bending, we can measure the masses of objects by observing how strongly they distort light paths.

Figure 22.9 shows a striking example of how a cluster of galaxies can act as a gravitational lens. Many of the yellow elliptical galaxies concentrated toward the center of the picture belong to the cluster, but at least one of the galaxies pictured does not. At several positions on various sides of the central clump of yellow galaxies you will notice multiple images of the same blue galaxy. Each one of these

images, whose sizes differ, looks like a distorted oval with an off-center smudge.

The blue galaxy seen in these multiple images lies almost directly behind the center of the cluster, at a much greater distance. Multiple images arise because photons traveling from this blue galaxy to Earth do not follow straight paths. Instead, the cluster's gravity bends their paths as they pass through the cluster so that light from this galaxy arrives at Earth from a few slightly different directions. Each alternative path produces a separate, distorted image of the blue galaxy, making this one galaxy look like several galaxies (Figure 22.10).

Multiple images of a gravitationally lensed galaxy are rare. They occur only when a distant galaxy lies directly behind the lensing cluster. However, single images of gravitationally lensed galaxies behind clusters are quite common. Figure 22.11 shows a more typical example. This picture shows numerous normal-looking galaxies and several arc-shaped galaxies. The oddly curved galaxies are not members of the cluster, nor are they really curved. They are normal galaxies lying far beyond the cluster whose images have been distorted by the cluster's gravity.

Careful analyses of the distorted images in pictures of clusters like these enable us to weigh the clusters without

Figure 22.9 Hubble Space Telescope picture of a galaxy cluster acting as a gravitational lens. The yellow elliptical galaxies are cluster members, and the odd, blue ovals are multiple images of a single galaxy that lies almost directly behind the cluster's center. (The picture shows a region about 1.4 million light-years across.)

VIS

| Mathematical Insight | **22.2** | **Masses of Clusters** |

The first two of the three ways described in the text for measuring the masses of galaxy clusters rely on the orbital velocity law from Mathematical Insight 19.1 to find the mass M_r within a distance r of the galactic center:

$$M_r = \frac{r \times v^2}{G}$$

where v is the velocity of objects moving under the influence of gravity. This formula can also be applied to clusters of galaxies.

In the case where we measure the orbital speeds of galaxies directly, we simply use the average speed of the galaxies and the cluster radius in the orbital velocity law. In the case where we measure the temperature of hot, intracluster gas from its X-ray emission, we must first convert the gas temperature into an average speed for the gas particles. Although we will not present a derivation here, we can make this conversion with the following formula, which gives the approximate average speeds of the hydrogen nuclei in a gas of temperature T:

$$\text{average speed of hydrogen nuclei} \approx 100 \, \frac{\text{m}}{\text{s}} \times \sqrt{T} \quad (T \text{ in Kelvin})$$

Once we find this speed, we can use it as v in the orbital velocity law.

Example: Suppose a cluster has a radius of 3.3 million light-years (3.1×10^{22} meters). Its galaxies orbit with an average speed of approximately 1,000 km/s (1×10^6 m/s), and its intracluster gas has a temperature of 9×10^7 K. Find the cluster's mass from both the galaxy speeds and the gas temperature. Do the results agree?

Solution: First, we find the mass from the galaxy speeds. We set $r = 3.1 \times 10^{22}$ m for the cluster radius, set $v \approx 1 \times 10^6$ m/s for the average speed of the galaxies, and assign the gravitational constant G its usual value. Plugging these values into the orbital velocity law, we find:

$$\text{cluster mass} = M_r \approx \frac{(3.1 \times 10^{22} \text{ m}) \times \left(1 \times 10^6 \, \frac{\text{m}}{\text{s}}\right)^2}{6.67 \times 10^{-11} \, \frac{\text{m}^3}{\text{kg} \times \text{s}^2}}$$

$$= 4.6 \times 10^{44} \text{ kg}$$

To find the mass from the temperature of the X ray–emitting gas, we first use the given formula to find the average speeds of the hydrogen nuclei in the gas:

$$\text{average speed} \approx 100 \, \frac{\text{m}}{\text{s}} \times \sqrt{T} = 100 \, \frac{\text{m}}{\text{s}} \times \sqrt{9 \times 10^7}$$

$$= 9.5 \times 10^5 \, \frac{\text{m}}{\text{s}}$$

Plugging this velocity into the orbital velocity law, we find:

$$\text{cluster mass} = M_r \approx \frac{(3.1 \times 10^{22} \text{ m}) \times \left(9.5 \times 10^5 \, \frac{\text{m}}{\text{s}}\right)^2}{6.67 \times 10^{-11} \, \frac{\text{m}^3}{\text{kg} \times \text{s}^2}}$$

$$= 4.2 \times 10^{44} \text{ kg}$$

Note that the two methods have given two results that agree fairly well: 4.6×10^{44} kg and 4.2×10^{44} kg. Thus, we can be confident that we are in the correct range for the actual mass of the cluster. Taking 4.4×10^{44} kg as an intermediate value and recalling that the Sun's mass is 2.0×10^{30} kg, we find:

$$\text{cluster mass} \approx (4.4 \times 10^{44} \text{ kg}) \times \frac{1 \text{ solar mass}}{2.0 \times 10^{30} \text{ kg}}$$

$$= 2.2 \times 10^{14} \text{ solar masses}$$

resorting to the orbital velocity law. Instead, Einstein's theory of general relativity tells us how massive these clusters must be to generate the observed distortions. It is reassuring that cluster masses derived in this way generally agree with those derived from galaxy velocities and X-ray tempera-tures. The three different methods all indicate that clusters of galaxies hold very substantial amounts of dark matter.

THINK ABOUT IT

Should the fact that we have three different ways of measuring cluster masses give us greater confidence that we really do understand gravity and that dark matter really does exist? Why or why not?

22.4 Dark Matter: Ordinary or Extraordinary?

We do not yet know what the dark matter in galaxies and clusters of galaxies is composed of, but our educated guesses fit into two basic categories. First, some or all of the dark matter could be *ordinary*, made of protons, neutrons, and electrons. In that case, the only unusual thing about dark matter is that it is dim. Otherwise, it is made of the same stuff as all the "bright matter" that we can see. The second possibility is that some or all of the dark matter is *extraordinary*, made of particles that we have yet to discover.

A bit of terminology will be useful. Protons and neutrons belong to a category of particles called **baryons**, so ordinary matter is sometimes called **baryonic matter**. (Technically, a baryon is a particle made from three quarks [Section S4.2].) By extension, extraordinary matter is called **nonbaryonic matter**.

Ordinary Dark Matter: MACHOs

Matter need not be extraordinary to be dark. In astronomy, "dark" merely means not as bright as a normal star and therefore not visible across vast distances of space. Your body is dark matter. Everything you own is dark matter. Earth and the rest of the planets are dark matter as well.

Figure 22.10 A cluster's powerful gravity bends light paths from background galaxies to Earth. If light can arrive from several differ-ent directions, we see multiple images of the same galaxy.

Labels in figure: image of galaxy — real galaxy — image of galaxy — cluster — Earth — View from Earth: the same galaxy appears at three positions in the sky.

Figure 22.11 Hubble Space Telescope picture of the cluster Abell 2218. The thin, elongated galaxies around the main clump of galaxies on the left side of the picture are the images of background galaxies distorted by the cluster's gravity. By measuring these distortions, astronomers can determine the total amount of mass in the cluster. Mea-surements of dark matter that rely on a cluster's light-bending properties generally agree with measurements determined from X-ray observations and the radial velocities of a cluster's galaxies. (The region pictured is about 1.4 million light-years from side to side.)

VIS

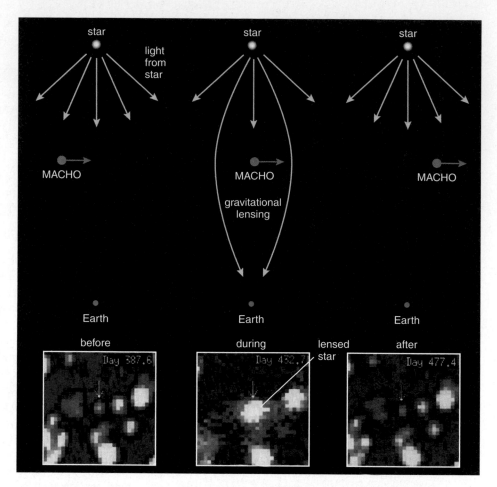

Figure 22.12 A lensing event in which a massive compact halo object (MACHO) passing in front of a more distant star temporarily makes the star appear brighter, revealing the presence of the dark MACHO. Multiple images of the star are possible, but they are too close together to resolve with current telescopes. Monitoring of events like this one show that our galaxy's halo contains MACHOs but that they do not comprise the majority of the galaxy's dark matter.

The process of star formation itself leaves behind dark matter in the form of *brown dwarfs,* those dim "failed stars" that are not quite massive enough to sustain nuclear fusion [Section 17.2]. Some scientists believe that trillions of faint red stars, brown dwarfs, and Jupiter-size objects left over from the Milky Way's formation still roam our galaxy's halo, providing much of its mass. They fancifully term these objects **MACHOs,** for *massive compact halo objects.*

Because MACHOs are too faint for us to see directly, astronomers who wish to detect them must resort to clever techniques. One innovative way to detect faint starlike objects in the Milky Way's halo takes advantage of gravitational lensing. If trillions of these MACHOs really exist, every once in a while a MACHO should drift across our line of sight to a more distant star. When a MACHO lies almost directly between us and the farther star, the MACHO's gravity will focus the star's light directly toward the Earth. The distant star will appear much brighter than usual for several days or weeks as the MACHO passes in front of it (Figure 22.12). We cannot see the MACHO itself, but the duration of the lensing event reveals its mass.

Gravitational lensing events such as these are rare, happening to about one star in a million each year. To detect MACHO lensing events, we therefore must monitor huge numbers of stars. Current large-scale monitoring projects now record numerous lensing events annually. These events demonstrate that MACHOs do indeed populate our galaxy's halo, but probably not in large enough numbers to account for all the Milky Way's dark matter. Similar measurements rule out black holes with masses like our Sun's. Something else lurks unseen in the outer reaches of our galaxy. Maybe it is also normal matter made from baryons, but maybe not.

Extraordinary Dark Matter: WIMPs

A more exotic possibility is that most of the dark matter in galaxies and clusters of galaxies is not made of ordinary, baryonic matter at all. Let's begin to explore this possibility by taking another look at those nonbaryonic particles we discussed in Section 15.3: neutrinos. These unusual particles are dark by their very nature, because they have no electrical charge and hence cannot emit electromagnetic radiation of any kind. Moreover, they are never bound together with charged particles in the way that neutrons are bound in atomic nuclei, so their presence cannot be revealed by associated light-emitting particles.

In fact, particles like neutrinos interact with other forms of matter through only two of the four forces: gravity and the *weak force* [Section S4.2]. For this reason, they are said to be *weakly interacting particles.* If you recall that trillions of neutrinos from the Sun are passing through your body at this very moment without doing any damage, you'll see why the name *weakly interacting* fits well.

The dark matter in galaxies cannot be made of neutrinos, because these tiny particles travel through the universe at enormous speeds and can easily escape a galaxy's gravitational pull. (However, neutrinos do make up a small amount of the dark matter outside galaxies.) What if other weakly interacting particles exist that are similar to neutrinos but considerably heavier? They too would evade direct detection, but they would move more slowly and could collect into galaxies, adding mass without adding light. Such hypothetical particles are called **WIMPs,** for *weakly interacting massive particles.* (Weakly interacting particles that are slow-moving enough to collect into galaxies are sometimes called *cold dark matter* to set them apart from faster-moving *hot dark matter* particles such as neutrinos.) WIMPs could make up most of our galaxy's mass, but they would be completely invisible in all wavelengths of light.

It might surprise you that scientists would suspect the universe is filled with particles they haven't yet discovered. However, WIMPs could also explain why dark matter doesn't behave like the visible matter in galaxies. During the early stages of the Milky Way's formation, the ordinary (baryonic) matter is thought to have settled toward our galaxy's center and then flattened into a disk [Section 21.2]. Meanwhile, the dark matter must have stayed where it was, out in the galaxy's halo. This resistance of the dark matter to settling is exactly what we would expect from weakly interacting particles.

Because WIMPs do not emit electromagnetic radiation, they cannot radiate away their energy. They are therefore stuck orbiting out at large distances and cannot collapse with the rest of the protogalactic cloud. By itself, the inability of dark matter to settle into the luminous regions of galaxies does not prove that dark matter is extraordinary and nonbaryonic. However, as we'll discuss in the next chapter, some other reasons lead us to believe that baryons represent only a minority of the universe's mass and hence that WIMPs really exist.

What do you think of the idea that much of the universe is made of as-yet-undiscovered particles? Can you think of other instances in the history of science in which the existence of something was predicted before it was discovered?

22.5 Structure Formation

Dark matter remains enigmatic, but every year we are learning more about its role in the universe. Because galaxies and clusters of galaxies seem to contain much more dark matter than luminous matter, we believe that dark matter's gravitational pull must be the primary force holding these structures together. Thus, we strongly suspect that the gravitational attraction of dark matter is what pulled galaxies and clusters together in the first place.

Growth of Structure

Stars, galaxies, and clusters of galaxies are all *gravitationally bound systems*—their gravity is strong enough to hold them together. In most of the gravitationally bound systems we have discussed so far, gravity has completely overwhelmed the expansion of the universe. That is, while the universe as a whole is expanding, space is *not* expanding within our solar system, our galaxy, or our Local Group of galaxies.

Our best guess at how galaxies formed, briefly outlined in Section 21.2, envisions them growing from slight density enhancements that were present in the very early universe. During the first few million years after the Big Bang, the universe expanded everywhere. Gradually, the stronger gravity in regions of enhanced density pulled in matter until these regions stopped expanding and became protogalactic clouds—even as the universe as a whole continued (and still continues) to expand.

If dark matter is indeed the most common form of mass in galaxies, it must have provided most of the gravitational attraction responsible for creating the protogalactic clouds. The hydrogen and helium gas in the protogalactic clouds collapsed inward and gave birth to stars, while weakly interacting dark matter remained in the outskirts because of its inability to radiate away its orbital energy. According to this model, the luminous matter in each galaxy must still be nestled inside the larger cocoon of dark matter that initiated the galaxy's formation, just as observational evidence seems to suggest.

The formation of galaxy clusters probably echoes the formation of galaxies. Early on, all the galaxies that will eventually constitute a cluster are flying apart with the expansion of the universe, but the gravity of the dark matter associated with the cluster eventually reverses the trajectories of these galaxies. The galaxies ultimately fall back inward and start orbiting each other randomly, like the stars in the halo of our galaxy.

Some clusters of galaxies apparently have not yet finished forming: Their immense gravity continues to draw in new members. For example, the nearby Virgo Cluster of galaxies appears to be tugging on our own Local Group. Right now the Local Group is still moving away from the Virgo Cluster, but not as quickly as a simple application of Hubble's law would predict. Hubble's law lets us calculate the speed at which universal expansion carries us away from the Virgo Cluster [Section 20.4], but it does not account for any gravitational effects. The discrepancy between a straight application of Hubble's law and our actual velocity away from the Virgo Cluster is about 400 km/s. That is, we are moving away from the Virgo Cluster 400 km/s more slowly than we would be as a result of the expansion of the universe alone. Astronomers call this 400-km/s deviation from Hubble's law a *peculiar velocity*, but it's really not so peculiar. It is simply the effect of the gravitational attraction pulling us back toward Virgo against the flow of universal expansion.

The overall effect is rather like that of swimming upstream against a strong current. You move "upstream" relative to other objects floating in the current, but you're still headed "downstream" relative to the shore because of the strong current. The speed of the current is like the speed of expansion of the universe, and your swimming speed "upstream" is your peculiar velocity.

Just as Earth's gravity slows a rising baseball and eventually turns it around and pulls it back toward the ground, the gravitational tug of the Virgo Cluster may eventually turn the galaxies of our Local Group around and pull them into the cluster. Similar processes are taking place on the outskirts of other large clusters of galaxies, where we see many galaxies with large peculiar velocities carrying them in the direction of the cluster. Their peculiar velocities indicate that the cluster's gravity is pulling on them. Eventually, some or perhaps all of these galaxies will fall into the cluster. Thus, many clusters are still attracting galaxies, adding to the hundreds they already contain.

On even larger scales, clusters themselves have surprisingly large peculiar velocities, hinting that they are parts of even bigger gravitationally bound systems, called **superclusters,** that are just beginning to form (Figure 22.13).

Large-Scale Structures

Beyond about 300 million light-years, we can no longer measure distances accurately enough to determine peculiar velocities. We can estimate the distances to faraway galaxies only by measuring their redshifts and applying Hubble's law. Today we have redshift measurements for many thousands of distant galaxies. When we convert the redshifts to distance estimates, we can create three-dimensional maps of the universe. Such maps have been made for only a few "slices" of the sky so far, but they are already revealing **large-scale structures** much vaster than clusters of galaxies.

One of the most famous depictions of large-scale structure is the "slice of the universe" pictured in Figure 22.14a. Many astronomers collaborated to measure the redshifts of all the galaxies they could find in this particular slice of the sky. This project dramatically revealed the complex structure of our corner of the universe. Each dot in such pictures represents an entire galaxy of stars. The arrangement of the dots reveals huge sheets of galaxies spanning many millions of light-years (Figure 22.14b,c). Clusters of galaxies are located at the intersections of these sheets. Between the sheets of galaxies lie giant empty regions called **voids**.

Some of the structures we see in the universe are amazingly large. The so-called *Great Wall* of galaxies stretches

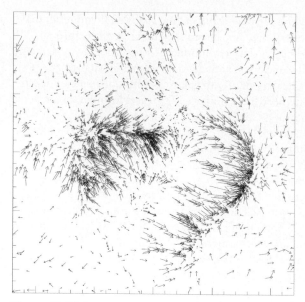

Figure 22.13 Peculiar velocities of galaxies flowing into superclusters. Each arrow shows the peculiar velocity of a galaxy inferred from a combination of observations and modeling. The Milky Way is at the center of the picture, and not all galaxies are shown. The area pictured is about 600 million light-years from side to side. Note how the galaxies tend to flow into regions where the density of galaxies is already high. These vast, high-density regions are probably superclusters in the process of formation.

across an expanse measuring some 180 million light-years side to side. Immense structures such as these apparently have not yet collapsed into randomly orbiting, gravitationally bound systems. They haven't had enough time. The universe may still be growing structures on ever larger scales even though it is billions of years old (Figure 22.15).

Most astronomers believe that all these large-scale structures grew from slight density enhancements in the early universe, just as galaxies did. Galaxies, clusters, superclusters, and the Great Wall probably all started as mildly high-density regions of different sizes. The voids in the distribution of galaxies probably started as mildly low-density regions.

If this picture of structure formation is correct, then the structures we see in today's universe mirror the original distribution of dark matter very early in time. Supercomputer models of structure formation in the universe can now simulate the growth of galaxies, clusters, and larger structures from tiny density enhancements as the universe evolves (Figure 22.16a). The results of these models look remarkably similar to the slices of the universe in Figure 22.14, bolstering our confidence in this scenario (Figure 22.16b). However, the models do not tell us *why* the universe started with these slight density enhancements—that is a topic for the next chapter. Nevertheless, it seems increasingly clear that these "lumps" in the early universe were the seeds of all the marvelous structures we see today.

Figure 22.14 These "slices" of the universe map out the galaxies in very thin, fanlike swaths of space. The tip of each slice corresponds to the location of Earth, and each dot represents a galaxy. Galaxies are not scattered randomly but instead trace out long strings and sheets surrounded by huge voids containing very few galaxies.

a This slice of the universe, extending to a distance of 700 million light-years, was first mapped in the 1980s. The sheet of galaxies that forms the arms of the stick figure at the center is known as the Great Wall and was one of the first very large-scale structures discovered. Large voids can be seen underneath the arms of the stick figure. (Tick marks at lower right indicate *recession* velocities of the surveyed galaxies.)

b The upper swath is similar to (a) but thicker and thus contains more galaxies. The lower swath points in a different direction.

c The galaxies in these two slices of the sky extend much deeper into space—up to 4 billion light-years. Many voids, walls, and strings of galaxies hundreds of millions of light-years in size are evident as tiny details in this image. However, the distribution of galaxies on scales larger than a billion light-years is nearly uniform.

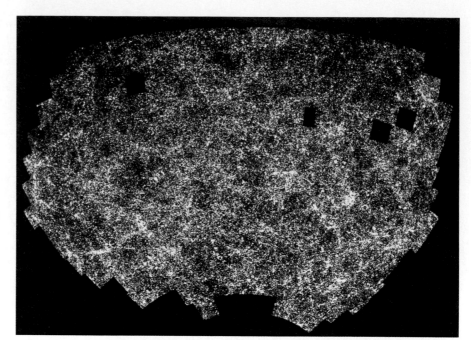

Figure 22.15 This picture shows the positions of over 3 million galaxies spread across 15% of the sky. Brighter areas contain more galaxies than darker areas, and the black rectangles are regions for which there is no data. Note that the universe is more uniform on large scales than on small scales.

Figure 22.16 Frames from supercomputer simulations of structure formation.

▶ **a** These six frames depict the development of a region the size of our Local Group. As the age of the universe progresses from about 50 million years (cosmological redshift $z = 50$) to the present ($z = 0$), the high-density portions of the universe, shown in orange, continue to attract more and more matter.

▲ **b** This frame shows how the present-day universe appears in a large-scale simulation. Note the similarity of the filamentary structures in this simulation to the slices of the universe in Figure 22.14.

22.6 The Universe's Fate

Some say the world will end in fire,
Some say in ice.
From what I've tasted of desire
I hold with those who favor fire.
But if it had to perish twice,
I think I know enough of hate
To say that for destruction ice
Is also great
And would suffice.

Robert Frost

We now arrive at one of the ultimate questions in astronomy: How will the universe end? Edwin Hubble's work established that the galaxies in the universe are rapidly flying away from one another [Section 20.3], but the gravitational pull of each galaxy on every other galaxy acts to slow the expansion. These facts suggest two possible fates for the universe. If gravity is strong enough, the expansion will someday halt and the universe will begin collapsing, eventually ending in a cataclysmic crunch. Alternatively, if the expansion can overcome the pull of gravity, the universe will continue to expand forever, growing ever colder as its galaxies grow ever farther apart. The fate of the universe thus boils down to a simple question: Does the expanding universe have enough kinetic energy to escape its own gravitational pull?

Density: Not Quite Critical

Hubble's constant essentially tells us the current kinetic energy of the universe, but we do not yet know the overall strength of the universe's gravitational pull. The strength of this pull depends on the density of matter in the universe: The greater the density, the greater the overall strength of gravity and the higher the likelihood that gravity will someday halt the expansion.

Precise calculations show that gravity can win out over expansion if the current density of the universe exceeds a seemingly minuscule 10^{-29} gram per cubic centimeter, which is roughly equivalent to a few hydrogen atoms in a volume the size of a closet. The precise density marking the dividing line between eternal expansion and eventual collapse is called the **critical density**.

Observations of the luminous matter in galaxies show that the mass contained in stars falls far short of the critical density. The visible parts of galaxies contribute less than 1% of the matter density needed to halt the universe's expansion. The fate of the universe thus rests with the dark matter. Is there enough dark matter to halt the expansion of the universe?

Recall that we can estimate the amount of dark matter by looking at the mass-to-light ratio. In order for dark matter to contribute enough mass for the universe to have the critical density, the average mass-to-light ratio throughout the universe would have to be approximately 1,000 solar masses per solar luminosity. Clusters of galaxies have mass-to-light ratios of a few hundred solar masses per solar luminosity, still a few times less than the ratio needed to halt the expansion. Thus, if the proportion of dark matter in the universe at large is similar to that in clusters, the universe will expand forever. For gravity to reverse the expansion and pull the universe back together, even more dark matter would have to lie beyond the boundaries of clusters.

If large-scale structures really do contain a higher proportion of dark matter than do clusters, the influence of that extra dark matter should show up in the velocities of galaxies near those large-scale structures: Larger amounts of dark matter should cause greater deviations from Hubble's law. As of 2003, however, most studies of galaxy velocities hold the line near the value we infer from clusters, which is about 30% of the critical density required to reverse the expansion. If that is the case, the universe will continue to expand forever.

Mysterious Acceleration

In the past few years, observations of distant white dwarf supernovae have enabled us to probe the fate of the universe in an entirely new way. Because white dwarf supernovae are such good standard candles [Section 20.3], we can use them to determine whether gravity has been slowing the universe's expansion, as it must if the universe is destined to end in a cataclysmic crunch. However, the astronomers who set out to measure gravity's influence over the universe by observing supernovae discovered something completely unexpected.

Instead of slowing because of gravity, the expansion of the universe appears to be speeding up, suggesting that some mysterious repulsive force is pushing all the universe's galaxies apart. This discovery, if it holds up to further scrutiny, has far-reaching implications for both the fate of the universe and our understanding of the forces that govern its behavior on large scales. To understand the evidence for an accelerating expansion, we must become more familiar with the possible fates of the universe.

Four Expansion Patterns Astronomers subdivide the two general possibilities for the fate of the universe, expanding forever or someday collapsing, into four broad categories. Each represents a particular pattern of change in the future expansion rate (Figure 22.17). We will call these four possible expansion patterns *recollapsing, critical, coasting,* and *accelerating.*

The first three possibilities assume that gravity is the only force that affects the expansion rate of the universe:

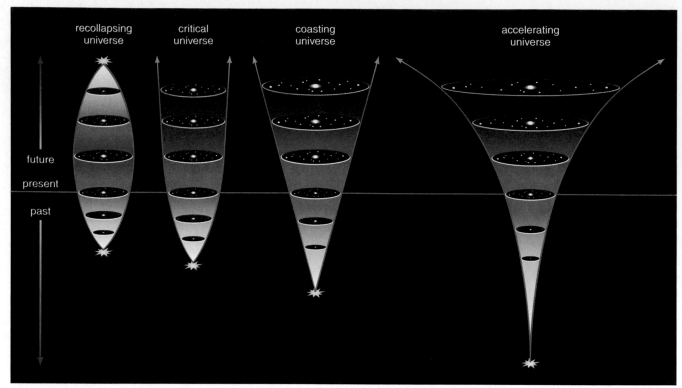

Figure 22.17 Four models for the fate of the universe. In each case, a region of the universe is shown expanding from a very small size in the past to its present-day size. The fate of the universe in each model depends on whether gravity can overcome the expansion and overpower any repulsive force that might be present. Gravity in a *recollapsing universe* is strong enough to halt and reverse the expansion, leading to a "Big Crunch" sometime in the future. Gravity in a *critical universe* slows the expansion but can't quite halt it. Gravity in a *coasting universe* is too weak to slow the expansion, so the universe always expands at the same rate. The expansion of an *accelerating universe* continually speeds up due to a repulsive force that overpowers gravity.

- If the matter density of the universe is *larger* than the critical density, the collective gravity of all its matter will eventually halt the universe's expansion and reverse it. The galaxies will come crashing back together, and the entire universe will end in a fiery "Big Crunch." If this is the fate of our universe, then we live in a **recollapsing universe**. (The overall geometry of a recollapsing universe closes upon itself like the surface of a sphere [Section S3.4], which is why this type of universe is sometimes called a *closed universe.*)

- If the matter density of the universe *equals* the critical density, the collective gravity of all its matter is exactly the amount needed to balance the expansion. The universe will never collapse but will expand more and more slowly as time progresses. If this is the fate of our universe, then we live in a **critical universe**. (Mathematically speaking, a critical universe stops expanding after infinite time, and its overall geometry is flat like the surface of a table [Section S3.4]. Thus, a critical universe is one example of what astronomers call a *flat universe.*)

- If the matter density of the universe is *smaller* than the critical density, the collective gravity of all its matter cannot halt the expansion. The universe will keep expanding forever, with little change in its rate of expansion. If this is the fate of our universe, then we live in a **coasting universe**. (A coasting universe is sometimes called an *open universe,* because its overall geometry is more like the open surface of a saddle than like the closed surface of a sphere [Section S3.4].)

Because observations of distant supernovae suggest that a repulsive force opposes gravity on very large scales, astronomers are now seriously considering a fourth possibility:

- If a repulsive force causes the expansion of the universe to accelerate with time, then we live in an **accelerating universe**. Its galaxies will recede from one another increasingly faster, and it will become cold and dark more quickly than a coasting universe. (Depending on the strength of gravity relative to the repulsive force, the overall geometry of an accelerating universe could be flat, open, or closed. However, most astronomers favor a flat geometry for this case.)

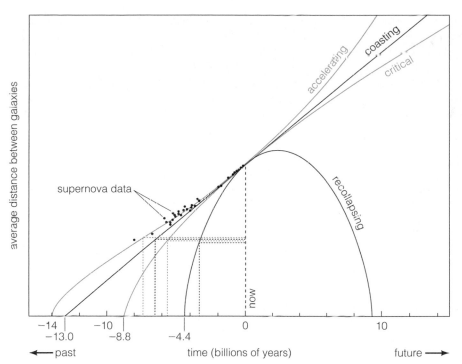

Figure 22.18 Relationship between look-back time and expansion rate in four models for the expansion of the universe. Each curve shows how the average distance between galaxies changes with time. A rising curve means that the universe is expanding, and a falling curve means that the universe is contracting. All the curves are aligned at the time labeled "now," where the expansion rate and average distance between galaxies are the same for all the models. The look-back time to the birth of the universe, which is the moment at which the average distance between galaxies is essentially zero, depends on how strongly gravity affects the expansion rate. If gravity is strong enough to slow the universe's expansion, as in the critical and recollapsing universe models, the look-back time is relatively short—10 billion years or less. In the coasting and accelerating universe models, the pull of gravity is relatively weak, and the lookback time to the birth of the universe is closer to 14 billion years. Observations of distant white dwarf supernovae suggest that the universe is indeed accelerating and is destined to expand forever.

THINK ABOUT IT

Do you think that one of the potential fates of the universe is preferable to the others? If so, why? If not, why not?

Figure 22.18 illustrates how the average distance between galaxies changes with time for each possibility. The lines for the accelerating, coasting, and critical universes always continue upward as time increases, because in these cases the universe is always expanding. The steeper the slope, the faster the expansion. In the recollapsing case, the line begins on an upward slope but eventually turns around and declines as the universe contracts. All the lines pass through the same point and have the same slope at the moment labeled "now," because the current separation between galaxies and the current expansion rate in each case must agree with observations of the present-day universe.

Note that the age for the universe that we infer from its expansion rate differs in each case. A recollapsing universe requires the least amount of time to arrive at the current

SPECIAL TOPIC   Einstein's Greatest Blunder

Shortly after Einstein completed his general theory of relativity in 1915, he found that it predicted that the universe could not be standing still: The mutual gravitational attraction of all the matter would make the universe collapse. Because Einstein thought at the time that the universe should be eternal and static, he decided to alter his equations. In essence, he inserted a "fudge factor" called the *cosmological constant* that acted as a repulsive force to counteract the attractive force of gravity.

Had he not been so convinced that the universe should be standing still, Einstein might instead have come up with the correct explanation for why the universe is not collapsing: because it is still expanding from the event of its birth. After Hubble discovered that the universe is expanding, Einstein called his invention of the cosmological constant the greatest blunder of his career.

More recently, astronomers have begun to take the idea of a cosmological constant more seriously. In the mid-1990s, a few observations suggested that the ages of the oldest stars are slightly older than the age of the universe derived from Hubble's constant under the assumption that gravity is the only force affecting the universe's expansion. Clearly, stars cannot be older than the universe. If these observations were being interpreted correctly, the universe had to be older than Hubble's constant implies. If the expansion rate has accelerated, so that the universe is expanding faster today than it was in the past, then the age of the universe would be greater than that ordinarily found from Hubble's constant (see Figure 22.18). What could cause the expansion of the universe to accelerate over time? The repulsive force represented by a cosmological constant, of course.

Further study of the troubling observations has shown that the oldest stars probably are *not* older than the age of the universe derived from Hubble's constant. However, measurements of distances to high-redshift galaxies using white dwarf supernovae as standard candles now suggest that the expansion *is* accelerating. A cosmological constant could account for this startling finding, but we'll need more observations before we can be sure it is correct. Einstein's greatest blunder, it seems, just won't go away.

separation between galaxies—the example in Figure 22.18 goes from zero separation to the current separation in less than 5 billion years. The cases for which gravity is less important require more time to achieve the current separation between galaxies. The ages we would infer from the examples in Figure 22.18 are 8.8 billion years for a critical universe, 13 billion years for a coasting universe, and around 14 billion years for an accelerating universe.

Evidence for Acceleration This relationship between the age of the universe and its expansion pattern enables us to determine the expansion pattern from supernova observations. To understand how, imagine observing a distant white dwarf supernova with spectral lines redshifted to twice their normal wavelengths. The expansion of the universe stretches light waves by the same amount that it increases the average distance between galaxies [Section 20.4]. Thus, the light we are observing from that supernova must have left when the average distance between galaxies was only half its current value. Because white dwarf supernovae are such excellent standard candles, we can also determine the actual distance—and thus the lookback time [Section 20.4]—to the supernova from its apparent brightness.

In principle, knowing both a supernova's redshift and its lookback time enables us to determine the universe's past expansion rate. The horizontal dotted lines in Figure 22.18 indicate the average distance between galaxies at the time of a particular supernova, and the vertical dotted lines indicate the corresponding lookback time for each of the illustrated expansion patterns. Note that the lookback times are smallest for models in which the universe grows most quickly to its present size. Thus, we expect a supernova with a given redshift to be somewhat closer, and therefore somewhat brighter, if the universe is recollapsing or critical than if it is coasting. Conversely, if the supernova turns out to be somewhat farther and dimmer than we would expect in a coasting universe, then we would conclude that the expansion is accelerating.

In practice, observations of such distant supernovae are still very difficult, but the results to date, indicated by the dots in Figure 22.18, favor an accelerating universe. White dwarf supernovae with large redshifts appear to be somewhat dimmer than we would expect if the universe were coasting. These observations are consistent with an accelerating universe and highly inconsistent with either a critical or a recollapsing universe.

Exactly why the universe might be accelerating remains a deep mystery. No known force would act to push the universe's galaxies apart, and an enormous amount of energy would be required. Because the source of this repulsive energy remains unknown, astronomers sometimes refer to it as *dark energy*. Terms such as *quintessence* and *cosmological constant* are also frequently used. If dark energy really exists, it is the most prevalent form of energy in the universe, outstripping the total mass-energy of all the matter in the universe—including the dark matter. Thus, astronomers are currently working very hard to confirm its existence. Whatever dark energy turns out to be, it now seems likely that the universe is indeed doomed to expand forever, its galaxies receding ever more quickly into an icy, empty future. However, forever is a very long time. The universe may have other surprises—like the mysterious acceleration—still be waiting to be discovered.

This is the way the world ends
This is the way the world ends
This is the way the world ends
Not with a bang but a whimper.

From The Hollow Men *by T. S. Eliot*

THE BIG PICTURE

Putting Chapter 22 into Context

In this chapter, we have found that there may be much more to the universe than meets the eye. Dark matter too dim for us to see seems to far outweigh the stars. Here are some key "big picture" points to remember about this mysterious matter:

- Either dark matter exists, or we do not understand how gravity operates across galaxy-size distances. We have many reasons to have confidence in our understanding of gravity, so the majority of astronomers believe that dark matter is real.

- Measurements of the mass and luminosity of galaxies and galaxy clusters indicate that they contain far more mass in dark matter than in stars.

- Despite the fact that dark matter is by far the most abundant form of mass in the universe, we still have little idea what it is.

- Superclusters, walls, and voids much larger than clusters of galaxies extend many millions of light-years across the universe. Each of these structures probably began as a very slight enhancement in the density of dark matter early in time, and these enormous structures are still in the process of forming.

- Dark matter holds the key to the fate of the universe. If there is enough of it, its gravity will cause the expansion of the universe to halt and reverse someday. Current indications are that there is not enough dark matter to halt the expansion, in which case the universe will continue to expand forever. Recent observations of an accelerating expansion only reinforce the suggestion that the expansion will never cease.

22.1 The Mystery of Dark Matter

- *What is the evidence for dark matter?* We infer the existence of dark matter from its gravitational effects on visible matter. The evidence for its existence is overwhelming—*if* we correctly understand the theory of gravity.

- *How does the distribution of dark matter compare to the distribution of luminous matter in spiral galaxies?* The luminous matter is concentrated in the disk, while the dark matter is distributed throughout the spherical halo and far beyond the visible light from the disk.

22.2 Dark Matter in Galaxies

- *How do we determine the distribution of mass in distant galaxies?* We can determine the mass distribution of a spiral galaxy from its rotation curve. We can determine the mass distribution of an elliptical galaxy from the average orbital speeds of its stars at different distances from the center, as measured from the broadening of the galaxy's spectral lines.

- *How does a galaxy's mass-to-light ratio tell us how much dark matter it contains?* The mass-to-light ratio tells us how many solar masses of matter the galaxy contains for each solar luminosity of light output. Because we can estimate the mass-to-light ratio the galaxy would have if it were made only of stars, we know that any excess mass must be dark matter.

- *What have we learned about galaxies from their mass-to-light ratios?* The mass-to-light ratios in the outer reaches of galaxies are particularly high, telling us that stars alone cannot account for the mass. Thus, we conclude that galaxies contain large amounts of dark matter.

22.3 Dark Matter in Clusters

- *What are the three independent ways to measure the total mass of a cluster of galaxies?* We can estimate a cluster's mass (1) from the orbital speeds and positions of its galaxies; (2) from the temperature of its hot, intracluster medium, which we can measure with X-ray observations; and (3) from observations of its effects on the appearance of more distant galaxies distorted by gravitational lensing.

- *What have we learned about dark matter in galaxy clusters?* All three methods of measuring cluster masses agree, and they imply that clusters contain far more mass in dark matter than in stars.

22.4 Dark Matter: Ordinary or Extraordinary?

- *What do we mean when we ask whether dark matter is ordinary or extraordinary matter?* Ordinary matter is often called baryonic matter because most of its mass comes from protons and neutrons, which are both baryons. However, the baryonic matter we have found does not account for all the dark matter. Much of the dark matter may be made of nonbaryonic particles that have yet to be discovered. This form of matter would be very different from the matter that we encounter in daily life and hence is considered extraordinary.

- *What are MACHOs, and can they account for all the dark matter?* MACHOs (massive compact halo objects) are ordinary objects—such as dim stars, brown dwarfs, and planet-size bodies—that might populate the galactic halo without being visible to our telescopes. MACHOs have been detected through gravitational lensing, but not in large enough numbers to account for all the dark matter inferred from gravitational effects.

- *What are WIMPs, and can they account for all the dark matter?* WIMPs (weakly interacting massive particles) are undiscovered particles of extraordinary (nonbaryonic) matter that do not interact with light (instead interacting only through the weak force and gravity). WIMPs are the leading candidate for dark matter, even though we have not yet discovered such particles.

22.5 Structure Formation

- *How does structure appear to be growing in the universe?* All the structure we see today probably grew from regions of slightly enhanced density in the early universe. Gravity in these higher-density regions drew matter together to form galaxies and later drew those galaxies together to form clusters. Some clusters of galaxies appear to be still in the process of formation, suggesting that gravity continues to draw matter into large-scale structures. Likewise, clusters of galaxies appear to be part of even larger structures, called superclusters, that are just beginning to form.

- *What does the universe look like on very large scales?* Galaxies appear to be distributed in gigantic chains and sheets that surround great voids. These giant structures trace their origin directly back to regions of slightly enhanced density early in time.

continued ▶

22.6 The Universe's Fate

- *What is the critical density?* The critical density is the average matter density the universe must have for the strength of gravity to eventually halt the expansion of the universe (assuming today's expansion rate). Although there may yet be dark matter unaccounted for, the overall matter density of the universe appears to be only about 30% of the critical density.

- *What are the four general models for the future expansion of the universe, and which model is currently favored?* (1) Recollapsing universe: The expansion will someday halt and reverse. (2) Critical universe: The universe will never collapse but will expand more and more slowly with time. (3) Coasting universe: The universe will continue to expand forever, with little change in the rate of expansion. (4) Accelerating universe: The expansion of the universe will accelerate with time. Recent observations favor the accelerating universe model.

- *Do we know what might be causing the universe to accelerate?* No. Although people give names to the mysterious force that could be causing acceleration—such as dark energy, quintessence, or cosmological constant—no one really knows what it is.

? True Statements?

Decide whether each of the following statements is true and explain why it is or is not.

1. Astronomers now believe that most of any galaxy's mass lies beyond the portions of the galaxy that we can see.

2. If our galaxy had less dark matter, its mass-to-light ratio would be lower.

3. A cluster of galaxies is held together by the mutual gravitational attraction of all the stars in the cluster's galaxies.

4. We can estimate the total mass of a cluster of galaxies by studying the distorted images of galaxies whose light passes through the cluster.

5. Clusters of galaxies are the largest structures that we have so far detected in the universe.

6. Dark matter gets its name from the fact that it emits no detectable light.

7. The primary evidence for an accelerating universe comes from observations of young stars in the Milky Way.

8. There is no doubt remaining among astronomers that the fate of the universe is to expand forever.

Problems

9. *Rotation Curves.* Briefly describe how we construct rotation curves for spiral galaxies and how these curves lead us to conclude that spiral galaxies contain dark matter.

10. *Dark Matter in Ellipticals.* Briefly describe how we can infer the average orbital speeds of stars in an elliptical galaxy from the widths of its spectral lines and how this allows us to weigh elliptical galaxies.

11. *Mass-to-Light Ratios.* What are typical mass-to-light ratios for inner regions of spiral galaxies? For inner regions of elliptical galaxies? How do these mass-to-light ratios compare to the mass-to-light ratios we find when we look farther from a galaxy's center? What does this tell us about dark matter in galaxies?

12. *Gravitational Lensing.* What is gravitational lensing? Why does it occur? How can we use it to estimate the masses of lensing objects?

13. *Weakly Interacting Particles.* Explain what we mean when we say that a neutrino is a weakly interacting particle. Why can't the dark matter in galaxies be made of neutrinos?

14. *Peculiar Velocities.* What are peculiar velocities? What causes them, and what do they tell us about the formation of clusters and superclusters?

15. *Fate and Dark Matter.* Explain how the total amount of dark matter in the universe holds the key to its final fate. According to current evidence about the amount of dark matter, what is the fate of the universe?

16. *Strange Acceleration.* What do we mean by an accelerating expansion, and what evidence suggests that such an acceleration may be occurring? If the acceleration is real, what is the fate of the universe? Explain.

17. *Strange Mass-to-Light Ratio.* Suppose you discovered a galaxy with a mass-to-light ratio of 0.1 solar mass per solar luminosity. Would you be surprised? Explain why or why not. What would this measurement say about the nature of the stars in this galaxy?

*18. *Weighing a Spiral Galaxy.* Figure 22.4 shows the rotation curve for a spiral galaxy called NGC 7541.

   a. Use the orbital velocity law to determine the mass (in solar masses) of NGC 7541 enclosed within a radius of 30,000 light-years from its center. (*Hint:* 1 light-year = 9.461×10^{15} m.)

   b. Use the orbital velocity law to determine the mass of NGC 7541 enclosed within a radius of 60,000 light-years from its center.

   c. Based on your answers to parts (a) and (b), what can you conclude about the distribution of mass in this galaxy?

*19. *Weighing a Cluster of Galaxies.* Suppose a cluster of galaxies has a radius of about 6.7 million light-years (6.2×10^{22} m) and an intracluster medium with a temperature of 8×10^7 K. Estimate the mass of the cluster. Give your answer in both kilograms and solar masses.

*20. *How Many MACHOs?* Suppose the rotation of a highly simplified galaxy whose stars are all identical to the Sun (1 solar mass per solar luminosity) shows that its overall mass-to-light ratio is equal to 30 solar masses per solar luminosity.

   a. What is the ratio of dark matter to luminous matter in this galaxy?

   b. Suppose all the dark matter consists of MACHOs similar to Jupiter, each with a mass of 0.001 solar mass. How many of these MACHOs must the galaxy contain for each ordinary star? Explain.

Discussion Questions

21. *Dark Matter or Revised Gravity.* One possible explanation for the large mass-to-light ratios we measure in galaxies and clusters is that we are currently using the wrong law of gravity to measure the masses of very large objects. If we really do misunderstand gravity, then many fundamental theories of physics, including Einstein's theory of general relativity, will need to be revised. Which explanation for these mass-to-light ratios do you find more appealing, dark matter or revised gravity? Explain why. Why do you suppose most astronomers find dark matter more appealing?

22. *Our Fate.* Scientists, philosophers, and poets alike have speculated on the fate of the universe. How would you prefer the universe as we know it to end, in a "Big Crunch" or through eternal expansion? Explain the reasons behind your preference.

For a complete list of media resources available, go to www.astronomyplace.com and choose Chapter 22 from the pull-down menu.

 Astronomy Place Web Tutorials

Tutorial Review of Key Concepts

Use the following interactive **Tutorials** at www.astronomyplace. com to review key concepts from this chapter.

Detecting Dark Matter in a Spiral Galaxy Tutorial

Lesson 1 Introduction to Rotation Curves

Lesson 2 Determining the Mass Distribution

Lesson 3 Where Is the Dark Matter?

Fate of the Universe Tutorial

Lesson 1 The Role of Gravity

Lesson 2 The Role of Dark Energy

Lesson 3 Fate and History of the Universe

Supplementary Tutorial Exercises

Use the interactive **Tutorial Lessons** to explore the following questions.

Detecting Dark Matter in Spiral Galaxies Tutorial, Lesson 3

1. Explain how stellar rotation curves suggest the existence of dark matter.

2. How does the amount of dark matter change as you move away from the center of a galaxy?

Fate of the Universe Tutorial, Lesson 1

1. If gravity were the only factor affecting the fate of the universe, what would the possible fates be?

2. How would the mass density of the universe determine which fate would be in store for the universe?

Fate of the Universe Tutorial, Lesson 2

1. What does dark energy do to the expansion of the universe?

2. How would the existence of dark energy change the fate of the universe from what it would be if only gravity played a role in its fate?

Fate of the Universe Tutorial, Lesson 3

1. How would the existence of dark energy change estimates of the age of the universe?

2. What evidence suggests that dark energy really exists?

Movies

Check out the following narrated and animated short documentary available on www.astronomyplace.com for a helpful review of key ideas covered in this chapter.

From the Big Bang to the Galaxies Movie

Web Projects

Take advantage of the useful Web links on www.astronomyplace. com to assist you with the following projects.

1. *Gravitational Lenses.* Gravitational lensing occurs in numerous astronomical situations. Compile a catalog of examples from the Web. Try to find pictures of lensed stars, quasars, and galaxies. Give a one-paragraph explanation of what is going on in each picture.

2. *Accelerating Universe.* Search for the most recent information about possible acceleration of the expansion of the universe. Write a one- to three-page report on your findings.

3. *The Nature of Dark Matter.* Find and study recent reports on new ideas about the possible nature of dark matter. Write a one- to three-page report that summarizes the latest ideas about what dark matter is made of.

23 The Beginning of Time

Somewhere, something incredible is waiting to be known.

Carl Sagan

The universe has been expanding for about 14 billion years. During that time, matter collected into galaxies. Stars formed in those galaxies, producing heavy elements that were recycled into later generations of stars. One of these late-coming stars formed about 4.6 billion years ago, in a remote corner of a galaxy called the Milky Way. This star was born with a host of planets that formed in a flattened disk surrounding it. One of these planets soon became covered with life that gradually evolved into ever more complex forms. Today, the most advanced life-form on this planet, human beings, can look back on this series of events and marvel at how the universe created conditions suitable for life.

To this point in the book, we have discussed how the matter in the early universe collected into galaxies, stars, planets, and ultimately people. We have seen that we are star stuff, and galaxy stuff as well. More to the point, we now recognize that we are the product of some 14 billion years of cosmic evolution. However, we have not yet answered one big question: Where did the *matter itself* come from?

To answer this ultimate question, we must go beyond the most distant galaxies and even beyond what we can see near the horizon of the universe. We must go back not only to the origins of matter and energy but to the beginning of time itself.

 Hubble's Law Tutorial, Lessons 1–3

23.1 Running the Expansion Backward

Is it really possible to study the origin of the entire universe? Not long ago, questions about creation were considered unfit for scientific study. That attitude began to change with Hubble's discovery that the universe is expanding. This discovery led to the insight that all things very likely sprang into being at a single moment in time, in an event that we have come to call the *Big Bang*. Today, powerful telescopes allow us to view how galaxies have changed over the past 14 billion years, and at great distances we see young galaxies still in the process of forming [Section 21.1]. These observations confirm that the universe is gradually aging, just as we should expect if the entire universe really was born some 14 billion years ago.

Unfortunately, we cannot see back to the very beginning of time. Light from the most distant galaxies shows us what the universe looked like when it was one or two billion years old. Beyond these galaxies, we have not yet found any objects shining brightly enough for us to see them. Ultimately, we face an even more fundamental problem. The universe is filled with a faint glow of radiation that appears to be the remnant heat of the Big Bang. This faint glow comes from a time when the universe was about 380,000 years old, and it prevents us from seeing directly to earlier times. Thus, just as we must rely on theoretical modeling to determine what Earth or the Sun is like on the inside, we must use modeling to investigate what happened in the universe in its first 380,000 years.

Fortunately, calculating conditions in the early universe is straightforward if we consider some fundamental principles. We know that the expanding universe is cooling and becoming less dense as it grows. Therefore, the universe must have been hotter and denser in the past. Calculating exactly how hot and dense the universe must have been when it was more compressed is similar to calculating the temperatures and densities in a car engine as the pistons compress the fuel mix, except that the conditions become much more extreme. Figure 23.1 shows just how

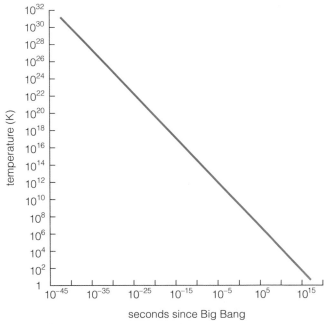

Figure 23.1 Temperature of the universe from the Big Bang to the present (10^{10} years $\approx 3 \times 10^{17}$ seconds). The universe cools as it expands. We can use the laws of physics to figure out how hot the universe had to be in the past in order to reach its present temperature.

hot the universe was during its earliest moments, according to such calculations. To understand these early moments, we need to know how matter and energy behave under such extraordinarily hot conditions.

Recent advances in physics have enabled us to predict the behavior of matter and energy under conditions far more extreme than in the centers of the hottest stars. When we apply our understanding of modern physics to the early universe, we find something so astonishing that it almost sounds preposterous. Today, we may understand the behavior of matter and energy well enough to describe what was happening in the universe just *one ten-billionth* (10^{-10}) of a second after the Big Bang. And that's not all. While our understanding of physics is less certain under the more extreme conditions that prevailed even earlier, we have some ideas about what the universe was like when it was a mere 10^{-38} second old, and perhaps a glimmer of what it was like at the age of just 10^{-43} second. These tiny fractions of a second are so small that, for all practical purposes, we are studying the very moment of creation.

23.2 A Scientific History of the Universe

The **Big Bang theory**—the scientific theory of the universe's earliest moments—presumes that all we see today, from Earth to the cosmic horizon, began as an incredibly tiny, hot, and dense collection of matter and radiation. It describes how expansion and cooling of this unimaginably intense mixture of particles and photons could have led to the present universe of stars and galaxies, and it explains several aspects of today's universe with impressive accuracy.

Our ideas about the earliest moments of the universe are quite speculative because they depend on aspects of physics that we do not yet fully comprehend. However, the evidence supporting this theory grows stronger and stronger as we focus on progressively later moments in time. We will discuss the evidence supporting the Big Bang theory later in this chapter. First, in order to help you understand the significance of the evidence, we'll examine the story of creation according to this theory.

Figure 23.2 summarizes the story by dividing the overall history of the universe into a series of *eras*, or time periods. Each era is distinguished from the next by some major change in the conditions of the universe. You'll find it easiest to keep track of the various eras if you refer back to this figure as we discuss each era in detail. (In more specialized books on cosmology, these eras are subdivided into many more eras, providing additional details about the early universe not covered here.)

Before we delve into the story of creation, it's important to be aware of a key aspect of the early universe: It was so hot that photons could transform themselves into matter, and vice versa, in accordance with Einstein's formula

$E = mc^2$ [Section 4.2]. Today, physicists can study many such reactions by reproducing them in their laboratories, but reactions that create and destroy matter are now relatively rare in the universe at large.

One example of such a reaction is the creation or destruction of an *electron–antielectron pair* [Section S4.5]. When two photons collide with a total energy greater than twice the mass-energy of an electron (i.e., the electron's mass times c^2), they can create two brand-new particles: a negatively charged electron and its positively charged twin, the *antielectron,* also known as a *positron* (Figure 23.3). The electron is a particle of **matter**, and the antielectron is a particle of **antimatter**. The reaction that creates an electron–antielectron pair also runs in reverse. When an electron and an antielectron meet, they *annihilate* each other totally, transforming all their mass-energy back into photon energy.

Similar reactions can produce or destroy any particle–antiparticle pair, such as a proton and antiproton or a neutron and antineutron. The early universe therefore was filled with an extremely dynamic blend of photons, matter, and antimatter, converting furiously back and forth. Despite all these vigorous reactions, describing conditions in the early universe is straightforward, because we can calculate the proportions of the various forms of radiation and matter from the universe's temperature and density at any time.

The First Instant

The scientific story of creation begins when the universe was an incomprehensibly tiny 10^{-43} second old. This instant in time is called the **Planck time** after physicist Max Planck, one of the founders of the science of quantum mechanics. According to the laws of quantum mechanics, there must have been substantial energy fluctuations from point to point in the very early universe. Because energy and mass are equivalent, Einstein's theory of general relativity tells us that these energy fluctuations must have generated a rapidly changing gravitational field that randomly warped space and time.

Prior to the Planck time, in the **Planck era**, these random energy fluctuations were so large that our current theories are powerless to describe what might have been happening. The problem is that we do not yet have a theory that links quantum mechanics and general relativity. In a sense, quantum mechanics is our successful theory of the very small (subatomic particles), and general relativity is our successful theory of the very big (the large-scale structure of spacetime). Perhaps someday we will be able to merge these theories of the very small and the very big into a single "theory of everything." Until that happens, science cannot describe the universe before the Planck time.

The GUT Era

Understanding the transition that marked the beginning of the next era requires thinking in terms of the *forces* that

Time Since Big Bang

present

Era of Galaxies

1 billion years

Era of Atoms

380,000 years

Era of Nuclei

3 minutes

Era of Nucleosynthesis

0.001 second

Particle Era

10^{-10} second

Electroweak Era

10^{-38} second

GUT Era

10^{-43} second

Planck Era

Major Events Since Big Bang

Humans observe the cosmos.

stars, galaxies, and clusters of galaxies (made of atoms and plasma)

First galaxies form.

atoms and plasma (stars begin to form)

Atoms form; photons fly free and become microwave background.

plasma of hydrogen and helium nuclei plus electrons

Fusion ceases; normal matter is 75% hydrogen, 25% helium, by mass.

protons, neutrons, electrons, neutrinos (antimatter rare)

elementary particles (antimatter common)

Matter annihilates antimatter.

Electromagnetic and weak forces become distinct.

elementary particles

Strong force becomes distinct, perhaps causing inflation of universe.

elementary particles?

????

Key

neutron
proton

electron
neutrino

antiproton
antineutron

antielectrons

quarks

Figure 23.2 This diagram summarizes the eras of the universe. The names of the eras and their ending times are indicated on the left, and the state of matter during each era and the events marking the end of each era are indicated on the right.

operate in the universe. Everything that happens in the universe today is governed by four distinct forces: *gravity, electromagnetism,* the *strong force,* and the *weak force* [Section S4.2]. Gravity dominates large-scale action, and electromagnetism dominates chemical and biological reactions. The strong and weak forces determine what happens within atomic nuclei: The strong force binds nuclei together, and the weak force mediates nuclear reactions such as fission or fusion.

These forces were not so distinct in the early universe. The four forces that act so differently today turn out to be separate aspects of a smaller number of more fundamental

forces, probably only one or two (Figure 23.4). As an analogy, think about ice, liquid water, and water vapor. These three substances are quite different from one another in appearance and behavior, yet they are just different phases of the single substance H_2O. In a similar way, physicists have shown that the electromagnetic and weak forces lose their separate identities under conditions of very high temperature or energy and merge together into a single *electroweak force.* Physicists also believe that the electroweak force and the strong force ultimately lose their separate identities at even higher energies.

The theories that predict this merger of the electroweak and strong forces are called **grand unified theories**, or **GUTs** for short. Their merger is usually referred to as the *GUT force*. Many physicists believe that at even higher energies the GUT force and gravity merge into a single "super force" that governs the behavior of everything. (Among the names you may hear for theories linking all four forces are *supersymmetry, superstrings,* and *supergravity*.)

Calculations from general relativity and quantum mechanics suggest that this unified "super force" may have reigned in the universe during the Planck era. If so, the Planck time (10^{-43} s) marks the instant when gravity became distinct from the other three forces, which were still merged as the GUT force. By analogy to ice crystals forming as a liquid cools, we say that gravity "froze out" at the Planck time. The universe subsequently entered the **GUT era**, when two forces operated in the universe: gravity and the GUT force.

The GUT era is thought to have lasted but a tiny fraction of a second, coming to an end when the universe had cooled to 10^{29} K at an age of about 10^{-38} second. Grand unified theories predict that the strong force froze out from the GUT force at this point, leaving three forces operating in the universe: gravity, the strong force, and the electroweak force. As we will discuss later, it now seems possible that this freezing out of the strong force released an enormous amount of energy. This energy release would have caused the universe to undergo a sudden and dramatic expansion that we call **inflation**. In the space of a mere 10^{-36} second, pieces of the universe the size of an atomic nucleus may have grown to the size of our solar system. Inflation sounds bizarre, but it might actually explain some puzzling features of today's universe.

The Electroweak Era

The end of the GUT era marks the beginning of the **electroweak era**, so named because the electromagnetic and weak

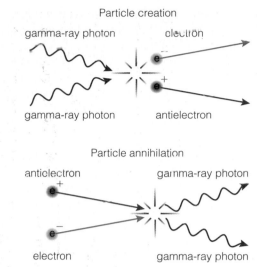

Figure 23.3 Electron–antielectron creation and annihilation. Reactions like these constantly converted photons to particles and vice versa in the early universe.

forces were still unified in the electroweak force. Intense radiation filled all of space, as it had since the Planck era, spontaneously producing matter and antimatter particles of many different sorts that immediately reverted back to energy. The universe continued to expand and cool throughout the electroweak era, dropping to a temperature of 10^{15} K when it reached an age of 10^{-10} second. This temperature is still 100 million times hotter than the temperature in the core of the Sun, but it was low enough for the electromagnetic and weak forces to freeze out from the electroweak force. After this instant (10^{-10} s), all four forces were forever distinct in the universe.

The end of the electroweak era marks an important transition not only in the universe, but also in human understanding of the universe. The theory that unified the weak and electromagnetic forces, developed in the 1970s, predicted the emergence of new types of particles (called

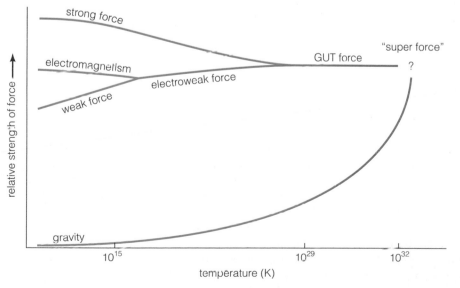

Figure 23.4 The four forces are distinct at low temperatures but may merge at very high temperatures.

the W and Z bosons, or *weak bosons*) at temperatures above the 10^{15} K that pervaded the universe when it was 10^{-10} second old. In 1983, experiments performed in a huge particle accelerator near the French/Swiss border (CERN [Section S4.2]) reached energies equivalent to such high temperatures for the first time. The new particles showed up just as predicted, produced from the extremely high energy in accord with $E = mc^2$.

Thus, we have direct experimental evidence concerning the conditions in the universe at the end of the electroweak era. We do *not* have any direct experimental evidence of conditions prior to that time. Our theories concerning the earlier parts of the electroweak era and the GUT era consequently are much more speculative than our theories describing the universe from the end of the electroweak era to the present.

The Particle Era

Spontaneous creation and annihilation of particles continued throughout the next era, which we call the **particle era** because tiny particles of matter were as numerous as photons. These particles included electrons, neutrinos, and *quarks*—the building blocks of heavier particles such as protons, antiprotons, neutrons, and antineutrons [Section S4.2]. Near the end of the particle era, when the universe was about 0.0001 second old, the temperature grew too cool for quarks to exist on their own, and quarks combined in groups of three to form protons and neutrons. The particle era came to an end when the universe reached an age of 1 millisecond (0.001 s) and the temperature had fallen to 10^{12} K. At this point, it was no longer hot enough to spontaneously produce protons and antiprotons (or neutrons and antineutrons) from pure energy.

If the universe had contained equal numbers of protons and antiprotons (or neutrons and antineutrons) at the end of the particle era, all of the pairs would have annihilated each other, creating photons and leaving essentially no matter in the universe. In today's universe, photons outnumber protons by over a billion to one. We conclude that protons must have slightly outnumbered antiprotons at the end of the particle era. For every billion antiprotons, there were about a billion and one protons. Thus, for each 1 billion protons and antiprotons that annihilated each other, creating photons, a single proton was left over. This seemingly slight excess of ordinary matter over antimatter makes up all the ordinary matter in the present-day universe. Some of those protons (and neutrons) left over from when the universe was 0.001 second old are the very ones that make up our bodies.

The Era of Nucleosynthesis

So far, everything we have discussed occurred within the first 0.001 second of the universe's existence—a time span shorter than the time it takes you to blink an eye. At this point, the protons and neutrons left over after the annihilation of antimatter attempted to fuse into heavier nuclei. However, the heat of the universe remained so high that most nuclei broke apart as fast as they formed. This dance of fusion and breakup marked the **era of nucleosynthesis**.

The era of nucleosynthesis ended when the universe was about 3 minutes old. By this time, the density in the expanding universe had dropped so much that fusion no longer occurred, even though the temperature was still about a billion Kelvin (10^9 K)—much hotter than the temperature at the center of the Sun today. When fusion ceased, about 75% of the mass of the ordinary (baryonic) matter in the universe remained as individual protons, or hydrogen nuclei. The other 25% of this mass had fused into helium nuclei, with trace amounts of deuterium (hydrogen with a neutron) and lithium. Except for the relatively small amount of matter that stars later forged into heavier elements, the chemical composition of the universe remains the same today.

The Era of Nuclei

At the end of the era of nucleosynthesis, the universe consisted of a very hot plasma of hydrogen nuclei, helium nuclei, and free electrons. This basic picture held for about the next 380,000 years as the universe continued to expand and cool. The fully ionized nuclei moved independently of electrons during this period (rather than being bound with electrons in neutral atoms), which we call the **era of nuclei**. Throughout this era, photons bounced rapidly from one electron to the next, just as they do deep inside the Sun today, never managing to travel far between collisions. Any time a nucleus managed to capture an electron to form a complete atom, one of the photons quickly ionized it.

The era of nuclei came to an end when the expanding universe was about 380,000 years old. At this point the temperature had fallen to about 3,000 K—roughly half the temperature of the Sun's surface today. Hydrogen and helium nuclei finally captured electrons for good, forming stable, neutral atoms for the first time. With electrons now bound into atoms, the universe suddenly became transparent, as if a thick fog had suddenly lifted. Photons, formerly trapped among the electrons, began to stream freely across the universe. We still see these photons today as the *cosmic microwave background*, which we will discuss shortly.

The Era of Atoms and the Era of Galaxies

We've already discussed the rest of the story in earlier chapters. The end of the era of nuclei marked the beginning of the **era of atoms**, when the universe consisted of a mixture of neutral atoms and plasma. Thanks to the slight density enhancements present in the universe at this time and the gravitational attraction of dark matter, the atoms and plasma slowly assembled into protogalactic clouds. Stars formed in these clouds, transforming the gas clouds into galaxies. The first full-fledged galaxies had formed by the time the universe was about 1 billion years old, beginning what we call the **era of galaxies**.

The era of galaxies continues to this day. Generation after generation of star formation in galaxies steadily builds elements heavier than helium and incorporates them into new star systems. Some of these star systems develop planets, and on at least one of these planets life burst into being a few billion years ago. Now here we are, thinking about it all. Describing both the follies and the achievements of the human race, Carl Sagan once said, "These are the things that hydrogen atoms do—given 15 billion years of cosmic evolution."

23.3 Evidence for the Big Bang

Like any scientific theory, the Big Bang theory is a model of nature designed to explain a set of observations. If it is close to the truth, it should be able to make predictions about the real universe that we can verify through more observations or experiments. The Big Bang model has gained wide scientific acceptance for two key reasons:

- The Big Bang model predicts that the radiation that began to stream across the universe at the end of the era of nuclei should still be present today. Sure enough, we find that the universe is filled with what we call the **cosmic microwave background**. Its characteristics precisely match what we expect according to the Big Bang model.

- The Big Bang model predicts that some of the original hydrogen in the universe should have fused into helium during the era of nucleosynthesis. Observations of the actual helium content of the universe closely match the amount of helium predicted by the Big Bang model. Fusion of hydrogen to helium in stars could have produced only about 10% of the observed helium.

The Cosmic Microwave Background

The first major piece of evidence supporting the Big Bang theory arrived in 1965. Arno Penzias and Robert Wilson, two physicists working at Bell Laboratories in New Jersey, were calibrating a sensitive microwave antenna designed for satellite communications (Figure 23.5). (*Microwaves* fall within the radio portion of the electromagnetic spectrum.) Much to their chagrin, they kept finding unexpected "noise" in every measurement they made with the antenna.

Fearing that they were doing something wrong, they worked frantically to discover and eliminate all possible sources of background noise. They even climbed up on their antenna to scrape off pigeon droppings, on the off-chance

Figure 23.5 Arno Penzias and Robert Wilson, discoverers of the cosmic microwave background, with the Bell Labs microwave antenna.

that these were somehow causing the noise. No matter what they did, the microwave noise wouldn't go away. The noise was the same no matter where they pointed their antenna, indicating that it came from all directions in the sky and ruling out the possibility that it came from any particular astronomical object or from any place on Earth. Embarrassed by their inability to explain the noise, Penzias and Wilson prepared to "bury" their discovery at the end of a long scientific paper about their antenna.

Meanwhile, physicists at nearby Princeton University were busy calculating the expected characteristics of the radiation left over from the heat of the Big Bang. They concluded that, if the Big Bang had really occurred, this radiation should be permeating the entire universe and should be detectable with a microwave antenna. On a fateful airplane trip home from an astronomical meeting, Penzias sat next to an astronomer who told him of the Princeton calculations. The Princeton group soon met with Penzias and Wilson to compare notes. The "noise" in the Bell Labs antenna was not an embarrassment after all. Instead, it was the cosmic microwave background—and the first strong evidence that the Big Bang had really happened. Penzias and Wilson received the 1978 Nobel Prize in physics for their discovery of the cosmic microwave background.*

The cosmic microwave background consists of photons arriving at Earth directly from the end of the era of nuclei,

---

*The dramatic story of the discovery of the cosmic microwave background is told in greater detail, along with much more scientific history, in Timothy Ferris, *The Red Limit* (New York: Quill, 1983).

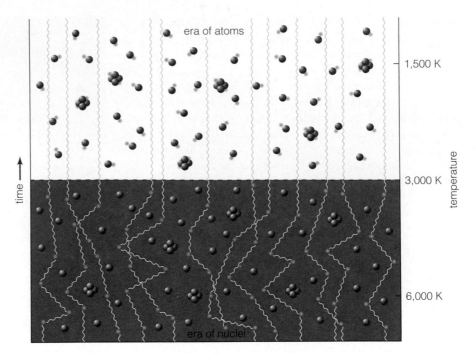

Figure 23.6 Photons frequently collided with free electrons during the era of nuclei and thus could travel freely only after electrons became bound into atoms. This transition was something like the transition from a dense fog to clear air. The photons released at the end of the era of nuclei, when the universe was about 380,000 years old, make up the cosmic microwave background. Precise measurements of these microwaves tell us what the universe was like at this moment in time.

era of atoms

1,500 K

3,000 K

temperature

6,000 K

era of nuclei

time

when the universe was about 380,000 years old. Because neutral atoms could remain stable for the first time, they captured most of the electrons in the universe. With no more free electrons to block them, the photons from that epoch have flown unobstructed through the universe ever since (Figure 23.6). Thus, when we observe the cosmic microwave background, we essentially are seeing back to a time when the universe was only 380,000 years old. In that sense, we are seeing light from the most distant regions of the observable universe—only 380,000 light-years from our cosmological horizon [Section 20.4]. Surprisingly, it does not take a particularly powerful telescope to "see" this radiation. In fact, you can pick it up with an ordinary television antenna. If you set an antenna-fed television (i.e., *not* cable or satellite TV) to a channel for which there is no local station, you will see a screen full of static "snow." About 1% of this static is due to photons in the cosmic microwave background. Try it. If your friends ask why you are watching nothing, tell them that you are actually watching the most incredible sight ever seen on a television screen: the Big Bang, or at least as close to it as we'll ever get.

The cosmic microwave background came from the heat of the universe itself and therefore should have an essentially perfect thermal radiation spectrum [Section 6.4]. When this radiation first broke free 380,000 years after the Big Bang, the temperature of the universe was about 3,000 K, not too different from that of a red giant star's surface. Thus, the spectrum of the cosmic microwave background originally peaked in visible light, just like the thermal radiation from a red star, with wavelengths of a few hundred nanometers. However, the universe has expanded by a factor of about 1,000 since that time, stretching the wavelengths of these photons by the same amount [Section 20.4]. Their wavelengths therefore should have shifted to about a millimeter, squarely in the microwave portion

of the spectrum and corresponding to a temperature of a few degrees above absolute zero.

In the early 1990s, a NASA satellite called the *Cosmic Background Explorer (COBE)* was launched to test these ideas about the cosmic microwave background. The results were a stunning success for the Big Bang theory. As shown in Figure 23.7, the cosmic microwave background does indeed have a perfect thermal radiation spectrum, with a peak corresponding to a temperature of 2.73 K. In a very real sense, the temperature of the night sky is a frigid 3 degrees above absolute zero.

THINK ABOUT IT

Suppose the cosmic microwave background did not really come from the heat of the universe itself but instead came from many individual stars and galaxies. Explain why, in that case, we would not expect it to have a perfect thermal radiation spectrum. How does the spectrum of the cosmic microwave background lend support to the Big Bang theory?

COBE achieved an even greater success mapping the temperature of the cosmic microwave background in all directions. It was already known that the cosmic microwave background is extraordinarily uniform throughout the universe. Conditions in the early universe must have been extremely uniform to produce such a smooth radiation field. For a time, this uniformity was considered a strike against the Big Bang theory because, as we discussed in Chapters 21 and 22, the universe must have contained some regions of enhanced density in order to explain the formation of galaxies. The COBE measurements restored confidence in the Big Bang theory because they showed that the cosmic microwave background is *not quite* perfectly uniform. Instead, its temperature varies very slightly from one

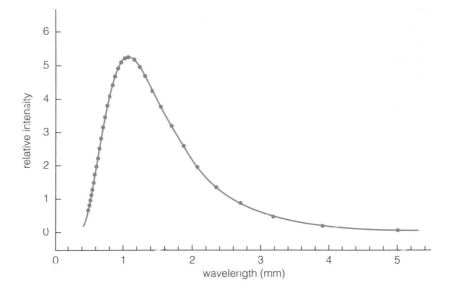

Figure 23.7 Spectrum of the cosmic microwave background from COBE. A theoretically calculated thermal radiation spectrum (smooth curve) for a temperature of 2.73 K perfectly fits the data (dots). This excellent fit is important evidence in favor of the Big Bang theory.

place to another by a few parts in 100,000 (Figure 23.8).* These variations in temperature indicate that the density of the early universe really did differ slightly from place to place. The seeds of structure formation were indeed present during the era of nuclei.

COBE's discovery of density enhancements bolstered the idea that some of the dark matter consists of WIMPs (weakly interacting massive particles [Section 22.4]) that we have not yet identified and that the gravity of this dark matter drove the formation of structure in the universe. Regions of enhanced density can grow into galaxies because the extra gravity in these regions draws matter together even while the rest of the universe expands. The greater the density enhancements, the faster matter should have collected into galaxies.

Detailed calculations show that, to explain the fact that galaxies formed within a few billion years, the density en-

---

*Earth's motion (e.g., orbit of Sun, rotation of galaxy, etc.) means that we are moving relative to the cosmic background radiation. Thus, we see a slight blueshift (about 0.12%) in the direction we're moving and a slight redshift in the opposite direction. We must first subtract these effects before analyzing the temperature of the background radiation.

hancements at the end of the era of nuclei must have been significantly greater than the few parts in 100,000 suggested by the temperature variations in the cosmic microwave background. Because WIMPs are weakly interacting and do not interact with photons, we do not expect them to influence the temperature of the cosmic microwave background directly. However, the gravity of the WIMPs can collect ordinary baryonic matter into clumps that *do* interact with photons. Thus, the small density enhancements detected by COBE may actually echo the much larger density enhancements made up of WIMPs. Careful modeling of the COBE temperature variations shows that they are consistent with dark-matter density enhancements large enough to account for the structure we see in the universe today.

Synthesis of Helium

The discovery of the cosmic microwave background in 1965 quickly solved another long-standing astronomical problem: the origin of cosmic helium. Everywhere in the universe, about one-quarter of the mass of ordinary matter (i.e., not dark matter) is helium. The Milky Way's helium fraction is about 28%, and no galaxy has a helium fraction

---

SPECIAL TOPIC The Steady State Universe

One of the cleverest alternatives to the Big Bang theory, developed in the late 1940s, was called the *steady state theory* of the universe. This theory accepted the fact that the universe is expanding but rejected the idea of a Big Bang and instead postulated that the universe is infinitely old. This idea may seem paradoxical at first: If the universe has been expanding forever, shouldn't every galaxy be infinitely far away from every other galaxy? The steady state theory suggested that new galaxies continually form in the gaps that open up as the universe expands, thereby keeping the same average distance between galaxies at all times. In a sense, the steady state theory said that the creation of the universe is an ongoing

and eternal process rather than having happened all at once with a Big Bang.

Two key discoveries caused the steady state theory to lose favor. First, the 1965 discovery of the cosmic microwave background matched a prediction of the Big Bang theory but could not be adequately explained by the steady state theory. Second, the steady state theory predicts that the universe should look about the same at all times, which is inconsistent with observations showing that galaxies at great distances look younger than nearby galaxies. As a result of these predictive failures, the steady state theory is no longer taken seriously by most astronomers.

Figure 23.8 This all-sky map shows temperature differences in the cosmic microwave background measured by COBE. The background temperature is about 2.73 K everywhere, but the brighter regions of this picture are slightly less than 0.0001 K hotter than the darker regions—indicating that the early universe was very slightly lumpy at the end of the era of nuclei. We are essentially seeing what the universe was like at the surface marked "380,000 years" in Figure 23.2. Gravity later drew matter toward the centers of these lumps, forming the structures we see in the universe today. (The central strip of this map, which corresponds to the disk of the Milky Way, has been masked out because the brightness differences there stem primarily from radio noise in the Milky Way.)

radio

lower than 25%. A small proportion of this helium comes from hydrogen fusion in stars, but most does not. The majority of the helium in the universe must already have been present in the protogalactic clouds that preceded the formation of galaxies. In other words, the universe itself must once have been hot enough to fuse hydrogen into helium. The current microwave background temperature of 2.73 K tells us precisely how hot the universe was in the distant past and exactly how much helium it should have made. The result—25% helium—is another impressive success of the Big Bang theory.

A helium nucleus contains two protons and two neutrons, so we need to understand what protons and neutrons were doing during the era of nucleosynthesis in order to see why the primordial fraction of helium was 25%. Early in this era, when the universe's temperature was 10^{11} K, nuclear reactions could convert protons into neutrons, and vice versa. For example, a proton and an electron could combine to form a neutron and a neutrino. This reaction is one of several that change protons into neutrons. All such proton–neutron conversion reactions involve neutrinos or antineutrinos, which experience only the weak force. As long as the universe remained hotter than 10^{11} K, these reactions kept the numbers of protons and neutrons nearly equal.

As the universe cooled from 10^{11} K to 10^{10} K, neutron–proton conversion reactions began to favor protons. Neu-

trons are slightly heavier than protons, and therefore reactions that convert protons to neutrons require energy to proceed (in accordance with $E = mc^2$). Below 10^{11} K, the required energy for neutron production was no longer readily available, so the rate of these reactions slowed. In contrast, reactions that convert neutrons into protons release energy and thus are unhindered by cooler temperatures. At the time the temperature of the universe fell to 10^{10} K, protons began to outnumber neutrons because the conversion reactions ran only in one direction. Neutrons changed into protons, but the protons didn't change back.

At 10^{10} K, the universe was still hot and dense enough for nuclear fusion to operate. Protons and neutrons constantly combined to form *deuterium*—the rare form of hydrogen nuclei that contains a neutron in addition to a proton—and deuterium nuclei fused to form helium. However, during the early part of the era of nucleosynthesis, the helium nuclei were almost immediately blasted apart by one of the many gamma rays that filled the universe.

Fusion finally created long-lasting helium nuclei when the universe was about 1 minute old, at which time the destructive gamma rays were gone (Figure 23.9). Calculations show that the proton-to-neutron ratio at this time should have been about 7 to 1. Moreover, almost all the available neutrons should have become incorporated into nuclei of helium-4. Figure 23.10 shows that, based on the 7-to-1 ratio of protons to neutrons, the universe should

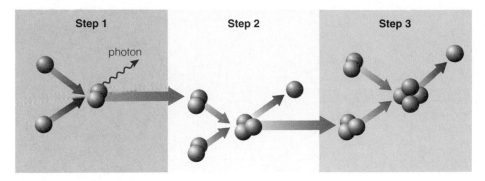

Figure 23.9 During the era of nucleo-synthesis, virtually all the neutrons in the universe fused with protons to form helium-4. This figure illustrates one of several possible reaction pathways. In step 1, a neutron and a proton fuse to form a deuterium nucleus, releasing a photon. In step 2, two deuterium nuclei fuse to make hydrogen-3, releasing a proton. In step 3, the hydrogen-3 nucleus fuses with deuterium to create helium-4, releasing one of the neutrons.

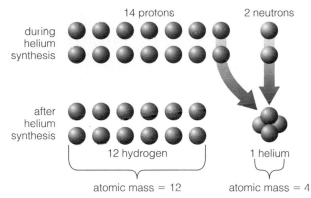

Figure 23.10 During helium synthesis, protons outnumbered neutrons 7 to 1, which is the same as 14 to 2. The result was 12 hydrogen nuclei (individual protons) for each helium nucleus. Thus, the hydrogen-to-helium mass ratio is 12 to 4, which is the same as 75% to 25%. The agreement between this prediction and the observed abundance of helium is important evidence in favor of the Big Bang theory.

have had a composition of 75% hydrogen and 25% helium by mass at the end of the era of nucleosynthesis.

Thus, the Big Bang theory makes a very concrete prediction about the chemical composition of the universe: It should be 75% hydrogen and 25% helium by mass. The fact that observations confirm this predicted ratio of hydrogen to helium is another striking success of the Big Bang theory.

THINK ABOUT IT

Briefly explain why it should not be surprising that some galaxies contain a little more than 25% helium, but it would be very surprising if some galaxies contained less. (*Hint.* Think about how the relative amounts of hydrogen and helium in the universe are affected by fusion in stars.)

Synthesis of Other Light Nuclei

Why didn't the Big Bang produce heavier elements? By the time stable helium nuclei formed, when the universe was about a minute old, the temperature and density of the rapidly expanding universe had already dropped too far for a process like carbon production (three helium nuclei fusing into carbon [Section 17.3]) to occur. Reactions be-

tween protons, deuterium nuclei, and helium were still possible, but most of these reactions led nowhere. In particular, fusing two helium-4 nuclei results in a nucleus that is unstable and falls apart in a fraction of a second, as does fusing a proton to a helium-4 nucleus.

A few reactions involving hydrogen-3 (also known as *tritium*) or helium-3 can create long-lasting nuclei. For example, fusing helium-4 and hydrogen-3 produces lithium-7. However, the contributions of these reactions to the overall composition of the universe were minor because hydrogen-3 and helium-3 were so rare. Models of element production in the early universe show that, before the cooling of the universe shut off fusion entirely, such reactions generated only trace amounts of lithium, the next lightest element after helium. Thus, aside from hydrogen, helium, and lithium, all other elements were forged much later in the nuclear furnaces of stars. (Beryllium and boron, which are heavier than lithium but lighter than carbon, were created later when high-energy particles broke apart heavier nuclei that formed in stars.)

The Density of Ordinary Matter

Calculations made with the Big Bang model allow scientists to estimate the density of ordinary (baryonic) matter in the universe from the observed amount of deuterium in the universe today. The fact that some deuterium nuclei still exist in the universe indicates that the fusing of neutrons into helium nuclei in the early universe was incomplete. The amount of deuterium—roughly one hydrogen atom in 40,000 contains a deuterium nucleus—tells us about the density of protons and neutrons (baryons) during the era of nucleosynthesis. If this density had been higher in the early universe, protons and neutrons would have fused more efficiently into helium, and less deuterium would have been left over.

The results of such calculations show that the density of ordinary (baryonic) matter in the universe is about 4% of the critical density (Figure 23.11). (Recall that the critical density is the density required if the expansion of the universe is to stop and reverse someday [Section 22.6].) Similar calculations based on the observed abundance of lithium lead to the same conclusion, adding to our confidence in the Big Bang model and implying something very

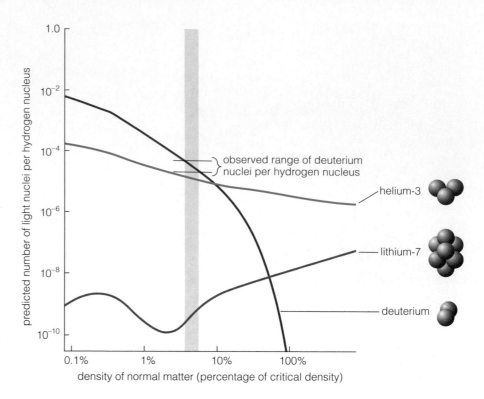

Figure 23.11 Predicted number of light nuclei created per hydrogen nucleus at the end of the era of nuclei. The predicted amounts of light nuclei like deuterium, helium-3, and lithium-7 depend on the total density of normal matter in the universe. If the density of normal matter were equal to the critical density, then virtually all the deuterium formed during the era of nuclei would have fused into helium, leaving almost none left over. Observations show that the universe contains about 2.5×10^{-5} deuterium nucleus for each hydrogen nucleus. Models of fusion processes during the era of nuclei indicate that this amount of deuterium is created if the density of normal matter in the universe is about 4% of the critical density.

interesting about the fate of the universe: Unless extraordinary (nonbaryonic) matter outweighs ordinary matter by at least a factor of 20, the universe cannot recollapse. Moreover, if the universe's density is close to 30% of the critical density—as appears to be the case—then extraordinary (nonbaryonic) matter such as WIMPs probably constitutes the majority of the universe's mass.

The ideas discussed above point to a rather amazing fact: Although we have yet to discover any WIMPs, we suspect they dominate the total mass of the universe. Briefly explain how this is possible, and comment on how confident we can be that weakly interacting particles make up the bulk of dark matter.

23.4 Inflation

The Big Bang model relies heavily on our knowledge of particle physics, which has been tested to temperatures of 10^{15} K, corresponding to the end of the electroweak era. Our knowledge of earlier times rests on a weaker foundation because we are less certain of the physical laws at work. In fact, the best laboratory for studying the laws of physics at such high temperatures is the Big Bang itself.

Different guesses as to how matter might behave at such high energies predict different outcomes for the universe we see today. If a particular model predicts that our universe should look different from the way it really does, then that model must be wrong. On the other hand, if a new model of particle physics explains some previously

unexplained aspects of the universe, then it may be on the right track. The grand unified theories of particle physics discussed in Section 23.2 have not yet been proved, but many scientists believe these theories are on the right track for just this reason.

Prior to the early 1980s, scientists had identified several major aspects of our actual universe that were unexplained by the Big Bang model. Three of the most pressing unanswered questions were the following:

● *Where does structure come from?* According to the Big Bang theory, the process of galaxy formation required the density of the early universe to differ slightly from place to place. Otherwise, the pull of gravity would have been equally balanced everywhere, prohibiting the collapse of protogalactic clouds. The subtle temperature differences seen in the cosmic microwave background tell us that regions of enhanced density did indeed exist at the end of the era of nuclei, when the universe was 380,000 years old. However, before inflation was added to the model, the Big Bang theory could not explain where these density enhancements came from.

● *Why is the large-scale universe so smooth?* Not only is the slight lumpiness of the early universe a problem, but its large-scale smoothness is also a problem. Observations of the cosmic microwave background show that, overall, the density of the universe at the end of the era of nuclei varied from place to place by no more than about 0.01%. Before the 1980s, the Big Bang theory could not explain why distant reaches of the universe look so similar.

- *Why is the density of the universe almost critical?* The density of matter in the universe is about 30% of the critical density [Section 22.6]—remarkably close to the critical density considering that the Big Bang theory before 1980 didn't say anything about what the density should be. Thus the Big Bang theory had no way to explain this coincidence. Why isn't the matter density 1,000 times the critical density or 0.0000001 times the critical density?

Physicist Alan Guth realized in 1981 that grand unified theories could potentially answer all three questions. These theories predict that the separation of the strong force from the GUT force should have released enormous energy, causing the universe to expand dramatically, perhaps by as much as 10^{30} times in less than 10^{-36} second. This dramatic expansion is what we call *inflation*. While it sounds outrageous, it may have shaped the way the universe looks today.

Structure: Giant Quantum Fluctuations

To understand how inflation solves the structure problem, we need to recognize a special feature of energy fields. Laboratory-tested principles of quantum mechanics (especially the uncertainty principle [Section S4.3]) require that the energy fields at any point in space be always fluctuating. Thus, the distribution of energy through space on very small scales is slightly irregular, even in a complete vacuum. These tiny quantum "ripples" can be characterized by a wavelength that corresponds roughly to their size. In principle, quantum ripples in the very early universe could have been the seeds for density enhancements that later grew into galaxies. However, the wavelengths of the original ripples were far too small to explain density enhancements like those we see imprinted on the cosmic microwave background.

Inflation would have dramatically altered these quantum fluctuations. The fantastic growth of the universe during the period of inflation would have stretched tiny ripples in space to enormous wavelengths (Figure 23.12). Ultimately, inflation could have caused these quantum ripples to grow large enough to become the density enhancements that later formed large structures in the universe. Amazingly, the structure of today's universe may have started as tiny quantum fluctuations just before the period of inflation.

Although inflation goes a long way toward solving the structure problem, it is not entirely solved. Models invoking inflation make reasonable predictions for how the wavelengths of the quantum ripples relate to their amplitudes, but they do not predict how large these amplitudes should be. This problem underscores the speculative nature of the idea of inflation. With improved models, scientists hope to be able to make more definitive predictions that can be tested observationally.

Smoothness: Equalizing Temperatures and Densities

Before we look at how inflation answers the smoothness problem, let's look a little closer at the problem itself. At

size of ripple before inflation = size of atomic nucleus

size of ripple after inflation = size of solar system

Figure 23.12 During inflation, ripples in spacetime would have stretched by a factor of perhaps 10^{30}. The peaks of these ripples then would have become the density enhancements that produced all the structure we see in the universe today.

first, the idea that different parts of the universe should be very similar shortly after the Big Bang may seem quite natural, but on further inspection the smoothness of the universe becomes difficult to explain.

Imagine observing the cosmic microwave background in a certain part of the sky. You are seeing that region as it was only 380,000 years after it formed. The microwaves coming from it have taken close to 14 billion years—almost the entire age of the universe—to travel all the way to Earth. Now imagine turning around and looking at the background radiation coming from the opposite direction. You are also seeing this region at an age of 380,000 years, and it looks virtually identical. The two microwave-emitting regions are billions of light-years apart, but we are seeing them as they were when they were only 380,000 years old. They can't possibly have had communication with each other then (Figure 23.13). A signal traveling at the speed of light from one to the other would barely have started its journey. So how do they come to be exactly the same temperature?

Inflation solves this problem because it says that the entire observable universe was less than 10^{-38} light-second across before the episode of enormous expansion. Radiation signals traveling at the speed of light would have had time to equalize all the temperatures and densities in this region. Inflation then pushed these equalized regions to much greater distances, far out of contact with one another (Figure 23.14). Like criminals getting their stories straight before being locked in separate jail cells, all parts of the observable universe came to the same temperature and density before inflation spread them far apart. Thus, inflation accounts for how vastly separated regions can look so similar.

This explanation of why the universe appears so uniform at first may appear paradoxical. It seems to require pieces of the observable universe to move faster than the

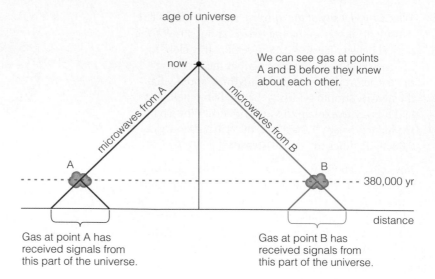

Figure 23.13 Light left the microwave-emitting regions we see on opposite sides of the universe long before they could have communicated with each other and equalized their temperatures, yet their temperatures are virtually identical. Without inflation, the similar temperatures are a puzzle.

(labels in figure)
age of universe
now
We can see gas at points A and B before they knew about each other.
microwaves from A
microwaves from B
A
B
380,000 yr
distance
Gas at point A has received signals from this part of the universe.
Gas at point B has received signals from this part of the universe.

speed of light during inflation, yet we know that nothing can move faster than the speed of light [Section S2.2]. To understand why inflation does not violate this universal speed limit, remember that the expansion of the universe is really the expansion of *space itself*. Objects with large cosmological redshifts aren't moving at high speed so much as they are riding along with the expansion of the universe.

The gap between two objects increases because the space between them is expanding. Photons can easily cross this gap before inflation, but during inflation the gaps expand much faster than photons can cross them, and communication between the two objects ceases. One object

never detects the other apparently moving faster than light speed because there is no communication or information exchange. Communication between these objects can resume only long after inflation, when photons have had time to cross the huge chasm in space that has opened.

Density: Balancing the Universe

The third question answered by inflation asks why the matter density of the universe is so close to the critical density. Another way to say that the universe's density is close to critical is to say that the overall geometry of the universe

Figure 23.14 Before inflation, regions A and B were near enough to communicate and equalize their temperatures. Inflation then pushed them far apart. Today, we can see both A and B, but they are too far apart to see each other.

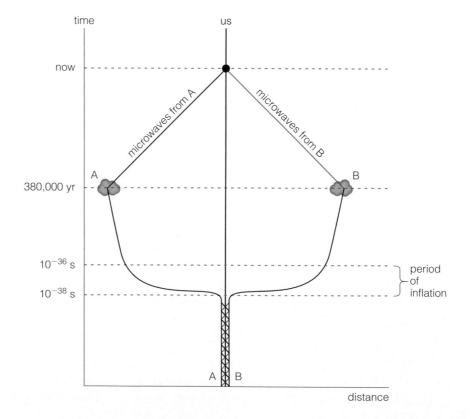

(labels in figure)
time
us
now
microwaves from A
microwaves from B
A
B
380,000 yr
10^{-36} s
10^{-38} s
period of inflation
A B
distance

is remarkably "flat" [Section 22.6]. If the density of the universe exactly equals the critical density, then the kinetic energy of expansion precisely balances the universe's overall gravitational pull.

According to Einstein's theory of general relativity, an imbalance in these energies causes curvature of spacetime, and deviations from precise balance grow more severe as the universe evolves. For example, if the universe had been 10% denser at the end of the era of nuclei, it would have collapsed long ago. On the other hand, if it had been 10% less dense at that time, galaxies would never have formed before expansion spread all the matter too thin. Thus, the universe had to start out remarkably balanced to be even remotely close to flat today.

Inflation balances the universe by ensuring that the mass-energy providing the universe's gravity precisely matches its kinetic energy of expansion. Once this balance is set, it should never waver. In terms of Einstein's theory, the effect of inflation on spacetime curvature is similar to the flattening of a balloon's surface when you blow into it (Figure 23.15). The flattening of space during the period of inflation would have been so enormous that it would virtually have eliminated any curvature the universe might have had previously.

The fine balance created by inflation turns out to be both a success and a potential pitfall of the inflation theory. On one hand, it explains how the universe managed to stay balanced long enough to bring forth galaxies. However, our studies of dark matter suggest that the current universe might be slightly out of balance. The best current estimates put the actual density of the universe at only about 30% of the critical density [Section 22.6], in which case the universe has about three times more kinetic energy than gravity can overcome.

Inflation theory has difficulty explaining this imbalance, but a loophole in Einstein's theory may solve the problem. Energy associated with a large-scale repulsive force can compensate for the shortfall in the matter density, making the present-day universe flatter than it would otherwise be. Remarkably, the supernova measurements indicating that our universe's expansion is accelerating [Section 22.6] also show that the strength of the implied repulsive force is about right to render the universe perfectly flat, just as inflation predicts. In other words, the total amount of matter plus energy may be precisely equal to the critical density.

Testing for Inflation

Verifying that the universe underwent an episode of inflation is difficult, because this inflationary episode would have happened extremely early in the history of the universe—much earlier than we can observe directly. Nevertheless, we can test the idea of inflation by exploring whether its predictions are consistent with our observations of the universe at later times. Such observations cannot prove the theory of inflation, but if they are inconsistent with the theory, then inflation probably never happened. Scientists are only beginning to make observations that test inflation, but the findings to date are consistent with the idea that an early inflationary episode smoothed and flattened the universe while planting the seeds of structure formation.

The earliest moment we can directly observe is the end of the era of nuclei, when photons began to stream freely across the universe. These photons ultimately became what we now observe as the cosmic microwave background. COBE was the first telescope to reveal the subtle temperature differences created by the density enhancements present at that moment. Many telescopes are following in COBE's footsteps. Because theoretical models of inflation make specific predictions about the nature of these density enhancements, better observations of temperature differences in the cosmic microwave background are subjecting the idea of inflation to ever more stringent tests.

A ground-breaking microwave map of the entire sky taken by NASA's Wilkinson Microwave Anisotropy Probe (WMAP) and released in February 2003 (Figure 23.16a) stands as the strongest test of inflation to date—and inflation has impressively passed the test. Comparing this map with Figure 23.8, we can see that WMAP detects ripples in the temperature of the early universe far subtler than those found by COBE.

Careful study of the temperature differences observed by WMAP and other microwave telescopes can tell us a great deal about both conditions in the early universe and the geometry of the universe as a whole (Figure 23.16b).

Figure 23.15 As a balloon expands, its surface seems increasingly flat to an ant crawling along it. Inflation is thought to have made the universe flat in a similar way.

radio

a An all-sky map of patterns in the cosmic microwave background measured by WMAP. The shades of color represent tiny temperature differences in the early universe when it was about 380,000 years old. Because these temperature differences depend on the density enhancements that later turned into galaxies, clusters of galaxies, and larger-scale structures, careful study of the temperature patterns reveals the "genetic code" imprinted on the universe in its earliest moments.

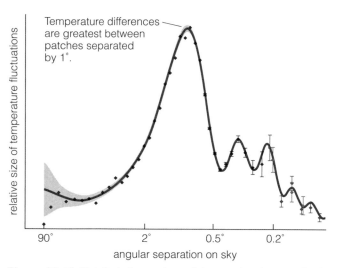

Temperature differences are greatest between patches separated by 1°.

relative size of temperature fluctuations

angular separation on sky

90° 2° 0.5° 0.2°

b This graph shows how scientists use temperature differences in the cosmic microwave background to test models of the universe. Black data points indicate typical temperature differences between patches separated by a particular angle in the map shown in (**a**). (Blue and green points represent data from other telescopes.) Complex calculations show that if the universe is geometrically flat we should observe the greatest temperature differences between patches separated by about a degree. The fact that the observed temperature differences peak at separations of 1° indicates that the universe is indeed flat, in accord with inflation. A detailed comparison of various models with the data tells us even more. The red line shows predictions of a model that relies on inflation to produce a universe whose normal matter density, dark matter density, and expansion rate are all similar to their observed values. Close agreement between the data points and the model indicates that the "genetic code" predicted by inflation is very similar to the true "genetic code" of our universe.

Figure 23.16 Detailed observations of the cosmic microwave background support the idea that inflation made the universe flat while planting the seeds for structure formation.

For example, complex calculations show that the largest temperature differences in the cosmic microwave background should typically be between patches of sky separated by about 1° if the overall geometry of the universe is flat. (Similar calculations show that this angular separation would be smaller than 1° if the universe were open and curved like a saddle, and larger than 1° if the universe were closed and curved like a sphere [Section 22.6].) The strongest temperature differences are indeed observed at angular separations of 1°, indicating that the universe is geometrically flat, as predicted by inflation. In fact, the overall pattern of temperature differences in the cosmic microwave background revealed by WMAP agrees very well with the predictions of

models based on inflation, which is why these results lend support to the idea that inflation really occurred.

In a certain sense, these new observations of the cosmic microwave background are revealing the "genetic code" of the universe. The characteristics of the density enhancements governed the subsequent growth of galaxies, clusters of galaxies, and the universe itself. To the extent that we have been able to read this code, it aligns reassuringly well with the universe we observe about us at the present time. The line in Figure 23.16b shows the inflation-based model that best fits the microwave observations. The "genetic code" according to this model corresponds to a universe with the following features:

- The overall geometry is flat.

- The total matter density is 27% of the critical density, in agreement with what we infer from the mass-to-light ratio in clusters of galaxies [Section 22.6]

- The density of baryonic matter is 4.4% of the critical density, in agreement with observations of deuterium in the universe [Section 23.3].

- The combination of a flat geometry and a matter density lower than the critical density implies the existence of a repulsive "dark energy" that currently accelerates the expansion, in agreement with observations of distant supernovae [Section 22.6].

- The universe's age should be about 13.7 billion years at the current microwave temperature of 2.73 K, in agreement with what we infer from Hubble's constant [Section 20.4] and the ages of the oldest stars [Section 16.6].

This close correspondence between the genetic code inherent in the universe at an age of 380,000 years and our observations of the present-day universe, some 14 billion years later, is persuasive evidence in favor of the Big Bang in general and inflation in particular.

The Bottom Line

All things considered, inflation does a remarkable job of addressing the problems inherent in the standard Big Bang theory. Many astronomers and physicists believe that some process akin to inflation did affect the early universe, but the details of the interaction between high-energy particle physics and the evolving universe remain unclear. If these details can be worked out successfully, we face an amazing prospect—a breakthrough in our understanding of the very smallest particles, achieved by studying the universe on the largest observable scales.

SPECIAL TOPIC   The End of Time

According to the Big Bang theory, time and space have a beginning in the Big Bang. Do they also have an end? If we live in a re-collapsing universe, the answer seems to be a clear yes. Sometime in the distant future, the universal expansion will cease and reverse, and the universe will eventually come to an end in a fiery *Big Crunch*. However, it appears more likely that the universe will continue to expand forever.

In a critical, coasting, or accelerating universe, the end will come much more gradually. The star–gas–star cycle in galaxies cannot continue forever, because not all the material is recycled. With each generation of stars, more mass becomes locked up in planets, brown dwarfs, white dwarfs, neutron stars, and black holes. Eventually, about a trillion years from now, even the longest-lived stars will burn out, and the galaxies will fade into darkness.

At this point, the only new action in the universe will occur on the rare occasions when two objects—such as two brown dwarfs or two white dwarfs —collide within a galaxy. The vast distances separating star systems in galaxies make such collisions extremely rare. For example, the probability of our Sun (or the white dwarf that it will become) colliding with another star is so small that it would be expected to happen only once in a quadrillion (10^{15}) years. Forever is a long time, however, and even low-probability events will eventually happen many times. If a star system experiences a collision once in a quadrillion years, it will experience about 100 collisions in 100 quadrillion (10^{17}) years. By the time the universe reaches an age of 10^{20} years, star systems will have suffered an average of 100,000 collisions each, making a time-lapse history of any galaxy look like a cosmic game of pinball.

These multiple collisions will severely disrupt galaxies. As in any gravitational encounter, some objects lose energy in such collisions and some gain energy. Objects that gain enough energy will be flung into intergalactic space, where they will drift ever farther from all other objects with the expansion of the universe. Objects that lose energy will eventually fall to the galactic center,

forming a gigantic black hole. The remains of the universe will consist of black holes with masses as great as a trillion solar masses widely separated from a few scattered planets, brown dwarfs, and stellar corpses. If Earth somehow survives, it will be a frozen chunk of rock in the darkness of the expanding universe, billions of light-years from any other solid object.

If grand unified theories are correct, Earth still cannot last forever. These theories predict that protons will eventually fall apart. The predicted lifetime of protons is extremely long: a half-life of at least 10^{32} years. However, if protons really do decay, then by the time the universe is 10^{40} years old Earth and all other atomic matter will have disintegrated into radiation and subatomic particles, such as electrons and neutrinos.

The final phase is predicted to come with the disintegration of the giant, black-hole corpses of galaxies. Black holes are predicted to slowly evaporate through the process of *Hawking radiation*, finally disappearing in brilliant bursts of radiation [Section S4.5]. The largest black holes last longest, but even trillion-solar-mass black holes will evaporate sometime after the universe reaches an age of 10^{100} years. From then on, the universe will consist only of individual photons and subatomic particles, each separated by enormous distances from the others. Nothing new will ever happen, and no events will ever occur that would allow an omniscient observer to distinguish past from future. In a sense, the universe will finally have reached the end of time.

Lest any of this sound depressing, keep in mind that 10^{100} years is an indescribably long time. As an example, imagine that you wanted to write on a piece of paper a number that consisted of a 1 followed by 10^{100} zeros (i.e., the number $10^{10^{100}}$). It sounds easy, but a piece of paper large enough to fit all those zeros *would not fit in the observable universe* today. If that still does not alleviate your concerns, you may be glad to know that a few creative thinkers speculate about ways in which the universe might undergo rebirth, even after the end of time.

23.5 Did the Big Bang Really Happen?

You might occasionally read an article in a newspaper or a magazine questioning whether the Big Bang really happened. We may never be able to prove with absolute certainty that the Big Bang theory is correct. However, no one has come up with any other model of the universe that so successfully explains so much of what we see. As we have discussed, the Big Bang model makes at least two specific predictions that we have observationally verified: the characteristics of the cosmic microwave background and the composition of the universe. It also explains quite naturally many other features of the universe. So far, at least, we know of nothing that is absolutely inconsistent with the Big Bang model.

The Big Bang theory's very success has also made it a target for respected scientists, skeptical nonscientists, and crackpots alike. The nature of scientific work requires that we test established wisdom to make sure it is valid. A sound scientific disproof of the Big Bang theory would be a discovery of great importance. However, stories touted in the news media as disproofs of the Big Bang usually turn out to be disagreements over details rather than fundamental problems that threaten to bring down the whole theory.

Yet scientists must keep refining the theory and tracking down disagreements, because once in a while a small disagreement blossoms into a full-blown scientific revolution.

You don't need to believe all you have read without question. The next time you are musing on the universe's origins, try an experiment for yourself. Go outside on a clear night and look at the sky. Notice how dark it is, and then ask yourself *why* it is dark. If the universe were infinite, unchanging, and everywhere the same, then the entire night sky would blaze as brightly as the Sun. Johannes Kepler [Section 3.4] was perhaps the first person to realize that the night sky in such a universe should be bright. This realization is now called **Olbers' paradox** after Heinrich Olbers, a German astronomer of the 1800s who independently reached the same conclusion.

To understand how Olbers' paradox comes about, imagine that you are in a dense forest on a flat plain. If you look in any direction, you'll likely see a tree. If the forest is small, you might be able to see through some gaps in the trees to the open plains, but larger forests have fewer gaps (Figure 23.17). An infinite forest would have no gaps at all—a tree trunk would block your view along any line of sight.

The universe is like a forest of stars in this respect. In an unchanging universe with an infinite number of stars, we would see a star in every direction, making every point in the sky as bright as the Sun's surface. Obscuring dust doesn't change this conclusion. The intense starlight would

Figure 23.17 Olbers' paradox is similar to the view through a forest. In a large forest (left), you'll see trees no matter where you look. In a small forest (right), you can see open spaces beyond the trees. Similarly, in an unchanging universe with an infinite number of stars, we would expect to see stars everywhere, making the sky bright even in daytime. Because the night sky is dark, the universe must either have a finite number of stars or it must change in a way that prevents us from seeing an infinite number of them.

heat the dust over time until it too glowed like the Sun or evaporated away.

There are only two ways out of this dilemma. Either the universe has a finite number of stars, in which case we would not see a star in every direction, or it changes over time in some way that prevents us from seeing an infinite number of stars. For several centuries after Kepler first recognized the dilemma, astronomers leaned toward the first option. Kepler himself preferred to believe that the universe had a finite number of stars because he thought it had to be finite in space, with some kind of dark wall surrounding everything. Astronomers in the early twentieth century preferred to believe that the universe was infinite in space but that we lived inside a finite collection of stars. They thought of the Milky Way as an island floating in a vast black void. However, subsequent observations showed that galaxies fill all of space more or less uniformly. We are therefore left with the second option: The universe changes over time.

The Big Bang theory solves Olbers' paradox in a particularly simple way. It tells us that we can see only a finite number of stars because the universe began at a particular moment. While the universe may contain an infinite number of stars, we can see only those that lie within our cosmological horizon [Section 20.4]. There are other ways in which the universe could change over time and prevent us from seeing an infinite number of stars, so Olbers' paradox does not *prove* that the universe began with a Big Bang. However, we must have some explanation for why the sky is dark at night, and no explanation besides the Big Bang

also explains so many other observed properties of the universe so well.

THE BIG PICTURE

Putting Chapter 23 into Context

Our "big picture" is now about as complete as it gets. We've discussed the universe from Earth outward, and from the beginning to the end. When you think back on this chapter, keep in mind the following ideas:

- Predicting conditions in the early universe is straightforward. The only real question is how matter and energy behave under such extreme conditions.

- Our current understanding of physics allows us to reconstruct the conditions that prevailed in the universe all the way back to the first 10^{-10} second. Our understanding is less certain back to 10^{-38} second. Beyond 10^{-43} second, we run up against the present limits of human knowledge.

- Although it may sound strange to talk about the universe during its first fraction of a second, our ideas about the Big Bang rest on a solid foundation of observational, experimental, and theoretical evidence. We cannot say with absolute certainty that the Big Bang really happened, but no other model ever proposed has so successfully explained how our universe came to be as it is.

SUMMARY OF KEY CONCEPTS

23.1 Running the Expansion Backward

- *Why do we think we understand the conditions that existed in the early universe?* If we imagine running the expansion backward, we can use fundamental physical principles to predict the temperature and density of the universe at various times in the past. We can then study how matter behaves at various temperatures and densities with laboratory experiments, including experiments in particle accelerators. Today, we have experimental evidence of conditions presumed to exist as early as one ten-billionth of a second after the Big Bang.

23.2 A Scientific History of the Universe

- *What was the universe made of during its earliest moments?* The early universe was filled with radiation and elementary particles. Because reactions involving matter and antimatter can turn matter

into radiation and vice versa, matter and energy were continually converted back and forth during the first few moments, but the total amount of mass-energy remained constant.

- *What key events define the various eras of the universe since the Big Bang?* Gravity became distinct from the other three forces at the end of the Planck era. Inflation may have occurred at the end of the GUT era. Electromagnetism and the weak force became distinct at the end of the electroweak era. Matter particles annihilated all the antimatter particles at the end of the particle era. Fusion of protons and neutrons into helium ceased at the end of the era of nucleosynthesis. Hydrogen nuclei captured all the free electrons, forming hydrogen atoms at the end of the era of nuclei. Galaxies began to form at the end of the era of atoms. The era of galaxies continues to this day.

continued ▶

23.3 Evidence for the Big Bang

- *What two key lines of evidence support the Big Bang model?* The Big Bang model predicts (1) the existence and characteristics of the cosmic microwave background and (2) the expected helium abundance in the cosmos. Both predictions agree with observations.

- *What is the cosmic microwave background?* It is microwave radiation that comes from all directions in space and thus seems to be the radiation of the universe itself. Its spectrum matches the characteristics expected of the radiation released at the end of the era of nuclei, providing spectacular confirmation of a key prediction of the Big Bang theory.

- *How does the Big Bang theory predict the helium abundance of the universe?* The theory predicts the temperature and density of the early universe. We can use these conditions to predict how many neutrons were in the universe when they fused with protons to make helium. The prediction matches observations of the cosmic helium abundance.

- *Why is the abundance of cosmic deuterium evidence in favor of dark matter being made of WIMPs?* We can use the abundance of deuterium to estimate the density of ordinary (baryonic) matter in the universe. Because this density turns out to be smaller than the overall matter density of the universe, we conclude that at least some—and probably most—of the matter in the universe must be extraordinary (nonbaryonic) matter, such as WIMPs.

23.4 Inflation

- *What do we mean by "inflation" in the early universe?* Inflation refers to a dramatic growth in the size of the universe, thought to have been driven by energy released as the strong force froze out at the end of the GUT era. The universe is thought to have grown by a factor of 10^{30} in less than 10^{-36} second.

- *What three questions were left unanswered by the Big Bang model prior to the inclusion of inflation in the model?* (1) Where does structure come from? (2) Why is the large-scale universe so smooth? (3) Why is the density of the universe almost critical?

- *How does inflation answer these three questions?* (1) Structure arose because the enormous growth of inflation stretched random quantum fluctuations into seeds for the density enhancements around which structure formed. (2) The universe is smooth on large scales because, prior to inflation, everything we can observe today was close enough together for temperatures and densities to equalize. (3) The growth that occurred with inflation ensures that the universe is geometrically flat, with a matter density close to the critical density for most of its history.

- *Is the overall density of mass-energy in the universe really equal to the critical density, as predicted by inflation?* The total density of matter in the universe appears to be about 30% of the critical density. However, detailed studies of the cosmic microwave background suggest that the universe is indeed flat, as predicted by inflation. In that case, the remaining density of mass plus energy needed to reach the critical density must exist in the universe in the form of energy—presumably as the mysterious energy driving the acceleration of the universe's expansion.

23.5 Did the Big Bang Really Happen?

- *What is Olbers' paradox?* If the universe were infinite, unchanging, and filled with stars, the sky would be everywhere as bright as the surface of the Sun, and it would not be dark at night.

- *How does the Big Bang theory solve Olbers' paradox?* The Big Bang theory tells us that the sky is dark at night because the universe changes over time. Because the universe has a finite age, we can see only a finite number of stars in the sky, which is why the sky is not everywhere bright.

❓ True Statements?

Decide whether each of the following statements is true and explain why it is or is not.

1. According to the Big Bang theory, the universe's temperature was greater than 10 billion (10^{10}) K when the universe was less than one ten-billionth (10^{-10}) of a second old.

2. Although the universe today appears to be made mostly of matter and not antimatter, the Big Bang theory suggests that the early universe had nearly equal amounts of matter and antimatter.

3. According to the Big Bang theory, the cosmic microwave background was created when energetic photons ionized the neutral hydrogen atoms that originally filled the universe.

4. While the existence of the cosmic microwave background is consistent with the Big Bang theory, it is also easily explained by assuming that it comes from individual stars and galaxies.

5. According to the Big Bang theory, most of the helium in the universe was created by nuclear fusion in the cores of stars.

6. The theory of inflation suggests that the structure in the universe today may have originated as tiny quantum fluctuations.

7. Within the next decade, observations of the cosmic microwave background will prove definitively whether inflation really occurred in the early universe.

8. The fact that the night sky is dark tells us that the universe cannot be infinite, unchanging, and everywhere the same.

Problems

9. *Life Story of a Proton.* Tell the life story of a proton from its formation shortly after the Big Bang to its presence in the nucleus of an oxygen atom you have just inhaled. Your story should be creative and imaginative, but it should also demonstrate your scientific understanding of as many stages in the proton's life as possible. You can draw on material from the entire book, and your story should be three to five pages long.

10. *The Big Bang.* Explain what we mean by the Big Bang theory.

11. *Antimatter.* What is antimatter? How were particle–antiparticle pairs created in the early universe? How were they destroyed?

12. *The Earliest Moment.* What are the Planck era and the Planck time? Why can't our current theories describe the history of the universe during the Planck era?

13. *Primordial Nucleosynthesis.* How long did the era of nucleosynthesis last? Explain why this era was so important in determining the chemical composition of the universe forever after.

14. *Origin of the Cosmic Microwave Background.* Briefly explain why radiation was trapped for 380,000 years during the era of nuclei and why the cosmic microwave background broke free at the end of this era.

15. *Later Eras.* What do we mean by the era of atoms and the era of galaxies? What era do we live in? Explain.

16. *Observations of the Cosmic Microwave Background.* Briefly describe how the cosmic microwave background was discovered. Explain why the Big Bang theory predicts that the cosmic microwave background has a perfect thermal radiation spectrum. How did the Cosmic Background Explorer (COBE) support this prediction of the Big Bang theory? What is the temperature of the cosmic microwave background?

*17. *Properties of the Cosmic Microwave Background.* The cosmic microwave background first began to stream freely across the universe at the end of the era of nuclei, when the universe was 380,000 years old and its temperature was about 3,000 K. Since that time, the universe has expanded by a factor of about 1,000.

   a. Using Wien's law (see Mathematical Insight 6.2), calculate the wavelength of maximum intensity for thermal radiation with a temperature of 3,000 K.

   b. The observed temperature of the cosmic microwave background today is about 2.73 K. Use Wien's law to calculate the wavelength of maximum intensity for the cosmic microwave background today.

   c. How much longer are the wavelengths of the cosmic microwave background today than they were at the time the radiation was first released? Is this change in wavelength consistent with the idea that the universe has expanded by a factor of about 1,000 since the end of the era of nuclei? Explain. (*Hint:* Look back at the discussion of cosmological redshifts in Chapter 20.)

18. *Testing Inflation.* Explain how inflation is being tested by astronomers.

*19. 10^{100} *Years.* In the box "The End of Time," we found that the final stage in the history of a critical, coasting, or accelerating universe will come about 10^{100} years from now. Such a large number is easy to write but difficult to understand. This problem investigates some of the incredible properties of very large numbers.

a. The current age of the universe is around 10^{10} years. How much longer is a trillion years than this current age? How much longer is 10^{15} years? 10^{20} years?

b. Suppose protons decay with a half-life of 10^{32} years. When will the number of remaining protons be half the current amount? When will it be a quarter of its current amount? How many half-lives will have gone by when the universe reaches an age of 10^{34} years? What fraction of the original protons will remain at this time? Based on your answers, is it reasonable to conclude that *all* protons in today's universe will be gone by the time the universe is 10^{40} years old? Explain.

c. Suppose you were trying to write 10^{100} zeros on a piece of paper and could write microscopically, so that each zero (including the thickness of the pencil mark) occupied a volume of 1 cubic micrometer—about the size of a bacterium. Could 10^{100} zeros of this size fit in the observable universe? Explain. (*Hints:* Calculate the volume of the observable universe in cubic micrometers by assuming it is a sphere with a radius of 15 billion light-years. The volume of a sphere is $4/3 \times \pi \times$ (radius)^3; 1 light-year $\approx 10^{15}$ meters; 1 cubic meter $= 10^{18}$ cubic micrometers.)

Discussion Questions

20. *The Moment of Creation.* You've probably noticed that, in discussing the Big Bang theory, we never quite talk about the first instant. Even our most speculative theories at present take us back only to within 10^{-43} second of creation. Do you think it will *ever* be possible for science to consider the moment of creation itself? Will we ever be able to answer questions such as *why* the Big Bang happened? Defend your opinions.

21. *The Big Bang.* How convincing do you find the evidence for the Big Bang model of the universe's origin? What are the strengths of the theory? What does it fail to explain? Overall, do *you* think the Big Bang really happened? Defend your opinion.

22. *The End of Time.* According to our current understanding, the universe as we know it will eventually come to an end—either in a fiery "Big Crunch" or in a slow death as all the stars eventually die. However, some people speculate that the universe might then undergo some type of rebirth. In the case of a recollapsing universe, the rebirth might consist of a new Big Bang. Some people even speculate that the universe might repeatedly oscillate through cycles of Big Bangs and Big Crunches. Discuss some ideas about how the universe might undergo rebirth if it continues to expand forever. (*Note:* You may wish to read Isaac Asimov's short story *The Last Question,* published in 1956.)

23. *Forever.* If you could live forever, would you choose to do so? Suppose the universe will keep expanding forever, as the bulk of observations currently suggest. How would you satisfy your energy needs as time goes on? How would you grow food? What would you do with all that time?

MEDIA EXPLORATIONS

For a complete list of media resources available, go to www.astronomyplace.com and choose Chapter 23 from the pull-down menu.

 Astronomy Place Web Tutorials

Tutorial Review of Key Concepts

Use the following interactive **Tutorial** at www.astronomyplace. com to review key concepts from this chapter.

Hubble's Law Tutorial

Lesson 1 Hubble's Law

Lesson 2 The Expansion of the Universe

Lesson 3 The Age of the Universe

Supplementary Tutorial Exercises

Use the interactive **Tutorial Lesson** to explore the following questions.

Hubble's Law Tutorial, Lesson 2

1. How does the observed rate of expansion allow us to determine distances between galaxies in the distant past?

2. Suppose we have an uncertainty of 10% in the rate of expansion. Would this cause significant changes in what we conclude about temperatures in the early universe? Why or why not?

Movies

Check out the following narrated and animated short documentary available on www.astronomyplace.com for a helpful review of key ideas covered in this chapter.

From the Big Bang to the Galaxies Movie

Web Projects

Take advantage of the useful Web links on www.astronomyplace. com to assist you with the following projects.

1. *New Tests of the Big Bang Theory.* The Cosmic Background Explorer (COBE) satellite provided striking confirmation of several predictions of the Big Bang theory. New satellites will test the Big Bang theory further, primarily by observing the subtle variations in the cosmic microwave background with a much higher sensitivity than COBE. Use the Web to gather pictures and information about the COBE mission and its successors, such as WMAP or Planck. Write a one-to two-page report about the strength of the evidence compiled by COBE and WMAP and how much more we might learn from upcoming missions.

2. *Decay of the Proton.* One of the most startling predictions of grand unified theories is that protons will eventually decay, albeit with a half-life of more than 10^{32} years. If this is true, it may be possible to observe an occasional proton decay, despite the extraordinarily long half-life. Several experiments to look for proton decay are under way or being planned. Find out about one or more of these experiments, and write a one- to two-page summary in which you describe the experiment(s), any results to date, and what these results (or potential results) mean to the grand unified theories.

3. *New Ideas in Inflation.* The idea of inflation solves many of the puzzles associated with the standard Big Bang theory, but we are still a long way from finding evidence confirming that inflation really occurred. Find recent articles that discuss some of the latest ideas about inflation and how we might test these ideas. Write a two- to three-page summary of your findings.

24 Life Beyond Earth

Prospects for Microbes, Civilizations, and Interstellar Travel

We, this people, on a small and lonely planet
Travelling through casual space
Past aloof stars, across the way of indifferent suns
To a destination where all signs tell us
It is possible and imperative that we learn
A brave and startling truth.

Maya Angelou
Excerpted from **A Brave and Startling Truth**

We have nearly completed our survey of the cosmos and our place within it. We have explored the nature of modern astronomy and the tools and methods we use to learn about the universe. We have discussed our own solar system in some depth, studying the other planets and what they teach us about our own. We have investigated the wide variety of structures in the universe and the history of the universe from the beginning of time to the present. We have even considered the future of our Sun and of the entire universe.

However, we have not yet discussed one of the most profound questions of all: Are we alone? The universe seems to be filled with worlds beyond imagination—more than 100 billion star systems in our galaxy alone, and some 100 billion galaxies in the observable universe—but we do not yet know whether any other world has ever been home to life.

In this chapter, we will discuss the possibility of life beyond Earth. We'll begin by discussing scientific interest in this topic. We'll consider the possibility of finding microbial life elsewhere in our solar system or beyond, and we'll examine efforts to search for extraterrestrial intelligence in other star systems (SETI). Finally, we'll explore how we might someday travel the great expanses of interstellar space and discuss the astonishing implications of the question of whether anyone else has already achieved this ability.

24.1 The Possibility of Life Beyond Earth

It may seem that aliens are everywhere. Television starships like *Enterprise* or *Voyager* are on constant prowl throughout the galaxy, seeking out new life and hoping it speaks English. In *Star Wars*, aliens from many planets gather at bars to share drinks and stories—and presumably to marvel at the fact that they all share a level of technology more similar than that shared by different nations on Earth. Closer to home, supermarket tabloids routinely carry headlines about the latest alien atrocities or about alien corpses hidden by the government at "area 51."

Despite their media popularity, any alien visitors have been sadly negligent in leaving scientific evidence of their trespass. Decades of scientific observation and study have not turned up a single piece of undeniable evidence that aliens have been here (see the box "Are Aliens Already Here?" on page 744). Scientists therefore are deeply skeptical of claims of intelligent visitors from outer space. Nevertheless, scientific interest in aliens of other types is on the rise, and the possibility of either microbial or intelligent life elsewhere is a hot topic of current research.

A Brief History of Ideas About Life Beyond Earth

Interest in life beyond Earth goes far back in human history. Many ancient cultures imagined beings living among the constellations, often treating such beings as gods. The ancient Greeks took such ideas a step further, debating whether other worlds like Earth and other beings like us might exist. Until quite recently, however, these ideas remained purely speculative, because we had no way to study the question of life beyond Earth scientifically.

The character of the debate began to change with the Copernican revolution [Section 3.4]. Once it became clear that the Moon and the planets really were other *worlds*, not mere lights in the sky, the idea of life beyond Earth became scientifically reasonable. Numerous scientists and philosophers of the seventeenth and eighteenth centuries considered it a near certainty that other worlds were home to life, including intelligent beings. Kepler suggested that the Moon was inhabited. William Herschel, discoverer of Uranus, spoke to fellow scientists about the physical conditions affecting the inhabitants of Mars. Indeed, Herschel imagined not only that all the planets were inhabited, but that the Sun harbored life as well.

By the beginning of the twentieth century, Percival Lowell's claims of Martian canals [Section 10.5] led many among the public to believe that Mars was home to an advanced but dying civilization. In 1938, a famous radio broadcast of H. G. Wells's novel *The War of the Worlds* created a panic when many listeners thought Earth really was under Martian attack (Figure 24.1). Venus, too, was often imagined as a home to life. Because Venus is closer to the Sun than Earth, some people guessed that beneath its clouds we would find a tropical paradise.

Hopes of finding life on our planetary neighbors were dashed by the bleak images of Mars returned by spacecraft and the discovery of the runaway greenhouse effect on Venus. For a few decades, scientific interest in extraterrestrial life waned. Many scientists began to think that Earth might be unique in having conditions for life, at least within our solar system.

New Discoveries and New Understanding

The pendulum has recently begun to swing back with a renewal of scientific interest in the possibility of life elsewhere. This new interest has been spurred by several important advances in our scientific understanding of the universe.

We now understand the conditions on other worlds within our solar system much better than we did a few decades ago. We have found that at least some worlds may have conditions—such as the presence of liquid water—that might allow life to survive. Looking more widely at the universe, discoveries about the lives of stars have taught us that planet formation often should be part of the star formation process. This theoretical understanding, along with recent discoveries of numerous extrasolar planets [Section 9.6], suggests that planets may be quite common in the universe. If planetary systems are common, then our galaxy alone might have billions of **habitable planets**—planets that are at least potentially capable of supporting life.

Of course, there's a great difference between finding planets that are *potential* homes for life and finding planets that actually have life. If life can arise only under extremely rare conditions, then it might be extremely rare even if habitable planets are common. However, three recent developments in biology and in the study of the origin of life on Earth may suggest that life can arise and survive with relative ease (Figure 24.2):

- We have learned that organic molecules—the building blocks of life on Earth—form easily and naturally under a wide range of conditions. We find organic molecules in places as diverse as meteorites that once inhabited the asteroid belt, comets still orbiting the Sun, the atmospheres of the jovian planets (Jupiter, Saturn, Uranus, and Neptune), and even clouds of gas between the stars. Moreover, laboratory experiments suggest that the chemical constituents that were common on the early Earth combine readily into complex organic molecules [Section 14.5]. Thus, the building blocks of life ought to be present on many planets and moons, both within our own solar system and beyond.

- Geological evidence suggests that life on Earth arose almost as early as conditions allowed. Study of ancient rocks shows that life was almost certainly widespread by 3.5 billion years ago, and some evidence pushes this date to earlier than 3.85 billion years ago [Section 14.5]. Moreover, the many large impacts that occurred during the heavy bombardment early in the solar system's history probably precluded life from taking permanent hold on Earth much before about 4.0 billion years ago. Thus, life arose on Earth within a period of no more than a few hundred million years, and possibly within a much shorter period. The relatively quick appearance of life on Earth suggests that the processes that led to life were not that difficult, because a difficult pro-

Figure 24.1 This front-page story from *The New York Times* describes the panic caused by a 1938 broadcast of *The War of the Worlds*.

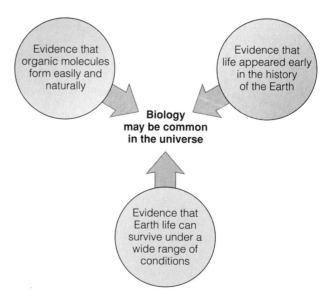

Figure 24.2 Three important lines of evidence suggest that biology may be quite common in the universe.

cess probably would have required much more time. If so, it's reasonable to suppose that the same processes have occurred over similar time periods on many other worlds.

- Studies of living organisms show that life can survive under a much broader range of conditions than we realized a few decades ago. For example, we now know that life exists in extremely hot water near deep-sea volcanic vents (see Figure 14.23), in the frigid conditions of Antarctica, and inside rocks buried a kilometer or more beneath Earth's surface. These examples suggest that the range of "right" conditions for life may be quite broad, in which case it might be possible to find life even on planets that are quite different in character from Earth.

In summary, we have no reason to think that life ought to be rare and several reasons to expect it to be common. If life is indeed common, studying it will give us new insights into life on Earth, even if we don't find other intelligent civilizations. These enticing prospects have captured the interest of scientists from many disciplines and many nations, giving birth to a new science devoted to the study of life in the universe. Sometimes called *astrobiology* or *exobiology*, this new science explores the origin of life, the conditions under which life is possible, and the prospects for finding life beyond Earth.

THINK ABOUT IT

The conclusion that life may be common in the universe rests on all three lines of evidence described above. How strong do you consider each line of evidence? Which line represents the "weakest link"? Overall, do you believe the current evidence is strong enough to warrant optimism about finding life elsewhere? Defend your opinion.

24.2 Life in the Solar System

If it's reasonable to assume that life may be common in the universe, then the easiest place to start a search is within our own solar system. Several worlds in our solar system are candidates for being habitable, but we don't yet know for sure whether any of them really have conditions under which life could have arisen and survived. The most likely candidates are Mars and a few of the moons that orbit the jovian planets, most notably Europa [Section 12.5].

Before we discuss the possibilities for life on these worlds, it's important for us to remember what we are looking for. Any life that exists elsewhere in our solar system is unlikely to be anything with which you could carry on a conversation. We now have sufficiently detailed images of Mars to be quite confident that no alien civilizations have ever existed there. The same holds true for every other candidate for harboring life in our solar system. Thus, the search for life in our solar system is primar-

ily a search for microbial life, not for large or intelligent beings.

In addition, our discussion will be based on the assumption that life elsewhere would be at least a little bit like life on Earth. For example, we will assume that life requires both liquid water and energy either from the Sun or from the world's internal heat. These requirements may seem rather limiting, especially since science fiction writers have imagined all sorts of bizarre life-forms existing under conditions far outside those we'll consider here. Nevertheless, a search for life must start somewhere, and it makes sense to begin by looking for "life as we know it." If this initial search turns out to be too narrow, we can always expand it in the future.

THINK ABOUT IT

Do you think that liquid water (or some other liquid) really is required for life? Why or why not?

Life on Mars

We now know that Percival Lowell's visions of a Martian civilization were mistaken, but more recent and real discoveries make Mars a good candidate for past or present microbial life. Two major factors explain scientific interest in the possibility of life on Mars:

1. Although the surface of Mars is dry and frozen today, early in its history Mars may have had surface conditions quite similar to those under which life arose on the early Earth. Numerous geological features of Mars appear to have been carved in the distant past by running water [Section 10.5], indicating that Mars had one or more periods during which it was warm and wet. The young Mars also had all the chemical ingredients needed for life, as well as energy both from sunlight and from now-dormant volcanoes. If early Mars really was similar to early Earth, then it is reasonable to imagine that life could have arisen there as well.

2. Recent observations suggest that Mars has significant amounts of subsurface water ice (see Figure 10.31). Because Mars may still have some volcanic heat, pockets of underground liquid water may exist. If so, life might still survive on Mars today, perhaps looking much like microbes that live deep underground on Earth.

Mars is not only the best candidate in our solar system for life beyond Earth but also the only place where we've begun an actual search for life. Our first attempt to search for life on Mars came with the Viking missions to Mars in the 1970s. More recently, scientists have debated the origin of odd structures found in meteorites from Mars. Let's briefly examine what we've learned from these studies.

The Viking Experiments Two Viking landers arrived on the surface of Mars in 1976. Each was equipped with a robotic arm for scooping up soil samples, which were fed into several on-board, robotically controlled experiments. The robotic arms even pushed aside rocks to get at shaded soil less likely to have been sterilized by ultraviolet light from the Sun (Figure 24.3).

Three of the Viking experiments were designed expressly to look for signs of life. None of these experiments could actually "see" life but rather looked for chemical changes that could be attributed to the respiration or metabolism of living organisms. Although all three experiments gave results that initially seemed consistent with life, further study suggested that chemical reactions could have produced the same results. Moreover, a fourth experiment, which analyzed the content of Martian soil, found no measurable level of organic molecules—the opposite of what we would expect if life were present. As a result, most scientists have concluded that the Viking results were inconsistent with the presence of life.

Despite the negative results from Viking, life still could exist on Mars. The Viking landers sampled only two rather bland locations on the planet and tested soils only very

near the surface. Life might be hiding elsewhere or underground on Mars, or life may have existed in the past and become extinct.

The Debate over Martian Meteorites More recently, some scientists claim to have found evidence of life in a Martian meteorite—one of the couple dozen known meteorites whose chemical composition suggests they came from Mars [Section 13.3]. The particular meteorite in question, designated ALH84001, was found in Antarctica in 1984 (Figure 24.4). Careful study of the meteorite shows that it landed in Antarctica about 13,000 years ago, following a 16-million-year journey through space after being blasted from Mars by an impact. The rock itself dates to 4.5 billion years ago, indicating that it solidified shortly after Mars formed and therefore resided on Mars throughout the times when Mars may have been warmer and wetter.

Painstaking analysis of the meteorite reveals several lines of evidence that could indicate the past presence of life on Mars. For example, the rock contains layered carbonate minerals and a type of complex organic molecule (polycyclic aromatic hydrocarbons, or PAHs) that are associated with life when found in Earth rocks. The rocks also contain microscopic chains of magnetite crystals that look quite similar to chains made in Earth rocks by living bacteria (Figure 24.5).

Most intriguingly, highly magnified images reveal rod-shaped structures that look much like recently discovered nanobacteria on Earth (Figure 24.6). The terrestrial nanobacteria are about a hundred times smaller than ordinary bacteria, and some biologists question whether they truly are living organisms. However, they appear to contain DNA, which suggests that they are indeed a form of life. If so, could the similar-looking structures in the Martian meteorite be fossil life from Mars?

It's possible, but each of the tantalizing hints of Martian life can also be explained in a nonbiological way. Subsequent studies have shown that chemical and geological

a This photograph shows a working model of the Viking landers, identical to those that landed on Mars in 1976. It is on display at the National Air and Space Museum in Washington, D.C.

b This pair of before-and-after photos, transmitted back to Earth by the *Viking 2* lander, shows where the robotic arm pushed away a small rock on the Martian surface.

Figure 24.3 Two Viking landers searched for life on Mars in the 1970s.

Figure 24.4 The Martian meteorite ALH84001, before it was cut open for detailed study. The small block shown for scale to the lower right measures 1 cubic centimeter, about the size of a typical sugar cube.

a Microscopic chains of magnetite crystals produced by bacteria on Earth.

b Similar chains of magnetite crystals found in the carbonate globules of Martian meteorite ALH84001. The similarity to the terrestrial chains has been cited as evidence of life on Mars.

Figure 24.5 Mineral evidence of life on Mars?

processes can produce structures very similar to those found in the Martian meteorite. In addition, terrestrial bacteria have been found living inside the meteorite, indicating that it was contaminated by Earth life during the 13,000 years it resided in Antarctica. This contamination may explain the presence of the complex molecules found in the rock.

On balance, most scientists now doubt that the Martian meteorite shows true evidence of Martian life. Never-theless, studies of ALH84001 and other Martian meteorites are continuing, and they may yet turn up surprises.

The Martian meteorites remind us of the possibility that living organisms have traveled between the planets by hitching rides on rocks blasted skyward by impacts. Suppose we someday discover living organisms on Mars. How will we be able to tell whether these organisms arose independently of Earth life or share a common ancestor with life on Earth?

Continuing the Search on Mars The debate over the Martian meteorite shows that we have at least some potential to search for life on Mars without ever leaving home. However, space missions offer by far our best hope of learning whether Mars ever had life. If all goes well, one new search for life will be getting under way by the time you read this book. The British-built lander *Beagle 2* is scheduled to reach the Martian surface in December 2003. Robotic equipment on the lander will drill into the surface and conduct experiments to search for life.

Within a decade or so, NASA hopes to launch a mission to Mars that will bring back surface samples for study in laboratories on Earth. Later, we may send humans to Mars, where they could search for fossils or living organisms in deep canyons like Valles Marineris, in ancient valley bottoms and dried-up lake beds, or in underground pockets of water near not-quite-dead volcanoes. The search will not be easy, but we should eventually learn whether life has ever existed on Mars.

Life on Jovian Moons

After Mars, the next most likely candidates for life in our solar system are some of the moons of the jovian planets—especially Jupiter's moons Europa, Ganymede, and Callisto and Saturn's moon Titan (Figure 24.7). These moons are

a This photo shows rod-shaped structures found in the carbonate globules of ALH84001. They measure about 100 nanometers in length and are as small as 10–20 nanometers in width.

b This photo shows terrestrial nanobacteria in a sample of volcanic rock from Sicily. They are close in size to the structures seen in ALH84001. The scale bar at the bottom is 1 micrometer, or 1,000 nanometers.

Figure 24.6 Does Martian meteorite ALH84001 contain fossils of Martian organisms?

Europa   Ganymede   Callisto   Titan

Figure 24.7 These moons are all candidates for life in our solar system. Europa, Ganymede, and Callisto are moons of Jupiter. Titan is a moon of Saturn.

all large enough that they would be considered planets if they orbited the Sun independently.

The strongest of these candidates for harboring life is Jupiter's moon Europa, on which tidal heating probably creates a deep ocean beneath an icy crust (although we are not yet certain that the ocean exists) [Section 12.5]. The ice and rock from which Europa formed undoubtedly included the necessary chemical ingredients for life, and Europa's internal heating (primarily due to tidal heating) is strong enough to power volcanic vents on the sea bottom. Thus, it's easy to imagine places on Europa's ocean floor that look much like deep-sea vents on Earth. Because many biologists suspect that life on Earth first arose near such undersea volcanic vents [Section 14.5], Europa would seem to have everything needed for an origin of life.

The possibility of life on Europa is especially interesting because, unlike any potential life on Mars, it would not necessarily have to be microscopic. After all, the several kilometers of surface ice that hide the ocean (if it exists) could also hide large creatures swimming within it. However, potential energy sources for life on Europa are far more limited than the energy sources for life on Earth (mainly because sunlight could not fuel photosynthesis in the subsurface ocean). As a result, most scientists suspect that any life that might exist on Europa would probably be quite small and primitive.

Some evidence suggests that Jupiter's moons Ganymede and Callisto may also have subsurface oceans. However, these moons lack significant tidal heating and therefore probably have even less energy for life than Europa. If they have life at all, it is almost certainly small and primitive. Nevertheless, Europa, Ganymede, and Callisto offer the astonishing possibility that Jupiter alone could be orbited by more worlds with life than we find in all the rest of the solar system.

Another enticing place to look for life is on Saturn's moon Titan. Titan's surface is far too cold for liquid water, but it may have lakes or oceans of liquid ethane and methane on its surface [Section 12.5]. Although we usually think of liquid water as a requirement for life, perhaps other liquids could play a similar role by acting as a medium in which organisms could transport chemicals needed for metabolism and facilitating chemical reactions needed for life. Most biologists consider life based on other liquids an unlikely possibility. Titan also appears to have many other organic molecules on its surface, and, at the very least, Titan probably offers an incredible natural laboratory of interesting organic chemistry.

Moreover, we can't completely rule out water-based life on Titan, even if it seems unlikely. While liquid water does not exist on Titan's surface today, pockets of liquid water may have existed in the past following impacts of asteroids or comets. The heat of a large impact could have melted water ice and kept it liquid for as long as a few thousand years. If water-based life arose in such temporary pockets of water, it might have found a way to survive as the water froze, perhaps by migrating deep underground where Titan's internal heat might still keep some water liquid.

Life in Other Places in Our Solar System

Some people have speculated about finding life elsewhere in the solar system, but the prospects for success fall rapidly after Mars and the four moons we've discussed. The primary reason for pessimism is that we believe that water or some other liquid is required to support life.

The Moon and Mercury have very little water at all, largely because of the way both were affected by giant impacts, which would have vaporized water and other ices that may once have been present [Section 9.4]. The runaway greenhouse effect on Venus caused it to lose most or all of the water it once had [Section 11.6], and in any case Venus seems far too hot to support life. Many moons of the outer solar system are ice-rich, but aside from Europa, Ganymede, Callisto, and Titan, they lack any source of heat that could allow these ices to melt into liquid form. Asteroids, comets, and Pluto similarly must remain in a perpetual state of deep freeze.

That leaves the jovian planets themselves as potential candidates for life. Because these planets are cold at their cloudtops but hot inside, each must have a narrow region in which temperatures are "just right" for liquid water. Indeed, Jupiter and Saturn each have layers of clouds made from droplets of liquid water [Section 12.3]. As a result, some scientists have speculated about life floating at these "just right" depths in the jovian planet atmospheres. Unfortunately, this idea appears to have a fatal problem: strong vertical winds.

Consider the case of Jupiter (similar ideas probably apply to Saturn, Uranus, and Neptune). Vertical winds would quickly carry any complex organic molecules that might form in Jupiter's atmosphere to depths at which the heat would destroy them, making it difficult to see how any life could arise indigenously. We might imagine microbes reaching Jupiter from elsewhere on meteorites, but again the vertical winds make their survival seem impossible. Such microbes would quickly be thrown onto a nonstop elevator ride between some cloud layers that are unbearably cold and others that are insufferably hot.

The only plausible way that life could survive in Jupiter's atmosphere is if it somehow had a buoyancy that allowed it to stay at the right altitude while the vertical winds rushed by it. However, such buoyancy would require large gas-filled sacs, making the organisms themselves enormous. Given that we cannot envision a way for microbes to survive, we can't imagine any way for large, buoyant organisms to evolve in the first place. They might survive if they arrived on Jupiter from elsewhere, but we know of no way such large organisms could survive a journey through space, even if they existed on other worlds.

24.3 Life Around Other Stars

We already know of planets orbiting more than 100 stars besides our Sun, and it's quite likely that billions of planetary systems inhabit our galaxy alone. These numbers might immediately make prospects for life elsewhere seem quite good, but numbers alone don't tell the whole story. In this section, we'll consider the prospects for life on worlds orbiting other stars.

Before we begin, we must distinguish between *surface* life like that on Earth and *subsurface* life like that we envision as a possibility on Mars or Europa. Although life is life and the discovery of any life elsewhere would be important, there are practical considerations. While large telescopes could in principle allow us to discover surface life on distant planets, no foreseeable technology will allow us to find life that is hidden deep underground in other star systems (unless the subsurface life has a noticeable effect on the planet's atmosphere). We therefore will focus on the search for life on planets with habitable surfaces—surfaces with temperatures and pressures that could allow liquid water to exist.

Which Stars Would Make Good Suns?

Before we consider planets themselves, it's useful to ask how many stars have any chance of having planets with life. In other words, which stars would make good "Suns," providing heat and light to the surfaces of terrestrial planets that happen to orbit them?

The first requirement for a star to have life-bearing worlds is that it be old enough that life could have arisen. More massive stars live shorter lives, and the most massive stars live no more than a few million years [Section 16.5]. Given that life on Earth did not arise for hundreds of millions of years after our solar system was born, we can rule out any star with more than a few times the mass of our Sun. However, because less massive stars are far more common than more massive stars, the lifetime constraint rules out only about 1% of all stars.

A second requirement is that the star allow planets to have stable orbits. About half of all stars are in binary or multiple star systems, in which stable planetary orbits are less likely than around single stars. If life is not possible in such systems, then we can rule out half the stars in our galaxy as potential homes to life. Of course, the other half—still some 100 billion stars or more—remain possible homes for life. Moreover, under some circumstances stable planetary orbits are possible in multiple star systems, so we shouldn't entirely rule out life in such systems.

A third constraint on the likelihood of finding habitable planets is the size of a star's **habitable zone**—the region in which a terrestrial planet of the right size could have a surface temperature that might allow for liquid water and life. Figure 24.8 shows the approximate sizes, to scale, of the habitable zones around our Sun, around a star with about half the mass of the Sun (spectral type K), and around a star with about $\frac{1}{10}$ the mass of our Sun (spectral type M). Although habitable planets seem possible in all three cases, the smaller size of the habitable zones around the less massive stars makes it less likely that suitable planets would have formed in these regions.

All in all, it seems that the vast majority of stars are at least potentially capable of having life-bearing planets. Moreover, even very conservative assumptions leave billions of possibilities. For example, limiting the search for habitable planets to stars very similar to our Sun (that is, spectral type G) would still mean billions of potential other Suns in the Milky Way.

Finding Habitable Planets

All the extrasolar planets found to date are closer in size to Jupiter than to Earth. Like Jupiter, they probably lack solid surfaces and cannot harbor life, although it is conceivable that they could be orbited by moons that have life. If we hope to find habitable planets around other stars, we need ways of identifying planets (or moons) the size of Earth or smaller.

Finding Earth-size planets is a daunting technological challenge. To see why, think back to the scale of the solar system we discussed in Chapter 1. On that scale, Earth is the

Figure 24.8 The approximate habitable zones around our Sun, a star with half the mass of the Sun (spectral type K), and a star with $\frac{1}{10}$ the mass of the Sun (spectral type M), shown to scale. The habitable zone becomes increasingly smaller and closer-in for stars of lower mass and luminosity.

size of a pinhead orbiting just 15 meters from a grapefruit-size Sun, and even the nearest stars are thousands of kilometers away (see Figure 1.7). Thus, looking for an Earth-like planet around a nearby star is like standing on the east coast of the United States and looking for a pinhead on the west coast. Moreover, even with a telescope powerful enough to detect such a tiny planet, its light is likely to be drowned out by the billions-of-times-brighter light of the star it orbits.

Nevertheless, advancing technology should soon put Earth-like planets within our telescopic reach. The Kepler mission, scheduled for launch in 2007, will look for planets that happen to orbit edge-on as seen from Earth [Section 9.6]. Because these planets periodically cross in front of their stars, Kepler will look for telltale dips in their stars' brightness during such eclipses (called *transits*). Scientists hope that Kepler will detect hundreds of Earth-size planets.

The orbital properties of planets detected by Kepler will tell us whether they lie within their stars' habitable zones. We will still need actual images or spectra to determine whether the planets really are habitable or have life. Scientists are actively working on technologies that may provide such data. If all goes well, within about a decade NASA hopes to launch the *Terrestrial Planet Finder (TPF)*, an orbiting interferometer (several telescopes working together as one [Section 7.5]). It will be capable of obtaining low-resolution spectra and crude images (a few pixels) of Earth-like planets around nearby stars (Figure 24.9).

Astronomers hope to deploy even more powerful interferometers in later decades, either in space or on the Moon. Within the lifetimes of today's college students, we could conceivably have optical interferometers with dozens of telescopes spread across hundreds of kilometers. Such telescopes will be able not only to detect Earth-like planets around nearby stars but also to obtain fairly clear images and high-resolution spectra of those planets.

Signatures of Life

The images from future telescopes may tell us whether the planets have continents and oceans like Earth and perhaps will even allow us to monitor seasonal changes. The spectra from future telescopes should prove even more important to the search for life. Moderate-resolution infrared spectra can reveal the presence and abundance of many atmospheric gases, including carbon dioxide, ozone, methane, and water vapor (Figure 24.10). Careful analysis of atmospheric makeup might tell us whether a planet has life.

On Earth, for example, the large abundance of oxygen is a direct result of photosynthetic life [Section 14.5]. Abundant oxygen in the atmosphere of a distant world might similarly indicate the presence of life, since we know of no non-biological way to produce an oxygen abundance as high as Earth's. Other evidence might come from the ratio of oxygen to the other detected gases. Scientists involved in the search for life in other planetary systems are working to improve our understanding of how life influences atmospheric chemistry so we can recognize the particular gas combinations that are unmistakable signatures of life.

Are Earth-like Planets Rare or Common?

We will not know for certain whether Earth-like planets exist or how common such planets may be until we survey many star systems with telescopes capable of detecting

Figure 24.9 This artist's rendering shows one possible configuration for NASA's planned Terrestrial Planet Finder (TPF). It consists of four separate telescopes flying freely in space. The central spacecraft performs the interferometry by combining the light from all four telescopes.

such small planets. If we are going to expend effort searching for habitable planets and life, however, it would be nice to think that the search will be fruitful.

Most scientists expect Earth-like planets to be common. Billions of stars have at least moderate-size habitable zones, and our understanding of planet formation tells us that rocky, terrestrial planets should form quite easily within these zones. However, a few scientists have questioned whether the picture is really this simple. Instead, they propose that Earth has been the beneficiary of several rare kinds of planetary luck. According to this idea, sometimes called the "rare Earth hypothesis," the specific circumstances that have allowed life on Earth to survive and evolve into complex forms (such as oak trees and people)

might be so rare that ours could be the only planet in the galaxy that harbors anything but the simplest life. Let's briefly examine some of the key issues involved in the rare Earth hypothesis.

Are There Enough Heavy Elements? We have assumed that, in general, both terrestrial (rocky) and jovian (gaseous) planets can form around any star. However, the fraction of heavy elements (elements other than hydrogen and helium) varies among different stars from less than 0.1% among the old stars in globular clusters to about 2% among stars like our Sun. Could it be that Earth-like planets can form only around stars with relatively high heavy-element abundances?

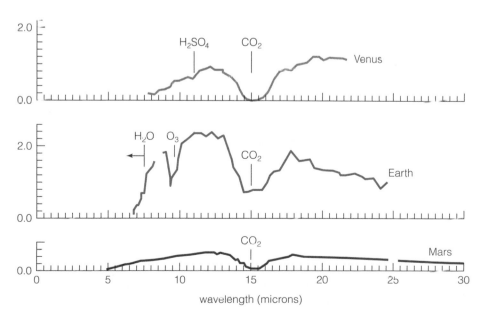

Figure 24.10 The infrared spectra of Venus, Earth, and Mars, as they might be seen from afar, showing absorption features that point to the presence of carbon dioxide (CO_2), ozone (O_3), and sulfuric acid (H_2SO_4) in their atmospheres. While carbon dioxide is present in all three spectra, only our own planet has appreciable oxygen (and hence ozone)—a product of photosynthesis. If we could make similar spectral analyses of distant planets, we might detect atmospheric gases that would indicate life.

We don't yet know. It's not merely a question of having enough raw material. Earth's mass is less than 1/100,000 of the mass of the Sun, so even a very small heavy-element abundance could be enough to make one or more Earth-like planets. However, we don't know the fine details of how terrestrial planets are made. It could be that a relatively high heavy-element abundance is important to the formation process. If so, a significant proportion of the stars in our galaxy may be incapable of having Earth-like planets. Earth-like planets would then be rarer than we would expect otherwise. Still, this constraint by itself would not make Earth-like planets very rare, because billions of stars have Sun-like heavy-element abundances.

Do Impacts Pummel Most Earth-Size Planets?
Another issue raised by rare Earth proponents concerns the impact rates on planets in other star systems. We have seen that Earth was probably subjected to numerous large impacts by asteroids and comets—some large enough to vaporize the oceans and sterilize the planet—during the heavy bombardment that went on during the first half-billion years after our planet's birth [Section 14.5]. In our solar system, the impact rate lessened dramatically after that. Might the impact rate remain high much longer in other planetary systems?

The most numerous small objects in our solar system are the trillion or so comets of the distant Oort cloud [Section 13.4]. Fortunately for us, nearly all of these myriad objects are essentially out of reach, posing no threat to our planet. However, the reason why they are out of reach can be traced directly to Jupiter.

The vast supply of comets in the Oort cloud is thought to have formed in the region of the solar system where the jovian planets were born. They were then "kicked" into their current, more distant orbits by close encounters with jovian planets—and with Jupiter in particular. Thus, if Jupiter did not exist, many of the comets might have remained in regions of the solar system where they could pose a danger to Earth. In that case, the heavy bombardment might never have ended, and huge impacts would continue to this day. From this viewpoint, our existence on Earth has been possible only because of the "luck" of having Jupiter as a planetary neighbor.

The primary question in this case is just how "lucky" this situation might be. Our discoveries of extrasolar planets so far suggest that Jupiter-size planets are in fact quite common. However, we've also found that many large planets migrate inward [Section 9.6], perhaps disrupting terrestrial planet orbits along the way. Whether having a Jupiter in the right place for ejecting comets is lucky or typical remains an open question.

The story of life on Earth is replete with disasters that served to stress terrestrial species, resulting in the rapid evolution of new, more complex organisms. For example, the impact thought to have led to the demise of the dinosaurs also led to their replacement by mammals. Do you think it's possible that a higher rate of impacts might have been *good* rather than bad for life on Earth? Explain.

Are Stable Climates Rare?
Another issue affecting the rarity of Earth-like planets concerns climate stability. Earth's climate has been stable enough for liquid water to exist throughout the past 4 billion years. This climate stability has almost certainly played a major role in allowing complex life to evolve on our planet. If our planet had frozen over like Mars or overheated like Venus, we would not be here today. Advocates of the rare Earth hypothesis point to at least two pieces of "luck" related to Earth's stable climate.

The first piece of "luck" concerns the existence of plate tectonics. As we discussed in Chapter 14, plate tectonics plays a major role in regulating Earth's climate through the carbon dioxide cycle. Plate tectonics probably was not necessary to the origin of life, but it seems to have been very important in keeping the climate stable enough for the subsequent evolution of plants and animals. But are we really "lucky" to have plate tectonics, or should this geological process be common on similar-size planets elsewhere? We do not yet know. The lack of plate tectonics on Venus, which is quite similar in size to Earth, might seem to argue for plate tectonics being rare. On the other hand, some scientists suspect that the lack of plate tectonics on Venus can be traced to its runaway greenhouse effect [Section 14.4], which occurred because Venus is not quite far enough from the Sun to be within the Sun's habitable zone. In that case, it's possible that any Earth-size planet within a star's habitable zone would have plate tectonics.

The second piece of "luck" in climate stability concerns the existence of Earth's relatively large Moon. The Moon helps keep Earth's axis tilt stable through time, which in turn keeps the climate stable because changes in axis tilt can lead to climate changes [Section 11.4]. If the Moon did not exist, our spin axis would be subject to large swings in its tilt over periods of tens to hundreds of thousands of years. This would cause deeper ice ages and more intense periods of warmth. Given that the Moon formed as a result of a random, giant impact [Section 9.4], we might seem to be very lucky to have the Moon and the climate stability it brings.

Again, however, there are other ways to look at the issue. Changes in axis tilt might warm or cool different parts of the planet dramatically, but the changes would probably occur slowly enough for life to adapt or migrate as the climate changed. In addition, our Moon's presumed formation in a random giant impact does not necessarily mean that large moons will be rare. At least a few giant impacts should be expected in any planetary system. Indeed, Earth may not be the only planet in our own solar system that ended up with a large moon through a giant impact. Pluto's moon (Charon) may have formed in the same way [Section 13.5]. Thus, while luck was certainly involved in Earth's having a large moon, it might not be a very rare kind of luck.

The Bottom Line The bottom line is that while the rare Earth hypothesis offers some intriguing arguments, it is too early to say whether any of them will hold up over time. For each potential argument that Earth has been lucky, we've seen counterarguments suggesting otherwise. There's no doubt that our solar system and our world have "personality"—they exhibit properties that might be found only occasionally in other star systems—but were such properties truly essential for our existence or merely a help?

Our solar system has no properties obviously essential to complex or even intelligent life that other star systems would never have. Indeed, it may be that we have missed out on some helpful phenomena that could have sped evolution on Earth. We might be less lucky than we recognize, and creatures on other worlds might regard the nature of our planet with disappointment. Until we learn much more about other planets in the universe, we cannot know whether Earth-like planets and complex life are common or rare.

24.4 The Search for Extraterrestrial Intelligence

So far, we have focused on search strategies for microbial or other nonintelligent life. However, if intelligent beings and civilizations exist elsewhere, we might be able to find them with a completely different type of search strategy. Instead of searching for hard-to-find spectroscopic signs of life, we might simply listen for signals that intelligent beings are sending into interstellar space, either in deliberate attempts to contact other civilizations or as a means of communicating among themselves. The search for signals from other civilizations is generally known as the **search for extraterrestrial intelligence**, or **SETI** for short.

How Many Civilizations Are Out There?

Given that we do not even know how common microbial life might be, we cannot reliably estimate the number of advanced civilizations that might exist. Nevertheless, for the purposes of planning a search for extraterrestrial intelligence, it is useful to have an organized way of thinking about the number of civilizations that might be out there. To keep our discussion simple, let's consider only the number of potential civilizations in our own galaxy. We can always extend our estimate to the rest of the universe by simply multiplying the result we find for our galaxy by 100 billion, the approximate number of galaxies in our universe.

The Drake Equation In principle, we could estimate the number of civilizations in the Milky Way Galaxy if we knew just a few basic facts. First, we'd need to know the number of habitable planets in the galaxy— that is, the number of planets on which life could potentially have arisen and survived. We'll call this number N_{HP} (for "number of habitable planets").

Second, we'd need to know what fraction of these habitable planets actually have life on them. We'll use the term f_{life} to stand for this fraction. For example, $f_{life} = 1$ would mean that all habitable planets have life, while $f_{life} = 1/1,000,000$ would mean that only 1 in 1 million habitable planets has life. Multiplying the number of habitable planets by the fraction that have life—$N_{HP} \times f_{life}$—would tell us the number of life-bearing planets in the galaxy.

Third, we'd need to know the fraction of the life-bearing planets upon which a civilization capable of interstellar communication *has at some time* arisen. Let's call this fraction f_{civ}. For example, $f_{civ} = 1/1,000$ would mean that a civilization has existed on 1 out of 1,000 planets with life, while the other 999 out of 1,000 have not had a species intelligent and industrious enough to build radio transmitters, high-powered lasers, or other devices for interstellar conversation. When we multiply this term by the first two terms to form the product $N_{HP} \times f_{life} \times f_{civ}$, we get the total number of planets upon which intelligent beings have evolved and developed a communicating civilization at some time in the galaxy's history.

Finally, we need to know the fraction of all these civilizations that exist *now*, as opposed to thousands, millions, or billions of years in the past. We will call this fraction f_{now}. After all, we can hope to contact only civilizations that are broadcasting signals we could receive now (assuming we take into account the light-travel time for signals from other stars). Because the product of the first three terms told us the total number of civilizations that have ever arisen in the galaxy, multiplying by f_{now} tells us how many civilizations we could potentially make contact with today. Thus, we have the following simple formula for the number of civilizations now inhabiting the Milky Way Galaxy (Figure 24.11):

$$\text{Number of civilizations} = N_{HP} \times f_{life} \times f_{civ} \times f_{now}$$

This simple formula is a variation of an equation first expressed in 1961 by astronomer Frank Drake, one of the pioneers in efforts to search for other civilizations. Now known as the **Drake equation**, in principle it gives us a simple way to calculate the number of civilizations capable of interstellar communication that are *currently* sharing the Milky Way Galaxy with us. Of course, we cannot actually perform the calculation until we learn the values of all its terms. Unfortunately, we don't yet know the value of any of them.

Try the following sample numbers in the Drake equation. Suppose that 1,000 habitable planets are in our galaxy, that 1 in 10 habitable planets has life, that 1 in 4 planets with life has at some point had an intelligent civilization, and that 1 in 5 civilizations that have ever existed is in existence now. How many civilizations would exist at present? Explain.

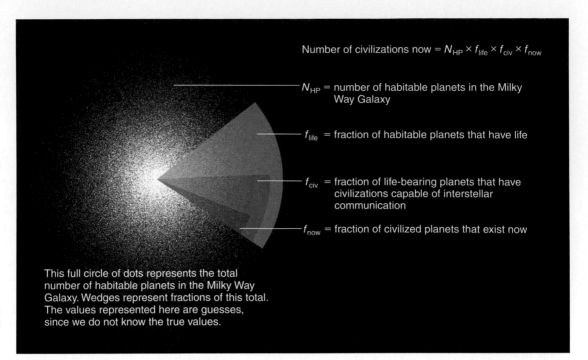

$$\text{Number of civilizations now} = N_{HP} \times f_{life} \times f_{civ} \times f_{now}$$

N_{HP} = number of habitable planets in the Milky Way Galaxy

f_{life} = fraction of habitable planets that have life

f_{civ} = fraction of life-bearing planets that have civilizations capable of interstellar communication

f_{now} = fraction of civilized planets that exist now

This full circle of dots represents the total number of habitable planets in the Milky Way Galaxy. Wedges represent fractions of this total. The values represented here are guesses, since we do not know the true values.

Figure 24.11 This diagram summarizes how the factors in the Drake equation lead to an estimate of the number of civilizations now in our galaxy. Keep in mind that we don't know the actual value of any of the terms, so we don't know the actual size of any of the wedges represented in the diagram.

Values for the Drake Equation The only term in the Drake equation for which we can make even a reasonably educated guess is the number of habitable planets, N_{HP}. As we discussed earlier, our current understanding of star system formation suggests that Earth-like planets ought to be common. Unless some of the "rare Earth" ideas prove to be correct, it seems entirely reasonable to suppose that 100 billion or more habitable planets could be orbiting among the several hundred billion stars in the Milky Way Galaxy. Nevertheless, the actual number may be far smaller. We won't know for sure until we begin to gather data about habitable planets over the next decade.

The rest of the formula presents more difficulty. For the moment, we have no rational way to estimate the fraction f_{life} of habitable planets upon which life actually arose. The problem is that we cannot generalize when we have only one example to study—our own Earth. Still, we are not completely without guidance. The fact that life arose rapidly on Earth suggests that the origin of life was fairly "easy," in which case we might expect most or all habitable planets to also have life, making the fraction f_{life} close to 1. However, until we have solid evidence that life arose anywhere else, such as on Mars, it is also possible that Earth might really have been very lucky. In that case, f_{life} might be so close to zero that life has never arisen on any other planet in our galaxy.

Similarly, we have little basis on which to guess the fraction f_{civ} of life-bearing planets that eventually develop a civilization capable of interstellar communication. On one hand, life flourished on Earth for almost 4 billion years before the rise of humans, and this fact might suggest that producing a civilization is very difficult even when life is present. On the other hand, roughly half the stars in the Milky Way are older than our Sun, so evolution has had

plenty of time to work on numerous planets. Any evolutionary drive toward intelligence might inevitably lead to huge numbers of civilizations, even if it takes billions of years on any given world. The value of the fraction f_{civ} comes down to the question of whether the 4 billion years it took for humans to evolve on Earth is typical, fast, or slow. If it was unusually fast and the typical time needed to develop complex species is much longer, then most life-bearing planets may be covered with nothing more advanced than bacteria.

The final term in the equation, f_{now}, is particularly interesting because it is related to the survivability of civilizations. Consider our own example. In the roughly 12 billion years during which our galaxy has existed, we have been capable of interstellar communication via radio for only about 60 years. Thus, if we were to destroy ourselves tomorrow (saving you the unpleasantness of a final exam), then other civilizations could have received signals from us during only 60 years out of the galaxy's 12-billion-year existence, equivalent to 1 part in 200 million of the galaxy's history.

If such a short technological lifetime is typical of civilizations, then f_{now} would be only 1/200,000,000, and some 200 million civilization-bearing planets would need to have existed at one time or another in the Milky Way in order for us to have a good chance of finding another civilization out there now. However, we'd expect f_{now} to be so small only if we are on the brink of self-destruction—after all, the fraction will grow larger for as long as our civilization survives. Thus, if civilizations are at all common, the key factor in whether any are out there now is their survivability. If most civilizations self-destruct shortly after achieving the technology for interstellar communication, then we are almost certainly alone in the galaxy at present. But if most survive and thrive for thousands or millions of years, the Milky Way may be brimming with civilizations—most of them far more advanced than our own.

Will Aliens Be Like Us?

When we imagine contact with other civilizations, we are making several important but usually unstated assumptions. For example, besides assuming that other intelligent beings may exist elsewhere, we are also assuming that their sociology will drive them to develop science and technology much as we have. Moreover, if we hope to make contact, then we must also hope that they share our innate curiosity and desire to explore the cosmos. Are these assumptions reasonable?

No one knows, but a fundamental assumption in nearly all of science today is that we are not "special" in any particular way. We live on a fairly typical planet orbiting an ordinary star in a normal galaxy, and we assume that living creatures elsewhere—whether they prove to be rare or common—would be subjected to evolutionary pressures quite similar to those that have operated on Earth. Thus, while the specifics of evolution might play out differently on different worlds, we have no reason to assume that we are anything but "average" among the types of creatures that ultimately evolve.

Most movies about aliens assume not only that they act somewhat like us, but also that they resemble us physically. For example, movie aliens often have two eyes, two arms with fingered hands, and two legs and are of two sexes. Do you think such movie aliens are realistic? Why or why not?

If we are indeed typical of intelligent species and if a lot of other intelligent species are out there (a very big *if*), then at least some of them ought to have the same interest in extraterrestrial communication that we have. In that case, we have a chance of making contact with them—especially if they are trying to make it easy for us to discover their presence.

SETI Strategies

The basic idea behind SETI efforts is that we might receive signals from other civilizations using technology available to us today. Based on our current understanding of physics, it seems likely that even very advanced civilizations would communicate much as we do—by encoding signals in radio waves or other forms of light. Most SETI researchers use large radio telescopes to search for alien radio signals (Figure 24.12). A few researchers are beginning to check other

Figure 24.12 The 64-meter Parkes radio telescope in New South Wales, Australia. A SETI experiment "piggybacks" on this telescope while it is engaged in other astronomical research.

parts of the electromagnetic spectrum. For example, some scientists use visible light telescopes to search for communications encoded as laser pulses. Of course, advanced civilizations may well have invented communication technologies that we cannot even imagine. In that case, SETI efforts will not detect them.

A good way to think about our chances of picking up an alien signal is to imagine what aliens would need to do to pick up signals from us. We have been sending relatively high-power transmissions into space since about the 1950s in the form of television broadcasts. Thus, in principle, anyone within about 50 light-years of Earth could watch our old television shows (perhaps a frightening thought). However, in order to detect our broadcasts, they would need far larger and more sensitive radio telescopes than we have today. If their technology were at the same level as ours, they could receive a signal from us only if we deliberately broadcast an unusually high-powered transmission.

To date, humans have made only a few attempts to broadcast our existence in this way. The most powerful of these occasional transmissions was made in 1974 and lasted only 3 minutes (Figure 24.13). The powerful planetary radar transmitter on the Arecibo radio telescope was fired up and used to send a simple pictorial message to the globular cluster M 13. This target was chosen in part be-cause it contains a few hundred thousand stars, seemingly offering a good chance that at least one has a civilization around it. However, M 13 is about 21,000 light-years from Earth, so it will take some 21,000 years for our signal to get there and another 21,000 years for any response to make its way back to Earth.

Several SETI projects under way or in development would be capable of detecting signals like the one we broadcast from Arecibo if they came from civilizations within a few hundred light-years. However, we could detect the signal only if we had the receiver tuned to the frequency of the broadcast—just as you can listen to your favorite radio station only by calling up the correct frequency on your radio dial. What radio frequency would aliens use? In the past, some astronomers made guesses about popular alien frequencies based on things such as frequencies emitted by common molecules in interstellar space. Today, SETI efforts generally seek to bypass this question by scanning millions of frequency bands simultaneously. Thus, if anyone nearby is deliberately broadcasting on an ongoing basis, we have a good chance of detecting the signals.

SETI efforts are often controversial, largely because of their uncertain chances of success and their need for large, expensive telescopes. As a consequence, nearly all SETI research is currently funded by private rather than govern-

Figure 24.13 In 1974, a short message was broadcast to the globular cluster M 13 using the Arecibo radio telescope.

▼ **a** The Arecibo radio telescope in Puerto Rico is the world's largest single radio dish, with a diameter of 305 meters (1,000 feet).

▶ **b** The message consisted of 1,679 bits, and each bit was represented by one of two radio frequencies. The bits make up a rectangular grid with 73 rows and 23 columns (each of these numbers is a prime number, which hopefully will enable any alien recipients to guess the layout of the grid). The resulting graphic represents the Arecibo radio dish, our solar system, a human stick figure, and a schematic of DNA and the eight simple molecules used in its construction. The colors are shown only to make the components clearer. The actual picture was sent in "black and white."

ment money. Nevertheless, SETI research offers several important benefits to more general science efforts. For example, SETI experiments that scan the sky at many different frequencies sometimes make unexpected discoveries that have nothing to do with aliens but are scientifically useful. In addition, technology developed for SETI has found applications in other areas. For SETI supporters, such spin-offs justify the ongoing costs of the effort, even though we don't know if or when we will ever discover another civilization.

THINK ABOUT IT

SETI supporters argue that contact with an extraterrestrial intelligence would be one of the most important discoveries in human history. Do you agree? Defend your opinion.

24.5 Interstellar Travel

So far, we have discussed ways of detecting distant life and civilizations without ever leaving the comfort of our own planet. But could we ever actually visit other worlds in other star systems?

In many science fiction movies, our descendants travel among the stars as routinely as we jet about the Earth in airplanes. They race around the galaxy in starships of all sizes and shapes, circumventing nature's prohibition on faster-than-light travel by entering hyperspace, wormholes, or warp drive. They witness incredible cosmic phenomena firsthand, such as stars and planets in all stages of development, accretion disks around white dwarfs and neutron stars, and the distortion of spacetime near black holes. Along the way they encounter numerous alien species, most of which look and act a lot like us.

Unfortunately, real interstellar travel is likely to be limited by the speed of light. Journeys even to nearby stars will require tremendous patience, as well as tremendous technological advances. In this section, we'll investigate prospects for interstellar travel by human beings. Then, in the book's final section, we'll see how these ideas lead to an unsettling paradox concerning the presence or absence of other interstellar travelers.

Starships: Distant Dream or Near-Reality?

Traveling by foot, early humans could sustain speeds of no more than a few kilometers per hour. Today, interplanetary spacecraft travel through the solar system at speeds of a few *tens of thousands* of kilometers per hour—10,000 times faster than the highest speed our ancestors could achieve.

These speeds may sound fast, but they are still slow by interstellar standards. Four interplanetary probes—*Pioneers 10* and *11*, and *Voyagers 1* and *2*—are on their way out of the solar system. It will take them more than 10,000 years to cover each light-year of distance, and their trajectories will not take them on close passes of any nearby stars. Nevertheless, these spacecraft should suffer little damage during their journeys and are likely to remain almost as good as new for millions of years. Each carries a greeting from Earth, just in case someone comes across one of them someday (Figure 24.14).

If we want to make interstellar journeys within human lifetimes, we will need starships that travel at speeds close to the speed of light. Light travels extremely fast—300,000 kilometers per second—so the required leap from our current technology is enormous. Indeed, current spacecraft speeds are less than 1/10,000 the speed of light, meaning

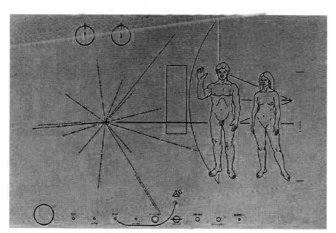

a The Pioneer plaque, about the size of an automobile license plate, shows human figures in front of the spacecraft (which provides them scale) as well as a diagram of our solar system (bottom). The spacecraft's origin, the third planet from the Sun, is schematically indicated, as is the Sun's position relative to nearby pulsars (the "prickly" graphic to the left). The periods of these pulsars, indicated in binary code, will help anyone reading the plaque to determine when the craft was launched, because pulsars slow over time. The radio frequency emitted naturally by neutral hydrogen, also indicated on the plaque, provides the unit of time for the pulsar data.

b *Voyagers 1* and *2* carry a phonograph record—a 12-inch gold-plated copper disk containing music, greetings, and images from Earth.

Figure 24.14 Messages aboard the Pioneer and Voyager spacecraft, which are bound for the stars.

that starships will need to go 10,000 times faster. From this standpoint, we are as far from interstellar flight as cavemen were from the space age.

From another standpoint, however, starflight might be just around the corner. Nearly all of the increase in speed we have achieved over our ancestors has occurred within the past century or so, with the advent of flight and the dawn of the space age. If we can achieve a comparable speed increase over the next hundred years, our great-grandchildren may be building starships.

The Challenge of Interstellar Travel

Regardless of how far in the future interstellar travel might be, we will have to overcome huge technological and social hurdles to achieve it. On the technological side, we will need entirely new types of engines to reach speeds close to the speed of light. Perhaps more important, we'll need vast new sources of energy. The energy needed to accelerate a single ship the size of *Star Trek*'s *Enterprise* to just half the speed of light would be more than 2,000 times the total annual energy use of the world today. In addition, fast-moving starships will require new types of shielding to protect crew members from instant death. As a starship travels through interstellar gas at near-light speed, ordinary atoms and ions will hit it like a flood of high-energy cosmic rays.

The social hurdles may be even more challenging. Building starships will not be something we can do in a garage. We will almost certainly need construction facilities in space as well as the ability to mine resources from the Moon or nearby asteroids. Starship development thus will depend on vast political will and huge budgets. It probably cannot occur without far greater international trust and cooperation than we enjoy today.

The starship crew will face other social hurdles. Einstein's theory of relativity offers high-speed travelers a "ticket to the stars," in that they will be able to make journeys across many light-years of space in relatively short amounts of time [Section S2.6]. For example, in a ship traveling at an average speed of 99.9% of the speed of light, the 50-light-year round-trip to the star Vega would take the travelers aboard only about 2 years. But more than 50 years would have passed on Earth while they were gone. Family and friends would be older or deceased, new technologies might have made their knowledge and skills obsolete, and many political and social changes would have occurred in their absence. Clearly, the crew will find it difficult to adjust to these changes.

Starship Design

Current spacecraft, including both robotic probes like those of the Pioneer and Voyager missions and crew-carrying spacecraft such as the Space Shuttle, are launched into space by *chemical rockets*. Their engines use energy from chemical reactions to drive hot gas out the back of the rocket, which causes the rocket to accelerate forward.

Chemical rockets will never be practical for interstellar travel because a limiting process prevents them from reaching very high speeds. Achieving higher speeds requires more fuel, but the added weight of additional fuel makes increasing the rocket's speed more difficult. As propulsion systems even for relatively small robotic spacecraft, chemical rockets cannot exceed speeds of about 0.001 of the speed of light—which means the journey to Alpha Centauri would take more than 4,000 years. For the much larger starships needed to support human crews, chemical rockets are completely out of the question. Fortunately, other technologies may allow much faster travel.

Nuclear Rocket Propulsion One way to improve rocket efficiency is to use nuclear power rather than chemical power. Fusing a kilogram of hydrogen generates more than a million times more energy than chemical reactions involving a kilogram of hydrogen and oxygen.

Scientists and engineers interested in interstellar travel have produced at least two design proposals for nuclear-powered rockets. One design, developed in the 1960s under the name Project Orion, envisions powering a rocket with repeated detonations of relatively small hydrogen bombs. Each explosion would take place a few tens of meters behind the spaceship and would propel the ship forward as the vaporized debris impacted a "pusher plate" on the back of the spacecraft (Figure 24.15). In principle, we could build an Orion spacecraft with existing technology, though it would be very expensive and would require an exception to the international treaty banning nuclear detonations in space.

A more futuristic idea—one which requires that we have the technology for controlled nuclear fusion reactors—was developed in the 1970s under the name Project Daedalus (Figure 24.16). The Daedalus design imagines shooting frozen pellets of deuterium and helium-3 into a reaction chamber, where they would undergo fusion and generate the energy to propel the starship.

Nuclear-powered rockets are undoubtedly feasible in some form. Still, at best they would achieve speeds of about 10% of the speed of light. Journeys to nearby stars would be possible but would take decades.

THINK ABOUT IT

From the sixteenth to the nineteenth century, many people left their homes in Europe on one-way journeys to the "new world" of America. In the future, similar one-way trips to colonies on Earth-like planets around other stars might be possible. If offered the opportunity, would you go? Do you know anyone who would? Why or why not?

Matter–Antimatter Rocket Engines In principle, we have an even more powerful energy source for rocket engines than fusion: matter–antimatter annihilation [Section S4.2]. Whereas fusion converts less than 1% of the mass of atomic nuclei into energy, matter–antimatter annihilation converts *all* the annihilated mass into energy. Starships with matter–

Figure 24.15 Artist's conception of the Project Orion starship, showing one of the small hydrogen-bomb detonations that propel it. Debris from the detonation impacts the flat disk, called the pusher plate, at the back of the spaceship. The central sections (enclosed in a lattice) hold the bombs, and the front sections house the crew.

antimatter engines could probably reach speeds of 90% or more of the speed of light. At these speeds, the slowing of time predicted by relativity becomes noticeable, putting many nearby stars within a few years' journey for the crew members.

However, while matter–antimatter engines may someday be the propulsion system of choice for starships, obtaining and storing the fuel will be very difficult. As far as we know, no natural reservoirs of antimatter exist, because antimatter that is produced in our universe (by numerous high-energy cosmic processes) is quickly annihilated when it comes into contact with ordinary matter. Thus, we would have to manufacture the antimatter, as physicists now do in high-energy particle accelerators.

Unfortunately, current worldwide production of antimatter amounts to only a few billionths of a gram per year.

With current technology, manufacturing 1 ton of antimatter—far less than would be needed for an interstellar trip—would require more energy than humanity has used in all of history. Moreover, even if we could make the antimatter, we don't yet know of a good way to store it aboard a rocket, because it would have to be kept in some type of container in which it never touched any ordinary matter within or beyond the walls.

Sails and Beamed Energy Propulsion Although nuclear or matter–antimatter rockets would be enormously more efficient than chemical rockets, their speeds are still ultimately limited by the weight of their fuel. More fuel means more weight, making it more difficult to reach higher speeds. As a result, some people have considered spacecraft that don't need to carry their fuel with them. At least two such

Figure 24.16 Artist's conception of a robotic Project Daedalus starship. The front section (right) holds the scientific instruments. The large spheres hold the fuel pellets for the central fusion reactor.

types of spacecraft are technologically within our reach, at least in principle.

One idea is to use sunlight as power. Large, thin, highly reflective *solar sails* could be pushed by the pressure exerted by sunlight. This pressure is so slight that we normally don't notice it, but in the vacuum of space it could gradually accelerate a spacecraft to impressive speeds. In the not-too-distant future, solar sailing may well prove to be an inexpensive way of navigating within the solar system. It may even be useful for interstellar travel (Figure 24.17). Although the push from the Sun slowly fades with distance, a solar sailing ship that started near the Sun might achieve speeds of a few percent of light speed. It could then coast to neighboring stars in less than a century.

We might achieve even higher speeds by shining powerful lasers at the spacecraft sails. Such *beamed energy propulsion* could accelerate ships to substantial fractions of the speed of light—but only if we could build enormous lasers. For example, accelerating a ship to half the speed of light within a few years would require a laser that uses more than 1,000 times more power than all current human power consumption. As a result, starships propelled by beamed energy probably remain far in the future.

Interstellar Ramjets

Another idea for avoiding the weight of fuel is to design a starship that collects fuel as it goes. An *interstellar ramjet* would use a giant scoop to collect interstellar hydrogen, which it would then use as fuel for its nuclear fusion engines (Figure 24.18).

Of course, this idea presents practical difficulties. The typical density of interstellar gas is only a few atoms per cubic centimeter, so the scoop would need to be enormous. As Carl Sagan said, we are talking about "spaceships the size of worlds." Nevertheless, an interstellar ramjet could in principle accelerate continuously, thereby achieving speeds arbitrarily close to the speed of light.

Imagine an interstellar ramjet that accelerated at 1g for half its journey and then turned around and decelerated at 1g until it reached its destination. The crew would find the trip quite comfortable, experiencing Earth-like gravity the entire way. During most of the journey, the ship would be traveling relative to Earth (and to the destination) at a speed very close to the speed of light, so time on the ship would pass very slowly compared to time on Earth. Moreover, longer trips would mean top speeds closer to the speed of light and therefore more extreme effects on time.

For example, such a ship could make a trip to a star 500 light-years away in only about 12 years of ship time. (You can calculate this and the other travel times for the crew using the formula given in problem 18 in Chapter S3.) It could travel the 28,000 light-years to the center of the Milky Way Galaxy, where the crew could observe firsthand the mysterious galactic center [Section 19.5], in only about 21 years of ship time. It could make the trip to a star system in the Andromeda Galaxy in about 29 years of ship time. Thus, the crew could go to the Andromeda Galaxy, spend 2 years studying one of its star systems and taking pictures of the Milky Way Galaxy to bring home, and return to Earth in just 60 years of ship time. However, because the Andromeda Galaxy is 2.5 million light-years away, the crew would find that 5 million years had passed on Earth by the time they returned home.

Science Fiction

If you are a science fiction fan, this discussion of interstellar travel may be depressing. Interstellar tourism and commerce seem out of the question, even with interstellar ramjets that reach speeds very close to the speed of light. If we are ever to travel about the galaxy the way we now travel about the Earth, we will need spacecraft that can get us from here to there at speeds much faster than the speed of light.

The theory of relativity leaves little hope that we can ever find a way to travel *through* space faster than the speed

Figure 24.17 Artist's conception of a spaceship propelled by a solar sail, shown as it approaches a forming planet in a young solar system. The sail is many kilometers across. The scientific payload is at the central meeting point of the four scaffoldlike structures.

Figure 24.18 Artist's conception of a spaceship powered by an interstellar ramjet. The giant scoop in the front (left) collects interstellar hydrogen for use as fusion fuel.

of light, but science fiction writers have imagined all kinds of novel shortcuts that don't necessarily violate relativity or any other known laws of physics. Some of these ideas go by names such as hyperspace, wormholes, and warp drive [Section S3.6]. The bottom line is that our present knowledge does not allow us to say whether any of these technologies are possible. Science fiction writers therefore can hope that we'll someday find a way to travel conveniently among the stars, but many scientists doubt that such rapid transit is possible.

Poised on the Brink

Interstellar travel will be difficult to achieve, but it is by no means impossible. Indeed, considered in the context of past technological development, it seems almost inevitable that we will eventually achieve the ability to build starships. With sufficient social and political will, our current technology could allow us to send small groups of colonists on decades-long journeys to the stars before this century is out. Future technologies might make interstellar travel even faster.

Thus, we seem bound to become interstellar travelers unless we choose otherwise. Such a choice might be deliberate. For example, many people argue that money for space exploration would be better spent here on Earth, and others argue against space colonization on philosophical grounds. Or it might be a consequence of choices that inadvertently lead us to catastrophe, such as nuclear war or a disaster brought on by overpopulation, epidemic disease, or global warming.

If an alien civilization were watching us, they might well conclude that we are poised on the brink of the most significant turning point in human history. If we choose poorly and destroy ourselves, all our achievements in science, art, and philosophy will be lost forever. But if we sur-

vive and choose to continue the exploration of space, we may be embarking on a path that will take us to the stars.

While we could certainly destroy our civilization today, some people argue that a civilization with colonies spread among many different star systems would be essentially "extinction proof." They claim that, because of the long travel times between star systems, no single event (such as war, disease, or environmental damage) could wipe out all the colonies. Do you agree that attaining large-scale interstellar travel would assure the long-term survival of the human species? Defend your opinion.

24.6 A Paradox: Where Are the Aliens?

Imagine that we survive and become interstellar travelers and that we begin colonizing habitable planets around nearby stars. As the colonies grow at each new location, some of the people may decide to set out for other star systems. Even if our starships traveled at relatively low speeds—say, a few percent of the speed of light—we could have dozens of outposts around nearby stars within a few centuries. In 10,000 years, our descendants would be spread among stars within a few hundred light-years of Earth. In a few million years, we could have outposts throughout the Milky Way Galaxy. We will have become a true galactic civilization.

Now, if we take the idea that *we* could develop a galactic civilization within a few million years and combine it with the reasonable (though unproved) idea that civilizations ought to be common, we are led to an astonishing conclusion: Someone else should already have created a galactic civilization. In fact, it should have been done a long time ago.

In this chapter, we have discussed contact with intelligent aliens as a possibility, not a reality. However, public opinion polls suggest that up to half the American public believes that aliens are already visiting us. What can science say about this remarkable notion?

The bulk of the claimed evidence for alien visitation consists of sightings of UFOs—unidentified flying objects. Many thousands of UFOs are reported each year, and no one doubts that unidentified objects are being seen. The question is whether they are alien spacecraft.

Aliens have long been a staple of science fiction, but modern interest in UFOs began with a widely reported sighting in 1947. While flying a private plane near Mount Rainier in Washington State, businessman Kenneth Arnold saw nine mysterious objects streaking across the sky. He told a reporter that the objects "flew erratic, like a saucer if you skip it across the water." (In fact, he may have seen meteors skipping across the atmosphere, though no one knows for sure.) He did *not* say that the objects were saucer-shaped, but the reporter nevertheless wrote up Arnold's experience as a sighting of "flying saucers." The story was front-page news throughout America, and within a decade "flying saucers" had invaded popular culture, if not our planet.

The flying saucer reports also interested the U.S. Air Force, largely out of concern that the UFOs might represent new types of aircraft developed by the Soviet Union. For two decades, the air force hired teams of academics to study UFO reports. In the overwhelming majority of cases, these experts were able to specify a plausible identification of the UFO. The explanations included bright stars and planets, aircraft and gliders, rocket launches, balloons, birds, ball lightning, meteors, atmospheric phenomena, and the occasional hoax. For a minority of the sightings, the investigators could not deduce what was seen, but their overall conclusion was that there was no reason to believe the UFOs were either highly advanced Soviet craft or visitors from other worlds. The air force ultimately dropped its investigations of the UFO phenomenon.

Believers discounted the air force denials and continued to gather "evidence" of alien visitation. None of this evidence has ever withstood close scrutiny. Photographs and film clips are nearly always too fuzzy to clearly show alien spacecraft, except in cases that are obviously faked. UFO witnesses are frequently credible (they include seasoned pilots), but generally there are several possible explanations for what they've seen besides alien spacecraft. Crop circles (which gained popularity with the Mel Gibson movie *Signs*) are easily made by pranksters. Stories of alien abductions are dramatic but cannot be verified. Many psychologists believe they may simply reflect experiences of *sleep paralysis*, which can occur during REM (rapid eye movement) sleep and affects about half of all people at some time. Pieces of metal that "UFO experts" say could not have been made by humans have turned out to be pieces of cars or refrigerators.

The most famous claim of physical evidence of alien UFOs comes from an incident that occurred in 1947 near Roswell, New Mexico. Just a few weeks after the nationwide coverage of Kenneth Arnold's "flying saucers," a rancher reported crash remnants in a pasture. Military personnel drove out to the ranch, picked up the debris, and explained to the local papers that they had recovered the remains of a "flying disk." However, the story quickly changed. Only a day later, an air force officer held a press conference in which he stated that the debris was merely a crashed weather balloon. This denial successfully buried the story until 1978, when UFO investigator Stanton Friedman began looking into the events at Roswell. Friedman claimed that the debris was from a spacecraft and that alien occupants had been picked up as well.

The Roswell incident quickly became part of modern folklore, but it doesn't seem to deserve much credence. Friedman based his claims on interviews he conducted more than three decades after the event. The witness testimonies were inconsistent. Some supposed witnesses to the crashed "saucer" had originally claimed not to have seen it. Others were caught in flat-out lies. A famed film of autopsies conducted on the alien bodies has been labeled a hoax even by the network that aired it. Moreover, declassified military records show that what crashed in Roswell was a top-secret, balloon-borne device designed to detect Soviet nuclear tests—which explains why the air force didn't want the truth to be made public at the time.

Champions of alien visitation generally explain away the lack of clear evidence in one of two ways: government cover-ups or a failure of the mainstream scientific community to take the relevant phenomena seriously. Neither explanation seems particularly compelling.

It's certainly conceivable that a secretive government might *try* to put the lid on evidence of alien visits, though the motivation for doing so is unclear. The usual explanations are that the public couldn't handle the news and that the government is taking secret advantage of the alien materials to design new military hardware (via "reverse-engineering"). Both explanations are silly. Half the population already believes in alien visitors and would hardly be shocked if newspapers announced that aliens were stacked up in government warehouses. As for reverse-engineering extraterrestrial spacecraft, we should keep in mind how difficult it is to travel from star to star. Any society that could do so routinely would be technologically far beyond our own. Reverse-engineering their spaceships is as unlikely as expecting Neanderthals to construct personal computers just because a laptop somehow landed in their cave. In addition, while a government might successfully hide evidence for a short time, does it really seem possible that evidence could remain secret for decades (more than five decades, in the case of the Roswell claims)? And unless the aliens landed only in the United States, can we seriously believe that *every* government has cooperated in hiding the evidence?

Alleged disinterest on the part of the scientific community is an equally unimpressive claim. Scientists are constantly competing with one another to be the first with a great discovery, and clear evidence of alien visitors would certainly rank high on the all-time list. Countless researchers would work evenings and weekends, without pay, if they thought they could make such a discovery. The fact that few scientists are engaged in such study reflects not a lack of interest, but a lack of evidence worthy of study.

Of course, absence of evidence is not evidence of absence. Most scientists are open to the possibility that we might someday find evidence of alien visits, and many would welcome aliens with open arms. So far, however, we have no hard evidence to support the belief that aliens are already here.

For argument's sake, suppose civilizations arise around one in a million stars. In this case, some 100,000 civilizations should have arisen in the Milky Way. Further, suppose civilizations typically arise when their stars are 5 billion years old. Given that the galaxy is some 12 billion years old, the first of these 100,000 civilizations would have arisen at least 7 billion years ago. Others would have arisen, on average, every 70,000 years. Under these assumptions, the youngest civilization besides ourselves would be some 70,000 years ahead of us technologically, and most would be millions or billions of years ahead of us.

Thus, we encounter a strange paradox: Plausible arguments suggest that a galactic civilization should already exist, yet we have so far found no evidence of such a civilization. This paradox is often called *Fermi's paradox,* after the Nobel Prize–winning physicist Enrico Fermi. During a 1950 conversation with other scientists about the possibility of extraterrestrial intelligence, Fermi responded to speculations by asking, "So where is everybody?"

This paradox has many possible solutions, but broadly speaking we can group them into three categories:

1. We are alone. There is no galactic civilization because civilizations are extremely rare—so rare that we are the first to have arisen on the galactic scene.

2. Civilizations are common, but no one has colonized the galaxy. There are at least three possible reasons why this might be the case. Perhaps interstellar travel is much harder or vastly more expensive than we have guessed, and civilizations are unable to venture far from their home worlds. Perhaps the desire to explore is unusual, and other societies either never leave their home star systems or stop exploring before they've colonized much of the galaxy. Most ominously, perhaps many civilizations have arisen, but they have all destroyed themselves before achieving the ability to colonize the stars.

3. There *is* a galactic civilization, but it has deliberately avoided revealing its existence to us.

We do not know which, if any, of these explanations is the correct solution to the question "Where is everybody?" However, each category of solution has astonishing implications for our own species.

Consider the first solution—that we are alone. If this is true, then our civilization is a remarkable achievement. It implies that through all of cosmic evolution, among countless star systems, we are the first piece of the universe ever to know that the rest of the universe exists. Through us, the universe has attained self-awareness. Some philosophers and many religions argue that the ultimate purpose of life is to become truly self-aware. If so, and if we are alone, then the destruction of our civilization and the loss of our scientific knowledge would represent an inglorious end to something that took the universe some 14 billion years to achieve. From this point of view, humanity becomes all the more precious, and the collapse of our civilization would be all the more tragic. Knowing this to be the case might help us learn to put petty bickering and wars behind us so that we might preserve all that is great about our species.

The second category of solutions has much more terrifying implications. If thousands of civilizations before us have all failed to achieve interstellar travel on a large scale, what hope do we have? Unless we somehow think differently than all previous civilizations, this solution says that we will never go far in space. Because we have always explored when the opportunity arose, this solution almost inevitably leads to the conclusion that failure will come about because we destroy ourselves. We can only hope that this answer is wrong.

The third solution is perhaps the most intriguing. It says that we are newcomers on the scene of a galactic civilization that has existed for millions or billions of years before us. Perhaps this civilization is deliberately leaving us alone for the time being and will invite us to join it when we prove ourselves worthy. If so, our entire species may be on the verge of beginning a journey every bit as incredible as that of a baby emerging from the womb and coming into the world.

No matter what the answer turns out to be, learning it is sure to mark a turning point in the brief history of our species. Moreover, this turning point is likely to be reached within the next few decades or centuries. We already have the ability to destroy our civilization. If we do so, then our fate is sealed. But if we survive long enough to develop technology that can take us to the stars, the possibilities seem almost limitless.

THE BIG PICTURE

Putting Chapter 24 into Context

Throughout our study of astronomy, we have taken the "big picture" view of trying to understand how we fit into the universe. Here, at last, we have returned to Earth and examined the role of our own generation in the big picture of human history. Tens of thousands of past human generations have walked this Earth. Ours is the first generation with the technology to study the far reaches of our universe, to search for life elsewhere, and to travel beyond our home planet. It is up to us to decide whether we will use this technology to advance our species or to destroy it.

Imagine for a moment the grand view, a gaze across the centuries and millennia from this moment forward. Picture our descendants living among the stars, having created or joined a great galactic civilization. They will have the privilege of experiencing ideas, worlds, and discoveries far beyond our wildest imagination. Perhaps, in their history lessons, they will learn of our generation—the generation that history placed at the turning point and that managed to steer its way past the dangers of self-destruction and onto the path to the stars.

24.1 The Possibility of Life Beyond Earth

- *Why do many scientists now think that it's reasonable to look for life on other worlds?* Discoveries in astronomy and planetary science suggest that planetary systems are common and that we can reasonably expect to find many habitable worlds. Meanwhile, discoveries in biology suggest that life can survive in a wide range of environments and may arise relatively easily under conditions that ought to exist on many habitable planets.

24.2 Life in the Solar System

- *Why does Mars seem a good candidate for life?* Mars apparently was warm and wet during at least some periods in its distant past, conditions that may have been conducive to an origin of life. It still has significant amounts of frozen water and might have some pockets of liquid water underground.

- *What evidence have we collected so far concerning life on Mars?* We do not now have any clear evidence of life on Mars. The Viking landers conducted experiments on the Martian surface, but the overall results of these experiments do not seem consistent with the presence of life in the samples studied. One Martian meteorite shows several intriguing lines of evidence of life, but each can also be explained in nonbiological ways.

- *Which outer solar system moons seem to be candidates for life, and why?* Europa probably has a deep, subsurface ocean of liquid water, and Ganymede and Callisto might have oceans as well. If so, life may have arisen and survived in these oceans. Titan may have other liquids on its surface, though it is too cold for liquid water. Perhaps life can survive in these other liquids, or perhaps Titan has liquid water deep underground.

24.3 Life Around Other Stars

- *What do we mean by a star's habitable zone?* A star's habitable zone extends over distances from the star at which a suitable-size terrestrial planet could have a surface temperature that might allow for oceans and life.

- *Have we discovered habitable planets around other stars?* No; our current technology is not quite up to the task. However, upcoming missions should soon tell us whether terrestrial planets exist within the habitable zones of nearby stars, and missions one or two decades away may tell us whether these planets are habitable and perhaps even whether they have life.

- *Are Earth-like planets rare or common?* We don't know. Arguments can be made on both sides of the

question, and we lack the data to determine their validity at present.

24.4 The Search for Extraterrestrial Intelligence

- *What is the Drake equation, and how is it useful?* The Drake equation (in a modified form) says that the number of civilizations in the Milky Way Galaxy is $N_{HP} \times f_{life} \times f_{civ} \times f_{now}$, where N_{HP} is the number of habitable planets in the galaxy, f_{life} is the fraction of habitable planets that actually have life on them, f_{civ} is the fraction of life-bearing planets upon which a civilization capable of interstellar communication has at some time arisen, and f_{now} is the fraction of all these civilizations that exist now. Although we do not know the value of any of these terms, the equation helps us organize our thinking as we consider the search for extraterrestrial intelligence.

- *What is SETI?* SETI, the search for extraterrestrial intelligence, generally refers to efforts to detect signals—such as radio or laser communications—coming from civilizations on other worlds.

24.5 Interstellar Travel

- *Why is interstellar travel difficult?* The technological requirements for engines, the enormous energy demands, and social considerations all make interstellar travel a difficult undertaking. In addition, the limitation of travel at speeds less than the speed of light means that journeys will always take a long time as seen by people on Earth, although at speeds close to the speed of light the journeys may be much shorter for the travelers.

- *Will we ever achieve interstellar travel?* Some technologies that could make interstellar travel possible, such as some method of nuclear rocket propulsion or the use of solar sails, are already within our reach, at least in principle. Thus, whether we ever achieve interstellar travel is primarily a question of political will and budgets.

24.6 A Paradox: Where Are the Aliens?

- *In what way is it surprising that we have not yet discovered alien civilizations?* Given that we are already capable in principle of colonizing the galaxy in a few million years and that the galaxy was around for at least 7 billion years before Earth was even born, it seems that someone should have colonized the galaxy long ago.

- *Why are the potential solutions to the paradox "Where are the aliens?" so profound?* Every category of possible solutions to the paradox has astonishing implications for our species and our place in the universe.

Fantasy or Science Fiction?

Each of the following describes some futuristic scenario that, while perhaps common and entertaining, may or may not be plausible. In each case, decide whether the scenario is plausible according to our present understanding of science or whether it is unlikely to be possible. Explain your reasoning.

1. The first human explorers on Mars discover that the surface is littered with the ruins of an ancient civilization, including remnants of tall buildings and temples.

2. The first human explorers on Mars drill a hole into a Martian volcano to collect a sample of soil from several meters underground. Upon analysis of the soil, they discover that it holds living microbes resembling terrestrial bacteria but with a different biochemistry.

3. In 2020, a spacecraft lands on Europa and melts its way through the ice into the Europan ocean. It finds numerous strange, living microbes, along with a few larger organisms that feed on the microbes.

4. It's the year 2075. A giant telescope on the Moon, consisting of hundreds of small telescopes linked together across a distance of 500 kilometers, has just captured a series of images of a planet around a distant star that clearly show seasonal changes in vegetation.

5. A century from now, after completing a careful study of planets around stars within 100 light-years of Earth, we've discovered that the most diverse life exists on a planet orbiting a young star that formed just 100 million years ago.

6. In 2030, a brilliant teenager discovers a way to build a rocket that burns coal as its fuel and can travel at half the speed of light.

7. In the year 2750, we receive a signal from a civilization around a nearby star telling us that the *Voyager 2* spacecraft recently crash-landed on its planet.

8. Crew members of the matter–antimatter spacecraft *Star Apollo*, which left Earth in the year 2165, return to Earth in the year 2450, looking only a few years older than when they left.

9. By traveling through a wormhole apparently constructed by an advanced civilization, future explorers can journey from our solar system to a star system near the center of the galaxy in just a few hours.

10. Aliens from a distant star system invade Earth with intent to destroy us and occupy our planet, but we successfully fight them off with a great effort by our best scientists and engineers.

11. The galaxy is divided into a series of empires, each having arisen from a different civilization, that hold each other at bay through the threat of military action.

12. A single, great galactic civilization exists. It originated on a single planet long ago but is now made up of beings from many different planets, each of which was assimilated into the galactic culture in turn.

Problems

13. *Most Likely to Have Life.* Suppose you were asked to vote in a contest to name the world in our solar system (besides Earth) "most likely to have life." Which world would you cast your vote for? Explain and defend your choice in a one-page essay.

14. *Are Earth-like Planets Common?* Based on what you have learned in this book, form an opinion as to whether you think Earth-like planets will ultimately prove to be rare, common, or something in between. Write a one- to two-page essay explaining and defending your opinion.

15. *Aliens in the Movies.* Choose a science fiction movie (or television show) that involves an alien species. Do you think aliens like this could really exist? Do you think they are portrayed in a realistic way? Write a one- to two-page critical review of the movie, focusing primarily on the question of how well the movie addresses the aliens in light of current scientific knowledge.

16. *Solution to the Fermi Paradox.* Among the various possible solutions to the question "Where are the aliens?" which do you think is most likely? (If you have no opinion on their likelihood, which do you like best?) Write a one- to two-page essay in which you explain why you favor this solution.

Discussion Questions

17. *Funding the Search for Life.* Imagine that you are a member of Congress, so that your job includes deciding how much government funding goes to research in different areas of science. How much would you allot to the search for life in the universe compared to the amount allotted to research in other areas of astronomy and planetary science? Why?

18. *Conducting the Search.* Given the large number of possible places to look for life, how would you prioritize the search? For example, how would you prioritize the search for life on other worlds in our own solar system, and how would you come up with a search strategy for other star systems? Explain your priorities and strategies clearly.

19. *Distant Dream or Near-Reality?* Considering all the issues surrounding interstellar flight, when (if ever) do you think we are likely to begin traveling among the stars? Why?

20. *Where Are the Aliens?* Consider the paradox concerning the question of why we do not yet have evidence of a galactic civilization. What do *you* think is the solution to this paradox? Why?

21. *The Turning Point.* Discuss the idea that our generation has acquired a greater responsibility for the future than any previous generation. Do you agree with this assessment? If so, how should we deal with this responsibility? Defend your opinions.

For a complete list of media resources available, go to www.astronomyplace.com and choose Chapter 24 from the pull-down menu.

Astronomy Place Web Tutorials

Tutorial Review of Key Concepts

Use the following interactive **Tutorial** at www.astronomyplace.com to review key concepts from this chapter.

Detecting Extrasolar Planets Tutorial

Lesson 1 Taking a Picture of a Planet

Lesson 2 Stars' Wobbles and Properties of Planets

Lesson 3 Planetary Transits

Supplementary Tutorial Exercises

Use the interactive **Tutorial Lesson** to explore the following questions.

Detecting Extrasolar Planets Tutorial, Lessons 1–3

1. Give two reasons why visual detection of planets orbiting other stars is extremely difficult.

2. As you move away from two objects what happens to the apparent angle between them?

3. How *can* we detect extrasolar planets?

4. How might transits allow us to detect Earth-size planets around other stars?

Movies

Check out the following narrated and animated short documentary available on www.astronomyplace.com for a helpful review of key ideas covered in this chapter.

The Search for Extraterrestrial Life Movie

Web Projects

Take advantage of the useful Web links on www.astronomyplace.com to assist you with the following projects.

1. *Astrobiology News.* Go to NASA's Astrobiology home page and read some of the recent news from the search for life in the universe. Choose one recent news article, and write a one- to two-page summary of the research and how it relates to the question of life in the universe in general.

2. *Martian Meteorites.* Find information about the latest discoveries concerning Martian meteorites and whether the meteorites contain evidence of life. Choose one recent discovery that seems important, and write a short summary of how you think it alters the debate about the habitability of Mars or about life on Mars.

3. *The Search for Extraterrestrial Intelligence.* Go to the home page for the SETI Institute. Learn more about how SETI is funded and carried out. In one page or less, describe the SETI Institute and its work.

4. *Starship Design.* Find more details about one of the proposals for starship propulsion or design discussed in this chapter. How would such a starship actually be built? What new technologies would be needed, and what existing technologies could be applied? Summarize your findings in a one- to two-page report.

5. *Advanced Spacecraft Technologies.* NASA supports many efforts to incorporate new technologies into spaceships. Although few of them reach the level of being suitable for interstellar colonization, most are innovative and fascinating. Learn about one such NASA project, and write a short summary of your findings.

Appendixes

A Useful Numbers

Astronomical Distances

1 AU $\approx 1.496 \times 10^8$ km [p. 17]

1 light-year $\approx 9.46 \times 10^{12}$ km [p. 9]

1 parsec (pc) $\approx 3.09 \times 10^{13}$ km ≈ 3.26 light-years [p. 525]

1 kiloparsec (kpc) = 1,000 pc $\approx 3.26 \times 10^3$ light-years

1 megaparsec (Mpc) = 10^6 pc $\approx 3.26 \times 10^6$ light-years

Astronomical Times

1 solar day (average) = 24^{h} [p. 87]

1 sidereal day $\approx 23^{\text{h}} 56^{\text{m}} 4.09^{\text{s}}$ [p. 87]

1 synodic month (average) ≈ 29.53 solar days [p. 88]

1 sidereal month (average) ≈ 27.32 solar days [p. 88]

1 tropical year ≈ 365.242 solar days [p. 89]

1 sidereal year ≈ 365.256 solar days [p. 89]

Universal Constants

Speed of light [p. 9]: $\qquad c = 3 \times 10^5$ km/s $= 3 \times 10^8$ m/s

Gravitational constant [p. 138]: $\qquad G = 6.67 \times 10^{-11} \dfrac{\text{m}^3}{\text{kg} \times \text{s}^2}$

Planck's constant [p. 157]: $\qquad h = 6.626 \times 10^{-34}$ joule $\times$ s

Stefan–Boltzmann constant [p. 162]: $\qquad \sigma = 5.7 \times 10^{-8} \dfrac{\text{watt}}{\text{m}^2 \times \text{Kelvin}^4}$

mass of a proton: $\qquad m_{\text{p}} = 1.67 \times 10^{-27}$ kg

mass of an electron: $\qquad m_{\text{e}} = 9.1 \times 10^{-31}$ kg

Useful Sun and Earth Reference Values

Mass of the Sun: $1 M_{\text{Sun}} \approx 2 \times 10^{30}$ kg

Radius of the Sun: $1 R_{\text{Sun}} \approx 696{,}000$ km

Luminosity of the Sun: $1 L_{\text{Sun}} \approx 3.8 \times 10^{26}$ watts

Mass of the Earth: $1 M_{\text{Earth}} \approx 5.97 \times 10^{24}$ kg

Radius (equatorial) of the Earth: $1 R_{\text{Earth}} \approx 6{,}378$ km

Acceleration of gravity on Earth: $g = 9.8$ m/s^2

Escape velocity from surface of Earth: $v_{\text{escape}} = 11$ km/s $= 11{,}000$ m/s

Energy and Power Units

Basic unit of energy [p. 117]: 1 joule $= 1 \dfrac{\text{kg} \times \text{m}^2}{\text{s}^2}$

Basic unit of power [p. 153]: 1 watt = 1 joule/s

Electron-volt [p. 124]: 1 eV $= 1.60 \times 10^{-19}$ joule

B ⬤ Useful Formulas

Universal law of gravitation for the force between objects of mass M_1 and M_2, distance d between their centers [p. 138]:

$$F = G\frac{M_1 M_2}{d^2}$$

Newton's version of Kepler's third law; p and a are period and semimajor axis, respectively, of either orbiting mass [p. 140]:

$$p^2 = \frac{4\pi^2}{G(M_1 + M_2)a^3}$$

Escape velocity at distance R from center of object of mass M [p. 146]:

$$v_{escape} = \sqrt{\frac{2GM}{R}}$$

Relationship between a photon's wavelength (λ), frequency (f), and the speed of light (c) [p. 157]:

$$\lambda \times f = c$$

Energy of a photon of wavelength λ or frequency f [p. 157]:

$$E = hf = \frac{hc}{\lambda}$$

Stefan–Boltzmann law for thermal radiation at temperature T (in Kelvin) [p. 162]:

$$\text{emitted power per unit area} = \sigma T^4$$

Wien's law for the peak wavelength (λ_{max}) thermal radiation at temperature T (in Kelvin) [p. 162]:

$$\lambda_{max} = \frac{2{,}900{,}000}{T} \text{ nm}$$

Doppler shift (radial velocity is positive if the object is moving away from us and negative if it is moving toward us) [p. 166]:

$$\frac{\text{radial velocity}}{\text{speed of light}} = \frac{\text{shifted wavelength} - \text{rest wavelength}}{\text{rest wavelength}}$$

Angular separation (α) of two points with an actual separation s, viewed from a distance d (assuming d is much larger than s) [p. 175]:

$$\alpha = \frac{s}{2\pi d} \times 360°$$

Luminosity–distance formula [p. 524]:

$$\text{apparent brightness} = \frac{\text{luminosity}}{4\pi d^2}$$
(where d is the distance to the object)

Parallax formula (distance d to a star with parallax angle p in arcseconds) [p. 525]:

$$d \text{ (in parsecs)} = \frac{1}{p \text{ (in arcseconds)}}$$

The orbital velocity law [p. 613], to find the mass M_r contained within the circular orbit of radius r for an object moving at speed v:

$$M_r = \frac{r \times v^2}{G}$$

C A Few Mathematical Skills

This appendix reviews the following mathematical skills: powers of 10, scientific notation, working with units, the metric system, and finding a ratio. You should refer to this appendix as needed while studying the textbook, particularly if you are having difficulty with the Mathematical Insights.

C.1 Powers of 10

Powers of 10 simply indicate how many times to multiply 10 by itself. For example:

$$10^2 = 10 \times 10 = 100$$

$$10^6 = 10 \times 10 \times 10 \times 10 \times 10 \times 10 = 1,000,000$$

Negative powers are the reciprocals of the corresponding positive powers. For example:

$$10^{-2} = \frac{1}{10^2} = \frac{1}{100} = 0.01$$

$$10^{-6} = \frac{1}{10^6} = \frac{1}{1,000,000} = 0.000001$$

Table C.1 lists powers of 10 from 10^{-12} to 10^{12}. Note that powers of 10 follow two basic rules:

1. A positive exponent tells how many zeros follow the 1. For example, 10^0 is a 1 followed by no zeros, and 10^8 is a 1 followed by eight zeros.

2. A negative exponent tells how many places are to the right of the decimal point, including the 1. For example, $10^{-1} = 0.1$ has one place to the right of the decimal point; $10^{-6} = 0.000001$ has six places to the right of the decimal point.

Multiplying and Dividing Powers of 10

Multiplying powers of 10 simply requires adding exponents, as the following examples show:

$$10^4 \times 10^7 = \underbrace{10,000}_{10^4} \times \underbrace{10,000,000}_{10^7} = \underbrace{100,000,000,000}_{10^{4+7} = 10^{11}} = 10^{11}$$

$$10^5 \times 10^{-3} = \underbrace{100,000}_{10^5} \times \underbrace{0.001}_{10^{-3}} = \underbrace{100}_{10^{5+(-3)} = 10^2} = 10^2$$

$$10^{-8} \times 10^{-5} = \underbrace{0.00000001}_{10^{-8}} \times \underbrace{0.00001}_{10^{-5}} = \underbrace{0.0000000000001}_{10^{-8+(-5)} = 10^{-13}} = 10^{-13}$$

Table C.1 Powers of 10

Zero and Positive Powers			Negative Powers		
Power	**Value**	**Name**	**Power**	**Value**	**Name**
10^0	1	One			
10^1	10	Ten	10^{-1}	0.1	Tenth
10^2	100	Hundred	10^{-2}	0.01	Hundredth
10^3	1,000	Thousand	10^{-3}	0.001	Thousandth
10^4	10,000	Ten thousand	10^{-4}	0.0001	Ten thousandth
10^5	100,000	Hundred thousand	10^{-5}	0.00001	Hundred thousandth
10^6	1,000,000	Million	10^{-6}	0.000001	Millionth
10^7	10,000,000	Ten million	10^{-7}	0.0000001	Ten millionth
10^8	100,000,000	Hundred million	10^{-8}	0.00000001	Hundred millionth
10^9	1,000,000,000	Billion	10^{-9}	0.000000001	Billionth
10^{10}	10,000,000,000	Ten billion	10^{-10}	0.0000000001	Ten billionth
10^{11}	100,000,000,000	Hundred billion	10^{-11}	0.00000000001	Hundred billionth
10^{12}	1,000,000,000,000	Trillion	10^{-12}	0.000000000001	Trillionth

Dividing powers of 10 requires subtracting exponents, as in the following examples:

$$\frac{10^5}{10^3} = \underbrace{100,000}_{10^5} \div \underbrace{1,000}_{10^3} = \underbrace{100}_{10^{5-3} = 10^2} = 10^2$$

$$\frac{10^3}{10^7} = \underbrace{1,000}_{10^3} \div \underbrace{10,000,000}_{10^7} = \underbrace{0.0001}_{10^{3-7} = 10^{-4}} = 10^{-4}$$

$$\frac{10^{-4}}{10^{-6}} = \underbrace{0.0001}_{10^{-4}} \div \underbrace{0.000001}_{10^{-6}} = \underbrace{100}_{10^{-4-(-6)} = 10^2} = 10^2$$

Powers of Powers of 10

We can use the multiplication and division rules to raise powers of 10 to other powers or to take roots. For example:

$$(10^4)^3 = 10^4 \times 10^4 \times 10^4 = 10^{4+4+4} = 10^{12}$$

Note that we can get the same end result by simply multiplying the two powers:

$$(10^4)^3 = 10^{4 \times 3} = 10^{12}$$

Because taking a root is the same as raising to a fractional power (e.g., the square root is the same as the 1/2 power, the cube root is the same as the 1/3 power, etc.), we can use the same procedure for roots, as in the following example:

$$\sqrt{10^4} = (10^4)^{1/2} = 10^{4 \times (1/2)} = 10^2$$

Adding and Subtracting Powers of 10

Unlike with multiplication and division, there is no shortcut for adding or subtracting powers of 10. The values must be written in longhand notation. For example:

$$10^6 + 10^2 = 1,000,000 + 100 = 1,000,100$$

$$10^8 + 10^{-3} = 100,000,000 + 0.001 = 100,000,000.001$$

$$10^7 - 10^3 = 10,000,000 - 1,000 = 9,999,000$$

Summary

We can summarize our findings using n and m to represent any numbers:

- To *multiply* powers of 10, *add* exponents: $10^n \times 10^m = 10^{n+m}$

- To *divide* powers of 10, *subtract* exponents: $\dfrac{10^n}{10^m} = 10^{n-m}$

- To *raise* powers of 10 to other powers, multiply exponents: $(10^n)^m = 10^{n \times m}$

C.2 Scientific Notation

When we are dealing with large or small numbers, it's generally easier to write them with powers of 10. For example, it's much easier to write the number 6,000,000,000,000 as 6×10^{12}. This format, in which a number *between* 1 and 10 is multiplied by a power of 10, is called **scientific notation**.

Converting a Number to Scientific Notation

We can convert numbers written in ordinary notation to scientific notation with a simple two-step process:

1. Move the decimal point to come after the *first* nonzero digit.

2. The number of places the decimal point moves tells you the power of 10; the power is *positive* if the decimal point moves to the left and *negative* if it moves to the right.

Examples:

$$3{,}042 \xrightarrow[\text{3 places to left}]{\text{decimal needs to move}} 3.042 \times 10^3$$

$$0.00012 \xrightarrow[\text{4 places to right}]{\text{decimal needs to move}} 1.2 \times 10^{-4}$$

$$226 \times 10^2 \xrightarrow[\text{2 places to left}]{\text{decimal needs to move}} (2.26 \times 10^2) \times 10^2 = 2.26 \times 10^4$$

Converting a Number from Scientific Notation

We can convert numbers written in scientific notation to ordinary notation by the reverse process:

1. The power of 10 indicates how many places to move the decimal point; move it to the *right* if the power of 10 is positive and to the *left* if it is negative.

2. If moving the decimal point creates any open places, fill them with zeros.

Examples:

$$4.01 \times 10^2 \xrightarrow[\text{2 places to right}]{\text{move decimal}} 401$$

$$3.6 \times 10^6 \xrightarrow[\text{6 places to right}]{\text{move decimal}} 3{,}600{,}000$$

$$5.7 \times 10^{-3} \xrightarrow[\text{3 places to left}]{\text{move decimal}} 0.0057$$

Multiplying or Dividing Numbers in Scientific Notation

Multiplying or dividing numbers in scientific notation simply requires operating on the powers of 10 and the other parts of the number separately.

Examples:

$$(6 \times 10^2) \times (4 \times 10^5) = (6 \times 4) \times (10^2 \times 10^5) = 24 \times 10^7 = (2.4 \times 10^1) \times 10^7 = 2.4 \times 10^8$$

$$\frac{4.2 \times 10^{-2}}{8.4 \times 10^{-5}} = \frac{4.2}{8.4} \times \frac{10^{-2}}{10^{-5}} = 0.5 \times 10^{-2-(-5)} = 0.5 \times 10^3 = (5 \times 10^{-1}) \times 10^3 = 5 \times 10^2$$

Note that, in both these examples, we first found an answer in which the number multiplied by a power of 10 was *not* between 1 and 10. We therefore followed the procedure for converting the final answer to scientific notation.

Addition and Subtraction with Scientific Notation

In general, we must write numbers in ordinary notation before adding or subtracting.

Examples:

$$(3 \times 10^6) + (5 \times 10^2) = 3,000,000 + 500 = 3,000,500 = 3.0005 \times 10^6$$

$$(4.6 \times 10^9) - (5 \times 10^8) = 4,600,000,000 - 500,000,000 = 4,100,000,000 = 4.1 \times 10^9$$

When both numbers have the *same* power of 10, we can factor out the power of 10 first.

Examples:

$$(7 \times 10^{10}) + (4 \times 10^{10}) = (7 + 4) \times 10^{10} = 11 \times 10^{10} = 1.1 \times 10^{11}$$

$$(2.3 \times 10^{-22}) - (1.6 \times 10^{-22}) = (2.3 - 1.6) \times 10^{-22} = 0.7 \times 10^{-22} = 7.0 \times 10^{-23}$$

C.3 Working with Units

Showing the units of a problem as you solve it usually makes the work much easier and also provides a useful way of checking your work. If an answer does not come out with the units you expect, you probably did something wrong. In general, working with units is very similar to working with numbers, as the following guidelines and examples show.

Five Guidelines for Working with Units

Before you begin any problem, think ahead and identify the units you expect for the final answer. Then operate on the units along with the numbers as you solve the problem. The following five guidelines may be helpful when you are working with units:

1. Mathematically, it doesn't matter whether a unit is singular (e.g., meter) or plural (e.g., meters); we can use the same abbreviation (e.g., m) for both.

2. You cannot add or subtract numbers unless they have the *same* units. For example, 5 apples + 3 apples = 8 apples, but the expression 5 apples + 3 oranges cannot be simplified further.

3. You *can* multiply units, divide units, or raise units to powers. Look for key words that tell you what to do.

   ■ *Per* suggests division. For example, we write a speed of 100 kilometers per hour as:

   $$100 \, \frac{\text{km}}{\text{hr}} \quad \text{or} \quad 100 \, \frac{\text{km}}{1 \, \text{hr}}$$

- *Of* suggests multiplication. For example, if you launch a 50-kg space probe at a launch cost *of* $10,000 per kilogram, the total cost is

$$50 \ \text{kg} \times \frac{\$10,000}{\text{kg}} = \$500,000$$

- *Square* suggests raising to the second power. For example, we write an area of 75 square meters as 75 m^2.
- *Cube* suggests raising to the third power. For example, we write a volume of 12 cubic centimeters as 12 cm^3.

4. Often the number you are given is not in the units you wish to work with. For example, you may be given that the speed of light is 300,000 km/s but need it in units of m/s for a particular problem. To convert the units, simply multiply the given number by a *conversion factor:* a fraction in which the numerator (top of the fraction) and denominator (bottom of the fraction) are equal, so that the value of the fraction is 1; the number in the denominator must have the units that you wish to change. In the case of changing the speed of light from units of km/s to m/s, you need a conversion factor for kilometers to meters. Thus, the conversion factor is:

$$\frac{1,000 \ \text{m}}{1 \ \text{km}}$$

Note that this conversion factor is equal to 1, since 1,000 meters and 1 kilometer are equal, and that the units to be changed (km) appear in the denominator. We can now convert the speed of light from units of km/s to m/s simply by multiplying by this conversion factor:

$$\underbrace{300,000 \ \frac{\text{km}}{\text{s}}}_{\substack{\text{speed of light} \\ \text{in km/s}}} \times \underbrace{\frac{1,000 \ \text{m}}{1 \ \text{km}}}_{\substack{\text{conversion from} \\ \text{km to m}}} = \underbrace{3 \times 10^8 \ \frac{\text{m}}{\text{s}}}_{\substack{\text{speed of light} \\ \text{in m/s}}}$$

Note that the units of km cancel, leaving the answer in units of m/s.

5. It's easier to work with units if you replace division with multiplication by the reciprocal. For example, suppose you want to know how many minutes are represented by 300 seconds. We can find the answer by dividing 300 seconds by 60 seconds per minute:

$$300 \ \text{s} \div 60 \ \frac{\text{s}}{\text{min}}$$

However, it is easier to see the unit cancellations if we rewrite this expression by replacing the division with multiplication by the reciprocal (this process is easy to remember as "invert and multiply"):

$$300 \ \text{s} \div 60 \ \frac{\text{s}}{\text{min}} = 300 \ \text{s} \times \underbrace{\frac{1 \ \text{min}}{60 \ \text{s}}}_{\substack{\text{invert} \\ \text{and multiply}}} = 5 \ \text{min}$$

We now see that the units of seconds (s) cancel in the numerator of the first term and the denominator of the second term, leaving the answer in units of minutes.

More Examples of Working with Units

Example 1. How many seconds are there in 1 day?

Solution: We can answer the question by setting up a *chain* of unit conversions in which we start with 1 *day* and end up with *seconds*. We use the facts that there are 24 hours per day (24 hr/day), 60 minutes per hour (60 min/hr), and 60 seconds per minute (60 s/min):

$$1 \, \cancel{day} \times \underbrace{\frac{24 \, \cancel{hr}}{\cancel{day}}}_{\substack{\text{conversion} \\ \text{from} \\ \text{day to hr}}} \times \underbrace{\frac{60 \, \cancel{min}}{\cancel{hr}}}_{\substack{\text{conversion} \\ \text{from} \\ \text{hr to min}}} \times \underbrace{\frac{60 \, s}{\cancel{min}}}_{\substack{\text{conversion} \\ \text{from} \\ \text{min to s}}} = 86{,}400 \, s$$

Note that all the units cancel except *seconds*, which is what we want for the answer. There are 86,400 seconds in 1 day.

Example 2. Convert a distance of 10^8 cm to km.

Solution: The easiest way to make this conversion is in two steps, since we know that there are 100 centimeters per meter (100 cm/m) and 1,000 meters per kilometer (1,000 m/km):

$$\underbrace{10^8 \, cm}_{\substack{\text{starting} \\ \text{value}}} \times \underbrace{\frac{1 \, m}{100 \, cm}}_{\substack{\text{conversion} \\ \text{from} \\ \text{cm to m}}} \times \underbrace{\frac{1 \, km}{1{,}000 \, m}}_{\substack{\text{conversion} \\ \text{from} \\ \text{m to km}}} = 10^8 \, \cancel{cm} \times \frac{1 \, \cancel{m}}{10^2 \, \cancel{cm}} \times \frac{1 \, km}{10^3 \, \cancel{m}} = 10^3 \, km$$

Alternatively, if we recognize that the number of kilometers should be smaller than the number of centimeters (because kilometers are larger), we might decide to do this conversion by dividing as follows:

$$10^8 \, cm \div \frac{100 \, cm}{m} \div \frac{1{,}000 \, m}{km}$$

In this case, before carrying out the calculation, we replace each division with multiplication by the reciprocal:

$$10^8 \, cm \div \frac{100 \, cm}{m} \div \frac{1{,}000 \, m}{km} = 10^8 \, cm \times \frac{1 \, m}{100 \, cm} \times \frac{1 \, km}{1{,}000 \, m}$$

$$= 10^8 \, \cancel{cm} \times \frac{1 \, \cancel{m}}{10^2 \, \cancel{cm}} \times \frac{1 \, km}{10^3 \, \cancel{m}}$$

$$= 10^3 \, km$$

Note that we again get the answer that 10^8 cm is the same as 10^3 km, or 1,000 km.

Example 3. Suppose you accelerate at 9.8 m/s² for 4 seconds, starting from rest. How fast will you be going?

Solution: The question asked "how fast?" so we expect to end up with a speed. Therefore, we multiply the acceleration by the amount of time you accelerated:

$$9.8 \, \frac{m}{s^2} \times 4 \, s = (9.8 \times 4) \, \frac{m \times \cancel{s}}{s^{\cancel{2}}} = 39.2 \, \frac{m}{s}$$

Note that the units end up as a speed, showing that you will be traveling 39.2 m/s after 4 seconds of acceleration at 9.8 m/s².

Example 4. A reservoir is 2 km long and 3 km wide. Calculate its area, in both square kilometers and square meters.

Solution: We find its area by multiplying its length and width:

$$2 \text{ km} \times 3 \text{ km} = 6 \text{ km}^2$$

Next we need to convert this area of 6 km² to square meters, using the fact that there are 1,000 meters per kilometer (1,000 m/km). Note that we must square the term 1,000 m/km when converting from km² to m²:

$$6 \text{ km}^2 \times \left(1{,}000 \, \frac{\text{m}}{\text{km}}\right)^2 = 6 \text{ km}^2 \times 1{,}000^2 \, \frac{\text{m}^2}{\text{km}^2} = 6 \, \cancel{\text{km}^2} \times 1{,}000{,}000 \, \frac{\text{m}^2}{\cancel{\text{km}^2}}$$

$$= 6{,}000{,}000 \text{ m}^2$$

The reservoir area is 6 km², which is the same as 6 million m².

C.4 The Metric System (SI)

The modern version of the metric system, known as *Système Internationale d'Unites* (French for "International System of Units") or **SI**, was formally established in 1960. Today, it is the primary measurement system in nearly every country in the world with the exception of the United States. Even in the United States, it is the system of choice for science and international commerce.

The basic units of length, mass, and time in the SI are:

■ The **meter** for length, abbreviated m

■ The **kilogram** for mass, abbreviated kg

■ The **second** for time, abbreviated s

Multiples of metric units are formed by powers of 10, using a prefix to indicate the power. For example, *kilo* means 10^3 (1,000), so a kilometer is 1,000 meters; a microgram is 0.000001 gram, because *micro* means 10^{-6}, or one millionth. Some of the more common prefixes are listed in Table C.2.

Metric Conversions

Table C.3 lists conversions between metric units and units used commonly in the United States. Note that the conversions between kilograms and pounds are valid only on Earth, because they depend on the strength of gravity.

Table C.2 SI (Metric) Prefixes

	Small Values			Large Values	
Prefix	**Abbreviation**	**Value**	**Prefix**	**Abbreviation**	**Value**
Deci	d	10^{-1}	Deca	da	10^{1}
Centi	c	10^{-2}	Hecto	h	10^{2}
Milli	m	10^{-3}	Kilo	k	10^{3}
Micro	μ	10^{-6}	Mega	M	10^{6}
Nano	n	10^{-9}	Giga	G	10^{9}
Pico	p	10^{-12}	Tera	T	10^{12}

Table C.3 Metric Conversions

To Metric	From Metric
1 inch = 2.540 cm	1 cm = 0.3937 inch
1 foot = 0.3048 m	1 m = 3.28 feet
1 yard = 0.9144 m	1 m = 1.094 yards
1 mile = 1.6093 km	1 km = 0.6214 mile
1 pound = 0.4536 kg	1 kg = 2.205 pounds

Example 1. International athletic competitions generally use metric distances. Compare the length of a 100-meter race to that of a 100-yard race.

Solution: Table C.3 shows that 1 m = 1.094 yd, so 100 m is 109.4 yd. Note that 100 meters is almost 110 yards; a good "rule of thumb" to remember is that distances in meters are about 10% longer than the corresponding number of yards.

Example 2. How many square kilometers are in 1 square mile?

Solution: We use the square of the miles-to-kilometers conversion factor:

$$(1 \text{ mi}^2) \times \left(\frac{1.6093 \text{ km}}{1 \text{ mi}}\right)^2 = (1 \text{ mi}^2) \times \left(1.6093^2 \frac{\text{km}^2}{\text{mi}^2}\right) = 2.5898 \text{ km}^2$$

Therefore, 1 square mile is 2.5898 square kilometers.

C.5 Finding a Ratio

Suppose you want to compare two quantities, such as the average density of the Earth and the average density of Jupiter. The way we do such a comparison is by dividing, which tells us the *ratio* of the two quantities. In this case, the Earth's average density is 5.52 grams/cm^3 and Jupiter's average density is 1.33 grams/cm^3 (see Table 11.1), so the ratio is:

$$\frac{\text{average density of Earth}}{\text{average density of Jupiter}} = \frac{5.52 \text{ g/cm}^3}{1.33 \text{ g/cm}^3} = 4.15$$

Notice how the units cancel on both the top and bottom of the fraction. We can state our result in two equivalent ways:

■ The ratio of the Earth's average density to Jupiter's average density is 4.15.

■ The Earth's average density is 4.15 times Jupiter's average density.

Sometimes, the quantities that you want to compare may each involve an equation. In such cases, you could, of course, find the ratio by first calculating each of the two quantities individually and then dividing. However, it is much easier if you first express the ratio as a fraction, putting the equation for one quantity on top and the other on the bottom. Some of the terms in the equation may then cancel out, making any calculations much easier.

Example 1. Compare the kinetic energy of a car traveling at 100 km/hr to that of a car traveling at 50 km/hr.

Solution: We do the comparison by finding the ratio of the two kinetic energies, recalling that the formula for kinetic energy is $1/2 \, mv^2$. Since we are not told the mass of the car, you might at first think that we don't have enough information to find the ratio. However, notice what happens when we put the equations for each kinetic energy into the ratio, calling the two speeds v_1 and v_2:

$$\frac{\text{K.E. car at } v_1}{\text{K.E. car at } v_2} = \frac{\frac{1}{2} m_{car} v_1^2}{\frac{1}{2} m_{car} v_2^2} = \frac{v_1^2}{v_2^2} = \left(\frac{v_1}{v_2}\right)^2$$

All the terms cancel except those with the two speeds, leaving us with a very simple formula for the ratio. Now we put in 100 km/hr for v_1 and 50 km/hr for v_2:

$$\frac{\text{K.E. car at } 100 \text{ km/hr}}{\text{K.E. car at } 50 \text{ km/hr}} = \left(\frac{100 \text{ km/hr}}{50 \text{ km/hr}}\right)^2 = 2^2 = 4$$

The ratio of the car's kinetic energies at 100 km/hr and 50 km/hr is 4. That is, the car has four times as much kinetic energy at 100 km/hr as it has at 50 km/hr.

Example 2. Compare the strength of gravity between the Earth and the Sun to the strength of gravity between the Earth and the Moon.

Solution: We do the comparison by taking the ratio of the Earth–Sun gravity to the Earth–Moon gravity. In this case, each quantity is found from the equation of Newton's law of gravity. (See Section 5.3.) Thus, the ratio is:

$$\frac{\text{Earth–Sun gravity}}{\text{Earth–Moon gravity}} = \frac{\cancel{G}\frac{\cancel{M_{\text{Earth}}}M_{\text{Sun}}}{(d_{\text{Earth–Sun}})^2}}{\cancel{G}\frac{\cancel{M_{\text{Earth}}}M_{\text{Moon}}}{(d_{\text{Earth–Moon}})^2}} = \frac{M_{\text{Sun}}}{(d_{\text{Earth–Sun}})^2} \times \frac{(d_{\text{Earth–Moon}})^2}{M_{\text{Moon}}}$$

Note how all but four of the terms cancel; the last step comes from replacing the division with multiplication by the reciprocal (the "invert and multiply" rule for division). We can simplify the work further by rearranging the terms so that we have the masses and distances together:

$$\frac{\text{Earth–Sun gravity}}{\text{Earth–Moon gravity}} = \frac{M_{\text{Sun}}}{M_{\text{Moon}}} \times \frac{(d_{\text{Earth–Moon}})^2}{(d_{\text{Earth–Sun}})^2}$$

Now it is just a matter of looking up the numbers (see Appendix E) and calculating:

$$\frac{\text{Earth–Sun gravity}}{\text{Earth–Moon gravity}} = \frac{1.99 \times 10^{30}\ \cancel{kg}}{7.35 \times 10^{22}\ \cancel{kg}} \times \frac{(384.4 \times 10^3\ \cancel{km})^2}{(149.6 \times 10^6\ \cancel{km})^2} = 179$$

In other words, the Earth–Sun gravity is 179 times stronger than the Earth–Moon gravity.

D The Periodic Table of the Elements

Key

12	— Atomic number
Mg	— Element's symbol
Magnesium	— Element's name
24.305	— Atomic mass*

*Atomic masses are fractions because they represent a weighted average of atomic masses of different isotopes— in proportion to the abundance of each isotope on Earth.

1 **H** Hydrogen 1.00794																	2 **He** Helium 4.003
3 **Li** Lithium 6.941	4 **Be** Beryllium 9.01218											5 **B** Boron 10.81	6 **C** Carbon 12.011	7 **N** Nitrogen 14.007	8 **O** Oxygen 15.999	9 **F** Fluorine 18.988	10 **Ne** Neon 20.179
11 **Na** Sodium 22.990	12 **Mg** Magnesium 24.305											13 **Al** Aluminum 26.98	14 **Si** Silicon 28.086	15 **P** Phosphorus 30.974	16 **S** Sulfur 32.06	17 **Cl** Chlorine 35.453	18 **Ar** Argon 39.948
19 **K** Potassium 39.098	20 **Ca** Calcium 40.08	21 **Sc** Scandium 44.956	22 **Ti** Titanium 47.88	23 **V** Vanadium 50.94	24 **Cr** Chromium 51.996	25 **Mn** Manganese 54.938	26 **Fe** Iron 55.847	27 **Co** Cobalt 58.9332	28 **Ni** Nickel 58.69	29 **Cu** Copper 63.546	30 **Zn** Zinc 65.39	31 **Ga** Gallium 69.72	32 **Ge** Germanium 72.59	33 **As** Arsenic 74.922	34 **Se** Selenium 78.96	35 **Br** Bromine 79.904	36 **Fr** Krypton 83.80
37 **Rb** Rubidium 85.468	38 **Sr** Strontium 87.62	39 **Y** Yttrium 88.9059	40 **Zr** Zirconium 91.224	41 **Nb** Niobium 92.91	42 **Mo** Molybdenum 95.94	43 **Tc** Technetium (98)	44 **Ru** Ruthenium 101.07	45 **Rh** Rhodium 102.906	46 **Pd** Palladium 106.42	47 **Ag** Silver 107.868	48 **Cd** Cadmium 112.41	49 **In** Indium 114.82	50 **Sn** Tin 118.71	51 **Sb** Antimony 121.75	52 **Te** Tellurium 127.60	53 **I** Iodine 126.905	54 **Xe** Xenon 131.29
55 **Cs** Cesium 132.91	56 **Ba** Barium 137.34		72 **Hf** Hafnium 178.49	73 **Ta** Tantalum 180.95	74 **W** Tungsten 183.85	75 **Re** Rhenium 186.207	76 **Os** Osmium 190.2	77 **Ir** Iridium 192.22	78 **Pt** Platinum 195.08	79 **Au** Gold 196.967	80 **Hg** Mercury 200.59	81 **Ti** Thallium 204.383	82 **Pb** Lead 207.2	83 **Bi** Bismuth 208.98	84 **Po** Polonium (209)	85 **At** Astatine (210)	86 **Rn** Radon (222)
87 **Fr** Francium (223)	88 **Ra** Radium 226.0254		104 **Rf** Rutherfordium (261)	105 **Db** Dubnium (262)	106 **Sg** Seaborgium (263)	107 **Bh** Bohrium (262)	108 **Hs** Hassium (265)	109 **Mt** Meitnerium (266)	110 **Uun** Ununnilium (269)	111 **Uuu** Unununium (272)	112 **Uub** Ununbium (277)						

Lanthanide Series

57 **La** Lanthanum 138.906	58 **Ce** Cerium 140.12	59 **Pr** Praseodymium 140.908	60 **Nd** Neodymium 144.24	61 **Pm** Promethium (145)	62 **Sm** Samarium 150.36	63 **Eu** Europium 151.96	64 **Gd** Gadolinium 157.25	65 **Tb** Terbium 158.925	66 **Dy** Dysprosium 162.50	67 **Ho** Holmium 164.93	68 **Er** Erbium 167.26	69 **Tm** Thulium 168.934	70 **Yb** Ytterbium 173.04	71 **Lu** Lutetium 174.967

Actinide Series

89 **Ac** Actinium 227.028	90 **Th** Thorium 232.038	91 **Pa** Protactinium 231.036	92 **U** Uranium 238.029	93 **Np** Neptunium 237.048	94 **Pu** Plutonium (244)	95 **Am** Americium (243)	96 **Cm** Curium (247)	97 **Bk** Berkelium (247)	98 **Cf** Californium (251)	99 **Es** Einsteinium (252)	100 **Fm** Fermium (257)	101 **Md** Mendelevium (258)	102 **No** Nobelium (259)	103 **Lr** Lawrencium (260)

E Planetary Data

Table E.1 Physical Properties of the Sun and Planets

Name	Radius (Eq[a]) (km)	Radius (Eq) (Earth units)	Mass (kg)	Mass (Earth units)	Average Density (g/cm^3)	Surface Gravity (Earth = 1)
Sun	695,000	109	1.99×10^{30}	333,000	1.41	27.5
Mercury	2,440	0.382	3.30×10^{23}	0.055	5.43	0.38
Venus	6,051	0.949	4.87×10^{24}	0.815	5.25	0.91
Earth	6,378	1.00	5.97×10^{24}	1.00	5.52	1.00
Mars	3,397	0.533	6.42×10^{23}	0.107	3.93	0.38
Jupiter	71,492	11.19	1.90×10^{27}	317.9	1.33	2.53
Saturn	60,268	9.46	5.69×10^{26}	95.18	0.70	1.07
Uranus	25,559	3.98	8.66×10^{25}	14.54	1.22	0.91
Neptune	24,764	3.81	1.03×10^{26}	17.13	1.64	1.14
Pluto	1,160	0.181	1.31×10^{22}	0.0022	2.05	0.07

[a]Eq − equatorial.

Table E.2 Orbital Properties of the Sun and Planets

Name	Distance from Sun[a] (AU)	Distance from Sun[a] (10^6 km)	Orbital Period (years)	Orbital Inclination[b] (degrees)	Orbital Eccentricity	Sidereal Rotation Period (Earth days)[c]	Axis Tilt (degrees)
Sun	—	—	—	—	—	25.4	7.25
Mercury	0.387	57.9	0.2409	7.00	0.206	58.6	0.0
Venus	0.723	108.2	0.6152	3.39	0.007	−243.0	177.3
Earth	1.00	149.6	1.0	0.00	0.017	0.9973	23.45
Mars	1.524	227.9	1.881	1.85	0.093	1.026	25.2
Jupiter	5.203	778.3	11.86	1.31	0.048	0.41	3.08
Saturn	9.539	1,427	29.42	2.48	0.056	0.44	26.73
Uranus	19.19	2,870	84.01	0.77	0.046	−0.72	97.92
Neptune	30.06	4,497	163.7	1.77	0.010	0.67	28.8
Pluto	39.54	5,916	248.0	17.14	0.248	−6.39	119.6

[a]Semimajor axis of the orbit.

[b]With respect to the ecliptic.

[c]A negative sign indicates rotation is backward relative to other planets.

Table E.3 Satellites of the Solar System (as of 2003)[a]

Planet Satellite	Radius or Dimensions[b] (km)	Distance from Planet (10³ km)	Orbital Period[c] (Earth days)	Mass[d] (kg)	Density[d] (g/cm³)	Notes About the Satellites
Earth						**Earth**
Moon	1,738	384.4	27.322	7.349×10^{22}	3.34	*Moon:* Probably formed in giant impact.
Mars						**Mars**
Phobos	13×11×9	9.38	0.319	1.3×10^{16}	2.2	*Phobos, Deimos:* Probable captured asteroids.
Deimos	8×6×5	23.5	1.263	1.8×10^{15}	1.7	
Jupiter						**Jupiter**
Small inner moons (4 moons)	10 to 135×82×75	128–222	0.295–0.6745	—	—	*Metis, Adrastea, Amalthea, Thebe:* Small moonlets within and near Jupiter's ring system.
Io	1,821	421.6	1.769	8.933×10^{22}	3.57	*Io:* Most volcanically active object in the solar system.
Europa	1,565	670.9	3.551	4.797×10^{22}	2.97	*Europa:* Possible oceans under icy crust.
Ganymede	2,634	1,070.0	7.155	1.482×10^{23}	1.94	*Ganymede:* Largest satellite in solar system; unusual ice geology.
Callisto	2,403	1,883.0	16.689	1.076×10^{23}	1.86	*Callisto:* Cratered iceball.
Irregular group 1 (7 moons)	4–85	7,500–17,100	30–457	—	—	*Themisto, Leda, Himalia, Lysithea, Elara, and 2 others:* Probable captured moons with inclined orbits.
Irregular group 2 (46 moons)	1–30	18,300–23,100	−854 to −901 / −504 to 1,312	—	—	*Ananke, Carme, Pasiphae, Sinope, and 42 others:* Probable captured moons in inclined backward orbits.
Saturn						**Saturn**
Small inner moons (6)	10 to 97×95×77	134–151	0.574–0.695	—	—	*Pan, Atlas, Prometheus, Pandora, Epimetheus, Janus:* Small moonlets within and near Saturn's ring system.
Mimas	199	185.52	0.942	3.70×10^{19}	1.17	*Mimas, Enceladus, Tethys:* Small and medium-size iceballs, many with interesting geology.
Enceladus	249	238.02	1.370	1.2×10^{20}	1.24	
Tethys	530	294.66	1.888	6.17×10^{20}	1.26	
Calypso	15×8×8	294.66	1.888	4×10^{15}	—	*Calypso, Telesto:* Small moonlets sharing Tethys's orbit.
Telesto	15×13×8	294.67	1.888	6×10^{15}	—	
Dione	559	377.4	2.737	1.08×10^{21}	1.44	*Dione:* Medium-size iceball, with interesting geology.
Helene	18×?×15	377.4	2.737	1.6×10^{16}	—	*Helene:* Small moonlet sharing Dione's orbit.
Rhea	764	527.04	4.518	2.31×10^{21}	1.33	*Rhea:* Medium-size iceball, with interesting geology.
Titan	2,575	1,221.85	15.945	1.3455×10^{23}	1.88	*Titan:* Dense atmosphere shrouds surface; ongoing geological activity possible.
Hyperion	180×140×112	1,481.1	21.277	2.8×10^{19}	—	*Hyperion:* Only satellite known not to rotate synchronously.

Moon	Radius (km)	Dimensions (a×b×c)	Distance (10³ km)	Period (days)	Mass (kg)	Density (g/cm³)
Iapetus	718		3,561.3	79.331	1.59×10^{21}	1.21
Phoebe	110		12,952	−550.4	1×10^{19}	—
Irregular group 1 (4 moons)	7–22		11,400–17,100	453–829	—	—
Irregular group 2 (3 moons)	5–15		17,400–18,000	854–901	—	—
Irregular group 3 (5 moons)	4–10		15,600–23,400	−723 to −1,325	—	—
Uranus						
Small inner moons (11 moons)		10 to 97×95×77	134–151	0.574–0.695	—	—
Miranda	236		129.8	1.413	6.6×10^{19}	1.26
Ariel	579		191.2	2.520	1.35×10^{21}	1.65
Umbriel	584.7		266.0	4.144	1.17×10^{21}	1.44
Titania	788.9		435.8	8.706	3.52×10^{21}	1.59
Oberon	761.4		582.6	13.463	3.01×10^{21}	1.50
Irregular group (5 moons)	???–60		7,170–25,000	580–2,280	—	—
Neptune						
Small inner moons (5 moons)		29 to 104×?×89	48–74	0.296–0.554	—	—
Proteus		218×208×201	117.6	1.121	6×10^{19}	—
Triton	1,352.6		354.59	−5.875	2.14×10^{22}	2.0
Nereid	170		5,588.6	360.125	3.1×10^{19}	—
Irregulars	15–20		20,200–21,900	2,520–2,870	—	—
Pluto						
Charon	635		19.6	6.38718	1.56×10^{21}	1.6

Iapetus: Bright and dark hemispheres show greatest contrast in the solar system.

Phoebe: Very dark; material ejected from Phoebe may coat one side of Iapetus.

2000 S2, S3, S5, S6: Probable captured moons with highly inclined orbits.

2000 S4, S10, S11: Probable captured moons in inclined orbits.

2000 S1, S7, S8, S9, S12, and 2003 S1: Probable captured moons in inclined backward orbits.

Uranus

Cordelia, Ophelia, Bianca, Cressida, Desdemona, Juliet, Portia, Rosalind, Belinda, Puck, 1986 U10: Small moonlets within and near Uranus's ring system.

Miranda, Ariel, Umbriel, Titania, Oberon: Small and medium-size iceballs, with some interesting geology.

Caliban, Sycorax, Stephano, Prospero, Setebos: Too recently discovered for accurate determination of their properties; several in backward orbits.

Neptune

Naiad, Thalassa, Despina, Galatea, Larissa: Small moonlets within and near Neptune's ring system.

Triton: Probable captured Kuiper belt object—largest captured object in solar system.

Nereid: Small, icy moon; very little known.

2002 N1, N2, N3: Possible captured moons in inclined or backward orbit.

Pluto

Charon: Unusually large compared to its planet; may have formed in giant impact.

[a]Note: Authorities differ substantially on many of the values in this table.

[b]a × b × c values for the Dimensions are the approximate lengths of the axes (center to edge) for irregular moons.

[c]Negative sign indicates backward orbit.

[d]Masses and densities are most accurate for those satellites visited by a spacecraft on a flyby. Masses for the smallest moons have not been measured but can be estimated from the radius and an assumed density.

F Stellar Data

Table F.1 Stars Within 12 Light-Years

Star	Distance (ly)	Spectral Type		RA h	RA m	Dec °	Dec '	Luminosity (L/L_{Sun})
Sun	0.000016	G2	V	—	—	—	—	1.0
Proxima Centauri	4.2	M5.5	V	14	30	−62	41	0.0006
α Centauri A	4.4	G2	V	14	40	−60	50	1.6
α Centauri B	4.4	K0	V	14	40	−60	50	0.53
Barnard's Star	6.0	M4	V	17	58	+04	42	0.005
Wolf 359	7.8	M6	V	10	56	+07	01	0.0008
Lalande 21185	8.3	M2	V	11	03	+35	58	0.03
Sirius A	8.6	A1	V	06	45	−16	42	26.0
Sirius B	8.6	DA2	—	06	45	−16	42	0.002
Luyten 726-8A	8.7	M5.5	V	01	39	−17	57	0.0009
Luyten 726-8B	8.7	M6	V	01	39	−17	57	0.0006
Ross 154	9.7	M3.5	V	18	50	−23	50	0.004
Ross 248	10.3	M5.5	V	23	42	+44	11	0.001
ε Eridani	10.5	K2	V	03	33	−09	28	0.37
Lacaille 9352	10.7	M1.5	V	23	06	−35	51	0.05
Ross 128	10.9	M4	V	11	48	+00	49	0.003
EZ Aquarii A	11.3	M5	V	22	39	−15	18	0.0006
EZ Aquarii B	11.3	M6	V	22	39	−15	18	0.0004
EZ Aquarii C	11.3	M6.5	V	22	39	−15	18	0.0003
61 Cygni A	11.4	K5	V	21	07	+38	42	0.15
61 Cygni B	11.4	K7	V	21	07	+38	42	0.09
Procyon A	11.4	F5	IV–V	07	39	+05	14	7.4
Procyon B	11.4	DA	—	07	39	+05	14	0.0005
Gliese 725 A	11.4	M3	V	18	43	+59	38	0.02
Gliese 725 B	11.4	M3.5	V	18	43	+59	38	0.01
Gliese 15 A	11.6	M1.5	V	00	18	+44	01	0.03
Gliese 15 B	11.6	M3.5	V	00	18	+44	01	0.003
DX Cancri	11.8	M6.5	V	08	30	+26	47	0.0003
ε Indi	11.8	K5	V	22	03	−56	45	0.26
τ Ceti	11.9	G8	V	01	44	−15	57	0.59
GJ 1061	11.9	M5.5	V	03	36	−44	31	0.0009

Note: These data were provided by the RECONS project, courtesy of Dr. Todd Henry. The luminosities are all total (bolometric) luminosities. The DA stellar types are white dwarfs. The coordinates are for the year 2000.

Table F.2 Twenty Brightest Stars

Star	Constellation	RA h	RA m	Dec °	Dec ′	Distance (ly)	Spectral Type		Apparent Magnitude	Luminosity (L/L$_{Sun}$)
Sirius	Canis Major	6	45	−16	42	8.6	A1	V	−1.46	26
Canopus	Carina	6	24	−52	41	313	F0	Ib–II	−0.72	13,000
α Centauri	Centaurus	14	40	−60	50	4.4	G2	V	−0.01	1.6
							K0	V	1.3	0.53
Arcturus	Boötes	14	16	+19	11	37	K2	III	−0.06	170
Vega	Lyra	18	37	+38	47	25	A0	V	0.04	60
Capella	Auriga	5	17	+46	00	42	G0	III	0.75	70
							G8	III	0.85	77
Rigel	Orion	5	15	−08	12	772	B8	Ia	0.14	70,000
Procyon	Canis Minor	7	39	+05	14	11.4	F5	IV–V	0.37	7.4
Betelgeuse	Orion	5	55	+07	24	427	M2	Iab	0.41	38,000
Achernar	Eridanus	1	38	−57	15	144	B5	V	0.51	3,600
Hadar	Centaurus	14	04	−60	22	525	B1	III	0.63	100,000
Altair	Aquila	19	51	+08	52	17	A7	IV–V	0.77	10.5
Acrux	Crux	12	27	−63	06	321	B1	IV	1.39	22,000
							B3	V	1.9	7,500
Aldebaran	Taurus	4	36	+16	30	65	K5	III	0.86	350
Spica	Virgo	13	25	−11	09	260	B1	V	0.91	23,000
Antares	Scorpio	16	29	−26	26	604	M1	Ib	0.92	38,000
Pollux	Gemini	7	45	+28	01	34	K0	III	1.16	45
Fomalhaut	Piscis Austrinus	22	58	−29	37	25	A3	V	1.19	18
Deneb	Cygnus	20	41	+45	16	2,500	A2	Ia	1.26	170,000
β Crucis	Crux	12	48	−59	40	352	B0.5	IV	1.28	37,000

Note: Three of the stars on this list, Capella, α Centauri, and Acrux, are binary systems with members of comparable brightness. They are counted as single stars because that is how they appear to the naked eye. All the luminosities given are total (bolometric) luminosities. The coordinates are for the year 2000.

G Galaxy Data

Table G.1 Galaxies of the Local Group

Galaxy Name	Distance (millions of ly)	Type[a]	RA h	RA m	Dec °	Dec ′	Luminosity (millions of L$_{Sun}$)
Milky Way	—	Sbc	—	—	—	—	15,000
WLM	3.0	Irr	00	02	−15	30	50
NGC 55	4.8	Irr	00	15	−39	13	1,300
IC 10	2.7	dIrr	00	20	+59	18	160
NGC 147	2.4	dE	00	33	+48	30	131
And III	2.5	dE	00	35	+36	30	1.1
NGC 185	2.0	dE	00	39	+48	20	120
NGC 205	2.7	E	00	40	+41	41	370
M 32	2.6	E	00	43	+40	52	380
M 31	2.5	Sb	00	43	+41	16	21,000
And I	2.6	dE	00	46	+38	00	4.7
SMC	0.19	Irr	00	53	−72	50	230
Sculptor	0.26	dE	01	00	−33	42	2.2
LGS 3	2.6	dIrr	01	04	+21	53	1.3
IC 1613	2.3	Irr	01	05	+02	08	64
And II	1.7	dE	01	16	+33	26	2.4
M 33	2.7	Sc	01	34	+30	40	2,800
Phoenix	1.5	dIrr	01	51	−44	27	0.9
Fornax	0.45	dE	02	40	−34	27	15.5
EGB0427+63	4.3	dIrr	04	32	+63	36	9.1
LMC	0.16	Irr	05	24	−69	45	1,300
Carina	0.33	dE	06	42	−50	58	0.4
Leo A	2.2	dIrr	09	59	+30	45	3.0
Sextans B	4.4	dIrr	10	00	+05	20	41
NGC 3109	4.1	Irr	10	03	−26	09	160
Antlia	4.0	dIrr	10	04	−27	19	1.7
Leo I	0.82	dE	10	08	+12	18	4.8
Sextans A	4.7	dIrr	10	11	−04	42	56
Sextans	0.28	dE	10	13	−01	37	0.5
Leo II	0.67	dE	11	13	+22	09	0.6
GR 8	5.2	dIrr	12	59	+14	13	3.4
Ursa Minor	0.22	dE	15	09	+67	13	0.3
Draco	2.7	dE	17	20	+57	55	0.3
Sagittarius	0.08	dE	18	55	−30	29	18
SagDIG	3.5	dIrr	19	30	−17	41	6.8
NGC 6822	1.6	Irr	19	45	−14	48	94
DDO 210	2.6	dIrr	20	47	−12	51	0.8
IC 5152	5.2	dIrr	22	03	−51	18	70
Tucana	2.9	dE	22	42	−64	25	0.5
UKS2323-326	4.3	dE	23	26	−32	23	5.2
Pegasus	3.1	dIrr	23	29	+14	45	12

[a]Types beginning with S are spiral galaxies classified according to Hubble's system (see Chapter 19). Type E galaxies are elliptical or spheroidal. Type Irr galaxies are irregular. The prefix d denotes a dwarf galaxy.

Table G.2 Nearby Galaxies in the Messier Catalog[a,b]

Galaxy Name (M / NGC)[c]	RA h	m	Dec °	′	RV_hel[d]	RV_gal[e]	Type[f]	Nickname[c]
M 31 / NGC 224	00	43	+41	16	-300 ± 4	-122	Spiral	Andromeda
M 32 / NGC 221	00	43	+40	52	-145 ± 2	32	Elliptical	
M 33 / NGC 598	01	34	+30	40	-179 ± 3	-44	Spiral	Triangulum
M 49 / NGC 4472	12	30	+08	00	997 ± 7	929	Elliptical/ Lenticular/Seyfert	
M 51 / NGC 5194	13	30	+47	12	463 ± 3	550	Spiral/Interacting	Whirlpool
M 58 / NGC 4579	12	38	+11	49	$1,519 \pm 6$	1,468	Spiral/Seyfert	
M 59 / NGC 4621	12	42	+11	39	410 ± 6	361	Elliptical	
M 60 / NGC 4649	12	44	+11	33	$1,117 \pm 6$	1,068	Elliptical	
M 61 / NGC 4303	12	22	+04	28	$1,566 \pm 2$	1,483	Spiral/Seyfert	
M 63 / NGC 5055	13	16	+42	02	504 ± 4	570	Spiral	Sunflower
M 64 / NGC 4826	12	57	+21	41	408 ± 4	400	Spiral/Seyfert	Black Eye
M 65 / NGC 3623	11	19	+13	06	807 ± 3	723	Spiral	
M 66 / NGC 3627	11	20	+12	59	727 ± 3	643	Spiral/Seyfert	
M 74 / NGC 628	01	37	+15	47	657 ± 1	754	Spiral	
M 77 / NGC 1068	02	43	-00	01	$1,137 \pm 3$	1,146	Spiral/Seyfert	
M 81 / NGC 3031	09	56	+69	04	-34 ± 4	73	Spiral/Seyfert	
M 82 / NGC 3034	09	56	+69	41	203 ± 4	312	Irregular/Starburst	
M 83 / NGC 5236	13	37	-29	52	516 ± 4	385	Spiral/Starburst	
M 84 / NGC 4374	12	25	+12	53	$1,060 \pm 6$	1,005	Elliptical	
M 85 / NGC 4382	12	25	+18	11	729 ± 2	692	Spiral	
M 86 / NGC 4406	12	26	+12	57	-244 ± 5	-298	Elliptical/ Lenticular	
M 87 / NGC 4486	12	30	+12	23	$1,307 \pm 7$	1,254	Elliptical/Central Dominant/Seyfert	Virgo A
M 88 / NGC 4501	12	32	+14	25	$2,281 \pm 3$	2,235	Spiral/Seyfert	
M 89 / NGC 4552	12	36	+12	33	340 ± 4	290	Elliptical	
M 90 / NGC 4569	12	37	+13	10	-235 ± 4	-282	Spiral/Seyfert	
M 91 / NGC 4548	12	35	+14	30	486 ± 4	442	Spiral/Seyfert	
M 94 / NGC 4736	12	51	+41	07	308 ± 1	360	Spiral	
M 95 / NGC 3351	10	44	+11	42	778 ± 4	677	Spiral/Starburst	
M 96 / NGC 3368	10	47	+11	49	897 ± 4	797	Spiral/Seyfert	
M 98 / NGC 4192	12	14	+14	54	-142 ± 4	-195	Spiral/Seyfert	
M 99 / NGC 4254	12	19	+14	25	$2,407 \pm 3$	2,354	Spiral	
M 100 / NGC 4321	12	23	+15	49	$1,571 \pm 1$	1,525	Spiral	
M 101 / NGC 5457	14	03	+54	21	241 ± 2	360	Spiral	
M 104 / NGC 4594	12	40	-11	37	$1,024 \pm 5$	904	Spiral/Seyfert	Sombrero
M 105 / NGC 3379	10	48	+12	35	911 ± 2	814	Elliptical	
M 106 / NGC 4258	12	19	+47	18	448 ± 3	507	Spiral/Seyfert	
M 108 / NGC 3556	11	09	+55	57	695 ± 3	765	Spiral	
M 109 / NGC 3992	11	55	+53	39	$1,048 \pm 4$	1,121	Spiral	
M 110 / NGC 205	00	38	+41	25	-241 ± 3	-61	Elliptical	

[a]Galaxies identified in the catalog published by Charles Messier in 1781; these galaxies are relatively easy to observe with small telescopes.

[b]Data obtained from NED: NASA/IPAC Extragalactic Database (http://ned.ipac.caltech.edu). The original Messier list of galaxies was obtained from SED, and the list data were updated to 2001 and M 102 was dropped.

[c]The galaxies are identified by the Messier number (M followed by a number) and by their NGC numbers, which come from the *New General Catalog* published in 1888.

[d]Radial velocity in km/s, with respect to the Sun (heliocentric). Positive values mean motion away from the Sun, and negative values are toward the Sun.

[e]Radial velocity in km/s, with respect to the Milky Way Galaxy, calculated from the RV_hel values with a correction for the Sun's motion around the galactic center.

[f]Galaxies are first listed by their primary type (spiral, elliptical, or irregular) and then by any other special categories that apply (see Chapters 19 and 20).

Table G.3 Nearby, X-ray Bright Clusters of Galaxies

Cluster Name	Redshift	Distance[a] (billions of ly)	Temperature of Intracluster Medium (millions of K)	Average Orbital Velocity of Galaxies[b] (km/sec)	Cluster Mass[c] (10^{15} M_{Sun})
Abell 2142	0.0907	1.20	101. ± 2	1,132 ± 110	1.8
Abell 2029	0.0766	1.07	100. ± 3	1,164 ± 98	1.8
Abell 401	0.0737	1.03	95.2 ± 5	1,152 ± 86	1.6
Coma	0.0233	0.34	95.1 ± 1	821 ± 49	1.6
Abell 754	0.0539	0.77	93.3 ± 3	662 ± 77	1.6
Abell 2256	0.0589	0.83	87.0 ± 2	1,348 ± 86	1.4
Abell 399	0.0718	1.01	81.7 ± 7	1,116 ± 89	1.3
Abell 3571	0.0395	0.57	81.1 ± 3	1,045 ± 109	1.3
Abell 478	0.0882	1.22	78.9 ± 2	904 ± 281	1.2
Abell 3667	0.0566	0.80	78.5 ± 6	971 ± 62	1.2
Abell 3266	0.0599	0.85	78.2 ± 5	1,107 ± 82	1.2
Abell 1651a	0.0846	1.17	73.1 ± 6	685 ± 129	1.2
Abell 85	0.0560	0.80	70.9 ± 2	969 ± 95	1.2
Abell 119	0.0438	0.63	65.6 ± 5	679 ± 106	0.94
Abell 3558	0.0480	0.69	65.3 ± 2	977 ± 39	0.94
Abell 1795	0.0632	0.89	62.9 ± 2	834 ± 85	0.88
Abell 2199	0.0314	0.46	52.7 ± 1	801 ± 92	0.68
Abell 2147	0.0353	0.51	51.1 ± 4	821 ± 68	0.65
Abell 3562	0.0478	0.68	45.7 ± 8	736 ± 49	0.55
Abell 496	0.0325	0.47	45.3 ± 1	687 ± 89	0.54
Centaurus	0.0103	0.15	42.2 ± 1	863 ± 34	0.49
Abell 1367	0.0213	0.31	41.3 ± 2	822 ± 69	0.47
Hydra	0.0126	0.19	38.0 ± 1	610 ± 52	0.42
C0336	0.0349	0.50	37.4 ± 1	650 ± 170	0.41
Virgo	0.0038	0.06	25.7 ± 0.5	632 ± 41	0.23

Note: This table lists the 25 brightest clusters of galaxies in the X-ray sky from a catalog by J. P. Henry (2000).

[a]Cluster distances were computed using a value for Hubble's constant of 65 km/sec/Mpc.

[b]The average orbital velocities of galaxies given in this column are the velocity dispersions of the clusters' galaxies.

[c]This column gives each cluster's mass within the largest radius at which the intracluster medium can be in gravitational equilibrium. Because our estimates of that radius depend on Hubble's constant, these masses are inversely proportional to Hubble's constant, which we have assumed to be 65 km/s/Mpc.

H Selected Astronomical Web Sites

The Web contains a vast amount of astronomical information. For all your astronomical Web surfing, the best starting point is the Web site for this textbook:

Astronomy Place
www.astronomyplace.com

The following are some other sites that may be of particular use. In case any of the links change, you can always find live links to these sites, and many more, on the Astronomy Place Web site.

Key Mission Sites

The following table lists the Web pages for major current astronomy missions.

Site	Description	Web Address
NASA's Office of Space Science Missions Page	**Direct links to all past, present, and planned NASA space science missions**	**http://spacescience.nasa.gov/missions**
Cassini/Huygens	Mission scheduled to arrive at Saturn in 2004	http://saturn.jpl.nasa.gov/index.ctm
Chandra X-Ray Observatory	Latest discoveries, educational activities, and other information from the Chandra X-Ray Observatory	http://chandra.harvard.edu
Far Ultraviolet Spectroscopic Explorer (FUSE)	Ultraviolet observatory in space	http://fuse.pha.jhu.edu
Galileo	Mission orbiting Jupiter	http://www.jpl.nasa.gov/galileo
Hubble Space Telescope	Latest discoveries, educational activities, and other information from the Hubble Space Telescope	http://hubble.stsci.edu
Mars Exploration Program	Information on current and planned Mars missions	http://mars.jpl.nasa.gov
Microwave Anisotropy Probe (MAP)	Mission to study the cosmic microwave background	http://map.gsfc.nasa.gov
Space Infrared Telescope Facility (SIRTF)	Infrared observatory scheduled for launch in 2002	http://sirtf.caltech.edu
Stratospheric Observatory for Infrared Astronomy (SOFIA)	Airborne observatory scheduled to begin flights in 2002	http://sofia.arc.nasa.gov

Key Observatory Sites

The following table lists the Web pages leading to major ground-based observatories.

Site	Description	Web Address
World's Largest Optical Telescopes	**Direct links to most of the world's major optical observatories**	**http://www.seds.org/billa/ bigeyes.html**
Arecibo Observatory (Puerto Rico)	World's largest single-dish radio telescope	http://www.naic.edu
Cerro Tololo Inter-American Observatory	Links to major observatories on site in Cerro Tololo, Chile	http://www.ctio.noao.edu
European Southern Observatory	Links to European telescope projects in Chile, including the Very Large Telescope	http://www.eso.org
Mauna Kea Observatories	Links to major observatories in Hawaii, including Keck, Gemini, Subaru, CFHT, and others	http://www.ifa.hawaii.edu/mko
Mt. Palomar Observatory	Powerful telescope near San Diego	http://www.astro.caltech.edu/ palomarpublic
National Optical Astronomy Observatory	Home page for United States national observatories in Arizona, Hawaii, and Chile	http://www.noao.edu
National Radio Astronomy Observatory	Home page for United States national radio observatories, including the Very Large Array (VLA)	http://www.nrao.edu

More Astronomical Web Sites

The following Web sites are some of the authors' favorites among many other non-commercial resources for astronomy.

Site	Description	Web Address
The Astronomy Place	**Don't forget to start here for all your astronomical Web surfing.**	**http://www.astronomyplace.com**
American Association of Variable Star Observers (AAVSO)	One of the largest organizations of amateur astronomers in the world. Check this site if you are interested in serious amateur astronomy.	http://www.aavso.org
Astronomical Society of the Pacific	An organization for both professional astronomers and the general public, devoted largely to astronomy education.	http://www.astrosociety.org
Astronomy Picture of the Day	An archive of beautiful pictures, updated daily.	http://antwrp.gsfc.nasa.gov/apod
AstroWeb	Listing of major resources for astronomy on the Web.	http://www.stsci.edu/astroweb/astronomy.html
Canadian Space Agency	Home page for Canada's space program.	http://www.space.gc.ca
European Space Agency (ESA)	Home page for this international agency.	http://www.esa.int
The Extrasolar Planets Encyclopedia	Information about the search for and discoveries of extrasolar planets.	http://cfa-www.harvard.edu/planets
NASA Home Page	Learn almost anything you want about NASA.	http://www.nasa.gov
NASA Science News	Read the latest news from NASA; has option to subscribe to e-mail notices of news releases.	http://science.nasa.gov
The Nine Planets (University of Arizona)	A multimedia tour of the solar system.	http://www.nineplanets.org
The Planetary Society	Has more than 100,000 members who are interested in planetary exploration and the search for life in the universe.	http://planetary.org
The SETI Institute	Devoted to the search for other civilizations.	http://www.seti.org
Voyage Scale Model Solar System	Take a virtual tour of the Voyage Scale Model Solar System.	http://www.voyageonline.org

Constellation Names (English equivalent in parentheses)

Andromeda (The Chained Princess)
Antlia (The Air Pump)
Apus (The Bird of Paradise)
Aquarius (The Water Bearer)
Aquila (The Eagle)
Ara (The Altar)
Aries (The Ram)
Auriga (The Charioteer)
Boötes (The Herdsman)
Caelum (The Chisel)

Camelopardalis (The Giraffe)
Cancer (The Crab)
Canes Venatici (The Hunting Dogs)
Canis Major (The Great Dog)
Canis Minor (The Little Dog)
Capricornus (The Sea Goat)
Carina (The Keel)
Cassiopeia (The Queen)
Centaurus (The Centaur)
Cepheus (The King)
Cetus (The Whale)

Chamaeleon (The Chameleon)
Circinus (The Drawing Compass)
Columba (The Dove)
Coma Berenices (Berenice's Hair)
Corona Australis (The Southern Crown)
Corona Borealis (The Northern Crown)
Corvus (The Crow)
Crater (The Cup)
Crux (The Southern Cross)

Cygnus (The Swan)
Delphinus (The Dolphin)
Dorado (The Goldfish)
Draco (The Dragon)
Equuleus (The Little Horse)
Eridanus (The River)
Fornax (The Furnace)
Gemini (The Twins)
Grus (The Crane)
Hercules
Horologium (The Clock)
Hydra (The Sea Serpent)
Hydrus (The Water Snake)
Indus (The Indian)
Lacerta (The Lizard)
Leo (The Lion)
Leo Minor (The Little Lion)
Lepus (The Hare)
Libra (The Scales)
Lupus (The Wolf)
Lynx (The Lynx)
Lyra (The Lyre)
Mensa (The Table)
Microscopium (The Microscope)

Monoceros (The Unicorn)
Musca (The Fly)
Norma (The Level)
Octans (The Octant)
Ophiuchus (The Serpent Bearer)
Orion (The Hunter)
Pavo (The Peacock)
Pegasus (The Winged Horse)
Perseus (The Hero)
Phoenix (The Phoenix)
Pictor (The Painter's Easel)
Pisces (The Fish)
Piscis Austrinus (The Southern
  Fish)
Puppis (The Stern)
Pyxis (The Compass)
Reticulum (The Reticle)
Sagitta (The Arrow)
Sagittarius (The Archer)
Scorpius (The Scorpion)
Sculptor (The Sculptor)
Scutum (The Shield)
Serpens (The Serpent)
Sextans (The Sextant)

Taurus (The Bull)
Telescopium (The Telescope)
Triangulum (The Triangle)
Triangulum Australe (Southern
  Triangle)
Tucana (The Toucan)
Ursa Major (The Great Bear)
Ursa Minor (The Little Bear)
Vela (The Sail)
Virgo (The Virgin)
Volans (The Flying Fish)
Vulpecula (The Fox)

Constellation Locations

Each of the charts on these pages shows
half of the celestial sphere in projection, so you can use them to learn the
approximate locations of the constellations. The grid lines are marked by right
ascension and declination [Section S1.4].

J Star Charts

How to use the star charts:

Check the times and dates under each chart to find the best one for you. Take it outdoors within an hour or so of the time listed for your date. Bring a dim flashlight to help you read it.

On each chart, the round outside edge represents the horizon all around you. Compass directions around the horizon are marked in yellow. Turn the chart around so the edge marked with the direction you're facing (for example, north, southeast) is down. The stars above this horizon now match the stars you are facing. Ignore the rest until you turn to look in a different direction.

The center of the chart represents the sky overhead, so a star plotted on the chart halfway from the edge to the center can be found in the sky halfway from the horizon to straight up.

The charts are drawn for 40°N latitude (for example, Denver, New York, Madrid). If you live far south of there, stars in the southern part of your sky will appear higher than on the chart and stars in the north will be lower. If you live far north of there, the reverse is true.

Jan–March
© Sky Publishing Corp.

©1999 *Sky & Telescope*

Use this chart January, February, and March.

Early January — 1 A.M.   Early February — 11 P.M.   Early March — 9 P.M.
Late January — Midnight   Late February — 10 P.M.   Late March — Dusk

Apr–June
© Sky Publishing Corp.

©1999 *Sky & Telescope*

Use this chart April, May, and June.

Early April — 3 A.M.*   Early May — 1 A.M.*   Early June — 11 P.M.*
Late April — 2 A.M.*   Late May — Midnight*   Late June — Dusk

*Daylight Saving Time

© 1999 *Sky & Telescope*

Use this chart July, August, and September.

Early July — 1 A.M.* Early August — 11 P.M.* Early September — 9 P.M.*
Late July — Midnight* Late August — 10 P.M.* Late September — Dusk

*Daylight Saving Time

Oct.–Dec.
©Sky Publishing Corp.

©1999 *Sky & Telescope*

Use this chart October, November, and December.

Early October — 1 A.M.*   Early November — 10 P.M.   Early December — 8 P.M.
Late October — Midnight*   Late November — 9 P.M.   Late December — 7 P.M.

*Daylight Saving Time

Glossary

21-cm line A spectral line from atomic hydrogen with wavelength 21 cm (in the radio portion of the spectrum).

absolute magnitude A measure of an object's luminosity; defined to be the apparent magnitude the object would have if it were located exactly 10 parsecs away.

absolute zero The coldest possible temperature, which is 0 K.

absorption (of light) The process by which matter absorbs radiative energy.

absorption-line spectrum A spectrum that contains absorption lines.

accelerating universe The possible fate of our universe in which a repulsive force (*see* cosmological constant) causes the expansion of the universe to accelerate with time. Its galaxies will recede from one another increasingly faster, and it will become cold and dark more quickly than a coasting universe.

acceleration The rate at which an object's velocity changes. Its standard units are m/s^2.

acceleration of gravity The acceleration of a falling object. On Earth, the acceleration of gravity, designated by g, is 9.8 m/s^2.

accretion The process by which small objects gather together to make larger objects.

accretion disk A rapidly rotating disk of material that gradually falls inward as it orbits a starlike object (e.g., white dwarf, neutron star, or black hole).

active galactic nuclei The unusually luminous centers of some galaxies, thought to be powered by accretion onto supermassive black holes. Quasars are the brightest type of active galactic nuclei; radio galaxies also contain active galactic nuclei.

active galaxy A term sometimes used to describe a galaxy that contains an *active galactic nucleus.*

adaptive optics A technique in which telescope mirrors flex rapidly to compensate for the bending of starlight caused by atmospheric turbulence.

albedo Describes the fraction of sunlight reflected by a surface; albedo = 0 means no reflection at all (a perfectly black surface); albedo = 1 means all light is reflected (a perfectly white surface).

altitude (above horizon) The angular distance between the horizon and an object in the sky.

amino acids The building blocks of proteins.

analemma The figure-8 path traced by the Sun over the course of a year when viewed at the same place and the same time each day; represents the discrepancies between apparent and mean solar time.

Andromeda Galaxy (M 31; the Great Galaxy in Andromeda) The nearest large spiral galaxy to the Milky Way.

angular momentum Momentum attributable to rotation or revolution. The angular momentum of an object moving in a circle of radius r is the product $m \times v \times r$.

angular resolution (of a telescope) The smallest angular separation that two pointlike objects can have and still be seen as distinct points of light (rather than as a single point of light).

angular size (or **angular distance**) A measure of the angle formed by extending imaginary lines outward from our eyes to span an object (or between two objects).

annihilation *See* matter–antimatter annihilation

annular solar eclipse A solar eclipse during which the Moon is directly in front of the Sun but its angular size is not large enough to fully block the Sun; thus, a ring (or *annulus*) of sunlight is still visible around the Moon's disk.

Antarctic Circle The circle on the Earth with latitude 66.5°S.

antielectron *See* positron

antimatter Refers to any particle with the same mass as a particle of ordinary matter but whose other basic properties, such as electrical charge, are precisely opposite.

aphelion The point at which an object orbiting the Sun is farthest from the Sun.

apogee The point at which an object orbiting the Earth is farthest from the Earth.

apparent brightness The amount of light reaching us *per unit area* from a luminous object; often measured in units of watts/m^2.

apparent magnitude A measure of the apparent brightness of an object in the sky, based on the ancient system developed by Hipparchus.

apparent retrograde motion Refers to the apparent motion of a planet, as viewed from Earth, during the period of a few weeks or months when it moves westward relative to the stars in our sky.

apparent solar time Time measured by the actual position of the Sun in your local sky; defined so that noon is when the Sun is *on* the meridian.

arcminutes (or **minutes of arc**) One arcminute is 1/60 of 1°.

arcseconds (or **seconds of arc**) One arcsecond is 1/60 of an arcminute, or 1/3,600 of 1°.

Arctic Circle The circle on the Earth with latitude 66.5°N.

asteroid A relatively small and rocky object that orbits a star; asteroids are sometimes called *minor planets* because they are similar to planets but smaller.

asteroid belt The region of our solar system between the orbits of Mars and Jupiter in which asteroids are heavily concentrated.

astrobiology The study of life on Earth and beyond; emphasizes research into questions of the origin of life, the conditions under which life can survive, and the search for life beyond Earth.

astronomical unit (AU) The average distance (semimajor axis) of the Earth from the Sun, which is about 150 million km.

atmospheric pressure The surface pressure resulting from the overlying weight of an atmosphere.

atomic mass The combined number of protons and neutrons in an atom.

atomic number The number of protons in an atom.

atoms Consist of a nucleus made from protons and neutrons surrounded by a cloud of electrons.

aurora Dancing lights in the sky caused by charged particles entering our atmosphere; called the *aurora borealis* in the Northern Hemisphere and the *aurora australis* in the Southern Hemisphere.

autumnal equinox *See* fall equinox

azimuth (usually called *direction* in this book) Direction around the horizon from due north, measured clockwise in degrees. E.g., the azimuth of due north is 0°, due east is 90°, due south is 180°, and due west is 270°.

bar The standard unit of pressure, approximately equal to the Earth's atmospheric pressure at sea level.

baryonic matter Refers to ordinary matter made from atoms (because the nuclei of atoms contain protons and neutrons, which are both baryons).

baryons Particles, including protons and neutrons, that are made from three quarks.

basalt A type of volcanic rock that makes a low-viscosity lava when molten.

belts (on a jovian planet) Dark bands of sinking air that encircle a jovian planet at a particular set of latitudes.

Big Bang The event that gave birth to the universe.

Big Crunch If gravity ever reverses the universal expansion, the universe will someday begin to collapse and presumably end in a Big Crunch.

binary star system A star system that contains two stars.

biosphere Refers to the "layer" of life on Earth.

BL Lac objects The name given to a class of active galactic nuclei that probably represent the centers of radio galaxies whose jets happen to be pointed directly at us.

blackbody radiation *See* thermal radiation

black hole A bottomless pit in spacetime. Nothing can escape from within a black hole, and we can never again detect or observe an object that falls into a black hole.

black smokers Structures around seafloor volcanic vents that support a wide variety of life.

blueshift A Doppler shift in which spectral features are shifted to shorter wavelengths, caused when an object is moving toward the observer.

bosons Particles, such as photons, to which the exclusion principle does not apply.

bound orbits Orbits on which an object travels repeatedly around another object; bound orbits are elliptical in shape.

brown dwarf An object too small to become an ordinary star because electron degeneracy pressure halts its gravitational collapse before fusion becomes self-sustaining; brown dwarfs have mass less than $0.08M_{Sun}$.

bubble (interstellar) The surface of a bubble is an expanding shell of hot, ionized gas driven by stellar winds or supernovae; inside the bubble, the gas is very hot and has very low density.

bulge (of a spiral galaxy) The central portion of a spiral galaxy that is roughly spherical (or football shaped) and bulges above and below the plane of the galactic disk.

Cambrian explosion The dramatic diversification of life on Earth that occurred between about 540 and 500 million years ago.

carbon stars Stars whose atmospheres are especially carbon-rich, thought to be near the ends of their lives; carbon stars are the primary sources of carbon in the universe.

carbonate rock A carbon-rich rock, such as limestone, that forms underwater from chemical reactions between sediments and carbon dioxide. On Earth, most of the outgassed carbon dioxide currently resides in carbonate rocks.

carbonate–dioxide cycle The process that cycles carbon dioxide between the Earth's atmosphere and surface rocks.

Cassini division A large, dark gap in Saturn's rings, visible through small telescopes on Earth.

CCD (charge coupled device) A type of electronic light detector that has largely replaced photographic film in astronomical research.

celestial coordinates The coordinates of right ascension and declination that fix an object's position on the celestial sphere.

celestial equator (CE) The extension of the Earth's equator onto the celestial sphere.

celestial navigation Navigation on the surface of the Earth accomplished by observations of the Sun and stars.

celestial sphere The imaginary sphere on which objects in the sky appear to reside when observed from Earth.

Celsius (temperature scale) The temperature scale commonly used in daily activity internationally. Defined so that, on Earth's surface, water freezes at 0°C and boils at 100°C.

central dominant galaxy A giant elliptical galaxy found at the center of a dense cluster of galaxies, apparently formed by the merger of several individual galaxies.

Cepheid *See* Cepheid variable

Cepheid variable A particularly luminous type of pulsating variable star that follows a period–luminosity relation and hence is very useful for measuring cosmic distances.

Chandrasekhar limit *See* white dwarf limit

charged particle belts Zones in which ions and electrons accumulate and encircle a planet.

chemical enrichment The process by which the abundance of heavy elements (heavier than helium) in the interstellar medium gradually increases over time as these elements are produced by stars and released into space.

chromosphere The layer of the Sun's atmosphere below the corona; most of the Sun's ultraviolet light is emitted from this region, in which the temperature is about 10,000 K.

circulation cells (also called *Hadley cells*) Large-scale cells (similar to convection cells) in a planet's atmosphere that transport heat between the equator and the poles.

circumpolar star A star that always remains above the horizon for a particular latitude.

climate Describes the long-term average of weather.

close binary A binary star system in which the two stars are very close together.

closed universe The universe is closed if its average density is greater than the critical density, in which case spacetime must curve back on itself to the point where its overall shape is analogous to that of the surface of a sphere. In the absence of a repulsive force (*see* cosmological constant), a closed universe would someday stop expanding and begin to contract.

cluster of galaxies A collection of a few dozen or more galaxies bound together by gravity; smaller collections of galaxies are simply called *groups*.

cluster of stars A group of anywhere from several hundred to a million or so stars; star clusters come in two types—open clusters and globular clusters.

CNO cycle The cycle of reactions by which intermediate- and high-mass stars fuse hydrogen into helium.

coasting universe The possible fate of our universe in which the mass density of the universe is *smaller* than the critical density, so that the collective gravity of all matter cannot halt the expansion. In the absence of a repulsive force (*see* cosmological constant), such a universe would keep expanding forever with little change in its rate of expansion.

coma (of a comet) The dusty atmosphere of a comet created by sublimation of ices in the nucleus when the comet is near the Sun.

comet A relatively small, icy object that orbits a star.

comparative planetology The study of the solar system by examining and understanding the similarities and differences among worlds.

compound (chemical) A substance made from molecules consisting of two or more atoms with different atomic numbers.

condensates Solid or liquid particles that condense from a cloud of gas.

condensation The formation of solid or liquid particles from a cloud of gas.

conduction (of energy) The process by which thermal energy is transferred by direct contact from warm material to cooler material.

conjunction (of a planet with the Sun) When a planet and the Sun line up in the sky.

conservation of angular momentum (law of) The principle that, in the absence of net torque (twisting force), the total angular momentum of a system remains constant.

conservation of energy (law of) The principle that energy (including mass-energy) can be neither created nor destroyed, but can only change from one form to another.

conservation of momentum (law of) The principle that, in the absence of net force, the total momentum of a system remains constant.

constellation A region of the sky; 88 official constellations cover the celestial sphere.

convection The energy transport process in which warm material expands and rises, while cooler material contracts and falls.

convection cell An individual small region of convecting material.

convection zone (of a star) A region in which energy is transported outward by convection.

core (of a planet) The dense central region of a planet that has undergone differentiation.

core (of a star) The central region of a star, in which nuclear fusion can occur.

Coriolis effect Causes air or objects moving on a rotating planet to deviate from straight-line trajectories.

corona (solar) The tenuous uppermost layer of the Sun's atmosphere; most of the Sun's

X rays are emitted from this region, in which the temperature is about 1 million K.

coronal holes Regions of the corona that barely show up in X-ray images because they are nearly devoid of hot coronal gas.

cosmic microwave background The remnant radiation from the Big Bang, which we detect using radio telescopes sensitive to microwaves (which are short-wavelength radio waves).

cosmic rays Particles such as electrons, protons, and atomic nuclei that zip through interstellar space at close to the speed of light.

cosmological constant The name given to a term in Einstein's equations of general relativity. If it is not zero, then it represents a repulsive force or a type of energy (sometimes called *dark energy* or *quintessence*) that might cause the expansion of the universe to accelerate with time.

cosmological horizon The boundary of our observable universe, which is where the lookback time is equal to the age of the universe. Beyond this boundary in spacetime, we cannot see anything at all.

cosmological redshift Refers to the redshifts we see from distant galaxies, caused by the fact that expansion of the universe stretches all the photons within it to longer, redder wavelengths.

critical density The precise average density for the entire universe that marks the dividing line between a recollapsing universe and one that will expand forever.

critical universe The possible fate of our universe in which the mass density of the universe *equals* the critical density. The universe will never collapse, but in the absence of a repulsive force it will expand more and more slowly as time progresses. *See* cosmological constant

crust (of a planet) The low-density surface layer of a planet that has undergone differentiation.

cycles per second Units of frequency for a wave; describes the number of peaks (or troughs) of a wave that pass by a given point each second. Equivalent to *hertz*.

dark energy Name sometimes given to energy that could be causing the expansion of the universe to accelerate. *See* cosmological constant.

dark matter Matter that we infer to exist from its gravitational effects but from which we have not detected any light; dark matter apparently dominates the total mass of the universe.

daylight saving time Standard time plus 1 hour, so that the Sun appears on the meridian around 1 P.M. rather than around noon.

declination (dec) Analogous to latitude, but on the celestial sphere; it is the angular north-south distance between the celestial equator and a location on the celestial sphere.

degeneracy pressure A type of pressure unrelated to an object's temperature, which arises when electrons (electron degeneracy pressure) or neutrons (neutron degeneracy pressure) are

packed so tightly that the exclusion and uncertainty principles come into play.

degenerate object An object in which degeneracy pressure is the primary pressure pushing back against gravity, such as a brown dwarf, white dwarf, or neutron star.

deuterium A form of hydrogen in which the nucleus contains a proton and a neutron, rather than only a proton (as is the case for most hydrogen nuclei).

differential rotation Describes the rotation of an object in which the equator rotates at a different rate than the poles.

differentiation The process in which gravity separates materials according to density, with high-density materials sinking and low-density materials rising.

diffraction grating A finely etched surface that can split light into a spectrum.

diffraction limit The angular resolution that a telescope could achieve if it were limited only by the interference of light waves; it is smaller (i.e., better angular resolution) for larger telescopes.

dimension (mathematical) Describes the number of independent directions in which movement is possible; e.g., the surface of the Earth is two-dimensional because only two independent directions of motion are possible (north-south and east-west).

direction (in local sky) One of the two coordinates (the other is altitude) needed to pinpoint an object in the local sky. It is the direction, such as north, south, east, or west, in which you must face to see the object. *See also* azimuth

disk component (of a galaxy) The portion of a spiral galaxy that looks like a disk and contains an interstellar medium with cool gas and dust; stars of many ages are found in the disk component.

disk population Refers to stars that orbit within the disk of a spiral galaxy. Sometimes called Population I.

Doppler effect (shift) The effect that shifts the wavelengths of spectral features in objects that are moving toward or away from the observer.

down quark One of the two quark types (the other is the up quark) found in ordinary protons and neutrons. Has a charge of $-\frac{1}{3}$.

dust (or **dust grains**) Tiny solid flecks of material; in astronomy, we often discuss interplanetary dust (found within a star system) or interstellar dust (found between the stars in a galaxy). *See also* interstellar dust grains

dust tail (of a comet) One of two tails seen when a comet passes near the Sun (the other is the plasma tail); composed of small solid particles pushed away from the Sun by the radiation pressure of sunlight.

dwarf elliptical galaxy A small elliptical galaxy with less than about a billion stars.

Earth-orbiters (spacecraft) Spacecraft designed to study the Earth or the universe from Earth orbit.

eccentricity A measure of how much an ellipse deviates from a perfect circle; defined as the center-to-focus distance divided by the length of the semimajor axis.

eclipse Occurs when one astronomical object casts a shadow on another or crosses our line of sight to the other object.

eclipse seasons Periods during which lunar and solar eclipses can occur because the nodes of the Moon's orbit are nearly aligned with the Earth and Sun.

eclipsing binary A binary star system in which the two stars happen to be orbiting in the plane of our line of sight, so that each star will periodically eclipse the other.

ecliptic The Sun's apparent annual path among the constellations.

ecliptic plane The plane of the Earth's orbit around the Sun.

ejecta (from an impact) Debris ejected by the blast of an impact.

electromagnetic field An abstract concept used to describe how a charged particle would affect other charged particles at a distance.

electromagnetic force One of the four fundamental forces; it is the force that dominates atomic and molecular interactions.

electromagnetic radiation Another name for light of all types, from radio waves through gamma rays.

electromagnetic spectrum The complete spectrum of light, including radio waves, infrared, visible light, ultraviolet light, X rays, and gamma rays.

electromagnetic wave A synonym for light, which consists of waves of electric and magnetic fields.

electron degeneracy pressure Degeneracy pressure exerted by electrons, as in brown dwarfs and white dwarfs.

electrons Fundamental particles with negative electric charge; the distribution of electrons in an atom gives the atom its size.

electron-volt (eV) A unit of energy equivalent to 1.60×10^{-19} joule.

electroweak era The era of the universe during which only three forces operated (gravity, strong force, and electroweak force), lasting from 10^{-38} second to 10^{-10} second after the Big Bang.

electroweak force The force that exists at high energies when the electromagnetic force and the weak force exist as a single force.

element (chemical) A substance made from individual atoms of a particular atomic number.

ellipse A type of oval that happens to be the shape of bound orbits. An ellipse can be drawn by moving a pencil along a string whose ends are tied to two tacks; the locations of the tacks are the foci (singular, focus) of the ellipse.

elliptical galaxies Galaxies that appear rounded in shape, often longer in one direction, like a football. They have no disks and contain very little cool gas and dust compared to spiral galaxies, though they often contain very hot, ionized gas.

elongation (greatest) For Mercury or Venus, the point at which it appears farthest from the Sun in our sky.

emission (of light) The process by which matter emits energy in the form of light.

emission-line spectrum A spectrum that contains emission lines.

emission nebula Another name for an ionization nebula. *See also* ionization nebula

energy Broadly speaking, energy is what can make matter move. The three basic types of energy are kinetic, potential, and radiative.

equation of time Describes the discrepancies between apparent and mean solar time.

equivalence principle The fundamental starting point for general relativity, which states that the effects of gravity are exactly equivalent to the effects of acceleration.

era of atoms The era of the universe lasting from about 500,000 years to about 1 billion years after the Big Bang, during which it was cool enough for neutral atoms to form.

era of galaxies The present era of the universe, which began with the formation of galaxies when the universe was about 1 billion years old.

era of nuclei The era of the universe lasting from about 3 minutes to about 500,000 years after the Big Bang, during which matter in the universe was fully ionized and opaque to light. The cosmic background radiation was released at the end of this era.

era of nucleosynthesis The era of the universe lasting from about 0.001 second to about 3 minutes after the Big Bang, by the end of which virtually all of the neutrons and about one-seventh of the protons in the universe had fused into helium.

erosion The wearing down or building up of geological features by wind, water, ice, and other phenomena of planetary weather.

eruption The process of releasing hot lava on the planet's surface.

escape velocity The speed necessary for an object to completely escape the gravity of a large body such as a moon, planet, or star.

evaporation The process by which atoms or molecules escape into the gas phase from a liquid.

event Any particular point along a worldline represents a particular event; all observers will agree on the reality of an event but may disagree about its time and location.

event horizon The boundary that marks the "point of no return" between a black hole and the outside universe; events that occur within the event horizon can have no influence on our observable universe.

exchange particle According to the standard model of physics, each of the four fundamental forces is transmitted by the transfer of particular types of exchange particles.

excited state (of an atom) Any arrangement of electrons in an atom that has more energy than the ground state.

exclusion principle The law of quantum mechanics that states that two fermions cannot occupy the same quantum state at the same time.

exosphere The hot, outer layer of an atmosphere, where the atmosphere "fades away" to space.

exposure time The amount of time for which light is collected to make a single image.

extrasolar planet A planet orbiting a star other than our Sun.

Fahrenheit (temperature scale) The temperature scale commonly used in daily activity in the United States. Defined so that, on Earth's surface, water freezes at 32°F and boils at 212°F.

fall equinox (autumnal equinox) Refers both to the point in Virgo on the celestial sphere where the ecliptic crosses the celestial equator and to the moment in time when the Sun appears at that point each year (around September 21).

false-color image An image displayed in colors that are *not* the true, visible-light colors of an object.

fault (geological) A place where rocks slip sideways relative to one another.

feedback relationships Processes in which one property amplifies (positive feedback) or counteracts (negative feedback) the behavior of properties.

fermions Particles, such as electrons, neutrons, and protons, that obey the exclusion principle.

Fermi's paradox The question posed by Enrico Fermi about extraterrestrial intelligence—"So where is everybody?"—which asks why we have not observed other civilizations even though simple arguments would suggest that some ought to have spread throughout the galaxy by now.

field An abstract concept used to describe how a particle would interact with a force. For example, the idea of a *gravitational field* describes how a particle would react to the local strength of gravity, and the idea of an *electromagnetic field* describes how a charged particle would respond to forces from other charged particles.

filter (for light) A material that transmits only particular wavelengths of light.

fireball A particularly bright meteor.

flare star A small, spectral type M star that displays particularly strong flares on its surface.

flat (or Euclidean) geometry Refers to any case in which the rules of geometry for a flat plane hold, such as that the shortest distance between two points is a straight line.

flat universe A universe in which the overall geometry of spacetime is flat (Euclidean), as would be the case if the density of the universe is equal to the critical density.

flybys (spacecraft) Spacecraft that fly past a target object (such as a planet), usually just

once, as opposed to entering a bound orbit of the object.

focal plane The place where an image created by a lens or mirror is in focus.

focus (of a lens or mirror) The point at which rays of light that were initially parallel (such as light from a distant star) converge.

force Anything that can cause a change in momentum.

formation properties (of planets) In this book, for the purpose of understanding geological processes, planets are defined to be born with four formation properties: size (mass and radius), distance from the Sun, composition, and rotation rate.

frame of reference (in relativity) Two (or more) objects share the same frame of reference if they are *not* moving relative to each other.

free-fall Refers to conditions in which an object is falling without resistance; objects are weightless when in free-fall.

free-float frame A frame of reference in which all objects are weightless and hence float freely.

frequency Describes the rate at which peaks of a wave pass by a point; measured in units of 1/s, often called *cycles per second* or *hertz*.

frost line The boundary in the solar nebula beyond which ices could condense; only metals and rocks could condense within the frost line.

fundamental forces There are four known fundamental forces in nature: gravity, the electromagnetic force, the strong force, and the weak force.

fundamental particles Subatomic particles that cannot be divided into anything smaller.

galactic cannibalism The term sometimes used to describe the process by which large galaxies merge with other galaxies in collisions. *Central dominant galaxies* are products of galactic cannibalism.

galactic disk (of a spiral galaxy) *See* disk component

galactic fountain Refers to a model for the cycling of gas in the Milky Way Galaxy in which fountains of hot, ionized gas rise from the disk into the halo and then cool and form clouds as they sink back into the disk.

galactic wind A wind of low-density but extremely hot gas flowing out from a starburst galaxy, created by the combined energy of many supernovae.

galaxy A huge collection of anywhere from a few hundred million to more than a trillion stars, all bound together by gravity.

galaxy cluster *See* cluster of galaxies

galaxy evolution The formation and development of galaxies.

Galilean moons The four moons of Jupiter that were discovered by Galileo: Io, Europa, Ganymede, and Callisto.

gamma-ray burst A sudden burst of gamma rays from deep space; such bursts apparently come from distant galaxies, but their precise mechanism is unknown.

gamma rays Light with very short wavelengths (and hence high frequencies)—shorter than those of X rays.

gap moons Tiny moons located within a gap in a planet's ring system. The gravity of a gap moon helps clear the gap.

gas phase The phase of matter in which atoms or molecules can move essentially independently of one another.

gas pressure Describes the force (per unit area) pushing on any object due to surrounding gas. *See also* pressure

genetic code The "language" that living cells use to read the instructions chemically encoded in DNA.

geocentric universe (ancient belief in) The idea that the Earth is the center of the entire universe.

geological controlling factors In this book, for the purpose of understanding geological processes, geology is considered to be influenced primarily by four geological controlling factors: surface gravity, internal temperature, surface temperature, and the presence (and extent) of an atmosphere.

geological processes The four basic geological processes are impact cratering, volcanism, tectonics, and erosion.

geology The study of surface features (on a moon, planet, or asteroid) and the processes that create them.

giant molecular cloud A very large cloud of cold, dense interstellar gas, typically containing up to a million solar masses worth of material. *See also* molecular clouds

giants (luminosity class III) Stars that appear just below the supergiants on the H–R diagram because they are somewhat smaller in radius and lower in luminosity.

global positioning system (GPS) A system of navigation by satellites orbiting the Earth.

global wind patterns (or **global circulation**) Wind patterns that remain fixed on a global scale, determined by the combination of surface heating and the planet's rotation.

globular cluster A spherically shaped cluster of up to a million or more stars; globular clusters are found primarily in the halos of galaxies and contain only very old stars.

gluons The exchange particles for the strong force.

grand unified theory (GUT) A theory that unifies three of the four fundamental forces—the strong force, the weak force, and the electromagnetic force (but not gravity)—in a single model.

granulation (on the Sun) The bubbling pattern visible in the photosphere, produced by the underlying convection.

gravitation (law of) *See* universal law of gravitation

gravitational constant The experimentally measured constant G that appears in the law of universal gravitation;

$$G = 6.67 \times 10^{-11} \frac{\text{m}^3}{\text{kg} \times \text{s}^2}.$$

gravitational contraction The process in which gravity causes an object to contract, thereby converting gravitational potential energy into thermal energy.

gravitational encounter Occurs when two (or more) objects pass near enough so that each can feel the effects of the other's gravity and can therefore exchange energy.

gravitational equilibrium Describes a state of balance in which the force of gravity pulling inward is precisely counteracted by pressure pushing outward.

gravitational lensing The magnification or distortion (into arcs, rings, or multiple images) of an image caused by light bending through a gravitational field, as predicted by Einstein's general theory of relativity.

gravitational redshift A redshift caused by the fact that time runs slow in gravitational fields.

gravitational time dilation The slowing of time that occurs in a gravitational field, as predicted by Einstein's general theory of relativity.

gravitational waves Predicted by Einstein's general theory of relativity, these waves travel at the speed of light and transmit distortions of space through the universe. Although not yet observed directly, we have strong indirect evidence that they exist.

gravitationally bound system Any system of objects, such as a star system or a galaxy, that is held together by gravity.

gravitons The exchange particles for the force of gravity.

gravity One of the four fundamental forces; it is the force that dominates on large scales.

grazing incidence (in telescopes) Reflections in which light grazes a mirror surface and is deflected at a small angle; commonly used to focus high-energy ultraviolet light and X rays.

great circle A circle on the surface of a sphere whose center is at the center of the sphere.

Great Red Spot A large, high-pressure storm on Jupiter.

greenhouse effect The process by which greenhouse gases in an atmosphere make a planet's surface temperature warmer than it would be in the absence of an atmosphere.

greenhouse gases Gases, such as carbon dioxide, water vapor, and methane, that are particularly good absorbers of infrared light but are transparent to visible light.

Gregorian calendar Our modern calendar, introduced by Pope Gregory in 1582.

ground state (of an atom) The lowest possible energy state of the electrons in an atom.

group (of galaxies) A few to a few dozen galaxies bound together by gravity. *See also* cluster of galaxies

GUT era The era of the universe during which only two forces operated (gravity and the grand-unified-theory or GUT force), lasting from 10^{-43} second to 10^{-38} second after the Big Bang.

GUT force The proposed force that exists at very high energies when the strong force, the weak force, and the electromagnetic force (but not gravity) all act as one.

H II region Another name for an ionization nebula. *See* ionization nebula

habitable zone The region around a star in which planets could potentially have surface temperatures at which liquid water could exist.

Hadley cells *See* circulation cells

half-life The time it takes for half of the nuclei in a given quantity of a radioactive substance to decay.

halo (of a galaxy) The spherical region surrounding the disk of a spiral galaxy.

Hawking radiation Radiation predicted to arise from the evaporation of black holes.

heavy elements In astronomy, *heavy elements* generally refers to all elements *except* hydrogen and helium.

helium-capture reactions Fusion reactions that fuse a helium nucleus into some other nucleus; such reactions can fuse carbon into oxygen, oxygen into neon, neon into magnesium, and so on.

helium flash The event that marks the sudden onset of helium fusion in the previously inert helium core of a low-mass star.

helium fusion The fusion of three helium nuclei into one carbon nucleus; also called the *triple-alpha reaction*.

hertz (Hz) The standard unit of frequency for light waves; equivalent to units of 1/s.

Hertzsprung–Russell (H–R) diagram A graph plotting individual stars as points, with stellar luminosity on the vertical axis and spectral type (or surface temperature) on the horizontal axis.

high-mass stars Stars born with masses above about $8M_{\text{Sun}}$; these stars will end their lives by exploding as supernovae.

horizon A boundary that divides what we can see from what we cannot see.

horizontal branch The horizontal line of stars that represents helium-burning stars on an H–R diagram for a cluster of stars.

horoscope A predictive chart made by an astrologer; in scientific studies, horoscopes have never been found to have any validity as predictive tools.

hot spot (geological) A place within a plate of the lithosphere where a localized plume of hot mantle material rises.

hour angle (HA) The angle or time (measured in hours) since an object was last on the meridian in the local sky. Defined to be 0 hours for objects that *are* on the meridian.

Hubble's constant A number that expresses the current rate of expansion of the universe; designated H_0, it is usually stated in units of km/s/Mpc. The reciprocal of Hubble's constant is the age the universe would have *if* the expansion rate had never changed.

Hubble's law Mathematically expresses the idea that more distant galaxies move away from us faster; its formula is $v = H_0 \times d$, where v is a galaxy's speed away from us, d is its distance, and H_0 is Hubble's constant.

hydrogen compounds Compounds that contain hydrogen and were common in the solar nebula, such as water (H_2O), ammonia (NH_3), and methane (CH_4).

hydrogen-shell burning Hydrogen fusion that occurs in a shell surrounding a stellar core.

hydrosphere Refers to the "layer" of water on the Earth consisting of oceans, lakes, rivers, ice caps, and other liquid water and ice.

hydrostatic equilibrium *See* gravitational equilibrium

hyperbola The precise mathematical shape of one type of unbound orbit (the other is a parabola) allowed under the force of gravity; at great distances from the attracting object, a hyperbolic path looks like a straight line.

hypernova A term sometimes used to describe a supernova (explosion) of a star so massive that it leaves a black hole behind.

hyperspace Any space with more than three dimensions.

hypothesis A tentative model proposed to explain some set of observed facts, but which has not yet been rigorously tested and confirmed.

ices (in solar system theory) Materials that are solid only at low temperatures, such as the hydrogen compounds water, ammonia, and methane.

image A picture of an object made by focusing light.

imaging (in astronomical research) The process of obtaining pictures of astronomical objects.

impact The collision of a small body (such as an asteroid or comet) with a larger object (such as a planet or moon).

impact basin A very large impact crater often filled by a lava flow.

impact crater A bowl-shaped depression left by the impact of an object that strikes a planetary surface (as opposed to burning up in the atmosphere).

impact cratering The excavation of bowl-shaped depressions (*impact craters*) by asteroids or comets striking a planet's surface.

impactor The object responsible for an impact.

inflation (of the universe) A sudden and dramatic expansion of the universe thought to have occurred at the end of the GUT era.

infrared light Light with wavelengths that fall in the portion of the electromagnetic spectrum between radio waves and visible light.

inner solar system Generally considered to encompass the region of our solar system out to about the orbit of Mars.

intensity (of light) A measure of the amount of energy coming from light of specific wavelength in the spectrum of an object.

interferometry A telescopic technique in which two or more telescopes are used in tandem to produce much better angular

resolution than the telescopes could achieve individually.

intermediate-mass stars Stars born with masses between about $2-8M_{Sun}$; these stars end their lives by ejecting a planetary nebula and becoming a white dwarf.

interstellar cloud A cloud of gas and dust between the stars.

interstellar dust grains Tiny solid flecks of carbon and silicon minerals found in cool interstellar clouds; they resemble particles of smoke and form in the winds of red giant stars.

interstellar medium Refers to gas and dust that fills the space between stars in a galaxy.

interstellar ramjet A hypothesized type of spaceship that uses a giant scoop to sweep up interstellar gas for use in a nuclear fusion engine.

intracluster medium Hot, X-ray-emitting gas found between the galaxies within a cluster of galaxies.

inverse square law Any quantity that decreases with the square of the distance between two objects is said to follow an inverse square law.

inversion (atmospheric) A local weather condition in which air is colder near the surface than higher up in the troposphere—the opposite of the usual condition, in which the troposphere is warmer at the bottom.

Io torus A donut-shaped charged-particle belt around Jupiter that approximately traces Io's orbit.

ionization The process of stripping an electron from an atom.

ionization nebula A colorful, wispy cloud of gas that glows because neighboring hot stars irradiate it with ultraviolet photons that can ionize hydrogen atoms.

ionosphere A portion of the thermosphere in which ions are particularly common (due to ionization by X rays from the Sun).

ions Atoms with a positive or negative electrical charge.

irregular galaxies Galaxies that look neither spiral nor elliptical.

isotopes Each different isotope of an element has the *same* number of protons but a *different* number of neutrons.

jets High-speed streams of gas ejected from an object into space.

joule The international unit of energy, equivalent to about 1/4,000 of a Calorie.

jovian nebulae The clouds of gas that swirled around the jovian planets, from which the moons formed.

jovian planets Giant gaseous planets similar in overall composition to Jupiter.

Julian calendar The calendar introduced in 46 B.C. by Julius Caesar and used until it was replaced by the Gregorian calendar.

Kelvin (temperature scale) The most commonly used temperature scale in science, defined such that absolute zero is 0 K and water freezes at 273.15 K.

Kepler's first law States that the orbit of each planet about the Sun is an ellipse with the Sun at one focus.

Kepler's laws of planetary motion Three laws discovered by Kepler that describe the motion of the planets around the Sun.

Kepler's second law States that, as a planet moves around its orbit, it sweeps out equal areas in equal times. This tells us that a planet moves faster when it is closer to the Sun (near perihelion) than when it is farther from the Sun (near aphelion) in its orbit.

Kepler's third law States that the square of a planet's orbital period is proportional to the cube of its average distance from the Sun (semi-major axis), which tells us that more distant planets move more slowly in their orbits. In its original form, written $p^2 = a^3$. *See also* Newton's version of Kepler's third law

kinetic energy Energy of motion, given by the formula $\frac{1}{2}mv^2$.

Kirchhoff's laws A set of rules that summarizes the conditions under which objects produce thermal, absorption line, or emission line spectra. In brief: (1) An opaque object produces thermal radiation. (2) An absorption line spectrum occurs when thermal radiation passes through a thin gas that is cooler than the object emitting the thermal radiation. (3) An emission line spectrum occurs when we view a cloud of gas that is warmer than any background source of light.

Kuiper belt The comet-rich region of our solar system that spans distances of about 30–100 AU from the Sun; Kuiper belt comets have orbits that lie fairly close to the plane of planetary orbits and travel around the Sun in the same direction as the planets.

Large Magellanic Cloud One of two small, irregular galaxies (the other is the Small Magellanic Cloud) located about 150,000 light-years away; it probably orbits the Milky Way Galaxy.

large-scale structure (of the universe) Generally refers to structure of the universe on size scales larger than that of clusters of galaxies.

latitude The angular north-south distance between the Earth's equator and a location on the Earth's surface.

leap year A calendar year with 366 rather than 365 days; our current calendar (the Gregorian calendar) has a leap year every 4 years (by adding February 29) except in century years that are not divisible by 400.

length contraction Refers to the effect in which you observe lengths to be shortened in reference frames moving relative to you.

lenticular galaxies Galaxies that look lens-shaped when seen edge-on, resembling spiral galaxies without arms. They tend to have less cool gas than normal spiral galaxies but more gas than elliptical galaxies.

leptons Fermions *not* made from quarks, such as electrons and neutrinos.

life track A track drawn on an H–R diagram to represent the changes in a star's surface temperature and luminosity during its life; also called an *evolutionary track*.

light-collecting area (of a telescope) The area of the primary mirror or lens that collects light in a telescope.

light curve A graph of an object's intensity against time.

light gases (in solar system theory) Refers to hydrogen and helium, which never condense under solar nebula conditions.

light pollution Human-made light that hinders astronomical observations.

light-year The distance that light can travel in 1 year, which is 9.46 trillion km.

liquid phase The phase of matter in which atoms or molecules are held together but move relatively freely.

lithosphere The relatively rigid outer layer of a planet; generally encompasses the crust and the uppermost portion of the mantle.

Local Bubble (interstellar) The bubble of hot gas in which our Sun and other nearby stars apparently reside. *See also* bubble (interstellar)

Local Group The group of more than 30 galaxies to which the Milky Way Galaxy belongs.

local sidereal time (LST) Sidereal time for a particular location, defined according to the position of the spring equinox in the local sky. More formally, the local sidereal time at any moment is defined to be the hour angle of the spring equinox.

local sky The sky as viewed from a particular location on Earth (or another solid object). Objects in the local sky are pinpointed by the coordinates of *altitude* and *direction* (or *azimuth*).

Local Supercluster The supercluster of galaxies to which the Local Group belongs.

longitude The angular east-west distance between the prime meridian (which passes through Greenwich) and a location on the Earth's surface.

lookback time Refers to the amount of time since the light we see from a distant object was emitted. I.e., if an object has a lookback time of 400 million years, we are seeing it as it looked 400 million years ago.

low-mass stars Stars born with masses less than about $2M_{Sun}$; these stars end their lives by ejecting a planetary nebula and becoming a white dwarf.

luminosity The total power output of an object, usually measured in watts or in units of solar luminosities ($L_{Sun} = 3.8 \times 10^{26}$ watts).

luminosity class Describes the region of the H–R diagram in which a star falls. Luminosity class I represents supergiants, III represents giants, and V represents main-sequence stars; luminosity classes II and IV are intermediate to the others.

luminosity–distance formula The formula that relates apparent brightness, luminosity, and distance:

$$\text{apparent brightness} = \frac{\text{luminosity}}{4\pi \times (\text{distance})^2}$$

lunar eclipse Occurs when the Moon passes through the Earth's shadow, which can occur only at full moon; may be total, partial, or penumbral.

lunar maria The regions of the Moon that look smooth from Earth and actually are impact basins.

lunar month *See* synodic month

lunar phase Describes the appearance of the Moon as seen from Earth.

MACHOs Stands for *massive compact halo objects* and represents one possible form of dark matter in which the dark objects are relatively large, like planets or brown dwarfs.

magma Underground molten rock.

magnetic braking The process by which a star's rotation slows as its magnetic field transfers its angular momentum to the surrounding nebula.

magnetic field Describes the region surrounding a magnet in which it can affect other magnets or charged particles in its vicinity.

magnetic-field lines Lines that represent how the needles on a series of compasses would point if they were laid out in a magnetic field.

magnetosphere The region surrounding a planet in which charged particles are trapped by the planet's magnetic field.

magnitude system A system of describing stellar brightness by using numbers, called *magnitudes*, based on an ancient Greek way of describing the brightnesses of stars in the sky. This system uses *apparent magnitude* to describe a star's apparent brightness and *absolute magnitude* to describe a star's luminosity.

main sequence (luminosity class V) The prominent line of points running from the upper left to the lower right on an H–R diagram; main-sequence stars shine by fusing hydrogen in their cores.

main-sequence fitting A method for measuring the distance to a cluster of stars by comparing the apparent brightness of the cluster's main sequence with the standard main sequence.

main-sequence lifetime The length of time for which a star of a particular mass can shine by fusing hydrogen into helium in its core.

main-sequence turnoff A method for measuring the age of a cluster of stars from the point on its H–R diagram where its stars turn off from the main sequence; the age of the cluster is equal to the main-sequence lifetime of stars at the main-sequence turnoff point.

mantle (of a planet) The rocky layer that lies between a planet's core and crust.

Martian meteorite This term is used to describe meteorites found on Earth that are thought to have originated on Mars.

mass A measure of the amount of matter in an object.

mass-energy The potential energy of mass, which has an amount $E = mc^2$.

mass exchange (in close binary star systems) The process in which tidal forces cause matter to spill from one star to a companion star in a close binary system.

mass extinction An event in which a large fraction of the species living on Earth go extinct, such as the event in which the dinosaurs died out about 65 million years ago.

mass increase (in relativity) Refers to the effect in which an object moving past you seems to have a mass greater than its rest mass.

mass-to-light ratio The mass of an object divided by its luminosity, usually stated in units of solar masses per solar luminosity. Objects with high mass-to-light ratios must contain substantial quantities of dark matter.

massive-star supernova A supernova that occurs when a massive star dies, initiated by the catastrophic collapse of its iron core; often called a Type II supernova.

matter–antimatter annihilation Occurs when a particle of matter and a particle of antimatter meet and convert all of their mass-energy to photons.

mean solar time Time measured by the average position of the Sun in your local sky over the course of the year.

meridian A half-circle extending from your horizon (altitude 0°) due south, through your zenith, to your horizon due north.

metallic hydrogen Hydrogen that is so compressed that the hydrogen atoms all share electrons and thereby take on properties of metals, such as conducting electricity. Occurs only under very high pressure conditions, such as that found deep within Jupiter.

metals (in solar system theory) Elements, such as nickel, iron, and aluminum, that condense at fairly high temperatures.

meteor A flash of light caused when a particle from space burns up in our atmosphere.

meteor shower A period during which many more meteors than usual can be seen.

meteorite A rock from space that lands on Earth.

Metonic cycle The 19-year period, discovered by the Babylonian astronomer Meton, over which the lunar phases occur on the same dates.

microwaves Light with wavelengths in the range of micrometers to millimeters. Microwaves are generally considered to be a subset of the radio wave portion of the electromagnetic spectrum.

mid-ocean ridges (on Earth) Long ridges of undersea volcanoes, along which mantle material erupts onto the ocean floor and pushes apart the existing seafloor on either side. These ridges are essentially the source of new seafloor crust, which then makes its way along the ocean bottom for millions of years before returning to the mantle at a subduction zone.

Milky Way Used both as the name of our galaxy and to refer to the band of light we see in the sky when we look into the plane of the Milky Way Galaxy.

millisecond pulsars Pulsars with rotation periods of a few thousandths of a second.

model (scientific) A representation of some aspect of nature that can be used to explain and predict real phenomena without invoking myth, magic, or the supernatural.

molecular bands The tightly bunched lines in an object's spectrum that are produced by molecules.

molecular clouds Cool, dense interstellar clouds in which the low temperatures allow hydrogen atoms to pair up into hydrogen molecules (H_2).

molecular dissociation The process by which a molecule splits into its component atoms.

molecule Technically the smallest unit of a chemical element or compound; in this text, the term refers only to combinations of two or more atoms held together by chemical bonds.

momentum The product of an object's mass and velocity.

moon An object that orbits a planet.

mutations Errors in the copying process when a living cell replicates itself.

natural selection The process by which mutations that make an organism better able to survive get passed on to future generations.

neap tides The lower-than-average tides on Earth that occur at first- and third-quarter moon, when the tidal forces from the Sun and Moon oppose one another.

nebula A cloud of gas in space, usually one that is glowing.

nebular capture The process by which icy planetesimals capture hydrogen and helium gas to form jovian planets.

nebular theory The detailed theory that describes how our solar system formed from a cloud of interstellar gas and dust.

net force The overall force to which an object responds; the net force is equal to the rate of change in the object's momentum, or equivalently to the object's mass × acceleration.

neutrino A type of fundamental particle that has extremely low mass and responds only to the weak force; neutrinos are leptons and come in three types—electron neutrinos, mu neutrinos, and tau neutrinos.

neutron degeneracy pressure Degeneracy pressure exerted by neutrons, as in neutron stars.

neutron star The compact corpse of a high-mass star left over after a supernova; typically contains a mass comparable to the mass of the Sun in a volume just a few kilometers in radius.

neutrons Particles with no electrical charge found in atomic nuclei, built from three quarks.

newton The standard unit of force in the metric system:

$$1 \text{ newton} = 1 \frac{\text{kg} \times \text{m}}{\text{s}^2}$$

Newton's first law of motion States that, in the absence of a net force, an object moves with constant velocity.

Newton's laws of motion Three basic laws that describe how objects respond to forces.

Newton's second law of motion States how a net force affects an object's motion. Specifically: force = rate of change in momentum, or force = mass × acceleration.

Newton's third law of motion States that, for any force, there is always an equal and opposite reaction force.

Newton's universal law of gravitation *See* universal law of gravitation

Newton's version of Kepler's third law This generalization of Kepler's third law can be used to calculate the masses of orbiting objects from measurements of orbital period and distance. Usually written as:

$$p^2 = \frac{4\pi^2}{G(M_1 + M_2)}a^3$$

nodes (of Moon's orbit) The two points in the Moon's orbit where it crosses the ecliptic plane.

nonbaryonic matter Refers to exotic matter that is not part of the normal composition of atoms, such as neutrinos or the hypothetical WIMPs.

nonscience As defined in this book, nonscience is any way of searching for knowledge that makes no claim to follow the scientific method, such as seeking knowledge through intuition, tradition, or faith.

north celestial pole (NCP) The point on the celestial sphere directly above the Earth's North Pole.

nova The dramatic brightening of a star that lasts for a few weeks and then subsides; occurs when a burst of hydrogen fusion ignites in a shell on the surface of an accreting white dwarf in a binary star system.

nuclear fission The process in which a larger nucleus splits into two (or more) smaller particles.

nuclear fusion The process in which two (or more) smaller nuclei slam together and make one larger nucleus.

nucleus (of an atom) The compact center of an atom made from protons and neutrons.

nucleus (of a comet) The solid portion of a comet, and the only portion that exists when the comet is far from the Sun.

observable universe The portion of the entire universe that, at least in principle, can be seen from Earth.

Occam's razor A principle often used in science, holding that scientists should prefer the simpler of two models that agree equally well with observations. Named after the medieval scholar William of Occam (1285–1349).

Olbers' paradox Asks the question of how the night sky can be dark if the universe is infinite and full of stars.

Oort cloud A huge, spherical region centered on the Sun, extending perhaps halfway to the nearest stars, in which trillions of comets orbit the Sun with random inclinations, orbital directions, and eccentricities.

opacity A measure of how much light a material absorbs compared to how much it transmits; materials with higher opacity absorb more light.

opaque (material) Describes a material that absorbs light.

open cluster A cluster of up to several thousand stars; open clusters are found only in the disks of galaxies and often contain young stars.

open universe The universe is open if its average density is less than the critical density, in which case spacetime has an overall shape analogous to the surface of a saddle.

opposition The point at which a planet appears opposite the Sun in our sky.

optical quality Describes the ability of a lens, mirror, or telescope to obtain clear and properly focused images.

orbital resonance Describes any situation in which one object's orbital period is a simple ratio of another object's period, such as 1/2, 1/4, or 5/3. In such cases, the two objects periodically line up with each other, and the extra gravitational attractions at these times can affect the objects' orbits.

orbital velocity law This law (a variation on Newton's version of Kepler's third law) allows us to use a star's orbital speed and distance from the galactic center to determine the total mass of the galaxy contained *within* the star's orbit. Mathematically, it is written:

$$M_r = \frac{r \times v^2}{G}$$

where M_r is the mass contained within the star's orbit, r is the star's distance from the galactic center, v is the star's orbital velocity, and G is the gravitational constant.

orbiters (of other worlds) Spacecraft that go into orbit of another world for long-term study.

outer solar system Generally considered to encompass the region of our solar system beginning at about the orbit of Jupiter.

outgassing The process of releasing gases from a planetary interior, usually through volcanic eruptions.

oxidation Refers to chemical reactions, often with the surface of a planet, that remove oxygen from the atmosphere.

ozone The molecule O_3, which is a particularly good absorber of ultraviolet light.

ozone depletion Refers to the declining levels of atmospheric ozone found worldwide on Earth, especially in Antarctica, in recent years.

ozone hole A place where the concentration of ozone in the stratosphere is dramatically lower than is the norm.

pair production The process in which a concentration of energy spontaneously turns into a particle and its antiparticle.

parabola The precise mathematical shape of a special type of unbound orbit allowed under the force of gravity; if an object in a parabolic orbit loses only a tiny amount of energy, it will become bound.

paradigm (in science) Refers to general patterns of thought that tend to shape scientific beliefs during a particular time period.

paradox A situation that, at least at first, seems to violate common sense or contradict itself. Resolving paradoxes often leads to deeper understanding.

parallax The apparent shifting of an object against the background, due to viewing it from different positions. *See also* stellar parallax

parallax angle Half of a star's annual back-and-forth shift due to stellar parallax; related to the star's distance according to the formula

$$\text{distance in parsecs} = \frac{1}{p}$$

where p is the parallax angle in arcseconds.

parsec (pc) Approximately equal to 3.26 light-years; it is the distance to an object with a parallax angle of 1 arcsecond.

partial lunar eclipse A lunar eclipse in which the Moon becomes only partially covered by the Earth's umbral shadow.

partial solar eclipse A solar eclipse during which the Sun becomes only partially blocked by the disk of the Moon.

particle accelerator A machine designed to accelerate subatomic particles to high speeds in order to create new particles or to test fundamental theories of physics.

particle era The era of the universe lasting from 10^{-10} second to 0.001 second after the Big Bang, during which subatomic particles were continually created and destroyed and ending when matter annihilated antimatter.

peculiar velocity (of a galaxy) The component of a galaxy's velocity relative to the Milky Way that deviates from the velocity expected by Hubble's law.

penumbra The lighter, outlying regions of a shadow.

penumbral (lunar) eclipse A lunar eclipse in which the Moon passes only within the Earth's penumbral shadow and does not fall within the umbra.

perigee The point at which an object orbiting the Earth is nearest to the Earth.

perihelion The point at which an object orbiting the Sun is closest to the Sun.

period–luminosity relation The relation that describes how the luminosity of a Cepheid variable star is related to the period between peaks in its brightness; the longer the period, the more luminous the star.

phase (of matter) Describes the way in which atoms or molecules are held together; the common phases are solid, liquid, and gas.

photon An individual particle of light, characterized by a wavelength and a frequency.

photosphere The visible surface of the Sun, where the temperature averages just under 6,000 K.

pixel An individual "picture element" on a CCD.

Planck era The era of the universe prior to the Planck time.

Planck time The time when the universe was 10^{-43} second old, before which random energy fluctuations were so large that our current theories are powerless to describe what might have been happening.

Planck's constant A universal constant, abbreviated h, with value $h = 6.626 \times 10^{-43}$ joule $\times$ s.

planet An object that orbits a star and that, while much smaller than a star, is relatively large in size; there is no "official" minimum size for a planet, but the nine planets in our solar system all are at least 2,000 km in diameter.

planetary nebula The glowing cloud of gas ejected from a low-mass star at the end of its life.

planetesimals The building blocks of planets, formed by accretion in the solar nebula.

plasma A gas consisting of ions and electrons.

plasma tail (of a comet) One of two tails seen when a comet passes near the Sun (the other is the dust tail); composed of ionized gas blown away from the Sun by the solar wind.

plate tectonics The geological process in which plates are moved around by stresses in a planet's mantle.

plates (on a planet) Pieces of a lithosphere that apparently float upon the denser mantle below.

Population I *See* disk population

Population II *See* spheroidal population

positron The antimatter equivalent of an electron. It is identical to an electron in virtually all respects, except it has a positive rather than a negative electrical charge.

potential energy Energy stored for later conversion into kinetic energy; includes gravitational potential energy, electrical potential energy, and chemical potential energy.

power The rate of energy usage, usually measured in watts (1 watt = 1 joule/s).

precession The gradual wobble of the axis of a rotating object around a vertical line.

pressure Describes the force (per unit area) pushing on an object. In astronomy, we are generally interested in pressure applied by surrounding gas (or plasma). Ordinarily, such pressure is related to the temperature of the gas (*see* thermal pressure). In objects such as white dwarfs and neutron stars, pressure may arise from a quantum effect (*see* degeneracy pressure). Light can also exert pressure. (*See* radiation pressure.)

primary mirror The large, light-collecting mirror of a reflecting telescope.

prime focus (of a reflecting telescope) The first point at which light focuses after bouncing off the primary mirror; located in front of the primary mirror.

prime meridian The meridian of longitude that passes through Greenwich, England, defined to be longitude 0°.

primitive meteorites Meteorites that formed at the same time as the solar system itself, about 4.6 billion years ago.

processed meteorites Meteorites that apparently once were part of a larger object that "processed" the original material of the solar nebula into another form.

protogalactic cloud A huge, collapsing cloud of intergalactic gas from which an individual galaxy formed.

proton–proton chain The chain of reactions by which low-mass stars (including the Sun) fuse hydrogen into helium.

protons Particles found in atomic nuclei with positive electrical charge, built from three quarks.

protoplanetary disk A disk of material surrounding a young star (or protostar) that may eventually form planets.

protostar A forming star that has not yet reached the point where sustained fusion can occur in its core.

protostellar disk A disk of material surrounding a protostar; essentially the same as a protoplanetary disk, but may not necessarily lead to planet formation.

protostellar wind The relatively strong wind from a protostar.

protosun The central object in the forming solar system that eventually became the Sun.

pseudoscience Something that purports to be science or may appear to be scientific but that does not adhere to the testing and verification requirements of the scientific method.

pulsar A neutron star from which we see rapid pulses of radiation as it rotates.

pulsating variable stars Stars that alternately grow brighter and dimmer as their outer layers expand and contract in size.

quantum mechanics The branch of physics that deals with the very small, including molecules, atoms, and fundamental particles.

quantum state Refers to the complete description of the state of a subatomic particle, including its location, momentum, orbital angular momentum, and spin, to the extent allowed by the uncertainty principle.

quantum tunneling The process in which, thanks to the uncertainty principle, an electron or other subatomic particle appears on the other side of a barrier that it does not have the energy to overcome in a normal way.

quarks The building blocks of protons and neutrons, quarks are one of the two basic types of fermions (leptons are the other).

quasar The brightest type of active galactic nucleus.

radar ranging A method of measuring distances within the solar system by bouncing radio waves off planets.

radial motion The component of an object's motion directed toward or away from us.

radial velocity The portion of any object's total velocity that is directed toward or away from us. This part of the velocity is the only part that we can measure with the Doppler effect.

radiation pressure Pressure exerted by photons of light.

radiation zone (of a star) A region of the interior in which energy is transported primarily by radiative diffusion.

radiative diffusion The process by which photons gradually migrate from a hot region (such as the solar core) to a cooler region (such as the solar surface).

radiative energy Energy carried by light; the energy of a photon is Planck's constant times its frequency, or $h \times f$.

radio galaxy A galaxy that emits unusually large quantities of radio waves; thought to contain an active galactic nucleus powered by a supermassive black hole.

radio lobes The huge regions of radio emission found on either side of radio galaxies. The lobes apparently contain plasma ejected by powerful jets from the galactic center.

radio waves Light with very long wavelengths (and hence low frequencies)—longer than those of infrared light.

radioactive dating The process of determining the age of a rock (i.e., the time since it solidified) by comparing the present amount of a radioactive substance to the amount of its decay product.

radioactive element (or **radioactive isotope**) A substance whose nucleus tends to fall apart spontaneously.

recession velocity (of a galaxy) The speed at which a distant galaxy is moving away from us due to the expansion of the universe.

recollapsing universe The possible fate of our universe in which the collective gravity of all its matter eventually halts and reverses the expansion. The galaxies will come crashing back together, and the universe will end in a fiery Big Crunch.

red giant A giant star that is red in color.

red-giant winds The relatively dense but slow winds from red giant stars.

redshift (Doppler) A Doppler shift in which spectral features are shifted to longer wavelengths, caused when an object is moving away from the observer.

reflecting telescope A telescope that uses mirrors to focus light.

reflection (of light) The process by which matter changes the direction of light.

reflection nebula A nebula that we see as a result of starlight reflected from interstellar dust grains. Reflection nebulae tend to have blue and black tints.

refracting telescope A telescope that uses lenses to focus light.

resonance *See* orbital resonance

rest wavelength The wavelength of a spectral feature in the absence of any Doppler shift or gravitational redshift.

retrograde motion Motion that is backward compared to the norm; e.g., we see Mars in apparent retrograde motion during the periods of time when it moves westward, rather than the more common eastward, relative to the stars.

revolution The orbital motion of one object around another.

right ascension (RA) Analogous to longitude, but on the celestial sphere; it is the angular east-west distance between the vernal equinox and a location on the celestial sphere.

rings (planetary) Consist of numerous small particles orbiting a planet within its Roche zone.

Roche tidal zone The region within two to three planetary radii (of any planet) in which the tidal forces tugging an object apart become comparable to the gravitational forces holding it together; planetary rings are always found within the Roche tidal zone.

rocks (in solar system theory) Material common on the surface of the Earth, such as silicon-based minerals, that are solid at temperatures and pressures found on Earth but typically melt or vaporize at temperatures of 500–1,300 K.

rotation The spinning of an object around its axis.

rotation curve A graph that plots rotational (or orbital) velocity against distance from the center for any object or set of objects.

runaway greenhouse effect A positive feedback cycle in which heating caused by the greenhouse effect causes more greenhouse gases to enter the atmosphere, which further enhances the greenhouse effect.

saddle-shaped (or **hyperbolic**) **geometry** Refers to any case in which the rules of geometry for a saddle-shaped surface hold, such as that two lines that begin parallel eventually diverge.

Sagittarius Dwarf A small, dwarf elliptical galaxy that is currently passing through the disk of the Milky Way Galaxy.

saros cycle The period over which the basic pattern of eclipses repeats, which is about 18 years $11\frac{1}{3}$ days.

satellite Any object orbiting another object.

scattered light Light that is reflected into random directions.

Schwarzschild radius A measure of the size of the event horizon of a black hole.

science The search for knowledge that can be used to explain or predict natural phenomena in a way that can be confirmed by rigorous observations or experiments.

scientific method An organized approach to explaining observed facts through science.

scientific theory A model of some aspect of nature that has been rigorously tested and has passed all tests to date.

secondary mirror A small mirror in a reflecting telescope, used to reflect light gathered by the primary mirror toward an eyepiece or instrument.

sedimentary rock A rock that formed from sediments created and deposited by erosional processes.

seismic waves Earthquake-induced vibrations that propagate through a planet.

semimajor axis Half the distance across the long axis of an ellipse; in this text, it is usually referred to as the *average* distance of an orbiting object, abbreviated a in the formula for Kepler's third law.

SETI (search for extraterrestrial intelligence) The name given to observing projects designed to search for signs of intelligent life beyond Earth.

Seyfert galaxies The name given to a class of galaxies found relatively nearby and that have nuclei much like those of quasars, except that they are less luminous.

shepherd moons Tiny moons within a planet's ring system that help force particles into a narrow ring. A variation on *gap moons*.

shield volcano A shallow-sloped volcano made from the flow of low-viscosity basaltic lava.

shock wave A wave of pressure generated by gas moving faster than the speed of sound.

sidereal day The time of 23 hours 56 minutes 4.09 seconds between successive appearances of any particular star on the meridian; essentially the true rotation period of the Earth.

sidereal month About $27\frac{1}{4}$ days, the time required for the Moon to orbit the Earth once (as measured against the stars).

sidereal period (of a planet) A planet's actual orbital period around the Sun.

sidereal time Time measured according to the position of stars in the sky rather than the position of the Sun in the sky. *See also* local sidereal time

sidereal year The time required for the Earth to complete exactly one orbit as measured against the stars; about 20 minutes longer than the tropical year on which our calendar is based.

silicate rock A silicon-rich rock.

singularity The place at the center of a black hole where, in principle, gravity crushes all matter to an infinitely tiny and dense point.

Small Magellanic Cloud One of two small, irregular galaxies (the other is the Large Magellanic Cloud) located about 150,000 light-years away; it probably orbits the Milky Way Galaxy.

snowball Earth Name given to a hypothesis suggesting that, some 600–700 million years ago, the Earth experienced a period in which it became cold enough for glaciers to exist worldwide, even in equatorial regions.

solar activity Refers to short-lived phenomena on the Sun, including the emergence and disappearance of individual sunspots, prominences, and flares; sometimes called *solar weather*.

solar circle The Sun's orbital path around the galaxy, which has a radius of about 28,000 light-years.

solar day Twenty-four hours, which is the average time between appearances of the Sun on the meridian.

solar eclipse Occurs when the Moon's shadow falls on the Earth, which can occur only at new moon; may be total, partial, or annular.

solar flares Huge and sudden releases of energy on the solar surface, probably caused when energy stored in magnetic fields is suddenly released.

solar luminosity The luminosity of the Sun, which is approximately 4×10^{26} watts.

solar maximum The time during each sunspot cycle at which the number of sunspots is the greatest.

solar minimum The time during each sunspot cycle at which the number of sunspots is the smallest.

solar nebula The piece of interstellar cloud from which our own solar system formed.

solar neutrino problem Refers to the disagreement between the predicted and observed number of neutrinos coming from the Sun.

solar prominences Vaulted loops of hot gas that rise above the Sun's surface and follow magnetic-field lines.

solar sail A large, highly reflective (and thin, to minimize mass) piece of material that can "sail" through space using pressure exerted by sunlight.

solar system (or **star system**) Consists of a star (sometimes more than one star) and all the objects that orbit it.

solar wind A stream of charged particles ejected from the Sun.

solid phase The phase of matter in which atoms or molecules are held rigidly in place.

sound wave A wave of alternately rising and falling pressure.

south celestial pole (SCP) The point on the celestial sphere directly above the Earth's South Pole.

spacetime The inseparable, four-dimensional combination of space and time.

spacetime diagram A graph that plots a spatial dimension on one axis and time on another axis.

spectral lines Bright or dark lines that appear in an object's spectrum, which we can see when we pass the object's light through a prismlike device that spreads out the light like a rainbow.

spectral resolution Describes the degree of detail that can be seen in a spectrum; the higher the spectral resolution, the more detail we can see.

spectral type A way of classifying a star by the lines that appear in its spectrum; it is related to surface temperature. The basic spectral types are designated by a letter (OBAFGKM, with O for the hottest stars and M for the coolest) and are subdivided with numbers from 0 through 9.

spectroscopic binary A binary star system whose binary nature is revealed because we detect the spectral lines of one or both stars alternately becoming blueshifted and redshifted as the stars orbit each other.

spectroscopy (in astronomical research) The process of obtaining spectra from astronomical objects.

spectrum (of light) *See* electromagnetic spectrum

speed The rate at which an object moves. Its units are distance divided by time, such as m/s or km/hr.

speed of light The speed at which light travels, which is about 300,000 km/s.

spherical geometry Refers to any case in which the rules of geometry for the surface of a sphere hold, such as that lines that begin parallel eventually meet.

spheroidal component (of a galaxy) The portion of any galaxy that is spherical (or football-like) in shape and contains very little cool gas; generally contains only very old stars. Elliptical galaxies have only a spheroidal component, while spiral galaxies also have a disk component.

spheroidal galaxy Another name for an elliptical galaxy.

spheroidal population Refers to stars that orbit within the spheroidal component of a galaxy. Thus, elliptical galaxies have only a spheroidal population (they lack a disk population), while spiral galaxies have spheroidal population stars in their bulges and halos. Sometimes called Population II.

spin (quantum) *See* spin angular momentum

spin angular momentum Often simply called *spin*, it refers to the inherent angular momentum of a fundamental particle.

spiral arms The bright, prominent arms, usually in a spiral pattern, found in most spiral galaxies.

spiral density waves Gravitationally driven waves of enhanced density that move through a spiral galaxy and are responsible for maintaining its spiral arms.

spiral galaxies Galaxies that look like flat, white disks with yellowish bulges at their centers. The disks are filled with cool gas and dust, interspersed with hotter ionized gas, and usually display beautiful spiral arms.

spreading centers (geological) Places where hot mantle material rises upward between plates and then spreads sideways creating new seafloor crust.

spring equinox (vernal equinox) Refers both to the point in Pisces on the celestial sphere where the ecliptic crosses the celestial equator and to the moment in time when the Sun appears at that point each year (around March 21).

spring tides The higher-than-average tides on Earth that occur at new and full moon, when the tidal forces from the Sun and Moon both act along the same line.

standard candle An object for which we have some means of knowing its true luminosity, so that we can use its apparent brightness to determine its distance with the luminosity–distance formula.

standard model (of physics) The current theoretical model that describes the fundamental particles and forces in nature.

standard time Time measured according to the internationally recognized time zones.

star A large, glowing ball of gas that generates energy through nuclear fusion in its core. The term *star* is sometimes applied to objects that are in the process of becoming true stars (e.g., protostars) and to the remains of stars that have died (e.g., neutron stars).

star cluster *See* cluster of stars

starburst galaxy A galaxy in which stars are forming at an unusually high rate.

state (quantum) *See* quantum state

steady state theory A now-discredited theory that held that the universe had no beginning and looks about the same at all times.

Stefan–Boltzmann constant constant that appears in the laws of thermal radiation, with value

$$\sigma = 5.7 \times 10^{-8} \frac{\text{watt}}{\text{m}^2 \times \text{Kelvin}^4}.$$

stellar evolution The formation and development of stars.

stellar parallax The apparent shift in the position of a nearby star (relative to distant objects) that occurs as we view the star from different positions in the Earth's orbit of the Sun each year.

stellar wind A stream of charged particles ejected from the surface of a star.

stratosphere An intermediate-altitude layer of the atmosphere that is warmed by the absorption of ultraviolet light from the Sun.

stratovolcano A steep-sided volcano made from viscous lavas that can't flow very far before solidifying.

stromatolites Large bacterial "colonies."

strong force One of the four fundamental forces; it is the force that holds atomic nuclei together.

subduction (of tectonic plates) The process in which one plate slides under another.

subduction zones Places where one plate slides under another.

subgiant A star that is between being a main-sequence star and being a giant; subgiants have inert helium cores and hydrogen-burning shells.

sublimation The process by which atoms or molecules escape into the gas phase from a solid.

summer solstice Refers both to the point on the celestial sphere where the ecliptic is farthest north of the celestial equator and to the moment in time when the Sun appears at that point each year (around June 21).

sunspot cycle The period of about 11 years over which the number of sunspots on the Sun rises and falls.

sunspots Blotches on the surface of the Sun that appear darker than surrounding regions.

superbubble Essentially a giant interstellar bubble, formed when the shock waves of many individual bubbles merge to form a single, giant shock wave.

supercluster Superclusters consist of many clusters of galaxies, groups of galaxies, and individual galaxies and are the largest known structures in the universe.

supergiants (luminosity class I) The very large and very bright stars that appear at the top of an H–R diagram.

supermassive black hole Giant black hole, with a mass millions to billions of times that of our Sun, thought to reside in the centers of many galaxies and to power active galactic nuclei.

supernova The explosion of a star.

Supernova 1987A A supernova witnessed on Earth in 1987; it was the nearest supernova seen in nearly 400 years and helped astronomers refine theories of supernovae.

supernova remnant A glowing, expanding cloud of debris from a supernova explosion.

synchronous rotation Describes the rotation of an object that always shows the same face to an object that it is orbiting because its rotation period and orbital period are equal.

synchrotron radiation A type of radio emission that occurs when electrons moving at nearly the speed of light spiral around magnetic field lines.

synodic month (or **lunar month**) The time required for a complete cycle of lunar phases, which averages about $29\frac{1}{2}$ days.

synodic period (of a planet) The time between successive alignments of a planet and the Sun in our sky; measured from opposition to opposition for a planet beyond Earth's orbit, or from superior conjunction to superior conjunction for Mercury and Venus.

tangential motion The component of an object's motion directed across our line of sight.

tangential velocity The portion of any object's total velocity that is directed across (perpendicular to) our line-of-sight. This part of the velocity cannot be measured with the Doppler effect. It can be measured only by observing the object's gradual motion across our sky.

tectonics The disruption of a planet's surface by internal stresses.

temperature A measure of the average kinetic energy of particles in a substance.

terrestrial planets Rocky planets similar in overall composition to Earth.

theories of relativity (*special* and *general*) Einstein's theories that describe the nature of space, time, and gravity.

thermal emitter An object that produces a thermal radiation spectrum; sometimes called a "blackbody."

thermal energy Represents the collective kinetic energy, as measured by temperature, of the many individual particles moving within a substance.

thermal escape The process in which atoms or molecules in a planet's exosphere move fast enough to escape into space.

thermal pressure The ordinary pressure in a gas arising from motions of particles that can be attributed to the object's temperature.

thermal pulses The predicted upward spikes in the rate of helium fusion, occurring every few thousand years, that occur near the end of a low-mass star's life.

thermal radiation The spectrum of radiation produced by an opaque object that depends only on the object's temperature; sometimes called "blackbody radiation."

thermosphere A high, hot X-ray-absorbing layer of an atmosphere, just below the exosphere.

tidal force A force that is caused when the gravity pulling on one side of an object is larger than that on the other side, causing the object to stretch.

tidal friction Friction within an object that is caused by a tidal force.

tidal heating A source of internal heating created by tidal friction. It is particularly important for satellites with eccentric orbits such as Io and Europa.

time dilation Refers to the effect in which you observe time running slower in reference frames moving relative to you.

timing (in astronomical research) The process of tracking how the light intensity from an astronomical object varies with time.

torque A twisting force that can cause a change in an object's angular momentum.

total apparent brightness *See* apparent brightness. We sometimes say "total apparent brightness" to distinguish it from wavelength-specific measures such as the apparent brightness measured in visible light.

total luminosity *See* luminosity. We sometimes say "total luminosity" to distinguish it from wavelength-specific measures such as the luminosity emitted in visible light or the X-ray luminosity.

total lunar eclipse A lunar eclipse in which the Moon becomes fully covered by the Earth's umbral shadow.

total solar eclipse A solar eclipse during which the Sun becomes fully blocked by the disk of the Moon.

totality (eclipse) The portion of either a total lunar eclipse during which the Moon is fully within the Earth's umbral shadow or a total solar eclipse during which the Sun's disk is fully blocked by the Moon.

transmission (of light) The process in which light passes through matter without being absorbed.

transparent (material) Describes a material that transmits light.

triple-alpha reaction *See* helium fusion

Trojan asteroids Asteroids found within two stable zones that share Jupiter's orbit but lie 60° ahead of and behind Jupiter.

tropic of Cancer The circle on the Earth with latitude 23.5°N. It is the northernmost latitude at which the Sun ever passes directly overhead (at noon on the summer solstice).

tropic of Capricorn The circle on the Earth with latitude 23.5°S. It is the southernmost latitude at which the Sun ever passes directly overhead (at noon on the winter solstice).

tropical year The time from one spring equinox to the next, on which our calendar is based.

troposphere The lowest atmospheric layer, in which convection and weather occur.

Tully–Fisher relation A relationship among spiral galaxies showing that the faster a spiral galaxy's rotation speed, the more luminous it is; it is important because it allows us to determine the distance to a spiral galaxy once we measure its rotation rate and apply the luminosity–distance formula.

turbulence Rapid and random motion.

ultraviolet light Light with wavelengths that fall in the portion of the electromagnetic spectrum between visible light and X rays.

umbra The dark central region of a shadow.

unbound orbits Orbits on which an object comes in toward a large body only once, never to return; unbound orbits may be parabolic or hyperbolic in shape.

uncertainty principle The law of quantum mechanics that states that we can never know both a particle's position and its momentum, or both its energy and the time it has the energy, with absolute precision.

universal law of gravitation The law expressing the force of gravity (F_g) between two objects, given by the formula

$$F_g = G\frac{M_1 M_2}{d^2}$$

$$(G = 6.67 \times 10^{-11} \frac{m^3}{kg \times s^2}).$$

universal time (UT) Standard time in Greenwich (or anywhere on the prime meridian).

universe The sum total of all matter and energy.

up quark One of the two quark types (the other is the down quark) found in ordinary protons and neutrons. Has a charge of $+\frac{2}{3}$.

velocity The combination of speed and direction of motion; it can be stated as a speed in a particular direction, such as 100 km/hr due north.

vernal equinox *See* spring equinox

virtual particles Particles that "pop" in and out of existence so rapidly that, according to the uncertainty principle, they cannot be directly detected.

viscosity Describes the "thickness" of a liquid in terms of how rapidly it flows; low-viscosity liquids flow quickly (e.g., water), while high-viscosity liquids flow slowly (e.g., molasses).

visible light The light our eyes can see, ranging in wavelength from about 400 to 700 nm.

visual binary A binary star system in which we can resolve both stars through a telescope.

voids Huge volumes of space between superclusters that appear to contain very little matter.

volatiles Refers to substances, such as water, carbon dioxide, and methane, that are usually found as gases, liquids, or surface ices on the terrestrial worlds.

volcanism The eruption of molten rock, or lava, from a planet's interior onto its surface.

wavelength The distance between adjacent peaks (or troughs) of a wave.

weak bosons The exchange particles for the weak force.

weak force One of the four fundamental forces; it is the force that mediates nuclear reactions; also the only force besides gravity felt by weakly interacting particles.

weakly interacting particles Particles, such as neutrinos and WIMPs, that respond only to the weak force and gravity; that is, they do not feel the strong force or the electromagnetic force.

weather Describes the ever-varying combination of winds, clouds, temperature, and pressure in a planet's troposphere.

weight The net force that an object applies to its surroundings; in the case of a stationary body on the surface of the Earth, weight = mass $\times$ acceleration of gravity.

weightless A weight of zero, as occurs during free-fall.

white-dwarf limit (also called the *Chandrasekhar limit*) The maximum possible mass for a white dwarf, which is about $1.4M_{Sun}$.

white dwarf supernova A supernova that occurs when an accreting white dwarf reaches the white-dwarf limit, ignites runaway carbon fusion, and explodes like a bomb; often called a *Type Ia supernova.*

white dwarfs The hot, compact corpses of low-mass stars, typically with a mass similar to the Sun compressed to a volume the size of the Earth.

WIMPs Stands for *weakly interacting massive particles* and represents a possible form of dark matter consisting of subatomic particles that are dark because they do not respond to the electromagnetic force.

winter solstice Refers both to the point on the celestial sphere where the ecliptic is farthest south of the celestial equator and to the moment in time when the Sun appears at that point each year (around December 21).

worldline A line that represents an object on a spacetime diagram.

wormholes The name given to hypothetical tunnels through hyperspace that might connect two distant places in our universe.

X rays Light with wavelengths that fall in the portion of the electromagnetic spectrum between ultraviolet light and gamma rays.

X-ray binary A binary star system that emits substantial amounts of X rays, thought to be from an accretion disk around a neutron star or black hole.

X-ray burster An object that emits a burst of X rays every few hours to every few days; each burst lasts a few seconds and is thought to be caused by helium fusion on the surface of an accreting neutron star in a binary system.

Zeeman effect The splitting of spectral lines by a magnetic field.

zenith The point directly overhead, which has an altitude of 90°.

zodiac The constellations on the celestial sphere through which the ecliptic passes.

zones (on a jovian planet) Bright bands of rising air that encircle a jovian planet at a particular set of latitudes.

Credits and Acknowledgments

ILLUSTRATIONS BY JOE BERGERON: Figures 1.1, 1.3, 1.16, 1.17, 1.18, 2.5, 9.2, 9.4, 9.8, 9.9, 9.10, 12.26, 1318.g, 17.4, 17.25, 17.26, 18.3, 18.9.b, 18.14.b, 19.1.a, 19.18, 21.4, 21.23, 21.24, 23.2. ©Joe Bergeron 24.15, 24.16, 24.17, 24.18

John and Judy Waller, the principal artists for this project, created many new figures and updated, redesigned, and adapted many others using elements created by artists who worked on previous editions.

Part Opener I: ©Roger Ressmeyer/CORBIS

Part Opener II: "An Expanding Bubble in Space": NASA, Donald Walter (South Carolina State University, Paul Scowen and Brian Moore (Arizona State University)

Part Opener III: "Valles Marineris—Point Perspective": Jody Swann, Tammy Becker, and Alfred McEwen of U.S. Geological Survey in Flagstaff, Arizona

Part Opener IV: "Disorder in Stephan's Quintet": Jane C. Charlton (Penn State) et al., HST, ESA, NASA

Part Opener V: "Hubble Peeks into a Stellar Nursery in a Nearby Galaxy": Hubble Heritage Team (AURA/STScI/NASA)

Part Opener VI: "A Galaxy on the Edge": Hubble Heritage Team (AURA/STScI/NASA)

CHAPTER 1 Opening Photo: *Niescja Turner and Carter Emmart*

1.2 NASA **1.4 (a)** ©Anglo-Australian Observatory. Photography by David Malin **1.4 (b)** Andrea Dupree (Harvard-Smithsonian CFA), Ronald Gilliland (STScI), ESA, and NASA **1.6** Jerry Lodrigus **1.7** ©Jeff Bennett **1.8 (a)** Stan Maddock **1.9** NASA **1.10** Akira Fujii **1.11** ©Jeff Bennett **1.12 (Cambrian explosion and pyramid)** Photos by Corel; **(Dinosaurs extinct)** Quade Paul, fiVth.com; **(Rise of dinosaurs)** ©John Eastcott & Eva Momatiuk/Photo Researchers, Inc.; **(Agriculture arises)** Blakeley Kim; **(Earth)** NASA

CHAPTER 2 Opening Photo: ©*David Nunuk*

2.4 Gordon Garradd **2.9** ©Richard Tauber Photography, San Francisco **2.10 (b)** ©Dennis diCicco **2.12** ©David Nunuk **2.16** ©Dennis diCicco **2.17** ©Husmo-foto **2.18** ©Tom Van Sant/GeoSphere **2.21 (Moons)** Akira Fujii; **(Earth)** NASA **2.23** Photo by John Q. Waller, art by John and Judy Waller **2.25 (top, bottom)** Akira Fujii; **(center)** Dennis diCicco **2.26, 2.27** Akira Fujii **2.28** Adapted from eclipse map by

Fred Espenak, NASA/GSF: (http://sunearth.gsfc. nasa.gov/ eclipse/eclipse.html) **2.29** ©2002 Jerry Lodrigus

CHAPTER 3 Opening Photo: ©*1987 by Margaret R. Curtis*

3.2 ©Michael Yamashita/CORBIS **3.3 (a)** ©N. Pecnik/Visuals Unlimited **3.4** ©Kenneth Garrett **3.5 (a)** ©Wm. E. Woolam/Southwest Parks; **(b)** ©Richard A. Cooke, III/Stone/Getty Images **3.6** ©1987 by Margaret R. Curtis **3.7** ©Richard A. Cooke, III **3.8** ©Loren McIntyre/Woodfin Camp & Assoc. **3.9** ©Jeff Henry/Peter Arnold, Inc. **3.10** ©Oliver Strewe/Wave Productions Pty, Ltd. **3.11** Werner Forman Archive/Art Resource, NY **3.14** © Bettmann/CORBIS **3.16 (a, b)** Courtesy of Carl Sagan Productions, Inc. From *Cosmos* (Random House), ©1980 Carl Sagan **Page 70** Giraudon/Art Resource, NY **Page 71** Archive Photos/Getty Images **3.17** The Granger Collection, NY **Page 72** ©Erich Lessing/Art Resource, NY **Page 74** ©Bettmann/CORBIS **3.22** ©Jerry Lodrigus **Page 76** Courtesy of Science Museum of Virginia

CHAPTER S1 Opening Photo: ©*Husmo-foto*

S1.5 Image from TRACE (Transition Region and Coronal Explorer), a mission of the Stanford-Lockheed Institute for Space Research (a joint program of the Lockheed-Martin Advanced Technology Center's Solar and Astrophysics Laboratory and Stanford's Solar Observatories Group), and part of the NASA Small Explorer program **S1.6** ©Bernd Wittich/Visuals Unlimited **S1.7** NASA/MSFC **S1.24 (a, b)** ©Bettmann/CORBIS; **(c)** The Granger Collection, NY; **(d)** ©Science VU/Visuals Unlimited

CHAPTER 4 Opening Photo: *Hubble Heritage Team/NASA/AURA/STScI*

4.1 (top) ©EyeWire/Getty Images; **(left)** ©Alvis Upitis/The Image Bank/Getty Images; **(right)** ©Fred Dana/CORBIS **4.2** ©Dimitri Iundt/Stone/Getty Images **4.5** Courtesy U.S. Department of Energy

CHAPTER 5 Opening Photo: *NASA*

Page 134 ©Bettmann/CORBIS **5.6 (a)** NASA; **(b)** ©Duomo/CORBIS; **(c)** NASA **5.14** NASA **5.16 (both)** ©Bill Bachmann/Gnass Photo Images

CHAPTER 6 Opening Photo: Dr. N. A. Sharp, NOAO/NSO/Kitt Peak FTS/AURA/NSF **6.1** ©Runk/Schoenberger, Grant Heilman Photography

CHAPTER 7 Opening Photo: ©*Joel Gordon Photography*

7.6 (b) Yerkes Observatory **7.7 (b)** NOAO/AURA/ NSF **7.9 (a)** ©Richard Wainscoat; **(b)** Russ Underwood (W. M. Keck Observatory) **7.11** Mark Voit (STScI) **7.12** ©Anglo-Australian Observatory, photography by David Malin **7.14** NASA/CXC/ SAO **7.16** ©Richard Wainscoat **7.17** NASA/ Ames Research Center **7.18 (a, b)** Canada-France-Hawaii Telescope Corporation, Hawaii **7.19 (a)** NASA; **(b)** Don Foley/National Geographic Image Collection **7.20** Jodi Schoemer **7.21 (c)** Eastman-Kodak **7.22** NASA **7.23** David Parker, 1997/Science Library. *The Arecibo Observatory is part of the National Astronomy and Ionosphere Center, which is operated by Cornell Univ. under a cooperative agreement with the National Science Foundation* **7.25** ©Joel Gordon Photography **7.26** NASA/JPL/Caltech

CHAPTER 8 Opening Photo: *NASA*

8.3 NEAR Project, Johns Hopkins University/APL and NASA **8.4** Niescja Turner and Carter Emmart **8.7 Sun (a)** Big Bear Solar Observatory New Jersey Institute of Technology and NASA Marshall Space Flight Center; **(b)** courtesy of SOHO. SOHO is a project of international cooperation between ESA and NASA **8.8 Mercury (left)** NASA/USGS; **(right)** From the Voyage scale model solar system, developed by Challenger Center for Space Science Education, the Smithsonian Institution, and NASA. Image created by ARC Science Simulations, ©2001 **8.9 Venus (left)** NASA, courtesy of NSSDC; **(right)** From the Voyage scale model solar system, developed by Challenger Center for Space Science Education, the Smithsonian Institution, and NASA. ©2001 David P. Anderson, Southern Methodist University **8.10 Earth (a)** From the Voyage scale model solar system, developed by Challenger Center for Space Science Education, the Smithsonian Institution, and NASA. Image created by ARC Science Simulations, ©2001; **(b)** NASA **8.11 Mars (top)** NASA/USGS, courtesy of NSSDC; **(bottom)** NASA **8.12 Jupiter** From the Voyage scale model solar system, developed by Challenger Center for Space Science Education, the Smithsonian Institution, and NASA. Image created by ARC Science Simulations, ©2001 **8.13 Saturn** From the Voyage scale model solar system, developed by Challenger Center for Space Science Education, the Smithsonian Institution, and NASA. Image created by ARC Science Simulations, ©2001 **8.14 Uranus** From the Voyage scale model solar system, developed by Challenger Center for Space Science Education, the Smithsonian Institution, and NASA. Image created by ARC Science Simulations, ©2001 **8.15 Neptune** From the Voyage scale model solar system, developed by Challenger Center for Space Science Education, the Smithsonian Institution, and NASA. Image created by ARC Science Simulations, ©2001 **8.16 Pluto and Charon** Dr. R. Albrecht

(ESA/ESO) and NASA **8.18 (inset)** NASA/JPL **8.19 (b)** NASA/JPL

CHAPTER 9 Opening Illustration: *Joe Bergeron/ John and Judy Waller*

9.1 C. R. O'Dell/Rice University and NASA **9.3 (a)** NASA/STScI, courtesy of Alfred Schultz and Helen Hart; **(b)** Courtesy of John Bally and Nathan Smith, CASA, Univ. of Colorado; **(c)** M. J. McCaughrean (MPIA), C. R. O'Dell (Rice Univ.), and NASA; **(d)** NASA, M. Clampin (STScI), H. Ford (JHU), G. Illingworth (UCO-Lick), J. Krist (STScI), D. Ardila (JHU), D. Golimowski (JHU), the ACS Science Team and ESA **9.7** Meteorite specimen courtesy of Robert Haag Meteorites **9.11 (a, b)** NASA/JPL, courtesy NSSDC **9.12** ©William K. Hartmann **9.18** Data adapted from STARE project: (www.hao.ucar.edu/public/ research/ stare/hd209458.html)

CHAPTER 10 Opening Photo: ©*Dr. Marco Fulle/ Osservatorio Astronomico*

10.1 (*Moon*-globe) Akira Fujii; **(others)** NASA **10.3 (both)** ©Roger Ressmeyer/CORBIS **10.6 (a)** ©Jules Bucher/Photo Researchers, Inc. **10.7** ©Don Davis **10.8** NASA and Lunar Planetary Institute **10.9 (a)** NASA, from the Apollo 16 crew's mapping camera; **(b)** NASA/USGS **10.10 (a)** NASA/JPL (Viking Orbiter); **(b)** NASA/USGS; **(c)** NASA/JPL/Malin Space Science Systems **10.11 (b)** ©Paul Chesley/Stone/Getty Images **10.12 (a)** NASA, courtesy of LPL; **(b)** NASA/JPL; **(c)** ©Craig Tuttle/CORBIS **10.13 (a)** USGS Photo Library, Denver, Colorado; **(b)** ©D. Cavagnaro/Visuals Unlimited **10.14 (left)** NASA/ USGS; **(right)** NASA/JPL (Magellan Mission) **10.15 (a)** ©Gene Ahrens/Bruce Coleman, Inc.; **(b)** ©J. Messerschmidt/Bruce Coleman, Inc.; **(c)** Craig Aurness/CORBIS; **(d)** ©C. C. Lockwood/ DDB Stock Photo **10.17 (a)** Akira Fujii; **(b)** NASA/USGS (Mariner 10 Mission) **10.18** NASA **10.19, 10.20 (a, b)** NASA **10.21 (a, b)** NASA/JPL; **(c)** Mark Robinson/Northwestern University **10.22 (b)** NASA, courtesy of Mark Robinson **10.23** Lowell Observatory Photographs **10.24** NASA/Mars Global Surveyor/NSSDC **10.25, 10.26** NASA/USGS **Page 280 (a)** Viking Project, NASA, NSSDC; **(2-top)** Viking Project, NASA, NSSDC; **(2-bottom)** Mar Global Surveyor, MLS, NASA; **(3)** Viking Orbiter, MLS, NASA **10.27** NASA/JPL **10.28 (a)** USGS; **(b)** NASA/JPL/ Maline Space Science Systems; **(c)** Dr. David E. Smith, NASA, and MOLA Science Team; **(d)** R. P. Irwin III and G. A. Franz, National Air and Space Museum, Smithsonian Institution **10.29 (a, b)** NASA/JPL **10.30 (left)** NASA/JPL/MGS/MSSS; **(right)** NASA/JPL/MSSS/Philip Christensen **10.31** NASA/JPL **10.32** NASA, the Magellan project. Additional processing by John Weiss **10.33 (a, b)** NASA/JPL **10.34 (a, b)** NASA/JPL **10.35** NASA, courtesy of NSSDC

CHAPTER 11 Opening Photo: *NASA*

11.1 (*Mercury*-globe) NASA/courtesy NSSDC, (*Mercury*-surface) ©Don Davis; (*Venus*-globe) NASA/JPL, (*Venus*-surface) From the Voyage scale model solar system, developed by Challenger Center for Space Science Education, the Smithsonian Institution, and NASA. ©2001 David P. Anderson, Southern Methodist University; (*Earth*-globe) NASA, (*Earth*-surface) ©Barrie Rokeach; (*Moon*-globe) Akira Fujii, (*Moon*-surface) Artis Planetarium/ The Netherlands;

(*Mars*-globe) NASA with enhancement by Calvin Hamilton, (*Mars*-surface) NASA/JPL/Caltech **11.2, 11.4** NASA **11.11 (a-inset)** Dr. L. A. Frank, Univ. of Iowa; **(b)** ©Daniel Hershman; **(c)** NASA **11.12, 11.13** ©Tom Van Sant/The Geosphere Project/CORBIS **11.15 (a)** ©Tom Van Sant/The Geosphere Project/CORBIS; **(b)** GOES-10 satellite image. Colorization by ARC Science Simulations, ©1998 **11.16** ©Tom Van Sant/The Geosphere Project/CORBIS **11.18** NASA/JPL/MSSS **11.22 (a)** D. Potter, T. Morgan, and R. Killen; **(b)** M. Mendillo, J. Baumgartner, and J. Wilson of Boston University **11.24** NASA, P. James (Univ. Toledo), T. Clancy (Space Science Inst.), S. Lee (Univ. Colorado), NSSDC **11.25** NASA/JPL and Malin Space Science Systems **11.26** J. Bell (Cornell), M. Wolff (Space Science Inst.), Hubble Heritage Team (STScI/AURA), and NASA **11.27** NASA/ JPL **11.28** NASA, courtesy NSSDC

CHAPTER 12 Opening Illustration: *Joe Bergeron*

12.1 NASA **12.3 (b)** NASA/JPL **12.6** ©Michael Carroll **12.8 (a)** STScI and R. Beebe and A. Simon (NMSU); **(b)** NASA Infrared Telescope Facility and Glenn Orton **12.11** (*Uranus*) E. Karkoscha (Univ. of Arizona and NASA); **(others)** NASA/JPL **12.13 (b)** J. Clarke (Univ. of Michigan) and NASA **12.14** John Spencer/Lowell Observatory **12.15** Data provided by Fran Bagenal **12.16** Courtesy Tim Parker/JPL **12.17** NASA/JPL **12.18** NASA/JPL/Galileo **12.19** PIRL/Lunar & Planetary Laboratory (Univ. of Arizona) and NASA **12.20 (a-left)** NASA/JPL; **(a-right)** University of Arizona and NASA; **(b)** Univ. of Arizona and NASA **12.21** NASA/JPL/Ames Research Center **12.23** DLR/NASA/JPL **12.25 (a, b)** NASA/ JPL/Arizona State University **12.27 (all)** NASA/ JPL/Galileo **12.28, 12.29 (a)** NASA/JPL/Galileo/ DLR (German Aerospace Center)/NSSDC; **12.29 (b)** NASA/JPL and Arizona State University **12.30** NASA/JPL/Galileo/NSSDC **12.31 (a)** M. E. Brown, A. H. Bouchez, C. A. Griffith; **(b)** ©Don Davis **12.32, 12.33** NASA/JPL **12.34 (a)** NASA/ USGS; **(b)** NASA/JPL **12.35 (a)** Jody Swann, Tammy Becker, and Alfred McEwen of U.S. Geological Survey in Flagstaff, Arizona; **(b)** NASA/JPL **12.36 (a)** S. Larson (Univ. of Arizona/LPL); **(b)** NASA/JPL; **(c)** ©William K. Hartmann **12.37 (a)** NASA and computer enhanced by Calvin J. Hamilton; **(b)** NASA/JPL, courtesy of M. Showalter (Plantary Data Systems Ring Node) **12.38** NASA/JPL **12.39** NASA/JPL/Caltech **12.40** (*Uranus*) Erich Karkoschka (Univ. of Arizona/LPL) and NASA; **(others)** NASA/JPL

CHAPTER 13 Opening Photo: *StockTrek/Photodisc/ Getty Images*

13.1 3 Discovery photograph made January 7, 1976 by Eleanor F. Helin/JPL (Helin then associated with Caltech/Palomar Observatory) **13.2 (a)** NASA/USGS; **(b)** NASA/JPL; **(c, d)** Johns Hopkins Univ./APL/NASA **13.3, 13.4** Data provided by Dave Tholen (Univ. of Hawaii) **13.5** Laird M. Close (Univ. of Arizona) and Wm. J. Merline (Southwest Research Institute, Boulder, Colorado) **13.6 (b)** Stephen Ostro et al. (JPL, Arecibo Radio Telescope, NSF, and NASA) **13.7 (a)** Walt Radomski (nyrockman.com); **(b)** PEANUTS reprinted by permission of United Feature Syndicate, Inc. **13.8** Meteorite specimens courtesy of Robert Haag Meteorites **13.9 (a)** ©Peter Ceravolo; **(b)** Tony and Daphne Hallas **13.10** LASCO, SOHO Consortium, NRL, ESA, and NASA **13.11 (left)** Astuo Kuboniwa, March 9,

1997, 19:25:00—19:45:out, BISTAR Astronomical Observatory, Japan, Telescope: d=125mm, f=500mm refractor, Film: Ektacrome E100S; **(right)** Halley Multicolour Camera Team, Giotto, ESA, ©MPAE **13.15** Dr. R. Albrecht, ESA/ESO, and NASA **13.16** Eliot Young/ Southwest Research Institute **13.17 (a)** Hal Weaver and T. E. Smith (STScI) and NASA; **(b)** Courtesy of Paul Schenk (Lunar and Planetary Institute) **13.18 (a)** NASA/JPL/Caltech; **(b)** HST Jupiter Imaging Science Team; **(c)** MSSO, ANU/ Science Library/ Photo Researchers, Inc.; **(d)** Courtesy of Richard Wainscoat et al (Univ. of Hawaii); **(e)** H. Hammel (MIT) and NASA; **(f)** HST Comet Team and NASA **13.19** ©Vic and Jen Winters/Icstars Astronomy **13.20 (top)** NASA, computer enhancement by Calvin J. Hamilton; **(center)** Geological Survey of Canada; **(bottom)** Brad Snowder **13.21** Kirk Johnson/Denver Museum of Natural History **13.22** Image courtesy of Dr. Virgil Sharpton, Univ. of Alaska—Fairbanks **13.23 (all)** Quade Paul/fiVth.com **13.24** ©TASS/Sovfoto

CHAPTER 14 Opening Photo: *StockTrek/Photodisc/ Getty Images*

14.1 NASA **14.2** Digital image by Dr. Peter W. Sloss (NOAA/NGDC) **14.6** NASA/SRTM, additional processing by John R. Spencer, Lowell Observatory **14.7** Earth Satellite Corp./SPL/Photo Researchers, Inc. **14.8** NASA **14.10 (a)** ©Roger Ressmeyer/CORBIS; **(b)** Mike Yamashita/Woodfin Camp & Assoc. **14.11 (clockwise from top)** ©Philip Rosenberg; ©Philip Rosenberg; University of Hawaii; R. Shallenberger/Midway Atol National Wildlife Refuge **14.12** ©Charles Mauzy/ CORBIS **14.14** NASA **14.19** ©Jeff Greenberg/ Visuals Unlimited **14.20** ©James L. Amos/Photo Researchers, Inc. **14.21** ©Nih R. Feldman/Visuals Unlimited **14.23 (a)** Woods Hole Oceanographic Institute; **(b)** ©Barrie Rokeach **14.25 (a)** Biological Photo Services; **(b, c)** ©S. M. Awramik, Univ. of California/Biological Photo Service **14.26** Biological Photo Services **14.27** ©Ken Lucas/ Visuals Unlimited **14.31** NASA/GSFC (Science Visualization Studio) **14.32** ©Kjell Sandved/ Photo Researchers, Inc.

CHAPTER S2 Opening Photo: ©*Bettmann/ CORBIS*

CHAPTER S3 Opening Photo: *NASA, Andrew Fruchter and the ERO Team (Sylvia Baggett/ STScI, Richard Hook/ST-ECF, Zoltan Levay/STScI*

S3.20 (a) ESA/NASA; **(b)** HST/MIRLIN/Lindsay King (Univ. of Manchester)

CHAPTER S4 Opening Photo: *Fermilab Visual Media Service, U.S. Department of Energy*

S4.1 Fermilab Visual Media Service **S4.5** ©Harold & Esther Edgerton Foundation, 1998. Palm Press, Inc.

CHAPTER 15 Opening Photo: *Solar and Heliospheric Observory (SOHO). SOHO is a project of international cooperation between ESA and NASA*

15.1 Photo by Corel **15.3** National Optical Astronomy Observatories/NSO, Sacramento Peak **15.9** National Optical Astronomy Observatories **15.11** Courtesy of Brookhaven National Laboratory **15.12 (a, b)** ICRR (Institute for Cosmic Ray Research), Univ. of Tokyo **15.14 (b)** Royal Swedish Academy of Sciences **15.15 (a-both)** Royal

Swedish Academy of Sciences; **(b)** NOAO/National Solar Observatory **15.17 (b)** Image from TRACE (Transition Region and Coronal Explorer), a mission of the Stanford-Lockheed Institute for Space Research (a joint program of the Lockheed-Martin Advanced Technology Center's Solar and Astrophysics laboratory and Stanford's Solar Observatories Group), and part of the NASA Small Explorer program **15.18** Courtesy of SOHO. SOHO is a project of international cooperation between ESA and NASA **15.19** Image from TRACE (Transition Region and Coronal Explorer), a mission of the Stanford-Lockheed Institute for Space Research (a joint program of the Lockheed-Martin Advanced Technology Center's Solar and Astrophysics laboratory and Stanford's Solar Observatories Group), and part of the NASA Small Explorer program **15.20** Courtesy of B. Haisch and G, Slater (Lockheed Palo Alto Research Laboratory) **15.21** Data adapted from Marshall Space Flight Center: (http://science.msfc.nasa.gov/ssl/pad/solar) **15.23 (b)** ©Hinrich Baesemann (www.polarfoto.com)

CHAPTER 16 Opening Photo: *Hubble Heritage Team (AURA/STScI/NASA)*

16.4 Hubble Heritage Team (AURA/STScI/NASA) **Table 16.1 (spectrum)** Roger Bell and Mike Briley, University of Maryland **16.5, Page 529** Harvard College Observatory **16.6 (photo-center)** Lowell Observatory **16.14** ©Anglo-Australian Observatory/Royal Observatory Edinburgh, photography by David Malin **16.15** Hubble Heritage Team (AURA/STScI/NASA)

CHAPTER 17 Opening Photo: *European Southern Observatory*

17.1 ©Anglo-Australian Observatory, photograph by David Malin **17.2** Quade Paul **17.3** IPAC (Infrared Processing and Analysis Center) and Caltech/JPL **17.5 (b, c)** C. Burrows and J. Morse (STScI), J. Hester (Arizona State Univ.), and NASA **17.14 (a)** Nordic Optical Telescope, La Palma; **(b)** NASA, Andrew Fruchter and the ERO Team (Sylvia Baggett/STScI, Richard Hook/ST-ECF, Zoltan Levay/STScI; **(c)** NASA and the Hubble Heritage Team (STScI/AURA); **(d)** R. Sahai, J. Trauger (JPL), the WFPC2 Science Team, and NASA **17.23** European Southern Observatory **17.24 (both)** ©Anglo-Australian Observatory, photography by David Malin

CHAPTER 18 Opening Photo: *NASA, McGill, and V. Kaspi et al.*

18.1 NASA/SAO/CXO **18.4 (b)** M. Shara, B. Williams, and D. Zurek (STScI); R. Gilmozzi (ESO); D. Prialnik (Tel Aviv Univ.); and NASA **18.6** NASA, McGill, and V. Kaspi et al. **18.8** European Southern Observatory **Pages 588 (Bell)** Copyright is held by CWP/Regents of the Univ. of California: www.physics.ucla.edu ~cwp/Phase2/ Burnell,_Jocelyn_Bell @841234567.html]; **(Oppenheimer)** ©CORBIS; **(others)** Courtesy of Astronomical Institute of Bonn University **18.16** ©Anglo-Australian Observatory, photography by David Malin

CHAPTER 19 Opening Photo: *©Axel Mellinger*

19.2 (counterclockwise from lower right) D. Malin (from Fig. 19.4); S. L. Snowden (from Fig. 19.7a); NASA/NSSDC (from Fig. 9.13a); J. Bally (from Fig. 19.10); J. Hester and P. Scowen (from Fig. 19.12); D. Malin (from Fig. 19.14) **19.3** NASA and the Hubble Heritage Team (STScI/AURA) **19.4** ©Anglo-Australian Observatory, photography by David Malin **19.5** NSA/CXC/SAO **19.6 (a)** Tony and Daphne Hallas; **(b)** Jeff Hester (Arizona State Univ.) and NASA **19.7 (a)** J. Keuhane, B. Koralesky, M. Anderson, L. Rudnick (Univ. of Minnesota) and R. Perley (NRAO); **(b)** NASA/CXO/SAO/Rutgers and J. Hughes **19.9** Simulation by D. Strickland and I. Stevens **19.10, 19.11** John Bally (Univ. of Colorado) **19.12** Jeff Hester and Paul Scowen (Arizona State Univ.) and NASA **19.13** Courtesy of NASA's National Space Science Data Center at the Goddard Space Flight Center **19.14, 19.16, 19.17** ©Anglo-Australian Observatory/Royal Observatory Edinburgh, photography by David Malin **19.20** Hubble Heritage Team (AURA/STScI/NASA); Acknowledgments: N. Scoville (Caltech) and T. Rector (NOAO) **19.21 (photo)** J. Trauger, JPL, and NASA **19.22 (photo)** Hubble Heritage Team (AURA/STScI/ NASA); Acknowledgments: N. Scoville (Caltech) and T. Rector (NOAO) **19.23 (a)** E. Kopan (IPAC/Caltech); **(b)** Produced by Navy Research Laboratory, data from NRAO/VLA; **(c)** NRAO/VLA, F. Zadeh et al.; **(d)** NRAO/AUI (D. A. Roberts, F. Yusef-Zadeh, W. M. Goss); **(e)** Courtesy of European Southern Observatory **19.24** Courtesy of European Southern Observatory **19.25** NASA/MIT/F. Baganoff et al.

CHAPTER 20 Opening Photo: *NASA, H. Ford (JHU), G. Illingworth (UCO-Lick), M. Clampin (STScI), G. Hartig (STScI), the ACS Science Team, and ESA*

20.1 ©Sky Publishing Corporation, reproduced with permission **20.2 (a)** ©Anglo-Australian Observatory, CCD image by Steve Lee and David Malin; **(b)** Hubble Heritage Team (AURA/STScI/ NASA); **(c)** ©IAC photo from plates taken with the Isaac Newton telescope, photography by David Malin **20.3** ©Anglo-Australian Observatory, photography by David Malin **20.4** ©Anglo-Australian Observatory, CCD image by Steve Lee and David Malin **20.5** M. Carollo (Swiss Federal Institute of Technology, Zurich) **20.6** Hubble Heritage Team (AURA/STScI/NASA) **20.7** NASA, N. Benitez (JHU), T. Broadhurst (The Hebrew University), H. Ford (JHU), M. Clampin (STScI), G. Hartig (STScI), G. Illingworth (UCO/Lick Observatory), the ACS Science Team and ESA **20.8 (a, b)** ©Anglo-Australian Observatory, photography by David Malin **20.9 (a, b)** ©Anglo-Australian Observatory/ Royal Observatory Edinburgh, photography by David Malin; **(c)** ©Anglo-Australian Observatory, photography by David Malin **20.10 (clockwise from bottom right)** David and Christine Smith/Adam Block/ NOAO/AURA/NSF; Daryl Seibel/Adam Block/ NOAO/AURA/NSF; Bob Birket and John Evelan/ Adam Block/NOAO/AURA/NSF; Courtesy Zsolt Frei and James E. Gunn; Courtesy Zsolt Frei and James E. Gunn; Courtesy Zsolt Frei and James E. Gunn; Courtesy Zsolt Frei and James E. Gunn; Rick Barry/Adam Block/NOAO/AURA/NSF; Michael Chase/Adam Block/NOAO/AURA/NSF; Tom Boemer and David Young/Adam Block/ NOAO/AURA/NSF **20.13 (a)** The Observatories of the Carnegie Institution of Washington; **(b)** NASA/STScI **20.14, 20.16** The Observatories of the Carnegie Institution of Washington **20.17 (a)** J. Trauger, JPL, and NASA; **(b)** Dr. W. L. Freedman (The Observatories of the Carnegie Institution of Washington) and NASA **20.19** Megan Donahue (STScI)

CHAPTER 21 Opening Photo: *Subaru Telescope, National Astronomical Observatory of Japan, © 2000, NAOJ*

21.1 Robert Williams and the HDF Team (STScI) and NASA **21.2** Courtesy of European Southern Observatory **21.5, 21.6** Quade Paul **21.7** Hyron Spinard, Univ. of California at Berkely et al. **21.8 (a)** Brad Whitmore (STScI) and NASA; **(b)** W. C. Keel (Univ. of Alabama) **21.9** Robert Williams and the HDF Team (STScI) and NASA **21.10** Frank Summers/American Museum of Natural History **21.11** ©Anglo-Australian Observatory, photography by David Malin **21.12** Dr. Michael J. West/ Univ. of Hawaii—Hilo **21.13 (a)** R. Thompson, M. Rieke, G. Schneider (Univ. of Arizona); N. Scoville (Caltech); and NASA **21.14 (a)** NASA/SAO/G. Fabbiano et al.; **(b)** Subaru Telescope, National Astronomical Observatory of Japan. ©2000 NAOJ **21.15** NASA/UCSB/C. Martinetal and NOAO/KPNO/C. Martin **21.16** Hubble Heritage Team (AURA/STScI/NASA) **21.19** John Bahcall/Institute for Advanced Study and NASA **21.20** NRAO/AUI **21.21 (a–c)** Alan Bridle (NRAO/AUI); **(d)** NRAO/AUI **21.22** John Biretta (STScI) **21.25** Rerendered after STScI illustration **21.26** H. Ford (STScI/Johns Hopkins Univ.); L. Dressel, R. Harms, A. Kochhar (Applied Research Corp.); Z. Tsvetanov, A. Davidsen, G. Kriss (Johns Hopkins Univ.); R. Bohlin, G. Hartig (STSci); B. Margon (Univ. of Washington-Seattle); and NASA **21.27** Emiko-Rose Koike/fiVth.com

CHAPTER 22 Opening Photo: *Ben Moore, Department of Physics, Durham University, Durham City, England*

22.3 (photo) ©Anglo-Australian Observatory, photography by David Malin **22.6** Courtesy of Caltech **22.8 (a)** Omar Lopez-Cruz and Iam Shelton; **(b)** Steve Snowden, NASA/GSFC/USRA **22.9** W. N. Colley and E. Turner (Princeton Univ.), J. A. Tyson (Bell Labs, Lucent Technologies), and NASA **22.11** A. Fruchter, the ERO Team (STScI, ST-ECF), and NASA **22.12 (inset photos)** Charles Alcock/Lawrence Livermore National Laboratory **22.13** Michael Strauss, Princeton Univ. **22.14 (a, b)** Harvard-Smithsonian Center for Astrophysics; **(c)** Emiko-Rose Koike/fiVth.com **22.15** S. Maddox, G. Efstathiou, and W. Sutherland, MNRAS, 242, 43P **22.16 (a, b)** Ben Moore, Dept. of Physics, Durham Univ., Durham City, England **22.17** Rerendered after STScI illustration

CHAPTER 23 Opening Photo: *E. Bunn/University of Richmond*

23.5 ©Roger Ressmeyer/CORBIS **23.8** E. Bunn/ University of Richmond **23.16 (a)** E. Bunn/University of Richmond **23.17 (left)** John Kieffer/Peter Arnold, Inc.; **(right)** Joel Gordon Photography

CHAPTER 24 Opening Illustration: *Joe Bergeron*

24.3 (a) NASA/NSSDC; **(b)** NASA/JPL **24.4** NASA, courtesy of Johnson Space Center **24.5** NASA/JPL **24.6 (a)** NASA; **(b)** Photo by R. L. Folk and F. L. Lynch **24.7** NASA/JPL **24.9** NASA/JPL/NSSDC **24.12** © Seth Shostak **24.13 (a)** David Parker, 1997/Science Library. *The Arecibo Observatory is part of the National Astronomy and Ionosphere Center, which is operated by Cornell Univ. under a cooperative agreement with the National Science Foundation* **24.14 (a, b)** NASA/JPL

Index

Page references in italics refer to figures. Page references preceded by a "t" refer to tables. Page references followed by an "n" refer to footnotes. Page references in boldface refer to main discussions.

A

AB Auriga, *230*
Abell 1689 galaxy cluster, *630*
Abell 2218 galaxy cluster, *687*
Abell 3827 galaxy cluster, *661*
Aberration of starlight, 76
Absolute magnitudes, 526
Absolute zero, 116, *116*
Absorption of light
  absorption line spectra, 160, *163*, **163–164**, *164*
  in the atmosphere, 302–303, *303*
  definition, 153
  spectral sequence, 528–529, t 528–529
  by thin gases, *159*, 159–161, *160*
Accelerating universe, *694*, 694–695, *695*
Acceleration
  around a curve, 135, *136*
  definition, 130, *131*
  equivalence principle, *455*, **455–457**, *456*, *457*
  of expansion of the universe, 693, 695–696
  gravitational, **130–131**, *131*, 146–147, A–2
  from net force, 132
  weight and, *132*, 132–133
Accretion
  around white dwarfs, 576–578, *577*
  from black holes, 667, 670, *670*
  internal heat from, 259–260, *260*
  planetesimals from, 233, *234*
Accretion disks
  around neutron stars, *582*, 582–583
  around supermassive black holes, 667, 670, 671
  around white dwarfs, 576–578, *577*
  quasars from, 665
Achondrites, 376
Acrux, A–19
Active galactic nuclei, **664–671**
  photographs of, *664*, *666*, *667*, *669*
  quasars, *664*, **664–666**, *665*, *667*
  radio galaxies, **666–667**, *667*, *668*, *669*
  Seyfert galaxies, 665
  supermassive black holes, 586, 665, **667–671**, *670*, *671*, *672*
Active galaxies, 664

Adams, Douglas, 477
Adams, John, 217
Adaptive optics, 185, *185*
Adrastea (moon of Jupiter), A–16
Advanced nuclear burning, 560–562, *561*
Aerobraking, 219
Aerosols, 314
African astronomy, ancient, 57–58, *58*
Age
  of Earth, 240–243, *242*
  of galaxies, 654, 656, *658*
  from impact craters, *264*, 264–265, *265*
  lookback time, 8–9, *10*, *644*, 644–645
  of Moon, 243, 264
  of solar system, 240–243, *242*
  of stars in galaxies, 654–655, *655*
  of Sun, 6, 16
  universal expansion and, 641–643
  of universe, 243, 539, 643, *695*, *695*, 717
Albedo, 301
Alcor, *530*
Aldebaran, 541, A–19
Aldrin, Buzz, *13*
Aleutian Islands, 404
Alexander the Great, 69
Algol, 531, 566, *567*
Algol paradox, 566
ALH84001 meteorite, *728*, 728–729
Alien visitations, 742
Al-Mamun, 69
Alpha Centauri
  Alpha Centauri A, t 528–529, A–18, A–19
  Alpha Centauri B, A–18
  distance to, *12*
  location of, *13*
  luminosity, 522–524, 541
Alphonso X, 70
Altair, A–19
Altitude, 31, *31*, 33–34, *35*, 96
Alvarez, Luis and Walter, 387
Amalthea (moon of Jupiter), A–16
Amerigo Vespucci, *106*
Ammonia, on Jupiter, 337–339, *338*, *339*
Ammonium hydrosulfide, 338, 340, *341*
Amplitude, 156
Analemma, *37*, 92, *92*
Ananke (moon of Jupiter), A–16
Anaxagoras, *67*, 374, 497
Anaximander, 66, *67*, 421
Andromeda Galaxy, 635–636, 684
Angelou, Maya, 725
Angle of reflection, *154*
Angstroms, 156

Angular measures, 31, *31*
Angular momentum, **137–138**
  elliptical orbits and, 139
  precession of axis and, 40
  principle of, *137*, 137–138, *138*
  in protogalactic clouds, 656–657, *657*
  in protostellar disks, 548, *548*
  solar nebulae collapse, *228*, 228
  solar system formation, 235–236, *236*
  spin angular momentum, 478n, **478–479**, *479*, 485–486
Angular resolution, 174, 177, 180n, **180–181**, *189*, 190
Angular separation, 175, *175*, A–3
Annihilation, **480**, *480*, 502, 703, *704*, *705*
Annular eclipse, 46, *46*
Antarctica, meteorites in, 375, *728*, **728–729**
Antarctica, ozone hole over, 425, *425*
Antares, 162, 541, A–19
Ante-meridiem (a.m.), 90
Antennae (NGC4038/4039), *659*
Antielectrons (positrons), *480*, **480**, *502*, 502–503, 703
Antileptons, 480
Antimatter
  in the early universe, 703, *704*, *705*
  in hydrogen fusion, 502–503
  matter-antimatter engines, 493, 740–741
  pair production and annihilation, *480*, **480**, 502, 703, *704*, *705*
  positrons, 480, *480*, *502*, 502–503, 703
  virtual particles, 489–490, *490*
Antiquarks, 480
Aperture, 174
Aphelion, *17*, 72–73, *73*
Aphrodite Terra (Venus), *285*
Apollo missions, *13*, 220, 274, *275*
Apollonius, *67*, 68
Appalachian Mountains, 269, 404
Apparent brightness, **522–524**, *523*, 526, 530, *530*, 632–633. *See also* Luminosity
Apparent magnitudes, 526
Apparent retrograde motion, **49–50**, *50*, *51*, 68
Apparent solar time, 90–91, *92*
Apparent weight, 132n
Arabic science, early, 69, 69–70
Arches National Park (Utah), *34*
Arcminutes/arcseconds, 31, *31*
Arctic/Antarctic Circles, 37, *38*, 104, *104*
Arcturus, t 528–529, 613, A–19
Arecibo radio telescope, *188*, *189*, *738*, *738*, A–24

Arguments, logical, 79
Ariel (moon of Uranus), 214, *345*, 355, A–17
Aristarchus, 50–51, *67*, 70
Aristotle
  cosmic model of, 68, 74
  ideas of, changed by Newton, 134
  life of, **66**, *67*
  on speed of light, 445
  on a spherical Earth, 32
Armstrong, Neil, *13*
Arp 220 starburst galaxy, *662*
Artificial stars, 185
Ashen light, 44
Asteroid belt, **369–372**
  effect of Jupiter on, 392, *393*
  formation, 237, 371–372
  misconception about, 371
  overview, 203, *205*, *206*, 370
Asteroids, **369–373**
  appearance, 203, *204*
  captured moons from, 238, *239*
  composition, 373
  current impact danger from, 391–392, *392*
  definition, 4, 369
  formation, 236–237, 368
  general characteristics, 369–370
  impacts and mass extinctions, **387–391**, *388*, *389*, *390*
  masses and densities, *372*, 373
  missions to, t 221
  naming of, 368–369
  orbits, 368, 369
  overview, 203, *205*
  photographs, *369*
  properties, t 200–201, **372–373**, *373*
  Trojan, 369–370
Astrobiology, 726–727. *See also* Life outside the Earth
Astrolabe, 105, *106*
Astrology, 40, 80–82
Astronauts, 118, 141, *141*
Astronomical units (AU), 4, 17, *17*, A–2
Astronomy
  ancient Greek, **64–70**, *65*, *67*, *68*
  ancient observation of planets and stars, 62–64, *63*, *64*
  ancient timekeeping, **58–60**, 59n, *59*, *60*
  ancient weather prediction and, 57–58, *58*
  astrology and, 40, 80–82
  Copernican revolution, *70*, **70–75**, *71*, *72*, *74*
  history of spectral sequence, 528–529
  as a science, 64
  Web sites, A–23 to A–25